T0399672

DEVELOPMENTS IN MARITIME TECHNOLOGY AND ENGINEERING

Proceedings in Marine Technology and Ocean Engineering

BOOK SERIES EDITOR

Carlos Guedes Soares

EDITORIAL BOARD MEMBERS

R. Ajit Shenoi, Enrico Rizzuto, Fenando Lopez-Peña, Jani Romanov, Joško Parunov

ABOUT THE SERIES

The 'Proceedings in Marine Technology and Ocean Engineering' series is devoted to the publication of proceedings of peer-reviewed international conferences dealing with various aspects of 'Marine Technology and Ocean Engineering'. The Series includes the proceedings of the following conferences: the International Maritime Association of the Mediterranean (IMAM) Conferences, the Marine Structures (MARSTRUCT) Conferences, the Renewable Energies Offshore (RENEW) Conferences and the Maritime Technology and Engineering (MARTECH) Conferences. The 'Marine Technology and Ocean Engineering' series is also open to new conferences that cover topics on the sustainable exploration and exploitation of marine resources in various fields, such as maritime transport and ports, usage of the ocean including coastal areas, nautical activities, the exploration and exploitation of mineral resources, the protection of the marine environment and its resources, and risk analysis, safety and reliability. The aim of the series is to stimulate advanced education and training through the wide dissemination of the results of scientific research.

BOOKS IN THE SERIES

Volume 1: Advances in Renewable Energies Offshore, 2019, C. Guedes Soares (Ed.).
Volume 2: Trends in the Analysis and Design of Marine Structures, 2019, J. Parunov and C. Guedes Soares (Eds.).
Volume 3: Sustainable Development and Innovations in Marine Technologies, 2020, P. Georgiev and C. Guedes Soares (Eds.).
Volume 4: Developments in the Collision and Grounding of Ships and Offshore Structures, 2020, C. Guedes Soares (Ed.).
Volume 5: Developments in Renewable Energies Offshore, 2021, C. Guedes Soares (Ed.).
Volume 6: Developments in Maritime Technology and Engineering, 2021, C. Guedes Soares and T.A. Santos (Eds.)
Volume 7: Developments in the Analysis and Design of Marine Structures, 2021, J. Amdahl and C. Guedes Soares (Ed.).

Proceedings in Marine Technology and Ocean Engineering (Print): ISSN: 2638-647X
Proceedings in Marine Technology and Ocean Engineering (Online): eISSN: 2638-6461

PROCEEDINGS OF THE 5TH INTERNATIONAL CONFERENCE ON MARITIME TECHNOLOGY
AND ENGINEERING (MARTECH 2020), LISBON, PORTUGAL, 16 – 19 NOVEMBER 2020

Developments in Maritime Technology and Engineering

Celebrating 40 years of teaching in Naval Architecture
and Ocean Engineering in Portugal
and the 25th anniversary of CENTEC

Volume 1

Editors

C. Guedes Soares
*Centre for Marine Technology and Ocean Engineering (CENTEC), Instituto Superior Técnico,
Universidade de Lisboa, Portugal*

T.A. Santos
Ordem dos Engenheiros, Portugal

CRC Press
Taylor & Francis Group
Boca Raton   London   New York

CRC Press is an imprint of the
Taylor & Francis Group, an **informa** business

A BALKEMA BOOK

CRC Press/Balkema is an imprint of the Taylor & Francis Group, an informa business

© 2021 the Author(s)

Typeset by Integra Software Services Pvt. Ltd., Pondicherry, India

The right of the Fifth International Conference on Maritime Technology and Engineering (MARTECH 2020) to be identified as author[/s] of this work has been asserted by him/her/them in accordance with sections 77 and 78 of the Copyright, Designs and Patents Act 1988.

Library of Congress Cataloging-in-Publication Data
A catalog record has been requested for this book

Published by: CRC Press/Balkema
              Schipholweg 107C, 2316 XC Leiden, The Netherlands
              e-mail: enquiries@taylorandfrancis.com
              www.routledge.com – www.taylorandfrancis.com

ISBN: 978-0-367-77374-8 (Set Hbk)
ISBN: 978-1-032-10699-1 (Set Pbk)
ISBN: 978-1-003-17107-2 (eBook)
DOI: 10.1201/9781003171072

Volume 1:
ISBN: 978-0-367-77376-2 (Hbk)
ISBN: 978-1-032-10693-9 (Pbk)
ISBN: 978-1-003-21658-2 (eBook)
DOI: 10.1201/9781003216582

Volume 2:
ISBN: 978-0-367-77377-9 (Hbk)
ISBN: 978-1-032-10696-0 (Pbk)
ISBN: 978-1-003-21659-9 (eBook)
DOI: 10.1201/9781003216599

Developments in Maritime Technology and Engineering – Guedes Soares & Santos (eds)
© 2021 Copyright the Author(s), ISBN 978-0-367-77376-2

Table of contents

Ship design

Ship structures - Ultimate strength

Ship structures - Composites

Preface

Since 1987, the Naval Architecture and Marine Engineering branch of the Portuguese Association of Engineers (Ordem dos Engenheiros) and the Centre for Marine Technology and Ocean Engineering (CENTEC) of the Instituto Superior Técnico (IST), Technical University of Lisbon, (now University of Lisbon) have been organizing national conferences on Naval Architecture and Marine Engineering. Initially, they were organised annually and later became biannual events.

These meetings had the objective of bringing together Portuguese professionals allowing them to present and discuss the ongoing technical activities. The meetings have been typically attended by 150 to 200 participants and the number of papers presented at each meeting was in the order of 30 in the beginning and 50 at later events.

At the same time as the conferences have become more mature, the international contacts have also increased and the industry became more international so that the fact that the conference was in Portuguese started to hinder its further development with wider participation. Therefore, a decision was made to experiment with having also papers in English, mixed with the usual papers in Portuguese. This was first implemented in the First International Conference of Maritime Technology and Engineering (MARTECH 2011), which was organized in the year that Instituto Superior Técnico completed 100 years. Subsequently, three more MARTECH conferences have been organized, namely in 2014, 2016 and 2018, always with a broad scope.

In this Fifth International Conference of Maritime Technology and Engineering (MARTECH 2020), some special events have marked it as a special one. To start with, the Conference is associated with the commemorations of the 40th anniversary of the teaching of Naval Architecture at IST, which has gone through different study plans and phases of development. The Centre for Marine Technology and Ocean Engineering (CENTEC), which was created some years later, is commemorating its 25th anniversary, making this a joint commemoration.

The other special event that marked this Conference was the appearance of COVID-19, which started having a marked spread in Portugal in March 2020, leading to a postponement of the Conference from the planned date of May to November 2020. Against initial hopes, the situation was still very serious in October and thus the Conference was held mostly online with a small presence at IST in Lisbon.

Running the Conference online was a challenge, because of the little experience accumulated with this type of events and, of the difficulties in dealing with the time difference for the authors from Asia. Finally, a compromise solution was found, allowing the authors to present their papers and to follow most of the sessions in the morning and early afternoon in Europe.

The postponement of the Conference allowed a few more abstracts and papers to be accepted, so in the end, around 285 abstracts have been received and after the review process, about 180 papers were finally accepted and are included in this book. Without the presence of participants, the book finished up being produced and distributed after the Conference, which is not the usual procedure.

The Scientific Committee had a major role in the review process of the papers although several other anonymous reviewers have also contributed and deserve our thanks for the detailed comments provided to the authors allowing them to improve their papers. Participation is coming from research and industry from almost every continent, which is also a demonstration of the wide geographical reach of the conference.

The contents of the present books are organized in the main subject areas corresponding to the sessions in the Conference and within each group, the papers are listed by the alphabetic order of the authors.

We want to thank all contributors for their efforts and the sponsors for their support and we hope that this Conference will be continued and improved in the future.

C. Guedes Soares & T.A. Santos

Conference chairmen

Carlos Guedes Soares, *IST, Universidade de Lisboa, Portugal*
Pedro Ponte, *Ordem dos Engenheiros, Portugal*

Organizing committee

Tiago A. Santos, *Ordem dos Engenheiros, Portugal*
Ângelo Teixeira, *IST, Universidade de Lisboa, Portugal*
Dina Dimas, *Ordem dos Engenheiros, Portugal*
Manuel Ventura, *IST, Universidade de Lisboa, Portugal*
José Manuel Cruz, *Ordem dos Engenheiros, Portugal*
José Gordo, *IST, Universidade de Lisboa, Portugal*
Francisco C. Salvado, *Ordem dos Engenheiros, Portugal*
José Varela, *IST, Universidade de Lisboa, Portugal*
Paulo Viana, *Ordem dos Engenheiros, Portugal*

Scientific committee

Felice Arena, *Mediterranea Univ. of Reggio Calabria, Italy*
Ermina Begovic, *UNINA, Italy*
Kostas Belibassakis, *NTUA, Greece*
Marco Biot, *Università degli Studi di Trieste, Italy*
Elzbieta Bitner-Gregerseon, *DNVGL, Norway*
Evangelos Boulougouris, *University of Strathclyde, UK*
Rui Carlos Botter, *University of São Paulo, Brazil*
Sofia Caires, *Deltares, The Netherlands*
Nian Zhong Chen, *Tianjin University, China*
Ranadev Datta, *IIT Kharagpur, India*
Vicente Díaz Casas, *Universidad A Coruña, Spain*
Leonard Domnişoru, *Univ. Dunarea de Jos Galati, Romania*
Soren Ehlers, *Technische Universität Hamburg, Germany*
Saad Eldeen, *Port Said University, Egypt*
Bettar Ould el Moctar, *Univ. of Duisburg-Essen, Germany*
Segen F. Estefen, *UFRJ, Brazil*
Selma Ergin, *Istanbul Technical University, Turkey*
Massimo Figari, *University of Genova, Italy*
Thor I. Fossen, *NTNU, Norway*
Yordan Garbatov, *Instituto Superior Técnico, Portugal*

Sergio Garcia, *University of Cantabria, Spain*
Lorena García Alonso, *Univ. de Oviedo, Spain*
Peter Georgiev, *TU Varna, Bulgaria*
Hercules Haralambides, *Erasmus Univ. Rotterdam, The Netherlands*
Spyros Hirdaris, *Aalto University, Finland*
Zhiqiang Hu, *Newcastle University, UK*
Chunyan Ji, *Jiangsu Univ. Science & Technology, China*
Xiaoli Jiang, *TUDelft, The Netherlands*
Jean-Marc Laurens, *ENSTA Bretagne, France*
Debabrata Karmakar, *National Inst. Techn. Karnataka, India*
Faisal Khan, *Memorial University, Canada*
Pentti Kujala, *Aalto University, Finland*
Xavier Martínez, *Uni. Politécnica de Catalunya, Spain*
Alba Martínez-López, *Univ. Las Palmas Gran Canária, Spain*
Marcelo Ramos Martins, *University of São Paulo, Brazil*
Jakub Montewka, *Gdynia Maritime University, Poland*
Muk Chen Ong, *University of Stavanger, Norway*
Josko Parunov, *University of Zagreb, Croatia*
Apostolos Papanikolaou, *HSVA, Germany*
Preben T. Pedersen, *DTU, Denmark*
L. Prasad Perera, *Arctic University of Norway, Norway*
Jasna Prpić-Oršić, *University of Rijeka, Croatia*
Harilaos N. Psaraftis, *DTU, Denmark*
Suresh Rajendran, *IIT Madras, India*
Huilong Ren, *Harbin Engineering University, China*
Jonas W. Ringsberg, *Chalmers Univ. of Technology, Sweden*
Liliana Rusu, *University Dunarea de Jos Galati, Romania*
António Souto-Iglesias, *Univ. Politécnica de Madrid, Spain*
Maciej Taczala, *West Pomeranian University, Poland*
Wiesław Tarełko, *Gdansk University of Technology, Poland*
Michele Viviani, *University of Genova, Italy*
Deyu Wang, *Shanghai Jiao Tong University, China*
Jin Wang, *Liverpool John Moores University, UK*
Xinping Yan, *Wuhan University of Technology, China*
Shengming Zhang, *Lloyds Register, UK*
Peilin Zhou, *University of Strathclyde, UK*
Xueqian Zhou, *Harbin Engineering University, China*

Technical programme & Conference secretariat

Sandra Ponce, IST, *Universidade de Lisboa, Portugal*
Maria de Fátima Pina, *IST, Universidade de Lisboa, Portugal*
Sónia Vicente, *IST, Universidade de Lisboa, Portugal*
Bárbara Azevedo, *IST, Universidade de Lisboa, Portugal*
Mina Abbasi, *IST, Universidade de Lisboa, Portugal*

Sponsors

Keynote lecture

Developments in Maritime Technology and Engineering – Guedes Soares & Santos (eds)
© 2021 Copyright the Author(s), ISBN 978-0-367-77376-2

Forty years of teaching and research in Naval Architecture and Ocean Engineering in Portugal

C. Guedes Soares

Centre for Marine Technology and Ocean Engineering (CENTEC), Instituto Superior Técnico, Universidade de Lisboa, Portugal

ABSTRACT: This paper provides an overview of how the teaching of Naval Architecture and Ocean Engineering has evolved at Instituto Superior Técnico since 1980. It describes the main changes in the curricula that occurred in 1988, 1998, 2007, 2017 and the planned study plan to be initiated in 2021/2022. At the same time, the evolution of the research activity of the academic staff is also described, providing an overview of the main research projects and achievements in that period. The evolution of the research centre CENTEC, which was created in the academic year of 1994/1995, and is now commemorating its 25th anniversary, is also described. A description of the research output is given, as well as information about the national and international evaluation of its performance.

1 INTRODUCTION

The MARTECH international conferences evolved from a series of Portuguese biannual conferences organised by the Engineering Faculty (IST, Instituto Superior Técnico) of the University of Lisbon and the Portuguese Engineering Association of Engineers (Ordem dos Engenheiros). Some of these conferences have been associated with special celebrations of significant Portuguese institutions or companies in the field, and on this occasion, the conference is associated with the celebrations of the 40th anniversary of the teaching of Naval Architecture and Ocean Engineering in Portugal and the related 25th anniversary of the Centre for Marine Technology and Ocean Engineering (CENTEC).

Therefore, it is appropriate to present an overview of the evolution of teaching and the associated research activity, which has been particularly important for the industry in Portugal and Europe in general, as a significant part of the graduates are working in various companies and institutions spread throughout Europe.

Teaching in Portugal started in 1976 with post-graduation courses, initially of one year and later of three semesters. These courses were initiated as a result of the planned expansion of the shipbuilding activity that was expected to occur in Portugal, with an increased need for engineers specialised in this area.

During the 70s, Portugal had a very large shipyard in Lisbon, Lisnave, which employed up to 10,000 workers undertaking ship repair and major conversions. There was also another shipyard in Setúbal, about 60 km from Lisbon, Setenave, building large tankers. In the early 70s, plans have been made to expand it with large docks to be able to build very large tankers, and for this planned expansion, the need to have enough engineers graduated in naval architecture and marine engineering was identified and the University of Lisbon was asked to respond.

The initial post-graduation courses have been organised directly under the aegis of the University of Lisbon, in which existing professional engineers practising in the country ensured the teaching. At the time, the largest group of these specialists were the "Naval Constructors" of the Portuguese Navy, who took up the leading role in this educational process as Adjunct Professors.

After three series of post-graduation courses, it was felt that it would make more sense to have a normal educational programme at the Engineering Faculty and thus in 1980, instead of starting the 4th post-graduation course, a regular 5-year programme started at IST.

2 TEACHING OF NAVAL ARCHITECTURE AT IST

In the academic year of 1980/1981, a new course started at IST under the designation of Shipbuilding Engineering. This was a full 5-year course that included two years of general engineering subjects common with other courses such as Mechanical Engineering. More specific subjects were taught in the last three years, including an expanded version

DOI: 10.1201/9781003216582-1

of the subjects taught in the three semesters of the earlier post-graduation course.

At that time, Prof. Luciano Faria was the leading person responsible for the post-graduation courses and the initiation of teaching at IST. The group of Adjunct Professors from the Portuguese Navy was led by C. Caldeira Saraiva in the post-graduation courses and by Rogério d'Oliveira at IST.

Starting a new educational programme at IST required the training of new teaching staff, as this was the first programme in Portugal, and it was not possible to hire specialists who graduated from other Portuguese Universities. Thus, an agreement was made with the Norwegian Institute of Technology at the University of Trondheim in Norway to provide a PhD programme for four new staff members who would be the first teaching staff of the new course.

Two positions were announced in 1980 and two more in 1981, but surprisingly only the author has been selected and sent to Norway from 1980 to 1984 to complete the PhD programme. Therefore, the build-up of the specific teaching staff took longer than initially planned as new staff had to be hired from the newly formed engineers. At that time, the teaching was totally in Portuguese, and internationalisation was not the standard approach in Portugal as in many European countries.

The teaching programme has evolved gradually to increase the number of specialised Naval Architecture subjects to make the nature of the programme more specific as the appropriate teaching capability became available.

The first curricular change was in 1988, when the course changed its designation from "Shipbuilding Engineering" to the classical designation of "Naval Architecture and Marine Engineering", or "Engenharia Naval", in Portuguese. This change has corresponded to increasing the number of subjects that strengthened the capability of designing ships, widening the earlier nature of the course, where shipbuilding technology had a larger relative weight.

The next curricular change was made in 1998 when the scope of the course was widened and was organised in two specialisation profiles. One corresponded to the traditional education that was ongoing and was denoted as "Ship Design and Shipbuilding", and a new one was created under the name of "Maritime Transportation and Ports".

The rationale behind this change was the recognition of the evolution that the shipbuilding and ship repair industry was undergoing in Europe, with the closing of several major shipyards and the transfer of industrial activities to countries in Asia. This transfer of activity has also occurred in other industries and, as a result, an increasingly large number of products started being produced in Asia, which required them to be transported to Europe for final consumption, increasing the role of maritime transportation in European countries. Therefore, the professional education of naval architects had to be adapted to the new conditions of the employment market, providing them with education in a neighbouring field where they could perform their professional activity.

The approach adopted was to maintain the main educational subjects common to both specialities to ensure the essential knowledge required by a naval architect to fulfil the main professional activities. The specialist subjects were limited to provide more in-depth knowledge of the more specific aspects of each of the specialisations.

In 2007, almost keeping the same ten-year period of curricular adjustment, a new reformulation of teaching was undertaken, this time as a consequence of the Bologna Agreement, which required uniformisation of the degrees in Europe. Thus, continental Europe, which had engineering degrees based on five years of education, decided to adjust the degrees to a system mostly existing in the UK of a first Bachelor degree of 3 years and a Master degree of 2 additional years.

In Portugal, the Universities reacted to these imposed changes by defining "Integrated Master degrees", in which the Bachelor and the Master programme were "integrated", and in that way, in practice, they continued having a 5-year programme. This "smart" way of dealing with imposed changes was not unique in Portugal as it was also adopted in other countries in Europe.

However, in Portugal, this approach was adopted only for the courses of a large number of students. In most countries, the naval architecture courses are almost always much smaller in the number of students than other more generic engineering specialities like mechanical, civil, electrotechnical and chemical, for example. For the courses with a small number of students like Naval Architecture and Marine Engineering, IST has chosen to have two separate cycles of Bachelor (Licenciado) and Master (Mestre).

In passing, it is interesting to note that the educational changes in Portugal have created a major problem that has not yet been solved, many years after the changes were made. The existing 5-year engineering programme in Portugal lead to a degree designated as "Licenciado". When creating a system in which the first degree was three years, the same designation of "Licenciado" was maintained. Thus, suddenly the same title designated an education of 5 years and another of 3 years.

The system that existed in Portugal and some other European countries before 2007 was to have a Master degree of about two years after the 5-year educational programme. Here also, the changes imposed by the Bologna Agreement lead to the title of a Master representing seven years or five years of education depending on whether it was before or after 2007.

While the problem associated with the designation of the degrees has not been changed, a recent law from the Ministry of Higher Education has dictated the end of the "Integrated Master" courses, so from the academic year of 2021/2022, all courses in Portugal will be functioning in strict 3+2-year cycles.

With the change of the study plan of 2007, there was the idea that the initial study cycle should have more generic education, leaving the specific subjects for the Master degree. This led to a reduction in the number of subjects specific to Naval Architecture in the first three years, and they were moved to the Master study plan. This unfortunate situation has been maintained until the present, more realistic, policy change, which allowed the return to the earlier situation in which the first three years will have a more substantial component of specific courses.

The Bachelor programme that will start functioning in 2021/2022 has the first year with mathematics and sciences, the second year with general engineering subjects and a third-year with Naval Architecture and Ocean Engineering subjects.

In 2017, a more limited change in the study plan has been implemented by creating a third specialisation profile, this one on Ocean Systems. This was a result of the interest and support of the Portuguese oil and gas company GALP, which have in more recent years become more involved in oil and gas production and exploration through joint ventures in Brasil, Mozambique and other locations worldwide. To support these activities, the company needed naval architects better prepared to deal with the specific aspects of the oil and gas industry and have supported several scholarships for students of this new specialisation.

With the addition of this new specialisation, the designation of the course was changed from Naval Architecture and Marine Engineering to Naval Architecture and Ocean Engineering, covering now the complete scope of activities where traditionally the graduates find work. This last addition to the study plan allowed it to reach the global scope of education that is found in several other countries. In Portugal, this educational programme continues to be the only one offered at any University.

The teaching of the Master course has been given for many years in English, while the Bachelor course maintained teaching in Portuguese, although starting in 2021/2022, subjects in the Bachelor programme will also be taught in English. The environment in the Master programme has been very much international with a moderate percentage of regular students from other countries and with 30 to 40 exchange students that every year take part in the study programme. The main number of exchange students come from Europe, being funded by the European ERASMUS programme, but several other international bilateral exchange programmes also bring students from other countries such as Brazil, India, China, Japan and Korea.

The third cycle of studies leads to Doctoral degrees, which has been very relevant to the built-up of the teaching staff of the Bachelor and Master programmes and also to shape the research produced at IST in this field. The PhD studies have started in the 1990s, with the traditional format based exclusively on research and in the early 2000s, following a change in all research fields at IST, Doctoral programmes were created with a full year of course work before the research component was conducted. This change has been particularly important after 2007 when the normal education period of 7 years before starting the PhD programme was reduced to 5 years. Thus, the reduction of 1 year of course work undertaken in the Master programmes after the 5-year Engineering degree was compensated by the 1-year taught programme on the Doctoral Programmes.

Several of the PhD graduates became teaching staff at IST, having produced thesis on a diversified range of subjects (Guedes Soares & Garbatov 1996; Fonseca & Guedes Soares, 1998; Santos & Guedes Soares 2008; Teixeira & Guedes Soares 2009; Gordo & Guedes Soares 2009; Ventura & Guedes Soares 2012; Ribeiro e Silva & Guedes Soares, 2013; Wang & Guedes Soares 2016a,b; Vettor & Guedes Soares 2016; Chen & Guedes Soares 2016; Campos et al. 2019).

Several other PhD graduates are now teaching staff at different Universities around the world, such as China (6), Egypt (5), Croatia, India, Germany, Japan, Norway, Peru and Romania, while others are working in the industry also in different countries such as Brasil, Italy, Netherlands, Norway, Poland and Portugal.

The PhD programme has been designated as Naval Architecture and Marine Engineering, but since 2017 it became Naval Architecture and Ocean Engineering, also aligning with the designations of the two earlier cycles of Bachelor and Master.

Some joint PhD degrees have been awarded with other Universities such as Universidade Federal do Rio de Janeiro, University of Zagreb, Universita Mediterranea de Reggio Calabria. Joint supervision of PhD's have been carried out with Universidade de Las Palmas de Gran Canaria, University of Rijeka, Universidade Federal do Rio de Janeiro, Universiti Teknologi Malaysia, Amirkabir University of Technology, Wuhan University of Technology, Ocean University of China, Shanghai Jiao Tong University, Dalian Maritime University and Northwestern Polytechnical University.

3 RESEARCH

At the same time, as the teaching staff was trained and increased in number, the research activity has also expanded, as it is indispensable for high-quality university teaching. During the '80s, the volume of research produced was low as it resulted from one single researcher. A significant jump in the volume of research has occurred during the '90s as a consequence of being the Coordinator of two important EU projects.

One was Reliability Methods for Ship Structural Design (SHIPREL), which ran from 1991 to 1995 and was funded by the industrial based EU

Programme BRITE-EURAM. This project was conducted with three major Classification Societies Bureau Veritas, Germanischer Lloyds and Registro Italiano Navale, and the Technical University of Denmark. Among the various advances achieved, it proposed a reliability-based design format based on the ultimate hull strength (Guedes Soares et al. 1996), which was adopted ten years later by the Common Structural Rules of the Classification Societies. The earlier midship section design requirement was based on elastic stresses and the first yield concept, specifying the minimum acceptable section modulus.

The other project was Probabilistic Methodology for Coastal Site Investigation Based on Stochastic Modelling of Waves and Current (WAVEMOD), which ran from 1993 to 1996 and was financed by the EU Marine Science and Technology (MAST) Programme. This project allowed the development of several probabilistic formulations of the variability of waves and currents, which are essential input information for reliability models (Guedes Soares, 2000).

This project was followed by the coordination of the project Hindcast of Dynamic Processes of the Ocean and Coastal Areas of Europe (HIPOCAS), which ran from 2000 to 2004 and was financed by the EU Programme Energy, Environment and Sustainable Development (EESD). This was a major project that produced hindcasts of about 40 years of the seas around Europe (Guedes Soares et al. 2002; Guedes Soares, 2008). It was funded in the same call as another project in which the EU funded ECMWF to conduct a global hindcast that led to the well-known ERA40 database. A comparison of these databases showed a good agreement for waves up to moderate significant wave heights (Campos and Guedes Soares, 2016), which is the limit that present-day hindcasting can be trusted, as significant difficulties still exist with the modelling of extreme sea states and new approaches continue being proposed to deal with this problem (Campos et al. 2018).

Another project that was coordinated almost at the same time was Freak Wave Generation in the Ocean (FREAK WAVES), which run from 2002 to 2005, financed by the EU Programme INTAS. This was a project with a small number of partners, but it reached the interesting conclusion at the time that freak waves tend to be generated when there are sudden significant changes in the shape of the wave spectra (Lopatoukhin et al., 2005).

The follow-up of SHIPREL in the 2000s were two large network projects. The first one was Safety and Reliability of Industrial Products, Systems and Structures (SAFERELNET), which ran from 2001 to 2005 and was financed by the industry-based EU Programme GROWTH. This project had 69 partners, collecting thus the contributions of a significant number of European groups, which produced many relevant papers and a book with the main outcomes (Guedes Soares, 2010).

The next project coordinated by CENTEC was the Network of Excellence on Marine Structures (MARSTRUCT), which ran from 2004 to 2010 and was funded by the EU Programme on Sustainable Development (SUSTDEV). This project involved 33 partners, including almost all groups in Europe that worked with Marine Structures. It has been a very successful project that produced a very large number of papers (more than 400). This project also initiated a series of Conferences, the first two of which were held during the project (Guedes Soares & Das 2007, 2009). This was also a major outcome of the project as this has generated a series of biannual conferences that have continued up to the present days (Guedes Soares & Fricke, 2011; Guedes Soares & Romanoff, 2013; Guedes Soares & Shenoi, 2015; Guedes Soares & Garbatov, 2017; Parunov & Guedes Soares, 2019).

Another interesting outcome is the MARSTRUCT Virtual Institute (http://www.marstruct-vi.com), which is an Association of the groups that were involved in the project, aiming at the continuation of the cooperation. In addition to the collaboration in the organisation of the biannual MARSTRUCT Conferences, it also conducts benchmark studies (e.g. Ringsberg et al. 2018; Parunov et al. 2020).

Another coordinated project that made the transition from the 2000s to the 2010s was Advanced Ship Design for Pollution Prevention (ASDEPP), which ran from 2006 to 2010 funded by the EU TEMPUS Programme. This project has organised and conducted several PhD courses and produced a book as one of the outcomes (Guedes Soares & Parunov, 2010).

At the same time as this series of projects had a major impact on the development of the research performed at IST, as a result of the number of researchers they allowed to hire, there were another series of important projects that had a similar level of involvement of the IST research group. For a relatively long period, the EU wanted industrially oriented projects to be led by a company, and therefore the strategy adopted in various projects was to have a company as the Administrative Coordinator and IST as the Technical Coordinator. This allowed IST to have in practice major participation in those projects, which also contributed to shaping the overall research profile at IST.

The first of these projects was Advanced Method to Predict Wave Induced Loads for High-Speed Ships (WAVELOADS), which run from 1998 to 2001, funded by the EU Programme BRITE-EURAM and coordinated by Germanischer Lloyds (Schellin et al., 2003).

The second project was Reliability-Based Structural Design of FPSO Systems (REBASDO), which ran from 2001 to 2003, funded by the EU Programme on Energy, Environment and Sustainable Development (EESD) coordinated by Shell International Exploration and Production. This project led to advances in reliability formulations (Garbatov et al. 2004) in addition to producing high-quality experimental results (Skourup et al. 2004) and improvements in wave descriptions (Ewans et al. 2006).

The next project, also coordinated by Shell, was Safe Offloading from Floating LNG Platforms (SAFEOFFLOAD), which run from 2006 to 2009, funded by the EU Programme Sustainable Surface Transport (SUST). This project dealt with the off-loading of LNG from platforms to shutter LNG tankers and led to several interesting hydrodynamic studies (Guedes Soares et al., 2015).

Another project with some time of overlap with that one was Decision Support System for Ship Operation in Rough Weather (HANDLING WAVES), which was held from 2007 to 2010, funded by the EU Programme Sustainable Surface Transport (SUST) and coordinated by Registro Italiano Navale. This project produced interesting experimental results (Rajendran et al., 2011) and led to developing a decision support system that was installed in one ship of Grimaldi Lines (Perera et al., 2012).

Another project undertaken from 2000 to 2003 was Rogue-Waves - Forecast and Impact on Marine Structures (MAXWAVE), which was funded by the EU Energy, Environment and Sustainable Development Programme. This project was coordinated by GKSS in Germany who was responsible for the group of partners dealing with wave modelling, while IST was responsible for the group of partners dealing with wave-induced responses. This was a very important project that made significant progress towards the description of abnormal or rogue waves (Guedes Soares et al., 2003) and the ship (Guedes Soares et al. 2008) and offshore structures (Guedes Soares et al. 2006) responses to them. This project recognised the important contribution of Prof Douglas Faulkner (Guedes Soares & Das 2008) to this field and invited him to be a consultant to the project.

The follow-up project, with technical coordination, of IST was Design for Ship Safety in Extreme Seas (EXTREME SEAS), which ran from 2009 to 2013, funded by the EU Sustainable Surface Transport Programme and was coordinated by Det Norske Veritas. This was also an important project with advances in modelling abnormal and extreme waves (Zhang et al., 2014) and responses to extreme waves represented by advanced models (Klein et al., 2016; Wang et al., 2016). Important experimental results were also obtained, allowing, for example, the analysis of pressures in the bow flare and stern of a containership (Wang and Guedes Soares 2016a, b).

In the mid-2010s, some new research areas have been identified, and among them, renewable energies offshore was one of the most important. As a consequence, the project being coordinated presently is Adaptation and implementation of floating wind energy conversion technology for the Atlantic region (ARCWIND), which runs from 2017 to 2021

and is funded by the INTERREG Atlantic Area Programme. It reflects the new priority area of renewable energies offshore that was adopted more recently in CENTEC. This ongoing project already allowed the identification of the status of present-day wind farms and the development of a siting approach (Diaz & Guedes Soares, 2020a, b), the establishment of a cost assessment procedure (Castro-Santos et al. 2020), the identification of the relevant maintenance policies (Kang et al. 2019) and even the design of a new platform concept (Uzunoglu & Guedes Soares, 2020).

The involvement in these projects has had a significant influence on the research conducted during the time span. However, at the same time, there has been the involvement in about 95 other EU funded projects as partners, to some of which IST made substantial contribution and others with a small contribution. They were 32 projects finishing in the decade of 1990, 40 in the 2000s and 22 in the 2010s. The EU projects had a tendency of becoming larger with time, with typical projects in 2010 having 3 to 4 times the budget of the ones in the 1990s.

Another source of project funding has been the national funding agencies. These have been projects mostly with one institution, and a few are a collaboration between two Portuguese institutions. The contribution of national projects has been relatively small in the 1900s, but since 2000 it became more significant. There have been 16, 25 and 28 projects respectively in the three decades of 1990, 2000 and 2010.

4 CENTRE FOR MARINE TECHNOLOGY AND OCEAN ENGINEERING (CENTEC)

Until the early 1990's the research activity in Portugal was made through a National Institute of Scientific Research (INIC), which had centres in various Universities, a system that resembles what still exists in France and Italy. However, INIC was dissolved at that time, and the research was transferred to the Universities, which created their own research centres.

So in 1994, the Unit of Marine Technology and Engineering (UETN) was created as a joint initiative of the author with one colleague from Mechanical Engineering and one colleague from Statistics. At that time, there was only one PhD in the field of Naval Architecture in the group, and 3 PhDs was the minimum number necessary to create a Research Unit.

When UETN was created, the group was already coordinating the two EU projects mentioned in the previous section, and thus there was already a critical mass of young researchers. The total number of researches of UETN has been relatively stable at around 55, but the number of researchers with PhD has increased from 3 to almost 20. Another interesting evolution is the internationalisation of the group, which went from a situation of a national group in 1994 to a group with about 35% of foreign researchers in 2006 (Figure 1).

By 2007, there were already about 55 researchers, 20 of which with a PhD degree and this was already an appropriate size to change the Research Unit into a Research Centre, which was done at an occasion that IST changed its own Statutes. Then the Centre for Marine Technology and Engineering was created and organised in 4 research groups:

- Marine Environment,
- Marine Dynamics and Hydrodynamics
- Marine Structures
- Safety, Reliability and Maintenance

The research groups represent the main scientific areas in which the members are active, and they are further subdivided into research lines that represent specialised topics in which there is a relatively permanent research activity and in which there is a minimum of two active PhD researchers.

The evolution of CENTEC up to the present day has witnessed an increase in size from 55 to an average level of about 100 researchers, which has been maintained in the last ten years. The other evolution was the number of PhD researchers, which increased from 20 to about 50 or 39% to about 45%. The number of foreign researchers has also had a similar evolution (Figure 2).

In 2014, an external evaluation of the research centres in Portugal was conducted under the leadership of the Portuguese Foundation for Science and Technology. This led to a review of the research activity and its projection for the future, and the name was readjusted to become Centre for Marine Technology and Ocean Engineering, giving a more accurate reflection of the research in Ocean Engineering that had been ongoing since the early '90s.

At this time, Maritime Transportation, which had been defined as one priority area for research, was developing but still did not have a critical mass of researchers, so to have consistency in the size of the groups, the fourth group was renamed as Safety and Logistics of Maritime Transportation, becoming, in fact, the junction of two subgroups.

So the present organisation of CENTEC's Groups in research lines is as follows:

Marine Environment Group

- Wave Spectral Models and Time Series Models
- Probabilistic Models of Wave Parameters
- Wave Modelling and Hindcasting
- Circulation and Oil Spill Modelling
- Oceanographic Instrumentation

Marine Dynamics and Hydrodynamics

- Dynamics of Moored Floaters
- Non-linear Motions and Loads
- Ship Manoeuvring and Control
- Computational Fluid Dynamics
- 3D Virtual Environments in Ship Dynamics
- Full-Scale Trials and Model Tests

Marine Structures

- Ultimate Strength
- Fatigue Strength
- Impact Strength
- Structures in Composite Materials
- Geometric Modelling of Ship Structures
- Offshore and Subsea Structures
- Experimental Analysis

Safety and Logistics of Maritime Transportation

- Structural Safety
- Reliability-Based Structural Maintenance
- System Reliability and Availability
- Maritime Safety and Human Factors
- Industrial and Occupational Safety
- Logistics of Maritime Transportation and Port Operations

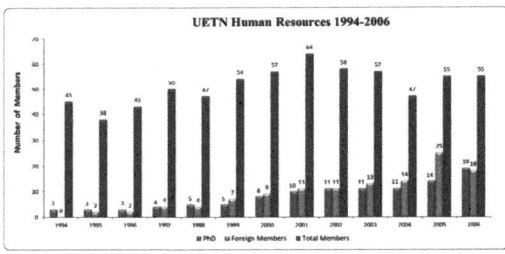

Figure 1. Evolution of human resources of UETN.

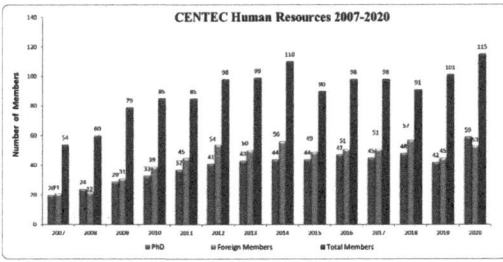

Figure 2. Evolution of human resources of CENTEC.

5 DISSEMINATION OF RESEARCH

Scientific publications are the end product of any research activity, and thus they are one important objective of any researcher. They have been an important index of the activity of CENTEC, and the productivity has changed over the years, as can be observed in Figure 3, which shows the output in each period of five years, separating the papers published in international journals from the ones presented in Conferences and published as book chapters or in proceedings.

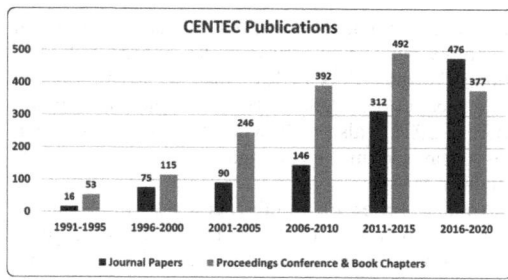

Figure 3. Evolution of CENTEC Publications.

An important difference can be observed between the first ten years and the recent years. Despite showing a permanent increase in the total number of publications, the increase was sharper after the first ten years. When correlating this with the number of researchers, it can be observed that the total number did not change much in the period, but the number of PhD researchers showed a significant increase around 2000, continuing to increase after that.

From 2000 to 2010, there was a significant increase in journal papers, with the ratio to conference papers being almost constant at 0.37, but in the following five years this changed to 0.63, and in the last five years the journal papers were even more than the conference papers, i.e. the change was to 1.26. This resulted from a deliberate policy to reduce the participation in Conferences in relation to the submission of papers in Journals.

The public acceptance and recognition of the results can be related to the number of citations that the papers have collected, which are now about 20,000 in the Web of Science and 29,000 in SCOPUS. The evolution of the number of citations in the Web of Science is shown in Figure 4. It can be observed that during the initial six years, the number of citations was moderate (average of 445/year), but a marked increase to high values can be observed, in particular in the last six years (average of 2350/year).

Some of the papers have even gained high public acceptance, becoming highly cited papers in the Web

of Science. In the current year, the following papers are on this list: Mantari et al. (2012a), Rusu & Guedes Soares (2012a), Chojaczyk et al. (2015), Gaspar et al. (2017), Wu et al. (2017, 2019). In earlier years other papers have also been highly cited: Rusu & Guedes Soares (2012b), Silva et al. (2013), Gonçalves et al (2014), Mantari & Guedes Soares (2014), Mantari et al (2012b) and Wu et al. (2018).

Other papers have won distinctions and awards such as the significant papers in the Journal of Ship Research (Saad-Eldeen et al, 2011; Sutulo et al. 2012; Teixeira et al, 2013; Corak et al. 2015), the best paper award of the Ship and Offshore Structures Journal (Saad-Eldeen et al. 2013) and of the International Journal of Maritime Engineering (Cubells et al. 2014; Zhou et al. 2016) the top paper award of the Renewable Energy Journal (Rusu & Guedes Soares 2012b) and the 10th Anniversary best paper of the journal Energies (Silva et al. 2013).

It may also be of interest to identify the main scientific area that has been covered by the publications, both in journals and in some conferences that are indexed in Web of Science. This identification is based on the results presented in the Web of Science, which is as follows:

Engineering Marine – 24%
Engineering Ocean – 17%
Engineering Civil – 15%
Oceanography – 10%
Engineering Mechanical – 8%
Transportation Sci & Techn – 3%
Mechanics – 3%
Energy & Fuels – 2%
Green Sustainable Sci. Tecn – 2%
Engineering Industrial – 1%
Operations Research – 1%
Eng Multidisciplinary – 1%
Materials Sci Composites – 1%
Others - 12%

Another interesting aspect that can be derived from the analysis of the publications is the identification of the international collaborations that they reflect. Again using the information from the Web of Science, the number of papers in collaboration with the various foreign institutions are as follows:

Wuhan University of Technology - 60
University of Zagreb – 47
Univ. Federal do Rio de Janeiro – 29
Jiangsu Univ Sci. Tech. - 27
Norwegian Univ Sci & Technology – 18
Harbin Engineering University – 17
Ocean University of China - 15
Univ las Palmas de Gran Canaria – 14
Shanghai Jiao Tong Univ – 13
Univ of Rijeka – 12
Univ Med. di Reggio Calabria – 11
SINTEF – 11
American Bureau Shipping – 10
Amirkabir Univ of Technology – 10
Gdansk Univ Technology – 10
Bulgarian Academy Sciences – 8

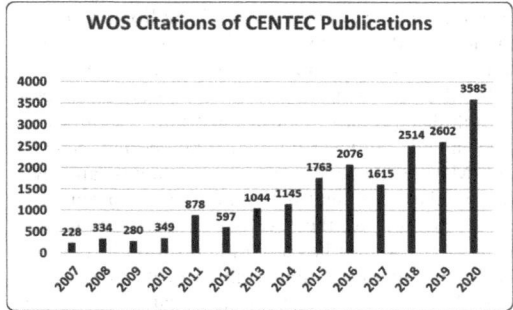

Figure 4. Evolution of WoS Citations of CENTEC Publications.

Delft Univ Technology – 7
Lloyds Register EMEA – 7
Universidade da Coruna – 7
University of Turin – 7
Nat Tech Univ Athens – 6
Tech Univ Berlin – 6
DNVGL – 5
Aalto Univ. – 5
Kuwait Univ – 5
Pol. Univ Milan – 5
Tech. Univ Varna – 5
University of Utah - 5
Hamburg Univ Tech – 5
Indian Inst Techn – 5
using 5 as the cut-off.

On the 15th anniversary of CENTEC, a commemoration book has been published mostly with state of the art review papers by CENTEC members and by several foreign collaborators at the time (Guedes Soares et al. 2011). This has been a very important source of information about the work performed in the period just before its publication.

The promotion of the dissemination of results through conferences has been considered an important activity, and there has been a significant engagement in organising conferences. There has been a direct involvement in the organisation of the yearly Offshore Mechanics and Arctic Engineering (OMAE) Conferences by coordinating Symposium 2 of that conference, which initially was the Safety and Reliability Symposium and later became the Structures, Safety and Reliability Symposium. In addition, the OMAE Conference was organised in Lisbon in 1998 and Estoril in 2008. During OMAE2018, which was held in Madrid, a special Symposium on Marine Technology and Ocean Engineering Honoring Prof. Carlos Guedes Soares was organised with about 120 papers (Garbatov, 2020). In the previous year, the Honouring Symposium of Prof Torgeir Moan had been held (Guedes Soares, 2019a) in Trondheim.

The European Safety and Reliability (ESREL) Conferences are promoted by ESRA, the European Safety and Reliability Association to whose creation the author made a significant contribution. The ESREL Conference has been organised in Lisbon in 1997 (Guedes Soares, 1997). and in Estoril in 2006 (Guedes Soares, & Zio 2006). There has also been cooperation in some of the Conferences organised in other years (Martorell et al., 2009; Bris et al., 2010; Berenguer et al., 2011).

A short series of biannual national conferences associated with the Portuguese ESRA chapter was conducted on Analysis and Management of Risk, Safety and Reliability from 2005 to 2012, leading to 4 sets of proceedings in Portuguese.

The Congress of the International Maritime Association of the Mediterranean (IMAM) was organised in Lisbon in 2005 (Guedes Soares et al. 2005), which was the first time the proceedings were published in book format. This was done in most of the following Congresses (Guedes Soares & Kolev, 2008; Rizzuto & Guedes Soares, 2012; Guedes Soares & Lopez Peña, 2014; Georgiev & Guedes Soares, 2020), also following the conference that was organised in Lisbon again (Guedes Soares & Teixeira 2018).

Another series of Conferences was the biannual series jointly organised by IST and the Portuguese Association of Engineers, which started in 1987, and initially, yearly Conferences were organised, but after some years, it stabilised in biannual conferences. In total, 16 books were edited in Portuguese with the proceedings of those conferences. In 2011 those conferences became international, leading to the Maritime Engineering & Technology (MARTECH) series, which have been organised biannually (Guedes Soares et al. 2012, Guedes Soares & Santos, 2015, 2016, 2018).

With the interest in renewable energies offshore and the priority that was given to research in this subject, a new series of Conferences on Renewable Energies Offshore (RENEW) has been initiated at IST in 2014 and has continued biannually (Guedes Soares, 2015, 2016, 2019, 2021).

6 NATIONAL AND INTERNATIONAL EVALUATION

The evaluation of work by peers is the normal process in academic environments, and thus, while one can describe the research activities developed, their significance can only be independently ascertained by the opinion of others. Therefore, it is worth mentioning some of the evaluations conducted on a national and international scale.

The Portuguese Foundation for Science and Technology (FCT) is responsible for funding the research centres directly and indirectly through competitive calls of research projects, research contracts and scholarships. Periodically FCT promotes the external evaluation of research centres, which is carried out by panels of foreign scientists.

In the evaluation conducted in 2014, quantitative measures have been used to characterise the performance of the centres in which bibliographic production and citation levels were taken into consideration across the various scientific areas.

In that evaluation, CENTEC was considered a multidisciplinary centre covering the areas of Mechanical Engineering and Engineering Science and Marine Sciences and Technologies and was evaluated as Excellent with a total of 24/25 points.

A selection of the Evaluation Panel comments reads as: "The unit has a unique position in Portugal and is one of the leading centres in its field worldwide. It has contributed very significantly to the high international ranking of IST. The Centre demonstrates a high scientific impact and is highly relevant to economic development. It is clear that the Centre is well-run and has clear strategic goals. The site visit confirmed that the culture of the Centre is forward-thinking and supportive of its members."

CENTEC was classified 1st among the 14 Centres of Portugal in Mechanical Engineering & Engineering Science and 3rd among the 63 Centres of Portugal in Engineering Sciences.

CENTEC was classified 2nd among the 6 Centres of Portugal in Marine Sciences & Technologies and 2nd among the 45 Centres of Portugal in Natural & Environmental Sciences.

In 2018, a new national evaluation exercise was conducted, CENTEC was again evaluated as Excellent although in a more qualitative way.

On an international level, the evaluations are mostly quantitative and objective. The first contact with results of international rankings was at a Workshop at NTNU commemorating the 70th birthday of Profs Faltinsen and Moan, where the Rector of the University announced the results of a study conducted by the Center for Measuring University Performance in the USA and the International Institute for Software Technology of The United Nations University, covering 250 disciplines, one of each Ocean Engineering.

The ranking criteria were the volume of total publications and their impact, based on SCOPUS in 2008-2011. Out of the 49 universities listed, the Technical University of Lisbon appeared in 2nd place after the Norwegian University of Science and Technology.

The next evaluation was the Shanghai Ranking of World Universities, which normally presented the ranking of universities for the broad area of Engineering, but has since 2017 started presenting results for the various branches of engineering. The ranking criteria was also based on publications but now using publications indexed in the Web of Science.

In the area of Marine & Ocean Engineering, the University of Lisbon has appeared in 3rd place, immediately after the Norwegian University of Science and Technology, while in the first place was the Shanghai Jiao Tong University. In the following years, the University of Lisbon maintained 2nd place in Europe consistently, although moving to 5th place internationally, while Chinese Universities, with their much larger manpower dimension, were moving up the ranking.

The most recent result is the Stanford World Ranking of Scientists (Ioannidis et al. 2020), which prepared the ranking based on the published work indexed in Scopus and made available the list of the 2% highest ranking scientists. It was interesting to note that CENTEC had two individual researchers among the 385 Portuguese scientists from all scientific areas on the list (1st Guedes Soares; 333rd Garbatov). In the area of Civil Engineering, which in this database includes the papers of Naval Architecture and Ocean Engineering, the same two individuals are included (1st Guedes Soares; 438th Garbatov) among the 900 scientists listed, which

were selected from a universe of 42,000 authors in that field.

7 CONCLUSIONS

An overview has been presented on the evolution of teaching and research in Naval Architecture and Ocean Engineering at IST, which is still the only University in Portugal dealing with this subject area. This has evolved from an initial phase in which the main concern was shipbuilding to progressively dealing with ship design, maritime transportation and ports, and ocean engineering. The simultaneous evolution of research was described as well as the development of CENTEC, the research centre through which the academic staff conducts its research and where the research students conduct their research towards their PhD degrees.

In general, one can say that teaching and research has gained international recognition and is in a leading position internationally.

ACKNOWLEDGEMENTS

The author acknowledges the dedication and hard work of the academic, research and administrative staff who have contributed to the development of Naval Architecture and Ocean Engineering at IST during the years. The students' involvement and dedication in the study and research, as well as in their professional work after graduation, have contributed to the continuous development of teaching and research and to the international recognition of the education provided.

This work contributes to the Strategic Research Plan of the Centre for Marine Technology and Ocean Engineering (CENTEC), which is financed by the Portuguese Foundation for Science and Technology under contract UIDB/UIDP/00134/2020.

REFERENCES

Berenguer, C., Grall A. & Guedes Soares C. (Eds.) 2012. *Advances in Safety, Reliability and Risk Management.* London, UK: Taylor & Francis Group.

Bris, R., Guedes Soares C. & Martorell S. (Eds.) 2010. Proceedings of the European Safety and Reliability Conference London, U.K. Taylor & Francis Group.

Campos, R.M.; Alves, J.H.G.M.; Guedes Soares, C.; Guimarães, L.G. & Parente, C.E. 2018. Extreme wind-wave modeling and analysis in the South Atlantic Ocean. *Ocean Modelling.* 124:75–93.

Campos, R.M. & Guedes Soares C. 2016. Comparison of HIPOCAS and ERA wind and wave reanalysis in the North Atlantic Ocean. *Ocean Engineering.* 112:320–334.

Campos, R.M.; Guedes Soares, C.; Alves, J.H.G.M.; Parente, L.G. & Guimarães, L.G. 2019. Regional Long-Term Extreme Wave Analysis using Hindcast

Data from the South Atlantic Ocean. *Ocean Engineering*, 179:202–212.

Castro-Santos, L.; Silva, D.; Bento, A.R.; Salvação, N. & Guedes Soares, C. 2020. Economic feasibility of floating offshore wind farms in Portugal. *Ocean Engineering*. 207:107393.

Chen, B.Q. & Guedes Soares, C. 2016. Effects of plate configurations on the weld induced deformations and strength of fillet-welded plates. *Marine Structures*. 50:243–259.

Chojaczyk, A.A.; Teixeira, A.P.; Neves, L.C.; Cardoso, J.B. & Guedes Soares, C. 2015. Review and application of Artificial Neural Networks models in reliability analysis of steel structures. *Structural Safety*. 52:78–89.

Corak, M.; Parunov, J. & Guedes Soares, C. 2015. Probabilistic load combination factors of wave and whipping bending moments. *Journal of Ship Research*. 59(1):11–30.

Cubells, A.; Garbatov, Y. & Guedes Soares, C. 2014. Photogrammetry measurements of initial imperfections for the ultimate strength assessment of plates. *International Journal of Maritime Engineering*. 156 (Part A4):A-291 - A-302.

Diaz, H.M. & Guedes Soares, C. 2020a. Review of the current status, technology and future trends of offshore wind farms. *Ocean Engineering*. 209: 107381.

Diaz, H.M. & Guedes Soares, C. 2020b. An integrated GIS approach for site selection of floating offshore wind farms in the Atlantic Continental European coastline. *Renewable and Sustainable Energy Reviews*. 134: 110328

Ewans, K.C.; Bitner-Gregersen, E. & Guedes Soares, C. 2006. Estimation of Wind-Sea and Swell Components in a Bimodal Sea State. *Journal of Offshore Mechanics and Arctic Engineering*. 128(4):265–270.

Fonseca, N. & Guedes Soares, C. 1998. Time-Domain Analysis of Large-Amplitude Vertical Ship Motions and Wave Loads. *Journal of Ship Research*. 42(2):139–153.

Garbatov, Y. 2020. Special Issue: Carlos Guedes Soares Honoring Symposium. *Journal of Offshore Mechanics and Arctic Engineering*. 142:030301.

Garbatov, Y.; Teixeira, A.P. & Guedes Soares, C. 2004. Fatigue Reliability Assessment of a Converted FPSO Hull. *Proceedings of the OMAE Specialty Conference on Integrity of Floating Production, Storage & Offloading (FPSO) Systems;* Houston, TX., ASME paper OMAE2004-FPSO'0035.

Gaspar, B.; Teixeira, A.P. & Guedes Soares, C. 2017. Adaptive surrogate model with active refinement combining Kriging and a trust region method. *Reliability Engineering and System Safety*. 165:277–291.

Georgiev, P. & Guedes Soares C. (Eds.) 2020. *Sustainable Development and Innovations in Marine Technologies*. Taylor & Francis.

Goncalves, M.; Martinho, P. & Guedes Soares, C. 2014. Wave energy conditions in the western French coast. *Renewable Energy*. 62:155–163.

Gordo, J.M. & Guedes Soares, C. 2009. Tests on Ultimate Strength Hull Box Girders Made of High Tensile Steel. *Marine Structures*. 22(4):770–790.

Guedes Soares, C. (Ed.) 1997. *Advances in Safety and Reliability*. London: Pergamon.

Guedes Soares, C. 2000. Probabilistic Based Models for Coastal Studies. *Coastal Engineering*. 40(4):279–283.

Guedes Soares, C. 2008. Hindcast of Dynamic Processes of the Ocean and Coastal Areas of Europe. *Coastal Engineering*. 55(11):825–826.

Guedes Soares, C. (Ed.) 2010. *Safety and Reliability of Industrial Products, Systems and Structures (SAFEREL-NET)*. London, U. K.: Taylor & Francis Group.

Guedes Soares, C. (Ed.), 2015, *Renewable Energies Offshore*, Taylor & Francis Group, London, UK.

Guedes Soares, C. (Ed.), 2016. *Progress in Renewable Energies*, Taylor & Francis Group, London, UK.

Guedes Soares, C. (Ed.) 2019. *Advances in Renewable Energies Offshore*, Taylor & Francis Group, London, UK.

Guedes Soares, C. 2019a Special Issue Honoring Prof. Torgeir Moan. *Journal of Offshore Mechanics and Arctic Engineering*. 141(3):030301

Guedes Soares, C. (Ed.) 2021. *Developments in Renewable Energies Offshore*, Taylor & Francis Group, London, UK.

Guedes Soares, C.; Cherneva, Z. & Antão, E. 2003. Characteristics of Abnormal Waves in North Sea Storm Sea States. *Applied Ocean Research*. 25(6):337–344.

Guedes Soares, C. & Das P.K. (Eds.) 2007. *Advancements in Marine Structures*. London, U.K.: Taylor & Francis Group.

Guedes Soares, C & Das P.K. (Eds.) 2009. *Analysis and Design of Marine Structures*. London, U.K.: Taylor & Francis Group.

Guedes Soares, C. & Das, P.K. 2008. Special Issue Douglas Faulkner Honouring Symposium. *Journal of Offshore Mechanics and Arctic Engineering*. 130:020201.

Guedes Soares, C. Dejhalla R. & Pavletic D. (Eds.) 2015. *Towards Green Marine Technology and Transport*. London, UK: Taylor & Francis Group.

Guedes Soares, C., Eatock-Taylor, R. & Ewans, K.C. 2015. Safe offloading from floating LNG platforms. *Applied Ocean Research*, 51, 252–254.

Guedes Soares, C.; Fonseca, N. & Pascoal, R. 2008. Abnormal Wave Induced Load Effects in Ship Structures. *Journal of Ship Research*. 52(1):30–44.

Guedes Soares, C.; Fonseca, N.; Pascoal, R.; Clauss, G.F.; Schmittner, C.E. & Hennig, J. 2006. Analysis of Design Wave Loads on a FPSO Accounting for Abnormal Waves. *Journal of Offshore Mechanics and Arctic Engineering*. 128(3):241–247.

Guedes Soares, C. & Fricke W. (Eds.) 2011. *Advances in Marine Structures*. London, U.K.: Taylor & Francis Group.

Guedes Soares, C. & Garbatov, Y. 1996. Fatigue Reliability of the Ship Hull Girder Accounting for Inspection and Repair. *Reliability Engineering and System Safety*. 51 (3):341–351.

Guedes Soares, C. & Garbatov Y. (Eds.) 2017. *Progress in the Analysis and Design of Marine Structures*. London, UK: Taylor & Francis Group.

Guedes Soares, C. Garbatov Y. & Fonseca N. (Eds.) 2005. *Maritime Transportation and Exploitation of Ocean and Coastal Resources*, London, U. K.: Francis and Taylor Group.

Guedes Soares, C., Garbatov, Y., Fonseca, N. & Teixeira, A.P. (Eds.), 2011. *Marine Technology and Engineering*, Taylor & Francis Group, London, UK.

Guedes Soares, C., Garbatov, Y., Sutulo, S. & Santos, T.A. (Eds.), 2012. *Maritime Engineering and Technology*, Taylor & Francis Group, London, UK.

Guedes Soares, C. & Kolev P., (Eds.) 2008. *Maritime Industry, Ocean Engineering and Coastal Resources*. London, U. K.: Taylor & Francis Group.

Guedes Soares, C. & Lopez Peña F., (Eds.) 2014. *Developments in Maritime Transportation and Exploitation of Sea Resources.* Taylor & Francis Group, London, UK.

Guedes Soares, C. & Parunov J. (Eds.) 2010. *Advanced Ship Design for Pollution Prevention.* London, U.K.: Taylor & Francis Group.

Guedes Soares, C. & Romanoff, J. (Eds.) 2013. *Analysis and Design of Marine Structures.* Taylor & Francis, Group.

Guedes Soares, C. and Santos T.A. (Eds.), 2015. *Maritime Technology and Engineering*, Taylor & Francis Group, London, UK.

Guedes Soares, C. & Santos T.A. (Eds.), 2016. *Maritime Technology and Engineering 3*, Taylor & Francis Group, London, UK.

Guedes Soares, C. & Santos T.A. (Eds.) 2018. *Progress in Maritime Technology and Engineering*, Taylor & Francis Group, London, UK,

Guedes Soares, C. & Shenoi R.A. (Eds.) 2015. *Analysis and Design of Marine Structures.* London, UK: Taylor & Francis Group.

Guedes Soares, C. & Teixeira A.P. (Eds.) 2018. *Maritime Transportation and Harvesting of Sea Resources.* Taylor & Francis.

Guedes Soares, C. & Zio E. (Eds.) 2006. *Safety and Reliability for Managing Risk.* London, U.K.: Taylor & Francis Group.

Guedes Soares, C.; Dogliani, M.; Ostergaard, C.; Parmentier, G. & Pedersen, P.T. 1996. Reliability-Based Ship Structural Design. *Transactions of the Society of Naval Architects and Marine Engineers* (SNAME). 104: 357–389.

Guedes Soares, C.; Weisse, R.; Alvarez, E. & Carretero, J. C. 2002. A 40 Years Hindcast of Wind, Sea Level and Waves in European Waters. *Proceedings of the 21st International Conference on Offshore Mechanics and Arctic Engineering (OMAE 2002)*; Oslo, Norway. New York, USA: ASME paper OMAE2002–28604.

Ioannidis, J.P.A.; Boyack, K.W. & Baas, J. (2020) Updated science-wide author databases of standardized citation indicators. *PLoS Biol* 18(10): e3000918. doi: 10.1371/journal.pbio.3000918

Kang, J.C.; Sobral, J. & Guedes Soares, C. 2019. Review of condition-based maintenance strategies for offshore wind energy. *Journal of Marine Science and Application*; 18(1):1–16.

Klein M.; Clauss, G.F.; Rajendran, S.; Guedes Soares, C. & Onorato, M. 2016. Peregrine breathers as design waves for wave-structure interaction. *Ocean Engineering.* 128: 199–212.

Lopatoukhin, L.; Boukhanovsky, A. & Guedes Soares, C. 2005. Forecasting and Hindcasting the Probability of Freak Waves Occurrence. Guedes Soares, C., Garbatov Y. & Fonseca N., (Eds.). *Maritime Transportation and Exploitation of Ocean and Coastal Resources.* London, U. K.: Francis & Taylor Group; 1075–1080.

Mantari, J.L. & Guedes Soares, C. 2014. Optimized sinusoidal higher order shear deformation theory for the analysis of functionally graded plates and shells. *Composites Part B.* 56:126–136.

Mantari, J.L.; Oktem, A.S. & Guedes Soares, C. 2012a. A new trigonometric shear deformation theory for isotropic, laminated composite and sandwich plates. *International Journal of Solids and Structures;* 49(1):43–53.

Mantari, J.L.; Oktem, A.S. & Guedes Soares, C. 2012b. A new higher order shear deformation theory for sandwich and composite laminated plates. *Composites Part: B.* 43(3):1489–1499.

Martorell, S. Guedes Soares C. & Barnett J. (Eds.). 2009. *Safety, Reliability and Risk Analysis: Theory, Methods and Applications.* London, UK: Taylor & Francis Group.

Parunov, J. & Guedes Soares C. (Eds.) 2019. *Trends in Analysis and Design of Marine Structures.* London, UK: Taylor & Francis Group.

Parunov, J.; Corak, M.; Guedes Soares, C.; Jafaryeganeh, H.; Kalske, S.; Lee, Y.W.; Liu, S.; Papanikolaou, A.; Prentice, D.; Prpic-Oršic, J.; Ruponen, P. & Vitali, N. 2020. Benchmark study and uncertainty assessment of numerical predictions of global wave loads on damaged ships. *Ocean Engineering.* 197:106876.

Perera, L.P.; Rodrigues, J.M.; Pascoal, R. & Guedes Soares, C. 2012. Development of an onboard decision support system for ship navigation under rough weather conditions. Rizzuto, E. & Guedes Soares C., (Eds.) *Sustainable Maritime Transportation and Exploitation of Sea Resources.* Taylor and Francis Group; 837–844.

Rajendran, S.; Fonseca, N.; Guedes Soares, C.; Clauss, G. F. & Klein, M. 2011. Time Domain Comparison with Experiments for Ship Motions and Structural Loads of a Containership in Abnormal Waves. *Proceedings of the 30th International Conference on Ocean, Offshore and Arctic Engineering (OMAE 2011)*; Rotterdam, The Netherlands. New York, USA: ASME paper OMAE2011–50316.

Ribeiro e Silva, S. & Guedes Soares, C. 2013. Prediction of parametric rolling in waves with a time domain non-linear strip theory model. *Ocean Engineering.* 72:453–469.

Ringsberg, J.W.; Amdahl, J.; Chen, B.Q.; Cho, S.R.; Ehlers, S.; Hu, ZQ.; Kõrgesaar, M.; Liu, B.; Nicklas, K.; Parunov, J.; Samuelides, M.; Guedes Soares, C.; Tabri, K.; Quinton, B. W.; Yamada, Y. & Zhang, SM. 2018. MARSTRUCT benchmark study on collision simulations. *Marine Structures.* 59:142–157.

Rizzuto, E. & Guedes Soares, C. (Eds.) 2012. *Sustainable Maritime Transportation and Exploitation of Sea Resources.* London, UK: Taylor and Francis Group.

Rusu, E. & Guedes Soares, C. 2012a. Wave Energy Pattern around the Madeira Islands. *Energy.* 45(1):771–785.

Rusu, L. & Guedes Soares, C. 2012b. Wave Energy Assessments in the Azores Islands. *Renewable Energy.* 45:183–196.

Saad-Eldeen, S.; Garbatov, Y., & Guedes Soares, C. 2011. Corrosion Dependent Ultimate Strength Assessment of Aged Box Girders Based on Experimental Results. *Journal of Ship Research.* 55(4):289–300.

Saad-Eldeen, S.; Garbatov, Y., & Guedes Soares, C. 2013. Experimental assessment of corroded steel box-girders subjected to uniform bending, *Ships & Offshore Structures.* 8(6):653–662.

Santos, T.A. & Guedes Soares, C. 2008. Study of Damaged Ship Motions Taking Into Account Floodwater Dynamics. *Journal of Marine Science and Technology.* 13(3):291–307.

Schellin, T.E.; Beiersdorf, C.; Chen, X.-B.; Fonseca, N.; Guedes Soares, C.; Loureiro, A.M.; Papanikolaou, A.; de Lucas, A. P. & Ponce Gomez, J.M. 2003. Numerical and Experimental Investigation to Evaluate Wave Induced Design Loads for Fast Ships. *Transactions of the Society of Naval Architects and Marine Engineers (SNAME);* 431–461.

Silva, D.; Rusu, E. & Guedes Soares, C. 2013. Evaluation of various technologies for wave energy conversion in the Portuguese nearshore. *Energies*. 6(3):1344–1364.

Skourup, J.; Sterndorff, M.J.; Smith, S.F.; Cheng, X.; Ahilan, R.V.; Guedes Soares, C. & Pascoal, R. 2004. Model Tests with an FPSO in Design Environmental Conditions. *Proceedings of the 23rd International Conference on Offshore Mechanics and Arctic Engineering (OMAE 2004)*; Vancouver, Canada. ASME paper OMAE2004-51618.

Sutulo, S.; Guedes Soares, C., & Otzen, J. F. 2012. Validation of Potential-Flow Estimation of Interaction Forces Acting upon Ship Hulls in Parallel Motion. *Journal of Ship Research*. 56(3):129–145.

Teixeira, A.P. & Guedes Soares, C. 2009. Reliability Analysis of a Tanker Subjected to Combined Sea States. *Probabilistic Engineering Mechanics*. 24(4):493–503.

Teixeira, A. P.; Guedes Soares, C.; Chen, N.-Z., & Wang, G. 2013. Uncertainty analysis of load combination factors for global longitudinal bending moments of double hull tankers. *Journal of Ship Research*. 57(1):42–58.

Uzunoglu, E. & Guedes Soares, C. 2020. Hydrodynamic design of a free-float capable tension leg platform for a 10 MW wind turbine. *Ocean Engineering*. 197:106888

Ventura, M. & Guedes Soares, C. 2012. Surface Intersection in Geometric Modeling of Ships' Hulls. *Journal of Marine Science and Technology*. 17(1):114–124.

Vettor, R. & Guedes Soares, C. 2016. Development of a ship weather routing system. *Ocean Engineering*. 123:1–14.

Wang, S. & Guedes Soares, C. 2016a. Experimental and numerical study of the slamming load on the bow of a chemical tanker in irregular waves. *Ocean Engineering*. 111:369–383.

Wang, S. & Guedes Soares, C. 2016b. Stern slamming of a chemical tanker in irregular head waves. *Ocean Engineering*. 122:322–332

Wang, S.; Zhang, HD. & Guedes Soares, C. 2016. Slamming occurrence for a chemical tanker advancing in extreme waves modelled with the nonlinear Schrodinger equation. *Ocean Engineering*. 119:135–142.

Wu, B.; Yan, X.P.; Wang, Y. & Guedes Soares, C. 2017. An evidential reasoning-based CREAM to human reliability analysis in maritime accident process. *Risk Analysis*. 37 (10):1936–1957.

Wu, B.; Yip, T.L.; Yan, X.P. & Guedes Soares, C. 2019. Fuzzy logic based approach to define risk factors for ship-bridge collision alert system. *Ocean Engineering*, 187:106152

Wu, B.; Zong, L.K.; Yan, X.P. & Guedes Soares, C. 2018. Incorporating evidential reasoning and TOPSIS into group decision-making under uncertainty for handling ships without command. *Ocean Engineering*. 164:590–603

Zhang, HD.; Cherneva, Z.; Guedes Soares, C. & Onorato, M. 2014. Modeling Extreme Wave Heights from Laboratory Experiments with the Nonlinear Schrödinger Equation. *Natural Hazards and Earth System Sciences*. 14(4):959–968.

Zhou, XQ.; Sutulo, S., & Guedes Soares, C. 2016. Ship-Ship hydrodynamic interaction in confined waters with complex boundaries by a o Panelled Moving Patch Method. *International Journal of Maritime Engineering*. 158(Part A1):A–21–A30.

Maritime transportation and ports

Developments in Maritime Technology and Engineering – Guedes Soares & Santos (eds)
© 2021 Copyright the Author(s), ISBN 978-0-367-77376-2

Life cycle assessment of a Ro-Ro ship for operation phase

D.D. Çiçek & M.C. Sağır
Hydraulics Laboratory, Civil Engineering Department, Middle East Technical University, Ankara, Turkey

E. Oğuz
Hydraulics Laboratory and Center for Wind Energy Research (METUWIND), Civil Engineering Department, Middle East Technical University, Ankara, Turkey

ABSTRACT: The aim of this research is to provide insight about the environmental burdens of a conventional Ro-Ro type ferry currently in use for policy makers, ship owners and the public. Since Turkey is currently in the transition phase of implementing the International Maritime Organization (IMO) regulations (following MARPOL Annex VI Rule 14: 0.5 % of sulfur content), the findings of this paper will aid in providing new information for establishing future decision-making processes. The life cycle assessment (LCA) approach is used to determine the long-term effects of the case ship focusing on the operation phase. Since the implementing new technologies to reduce the environmental burdens and increase the fuel efficiency of the Ro-Ro type ships will be important in the directorial processes for the shipping industry due to upcoming policies of Turkish Government, the findings of this research were also compared to a hybrid ship which is operating in Scotland.

1 INTRODUCTION

Today's world is facing problems arising from excessive fossil fuel consumption. Fossil fuel consumption has increased the concentration of greenhouse gasses (GHG) in the atmosphere over the years. In fact, the concentrations of carbon dioxide, methane, and nitrous oxides have exceeded the pre-industrial levels respectively by about 40%, 150%, and 20%, due to human activities(IPCC 2014). Meanwhile, the energy demand and energy consumption of the world has drastically increased in recent years. In 2018, global energy consumption increased by 2.9% and the power sector's contribution of carbon emissions increased by 2.7% in 2018(BP 2019).

The increased concentration of GHG in the atmosphere is the main reason for global warming and for the deposition of GHG in sea water and the earth; the latter of which is one of the leading contributors of acidification and eutrophication. The global economy and millions of people suffer from the side effects caused by global warming as well as other issues caused by the onset of GHG emissions.

The 2014 California drought presents a direct example of the arising problem. In 2014, California experienced the highest annual average temperatures with 10 degrees between November to April (Agha-Kouchak et al. 2014). The drought caused a loss of 2.2 billion US dollars to the state of California (Howitt et al. 2014). This climate disaster shows only a small fraction of the total economic loss due to climate change. Between 2017 and 2018, the damage of the weather and climate disasters resulted in $411.3 billion in economic loss to the US economy (NCEI 2018).

There are also indirect effects of the GHG emissions. Doney, et al. (2009) conducted inclusive research on ocean acidification and explained the underlying mechanism of the acidification and impact on calcifying organisms. Due to the excessive amount of dissolved CO_2, oceans have become increasingly acidic environments for calcifying organisms. The decrease in pH causes harmful effects on the skeletal formation of marine organisms including plankton, corals and coralline algae (Doney et al. 2009). Moreover, the Great Barrier Reef was reduced by 21% between 1988-2003(Cooper et al. 2008).

Air pollution is another important issue that arises from burning fossil fuels. Apart from GHG emissions, burning processes release particulate matter (PM) or gaseous precursors of particulate matter. Exposure to ambient air pollution is directly linked to mortality and morbidity and it increases the risk of cardiovascular and respiratory diseases and lung cancer (Brook et al. 2010, Cohen et al. 2017, Health Organization & Office for Europe 2013, US EPA 1979). Moreover, the analysis of data from the Global Burden of Diseases Study 2015 showed that long term exposure to ambient PM lead to 4.2 million deaths and 103.1 lost years of healthy life in 2015 and was the fifth-ranking global mortality risk factor in 2015 (Cohen et al. 2017).

DOI: 10.1201/9781003216582-2

Emerging consequences of climate change caused by the adverse effects of GHG emissions has garnered public attention and forced international authorities and governments to act against climate change. The Kyoto Protocol and Paris Agreement are accepted by most countries in the world which established limiting targets for global GHG emissions. Consequently, most governments and international agencies set regulations to limit GHG emissions. For instance, The European Commission set exorbitant targets to reach until 2030. The two key points of this framework are cutting GHG emissions by 40% from 1990's level and increasing energy efficiency by 32.5% (EC 2019).

Most parties of the Paris Agreement consider the transport sector to be one of the key points in achieving the target emissions as the transport sector produces 23% of global CO_2 emissions (International Energy Agency 2017). In fact, 81% of Nationally Determined Contributions (NDCs) mentioned the transport sector as part of their effort to reduce CO_2 while 61% of NDCs suggested measures to be implemented in the transport sector to reduce CO_2 emissions (International Transport Forum 2018).

Established emission targets forced the transportation sector to increase fuel efficiency to reach the target emission values. Most of the automobile industry focused on hybrid and electric solutions to achieve greener vehicles. Market shares of Electric Vehicles (EVs) has escalated in recent years. The number of electric cars surpassed 5.1 million in 2018, an increase of 2 million when compared to that of 2017 (International Energy Agency 2019).

Another key industry to realize target emissions is the shipping industry. Since shipping transfers 80-90% of the volume of international trade (UNCTAD 2018), it is mainly based on marine transportation. Due to its sheer size, according to The Third IMO GHG Study 2014, International Shipping emitted 2.2% of the global CO2 emissions in 2012 and by 2050, International Maritime Organization (IMO) estimated that CO_2 emissions from international shipping may show an increase between 50% and 250% (Ebi et al. 2014). Moreover, as shown in Figure 1,

Maritime transportation holds 13.6% of total transport greenhouse gas emissions in the EU (EEA 2018).

In order to forestall the increase in the GHG emissions of global shipping, the International Maritime Organization (IMO) established forceful environmental regulations. (Julian 2000). IMO also set two new measures for the shipping sector which are the Energy Efficiency Design Index (EEDI) for newly built ships and the Ship Energy Efficiency Management Plan (SEEMP) for all ships with the MARPOL Annex VI (IMO 2015). Although a number of research and development studies are funded by several organizations, the shipping industry still needs to adopt new greener solutions to decrease the environmental burdens of conventional ships.

The aim of this paper is to focus on the environmental impact of an in-service Ro-Ro type ship in Turkey by applying a bottom up approach and Life Cycle Assessment (LCA) methodology. In recent decades, the transportation need for passengers and goods has increased drastically in Turkey due to an increase in population. Turkey could not prevent the environmental effects caused by the fossil fuel dependent transportation sector while meeting the increased demand for transportation.

The data provided by the European Environment Agency (EEA) for changes in total greenhouse gas emissions from transport between 1990-2016 in Figure 2, clearly indicates that Turkey has the highest increase in emission rates over these years (EEA, 2018b). This increase mainly resulted from the late implementation of regulations concerning GHG emissions.

Inland waterway transportation holds an important role in Turkey's transportation network. In 2017, Ro-Ro type ships carried 12,638,289 cars with an increase of 82% from 2005(DTGM 2017). However, Turkey has still not been able to implement and integrate the MARPOL regulations into Turkish shipping policies. Yet, Turkey has taken the necessary first steps towards the implementation of the MARPOL regulations. In fact, the Chamber of Shipping has announced that all

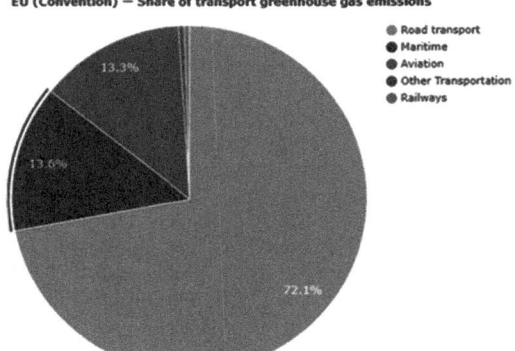

Figure 1. Share of transport GHG emissions in EU.

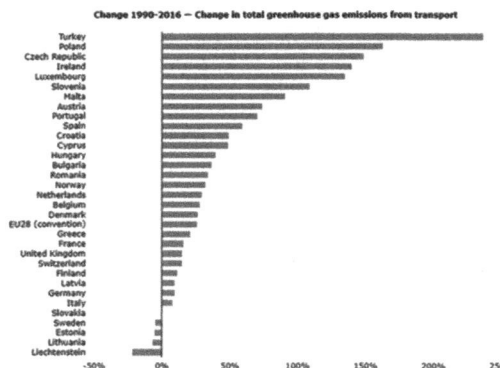

Figure 2. Change in total GHG emissions from the transportation sector between 1990-2016.

ships that operate in Turkey's jurisdiction should utilize fuel that has a maximum 0.5% sulfur content following MARPOL Annex VI Rule 14 regulation starting on January 1st, 2020, as decreed by International Maritime Organization (SALİHOĞLU 2020). The projected SO_2 content of marine fuels according to MARPOL Annex VI Rule 14 is shown in Figure 3

This paper is organized as follows. Section 1.1 gives a brief literature review on LCA applications for many engineering purposes. Afterwards, LCA methodology and LCA model of the case ship is presented in Section 2. Then, the results from this work are discussed in detail in Section 3. Finally, a brief summary of the main results from this work are provided in Section 4 and suggestions are made for future study.

1.1 Literature review

LCA methodology is a convenient tool for investigating the environmental burdens of any process or good. Generally, LCA study examines the life cycle of a product or process from the extraction of raw material, that is needed for the production steps of the product, to the manufacturing process, the use phase, and the disposition or recycling phase.

Many industries have been integrating or considering the integration of LCA methodology into their production plans especially in the EU since, the new regulations and the certification programs force many industries to reduce the environmental effects of both the production phase and the operation phase of goods.

Eco-design principles, which take into account not just the cost of production but also the environmental burdens of the goods, are getting more popular among manufacturers. To illustrate, Favi et al. (2019) taught eco-design principles by using LCA methodology to an Italian coffee machine producer's employee. Together, they made a few optimizations and replacements for an existing project and they managed to reduce Global Warming Potential (GWP) by 25% and energy consumption by 35% while keeping the cost in allowance of plus 5%. Vinci et al. (2019) used LCA methodology and eco-design principles to improve the efficiency of the glass manufacturing process. By re-using and re-directing the heated air in the furnace, they reduced costs by 20% and GHG emission by 5%. These two examples

show the effectiveness of the application of LCA methodology in completely different industries.

LCA methodology can also be used for evaluations of the energy efficiency of different industrial solutions. Chàfer et al (2019) questioned the efficiency of pneumatic municipal waste collection systems in terms of reduction in adverse environmental effects. They compared this system with traditional truck collection by adopting LCA methodology. They concluded that the environmental burdens of this system mainly depend on the source of electricity that is used in the operation phase.

Similarly, the benefits of the LCA methodology can be used by the transportation sector in which the interest in LCA applications has increased over the recent years. The automobile industry frequently applies LCA approach to analyze the different possible solutions for achieving greener vehicles. For instance, Dhingra & Das (2014) analyzed the environmental impacts of two new trends to reduce fuel consumption in the automotive sector, namely downsized engines and lightweight engines. Their findings indicated that both downsized and lightweight engines reduce fuel consumption. The authors recommended to use a combination of light weighting and downsizing in order to improve economic feasibility and fuel saving. Benajes et al. (2020) compared conventional diesel car with different types of hybrid cars, namely Mild (MHEV), Full (FHEV) and Plug-in (PHEV) hybrid electric vehicles, by considering different usage scenarios. Their findings showed that plug-in hybrid car reduces the carbon dioxide emission by 30% compared to that of the conventional one. The plug-in hybrid car also accomplished the ultra-low NO_x emission target.

The shipping industry has also started to use LCA methodology to evaluate and to improve the energy efficiency of the existing and new ships. Wang et al. (2018) investigated the optimal interval of re-coating and steel renewal for a ship hull by applying LCA methodology. The LCA study revealed the importance of an optimal hull maintenance strategy to reduce ship's cost and emissions during the operation stage. Porras et al. (2018) integrated LCA approach into retrofitting design stage of ships in order to improve effectiveness of adopted new technologies such as scrubbers. These examples show that by the integration of LCA methodology into the maintenance strategies of the ship operators, both the environmental impacts and the fuel consumption of an already existing ship can also be reduced. Due to emission limiting regulations of the IMO, the shipping industry has started to seek new greener propulsion systems in order to decrease the emissions of new ships.

LCA methodology has been proven to be a convenient tool for measuring the superiority of the different engineering solutions in terms of environmental impacts. Hence, the researchers commonly use LCA methodology to analyze the effectiveness of different propulsion systems and different fuels in reducing GHG emissions. Two extensive studies,

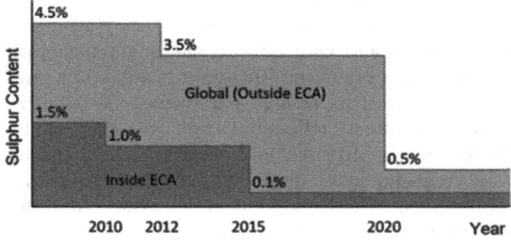

Figure 3. Projected SO_2 content of marine fuels (Hwang et al. 2019).

J. Ling-Chin & Roskilly, (2016) show the environmental benefits of adopting hybrid engines for the shipping industry by implementing LCA methodology. Jeong et al (2018) used LCA methodology to determine the optimal propulsion system for marine vessels. A hybrid engine, a diesel engine and a diesel marine engine were compared in the scope of their study and they concluded that the hybrid engine is far more superior to others in terms of environmental and cost impacts. Laugen (2014) conducted a study in order to compare and assess the environmental effects of LNG and HFO as marine fuels. The research showed that LNG is more environmentally friendly than HFO in terms of GWP. With a reduction in the methane slip in the supply chain, environmental damage of using LNG can be improved.

The abovementioned studies prove the effectiveness of the LCA applications. LCA studies help engineers and managers to understand the holistic environmental effects of a good or a process. By adopting LCA methodology and eco-design principles, reduction in both cost and emissions can be achieved. For these reasons, the popularity of LCA studies has increased. Due to increased demand and interest on life cycle assessment studies, software that is specialized on the subject of LCA have been developed. EU funded many research projects to develop a virtual prototyping system for ships that is combined with LCA approach(EU Commission, HOLISHIP, LINCOLN 2016, Volbeda(FERG) 2017). LCA study can be conducted with homemade tools which are developed on special industrial processes or it can be done by use of comprehensive programs such as GaBi, SimaPro and COMPASS. These programs provide large databases to users about different industrial processes. GaBi and SimaPro are the most commonly used software in this area. Generally, both of these programs give the same results with reasonable differences (Herrmann & Moltesen 2015, Mojzes et al. 2012) However, in this study, GaBi is chosen because it provides a large database and friendly user interface. A homemade MATLAB tool is developed in order to assess the environmental effects of the case ship.

2 METHODOLOGY

The above-mentioned side effects of the environmental burdens of the shipping industry and the obligation to comply with regulations for both international and national shipping has brought to attention the necessity of evaluating emission rates for ships. Life cycle assessment methodology became a prominent tool for forecasting the environmental burdens. In order to create a global standardized process for LCA study, International Organization for Standardization (ISO) established international standards under the names of ISO 14040 and ISO 14044 (ISO 14043, 2002).

According to these standards, LCA is divided into four stages; the definition of goal and scope, inventory analysis, impact assessment and interpretation. The definition of goal and scope include the aim of life cycle assessment and the steps of the study which are referred as system boundary. Inventory analysis phase includes inputs and outputs of the system boundaries. The impact assessment phase evaluates the output of inventory analysis phase in terms of the global warming potential (GWP), the acidification potential (AP) and the eutrophication potential (EP). The last phase is a result of the inventory and impact assessment analysis. Applying LCA methodology helps ship owners, ship builders, and policy makers to make sustainable decisions both for the environment and resource management.

In general, life cycle assessment is a method which evaluates the environmental impact of a product in its whole life span. From the cradle to grave, ships' life span has four main stages; construction, operation, maintenance and dismantling. The construction phase includes hull construction and battery installation. The operation phase consists of transition, maneuvering and slip parts. The maintenance parts include hull and machinery maintenance. However, the most influenced stage on GHG emissions is the operation phase (Kameyama et al. 2005, Michihiro et al. 2007, Kjær et al. 2015, kavli et al. 2017, Wang et al. 2018). Therefore, this study is focused on the operation stage of the case ship.

2.1 Goal and scope of the study

In this study, the aim is to adopt LCA methodology in order to evaluate the environmental effects of in-service conventional Ro-Ro type ships in Turkey. This paper also attempts to assist policy makers and ship owners by providing a quantified environmental impact assessment of a Ro-Ro type ship Boundary setting and data quality requirement.

As stated in section 2.1, this study will only cover the operation phase of the life cycle of the case ship and the applied LCA methodology is based on a bottom-up approach. The bottom-up approach simplifies the system boundary investigation process and data collection process. It divides the whole life cycle of a process or production into sub-components which are referred to as unit processes. While adopting bottom-up approach into an LCA study, one should take into consideration the most influential process in the life cycle of a good or process in terms of its desired parameters. By investigating the most influential process, this study attempts to simplify system boundary setting and to avoid time and cost requirements for the data collection process. In order to achieve necessary simplification, some assumptions are made.

- Manufacturing process is not considered.
- Contribution of maintenance parts are negligible.

- Dismantling parts are not considered as a part of the life cycle assessment.
- It is assumed that the case ship will operate on the same route during its whole life span.
- Transportation of fuel for the ship is neglected.

2.1.1 Case ship description

In this research, a conventional Ro-Ro type ship is chosen, namely Alınteri 17, as a case ship. Alınteri 17 is a conventional diesel engine driven short route ferry which operates between Kilitbahir and Canak-kale. In Table 1, Technical details of the case ship are shown. In Figure 4, the route of the case ship is shown.

Table 1.   Specifications of the case ship.

Name	Alınteri 17
Gross weight	415 tons
Length	49,5 m
Breadth	13,5 m
Depth	3 m
Power	500 BHP-710 BHP-400 BHP
Built year	1993
Working Hour (Daily/ Weekly)	4 h/28 h

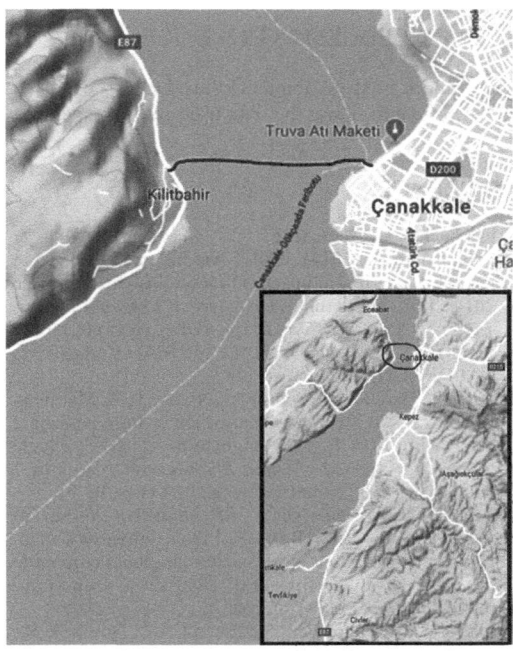

Figure 4. Route of the case ship.

2.2 Life cycle inventory assessment

LCA model is conducted in GaBi and in the home-made MATLAB tool. In the GaBi model, the database of GaBi was used in terms of specified boundary settings and assumptions. In the MATLAB tool, emissions factors, that is provided by Environment Protection Agency (EPA), was used to compute total CO_2 emissions from different fuels and CML's global warming potential factors was used to compute CO2-equivalent. For estimating acidification potential (AP) and eutrophication potential (EP) a combination of EPA's database and EEA's (European Environment Agency) database is used and like CO_2 -equivalent calculation CML's acidification potential factors and eutrophication potential factors used to compute SO_2-equivalent and PO_4-eqguivalent.(CML - Department of Industrial Ecology, n.d.). In GaBi, environmental effects of fuel production at refinery is automatically considered, however, in our MATLAB tool, this part is neglected.

3   RESULTS

In this section, the results of the LCA model is presented by using GaBi and the homemade MATLAB tool. The Environmental impacts of Alınteri 17 are visualized and simplified by converting different emissions into CO_2-Equivalent, SO_2-Equivalent and PO_4-Equivalent. GWP 100, AP and EP are taken as impact categories in the process of the LCA model.

Results of the LCA models is presented in Table 2, Table 3. The differences between the GaBi outputs and the MATLAB outputs are negligible since the difference between the results are less than 5%. However, determining the accurate result is not a simple task because in the GaBi model all the pre-defined ships are specified as having some type of mix-technology. When the GaBi emission data is processed in Excel or MATLAB, the results are greater than the GaBi results. Findings indicated that GaBi

Table 2.   Result of the MATLAB model.

Categories	kg CO_2-Equi.	kg SO_2-Equi.	kg PO_4-Equi.
GWP 100	2.11×10^7	-	-
AP	-	1.46×10^5	-
EP	-	-	2.96×10^4

Table 3.   Result of the GaBi model.

Categories	kg CO_2-Equi.	kg SO_2-Equi.	kg PO_4-Equi.
GWP 100	2.02×10^7	-	-
AP	-	1.33×10^5	-
EP	-	-	3.16×10^4

considers ships with some emission reducing technologies to provide users a mean environmental impact. However, both IMO and EEA recommend using Tier I procedure since the engine profile within the fleet was not available and the case ship was built in 1993. Therefore, in the MATLAB code, default emission factors that are based on fuel consumption is used. Also, the case ship uses diesel with a 2% sulfur content as fuel but, the maximum sulfur content of the pre-defined fuels is 0.08% in GaBi. Since, the sulfur content in MATLAB tool is taken as 2%, the MATLAB results of AP are more reliable.

3.1 Comparison of Alınteri 17 with a hybrid ship

Wang et al. (2018) conducted an LCA study on a hybrid ship that operates in Scotland. As seen in Table 1, Table 4, their length, breadth, and gross weights are close enough to make a valid comparison to analyze the benefits of replacing Alınteri 17 with a hybrid alternative. Since, the working hour of the ships are different, in order to make more accurate comparisons, LCA model of Alınteri 17 is modified to simulate the same conditions as the hybrid ship. For this reason, 10 working hours is used daily in the modified LCA model since the studied hybrid ship also had 10 daily working hours. The inorganic emissions to air during the operation phases of the ships are presented in, in terms of CO_2, CO, NO_x, SO_2.

When the same working hours are considered, Alınteri 17 has 4-5 times higher inorganic emissions to air compared to the hybrid alternative in terms of CO_2, CO and SO_2. Also, the NO_x emission of Alınteri 17 is 2 times higher than that of the hybrid one.

Wang et al. (2018) presented the results of their LCA models in terms of GWP 100 by considering the whole life span of the studied hybrid ship from cradle to grave. As mentioned in section 1.1, the researchers aimed to find the optimal hull coating interval for the hybrid ship. In the worst-case scenario, the environmental impact of the studied hybrid ship is determined around 1.55×10^7 kg CO_2-Equi. in terms of GWP 100. The environmental impact of Alınteri 17 in the modified LCA model is determined as 6×10^7 kg CO_2-Equi. in terms of GWP 100. Alınteri 17 has 4 times higher environmental effects at its operation

Table 4. The case ship description from the study of Wang et al. (2018).

Name	MV Hallaig
Gross weight	499 tons
Length	43,5 m
Breadth	12,5 m
Depth	3 m
Power	360 kW x 3
Built year	2012
Daily Working Hour	10 h

Table 5. Inorganic emissions to air during the operation phase.

Emission Flows	Inorganic emissions (kg)	
	Alınteri 17	The Hybrid Ship
CO_2	5.81×10^7	1.36×10^7
CO	1.87×10^5	3.1×10^4
NO_x	$7,21 \times 10^5$	3.36×10^5
SO_2	2.93×10^4	6.37×10^3

phase compared to the hybrid ship's environmental effect over the course of its whole life span.

4 DISCUSSION AND FUTURE RESEARCH

This paper investigated the environmental effects of an in-service Ro-Ro type ship in Turkey by using GaBi and MATLAB. The findings of this research presented a quantified data about the environmental impacts of the case ship. The comparison conducted with the hybrid ship indicate the importance of implementing the regulations set by IMO and the necessity for stricter regulations to be set. This study highlighted the significant importance of new greener propulsion technologies. Also, it is critical to carry out an extensive research considering a large fleet of ships in order to indicate the research impact.

ACKNOWLEDGEMENTS

The authors wish to thank Gestaş Deniz Ulaştırma Tic. A.Ş for providing the data used in this paper.

REFERENCES

AghaKouchak, A., Cheng, L., Mazdiyasni, O., & Farahmand, A. (2014). Global warming and changes in risk of concurrent climate extremes: Insights from the 2014 California drought. *Geophysical Research Letters*, *41*(24), 8847–8852. https://doi.org/10.1002/2014GL062308@10.1002/(ISSN)1944-8007.CALDROUGHT1.

Benajes, J., García, A., Monsalve-Serrano, J., & Martínez-Boggio, S. (2020). Emissions reduction from passenger cars with RCCI plug-in hybrid electric vehicle technology. *Applied Thermal Engineering*, *164*. https://doi.org/10.1016/j.applthermaleng.2019.114430.

BP. (2019). *CO2 emissions – BP Statistical Review of World Energy 2019*. Retrieved from https://www.bp.com/content/dam/bp/business-sites/en/global/corporate/pdfs/energy-economics/statistical-review/bp-stats-review-2019-co2-emissions.pdf.

Brook, R. D., Rajagopalan, S., Pope, C. A., Brook, J. R., Bhatnagar, A., Diez-Roux, A. V, … Kaufman, J. D. (2010). Particulate matter air pollution and cardiovascular disease: An update to the scientific statement from the american heart association. *Circulation*, Vol. 121, pp. 2331–2378. https://doi.org/10.1161/CIR.0b013e3181dbece1.

Chàfer, M., Sole-Mauri, F., Solé, A., Boer, D., & Cabeza, L. F. (2019). Life cycle assessment (LCA)of a pneumatic municipal waste collection system compared to traditional truck collection. Sensitivity study of the influence of the energy source. *Journal of Cleaner Production*. https://doi.org/10.1016/j.jclepro.2019.05.304.

CML - Department of Industrial Ecology. (n.d.). *CML-IA_aug_2016*. Retrieved from https://www.universiteitleiden.nl/en/research/research-output/science/cml-ia-characterisation-factors#downloads.

Cohen, A. J., Brauer, M., Burnett, R., Anderson, H. R., Frostad, J., Estep, K., … Forouzanfar, M. H. (2017). Estimates and 25-year trends of the global burden of disease attributable to ambient air pollution: an analysis of data from the Global Burden of Diseases Study 2015. *The Lancet*, *389*(10082), 1907–1918. https://doi.org/10.1016/S0140-6736(17)30505-6.

Commission, E. (n.d.-a). Lean innovative connected vessels | LINCOLN Project | H2020 | CORDIS | European Commission. Retrieved January 29, 2020, from https://cordis.europa.eu/project/id/727982.

Commission, E. (n.d.-b).Ship Lifecycle Software Solutions | SHIPLYS Project | H2020 | CORDIS | European Commission. Retrieved January 29, 2020, from https://cordis.europa.eu/project/id/690770.

Cooper, T. F., De'ath, G., Fabricius, K. E., & Lough, J. M. (2008). Declining coral calcification in massive Porites in two nearshore regions of the northern Great Barrier Reef. *Global Change Biology*, *14*(3), 529–538. https://doi.org/10.1111/j.1365-2486.2007.01520.x.

Dhingra, R., & Das, S. (2014). Life cycle energy and environmental evaluation of downsized vs. lightweight material automotive engines. *Journal of Cleaner Production*. https://doi.org/10.1016/j.jclepro.2014.08.107.

Doney, S. C., Fabry, V. J., Feely, R. A., & Kleypas, J. A. (2009). Ocean Acidification: The Other CO 2 Problem . *Annual Review of Marine Science*, *1*(1), 169–192. https://doi.org/10.1146/annurev.marine.010908.163834.

DTGM. (2017). *Deniz Ticareti 2017 İstatistikleri*. 1–114.

Ebi, K. L., Hallegatte, S., Kram, T., Arnell, N. W., Carter, T. R., Edmonds, J., … Luo, M. (2014). Third IMO Greenhouse Gas Study 2014. *International Maritime Organization (IMO)*, *122*(version 0.2), 327. https://doi.org/10.1007/s10584-013-0912-3.

EC. (2019). 2030 climate & energy framework. Retrieved December 22, 2019, from European Commission website: https://ec.europa.eu/clima/policies/strategies/2030_en.

EEA. (2018a). Greenhouse gas emissions from transport in Europe — European Environment Agency. *European Environmental Agency (EEA)*. Retrieved from https://www.eea.europa.eu/data-and-maps/indicators/transport-emissions-of-greenhouse-gases/transport-emissions-of-greenhouse-gases-12.

EEA. (2018b). Greenhouse gas emissions from transport in Europe — European Environment Agency. *European Environmental Agency (EEA)*. Retrieved from https://www.eea.europa.eu/data-and-maps/indicators/transport-emissions-of-greenhouse-gases/transport-emissions-of-greenhouse-gases-12.

Favi, C., Marconi, M., & Germani, M. (2019). Teaching eco-design by using LCA analysis of company's product portfolio: The case study of an Italian manufacturing firm. *Procedia CIRP*. https://doi.org/10.1016/j.procir.2019.01.032.

Health Organization, W., & Office for Europe, R. (2013). *HealtH effects of particulate matter*. Retrieved from http://www.euro.who.int/pubrequest.

Herrmann, I. T., & Moltesen, A. (2015). Does it matter which Life Cycle Assessment (LCA) tool you choose? - A comparative assessment of SimaPro and GaBi. *Journal of Cleaner Production*, *86*, 163–169. https://doi.org/10.1016/j.jclepro.2014.08.004.

HOLISHIP. (n.d.). Approach - Holiship. Retrieved January 29, 2020, from http://www.holiship.eu/approach/.

Howitt, R., Medellín-Azuara, J., Macewan, D., Lund, J., & Sumner, D. (2014). *Economic Analysis of the 2014 Drought for California Agriculture*. Retrieved from http://watershed.ucdavis.edu.

Hwang, S., Jeong, B., Jung, K., Kim, M., & Zhou, P. (2019). Life cycle assessment of lng fueled vessel in domestic services. *Journal of Marine Science and Engineering*, *7*(10), 1–24. https://doi.org/10.3390/jmse7100359.

IMO. (2015). *GHG emissions*. https://doi.org/10.1787/9789264227385-graph2-en.

International Energy Agency. (2017). *Tracking Clean Energy Progress 2017: Informing Energy Sector Transformations*. 116. Retrieved from www.iea.org/etp/tracking.

International Energy Agengy. (2019). Global EV Outlook 2019. In *Global EV Outlook 2019*. 10.1787/35fb60bd-en.

International Transport Forum. (2018). *Transport CO2 and the Paris Climate Agreement - Reviewing the Impact of Nationally Determined Contributions*. Retrieved from https://www.itf-oecd.org/transport-co2-paris-climate-agreement-ndcs.

IPCC. (2014). Climate Change 2014. In *Climate Change 2014: Synthesis Report*. https://doi.org/10.1017/CBO9781107415324.

ISO 14043. (2002). *International Standard Environmental Management - Life Cycle Assessment - Life Cycle Interpretation*. 2002.

Jeong, B., Oguz, E., Wang, H., & Zhou, P. (2018). Multi-criteria decision-making for marine propulsion: Hybrid, diesel electric and diesel mechanical systems from cost-environment-risk perspectives. *Applied Energy*, *230* (April), 1065–1081. https://doi.org/10.1016/j.apenergy.2018.09.074.

Julian, M. (2000). MARPOL 73/78: the International Convention for the Prevention of Pollution from Ships. *Maritime Studies*, *2000*(113), 16–23. https://doi.org/10.1080/07266472.2000.10878605.

Kameyama, M, Hiraoka, K., Sakurai, A., Naruse, T., & Tauchi, H. (2005). Development of LCA Software for Ships and LCI Analysis based on Acutual Shipbuilding and Operation. *Proceedings of the 6th International Conference on Ecobalance*, 159–162. Retrieved from http://202.241.16.30/env/lca/Paper/pdf/44.pdf.

Kameyama, Michihiro, Hiraoka, K., & Tauchi, H. (2007). Study on Life Cycle Impact Assessment for Ships. *Relatório Geral Instituto de Pesquisa Marítima*, *7*(3), 133–143.

Kjær, L. L., Pagoropoulos, A., Hauschild, M., Birkved, M., Schmidt, J. H., & McAloone, T. C. (2015). From LCC to LCA using a hybrid Input Output model - A maritime case study. *Procedia CIRP*, *29*, 474–479. https://doi.org/10.1016/j.procir.2015.02.004.

Laugen, L. (2014). An Environmental Life Cycle Assessment of LNG and HFO as Marine Fuels. https://doi.org/10.1177/1475090211402136.

LINCOLN. (2016). *Summary and Structure*. https://doi.org/10.1016/s0921-9110(98)80003-7.

Ling-Chin, J., & Roskilly, A. P. (2016). Investigating a conventional and retrofit power plant on-board a Roll-on/Roll-off cargo ship from a sustainability perspective - A life cycle assessment case study. *Energy Conversion and Management*. https://doi.org/10.1016/j.enconman.2016.03.032.

Ling-Chin, Janie, & Roskilly, A. P. (2016). Investigating the implications of a new-build hybrid power system for Roll-on/Roll-off cargo ships from a sustainability perspective – A life cycle assessment case study. *Applied Energy*. https://doi.org/10.1016/j.apenergy.2016.08.065.

Mojzes, A., Foldesi, P., & Borocz, P. (2012). Define Cushion Curves for Environmental Firendly Packaging Foam. *Annals of Faculty Engineering Hunedoara*, (February), 113–118. https://doi.org/10.1002/pts.

NCEI. (2018). What's the Difference Between Weather and Climate? | News | National Centers for Environmental Information (NCEI). Retrieved December 17, 2019, from National Oceanic and Atmospheric Administration (NOAA) website: https://www.ncdc.noaa.gov/billions/summary-stats.

Peder Kavli, H., Oguz, E., & Tezdogan, T. (2017). A comparative study on the design of an environmentally friendly RoPax ferry using CFD. *Ocean Engineering*, *137*(January), 22–37. https://doi.org/10.1016/j.oceaneng.2017.03.043.

Porras, A., Herrera, L., Carneros, A., & Zanón, J. I. (2018). Lifecycle and virtual prototyping requirements for ship repair projects. *Maritime Transportation and Harvesting of Sea Resources*, 2, 953–959. https://doi.org/10.5281/ZENODO.1248740.

SALİHOĞLU, İ. (2020). *Sirküler No :765. 320*(74), 1–5. Retrieved from https://www.denizticaretodasi.org.tr/tr/sirkuler/gemilerde-dusuk-sulfurlu-yakit-kullanimi-hk-11188.

UNCTAD. (2018). 50 Years of Review of Maritime Transport, 1968–2018: Reflecting on the past, exploring the future. *50 Years of Review of Maritime Transport, 1968–2018: Reflecting on the Past, Exploring the Future*, (812), 86p. Retrieved from https://unctad.org/en/PublicationsLibrary/dtl2018d1_en.pdf.

US EPA, O. (1979). Health and Environmental Effects of Particulate Pollutants. In *Fine Particulate Pollution* (pp. 9–20). https://doi.org/10.1016/b978-0-08-023399-4.50008-9.

Vinci, G., D'Ascenzo, F., Esposito, A., Musarra, M., Rapa, M., & Rocchi, A. (2019). A sustainable innovation in the Italian glass production: LCA and Eco-Care matrix evaluation. *Journal of Cleaner Production*. https://doi.org/10.1016/j.jclepro.2019.03.124.

Volbeda(FERG), C. (2017). *Ship Lifecycle Software Solutions (SHIPLYS)*.

Wang, H., Oguz, E., Jeong, B., & Zhou, P. (2018). Life cycle cost and environmental impact analysis of ship hull maintenance strategies for a short route hybrid ferry. *Ocean Engineering*, *161*(April), 20–28. https://doi.org/10.1016/j.oceaneng.2018.04.084.

Developments in Maritime Technology and Engineering – Guedes Soares & Santos (eds)
© 2021 Copyright the Author(s), ISBN 978-0-367-77376-2

An analysis of tug's escorting capability in low visibility conditions based on chase model: A case study of Qingdao port

J.S. Dai, T.F. Wang & Y. Wang
Intelligent Transportation Systems Center (ITS Center), Wuhan University of Technology, Wuhan, China
National Engineering Research Center for Water Transportation Safety (WTS Center), Wuhan University of Technology, Wuhan, China

Z.X. Yu
Qingdao Port Group Co., Ltd. of Shandong Port Group

ABSTRACT: The safety of container ships navigating in and out of the ports in low visibility conditions has always been an important problem for the operations at ports. Worldwide, the surveillance devices are often disrupted by foggy weather, resulting in the performance degradation of monitoring ashore and onboard, this will cause lagged detection of abnormal events such as fishing boats cross the channel at will, which make severe threat to merchant ships. The role of tugs is not unique. In addition to towing container ships at docks, tugboats are often used to clear waterways and escort container fleets inbound and outbound. In order to analyze the escorting capability of tugs and provide guidance for port dispatching, this paper introduces a real ship experiment at sea, a collision avoidance steering simulation experiment, and the escort simulation experiment under six scenarios. The research results show that the escort effectiveness is qualified.

1 INTRODUCTION

Large ports suffer from fog is a long-standing problem in the shipping industry. Take Qingdao port as an example, there are about 100 days every year, the fog obstructs navigation within the port area. According to the current relevant regulations, the operation is almost stopped with visibility below 500m, which greatly affects the production.

In the occasions of foggy weather, the surveillance devices ashore and onboard will be significantly compromised. Besides, many fishing boats whose motions are largely driven by fish school and usually ignore the maritime traffic rules, which lead to collisions between fishing boats and merchant ships frequently, make severe threat to merchant ship, especially in low visibility conditions. These fishing vessels are termed as unidentified and unpredictable moving object (UUMO). According to statistics, there have been about 40 collision accidents between UUMO and merchant ship along the coast of Shandong province, accounting for 30% of traffic accidents in the waters.

At present, escorting fleet with tugs against UUMOs in the critical channel of Qingdao waters is the most common method, especially in low visibility conditions. However, the existing research on tugs' escorting capability and organization scheme is still very scarce. On the one

hand, it is difficult to define the effect of intercepting UUMO in the complex encounter situation, which brings difficulties in quantifying the escort scope. On the other hand, most of the studies focused on the single tug and less on evaluation of the joint escorting capability with multiple tugs. Under the condition of low visibility, the role of tug in port operation is becoming more and more important, so it is more urgent to carry out research on this accordingly.

The study on the escort scope of single tug and the joint escorting capability of multiple tugs under low visibility conditions is conducive to provide more sensible scheduling strategies for tugs, which can not only ensure the safety of container fleet, but also improve the operation efficiency of ports. This paper will establish a numerical model of the real-time threat caused by UUMO to members of the fleet based on the geometric parameters and dynamic parameters of the encounter scenario, meantime, estimate the escort range of single tug and multi-tug, in order to propose a more reasonable scheme.

Some efforts have been paid to investigate the vessel traffic organization in harbors. The vessel traffic organization belongs to the scope of channel service, which refers to the management of overall behaviors of ships in the designated area, with the purpose of ensuring navigation safety and improving

DOI: 10.1201/9781003216582-3

navigation efficiency. Fujii proposed the concept of ship domain firstly, and then Goodwin et al (1975) defined ship domain as the waters required by each ship to ensure navigation safety. On the basis of the research above, scholars proposed the theory of the ship's dynamic boundary further, Wang et al (2010) built a quaternion model of ship by taking the speed of ships into consideration.

Besides, Pedersen et al (2010) applied fuzzy mathematics to risk assessment for ship collision, and Goerlandt et al (2015) constructed the theoretical framework of ship collision risk, take probability distribution and data statistics into risk assessment, at the same time, Hyeong et al (2012) proposed a method to analyze ship collision risk based on DCPA and TCPA.

Davidson defined ship maneuverability as the ability to maintain or change course, Nomoto et al (1957) deduced the equation of steering motion according to the response relationship between the input of rudder angle and the output of steering motion, based on the Nomoto model, Ghorbani et al (2016) identified the steering and rolling power of ships, researched on the PID controller deeply.

It should be mentioned that scholars domestic and overseas have little research on the escort ability of tugs in different situations. The reasonable assumption is that collision threat can be eliminated if the tug intercept UUMO in time. The escort mission can be regarded as a problem of control actively, and the distance is a key factor that determines the interaction between tugs and UUMO. In the course of the chase, the tug will monitor the situation all around and adjust its speed and course by the dynamic parameter of UUMO. In this paper, ship maneuverability is taken into consideration, verify the escorting capability by analyzing the result of tug's work in various encounter scenarios.

2 SITUATIONAL MODEL

2.1 Encounter scenario construction

According to International Regulations for Preventing Collisions at Sea (COLREGS), the scene can be divided into three situations, includes meeting end on, crossing and overtaking when two ships meet and the risk of collision exist persistently, it is stipulated that one ship should go straight with the same speed and course, and the other ship also should take measures to avoid collision as soon as possible,

nevertheless, the ship who should go straight should also take measures if the collision cannot be avoided only by the behavior of the giving way vessel. In the case of crossing, the ship should give way to the vessel when the vessel on the starboard side of this ship, and should not turn left when the vessel on the port side of the ship, as circumstances permit.

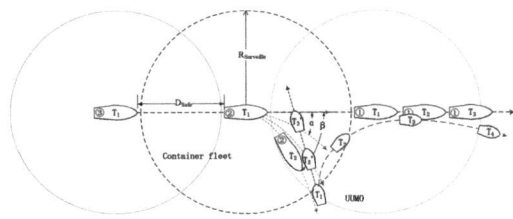

Figure 1. The typical crossing scenario of UUMO and container fleet.

Set up a typical crossing encounter scenario, as shown in Figure 1.

At T_1, the container ship number two (vessel$_2$) is in the middle of the fleet, keeping a safe distance (D $_{Safe}$) from the first (vessel$_1$) and the last (vessel$_3$), the monitoring scope of container ships is R (R $_{Surveille}$). When the vessel$_2$ detects a UUMO invading the dynamic boundary of the fleet from its starboard side, will turn right to avoid collision immediately. At the same time, the UUMO will also turn right to match the movements of vessel$_2$ to reduce the risk of collision, when the UUMO receives an early warning signal.

At T_2, both of the UUMO and the vessel$_2$ passed the closest point of approach, and the risk of collision declined gradually. However, the UUMO and the vessel$_1$ had formed a new trend of overtaking from T_1 to T_2.

From T_3 to T_4, the vessel$_1$ go straight with the same speed and course, and the UUMO keeps turning right to get away from the fleet, until the risk of collision vanishes completely.

2.2 Collision avoidance space-time analysis

The traffic situation in Qingdao port is complex, which is composed of various elements such as ship, route, weather flow and so on. In essence, the risk of traffic situation is measured by the spatiotemporal relationship of these elements.

It is necessary to consider the maneuverability of vessels when vessels meet on a small scope, because in this case, there is not enough time and space to avoid collision. In this paper, a simple K-T equation is adopted to analyze the ship maneuverability. Nomoto defines the equation as:

$$T\dot{r} + r = K\delta \qquad (1)$$

where T is the followability index of the ship, reflects the response speed of the steering action.

The followability is better when T is smaller. Besides, K is the cyclicity index of the ship, if K is larger, the steering angular velocity of the ship is faster, so it is easier to have a larger steering angle.

Since the research focus of this paper is not an accurate model of ship motion, the K-T equation only considers the ship maneuvering response under static water in the process of research, the effect of the drift angle is not considered.

In the typical crossing scenario shown in Figure 1, when the rudder angle changes from 0 to $\delta = \delta_0$, the course of the vessel changes. Based on the K-T equation, the angular acceleration and angular velocity of the vessel at any time can be obtained:

$$\dot{r} = (K/T)\delta_0 e^{-t/T} \tag{2}$$

$$r(t) = K\delta_0\left(1 - e^{-\frac{t}{T}}\right) \tag{3}$$

The change in course angle is given by:

$$\varphi = K\delta_0\left(t - T + Te^{-t/T}\right) \tag{4}$$

2.3 Traffic situation risk assessment

In order to help seaman to establish a more intuitive understanding of the collision risk and take effective measures to avoid timely under the urgent situation of the UUMO invading the fleet, this section provides a real-time assessment of the collision risk between vessels in close proximity, and takes the ship maneuverability and the actual environmental elements into consideration.

The motion parameters of vessels include the latitude and longitude, speed, course, turning angular velocity and so on; see Ma et al. (2019). Suppose the vessel takes the maximum steering angle within the safe range, the collision risk in real time can be identified by calculating the distance to closest point of approach (DCPA) and the time of closest point of approach (TCPA); see Sang et al. (2016). The relative distance index is introduced as an important supplement to the risk assessment to reduce the estimation deviation.

During the process of steering to avoid collision, the formula of collision risk between two vessels is:

$$R_i = \alpha \log\left(1 + \frac{D_S}{DCPA_i}\right) + \beta \log\left(1 + \frac{T_S}{TCPA_i}\right) + \gamma\frac{d_S}{d_i} \tag{5}$$

where d_i is the distance between container ship i and UUMO. α, β, γ are the weight values of the risk assessment, respectively. D_S, T_S, d_S are the safety thresholds of the risk assessment, respectively.

3 DYNAMIC ESCORT MODELING

3.1 Process of dynamic escort

The UUMO usually cannot perceive threats by means of optical equipment timely in low visibility conditions. Due to the characteristics of moving follow the fish school, it is difficult to model and analyze its trajectory according to its nonlinear and time-varying kinematic parameters.

Based on the consideration of the real situation Figure 2, it is assumed that when the UUMO enters the area where tug monitored, the tug will obtain information about the position, course and speed of the UUMO continuously, the UUMO has no response to the approach of tug and will take evasive action according to the instructions after the tug approached.

In view of the autonomy and maneuverability of tug in low visibility conditions, the process of escort can be seen as three steps in Figure 3.

Step1: After the tug has detected the movement of UUMO, estimate the risk of collision between UUMO and each container ship based on the geometric parameters of vessel movement such as DCPA, TCPA and relative distance, so as to facilitate the decision-making of escort actions.

Step2: Take the ship maneuverability into consideration, establish the mathematical model for the process of chasing, so as to determine the optimal action

Figure 2. The scenario of escorting capability test.

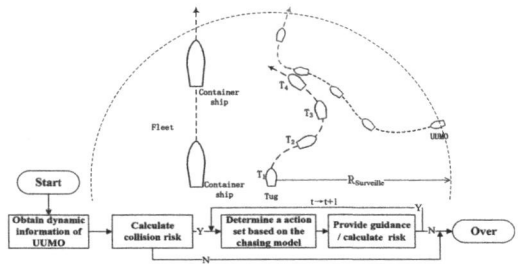

Figure 3. The process of dynamic escort.

set of the tug at each moment. Besides, the tug should also provide reasonable guidance to the UUMO when the tug is within the effective warning range.

Step3: Monitor the condition of UUMO continuously after the tug completed the guiding task, so as to determine whether there is still a risk of collision.

3.2 Chase model construction

In order to describe the chasing process of tug accurately, this section makes the following assumptions:

The channel is shrouded in fog, and the dynamic data (position, course, speed) of each vessel and UUMO, which in the monitoring range are known.

3.2.1 Coordinate system

In the fixed coordinate system, the X-axis points to the east and the Y-axis points to the north as shown in Figure 4. The dynamic parameters of tug and UUMO, are described by four variables respectively:

$$T = (X, Y, \theta, V); U = (x, y, \psi, v) \qquad (6)$$

Where $\theta \in (0, 2\pi)$; $\psi \in (0, 2\pi)$.

3.2.2 Course decision

The tug will adjust the course according to the relative distance between the tug and UUMO in the process of chasing timely. Select the target as point A ahead of the UUMO, that avoid to crash with UUMO. R_A and ϕ represent the safe distance and direction of point A relative to UUMO. Then the position of point A can be expressed as (x_A, y_A).

$$\begin{cases} x_A = x + R_A cos(\psi + \phi) \\ y_A = y + R_A sin(\psi + \phi) \end{cases} \qquad (7)$$

The dynamic parameters of UUMO at time t is $U(t) = (x, y, \psi, v)$, thus the steering angle of the tug relative to point A can be determined.

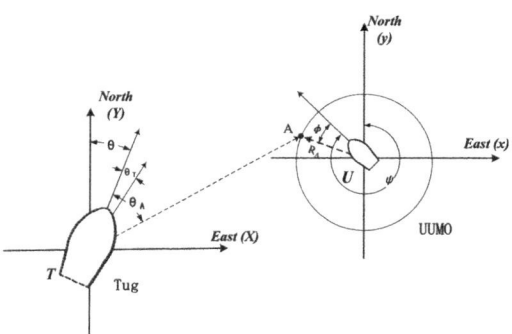

Figure 4. The fixed coordinate system.

$$\theta_A(t) = arctan \frac{x_A - X(t)}{y_A - Y(t)} - \theta \qquad (8)$$

On the premise of safety, the change in course angle (θ') caused by adjusting the rudder angle is often smaller than the desired steering angle (θ_A), therefore, it is necessary to correct the desired steering angle based on the K-T equation to improve the facticity of course decision, the actual steering angle of tug at time t can be expressed as θ_T.

$$\theta'(t) = K\delta_M \left(t - T + Te^{-t/T} \right) \qquad (9)$$

$$\theta_T(t) = \begin{cases} \theta_A(t) if \theta_A(t) < \theta'(t) \\ \theta'(t) if \theta'(t) < \theta_A(t) \end{cases} \qquad (10)$$

Where $\theta'(t)$ is the effective steering angle of tug and δ_M is the largest rudder angle.

3.2.3 Speed decision

In the initial stage of the chasing process, the tug will accelerate gradually to approach the UUMO for an intercept mission.

Besides, the tug will slow down in advance so that it can just match with the speed of the UUMO while the tug reached to the point A.

Limited by engine power, the maximum acceleration of tug is a, where $a = \Delta v/\Delta t$ and Δt is a time increment.

Set the r as the threshold of safe distance that the setpoint refers to the safety domain of the UUMO. The speed of the tug at t is $V_T(t)$, and the distance between the tug and point A is $d(T, A)$. In order to match with the speed of the UUMO, the tug decreases its velocity by its maximum acceleration during the time period $[t, t+T]$.

$$v = V_T(t) - aT \qquad (11)$$

$$\int (V_T(t) - a\tau)d\tau \leq d(T, A) - r \qquad (12)$$

Keeping the distance from the tug to the aiming location at least by r, Eq. (11) and (12) lead to

$$V_T(t) \leq \left(2a(d(T, A) - r) + v^2 \right)^{1/2} \qquad (13)$$

This condition defines an upper bound ω on the change of the speed between two steps, in order to avoid collision.

$$\omega = \frac{\left(2a(d(T, A) - r) + v^2 \right)^{1/2} - V_T(t-1)}{\Delta t} \qquad (14)$$

The speed decision can be expressed as $V_T(t+1)$.

If $d(T,A)>R$

$$V_T(t+1) = V_T(t) + sgn(\omega)\min(a,|\omega|) \qquad (15)$$

If $r<d(T,A)<R$

$$V_T(t+1) = V_T(t) - sgn(\xi)\min(a,|\xi|) \qquad (16)$$

Where ξ is the difference between $V_T(t)$ and v.
If $d(T,A)<r$

$$V_T(t+1) = V_T(t) - a \qquad (17)$$

3.3 Decision-making solving

The solution of escort decision provides guidance to tug, its essential is figure up the real-time speed and course of the tug. Construct the mathematical model of chasing movement to find the right way when the collision risk is higher than the safety threshold, which is essentially to find a optimal operation set in a series of actions, the risk is the standard that decide whether the tug should start to intercept or not.

The time period $[t, t+1]$ is taken as an example, Figure 5 shows the process of the escort decision-making.

Firstly, the tug will assess the collision risk between UUMO and each member of the container fleet at time t, according to the dynamic parameters of vessels ($M = [C_1, C_2, \cdots, C_n, T, U]$).

It is worth noting that the tug's position relative to point A is not only the key factor in determining the tug's course decision, but also the basis of speed decision-making. After that, evaluate the risk again at $t+1$

when the speed and course response to its manipulative behavior. Only if the risk satisfies the requirement, stop intercepting and back to the escorting position.

4 ESCORTING CAPABILITY ANALYSIS

The geographical features and resources of each port are not the same, result in the dispatching scheme for tugs is hard to be unified in a dense fog. This section constructs various typical scenes that the container fleet is invaded by UUMO in the passage based on the actual navigation environment of Qingdao port in low visibility conditions.

The escorting capability of tug in each scenario is calculated by the monitoring radius of tugs, the duration of intercepting process and back to the escorting position. The following assumptions are convenient for the study.

1. Consider each ship as a moving point instead of its size. The size of vessels can be ignored relative to the distance between ships.
2. The container ship should take no operation for direction change and speed change as far as possible, and the tug will clear up the waterway to ensure the navigation safety for the fleet.
3. The monitoring efficiency of imaging equipment in tugs is poor in low visibility conditions. The radar can only obtain the dynamic information (position, course, speed) within 2.5n mile.

There is no clear definition of various encounter scenarios in COLREGS. In this section, the typical scenes are divided into four types, according to the position and heading of the incoming ship, as shown in Figure 6.

The escorting capability of tugs is limited by the radar's monitoring range and its maneuvering capability. In order to balance the consumption of resource and the security of navigation to the greatest extent, the mission of escort is divided into four steps.

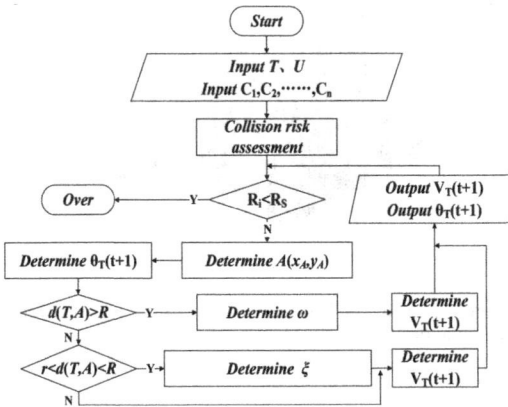

Figure 5. The process of the escort decision-making.

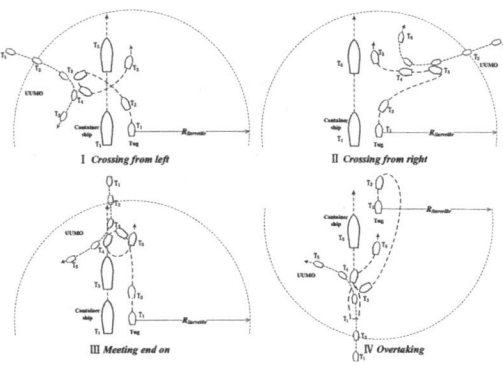

Figure 6. The typical scenes.

Step1 (T_1-T_2): Follow the fleet. In the course of following with the fleet, the tug will grasp the trend of traffic situation according to the dynamic information of each ship within the monitoring range, and provide the basis for the decision.

Step2 (T_2-T_3): Get close to UUMO. The tug will consider to chase the UUMO when the UUMO pose a threat on fleet. According to the chasing model and K-T equation, the optimal action set of the tug at each moment will be determined.

Step3 (T_3-T_4): Intercept the UUMO and give guidance. The tug will warn the UUMO of dangerous action when it is within the effective communication range of the UUMO, provide guidance to UUMO to get away from the fleet in the meantime. In a period of time, the tug and UUMO remain relatively still.

Step4 (T_4-T_5): Back to the escorting position. After the mission of intercept is completed, the tug should return to the escorting position as soon as possible to continually escort for the fleet. At the beginning of the return journey, the tug sails at the maximum speed, when there is only 100m from the homing position, the tug will slow down until just match the speed of fleet.

5 SIMULATIONS

In this section, several typical encounter scenarios will be designed, and the tug's escorting capability will be demonstrated by the simulation results of the time for chasing, intercepting and coming back.

5.1 *Encounter scenario simulation*

The initial parameters of each vessel are summarized in Table 1. It can be known that the UUMO will collide with C_2 after a period of sailing if all vessels maintain the same speed and course.

The vessels C_1, C_2, C_3 and UUMO are set as simulation objects. The center coordinate of the simulation interface is (0, 0), the X-axis direction is

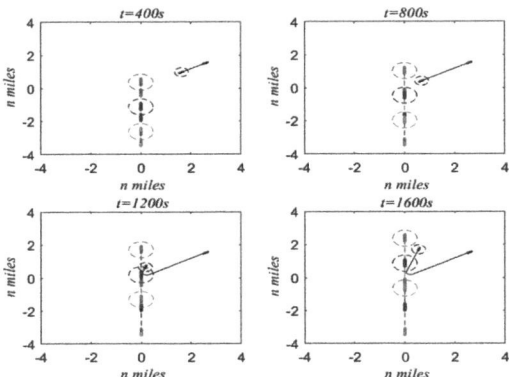

Figure 7. The trajectories of vessels.

from left to right and the Y-axis direction is from bottom to top.

Suppose that the merchant ship's monitoring effect on UUMO is poor, and the UUMO consider to turn right to avoid collision after sailing into the safety domain of the fleet. Figure 7 shows the trajectories of four vessels and the relative location at certain times, the closest distance from UUMO to C_2 in the process is 594.5m.

Although the collision is not exist, the reality is always unsatisfied. Analyze the risks caused by different manipulations at the moment the UUMO began to turn, Figure 8 shows that the risk will reach peak if the speed and course are adjusted to 7.9 and 243 respectively. Thus, it is easy to form an urgent situation or even a collision accident, if the UUMO fails to take right operations in time.

Table 1. Initial parameters in typical encounter scenario.

Vessel	Position /n mile	Speed /Kn	Course /°	Risk exist
C_1	(0, -0.3)	6	0	N
C_2	(0, -1.8)	6	0	Y
C_3	(0, -3.3)	6	0	Y
UUMO	(2.6, 1.53)	10	240	N

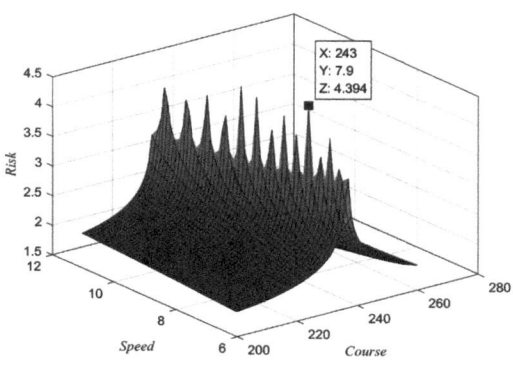

Figure 8. The collision risk to C_2 of different manipulations.

5.2 Escort process simulation

The vessels C_1, C_2, UUMO and Tug are set as simulation objects. The center coordinate of the simulation interface is (0, 0), the X-axis direction is from left to right and the Y-axis direction is from bottom to top. The initial course and speed of C_1 and C_2 are 0 and 6 respectively, and the initial positions are (0, -1.80), (0, -3.30) respectively.

The escorting position of the tug is set as the front of the starboard of the first container ship, with the transverse and longitudinal distance from C_1 are 1000m and 0.5n miles respectively.

In order to analyze the escorting capability of tugs, there are six typical encounter scenarios describe the situations that the UUMO intruding into the safety domain of the fleet from different directions in the speed of 12Kn.

The initial parameters and simulation results of escorting process are summarized in Table 2.

In order to simulate the motion state of vessels more precisely, the length and the scope of the safety domain of container ships are set to 300m and 900m respectively, the trajectories in different typical encounter scenarios are shown in Figure 9.

Simulation results represent that the tug makes right decisions and intercepts the UUMO out of the safety domain of the fleet within 600s in every typical encounter scenario, after then, it takes 1218s to return to escorting position in addition at most, about 20 minutes. Thus, a single tugboat can effectively escort the fleet composed of two

container ships to ensure the navigation safety in low visibility conditions.

6 CONCLUSIONS

In view of the actual navigation environment of Qingdao port in low visibility conditions, and put the UUNO under the assumption of irrational behavior, this paper proposes an analysis method of tug's escorting capability based on the chasing model and the maneuverability of ship, take the real-time collision risk as the standard for decision-making, the optimal set of operations is easy to obtain.

Several simulation experiments are carried out in different typical scenarios to verify the performance of the proposed method. All of the steering operations of vessels conform to the K-T equation. Simulation results represent that the tug can always intercept the UUMO out of the safety domain of the fleet within 600s and come back to the escorting position within 20 minutes.

A single tugboat can effectively escort the fleet composed of two container ships to ensure the navigation safety in pre-designed typical encounter scenarios. However, in the urgent situation that many UUMOs invade into the safety domain of the fleet at the same time, it is harder to reach a consensus on a convoy mission for multi-tugs if more than one tug is involved, this will be considered in future.

Table 2. The initial parameters and simulation results of escorting process.

Typical scenarios	Vessel	Initial parameters			The moment of tug's beginning (t)	The moment of UUMO's stopping (t)	The moment of tug's returning (t)
		Position	Speed	Course			
1	UUMO	(3.12, 1.91)	12Kn	240°	255	417	674
	Tug	(0.54, -1.30)	6Kn	0°			
2	UUMO	(3.12, -3.30)	12Kn	300°	0	550	1218
	Tug	(0.54, -1.30)	6Kn	0°			
3	UUMO	(0, -5.10)	12Kn	0°	511	465	1203
	Tug	(0.54, -1.30)	6Kn	0°			
4	UUMO	(-3.12, -3.30)	12Kn	60°	494	420	1130
	Tug	(0.54, -1.30)	6Kn	0°			
5	UUMO	(-3.12, 1.91)	12Kn	120°	413	452	823
	Tug	(0.54, -1.30)	6Kn	0°			
6	UUMO	(0, 3.60)	12Kn	180°	391	587	924
	Tug	(0.54, -1.30)	6 Kn	0°			

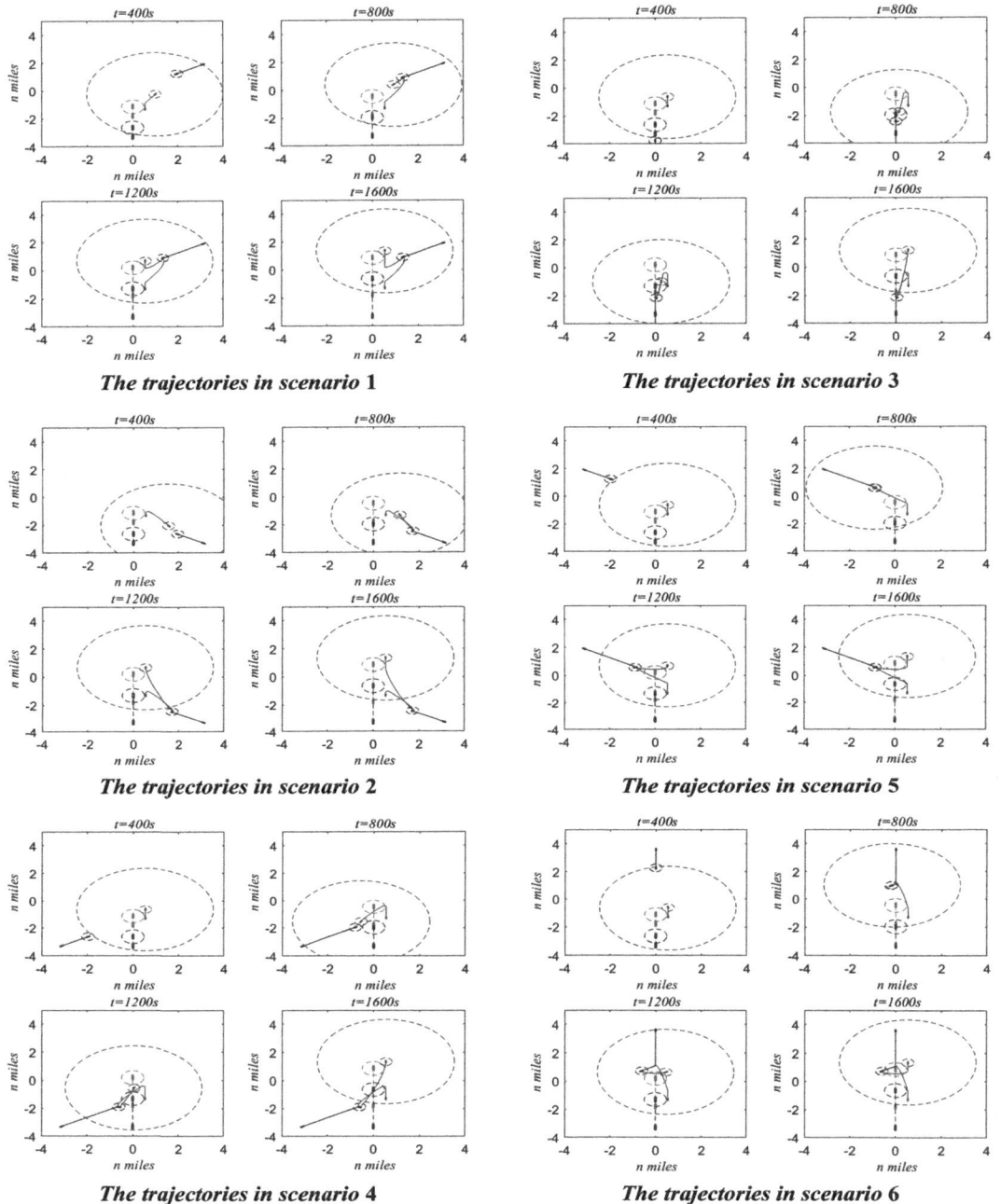

Figure 9. The trajectories in different typical encounter scenarios.

ACKNOWLEDGEMENTS

This work was supported by the Intelligent Transport Systems Research Center (ITSC), funded by the project "Feasibility study on navigation of container ships in and out of Qingdao port in foggy days" and "High-tech ship research project of the ministry of industry and information technology" (G18473CZ06/G18473CZZ05). All supports are gratefully acknowledged by the authors.

REFERENCES

Wang, N. 2010. An Intelligent Spatial Collision Risk Based on the Quaternion Ship Domain. Journal of Navigation 63(4): 733–749.

Goodwin, E.M. 1975. A statistical Study of Ship Domains. Journal of Navigation 28(3): 328–344.

Pedersen, P.T. 2010. Review and application of ship collision and grounding analysis procedures. Marine Structures 23

Goerlandt, F., Montewka, J., Kuzmin, V. & Kujala, P. 2015. A risk-informed ship collision alert system: Framework and application. Safety Science 77(1): 182–204.

Hyeong, J.A., Pyo, K.R. & Jun, Y.Y. 2012. A study on the collision avoidance of a ship using neural networks and fuzzy logic. Applied Ocean Research 37: 162–173.

Nomoto, K., Taguchi, T., Honda, K. & Hirano, S. 1957. On the steering qualities of ships. International Shipbuilding Progress 4(35): 354–370.

Sang, L.Z., Yan, X.P. & Wall, A. 2016. CPA calculation method based on AIS position prediction. Journal of Navigation 1(06): 1–18.

Ghorbani, M.T. 2016. Line of sight waypoint guidance for a container ship based on frequency domain identification of Nomoto model of vessel. Journal of Central South University 23(08): 1944–1953.

Ma, J., Liu, Q., Zhang, C.W., Liu, K.Z. & Zhang, Y. 2019. A method for extracting ship encounter situation based on spatio-temporal analysis of AIS data. China Safety Science Journal 29(05): 111–116.

Developments in Maritime Technology and Engineering – Guedes Soares & Santos (eds)

Digitalization of Iberia Maritime Transport: Preparation of a survey on the as-is status

D. Díaz Gutiérrez & P-L. Sanchez-Gonzalez

Higher Technical School of Naval Architecture and Ocean Engineering, Technical University of Madrid, Madrid, Spain

ABSTRACT: Seaborne trade keeps a growing path. This positive impact on the Maritime Transport industry is shadowed by the entry of other industry digital players like Amazon and Alibaba. In this context, digitalization is mandatory not just for outperforming but simply for survival. This paper presents the work that the Technical University of Madrid is developing to determine the status of digitalization of the different players in the maritime transport transportation chain to verify the state-of-the-art regarding digitalization of the Maritime Transport sector; find out the possible bottlenecks or barriers to this digitalization and identify possible solutions to sort out these obstacles. The focus of this paper is the development of a survey on Maritime Transportation companies. The main contribution of this paper is the description of the process for creating the tool that will help in the data capture and analysis of the as-is status of digitalization of Maritime Transportation companies in the Iberia Peninsula.

1 INTRODUCTION

Seaborne trade keeps a growing path, as can be seen when comparing the United Nations Conference on Trade and Development UNCTAD (2016) and UNCTAD (2019). Whereas in their 2016 report UNCTAD quoted a shipment expansion of 2.1 per cent and the surpass of the 10 billion tons milestone for the first time in world seaborne trade volumes, these figures moved to 2.7 per cent and 11 billion tons respectively in the 2019 report.

Being these numbers impressive and having a positive impact as of today on the Maritime Transport industry, threats are coming from other industries digital players. Not only UNCTAD (2018) quoted this threat coming from e-retailers. First line analysts like McKinsey (2017) cited this threat coming from tech giants like Amazon and Alibaba. Amazon has landed into logistics with its Prime Air cargo service and its recent acquisition of an ocean-freight-forwarder license in China; it has even reported interest in buying physical infrastructure assets such as Frankfurt's Hahn airport. Alibaba have announced a partnership with COSCO Shipping Co. to develop an integrated logistics platform for small and midsize enterprises.

In this context, digitalization is mandatory not just for outperforming but simply for survival. This move towards digitalization has started. Governments and Public Institutions have in their agendas actions to digitalize the sector, but these are moving slow. An example is the European Commission (EC) initiative "European Maritime Single Window environment" that aims to improve efficiency using digital information as well as contributing to the integration of the sector to the digital multimodal logistic chain. This initiative, launched in 2010, is declared by the same EC in 2016 as not achieving the objectives. EC also remarked the lack of enough progress at the European Union level in terms of harmonization in this topic. The Euro Parliament (2018) has launched a draft report on the Single Window proposal in October 2018 trying to unblock this situation.

In this same direction, the Spanish Government launched the SIMplification of Processes for a Logistic Enhancement initiative (also known as SIMPLE) within the 2017-2020 Transportation and Infrastructures Innovation Plan. As of today, it is still in its early stages and given the existing plans, will not be a reality before 2020.

Another example is the efforts done by the European Shortsea Network, that back in 2016 they were already demanding the global launch of the e-Manifest at European level. The European Maritime Safety Agency (EMSA) ran a pilot in 2019 with the objective of deploying it in Europe in the short/mid-term.

This paper describes part of a project that is being developed at the Technical University of Madrid (UPM in its Spanish acronym) that will help improving Maritime Transportation companies' operations by introducing digitalization on their operations.

The first step of the project has been the use of systematic literature review technique to assess the status

DOI: 10.1201/9781003216582-4

on digitalization of Maritime Transport. This work has already been published (Sanchez-Gonzalez et al. (2019)), and its main conclusion is that when looking to the relevant literature, Maritime Transport is moving towards digitalization at different speeds in the eight domains considered in that analysis (Autonomous Vehicles and Robotics; Artificial intelligence; Big data; Virtual, Augmented and Mixed Reality; the Internet of Things; Cloud and Edge Computing; Digital Security; and 3D Printing and Additive Engineering). This is also the case when crossing those domains impact with the three industrial sectors used for that work (Ship Design and Shipbuilding; Shipping; and Ports). While robotics has been investigated in detail, mostly on unmanned vehicles in Robotics and the use of Artificial Intelligence as a means of supporting vessels aids for navigation, there are domains practically unexplored (Cloud, Security, and 3DP have room for study).

The next step of this project is the assessment of digitalization on Maritime Transportation companies that operate in Iberia (Portugal and Spain), to find out the opportunities for increasing performance through digitalization.

This paper presents the work done to develop an instrument together with its statistical validation for providing agnostically answer to:

1. Verify the state-of-the-art regarding Digitalization of the Maritime Transport sector.
2. Find out the possible bottlenecks or barriers to this Digitalization.
3. Identify possible solutions to sort out these obstacles.

2 RESEARCH APPROACH AND METHODOLOGY

The methodology chosen for this research design is the development of a survey on Maritime Transportation companies. This technique has been chosen as the right approach for this investigation was a cross sectional design, following the classification of research design from Bryman et al. (2015):

- It is an analysis that will include more than one case so that the conclusions were not affected by the idiosyncrasies of one specific Maritime Transportation company.
- The analysis will be performed at one single point of time, so that optimal resources, such as time and funding, could be used efficiently for this research.
- Quantitative data will be used from data collected via questionnaires.
- This collected data will be examined to detect their patterns of association.

Using this technique, the information necessary for assessing the status of digitalization of Maritime Transportation companies will be gathered. This survey will comply with at least 2 of the 3 objectives defined by Babbie (1990):

- It will describe the situation "as-is" in terms of digitalization of the Maritime Transportation companies.
- It will provide a suitable explanation on the reasons why they have that level of digitalization (for example, given their Human Resources, or their current strategic intent).
- Initially it was not planned to be used as a search device to open new areas of investigation.

Surveys have been used as research tool of the maritime transport industry. Works are on a variety of matters, so the list below is not exhaustive, it tries to illustrate some areas of research where surveys have been used in this industry:

- Applications in the carrier selection process (Wong et al. (2008); Kent et al. (1999); van den Berg et al. (2015)).
- Analysis of environmental impacts (Hulskotte et al. (2010); Chung-Shan et al. (2013); Giziakis et al. (2012); Rehmatulla et al. (2015); Lun et al. (2015)).
- Studies on port selection process (Nazemzadeh et al. (2015); Mittal et al. (2016)).
- Use of internet and other Information Technology (IT) tools (Lu et al. (2006); Nikitakos et al. (2007); Triantafylli et al. (2010)).
- Works on Security Management (Yang et al. (2013); Oltedal et al. (2010); Chen et al. (2018)).

There is not much literature on the methodology for creating survey questions. There is literature on recommendations on how questions should be posed, on type of questions, and general guides on how to construct questions (Babbie (1990), Cohen et al. (2018)). But not real methodologies on how to develop questions for a survey. Therefore, the decision was to build one taking as starting point the optimal situation in terms of digitalization. In other words, the questions were developed based on the scenario of a Maritime Transportation company that has really embedded digitalization into all their processes. This method has helped in the identification of the key issues that were going to be identified as well as in the generation of the variables and indicators that were going to allow a fair comparison on the status of digitalization amongst Maritime Transportation companies. For example, in an ideal scenario a Maritime Transportation company must make use and correlate the different data coming from the sensors of their ships in order to anticipate failures. Several questions were developed to assess the status of such scenario to guarantee a legitimate answer that is not biased by a question too straight forward.

Simplicity and ease on answering has been one of the main concerns when developing the questionnaire. For this reason, it was decided the intensive use of dichotomous questions that should be answered with a simple "Yes" or "No" followed by a space where the respondent could explain his/her answer with additional details. This strategy also provided inputs that were used to identify the main issues blocking the progress of digitalization as well as improvements that could be applied to the operations of Maritime Transportation companies using digital technologies.

Questions were grouped on these chapters to facilitate answering:

- Digital strategy.
- Customer experience.
- Digital interaction.
- Operational digitalization.
- Data management.
- Cybersecurity
- Systems architecture.
- Organization and Human Resources.

One additional section was added named Digitalization Impact. In this section the Analytic Hierarchy Process (AHP) developed by Saaty (1980) has been used to make pair-wise comparisons between the eight domains used in the literature review from Sanchez-Gonzalez et al. (2019). The final objective is to determine which domain is expected to have a higher impact in the coming years, according to the answers received.

The questionnaire was completed with a group on instructions for answering correctly to the different questions. It also included a chapter explaining the fulfilment of the European Union (EU) General Data Protection Regulation (GDPR). The chapter on the fulfilment of GDPR was developed following the recommendations received from the Spanish Agency for Data Protection (hereinafter, AEPD).

For practical reasons the site of this research was the Maritime Transportation companies located in Iberian Peninsula. In addition, and due to the importance of the sector from a commercial perspective, the focus will be the Maritime Transportation companies dedicated to containers transportation. The companies of this research came from the list of Maritime Transportation companies from ANAVE (*Asociación de Navieros Españoles*, Spanish association of Maritime Transportation companies), as well as other coming from the CME (Spanish Maritime Cluster, organization that groups all the industries, services and economic activities from Spain related with the sea). Some others that operate in Iberian Peninsula and that have a big impact in the sector, like Maersk or CMC, will be part of the sample. Finally, given the low number of this population, the survey will be sent to the full population of container Maritime Transportation companies.

The content validation of the survey was done using the standard inter-judge validation process. This was done at three levels:

- Methodology: the questions were sent for validation to an expert on surveys methods.
- Digitalization: several digitalization experts performed the clarity and effectiveness of the questions used.

- Maritime Transportation: as with digitalization, questions were sent for validation to Maritime Transportation experts.

The method requires the agreement amongst judges on the validity and clarity of each of the questions. This method has been extensively used for surveys' content validation. The quantification of the agreement was done using the content validity ratio (CVR) developed by Lawshe (1975):

$$CVR = \frac{n_{e-N/2}}{N/2} \tag{1}$$

where n_e = number of judges indicating the question as "essential"; and N = total number of judges (in this work, N = 7).

Lawshe (1975) considered the values of CVR included in Table 1 as the ones necessary for item validation:

Table 1.   Minimum values of CVR.

Number of judges	CVR Min. Value
5 - 7	0.99
8	0.85
9	0.78
10	0.62
11	0.59
12	0.56
13	0.54
14	0.51
15	0.49
20	0.42
25	0.37
30	0.33
35	0.31
40	0.29

To assess the reliability and validity of the survey research the following classical methods will be used once received the answers:

- Cronbach's α: As per Martínez González et al. (2014) and since different questions or items are being used to measure the same theoretical concept, there should be consistency amongst them. And to asses this, the Cronbach's α test is the adequate one, since it measures the average correlation amongst variable from one scale and its scoring should be on the range of 0.7 or above (Bryman at al. (2015)).
- The following tests will be used to verify correlations:
  o Factor loading tests. They must confirm the adequacy of factor analysis in this case. This is especially relevant since the sample used for this

survey was not a large one. Factor loading above 0,80 is considered valid; otherwise factor analysis should not be considered, or the sample should be increased (Martínez González et al. (2014)).

o Bartlett´s sphericity tests. Given that the sample use is not a large one, it is adequate using this analysis (Martínez González et al. (2014)). This test assumes as null hypothesis that, though the sample used in this research has been built randomly, all the variables are independent and therefore, factorization of the correlation matrix is useless (Barbero García et al. (2011)). If the Bartlett's test comes to a low level of signification (or a value of the test over 0,05, which is the same) this null hypothesis should not be accepted, and the matrix shouldn't be factorized.

o Kaiser-Meyer-Olkin (KMO) test. This test generated a parameter that indicates the degree of correlation of one variable with the rest of the variables, being the result between 0 and 1. The criteria use to qualify the results is in Table 2:

Table 2. KMO Values.

KMO	Adequacy
> 0.90	Excellent
0.90 – 0.80	Good
0.79 – 0.70	Normal
0.69 – 0.60	Mediocre
< 0.60	Unacceptable

A low result of KMO indicates that factor analysis is not adequate for the case (Martínez González et al. (2014)).

3 RESULTS

The method applied for generating these questions resulted into a total of 62 questions that were sent to judges for validation purposes.

The number of questions that were finally included in the survey were 62. It may seem that there were not changes coming from the inter-judges' validation but that is not the case. Table 3 summarizes the outcome of the inter-judge's validation done on the initial survey instrument generated.

The most relevant change coming out of the validation was the complete change of Cybersecurity section. The initial 6 questions included in this section were not addressing the developments on cybersecurity that the IMO (International Maritime Organization) imposes to the Maritime Transportation companies through regulations like the IMS code (International Security Management).

In the Digital Interaction section, there were 3 questions that one of the judges considered not

Table 3. Questions per survey section.

Survey Section	Initial # quest.	Final # quest.
Digital strategy	4	2
Customer exper.	8	8
Digital interact.	7	7
Operat. Digit.	14	14
Data Managmt.	7	7
Cybersecurity	6	8
Syst. Arch.	10	8
Org & HHRR	3	4
Digit. Impact	1	1
Conclusions	2	3

essential. After discussing this point internally, the decision was to keep the question since it was finally judged essential for the purposes of the survey. This situation also happened in the Operational Digitalization section with 2 of the initial questions and in the Systems Architecture section, also with 2 of the questions. The rational for keeping these questions is that they make look irrelevant in a not digitalized context, but the to-be situation makes them important, a context were digitalization is a must.

The Organization and Human Resources section was increased in 1 question, recommended by the judges. Same happened with the "Conclusions" section.

4 CONCLUSIONS AND NEXT STEPS

As said in the introduction, this is just a small component of a research that UPM is developing on digitalization of Maritime Transportation companies. A research that is broken done in four components:

1. Maritime Transportation digitalization research state-of-the-art, already completed and published (Sanchez-Gonzalez et al. (2019)).
2. Maritime Transportation companies' digitalization as-is status.
3. Maritime Transportation companies Business Process Model (BPM), with the definition of the to-be BPM for these companies.
4. Maritime Transportation companies' digitalization. A set of digital applications will be defined with the information collected in the previous steps to enhance their performance.

The work described in this paper is part of the second workstream, that is based on the data collection via survey of the as-is status for these companies and their main threads and barriers for improving their status. The work done has allowed the creation of a valid instrument for the verification of the digitalization status of Maritime Transportation companies operating in the Iberian Peninsula.

This paper also contributes with the summarization on how a survey should be created and validated.

The team is now working on data capture coming from the survey sent. Results will be communicated once available.

REFERENCES

Babbie, E. 1990. *Survey Research Methods (2nd edition)*. California, USA. Wadsworth Publishing Company.

Barbero García M A, Vila Abad E & Holgado Tello F P. 2011. *Introducción básica al análisis factorial*. Madrid, Spain. Universidad Nacional de Educación a Distancia.

Bryman, A. & Bell, E. 2015. *Business Research Methods*. Oxford, UK. Oxford University Press.

Chen, C.; Chiang, Z.; Liu, Y.; Zeng, X. 2018. Critical success factors in marine safety management in shipping industry. *Journal of Coastal Research 2018, 83, 846-850.*

Chung-Shan, Y.; Chin-Shan, L.; Haider, J. J.; Marlow, P. B. 2013. The effect of green supply chain management on green performance and firm competitiveness in the context of container shipping in Taiwan. *Transportation Research Part E-Logistics and Transportation Review 2013, 55, 55-73.*

Cohen, L.; Manion, L. & Morrison, K. 2018. *Research Methods in Education*. Oxon, UK. Routledge.

European Parliament. 2018. Draft Report on the proposal for a Regulation of the European Parliament and of the Council establishing a European Maritime Single Window environment and repealing Directive 2010/65/EU. *Committee on Transport and Tourism, European Parliament.*

Giziakis, C.; Christodoulou, A. 2012. Environmental awareness and practice concerning maritime air emissions the case of the Greek shipping industry. *Maritime Policy & Management 2012, 39:3, 353-368.*

Hulskotte, J.H.J.; Denier van der Gon, H.A.C. 2010. Fuel consumption and associated emissions from seagoing ships at berth derived from an on-board survey. *Atmospheric Environment 2010, 44:9, 1229-1236.*

Kent, J. L.; Parker, R. S. 1999. International containership carrier selection criteria. Shippers/carriers differences. International *Journal of Physical Distribution & Logistics Management 1999, 29:6, 398-408.*

Lawshe 1975. A quantitative approach to content validity. *Personnel Psychology, 28: 563-575.*

Lu, C.-S.; Lai, K.-H.; Cheng, T. C. E. 2006. Adoption of internet services in liner shipping: An empirical study of shippers in Taiwan. *Transport Reviews 2006, 26:2, 189-206.*

Lun, Y.H. V.; Lai, K-H.; Wong, C. W.Y.; Cheng, T.C.E. 2015. Greening and performance relativity: An application in the shipping industry. *Computers & Operations Research 2015, 54, 295-301.*

Martínez González M A, Sánchez-Villegas A, Toledo Atucha E A, & Faulin Fajardo J. 2014. *Bioestadística amigable*. Barcelona, Spain. Elsevier España.

McKinsey and Company. 2017. Container shipping: The next 50 years. *McKinsey & Company.*

Mittal, N.; McClung, D. 2016. Shippers changing priorities in port selection decision – A survey analysis using Analytic Hierarchy Process (AHP). *Journal of the Transportation Research Forum 2016, 55(3), 65-81.*

Nazemzadeh, M.; Vanelslander, T. 2015. The container transport system: Selection criteria and business attractiveness for North-European ports. *Maritime Economics & Logistics 2015, 17(2), 221-245.*

Nikitakos, N.; Lambrou, M. A. 2007. Chapter 12 digital shipping: The Greek experience. *Research in Transportation Economics 2007, 21, 383-417.*

Oltedal, H.; Wadsworth, E. 2010. Risk perception in the Norwegian shipping industry and identification of influencing factors. *Maritime Policy & Management 2010, 37:6, 601-623.*

Rehmatulla, N.; Smith, T. 2015. Barriers to energy efficient and low carbon shipping. *Ocean Engineering 2015, 110-B, 102-112.*

Saaty, T.L. 1980. *The Analytic Hierarchy Process*. New York USA. McGraw Hill Company.

Sanchez-Gonzalez, P-L.; Diaz-Gutiérrez, D.; Leo, T. J.; Nuñez-Rivas, L. R. 2019. Towards a Digitalization of Maritime Transport? *Sensors 2019, 19(4), 926.*

Triantafylli, A. A.; Ballas, A. A. 2010. Management control systems and performance evidence from the Greek shipping industry. *Maritime Policy & Management 2010, 37:6, 625-660.*

UNCTAD (United Nations Conference on Trade and Development). 2016. Review of maritime transport 2016". *UNCTAD/RMT/2016.*

UNCTAD (United Nations Conference on Trade and Development). 2018. Review of maritime transport 2018". *UNCTAD/RMT/2018.*

UNCTAD (United Nations Conference on Trade and Development). 2019. Review of maritime transport 2019". *UNCTAD/RMT/2019.*

van den Berg, R.; de Langen, P. W. 2015. Assessing the intermodal value proposition of shipping lines: Attitudes of shippers and forwarders. *Maritime Economics & Logistics 2015, 17(1), 32-51.*

Wong, P. C.; Yan, H. & Bamford, C. 2008. Evaluation of factors for carrier selection in the China Pearl River delta. *Maritime Policy & Management 2008, 35:1, 27-52.*

Yang, C-C.; Wei, H-H. 2013. The effect of supply chain security management on security performance in container shipping operations. *Supply Chain Management 2013, 8(1), 74-85.*

Developments in Maritime Technology and Engineering – Guedes Soares & Santos (eds)
© 2021 Copyright the Author(s), ISBN 978-0-367-77376-2

Competitive strategic position analysis of ports of the Iberian Peninsula hosting car-carrier traffic

J. Esteve-Perez & J.E. Gutierrez-Romero

Department of Naval Technology, Universidad Politécnica de Cartagena (UPCT), Cartagena, Murcia, Spain

ABSTRACT: Spain and Portugal have an important automobile production industry, which is one of the most significant in Europe. In this regard, the ports of the Iberian Peninsula play a key role in the logistics of import and export of vehicles in Southern Europe. The analysis of Ro-Ro cargoes, in this particular case of vehicles, does not have a notable presence in the literature specialized in Maritime Transportation and Port Logistics. In the present work, a dynamic analysis of strategic competitive positions of Spanish and Portuguese ports with car-carrier vessel traffic is developed. For this purpose, time series of vehicles movements corresponding to the period from 2011 to 2018 are used. In addition, the seasonal pattern of this port traffic is analyzed to determine whether this maritime traffic has seasonality and the variables that influence the patterns. Moreover, forecasts are made for this port traffic through the application of simulations with Monte Carlo techniques.

1 INTRODUCTION

Roll-On/Roll-Off shipping can be divided into two main segments. On the one hand, the traffic of vehicles in passage regime and of vehicles that transport goods, which is mainly developed with ferry, Ro-Pax or Roll-on/Roll-off (Ro-Ro) ships. On the other hand, the traffic of new vehicles, which is fundamentally associated with the import and export of all types of new vehicles. This traffic is carried out with pure car carrier (PCC) or pure car-truck carrier (PCTC) ships, which are designed to provide fast loading and unloading of vehicles by means of ramps. This article focuses on the traffic of PCC/PCTC ships in ports of the Iberian Peninsula, i.e. in Portuguese and Spanish ports.

Maritime transport plays a key role in global vehicle logistics connecting vehicle manufacturers with the target customer markets. Global vehicle seaborne trade grew significantly during the period from 1996 to 2013, with an average annual growth rate of 5.98%. Additionally, global vehicle sea trade evolved from 8 million in 1996 to 21 million in 2016 (Clarkson Research Services, 2019).

The vehicle production industry is a key sector for the Iberian Peninsula because 3.12 million vehicles were manufactured during 2018 in the factories located in Spain and Portugal, the former, produced 2.82 million and, the latter, manufactured 0.29 million. Spain during 2018 was the ninth largest producer of vehicles worldwide. The world leader during 2018 was China with a production of 27.8 million vehicles. Regarding production in

Europe, Germany was the leader in 2018 with 5.12 million vehicles, Spain occupied the second position within the European vehicle manufacturers in 2018 (ANFAC, 2019).

The vast majority of the vehicles produced in Portugal and Spain are intended to exportation. For instance, the 97% and 81.7% of the Portuguese and Spanish production in 2018, respectively, was exported. The 91.15% of the Spanish exportation of vehicles was destined to Europe. Specifically, the main countries of destination of the vehicles were France, Germany, United Kingdom and Italy, which jointly accounted for 60.3% of exportation. The second region of destination for vehicle exports was America with a 4.25%; whose main destinations were the United States of America and Mexico.

Based on the importance of Portugal and Spain as producers of vehicles both at European and global level, it is of interest to develop a study about the traffic of PCC/PCTC ships in ports of the Iberian Peninsula, because of the majority of the production is intended to exportation and the importance that maritime transport has in global vehicle logistics. The transport of vehicles by sea constitutes an important segment of the maritime industry, however, there is a lack of works focusing on this topic in the literature specialized in maritime transport (Hall & Olivier, 2005; Kahveci & Nichols, 2006). In contrast, in the literature specialized in maritime transport, numerous works focused on container traffic (AlMarar & Cheaitou, 2018; Yang & Wong, 2016), liquid (González Laxe et al., 2014; Siddiqui et al., 2013) and solid bulk cargoes (Yin et al., 2017; Chen et al., 2010) can be

DOI: 10.1201/9781003216582-5

found. The objectives of the article are as follows: (1) to identify the main ports of the Iberian Peninsula with PCC/PCTC ship traffic, (2) to know the competitive positions of the ports of the Iberian Peninsula with PCC/PCTC ship traffic, (3) to determine whether this maritime traffic has seasonality, and (4) to carry out forecasts of vehicle traffic through the application of simulations with Monte Carlo techniques.

The article is structured as follows. Section 2 is devoted to literature review on the features associated with maritime transport of vehicles and vehicle traffic in ports of the Iberian Peninsula. Section 3 presents the analysis of competitive positions of the ports hosting PCC/PCTC ship traffic in the Iberian Peninsula. Section 4 includes the determination of seasonality patterns of vehicle traffic. In section 5 the Monte Carlo simulations for vehicle traffic are performed. Finally, in Section 6 the main conclusions of the work are shown.

2 LITERATURE REVIEW

2.1 *Features of maritime transport of vehicles*

Transport of vehicles by sea has been deeply influenced by the process of containerization, since both cargoes generated vessels designed specifically for the characteristics of a break bulk cargo, in this case, vehicles and containers, respectively. In this process of specialization of ships, collaborative relationships between shipping companies and vehicle companies were key to adapting new ships to the needs of the industry (Hall & Olivier, 2005). In this sense, the pioneering role of American companies in the development of containerization is well documented (Broeze, 2002), however, the key role of Japanese and European companies in innovation in car transport remains less known. Before to the specialization of ships, cars were transported in the holds of general cargo ships and were handled in the same way as other general cargoes (Kendall & Buckley, 2001).

Currently, most new cars and other types of vehicles are transported in what is often described as floating garages, formally known as PCC ships or PCTC ships. PCC ships appeared in the 1960s and gradually became the dominant means of transporting vehicles (MOL, 2019). Since the early 1960s, innovation in car transport has been associated with a small number of Japanese, Korean and Scandinavian shipping companies. This type of vessel, in addition to cars, also carries large and heavy loads, such as construction machinery and agricultural equipment, and trucks, which have their own commercial patterns.

Fundamentally, car transport by sea is an exclusive service for the automobile manufacturing industry. In fact, the global growth of the car carrier fleet was significantly promoted by the development of the automobile manufacturing industry. The steady growth of the global automobile manufacturing industry has caused the global maritime automobile trade to have grown steadily over a long period of time. PCC/PCTC have become the best means of car transport. The main shipping companies of PCC/PCTC ships cooperate closely with global car manufacturers and coordinate their development strategy to provide sufficient ship capacity and high quality service to meet their transport demand.

The most important vehicle shipping routes are from Asia to Europe and North America. Historically, the most important car exporters are located in the Far East, where China, Japan and South Korea are the main exporters of this region. In the case of Europe and North America their function is twofold, since they are both important exporters and importers of cars. As can be deduced from the main maritime routes of vehicle transport, vehicle trade has a global and highly globalized scope. These characteristics are positive for PCC/PCTC ship traffic because the demand of tons per mile increases, as well as offering greater business opportunities for car manufacturers and more options for consumers.

There are three ways for car carrier operators to participate in the car carrier market: (1) Major operators, large car carrier operating companies control the majority of PCC/PCTC capacity, either themselves or via subsidiary companies by owning or chartering vessels, for example, Hyundai Glovis, Hoegh Autoliners or Grimaldi; (2) Smaller operators, which provide some car carrier services on relatively lower-volume and/or emerging trades, such as, Abou Merhi shipping company; and, (3) Charter owners, the owners charter out vessels under time charter arrangement, for example, Cido Shipping and Ray Shipping.

The most prominent feature of the vehicle shipping industry is that long term stable strategic cooperation has to be established between car carrier companies and automobile manufacturers. In the meantime, the automobile manufacturers also have fixed car carrier companies to provide shipping logistics. This is a symbiotic relationship. On the one hand, the car carrier company can develop its fleet harmoniously according to the expansion and export plan of the automobile manufacturers. On the other hand, automobile manufacturers need car carrier companies to provide sufficient shipping space, reliable schedules and controllable cost of freight. This collaboration supports the development of global seaborne car transportation in a long run (Liu, 2014).

2.2 *Vehicle traffic in the ports of the Iberian Peninsula*

The variable used to identify the ports of the Iberian Peninsula with car-truck carrier ship traffic was the number of vehicles handled during the period 2011-2018. We could not analyze years prior to 2011 because the statistics for those years are not available. Specifically, the statistics referring to the following types of vehicles, cars, buses and trucks, were used. Among the 38 commercial ports situated on the Iberian Peninsula, six Spanish ports, Barcelona, Pasajes, Santander, Tarragona, Valencia and Vigo, and one

Table 1. Vehicle movements during the period 2011-2018 in the ports of the Iberian Peninsula with PCC/PCTC ship traffic.

Year	Vehicle movements by port (thousands)							
	Barcelona	Valencia	Santander	Vigo	Pasajes	Setúbal	Tarragona	Total
2011	550	390	319	363	255	175	65	**1941**
2012	557	287	271	311	198	136	43	**1666**
2013	575	371	316	400	201	125	39	**1901**
2014	656	459	376	368	223	149	59	**2142**
2015	750	623	459	391	245	169	120	**2589**
2016	813	710	490	425	249	172	135	**2822**
2017	752	747	493	430	235	224	200	**2857**
2018	731	795	486	415	255	274	192	**2874**
Total	**5383**	**4383**	**3211**	**3102**	**1859**	**1422**	**853**	

Portuguese port, Setúbal, are the ones that concentrated the highest PCC/PCTC ship traffic (Puertos del Estado, 2019; Port of Setúbal, 2019).

During the period from 2011 to 2018, the traffic of vehicles grew at an average annual rate of 6.4%, which means six consecutive years of growth from 2013 to 2018. In addition, since 2015 the milestone of 2.5 million vehicles imported and exported through the ports of the Iberian Peninsula was overcome, see Table 1. The port of Barcelona is the one that has moved the highest number of vehicles during the analyzed period, with 5.38 million, followed by the port of Valencia with 4.38 million. Additionally, the port of Barcelona from 2011 to 2017 was the main port moving vehicles, however, in 2018 the leadership was taken by the port of Valencia.

Regarding the export traffic, there is a direct link between each vehicle factory and the port through which it exports part of its production. That is, the factory is placed in the hinterland of the port that runs as a gate for vehicle exportation. Figure 1 shows the locations of vehicle factories on the Iberian Peninsula and the seven ports with PCC/PCTC ship traffic.

Specifically, the exportation of vehicles is made through the port closest to the vehicle factory. In particular, Ford uses the port of Valencia; the port of Vigo is employed by Citroën; the factories of Audi, Nissan and Seat employ Barcelona, the port of Santander is

used by Iveco, Nissan and Renault; the port of Tarragona is the gateway for the Opel factory in Zaragoza; the port of Pasajes is used by Mercedes and Volkswagen factories in Spain; and the Volkswagen vehicles manufactured in Portugal are exported through the port of Setúbal, see Figure 1.

3 COMPETITIVE POSITION ANALYSIS OF PORTS OF THE IBERIAN PENINSULA WITH CARTRUCK CARRIER SHIP TRAFFIC

3.1 *Methodology*

Fleisher and Bensoussan (2007) defined the competitive position of an organization as the position of an organization compared to its competitors in the same market or industry. Therefore, an analysis of competitive positions of a set of ports will enable port authority managers to gain insights into the structure of the traffic flows as well as their port's past performance in comparison with their competitors.

There are several methods to measure and identify the competitive position of a set of ports, among which the growth-market share matrix or Boston Consulting Group (BCG) matrix stands out. This matrix is divided into four quadrants, each of which

Figure 1. Map of the locations of vehicle factories and ports with PCC/PCTC ship traffic on the Iberian Peninsula.

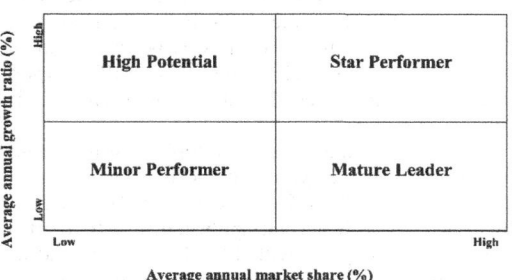

Figure 2. Growth-market share matrix applied to the port industry.

corresponds to a market position, see Figure 2. For the period analyzed, the border on the X axis is defined by the average market share and the border in the Y axis is defined by the average growth rate of all the ports analyzed.

Minor Performers and Star Performers respectively represent the least and the most auspicious ports. Star Performers represent dominant ports with high market share and high growth. However, this status might not be sustained long term (Haezendonck, 2001). Minor Performers represent ports with low growth rate less than average growth rate, small market share and overall port performances that are not good in comparison with the others. High Potential represents high growth but low market share ports; the ports that belong to this category have typically not achieved mature status yet and might move to stars' status if they can improve performance by remaining their high growth rate and gain more market share from their rivals over time. Lastly, the Mature Leaders within this matrix represent ports that have reached maturity and have well established positions with low growth rate and high market share. Through these four categories of the growth-market share matrix, the analyzed ports will be positioned within the matrix based on the vehicle traffic of each port.

The static growth-market share matrix is an analysis technique used to assess the current position of the ports through the analysis of the relationship between the growth rate and the market share. However, some authors, such as David (2000), have pointed out that a main drawback of the growth-market share matrix is that it simply provides a temporary position of a company at a specific point in time. That is, it does not show the dynamic position of the company over time. Therefore, in this work a portfolio analysis is carried out to address this issue. A portfolio analysis is based on the principles of the growth-market share matrix, but provides a dynamic view of the progress of port positions over a distinct span of time (Haezendonck, 2001). In this work, the period 2011-2018 was divided into two sub-periods, the first one consisting of the years from 2011 to 2014, and, the second, composed of the years from 2015 to 2018. Through this approach, it is possible to determine how the competitive positions of the ports have changed over time. In addition, the analysis was performed independently for the vehicle embarkation traffic (export flow) and the vehicle disembarkation traffic (import flow).

This methodology has been applied in numerous research devoted to container ports (Dang & Yeo, 2017; Liu et al., 2016; Shevchenko, 2013). However, there are few articles that apply the growth–market share matrix to ports with PCC/PCTC ship traffic. This work is pioneering in the application of this methodology to determine the competitive positions of ports of the Iberian Peninsula with PCC/PCTC ship traffic.

3.2 Results

Figure 3 shows the results of competitive positions for vehicle embarking traffic. In both analyzed periods there are ports located in each quadrant. The port of Vigo has maintained the position of Mature Leader during both periods, this position is fundamentally linked to the high market share it has. Barcelona has evolved from the position of Star Performer to Mature Leader, that change is associated with a decrease in the growth rate. The port of Valencia was located in the Star Performer quadrant during both periods. Two factors explain this position, on the one hand, the high growth rate registered during both periods and, on the other hand, the increase in market share during the period from 2015 to 2018. The most significant evolution corresponds to the port of Santander, which moved from Minor Performer to Star Performer quadrant. This evolution was due to the notable increase in both the market share and the growth rate. The increase in the growth rate registered by the ports of Setúbal and Tarragona during the period from 2015 to 2018 led to its evolution to the position of High Potential, especially in the case of Setúbal. However, their market shares decreased slightly. Finally, the port of Pasajes remained during both periods as a Minor Performer, registering a decrease of market share during the second period.

Figure 4 shows the results of the competitive positions for vehicle disembarking traffic. Barcelona and Santander during both periods have remained in the Mature Leader quadrant. Both ports during the period from 2015 to 2018 registered a decrease in market share. The port of Valencia has evolved from the position of Mature Leader to Star Performer mainly due to the growth registered during the 2015-2018 period. The port of Vigo during the periods analyzed maintained the same market share, although it increased its growth rate, which yielded to evolution from the Minor Performer quadrant to the High Potential quadrant. The port of Setúbal, evolved from the Minor Performer position to the High Potential position,

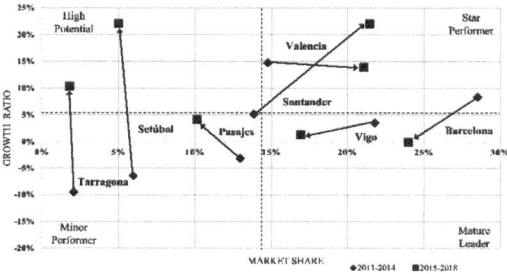

Figure 3. Matrix with portfolio analysis of port competitive positions for vehicle embarking traffic during the period from 2011 to 2018.

44

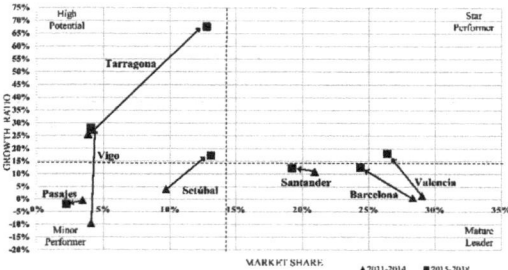

Figure 4. Matrix with portfolio analysis of port competitive positions for vehicle disembarking traffic during the period from 2011 to 2018.

mainly because it registered an increase in the market share. The port of Tarragona remained in the position of High Potential during the two periods, although with a significant growth in the market share during the period from 2015 to 2018. Finally, the port of Pasajes remained as a Minor Performer during the two periods analyzed, registering a drop in the market share during the 2015-2018 period.

4 ANALYSIS OF SEASONALITY PATTERNS OF VEHICLE MOVEMENTS

4.1 Methodology

In this section the seasonality pattern of vehicle traffic of each port is determined. The variable used to perform the analysis was the total vehicle movements. This seasonality analysis was conducted using a time series composed of 96 observations corresponding to each port's monthly registers for the period from 2011 to 2018. For each port, the time series follows a multiplicative model. In this type of time series, the seasonal component is measured by an index called the seasonal variation index (SVI), which is calculated with equation (1). This index, which is expressed as a percentage, represents the value fluctuation of the series with respect to the value of the annual average trend (Rey-Graña & Ramil-Díaz, 2007).

$$SVI(month\ i) = \left(\frac{\frac{1}{N-1} \sum \frac{y_{it}}{tc_{it}}}{1/12 \sum \left(\frac{1}{N-1} \sum \frac{y_{it}}{tc_{it}} \right)} \right) \cdot 100 \tag{1}$$

where y_{it} is the data in year t and month i; tc_{it} is the moving average in year t and month i; N is the number of years (specifically 8); and i varies from 1 to 12.

4.2 Results

Figures 5 to 11 show the representation of the seasonality patterns obtained for each port analyzed. The vehicle traffic in the ports of the Iberian Peninsula has a seasonal behavior. The peak-season months are identified by IVE values greater than 100, while the low-season months are identified with IVE values below 100. The seasonality patterns of the seven ports have certain similarities. For the seven ports, the month of August is the one with the lowest vehicle movements. Additionally, the months of January and December are also low-season months for all ports. Regarding the peak-season months, the seven ports concentrated the highest activity, mainly, during the period from February to July.

The seasonality patterns obtained are directly linked to the monthly vehicle production ratios in the different vehicle factories in the Iberian Peninsula. Through the comparison of the seasonality patterns obtained and the ratios of vehicle production, it is observed that the seasonality patterns of vehicle traffic are conditioned by the production of the

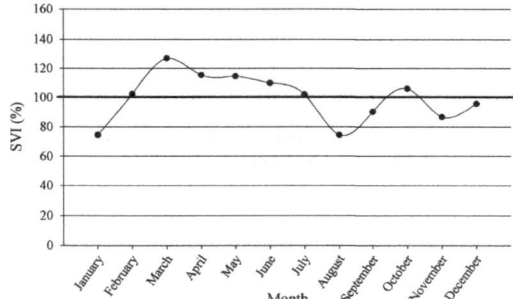

Figure 5. Seasonality pattern of vehicle traffic of the port of Barcelona during the period from 2011 to 2018.

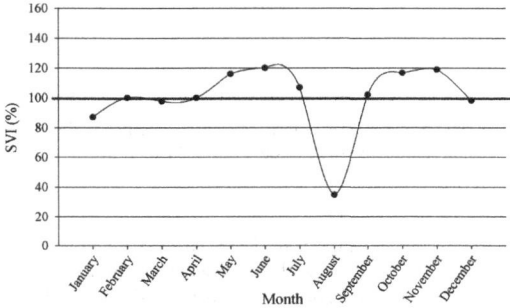

Figure 6. Seasonality pattern of vehicle traffic of the port of Pasajes during the period from 2011 to 2018.

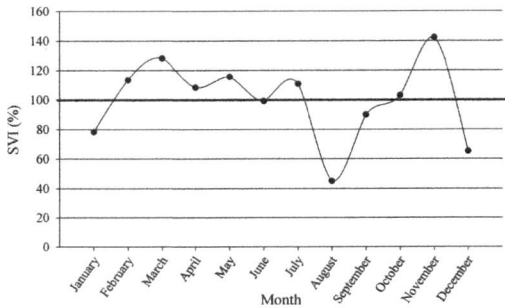

Figure 7. Seasonality pattern of vehicle traffic of the port of Santander during the period from 2011 to 2018.

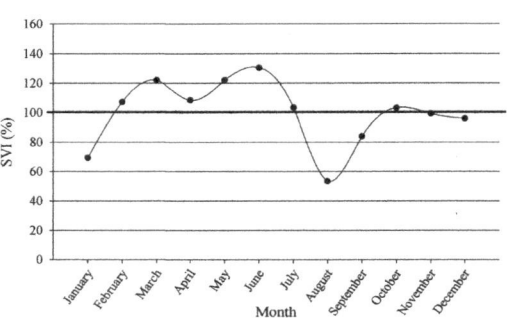

Figure 8. Seasonality pattern of vehicle traffic of the port of Setúbal during the period from 2011 to 2018.

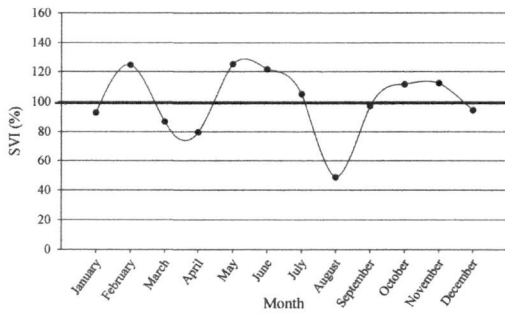

Figure 9. Seasonality pattern of vehicle traffic of the port of Tarragona during the period from 2011 to 2018.

factories. For instance, in 2018 August and December were the months with the lowest production, and the period from February to July had the highest concentration of production (ANFAC, 2019; ACAP, 2019).

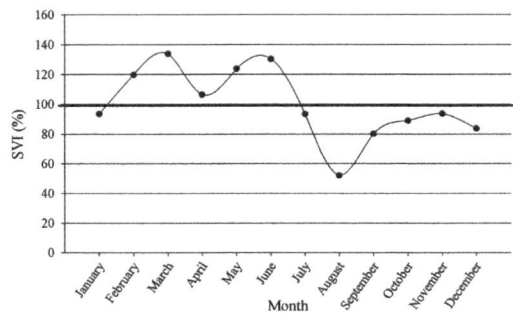

Figure 10. Seasonality pattern of vehicle traffic of the port of Valencia during the period from 2011 to 2018.

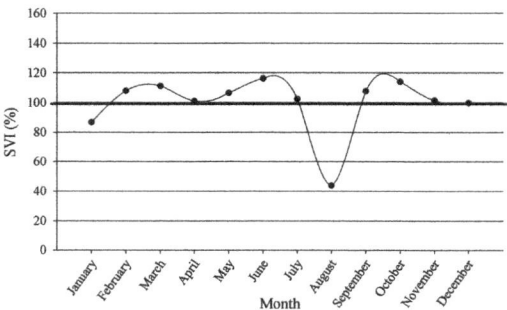

Figure 11. Seasonality pattern of vehicle traffic of the port of Vigo during the period from 2011 to 2018.

5 FORECASTS OF VEHICLE MOVEMENTS THROUGH MONTE CARLO TECHNIQUES

5.1 *Methodology*

With the aim of making forecasts about the behavior of the traffic of vehicles, simulations have been carried out with Monte Carlo techniques for each port analyzed. The result of the simulations is the average monthly vehicle movements that each port would handle. The method of Monte Carlo allows simulating a system through the construction of mathematical models that accurately represent the reality, and it is used when a solution in the form of an equation would be difficult or impossible to obtain. Monte Carlo method allows solving mathematical problems through the simulation of probabilistic variables (Gómez González, 2015).

Monte Carlo series simulations have numerous applications in several fields of the maritime industry. This method has been applied, among others, in works related to the assessment of damaged ship stability (Santos & Guedes Soares, 2005), the energy consumption of ships in port (Gutierrez-Romero et al., 2019), the evaluation of the quality of port services in container terminals (Yeo et al., 2015), and the optimization of container stowage (Zhao, 2018).

Monte Carlo simulations are executed using algorithms that generate stochastic values based on the Probability Density Function of the input data. The Probability Density Function explains the range of potential values of a given variable and the probability that different values represent the true value. The aim of these repeated simulations is to produce distributions that represent the probability of different estimates. The distributions associated with the most common Monte Carlo series are uniform, triangular, normal and lognormal.

Simulations were carried out for each port and divided by type of traffic, that is, embarking or disembarking vehicles, using as input data the database of vehicle movements during the period 2011-2018. Simulations were performed with calculation software obtaining approximately 10^6 random values per each port and traffic. The value of simulations was selected based on the values recommended in the Monte Carlo series specialized literature (McMurray et al., 2017). Additionally, the simulations were made with a confidence grade of 95%.

5.2 Results

Table 2 summarizes the results of the simulations carried out for each port and traffic. According to the results obtained, Barcelona would be the port with the highest embarking traffic with 40,331 average vehicle movements per month and Valencia would handle the highest disembarking traffic with 15,823 average vehicle movements per month. Additionally, both ports would be the leaders on the East coast of the Iberian Peninsula.

Furthermore, Vigo would be the port with the highest embarking traffic, accounting 30,212 average vehicle movements per month, and Setúbal the port

with the most important disembarking traffic, registering 6763 average vehicle movements per month, on the West coast of the Iberian Peninsula. Moreover, Santander would manage the highest embarking and disembarking traffic on the North coast of the Iberian Peninsula with 21,356 and 11,365 average vehicle movements per month, respectively.

6 CONCLUSIONS

The traffic of PCC/PCTC ships in the Iberian Peninsula is concentrated in seven ports, three located on the Eastern coast (Barcelona, Tarragona and Valencia), two situated on the Western coast (Setúbal and Vigo), and the remaining two located on the Northern coast (Pasajes and Santander). The Eastern coast concentrated the highest activity with a 52.5% of the total PCC/PCTC traffic of the Iberian Peninsula. The seven ports are directly linked to vehicle production factories, therefore, they act as the gateways for exporting the production of these factories. Moreover, these seven ports also registered vehicle disembarking traffic (import flow) because the port terminal requirements for exporting and importing vehicles are the same. The PCC/PCTC ship traffic has had a positive evolution in the Iberian Peninsula during the last eight years with an average annual growth rate of 6.4%.

The results of the matrix and portfolio analysis revealed that the seven ports analyzed were distributed within the four positions in both types of traffic, embarkation and disembarkation of vehicles. Vigo and Barcelona are the Mature Leader ports in the embarkation of vehicles. The ports of Santander and Valencia were located in the period 2015-2018 as Star Performers due to the remarkable growth that they registered in the embarkation of vehicles. Setúbal and Tarragona evolved to the quadrant of Minor Potential thus improving the position of Minor Performer that they had in the period 2011-2014. The port of Pasajes remained throughout the period analyzed as a Minor Performer, because it was the only port that did not change its competitive position. Regarding the competitive positions in the disembarking traffic, the evolution was slightly different. Barcelona is also positioned as a Mature Leader in this traffic, position that it is also occupied by the port of Santander. Valencia evolved from the position of Mature Leader to Star Performer as a result of the increasing in vehicle disembarking during the period 2015-2018. The ports of Setúbal and Vigo showed a positive evolution changing from Minor Performer to High Potential. The port of Tarragona, despite remaining throughout the period in the position of High Potential, it registered a remarkable increase of growth rate and market share. These changes placed it near the border of Star Performer quadrant. The port of Pasajes was located throughout the analysis period in the Minor Performer quadrant.

Table 2. Results of the Monte Carlo simulations for each port and traffic.

| Port | Interval for 95% of confidence grade (average monthly vehicle movements) | | | |
| | Embarking | | Disembarking | |
	Minimum	Maximum	Minimum	Maximum
Barcelona	40,160	40,502	14,854	14,689
Pasajes	18,767	18,914	1,799	1,820
Santander	21,250	21,461	11,295	11,435
Setúbal	9,167	9,264	6,664	6,861
Tarragona	21,396	21,606	11,294	11,436
Valencia	28,260	28,663	15,725	15,921
Vigo	30,103	30,320	2,926	2,981

The analysis of time series of vehicle movements showed that the PCC/PCTC ship traffic has seasonality. The seasonality of this maritime traffic is explained by the production patterns of the vehicle factories. The highest traffic was concentrated during the period from February to July. Whereas, August was notably the low-season month with the lowest activity for all ports. Moreover, January and December were low-season months, although with more activity than August. The low-season months obtained are associated with holiday periods of the vehicle manufacturing factories and their auxiliary industry.

The forecasts made through Monte Carlo techniques placed Barcelona and Valencia as the main ports for embarking and disembarking vehicles, respectively, on the Iberian Peninsula. In addition, the ports of Vigo and Setúbal would register the highest embarking and disembarking traffic of vehicles, respectively, on the Western coast of the Iberian Peninsula. Finally, the port of Santander would manage the highest embarking and disembarking movements of vehicles on the Northern coast of the Iberian Peninsula.

REFERENCES

ACAP (Associação Automóvel de Portugal). 2019. *Estadísticas Produção Automóvel Nacional*. Lisboa: ACAP.

AlMarar, A. & Cheaitou, A. 2018. Cargo flow, freight rate and speed optimisation of container liner services. *International Journal of Shipping and Transport Logistics* (IJSTL) 10(5/6): 533–566.

ANFAC (Asociación Española de Fabricantes de Automóviles y Camiones). 2019. *Informe anual 2018*. Madrid: ANFAC.

Broeze, F. 2002. *The Globalisation of the Oceans: Containerisation from the 1950s to the Present*. Liverpool: Liverpool University Press.

Chen, S., Frouws, K. & Van de Voorde, E. 2010. Technical changes and impacts on economic performance of dry bulk vessels. *Maritime Policy & Management* 37(3): 305–327.

Clarkson Research Services. 2019. *Global Seaborne Car Trade* [Online]. Available at: https://www.clarksons.com/services/broking/pctc/(Accessed: 4 September 2019).

Dang, V.L. & Yeo, G.T. 2017. A Competitive Strategic Position Analysis of Major Container Ports in Southeast Asia. *The Asian Journal of Shipping and Logistics* 33 (1): 19–25.

David, F. 2009. *Strategic management: concepts*. 12th edn. New Jersey: Pearson International Edition.

Fleisher, C.S. & Bensoussan, B.E. 2007. *Business and Competitive Analysis: Effective Application of New and Classic Methods*. Upper Saddle River: FT Press.

Gómez González, R. 2015. *A Methodology Proposal to Obtain Operational Parameter Forecasts and Operational Risk Assessment in Container Terminals*. PhD Thesis, Escuela de Ingenieros de Caminos, Canales y Puertos, Universidad Politécnica de Madrid.

González Laxe, F., Martín Bermúdez, F., Martín Palmero, F. & Novo-Corti, I. 2014. Sustainability at Spanish ports specialized in liquid bulk: evolution in times of crisis (2010–2015). *Maritime Policy & Management* 46(4): 491–507.

Gutierrez-Romero, J.E., Esteve-Pérez, J. & Zamora, B. 2019. Implementing Onshore Power Supply from renewable energy sources for requirements of ships at berth. *Applied Energy* 255 (2019) 113883. 10.1016/j.apenergy.2019.113883.

Haezendonck, E. 2001. *Essays on strategy analysis for seaports*. Antwerp: Garant Publishers.

Hall, P.V. & Olivier, D. 2005. Inter-firm relationships and shipping services: the case of car carriers and automobile importers to the United States. *Maritime Policy & Management* 32(3): 279–295.

Kahveci, E. & Nichols, T. 2006. *The Other Car Workers*. Basingstoke: Palgrave Macmillan.

Kendall, L.C. & Buckley, J.J. 2001. *The Business of Shipping*. 7th edn. Centreville, MD: Cornell Maritime Press.

Liu, M., Kang, S.H. & Ahn, W.C. 2016. Analysis of the Market Structure and Shift-effects in North China Ports. *The Asian Journal of Shipping and Logistics* 32(3): 179–186.

Liu, Y. 2014. *The prospects of development of the car carrier industry in China*. Malmo: World Maritime University.

McMurray, A., Pearson, T. & Casarim, F. 2017. *Guía para aplicar el enfoque de Monte Carlo*. Arlington: Winrock International.

MOL. 2019. *Part I: the Car Carrier Story "A reservoir of Innovative Technologies"* [Online]. Available at: https://www.mol.co.jp/ishin/en/carcarrier/past/(Accessed 28 August 2019).

Port of Setúbal. 2019. *Estatísticas Portuárias* [Online]. Available at: https://wwwportodesetubal.pt/setubal_e_sesimbra_estatisticas_portuarias.htm (Accessed 15 December 2019).

Puertos del Estado. 2019. *Estadística mensual de vehículos en régimen de mercancía*. Available at: http://www.puertos.es/es/estadisticas/Paginas/estadistica_mensual.aspx (Accessed 8 July 2019).

Rey-Graña, C. & Ramil-Díaz, M. 2007. *Introducción a la estadística descriptiva*. 2nd edn. A Coruña: Netbiblo.

Santos, T.A. & Guedes Soares, C. 2005. Monte Carlo simulation of damaged ship survivability. *Proceedings of the Institution of Mechanical Engineers Part M Journal of Engineering for the Maritime Environment* 219: 25–35.

Shevchenko, O. 2013. *Benchmarking of Russian ports case study from ports of Vladivostok and Nakhodka*. MSc dissertation, Faculty of Technology and Maritime Sciences, Vestfold University College.

Siddiqui, A., Verma, M. & Tulett, D. 2013. A periodic planning model for maritime transportation of crude oil. *EURO Journal on Transportation and Logistics* 2(4): 307–335.

Yang, C.C. & Wong, C.W.Y. 2016. Configurations of environmental management strategy: evidence from the container shipping service industry. *International Journal of Shipping and Transport Logistics* (IJSTL) 8(3): 334–356.

Yeo, G.T., Thai, V.V. & Roh, S.Y. 2015. An Analysis of Port Service Quality and Customer Satisfaction: The Case of Korean Container Ports. *The Asian Journal of Shipping and Logistics* 31(4): 437–447.

Yin, J., Luo, M. & Fan, L. 2017. Dynamics and interactions between spot and forward freights in the dry bulk shipping market. *Maritime Policy & Management* 44(2): 271–288.

Zhao, N., Guo, Y., Xiang, T., Xia, M., Shen, Y. & Mi, C. 2018. Container Ship Stowage Based on Monte Carlo Tree Search. *Journal of Coastal Research* 83: 540–547.

Developments in Maritime Technology and Engineering – Guedes Soares & Santos (eds)
© 2021 Copyright the Author(s), ISBN 978-0-367-77376-2

Shortcomings in cybersecurity education for seafarers

D. Heering, O.M. Maennel & A.N. Venables
Tallinn University of Technology, Tallinn, Estonia

ABSTRACT: Ships, ports, terminals and offshore facilities are increasingly becoming dependent on networked information and communication technology (ICT). Seafarers must be ready to cope with a growing number of cyber threats onboard ships with cybersecurity awareness playing an important role in emergency and crisis management. Unfortunately, current maritime education and training (MET) programmes do not provide enough information on cybersecurity to seafarers to be able to identify and mitigate the prevailing cyber threat landscape. This paper provides a structured survey of published maritime cybersecurity research and gives an overview of the role of the cybersecurity component in MET for seafarers. The results show that currently there are no requirements for MET institutions to include cybersecurity awareness or cyber hygiene practice in the curricula. Some areas for future research are also proposed.

1 INTRODUCTION

International trade is highly dependent on the shipping industry. Shipping provides an efficient and low-cost transportation of goods and it is estimated that over 90% of the world's trade is carried by sea (United Nations, n.d.). Maritime trade expanded by 2.7% to reach 11 billion tons in 2018 (UNCTAD, 2019). With these increasing volumes, the importance of maritime transportation to the world economy cannot be over-emphasized. The global economic inter-dependency among nations relies largely on the successful operation of the shipping industry. Since maritime accidents have a significant impact on the surrounding environments and heavily influences world trade, the safe and secure operation of today's modern shipping fleet is of utmost importance. This is ensured by various onboard and onshore maritime systems working together simultaneously.

The traditional physical security threats that may affect safe and secure ships' operation usually include piracy, smuggling, boarding, theft, stowaways and destruction. As these incidents have often been reported and continue to occur, they are well understood and the risks are known, enabling appropriate measures to be taken to mitigate these threats. One way this is achieved is through the structured education and training of seafarers. The International Convention on Standards of Training, Certification and Watchkeeping for Seafarers (STCW), 1978, as amended, sets the standards of competence for seafarers internationally (Cockcroft & Lameijer, 2012). The International Maritime Organization (IMO) has also developed a series of model courses. These provide suggested syllabi, course timetables and learning objectives to assist instructors develop training programs to meet the STCW Convention standards for seafarers.

As the maritime industry embraces the digital era, new technological developments allow shipowners to operate ships more safely, securely and economically by optimizing routes and taking ship movement dynamics into account (Ishii et al., 2010). Smart shipping initiatives support crews and improve the performance of fleets. With the introduction of new digital solutions, ships are now more reliant on the internet to continually share data between the ship and shore. Connection to the cyber domain has exposed crew members, cargo, information technology (IT) and operational technology (OT) systems to a range of different cyber threats. Operational technology is the hardware and software that is dedicated to detecting and causing changes in physical processes through direct monitoring and control of physical devices, processes and events (Gartner, n. d.). Traditionally IT and OT have been separated, but with the widespread introduction of the Internet onboard ships, they are increasingly connected and the lines between them has started to blur. This poses a significant risk as any disruption in the operation of OT systems may impact the ship's safety (BIMCO et al., 2018). Cybersecurity has become a difficult challenge for the maritime industry with its multidisciplinary nature. The strategies that most shipping companies have currently implemented are not able to efficiently counter and deter intrusions in the maritime cyber domain. The maritime industry globally has failed to make the cybersecurity a priority (Caponi & Belmont, 2015) and Estonian shipowners are no exception (Heering, 2017a).

DOI: 10.1201/9781003216582-6

2 CYBERSECURITY IN MARITIME DOMAIN

New technologies implemented in the shipping sector have impacted on the responsibilities, skills and training of seafarers. The employment of advanced equipment and working in more automated and integrated environments has increased demands on the maritime community. In order to keep the systems functioning safely and securely it is important to protect them from cyber threats. When seafarers are sailing in different geographical locations (e.g. high seas, coastal areas, ports, high dangers areas etc.) they face different challenges, which now also include cybersecurity. For example, in 2019 the Norwegian National Security Authority (NSM) together with the Norwegian Shipowner's Association and Maritime Authority published recommended measures on ICT security and usage of social media. These recommendations were published based on information received regarding a cyber campaign targeting several different sectors and companies around the world (Norwegian Maritime Authority, 2019). Norwegian authorities assessed that "all types of ships and shipowners' land-based infrastructure can be vulnerable to cyber incidents. Shipowners that operate in ISPS/MARSEC level 2 areas or higher should be aware of the situation".

Recent cyber incidents that have taken place in the shipping sector and which have been publicized have proven that the sector is not immune to cyber criminals, state-sponsored hackers or to the reckless behaviour of their own people, which have exploited vulnerabilities in their systems. Table 1 includes some latest known cyber incidents in the sector. This list does not reflect all known cyber incidents.

It has been successfully demonstrated that both IT and OT systems used onboard ships have

Table 1. Latest known prominent cyber incidents in the sector.

2017	The ransomware NotPetya was released by a suspected state actor. While the main targets of the attack were companies in Ukraine, the malware crossed the border within hours and affected the operations of computers around the world. Among the victims was the world's largest container ship and supply vessel operator, A.P. Møller–Mærsk A/S (Greenberg, 2018). This incident impacted operations at Maersk terminals around the world, caused delays and disruption that lasted weeks. According to the company's interim financial report for the second quarter of 2017, the financial impact of the incident was estimated at between $200m to $300m (Maersk, 2017).
2018	In 2018 several ransomware attacks affected normal operations of major ports around the world (Cimpanu, 2018; Drougkas et al., 2019). These included the Port of Barcelona on September 20th and the Port of San Diego declared on September 25th. However, both ports were able to continue normal operations and ships continued to access their facilities without impacts from the cybersecurity incident (Port Strategy, 2018).
2018	China Ocean Shipping Company (COSCO) reported a cyber-attack. Its fleet comprises 1317 ships with a capacity of 105.36 million DWT, ranking it as the biggest in the world (COSCO Shipping, 2020). According to the company's statement, telephone networks and e-mail system were impacted in U.S, Canada, Panama, Argentina, Brazil, Peru, Chile and Uruguay. The incident affected the carrier's ability to communicate with its ships, customers, and marine terminals (Gallagher, 2018).
2019	The crew of a ship that was bound for the ports of New York and New Jersey in February 2019 reported to the Coast Guard that they were experiencing a significant cyber incident impacting their shipboard network. A team of cyber experts concluded that although the malware had significantly degraded the functionality of the onboard computer system, essential ship control systems had not been impacted (U.S. Coast Guard, 2019a). The team also found that the ship was operating without any effective cybersecurity measures and that the critical ship control systems were exposed to significant vulnerabilities. As the ship's crew was aware of the cyber risks and of possible malware infection presented in the shipboard network, they didn't use onboard computers for personal purposes (e.g. reading e-mails, making online purchases). However, they continued using the network for updating electronic charts, manage cargo data and communicating with relevant institutions.
2019	The Maritime Transportation Security Act (MTSA) facility was involved in a ransomware intrusion case. According to the U.S. Coast Guard, the attack vector was likely a phishing e-mail sent to the operators at the MTSA facility (U.S. Coast Guard, 2019b). The Ryuk ransomware was activated by an employee who clicked on an embedded malicious link in the sent e-mail. This allowed for malware to access significant enterprise IT network files and encrypt them. It also infected the industrial control systems that monitored and controlled cargo transfer disrupting the entire corporate IT network, including camera and physical access systems.
2019	The Kuwait transportation and shipping industry experienced two cyber-attack campaigns against their IT systems between May and June 2019. The intelligence team at Palo Alto Networks believed that the tools used to carry out the attacks were created by the same developer as The Ryuk ransomware (Loock, 2019).
2016-2020	Maritime industry has seen the rise of GPS jamming and spoofing cases in different parts of the world (GPS spoofing in the Black Sea on 2016, GPS outage in South Korea in 2016, GPS jamming and spoofing during NATO exercises in 2017 and 2018, GPS disruptions in the Mediterranean Sea in 2018) (Dunn, 2020).

vulnerabilities that can be exploited or exposed unintentionally by crew members. Researchers from the University of Texas at Austin have also demonstrated the capability to take over control of a 65-m yacht in the Mediterranean Sea. Experiments have shown that the ship's along-track and cross-track positions can be modulated by feeding false positions to the ship's autopilot system, that are offset from the ship's true position. Besides this system-level effect of spoofing, specific navigation and collision avoidance instruments can also be affected. These include automatic radar plotting, the automatic identification system, the dead reckoning system built into the ship's electronic chart display and information system (ECDIS), and the ship's satellite communications. All can all generate hazardously misleading information during a GNSS spoofing attack (Bhatti & Humphreys, 2017). In January 2016 two U.S. Navy patrol boats were intercepted by Iranian military in Iranian territorial waters. Officially, it was stated, that the U.S. sailors made a navigational error that mistakenly took them into Iranian territorial waters (Schmidt & Cooper, 2016). However, it is also suspected that the patrol boats were guided in the wrong direction and into the Iranian waters by manipulating GPS signals (e.g. spoofing) (Psiaki & Humphreys, 2016).

3 CYBERSECURITY EDUCATION FOR SEAFARERS

We argue that the regulatory framework for seafarer's training is struggling to keep pace with the technological change taking place in the maritime industry. Current multilateral decision-making processes mean that STCW Convention revisions can take a long time to adopt and incorporate into the curriculum. At the same time, seafarers' skills are in urgent need for an upgrade. This means that both the MET institutions and seafarers need to be agile (Alfultis, 2018).

A number of maritime organizations have already taken steps to address this significant problem and the threat to the shipping industry. The International Maritime Organization (IMO) has issued guidelines for shipowners and authorities on maritime cyber risk management (MSC-FAL.1/Circ.3) and has also adopted Resolution MSC.428(98) – Maritime Cyber Risk Management in Safety Management Systems. This resolution encourages administrations to ensure that cyber risks are appropriately addressed in existing safety management systems (as defined in the ISM Code). This should be achieved no later than the first annual verification of the company's Document of Compliance after 1 January 2021 (International Maritime Organization, 2020).

In addition to the IMO there are a range of different guidelines from other industry organizations. This include the International Chamber of Shipping (ICS), INTERCARGO, INTERTANKO, Oil Companies International Marine Forum (OCIMF), the International Union of Marine Insurance (IUMI),

BIMCO, Department for Transport (UK), DNV GL and others.

Studies have shown that the main type of cyber breaches that organizations face (phishing, viruses and ransomware attacks) are related to human error and technical cybersecurity vulnerabilities (Klahr et al., 2017). Employees are perceived as the weak link in relation to the cybersecurity breaches (Canfield et al., 2016; Safa & Maple, 2016).

The IBM X-Force Threat Intelligence Index 2020 shows, that there was a 2000% increase in operational technology targeting incidents in 2019 (IBM X-Force Incident Response and Intelligence Services, 2020). This predicts a rising number of threat actors attacking industrial systems in the future. According to the report, the top three infection vectors in 2019 were phishing (31%), scan and exploit (30%) and stolen credentials (29%). The same report also shows that over 8.5 billion records were compromised in 2019 and that careless employees can be largely responsible for a rise of more than 200% compared to the records lost in 2018. The transportation sector was the third-most attacked in 2019, after financial services and retail sector. One of the reasons the cyber threats have spread around the world so rapidly in last few years is a lack of awareness on cybersecurity.

Although cybersecurity knowledge has increased somewhat in the maritime sector, partly also due to the publicly known cyber incidents, the results indicate that there is still a long way to go. Onboard personnel have a key role to play when operating IT and OT systems onboard. Therefore, it is of the utmost importance that the crew know how to operate the systems safely and securely, understand the risk, including cyber risks, and can identify and report suspected cyber incidents.

A 2018 Maritime Cybersecurity Survey revealed, that only 36% of the 126 respondents from maritime companies across the United States believed that their own companies were prepared enough in cybersecurity. 38% of the respondents reported that cyber attackers targeted their companies in the past year (Lee & Wogan, 2018).

Key findings from Crew Connectivity 2018 Survey Report show that the number of seafarers who have access to the internet is increasing with every year (Futurenautics Ltd, 2018). The survey comprised 5889 seafarers of which 75% confirmed that they have some form of Internet access while at sea. At the same time only 15% of seafarers disclosed that they had received any form of cybersecurity training. Of the training that did take place, most was provided by the crewing and manning companies before joining their next ship. Worryingly 47% of seafarers said that they had sailed on a ship that had been a target of a cyber-attack. The same survey also showed that 49% of the seafarers confessed that they were unaware of their employers' cyber policies, and 41% thought the responsibility lay with the ship's master. One Dutch ship owner declared during the Digital Ship Maritime Cyber Resilience Forum in Rotterdam in 2017 that all the recent incidents in his company could have been prevented from individual users being more alert (de Vleeschhouwer, 2017).

As new technologies, increasing automation and emerging threats from cyberspace are transforming seafarers' skills and responsibilities, it is important that future seafarers are supported. This requires skills training during their studies at the maritime education and training institutions including an increased focus on improving cyber resilience.

Although there is an abundance of research on cybersecurity and maritime safety and security, there is a lack of evidence highlighting the gaps and challenges of cybersecurity education for seafarers. The overall aim of this paper is to provide an overview of the research field, conclusions made to date and synthesize the collective knowledge of the field. This will provide the justification for recommended future research directions.

Table 2. List of keywords used in the literature review.

automatic identification system	maritime accidents
autonomous ship	maritime safety
bridge operations	maritime security
bridge procedures	maritime threat actors
cyber attacks	mitigation
cyber awareness	navigation
cyber risk management	cyber risk assessment
cyber security	satellite communication
cybersecurity	seafarer
cyber situational awareness	ship
cyber threats	shipping
ECDIS	simulator training
e-navigation	situational awareness
education and training	smart ship
global maritime distress and	spoofing
safety system	STCW
global positioning system	unmanned ship
human factor	vessel
jamming	vulnerabilities

4 METHODOLOGY

In this section we discuss the research methodology and limits of our survey. The method applied in this study was a comprehensive literature review. The review focused on published journal articles discussing cybersecurity in the maritime domain and those that take notice on the needs for future education and training of seafarers.

The literature search was conducted using a number of the databases These included IEEE, ScienceDirect, Springer, Google Scholar, ResearchGate, TransNav Journal, The Journal of Navigation, WMU Journal of Maritime Affairs, International Maritime Science Conference and the Digital Collection of Tallinn University of Technology Library. The focus was on the publications published within the last 7 years (2013-2020).

The keywords employed for the publication search are listed in Table 2. Keywords were used in search engines both separately and in combination in order to expand the results of the search. The scope of the survey includes the cyber threats and their impacts in maritime cybersecurity and the cybersecurity component in the maritime education and training (MET).

A selection process was also used to exclude papers that were not relevant to the current review. Papers only referring to maritime logistics, piracy, hijacking, financial risks, digital twin, cybersecurity in other domains, vessel traffic service, ship design, ship collisions, spatial planning and ship performance were excluded.

The search resulted in a range of papers covering cybersecurity threats in the shipping industry. These also included the challenges of big data in shipping, risks related to autonomous ships, vulnerabilities of the ship systems, the role of human error in ship accidents, maritime education and training, cyber situational awareness and cyber risk assessment methods.

Papers were excluded that were not related to the scope of the review and which were not accessible on the internet. A total of 88 unique papers were retrieved and analysed (Table 3.).

5 RESULTS

In this section, the authors provide a review of the literature and the role of cybersecurity training in the maritime education and training sector. Until recently, cybersecurity in the maritime domain has not been seen as a significant issue. The European Union Agency for Network and Information Security (ENISA) carried out a study on cybersecurity challenges in the maritime industry in 2011 (Cimpean et al., 2011). The aim of the study was to assist the sector to understand better its key cybersecurity risks. The target audience of the study included organizations, national authorities, government bodies and private companies that were involved in maritime activities. The following key findings were made:

- The awareness of cybersecurity is either at a very low level or even non-existent in the maritime sector. This observation was applicable at all layers, including government bodies, port authorities and maritime companies.
- ICT systems supporting maritime operations, from port management to ship communication, are generally highly complex and employ a variety of ICT technologies that also include very specific elements.

Table 3.   List of articles (sorted by publishing year).

Year	No of articles	Reference
2014	2	(Boyes, 2014), (Škrlec et al., 2014)
2015	3	(Rødseth & Burmeister, 2015), (Fitton et al., 2015), (DiRenzo et al., 2015)
2016	3	(Burke & Clott, 2016), (Jones et al., 2016), (Bolat et al., 2016)
2017	14	(Glomsvoll & Bonenberg, 2017), (Bhatti & Humphreys, 2017), (Batalden & Sydnes, 2017), (Garcia-Perez et al., 2017), (Hassani et al., 2017), (Kolev & Dimitrov, 2017), (Bhandari et al., 2017) (Tucci, 2017), (Bothur et al., 2017) (Y.-C. Lee et al., 2017), (Becmeur et al., 2017), (Bou-Harb et al., 2017), (Radmilo et al., 2017), (Botunac & Gržan, 2017)
2018	15	(Kessler et al., 2018), (Filić, 2018), (Ahvenjärvi, 2018), (Svilicic et al., 2018) (Hareide et al., 2018), (Zăgan et al., 2018), (Kimberly Tam & Jones, 2018a), (Vinnem & Utne, 2018), (Forbes, 2018) (Alfultis, 2018), (Mileski et al., 2018), (Lund, Hareide, et al., 2018), (Jacq et al., 2018), (Lund, Gulland, et al., 2018), (Kimberly Tam & Jones, 2018b)
2019	44	(Svilicic, Kamahara, Rooks, & Yano, 2019), (Lovell & Heering, 2019), (Nasaruddin & Emad, 2019), (Svilicic, Rudan, Jugović, et al., 2019), (Kitada & Baum-Talmor, 2019), (Greiman, 2019), (Baskar et al., 2019), (Bolmsten et al., 2019), (Kavallieratos et al., 2019), (Chia, 2019), (Mednikarov et al., 2019), (Oruc, 2019), (Kidd & Mccarthy, 2019), (Zăgan & Raicu, 2019), (Ahvenjärvi, Czarnowski, Kåla, et al., 2019), (Kimberly Tam & Jones, 2019a), (Kimberly Tam & Jones, 2019d), (Svilicic, Brčić, Žuškin, & Kalebić, 2019), (Svilicic, Rudan, Frančić, et al., 2019), (Svilicic, Kamahara, Celic, & Bolmsten, 2019), (Park et al., 2019), (Jacq et al., 2019), (Rana, 2019), (Hong et al., 2019), (Bolat & Kayişoğlu, 2019), (Kimberly Tam & Jones, 2019b), (Mraković & Vojinović, 2019), (Ahvenjärvi, Czarnowski, & Mogensen, 2019), (Sakar et al., 2019), (Voliotis & Filippopoulos, 2019), (Kaleem Awan & Ghamdi, 2019), (Lutzhoft et al., 2019), (K Tam et al., 2019), (Blagovest, 2019), (Heffner & Rødseth, 2019), (Hult et al., 2019), (Kimberly Tam & Jones, 2019c), (Said & Agamy, 2019), (Dimakopoulou et al., 2019), (Vidan et al., 2019), (Lušić et al., 2019), (Alop, 2019), (Daum, 2019)
2020	7	(Svilicic et al., 2020), (Alcaide & Llave, 2020), (Hynnekleiv et al., 2020), (Emad et al., 2020), (Caprolu et al., 2020), (Heering, 2020), (Kimberly Tam et al., 2020)

- In the current regulatory context for the maritime sector on global, regional and national levels, there is very little consideration given to cyber security. Most security related regulation only includes provisions relating to safety and physical security concepts.
- No holistic approach to maritime cyber risks exists. It was observed that maritime stakeholders are setting and managing cyber security expectations and measures in a rather ad hoc manner. Not all of the actual risks are being considered, such as the disruption of critical telecommunication means or the exposure of cargo information.

One of the high-level recommendations made in the ENISA study was the need for the development and implementation of awareness raising campaigns targeting the maritime sector and provision of appropriate cybersecurity training to relevant stakeholders (e.g. shipping companies, ship crews, port authorities etc.).

Categorization in chronological order allows to follow the developments in the maritime sector in regard to cybersecurity education and training. The reviewed papers were published between 2014 and 2020.

(Bloor & Sampson, 2009) describes the issues with the quality of seafarer training which were affecting maritime sector before 2010, and which are also still relevant today. The quality of the maritime education and training is not on the level that it should be. It varies widely, from clearly substandard to the highest international quality. Cybersecurity was not on the agenda at that time.

The importance of providing training on cybersecurity for shipping companies and crews was brought up by ENISA in 2011 (Cimpean et al., 2011). It pointed out the problems and shortcomings in maritime sector concerning cybersecurity. This was described at the beginning of this section. In (Rødseth & Burmeister, 2015) the authors discuss and describe the risk assessment method related to the MUNIN project; a feasibility study on an unmanned bulk carrier on an intercontinental voyage. A total of 65 main hazards were identified and then classified according to its consequence and the probability that it will happen. Risks related to cyber domain were not included.

In their work Fitton et al., 2015 address cyber operations in the maritime domain in three elements: information, technology and people. Ship crews who have been isolated from the rest of the world for many years, are now, due to the needs of a modern maritime business, are constantly connected to the Internet. It also means, that once unreachable individuals can now be targeted by the cyber criminals. In their paper the authors recommend for preventing, spotting and defending against cyber-attacks to educate, train and drill people, so that they could continue to operate under cyber-attack conditions. They also stress the importance of understanding social engineering attacks and recommend appropriate training to mitigate it.

But it is only in 2016-2017, when the importance of cybersecurity training for the shipping companies and personnel was emphasized more strongly by researchers, e.g. (Becmeur et al., 2017; Bothur et al., 2017; DiRenzo et al., 2015; Fruth & Teuteberg, 2017; Garcia-Perez et al., 2017; Heering, 2017b; Jones et al., 2016; Kolev & Dimitrov, 2017; Radmilo et al., 2017).

The case of A.P. Møller-Maersk in June 2017 (Greenberg, 2018) did finally raise the awareness of the vulnerabilities of shipping companies and ports to technological failure. This was followed by the increasing number of articles published from 2017 on the threats and vulnerabilities in shipping (Table 3.).

Articles published in 2018-2020 provide preliminary recommendations for maritime cybersecurity training (Ahvenjärvi, 2018; Alfultis, 2018; Daum, 2019; Hareide et al., 2018; Kidd & Mccarthy, 2019; Mileski et al., 2018; Kimberly Tam et al., 2020; Kimberly Tam & Jones, 2018b; Zăgan et al., 2018).

The vigilant seafarer onboard the ship is the most important security asset for the shipping company. (Hareide et al., 2018) emphasize that in addition to the need for a high degree of situational awareness in order to be able to make well informed navigation decisions, the navigators need to be also situationally aware of the status of the new IT systems and the limitations and possibilities they present. If one lacks system awareness, one would lack a vital part of the overall situational awareness and potentially present a risk factor rather than a risk reduction factor. So, in order to utilize the human capacity to be the strongest link in the maritime cybersecurity chain, it has to become a part of education and training in order to enhance the navigator's competence by increasing system awareness (Lund, Hareide, et al., 2018).

(Roolaid, 2018) hypothesized in his dissertation that maritime educational institutions have not given enough attention to specific cybersecurity education in deck officers training to ensure their ability to operate the ships safely. The author carried out research (surveys, interviews) among the European maritime educational institutions. The aim of the research was to find out which institutions are providing cybersecurity training for the seafarers. The author searched for the study programs, different courses and their learning outcomes related to cybersecurity. Information was received from about 35 MET institutions. The quantitative results of the mapping are:

– 2 institutions out of 35 provided specific cybersecurity education for ship's officers;
– 3 institutions out of 35 provided only general cybersecurity awareness education for ships' officers;
– 11 institutions out of 19 thought that it was necessary to teach cybersecurity to ships' officers;
– 7 institutions out of 19 thought that cybersecurity education will be necessary in the future;
– 1 institution out of 19 thought that cybersecurity education is not necessary for ship's officers.

The results of the survey confirm that even though the reports and industry guidelines recommend educating and training for the cadets, the collaboration between the shipping companies and educational institutions in Europe is still lacking the cybersecurity component.

According to Roolaid's research the main obstacles in providing cybersecurity training to the seafarers are: already excessive workload of the teachers, lack of study materials in native language, already filled curricula, and a lack of specific requirements in the STCW Code (Roolaid, 2018). The same issue has been brought up by (Kidd & Mccarthy, 2019). Roolaid recommends starting with a two-day cybersecurity course for ship officers. This would comprise teaching theoretical knowledge about cybersecurity in shipping and risk assessment training based on given scenarios. The practical part of the course could be based on a model-based framework for maritime cyber risk assessment created by the University of Plymouth (Kimberly Tam & Jones, 2019a).

Similar research was carried out by researchers within the project "Addressing Cyber Security in Maritime Education and Training" (CYMET) (Ahvenjärvi, Czarnowski, Kåla, et al., 2019). Ten different bachelor's degree programs on navigation in ten European maritime universities were analysed. None of the study programs included courses in maritime cybersecurity. Only two programs provided basic computer science with some elements of cybersecurity. The authors found the results to be unsatisfactory considering the importance of cybersecurity awareness and the need for proper cyber risk management on the ships.

An important part of understanding the cyber risks and in the detection of cyber incidents lies on forensic readiness of ships. This is covered by (Kimberly Tam & Jones, 2019d). The risks and forensic needs of ships are highly divergent from traditional systems (Jones et al., 2016). Currently, ship crews receive no training for recognising cyber-elements and there are no IMO requirements for cyber-related evidence to be stored. In their recommendations for improving forensic readiness in the maritime sector are steps for training staff, crew members and management to increase the cyber-incident awareness and for secure evidence handling.

(Alcaide & Llave, 2020; Bolat et al., 2016; Bolat & Kayişoğlu, 2019; Dimakopoulou et al., 2019; Heering, 2020; Kimberly Tam & Jones, 2019b, 2019c) have conducted surveys among shipping companies and maritime professionals. The purpose of these surveys is to appreciate the state of cybersecurity in the sector, understand the motivation of actions of the companies, increase the level of cyber awareness and understand the needs of the sector related to education and training of their personnel (on shore and onboard ships). The results show that seemingly cyber-secure maritime domain and ships exhibit vulnerabilities and critical components. The research carried out in Estonia (Heering, 2020) reveals that the cyber threats at sea are very real and inflict damage to the shipping operations. The main cyber incidents that were mentioned by the companies were: ship computers infecting with malware, phishing attacks, e-mail spoofing, GPS interference, ransomware in ship computers and network problems. According to the feedback from the companies the biggest cyber threats are from third parties (hackers, suppliers, passengers, port officials), company's own employees and crews, IT systems on ships and the procedures. The results also show that companies are interested in providing cyber hygiene and awareness training for their personnel and also would like to carry out cyber incident trainings onboard ships (drills).

In June 2018, the Tallinn University of Technology organised a Cyber Security Summer School. The main focus was on maritime cybersecurity. In (Lovell & Heering, 2019) the authors give an overview of the exercise developed and carried out. A novel method was used to present different approach to cybersecurity-related education and training of seafarers. This included simulator-based exercise on Wärtsilä bridge simulators and developing possible cyber-attack vectors via open-source intelligence (OSINT). The results were surprising: participants successfully developed cyber-attacks against opposing ships, were able to get hold of over 7000 usernames and passwords used by the employees and crews of NATO warships, track NATO ships using Snap Map, Twitter and other social media sites. Webcams in ports were accessed to use as intelligence gathering assets.

The maritime industry, being quite conservative has been behind other sectors in adopting new technologies. Now, when modern ships have reduced crews, the demand for adequately educated and trained professionals is on the rise. Several reviewed papers address the deficiencies and shortcoming in maritime education and training related to digital skills. The EU funded project SkillSea has been initiated with the aim of ensuring that maritime professionals possess key digital, green and soft management skills for the rapidly changing maritime labour market. In their latest report on current skills needed, the consortium addresses the main challenges the maritime shipping sector must face in nearest future (Zec et al., 2020). As the digital services are increasing, the digital skills of maritime professionals are becoming more and more important. This also includes skills required to maintain cybersecurity onboard ships or on shore. As already mentioned in this paper the STCW Convention does not make any reference to digital skills, including skills to ensure cybersecurity onboard ships. The threats and vulnerabilities described and presented in the articles in this review give completely a different view to maritime safety and security. They describe risks that didn't exist or weren't relevant in the maritime domain 10-15 years ago. They are now, but STCW Convention doesn't address them. This should change with the next revision of the Convention. The same conclusions and finding have also been made by (Ahvenjärvi, Czarnowski, & Mogensen, 2019; Ahvenjärvi, Czarnowski, Kåla, et al., 2019; Alfultis, 2018; Blagovest, 2019; Bolat et al., 2016; Bolat & Kayişoğlu, 2019; Bothur et al., 2017; Botunac & Gržan, 2017; Boyes, 2014; Burke & Clott, 2016; Emad et al., 2020; Fitton et al., 2015; Heering, 2020; Hong et al., 2019; Kidd & Mccarthy, 2019; Kolev & Dimitrov, 2017; Lovell & Heering, 2019; Lušić et al., 2019; Lutzhoft et al., 2019; Mednikarov et al., 2019; Nasaruddin & Emad, 2019; Sakar et al., 2019; Vidan et al., 2019).

The authors argue that including cybersecurity awareness training into the MET programmes of all specialities is essential. The next step would be specifying additional concrete skills and knowledge that are essential for different positions onboard the ship (bridge, engine room, etc.). These would be related to their duties and include the development of the framework for a holistic approach to increased cybersecurity awareness and competence on ships in order to avoid accidents caused by cyber incidents. This includes (i) theoretical and practical training in classrooms and simulated environments (Lovell & Heering, 2019; Kimberly Tam et al., 2020), (ii) demonstrations and experiments in special laboratories or platforms (Becmeur et al., 2017; K Tam et al., 2019; Zăgan et al., 2018) and (iii) development of bridge and operational procedures that assists the crew in identifying possible cyber threats and the immediate measures to contain them.

6 DISCUSSION

The authors speculate that in the near future shipowners will start requiring a proof or a certificate from crew members joining their ships of passing a course on basic cybersecurity or attending cyber awareness course. In the long-term these measures can play a significant role in reducing the risks of possible cyber incidents that may cause severe consequences and expenses for the company and to the environment.

Crew members should be able to understand the cyber threats that they are facing when using internet-connected and sophisticated equipment. They should also be able to question if the systems they are using, are properly updated and securely configured.

Baptiste Ossena, Global Product Leader of AGCS Hull and Marine Liabilities has said "As the use of new technologies on board vessels grows, we expect to see changes in both the risk profile of shipowners and the maritime loss environment in future. Insurers will have to deal with a growing number of more technical claims - such as cyber incidents or technological defects - in addition to traditional losses, such as collisions or groundings." (Allianz Global Corporate & Specialty, 2018).

It is only a question of time when the next big cyber-attack will hit the shipping industry. In the last few years there have been many wake-up calls. The criminal activity in the Port of Antwerp (Pol, 2015), the NotPetya malware in 2017 (Greenberg, 2018; Mimoso, 2017), the cyber-attack in shipbroking company Clarkson in 2017 (PLC, 2017), a cyber-attack on COSCO in 2018 (Mongelluzzo, 2018) and the cyber-attack on the Port of San Diego in 2018 (Cimpanu, 2018).

The industry has to be prepared for more damaging incidents in the future. This could involve a vessel carrying hazardous or polluting material, a passenger vessel with large numbers of tourists on board, or a loaded container vessel in a narrow passage. The cybersecurity company Naval Dome has reported that cyber-attacks on maritime OT systems have increased by 900% in last three years (SAFETY4SEA, 2020).

7 CONCLUSION AND FUTURE WORK

This paper reviewed the literature and research done on maritime cybersecurity with the focus on education and training for ships' crews. The results of this review indicate that there is a need for more research on cybersecurity education and training for seafarers. In most guidelines a proper cybersecurity education and cyber awareness development are seen as an important part of prevention and protection, but we conclude that education and awareness are important in all phases. The key to cyber-safer maritime operations is raising the awareness of seafarers of all

possible cyber threats, make them understand the challenges, prepare them to prevent cyber incidents on ships, train them to act properly if the problems arise, and also make them aware of their digital behaviour in cyber space in relation to cybersecurity. It's not just an issue for the crew on board, but also for office personnel and third parties, who have access to their systems.

The authors propose to introduce cybersecurity education in the maritime education and training of seafarers without delay by adopting and integrating the best practices and tools implemented and used in other sectors. The need for additional cybersecurity training has been discussed and emphasized in many research papers and reports. The review of the literature show, that although the importance of cybersecurity has increased in the maritime domain, the role and importance of maritime education and training has been so far undervalued. The training programmes should be tailored to each profession and rank onboard the ship.

Cybersecurity cannot be handled exclusively by the company IT department alone. Future research has to also investigate the cybersecurity needs for different ship crew positions (e.g. navigating officer, electrical engineer, engineer, master, ratings etc.). Currently published research provides a good overview of existing cyber threats and vulnerabilities in the maritime domain. This research can be used when updating existing educational programmes for seafarers with a cybersecurity component.

The future work and research will focus on the development of a blueprint for maritime cybersecurity courses for deck officers taking into account the guidelines and workbook published by BIMCO and ICS (BIMCO et al., 2018; BIMCO & ICS, 2019), IMO guidelines MSC-FAL.1/Circ.3 and IMO resolution MSC.428(98) (International Maritime Organization, 2017b, 2017a) and testing it in the maritime educational institution.

REFERENCES

Ahvenjärvi, S. (2018). Addressing cyber security in training of the mariner of the future - the CYMET project. *International Symposium on Integrated Ship's Information Systems & Marine Traffic Engineering Conference.* https://www.dgon-isis.org/index.php?id=46

Ahvenjärvi, S., Czarnowski, I., Kåla, J., Kyster, A., Meyer, I., Mogensen, J., & Szyman, P. (2019). Safe information exchange on board of the ship. *TransNav, 13*(1), 165–171. https://doi.org/10.12716/1001.13.01.17

Ahvenjärvi, S., Czarnowski, I., & Mogensen, J. (2019). Joint production of web-learning material by IAMU member universities. *20th Commemorative Annual General Assembly, AGA 2019 - Proceedings of the International Association of Maritime Universities Conference, IAMUC 2019*, 175–181.

Alcaide, J. I., & Llave, R. G. (2020). Critical infrastructures cybersecurity and the maritime sector. *Transportation Research Procedia.* https://doi.org/10.1016/j.trpro.2020.03.058

Alfultis, M. A. (2018). Educating the future maritime workforce in a sea of constant disrupters and change. *AGA 2018-19th Annual General Assembly (AGA) of the International Association of Maritime Universities (IAMU)*, 87–93.

Allianz Global Corporate & Specialty. (2018). *Safety and Shipping Review 2018*. 25. https://www.agcs.allianz.com/assets/PDFs/Reports/AGCS_Safety_Shipping_Review_2018.pdf

Alop, A. (2019). The Main Challenges and Barriers to the Successful "Smart Shipping." *The International Journal on Marine Navigation and Safety of Sea Transportation*, *13*(3). https://doi.org/10.12716/1001.13.03.05

Baskar, K., Kala, N., & Balakrishnan, M. (2019). Cyber Preparedness in Maritime Industry. *International Journal of Scientific and Technical Advancements*.

Becmeur, T., Boudvin, X., Brosset, D., Héno, G., Merien, T., Jacq, O., Kermarrec, Y., & Sultan, B. (2017). A Platform for Raising Awareness on Cyber Security in a Maritime Context. *Proceedings - 2017 International Conference on Computational Science and Computational Intelligence, CSCI 2017*. https://doi.org/10.1109/CSCI.2017.17

Bhandari, R., Mohanty, S. S., & Wylie, J. (2017). Cyber security the unknown threat at sea. *18th Annual General Assembly of the International Association of Maritime Universities - Global Perspectives in MET: Towards Sustainable, Green and Integrated Maritime Transport, IAMU 2017*.

Bhatti, J., & Humphreys, T. E. (2017). Hostile Control of Ships via False GPS Signals: Demonstration and Detection. *Navigation, Journal of the Institute of Navigation*, *64*(1), 51–66. https://doi.org/10.1002/navi.183

BIMCO, CLIA, ICS, INTERCARGO, INTERMANAGER, INTERTANKO, IUMI, OCIMF, & World Shipping Council. (2018). *The Guidelines on Cyber Security onboard Ships*. https://www.bimco.org/products/publications/free/cyber-security

BIMCO, & ICS. (2019). *Cyber Security Workbook for On Board Ship Use. 1st Edition 2019*. https://www.witherbyseamanship.com/cyber-security-workbook-for-on-board-ship-use-1st-edition-2019.html

Blagovest, B. (2019). Maritime education development for environment protection behaviour in the autonomous ships era. *Scientific Bulletin of Naval Academy*, *22*, 21–27. https://doi.org/10.21279/1454-864X-19-I1-003

Bloor, M., & Sampson, H. (2009). Regulatory enforcement of labour standards in an outsourcing globalized industry: The case of the shipping industry. *Work, Employment and Society*, *23*(4), 711–726. https://doi.org/10.1177/0950017009344915

Bolat, P., & Kayişoğlu, G. (2019). Antecedents and Consequences of Cybersecurity Awareness: A Case Study for Turkish Maritime Sector. *Journal of ETA Maritime Science*. https://doi.org/10.5505/jems.2019.85057

Bolat, P., Yüksel, G., & Uygur, S. (2016). A Study for Understanding Cyber Security Awareness Among Turkish Seafarers. *SECOND GLOBAL CONFERENCE ON INNOVATION IN MARINE TECHNOLOGY AND THE FUTURE OF MARITIME TRANSPORTATION*. https://doi.org/10.1007/s11628-013-0202-1

Bolmsten, J., Kasepõld, K., Heering, D., Kaizer, A., Ziemska, M., Alop, A., Chesnokova, M., Sköld, D., & Olena, S. (2019). Maritime Innovation Management - A concept of an innovative course for young maritime professionals. *Proceedings of the International Association of Maritime Universities Conference*, 268–274.

http://iamu-edu.org/wp-content/uploads/2019/11/IAMUC2019_Proceedings-1.pdf

Bothur, D., Zheng, G., & Valli, C. (2017). A critical analysis of security vulnerabilities and countermeasures in a smart ship system. *Proceedings of the 15th Australian Information Security Management Conference, AISM 2017*, 81–87.

Botunac, I., & Gržan, M. (2017). Analysis of software threats to the automatic identification system. *Brodogradnja*, *68*(1), 97–105. https://doi.org/10.21278/brod68106

Bou-Harb, E., Kaisar, E. I., & Austin, M. (2017). On the impact of empirical attack models targeting marine transportation. *5th IEEE International Conference on Models and Technologies for Intelligent Transportation Systems, MT-ITS 2017 - Proceedings*. https://doi.org/10.1109/MTITS.2017.8005665

Boyes, H. A. (2014). Maritime Cyber Security – Securing the Digital Seaways. *Engineering & Technology Reference, April*, 56–63. https://doi.org/10.1049/etr.2014.0009

Burke, R., & Clott, C. (2016). Technology, collaboration, and the future of maritime education. *RINA, Royal Institution of Naval Architects - International Conference on Education and Professional Development of Engineers in the Maritime Industry, EPD 2016, September*.

Canfield, C. I., Fischhoff, B., & Davis, A. (2016). Quantifying Phishing Susceptibility for Detection and Behavior Decisions. *Human Factors*. https://doi.org/10.1177/0018720816665025

Caponi, S. L., & Belmont, K. B. (2015). Maritime Cybersecurity: A Growing Threat Goes Unanswered. *Intellectual Property & Technology Law Journal; Clifton*. https://doi.org/10.1093/ser/mwy024

Caprolu, M., Di Pietro, R., Raponi, S., Sciancalepore, S., & Tedeschi, P. (2020). Vessels Cybersecurity: Issues, Challenges, and the Road Ahead. *ArXiv*. http://arxiv.org/abs/2003.01991

Chia, R. (2019). The Need for Ethical Hacking in the Maritime Industry. *The Society of Naval Architects and Marine Engineers, Singapore*, *38*, 108–121.

Cimpanu, C. (2018). *Port of San Diego suffers cyber-attack, second port in a week after Barcelona*. ZDNet. https://www.zdnet.com/article/port-of-san-diego-suffers-cyber-attack-second-port-in-a-week-after-barcelona/

Cimpean, D., Meire, J., Bouckaert, V., Stijn, V. C., Pelle, A., & Hellebooge, L. (2011). *Analysis of cyber security aspects in the maritime sector*. https://www.enisa.europa.eu/publications/cyber-security-aspects-in-the-maritime-sector-1

Cockcroft, A. N., & Lameijer, J. N. F. (2012). International convention on standards of training, certification and watchkeeping for seafarers, 1978, as amended. In *A Guide to the Collision Avoidance Rules (Seventh Edition)*. https://doi.org/10.1371/journal.pone.0029637

COSCO Shipping. (2020). *COSCO Shipping, Group profile*. COSCO SHIPPING Group. http://en.coscocs.com/col/col6918/index.html

Daum, O. (2019). Cyber security in the maritime sector. *Journal of Maritime Law and Commerce*.

de Vleeschhouwer, S. (2017). *Safety of data The risks of cyber security in the maritime sector*. https://maritimetechnology.nl/media/NMT_Safety-of-data-The-risks-of-cyber-security-in-the-maritime-sector.pdf

Dimakopoulou, A., Nikitakos, N., Dagkinis, I., Lilas, T. E., Papachristos, D. A., & Papoutsidakis, M. (2019). The

New Cyber Security Framework in Shipping Industy. *Journal of Multidisciplinary Engineering Science and Technology*, *6*(12), 11227–11233.

DiRenzo, J., Goward, D. A., & Roberts, F. S. (2015). The little-known challenge of maritime cyber security. *6th International Conference on Information, Intelligence, Systems and Applications (IISA)*, 1–5. https://doi.org/10.1109/IISA.2015.7388071

Drougkas, A., Sarri, A., Kyranoudi, P., & Zisi, A. (2019). *Port Cybersecurity. Good practices for cybersecurity in the maritime sector*. https://doi.org/10.2824/328515

Dunn, K. (2020, January 22). *Mysterious GPS outages are wracking the shipping industry*. Fortune. https://fortune.com/longform/gps-outages-maritime-shipping-industry/

Emad, G. R., Khabir, M., & Shahbakhsh, M. (2020). Shipping 4. 0 and Training Seafarers for the Future Autonomous and Unmanned Ships. *Marine Industries Conference, January*.

Fitton, O., Prince, D., Germond, B., & Lacy, M. (2015). *The Future of Maritime Cyber Security*. Lancaster University. http://www.research.lancs.ac.uk/portal/en/publications/the-future-of-maritime-cyber-security(d6a02f20-3125-4337-b189-e8420ca71316).html

Forbes, V. L. (2018). *The Global Maritime Industry Remains Unprepared for Future Cybersecurity Challenges. December 2014*.

Fruth, M., & Teuteberg, F. (2017). Digitization in maritime logistics—What is there and what is missing? *Cogent Business and Management*. https://doi.org/10.1080/23311975.2017.1411066

Futurenautics Ltd. (2018). *Crew Connectivity 2018 Survey Report*. https://knect365.com/shipping/article/37c4946d-cae7-4749-98dd-a4bc1f5b11e8/crew-connectivity-2018-survey-results

Gallagher, J. (2018). *Cyber attack hits COSCO Shipping*. Safety at Sea. https://safetyatsea.net/news/2018/cyber-attack-hits-cosco-shipping/

Garcia-Perez, A., Thurlbeck, M., & How, E. (2017). Towards cyber security readiness in the Maritime industry : A knowledge-based approach. *Semantic Scholar*, 1–7. https://www.semanticscholar.org/paper/Towards-cyber-security-readiness-in-the-Maritime-%3A/0bca56d7f4c56899540d3ee9180ee6c8557a813b

Gartner. (n.d.). *Operational Technology (ot)*. Gartner Glossary. Retrieved February 19, 2020, from https://www.gartner.com/en/information-technology/glossary/operational-technology-ot

Glomsvoll, O., & Bonenberg, L. K. (2017). GNSS jamming resilience for close to shore navigation in the Northern Sea. *The Journal of Navigation*, *70*, 33–48. https://doi.org/10.1017/S0373463316000473

Greenberg, A. (2018). *The Untold Story of NotPetya, the Most Devastating Cyberattack in History*. Wired. https://www.wired.com/story/notpetya-cyberattack-ukraine-russia-code-crashed-the-world/

Greiman, V. (2019). Navigating the cyber sea: Dangerous atolls ahead. *14th International Conference on Cyber Warfare and Security, ICCWS 2019*.

Hareide, O. S., Josok, O., Lund, M. S., Ostnes, R., & Helkala, K. (2018). Enhancing Navigator Competence by Demonstrating Maritime Cyber Security. *Journal of Navigation*, *71*(5), 1025–1039. https://doi.org/10.1017/S0373463318000164

Hassani, V., Crasta, N., & Pascoal, A. M. (2017). Cyber security issues in navigation systems of marine vessels from a control perspective. *Proceedings of the International Conference on Offshore Mechanics and Arctic Engineering - OMAE*. https://doi.org/10.1115/OMAE201761771

Heering, D. (2017a). *Ensuring Cyber Security in Shipping with Reference to Estonian Shipowners and Proposals for Risk Mitigation* [Tallinn University of Technology]. https://digi.lib.ttu.ee/i/file.php?DLID=8512&t=1

Heering, D. (2017b). *Küberturvalisuse tagamine laevanduses eesti laevaomanike näitel ning ettepanekud riskide maandamiseks*. https://digikogu.taltech.ee/et/Item/7bb85829-2c56-4c8e-9895-f955385f627b

Heering, D. (2020). Ensuring Cybersecurity in Shipping: Reference to Estonian Shipowners. *TransNav, the International Journal on Marine Navigation and Safety of Sea Transportation*, *14*(2), 271–278. https://doi.org/10.12716/1001.14.02.01

Heffner, K., & Rødseth, Ø. J. (2019). Enabling Technologies for Maritime Autonomous Surface Ships. *Journal of Physics: Conference Series*, *1357*, 12021. https://doi.org/10.1088/1742-6596/1357/1/012021

Hong, J.-H., Lee, C.-H. L., & Yun, G. (2019). A Study on the New Education and Training Scheme for Developing Seafarers in Seafarer 4. 0 - Focusing on the MASS. *Journal of the Korean Society of Marine Environment and Safety*, *25*(6), 726–734. https://doi.org/https://doi.org/10.7837/kosomes.2019.25.6.726

Hult, C., Praetorius, G., & Sandberg, C. (2019). On the Future of Maritime Transport – Discussing Terminology and Timeframes. *TransNav, the International Journal on Marine Navigation and Safety of Sea Transportation*, *13*(2), 269–273. https://doi.org/10.12716/1001.13.02.01

Hynnekleiv, A., Lutzhoft, M., & Earthy, J. V. (2020). Towards an ecosystem of skills in the future maritime industry. *Human Factors, February*.

IBM X-Force Incident Response and Intelligence Services. (2020). *X-Force Threat Intelligence Index 2020*. https://www.ibm.com/security/data-breach/threat-intelligence

International Maritime Organization. (2017a). *Guidelines on maritime cyber risk management*.

International Maritime Organization. (2017b). *Resolution MSC.428(98) Maritime cyber risk management in safety management systems*.

International Maritime Organization. (2020). *Maritime cyber risk*. International Maritime Organization. http://www.imo.org/en/OurWork/Security/Guide_to_Maritime_Security/Pages/Cyber-security.aspx

Ishii, E., Kobayashi, E., Mizunoe, T., & Maki, A. (2010). Proposal of new-generation route optimization technique for an oceangoing vessel. *OCEANS'10 IEEE Sydney, OCEANSSYD 2010*. https://doi.org/10.1109/OCEANSSYD.2010.5603624

Jacq, O., Boudvin, X., Brosset, D., Kermarrec, Y., & Simonin, J. (2018). Detecting and Hunting Cyberthreats in a Maritime Environment : Specification and Experimentation of a Maritime Cybersecurity Operations Centre. *2018 2nd Cyber Security in Networking Conference (CSNet)*, 1–8.

Jacq, O., Merino, P., Brosset, D., Simonin, J., Kermarrec, Y., Giraud, M., Lab-sticc, I. M. T. A., & Cedex, F.-B. (2019). *Maritime Cyber Situational Awareness Elaboration for Unmanned Vehicles*.

Jones, K. D., Tam, K., & Papadaki, M. (2016). Threats and Impacts in Maritime Cyber Security. *Engineering & Technology Reference*, 1–12. https://doi.org/10.1049/etr.2015.0123.Published

Kaleem Awan, M. S., & Ghamdi, M. A. A. (2019). Understanding the vulnerabilities in digital components of an integrated bridge system (IBS). *Journal of Marine Science and Engineering.* https://doi.org/10.3390/jmse7100350

Kavallieratos, G., Katsikas, S., & Gkioulos, V. (2019). Cyber-attacks against the autonomous ship. In *Lecture Notes in Computer* Science *(including subseries Lecture Notes in Artificial* Intelligence *and* Lecture Notes in Bioinformatics): Vol. 11387 LNCS. https://doi.org/10.1007/978-3-030-12786-2_2

Kessler, G. C., Craiger, P., & Haass, J. C. (2018). A Taxonomy Framework for Maritime Cybersecurity: A Demonstration Using the Automatic Identification System. *TransNav, the International Journal on Marine Navigation and Safety of Sea Transportation, 12*(3), 429–437. https://doi.org/10.12716/1001.12.03.01

Kidd, R., & Mccarthy, E. (2019). Maritime Education in the Age of Autonomy. *WIT Transactions on The Built Environment, 187*, 221–230. https://doi.org/10.2495/mt190201

Kitada, M., Baldauf, M., Mannov, A., Svendsen, P. A., Baumler, R., Schröder-Hinrichs, J. U., Dalaklis, D., Fonseca, T., Shi, X., & Lagdami, K. (2019). Command of vessels in the era of digitalization. In *Advances in Human* Factors, *Business* Management *and* Society. AHFE 2018. Advances in *Intelligent* Systems *and* Computing (Vol. 783). Springer, Cham. https://doi.org/10.1007/978-3-319-94709-9_32

Kitada, M., & Baum-Talmor, P. (2019). Maritime digitisation and its impact on seafarers' employment from a career perspective. *20th Commemorative Annual General Assembly, AGA 2019 - Proceedings of the International Association of Maritime Universities Conference, IAMUC 2019, November*, 259–267.

Klahr, R., Shah, J. N., Sheriffs, P., Rossington, T., & Pestell, G. (2017). Cyber Security Breaches Survey 2017: Main report. *UK Government.* https://doi.org/10.13140/RG.2.1.4332.6324

Kolev, K., & Dimitrov, N. (2017). Cyber threat in maritime industry-Situational awareness and educational aspect. *18th Annual General Assembly of the International Association of Maritime Universities, 1*, 352–360.

Lee, A., & Wogan, H. (2018). *Jones Walker LLP 2018 Maritime Cybersecurity Survey.* https://sites-communications.joneswalker.com/38/990/landing-pages/2018-maritime-cybersecurity-survey-landing-page-only-(v1).asp

Lee, Y.-C., Park, S.-K., Lee, W.-K., & Kang, J. (2017). Improving cyber security awareness in maritime transport: A way forward. *Journal of the Korean Society of Marine Engineering, 41*(8), 738–745. https://doi.org/10.5916/jkosme.2017.41.8.738

Loock, J. (2019). *Two Major Cyberattacks Have Targetted Kuwait Transportation and Shipping Industry This Year.* Maritime Security Review. http://www.marsecreview.com/2019/10/two-major-cyberattacks-have-targetted-kuwait-transportation-and-shipping-industry-this-year/

Lovell, K. N., & Heering, D. (2019). Exercise Neptune: Maritime Cybersecurity Training Using the Navigational Simulators. *5th Interdisciplinary Cyber Research Conference (ICR2019), June*, 34–37.

Lund, M. S., Gulland, J. E., Hareide, O. S., Josok, eyvind, & Weum, K. O. C. (2018). Integrity of Integrated Navigation Systems. *2018 IEEE Conference on Communications and Network Security (CNS)*, 1–5. https://doi.org/10.1109/CNS.2018.8433151

Lund, M. S., Hareide, O. S., & Jøsok, Ø. (2018). An Attack on an Integrated Navigation System. *Sjøkrigsskolen, 3*(2), 149–163. https://doi.org/10.21339/2464-353x.3.2.149

Lušić, Z., Bakota, M., Čorić, M., & Skoko, I. (2019). Seafarer market – challenges for the future. *Transactions on Maritime Science, 8*(1), 62–74. https://doi.org/10.7225/toms.v08.n01.007

Lutzhoft, M., Hynnekleiv, A., Earthy, J. V., & Petersen, E. S. (2019). Human-centred maritime autonomy-An ethnography of the future. *Journal of Physics: Conference Series.* https://doi.org/10.1088/1742-6596/1357/1/012032

Maersk. (2017). *Interim Report Q2 2017.* https://investor.maersk.com/news-releases/news-release-details/interim-report-q2-2017

Mednikarov, B., Kalinov, K., Kanev, D., Madjarova, T., & Lutzkanova, S. (2019). Current trends in the maritime profession and their implications for the maritime education. In B. Svilicic, Y. Mori, & S. Matsuzaki (Eds.), *Proceedings of the International Association of Maritime Universities Conference* (pp. 275–286). International Association of Maritime Universities. http://iamu-edu.org/wp-content/uploads/2019/11/IAMUC2019_Proceedings-1.pdf

Mileski, J., Clott, C., & Galvao, C. B. (2018). Cyberattacks on ships: a wicked problem approach. *Maritime Business Review, 3*(4), 414–430. https://doi.org/10.1108/mabr-08-2018-0026

Mimoso, M. (2017). *Maersk Shipping Reports $300M Loss Stemming from NotPetya Attack.* ThreatPost - The Kaspersky Lab Security News Service. https://doi.org/10.1177/1077546308094431

Mongelluzzo, B. (2018). *Cosco's pre-cyber attack efforts protected network.* JOC.Com Magazine, Maritime News. https://www.joc.com/maritime-news/container-lines/cosco/cosco's-pre-cyber-attack-efforts-protected-network_20180730.html

Mraković, I., & Vojinović, R. (2019). Maritime cyber security analysis – How to reduce threats? *Transactions on Maritime Science, 8*(1), 132–139. https://doi.org/10.7225/toms.v08.n01.013

Nasaruddin, M. M., & Emad, G. R. (2019). Preparing maritime professionals for their future roles in a digitalized era: Bridging the blockchain skills gap in maritime education and training. *20th Commemorative Annual General Assembly, AGA 2019 - Proceedings of the International Association of Maritime Universities Conference, IAMUC 2019.*

Norwegian Maritime Authority. (2019). *Maritime cyber risks.* Norwegian Maritime Authority. https://www.sdir.no/en/news/news-from-the-nma/cyber-risk-in-the-maritime-sector/

Oruc, A. (2019). Tanker Industry is More Ready against Cyber Threats. *International Conference on Marine Engineering and Technology (ICMET), November.* https://doi.org/10.24868/icmet.oman.2019.030

Park, C., Shi, W., Zhang, W., Kontovas, C., & Chang, C. H. (2019). Cybersecurity in the maritime industry: A literature review. *20th Commemorative Annual General Assembly, AGA 2019 - Proceedings of the International Association of Maritime Universities Conference, IAMUC 2019*, 79–86.

PLC, C. (2017). *Notice of cyber security incident.* https://www.clarksons.com/news/notice-of-cyber-security-incident/

Pol, W. van de. (2015). *Gehackte haven, cokesmokkel 2.0 (#1)*. Crimesite. https://www.crimesite.nl/gehackte-haven-cokesmokkel-2-0-1/

Port Strategy. (2018). *San Diego cyber-attack included ransom note*. https://www.portstrategy.com/news101/world/americas/cyber-attack-on-san-diego-included-ransom-note

Psiaki, M. L., & Humphreys, T. E. (2016). Attackers can spoof navigation signals without our knowledge. Here's how to fight back GPS lies. *IEEE Spectrum*, *53*(8), 26–53. https://doi.org/10.1109/MSPEC.2016.7524168

Radmilo, I., Gudelj, A., & Ristov, P. (2017). Information Security in Maritime Domain. *International Maritime Science Conference*, 76–93.

Rana, A. (2019). Commercial Maritime and Cyber Risk Management. *Safety & Defense*, *5*(1), 46–48. https://doi.org/10.37105/sd.42

Rødseth, Ø. J., & Burmeister, H.-C. (2015). Risk Assessment for an Unmanned Merchant Ship. *The International Journal on Marine Navigation and Safety of Sea Transportation*, *9*(3), 357–364. https://doi.org/10.12716/1001.09.03.08

Roolaid, L. (2018). *Küberturbe haridus laevaohvitseride väljaõppes ning soovitused selle korraldamiseks* [Tallinn University of Technology]. https://digi.lib.ttu.ee/i/file.php?DLID=10538&t=1

Safa, N. S., & Maple, C. (2016). Human errors in the information security realm – and how to fix them. *Computer Fraud and Security*. https://doi.org/10.1016/S1361-3723(16)30073-2

SAFETY4SEA. (2020). *Cyber attacks on maritime OT systems increased 900% in last three years*. SAFETY4SEA. https://safety4sea.com/cyber-attacks-on-maritime-ot-systems-increased-900-in-last-three-years/

Said, K., & Agamy, M. (2019). The impact of cybersecurity on the future of Autonomous ships. *International Journal of Recent Research in Interdisciplinary Sciences*, *6*(2), 10–15.

Sakar, C., Koseoglu, B., Buber, M., & Toz, A. C. (2019). Are The Ships Fully Secured Against The Cyber-Attacks? *Global Conference on Innovation in Marine Technology and the Future of Maritime Transportation*, 276–288.

Schmidt, M. S., & Cooper, H. (2016). *Defense Secretary Says U.S. Sailors Made Navigational Error Into Iranian Waters*. The New York Times. https://www.nytimes.com/2016/01/15/world/middleeast/us-navy-iran.html

Škrlec, Z., Bićanić, Z., & Tadić, J. (2014). Maritime Cyber Defense. *6th International Maritime Science Conference (IMSC 2014)*, *1*, 19.

Svilicic, B., Brčić, D., Žuškin, S., & Kalebić, D. (2019). Raising awareness on cyber security of ecdis. *TransNav*. https://doi.org/10.12716/1001.13.01.24

Svilicic, B., Celic, J., Kamahara, J., & Bolmsten, J. (2018). *A framework for cyber security risk assessment of ships*. 21–28.

Svilicic, B., Kamahara, J., Celic, J., & Bolmsten, J. (2019). Assessing ship cyber risks: a framework and case study of ECDIS security. *WMU Journal of Maritime Affairs*, *18*(3), 509–520. https://doi.org/10.1007/s13437-019-00183-x

Svilicic, B., Kamahara, J., Rooks, M., & Yano, Y. (2019). Maritime Cyber Risk Management: An Experimental Ship Assessment. *Journal of Navigation*. https://doi.org/10.1017/S0373463318001157

Svilicic, B., Rudan, I., Frančić, V., & Doričić, M. (2019). Shipboard ECDIS cyber security: Third-party component threats. *Pomorstvo*, *33*(2), 176–180. https://doi.org/10.31217/p.33.2.7

Svilicic, B., Rudan, I., Frančić, V., & Mohović, D. (2020). Towards a Cyber Secure Shipboard Radar. *Journal of Navigation*, *73*(3), 547–558. https://doi.org/10.1017/S0373463319000808

Svilicic, B., Rudan, I., Jugović, A., & Zec, D. (2019). A study on cyber security threats in a shipboard integrated navigational system. *Journal of Marine Science and Engineering*, *7*(10), 1–11. https://doi.org/10.3390/jmse7100364

Tam, K, Forshaw, K., & Jones, K. (2019). Cyber-SHIP: Developing Next Generation Maritime Cyber Research Capabilities. *International Conference on Marine Engineering and Technology*. https://doi.org/10.24868/icmet.oman.2019.005

Tam, Kimberly, & Jones, K. (2018a). Cyber-Risk Assessment for Autonomous Ships. *2018 International Conference on Cyber Security and Protection of Digital Services, Cyber Security 2018*. https://doi.org/10.1109/CyberSecPODS.2018.8560690

Tam, Kimberly, & Jones, K. (2019a). MaCRA: a model-based framework for maritime cyber-risk assessment. *WMU Journal of Maritime Affairs*. https://doi.org/10.1007/s13437-019-00162-2

Tam, Kimberly, & Jones, K. D. (2018b). Maritime cybersecurity policy: the scope and impact of evolving technology on international shipping. *Journal of Cyber Policy*, *3*(2), 147–164. https://doi.org/10.1080/23738871.2018.1513053

Tam, Kimberly, & Jones, K. D. (2019b). Situational Awareness : Examining Factors that Affect Cyber-Risks in the Maritime Sector. *International Journal on Cyber Situational Awareness*, *4*(1), 40–68. https://doi.org/10.22619/IJCSA.2019.100125

Tam, Kimberly, & Jones, K. D. (2019c). Factors Affecting Cyber Risk in Maritime. *2019 International Conference on Cyber Situational Awareness, Data Analytics And Assessment (Cyber SA)*, 1–8. https://doi.org/10.1109/CyberSA.2019.8899382

Tam, Kimberly, & Jones, K. D. (2019d). Forensic Readiness within the Maritime Sector. *2019 International Conference on Cyber Situational Awareness, Data Analytics And Assessment (Cyber SA)*, 1–4. https://doi.org/10.1109/CyberSA.2019.8899642

Tam, Kimberly, Moara-Nkwe, K., & Jones, K. (2020). The Use of Cyber Ranges in the Maritime Context: Assessing maritime-cyber risks, raising awareness, and providing training. *Maritime Technology and Research*, *3*(1). https://doi.org/10.33175/mtr.2021.241410

Tucci, A. E. (2017). Cyber Risks in the Marine Transportation System. In R. M. Clark & S. Hakim (Eds.), *Cyber-Physical Security* (pp. 113–131). Springer, Cham. https://doi.org/10.1007/978-3-319-32824-9_6

U.S. Coast Guard. (2019a). Cyber Incident Exposes Potential Vulnerabilities Onboard Commercial Vessels. In *Marine Safety Alert*. https://www.dco.uscg.mil/Our-Organization/Assistant-Commandant-for-Prevention-Policy-CG-5P/Inspections-Compliance-CG-5PC-/Office-of-Investigations-Casualty-Analysis/Safety-Alerts/

U.S. Coast Guard. (2019b). Cyberattack Impacts MTSA Facility Operations. In *Marine Safety Information Bulletin*. https://www.dco.uscg.mil/Portals/9/DCO Documents/5p/MSIB/2019/MSIB_10_19.pdf?ver=2019-12-23-134957-667

UNCTAD. (2019). Review of Maritime Transport 2019. In *Review of Maritime Transport*. UNCTAD. https://unctad.org/en/pages/PublicationWebflyer.aspx?publicationid=2563

United Nations. (n.d.). *IMO profile*. Business.Un.Org. Retrieved June 2, 2020, from https://business.un.org/en/entities/13

Vidan, P., Bukljaš, M., Pavić, I., & Vukša, S. (2019). Autonomous Systems & Ships - Training and Education on Maritime Faculties. *8th International Maritime Science Conference*. https://www.bib.irb.hr/1005959?rad=1005959

Vinnem, J. E., & Utne, I. B. (2018). Risk from cyberattacks on autonomous ships. *Safety and Reliability - Safe Societies in a Changing World - Proceedings of the 28th International European Safety and Reliability Conference, ESREL 2018*. https://doi.org/10.1201/9781351174664-188

Voliotis, A., & Filippopoulos, I. (2019). *An Integrated Maritime Cyber Security Policy Proposal. August*.

Zăgan, R., & Raicu, G. (2019). Understanding the OT cyber risk on board ship and ship stability. *Analele Universității "Dunărea de Jos" Din Galați. Fascicula XI, Construcții Navale/Annals of "Dunărea de Jos" of Galati, Fascicle XI, Shipbuilding, 42*, 81–90. https://doi.org/10.35219/annugalshipbuilding.2019.42.11

Zăgan, R., Raicu, G., Hanzu-Pazara, R., & Enache, S. (2018). Realities in Maritime Domain Regarding Cyber Security Concept. *Advanced Engineering Forum, 27*, 221–228. https://doi.org/10.4028/www.scientific.net/aef.27.221

Zec, D., Maglic, L., Šimić, H. M., & Gundić, A. (2020). *Current Skills Needs: Reality and Mapping*. https://www.skillsea.eu/index.php/news-events/spotlight/106-read-the-full-report-on-currents-skills-needs

Developments in Maritime Technology and Engineering – Guedes Soares & Santos (eds)

External costs in short sea shipping based intermodal transport chains

M.M. Ramalho, T.A. Santos & C. Guedes Soares
Centre for Marine Technology and Ocean Engineering (CENTEC), Instituto Superior Técnico, Universidade de Lisboa, Lisbon, Portugal

ABSTRACT: This paper presents a methodology for calculating external costs implied by intermodal transport chains, with particular focus on the specific features of its application to short sea shipping (SSS) using roll-on/roll-off (Ro-Ro) and container ships. An existing methodology for the economic assessment of SSS routes, is extended to comprise the external costs of transportation. A transport networks model is utilized to assess the full transportation costs and, therefore, the true competitiveness of different SSS Ro-Ro and container routes in comparison with road and rail alternatives. The methodologies have been applied in a case study dedicated to the European Atlantic Area that includes different routes, using all modes of freight transportation: road, rail, inland waterways and maritime, in different countries, regions, road types and travelling conditions. The case study shows that SSS is a suitable alternative in terms of CO2 emissions and external costs, but not so much for SOx, NOx and PM.

1 INTRODUCTION

The focus of the EU transport policy regarding its negative externalities has been the shifting of freight transport from the roads to other modes of transport. Freight transport (heavy good vehicles), which is responsible for more than three quarters of total externalities in the EU, contributes significantly to road degradation and air pollution, leading to increased maintenance costs and health related costs, including a significant number of fatalities arising from accidents.

The rail and waterborne modes of transport, which include short sea shipping (SSS) and inland waterways (IWW), may well assist the EU in attaining its policy objectives. However, rail and IWW networks of most peripheral countries in the EU are not as developed as in Central Europe and, in most cases, are not effectively interconnected with other regions of the EU. Simultaneously, many issues remain in terms of technical compatibility of railway lines and inland waterways in different countries. Therefore, for peripheral countries, SSS remains the most feasible alternative.

Increasing the utilization of SSS will depend on the relative performance of this mode of transportation in comparison with other modes. While it is known for being cost competitive, its transit time is generally much larger than for road transport. Also, SSS based routes often rely on integration of different transport modes, becoming less reliable when compared to the door to door road service. Furthermore, while shipping is recognisably green regarding CO_2 emissions, its performance is lagging in terms

of reducing air pollution, for example due to the much higher sulphur content of maritime fuels in comparison with road diesel. The environmental performance of shipping can be taken into account when making transport decisions based on the social (and not only private) costs. Hence, the internalisation of transport externalities is being discussed as a way of promoting sustainable multimodal transport in the EU, also within the framework of the European Year of Multimodality 2018.

Research works have already emerged on the comparison between short sea shipping and other transport modes services based on the calculation of external costs in Europe. The works of Sambracos & Maniati (2012), Tzannatos, et al. (2014), Vierth, et al. (2018) and Jiang, et al. (2010), just to mention a few, lead to the conclusion that the competitiveness of the sea alternative depends from several factors such as the world region, the transport infrastructure, the distance travelled and the type of ship to be used. This variety of factors leads to the need for a case-by-case analysis of the external costs.

Furthermore, the green label of shipping was originally questioned by Hjelle (2010) and Martínez-López (2015, 2016, 2019). While there is no doubt about the superior comparative efficiency of ships when fuel consumption is calculated per deadweight tonne (conclusions often based on bulk carriers), that might not be the case for SSS services based on container or Ro-Ro technologies when dividing fuel consumption per cargo tonne as these ships have a lower payload capacity, cargo utilization factors and operated at higher speeds.

DOI: 10.1201/9781003216582-7

Also, while in terms of CO_2 emissions shipping is accepted as an environmentally friendly transport mode that is not the case for other pollutant emissions when no emissions abatement technologies are applied (Hjelle & Fridell 2012). The case of sulphur content is especially concerning as road diesel has a much lower sulphur content than the very low sulphur fuel oil (VLSFO) used in shipping.

In order to further clarify these aspects, especially in the context of SSS and intermodal transport chains, this paper presents the application of a methodology for calculating emissions from the different transport modes used in intermodal transport chains. The methodology is explained in section 2, whereas results of a case study are shown in section 3. Conclusions are presented in section 4.

2 METHODOLOGY

A methodology has been developed and applied to an existing transportation network for calculation of the external costs and emissions in routes of interest. As a result of this, two computer programs, which use the 2014 and 2019 versions of the EU external costs handbook (see Korzhenevych *et al.* (2014) and van Essen *et al.* (2019)), were developed.

These codes use a pre-existing transport network, being only different by the input files with the costs coefficients data they work upon. Below, in general, the calculation is explained using the 2014 version of the handbook, but the code implementing the 2019 version is similar. Figure 1 shows the procedure for the calculations according with the 2019 version.

2.1 *Transport network model*

The network consists, currently, on 3018 nodes connected by 3828 links spread over the geographical region comprised between Portugal and Northern Europe, but including also Italy, Greece and Sweden. It is an intermodal transportation network accounting for four different modes of transportation: road, rail, inland waterways and short sea/maritime shipping. The nodes and links form an extensive database along with its relevant node-specific and link-specific characteristics for the calculation of both internal and external costs of transportation.

The specification of transport services is done by the user through paths. The user may specify as many paths as are necessary to describe alternative transportation services between a pair origin-destination. A path is a set of links used in succession and is specified using an ordered list of nodes. A computer

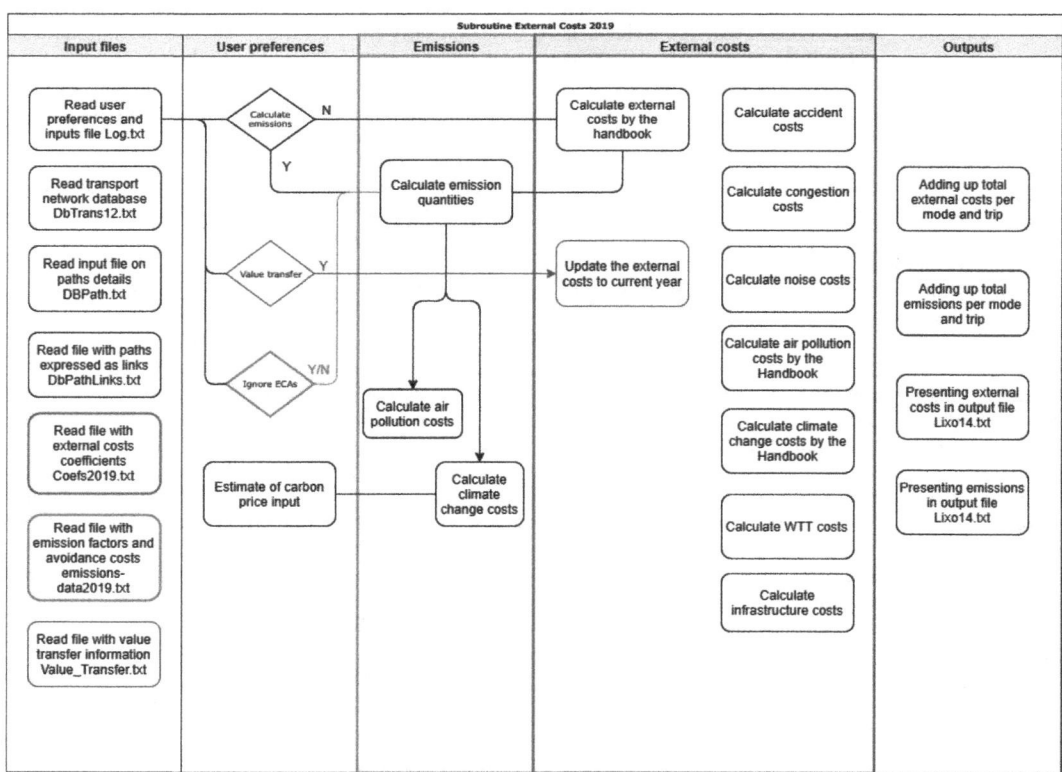

Figure 1. Procedure for calculating external costs according with 2019 handbook.

code will identify the succession of links connecting the different nodes included in this list and will also identify the type (road, rail, IWW, SSS) of each link.

For each path, there is also a number of variables specific of each transport mode. First, a variable takes the information on whether the trip takes place during the day or during the night. For the truck, the truck type, cargo capacity, gross weight, type of propulsion, EURO class, utilization factor, engine power and specific consumption and the number of axles is included. For ships (SSS) the ship type, deadweight, freight capacity (trailers), capacity utilization, type of fuel for main and auxiliary machinery, IMO emission standard, main and auxiliary engines speed rating and propulsion power and ship's design speed, are included. Similar definitions are included for trains and inland waterways vessels.

2.2 Method for costs calculation

As mentioned before, the external costs fall in the following categories: congestion, accidents, noise, air pollution, climate change, costs of up and downstream processes and marginal infrastructure costs. External costs are expressed as function of marginal cost coefficients (expressed in euros per ton, ton-km or vehicle-km). In this chapter, the vehicle will be considered as roughly equivalent to a road semi-trailer or a FEU container. Depending on the ship type (Ro-Ro or container ship), the cargo unit will then be of one or the other type.

The external costs are given by:

$$C_{Ext-ij} = C_{Cong-ij} + C_{Acc-ij} + C_{Noi-ij}$$
$$+ C_{Air-ij} + C_{CC-ij} + C_{Ud-ij} \qquad (1)$$
$$+ C_{Inf-ij}$$

where the different terms are self-explanatory (see also the Annex), but is must be taken in consideration that over a path between i and j, there will be, for example, air pollution costs arising from road, rail, IWW and maritime transportation. It may happen that the path involves only some of these transport modes, implying that some air pollution cost components will be inexistent. Also, Korzhenevych et al. (2014) indicates that for maritime transportation (IWW and SSS) there are no accident, congestion and noise external costs.

The general formulation for the external costs' calculation will be exemplified below for selected cases which serve to illustrate the general principles. Let us take again a path between an origin i and a destination j. Within this path there will be a number of links in succession, each one being denoted as k. These links may correspond to different modes of transportation. For example, the congestion cost borne in a given link k of road type, in € per vehicle, may be calculated by:

$$C_{cong-ij_{road-k}} = c_{m_{cong_{road}}} \times d_{ijk}/100 \qquad (2)$$

where d_{ijk} is the travelled distance within the link in km and the marginal coefficient $c_{m\_cong\_road}$, given in €ct/vkm, is taken from the tables provided in Korzhenevych et al. (2014). This coefficient is dependent on the type of vehicle, the link's region, road type and its congestion band:

$$c_{m\_cong\_road} = f(vehicle, region, road\ type, congestion\ band) \qquad (3)$$

The metropolitan regions in Korzhenevych et al. (2014) are considered to be referring to the suburban regions in the method's implementation. Also, the links labelled as motorways in urban areas are treated as urban main roads when choosing the most suitable cost coefficient. The congestion band of a road is defined by its volume (actual traffic flow v) to capacity (theoretical maximum traffic flow c) ratio, which may range from less than 0.25 to over 1.0, as categorized in the FORGE model (Department for Transport (2015)).

As it was not possible to find information enabling a characterization of the different types of road at a European level, only the free flow band and the higher bands are chosen as the most representative. In the implementation, congestion bands 1 to 3 are considered free flow (up to v/c equal to 0.75). The links in urban and suburban areas have been assigned category 4 (near congestion) while the others are categorized as free flow. A few notoriously congested motorways in urban areas where categorized as category 5 (over capacity).

While road transportation is carried out on an individual basis (on a truck), rail, IWW and maritime transportation is carried out collectively and the external costs need to be split by the individual cargo units. The procedure adopted for this split is now exemplified, for the case of a link of rail type. The congestion cost, in € per cargo unit, is calculated by:

$$C_{cong-ij_{rail-k}} = c_{m_{cong_{rail}}} \times d_{ijk} \times LF_{MP}/1000 \times TraCap_{ij} \times TraUt_{ij} \qquad (4)$$

where the cost coefficient $c_{m\_cong\_rail}$, given in €/1000tkm, is a coefficient from Korzhenevych et al. (2014) and depends on the country of transit:

$$c_{m\_cong\_rail} = f(country) \qquad (5)$$

When country specific values are not available, the EU average values are used. The unit costs per tonne kilometre are transformed into costs per train kilometre using the rail-specific load factors (LF_{MP}) in tonnes referred in Brons *et al.* (2013), actually based on the TREMOVE model (De Ceuster *et al.* 2004). The cost is divided by the train capacity in trailers $TraCap_{ij}$ multiplied by the train capacity utilization factor $TraUt_{ij}$ in order to obtain the values per carried trailer. These variables are included in the path definition.

The climate change cost borne in a given link k of maritime transport type, in € per cargo unit, is calculated by:

$$C_{cc-ij_{mar}-k} = c_{cc\_mar} \times d_{ijk}/SssCap_{ij} \times SssUt_{ij} \tag{6}$$

The cost coefficient c_{cc_mar} given in € per ship-km and depends on the type of ship, DWT and sea region:

$$c_{cc\_mar} = f(type\ of\ ship,\ DWT,\ sea\ region) \tag{7}$$

The type of ship and DWT must be in the path definition. Container ships, Ro-Ro and Ro-Pax ships are considered to fall into the general cargo vessel type. For bulk carriers, feeder size is considered as up to 15 kt DWT, handysize from 15kt to 40 kt and handymax greater than 40 kt. The cost is divided by the vessel's capacity in trailers $SssCap_{ij}$ multiplied by the capacity utilization factor $SssUt_{ij}$ in order to obtain the values per trailer carried onboard. These variables are defined in the path definition. The central value of the carbon price used in Korzhenevych *et al.* (2014) is 90€ and a recommendation is put on the update of the unit costs proportional to the variation of the estimates on the carbon price. In the most recent version of the handbook the recommended price is 100 €/ton CO_2.

2.3 *Activity-based emissions and external costs*

The lack of consistence between the different sources of air pollutant emissions related external costs factors, especially in the case of ships, supports in many studies the use of an activity based methodology to calculate those costs. This alternative methodology is based on an estimate of the actual quantity of pollutant emissions in a trip through the calculation of the vehicle's fuel consumption combined with emission factors. In the end, the damage costs per quantity of pollutant are used to evaluate the external costs. This can be done for all transport modes considering that either engine load or consumption characteristics and fuel are known, and emission factors or vehicle standards are available.

For the case of short sea shipping, first, the vessel's fuel consumption must be calculated, it should include main and auxiliary engine(s) consumption.

The vessel's main engine fuel consumption FC_{ME} per link, in g, can be calculated according with:

$$FC_{ME,ijk} = \frac{SFOC_{ME} \times P_{EF,ijk}}{s_{ijk}} \times d_{ijk} \tag{8}$$

where P_{EF} is the effective capacity of the main engine, in KW, $SFOC_{ME}$ is the specific fuel oil consumption of the main engine in g/kWh and s_{ijk} is the travel speed within the link in km/h.

The effective output at a given link speed s_{ijk}, in kW, can be estimated using the brake power-speed relation in (9), assuming an engine load factor at design speed with clean hull and in calm weather of 0.9. The factor 1.09 accounts for hull roughness and 1.15 for wave resistance in average conditions as per EcoTransIT World Initiative (EWI 2018) .P_{ME} is the nominal power of the main engine, in kW, and $SssSpeed_{ij}$ is the design speed of the vessel (without sea margin):

$$P_{EF,ijk} = 0.9 \times 1.09 \times 1.15 \times SssP_{MEij} \times \left(\frac{s_{ijk}}{SssSpeed_{ij} \times 1.852}\right)^3 \tag{9}$$

The main engine fuel oil consumption factors for engines as old as 2001 (IMO 2015) assumed in this model are 175 g/kWh, 185 g/kWh and 195 g/kWh, for, respectively, slow speed, medium speed and high speed engines.

Similarly, the fuel consumption in g of the auxiliary engines (and/or boilers) can be calculated as follows:

$$FC_{AE,ijk} = SFOC_{AE} \times P_{AE} \times \frac{d_{ijk}}{s_{ijk}} \tag{10}$$

depending on the time and engine load at sea P_{AE} in kW. No emissions in port are being accounted for. The engine load at sea is assumed to be 30 % of engine MCR according to Whall *et al.* (2002). For auxiliary engines and boilers, the specific fuel oil consumption factor, $SFOC_{AE}$, is 225 g/kWh for medium speed engines operating on either HFO or MDO.

Next, the amount of emissions per pollutant is obtained by multiplying the fuel consumption by the emission factors in g/g fuel (IMO 2015):

$$gCO_{2,ijk} - eq = FC_{ME,ijk} \times ef_{CO-eq_2ME} + FC_{AE,ijk} \times ef_{CO-eq_2AE} \tag{11}$$

$$gSO_{X,ijk} = FC_{ME,ijk} \times ef_{SO_XME} + FC_{AE,ijk} \times ef_{SO_XAE} \tag{12}$$

$$gNO_{X,ijk} = FC_{ME,ijk} \times ef_{NO_XME} + FC_{AE,ijk} \times ef_{NO_XAE} \tag{13}$$

$$gPM_{ijk} = FC_{ME,ijk} \times ef_{PMME} + FC_{AE,ijk} \times ef_{PMAE}` \tag{14}$$

The most important air pollutants impacting human health are particulate matter (PM) (mostly fine $PM_{2.5}$ from exhaust emissions), nitrogen oxides (NO_x) and sulphur dioxides (SO_2).

The most recent emission factors available for maritime transport come from (IMO 2015), differentiated by main engine and auxiliary engine, fuel type, engine rating and IMO tier for NO_x emissions. The sulphur content of HFO is assumed to be 2.51% and for MDO 0.1% according with the latest regulations. Tier 3 emissions factors are deducted from Tier 1 with an 80% reduction.

Finally, damage cost factors of exhaust emissions in €/ton of emitted pollutant differentiated per sea area are used to estimate air pollution external costs due to exhaust emissions from maritime transport $C_{air_{mar}ijk}$ € per trailer according with:

$$C_{air_{mar}ijk} = c_{ton\,SO_X} \times gSO_{X,ijk} + c_{ton\,NO_X} \times gNO_{x,ijk} + c_{ton\,PM_{ijk}} \times gPM_{ijk}/10^6 \times SssCap_{ij} \times SssUt_{ij} \tag{15}$$

For climate change costs, the amount of emitted CO_2 equivalent gases (GHG) is multiplied by the estimate of the carbon price to calculate the climate change external costs, given in € per trailer:

$$C_{CC_{mar}ijk} = c_{tonCO_2} \times gCO_{2,ijk} - eq/10^6 \times SssCap_{ij} \times SssUt_{ij} \tag{16}$$

For those alternative fuel technologies other than the use of MDO or HFO for which emissions factors are known, external cost reduction factors can be applied. For heavy-duty truck emissions, the limits of the standards introduced by Directive 88/77/EEC and following amendments are used. In the rail links, emission factors for line-haul locomotives are from (EEA 2016). Emission factors of SO_2, when not directly available, are estimated using the fuel's sulphur content and the molecular weight relation (U.S. Environmental Protection Agency (EPA) 2019).

3 NUMERICAL RESULTS

The methodology and numerical methods described above will now be applied to routes located within the European Atlantic Area. This choice is due to the fact that Portugal and Spain have significant volumes of trade in goods with central European countries, such as France, and could therefore benefit from the use of Ro-Ro based SSS services as an alternative to routes that use mainly the road (Santos & Guedes Soares 2017).

Accordingly, in this section a study of intermodal transport solutions in this corridor is reported, focusing mainly on the external costs implied by the different routes, while in Santos & Guedes Soares (2019) the same corridor was studied considering only the internal costs and the transit time of the various routes. Different transport scenarios are chosen involving all transport modes considered in the model, as represented in Figure 2, and its characteristics are presented in Table 1.

The objective of this case study is to include paths with all transport modes being tested. For the road links, a 5 axles 40t EURO 5 articulated truck is considered at 85% of loading capacity and having an engine with a power of 365 kW and a specific fuel oil consumption of 215 g/kWh. For the rail links, a diesel train with a capacity of 40 FEU's utilized at 75% is considered, having an installed power of 2240 Kw and a SFOC of 236 g/kWh. A vessel with 737 kW of installed power is considered for inland waterways links, having a capacity of 50 FEUs utilized at 60%. Two reference cargo ships are considered for maritime links: a containership[1] having a deadweight of 11 200 t, a capacity of 306 FEUs/trailers (14 tons) utilized at 70%, an installed power for the main engine of 8400 kW and two auxiliary generators at 500 kW each; a ro-ro cargo ship having a deadweight

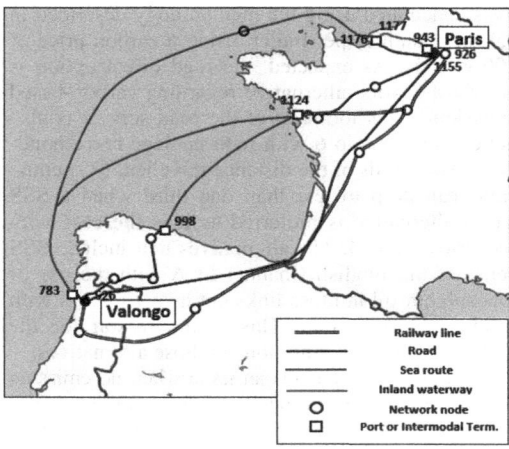

Figure 2. Transport network supporting the intermodal and unimodal routes in the corridor Valongo–Paris.

---

1 Reference vessel: WEC VERMEER IMO 9237371.

Table 1. Characteristics of the different scenarios studied in the corridor Valongo-Paris.

Identification	Name	Description	Distances	Speed average	Number of nodes
Scenario 1	Road direct	Long range road haulage	Road distance: 1628 km	Road: 80 km/h	54
Scenario 2	Rail	Long-haul freight train Valongo-Paris + Short range road haulage	Road distance: 47.5 km Railway distance: 1745 km	Road: 50 km/h Rail: 37 km/h	33
Scenario 3	SSS LeHavre (Ro-Ro)	Ro-Ro cargo ship Leixões-Le Havre + Medium range road haulage	Road distance: 246 km Sea distance: 836 km	Road: 80 km/h Sea: 15 kn	24
Scenario 4	IWW	Short range road haulage + Containership Matosinhos-Le Havre + Inland vessel Le Havre-Paris	Road distance: 38 km Inland distance: 343 km Sea distance: 841 km	Road: 80 km/h Inland: 10 kn Sea: 15 kn	21
Scenario 5	SSS St.Nazaire (Ro-Ro)	Medium range road haulage + Ro-Ro cargo ship Leixões-St. Nazaire	Road distance: 460 km Sea distance: 1086 km	Road: 80 km/h Sea: 15 kn	25
Scenario 6	SSS Gijon (Ro-Ro)	Long range road haulage + Ro-Ro cargo ship Gijon-St. Nazaire	Road distance: 988 km Sea distance: 482 km	Road: 80 km/h Sea: 15 kn	47
Scenario 7	SSS Gijon & Rail St. Nazaire	Medium range road haulage + Ro-Ro cargo ship Gijon-St. Nazaire + Freight train St. Nazaire-Paris	Road distance: 585.5 km Railway distance: 507 km Sea distance: 482 km	Road: 80 km/h Railway: 37 km/h Sea: 15 kn	45

of 13 535 t, a capacity of 239 FEUs/trailers utilized at 80%, an installed power for the main engine of 12000kW and 1270kW of auxiliary power.

Figure 3 shows the air pollutant and climate change emissions for the various transport alternatives, calculated using the methodology described in 3.4, in quantities per trailer, using a carbon price of 100 €/tCO$_2$. As expected, the road direct option is clearly the worst alternative regarding carbon-based emissions. This footprint of the road service is also noticed in scenario 6 with road haulage corresponding to two thirds of the distance travelled. CO$_2$ emissions can drop to less than one third when a SSS based alternative is preferred as it is the case with scenarios 3 and 4. The alternatives that include SSS consider the English Channel ECA with the use of low sulphur oil in those links but heavy fuel oil with sulphur 2.5% outside. This is the reason for the higher air pollutant emissions in those alternatives.

Although NO$_x$ remains an issue when no emission control options are considered, the least pollutant alternative is the SSS based route to Le Havre (Ro-Ro) represented by scenario 3. In scenario 4, while the maritime distance is the same as in scenario 3, NO$_x$ emissions are more than double. This is due to the inland waterways leg, which is using low sulphur oil but no emissions abatement technology, resulting in much higher external costs than when the leg Le Havre-Paris is done by road.

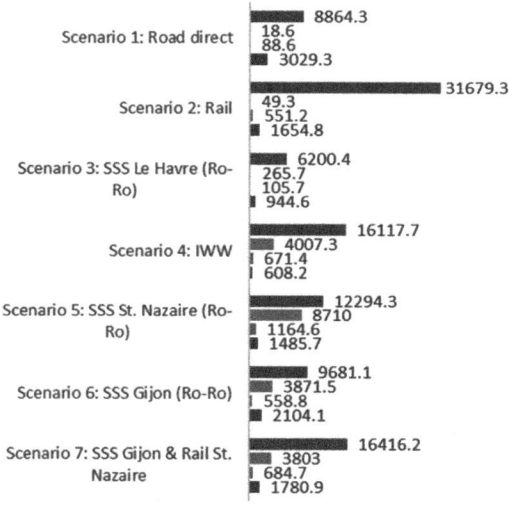

Figure 3. Air pollutant and climate change emissions for the different route alternatives Valongo-Paris.

The shares of external costs types in each route option are presented in Figure 4. For this corridor, emissions related external costs (air pollution and

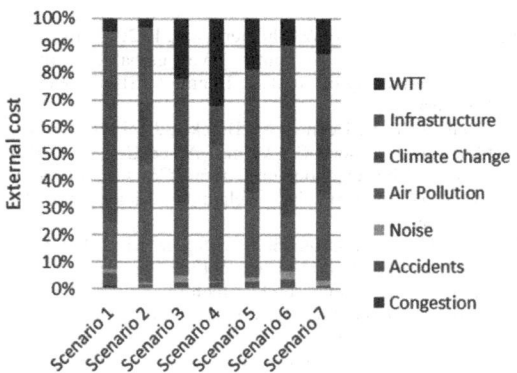

Figure 4. Distribution of external cost types in the transport scenarios Valongo-Paris.

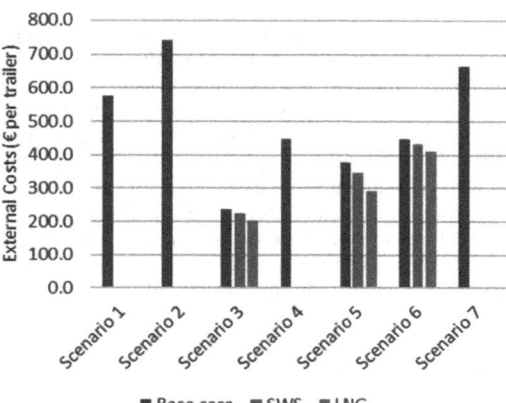

■ Base case ■ SWS ■ LNG

Figure 6. External costs per trailer comparison of the different routes Valongo-Paris – seawater scrubbing and LNG.

climate change) are responsible for more than 50% of the total external cost figures and should therefore be the main concern in the path for achieving freight transport long term sustainability. This conclusion is transversal to all transport modes.

The results on the external costs of this transport scenarios were added to its internal costs (Santos & Guedes Soares, 2017), resulting in Figure 5. This means that this figure shows the results of a full internalization of the external costs. Considering this, the unimodal solution becomes more attractive compared to scenario 6 using SSS contrary to what happens when only external costs are considered. Even so, the superiority in terms of total costs of SSS routes in this corridor is noticeable at least for scenarios making use of the ports of Le Havre and Saint Nazaire (the ones leading to less road distances).

Even without any emission control option considered and using H FO in the sea links outside the English Channel, SSS alternatives prove to be very competitive in this corridor. By 2020 the compliance with the sulphur limit of 0.5% should be mandatory in all sea areas and will further benefit the competitiveness of those (in the same travelling conditions). Let us then assume alternatively the use of a seawater

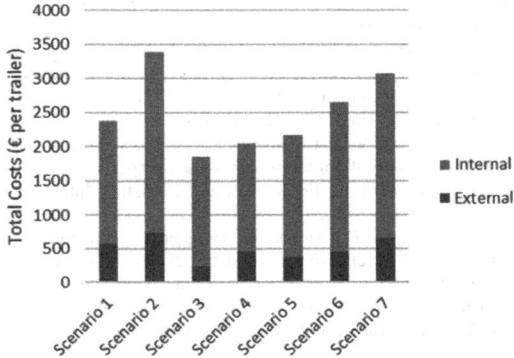

Figure 5. Total costs per trailer comparison of the different route alternatives Valongo-Paris.

scrubber and LNG/methanol in the maritime links in scenarios 3, 5 and 6, reducing air pollution costs in 58% and more than 90%, respectively. The comparison of total external costs when those emission abatement and fuel technologies are considered is in Figure 6. Because the road leg is dominant in terms of external costs share in those routes, the use of a scrubber represents in all scenarios less than a 10% reduction in total external costs, and the use of LNG at most 22% reduction in scenario 5.

4 CONCLUSIONS

A methodology for calculating emissions and external cost components was presented that uses a detailed database of transport networks, including both roads, rail, inland and sea routes. Results obtained may assist shipping companies, ports and national authorities in the development and promotion of more sustainable transport chains.

This methodology has been applied in a case study dedicated to the study of transport operations in the European Atlantic Area corridor (Portugal – Northern Europe), chosen due to its interest for the Portuguese freight market. Various scenarios were built for different land/sea routes and varying ship speeds and fuel technologies were chosen.

Regarding the performance of different transport modes compared within the same transport corridor, road transport is, as expected, responsible for most external costs of transport and loses its competitiveness over CO_2 emissions costs. In comparison, SSS services are greener with respect to climate change costs but not when other air pollutant emissions are considered. The change on air pollution and climate change costs with the use of alternative fuel technologies and when in slow steaming was demonstrated.

In this corridor, SSS base routes seem to be the most competitive alternatives at least while there are no changes in the road and rail sectors. In the future, with new engines and exhaust emissions cleaning technologies encouraged by regulations in force among all transport modes, these conclusions may well be changed. This means that a quick transition into green shipping could represent a significant advantage for the shipping sector. As shown, damage costs and the avoidance costs of carbon play a big role in the valuation of respectively, air pollution and climate change costs and need always to be updated along with the most recent estimates, before quantitative evaluations are carried out.

ACKNOWLEDGEMENTS

The research presented in this paper received support from the research project "SHORTSEACHAIN Evaluation of short sea shipping services integrated in supply chains", PTDC/ECI-TRA/28754/2017, financed by the Portuguese Foundation for Science and Technology (Fundação para a Ciência e Tecnologia - FCT). The authors would like to acknowledge also the comments of an anonymous reviewer. This work contributes to the Strategic Research Plan of the Centre for Marine Technology and Ocean Engineering (CENTEC), which is financed by the FCT under contract UIDB/UIDP/00134/2020

REFERENCES

Brons, M. & Christidis, P. 2013. External cost calculator for Marco Polo freight transport project proposals Call 2013 updated version, European Union Join Research Centre (JRC), Luxembourg: Publications Office of the European Union, Sevilla, Spain, ISBN 978-92-79-25820–6.

De Ceuster, G., Van Herbruggen, B., Logghe, S., Proost, S. 2004. TREMOVE 2.2 Model and Baseline Description. Report B4-3040/2002/342069/MAR/C.1, European Commission, DG ENV, Directorate C – Environment and Health.

Department of Transport 2005. National Transport Model FORGE - The Road Capacity & Costs Model, Research Report.

EcoTransIT World Initiative (EWI) 2018. Ecological Transport Information Tool for Worldwide Transports, Methodology and Data Update, s.n., Berne-Hannover-Heidelberg.

EEA 2016. EMEP/EEA Air pollutant emission inventory guidebook 2016 – Technical guidance to prepare national emission inventories, EEA Report 21/2016

Hjelle, H. M. 2010. Short Sea Shipping's Green Label at Risk, Transport Reviews, 30:5, pp. 617–640.

Hjelle, H.M. & Fridell, E. 2012. When is Short Sea Shipping environmentally competitive? Environmental Health - Emerging Issues and Practice, Jacques Oosthuizen, IntechOpen, DOI: 10.5772/38303.

IMO 2015. Third IMO GHG Study 2014 Executive Summary and Final Report, International Maritime Organization, London.

Jiang, L., Kronbak, J., & Christensen, L.P. 2010. External Costs of Maritime Shipping: A Voyage-based Methodology. University of Southern Denmark (SDU), Odense, Denmark.

Korzhenevych, A., Dehnen, N., Bröcker, J., Holtkamp, M., Meier, H., Gibson, G., Varma, A. & Cox, V. 2014. Update of the Handbook on External Costs of Transport, DG-MOVE. Brussels, Belgium, Ricardo-AEA/R/ED57769.

Martínez-López A., Caamaño, P., Míguez, M. 2016. Influence of external costs on the optimisation of container fleets by operating under motorways of the sea conditions. International Journal of Shipping and Transport Logistics, 8:6, pp. 653–686.

Martínez-López A., Caamaño P., Castro, L. 2015. Definition of optimal fleets for Sea Motorways: the case of France and Spain on the Atlantic coast. International Journal of Shipping and Transport Logistics, 7:1, pp. 89–113.

Martínez-López A., Caamaño P., Chica, M., Trujillo, L. 2019. Choice of propulsion plants for container vessels operating under Short Sea Shipping conditions in the European Union: An assessment focused on the environmental impact on the intermodal chains. Proceedings of the Institute of Mechanical Engineers, Part M, Journal of Engineering for the Maritime Environment, 233:2, pp. 653–669.

Sambracos, E., Maniati, M. 2012. Competitiveness between short sea shipping and road freight transport in mainland port connections; the case of two Greek ports, Maritime Policy & Management, 39:3, pp. 321–337.

Santos, T.A. & Guedes Soares, C. 2017. Modelling of Transportation Demand in Short Sea Shipping. Maritime Economics and Logistics, 19:4, pp. 695–722.

Santos, T.A., Guedes Soares, C. 2019. Short sea shipping and the promotion of multimodality in the European Atlantic Area. Proc. of the Conference of the International Association of Maritime Economists, Athens, Greece.

Tzannatos, E., Papadimitriou, S. & Katsouli, A. 2014. The cost of modal shift: a short sea shipping service compared to its road alternative in Greece, European Transport\Transporti Europei, 56:2.

U.S. Environmental Protection Agency (EPA) 2019. Regulatory Impact Analysis: Control of Emissions of Air Pollution from Category 3 Marine Diesel Engines, Office of Transportation and Air Quality, EPA-420-R-09-R–019.

van Essen, H., van Wijngaarden, L., Schroten, A., Sutter, D., Bieler, C., Maffii, S., Brambilla, M., Fiorello, D., Fermi, F., Parolin, R. & El Beyrouty, K. 2019. Handbook on the external costs of transport Version 2019, Publications Office of the European Union. Luxembourg, ISBN 978-92-79-96917–1.

Vierth, I., Sowa, V. & Cullinane, K. 2018. Evaluating the external costs of trailer transport: a comparison of sea and road, Maritime Economics and Logistics. http://dx.doi.org/10.1057/s41278-018-0099-7.

Whall, C., Cooper, D., Archer, K., Twigger, L., Thurston, N., Ockwell, D., McIntyre, A. & Ritchie, A. 2002. Quantification of emissions from ships associated with ship movements between ports in the European Community, Entec UK Limited, Report to the European Commission.

ANNEX

Abbreviations:
AE: Auxiliary Engine
DWT: Deadweight
ECA: Emission Control Area
ef: Emission Factor
GHG: Greenhouse Gases
HFO: Heavy Fuel Oil
IMO: International Maritime Organization
IWW: Inland Waterways
kn: knots
kt: kilotonnes
LF: Load Factor
LNG: Liquefied Natural Gas
MCR: Maximum Continuous Rating
MDO: Marine Diesel Oil
ME: Main Engine
RO-RO: Roll-On Roll-Off cargo ship
RO-PAX: Roll-on Roll Off passenger ship
SFOC: Specific Fuel Oil Consumption
SSS: Short Sea Shipping
tkm: tonne-kilometre
vkm: vehicle-kilometre

Symbology:

CO_2	Carbon Dioxide
SO_x/SO_2	Sulphur Oxides/Sulphur Dioxide
$PM/PM_{2.5}$	Particulate Matter/Fine Particulate Matter
NO_x	Nitrogen Oxides
FC_{ME}	Main engine fuel consumption
$SFOC_{ME}$ /$SFOC_{AE}$	Specific fuel oil consumption of the main engine/auxiliary engine
P_{ME}/P_{AE}	Nominal power of main engine/auxiliary engine
P_{EF}	Effective power of main engine
$TraCap$	Train capacity in number of trailers
$TraUt$	Train capacity utilization factor
$SssCap$	SSS vessel's capacity in number of trailers
$SssUt$	SSS vessel's freight capacity utilization factor
$SssSpeed$	SSS vessel's design speed in kn
d_{ijk}	Travelled distance within the link in km
s_{ijk}	Travel speed within the link in km/h
$C_{Cong-ij}$	External cost due to congestion
C_{Acc-ij}	External cost due to accidents
C_{Air-ij}	External cost due to air pollution
C_{Noi-ij}	External cost due to noise
C_{CC-ij}	External cost due to climate change
C_{Ud-ij}	External cost due to upstream and downstream activities
C_{Inf-ij}	External cost due to infrastructure costs

Review of waste handling plans in insular ports: Lessons from Las Palmas Port

A. Ruiz & A. Martínez-López
Department of Mechanical Engineering, University of Las Palmas de Gran Canaria, Spain

I. Pérez
Department of Applied Economics, University of Las Palmas de Gran Canaria, Spain

ABSTRACT: This paper examines the consequences of Directive (EU) 2019/883 implementation by assessing the performance of the sewage treatment plan (Annex IV of MARPOL) in Las Palmas Port from the social interest perspective. So, the accomplishment of the Cost Recovery System in Las Palmas is assessed by considering the effectiveness of the current protocol for the reception and treatment of Annex IV waste from the point of view of the on-shore side social benefits. This involves review of the mandatory fees for landing this waste in port, the cost of the sewage treatment, the environmental impact of the treated effluent from the vessels (environmental costs) and addressing technical difficulties in order to meet the normative standard required from the vessels' sewage in the framework of Las Palmas Port. Next, alternative waste treatment plans are discussed through a performance's comparison in terms of social benefits.

1 INTRODUCTION

In order to ensure fulfilment of the international maritime regulation (IMO MARPOL 73/78 Convention -International Convention for the Protection of Pollution from Ships, 1973, as amended by the 1978 Protocol-) about the reduction of pollution due to sea discharges from spills and waste generated on board, the Directive 2000/59/EC (PRF Directive), as amended by Directive 2015/2087/EC, states that all European Union ports must have adequate waste reception facilities (Port Waste Reception Facilities, hereinafter PRF) and perform the withdrawal service without causing unnecessary delays.

PRF Directive also incorporates a charging system based on the "the Polluter Pays Principle" (Cost Recovery System -CRS-), so that the charge for this service is fully financed (planning, collection and treatment). In this regard, Member States have autonomy to design their waste treatment plans, their PRFs and the management of the CRS. This heterogeneity in the member countries has generated complaints from users, especially due to the high cost per m³ that must be met by vessels in order to discharge Annex I and Annex IV of MARPOL wastes and the high volumes for treatment. In turn, the REFIT evaluation (COM 2016(168)) of the PRF Directive concluded several 'weak points' to be solved.

One of them was especially highlighted: The costs assumed to implement the Directive are not always proportionate to the benefits gained from complying with the Directive. This is especially true in ports with additional difficulties, which lead to the development of Waste Reception and Handling Plans (WRHPs) becoming a serious challenge. This is the case of European archipelagos, where the difficulties inherent to insularity (supplies, stockpiles, etc.) are often on top of a high level of port activity (perhaps because the port is a strategic location in certain kinds of traffic).

Las Palmas Port (Gran Canaria) provides a good example case in this point. As it is one of the main ports in Spain, the lack of a specific wastewater treatment plant for these waters forces the transfer of bilge waters to wastewater treatment plants located in the Iberian Peninsula to comply with the national discharge regulations.

Recently, Directive (EU) 2019/883 repealed the PRF Directive especially mentions the lack of transparency in the information included in the Waste Reception and Handling Plans (WRHPs). Article 5 states that Member States must ensure that this information about the adequate port reception facilities in their ports and the structure of the costs (Cost Recovery Systems description with waste management schemes) must be publicly available and easily accessible.

This paper contributes to this aim though the review of the mandatory fees for landing this waste in port, the cost of sewage treatment, the environmental impact of the treated effluent from the vessels (environmental costs), and specific technical difficulties to achieve the normative quality from the vessels sewage in the current scenario of the port of Las Palmas. Next,

DOI: 10.1201/9781003216582-8

alternative waste treatment plans (treatment routes) were analyzed through a performance comparison in terms of social benefits. Since it is well known that the regulatory framework is an effective instrument to ensure the Polluter Pays Principle (PPP), the results of this work are not only useful for port authorities, or waste operators, but also for European policy makers, who should consider the particularities of insular territories when modifying the Directives.

2 SEWASTE PLAN MANAGEMENT ONBOARD

According to MARPOL 73/78 (IMO, 2017) ship waste can be divided into six Annexes. This paper focuses on the quantitative analysis of Annex IV waste management. This mainly involves sanitary wastewater that can be divided into black water, or sewage, and grey water. The former includes discharge from toilets, urinals, medical rooms and spaces with live animals. On the other hand, grey waters come from the baths, showers, washing machines, and swimming pools, among others. The quality of sanitary wastewater is determined by the amount of certain substances and energy contained by the wastewater (Peric et al., 2016).

The revised Annex IV of MARPOL 73/78, states the treatment requirements for the discharge of these residues on the basis of the distance from the coast (Regulation 11). When the navigation area is 3 nautical miles from the coast, sewage discharge is only permitted when it is fully treated (Advanced Wastewater Treatment System-AWTS-) otherwise it must be kept on board (in wastewater tanks) until it can be discharged in port. In a navigation area between 3 and 12 nautical miles a partially treated discharge of wastewater (comminute and disinfected) is required, through an AWTS or a Marine Sanitation Device -MSD- (the latter uniquely treats black waters). Finally, when the shipping area is further than 12 nautical miles from the coast and the vessel speed is greater than 4 knots, previous treatment of sewage is not compulsory for discharge to be allowed. In any case, the maximum untreated discharge for Annex IV residues (a volume) is limited by Resolution MEPC.157(55) of the Marine Environment Protection Committee (MEPC).

Wastewater treatment plants are certified according to their effluent treatment performance. The Resolutions MEPC.159(55) and MEPC.227-(64) identify the effluent standards that must be met to obtain the Certificate of Type Approval by the Administration. It is notable that the maximum values for the most significant outputs coincide with those required by the land normative (Directives 91/271/CEE) for discharging the treated wastewaters into the sea, apart from nitrogenous concentration. Whereas for the land normative the maximum permitted nitrogenous concentration is 10 mg/L (Directives 91/271/CEE) for the discharge on-board, the limited concentration is 20 mg/L (MEPC.227(64)).

3 WRHP IN LAS PALMAS PORT: THE SEWASTE PLAN MANAGEMENT

Directive 2000/59/EC was incorporated into Spanish Law by Royal Decree 1381/2002 on port facilities for the reception of ship-generated waste and cargo residues. This details the procedure for the reception and treatment of waste in port, which is a service that each port has to offer.

Port reception service for ship-generated waste and cargo residues consists of the collection of ship-generated waste, its transfer to an authorized treatment facility and, if applicable, its storage, classification and pretreatment in the area approved by the competent authorities (Royal Legislative Decree 2/2011). In order to carry out this service, the owners of the waste reception facilities (waste operators) need a license from the Las Palmas Port Authority (APLP), which is a body that manages Las Palmas Port. In addition, they need an authorization from the competent environmental body and have to demonstrate a commitment to the treatment or disposal of the waste by a final manager. WRHP include, general requirements - aside from the economic-financial, technical and professional solvency to be met by the bidding companies -, the technical characteristics and conditions to be met by the waste reception operations and facilities, and the equipment to be provided by the service provider to collaborate in marine pollution control services.

With a few exceptions, in order to initiate the service, the captain of a ship heading to the Port of Las Palmas has to inform *Capitanía Marítima* (the Ministry of Public Works and Transport) and APLP of its intention to deliver or keep on board the waste generated, along with its maximum storage capacity. This notification is mandatory and *Capitanía Marítima* must verify that there is sufficient storage capacity for the accumulated waste on the ship, in addition to the matter that will be accumulated until it is unloaded.

If there are strong reasons to assume that the intended port of delivery, this is the next port of call, does not have adequate reception facilities or if that port is unknown, *Capitanía Marítima* has to order the discharge of the waste to avoid the risk of it being dumped at sea on the route between Las Palmas Port and the next port of call. The same procedure applies if the obligation to notify on board waste is breached and it verifies that the storage conditions are not adequate.

The collection can be carried out afloat, by barge, or from the side of a ship, through at anker truck or a boat. Once the waste has been removed, the MARPOL certificate is issued to the vessel and the waste is temporarily stored or receives primary treatment.

The waste collected must be treated and disposed of according to the current regulations. When such processes are not carried out by own equipment, waste operators should achieve an agreement with a duly authorized waste manager. In addition, if the final treatment plant is located outside the port, it is necessary to complete the Special Waste Transport Control and Monitoring Document, in relation to the

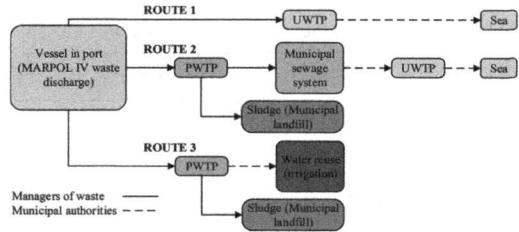

Figure 1. Possible routes for MARPOL IV wastewater according to the current Waste Handling Plan in Las Palmas Port.

applicable regulations of the Autonomous Community. This might be the case, for example, of Urban Wastewater Treatment Plants (UWTPs). Figure 1 shows the possible routes for Annex IV waste according to the current WRHP in Las Palmas Port. Currently, the most frequent route is the first (Route 1); while the second (Route 2) is also dependent on the Port Wastewater Treatment Plant (PWTP). On the other hand, Route 3 provides a possible new scenario due to the recent PWTP operation.

In general, this type of service is provided on a regular and continuous basis, it is operational 24 hours a day throughout the year, except in cases of force majeure and a permanent physical presence in port is not required. In the context of the particularities of Annex IV, a low number of operations are being carried out and a low quantity unloaded per operation in Las Palmas Port. In order to provide this service, waste operators have to pay various fees to APLP. Ships, for their part, have to pay a fee directly to the waste operator which depends on the volumes downloaded. These tariffs have to be public and the APLP has to verify that they do not exceed the maximum tariffs established in the Waste Reception and Handling Plan.

These charges include any expenditure or cost necessary for the provision of the service, emphasizing the payment for reception and treatment (63.0 €/m^3 by terrestrial means and 69.5 €/m^3 by barge).

4 ANNEX I AND ANNEX IV WASTES MANAGEMENT AFTER RECEPTION IN LAS PALMAS PORT (ON-SHORE PLAN)

4.1 Management regulation

Post-reception regulation is included in the WRHP in the ports managed by APLP. This regulation establishes that Annex IV waste must be delivered to a treatment plant, without specifying the plant's characteristics. The WRHP mentions that the emission characteristics in waste treatment processes must be in accordance with the current regulation in Spain and in the region. Quality and control requirements established by Directives (EU) 2000/60/CE and 2008/56/CE should be taken into consideration when discharges into water bodies are carried out.

Since the sewage features from Annex IV are different to urban wastewater, there is an additional difficulty to reach the normative requirements when the Annex IV residues are treated through UWTPs (that are specifically designed to treat urban wastewater).

4.2 Treatment problems

Usually, Annex IV wastewater is transported to UWTPs (see Figure 1) while Annex I waste is transported to the Iberian Peninsula to be treated after being centrifuged (an oil and hydrocarbon separation process). The UWTPs are made to treat wastewater with specific physical-chemical characteristics such as biological oxygen demand (BOD), chemical oxygen demand (COD), suspended solids (SS), concentrations of nitrogen, phosphorus, etc. Normally, the UWTPs undergo a biological treatment, among other things, to reduce the concentration of BOD, COD, SS, etc. (Christensen et al. 2015) to meet the discharge criteria according to the Directives 91/271/CEE and 98/15/CE. Biological treatments in UWTPs involve microorganisms (mainly bacteria), which transform biodegradable organic matter into 'simple products'. Predominantly anaerobic and anaerobic processes are used in UWTPs.

In principle, Annex IV waste should have similar characteristics to urban wastewater. But, the utilization of flushing toilets with seawater onboard in order to reduce the freshwater load makes Annex IV waste much more saline than urban wastewater (Jiang et al., 2019). Table 1 shows typical composition ranges of Annex IV waste (Zhu, He, and Wang, 2016; Chen, Shi, and Wang, 2019; Cai et al., 2019).

It is important to bear in mind that there is a wide variability in concentrations that are dependent on the sort of ships in port. The salinity range was changed from 5-10 to 1-10 g/L and the BOD range from 500-800 to 100-800 once these values were found in Las Palmas Port. It is well known that, effluents with high salt concentrations exert inhibitory/toxic effects on both aerobic and anaerobic microorganisms during biodegradation processes (Lefebvre and Moletta, 2006; Tomei et al., 2017). Hence, the discharge of Annex IV waste in UWTPs affect the correct functioning and separation performance of these types of plants. Consequently, a direct route for Annex IV waste from the vessel to the UWTPs (Route 1 in Figure 1) is not fully appropriate from an efficiency standpoint.

Table 1. Composition ranges of Annex IV wastewater.

Parameter	Range (mg/L)
COD	240 - 1,216
BOD$_5$	100 - 800
SS	450 - 750
E. coli (CFU/100 mL)	>100000
Salinity	1,000 - 10,000

The treatment of Annex I waste is a different issue. This oily wastewater has different physical-chemical characteristics in comparison with MARPOL IV waste. Usually, Annex I waste is treated in specific wastewater treatment plants and the discharge criteria to the sewage system has to be in line with the Spanish and regional regulation. Table 2 shows part of the maximum required values of outputs to permit the discharge of treated wastewaters to the sewage system in Gran Canaria. A similar table is required by the Spanish regulation.

4.3 *Possible on-shore solutions to wastewater treatment*

The recent building of a specific plant for treating Annex IV waste in port (PWTP) is a promising solution to the technical difficulties highlighted in previous paragraphs. This plant should be able to meet not only the quality criteria of the outputs mentioned in Table 2, but also to be able to treat saline wastewater. In this case, biological treatments are not recommended, consequently, physico-chemical treatment is generally required to remove the organic matter from such effluents. The main technologies that have been researched are coagulation–flocculation, direct flocculation and membrane separation processes (Lee, Robinson and Chong, 2014; Sharma and Sanghi, 2012). The reduction of conductivity or dissolved salt can be achieved by two kind of processes, thermal and membranes. Usually the thermal processes are discarded due to their high specific energy consumption in comparison with membrane separation techniques such as nanofiltration (NF) and reverse osmosis (RO) (Ghaffour, Missimer and Amy, 2013). If the salinity of the wastewater to be treated is close to seawater, seawater reverse osmosis (SWRO) membranes are required as NF and brackish water reverse osmosis (BWRO) membranes are designed to treat brackish water (salt concentration < 10 g/L). The SWRO and BWRO systems require a certain sort of pre-treatment in order to reduce operating problems such as

Table 3. Separation efficiency of microfiltration + RO.

Parameter	Value (%)
COD	80
BOD$_5$	83
SS	100
E. coli (CFU/100 mL)	100
Salinity	95

(Source: del Pino and Durham, 1999)

colloidal fouling etc. Microfiltration cartridge filters (~5 μm) and/or ultrafiltration membrane are usually used as a pre-treatment of SWRO and BWRO systems. It has been found that Annex IV waste has a salinity in a range of 1 and 10 g/L approximately. As can be seen, treating Annex IV waste to meet the criteria for discharge to the sewerage network in Gran Canaria is not a trivial matter.

A physico-chemical treatment and microfiltration of 100 μm treatment was tested in Las Palmas port using Annex IV waste. The COD, BOD and E. coli decreased by 70, 80 and 99.99% respectively while SS and salinity increased by 10 and 800 mg/L due to coagulant and flocculant dosing. A thinner microfiltration stage plus RO treatment would increase the separation efficiency in line with that shown in Table 3.

5 ETHODOLOGY FOR COST-BENEFIT ANALYSIS

Even though, the costs linked to the implementation of the Directive can be clearly established (mainly costs for Member States), not all benefits are so easily quantifiable (European Commission report, 2015). The market benefits (mainly, for waste operators) can directly be accounted for in monetary terms but the quantification of non-market benefits (the health and environmental benefits) requires specific economic evaluation methods.

In such an analysis the following items must be considered: costs to Member States, benefits for human health, environmental benefits, benefits for productive activities and costs by the treatment systems. The later are closely associated with the process route for the sewage handling plan (see Figure 1), they are mainly supported by the waste operators and by public wastewater treatment plants (local plants).

The method proposed is lied on the proportion of the costs incurred, the benefits achieved and the port fees charged for it. An imbalance between costs, benefits and charges should be carefully analyzed because, this directly affects the effectiveness of the CRS.

Table 2. Limiting values of treated wastewater to be discharged to the sewage system in Gran Canaria.

Parameter	Limiting value
BOD$_5$	1000 mg/L
pH	5.5 – 9.5
COD	1,600 mg/L
Conductivity	2,500 μS/cm
SS	1,000 mg/L
Oils and/or fats	100 mg/L

(Source: Council of Aguas de Gran Canaria, 2010)

6 CALCULATIONS FOR LAS PALMAS PORT

In order to evaluate the suitability of the CRS to the management of the Annex IV sewage in Las Palmas port, a first approach to the problem is tackled in this section through a preliminary cost-benefit analysis (see section 5).

6.1 Inputs from Las Palmas Port

This empirical application considers information for year 2018: 48,308.85 m³ of Annex IV discharged by 10,050 yearly calls with a standard fee 63 €/m³ for Annex IV management in port.

Regarding the assumed performance for the treatment plants involved in the possible routes for sewage handling in Las Palmas Port (see Figure 1), the initial values for the outputs in a standard discharge were taken from Table 1, and in turn, the output values after a sewage treatment in PWTP were assumed from the Table 3. Finally, the output values assumed for the UWTP treatment are those limits identified in the Directive 91/271/EEC. According to Figure 1, three different routes are possible for the handling plan of the sewage (Annex IV waste) in Las Palmas port: Route 1 (where only UWTP is involved); Route 2 (in which the sewage is treated by two plants: PWTP and UWTP); and Route 3 (only PWTP treatment is necessary). This leaves three different scenarios with different costs and benefits. In this domain, it is interesting to bear in mind that the PWTP treated water meets the requirements for its reuse in agrarian activities (irrigation of crops, according to the Spanish Regulation - Royal Legislative Decree 1620/2007); this destination for the effluent corresponds to Route 3 (see Figure 1).

6.2 Costs associated to the Annex IV management (Costs to Member States)

Only the costs for the port management of the Annex IV sewage have been considered in this regard due to the lack of accurate information. The Department of Sustainability, Safety and Security from Las Palmas Port invested 60% of their time in 2018 (1,736 working hours) to managing the MARPOL residues, and specifically 10% annually (173.6 hours) regardless of the route followed on-shore by the Annex IV discharges (see Figure 1). Assuming an average hourly wage costs for administrative tasks in 2018 of 14.40 €/h (II Collective agreement for workers of State Ports and Port authorities of Spain –BOE N9, 11 January, 2006) the cost results is 2,499.84 € in 2018.

6.3 Environmental benefits

The shadow prices method (Hernández-Sancho et al., 2010; Hernández-Sancho et al., 2015) was assumed to calculate the environmental benefits from the application of treatment systems for the sewage from Annex IV of MARPOL. In such a way, the environmental benefits are assessed through shadow prices associated with the undesirable outputs of the sewage (Hernández-Sancho et al., 2010). Even though the treated outputs which are involved in the water pollution are numerous, this study has focused on four: Nitrogen (N), Phosphorus (P), SS and organic matter. The latter is measured as BOD and COD.

Table 4 shows shadow prices of the previous pollutants for 2018 year in Spain (the initial values published by Hernández-Sancho et al., 2010 were updated through the CPI- Consumer Price Index for 2018 from 2004-).

The shadow prices along with the reference prices are highly dependent on the destination effluent. Routes 1 and 2 (see Figure 1) discharge their effluents into the sea. However, the Route 1 treatment (see Figure 1) is only suitable for discharges into the sea. Route 3 allows the possibility of reusing the water through a unique treatment plant (PWTP). For that reason, two possible destinations are considered for the discharges from the routes: the sea and irrigation (wetlands).

Table 5 shows the expected final values for the harmful outputs, by taking into account the route followed for the sewage treatment (see Figure 1).

Table 6 collects the environmental benefits calculated for the possible routes and the effluent destinations. Even though the overall results offer similar advantages through Route 1 and 2, where the sea is the final destination of the effluent, a significant improvement in environmental benefits is found through Route 3, where the effluent is reused (see Figure 1).

The role of treated water as an economic resource for irrigation in agriculture is especially significant in the Canary Islands, due to the lack of rivers and sweet water lakes. This leads to substantial environmental benefits through Route 3 (see Table 6) with a high reference price for this treated water (see Table 4) in comparison to that poured into the sea.

Table 4. Shadow Prices for the undesirable outputs 2018.

Destination of effluent		Sea	Wetlands/ irrigation
Reference price for the treated water (€/m³)		0.128	1.150
Shadow Prices for undesirable outputs (€/kg)	N	-5.902	-83.337
	P	-9.627	-132.176
	SS	-0.001	-0.013
	BOD	-0.006	-0.150
	COD	-0.013	-0.156

(Source: Hernández-Sancho et al., 2010; updated CPI Spain 2018)

Table 5. Minimum values reached for the possible routes in the sewage treatment in Las Palmas Port.

mg/L	Route 1	Route 2	Route 3
COD	125	43.68	43.68
BOD	25	13.6	13.6
SS	35	0	0
N	10	10	19.98
P	1	1	1.267
Salinity	—	290	290

Table 6. Environmental benefits obtained from the treatment of the Annex IV waste in Las Palmas port (2018).

Parameter (€/year)	Route 1 (Sea)	Route 2 (Sea)	Route3 (Wetland/ irrigation)
COD	333.75	378.76	4,620.89
BOD	103.77	106.93	2,502.24
SS	31.27	33.20	332.09
N	25,815.63	25,815.63	328,512.48
P	7,129.70	7,129.70	96,358.57
Overall (€)	33,414.14	33,464.24	432,326.29
(€/m³)	0.771	0.772	9.982

6.4 Human health benefits

Since this application case is not in the context of a developing country, despite the existence of the human health benefits in sewage treatment, this study assumes that these ones might be negligible in this work.

6.5 Benefits from productive activities (Industrial Benefits)

Aside from waste operators in port (Annex IV residues management), the agriculture sector, the fishing sector and tourism (with a high impact on the economy of the Canary Archipelago) will be directly affected by the sewage treatments and the possibilities of its reuse. The treated waters, with the sea as a final destination (Routes 1 and 2, see Table 9), obviously impact the fishing and recreational sector, whereas the treated waters for irrigation uses (wetland destination in Route 3) potentially benefit agriculture and tourism. These productive benefits have been quantified through references prices (see Table 4) in Table 7.

6.6 Costs by the treatment systems

According to the possible routes, the treatment plants involved in the processes are different and as a consequence of this, the treatment costs are also different (see Figure 1). Table 8 collects the costs associated with the plants provided by the owners (PWTP) and the public tariffs charged by the urban plant

(UWTP) to the waste operators. In this latter case, the UWTP manager in Las Palmas Port has provided estimated costs for the operation, of around 1 €/m³, however the operation costs for these plants (UWTP) found in the literature (Hernández-Sancho et al., 2010) are between 0.2-0.6 €/m³ (2018 values). In any case, these costs are very far from the required UWTP tariffs (see Table 8) for waste operators. Therefore, a significant margin between the real cost and the tariffs seems to exist. This involves significant market benefits for the UWTP.

7 COST-BENEFIT ANALYSIS. RESULTS

The costs and benefits that could be quantified, according to the above sections, are shown in Table 9 and some insights can be draw from the analysis: the total costs are higher than the total benefits when the final destination of the effluent is the sea (Routes 1 and 2) and the incomes by the port users are not considered. In addition, only Route 3 provides greater benefits than costs.

When the incomes associated with the port fees for MARPOL IV waste management are considered (port users in Table 9), significant extra-benefits arise. These extra-benefits are the difference between the total costs and the incomes from port fees (portusers). Due to the fact that the fees are directly charged by the waste operators for Annex IV management, the main beneficiaries are these stakeholders with an estimated margin of between 40 and

Table 7. Industrial benefits calculated from reference prices for 2018 in Las Palmas Port.

Industrial Benefits (€/year)	Route 1 (Sea)	Route 2 (Sea)	Route3 (Wetland/ irrigation)
	5,534.87	5,534.87	49,813.83

Table 8. Costs of the treatment systems of the sewage Annex IV in Las Palmas Port (2018).

	Route 1 (Sea)	Route 2 (Sea)	Route 3 (Wetland/irrigation)
PWTP (€/m³)	—	2.8	2.8
UWTP (€/m³)	16.5	16.5	—
Municipal Landfill (Sludge)		3.6	3.6
Total costs (€/m³)	16.5	22.9	6.4
Yearly costs (€)	714,596.02	991,772.66	277,176.64

78

Table 9. Cost-Benefit analysis for 2018 of Annex IV waste management in Las Palmas Port.

	Route 1 (Sea)		Route 2 (Sea)		Route 3 (Wetland/ Irrigation)	
	Yearly Costs (€)	€/m^3	Yearly Costs (€)	€/m^3	Yearly Costs (€)	€/m^3
CMS	2,499,84	0.06	2,499,84	0.06	2,499,84	0.06
CTS	714,596.03	16.5	991,772.67	22.9	277,176.64	6.4
Total Costs	717,095.87	16.5	994,272.51	22.9	279,676.48	6.4
EB	33,414.14	0.77	33,464.25	0.77	432,326.30	10.8
IB	5,534.87	0.13	5,534.87	0.13	49,813.84	1.15
Total Benefits	38,949.01	0.90	38,999.12	0.90	482,140.14	11.1
Port-users	2,728,457.5	63.0	2,728,457.5	63.0	2,728,457.5	63.0
Extra Benefits	2,011,361.7	46.4	1,734,185.0	40.0	2.448.781,1	56,5

CMS: Costs to Member State; CTS: Costs by Treatment Systems; EB: Environmental Benefits; IB: Industrial Benefits

56.5 €/m^3 (see Table 9). Nonetheless, these margins do not involve necessary large profits, since several items of their activity were not taken into account: capital costs (PWTP in Routes 2 and 3, see Figure 1), storage and transportation costs (for all routes), maintenance costs, etc. Likewise, the estimated difference between the public tariffs charged by the UWTP and the real costs provides margins of up to 15.5 €/m^3.

8 MAIN INSIGHTS AND CONCLUSIONS

This work has tackled the WRHP (Directive (EU) 2019/883) performance in terms of Annex IV management, from an analysis of the appropriateness of the CRS. the paper has employed a cost-benefit method based on the on-shore advantages to citizens. Conscious of the singularity of the Archipelagos, the analysis is carried out for Las Palmas Port where three different routes have been identified in the sewage management: Route 1 where only the UWTP acts, Route 2 where, aside from UWTP, the PWTP is involved, and finally Route 3 where only the latter plant acts to reuse the treated effluent.

Even though, the lack of accuracy in the necessary information has meant that important costs and benefits have not been taken into account, from the results obtained, we can affirm that the only route, which is able to offer social benefits that outweigh the costs, is Route 3. Unlike Route 3, the other routes do not offer higher benefits than costs. Nevertheless, Route 2, against Route 1(the most frequently used), provides an additional advantage because it avoids significant technical difficulties (deficiencies in the treatment process) in UWTP associated with the high salinity of Annex IV residues.

As one of the main findings, the costs invested in the most frequent Annex IV waste management in Las Palmas Port (Route 1) have significantly overtaken the benefits for citizens. In this regard it is especially noteworthy, the wide difference between the estimated costs for the treatment process in the UWTP and the tariffs charged to the waste operators.

Finally, paying attention to the suitability of the CRS for Annex IV waste, the port users costs due to the fees have proven to be very significant. This income not only broadly covers the costs assumed for Annex IV waste management, but also provides extra benefits for the market.

ACKNOWLEDGEMENTS

The authors wish to thank anonymous reviewers for their comments and suggestions that have improved the quality of this paper. This work has been supported by the 'Cátedra Marítimo-Portuaria' (University of Las Palmas de Gran Canaria, Mapfre Guanarteme foundation and ASTICAN shipyard)

REFERENCES

Cai, Yuhang, Xin Li, Asad A. Zaidi, Yue Shi, Kun Zhang, Ruizhe Feng, Aqiang Lin, and Chen Liu. (2019). "Effect of Hydraulic Retention Time on Pollutants Removal from Real Ship Sewage Treatment via a Pilot-Scale Air-Lift Multilevel Circulation Membrane Bioreactor." Chemosphere 236 (December): 124338.https://doi.org/ 10.1016/J.CHEMOSPHERE.2019.07.069.

Chen, Rongchang, Jing Shi, and Zheng Wang. (2019). "Method Study on Establishing of Ship Sewage Pollutants Discharging Inventory Based on {AIS}." {IOP} Conference Series: Earth and Environmental Science 237 (March): 22018. https://doi.org/10.1088/1755-1315/ 237/2/022018.

Christensen, Morten Lykkegaard, Kristian Keiding, Per Halkjær Nielsen, and Mads Koustrup Jørgensen. (2015). "Dewatering in Biological Wastewater Treatment: A Review." Water Research 82 (October): 14–24. https:// doi.org/10.1016/J.WATRES.2015.04.019.

Council of Aguas de Gran Canaria. (2010). Hydrological Plan of Gran Canaria. Gran Canaria (Spain).

European Comission report. (2015). "Ex-Post evaluation of Directive 2000/59/EC on port reception facilities for ship-generated waste and cargo residues".

Ghaffour, Noreddine, Thomas M. Missimer, and Gary L. Amy. (2013). "Technical Review and Evaluation of the Economics of Water Desalination: Current and Future Challenges for Better Water Supply Sustainability." Desalination 309 (January): 197–207. https://doi.org/10.1016/J.DESAL.2012.10.015.

Hernández-Sancho, F., Molinos-Senante, M. and Sala-Garrido, R. (2010). Economic valuation of environmental benefits from wastewater treatment processes: An empirical approach for Spain. Science of Total Environment, 408, pp. 953–57

Hernández-Sancho H., Lamizana-Diallo B., Mateo-Sagasta, J., Qadir M. (2015). Economic Valuation of Wastewater. The cost of action and the cost of no action.

Division of Environmental Policy Implementation. United Nations Environment Programme.

Jiang L., Chen X., Qin M., Cheng M., Wang Y., Zhou W. (2019) On-board saline black water treatment by bioaugmentation original marine bacteria with Pseudoalteromonas sp. SCSE709-6 and the associated microbial community. *Bioresource Technology.* 273(2019): 496–505

Lee, Chai Siah, John Robinson, and Mei Fong Chong. (2014). "A Review on Application of Flocculants in Wastewater Treatment." *Process Safety and Environmental Protection* 92 (6): 489–508. https://doi.org/10.1016/J.PSEP.2014.04.010.

Lefebvre, Olivier, and René Moletta. (2006). "Treatment of Organic Pollution in Industrial Saline Wastewater: A Literature Review." *Water Research* 40 (20): 3671–82. https://doi.org/10.1016/J.WATRES.2006.08.027.

Perić T., Komadina P., Račić N. (2016) Wastewater Pollution from Cruise Ships in the Adriatic Sea. Promet–Traffic&Transportation, Vol. 28, 2016, No. 4, 425–433

Pino, Manuel P. del, and Bruce Durham. (1999). "Wastewater Reuse through Dual-Membrane Processes: Opportunities for Sustainable Water Resources." Desalination 124 (1–3): 271–77. https://doi.org/10.1016/S00119164(99)00112-5.

Sharma, Sanjay K, and Rashmi Sanghi. (2012). Advances in Water Treatment and Pollution Prevention. Springer Science & Business Media.

Tomei, M. Concetta, Domenica Mosca Angelucci, Valentina Stazi, and Andrew J. Daugulis. (2017). "On the Applicability of a Hybrid Bioreactor Operated with Polymeric Tubing for the Biological Treatment of Saline Wastewater." Science of The Total Environment 599–600 (December): 1056–63. https://doi.org/10.1016/J.SCITOTENV.207.05.042.

Zhu, Linan, Hailing He, and Chunli Wang. (2016). "COD Removal Efficiency and Mechanism of HMBR in High Volumetric Loading for Ship Domestic Sewage Treatment." Water Science and Technology 74 (7):1509–17. https://doi.org/10.2166/wst.2016.271

Developments in Maritime Technology and Engineering – Guedes Soares & Santos (eds)
© 2021 Copyright the Author(s), ISBN 978-0-367-77376-2

Short sea shipping routes hinterland delimitation in the European Atlantic Area

T.A. Santos, J. Escabelado, P. Martins & C. Guedes Soares
Centre for Marine Technology and Ocean Engineering (CENTEC), Instituto Superior Técnico, Universidade de Lisboa, Lisbon, Portugal

ABSTRACT: This paper presents a numerical method to delimit the hinterland and foreland of short sea shipping (SSS) routes based on the utilization of Ro-Ro ships with an application to a route in the European Atlantic Area. A review of literature is carried out and the numerical method, based on transportation cost and time combined in a generalized transportation cost, is described. A case study is developed for a route between Portugal and Northern France in which short sea shipping is in competition with fully road-based transportation. The method is applied to carry out systematic calculations between pairs origin/destination, allowing the delimitation of the geographic boundaries of the route's hinterland and foreland. The effects of variations in the maritime freight rate, ship service speed and of perturbations in road transportation caused by border controls (necessary as result of the Covid-19 crisis) are evaluated. Conclusions are drawn regarding the competitiveness of SSS compared to road transportation.

1 INTRODUCTION

In late 2019 the European Commission (EC) presented its new flagship economic growth strategy, entitled "European Green Deal", the details of which may be found in EC (2019). This strategy seeks to tackle the issues of climate change and air pollution, turning these threats into opportunities for a renewed economic growth of the European Union (EU). In the wake of the Covid-19 pandemic, the EC has reiterated that the European Green Deal is still the utmost priority of the EU and should significantly contribute to the economic recovery. This strategy puts forward multimodality as a transport policy priority and indicates that this should replace a large portion of road transport in the EU. It also indicates that integration between Short Sea Shipping (SSS), rail freight corridors and inland waterways should be enhanced and this should be achieved also through a revision of the Combined Transport Directive, see EC (2017).

SSS has a long history in the EU and faces significant challenges, as discussed by Psaraftis and Thalis (2020). It is generally considered to be cost competitive but slow, unreliable and bureaucratic. Therefore, it remains underused even for coastal regions located near Motorways of the Seas, as it happens in the European Atlantic Area (https://www.atlanticarea.eu/), which congregates coastal 36 regions in five different European countries.

SSS has been regarded as effective when significant amounts of cargo need to be carried between regions located near the coastline, thus ensuring good accesses from locations of production and consumption to seaports. Therefore, the hinterland/foreland of SSS has long been regarded as being located mostly along the EU coastlines. It is interesting to note that the hinterland and foreland of ports are concepts long established in the literature, see for example Bird (1963) and Hayuth (1981). Notteboom and Rodrigue (2005) describe the ongoing process of port regionalization as a major contributor to the expansion of port hinterlands. Many authors have studied the hinterland of specific ports or within specific countries, including Santos and Guedes Soares (2017, 2019a), who studied the hinterlands of Portuguese ports. Therefore, these concepts have been used mainly in connection with ports rather than applied to shipping routes (short or deep sea in nature).

This paper will attempt to delimit hinterlands and forelands in the context of short sea shipping routes. The general idea is that a regular SSS service between two ports possesses a set of geographic regions for which the multimodal transport chains (SSS plus road haulage, i.e., pre and on carriage) are competitive when compared with road haulage (unimodal transport solution). The set of regions adjacent to the loading port will form the hinterland of the maritime service, whereas the set of regions adjacent to the unloading port will form the foreland of the maritime service. For regions outside these two sets, road haulage will be more competitive than the multimodal solution (which includes SSS) and these other regions

DOI: 10.1201/9781003216582-9

will form the hinterland/foreland for road haulage. It is fairly clear that the hinterland/foreland of SSS-based transport will lay along the coastlines, but the research question is: how do we establish the border between the hinterlands of SSS and road haulage? Putting this in another way, how far inland from the coastline extends the hinterland/foreland of SSS transport and which parameters determine this extent?

Having established the research questions to be answered, it is necessary to identify methods and tools enabling the characterization of these hinterland/foreland borders. This paper proposes the use of a transport network model and transport chain assessment tool to evaluate systematically transport chains, unimodal or multimodal, between multiple pairs origin/destination. These pairs origin/destination will be taken as, typically, major cities in NUTS (Nomenclature of Territorial Units for Statistics) regions. For each pair, transport cost and time will be calculated using the tool and the hinterlands/forelands of SSS and road haulage will be delimited based on the generalized transport cost. Santos et al. (2019b) studied a variety of different transport solutions, unimodal and multimodal, between a fixed pair origin/destination: Porto (Valongo) and Paris. This paper seeks to extend the analysis to multiple pairs in these two geographical regions, thus determining the hinterland/foreland of this SSS maritime service.

The remainder of this paper is organized in the following manner. Section 2 presents a brief review of the literature on SSS and transport chains analysis, forming the basis for section 3, which presents the numerical method adopted for the assessment of transport chains. Section 4 presents the application of the numerical method in a case study set in the European Atlantic Area, dealing with a SSS maritime service between Leixões and Le Havre. The hinterland and foreland of this SSS route are delimited for a range of different parameters (maritime freight rate, ship speed, border controls). Section 5 puts forward the main conclusions of the case study.

2 LITERATURE REVIEW

A number of models of transport networks have been presented in recent years, especially in southern Europe, allowing the evaluation of transport chains combining short sea shipping and road haulage. Carrese et al. (2020), Lupi et al. (2020) and Morales-Fusco et al. (2018) provide the details of such model. Out of this, Carrese et al. (2020) appears to present the more comprehensive model in its description of the full transport chains, but still resorts to average speeds over the entire networks, while Lupi et al. (2020) consider only the maritime routes network.

Santos et al. (2019b) studied transport chains between a fixed pair origin/destination, using as support a model of transport networks available at sea and on land along the Motorway of the Sea of Western Europe. This paper only considered internal

costs and transport time in the evaluation of the performance of the different transport chains, which were unimodal or intermodal, running between a particular pair origin/destination: Northern Portugal (Porto-Valongo) and Northern France (Paris). The transport solutions included using Ro-Ro cargo ships and container ships, sailing from Leixões to Le Havre, coupled with road haulage, inland waterways (river Seine) and rail freight corridors. Ro-Ro cargo ships coupled to road haulage (pre and on carriage) proved to be a competitive alternative to road transportation for this particular pair origin/destination.

Santos and Guedes Soares (2020) have also studied the impact of external costs in SSS services hinterland's, with the external costs calculated using the same coefficients given in the Marco Polo calculator, see Brons and Christidis (2013). Four routes were considered by the authors, from Leixões (Portugal) to Le Havre, Antwerp, Rotterdam and Hamburg, and in most cases a significant expansion of the hinterland was identified when the external costs, calculated using the Marco Polo approach, were included in the generalised cost (full internalisation). This paper did not use a model of the transport networks and only considered road transport and SSS, assuming a constant speed for the entire pre and on carriage, with road distances taken from commercial route mapping software. Ship speed was also considered constant. The current paper uses a model of the transport networks, entailing the possibility to consider different speeds in each type of road and in different parts of the ship's voyage. It also allows considering cargo handling and storage times in port terminals as well as associated handling and storage costs.

Another application of transport network models and of the numerical tool mentioned above is in the calculation of external costs occurring in door-to-door transport chains. Recently, Santos, Ramalho and Guedes Soares (2020) used such a tool to calculate the generalized transport cost and the external costs occurring in door-to-door transport chains, in three SSS routes: Leixões-Le Havre, Stockholm-Germany, Athens-Thessaloniki. The external costs were calculated using the EU handbook for the calculation of external costs, see van Essen (2019). It was identified that maritime transport is competitive with road transport in terms of external costs as long as ship's speed is kept moderate.

3 NUMERICAL METHOD

The transport chain assessment tool used in this paper builds upon a transport network model exemplified in Figure 1. Road and SSS networks are modelled using nodes (road junctions, seaport terminals) and links (roads, motorways, maritime routes). These may be located in different countries (mainland or islands). Nodes are characterized by, if applicable, time delays (representing dwell time in seaport terminals for example) and costs (cargo

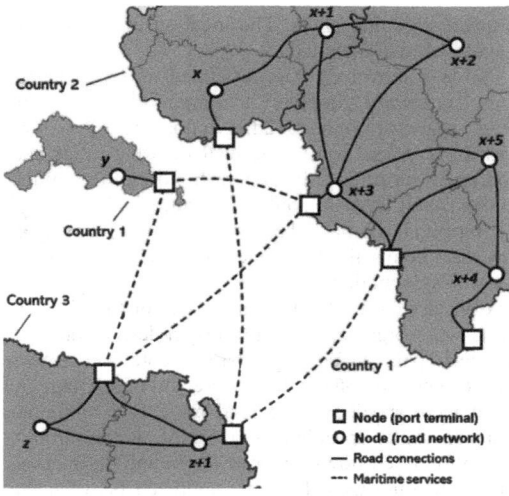

Figure 1. Transport network model.

The transportation time taken along path p of set k is then given by:

$$T_{TRkp} = T_{RD_{kp}} + T_{RL_{kp}} + T_{IW_{kp}} + T_{RR_{kp}} + T_{CC_{kp}} \quad (2)$$

where time taken in links of other types, namely rail, inland waterways and maritime (using Ro-Ro or container ships) is denoted as, respectively, T_{RL}, T_{IW}, T_{RR} and T_{CC}. The transportation time needs to be increased due to delays at certain nodes in the path. Let N be the number of nodes in the database representing the transport network. The time in nodes is user specified and can represent, for example, dwell time in a Ro-Ro terminal (waiting plus loading in the ship). Therefore, equation (3) defines the total time between an origin and a destination using a certain path as the sum of actual transportation time, given by equation (2), with the time taken in nodes j along the path:

$$T_{kp} = T_{TR_{kp}} + \sum_{j=1}^{N} \delta_{kp_j}.T_{Dw_j} \quad (3)$$

where δ_{kp_j} is a binary variable representing whether or not node j belongs to the path p of set k. T_{Dw_j} represents the average dwell time in node j. A similar procedure is used for calculating the total cost of freight transportation in each path. Firstly, the total distance travelled in path p of set k is determined. In the case of road transportation, the total distance travelled is given by:

$$D_{RD_{kp}} = \sum_{i=1}^{L} \delta_{l_i}.\delta_{kp_i}.\delta_{Rd_i}.D_{l_i} \quad (4)$$

The cost associated with such distance is then calculated by:

$$C_{RD_{kp}} = D_{RD_{kp}}.c_{RD} \quad (5)$$

where the cost coefficient c_{RD} is, itself, a function of the distance travelled by road. This coefficient is obtained by interpolation over a non-linear function of specific cost $f(D_{RD_{kp}})$ (monetary units per km for a cargo unit), specified according with market conditions. The same principle is applied to calculate the costs of other modes of transportation in the path, which are represented by $C_{RL_{kp}}$, $C_{IW_{kp}}$, $C_{RR_{kp}}$ and $C_{CC_{kp}}$. The total transportation cost is the sum of the costs associated with each mode of transport:

handling costs). Links are characterized by distance, type (road, motorway, maritime, rail, inland waterway), end nodes, congestion level and country. A database, covering the core and comprehensive road and seaport networks (as defined by the EU) for the region between Portugal and Northern Europe, has been developed. The overall goal of the transport chain assessment tool is not to optimize transport operations but to study the relative competitiveness of transport chains based on road transportation or on SSS.

In order to calculate transport time and cost, let us consider that the network database has L links. A sequence of links between an origin and a destination is called a path. Paths with a common origin and destination are organized in a set of paths k. All different paths in each set (eventually using different transport modes) represent possibilities for the transportation of cargo units between this origin and destination. Each path within a set is indicated using the index p. Furthermore, each link, i, corresponds to a mode of transport and, in addition, is characterized by a distance (length) D_{l_i} and an average speed S_{l_i}. The binary variable δ_{l_i} indicates whether the link i is operational or not, whereas the binary variable δ_{kp_i} indicates whether link i is used in path p of set k. A third binary variable, δ_{Rd_i}, identifies the type of link i as being of road or not.

Considering these definitions, the transportation time taken by a cargo unit along a given path, using road type links, is given by the following summation over all links existing in the database:

$$T_{RD_{kp}} = \sum_{i=1}^{L} \delta_{l_i}.\delta_{kp_i}.\delta_{Rd_i}.\frac{D_{l_i}}{S_{l_i}} \quad (1)$$

$$C_{TRkp} = C_{RD_{kp}} + C_{RL_{kp}} + C_{IW_{kp}} + C_{RR_{kp}} + C_{CC_{kp}} \quad (6)$$

while the total transportation cost is the summation of the cost incurred in the transportation operations plus costs associated with transfers between transport modes occurring in nodes of the network:

$$C_{kp} = C_{TRkp} + \sum_{j=1}^{N} \delta_{nj}\delta_{kp_j} \cdot (C_{u_j} + C_{l_j} + C_{s_j}) \quad (7)$$

where C_{u_j}, C_{l_j} and C_{s_j} represent the unloading, loading and storage costs in node j, δ_{nj} is a binary variable which indicates whether or not node j is active and δ_{kp_j} is a binary variable which indicates if node j is used in path p of the set of paths k. The total transportation time and cost on a given path, p, are combined to produce a generalized transportation cost (GTC), given by:

$$GTC_{kp} = C_{kp} + VoT.T_{kp} \quad (8)$$

where VoT represents the value of time for the cargo in euros per hour.

4   APPLICATION

Table 1 shows the NUTS 3 regions considered to be the origins of cargos and the NUTS 2 regions considered to be the destinations of cargos. Within each NUTS, the origin or destination is considered to be its capital city (in most cases the largest city).

Figure 2 shows the NUTS 2 regions across Northern France, Belgium, Luxembourg and Southern Netherlands considered in this study and the respective capitals (including also the port of Le Havre).

Figure 3 shows the NUTS 3 regions in Northern Portugal and Spain considered in this study and the respective capital cities. The location of the port of Leixões is also included.

Table 2 shows the costs assumed for cargo handling (landward implies loading/unloading from truck; seaward implies loading/unloading from ship) and the average dwell time in Ro-Ro terminals and container terminals. It is worth noting the much higher dwell times in container terminals than in Ro-Ro terminals (2 or 3 days, instead of 0.5 days), in any case below the free time conceded by terminals for storing containers.

Considering the NUTS 2 and 3 regions in Table 1, these were organized in pairs origin-destination and, for each pair, a fully road-based route and a SSS based route (complemented by pre and on carriage by road) were specified. The road route is a minimum distance route, specified as a set of nodes through which the truck flows between origin and destination. The SSS route uses the same approach between the origin in Portugal and the port of Leixões and the port of Le Havre and the final destination. Similarly, the SSS route is specified as a set of nodes that define accurately the ship's route in a realistic way (considering the most common waypoints between the two ports). These two routes are specified for each pair origin-destination and the calculations in section 3 are then run through the 14x26 pairs to determine the cost, time and generalized transportation cost.

Table 3 shows one example of numerical results, in this case for SSS routes between Guimarães and NUTS 2 in northern Europe. The maritime freight is considered to be 631.5€ and the ship's service speed is 15 knots (moderate speed). The cargo is carried in

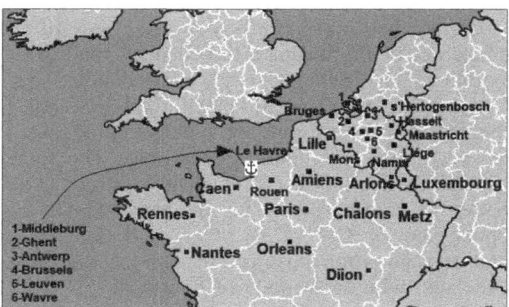

Figure 2. Geographical scope of Northern Europe region under study.

Figure 3. Geographical scope of Northern Portugal/Spain region under study.

Table 1. NUTS 3 and 2 codes (origins and destinations of cargo units) and its capitals.

Northern Portugal (NUTS 3)	Northern France, Belgium, Luxembourg and Southern Netherlands (NUTS 2)	
PT11A-Porto	FR23-Rouen	BE35-Namur
PT119-Guimarães	FR25-Caen	BE1-Brussels
PT112-Braga	FR22-Amiens	BE24-Leuven
PT11C-Amarante	FR10-Paris	BE32-Mons
PT111-Viana do Castelo	FR52-Rennes	BE21-Antwerp
PT11E-Bragança	FR24-Orleans	BE33-Liege
PT11D-Vila Real	FR3-Lille	BE22-Hasselt
PT11B-Chaves	FR51-Nantes	FR26-Dijon
PT16D-Aveiro	BE-25-Bruges	FR41-Metz
PT16G-Viseu	BE23-Ghent	BE34-Arlon
PT16J-Guarda	FR21-Chalons	NL34-Middleburg
PT16E-Coimbra	BE31-Wavre	NL41-s'Hertogenbosch
PT16F-Leiria	LU-Luxembourg	NL42-Maastricht
ES114-Pontevedra (Spain)		

Table 2. Seaport terminal cost and time parameters.

	Load/Unload Landward [€]	Load/Unload Seaward [€]	Dwell Time [h]	Free Time [h]	Storage Cost [€]
Ro-Ro terminal (Leixões)	25.0	25.0	12.0	48.0	2.0
Ro-Ro terminal (Le Havre)	25.0	25.0	12.0	48.0	2.0
Container terminal (Leixões)	0.0	142.2	72.0	120.0	1.79
Container terminal (Le Havre)	25.0	120.0	48.0	96.0	2.0

Table 3. Numerical results for the pair origin/destination Guimarães-Paris when using SSS (maritime freight rate 631.5 €, speed 15 knots).

	Road Cost [€]	SSS Cost [€]	Time on Road [h]	Time at sea [h]	Total time [h]	GTC [€]
Rouen	450	631.5	2.1	48.93	74.99	1693
Caen	448	631.5	2.0	48.93	74.92	1691
Amiens	541	631.5	3.1	48.93	75.99	1791
Paris	656	631.5	3.8	48.93	76.76	1910
Rennes	753	631.5	4.3	48.93	77.25	2011
Orleans	847	631.5	4.9	48.93	77.87	2109
Lille	829	631.5	4.8	48.93	77.7	2090
Nantes	948	631.5	5.5	48.93	79.17	2219
Brugges	997	631.5	5.9	48.93	79.54	2271
Ghent	981	631.5	5.7	48.93	79.4	2254
Middelburg	1095	631.5	6.8	48.93	80.43	2375
Chalons	968	631.5	5.6	48.93	79.31	2240
Wavre	1027	631.5	6.0	48.93	79.72	2302
Namur	1024	631.5	6.1	48.93	79.77	2299
Brussels	1015	631.5	6.0	48.93	79.63	2289
Leuven	1045	631.5	6.2	48.93	79.84	2321
Mons	882	631.5	5.1	48.93	78.01	2145

(Continued)

Table 3. (*Cont.*)

	Road Cost [€]	SSS Cost [€]	Time on Road [h]	Time at sea [h]	Total time [h]	GTC [€]
Antwerp	1078	631.5	6.4	48.93	80.08	2356
s'Hertogenbosch	1312	631.5	8.2	48.93	81.92	2602
Liege	1114	631.5	6.7	48.93	80.34	2393
Hasselt	1222	631.5	7.5	48.93	81.16	2507
Maastricht	1310	631.5	8.2	48.93	81.85	2600
Dijon	1273	631.5	7.9	48.93	81.58	2561
Metz	1235	631.5	7.6	48.93	81.26	2521
Arlon	1229	631.5	7.5	48.93	81.21	2514
Luxembourg	1277	631.5	7.9	48.93	81.59	2565

a Ro-Ro cargo ship, unaccompanied, in wheeled platforms or semi-trailers. Road cost represents the cost of pre and on carriage between Guimarães and Leixões and Le Havre and the final destination. Total time on road and sea are also shown and, when added to the 24 hours spent in terminals, give the total transport time. This is multiplied by the value of time for cargo (6.82 €/hour) and summed to the other costs to produce the generalized transport cost.

Table 3 shows a wide dispersion of generalized transport costs, meaning that there are NUTS 2 regions for which it is preferable to use SSS while there are other regions for which road based transportation presents lower costs. Figure 4 shows the NUTS 2 regions for which SSS is preferable and those for which road transport is more advantageous, when cargo units originate in Guimarães. The maritime freight rate and speed are kept unchanged in relation to Table 3. It may be seen that for this maritime freight and ship speed the only NUTS 2 regions preferring road transport are the southernmost ones in the geographical region considered: FR51, FR24 and FR26 (Nantes, Orleans, Dijon). This occurs because the road distance between Guimarães and these NUTS 2 is the smallest for this geographical region, making road haulage very competitive.

Figure 5 shows the NUTS 2 regions for which SSS is preferable and those for which road transport is more advantageous, when cargo units originate in Bragança (on top right in Figure 3). The difference in comparison with the previous figure is evident, evidencing that SSS is only preferable for three NUTS 2 regions, all located in the close vicinity of Le Havre. This is caused by the fact that Bragança is far away from Leixões (210 km) while Guimarães is only 58 km away.

Figure 6 shows graphically the generalized transport cost for NUTS 2 regions, which prefer SSS to road transport, when the cargo has as origin Guimarães (same results as Table 3). The maritime freight rate is set at 631.5 € per unit and the ship's speed is constant at 15 knots, except for short distances next to the ports, where it is lower. It may be seen that values range from 1800 € to 2600 €, depending on the distance from the port of Le Havre.

Figure 7 shows the generalized transport cost when using SSS from NUTS 3 regions in Portugal to Paris, keeping the maritime freight rate and ship speed unchanged. It may be seen that as the distance from the port of Leixões increases, the generalized transport cost also increases up to 200 € more.

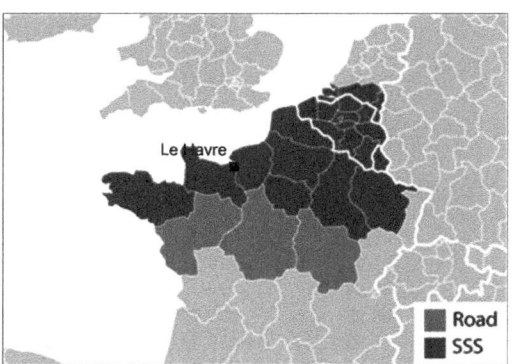

Figure 4. Regions preferring SSS or road transport across northern France-Belgium, for origin of cargo unit located in Guimarães (maritime freight rate 631.5 €, speed 15 knots).

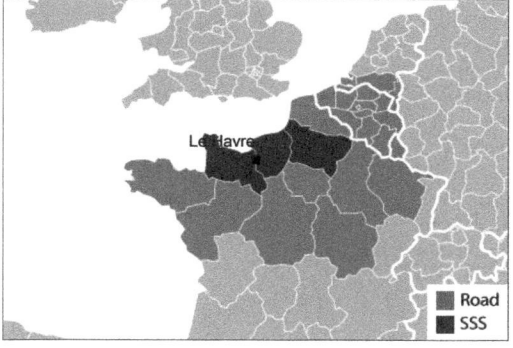

Figure 5. Regions preferring SSS or road transport across northern France-Belgium, for origin of cargo unit located in Bragança (maritime freight rate 631.5 €, speed 15 knots).

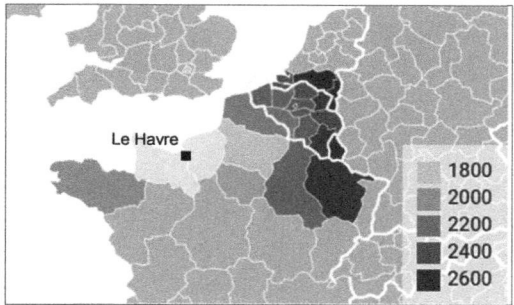

Figure 6. Generalized transport costs (in €) when using SSS for NUTS 2 regions of France-Belgium with origin in Guimarães (maritime freight rate 631.5 €, speed 15 knots).

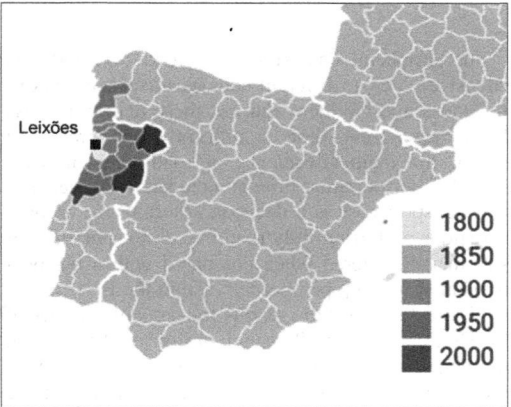

Figure 7. Generalized transport costs (in €) when using SSS from NUTS 3 regions of Portugal with destination to Paris (maritime freight rate 631.5 €, speed 15 knots).

Figure 4 showed that, for a given origin in Portugal, there are NUTS 2 regions for which it is preferable to use SSS while for others it is preferable to use the road. It is also useful to quantify, for each NUTS 2 region, how many Portuguese NUTS 3 regions prefer SSS to road transport, when the generalized transport cost is taken as the decision parameter. This indicator (number of regions preferring SSS), shown in Figure 8, points out those NUTS 2 regions for which a substantial number of origins in Portugal prefer SSS. This would constitute the "foreland" of this SSS service. The boundary of this "foreland" could certainly be put at the border of such NUTS 2 for which there is not a single NUTS 3 origin that prefers SSS to road transport.

Using this criterion, the foreland would be, for this freight rate and speed, the whole geographical area considered in this paper except for NUTS 2 regions around Nantes, Orleans, Dijon. In fact, all NUTS 2 regions except for Metz have 8 or more NUTS 3 in Portugal preferring this SSS service to the road alternative. The foreland seems to be the whole French coast

from Brittany to Pas-de-Calais, most of northern France, Belgium and parts of south Netherlands. However, it should be taken in consideration that in Belgium and Netherlands this SSS service would face competition of other SSS services calling in Antwerp and Rotterdam.

Figure 9 shows the same parameter, number of NUTS 3 origin regions preferring SSS, when the maritime freight rate is increased by 300 €. Such an increase might be dictated by the needs of the shipping company to cover increases in the ship's cost structure.

It is visible in Figure 9 a notable reduction in the number of NUTS 3 preferring SSS for most NUTS 2 of destination, as the regions located further away from the Portuguese coastline and, simultaneously, closer to destination by road, now find it more advantageous to use road transport. The foreland is now restricted to three NUTS 2 along the French coastline. A few NUTS 2 in Belgium and Netherlands show preference for SSS for a few NUTS 3 origin regions, but the numerical values for GTC are only marginally superior to those of road.

An important factor in the decision making process of shippers or logistics companies is the

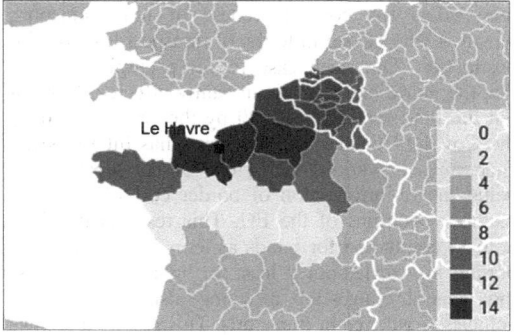

Figure 8. Number of NUTS 3 origin regions preferring SSS, for each NUTS 2 region in France-Belgium, when GTC is used as decision criterion (maritime freight rate 631.5 €, speed 15 knots).

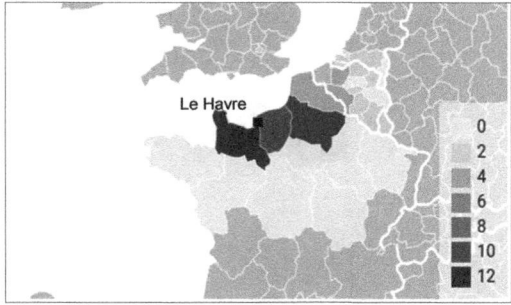

Figure 9. Number of NUTS 3 origin regions preferring SSS, for each NUTS 2 region in France-Belgium, when GTC is used as decision criterion (maritime freight rate 931.5€, speed 15 knots).

transport time between the origin and destination. This time depends on the ship's speed, truck speed and the dwell times in seaports or intermodal terminals. All these parameters present significant uncertainties. Therefore, the database includes standard average values, which can be promptly changed as required. It is useful to assess the effects in the number of NUTS 3 regions in Portugal preferring SSS, for various NUTS 2 in France-Belgium, when the ship's speed is increased to 18 knots. This increase in ship speed from 15 knots is feasible for most Ro-Ro cargo ships, that is, the installed power would be sufficient for most ships (design speed above 18 knots).

In the previous Figure, the maritime freight rate had been increased to 931.5 €, most certainly covering the extra fuel costs incurred when increasing the ship's operating speed to 18 knots. Figure 10 shows the number of NUTS 3 regions preferring SSS when ship's speed is increased to 18 knots, implying a significant reduction in transport time. It may be seen that for many NUTS 2 in Eastern Belgium (and FR10-Paris) there is a moderate increase in the number of NUTS 3 regions preferring SSS.

There are other circumstances, time related, that may affect the competitiveness of road transport, which is very strong in terms of transport time origin-destination. One such case is, for example, the fact that the world has been significantly affected throughout the first semester of 2020 by the crisis caused by the Covid-19 pandemic. This crisis has hit European economies in various ways, one of the most severe being the re-introduction of border controls between member countries of the EU. This restricts the freedom of circulation for goods as temporary controls at borders are carried out at least for light vehicles, causing general congestion affecting also heavy goods vehicles (road freight transport). This results in delays of several hours in each border crossing.

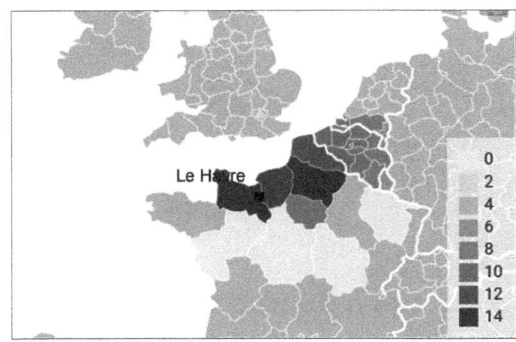

Figure 11. Number of NUTS 3 origin regions preferring SSS, for each NUTS 2 region in France-Belgium, when GTC is used as decision criterion (maritime freight rate 931.5€, speed 18 knots, border controls in force).

For peripheral countries using road transportation extensively across different countries to reach markets in Central and Northern Europe, these delays increase the transportation time significantly, compromising one of the main advantages of road transport over SSS. As transportation time is included in the generalized transport cost as a decision criterion, the number of regions preferring fully road-based transportation or SSS may be altered when compared with a business as usual scenario (open borders within the EU).

Figure 11 shows the results in terms of number of NUTS 3 regions in Portugal preferring SSS as an alternative to road transportation for each NUTS 2, when border controls are in force leading to 12 hours' delays in each border crossing (Portugal-Spain, Spain-France, France-Belgium, Belgium-Netherlands). For NUTS 2 in Belgium and Netherlands, border controls also impact transport time even when using SSS, as the trucks coming from Le Havre suffer delays in border crossings into these countries. However, when using full road-based transport from Portugal, three border controls have to be cleared, leading to a loss in the main advantage of road haulage: transport time. The results in this Figure should be compared with those shown in Figure 10 as they relate to the same maritime freight rate and ship speed. It is possible to see that SSS gains competitiveness due to delays of trucks at border crossings. For example, for Ille de France (FR10-Paris), the number of NUTS 3 regions for which SSS is the preferable option jumps from 3 to 9. Such improvements occur also for Brittany (Rennes), Champagne-Ardennes (Chalons), Belgium and southern Netherlands.

5 CONCLUSIONS

This paper has firstly presented a numerical method to delimit the hinterland and foreland of short sea shipping (SSS) routes based on the generalized

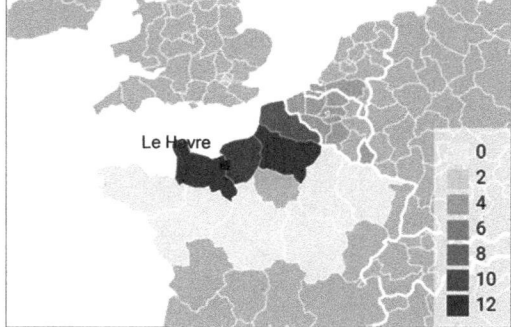

Figure 10. Number of NUTS 3 origin regions preferring SSS, for each NUTS 2 region in France-Belgium, when GTC is used as decision criterion (maritime freight rate 931.5€, speed 18 knots).

transport cost, which combines the two main parameters in transport decisions: cost and time. In this context, the hinterland and foreland of the route are the set of regions adjacent, respectively, to the loading port and the unloading port, for which transport chains based on SSS are competitive when compared with road transport chains. The method thus identifies the potential hinterland of a SSS route, but it should be recognized that other parameters, albeit less relevant than the above, such as the reliability, frequency and schedule of service also impact the route's competitiveness. A case study dedicated to a SSS route between Northern Portugal and Northern France, in the European Atlantic Area, was presented to illustrate the results of this approach.

The first conclusion is that the foreland of the SSS service in France depends on the location of the origin of cargo in Portugal. Cargo originating far away from the Portuguese coastline only finds it advantageous to use the ship if it is bound for NUTS 2 regions in France located along the coastline, not far from the port of Le Havre, implying a restricted foreland. Cargos originating along the Portuguese coastline may be competitively sent using SSS to a wide area in Northern France, Belgium and even the Netherlands, implying a much larger foreland. The numerical method provides results that can be used to identify these regions.

The numerical method was also used systematically for all possible pairs origin/destination, allowing a comprehensive evaluation of the foreland of the route. The foreland was characterized, for every NUTS 2 in Northern France, on the basis of the number of origin NUTS 3 regions preferring SSS to road transport. This evaluation was carried out for different maritime freight rates and ship speeds. Higher freight rates and lower speed clearly lead to smaller forelands. It is also evident from the results that forelands have no clear boundaries but rather fade away for NUTS 2 regions at longer distances from the coast.

Finally, as the numerical method is based on the calculation of transport cost and time over transport networks composed of nodes and links, it has been proven possible to model the effects of phenomena affecting transport performance such as border controls between EU countries, a situation that has recently occurred due to the Covid-19 pandemic. These border controls lead to significant delays in road transport, degrading its main advantage in comparison with SSS: shorter transport times. Results presented in this paper show how the numerical method indeed predicts an increase in the size of the foreland of the SSS route when border controls are in force.

The results presented show that SSS forelands and hinterlands are located primarily along coastlines. Further research will be directed at establishing whether the combination of SSS with other high capacity modes of transport (rail or IWW) may extend the foreland of SSS routes further inland, thus restricting the use of road transportation in line with the transport policy objectives included by the European Commission in the European Green Deal.

ACKNOWLEDGEMENTS

The research presented in this paper was conducted within the research project SHORTSEACHAIN - "Evaluation of short sea shipping services integrated in supply chains", financed by the Portuguese Foundation for Science and Technology (Fundação para a Ciência e Tecnologia - FCT), under contract PTDC/ECI-TRA/28754/2017. This work contributes to the Strategic Research Plan of the Centre for Marine Technology and Ocean Engineering (CENTEC), which is financed by FCT under contract UIDB/UIDP/00134/2020.

REFERENCES

Bird, J.H. 1963. Seaports and Seaport Terminals. Hutchinson University Library, London.

Brons, M., Christidis, P. 2013. External Cost Calculator for Marco Polo Freight Transport Project Proposals. Spain: European Union Joint Research Centre, Sevilla.

Carrese, S., Nigro, M., Petrelli, M., Renna, A. 2020. Identifying policies for intermodal logistics chains based on domestic Ro-Ro services, in *Integration of the Maritime Supply Chain: evolving from collaboration processed to maritime supply chain network*, Vanelslander, T., Sys, C. (Eds.), Elsevier, pp. 149–165.

EC 2017. Proposal for a directive amending Directive 92/106/EEC on the establishment of common rules for certain types of combined transport of goods between Member States, COM (2017) 648, Brussels, Belgium.

EC 2019. The European Green Deal, COM (2019) 640 final, Brussels, Belgium.

Hayuth, Y. 1981. Containerization and the load center concept. Econ. Geogr. 57 (2), pp. 160–175.

Lupi, M., Farina, A., Pratelli, A. 2020. The competitiveness of Motorways of the Sea: a case study in Italy, in *Short Sea Shipping in the Age of Sustainable Development and Information Technology*, Santos, T.A., Guedes Soares, C. (Eds.), pp. 199–223, Routledge, Taylor & Francis Group, London.

Morales-Fusco, P., Grau, M., Saurı, S. 2018. Effects of RoPax shipping line strategies on freight price and transporter's choice. Policy implications for promoting MoS. Transport Policy, vol. 67, pp. 67–76.

Notteboom, T., Rodrigue, J.-P., 2005. Port regionalization: towards a new phase in port development. Maritime Policy Management, 32 (3), 297–313.

Psaraftis, H., Zis, T. 2020. European policies for short sea shipping and intermodality, in *Short Sea Shipping in the Age of Sustainable Development and Information Technology*, Santos, T.A., Guedes Soares, C. (Eds.), pp. 3–21, Routledge, Taylor & Francis Group, London.

Santos, T.A., Guedes Soares, C. 2017. Development dynamics of the Portuguese range as a multi-port gateway system, Journal of Transport Geography, Vol. 60, April, pp. 178–188.

Santos, T.A., Guedes Soares, C. 2019a. Methodology for container terminal potential hinterland characterization in a multi-port system subject to a regionalization process, Journal of Transport Geography, Volume 75, pp. 132–146.

Santos, T.A., Martins, P., Guedes Soares, C. 2019b. Short sea shipping and the promotion of multimodality in the European Atlantic Area. Proc. of 27th Annual Conference of the International Association of Maritime Economists, Athens, Greece.

Santos, T.A., Guedes Soares, C. 2020. Assessment of transportation demand on alternative short sea shipping services considering external costs, in *Integration of the Maritime Supply Chain: evolving from collaboration processed to maritime supply chain network*, Vanelslander, T., Sys, C. (Eds.), Elsevier, pp. 13–45.

Santos, T.A., Ramalho, M., Guedes Soares, C. 2020. Sustainability in short sea shipping–based intermodal transport chains, in *Short Sea Shipping in the Age of Sustainable Development and Information Technology*, Santos, T.A., Guedes Soares, C. (Eds.), pp. 89–115, Routledge, Taylor & Francis Group, London.

van Essen, H., van Wijngaarden, L., Schroten, A., Sutter, D., Bieler, C., Maffi i, S., Brambilla, M., Fiorello, D., Fermi, F., Parolin, R., & El Beyrouty, K. 2019. Handbook on the External Costs of Transport, Version 2019. Luxembourg: Publications Office of the European Union.

Developments in Maritime Technology and Engineering – Guedes Soares & Santos (eds)
© 2021 Copyright the Author(s), ISBN 978-0-367-77376-2

Characterization of the cruise ship fleet calling in the port of Lisbon

T.A. Santos, P. Martins & C. Guedes Soares
Centre for Marine Technology and Ocean Engineering (CENTEC), Instituto Superior Técnico, Universidade de Lisboa, Lisbon, Portugal

ABSTRACT: This paper presents a study of the cruise ship fleet that called in the port of Lisbon in 2018, based on data provided by the port authority. This analysis is supported by a comprehensive database of cruise ships currently operating in the world fleet, which was developed to support the study. The information in the database is presented and its representativeness is assessed in comparison with other sources of information. The technical parameters, the characteristic ratios and the technical trends for this type of ship are identified. This world fleet is then compared with the subset of ships (fleet) that actually called in the port of Lisbon during the year of 2018. Relevant differences between this subset and the world fleet in terms of ship size (length, draught, tonnage), installed propulsion and auxiliary machinery power, passenger capacity, age and market segment are identified and analysed. An attempt is made to identify the profile of the typical cruise ship calling in Lisbon. Conclusions regarding implications for the port of Lisbon are drawn.

1 INTRODUCTION

The cruise shipping industry has been growing significantly in recent years, attracting 28.5 million passengers in 2018 while ten years before it carried only 17.2 million passengers, as per data from CLIA (2019a). The major regions for cruising are the Caribbean islands and the Mediterranean Sea, but other regions such as the Far East are booming. Portugal has also known some increase in cruise tourism, especially in the ports of Lisbon and Funchal. These ports have grown, respectively, from 407,504 and 435,821 passengers in 2009, to 577,603 and 537,851 passengers in 2018, implying a growing number of cruise ship calls. Correspondingly, port authorities in Lisbon, Funchal and Leixões have invested in new cruise terminal facilities aiming at improving the experience of cruise tourists.

The dynamics of cruise tourism worldwide presents different trends, as regards the way shipping companies and ports evolve, well reviewed in Pallis (2015). This dynamic is increasing cruise ship traffic in certain ports, and this fact has become very notorious, raising concerns over the impacts of this segment of tourism on cities. These impacts occur at many different levels, some are positive (revenue generated for port authorities, terminal operators and local businesses), but some others are negative (congestion, noise, air and water pollution). A limited number of studies has been dedicated to the evaluation of these impacts, especially in Portugal. At European scale level, the most notorious and controversial study has been one dedicated to air pollution,

see Transport and Environment (2019). In parallel, especially in Europe, the press has been flooded with news about the impacts of cruises.

Some of the news coming up in press and, in some cases, several findings in reports and studies, clearly lack a closer scrutiny. Therefore, it seems timely and appropriate to contribute to this public discussion with a technically rigorous characterization of ships calling in one of the two major cruise ports in Portugal: the port of Lisbon. This characterization is carried out for the year of 2018 and is based on data provided by the port of Lisbon authority (ship calls), see APL (2018), and on a database of technical characteristics of cruise ships developed to support this analysis. The objectives of this paper are to examine the characteristics of the world cruise ship fleet and to compare it with the cruise fleet that called the port of Lisbon in 2018. The research question under consideration in this paper is: *What is the typical profile (if any) of a cruise ship calling in Lisbon and how does it compare with the ships in the world cruise fleet?* The focus of the study is therefore largely on the examination of technical characteristics of the ships.

The paper, apart from this introduction, is organized in the following manner. Section 2 presents a literature review dedicated to papers on cruise ship technical characteristics and cruise ship design. Section 3 examines the world cruise ship fleet through a ship database developed for this study. Section 4 puts the ships calling in Lisbon in 2018 in the context of this database and attempts to answer the research question above, while section 5 summarizes the conclusions of the paper.

DOI: 10.1201/9781003216582-10

2 LITERATURE REVIEW

The literature on cruise ship design and on cruise ship technical characteristics is relatively scarce. Databases with technical information for the existing fleet are rare and available only with significant costs attached. Seatrade Cruise (2017) provides a comprehensive list of cruise ships in the world fleet, but technical data for each individual ship is very scarce. In any case, data in these sources does not allow an in-depth analysis of the fleet's characteristics.

Academic literature on cruise ships is focused on the evolution of the ship's technical characteristics and presents limited statistical information on the world's fleet. Often, papers are written by authors with experience on a specific shipyard and present data which reflects the focus of the activity of that specific shipyard rather than global trends.

However, studies such as those of Levander (2001), Landtman (2002) and Kvaerner Masa (2003), provide good overviews of developments in cruise ship design covering the period 1970-2002. These include statistics on main technical characteristics, main factors influencing ship design, the cost structure implied in the operation of these ships and main technological challenges. The importance of cabins and amenities onboard is discussed in detail, as well as the impact of the Panama Canal expansion on the design of ships.

Onoguchi *et al.* (2004) report the efforts put in the design of cruise ships built in Japan, concerning mainly the issues of smoke, noise and vibration reduction. These aspects are known to have significant impact in passenger comfort onboard and also on the environment, especially when in port, in the case of smoke. Levander (2010) applies the principles of goal based design to cruise ships, with particular focus on fire safety, damage survivability, evacuation and safe return to port. Vie (2014) provides an excellent overview of cruise ship design and construction, from the point of view of a major shipping company.

Several papers are dedicated to the discussion of different types of machinery and how these impact on fuel cost, flexibility and capital cost. Sippilä (2000) provides an overview of developments in the specific aspect of the machinery used for propulsion in large cruise ships while Levander (2006) discusses the different propulsion and auxiliary power installations in cruise ships, including the differences in fuel costs and emissions arising from different options in this matter. Sanneman (2004) discusses the specific aspects of gas turbine-electric propulsion systems in cruise ships.

Due to the significant number of passengers onboard, especially in more recent cruise ships, there are also some concerns with topics such as safety. This is focused primarily on damage stability, fire prevention and passenger evacuation. Vassalos (2016) provides a study dedicated to the issue of damage survivability, including methods to determine the probability of survival of a flooded ship. Luhmann (2016) discusses the impact of new damage stability regulations in the design of cruise ships.

Still dealing with safety, although mainly in port areas, Mileski *et al.* (2014) puts forward interesting data on the number of accidents and its general consequences, occurring in recent years, involving cruise ships. This information is used to derive societal risk levels and is of interest also in cruise ship design.

At a local scale, port authorities also provide some information on the cruise ships calling in their own ports. In this respect, the Port Authority of Lisbon is especially noteworthy, as it has published, for many years now, an annual review of cruise activity in this port, see APL (2018). This annual report includes an analysis of the characteristics of ships calling in this port, including also their names and a short number of main characteristics. Again, data presented in this document does not allow an in-depth analysis of the fleet's characteristics in Lisbon and a comparison with the world's fleet, which is the main objective of this paper.

3 ANALYSIS OF THE WORLD CRUISE SHIP FLEET

The analysis of the world cruise ship fleet should necessarily start with an examination of the size of this fleet. Figure 1 shows the evolution between 2000 and 2020 of the world cruise ship fleet in terms of number of ships and total number of berths fitted in these ships (a measure of carrying capacity). It may be seen that there has been a continuous growth in the number of ships since 2011. Even more noteworthy is the growth in number of berths, indicating that ships are becoming larger and larger.

Overall, even the economic crisis of 2009 appears to have had little effect in this industry, leading to a small and temporary decline in the number of ships, but not on the number of berths. In 2016 a total of 300 cruise ships were in operation while, in 2020, it is expected that a total of 363 ships will

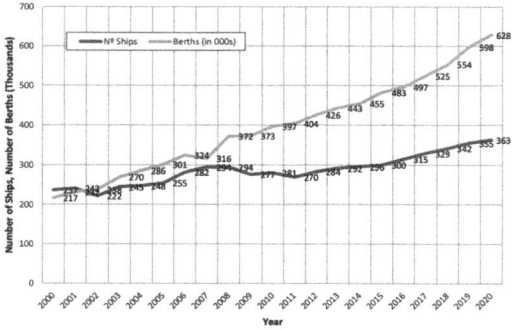

Figure 1. Evolution of the total number of ships and berths in the world cruise ship fleet (Source: own elaboration based on CLIA data).

be in operation worldwide and they will have a combined capacity of 628,000 berths.

3.1 *Cruise ship database validation*

A database of cruise ships in the world fleet has been developed using various sources, including specialized shipping publications (RINA significant ships of the year, Motorship, Ship Technology, Seatrade), specialized websites (Faktaomfartyg, Ferry-site, Scheepvartwest, Nedcruise, Globalcruiseship), industry organizations (CLIA), classification society's websites and the Equasis database. The information arising from these sources has been cross-checked as far as practicable and now includes data on a total of 205 cruise ships of all sizes and market segments. Henceforth, this database will be designated as the "IST Database".

This database includes a large number of technical characteristics for each ship, typically many more than are generally recorded in such databases, even the well-known World Fleet Register (Clarkson). Table 1 shows most of the technical characteristics included in the database. It must be mentioned that not all these characteristics, for every ship, are included in the database, as sources used do not provide a perfectly even and complete information. However, the database includes only those ships for which most of the relevant characteristics could be obtained. Furthermore, this database avoided the significant cost of actually acquiring it from consultancy or shipping research companies.

The world cruise ship fleet numbered in 2017, according with Seatrade Cruise (2017), a total of 294 ships. Therefore, the database developed for this study comprises about 2/3 of the world cruise ship fleet in 2016. It is believed that this sample is representative enough of the world fleet. In order to confirm this hypothesis a comparison is made with the characteristics of the fleet reported in Seatrade Cruise (2017), as far as possible, since this publication only reports a few main characteristics of the ships. Figure 2 shows the total gross tonnage of ships included in each size band in the existing fleet (as per Seatrade Cruise (2017)) and in the IST database, the conclusion being that the database now developed describes fairly well the world fleet.

Cruise ships may be categorized according with the quality of the ship's amenities and type of client experiences, into different market segments (five): contemporary, premium, luxury, budget, niche. Figure 3 shows the number of ships per market segment in the IST database and in Seatrade Cruise (2017), the matching between both being relatively good.

Figure 4 shows the percentage of ships in each passenger capacity range, in the case of the IST database and in Seatrade Cruise (2017).

Figure 5 shows the age distribution (in 2018) for ships in both databases.

Table 1. Technical characteristics included in the database.

IMO number	Gross and net tonnage
MMSI	Length over all
Name	Length between perpendiculars
Flag	Breadth
Operator	Depth
Sisterships	Summer draught
Year of delivery	Scantling draught
Shipyard	Displacement
Type of propulsion	Lightweight
Number and power of main engines	Fuel capacity
Number of propellers	Fresh water capacity
Propeller diameter	Maximum passenger capacity
Number and power of generators	Lower bed capacity
Number of boilers	Crew number
Number of bow thrusters	Newbuilding price
Service and maximum speed	

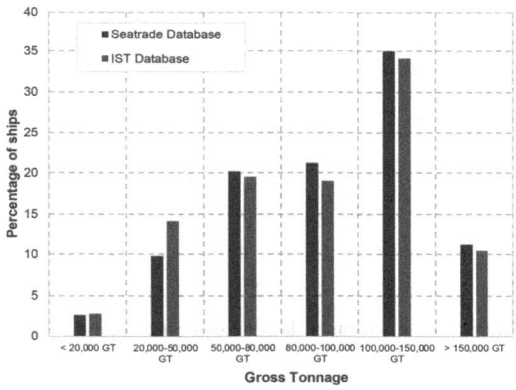

Figure 2. Comparison of the total gross tonnage of ships per size band.

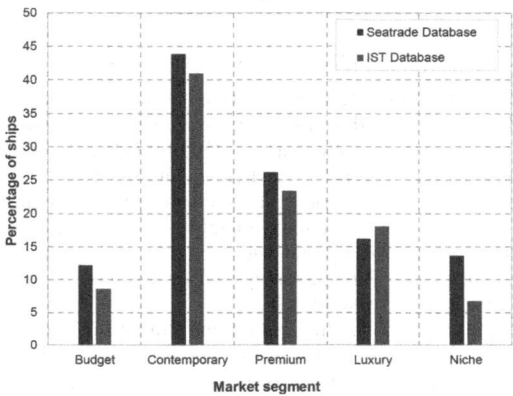

Figure 3. Comparison of the total number of ships per market segment.

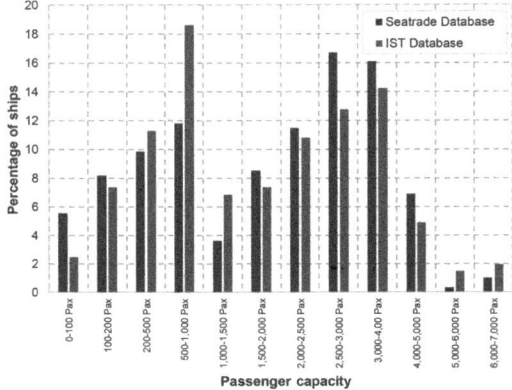

Figure 4. Comparison of the total number of ships per passenger capacity band.

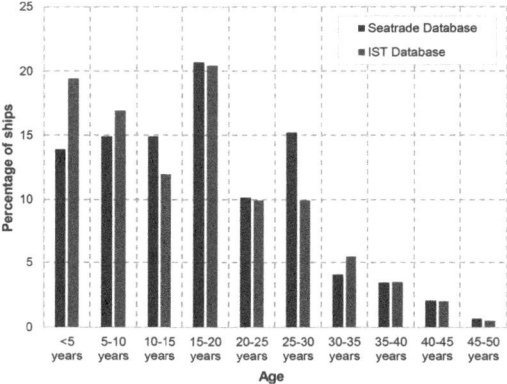

Figure 5. Comparison of the total number of ships per age band.

Considering the results shown in Figures 1 to 4 it is confirmed that the IST database reproduces well the characteristics of the world fleet and may be used to study the technical details of the ships.

3.2 Analysis of main dimensions and ratios

Figure 6 shows the gross tonnage and net tonnage of ships in the IST database. Ships have been built with gross tonnages of up to about 250,000 GT, although most have gross tonnages below 150,000 GT. It is also interesting to note a large cluster of ships with length over all of approximately 294m, the maximum permitted by the old locks of the Panama Canal. These Panamax ships generally have gross tonnages up to 100,000 GT.

Figure 7 shows the breadth and draught of cruise ships as a function of gross tonnage. Draught grows almost linearly with gross tonnage, staying always below 10m. Breadth presents several interesting features, one of them being the cluster of ships with a breadth of 32.3m, a restriction arising from the

previous Panama Canal locks. The other is the rapid growth of ship's breadth with the increase of gross tonnage, upwards of 100,000 GT.

Figure 8 shows the ratios length between perpendiculars to breadth and breadth to draught. Both show some tendency to rise as gross tonnage

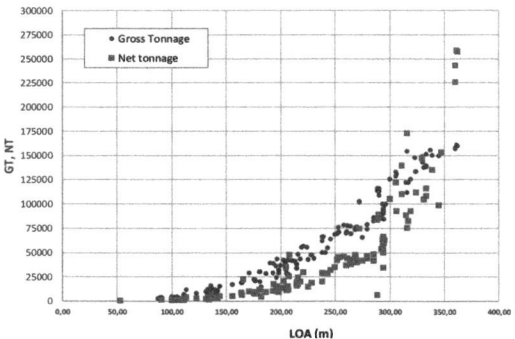

Figure 6. Gross and net tonnage of ships as a function of length over all of ships.

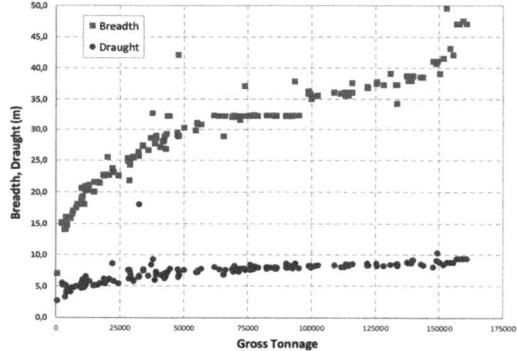

Figure 7. Breadth and draught of ships as a function of gross tonnage of ships.

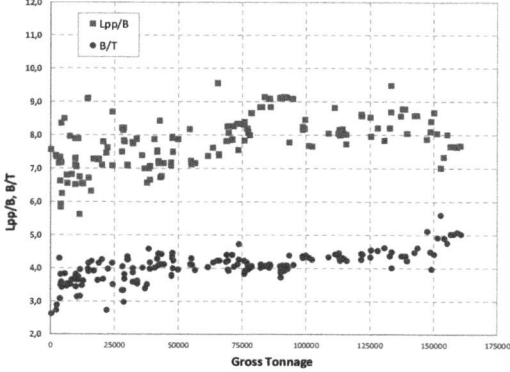

Figure 8. Length between perpendiculars to breadth and breadth to draught ratios as a function of gross tonnage.

increases: Lpp/B increases from 7.5 to 8.5 while B/T goes from less than 4.0 to 4.5.

3.3 Analysis of passenger capacity and crew number

Figure 9 shows the capacity of the ships in terms of maximum passenger capacity, lower berths capacity and crew number. Crew number is almost linearly proportional to gross tonnage. However, both passenger capacities seem to rise steeply with increases in gross tonnage for ships larger than 150,000 GT.

Figure 10 shows the ratio passengers per crew member, which rises almost linearly with gross tonnage. For smaller ships (below 50,000 GT) it is less than 2,5, while for larger ships it is about 3,5. The same figure shows the ratio between gross tonnage and the number of lower berths, which generally is comprised between 25 and 40. A significant number of smaller ships (below 50,000 GT) present much higher values (above 40). These values correspond to

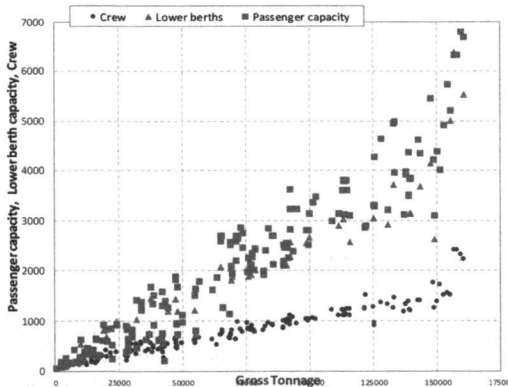

Figure 9. Passenger capacity, lower bed capacity and crew number as a function of the gross tonnage of ships.

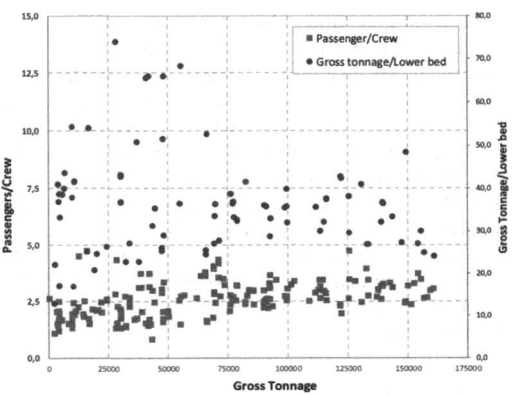

Figure 10. Passenger to crew and gross tonnage to lower bed ratios for ships categorized as per their market segment.

ships in the luxury market segment, where passenger density is generally much lower.

3.4 Analysis of propulsion and auxiliary machinery

Cruise ships may have diesel-mechanic or diesel-electric propulsion. Figure 11 shows the propulsion power (rated power of propulsion engines, either diesel or electric motors) installed on-board the ships, as a function of the gross tonnage. Each ship is identified according with its type of propulsion: diesel-electric or diesel-mechanic. It may be seen that ship's power increases with gross tonnage, as might be expected, but tends to stabilize at an average of 40,000 kW for ships above 100,000 GT. Very few ships have more than 50,000 kW of installed propulsion power. The ship with 86,000 kW is the *Queen Mary II*, capable of achieving almost 30 knots. The database comprises about 60 diesel-mechanic ships and 140 diesel-electric ships. Diesel-mechanic ships are almost all below 75,000 GT and 25,000 kW.

When ships are diesel-electric, the electric motors may actuate propulsion shafts and propellers or be fitted in azimuthing pods (azipods). Figure 12 shows the number of propulsion shafts and number of azipods. Propulsion shafts may be actuated by diesel engines, with one or two diesel engines per shaft (there is a significant number of diesel-mechanic ships with one shaft per pair of diesel engines), or by electric motors (always one per shaft).

As may be seen also in Figure 12, there might be 2 to 4 azipods (mostly manufactured by ABB). There are a couple of ships for which one of these pods is actually fixed, while the others are azimuthing. Overall, the vast majority (above 90%) of the ships either has two shaft lines (diesel-mechanic ships) or has 2 azipods or 2 electric motors (and shafts).

Figure 13 shows the service speed and maximum speed of ships. It may be seen that larger ships present higher speeds, in any case almost always below 25 knots. Service speed of larger ships is about 21-22 knots. Ships below 50,000 GT have, generally, service

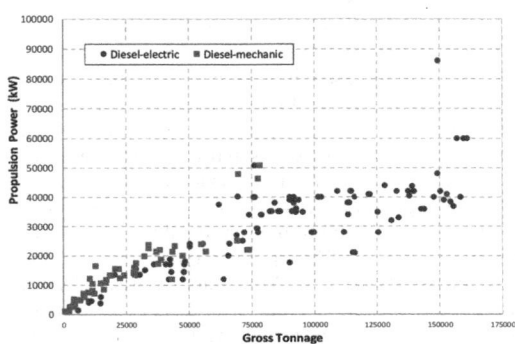

Figure 11. Propulsion power as a function of the gross tonnage of ships.

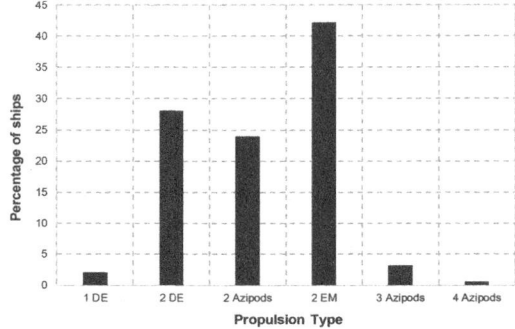

Figure 12. Type of propulsion installed in the ships (DE-diesel engines; EM-electric motors; in any case with shafts).

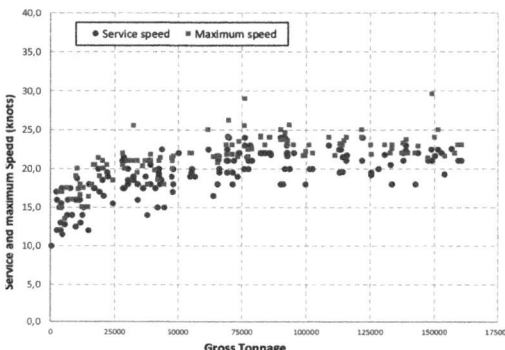

Figure 13. Service speed and maximum speed as a function of the gross tonnage of ships.

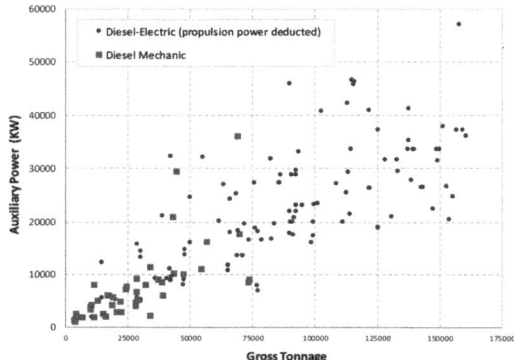

Figure 14. Auxiliary power as a function of the gross tonnage of ships.

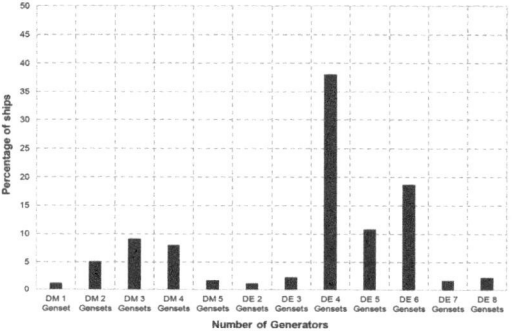

Figure 15. Number of propulsion and auxiliary power units for diesel-mechanic and diesel-electric ships.

speeds below 20 knots. There is a difference between service speed and maximum speed of about 2 knots, in most cases.

Figure 14 shows the auxiliary power installed onboard the ships, after deducting the propulsion power, in case the ship has diesel-electric propulsion. Each ship is identified according with its type of propulsion. The auxiliary power appears to grow linearly with the gross tonnage.

Regarding the types of main and auxiliary sources of power, it may be said that either diesel generators or gas turbines are used in this type of ship. Diesel-mechanic ships are all fitted with diesel engines for propulsion or auxiliary purposes. However, some diesel-electric ships (such as those of the Coral and Vista Classes and *Queen Mary II*) are fitted with a mix of diesel engines (in gensets) and gas turbines (generally the GE LM-2500 turbine) in combined diesel and gas installations (CODAG). Some other ships, such as those of the Radiance or Millennium Classes, feature also a steam turbine working with exhaust gases from the steam turbines, corresponding to a combined gas turbine electric and steam (COGES) installation.

Figure 15 shows, for diesel-mechanic and diesel-electric ships, the number of power units (diesel engines, gas turbines or steam turbines) used for auxiliary purposes. It must be taken into account that for diesel-electric ships, a significant part of the power generated is used for propulsion. The figure allows the conclusion that most diesel-mechanic ships have 2-4 gensets, while diesel-electric ships have generally 4-6 gensets. This increase is due to the need for generating also propulsion power.

Considering the significant propulsion power and auxiliary power installed on-board these ships, emissions have become a major concern for the public, but also for maritime administrations and port authorities. However, it is important to have in mind that many cruise ships have installed exhaust gas cleaning systems (EGCS), particulate filters, selective catalytic reduction (SCR) systems or are able to connect to shore power (cold ironing), as reported in CLIA (2019b). Additionally, a few ships (and particularly newbuildings) are now capable of using LNG or other alternative fuels.

3.5 Newbuilding yards and prices

Figure 16 shows the country and shipyard that builds the ships in the database. It may be seen that Italy,

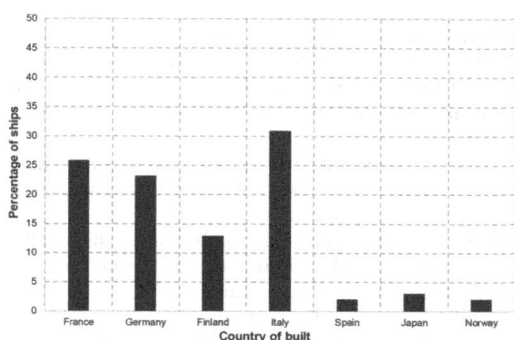

Figure 16. Country and shipyard of newbuilding.

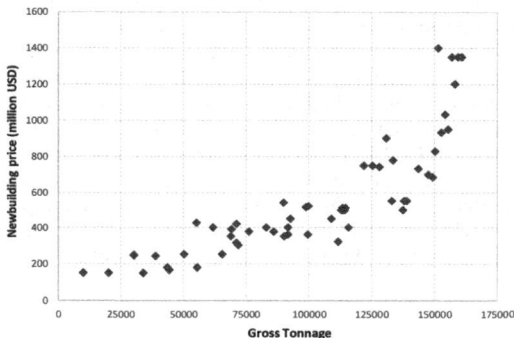

Figure 17. Published newbuilding price (in USD of the year of construction) as a function of the gross tonnage (Source: various specialized press).

France and Germany largely dominate this market, followed by Finland. Each of these countries is mostly represented by one shipyard (in some cases under different names), namely Fincantieri, Chantiers de L'Atlantique, Meyer-Werft and Wartsila. Contrary to other ship types, shipyards in the European Union have been largely successful in retaining the construction of these ships.

Figure 17 shows the newbuilding prices in USD reported in the specialized press for a number of ships, as function of the gross tonnage. Newbuilding price appears to increase linearly with gross tonnage, but upwards of 125,000 GT the increase becomes much steeper, with prices for the largest ships nearing 1.4 billion USD.

4 COMPARISON WITH CRUISE SHIP FLEET CALLING IN LISBON IN 2018

The database used to characterise the world cruise ship fleet in the previous section will now be used to characterise the subset of ships (fleet) calling in the port of Lisbon in 2018. Lisbon has been for many years one of the two major cruise ports in Portugal, the other being Funchal in the island of Madeira. The Port Authority of Lisbon provides in their website information on the number of cruise ships calling during the period of a year, as well as their names and number of calls per ship. This information is used to support the analysis presented below, in connection with the data presented in the preceding section.

Figure 19 shows the evolution of the number of cruise passengers in the port of Lisbon from 1996 onwards. In parallel, the number of ship calls is also shown. It is possible to see that the number of passengers has grown more than the number of ship calls, indicating mainly that the average size of the ships received in this port has increased significantly. In fact, data from the Port Authority of Lisbon (2006-2018) indicates that ship's occupation levels are consistently above 95%, implying that this is not the reason for the increase in number of passengers. Regarding the number of calls, it stands at 339 calls in 2018, but this corresponds actually to only 123 different ships, as many of them visit Lisbon several times during the course of the year.

Figure 19 shows the percentages of ships calling 1 up to 14 times during the year of 2018. The main conclusion is that about ¾ of the ships called in

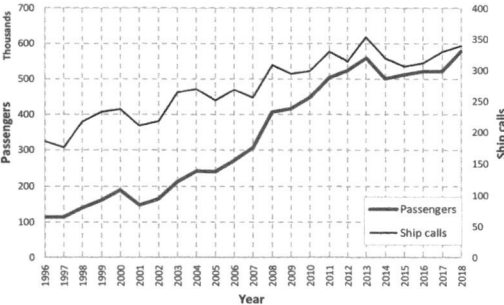

Figure 18. Number of cruise passengers and cruise ship calls in the port of Lisbon 1996-2018.

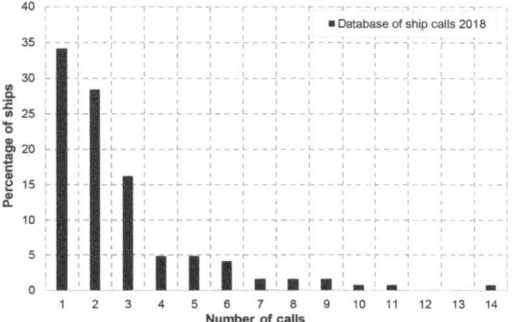

Figure 19. Number of calls per cruise ship in Lisbon in 2018.

97

Lisbon up to 3 times during 2018. Very few called more than 6 times in 2018. This implies a large variety of ships coming into Lisbon, indicating that the deployment of the ships changes significantly throughout the year. Seasonality, which is strong in Lisbon, probably plays a role in this situation.

Figure 20 shows the daily distribution of times of arrival and departure of cruise ships in the port of Lisbon in 2018. It may be seen that the arrivals are concentrated in the early hours of the morning, namely between 5AM and 9AM, while the departures occur mainly in the evening hours, between 16PM and 21PM. However, the distribution of departures is less concentrated, with some ships departing immediately after lunch and others in the early hours of the night.

Figure 21 shows the gross tonnage of ships calling in Lisbon in 2018 and of those in the world fleet. It is possible to note that ships are distributed mostly between 50,000 GT and 150,000 GT. The range of gross tonnage between 100,000 and 150,000 GT is the one concentrating more tonnage. Globally, the split of the ships throughout the ranges of gross tonnage is very similar, but with a slightly larger focus in Lisbon on ships with less than 100,000 GT.

Figure 22 shows the maximum passenger capacity of ships calling in Lisbon in 2018 and in the world fleet database. A larger emphasis on ships with small passenger capacities (up to 1500 passengers) is seen in Lisbon.

Figure 23 shows the market segment of ships calling in Lisbon in 2018 and in the world database. Distributions are very similar

Figure 24 shows the age distribution of ships calling in Lisbon in 2018 and in the world database. The distributions are again very similar.

Figure 25 shows the propulsion power of ships calling in Lisbon in 2018, showing some preponderance of ships (56% of total) with installed power between 10,000 and 30,000 kW. Ships with larger installed propulsion power are less common than in the world fleet, a consequence of the smaller number of ships above 100,000 GT than in the world fleet.

Table 2 shows a final comparison between the average characteristics of ships in the world fleet and in the fleet that called in Lisbon in 2018. It may be concluded that ships calling in Lisbon are slightly smaller (GT, LOA, passenger capacity) and slightly older (one year more). The installed propulsion and auxiliary power (DE stands for diesel-electric hips and DM for diesel-mechanic ships) are also slightly smaller, mainly due to smaller dimensions of ships.

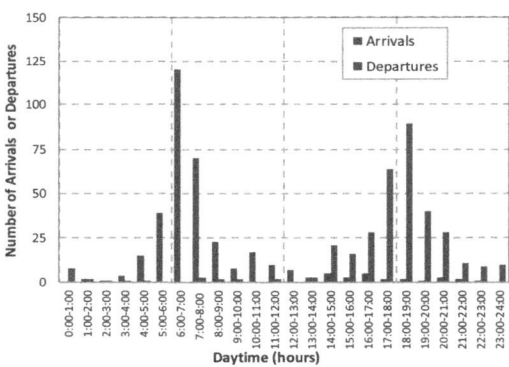

Figure 20. Distribution of times of arrival and departure in the port of Lisbon in 2018.

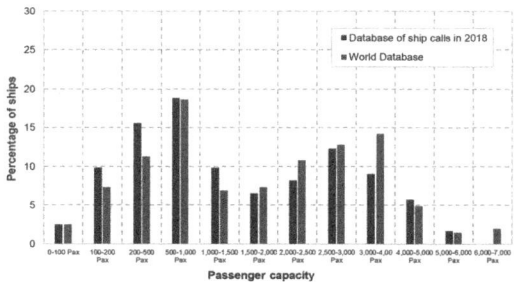

Figure 22. Passenger capacity of ships calling in Lisbon compared with the world fleet.

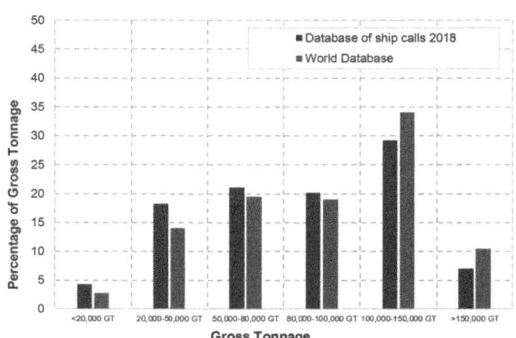

Figure 21. Gross tonnage of ships calling in Lisbon compared with world fleet database.

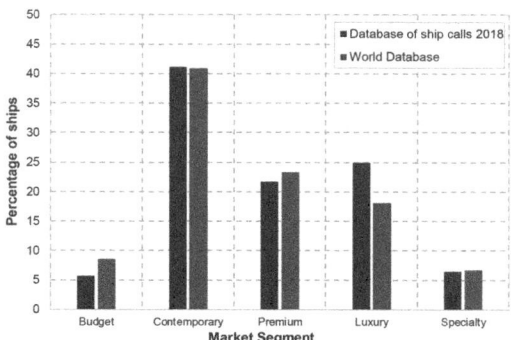

Figure 23. Market segment of the ships calling in Lisbon compared with the world fleet.

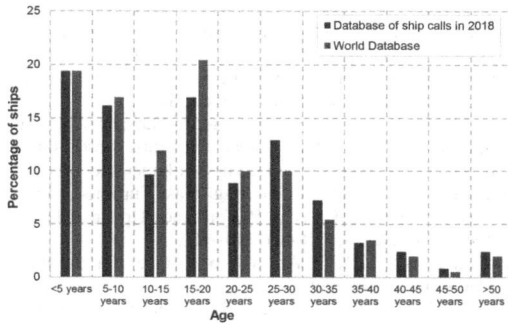

Figure 24. Age of the ships calling in Lisbon compared with the world fleet.

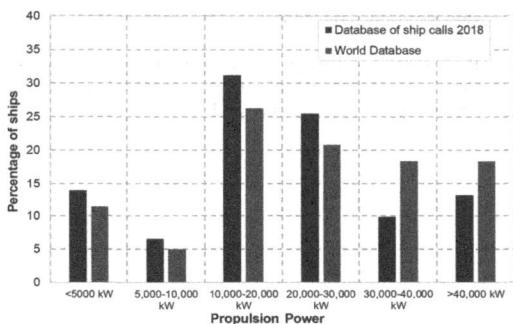

Figure 25. Propulsion power of ships calling in Lisbon compared with the world fleet.

Table 2. Comparison of average ship characteristics.

	World Fleet	Lisbon Fleet in 2018
Gross Tonnage	65146	55182
Length over all (m)	234.0	217.2
Passenger maximum capacity	1925	1615
Age (years)	17.1	18.1
Propulsion power DE-DM (kW)	25900-14267	24578-14211
Auxiliary Power DE-DM (kW)	49416-7191	43076-6597

5 CONCLUSIONS

This paper has presented a characterization of the world cruise ship fleet and has compared this fleet with the fleet calling in the port of Lisbon in 2018. This study has been based on a database containing technical information on cruise ships, developed using data provided in a number of technical publications and websites. The sample of ships in this database comprises about 2/3 of the world fleet in 2016 and was found, in the first part of this paper, to represent well the world fleet.

Regarding the world fleet, it is possible to conclude that ships have become larger and have now reached gross tonnages of 250,000 GT, lengths of 360m, breadths of 50m and draughts of 10m. They are mostly built in Europe, reaching extreme new-building costs, now nearing 1.4 billion USD. These dimensions have allowed the passenger capacity to rise to 6300 plus 2200 crew members, making space per person rather limited. However, ships in the luxury market segment have generally much more space per passenger (and crew members per passenger) than other ships. Two thirds of the ships in the world fleet are diesel-electric, with propulsion power up to 45,000 kW and auxiliary power up to 40,000 kW (with a few exceptions). Service speed is between 15 knots (smaller ships) and 22 knots (larger ships).

A database of cruise ships that called in Lisbon in 2018 has been developed and its analysis shows that this fleet is not significantly different from the world fleet, as ships of all sizes and market segments are present. Indeed, in 2018, 124 different ships called in Lisbon, which is a significant fraction of the world fleet. Most ships called 1-3 times, some called in this port 4-6 times. Typically, ships arrive in the early morning and depart in the evening. Although the Lisbon fleet is not very different from the world fleet, it is clear from the statistics that there are slightly more ships with lower gross tonnage, higher age, smaller passenger capacity and less propulsion power. Also, a larger proportion of luxury and specialty (niche) ships is present. Nevertheless, the overall averages for most parameters do not differ significantly from the world fleet. Accordingly, there appears not to exist a typical profile of cruise ship which calls in Lisbon.

ACKNOWLEDGEMENTS

This work was performed within the scope of the Strategic Research Plan of the Centre for Marine Technology and Ocean Engineering (CENTEC), which is financed by the Portuguese Foundation for Science and Technology (Fundação para a Ciência e Tecnologia) under contract UIDB/UIDP/00134/2020.

REFERENCES

APL (2018). Cruise ship traffic – activity report (in Portuguese), Lisbon Port Authority, Lisbon, Portugal.

Chen, J. M., Nijkamp, P. (2017). Itinerary planning: modelling cruise line's lengths of stay in ports. International Journal of Hospitality Management, 73, 2018, 55–63.

CLIA (2019a). Cruise trends & industry outlook. Washington D.C., USA.

CLIA (2019b). Environmental Technologies and Practices Report (11 September), Washington D.C., USA.

Kvaerner Masa (2003). Breaking the mold: the way ahead in cruise ship design. Seatrade (2003) Cruise Shipping Convention.

Landtman, M. (2002). Designing Cost Efficient Cruise Ship, Optimizing the Money Making Potential. Seatrade Cruise Shipping Convention.

Levander, K. (2001). New Challenges in Cruise Ship Design Optimizing the Panamax Size Ship. Seatrade Cruise Shipping Convention.

Levander, K. (2006). Continuing Economic Advantage for Diesel-Electric Cruise Ships, The Ship Power Supplier.

Levander, K. (2010). Goal based ship safety application in large cruise ship design. Proc. of the 5th International Conference on Collision and Grounding, 3-12, June 14th-16th, Espoo, Finland.

Luhmann, H. (2016). Damage stability of cruise ships. Meyer-Werft (retrieved from https://vsm.de on 23.10.2018).

Mileski, J. P., Wang, G., Beachman IV, L. L. (2014). Understanding the causes of recent cruise ship mishaps and disasters. Research in Transportation Business & Management, 13, 65–70.

Onoguchi, Y., Iwamoto, M., Senju, H., Kanaga, T., Terada, S. (2004). Debut of First Large Passenger Cruise Ship in Japan for Princess Cruises. Mitsubishi Heavy Industries, Technical Review, 41, 6.

Pallis, T. (2015). Cruise Shipping and urban development: State of the art of the industry and cruise ports. International Transport Forum Discussion Paper, No 14.

Sanneman, B. (2004). Pioneering Gas Turbine-Electric System in Cruise Ships: A Performance Update. Marine Technology, 41, 4, October, 161–166.

Seatrade Cruise (2017). An A-Z of Cruise Brands, their Fleets and Passengers.

Sippilä, H. (2000). Machinery in Large Cruise Ships. Marine Propulsion Conference.

Transport & Environment (2019). One Corporation to Pollute Them All - Luxury cruise air emissions in Europe, European Federation for Transport and Environment AISBL.

Vassalos, D. (2016). Damage survivability of cruise ships – Evidence and conjecture. Ocean Engineering, 121, 89–87.

Vie, R. (2014). The Design and Construction of a Modern Cruise Vessel. President's Day Lecture. Institute of Marine Engineering, Science and Technology.

Hinterland corridor management initiatives in the EU and the US: The role of ports

E. Sdoukopoulos & M. Boile
Department of Maritime Studies, University of Piraeus, Piraeus, Greece
Hellenic Institute of Transport, Centre for Research and Technology Hellas, Athens, Greece

ABSTRACT: With hinterland connectivity increasingly shaping now port competitiveness, port actors have intensified their regionalization efforts, establishing strong and functional links with the hinterland transport system. However, they did not stop there. For driving further efficiencies to the system, they have started to step more into the hinterland through vertical integration strategies and other targeted investments, gradually advancing their role from facilitators to entrepreneurs. Those business and market strategies need though to be efficiently complemented with others aiming to ensure institutional support and drive infrastructural developments into the hinterland. This is where hinterland corridor management initiatives can play an important role. Ports participating into such initiatives are provided with the opportunity to enhance their hinterland intelligence and stress out bottlenecks to be addressed that are beyond their reach but nevertheless can impose a considerable impact on their competitiveness. All the aforementioned strategies are being discussed in this paper, with emphasis on corridor management initiatives established in the world's two key trading regions i.e. Europe and the U.S. The structure and activities of those initiatives are outlined and valuable insights are provided on how these can be effectively exploited by ports for devising comprehensive hinterland strategies.

1 INTRODUCTION

Supply chain globalization has driven the exponential rise of international trade. Maritime transport serves as the backbone of the latter, since it accounts for nearly 80% of world's cargo by volume and 70% by value. Indeed, over the last three decades, international maritime transport volumes have been increasing at an average rate of 3%, with the exception of 2009 when the implications of the global financial crisis became apparent also to the maritime transport industry, among several other economic sectors that were impacted (UNCTAD, 2019a). Another exception though, and most probably with an even greater impact, is being expected in 2020 as a result of the global outbreak of Covid-19 (Baldwin & Tomiura, 2020).

This constant growth of international trade has been heavily supported by developing economies (i.e. mainly in East Asia), which still account today as major exporters of raw materials, although share increases in import volumes highlight that the role of these economies is gradually changing. Besides continuing to serve as major players in global manufacturing processes, they also account now as key areas of growing consumption.

The aforementioned considerations dully justify the majority of international trade volumes being concentrated on the east and westbound main-lane routes

connecting East Asia with its key trading partners namely Europe and North America (Figure 1). The ever increasing cargo volumes that those regions receive, are being accommodated via an extended and highly functional port network, efficiently connected with multimodal hinterland transport corridors providing cost-effective access to the main hinterland markets. The concentration of the latter presents however different regional patterns (Figure 2).

More specifically, the North American market is mainly concentrated along the continent's East and West Coast, with cross-cutting (mainly rail) land-bridges connecting major and regional maritime hubs located at each end. This proves not to be the case in Europe, where the hinterland market is centrally located with multiple transport corridors, organized in a radial type format, providing efficient connections with a high number of gateway ports that have been developed along key surrounding coastal regions.

Hinterland transport connectivity is indeed an important driver of port competitiveness, and actually the second most important one, according to Parola et al., 2017, after port costs. It is therefore of great interest to review how port actors can drive or influence hinterland transport developments and efficiency improvements, going beyond measures limited to their area (e.g. hinterland access rules,

DOI: 10.1201/9781003216582-11

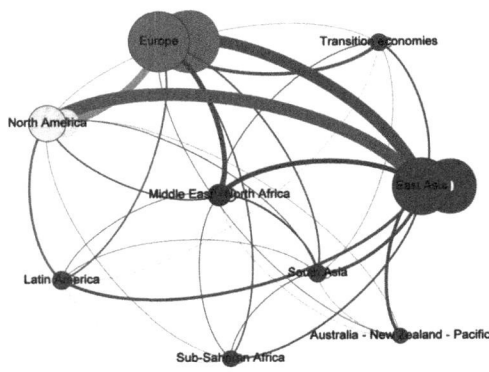

Figure 1. Trade flows across world's regions between 2015 and 2016 (UNCTAD, 2019b).

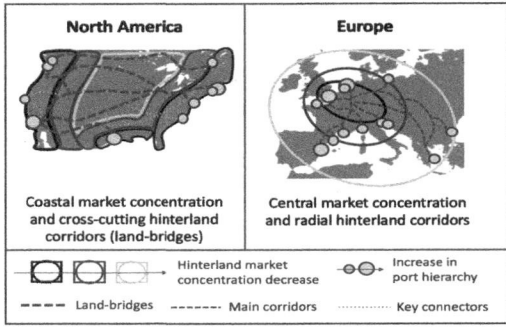

Figure 2. Regional patterns of hinterland market concentration and corridor formation (adapted from Lee et al., 2008).

etc.). Section 2 provides such a review, with emphasis on the relevant opportunities that corridor management initiatives can provide for ports. The latter are being discussed in more detail for both Europe and the U.S. in sections 3 and 4 respectively, enabling to draw some interesting conclusions and gain valuable insights in section 5, on how participation in such initiatives may form an integral part of the hinterland strategies of ports, complementing a number of other measures/actions that are being planned.

2 PORT STRATEGIES FOR ENHANCING HINTERLAND INFLUENCE

Sea-driven, as they are often called, hinterland developments are very limited compared to land-driven ones, since ports lack of the required institutional capacity. To this end, vertical integration strategies, cooperation schemes and coordination measures, as well as influence exercised through participation in hinterland corridor management initiatives, are often exploited by ports for driving, to the

extent possible, hinterland developments and efficiency improvements.

2.1 Vertical integration

Vertical integration is defined as the strategy of a company providing services in a part of the supply chain, investing (i.e. majority or minority) in another company that provides services in another part down the same chain.

Shipping lines are one of the most indicative examples. In the container industry, seven out of ten largest shipping lines in the world, are also involved into terminal operations managing approximately 188 terminals worldwide (e.g. Maersk through APM Terminals, MSC through TiL, CMA CGM through Terminal Link, etc.) (Zhu et al., 2019; Álvarez-SanJaime et al., 2013). In many cases, shipping lines have also expanded their strategy further down the chain, being heavily involved now in hinterland intermodal transport and logistics services (e.g. COSCO through PEARL, MSC through MEDLOG, etc.).

Terminal operating companies, mainly the global ones, have also broaden their scope, more actively participating now, through majority or minority investments, in intermodal transport and inland terminal operations (i.e. with the latter supporting the 'extended gates' concept) (Table 1) (De Langen, 2008; Franc & Van der Horst, 2010; Van den Berg & De Langen, 2011; Van den Berg & De Langen, 2015).

Through such a strategy, port actors can better control the efficiency of the hinterland system and drive developments, though to a certain extent depending on the multiplicity of the hinterland actors and their respective market shares.

Table 1. Vertical integration of selected leading global terminal operators (GTOs).

GTO	Intermodal transport operations	Inland terminal operations
COSCO Shipping Ports (CSP)	CSP Iberian Rail Services	CSP Iberian Zaragoza Rail Terminal
HPH	European Gateway Services, Synergy	DeCeTe Duisburg, Venlo, Willebroek
APM Terminals	APM Terminals Railway Spain	Depot tmZ Services
DP World	-	DP World Mannheim, DP World Germersheim, DP World Stuggart, swissterminal
Eurogate	Eurogate Intermodal, Contship Intermodal (Hannibal, Oceanogate)	Contship Intermodal (RHM)

(authors' own elaboration)

102

2.2 Cooperation and coordination

Effective cooperation and efficient coordination between port and hinterland actors is a long-lasting objective that lies at the very heart of every hinterland strategy of all port actors worldwide, and various measures have been deployed towards meeting this goal.

Unlike vertical integration strategies that have been mostly implemented by global terminal operating companies, though with a few exceptions mainly of corporatized authorities of large ports (e.g. Rotterdam, Marseille, Le Havre, etc.), port authorities have been, so far, more active in ensuring the establishment of strategic partnerships with hinterland actors (ESPO, 2011). However, the extent to which they will continue to act as conservators (i.e. fostering developments limited to the port domain) or facilitators (i.e. introducing incentives to stimulate hinterland developments), or will be undertaking a more active entrepreneur role (i.e. making targeted investments within the context of vertical integration) in the future, remains to be seen (Magnan & van der Horst, 2020).

Table 2 below provides an overview of the key measures that port authorities have implemented for enhancing cooperation with hinterland actors, and ensuring a better coordination of the relevant operations. As it has been stressed out before, emphasis is being provided on the role of ports as facilitators. For the measures included, a few indicative examples are also provided.

2.3 Hinterland corridor management initiatives

Another opportunity for ports driving or influencing hinterland developments, is their participation into the governance of specific corridors that provide efficient connections with key production or consumption areas. Those corridors can be of different scale and thus of purpose, with various relevant stakeholders constituting the respective governance body.

Local corridors are more specific in purpose, aiming most often to connect ports with key parts of the primary transport network. Regional corridors aim to provide efficient connections of ports with key areas in their captive hinterland, while extended cross-cutting corridors can allow ports to reach more distant hinterland markets, for which other ports are also competing. The port co-opetition dimension falls within the context of the latter, since adjacent and often competing ports have started to better realize the mutual benefits that they can gain if the performance of shared hinterland corridors is improved, enabling them to attract freight flows from other multi-port regions (Song, 2003).

Obviously, the criticality of certain corridors necessitated the establishment of a governance body, compared to others that just fall under the general responsibility of governmental institutions. Those bodies are mainly tasked within the supervision and effective management of the respective corridors, identifying any bottlenecks that may exist or areas where further improvements may be realized, and taking appropriate actions for addressing them so that the overall performance of the corridors is enhanced and trade flows can be better facilitated.

For getting a better perspective on the aforementioned considerations, a number of highly indicative examples that can be found within the European and U.S. context are presented in the two following sections. Corridors of different scale are taken into consideration, and insights on the role of ports within the respective governance bodies are being provided.

Table 2. Port authorities' hinterland cooperation and coordination measures.

Measure	Brief description	Indicative examples
Strategic partnerships with industry associations	Joint development of strategic initiatives for stimulating stakeholder collaboration, supporting the creation of new services (e.g. in intermodal transport & inland terminal operations)	Inlandlinks.eu, Rail Incubator
Acting as (neutral) mediator in commercial business relations	Facilitating the connection of different parties that can provide the necessary conditions for launching new services (e.g. shippers in a region providing enough volumes for introducing a new transport service)	HungaRo Express
Hinterland network promotional partnerships	Partnership of seaport(s) with inland terminals and possibly with other authorities, for promoting integrated multimodal solutions, improving the commercial offering of door-to-door logistics chain and increase multimodal traffic via a nationally and internationally recognized name	Medlink ports
Community Management	Implementation of Port and Cargo Community Systems (PCS & CCS) for facilitating information and data exchange between public and private stakeholders of the extended port community, hinterland requirements (i.e. modal split shares) in concession contracts, etc.	Portbase, DAKOSY, etc. (PCS), AP+ (CCS) Maasvla-kte 2 concession at Port of Rotterdam

(authors based on Magnan & van der Horst, 2020)

3 THE EUROPEAN CONTEXT

3.1 *The TEN-T core network corridors*

The development of an efficient transport network and the establishment of a common transport policy had been a fundamental part of the European Union (EU) since its establishment. To this end, with the Maastricht Treaty (1992), the Trans-European Transport Network (TEN-T) was also introduced, aiming to address existing gaps, remove any major bottlenecks and eliminate the technical and organizational barriers that could hinter the interoperability of the transport networks of the different Member States (European Commission, 2014).

The TEN-T core network comprises of nine hinterland and four maritime corridors (Figure 3). The former span both vertically and horizontally across the European context, linking hub ports (i.e. TEN-T core ports) with the main hinterland market of Europe, which is characterized by a high degree of centrality.

For each of the nine corridors, the European Commission (EC) nominated, in 2014, a European Coordinator considering the knowledge of the selected experts on issues related to transport and financing, as well as their experience of European institutions (three of them were replaced though in 2018). The Coordinators' mandate is to draw-up a work plan for the corridor under their responsibility, which outlines its current state and sets out the challenges to be addressed for its future development (European Commission, 2020).

For setting the main basis of these work plans, comprehensive corridor studies are being undertaken first. During this process, the Coordinators consult their corridor forum, which comprises of various working groups including one on ports. Port authorities participating in those groups have thus the chance to present the challenges they are facing, also with regard to their hinterlands among others, and propose proper solutions for addressing them whilst highlighting any areas where additional improvements can be realized. Considering also the feedback collected from the other

working groups, the corridor work plan is then prepared, which mainly comprises of (a) the corridor's current characteristics (i.e. infrastructural, operational and market), (b) critical issues (e.g. cross-border and capacity issues, interoperability and intermodality, hinterland transport bottlenecks, better integration of seaports, etc.), and (c) recommendations (i.e. suggested improvements) and future outlook. The work plan is then sent for approval to the Member States concerned, and once approved its implementation can be initiated. For supporting the latter, funding is available for all corridors through the Connecting Europe Facility (CEF) Programme (□ 24 billion for 2014-2020).

Very interestingly and for further supporting the successful implementation of the TEN-T core network corridors, Öberg et al., 2018 also discuss the need for complementary governance considering not the key but other stakeholder groups that are affected by the corridors' implementation (i.e. local authorities, infrastructure providers, private companies). The authors highlight the need for adopting an inclusive approach where all these stakeholders, who together form the extended corridor community, are involved in its implementation, receiving relevant information and updates, providing opinions and views in the context of a longlasting open dialogue and participate in relevant awareness raising events to be organized. This process is important for all key stakeholder groups, including port authorities, since all other actors adopting a corridorwide perspective, can contribute towards the development of additional joint initiatives, the intensification of cooperation efforts, and a greater level of transport and logistical integration being achieved.

3.2 *Regional corridor projects*

Besides the TEN-T core network corridors, no other transport corridors within the European context are being managed by a governance body. However, on a project basis, with funding received from various programmes, different stakeholders often come together in a partnership for analyzing regional corridors, which in most cases serve as connectors to the TEN-T core network corridors. Port authorities constitute an important part of such partnerships, sharing their views, and supporting the development of targeted action plans for further improving (a) the integration of ports with the hinterland corridors and (b) the overall efficiency of the latter, since the benefits to be derived in terms of port competitiveness are important.

The SETA - South East Transport Axis corridor is such a case, with the relevant project receiving funding from the South East Europe programme. SETA provides an efficient railway connection between the Baltic-Adriatic Corridor, the Mediterranean Corridor, the Orient/East-Med Corridor and the Rhine-Danube Corridor. Within the framework of the project, a corridor diagnosis was performed and a development plan was devised putting forward organizational and infrastructural measures to be implemented for reducing travel time and

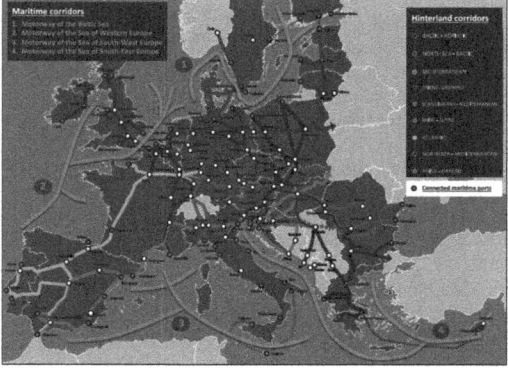

Figure 3. TEN-T core hinterland maritime corridors and connected seaports (European Commission, 2013).

eliminating capacity constraints, further boosting in that way economic development in the region. Although, as mentioned above, no governance body exists for managing the corridor, a Memorandum of Understanding (MoU) was signed between all project partners (on the port side, the port of Koper was part of the consortium) for continuing the established collaboration and implementing the planned measures (SETA Consortium, 2014).

Such informal forms of corridor development schemes prove to exist in other parts of Europe as well (e.g. the Adriatic – Ionian Transport Corridor receiving support from the Interreg ADRION funding programme). Although they account for a lower level of commitment, if the established agreement is respected by all parties, they can successfully drive corridor developments and upgrades, generating multiple benefits for a variety of stakeholders.

4 THE U.S. CONTEXT

As visualized in Figure 2, the setting of hinterland corridors in the U.S. is very different compared to the European context. The scale and spatial allocation of the main production centres, distribution hubs and consumption areas drove the development of a number of maritime gateways and hinterland land-bridges that facilitated, considering the great distances involved, the efficient expansion of the internal market, thus ensured easier access to natural resources. The liberalization and globalization of trade, along with the deregulation of transport in North America (i.e. North American Free Trade Agreement – NAFTA) resulted in a significant restructuring of the hinterland corridors so as to better cope with the changing business environment. Their restructuring was undertaken along three horizontal and four vertical axes (Figures 4 & 5).

4.1 Management initiatives of major hinterland corridors

Considering the high concentration of the main U.S. market along its East and West Coast as

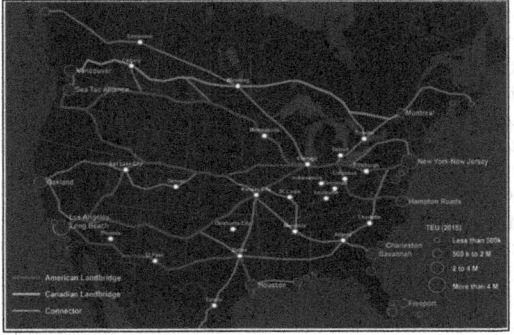

Figure 4. The North American land-bridges (Rodrigue, 2017).

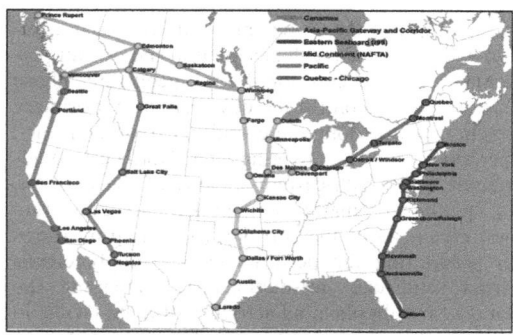

Figure 5. North American vertical corridors (Rodrigue, 2017).

stressed out before, both the I-95 and the Pacific corridor (Figure 5) are of great importance for facilitating trade flows and boosting the national economy.

The I-95 corridor benefits substantially from a Coalition that was established in 1993, as an advancement of an informal group of transportation professionals that existed for a few years till then, aiming mainly to effectively manage highway incidents. At its current form, the Coalition accounts for a partnership of more than 100 members including Federal Administrations, State and Local Departments of Transportation (DOTs), transportation and port authorities, transit, rail and motor vehicles agencies, and transportation industry associations. Originally, it concentrated more on highways and the application of intelligent transportation systems (ITS), but has now a broader scope encompassing all modes of transport and focusing on (a) travel information services, (b) coordinated incident management and safety, and (c) intermodal movement of freight and passengers. For each of these three focus areas, a relevant committee is established for facilitating stakeholder coordination and cooperation, and undertaking projects that can add further value to the corridor (The I-95 Corridor Coalition, n.d.).

With the I-95 corridor serving 15 of the largest 50 seaports in the U.S. (i.e. considering their total throughput), as well as a large number of smaller niche ports, port authorities as well as the Federal Maritime Administration (MARAD) have an important role to play within the Coalition, substantially contributing to the work of its Intermodal Movement of Freight and Passengers Committee.

In 2009, after the positive evaluation of its proposal, the Coalition received the designation of its waterways/waterside system as a Marine Highway Corridor (M-95), which allowed to get federal assistance for a number of targeted projects aiming to further develop and expand marine services, thus ensure their better integration with the hinterland transport system. Since then, the Committee conducted a number of additional projects aiming to further improve the aforementioned level of integration and support the enhancement of the overall performance of the corridor. More specifically, those projects focused on (a) the designation of a primary freight network along the corridor, (b) rail and truck operations studies covering the Northeast, Mid-Atlantic and South-eastern regions of the corridor, (c) monitoring major bottlenecks and estimating their effects along the corridor, (d) setting measures for monitoring the performance of supply chains across multistate jurisdictions, (e) assessing the status of freight planning activities of State DOTs, in response to the requirements of the 2015 Fixing America's Surface Transportation Act (FAST Act), and (f) setting-up a Freight Academy program and a commercial vehicle infrastructure integration library (The I-95 Corridor Coalition, n.d.).

On the U.S. West Coast, a similar partnership (i.e. the West Coast Corridor Coalition – WCCC) was established in 2003. It includes State DOTs, regional and local transportation agencies, ports, and related transportation organizations (both public and private) from Alaska all the way down to California. Within its structure, as was the case with the I-95 Corridor Coalition, thematic committees have been formed allowing for a better allocation of efforts and resources on a number of issues that need to be efficiently tackled. More specifically, the following three committees have been established: Goods Movement Committee, ITS/Environment Committee and Federal Relations Committee.

For setting a good basis and identifying, as a next step, projects that can add further value to the corridor, the Goods Movement Committee issued in 2008 the development of a corridor diagnostic study (Cambridge Systematics, 2008). The latter stressed out (a) the 7 major maritime and 5 air transport hubs that the corridor accommodates, as well as increases in demand resulting from the rapidly growing Trans-Pacific trade, (b) key regional bottlenecks to be addressed, and (c) key dimensions that should shape the approach of the Coalition for successfully addressing identified bottlenecks and challenges. More specifically, those dimensions include (i) partnering with federal government bodies for developing and implementing multistate planning and funding mechanisms that can actively contribute towards addressing system chokepoints, (ii) promoting, in a coordinated manner, innovative planning, funding and project development strategies aiming to relieve congestion at major metropolitan areas, (iii) developing freight investment models that incorporate market and economic principles while ensuring environmental sustainability, and (iv) developing new approaches in order to balance environmental protection and community interests with system expansion needs.

Through this Coalition, ports had a great chance to work closely with the partnership and define together appropriate measures to be implemented for successfully addressing port access and border crossing chokepoints that would enable a smoother flow of incoming and outgoing freight volumes. However, no track of the Coalition's activities can be found beyond 2012. If this is combined with an appropriation request that was agreed to be submitted by the Coalition in 2011 for continuing its work, most likely the Coalition as such, failed to progress any further activities that would add value to the West Coast freight transportation system. Since then however, a number of other complementary, to its scope, thematic initiatives prove to have emerged. More specifically, the West Coast Clean Transit Corridor Initiative aims to identify optimal locations for electric charging infrastructure to support truck operations along the I-5 corridor, while the West Coast Collaborative Medium & Heavy-duty Alternative Fuel Infrastructure Corridor Coalition aims to accelerate the modernization of the West Coast transportation corridors by deploying alternative fuel infrastructure for medium and heavy-duty vehicles and equipment in combination with other relevant investments.

For most of the other major U.S. transportation corridors (e.g. I-5, I-10, I-69, I-70, Heartland corridor, Cresent corridor, etc.) a coalition or a similar management initiative exists (Federal Highway Administration, 2018), while an overarching one has also been established (i.e. the Coalition for America's Gateways & Trade Corridors - CAGTC) taking a network-wide perspective but more with an awareness raising role. However, since those corridors mentioned above mostly extend, either horizontally or vertically, within the middle of the continent, and taking into account the coastal concentration that characterizes the main hinterland market in the U.S. as stated before, the respective management initiatives are of less relevance to the scope of this paper, and thus no further details are being provided herein. Relevant information can be found however in pertinent literature (Blank, 2006; Monios & Lambert, 2013; Sugawara, 2017).

4.2 Local corridor management initiatives

Besides national corridors in the U.S., the scale of which resembles or is even greater than that of international corridors in other contexts (e.g. TEN-T core network corridors in Europe), management initiatives for local corridors have also been established. However, they prove to be very limited in number and mostly relate to specific infrastructure projects that aim to enhance the access of ports into the hinterland. The Alameda Corridor Transportation Authority (ACTA) is the most distinctive case. It accounts for a joint powers authority established by the cities of Long Beach and Los Angeles, with representatives from each port, each city council and a representative of the Los Angeles County Metropolitan Transportation Authority constituting its Governing Board. ACTA financed the construction of the Alameda Corridor, which comprises of a series of bridges, underpasses, overpasses and street improvements that enabled to separate the freight trains serving the two ports from passenger trains and street traffic, greatly improving in that way the ports' access into the hinterland. Corridor operations started in 2002, with ACTA's main responsibility lying today in the collection of fees for paying the debt service on the outstanding revenue bonds (Alameda Corridor Transportation Authority, n.d.).

It is worth noting at this point that following the success of the Alameda Corridor, as of the start of 2008 and within the framework of the National Gateway project, similar improvements (i.e. bridges, tunnels, underpasses, overpasses, etc.) were also conducted at network level, for improving the railway connection of mid-Atlantic ports with key midwest distribution centers. The project was actively supported by all relevant corridor coalitions and facilitated the greater use of double-stack trains while intermodal terminals were also improved. To this end, the existing, at that time, rail and intermodal terminal capacity was considerably enlarged (Railway Technology, n.d.).

5 CONCLUSIONS AND RECOMMENDATIONS

The greater emphasis that ports are now placing on enhancing their links with the hinterland systems they are connected to is evident and this is actually taking place within a highly dynamic business environment characterized today by high levels of competition, industry consolidation and scale increases, as well as the fast penetration of new technologies and (big) data management solutions. Ports have to efficiently tackle both the relevant business and market dimension of this hinterland integration, as well as the infrastructural and institutional one, so that their global competitiveness is enhanced and additional freight flows are attracted.

On the business and market side, port authorities have mainly acted, until now, as facilitators. However, it seems that they will be transiting more and more into the entrepreneur role in the near future, considering the prospective benefits to be realized, better aligning in that way their hinterland strategy with that of global terminal operators. The latter have been heavily investing over the last decade on building an efficient and resilient port network at global level, but did not stop there. They strived to acquire better control over the supply chain and achieve efficiencies that existing, to that time, cooperation schemes had failed to provide. Through majority investments, they are now heavily involved in intermodal transport and terminal operations, ensuring higher levels of coordination that have led to capacity improvements and considerable cost and time savings. The aforementioned business and market strategies however, need to be efficiently

coupled with additional ones capable of addressing hinterland infrastructural and institutional aspects. This is where hinterland corridor management initiatives have an important role to play.

As discussed in the two previous sections, the world's key trading regions, which also account for the main areas of consumption at global level, benefit from hinterland corridor management initiatives that have been established for supporting trade facilitation through the efficient connection of seaports with the main hinterland markets. In both contexts, large-scale corridors are mostly considered, although management initiatives for regional or local corridors also exist in some form (e.g. informal forum, authority, etc.). In most cases, they account for a direct, long-term outcome of infrastructural projects.

Taking into consideration the diverse characteristics that each context presents, as well as the institutional differences that exist and have to be carefully accounted for, a different approach was followed with regard to the structure and governance of these initiatives in Europe and the U.S. More specifically, a more centralized structure has been established in Europe, with the European Commission acting, through the European Coordinators, as the main management body for the TEN-T core network hinterland corridors. Corridor-related stakeholders shape up the corridors' fora and are being regularly consulted by the Coordinators for running the corridors' diagnostic, identifying bottlenecks and issues to be addressed and devising, as a next step, appropriate work plans. The latter are then sent to Member States for approval, and once approved their implementation phase is initiated, exploiting available funding programs such as CEF. On the other hand, such initiatives in the U.S. prove to be more decentralized, with CAGTC that was mentioned before mainly acting as an awareness raising body. No overall management body proves to exist. Besides that, the processes followed for stakeholder engagement, bottleneck identification and project implementation are quite similar. However, both contexts could benefit from the greater formalization of regional and local corridor partnerships, adding in that way another management layer that may provide further value to the relevant stakeholders involved. Such a bottom-up approach can further strengthen stakeholder engagement and better shape in that way cooperation and coordination, avoiding however the risk of increasing management complexity since the latter will act as a disincentive. Of course, the efficient interaction of such initiatives with the existing broader ones needs to be carefully assessed and defined, so that the expected benefits are maximized.

Through such initiatives, ports are provided with the opportunity to further enhance their hinterland intelligence and stress out chokepoints to be addressed, which lie beyond their reach but nevertheless can impose a considerable impact on their competitiveness. The institutional support ensured through these initiatives is also important for driving hinterland developments in a timely manner.

A highly indicative and interesting case is that of the port of Piraeus in Greece, which has been experiencing tremendous growth over the last few years, ranking in 2019 as the first container port in the Mediterranean region and the fourth one in Europe. A considerable share of its container throughput accounts for transit cargo (i.e. for Pier I that is operated by the Piraeus Port Authority, this share was estimated to 19.2% in 2018), and COSCO (i.e. the port's major shareholder) has undertaken important actions for driving further growth (Hughes, 2019). Its investment, through a group company, on a new railway operator, should however be coupled with relevant infrastructure improvements undertaken in Greece and more importantly in the Balkans, so that rail access to the main central European market is enhanced. The East-MED TEN-T corridor forum can play an important role towards driving and efficiently coordinating such infrastructural developments.

The hinterland strategies of ports therefore, should cover and efficiently integrate all the aforementioned dimensions exploiting in a coordinated way all different opportunities that are available for ensuring that the ports are connected with a highly-efficient hinterland transport system, enabling them to cost-effectively reach and accommodate more distant hinterland markets. Of course, besides port authorities, several other actors will benefit from such a system further boosting in that way local, regional and national economies.

ACKNOWLEDGEMENTS

This work has been partly supported by the University of Piraeus Research Center and the Fulbright Student Program 'Visiting Research Students'.

REFERENCES

Alameda Corridor Transportation Authority (ACTA). n.d. *Governance*. http://www.acta.org/about/governance.asp

Álvarez-SanJaime, O., Cantos-Sánchez, P., Moner-Colonques, R. & Sempere-Monerri, J. 2013. Vertical integration and exclusivities in maritime freight transport. *Transportation Research Part E*, 51: 50–61.

Baldwin, R. & Tomiura, E. 2020. Thinking ahead about the trade impact of COVID-19. In: Baldwin R. and Weder di Mauro B. (Eds.) *Economics in the time of COVID-19*. London: CEPR Press.

Blank, S. 2006. North American Trade Corridors: An Initial Exploration. *Faculty Working Papers*, Paper 50. http://digitalcommons.pace.edu/lubinfaculty_workingpapers/50

Cambridge Systematics. 2008. *West Coast Corridor Coalition Trade and Transportation Study*. Final report prepared for West Coast Corridor Coalition, Austin, Texas.

De Langen, P.W. 2008. Ensuring hinterland access: The role of port authorities. OECD/ITF, Joint Transport Research Centre, Discussion Paper, 18. https://www.itf-oecd.org/ensuring-hinterland-access-role-port-authorities

European Commission. 2013. *Trans-European Transport Network – Maps of core network corridors*. http://ec.europa.eu/transport/infrastructure/tentec/tentec-portal/site/en/maps .html

European Commission. 2014. *The European Union explained: Transport*. Luxembourg: Publications Office of the European Union.

European Commission. 2020. *European Coordinators*. https://ec.europa.eu/transport/themes/infrastructure/ten-t-guidelines/european-coordinators/

European Sea Ports Organization - ESPO. 2011. *European Port Governance: Report of an inquiry into the current governance of European seaports*. Brussels: ESPO.

Federal Highway Administration (FHWA) of the U.S. Department of Transportation. 2018. *Major Corridor Coalitions*. https://ops.fhwa.dot.gov/freight/corridor_coal.htm

Franc, P. & Van der Horst, M. 2010. Understanding hinterland service integration by shipping lines and terminal operators: a theoretical and empirical analysis. *Journal of Transport Geography*, 18(4): 557–566.

Hughes, A. 2019. Game of leapfrog in the Mediterranean. PortStrategy. https://www.portstrategy.com/news101/world/europe/south-med-article

Lee, S-W., Song, D-W. & Ducruet, C. 2008. A tale of Asia's world ports: The spatial evolution in global port cities. *Geoforum*, 39(1): 372–385.

Magnan, M. & van der Horst, M. 2020. Involvement of port authorities in inland logistics markets: the cases of Rotterdam, Le Havre and Marseille. *Maritime Economics & Logistics*, 22: 102–123.

Monios, J. & Lambert, B. 2013. Intermodal freight corridor development in the United States. In Bergqvist, R., Wilmsmeier, G. & Cullinane, K. (Eds.) *Dry ports: A Global Perspective*. London: Ashgate.

Öberg, M., Nilsson, K.L. & Johansson, C.M. 2018. Complementary governance for sustainable development in transport: The European TEN-T core network corridors. *Case Studies on Transport Policy*, 6 (4): 674–682.

Parola, F., Risitano, M., Ferretti, M. & Panetti, E. 2017. The drivers of port competitiveness: a critical review. *Transport Reviews*, 37(1): 116–138.

Railway Technology. n.d. National Gateway Project. https://www.railway-technology.com/projects/national-gateway-project/

Rodrigue, J-P. 2017. *The Geography of Transport Systems* (4th ed.). New York: Routledge.

Song, D-W. 2003. Port co-opetition in concept and practice. *Maritime Policy & Management*, 30(1): 29–44.

South East Transport Axis (SETA) Consortium. 2014. SETA Corridor Development Plan. http://www.southeast-europe.net/document.cmt?id=683

Sugawara, J. 2017. Port and hinterland network: a case study of the Cresent Corridor intermodal freight program in the US. *Transportation Research Procedia*, 25: 916–927.

The I-95 Corridor Coalition. n.d. *The Coalition*. https://i95coalition.org/the-coalition-2/

United Nations Conference on Trade and Development. 2019a. *Review of Maritime Transport 2019*. https://unctad.org/en/pages/PublicationWebflyer.aspx?publicationid=2563

United Nations Conference on Trade and Development. 2019b. Key Statistics and Trends in International Trade 2018. https://unctad.org/en/pages/PublicationWebflyer.aspx?publicationid=2446

Van den Berg, R. & De Langen, P.W. 2011. Hinterland strategies of port authorities: A case study of the port of Barcelona. *Research in Transportation Economics*, 33 (1): 6–14.

Van den Berg, R. & De Langen, P.W. 2015. Towards an 'inland terminal centred' value proposition. *Maritime Policy & Management*, 42(5): 499–515.

Zhu, S., Zheng, S., Ge, Y-E., Fu, X., Sampaio, B. & Jiang, C. 2019. Vertical integration and its implications to port expansion. *Maritime Policy & Management*, 46 (8): 920–938.

Developments in Maritime Technology and Engineering – Guedes Soares & Santos (eds)
© 2021 Copyright the Author(s), ISBN 978-0-367-77376-2

Topological surfaces based advanced data analytics to support industrial digitalization in shipping

Lokukaluge P. Perera
UiT The Arctic University of Norway, Tromso, Norway

ABSTRACT: Advanced data analytics based on topological surfaces derived from ship performance and navigation data sets with possible commercial applications as a part of the industrial digitalization is discussed. That can be used to utilize ship performance and navigation data sets collected by various ocean-going vessels as a part of an advanced data analytics framework. These data analytics can be divided into several categories of digital models, descriptive, diagnostic, predictive and visual analytics. The digital models, a representation of topological surfaces, consist of linear or nonlinear data driven relationships of ship performance and navigation parameters. Since these parameters consist of various sensor measurements, the respective measurement space represents the dimensionality of the topological surfaces, i.e. the digital models. Therefore, the measurement space, i.e. the number of sensors and parameters, can play an important role in the topological surface development process and such an example is discussed in this study.

1 INTRODUCTION

1.1 *Ship intelligence*

Future vessels will be supported by smart systems with IoT (internet of things) to achieve the required integrity levels of ship navigation. Such ship systems can be divided into two main divisions: navigation and automation systems. The respective reliability conditions of both systems should be investigated to quantify the required integrity levels of ship navigation. Since autonomous and remote controlled facilities have been discussed for future vessels, the integrity levels of both systems, i.e. navigation and automation, should complement to each other. Furthermore, to achieve the required regulatory requirements of future ship navigation, additional decision support type facilities should be accommodated. Therefore, adequate vessel IoT should be onboard to collect required ship performance and navigation information as big data sets and that information can support the same decision supporting type facilities. One should note that big data can create a bridge between the decision support facilities and vessel navigation and automation systems to achieve the required integrity levels of future ship navigation.

1.2 *Big data in shipping*

The respective ship performance and navigation parameters collected by vessel IoT are often categorized as big data in shipping (Rodseth *et al.*, 2016). These data sets should be transformed into useful information, therefore that can facilitate decision support systems in future vessels (Perera, 2019), as discussed before. A general process that transforms big data into useful information facilitates decision support systems in ship navigation is presented in Figure 1. However, there are various challenges that can be encountered during this process are also noted in the same figure. These data sets should go through a data management layer that may consist of data collection, preparation, communication and storage type applications. As the next step, useful information from these data sets should be extracted and that can be done by the data analytics layer. However, the transformation of big data into useful information may be an indirect process due to the respective challenges. These challenges can be categorized into four groups: data volume, velocity, variety and veracity.

The volume and velocity issues relate to the size and communication speeds of ship performance and navigation data sets, respectively and various solutions under the data management layer have been developed to overcome the same by the shipping industry. The variety and veracity issues relate to the different forms and uncertainty of ship performance and navigation data sets, respectively and various solutions under advanced data analytics have been considered to overcome the same under various research studies (Perera & Mo, 2019).

One should note that different forms of ship performance and navigation data can introduce various modeling challenges, where conventional mathematical models may not be available to accommodate

DOI: 10.1201/9781003216582-12

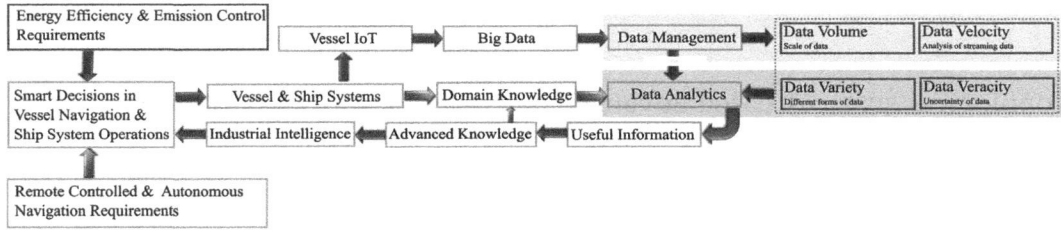

Figure 1. From ship performance and navigation data to decision support.

such data forms. On the other hand, the uncertainty or quality of ship performance and navigation data sets can introduce additional anomalies that can degrade the modeling process (Perera & Mo, 2018a). It is considered that the data variety and veracity issues can be the two biggest challenges in the ear of intelligent shipping. Therefore, appropriate solutions to overcome the same challenges are considered in this study as its main contribution.

1.3 *Smart decisions*

It is believed that advanced analytics can provide adequate solutions to data variety and veracity issues in shipping. These data analytics should support with the respective domain knowledge, where some fundamental data quality issues can be addressed, appropriately. That can be denoted as classifying each data point either as a good value, i.e. a proper data point, or a bad value, i.e. an erroneous data point, during the data handling process (Perera and Mo, 2017). However, such erroneous data points should further be classified either as a sensor fault event or system abnormal event.

Sensor faults may relate to the failure situations of vessel IoT and system abnormal events may relate to the degradation conditions of the vessel or ship systems. Such classifications should be done properly to either remove or recover erroneous data points by considering the respective domain knowledge. Since such fundamental issues in the data quality have been overlooked by the research community, the outcome can degrade the respective data analyses (Perera & Czachorowski, 2019). It is believed that the proposed data analytics, i.e. advanced data analytics framework, should have adequate features to overcome such issues (Perera & Mo, 2018b).

The same data analytics can be used to extract useful information from ship performance and navigation data sets. That information can be transformed into advanced knowledge in the shipping industry (see Figure 1). Such advanced knowledge in shipping can also improve the respective domain knowledge and support industrial intelligence. That can transform into smart decisions in vessel navigation and ship system operations. One should note that these smart decisions in shipping can be categorized into two

main groups: energy efficiency and emission control requirements and remote controlled and autonomous navigation requirements.

1.4 *Machine learning and artificial intelligence*

The focus of this study is to develop advanced data analytics, i.e. an advanced data analytics framework, for the shipping industry based on machine learning (ML) and artificial intelligence (AI) algorithms. ML can be considered as a subset of AI in the field of computer science. That may consist of various neural networks that have the ability to "learn" from the respective data sets. In general, neural networks are considered as a black box approach, since the internal structures of these networks have never been understood, adequately with respect to physical systems i.e. the theories in classical mechanics.

However, the recent developments in deep learning type applications make such networks no longer a black box type approach, rather one can observe the internal structure of these network as in the sense of data. i.e. the linear version of deep learning consists of linear autoencoders. The structure of a linear autoencoder can be a representation of singular values and vectors, i.e. building blocks of a multi-input multi-output (MIMO) physical system. Therefore, these autoencoders, i.e. neural networks, can be a representation of physical theories that have been developed under classical mechanics. The proposed data analytics is also based on a linear version of deep learning and that have been further illustrated under digital models of the advanced data analytics framework. Hence, classical mechanical models derived previously can also be accommodated into such autoencoders through the respective domain knowledge. That would consist of data driven mathematical models to capture ship performance and navigation conditions.

2 ADVANCED DATA ANALYTICS

2.1 *Framework structure*

The information flow process from ship performance and navigation data to smart decisions (i.e. in Figure 1) is further illustrated as an advanced data analytics framework with high-performance computing (HPC) facilities in Figure 2. Vessel IoT is connected to various ship systems (i.e. navigation and automation systems) and collects ship performance and navigation data in real-time. One should note that external conditions, i.e. weather and environmental conditions, can influence on vessel navigation and ship system operations, therefore adequate information on external condition should also be collected by vessel IoT. Hence, the same framework can be supported not only by ship performance and navigation data sets but also weather and AIS (Automatic identification system) data sets (Murray & Perera, 2018). These big data sets will go through a real-time data pre-processing step supported by advanced data analytics with appropriate APIs (application programming interface) under a HPC platform.

2.2 Data analytics

These analytics can be divided into several categories of digital models, descriptive, diagnostic, predictive and visual analytics. The digital models consist of data driven relationships of ship performance and navigation parameters, i.e. the structure of data sets that is the basis of the proposed digital models. One should note that the respective data sets should consist of vessel and ship system input and output parameters measured by vessel IoT. These parameter relationships can be either linear or nonlinear. The respective measurement space from vessel IoT represents the dimensionality of the digital models. On the other hand, this approach can be seen as a method of identifying the respective structure of the data set, i.e. that represents the same digital models. Hence, the measurement space, i.e. the number of parameters measured by vessel IoT, can play an important role in the digital model development process and that should be selected, appropriately to support the respective model objectives.

Digital models should be developed for specific applications, therefore that may not be a universal approach to satisfy every application in shipping. That represents an information model of the vessel and ship systems and can also be used towards vessel navigational and ship system operational decisions, as discussed before. One should note that such onboard digital models should be developed onshore and deployed into vessels, therefore the required computational power requirements of onboard vessels can be reduced.

Furthermore, descriptive analytics identifies various data anomalies and diagnostic analytics recovers/ removes such data anomalies in the same data sets. Both descriptive and diagnostic analytics can consist of sensor & DAQ fault detection & isolation, parameter reduction & error compression, and data visualization steps under the same framework. One should note that some data analytics can complement to each other, therefore clear boundaries among such analytics might not be visible. The predictive analytics forecasts vessel and ship system behavior with the

support of the proposed digital models. The visual analytics visualizes the respective ship performance and navigation information, where useful information can be extracted. One should note that some data anomalies can be visualized under visual analytics. Visual information can create advanced knowledge in shipping and that can lead to industrial intelligence, as discussed before. Both advanced knowledge and industrial intelligence support decision analytics, where appropriate key performance indicators (i.e. KPIs) for ship performance and navigation conditions can be derived. Furthermore, the respective KPIs can be a considerable part of visual analytics.

2.3 Data and error compression

The number of ship performance and navigation parameters measured by vessel IoT should be reduced while preserving its information, i.e. of the respective data sets. That can be done under a parameter reduction step, where a smaller sub-set of new parameters, a good representation of ship performance and navigation data set, can be derived (Perera & Mo, 2018(c)). One should note that this parameter reduction step can also reduce various data anomalies and that can be seen as an error compression approach. Therefore, a much smaller parameter set with higher data quality can be delivered by the respective communication and satellites networks within reduced costs. This process should also be supported by the respective digital models, where the structure of ship performance and navigation data set can be used to reduce its dimensionality as a part of the advanced data analytics framework as illustrated in Figure 2.

2.4 Online services

These advanced analytics can support appropriate online services, such as weather routing, safe ship handling and emission and fuel calculations under various APIs (see Figure 2). The respective APIs should be supported by HPC (High Performance Computing) platforms due to their computational power requirements and should consist of online data pre-processing steps. Hence, the outcome of such APIs can be used towards decision support facilities of energy efficiency, reliable vessel navigation and ship system operation conditions. That can also be a part of future ship navigation under autonomous and remote-controlled conditions. The reduced data sets transferred into shore-based data centers should be further analyzed and information should be extracted to support the respective online applications. That should consist of advanced analytics and execute as an online post-processing step, i.e. that consists of parameter expansion and data recovery, integrity verification and data regression steps (see Figure 2).

The parameter expansion and data recovery step can be used to derive the estimated ship performance and navigation parameters from the transferred data

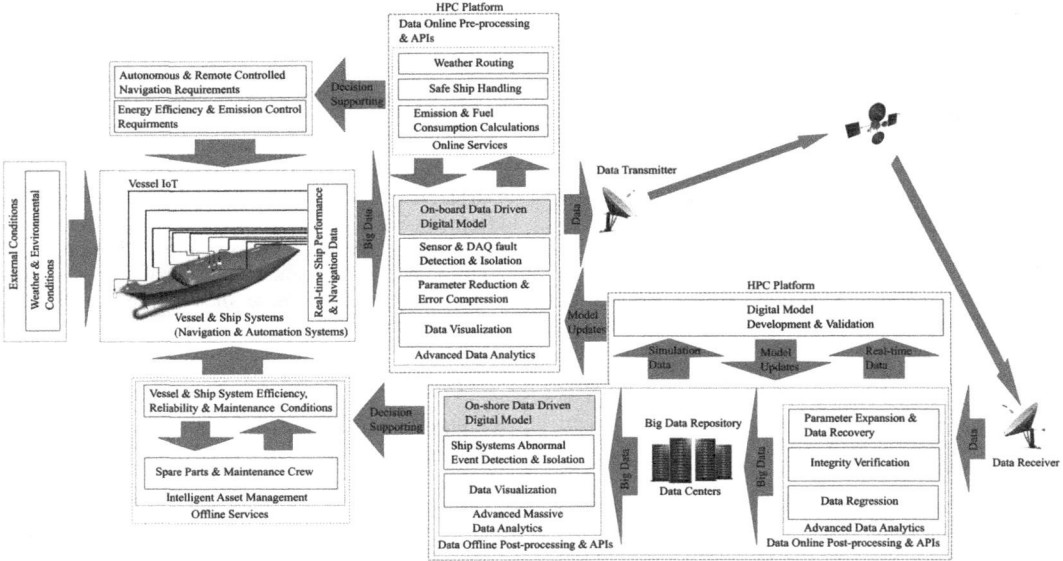

Figure 2. Advanced Data Analytics Framework.

set. The parameter expansion step can also recover some data anomalies, observed previously. This process is also supported by the respective digital models, where the structure of ship performance and navigation data is used to expand its dimensionality, as a part of both descriptive and diagnostic analytics.

The same ship performance and navigation data sets can be compared with external data sources, i.e. AIS and weather data, as an integrity verification step. Furthermore, the data regression step can be executed under the same framework and that can derive additional parameters, required for further analyses of ship performance and navigation conditions. The data regression step can also be a part of the respective digital models, where the accuracy of such models can further be improved. Furthermore, the same data sets can be included under big data repository for further analysis (see Figure 2).

These data centers can develop onshore data driven digital models, where adequate real-time data and simulations can be conducted to validate the same. In some situations, the required modifications into the digital models can also be introduced as a part of the validation step (see Figure 2) of data offline post-processing. Such approaches can consist of advanced massive data analytics and should be executed under offline conditions due to high computational power requirements. Hence, an onshore HPC environment is required to execute massive data analytics and store big data sets.

2.5 *Offline services*

Various offline services as decision support applications that relate to intelligent asset management-

based APIs can be offered for the shipping industry under the same framework. Vessel and ship system health monitoring under data visualization can also be an important part of such analytics. The respective reliability levels in ship systems can be observed under such offline services. Furthermore, the required parts and maintenance crew for ship systems under possible system degradation conditions can be arranged, earlier due to such services. Ship system abnormal event detection, isolation and elimination by appropriate operational and maintenance actions can be a part of such health monitoring APIs. Therefore, vessel energy efficiency as well as ship system availability can be improved by these services (Perera *et al.*, 2019).

3 TOPOLOGICAL SURFACES

3.1 *Data clustering*

The proposed digital models play an important role in the advanced data analytics framework, as discussed before. Such models are based on ML and AI algorithms and can be implemented as both onboard and onshore applications. In general, ML applications can be divided into two section: regression and classification applications. The classification and clustering approaches can be considered as similar applications. The clusters can have the dimensionality of a space that is known or somewhat fuzzy/undefined. i.e. the recent development of natural language process, the respective words are often been clustered into a space with an unknown dimensionality. The basic idea is that the words can be clustered into a space with unknown dimensionality and that can be seen as a classification approach (Collobert & Weston, 2008).

One should note that the breakthrough in ML came from data clustering type applications (Liu *et al.*, 2017). It has been shown that if any engineering problem can be transformed into a data classification problem, then an elegant solution can be found under deep learning type approaches. On the other hand, the success in data regression type applications, is yet to be investigated.

The concept of digital models proposed in this study is based on both approaches of data classification and regression. The respective ship performance and navigation data sets are clustered around a specific dimensional space, i.e. with respect to the model objectives, i.e. that should support the object of the digital model. In general, vessel navigation and ship system operational conditions, i.e. marine engine operational points, can be considered to derive these data clusters (Perera & Mo, 2016), initially.

Furthermore, data regression type approaches can be used in each data cluster, to curve-fit that into high dimensional surfaces. These high dimensional surfaces can be either linear or nonlinear. A linear high dimensional surface can be derived by calculating an appropriate vector structure, i.e. eigenvectors, that represents the respective linear parameter correlations within the respective data cluster. However, the derivation of a nonlinear high dimensional surface may not be a simple process, where complex topological surfaces should be considered. Such surfaces can create complex structures for ship performance and navigation data sets as a part of proposed digital models.

One should note that the respective data clusters represent normal as well as abnormal regions of vessel navigation and ship system operational conditions. That can further increase the complexity in the proposed topological surfaces. The domain knowledge in shipping should extensively be used to identify the normal and abnormal situations under such topological surfaces. One the other hand, there are various outliers that can detected beyond these topological surfaces. It is noted that such outliers, i.e. beyond these topological surfaces, relate to various data anomalies. Since various data anomalies can be detected under these topological surfaces, that approach can be used to improve the data quality. Hence, such topological surfaces can be useful for both online and offline services, since that can also provide adequate solutions to the data quality issues.

3.2 *Data structural shapes*

Each data cluster in digital models can have its own topological shape that can be categorized as data structural shapes. Therefore, identifying of the respective shape of each data cluster is an important step of developing digital models. One should note

that shapes of these data clusters relates to their covariance directions (i.e. that can be either linear or nonlinear), therefore identifying the covariance directions will quantify the shape of each data cluster.

The covariance directions of a data cluster can be calculated by the respective singular values and vectors in a linear topological surface, as mentioned before. The top singular vector represents the largest covariance direction in a data cluster and the next singular vector represents the second largest covariance direction that is orthogonal to the previous singular vector (Perera & Mo, 2017(b)). Each data cluster represents a unique vessel navigation and ship system operation situation, therefore these singular vectors represent the most important ship performance and navigation parameter relationships. On the other hand, these singular vectors derive a proper structure into the respective ship performance and navigation data sets. Therefore, this approach can also be seen as a way extract the ship performance and navigation information from the respective data sets under a linear topological surface.

3.3 *Linear topological surface*

Since a proper structure for each data cluster can be introduced by the singular vectors under a linear topological surface, the distribution structure of such clusters will create an overall structure for the respective data set. That same data structure becomes the digital model under the proposed advanced data analytics framework. This study proposes to consider this structure of data sets as the basis for the proposed digital models.

One should note that the respective data sets should consist of ship performance and navigation parameters measured by vessel IoT. The measurement space represents the dimensionality of the system models for vessel navigation and ship system operations. Therefore, selecting an appropriate measurement space, i.e. the number of sensors and parameters, can also play an important step in the development process of digital models. Therefore, appropriate ship performance and navigation parameters should be measured and included in the data sets to archive the respective model objectives. i.e. ship energy efficiency can be considered as one of the model objectives.

4 COMPUTATIONAL RESULTS

4.1 *Ship performance and navigation data sets*

Several topological surfaces are visualized to observe their structural shapes and that are derived from a ship performance and navigation data set of a selected vessel. The approximated vessel particulars are presented in Table 1. The data sets of this vessel consist of the following ship performance and navigation parameters: average (avg.) draft (m), speed through

Table 1. Vessel particulars.

Parameter	Particulars
Ship Type	General Cargo Carrier
Ship length	230 (m)
Ship beam	35 (m)
Gross tonnage	40000 (tons)
Deadweight (at max draft)	75000 (tons)
Main engine	2 Stroke engine with MCR 7500 (kW) at 100 (rpm)
Propeller	FPP with diameter 6 (m) and 4 blades.

water (STW) (Knots), main engine (ME) power (kW), shaft speed (rpm), main engine (ME) fuel consumption (Tons/day), trim (m), relative (rel.) wind speed (m/s) and direction (deg) and auxiliary (aux.) fuel consumption (Tons/day).

The respective topological surfaces of the same ship performance and navigation parameters, as Gaussian kernel density functions, are presented in Figure 3. The top-left plot of Figure 3 represents the data distribution of marine engine (ME) power vs. shaft speed and there are three data clusters, i.e. that relates to engine operational modes can be identified. One should note that the dense data regions are presented by higher topological surface contours. Therefore, such clusters

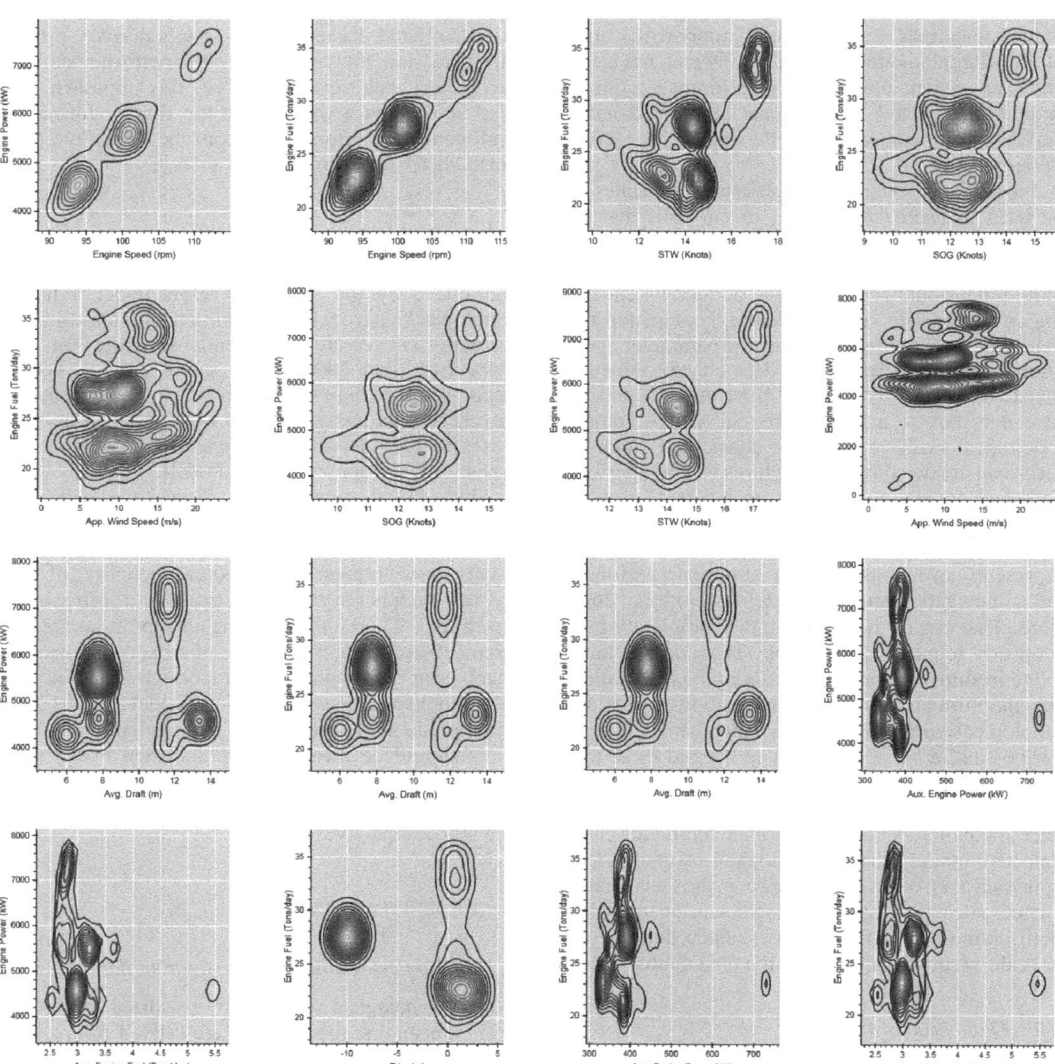

Figure 3. Topological surfaces for ship performance and navigation data.

represent the structural shape, i.e. topological surfaces, of engine speed-power data. These topological surface contours are a representation of various sectors of the ship performance and navigation data, i.e. that determine the overall structural shape of the ship performance and navigation data sets. Similarly, the respective ship performance and navigation parameters that relate to various topological surfaces are presented in the following plots of the same figures and that have various sizes and shapes. These topological surface sizes and shapes relate to various vessel navigational and ship system operational conditions.

5 CONCLUSIONS

Large scale data sets, so called big data, collected by ocean going vessels should be utilized, adequately to support industrial digitalization in the shipping industry. Even though there are various data analysis methods have been proposed by the research community (Nakatani *et al.* (2013), Osborne *et al.* (2007) and Trodden *et al.* (2015)), adequate analytics to facilitate these data sets and overcome the big data challenges are yet to be developed. Development of an advanced data analytics framework with possible commercial applications in shipping is presented and that can facilitate to overcome the respective big data challenges. The proposed digital models can play an important role in the proposed advanced data analytics framework, as presented in this study. However, various layers of ship performance and navigation data in an appropriate format should be included and that can create the most suitable structure of the digital models, i.e. that can cerate proper data structural shapes as topological surfaces for ocean going vessels. That outcome can introduce new types of data driven mathematical models into the shipping industry.

REFERENCES

Collobert, R. and Weston, J. 2008. A Unified Architecture for Natural Language Processing: Deep Neural Networks with Multitask Learning, *Proceedings of the 25th International Conference on Machine Learning.* (ICML '08) New York, USA: ACM: pp. 160–167.

Liu, S. Tang, J. Zhang, Z. and Gaudiot, J.L. 2017. Computer Architectures for Autonomous Driving, *Computer*, 50(8), pp. 18–25.

Murray, B. and Perera, L.P. 2018. A Data-Driven Approach to Vessel Trajectory Prediction for Safe Autonomous Ship Operations, *In Proceedings of the 13th International Conference on Digital Information Management (ICDIM)*: 240–247.

Perera, L.P. and Mo, B. 2016. Data Analysis on Marine Engine Operating Regions in relation to Ship Navigation *Journal of Ocean Engineering*, 128, pp. 163–172.

Perera, L.P. and Mo, B. 2017(a). Machine Intelligence based Data Handling Framework for Ship Energy Efficiency, *IEEE Transactions on Vehicular Technology*, 66(10), pp. 8659–8666.

Perera, L.P. and Mo, B. 2017(b). Digitalization of Sea going Vessels under High Dimensional Data Driven Models," *In Proceedings of the 36th International Conference on Ocean, Offshore and Arctic Engineering (OMAE 2017)*, Trondheim, Norway (OMAE2017-61011).

Perera, L.P. and Mo, B. 2018(a). An overview of Data Veracity Issues in Ship Performance and Navigation Monitoring, *In Proceedings of the 37th International Conference on Ocean, Offshore and Arctic Engineering (OMAE 2018)*, Madrid, Spain, (OMAE2018-77669).

Perera, L.P. and Mo, B. 2018(b). Ship Speed Power Performance under Relative Wind Profiles in relation to Sensor Fault Detection, *Journal of Ocean Engineering and Science*, 3(4): 355–366.

Perera, L.P. and Mo, B. 2018(c). Ship Performance and Navigation Data Compression and Communication under Autoencoder System Architecture, *Journal of Ocean Engineering and Science*, 3(2), pp. 133–143.

Perera, L.P. 2019. Deep Learning towards Autonomous Ship Navigation and Possible COLREGs Failures, *Journal of Offshore Mechanics and Arctic Engineering-Transactions of The ASME.*, (OMAE-19-1027), (DOI: 10.1115/1.4045372).

Perera, L.P. and Mo, B. 2019. Ship Performance and Navigation Information under High Dimensional Digital Models, *Journal of Marine Science and Technology*, (DOI: 10.1007/s00773-019-00632-5).

Perera, L.P. and Czachorowski, K. 2019. Decentralized System Intelligence in Data Driven Networks for Shipping Industrial Applications: Digital Models to Blockchain Technologies, *In Proceedings of the MTS/IEEE OCEANS 19*, Marseille, France, pp. 1–6.

Perera, L.P. Machado, M.M. Valland, A. and Manguinho, D.A.P. 2019. Failure Intensity of Offshore Power Plants under Varying Maintenance Policies, *Engineering Failure Analysis*, 97, pp. 434–453.

Nakatani, T. Miwa, T. Yamatani, N. Sasaya, K Okada, D. Kaneda, T. Kanayama, E. and Ura, E. 2013. Dynamics analysis and optimal control of a marine diesel engine, *in Control, Automation and Systems (ICCAS), 13th International Conference on*, pp. 1261–1265.

Osborne, P. D. Hericks, D. B and Cote, J. M. 2007. Full-Scale Measurements of High Speed Passenger Ferry Performance and Wake Signature," in *OCEANS 2007*, pp.1–10.

Trodden, D. G. Murphy, A. J. Pazouki, K. and Sargeant, J. 2015. Fuel usage data analysis for efficient shipping operations, *Ocean Engineering*, 110(B), pp. 75–84.

Rodseth, O.J., Perera, L.P., and Mo, B. 2016. Big data in shipping - Challenges and opportunities, *In Proceedings of the 15th International Conference on Computer Applications and Information Technology in the Maritime Industries (COMPIT 2016)*, Lecce, Italy, pp. 361–373.

Developments in Maritime Technology and Engineering – Guedes Soares & Santos (eds)
© 2021 Copyright the Author(s), ISBN 978-0-367-77376-2

A comparative life cycle assessment study on environmental performances between battery-powered and conventional marine vessels

H. Wang, E. Boulougouris, G. Theotokatos, A. Priftis, G. Shi & P. Zhou
Department of Naval Architecture, Ocean and Marine Engineering, University of Strathclyde, Glasgow, UK

ABSTRACT: International Maritime Organization (IMO) has initialized a new strategy on greenhouse gas (GHG) to reduce 50% of GHG from international shipping by 2050 comparing to 2008. Recently, solutions, like fuel cell and renewable energy, have been proposed by technology providers and ship designers. Battery-powered ship is a solution that can lead to a considerable reduction or even elimination of gaseous emissions, thus limiting fossil fuels consumptions. This paper aims to determine the benefits and drawbacks of marine battery propulsion system by conducting a comparative study using Life Cycle Assessment (LCA). It holistically evaluates the performances of the proposed system and the respective conventional one, applied on a 33-meter-long short-route ferry operating on the Thames. The results indicate the battery-powered system instead of marine engines can reduce the environmental impacts and GHG from ships, which fulfils the IMO requirements and proves the emission control potential of battery-powered systems.

1 INTRODUCTION

International shipping is one of the most efficient transportation means with the largest quantity of cargo transportation and the least air pollution released. According to IMO's report in 2014, the estimation of greenhouse gas emissions from international shipping accounted for only 2.2% of global anthropogenic carbon dioxide emissions. However, to achieve the temperature goal of Paris Agreement and the vision of United Nations' (Committee on Climate Change, 2019), GHG emission from marine sector should be highly addressed and urgent actions are also required to meet IMO's new targets comparing to 2008: 40% and 70% CO_2 reductions in 2030 and 2050 respectively and 50% GHG emission reduction in 2050 (IMO, 2018). There are a series of emission reduction technologies are under investigation and development, such as applications of low sulphur fuel oil, liquefied natural gas and renewable energy, but it is extremely challenging to realize this target from the perspective of time frame and technology maturity (Hwang et al., 2019). Due to the pace of the challenging target, most of the GHG emission reduction technologies are required to be developed rapidly such as the usage of liquefied natural gas as a marine fuel, the development of hydrogen engine, the application of the after-treatment system and all-electric ships. Among these GHG reduction methods, all-electric ships are one promising solution since all-electric cars are practically in-use which provides a solid fundamental for marine applications (Divya and Østergaard, 2009; Vicenzutti et al., 2015).

Recent years, the application of fossil fuels has raised significant concerns due to emission generation during combustion, such as SOx, NOx and CO_2, polluting the environment. The first two air pollutions are main contributors to acidification of the ambition such as the formation of acid rains. According to the revised MARPOL Annex VI, starting from 2020, the global sulphur limit will be reduced from the current 3.50% to 0.50%. From 2015, the limits for SOx and particulate matter released in the Emission Control Areas were reduced to 0.10%. Also, in the ECA region, Tier III NOx standard has become effective on ship build on or after 2016. Regarding CO_2 emission, it is more concerned as the largest contributor in global warming effect than its acidification impact. Although many higher global warming potential gases are existing, CO_2 has become the most significant GHG due to the substantial emission released from the burning of fossil fuels. According to the data collected by the World Bank, in 2014, the annual global CO_2 emission is more than 36 billion tonnes and the other substantial released GHGs are methane and NO_2 which are about 8 billion and 3 billion tonnes in 2012. Therefore, methods and technologies to reduce the carbon dioxide emission (and thus reduce GHG emission) release are attracting to all industries.

Electricity has been used centuries ago onboard ships and it supports the operation of marine systems and equipment, for instance, navigation, lighting, equipment control and cargo handling. However, the

DOI: 10.1201/9781003216582-13

majority of sources of ship propulsion power was still coming from marine engines which combust fossil fuels to supply mechanical power and electricity. To replace these engines to get rid of fossil fuels onboard ship, battery-powered ship concept which intends to use electricity from onshore power plants. The battery system has not only avoided emission release during shipping but also considered as a more stable system and the maintenance requirement of the system is lower than engines due to no moving parts in the battery system. The hybrid system, combining engines and batteries, and sometimes other renewable resources, has been investigated for many years (Dedes, 2013; Jeong et al., 2018; Ling-Chin and Roskilly, 2016).

Since electric vehicles have been discussed for many years, many research work has already evaluated the application of different batteries technologies to discover their environmental impacts. Sanfelix et al. evaluated both the environmental and economic performance of the lithium-Ion batteries in 2016 and discovered the manufacturing stage has the highest environmental load for all the considered impact categories (Sanfélix et al., 2016). In 2019, Dai's team has investigated the application of li-ion batteries on electric vehicles which also indicates the main environmental impact contributor of the batteries are from the material and energy use during production (Dai et al., 2019). Any other interesting research carried out by Zhao and You presented and compared the environmental impacts of four different types of cathode materials of Li-ion batteries and also conclude the most significant contributor to life cycle GHG emission is the cell production phase (Zhao and You, 2019). Matheys has assessed 5 different battery technologies for electric vehicles to indicate the most environmentally friendly one with the best overall environmental impact (Matheys et al., 2009). Similarly, the research carried out in the University of Oldenburg investigated four different stationary battery technologies on their cumulative energy demand and global warming potential. However, the results showed the usage of batteries dominates the life cycle impact which is different from others' research results (Hiremath et al., 2015; Mitavachan, 2014).

In the meantime, the applications of batteries on marine vessels are under development: Kluiters and his research team have tested the application of sodium/nickel chloride battery for ships back to 1999. Due to the limitation of the battery technologies, the advantages of sodium/nickel chloride battery was only proved to be better than sodium/ sulphur battery (Kluiters et al., 1999). In the last decade, many pieces of research on batteries technologies are focusing on their application on ships. Lan et al. have carried out one study to investigate the optimal size of a hybrid ship power system combining solar energy, diesel engine and battery system. The result from the research was to minimize the investment and operational costs and the carbon dioxide emissions release of the ship using

a hybrid system with the consideration of the solar irradiations and temperature along the voyage (Lan et al., 2015). Misyris' research team also developed an estimation algorithm to illustrate the battery state during the ship operation (Misyris et al., 2017).

To identify and evaluate the performance of the marine propulsion system, life cycle assessment is widely adopted which covers the material exploitation, manufacturing, operation, maintenance and scrapping stages. LCA has been adopted to evaluate the performance of Thai Island's diesel/PV/wind hybrid microgrid from the perspective of environmental sustainability (Smith et al., 2015). There is also application of LCA in automotive vehicle production such as the material selection, and forming process from scrapped materials (Delogu et al., 2016; Raugei et al., 2015, 2014). There are also many applications of LCA on battery system evaluations and all the previously mentioned researches on battery system in automobile industry applied LCA methodology to assess and compare their environmental impacts (Dunn et al., 2016).

However, although there are several pieces of research on the evaluation of battery technologies on marine vessels, it is still lack of investigation either from the point view of using a battery as an only power source or applying holistic analysis such LCA approach. It is necessary to determine the performance of a battery-powered vessel comparing to a conventional vessel from the perspective of environmental friendliness to address the advantages of electric vessels.

2 METHODOLOGY

ISO standard has provided comprehensive procedures to carry out an LCA analysis and basically, LCA method can be carried out by following four phases: a) Goal and scope definition, b) Life cycle inventory analysis, c) Life cycle impact analysis and d) Life cycle interpretation (ISO, 2006).

2.1 Goal and scope definition

In goal and scope definition phase, the target of this LCA investigation will be identified first which could be a product, a system; a technology or a strategy. Then the reasons behind the investigation should be clarified, e.g. most studies are aiming to find out the environmental impacts of the target and sometimes only one or several interesting impacts are under considerations. The reasons for the study sometimes can be determined according to the audiences and their interests so that the objective of the study could be adapted accordingly.

The study scope will be determined by identifying the product's main functions, tightly connected activities and their inputs and outputs, impact assessment method selection, and associated data collection, assumption and quality check. As a fact, an ideal LCA assessment should be boundless who has no defined scope and modelling every single activity during the target's life span. However, a boundless study requires support from many industries and it becomes time-consuming to collect data. Since the objective sometimes focuses inside one or a few industries, the boundary could be determined and help to restrict the never-ending data collection process. Furthermore, to restrict the scope is not only to accelerate the study but also due to the data availability, such as lack of data recording, difficulty to retrieve data from other industry and sometimes confidential issues. Therefore, at the beginning to the study, the goal and scope should be well defined and this definition phase will provide a focused point and an overall view of the LCA study and model which will help to determine the data required in next phase.

2.2 Life cycle inventory analysis

In this analysis, an inventory of emissions will be developed based on the LCA model and data requirement established in the goal and scope definition phase. In the marine industry or shipbuilding industry, data required are usually the vessel specification, operational profiles and maintenance plans, with which a preliminary design could be conducted. However, the LCA model will not be focused only inside the marine industry and the material exploitation, transportation, manufacturing and many other processes should be involved to develop a comprehensive LCA assessment. These data will be collected and based on relevant formulas, the emissions quantities could be eventually calculated which will be used to develop the emission inventory.

In this phase, the goal and scope of the LCA study could be refined because sometimes the data requirement is different from what was defined in the goal and scope definition phase. For example, the raw material exploitation data may be difficult to derive, some emission categories could be less interesting in different industry and sometimes more activities are found to be available during the data collection and calculation.

The output of this phase will be an emission inventory (LCI) covering a list of emissions with their quantities in the selected environmental categories. This inventory will be categorized and characterized in the next phase to determine the environmental impact of the target.

2.3 Life cycle impact analysis

This phase will be divided into three steps: a) selection: to select environmental categories; b) classification: to assign emissions from the inventory to different categories; c) characterization: convert/transform emissions assigned to the categories into the indicator of its group. The detailed processes are presented in Figure 1.

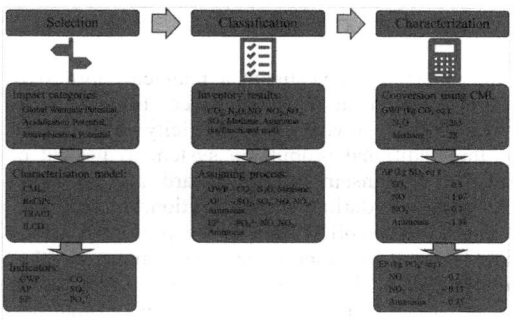

Figure 1. Processes of life cycle impact analysis.

In the selection step, impact categories will be selected based on the goal of the study. For each category selected, there will be an emission indicator which is usually the most common emission in the categories. For different characterization model, it has a different indicator selector. For CML model as an example, carbon dioxide is the indicator for global warming potential, sulphur dioxide for acidification potential, phosphate for eutrophication potential.

After selection, all the inventory results will be assigned to different impact categories. One emission could contribute more than one emission impact. An example is given in Figure 1 and example emissions are presented under different categories.

The last step will convert all the assigned emission in one category into the indicator based on the factors provided in the characterization model. Figure 1 provides an example of conversion factors using CML model.

This is also the phase of considering the sensitivity and uncertainty of data collected or factors involved. Sensitivity analysis is a way to investigate how variations in input values or assumptions affect the model outputs. When the data or factors varied significantly in the past, it will help to determine the impact of variation on the assessment results. Uncertainty treatment is used when limited knowledge is available to exactly describe the existing state or possible outcome. Under certain circumstances, data or factors will be based on assumptions, such as the engineer's judgement, as it is difficult to determine an exact value. Uncertainty treatment could apply possible distributions of the data or factor to find out the trust level.

After the phase of LCIA, the final step will be the interpretation of all phases.

2.4 Interpretation phase

The final phase of the LCA assessment is to deliver results obtained through the study based on the inventory and impact assessment. This phase will also conclude the LCA study and provide suggestions to the participants such as manufacturers, operators, and decision-makers.

3 CASE SHIP STUDY

This paper will investigate a planned short route ferry operated in Thames River, London. This ferry will be powered by electricity only instead of the traditional propulsion system to get rid of fossil fuel consumption on board and to avoid the emission during ship operation. In this section, the data collection will be presented which will be used to support the development of LCA model and the LCA assessment. The specification of the case ship is shown in Table 1 and the route of the vessel is presented in Figure 2. The route of the ferry is along the Thames, servicing between Westminster and Woolwich and stopping at London Eye, Embankment, Blackfriars, Bankside, London Bridge, Tower, Canary Wharf, Greenland (Surry Quays), Masthouse Terrance, Greenwich, North Greenwich. The service hours are about 18 hours per day and battery power system is considered to be applied. This study will identify the performance of the battery power system and determine its advantages comparing to traditional propulsion system (diesel engines).

Table 1. Case ship specification.

Main particulars			
LOA (m)	35	Block coefficient (-)	0.582
LBP (m)	32.538	Midship coefficient (-)	0.849
Breadth (m)	8	Prismatic coefficient (-)	0.69
Depth (m)	2.354	Waterplane coefficient (-)	0.842
Draught (m)	1	Geometric displacement (t)	75
No. of engines	2	Engine power (kW)	625
Battery type	Li-ion	Battery capacity (kWh)	650

3.1 Case scenarios

Two different propulsion systems will be under consideration so that two different case scenarios are developed: 1) application of diesel engines; 2) application of battery power system.

3.1.1 Case 1- Use a diesel mechanical system
This case scenario applies traditional propulsion system which equips with two diesel engines. The ferry will be operated 18 hours per day. A round trip will be 3.6 hours and the idling time at destinations will be about 0.35 hours (21 minutes).

3.1.2 Case 2- Use a battery power system
In this scenario, the diesel engine will be replaced by a battery power system with the same power loads. The usage of electricity will take the place of diesel oil so the emissions from ferry operation will be eliminated but the upstream emission from the electricity generation must be taken into account.

3.2 Preliminary analysis

Based on the data provided by Thames Clipper, the power requirement of a similar ferry operated in the same route is derived (Figure 3, Figure 4 and Figure 5). Based on the power requirement presented in Table 2, the weights and costs of different types of battery packs could be determined according to Table 3.

4 LIFE CYCLE ASSESSMENT

This section will present the LCA analysis on the short route ferry under two different scenarios. The approach from section 2 will be followed.

4.1 Goal and scope

The goal of this LCA study is to determine the environmental impacts of two different power

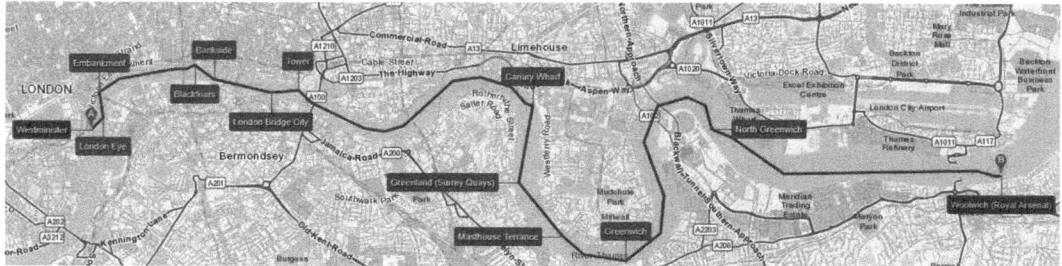

Figure 2. Case ferry route.

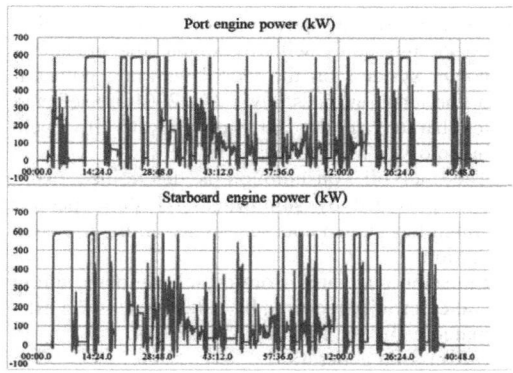

Figure 3. Port and starboard engine power outputs.

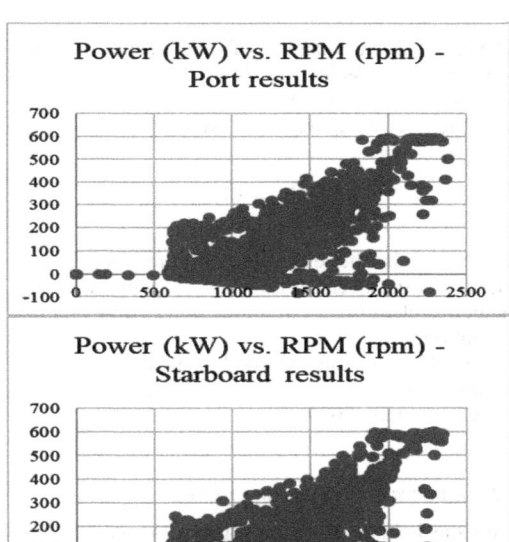

Figure 4. Engine power output under different RPM.

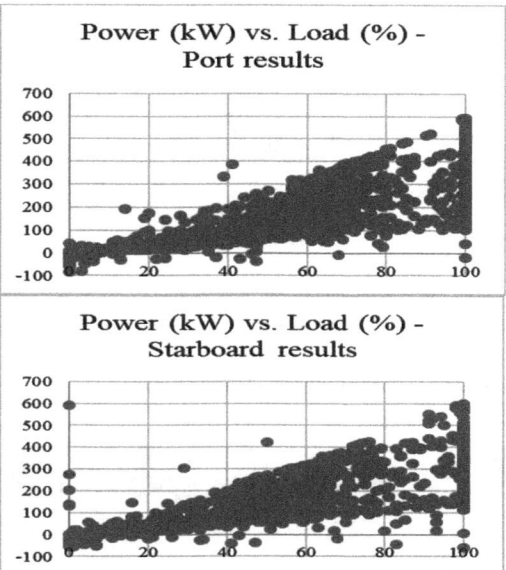

Figure 5. Engine power output under different loads.

Table 2. Engine energy requirement, fuel oil consumptions and emission releases.

Stbd. engine	Port engine	Total
Total propulsion energy (kWh)		
319.97	329.74	649.7
Total fuel (l)		
82.79	86.20	168.98
Total fuel energy (kWh)*		
834.65	869.01	1703.67
Efficiency (-)		
0.38	0.38	
Total CO_2 (t)		
0.27	0.28	0.55
NOx (kg)		
3.84	3.96	7.80

* Low heating value of fuel is 42700kJ/kg

Table 3. Battery types, costs and energy-weight relations (Galloway and Dustmann, 2003; Linden et al., 2002).

Type	Wh/kg	Cost ($/kWh)	Cost ($/kg)
Lead-acid	35	90	3.2
Vanadium Bromine	50	300	15
Silver Cadmium	70		
Zinc Bromine	70		
Sodium/Nickel-Chloride	115	110	12.7
Lithium-Ion	150	600	90

systems on marine vessels (ferry in this case). The benefits of applying battery power system will be derived by comparing with engine power system from the perspective of environmental impacts. Usually, the life phases of on-board power system include five phases (Figure 6): material exploitation, system manufacturing, utilization phase, maintenance periods and end of life. Under different phases, associated material, energy and emissions will be under consideration and shown in Figure 7 (for further study, cash flow will be considered). To retrieve the material, energy and emission flow, main activities under these phases should be identified and associated data are required for LCA study.

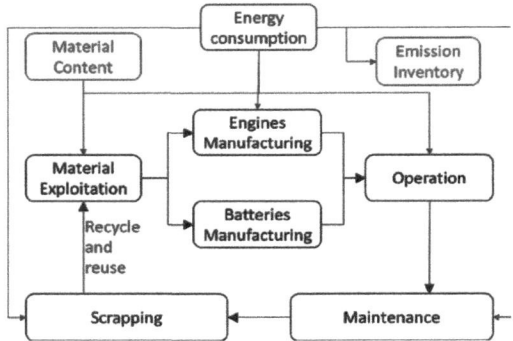

Figure 6. Life cycle stages and interactions.

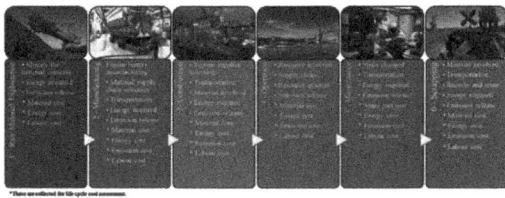

Figure 7. Flow chart of life cycle activities and data collection master checklist.

To simplify the method, a series of assumptions are made based on engineers' judgement and industry's practices:

A) The ship is planned to be built in the UK; the equipment of the propulsion system will be supplied by Wartsila and Corvus (Corvus, 2019; Wartsila, 2020);
B) The raw materials of battery and engines come from exploitation in the same area;
C) Costs and impacts of charging stations will not be considered in this study;
D) The assumption is made that all the rest parts of the vessel remain unchanged when replacing diesel engines with battery packs. Since the differences between the two cases are only the propulsion/power systems, the comparison and evaluation focus on these systems;

E) As the growing attention on global warming effect, this study will investigate the impact of power systems on the global warming potential of the ferry; in this study, CML 2001 are selected to characterise the emissions in the life cycle inventory (CML, 2016);
F) The fuel oil consumption is determined using data provided by shipowner and ship operator which has shown and determined in the previous section;
G) Emission generated from diesel engines running and electricity required by battery packs are estimated using the following equations:

d.1.) Emission generation

$$M_E = FOC \times CF \tag{1}$$

where, M_E is the quantity of one specific emission from fuel combustion; CF is the carbon conversion factor provided by IMO.

d.2.) Electricity required from battery packs

$$E_B = FOC \times LHV \times r \tag{2}$$

where, E_B is the electricity required for battery packs; r is the overall efficiency of using electricity including charging from shore power and onboard utilization.

H) OpenLCA is used to model the life cycle processes of the ship power system (Figure 8) (Ciroth et al., 2020).

4.2 Life cycle inventory assessment

With the help of software OpenLCA and its database, the following emission inventory is developed which are trailered to the emission of interest in Table 4.

4.3 Life cycle impact assessment results

This section will present the results of characterizing the emissions in LCI by applying CML

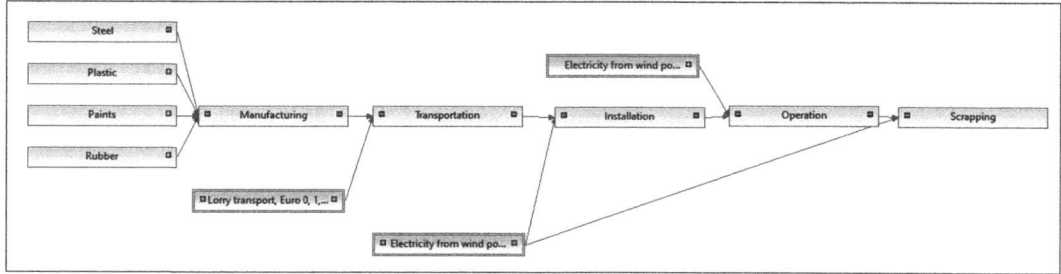

Figure 8. Life cycle modelling of the ship power system.

Table 4. Life cycle emission inventory.

Emission	Category	Result (kg)
Ammonia	Air	65.17
Carbon dioxide	Air	4.34E+7
Dinitrogen monoxide	Air	105.12
Methane	Air	44649.40
Nitrogen dioxide	Air	13234.07
Nitrogen monoxide	Air	0.63
Sulphur dioxide	Air	24733.96
Sulphur trioxide	Air	7.46E-08
Ammonia	Soil	0.010
Phosphate	Soil	0.0092
Ammonia	Water	1.87
Phosphate	Water	20.96
Ammonia	Water	2.75E-05
Phosphate	Water	0.00073

Table 6. GWP potential and reduction comparison.

GWP (kg CO_2 eq.)		Reduction Rate
Diesel oil	4.46E+07	Benchmark
Electricity (grid mix)	3.31E+07	25.78%
Electricity (hydro)	4.10E+05	99.08%
Electricity (wind)	3.11E+05	99.30%

of GHG emission has been avoided if using renewable energy as electricity source either comparing to using diesel engines or battery system supplied by grid mix electricity on the ferry.

5 CONCLUSION

This study investigated the performance of applying battery power system on a short ferry from the perspective of environmental protection especially to the global warming contributions. LCA approach is applied to holistically consider all the essential life processes and activities in the ferry's life span, covering the production of the system, construction, operational and scrapping phases. An LCA model is established in OpenLCA software to take all these phases and activities into account and an LCI inventory is established to cover all the emissions from the targeted emission categories. Eventually, after characterization of the LCI results, a comparative study was carried out to compare the result from different power system candidates. It was derived that comparing to the traditional power system with diesel engines (consuming fossil fuels), the ferry with battery system could achieve about one-fourth of GHG emissions reduction whose electricity supply is from grid mix. However, due to the challenging emission reduction target from IMO, two more considerations of electricity from two other sources are made: hydro and wind power. The reduction rate could be reached about 99% comparing to using diesel engines on the ferry which reduce the GWP from 45 thousand tons CO_2 eq. to 410 and 311 tons CO_2 eq. Hence this paper has developed an LCA model to estimate the emission from a selected ferry and the model also helped to determine the advantage of using battery power systems on the ferry from the point view of emission reduction. Recommendations are made to apply renewable energy for electricity generation to further dig the environmental protection potential of applying the battery system in the marine industry.

characterization model. The results are determined not only considering the emission mentioned in the inventory in this paper but also many other insignificant and low quantity emissions included in the Open-LCA database. The comparison of results between ferry with diesel engines and battery system is presented in Table 5. From the table, it is observed that with the battery system as the propulsion system on the ferry instead of diesel engines, the global warming potential is reduced by around 25%.

4.4 Detail discussion (sensitivity analysis)

Since the battery power ferry could reduce 25% of GHG emissions, it is still challenging to achieve the target of reducing 40% of carbon dioxide (CO_2) by 2030 and 50% of the GHG emission release by 2050. The main source of the GHG emissions from the life span of battery power ferry comes from the electricity generation process which, in this analysis, uses supply from grid mix in the UK. It is a mix of many different energy sources for electrical power generation which includes fossil fuel consumptions, such as hard coal, natural gas, fuel oil and so on. In this section, the impact of using different energy sources for electricity generation will be investigated.

Two energy sources of electricity are under consideration: hydropower and wind power and the results are compared with previous results (engine using diesel oil and battery using grid mix electricity). The comparison results are shown in Table 6. From the results, the observation is that around 99%

Table 5. Global warming potential comparison.

Global warming potential (GWP) (kg CO_2 eq.)	
Ferry with diesel engines	4.46E+07
Ferry with a battery system	3.31E+07

ACKNOWLEDGEMENTS

The authors wish to thank Thames Clipper for providing the data used in this paper. The authors also gratefully acknowledge that the research presented in this paper was partially generated as part of the TrAM project. TrAM has received funding from the European Union's Horizon 2020 research and innovation programme under grant agreement No 769303.

REFERENCES

Ciroth, D.A., Noi, C. Di, Lohse, T., Srocka, M., 2020. openLCA 1.10 Comprehensive User Manual.

CML, 2016. CML-IA Characterisation Factors - Leiden University [WWW Document]. Inst. Environ. Sci. URL https://www.universiteitleiden.nl/en/research/research-output/science/cml-ia-characterisation-factors (accessed 6. 9.20).

Committee on Climate Change, 2019. Net Zero: The UK's contribution to stopping global warming 277.

Corvus, 2019. Corvus Dolphin Energy.

Dai, Q., Kelly, J.C., Gaines, L., Wang, M., 2019. Life cycle analysis of lithium-ion batteries for automotive applications. Batteries 5. https://doi.org/10.3390/batteries5020048

Dedes, E.K., 2013. Investigation of Hybrid Systems for Diesel Powered Ships By. Soton 324.

Delogu, M., Zanchi, L., Maltese, S., Bonoli, A., Pierini, M., 2016. Environmental and economic life cycle assessment of a lightweight solution for an automotive component: A comparison between talc-filled and hollow glass microspheres-reinforced polymer composites. J. Clean. Prod. 139, 548–560. https://doi.org/10.1016/J.JCLEPRO.2016.08.079

Divya, K.C., Østergaard, J., 2009. Battery energy storage technology for power systems-An overview. Electr. Power Syst. Res. https://doi.org/10.1016/j.epsr.2008.09.017

Dunn, J.B., Gaines, L., Kelly, J.C., Gallagher, K.G., 2016. Life cycle analysis summary for automotive lithiumion battery production and recycling. REWAS 2016 Towar. Mater. Resour. Sustain. 73–79. https://doi.org/10.1007/978-3-319-48768-7_11

Galloway, R.C., Dustmann, C.H., 2003. ZEBRA Battery-Material Cost Availability and Recycling, in: EVS 20, 20th International Electric Vehicle Symposium and Exposition, Powering Sustainable Transportation, Conference Proceedings, Long Beach, US, Nov 15-19, 2003. pp. 1–9.

Hiremath, M., Derendorf, K., Vogt, T., 2015. Comparative life cycle assessment of battery storage systems for stationary applications. Environ. Sci. Technol. 49, 4825–4833. https://doi.org/10.1021/es504572q

Hwang, S., Jeong, B., Jung, K., Kim, M., Zhou, P., 2019. Life cycle assessment of lng fueled vessel in domestic services. J. Mar. Sci. Eng. 7, 1–25. https://doi.org/10.3390/jmse7100359

IMO, 2018. I:\MEPC\72\MEPC 72-17-ADD.1.docx 304, 1–11.

ISO, 2006. ISO 14040:2006 - Environmental management – Life cycle assessment – Principles and framework [WWW Document]. Int. Organ. Stand. Geneva, Switz. URL https://www.iso.org/standard/37456.html (accessed 7. 18.18).

Jeong, B., Wang, H., Oguz, E., Zhou, P., 2018. An effective framework for life cycle and cost assessment for marine vessels aiming to select optimal propulsion systems.

J. Clean. Prod. 187, 111–130. https://doi.org/10.1016/J.JCLEPRO.2018.03.184

Kluiters, E.C., Schmal, D., Ter Veen, W.R., Posthumus, K. J.C.M., 1999. Testing of a sodium/nickel chloride (ZEBRA) battery for electric propulsion of ships and vehicles. J. Power Sources 80, 261–264. https://doi.org/10.1016/S0378-7753(99)00075-0

Lan, H., Wen, S., Hong, Y.Y., Yu, D.C., Zhang, L., 2015. Optimal sizing of hybrid PV/diesel/battery in ship power system. Appl. Energy 158, 26–34. https://doi.org/10.1016/j.apenergy.2015.08.031

Linden, D., Reddy Editor, T.B., York, N., San, C., Lisbon, F., Madrid, L., City, M., New, M., San, D., Seoul, J., 2002. HANDBOOK OF BATTERIES.

Ling-Chin, J., Roskilly, A.P., 2016. Investigating the implications of a new-buildon/Roll-off cargo ships from a sustainability perspective– A life cycle assessment case study. Appl. Energy 181, 416–434. https://doi.org/10.1016/J.APENERGY.2016.08.065

Matheys, J., Timmermans, J.M., Van Mierlo, J., Meyer, S., Van Den Bossche, P., 2009. Comparison of the environmental impact of five electric vehicle battery technologies using LCA. Int. J. Sustain. Manuf. 1, 318–329. https://doi.org/10.1504/IJSM.2009.023977

Misyris, G.S., Marinopoulos, A., Doukas, D.I., Tengnér, T., Labridis, D.P., 2017. On battery state estimation algorithms for electric ship applications. Electr. Power Syst. Res. 151, 115–124. https://doi.org/10.1016/j.epsr.2017.05.009

Mitavachan, H., 2014. Comparative Life Cycle Assessment of Stationary Battery Storage Technologies for Balancing Fluctuations of Renewable Energy Sources. Thesis 103.

Raugei, M., El Fakir, O., Wang, L., Lin, J., Morrey, D., 2014. Life cycle assessment of the potential environmental benefits of a novel hot forming process in automotive manufacturing. J. Clean. Prod. 83, 80–86. https://doi.org/10.1016/J.JCLEPRO.2014.07.037

Raugei, M., Morrey, D., Hutchinson, A., Winfield, P., 2015. A coherent life cycle assessment of a range of light-weighting strategies for compact vehicles. J. Clean. Prod. 108, 1168–1176. https://doi.org/10.1016/J.JCLEPRO.2015.05.100

Sanfélix, J., de la Rúa, C., Schmidt, J.H., Messagie, M., Van Mierlo, J., 2016. Environmental and economic performance of an li-ion battery pack: A multiregional input-output approach. Energies 9. https://doi.org/10.3390/en9080584

Smith, C., Burrows, J., Scheier, E., Young, A., Smith, J., Young, T., Gheewala, S.H., 2015. Comparative Life Cycle Assessment of a Thai Island's diesel/PV/wind hybrid microgrid. Renew. Energy 80, 85–100. https://doi.org/10.1016/J.RENENE.2015.01.003

Vicenzutti, A., Bosich, D., Giadrossi, G., Sulligoi, G., 2015. The Role of Voltage Controls in Modern All-Electric Ships: Toward the all electric ship. IEEE Electrif. Mag. 3, 49–65. https://doi.org/10.1109/MELE.2015.2413437

Wartsila, 2020. TrAm project report.

Zhao, S., You, F., 2019. Comparative Life-Cycle Assessment of Li-Ion Batteries through Process-Based and Integrated Hybrid Approaches. ACS Sustain. Chem. Eng. 7, 5082–5094. https://doi.org/10.1021/acssuschemeng.8b05902

Maritime traffic

Developments in Maritime Technology and Engineering – Guedes Soares & Santos (eds)
© 2021 Copyright the Author(s), ISBN 978-0-367-77376-2

Behavior feature analysis on passenger ferry of Jiangsu Section in the Yangtze River based on AIS data

M.Y. Cai, J.F. Zhang & B. Wu
Intelligent Transportation Systems Center (ITS Center), Wuhan University of Technology, Wuhan, China
National Engineering Research Center for Water Transportation Safety (WTS Center), Wuhan University of
Technology, Wuhan, China

W.L. Tian
Maritime College, Beibu Gulf University Qinzhou, China

C. Guedes Soares
Centre for Marine Technology and Ocean Engineering (CENTEC), Instituto Superior Técnico, Universidade de
Lisboa, Lisbon, Portugal

ABSTRACT: There is high risk of collision between passenger ferries and other nearby ships in Inland River when the ferries are crossing the channel. Some serious accidents might happen and result in catastrophic fatalities. Therefore, researches on the behavior feature of ferries are performed in this paper. Most of the information related to ship's encountering can be extracted from AIS data, which provides with effective assistance to the investigation on the behavior features of ferries. As a result, the AIS data of ships in Jiangsu section of the Yangtze River is introduced, and the linear interpolation is applied after data preprocessing to obtain the navigation data of other ships near ferries at the same time. Based on this, some statistical analysis, including the encounter situation, ship type, size, navigation speed and course when the passenger ferries are encountered with target ships, are analyzed so that the detailed navigation features about the crossing behaviors can be identified in more detail. This study presents a theoretical basis for ensuring the navigation safety of passenger ferries and reducing the probability of accidents.

1 INTRODUCTION

Due to the advantages of large carrying capacity and low cost, water transportation always plays a significance role in transportation system. As one of the best inland waterways in China, the Yangtze River is known as the golden waterway. The development of transportation along the Yangtze River is promoted. While the shipping of the Yangtze River makes it more convenient to transport passengers and goods, there are also many safety issues that need to pay attention to. The traffic volume of ships in the Yangtze River is very large, and the navigational environment and channel conditions are relatively complicated, which makes it necessary to promote the safety of ships in the waterway as well. The encounter situations between ferries and the ships in the channel are complex and the risk of collision is supposed to be at a relative high level. Many injuries or fatalities occur throughout the world as a result of accidents involving passenger ferries (Lu & Yang, 2011). Due to that such

accidents have direct relationship with the safety of multiple passengers, the navigation safety of ferries has always been the top concerns of the Maritime Safety Administration (MSA).

When ferry ships are navigating in the Yangtze River periodically, they always sail across the waterway, which interacts with the regular ship traffic of other ships to some extent, and the probability of encountering with target ships would also increase (Uğurlu et al., 2018), and further increase the probability of collision. In order to ensure the navigation safety under such scenarios, it is necessary to investigate the characteristics of the navigation behaviour of the passenger ferries during crossing the waterways.

The employment of Automatic Identification System (AIS) provides with a solid foundation for researchers to conduct quantitative analysis on the motion patterns of ship traffic (Xu et al., 2019). However, in most cases the AIS data volume is very large, and there are also many errors. It is necessary to make some pre-processing, so as to enable the research be more efficient and more reliable.

DOI: 10.1201/9781003216582-14

Some effective methods have been applied to data analysis by some researchers in order to better deal with the AIS data. For example, Xu et al. (2015) applied linear interpolation to AIS data to reconstruct the linear segment of ship trajectory, and used spline interpolation to reconstruct the arc segment of ship trajectory. Sang et al. (2015) introduced a method to restore the trajectory of an inland waterway ship based on the AIS data according to three rules developed to identify and remove the inaccurate data and the navigational features of the inland waterway ship.

Zhang et al. (2018) introduced the Douglas-Peucker (D-P) algorithm into AIS trajectory data restoration, which can simplify AIS trajectory by extracting some characteristic points and compress redundant information as well as improve subsequent processing efficiency. Zhao and Shi (2018) also applied the improved D-P algorithm to fully compress the AIS big data on the premise of keeping the ship trajectory information that can improve the efficiency of data processing as well. Zhang et al. (2018) proposed a ship trajectory reconstruction model consisting of three steps including outlier removal, navigation state estimation and ship trajectory fitting, which can effectively fit various types of trajectories based on AIS data. Rong et al. (2020) presented a data mining approach which consists of identifying relevant waypoints along a route where significant changes in the ships' navigational behaviour are observed using trajectory compression and clustering algorithms.

The first step of performing collision risk is to identify the encounter situations from AIS data. It is illustrated that there is collision risk between the encounter ships if one ship is encountered with another. Therefore, the encounter can be understood as the situation with the distance and time to closest point of approach less than a threshold value (i.e. a distance under the absence of collision avoidance operations). The encounters between ships provides with sufficient samples so that the ship's behavior features and collision risk can be further analyzed. The identification of encounters is very important for risk assessment and collision avoidance. A lot of valuable information can be identified from the behavior features of ships when the encounter occurs. The ship encounter identification methods have been extensively studied. Ma et al. (2019) applied Support Vector Machine (SVM) to realize the automatic extraction of the encounter situation by constructing model to classify and identify the feature sequences of encounters. Zhang et al. (2015) proposed a Vessel Conflict Ranking Operator (VCRO) methods to evaluate the potential collision risk and safety of ship encounters based on the AIS data in the Gulf of Finland. Chen et al. (2020) proposed an improved Time Discretized non-linear Velocity Obstacle algorithm (TD-NLVO) to identify multi-ship encounters effectively. Besides the above, ship domain models with different shapes (Szlapczynski & Szlapczynska, 2017) are usually applied to identifying the encounter situations by judging whether the ship domain is

violated. For example, Rong et al. (2016) used ship domain as criterion to determine the near collision scenarios and a distinction is made between head-on, crossing and overtaking near collision scenarios. Silveira et al. (2013) applied an approach to calculates the number of collision candidates by estimating the future positions of ships and the distance between them, based on the positions, courses and speeds given by the AIS messages. Some parameters such as the Distance at Closest Point of Approach (DCPA), Time at Closest Point of Approach (TCPA) and relative distance between the ships can be well used in the analysis of the encounters (Rong et al., 2019).

Based on the comprehensive assessment methods and spatial-temporal data mining algorithms for ship encounter, the navigation condition on four ferries in the Jiangsu section of the Yangtze River during encountering with target vessels is studied in this paper through the analysis on AIS data. The behavior features of ferries including spatial and temporal distribution of encounters as well as the relative speed and relative course distribution in different scenarios of encounters are investigated.

The rest of the paper is organized as follows: The processing methods and principles of AIS data are present in Section 2. In Section 3, the behavior features of ferries when they are encountered with target ships are analyzed in more detail according to some specific cases and data processing results.

2 THE METHODS

2.1 *Data extraction*

According to the requirements by the International Convention for Safety of Life at Sea (SOLAS), most of the ships must be equipped with AIS on board except a few specific types of ships. Useful information can be broadcasted and received among the nearby ships as well as the base station, which is very useful for the navigating ships in identifying the nearby ships, and assists target tracking, and simplify information exchange (Huang et al., 2019). The data frequency of dynamic information is slightly different with different speed and angular speed of the ships. In general, the dynamic information of the ships with speed between 0 and 14 knots as well as over 14 knots while changing the heading is usually transmitted with an interval of 4 and 2 seconds, respectively, while the static data is broadcast every 6 minutes. The phenomenon reflects the characteristics of AIS big data which generates frequently and contains large data volume. Only through data processing and mining can valuable information be obtained. As present in Figure 1, the process of AIS data processing can be divided into three steps: data extraction, data pre-processing, and data analysis.

The data used in this study is obtained from the AIS database of the Yangtze River, which contains both the static and dynamic information of the ship,

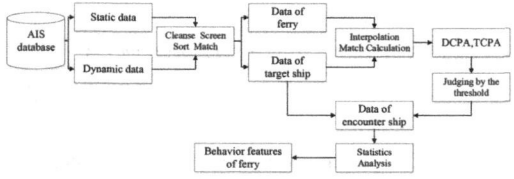

Figure 1. Flow chart of AIS data processing.

as well as other types of navigation data that is similar with the classification of the research by Chen et al. (2018). The static data mainly includes ship name, length and width of the ship, and ship type. The ship type is usually indicated in the form of a code. The relationship between the code and its corresponding type is shown in Table 1. The dynamic information includes the latitude and longitude of the ship, the speed over ground, the course over the ground, the true heading, and the timestamp of the AIS data. These data are always used by researchers e.g. Weng et al. (2019). There is one unique Maritime Mobile Service Identify (MMSI) for each ship, so MMSI is able to be used to identify different ships. The range of the studied water area should be constrained to be near the ferry ships for the AIS data obtained from the database, so as to ensure that the obtained ship information is within the desired area. It is necessary to obtain dynamic information such as the longitude, latitude, speed over ground, course over ground, true heading, and timestamp of the ships, as well as static information such as MMSI, ship type and length in order to prepare raw data for the analysis in the subsequent research.

2.2 Data pre-processing

There are many missing, repetitive and erroneous cases in the raw AIS data. Therefore, it is necessary to pre-process the data. First of all, the AIS data in the specified time and place should be screened out. Then, the data of static ship, that is, the ship whose

Table 1. A part of the code of ship type in AIS data.

code	ship type
30	fishing-boat
35	military ship
36	sailboat
37	yacht
4X	high speed vessel
50	pilot vessel
52	tugboat
6X	passenger boat
7X	cargo ship
8X	tanker
9X	other ship

speed is 0 or a very small value, should be removed from the sample. Moreover, those abnormal speed with very large values also need to be removed. Due to the fact that the influence of external factors in Inland River such as the wind and waves is relatively small, the value of the true heading and the course over ground of a ship are roughly the same. Therefore, during data pre-processing, course over ground is taken as the navigation course, and the true heading is taken as the navigation course when the course over ground of a ship is missing, wrong or zero. The navigation data of ships which will be used later can be more accurate and adequate.

In order to facilitate the subsequent calculation and analysis, the units of the ship data should be converted and unified. The units of time can be converted into seconds, and the units of speed can be converted into m/s.

Because the static and the dynamic data in the database are stored separately, the MMSI can be used to match the dynamic data and the static data (Zhou et al., 2019). After the matching process, each AIS data can include the information of speed, course, length, width of a ship at a certain time along with the latitude and longitude.

Passenger ferries and their related data can be obtained according to the code of ship type and MMSI. Then according to the longitude and latitude data as well as the time data of the ferries, nearby target ships and their related information can be further obtained. But navigation data of the ferries and target ships are not synchronization, that is to say that the time stamps of them are not the same as mentioned by Wu et al. (2016) which makes it difficult to conduct the next step of calculation. In view of that when the time interval of AIS data is short enough, the change of angular velocity and speed are slow and they are able to be approximately matched by linear function. The linear interpolation method can be used to obtain the data of ferries and target ships at the same moment. Meanwhile, the missing data can be interpolated.

Linear interpolation is used to estimate the function value of interpolation data points in the region according to the linear function of adjacent data points. The time interval of the acquired discrete AIS data is in general about 1 minute. However, in practice the time interval between the AIS data is much different due to some exterior influence brought by the reception coverage, data loss, as well as ship status, and the AIS radio station (Zhang et al., 2019). Thus, the intervals of time are set as 60 seconds to interpolate the new dynamic data. The speed, course, longitude and latitude of ferries at time T_{i-1} and T_{i+1} are $V_{A,Ti-1}$, $C_{A,Ti-1}$, $X_{A,Ti-1}$, $Y_{A,Ti-1}$ and $V_{A,Ti+1}$, $C_{A,Ti+1}$, $X_{A,Ti+1}$, $Y_{A,Ti+1}$, respectively. The corresponding parameters of the nearby targets ships at time T_{i-1} and T_{i+1} is $V_{B,Ti-1}$, $C_{B,Ti-1}$, $X_{B,Ti-1}$, $Y_{B,Ti-1}$ and $V_{B,Ti+1}$, $C_{B,Ti+1}$, $X_{B,Ti+1}$, $Y_{B,Ti+1}$, respectively. According to the principle of linear interpolation, the speed, course, longitude and latitude of the ferries at time T_i can be expressed as follows:

$$V_{A,T_i} = V_{A,T_{i-1}} + \frac{T_i - T_{i-1}}{T_{i+1} - T_{i-1}}(V_{A,T_{i+1}} - V_{A,T_{i-1}}) \quad (1)$$

$$C_{A,T_i} = C_{A,T_{i-1}} + \frac{T_i - T_{i-1}}{T_{i+1} - T_{i-1}}(C_{A,T_{i+1}} - C_{A,T_{i-1}}) \quad (2)$$

$$X_{A,T_i} = X_{A,T_{i-1}} + \frac{T_i - T_{i-1}}{T_{i+1} - T_{i-1}}(X_{A,T_{i+1}} - X_{A,T_{i-1}}) \quad (3)$$

$$Y_{A,T_i} = Y_{A,T_{i-1}} + \frac{T_i - T_{i-1}}{T_{i+1} - T_{i-1}}(Y_{A,T_{i+1}} - Y_{A,T_{i-1}}) \quad (4)$$

The speed, course, longitude and latitude of target ships at time T_i can be calculated in the same way. After the whole process of data pre-processing is completed, the dynamic data at time T_i can be replaced by the data at time T_i that is obtained from the interpolation operations, which can be used for further analysis.

2.3 Data analysis

The above data, which has been pre-processed previously is used for calculation, and the relative motion parameters such as relative speed, relative course and relative distance, Distance at Closest Point of Approach (DCPA) and Time at Closest Point of Approach (TCPA) are applied to measuring the relative motion condition and encounter situation between the ferries and target ships. These relative motion parameters were also used in Yoo (2018). In order to facilitate the analysis, the coordinate system with the ferry located at the origin can be constructed. At this time, it is assumed that the ferry is stationary, while the nearby target ships are sailing to it with relative course and relative speed. The relative course (C_r) of the target ship to the ferry at time T_i is:

$$C_r = \begin{cases} C_{A,T_i} - C_{B,T_i} & C_{A,T_i} \geq C_{B,T_i} \\ C_{A,T_i} - C_{B,T_i} + 2\pi & C_{A,T_i} < C_{B,T_i} \end{cases} \quad (5)$$

The relative speed (V_r) of the target ship to the ferry at time T_i is:

$$V_r = \sqrt{V_{A,T_i}^2 + V_{B,T_i}^2 - 2V_{A,T_i}V_{B,T_i}\cos C_r} \quad (6)$$

The relative distance (D_r) between the target ship and the ferry can be calculated according to the longitude and latitude of them by so-called Haversine formula:

$$hav(\frac{D_r}{R}) = hav(Y_2 - Y_1) + \cos(Y_1)\cos(Y_2)hav(X_2 - X_1) \quad (7)$$

where R is the radius of the earth, with a mean of 6371000 meters, X_1, Y_1 and X_2, Y_2 are the longitude and latitude of ferry and target ship, respectively.

$$hav(\alpha) = \sin^2(\frac{\alpha}{2}) = \frac{1 - \cos\alpha}{2} \quad (8)$$

The angle between D_r and C_r (θ) can be expressed as follows:

$$\gamma = \begin{cases} C_b + \pi & 0 \leq C_b \leq \pi \\ C_b - \pi & \pi < C_b \leq 2\pi \end{cases} \quad (9)$$

$$\theta = \begin{cases} C_r - \gamma & C_r \geq \gamma \\ C_r - \gamma + 2\pi & C_r < \gamma \end{cases} \quad (10)$$

where C_b is the azimuth angle between the ferry and the target ship.

The meaning of DCPA and TCPA is shown in Figure 2. The formula can be expressed as follows:

$$DCPA = \begin{cases} D_r\sin\theta & 0 < \theta < \frac{\pi}{2} \\ D_r\sin(\pi - \theta) & \frac{\pi}{2} < \theta < \pi \\ D_r\sin(\theta - \pi) & \pi < \theta < \frac{3\pi}{2} \\ D_r\sin(2\pi - \theta) & \frac{3\pi}{2} < \theta < 2\pi \\ 0 & \theta = 0, \frac{\pi}{2} \\ D_r & \theta = \pi, \frac{3\pi}{2} \end{cases} \quad (11)$$

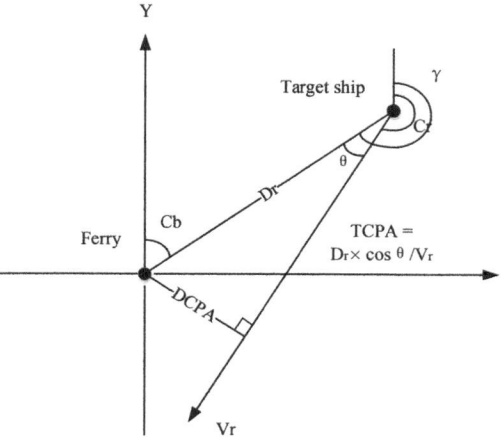

Figure 2. The schematic of TCPA and DCPA.

$$TCPA = \begin{cases} \frac{D_r \cos\theta}{V_r} & 0 < \theta < \frac{\pi}{2} \\ \frac{D_r \cos(\pi-\theta)}{V_r} & \frac{\pi}{2} < \theta < \pi \\ -\frac{D_r \cos(\theta-\pi)}{V_r} & \pi < \theta < \frac{3\pi}{2} \\ \frac{D_r \cos(2\pi-\theta)}{V_r} & \frac{3\pi}{2} < \theta < 2\pi \\ \frac{D_r}{V_r} & \theta = 0 \\ -\frac{D_r}{V_r} & \theta = \frac{\pi}{2} \\ 0 & \theta = \pi, \frac{3\pi}{2} \end{cases} \quad (12)$$

Whether the ferries will encounter with target ships can be determined through the range of DCPA and TCPA (Szlapczynski & Szlapczynska, 2017). After identifying the encounter data, it becomes possible to analyse the behaviour of ferries, and the navigation mode and rule of ferries can be investigated in more detail. The influencing factors that have impact on the behaviour of ferries can be also understood (Shu et al., 2017) accordingly.

The overall procedure of the AIS data preprocessing and analysis is composed with the following steps:

(1) Group the pre-processed data according to MMSI, and match the data of ferries with encounter ships which is at the same location and moment.
(2) Calculate the relative motion parameters of ferries and encounter ships, including relative speed, relative course, relative distance, DCPA and TCPA, etc.
(3) Set the thresholds of DCPA and TCPA according to vessel traffic flow and channel status, and determine whether the encounter between ferry and target ship happens accordingly.
(4) Take the encounter data as the identified encounter situation when its TCPA reaches the minimum.
(5) Analyse the data according to the identified encounter situations, and conclude the behaviour features of ferries such as spatial distribution, speed and course distribution as the research by Xiao et al. (2015).

3 CASE STUDIES

3.1 The study water area

The study water area in this paper is the Jiangsu section of the Yangtze River, which is close to the estuary of the river. The ship traffic flow in this water area is very high, which has much influence on the navigation safety of ferry ships. The operation of the ferries will increase the risk of collision in this region to some extent, which will not only affect navigation capacity, but also pose a potential threat to the safety of customers and crew in the passenger ferries. The longitude and latitude of research area is set as [118.5396°E,

121.5167°E] and [31.75°N, 32.33°N], and the AIS data used in the research is obtained according to the above range of latitude and longitude from 0:00:00 to 23:59:59 on July 26, 2019 from the AIS database.

After the process of removing the duplicate and erroneous data, matching the static data and dynamic data to the raw AIS data, 461142 samples of the AIS data which is obtained. The data of passenger ships whose type codes are between 60 and 69 is screened out, and further a total of 1238 of data of 4 passenger ferries is selected. Then, a total of 21586 sample of data of 621 target ships is identified according to the spatial distribution of passenger ferries. In the next step, linear interpolation is applied to the data of passenger ferries and target ships. The relative speed V_r, relative course C_r, relative distance D_r, DCPA and TCPA is obtained after the calculation, which is used in the further analysis (Chen et al., 2019).

3.2 Results and discussion

The width of the channel where the ferries are sailing is relatively small and there are numerous ships around the ferries. Therefore, the probability of encounter is extremely high (Altan & Otay, 2018). Based on some relevant researches, the thresholds of DCPA and TCPA for determining whether the encounters happen is set as follows:

$$\begin{cases} DCPA \leq 300m \\ -60s \leq TCPA \leq 180s \end{cases} \quad (13)$$

After further screening based on the above thresholds, 6842 samples of data are obtained, and then the ones with the minimum TCPA of each encounter is taken as the encounter data. A total of 1554 samples of encounter data is identified.

One of the obvious behaviour features of passenger ferries in the Jiangsu section of the Yangtze River is that they are sailing from one side to another across the waterway periodically, and they generally stop for a short time when they arrive at destination in order to let the passengers on board get off the ferry ship and wait for new passengers. More information about waiting time can be learnt from Jørgensen & Solvoll, (2018). It approximately takes half an hour to complete a whole voyage and the ferries stop and stay at the wharf for a short time. However, the period of anchoring is difficult to predict because the time interval of the AIS data is long while the standing time is short. When the passenger ferries are crossing the channel, the routes of navigation are not simply along the shortest straight line connecting the two docks on the either side of the river. The trajectories of ferries keep a large angle with the river bank in the middle of the channel, and the angle of the navigating direction with the river bank gradually reduces when the ferries are close to the shore so as to berth.

According to International Regulations for Preventing Collisions at Sea (COLREGs), a vessel shall be deemed to be overtaking when coming up with another vessel from a direction more than 22.5° abaft her beam. When two power-driven vessels are meeting on reciprocal or nearly reciprocal courses so as to involve risk of collision, they are deemed to be head-on.

When two power-driven vessels are crossing so as to involve risk of collision, they can be deemed to be crossing. These definitions can be appropriately applied to judging the encounter situation in Jiangsu section of the Yangtze River. The encounter situation can be divided into 3 categories according to relative course C_r between the ferries and target ships. As shown in the Figure 3, when C_r is in the range of $0°<C_r≤67.5°$ or $292.5°<C_r≤360°$, the situation is deemed to be overtaking. When C_r is in the range of $175°< C_r≤185°$, the situation is deemed to be head-on. When C_r is in the range of $67.5°<C_r≤175°$ or $185°< C_r≤292.5°$, the situation is deemed to crossing.

The distribution of the relative speed and relative course in one day when the four ferries are encountering with target ships is shown in Figure 4. The relative speed is obviously larger when the relative course is between 175° and 185° than that when the relative course is close to zero. The relative speed of head-on is relatively large, while relative speed of overtaking is relatively small.

According to the statistical results on the encounters, the proportion of the three types of encounter situations is presented in Figure 5.

When analysing the encounters between ferries and target ships, it is found that the probability of the encounter frequency between ferries and target ships is very high. Among the 621 target ships that cross the

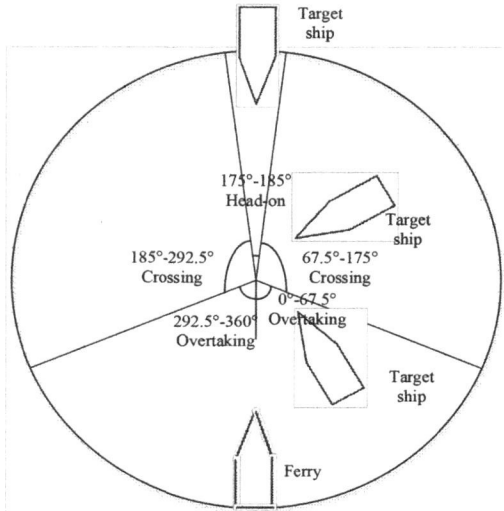

Figure 3. Diagram of encounter situation.

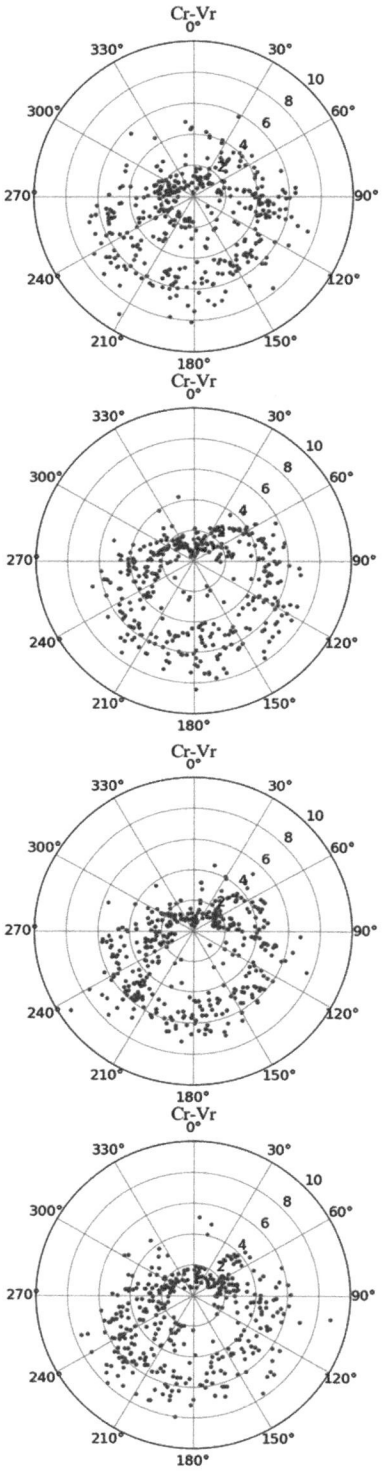

Figure 4. The distribution of the relative speed and relative course when the ferries encountering with target ships.

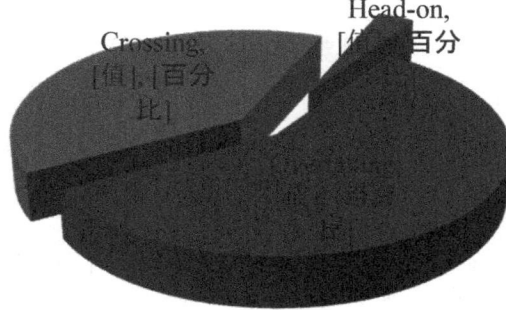

Figure 5. Encounter situations.

Figure 6. Ship length.

route of ferries, more than half of them were encountered with one ferry. The reason is that inland waterways are generally very narrow that the relative distance between the ferry and target ship is small. With respect to the encounter situations, head-on is the least which accounts for 3%. This is because the target ships tend to sail along the channel, while ferries can hardly sail against the channel. The risk of head-on is very high due to the high relative speed, so it should be avoided as much as possible. Overtaking appears the most, which accounts for 58%. Under this situation, as long as the relative distance and relative speed can be kept in a reasonable range, the safety of navigation of both ferries and nearby ships can be guaranteed. Crossing also occurs frequently, which account for 39%, because the route of ferries is cross with the route of target ships. The risk of this situation is also relatively high under the circumstance of that it is difficult for ship to steer flexibly in the inland waterway. The crews on ferries and other ships need to take effective measures to ensure that accidents can be avoided.

In terms of the length (L) of the encounter ships, they are divided into three categories as follows: L≤50m, 50m<L≤100m and L>100m. The distribution is present in Figure 6. The average length of them is 73 meters.

The types of the encounter ships can be divided into cargo ships, oil tankers, tugs and others according to their codes of ship types, and the corresponding distribution is shown in Figure 7.

In terms of the type and length of ships that are encountered with ferries, the majority of the ships are cargo ships and tankers, which account for 90% and 7%, respectively. Small and medium ships are 5 times as many as large ships, which is mainly determined by the condition of the waterway. There are many cargo ships in the Yangtze River because of a significance role it plays in the transportation of goods. The navigation capacity of inland waterway navigation is not suitable for large ships due to the narrow channel and complicated condition of navigation. Moreover, there are many bridges over the Yangtze River which have restriction to large ships. If there are more large ships or ships carrying dangerous goods, the risk of accident will increase apparently, which might result in enormous economic losses and casualties to human beings.

The time of encounters can be divided into 4 time periods in order to analyze the temporal distribution of encounters, which are 0:00 to 6:00, 6:00 to 12:00, 12:00 to 18:00, and 18:00 to 24:00. The distribution of the encounters in each time period is shown in Figure 8.

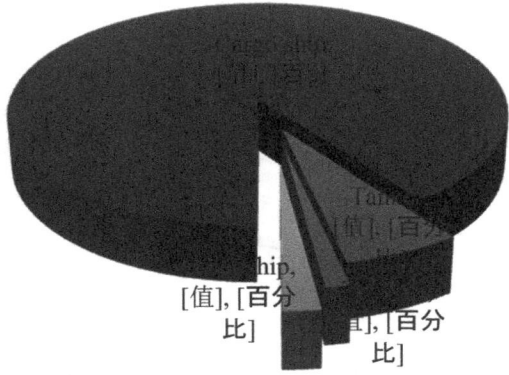

Figure 7. Ship Types.

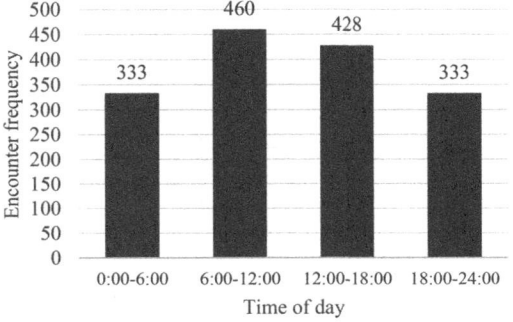

Figure 8. Temporal distribution of encounters.

135

With respect to the temporal distribution of encounters, there are more encounters from 6:00 to 18:00 than that during other time periods. It is mainly because the traffic flow of ships in the Jiangsu section of the Yangtze River at night is much less than other time periods, and the visibility is bad at night which can influence the judgement of the captain on the ship, so that the risk of navigation at night is relatively high.

The relative position and relative motion parameters between ferries and target ships under the three typical scenarios are present in Figure 9 and 10. The starting point in Figure 9 is

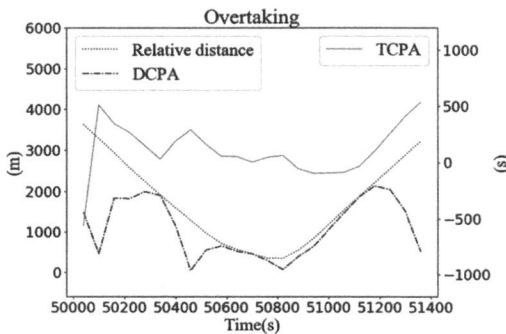

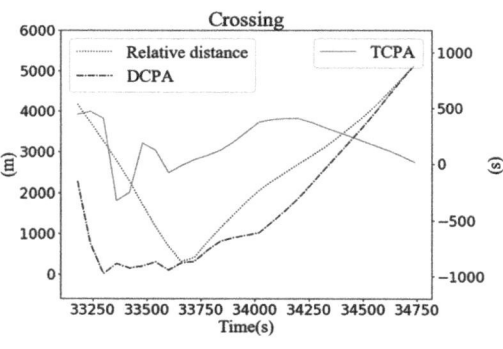

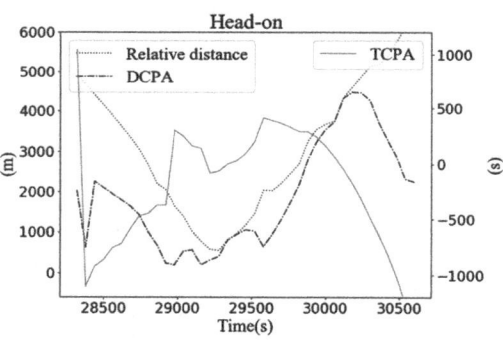

Figure 10. The DCPA, TCPA and relative distance when ferry encounters with target ship of 3 scenarios.

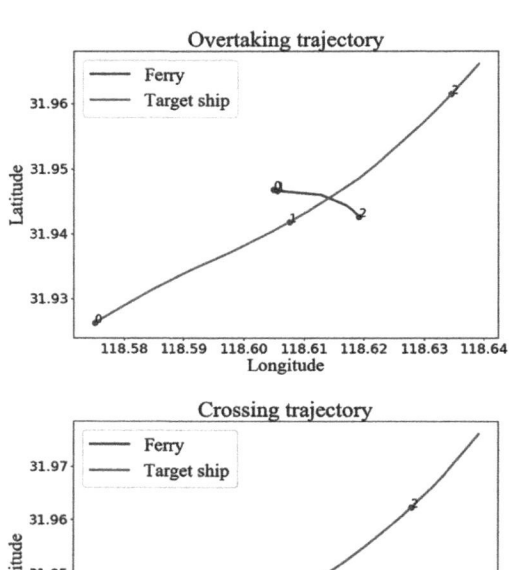

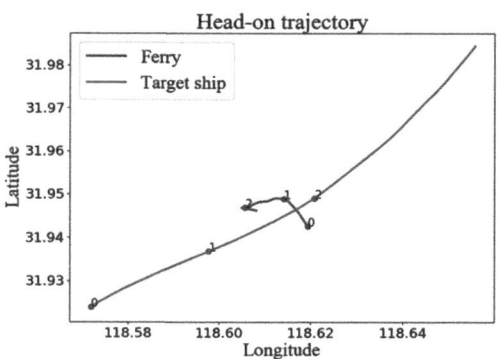

Figure 9. The relative location when ferry encounters with target ship of 3 scenarios.

represented by point 0, and the time intervals between two points are 600s. It can be seen from the trajectory of the three scenarios that they all stay in wharf for a short time. As shown in Figure 10, the distance between ferry and target ship reduces initially and then increases, and DCPA, TCPA changes with the relative motion between them and the evasive actions taken by the ferry or the target ship.

4 CONCLUSIONS

By means of data extraction, preprocessing and analysis on AIS data, the behavior features of passenger ferries when crossing the busy waterways are identified and investigated in this paper. According to the statistical analysis on the encounters between ferry and target ships in Jiangsu section of the Yangtze River, it is concluded that encounters happen more frequently between 6:00 to 18:00 during the day. The ferries are more frequently encountered with small and medium cargo ships, and the probability of overtaking and crossing account for the majority when ferry encounters with target ship.

By identifying these behavior features of ferry, the more valuable information and characteristics about passenger ferry is obtained, so that it can effectively promote the further study in terms of collision avoidance and ship safety domain. There will be more reference for the Officer On Watch (OOW) to judge the encounter situations and take the effective strategies to prevent collision. The research can also provide with a foundation for the development of navigational aids. And the navigation of ferry and other ship can be promoted, and the transportation efficiency of the area where the ferries navigate can also be improved.

In the future work, the volume of AIS data for research can be further expanded to include a longer time period and a wider area, so that the results can be more comprehensive and accurate. Moreover, the trajectory of the ferry can be restored in a better way. Moreover, modeling and simulation of the ship traffic can also be performed and the behavior features of the ferry affected by human and environment factors can be investigated in more depth.

ACKNOWLEDGEMENTS

The research was supported by National Key Technologies Research & Development Program (2017YFE0118000), Funds for International Cooperation and Exchange of the National Natural Science Foundation of China (51920105014), the European Union's Horizon 2020 research and innovation programme under the Marie Skłodowska-Curie grant agreement No 730888 (RESET), the National Science Foundation of China (NSFC) (41801375) and Technical Innovation Project of Hubei province (International Cooperation) (2018AHB003).

REFERENCES

Altan, Y. C., & Otay, E. N., 2018. Spatial mapping of encounter probability in congested waterways using AIS. Ocean Engineering, 164, 263–271.

Chen, P.F., Huang, Y.M., Papadimitriou, E., et al., 2020. An improved time discretized non-linear velocity obstacle method for multi-ship encounter detection. Ocean Engineering, 196.

Chen, P.F., Mou, J.M., van Gelder, P.H.A.J.M., 2019. Integration of individual encounter information into causation probability modelling of ship collision accidents. Safety Science, 120, 636–651.

Chen, Z.J., Xue, J., Wu, C.Z., et al., 2018. Classification of vessel motion pattern in inland waterways based on Automatic Identification System. Ocean Engineering, 161, 69–76.

Huang, J.C., Nieh, C.Y., Kuo, H.C., 2019. Risk assessment of ships maneuvering in an approaching channel based on AIS data. Ocean Engineering, 173, 399–414.

Jørgensen, F., & Solvoll, G., 2018. Determining optimal frequency at ferry crossings. Transport Policy, 63, 200–208.

Lu, C.S., & Yang, C.S., 2011. Safety climate and safety behavior in the passenger ferry context. Accident Analysis and Prevention, 43(1), 329–341.

Ma, J., Liu, Q., Zhang, C.W., et al., 2019. A Method for Extracting Ship Encounter Situation Based on Spatio-temporal Analysis of AIS Data. China Safety Science Journal, 29.

Rong, H., Teixeira, A.P., Guedes Soares, C., 2016. Assessment and characterization of near ship collision scenarios off the coast of Portugal. In: Guedes Soares, C., Santos, T.A., (Eds.), Maritime Technology and Engineering 3. Taylor & Francis Group, London, 871–878.

Rong, H., Teixeira, A.P., Guedes Soares, C., 2019. Risk of ship near collision scenarios off the coast of Portugal. In: Beer, M., Zio, E., (Eds.), 29th European Safety and Reliability Conference (ESREL) (Hannover, Germany).

Rong, H.; Teixeira, A. P., and Guedes Soares, C. 2020; Data mining approach to shipping route characterization and anomaly detection based on AIS data. Ocean Engineering. 198:106936.

Sang, L.Z., Alan, W., Yan, X.P., et al., 2015. A novel method for restoring the trajectory of the inland waterway ship by using AIS data. Ocean Engineering, 110, 183–194.

Shu, Y.Q., Daamen, W., Ligteringen, H., et al., 2017. Influence of external conditions and vessel encounters on vessel behavior in ports and waterways using Automatic Identification System data. Ocean Engineering, 131, 1–14.

Silveira, P. A. M., Teixeira, A. P., Guedes Soares, C., 2013. Use of AIS Data to Characterise Marine Traffic Patterns and Ship Collision Risk off the Coast of Portugal. Journal of Navigation, 66(6), 879–898.

Szlapczynski, R., & Szlapczynska, J., 2017. A method of determining and visualizing safe motion parameters of a ship navigating in restricted waters. Ocean Engineering, 129, 363–373.

Szlapczynski, R., & Szlapczynska, J., 2017. Review of ship safety domains: Models and applications. Ocean Engineering, 145, 277–289.

Uğurlu, Ö., Yıldız, S., Loughney, S., et al., 2018. Modified human factor analysis and classification system for passenger vessel accidents (HFACS-PV). Ocean Engineering, 161, 47–61.

Weng, J.X., Shi, K., Gan, X.F., et al., 2019. Ship emission estimation with high spatial-temporal resolution in the Yangtze River estuary using AIS data. Journal of Cleaner Production.

Wu, X., Mehta, A.L., Zaloom, V.A., et al., 2016. Analysis of waterway transportation in Southeast Texas waterway based on AIS data. Ocean Engineering, 121, 196–209.

Xiao, F.L., Han, L., Coen, G., et al., 2015. Comparison study on AIS data of ship traffic behavior. Ocean Engineering, 95, 84–93.

Xu, H.T., Rong, H., Guedes Soares, C., 2019. Use of AIS data for guidance and control of path-following autonomous vessels. Ocean Engineering, 194.

Xu, W.X., Chu, X.M., Liu, X.L., 2015. Key Technologies of Visual Analytics for Vessel Behaviors. *Journal of WUT (Information & Management Engineering)*, 37, 451–454.

Yoo, S.L., 2018. Near-miss density map for safe navigation of ships. *Ocean Engineering*, 163, 15–21.

Zhang, L.Y., Meng, Q., Fang Fwa, T., 2019. Big AIS data based spatial-temporal analyses of ship traffic in Singapore port waters. *Transportation Research Part E: Logistics and Transportation Review*, 129, 287–304.

Zhang, L.Y., Meng, Q., Xiao, Z., et al., 2018. A novel ship trajectory reconstruction approach using AIS data. *Ocean Engineering*, 159, 165–174.

Zhang, S.K., Shi, G.Y., Liu, Z.J., et al., 2018. Data-driven based automatic maritime routing from massive AIS trajectories in the face of disparity. *Ocean Engineering*, 155, 240–250.

Zhang, W.B., Goerlandt, F., Montewka, J., et al., 2015. A method for detecting possible near miss ship collision from AIS data. *Ocean Engineering*, 107, 60–69.

Zhao, L.B., Shi, G.Y., 2018. A method for simplifying ship trajectory based on improved Douglas-Peucker algorithm. *Ocean Engineering*, 166, 37–46.

Zhou, Y., Winnie, D., Tiedo, V., 2019. Ship classification based on ship behavior clustering from AIS data. *Ocean Engineering*, 175, 176–187.

Developments in Maritime Technology and Engineering – Guedes Soares & Santos (eds)
© 2021 Copyright the Author(s), ISBN 978-0-367-77376-2

A novel framework of real-time regional collision risk prediction based on RNN approach

Dapei Liu, Yao Cai, Xin Wang, Zihao Liu & Zhengjiang Liu
College of Navigation, Dalian Maritime University, Dalian, P.R. China

ABSTRACT: Regional collision risk assessment is important for traffic surveillance in maritime transportation. This study proposes a framework of real-time prediction for regional collision risk by combining density-based spatial clustering of applications with noise (DBSCAN) technique, Shapley value method and recurrent neural network (RNN). Firstly, the DBSCAN technique is applied to cluster vessels in specific sea area, then the regional collision risk is quantified by calculating the contribution of each vessel and each cluster with Shapley value method. Afterwards, the optimized RNN method is employed to predict the regional collision risk of specific seas in short time. At last, a case study is carried out with actual automatic identification system (AIS) data, the results show that the proposed framework is an effective tool for regional collision risk prediction.

1 INTRODUCTION

Maritime transport is the backbone of international trade and the global economy. In general, vessel collision may cause great loss of human lives and property, as well as severe environment pollution (Zhang et al. 2019). Particularly, surveillance plays an important role in preventing vessel collision accidents (Liu et al. 2019) and the collision risk is a major indicator for surveillance operators to judge the danger between meeting vessels (Zhen et al. 2017). However, VTS surveillance operators' burden get heavier with the shipping rapidly increasing. Hence, the optimized marine surveillance framework has been emerged as an important issue.

The prediction of regional collision risk of vessels in water areas is based on the study of regional collision risk. The regional collision risk refers to the vessel collision risk in a certain water area. Compared with the study of collision risk for single or multiple vessels, the researchers get less achievements in the study and application of regional collision risk.

Recently, Liu (2019) proposed a regional collision risk calculation model based on AIS data. The cooperative game theory is used to model the collision risk of regional collision risk in the framework, which stars from the information of single vessel's DCPA, TCPA, etc. Compared with the traditional model, the framework can obtain more accurate results of instantaneous regional collision risk, and avoids the influence of traffic flow in the water area. Therefore, in this paper, DCPA, TCPA and Ship Domain Overlapping Index (SDOI) are used as the vessel parameters for measuring regional collision risk of selected water area.

In recent years, researchers in maritime transportation worldwide have also carried out researches on the prediction of regional collision risk and achieved certain achievements. Nivoliantou at al. (2016) selected the major accidents in the Aegean waters from 2008 to 2012 as historical data, and carried out a Bayesian network-based forecasting study for the vessel's navigational environment risk in the Aegean waters. Fan et al. (2017) extracted the influencing factors from the data of 218 vessel accidents in the research water area in 2013, and combined with the Bayesian network to build a prediction model of collision accident levels in the Yangtze River waters. Kim et al. (2018) used a deep neural network called SETnet, which consisted with convolutional neural network (CNN) and a large amount of historical AIS data to make mid-term and long-term predictions of vessel traffic in congested port waters. Okazaki et al. (2018) used the Support Vector Machine (SVM) method in the study of collision risk prediction of vessels at the exit of the sea route, Fukuto & Imazu (2013) combined the vessel's course prediction work and the Obstacle Zone by Target (OZT) method to determine the collision probability area.

Although the prediction of regional collision risk has been obtained by various methods in previous studies, the prediction of regional collision risk still has certain limitations. Some models require a large amount of historical data of vessel accidents to build a database for predicting the risk of regional vessel collisions, which has a random effect on the real-time quantification and short-term prediction of vessel collision risks in water areas. In addition, the existing models have good effects on mid-term and long-term predictions of regional collision risk, but

DOI: 10.1201/9781003216582-15

there is no corresponding research on real-time and short-term predictions of regional collision risk. The real-time and short-term regional collision risk prediction in water areas with limited data is a shortcoming of current research.

In order to overcome this limitation, in this article, a prediction framework for regional collision risk is proposed. This framework uses limited non-accident data from selected water area to achieve the prediction of regional collision risk more accurately and effectively. Vessel parameters include speed, position, course, length; and the number of vessels entered the selected water area through the traffic lane in a specific period of time, are all extracted from AIS data. The DBSCAN algorithm, improved Shapely value method and recurrent neural network (RNN) divide the framework into two steps. Prediction of the regional collision risk in selected water areas can help the operators of maritime surveillance grasp the trend and value of the regional collision risk more accurately. In addition, in the water area under the charge of maritime surveillance department, the application of the framework can also clearly display the distribution of regional collision risk in future, so as to display areas that deserve special attention.

The following content was arranged as follows: In Section 2, the assessment of regional collision risk was introduced. Section 3 is a detail description of the optimized RNN approach for regional collision risk prediction. In Section 4, a case study to verify the feasibility and effectiveness of the framework is developed. Finally, conclusions, discussions and future perspectives are presented in Section 5.

2 THE ASSESSMENT OF REGIONAL COLLISION RISK

As an approach of expressing the collision risk of vessels in a certain area, the regional collision risk is the overall collision risk formed by all vessels in a certain area. Therefore, the contribution of all vessels in the area to the collision risk is used in the measurement of regional collision risk.

In a certain busy water area that meets the research requirements, plenty of vessels will appear at the same time. To calculate the regional collision risk in the water area at that moment, a clustering algorithm based on spatial density is used for data processing. In order to improve the calculation efficiency of regional collision risk and reduce the computational complexity, this study uses the spatial density clustering algorithm DBSCAN (Density-Based Spatial Clustering of Applications with Noise) to cluster vessels in selected water areas.

The clustering results of DBSCAN algorithm are relatively more accurate when there is no specified number of clustering results, and DBSCAN can find clusters of any shape in the data with noisy points.

This study used the DBSCAN clustering algorithm in the process of quantifying the regional collision risk in selected water area, which improved the efficiency and visualization of the quantization process, while reducing the amount of calculation and the complexity of the calculation. After clustering with DBSCAN, several clusters will be obtained in the area as shown in Figure 1.

The vessels in this area are divided into several clusters. Therefore, the cluster's collision risk is obtained by calculating the collision risk of every single vessel based on Eq. (1), and then the regional collision risk is calculated from the cluster's collision risk based on Eq. (2).

$$CRC = \sum_{i=1}^{n} CRi * Wi \qquad (1)$$

where CRC refers to collision risk of each cluster, CR_i refers to collision risk of every single vessel and W_i refers to every single vessel's contribution.

$$RCR = \sum_{j=1}^{m} CRC_j \times W_j \qquad (2)$$

where RCR refers to regional collision risk, CRC_j refers to collision risk of every single cluster and W_j refers to every single cluster's contribution.

There are n vessels in each cluster, and any vessel in the cluster and other vessels form n vessel pairs. The collision risk of each single vessel in the cluster can be determined by summing all collision risk of corresponding vessel pairs. The collision risk of all vessel pairs of every single vessel are summed and expressed as:

$$CRi = \frac{1}{n-1}(CR_{i1} + CR_{i2} + CR_{i3} + \cdots + CR_{in})$$
$$(3)$$

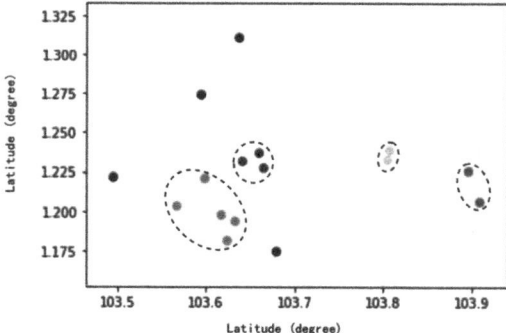

Figure 1. Vessel clusters.

140

where CR_i is the collision risk of Vessel i, CR_{ij}, j=1, 2, ..., n is the collision risk of vessel pairs which include Vessel i.

In this study, the analytical method was used to calculate the collision risk of the vessel pair. As a method to determine the collision risk through the vessel dynamic elements directly, the analytical method is more objective than the fuzzy logic method based on marine experts. Besides, it is simpler than the method based on artificial intelligence and the calculation results will unaffected by the previous training set. In order to overcome these limitations, this paper uses DCPA, TCPA and SDOI to calculate collision risk to improve the calculation accuracy.

Considering the simplicity and practical application of the model, this study directly uses the AIS data to obtain the essential parameters of vessels, including speed, course, position (longitude and latitude) and length are obtained by decoding and extracting AIS data.

In every vessel pair, a vessel was assigned as own vessel, the other one was target vessel. The vessel parameters including speed, course, longitude, latitude and length of own vessel and target vessel can be expressed as $(v_0, c_0, x_0, y_0, l_0)$ and $(v_t, c_t, x_t, y_t, l_t)$.

Then the relative distance (r), relative bearing (c_b), relative speed (v_r) and relative course (c_r) can be calculated as following:

$$r = \sqrt{(x_t - x_0)^2 + (y_t - y_0)^2} \qquad (4)$$

$$c_b = \begin{cases} \arctan\frac{x_t - x_0}{y_t - y_t} & \text{if } y_t > y_0 \\ \arctan\frac{x_t - x_0}{y_t - y_0} + \pi & \text{if } y_t \leq y_0 \end{cases} \qquad (5)$$

$$v_r = \sqrt{v_0^2 + v_t^2 - 2v_0v_t \cos(c_t - c_0)} \qquad (6)$$

$$c_r = \begin{cases} c_0 + p + \arccos\frac{v_0^2 + v_r^2 - v_t^2}{2v_0v_r}, \\ \qquad \text{if } \sin(c_0 + p - c_t) < 0 \\ c_0 + p - \arccos\frac{v_0^2 + v_r^2 - v_t^2}{2v_0v_r}, \\ \qquad \text{if } \sin(c_0 + p - c_t) \geq 0 \end{cases} \qquad (7)$$

Here in, DCPA, TCPA and SDOI can be calculated as follows:

$$DCPA = r \times |\sin(c_r - c_b - \pi)| \qquad (8)$$

$$TCPA = r \times \cos(c_r - c_b - \pi)/v_r \qquad (9)$$

$$SDOI = \frac{\sqrt{(x'_t - x'_0)^2 + (y'_t - y'_0)^2}}{10(l_0 + l_t)} \qquad (10)$$

The calculation of collision risk of vessel pair by DCPA, TCPA (Zhen, 2017) and SDOI (Liu, 2019) can be expressed in the form of negative exponential equations:

$$\begin{aligned} CR_{pair} = {}& a_{DCPA}\alpha_{DCPA} \exp(-\beta_{DCPA}) \\ & + a_{TCPA}\alpha_{TCPA} \exp(-\beta_{TCPA}) \\ & + a_{SDOI}\alpha_{SDOI} \exp(-\beta_{SDOI}) \end{aligned} \qquad (11)$$

where the sum of a_{DCPA}, a_{TCPA} and a_{SDOI} is 1 and can be set according to the actual situation of selected water area for better accuracy.

After the collision risk calculation of vessel pairs in the cluster is quantified, in order to obtain the collision risk of each cluster more accurately, this paper uses improved Shapely value to determine the summing weight of each vessel pair in every single cluster.

$$s_i = \sum_{\substack{G \subseteq N \\ i \in G}} \frac{(g-1)!(n-g)!}{n!} [A(G) - A(G - \{i\})] \qquad (12)$$

$$p_{fi} = CR_{fi} / \sum_{i=1}^{n} CR_{fi} \qquad (13)$$

$$s'_i = \sum_f s_f p_{ji} s_i / \left(\sum_{i=1}^{n} p_{fi} \times s_i \right) \qquad (14)$$

where S_i refers to the summing weight determined by Shapely value of vessel i, G is the group formed according to vessel i, g represents the number of vessels in group G, N represents the group of all vessels, n refers the vessel number of group N, $A(G)$ refers to the total vessel number of group G and A (G- $\{i\}$) refers to the total vessel number of group G without vessel i. S'_i means the summing weight determined by Improved Shapley value of vessel i, f refers to the influencing factor of collision risk, CR_{fi} refers to the collision risk of the factor f of the vessel i, p_{fi} refers to the weight of the factor for vessel i, and σ_f refers to the influence coefficient which can be determined by maritime experts.

The improved Shapely value method applied here can also be used in the quantification of the summing weight of clusters in Eq. (2), therefore, regional collision risk can be quantified more precise. After calculating regional collision risk at different time points, a series of regional collision risk at different time points in selected water area is obtained. This study uses a RNN to predict regional collision risk of selected water area in the following content as the second step of the framework.

3 THE RNN PREDICTION FRAMEWORK

3.1 Related work of RNN

Recurrent neural networks have always been an interesting and important part of neural networks. Researchers have studied and applied recurrent neural networks since the 1990s (Costa, 1999, Coulibaly & Anctil, 1999, Li, 1999, Liang, 1999, Giles, 1997). As an important method in the field of machine learning and artificial intelligence networks, recurrent neural networks can realize the prediction of sequence labeled data with time. Recurrent neural networks have been applied to a variety of problems, especially those involving ordered data processing. In terms of data prediction based on existing databases, RNN has shown great success, such as such as image processing (Karpathy, 2017, Mou, 2017), language processing (Kolbæk, 2017), prediction problems in different fields (Ma, 2017, Guo, 2017, Shi, 2018, Gao, 2018, Kong, 2019). RNNs contain recurrent connections that make them more powerful than traditional neural networks to model such sequence data. In the application of recurrent neural networks in speech recognition, recurrent neural networks contain network activations that recurrent from the previous time step as input networks to influence the prediction of the current time step. These activations are stored in the internal state of the network which can save long-term temporal context information in principle. This mechanism allows RNNs to take advantage of changing the context according to the history of input sequence dynamically instead of using static context like traditional neural networks. Such an advantage makes RNNs have better performances in prediction and sequence-labeled issues than traditional forward neural networks.

A simple recurrent neural network consists of an input layer, a hidden layer, and an output layer, where X_t is the input, Y_t is the output, H_t is the hidden layer, U_t is the weight of the input value, X_t, and V_t is the weight of the hidden layer H_t. In a recurrent neural network, the input information continuously loops through the structure. The simple structure of recurrent neural network is shown in Figure 2.

The structure of the unfolded recurrent neural network is shown in the Figure 3.

In Figure 3, $X_1,\ldots, X_t$ are inputs at different times, $Y_1,\ldots, Y_t$ are outputs at different times, $H_1,\ldots, H_t$ are hidden layers at different times in the network, and $U_1,\ldots, U_t$ are weights of input values at different times. $V_1,\ldots, V_t$ are the weights of the hidden layers at different times, $W_1,\ldots, W_{t-1}$ are the transfer weights of the hidden layers at different times. It can be seen that the inputs of the recurrent network include not only the current input data, but also the previous input information. The determination of the recurrent network at previous time steps will affect the subsequent determination of following time steps. These continuous massages are saved in the hidden

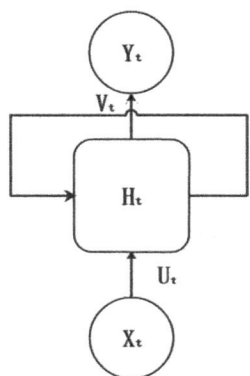

Figure 2. A simple structure of RNN.

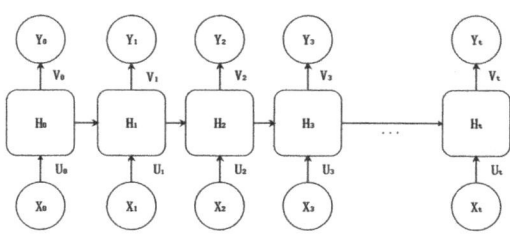

Figure 3. Unfolded structure of a simple RNN.

state of the recurrent network. This hidden state spans multiple time steps and is passed forward layer by layer, which has always affected the network's processing of each new example. At the same time, the hidden state has been constantly corrected. Therefore, the real-time input data and the lasted input data become two input sources of the recurrent neural network. The combination of them will determine how the recurrent neural network processes new data. In traditional feed-forward networks, neurons directly forward information, and the transmitted information will not contact the nodes that have passed through it again. While the recurrent network uses historical information to update the weights in the network. The value of the hidden layer H_t of the recurrent neural network depends not only on the current input X_t, but also on the value of the previous hidden layer H_{t-1}. The weight matrix W_{t-1} is the weight collected by the previous value of the hidden layer and used in the hidden layer of this time. In order to realize the prediction of the regional collision risk in selected water area, RNN was used in this study.

$$Y_t = gV_tH_t \tag{15}$$

$$H_t = f(U_tX_t + W_{t-1}H_{t-1}) \tag{16}$$

3.2 RNN model for regional collision risk prediction

Because the recurrent neural network has the advantage of updating the system weights based on historical input data, this study applies the recurrent neural network in the prediction of the regional collision risk of vessels at continuous time points. In this study, a recurrent neural network with a hidden layer structure is used as an input model with two sets of matrices as inputs and one matrix as output. The inputs include the number of vessels entered the selected water area through the traffic lane in different time periods and the regional collision risk value of the selected water area at different times. Among them, the number of vessels passing through the traffic separation lane in two consecutive time periods is used as a set of 2×2 input matrices, and the regional collision risk value in selected water area at two consecutive time points is used as another set of 2×2 size input matrix. A regional collision risk value at different time points in selected water area is used as the output data in this model, and the recurrent neural network is trained with a certain data sample. The learning results of the recurrent neural network on the samples are used as the basis for predicting the regional collision risk in selected water area.

$$H_t = W_{t-1}H_{t-1} + U_{t1}X_{t,1} + U_{t2}X_{t,2} + Y_{t-1} \quad (17)$$

$$H_{t-1} = W_{t-2}H_{t-2} + U_{(t-1)1}X_{t-1,1} + U_{(t-1)2}X_{t-1,2} + Y_{t-2} \quad (18)$$

$$\hat{Y}_{t-1} = W_{2t-1}H_{2t} + Y_t \quad (19)$$

$$F = (\hat{Y} - Y)/2 \quad (20)$$

The structure of the recurrent neural network used in the prediction model is shown in the Figure 4.

Among them, $X_{1-2t, 1}$, $X_{1-2t, 2}$ are input matrices, Y_{1-t} are 1×1 output matrix, H_{1-2t} are hidden layers, $U_{1-2t, 1}$, $U_{1-2t, 2}$, are the weights of different input matrices, and V_{1-t} are the weights of the hidden layer H_{2-2t}. Therefore, we can get the formula, and the regression

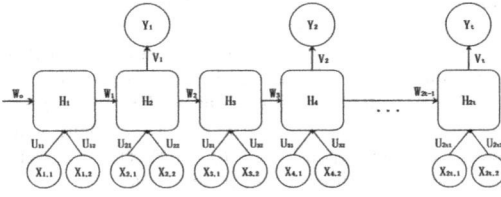

Figure 4. The structure of RNN prediction model.

equation about the output value. The difference between the regression equation of the output value and the output value is used to obtain a new equation shown as Eq. (20). When the new equation approaches to 0, it indicates that the recurrent neural network has achieved good results in learning the samples.

Construct the historical data obtained from Section. 2 as a training database for RNN to learn. By calculating the regional collision risk at different time points, the time distribution of regional collision risk in selected water area can be obtained as inputs with the number of entered vessels at different time periods, therefore, the RNN is used to predict regional collision risk of selected water area at following subsequent time point. The flow chart of this second-step framework is shown in Figure 5.

Finally, the prediction of regional collision risk at a future time point can be calculated in RNN, which helps maritime traffic surveillance department to grasp the regional collision risk in the future and allocate reasonable surveillance force more reasonable.

4 CASE STUDY

4.1 Data selection

To verify the validity of the prediction framework, a case study was performed at the western entrance of the Malacca Strait, which is the busiest water area of Singapore. The water area in this study is selected between 103.4 ° E to 103.6 ° E in longitude and 1.1 ° N to 1.2 ° N in latitude, as shown in Figure 6 and Figure 7.

This area is located at the end of a traffic separation scheme. In this area, the vessel enters the port and anchorage after passing through the traffic lane, or enters the traffic lane from the port and anchorage. Without traffic separation scheme's control, the increase in traffic density and more vessel intersections will lead to greater collision risks. The selected area does not contain anchorages, narrow channels and shallow water areas.

In this study, MMSI, longitude, latitude, speed, heading, and vessel length information were selected from the 27 kinds of dynamic and static information contained in AIS data. After processing the AIS data, they were used to calculate the real-time regional collision risk of selected water area. The selected AIS data is the AIS vessel data received in the selected water area from 1800 to 1900 on January 3, 2014.

4.2 Data optimizing

4.2.1 AIS data screening
The obtained AIS data is decoded and stored in the database, then carry out the work of pre-processing

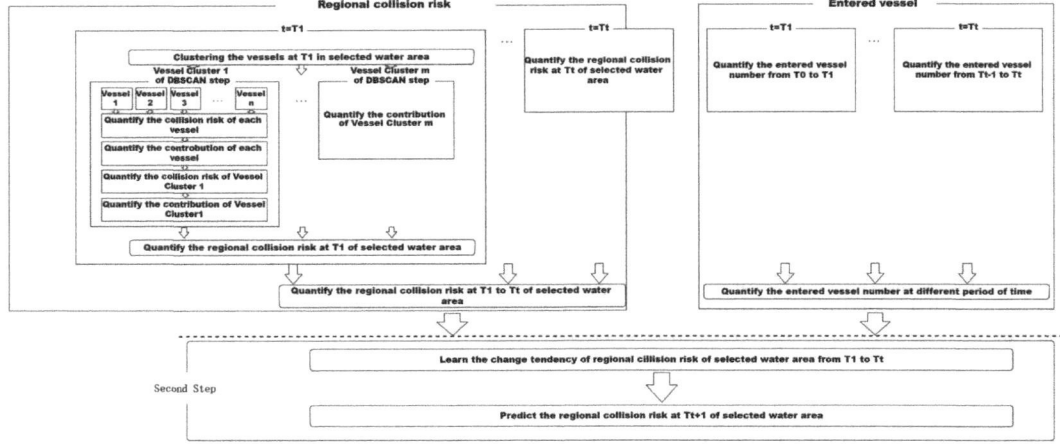

Figure 5. The flow chart of RNN prediction framework.

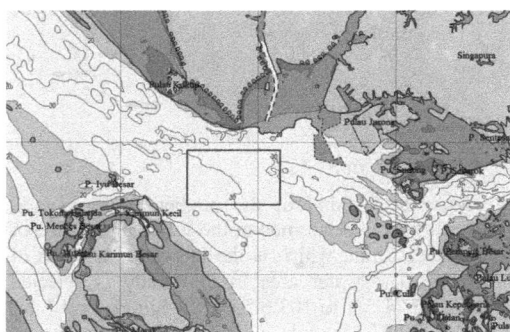

Figure 6. Selected water area-1.

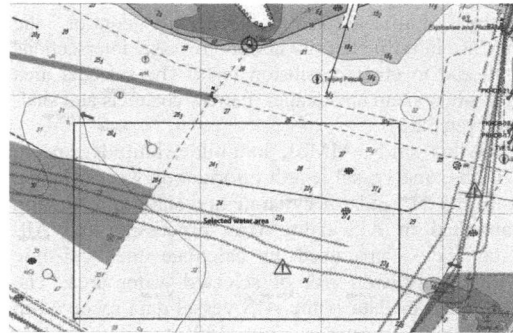

Figure 7. Selected water area-2.

and cleaning to the data, so as to obtain valid AIS data. The main work of the pre-processing is to filter the AIS data on longitude and latitude according to the selected water area position information. Delete the

data with MMSI of 0, and delete the AIS data where the position, speed or course exceeds a reasonable value.

4.2.2 AIS data processing

AIS information is sent discontinuously by different vessels at different time intervals. Because this study needs to calculate the regional collision risk at a specific time in selected water area, interpolation algorithm processing is performed on the filtered AIS data to obtain the different vessels' characteristic information at a specific time.

By collecting AIS data 3 minutes before and after 1800, 1810, …, 1900 time points, the distribution of vessels in selected water area at the time point was optimized and applied in the following study.

4.3 Prediction model application

After the processed AIS information database was established according to the above process, Spatial clustering work was carried out to selected water area at different time point according to database, the results were shown in Figure 8-13.

The regional collision risk at the time of 1800, 1810, …, 1900 in the selected water area was calculated by using the improved shapely value after clustering analysis at each time point, the regional collision values were given in Table 1. And the number of entered vessels through traffic lane to selected water area is shown in Table 2.

In the next steps, the regional collision risk at different time points and the number of vessels entering the selected water area from the traffic lane in different time periods (10 minutes before each time point) were used as one input parameters to construct a recurrent neural network training data set, see Table 3.

The above data was applied to the RNN used in this study, and the parameters' weights were

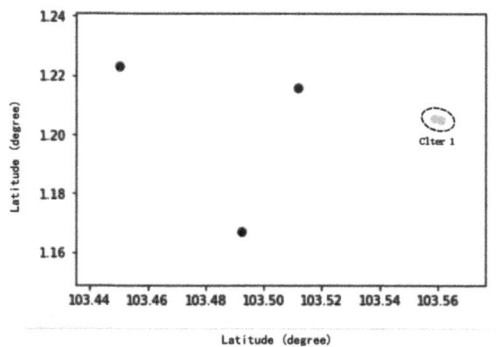

Figure 8. Result of DBSCAN at 1800.

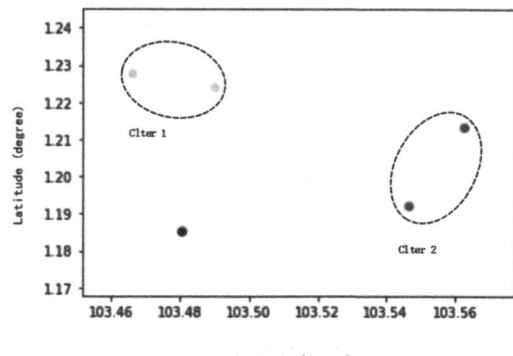

Figure 11. Result of DBSCAN at 1830.

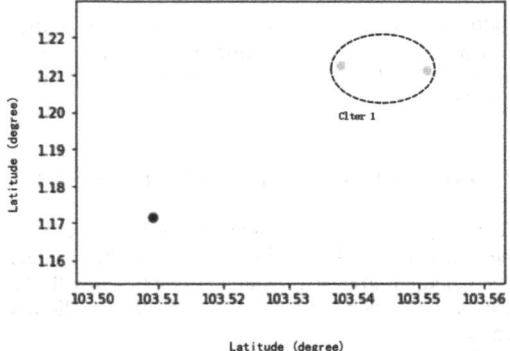

Figure 9. Result of DBSCAN at 1810.

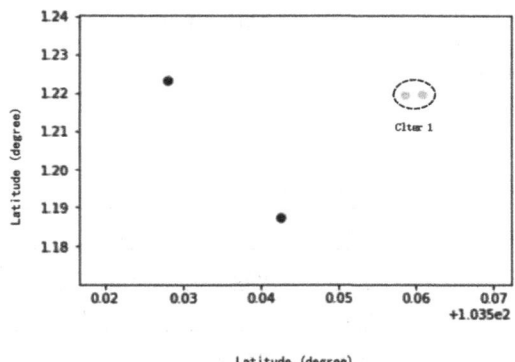

Figure 12. Result of DBSCAN at 1850.

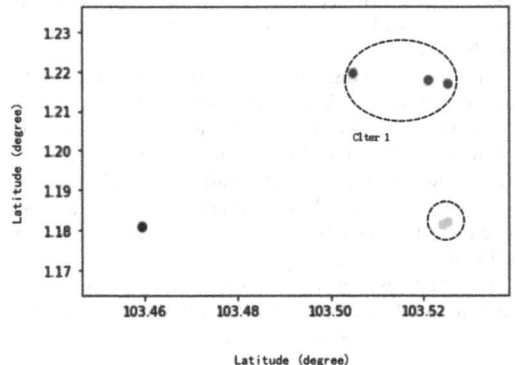

Figure 10. Result of DBSCAN at 1820.

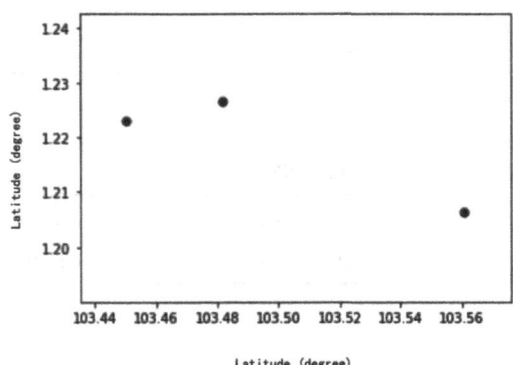

Figure 13. Result of DBSCAN at 1900.

gradually modified, set the number of learning times to 3000 rounds. The training process of RNN is shown as follows Figure 14

It can be illustrated from Figure 14 that the values of the parameter *mae* and parameter loss have changed from large to small, and have gradually stabilized.

The meaning of parameter loss in RNN is shown in Eq. 20, the meaning of *mae* here can be expressed in Eq. 21. The lower the value of *mae,* the better the

145

Table 1. Regional collision risk of selected water area.

Moment	1800	1810	1820	1830
Regional Collision Risk	0	0.93935	0.61104	0.49891
Moment	1840	1850	1900	-
Regional Collision Risk	0	0.83017	0	-

Table 2. The number of entered vessels.

Period	1800-1810	1810-1820	1820-1830
Entered Vessel Number	4	2	1
Period	1830-1840	1840-1850	1850-1900
Entered Vessel Number	0	3	0

Table 3. Input Set of RNN.

Input Set	Set 1,1		Set 1,2	
	0	0.93935	4	2
	Set 2,1		Set 2,2	
	0.939	0.61104	2	1
	Set 3,1		Set 3,2	
	0.611	0.49891	1	0
	Set 4,1		Set 4,2	
	0.499	0	0	3

Table 4. Output Set of RNN.

Output Set	Set1	Set2	Set3	Set4
	0.611	0.49891	0	0.83017

goodness of fit. The lower the value of loss, the better the predictive ability of a model.

$$mae = \frac{1}{n}\sum_{i=2}^{t}|Y_i - Y_{i-1}| \quad (21)$$

Finally, the regional collision risk value at 1840 and 1850 of selected water area, the number of entered vessel in 1840-1850 and 1850-

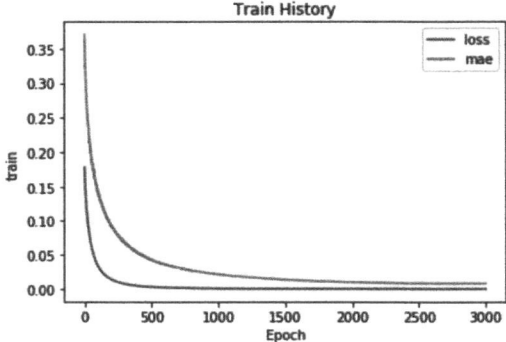

Figure 14. The training process of RNN.

1900 were applied in the previous trained prediction model, therefore, a value of 1900 was calculated. The value of regional risk value at 1900 is 0.11747.

5 VALIDATION OF PREDITION FRAMEWORK

To verify the validity of the proposed framework, the actual regional collision risk of selected waters calculated based on AIS data was compared with the results obtained from the RNN prediction framework.

This study uses the RNN regional collision risk prediction model to predict the regional collision risk at different time points in selected water area from 1800 to 1900 on January 3, 2014. The results obtained from prediction framework and actual value of regional collision risk obtained from historical AIS data are shown in Table 5 and Figure 15.

In the following study, the data set was constructed from AIS data of selected water area at different time points from 1800 to 1900 on 4 and 5 January 2014, and the number of vessels entered the selected waters through traffic lane in different time periods. The RNN prediction framework proposed in this study was used to predict the regional collision risk respectively.

The data set constructed based on the AIS historical data on 4 and 5 January 2014 is shown in Tables 6, 7, 8 and 9. The RNN prediction framework

Table 5. 03-Jan-14 Regional collision risk.

03-Jan-14					
Regional Collision Risk	1820	1830	1840	1850	1900
Actual Value	0.611	0.498	0	0.830	0
Prediction Value	0.634	0.339	0.081	0.886	0.117

146

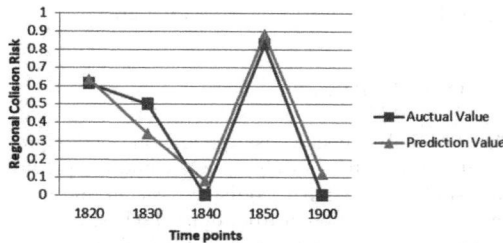

Figure 15. 03-Jan-2014 Regional collision risk.

Table 6. 04-Jan-14 Input Set.

Input Set	Set 1,1		Set1,2	
	0	0.13542	2	1
	Set 2,1		Set 2,2	
	0.13542	0	1	3
	Set 3,1		Set 3,2	
	0	0	3	0
	Set 4,1		Set 4,2	
	0	0.54425	0	3

Table 7. 04-Jan-14 Output Set.

Output Set	Set1	Set2	Set3	Set4
	0	0	0.54425	0

Table 8. 05-Jan-14 Input Set.

Input Set	Set 1,1		Set1,2	
	0	0.61407	1	1
	Set 2,1		Set 2,2	
	0.61407	0	1	2
	Set 3,1		Set 3,2	
	0	1	2	0
	Set 4,1		Set 4,2	
	1	0	0	1

proposed in this study was trained based on the data set, and the factor weights were gradually modified, set the number of learning times as 3000 rounds and 7000 rounds. The training process diagram of RNN

Table 9. 05-Jan-14 Output Set.

Output Set	Set1	Set2	Set3	Set4
	0	1	0	0

is shown in Figures 16 and 17, the parameter *mae* and parameter loss in each figure have changed from large to small, and have gradually stabilized.

In the next step, the prediction models obtained by this approach was used to predict the regional collision risk at 1900 on January 4 and 5, 2014. The predicted and true values for regional collision risk values are shown in Table 10 and 11, Figures 18 and 19.

In the prediction of regional collision risk at 1900 on 3, 4 and 5, January 2014, the prediction results of regional collision risk obtained through the RNN prediction framework is close to the actual value obtained based on AIS historical data. In addition, the prediction results of the prediction framework for the other time points are also closer to the actual values. Comparing the predicted values of regional collision risk with actual historical data, it can be seen from Figures 15, 18 and 19 that the predicted

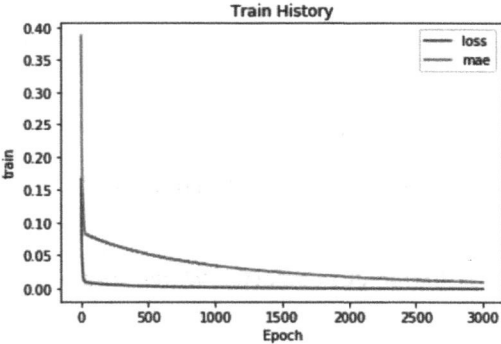

Figure 16. 04-Jan-2014 The training process of RNN.

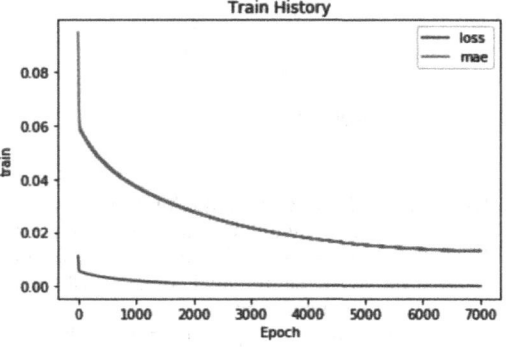

Figure 17. 05-Jan-2014 The training process of RNN.

147

Table 10. 04-Jan-14 Regional collision risk.

04-Jan-14					
Regional Collision Risk	1820	1830	1840	1850	1900
Actual Value	0	0	0.544	0	0
Prediction Value	0.014	0	0.648	0.079	0

Table 11. 05-Jan-14 Regional collision risk.

05-Jan-14					
Regional Collision Risk	1820	1830	1840	1850	1900
Actual Value	0	1	0	0	0
Prediction Value	0.262	0.872	0	0	0.069

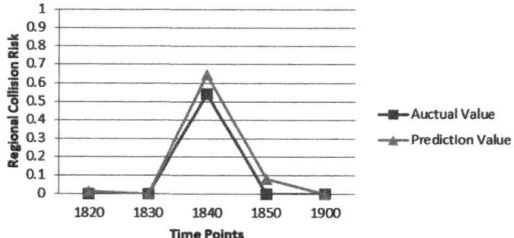

Figure 18. 04-Jan-2014 Regional collision risk.

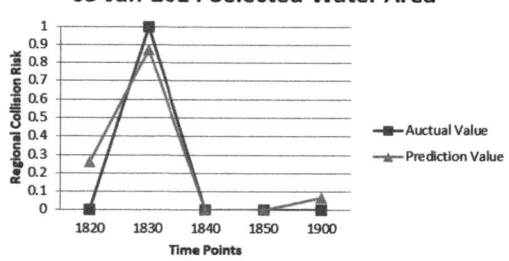

Figure 19. 05-Jan-2014 Regional collision risk.

values obtained from RNN prediction framework are close to the actual values of regional collision risk. The change tendency of predicted regional collision risk value and the actual regional collision risk value have reach a good agreement.

The results of previous application and validation all shown that the RNN prediction framework proposed in this study can effectively predict the regional collision risk in specific water area.

6 CONCLUSIONS

The regional collision risk prediction model proposed in this paper introduces RNN into the prediction study of regional collision risk in maritime transportation. Different from other models, this prediction framework starts from the calculation of the collision risk of a single vessel, uses real-time AIS data in selected water area and other information to predict the regional collision risk in future time point. In addition, the proposed framework can achieve more accurate real-time prediction of regional collision risk, without the limitation of construction of large database or large index set.

The limited size of training data and simple structure of RNN structure applied here can affect the predictions. In order to achieve more accurate prediction results, considering more factors, having more advanced structures and constructing larger training data, optimized RNN algorithms will also be applied in the future work.

ACKNOWLEDGEMENTS

This work is supported in part by the National Natural Science Foundation of China (Grant. 51909022, 61976033), the Natural Science Foundation of Liaoning Provence (Grant. 2019-BS-024), the Key Scientific Research Project of Ministry of Transport of China (Grant. 2019-ZD7-042), and the Fundamental Research Funds for the Central Universities (Grant. 3132019347).

REFERENCES

Costa M., Pasero E., Piglione F., Radasanu D.,1999. Short term load forecasting using a synchronously operated recurrent neural network. *Proceedings of the International Joint Conference on Neural Networks.*
Coulibaly P & Anctil F. 1999. Real-time short-term water inflows forecasting using recurrent neural networks. *Proceedings of the International Joint Conference on Neural Networks.*
Fan S., Sang L. Mao Z.,2017. The prediction of the collision incident level in the lower reaches of the Yangtze River based on the mutual information. *2017 4th International Conference on Transportation Information and Safety (ICTIS).*
Fukuto J. & Imazu H., 2013. Application of "Obstacle Zone by Target (OZT)" Algorithm for Collision Alarm. *The Jounal of Japan Insitute of Navigation.* 128(0), 49–54.
Gao M., Shi G., Li S., 2018. Online Prediction of Ship Behavior with Automatic Identification System Sensor Data Using Bidirectional Long Short-Term Memory Recurrent Neural Network. *Sensors* 18(12), 4211.
Giles C.L., Lawrence S., Tsoi A.C.,1997. Rule inference for financial prediction using recurrent neural networks.

IEEE Conference on Computational Intelligence for Financial Engineering, IEEE Press.

Guo L., Li N., Jia F., Lei Y., Lin J., 2017. A recurrent neural network based health indicator for remaining useful life prediction of bearings. *Neurocomputing.* 240 (2017) 98–109.

Karpathy A., Fei-Fei L., 2017. Deep Visual-Semantic Alignments for Generating Image Descriptions. *IEEE Transactions on Pattern Analysis and Machine Intelligence.* 39(4), 3128–3137.

Kim K. & Lee K.M. 2018. Deep Learning-Based Caution Area Traffic Prediction with Automatic Identification System Sensor Data. *Sensors* 18(9), 3172.

Kolbæk M., Yu D., Tan Z.H., Jensen J., 2017. Multitalker Speech Separation With Utterance-Level Permutation Invariant Training of Deep Recurrent Neural Networks. *IEEE/ACM Transactions on Audio, Speech, And Language Processing.* 25(10).

Kong W., Dong Z.Y., Jia Y., Hill D.J., Xu Y., Zhang Y., 2019. Short-Term Residential Load Forecasting Based on LSTM Recurrent Neural Network. *IEEE Transactions on Smart Grid, Vol.* 10(1).

Liang S.F., Su A.W.Y., Lin C.T., 1999. A new recurrent-network-based music synthesis method for Chinese plucked-string instruments-pipa and qin. *Proceedings of the International Joint Conference on Neural Networks.*

Li S., Wunsch D.C., O'Hair E., Giesselmann M.G., 1999. Wind turbine power estimation by neural networks with Kalman filter training on a SIMD parallel machine. *Proceedings of the International Joint Conference on Neural Networks.*

Liu Z.H., Wu Z.L., Zheng Z.Y., 2019. A cooperative game approach for assessing the collision risk in multi-vessel encountering. *Ocean Engineering* 187(1) 106–175.

Liu Z.H., Wu Z.L., Zheng Z.Y., 2019. A novel framework for regional collision risk identification based on AIS data. *Applied Ocean Research* 89 261–272.

Ma X., Dai Z., He Z., Ma J., Wang Y., 2017. Learning Traffic as Images: A Deep Convolutional Neural Network for Large-Scale Transportation Network Speed Prediction. *Sensors.* 17(4), 818.

Mou L., Ghamisi P., Z., 2017. Deep Recurrent Neural Networks for Hyperspectral Image Classification. *IEEE Transactions on Geoscience and Remote Sensing.* 55(7), 3639–3655.

Nivoliantou Z.S, Koromila I.A, Giannakopoulos T., 2016. Bayesian Network to Predict Environmental Risk of a Possible Ship Accident. *Proceedings of the 7th International Conference on PErvasive Technologies Related to Assistive Environments.* 44, 1–5.

Okazaki T., Terayama M., Nishizaki C., 2018. Feasibility Study for Predicting Collision Possibility Sea Area for Each Ship by Using Support Vector Machine. *2018 IEEE International Conference on Systems, Man, and Cybernetics.*

Shi Z., Xu M., Pan Q., Yan B., Zhang H., 2018. LSTM-based Flight Trajectory Prediction. *2018 International Joint Conference on Neural Networks.*

Zhen R., Riveiro M., Jin Y., 2017. A novel analytic framework of real-time multi-vessel collision risk assessment for maritime traffic surveillance. *Ocean Engineering.* 145, 492–501.

Zhang S., Villavicencioa R., Zhu L., Pedersen P.T., 2019. Ship collision damage assessment and validation with experiments and numerical simulations. *Marine Structures* 63 239–256.

Developments in Maritime Technology and Engineering – Guedes Soares & Santos (eds)
© 2021 Copyright the Author(s), ISBN 978-0-367-77376-2

Maneuvering and operational strategies using AIS data

A. Mujal-Colilles
Department of Nautical Studies and Engineering, Barcelona School of Nautical Studies, UPC-BarcelonaTech

C. Bagés
Department of Civil and Environmental Engineering, Civil Engineering School, UPC-BarcelonaTech

J. Fonollosa
Department of Automatic Control ESAII & Barcelona School of Nautical Studies, UPC-BarcelonaTech

ABSTRACT: Automatic Identification System (AIS) is a transmitter information system mandatory for marine traffic. Although it was originally designed to help vessels in avoiding collisions, AIS data can be used as a decision-making tool to assist harbor authorities. An exact knowledge of the temporal usage of the areas is relevant to harbor authorities to assist operational strategies, risk management, resource assignment and new infrastructures appraisal. Also, maneuvering patterns extracted from AIS data can be further used to estimate harbor basin bed morphodynamics as well as water quality estimations based on the fluxes generated by the maneuvers. The paper presents a preliminary methodology to obtain operational and maneuvering maps using decoded AIS messages with the objective of estimating a usage-index of the areas of the Port of Barcelona.

1 INTRODUCTION

Automatic Identification System (AIS) was included in the IMO SOLAS Agreement as a mandatory vessel information transmitter for vessels with gross tonage larger than 300 GT. Information contained in AIS system is sent from vessel stations using the VHF radio frequency and can be captured either by other vessels or by in-land stations. The information contained in the AIS data was mainly designed to prevent ship collisions (Silveira et al. 2013). However, it has already been used for different purposes, such as ship traffic analysis, (Aarsæther et al. 2009; Rong et al. 2019; Silveira et al. 2013), ship emissions from vessels (Li et al. 2018; Miola et al. 2011) and as a guidance for autonomous vessels (Xu et al. 2019).

From a harbor management point of view, AIS data can be useful to see time-dependent operational areas and maneuvering patterns in order to assist operational strategies, risk management, resource assignment and new infrastructures appraisal. As an example, this contribution presents a preliminary work to analyze raw AIS data in order to create operational maps based on traffic density and manoeuver analysis at the Port of Barcelona.

The Port of Barcelona is one of the main ports in the western Mediterranean Sea. It combines a cargo-oriented design and an in-port city concept since it is also an important port in cruise calling. According to the annual statistic reports of the Port of Barcelona, in 2019 there was an increase between 9% to 15% of

passengers and TEU's respectively, although the total calls only increased 0.7%, which means that larger and more powerful vessels are docking in it. Moreover, the Port of Barcelona is the first port in cruise passengers in the Mediterranean Sea.

2 METHODOLOGY

The AIS system provides a total of 27 messages, each containing specific information that can be automatically broadcasted by the vessel's information (e.g. speed, rate of turn, gpss position and error) or having a human input source from the crew (e.g. estimated time of arrival, ship dimensions). This leads to big amounts of data that should be accurately selected to be properly analyzed. As an order of magnitude, a single inland station ranging up to ~120NM receives can receive around 3MB/h of undecoded messages. However, the most useful information for the previously named purposes is the dynamic information contained in messages 1, 2, 3 and 18 and the static information yielded by messages 5 and 24B.

The Barcelona School of Nautical Studies (FNB-UPC) hosts a Class B-AIS (SeaTraceR AIS Class B Transponder S.287) receiver covering an area ranging from Barcelona to the Balearic Islands (~120NM). Raw data from the receiving system is decoded using a python script to access the specific information contained in each of the messages. Therefore, the

DOI: 10.1201/9781003216582-16

information can be manually selected from raw messages based on the fields contained in each message type. The details of the information contained in each messages can be found in (ITU-R 2014).

In the present contribution only messages of type 1-3, 5, 18 and 24B are decoded using 70% of the information contained. For instance, *Communication State, RAIM Flag*, from messages 1-2-3 and 18 and *AIS Version Indicator* or *Call Sign* from messages 5 and 24B are not considered. Once decoded from raw messages, the information is stored in ASCII files ready to be used for detailed analysis, each type of message in a separate file. As an example, one month of data in raw messages is ~1GB whereas the sum of the ASCII files is slightly larger. Afterwards, each type of messages is filtered following the instructions given in (ITU-R 2014) for bad data in each field, which represented a total of 2% of the initial data.

3 RESULTS AND DISCUSSION

First, we analyzed the time consistency of the messages using the field *Time Stamp*, t_s, present in dynamic messages (this is messages 1-2-3, m1-3 and messages 18, m18). This field contains the information relative to the second when the whole message was sent from the vessel station. In order to see the reliability of this data, and due to the fact that only the second of the sent message is included (missing the minute and hour), the raspberry-pi that receives the raw messages was programmed in order to add the local time when the messages were received.

Surprisingly, 15% of the dynamic data presented errors in t_s, meaning that the second when the message was received differs with the second when the message was sent, which is physically impossible. This discrepancy, which apparently does not seem to cause further errors, can become important when latitude-longitude (lat-lon) data is plotted because the error comes from a bad interrogation from base stations through messages 10-11. However, 80% of this messages with t_s errors are only 1s delayed and will be further considered as good data (see Figure 1).

Next, marine traffic according to ship type and cargo was obtained from AIS data and results are plotted in Figure 2. Passenger ships are described either as passenger type or as vessel-passenger type (Figure 2b) but, as detailed in the introduction this is the main traffic at the Port of Barcelona. There are no significant differences among days of the week when looking into weekdays histograms of ship types.

Operational maps are defined as the amount number of dynamic messages received over 1ha. Figure 3 shows the example of all Saturdays within one month.

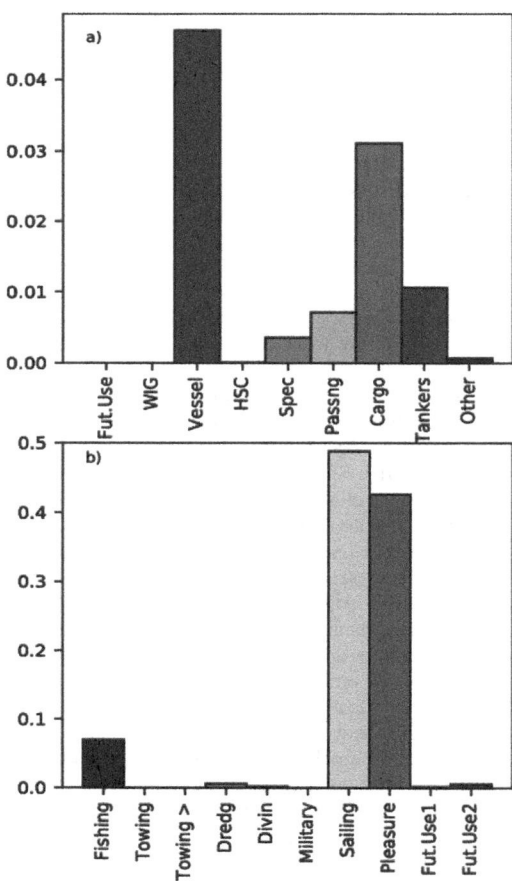

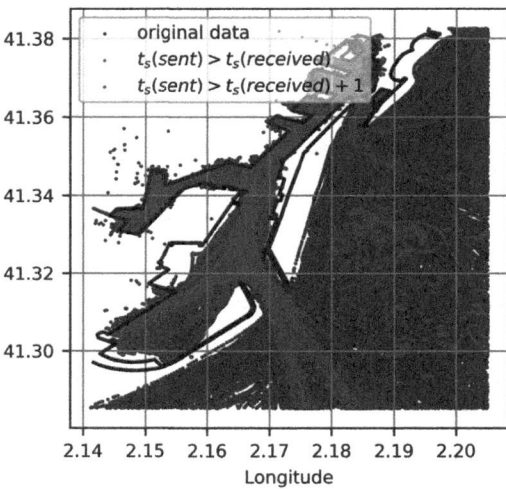

Figure 1. One month AIS messages of type 1-2-3 in blue; green points are messages with differences between the t_s received and t_s sent and red points are messages with differences between the t_s received and t_s sent larger than 1 s.

Figure 2. a) histogram of marine traffic in the Port of Barcelona during one month according to ship types; b) histogram of vessel ship type from Figure 2a.

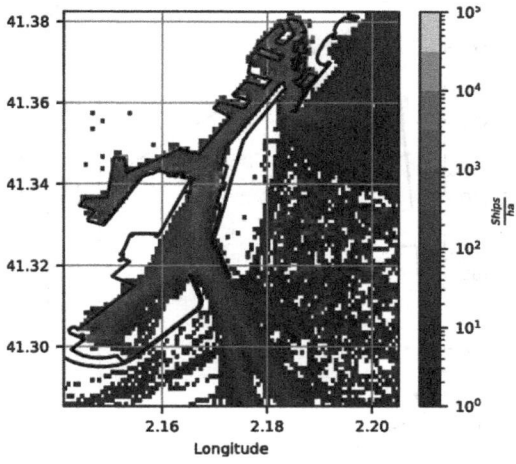

Figure 3. Operational map of the Port of Barcelona. Data is from all Saturdays of a month AIS data.

Also, Figure 3 is an example of bad lat-lon data that was not originally filtered but should not be taken into account. According to the analysis of the own signal sent and received by the FNB-UPC AIS transmitter, the error from lat-lon data is between 8 to 20 meters, but outliers plotted in Figure 3 are far beyond these limits.

Maps of spatial occupation of the Port of Barcelona can also be used to detect main trajectories if temporal-spatial maneuvers are plotted, like in Figure 4. In Figure 4 no vessels with a ship type classification as Fut. Use (Future Use), WIG (Wing-In-Ground) are detected. Surprisingly, there are also no HSC (High Speed Craft) either.

However, green dots in Figure 4 from Specific Vessels are hiding the rest of the data due to the high occupation of the port area. These include pilot vessels, tugs and search and rescue vessels. In particular, pilot vessels and tugs may sail throughout the port area covering it entirely. Therefore, if dynamic messages m1-3 and m18 from Specific Vessels are filtered from AIS data, manoeuvres and trajectories are easily identified as shown in Figure 5. This filter does not affect operational analysis.

From Figure 5 the specialization of harbor basins inside the Port of Barcelona can be concluded following the spatial representation of AIS data. North entrance is mainly used by recreational, local fishing and tourist sailing vessels, most of them carrying a Class B AIS transmitter. Also, cargo vessels use around 70% of the docking infrastructures of the inner side of the harbor coast, bounding tankers in a specific terminal. Passenger vessels, which include cruises use the eastern docking infrastructures. Light blue dots representing the vessels using the southern harbor basin are likely to be cargo and tanker ships being towed to these areas (assuming that crews have changed their ship type entry accordingly).

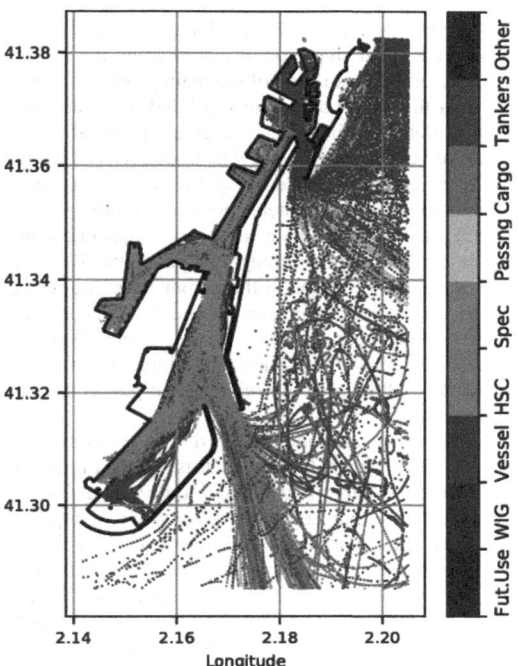

Figure 4. Trajectories of the ships according to ship type of AIS data plotted in Figure 3.

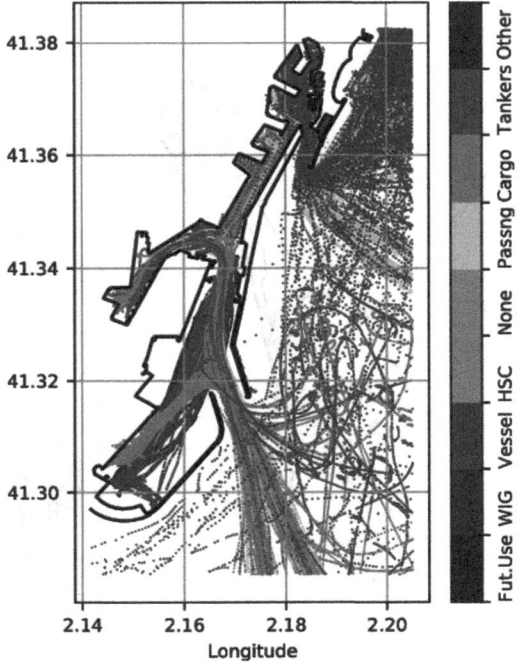

Figure 5. Trajectories of the ships according to ship type of AIS data plotted in Figure 3 with Specific Vessels filtered.

153

Both in Figure 4 and Figure 5 trajectories of ships classified as other ship types are masked by the rest of the trajectories. This is one of the main problems of the AIS data that is not automatically filled: the ship type classification of Other does not give any information about the vessel. However, if AIS lat-lon data of Other ship type vessels is plotted along with vessels engaged in dredging or underwater operations (ship type = Vessel) in Figure 6, Other ship type trajectories are almost coincident with dredging operations. But not only dredging operations in the harbor basin are being carried out in the south part of the Port. There are ongoing construction works in the South Dike to increase docking areas. Therefore, ships classified in AIS data as Other are only related to in-port construction works.

Maneuvers can be analyzed by zooming in a particular harbor basin. In this case, Figure 7 shows the docking and undocking maneuvers of a harbor basin that combines a terminal of regular ro-ro ferry lines and a dry bulk terminal. This figure also shows the spatial error induced by AIS lat-lon data and the need to filter specific data from Vessel type. AIS data also provides the speed, heading and rate of turn of all the points present in Figure 7 that can be very useful to compare with maneuvering models (Sutulo et al. 2019).

This harbor basin was particularly affected by a continuous sediment morphodynamic problem with important consequences on the stability of the infrastructures and reduction of the basin water depth,

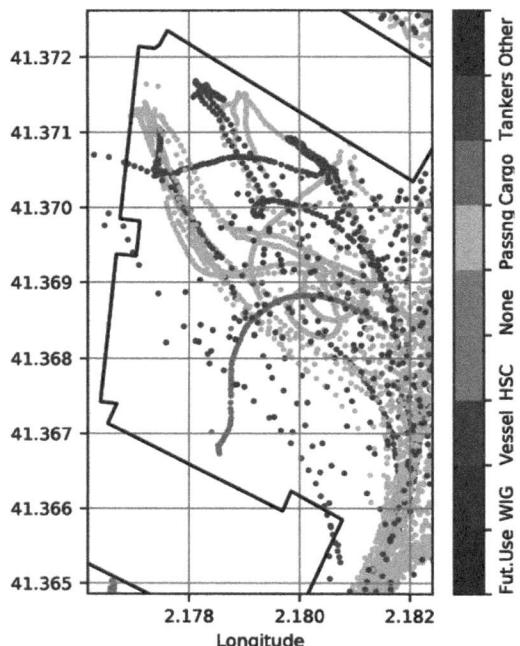

Figure 7. Manoeuvres of a particular harbor basin. Zoom in from Figure 5.

(Mujal-Colilles et al. 2017). The use of maneuvering maps can be helpful to predict areas of sediment accumulation when combining position maps with velocities, heading and rate of turn. For instance, in Figure 7, the lower use of the south and south-west berth indicates that the sediment eroded in the north berth can be deposited in the south area due to i) its low use and ii) the flux generated by vessels docking in the north berth not affecting this area.

4 CONCLUSIONS

It is presented a methodology to extract relevant information from AIS data of vessel stations. This information contains ship type, speed, rate of turn, gps position and error, estimated time of arrival and ship dimensions, among others. A smart visualization and AIS data management can be very relevant and beneficial for harbor authorities to, for example, plan arrival queues, estimated time of departure/arrival to foresee operational areas and maneuvering patterns to assist in operational strategies, risk management, resource assignment and new infrastructures appraisal.

The preliminary results presented in this contribution have been used to detect AIS data errors such as points located in dry areas, time inconsistencies and the use of Other ship type for dredging and construction works operations inside the Port. Moreover, the usage of berthing areas according to ship types can be clearly identified from AIS data.

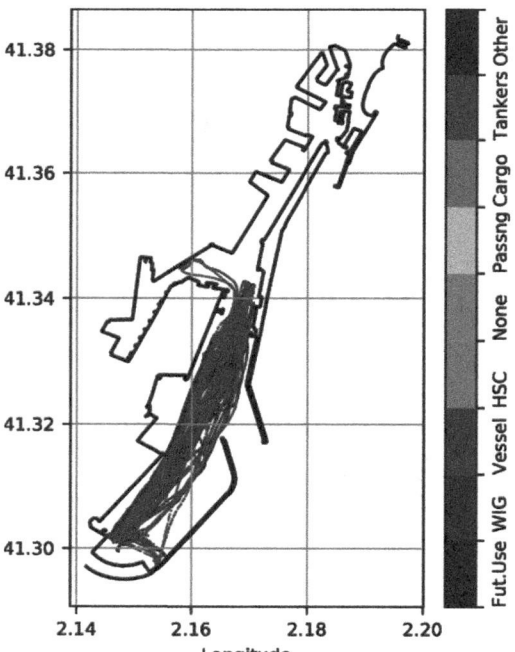

Figure 6. AIS lat-lon data of all Saturdays within one month. Only specific ship type vessels are plotted.

Further work will be performed in these lines including other fields from raw data as, for instance, position accuracy and navigational status.

REFERENCES

Aarsæther, K. & Moan, T. 2009, 'Estimating navigation patterns from AIS', *Journal of Navigation*, vol. 62, no. 4, pp. 587–607.

ITU-R 2014, *Technical characteristics for an automatic identification system using time division multiple access in the VHF maritime mobile frequency band*, International Comunication Union.

Li, C., Borken-Kleefeld, J., Zheng, J., Yuan, Z., Ou, J., Li, Y., Wang, Y. & Xu, Y. 2018, 'Decadal evolution of ship emissions in China from 2004 to 2013 by using an integrated AIS-based approach and projection to 2040', *Atmos. Chem. Phys*, vol. 18, pp. 6075–93.

Miola, A. & Ciuffo, B. 2011, 'Estimating air emissions from ships: Meta-analysis of modelling approaches and available data sources', *Atmospheric Environment*, vol. 45, no. 13, pp. 2242–51.

Mujal-Colilles, A., Gironella, X. & Sanchez-Arcilla, A. 2017, 'Erosion caused by propeller jets in a low energetic harbour basin', *Journal of Hydraulic Research*, vol. 55, no. 1.

Rong, H., Teixeira, A.P. & Guedes Soares, C. 2019, 'Ship trajectory uncertainty prediction based on a Gaussian Process model', *Ocean Engineering*, vol. 182, no. December 2018, pp. 499–511.

Silveira, P.A.M., Teixeira, A.P. & Guedes Soares, C. 2013, 'Use of AIS Data to Characterise Marine Traffic Patterns and Ship Collision Risk off the Coast of Portugal', *Journal of Navigation*, vol. 66, no. 6, pp. 879–98.

Sutulo, S. & Guedes Soares, C. 2019, 'On the application of empiric methods for prediction of ship manoeuvring properties and associated uncertainties', *Ocean Engineering*, vol. 186, no. September 2018, p. 106111.

Xu, H., Rong, H. & Guedes Soares, C. 2019, 'Use of AIS data for guidance and control of path-following autonomous vessels', *Ocean Engineering*, vol. 194, p. 106635.

Developments in Maritime Technology and Engineering – Guedes Soares & Santos (eds)
© 2021 Copyright the Author(s), ISBN 978-0-367-77376-2

Deep representation learning-based vessel trajectory clustering for situation awareness in ship navigation

Brian Murray & Lokukaluge Prasad Perera
Department of Technology and Safety, UiT The Arctic University of Norway, Tromsø, Norway

ABSTRACT: Vessel trajectory clustering using historical AIS data has been a popular research topic in recent years. However, few studies have investigated applying deep learning techniques. In this study, deep representaion learning is investigated for use in clustering historical AIS trajctories to provide insight into navigation patterns to support maritime situation awareness. A recurrent autoencoder and β -variational recurrent autoencoder are investigated to generate fixed size vector representations of the AIS trajectories. Subsequently, clustering is facilitated by applying the Hierarchical Density-Based Spatial Clustering of Applications with Noise algorithm to the representations. The method was tested on historical AIS data for a region surrounding Tromsø, Norway, with successful results. The results also indicate that the β -variational recurrent autoencoder was able to generate representations of the AIS trajectories that resulted in more compact clusters.

1 INTRODUCTION

Maritime situation awareness is an essential aspect of safe maritime operations. Recently, historical Automatic Identification System (AIS) data have been the subject of significant research to aid in intelligent navigation systems (Tu et al. 2017) that can support in providing maritime situation awareness. One area of interest has been identifying navigational patterns for various geographical regions based on historical ship behavior. Knowledge of such patterns can be useful for multiple purposes including anomaly detection, route prediction, path planning, collision avoidance and general maritime situation awareness for ship operators. Multiple studies have addressed identifying such patterns, where trajectory clustering is a central element. Clustering is a field of machine learning concerned with identifying groupings within a dataset. Aarsæther and Moan (2009), for instance, applied computer vision techniques to find groupings of trajectories, and Pallotta et al. (2013) introduced the Traffic Route Extraction and Anomaly Detection (TREAD) methodology. Murray and Perera (2020) also clustered trajectories to predict the future position of a vessel for collision avoidance purposes. Other relevant studies include Zhang et al. (2018) and Zhou et al. (2019).

Many of the studies in the literature apply machine learning to historical AIS data to achieve the desired effect. To effectively conduct clustering, a similarity metric must be calculated to compare the trajectories. This can be challenging as the trajectories may be of variable length. This is not conducive with standard clustering techniques that require data points to be vectors of equal length. Various techniques and metrics have been applied to overcome this challenge, e.g Zhou et al. (2019) which applied Dynamic Time Warping (DTW). These techniques attempt to generate a representation of the trajectories such that they can be compared against each other. Few studies, however, have investigated utilizing deep learning techniques in AIS trajectory clustering. Nguyen et al. (2018) introduced a multi-task deep learning framework based on a variational recurrent neural network trained on historical AIS data for a region. This framework can be used for multiple tasks including trajectory reconstruction and anomaly detection. However, the framework employs a 4-hot encoding of the data that reduces the resolution. Additionally, the architecture does not provide suitable trajectory representations for clustering. Yao et al. (2017), however, investigated AIS trajectory clustering based on deep representation learning. The outlined method gave good results in providing representations of the trajectories, with the clustering based on the deep representations outperforming other non-deep learning based metrics. The method, however, does not investigate more advanced deep learning architectures, and can, therefore, be further improved. Additionally, the study employs the k-means clustering algorithm, which likely reduces the clustering performance compared to non-parametric, density-based approaches that better handle clusters of varying shape and size.

In this study, deep representation learning is investigated to facilitate historical AIS trajectory clustering. A method is presented where deep representations of trajectories are generated for a given geographical region. A recurrent autoencoder is compared with

DOI: 10.1201/9781003216582-17

a more advanced architecture, the β-variational recurrent autoencoder. Using the representations from these architectures, an approach to cluster the trajectories is introduced using a density-based clustering approach that can adapt to a high number of clusters of variable density and shape. The method is applied to a test case, and the results indicate that deep learning can generate powerful representations that facilitate effective clustering. The focus of the study is on the ability of deep learning to generate meaningful representations, and as such aims to determine if such representations are appropriate for use in trajectory clustering.

1.1 Deep representation learning

One of the most powerful aspects of deep learning is its ability to learn meaningful features. Features in this case are a representation of the data, that in the case of deep learning, are learned via the training of the network. Most applications within deep learning deal with classification tasks, and in such cases these features are optimized such that they can be used to discriminate between classes. Deep learning can be split into two main groups, supervised and unsupervised learning. In supervised learning, class labels for each data point are available. In a classification setting, the accuracy of the network in correctly classifying the data is utilized to optimize the parameters in the network. Unsupervised learning on the other hand, deals with cases in which labels for the data are unavailable. In such cases, it is desirable to discover the underlying structure in the data. In machine learning, the task of finding underlying groupings in the data is known as clustering. Deep learning in and of itself does not present a method to cluster the data, but can be utilized in an unsupervised form to generate more meaningful representations, i.e. features, through which the structure of the data is more apparent. Using such new representations, conventional clustering techniques can be applied to the data.

A meaningful representation of the data should be one that preserves information. One architecture that can generate such representations is the autoencoder. The simplest form of an autoencoder is a multi-layer perceptron (Bourlard and Kamp 1988). However, alternative frameworks also exist that utilize either convolutional or recurrent layers. An autoecoder has as its objective to reconstruct the data input to the network. Autoencoders can be separated into two parts, an encoder and a decoder. The encoder produces a latent representation of the input, and the decoder then reconstructs the data form this latent representation.

It is desirable to learn meaningful representations in the latent space of the data. One approach to achieve this is to utilize undercomplete autoencoders (Goodfellow et al. 2016), i.e. where the latent space has a lower dimensionality than the input. Such architectures create a bottleneck in the latent space. In this manner, information must be compressed. By optimizing the network via the reconstruction loss, the network is forced

to learn a meaningful representation of the data that preserves the mutual information between the input data and their latent representations. This data compression can also be viewed as a form of dimensionality reduction. It has been shown that deep autoencoders have the ability to produce much better representations of the data than other dimensionality reduction techniques such as Principle Component Analysis (PCA) (Hinton and Salakhutdinov 2006).

1.2 Recurrent neural networks

Historical AIS trajectories are datasets of multivariate time series. Traditional autoencoders utilize architectures that require a fixed size input for each data point. In the case of trajectory data, each data point is a time series of variable length. As such, in this study it is suggested to utilize the power of recurrent neural networks (RNNs) (Rumelhart et al. 1986) to facilitate effective trajectory representations.

RNNs are designed to process sequences, and are additionally capable of handling sequences of variable length. The networks are ideal for time series as they are able to incorporate temporal information and dependencies. In this sense, they have a form of memory. An RNN can be thought of as an unfolded computational graph, where the recurrent operation is unfolded. This is visualized in Figure 1. Consider the sequence of length L: $\mathbf{x} = \{\mathbf{x}_0, \mathbf{x}_1, ..., \mathbf{x}_L\}$. An RNN cell takes the current state of the sequence, $\mathbf{x}_t$ as well as the previous hidden state $\mathbf{h}_{t-1}$ as input. The cell then outputs the current hidden state, $\mathbf{h}_t$. This process then repeats for all states in the sequence. The output of the cell, i.e. the current hidden state, is fed back into the same cell along with the next state in the sequence. In this manner, the operation is recurrent, as visualized by the red box in Figure 1. In this sense the same cell and parameters are shared between all operations.

The standard RNN, also known as the vanilla RNN, however, encounters challenges during training due to vanishing gradients during backpropagation (Bengio et al. 1994). This prevents the network from learning long-term dependencies. More advanced architectures have, therefore, been introduced to ameliorate the challenge of vanishing gradients. Gated recurrent architectures including Long Short-Term Memory (LSTM) (Hochreiter and Schmidhuber 1997) and Gated Recurrent Unit (GRU) (Cho et al. 2014, Chung et al. 2014) address the issue of vanishing gradients through the introduction of gates. In this

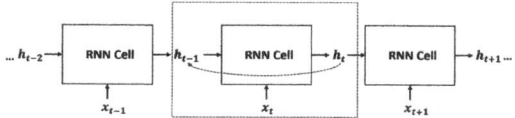

Figure 1. Unfolded RNN structure. The recurrent operation is illustrated within the red box.

manner, the networks are able to learn long-term dependencies in the data.

2 METHODOLOGY

In this section, the methodology utilized to facilitate clustering of historical AIS trajectories is presented. The overall architecture is two-fold, and is illustrated in Figure 2. In the first step, deep learning is utilized to facilitate representation generation. The architecture takes as input historical AIS trajectories for a given geographical region, and outputs a latent representation that can be further processed. In the second step, clustering is conducted on the latent trajectory representations to evaluate the ship traffic in the region. Two deep learning-based approaches are investigated in this study. The first is a recurrent autoencoder (RAE), and the second a β-variational recurrent autoencoder (β-VRAE). These approaches are further described in this section. Using the latent trajectory representations, the Hierarchical Density-Based Spatial Clustering of Applications with Noise algorithm is applied to cluster the trajectories. This study aims to investigate the ability of deep representation learning to facilitate effective clustering of vessel trajectories. As such, the study is limited to the methods described in this section.

2.1 Preprocessing

Prior to applying the representation generation step, a preprocessing of the the historical AIS data is conducted. Each unique vessel trajectory for the given

region is extracted and stored. Each trajectory is a sequence $\mathbf{x} = \{\mathbf{x}_0, \mathbf{x}_1, ..., \mathbf{x}_L\}$ that has been interpolated at one minute intervals. Each state $\mathbf{x}_t$ is a vector of spatio-temporal data, that includes the positional data x, y in UTM-coordinates, the speed over ground v, and the $x-$ and $y-$ components of the course over ground χ_x and χ_y in (1). Each of these values are normalized across the dataset to have values between -1 and 1.

$$\mathbf{x}_t = \left[x, y, v, \chi_x, \chi_y\right] \tag{1}$$

2.2 Recurrent autoencoder

In order to learn good representations of the trajectories, an undercomplete autoencoder structure is investigated. This introduces a bottleneck through which the network must learn the best representation to reconstruct the data. Given that AIS trajectories are sequences of data, recurrent neural networks provide the core architecture applicable to generate meaningful representations. Some of the most popular forms of recurrent neural networks are sequence to sequence models (Sutskever et al. 2014). These provide the basis for many natural language processing tasks such as translation (Cho et al. 2014). The basis of these models is to train an encoder-decoder model in a similar manner to an autoencoder. The encoder takes an input sequence, for instance a sentence in English, and encodes it to a fixed size vector. The decoder then takes the hidden representation of the sentence and generates a target sentence in another language, for instance Spanish. The input and target sequences can be of variable length as well.

If one, however, utilizes a sequence to sequence model to reconstruct its input, i.e. with the target sequence equal to the input sequence, the architecture functions as a recurrent autoencoder (RAE) (Srivastava et al. 2015). The structure of an RAE is visualized in Figure 3. In this case, the input sequence is run through an encoder recurrent neural network. The output of the

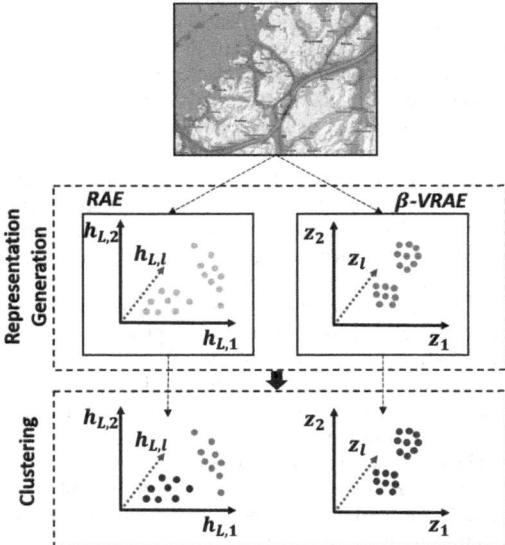

Figure 2. Overview of methodology. Map courtesy of Google Maps (2020).

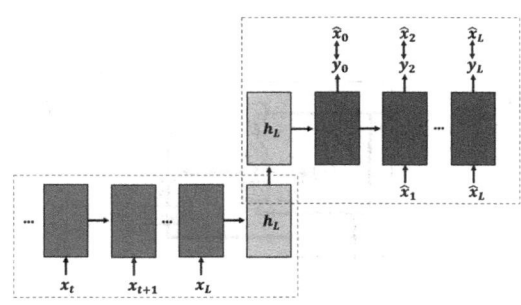

Figure 3. RAE architecture. The encoder is visualized in green, the final hidden state, i.e. latent representation, in yellow, and the decoder in orange.

final cell will have compressed the information in the sequence into a fixed size vector, i.e. the final hidden representation, $\mathbf{h}_L$. $\mathbf{h}_L$ is then fed into a decoder recurrent neural network that predicts one time step at a time. First, the initial state $\mathbf{x}_0$ is predicted using $\mathbf{h}_L$ as input. The following hidden state is then fed into the next cell along with the predicted initial state as input. Each predicted value, i.e. $\mathbf{y}_t$, is estimated using a fully connected layer that takes the current hidden state, $\mathbf{h}_t$ as input. This process then repeats for the remainder of the sequence. The network can be optimized by using the mean squared error as a loss function, where the error between the predicted and true states for each time step are calculated.

By training such an architecture, variable length AIS trajectories can be represented by a fixed size vector via their hidden representations, $\mathbf{h}_L$. This space is equivalent to the latent space for this architecture. When training the recurrent autoencoder, trajectories that have a higher degree of similarity will be in closer proximity to each other than dissimilar sequences in the latent space. Similarity measures can easily be evaluated based on distances in this space, and classical clustering techniques can be applied.

2.2.1 *Gated recurrent unit*

The more recent Gated Recurrent Unit (GRU) cell is a variant of the LSTM that reduces the number of parameters necessary to learn, and is, therefore, investigated for use in this study as the core architecture in the recurrent autoencoder. Each cell in Figure 3 can, therefore, be thought of as a GRU cell. The architecture of the GRU cell is illustrated in Figure 4. The cell takes in the previous hidden state, $\mathbf{h}_{t-1}$, and the current input state, $\mathbf{x}_t$. The reset gate (2) and the update gate (3) regulate what information should be retained or forgotten in the network. Each gate comprises a weight matrix, $\mathbf{W}$, and bias term, $\mathbf{b}$, that consist of parameters that are updated during the training of the network. The output of the operations are fed into sigmoid activation functions. These force the values between 0 and 1, and are multiplied with the various inputs using the Hadamard product, thereby regulating the amount of information that should be

passed on. In this manner, they function as gates, either allowing or preventing information from flowing. (4) calculates a new candidate vector for the hidden state via a hyperbolic tangent activation function. In (5), the hidden state is calculated and passed on to the next cell in the network. The hidden state, $\mathbf{h}_t$, can also used for a prediction, $\mathbf{y}_t$, via (6).

$$\mathbf{r}_t = \sigma(\mathbf{W}_{xr}\mathbf{x}_t + \mathbf{b}_{xr} + \mathbf{W}_{hr}\mathbf{h}_{t-1} + \mathbf{b}_r) \quad (2)$$

$$\mathbf{u}_t = \sigma(\mathbf{W}_{xu}\mathbf{x}_t + \mathbf{b}_{xu} + \mathbf{W}_{hu}\mathbf{h}_{t-1} + \mathbf{b}_{hu}) \quad (3)$$

$$\mathbf{n}_t = tanh(\mathbf{W}_{xn}\mathbf{x}_t + \mathbf{b}_{xn} + \mathbf{r}_t \odot (\mathbf{W}_{hn}\mathbf{h}_{t-1} + \mathbf{b}_{hn})) \quad (4)$$

$$\mathbf{h}_t = (1 - \mathbf{u}_t) \odot \mathbf{h}_{t-1} + \mathbf{u}_t \odot \mathbf{n}_t \quad (5)$$

$$\mathbf{y}_t = \mathbf{W}_{hy}\mathbf{h}_t + \mathbf{b}_{hy} \quad (6)$$

2.3 β-*variational recurrent autoencoder*

As previously mentioned, autoencoders can be powerful in generating meaningful representations of the data they are trained on. Typically, however, the latent space is sparsely populated. This is not an issue for tasks where compression is the goal of the autoencoder, as the scatted data indicate an ideal utilization of the latent space and often lead to better reconstruction results (Spinner et al. 2018). This, however, may be challenging for a clustering algorithm, as data points belonging to the same class may be scattered over a large region in the latent space. The variational autoencoder (Kingma and Welling 2014, Rezende et al. 2014) attempts to limit the chaos in the latent space. This is achieved by forcing latent variables, denoted $\mathbf{z}$, to become normally distributed. The main goal of the variational autoencoder is data generation, as they attempt the learn the underlying distribution of the data, $p(x)$, such that new data points can be generated by sampling from the distribution. However, the resulting latent representations of the data end up being more compact than for standard autoencoders, where similar data are more closely grouped. This may provide a better basis for a clustering algorithm.

The variational autoencoder is a probabilistic version of a traditional autoencoder. It is assumed that the data are generated by a random process utilizing a continuous random variable, $\mathbf{z}$. The general idea is that a value $\mathbf{z}^i$ is generated from a prior distribution $p_\theta(\mathbf{z})$, and a data point $\mathbf{x}^i$ is generated via some conditional distribution $p_\theta(\mathbf{x}|\mathbf{z})$.

The marginal likelihood $p_\theta(\mathbf{x})$ in (7) and posterior density $p_\theta(\mathbf{z}|\mathbf{x})$ in (8) are, however, intractable. As a result, the variational aspect of the autoencoder is introduced in that $p_\theta(\mathbf{z}|\mathbf{x})$ is replaced with an approximation, $q_\phi(\mathbf{z}|\mathbf{x})$. In the context of an autoencoder, the function $q_\phi(\mathbf{z}|\mathbf{x})$ can be thought of as

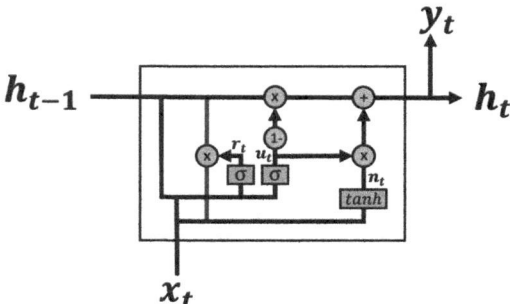

Figure 4. GRU cell. Illustration adapted from colah.github.io.

a probabilistic encoder that produces a distribution over the latent variable, $\mathbf{z}$. $p_\theta(\mathbf{x}|\mathbf{z})$ can, therefore, be thought of as the decoder, taking $\mathbf{z}$, and reconstructing the input, $\mathbf{x}$.

$$p_\theta(\mathbf{x}) = \int p_\theta(\mathbf{z})p_\theta(\mathbf{x}|\mathbf{z})d\mathbf{z} \qquad (7)$$

$$p_\theta(\mathbf{z}|\mathbf{x}) = \frac{p_\theta(\mathbf{x}|\mathbf{z})p_\theta(\mathbf{z})}{p_\theta(\mathbf{x})} \qquad (8)$$

It is assumed that the approximate posterior is a multivariate Gaussian with a diagonal covariance structure according to (11). A neural network is utilized to estimate the parameters of the distribution, i.e. the mean μ_z and standard deviation σ_z. The latent variable, $\mathbf{z}$, is then sampled from this distribution and decoded by $p_\theta(\mathbf{x}|\mathbf{z})$ to generate the reconstructed data, $\mathbf{x}$. A neural network is also used to model $p_\theta(\mathbf{x}|\mathbf{z})$. In Kingma and Welling (2014), a multi-layer perception network was suggested, but alternative architectures have also been proposed.

Fabius and van Amersfoort (2015) introduced the variational recurrent autoencoder (VRAE) by integrating an RNN architecture. As such, the encoder and decoder functions are assumed to be described by RNNs. In this study, a GRU architecture is utilized as the RNN cell. The encoder RNN produes the final hidden state $\mathbf{h}_L$, which compresses the information in the given AIS trajectory. The parameters of the normal distribution in (11) are subsequently calculated via fully connected layers in (9) and (10).

$$\mu_z = \mathbf{W}_\mu \mathbf{h}_L + \mathbf{b}_\mu \qquad (9)$$

$$\sigma_z = \mathbf{W}_\sigma \mathbf{h}_L + \mathbf{b}_\sigma \qquad (10)$$

$$q_\phi(\mathbf{z}|\mathbf{x}) \sim N\left(\mu_z, \sigma_z^2 \mathbf{I}\right) \qquad (11)$$

However, training such an architecture using backpropagation is not possible, as gradients are unable to flow through the sampling operation. The reparametrization trick is, therefore, utilized. Instead of sampling from (11), the latent vector $\mathbf{z}$ is calculated in (13). The sampling effect is achieved by sampling from a noise vector, $\mathbf{e}$, distributed according to (12). This allows for gradients to flow freely. The initial hidden state that is input to the decoder RNN is then calculated according to (14). Subsequently, the decoder RNN reconstructs the data in the same manner as for a conventional recurrent autoencoder. The overall architecture of the variational recurrent autoencoder is visualized in Figure 5.

$$\mathbf{e} \sim N(0, \mathbf{I}) \qquad (12)$$

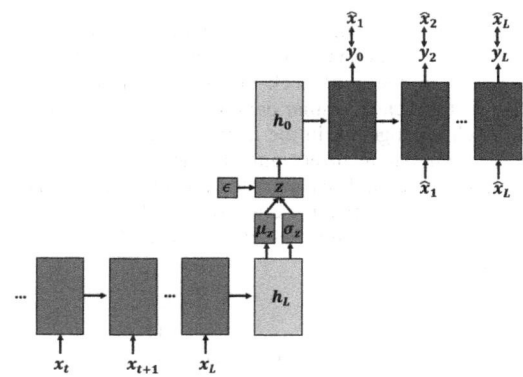

Figure 5. VRAE architecture. The encoder is visualized in green, the final and initial hidden states in yellow, the latent variable in blue and the decoder in orange.

$$\mathbf{z} = \mu_z + \sigma_z \odot \epsilon \qquad (13)$$

$$\mathbf{h}_0 = tanh(\mathbf{W}_{zh}\mathbf{z} + \mathbf{b}_{zh}) \qquad (14)$$

The approximation of the true posterior $p_\theta(\mathbf{z}|\mathbf{x})$ by $q_\phi(\mathbf{z}|\mathbf{x})$ is optimized by maximizing a lower bound on the log-likelihood, i.e. $log(p_\theta(\mathbf{x})) \geq \mathcal{L}(\theta, \phi; \mathbf{x}, \mathbf{z}, \beta)$. The lower bound is defined in (15). In this study, the network is, therefore, trained using a loss function defined by maximizing the lower bound in (15). The first term is the expectation of the decoder function under the approximation of the encoder function. This can be thought of as the likelihood of the reconstruction of the data. As a result, the lower bound is maximized by minimizing the reconstruction loss. In this study, the reconstruction loss is evaluated via the mean squared error.

The second term is the negative Kullback-Leibler (KL) divergence between the approximate posterior distribution and the prior. This means that maximizing the lower bound entails minimizing the KL-divergence. This can be thought of as forcing the distribution of the posterior to be close to the distribution of the prior. In effect, this is a regularization term that encourages the latent variables to have Gaussian distributions. In this study, this is evaluated by assuming the prior is distributed as $p_\theta(\mathbf{z}) \sim \mathcal{N}(0, \mathbf{I})$. This regularization term is weighted according to a factor β as seen in (15). For a variational autoencoder, a value of $\beta = 1$ is utilized.

Higgins et al. (2017) introduced the β-variational autoencoder (β-VRAE), often referred to as a disentangled autoencoder. In such an architecture, values of $\beta > 1$ are utilized when training the network. As such, the regulation term is weighted higher. This further encourages the network to learn compact Gaussian distributions. In this manner, the network is able to learn more ordered, i.e. disentangled, representations, since a stronger constraint is imposed on the bottleneck in the latent representation. In this sense, similar data points are encouraged to be closer together in the latent space. As a result, a β-variational recurrent autoencoder may generate more compact and disentangled representations that result in a more effective clustering of AIS trajectories. This architecture is, therefore, chosen for investigation in this study.

$$\mathcal{L}(\theta, \phi; \mathbf{x}, \mathbf{z}, \beta) = \mathbf{E}_{\mathbf{z} \sim q_\phi(\mathbf{z}|\mathbf{x})}[log(p_\theta(\mathbf{x}|\mathbf{z}))] \\ -\beta D_{KL}(q_\phi(\mathbf{z}|\mathbf{x})||p_\theta(\mathbf{z})) \quad (15)$$

2.4 Trajectory clustering

Trajectory clustering is facilitated via the representations generated in Sec. 2.2 and 2.3. All the trajectories for the region of interest will be run through a forward pass of the encoder for each architecture, resulting in either a hidden state, $\mathbf{h}$, or latent variable, $\mathbf{z}$, representation for each trajectory. Clustering can then be applied to the dataset in either the $\mathbf{h}$- or $\mathbf{z}$-space depending on the representation architecture chosen.

In the representation space, standard clustering techniques can be applied. However, due to the unsupervised nature of the problem, the number of clusters is unknown. This is due to the fact that there could be any number of trajectory routes, or groupings of similar trajectories in the data. A technique that can discover the most likely number of clusters is, therefore, necessary. Additionally, there may exist hundreds of trajectory clusters, and a method that can handle such a dataset must, therefore, be utilized. In this study the Hierarchical Density-Based Spatial Clustering of Applications with Noise (HDBSCAN) algorithm is investigated to facilitate trajectory clustering.

2.4.1 HDBSCAN algorithm

The HDBSCAN algorithm was introduced in Campello et al. (2013). Density-based clustering approaches, such as Density-Based Spatial Clustering of Applications with Noise (DBSCAN) (Ester et al. 1996), are powerful in that they provide a non-parametric clustering approach that is capable of handling clusters of variable shape, in addition to identifying noise in the data. However, DBSCAN requires hyperparameters that determine the sensitivity of the clustering to noise, and constrict the clusters to having similar densities. HDBSCAN introduces a hierarchical approach that allows for the discovery of clusters of varying density, and is, therefore, more powerful for the case of AIS trajectory clustering.

HDBSCAN begins by finding the core distance of each point d_c, i.e. the distance to the kth-nearest neighbor. The value of k is input as a hyper-

parameter. This functions as a local density estimate for the point. The algorithm then calculates a distance metric given by the mutual reachabliity distance d_m, where the distance between two points $\mathbf{x}^i$ and $\mathbf{x}^j$ is calculated in (16).

Using the mutual reachability distance metric, a minimum spanning tree is constructed, and subsequently converted into a hierarchy of connected components. An additional hyper-parameter is introduced that defines the minimum size of a cluster. Clusters in the hierarchy that do not have a size larger than this value are filtered out. It is then desirable to choose the clusters that have the greatest stability in the hierarchy. This choice, however, must be made under the constraint that once a cluster is selected, no cluster that is a descendant of it may be selected. This results in a clustering scheme where the most stable clusters are found, and all points not belonging to these clusters are labeled as noise. For further details, see HDBSCAN.

$$d_m(\mathbf{x}^i, \mathbf{x}^j) = max(d_c(\mathbf{x}^i), d_c(\mathbf{x}^j), ||\mathbf{x}^i - \mathbf{x}^j||_2) \quad (16)$$

3 RESULTS AND DISCUSSION

In this section, the power of deep representation learning to facilitate effective AIS trajectory clustering is investigated in a case study of the region surrounding the city of Tromsø, Norway. One year of historical AIS data from January 1st, 2017 to January 1st, 2018 was utilized. This corresponds to 81033 unique trajectories, each of varying length. Zero-padding was utilized on the trajectories to facilitate batch-training, where each trajectory was padded such that it had a length equal to that of the longest trajectory in the batch. Additionally, reversing the order of the input sequence when training recurrent autoencoder architectures has been found to make the optimization easier, as the model can start by looking at low range correlations seq2seq. Each trajectory, therefore, had its order reversed after padding before being fed to the encoder. The decoder, however, strives to reconstruct the trajectories in the forward direction.

The RNNs utilized in this study had a hidden size of 50 neurons. As a result, the dimensionality of the latent representation for the RAE was equal to the dimensionality of the hidden state, i.e. $\mathbf{h}_t \in IR^{50 \times 1}$. For the β-VRAE, the latent vector was chosen to have a dimensionality of 20, i.e. $\mathbf{z} \in IR^{20 \times 1}$. Furthermore, a value of $\beta = 20$ was utilized for the β-VRAE. A number of iterations using various hyperparameters were conducted, with good results for those presented in this study. PyTorch (Paszke et al. 2019) was utilized to implement the neural networks. All models were trained using the Adam optimizer (Kingma and Ba 2015).

3.1 Deep representation generation

In this section, the deep representations of the historical AIS trajectories are presented. In Figure 6(a), the final hidden representation, $\mathbf{h}_L$, representing the latent space of the RAE is illustrated. Each data point represents a unique trajectory in the dataset, facilitated by a forward pass of the encoder of the RAE. As such, the figure illustrates the distribution of the trajectories in the latent space. The latent space for the RAE has 50 dimensions, as the hidden state is 50-dimensional. The top two principle components of the 50 dimensional data are, therefore, visualized in the figure via PCA. The figure indicates that there are groupings of data discovered by the RAE. These should be easily identified when clustering in this space. However, the data do appear to be spread out significantly, which may prove difficult to cluster effectively. Nonetheless, it appears that the RAE is able to discover meaningful representations of the historical AIS data for the selected region in this case.

In Figure 6(b), the latent distribution of the β-VRAE is illustrated. The latent space in this case is 20-dimensional. In order to effectively visualize the data, the top two principle components are illustrated via PCA. Comparing the deep representations generated by the β-VRAE in Figure 6(b) to those from the RAE in Figure 6(a), it appears that the β-VRAE generated more compact clusters than the RAE. The data are spread out, but appear to be significantly more compact than that in Figure 6(a). The results indicate, therefore, that the β-VRAE is more effective in generating meaningful representations suitable for clustering, due to the increased grouping of data in the latent space.

3.2 Trajectory clustering

Utilizing the trajectory representations in Sec. 3.1, HDBSCAN was applied to the latent distributions of the RAE and β-VRAE with a minimum cluster size of 50. This was deemed a minimum size to be considered a significant cluster of ship trajectories. The implementation in McInnes et al. (2017) was utilized to facilitate the clustering. The results for the RAE and β-VRAE

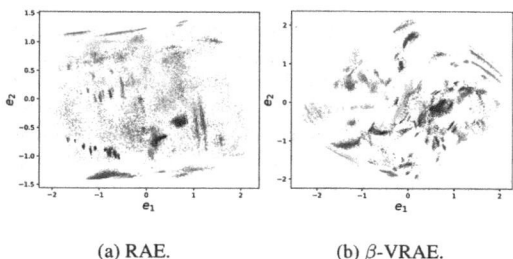

(a) RAE.   (b) β-VRAE.

Figure 7. Clustering results. The data are illustrated using the top two principle components of the latent data, e_1 and e_2.

are illustrated in Figure 7(a) and 7(b) respectively. The reader should note that multiple clusters appear to have similar colors, but are in fact separate. For the RAE, 103 clusters were discovered, and 111 for the β-VRAE. HDBSCAN appears to have effectively clustered the data, where clusters of varying density and shape were discovered. Furthermore, these architectures can be utilized to support anomaly detection, either via the identified noise data from HDBSCAN, or by further analysis of individual clusters. As indicated by the results in Sec. 3.1, however, it appears that representations generated using the β-VRAE resulted in a more effective clustering regime. The RAE generated clusters that were much less dense and defined, whilst the β-VRAE generated more compact and well defined clusters. As a result, the results indicate that a β-VRAE has superior performance with respect to generating deep representations to facilitate clustering.

Clustering using both the RAE and β-VRAE resulted in over 100 different clusters. Therefore, it is very difficult to visualize the clustered trajectories on a map. A subset consisting of 11 clusters from the β-VRAE results are presented in Figure 8 to illustrate the clustering performance. The reader should note that the colors do not match those for the clustering results in Figure 7. Given the

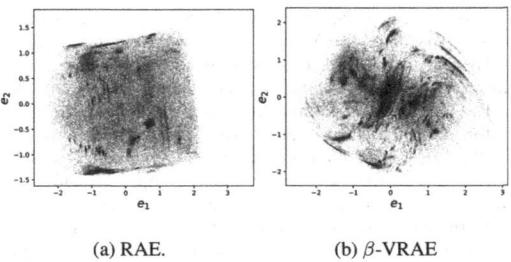

(a) RAE.   (b) β-VRAE

Figure 6. Latent distributions, i.e. deep representations. The data are illustrated using the top two principle components of the latent data, e_1 and e_2.

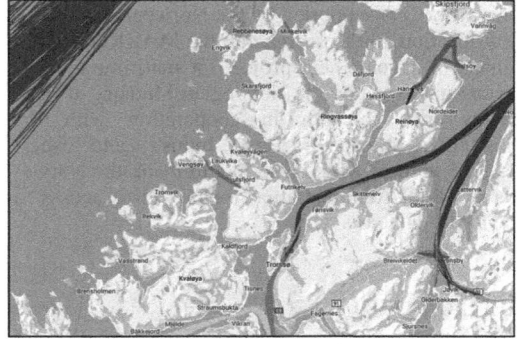

Figure 8. Subset of clustered trajectories for the region of Tromsø, Norway. Map courtesy of Google Maps (2020).

unsupervised nature of the problem, the clustering must be validated by a visual inspection of the raw trajectories on a map. The results indicate that the deep representations in conjunction with the HDBSCAN algorithm resulted in a successful clustering of the trajectories, with the β-VRAE having the best performance.

4 CONCLUSION

In this study, it has been shown that deep representation learning can provide powerful representations of historical AIS trajectories to facilitate vessel trajectory clustering. By utilizing recurrent autoencoder architectures, variable length trajectories can be encoded to a fixed size vector. A recurrent autoencoder and β-variational recurrent autoencoder were compared. Both architectures were found effective in generating groupings of similar trajectories. However, the β-variational recurrent autoencoder provided more compact representations. Using the region of Tromsø, Norway as a test case, the Hierarchical Density-Based Spatial Clustering of Applications with Noise algorithm was applied to the deep representations, providing a successful clustering of the data. Based on a visual inspection of the clusters plotted on a map, the trajectory clusters appear to be meaningful.

The method outlined in this study can be expanded to include more complex architectures such as stacked- and bi-directional recurrent autoencoders that likely will be able to capture more information in their deep representations. Nonetheless, the less complex models described in this study appear to provide a successful clustering of the trajectories for the tested region. Further work will investigate more complex deep learning architectures that can be used in conjunction with trajectory prediction algorithms for collision avoidance purposes.

ACKNOWLEDGMENTS

This work was supported by the Norwegian Ministry of Education and Research and the MARKOM-2020 project, a development project for maritime competence established by the Norwegian Ministry of Education and Research in cooperation with the Norwegian Ministry of Trade, Industry and Fisheries. The authors would also like to thank the Norwegian Coastal Administration for providing access to their AIS database.

REFERENCES

Aarsæther, K. G. & T. Moan (2009). Estimating Navigation Patterns from AIS. *Journal of Navigation 62*(04), 587–607.

Bengio, Y., P. Simard, & P. Frasconi (1994). Learning Long-Term Dependencies with Gradient Descent is Difficult. *IEEE Transactions on Neural Networks 5*(2), 157–166.

Bourlard, H. & Y. Kamp (1988). Auto-association by multilayer perceptrons and singular value decomposition. *Biological Cybernetics 59*(4-5), 291–294.

Campello, R. J. G. B., D. Moulavi, & J. Sander (2013). Density-based clustering based on hierarchical density estimates. In J. Pei, V. S. Tseng, L. Cao, H. Motoda, and G. Xu (Eds.), *Advances in Knowledge Discovery and Data Mining*, Berlin, Heidelberg, pp. 160–172. Springer Berlin Heidelberg.

Cho, K., B. van Merrienboer, D. Bahdanau, & Y. Bengio (2014). On the Properties of Neural Machine Translation: Encoder-Decoder Approaches.

Chung, J., C. Gulcehre, K. Cho, & Y. Bengio (2014). Empirical Evaluation of Gated Recurrent Neural Networks on Sequence Modeling. In *NIPS'2014 Deep Learning workshop*.

Ester, M., H.-P. Kriegel, J. Sander, & X. Xu (1996). A density-based algorithm for discovering clusters in large spatial databases with noise. In *Proceedings of the Second International Conference on Knowledge Discovery and Data Mining*, KDD96, pp. 226–231. AAAI Press.

Fabius, O. & J. R. van Amersfoort (2015). Variational Recurrent Auto-Encoders. In *Proceedings of the International Conference on Learning Representations (ICLR)*.

Goodfellow, I., Y. Bengio, & A. Courville (2016). *Deep Learning*. MIT Press.

Google Maps (2020). Map of Tromsø, Norway.

Higgins, I., L. Matthey, A. Pal, C. Burgess, X. Glorot, M. M. Botvinick, S. Mohamed, & A. Lerchner (2017). beta-vae: Learning basic visual concepts with a constrained variational framework. In *Proceedings of the International Conference on Learning Representations (ICLR)*.

Hinton, G. E. & R. R. Salakhutdinov (2006). Reducing the dimensionality of data with neural networks. *Science 313*(5786), 504–507.

Hochreiter, S. & J. Schmidhuber (1997). Long Short-Term Memory. *Neural Computation 9*(8), 1735–1780.

Kingma, D. P. & J. L. Ba (2015). Adam: A method for stochastic optimization. In *Proceedings of the International Conference on Learning Representations (ICLR)*.

Kingma, D. P. & M. Welling (2014). Auto-Encoding Variational Bayes. In *Proceedings of the International Conference on Learning Representations (ICLR)*.

McInnes, L., J. Healy, & S. Astels (2017). hdbscan: Hierarchical density based clustering. *The Journal of Open Source Software 2*(11), 205.

Murray, B. & L. P. Perera (2020). A Dual Linear Autoencoder Approach for Vessel Trajectory Prediction Using Historical AIS Data. *Ocean Engineering*.

Nguyen, D., R. Vadaine, G. Hajduch, R. Garello, & R. Fablet (2018). A multi-task deep learning architecture for maritime surveillance using ais data streams. In *2018 IEEE 5th International Conference on Data Science and Advanced Analytics (DSAA)*, pp. 331–340.

Pallotta, G., M. Vespe, & K. Bryan (2013). Vessel Pattern Knowledge Discovery from AIS Data: A Framework for Anomaly Detection and Route Prediction. *Entropy 15*(12), 2218–2245.

Paszke, A., S. Gross, F. Massa, A. Lerer, J. Bradbury, G. Chanan, T. Killeen, Z. Lin, N. Gimelshein, L. Antiga, A. Desmaison, A. Kopf, E. Yang, Z. DeVito,

M. Raison, A. Tejani, S. Chilamkurthy, B. Steiner, L. Fang, J. Bai, & S. Chintala (2019). Pytorch: An imperative style, high-performance deep learning library. In H. Wallach, H. Larochelle, A. Beygelzimer, F. Alché-Buc, E. Fox, and R. Garnett (Eds.), *Advances in Neural Information Processing Systems 32*, pp. 8024–8035. Curran Associates, Inc.

Rezende, D. J., S. Mohamed, & D. Wierstra (2014). Stochastic backpropagation and approximate inference in deep generative models. In *Proceedings of the 31st International Conference on International Conference on Machine Learning - Volume 32*, ICML'14.

Rumelhart, D. E., G. E. Hinton, & R. J. Williams (1986). Learning representations by back-propagating errors. *Nature 323*(6088), 533–536.

Spinner, T., J. Krner, J. Grtler, & O. Deussen (2018). Towards an interpretable latent space: an intuitive comparison of autoencoders with variational autoencoders. In *Proceedings of the Workshop on Visualization for AI Explainability 2018 (VISxAI)*.

Srivastava, N., E. Mansimov, & R. Salakhutdinov (2015, feb). Unsupervised Learning of Video Representations using LSTMs. *32nd International Conference on Machine Learning, ICML 2015 1*, 843–852.

Sutskever, I., O. Vinyals, & Q. V. Le (2014). Sequence to sequence learning with neural networks. In *Advances in neural information processing systems*, pp. 3104–3112.

Tu, E., G. Zhang, L. Rachmawati, E. Rajabally, & G.-B. Huang (2017). Exploiting AIS Data for Intelligent Maritime Navigation: A Comprehensive Survey From Data to Methodology. *IEEE Transactions on Intelligent Transportation Systems*, 1–24.

Yao, D., C. Zhang, Z. Zhu, J. Huang, & J. Bi (2017). Trajectory clustering via deep representation learning. In *Proceedings of the International Joint Conference on Neural Networks*, Volume 2017-May, pp. 3880–3887. Institute of Electrical and Electronics Engineers Inc.

Zhang, S.-k., G.-y. Shi, Z.-j. Liu, Z.-w. Zhao, & Z.-l. Wu (2018). Data-driven based automatic maritime routing from massive AIS trajectories in the face of disparity. *Ocean Engineering 155*, 240–250.

Zhou, Y., W. Daamen, T. Vellinga, & S. P. Hoogendoorn (2019). Ship classification based on ship behavior clustering from AIS data. *Ocean Engineering*, 176–187.

Developments in Maritime Technology and Engineering – Guedes Soares & Santos (eds)
© *2021 Copyright the Author(s), ISBN 978-0-367-77376-2*

A comparison of qualitative and quantitative models evaluating intelligent vessel safety

J. Montewka & K. Wróbel
JPP Marine, Szczecin, Poland
Gdynia Maritime University, Maritime Safety Group, Gdynia, Poland

M. Mąka, Ł. Nozdrzykowski & P. Banaś
JPP Marine, Szczecin, Poland
Maritime University of Szczecin, Szczecin, Poland

ABSTRACT: Intelligent vessels are being developed with an ultimate goal of reaching full autonomy in the future. Although detailed means of doing so are still being investigated, it is clear that they will be complex socio-technical systems. In order to analyze their safety, appropriate models mimicking their behaviour must be developed to include not only technical aspects but also liveware ones. These models can be of a qualitative or quantitative nature and their joint use can provide exhaustive results aiming at ensuring that intelligent merchant vessels are operated safely and efficiently. The hereby paper presents and discusses differences in results obtained by applying both types of models as well as benefits of using them. Herein, a system being currently under development is being investigated.

1 INTRODUCTION

With an apparent drive towards intelligent shipping, new solutions emerge. Their aim is to increase either safety, economic efficiency, or both. One of the biggest issues in this respect is to encompass the influence of human factors on the overall performance of shipping. To this end, an effort has been focused on numerous aspects including shore-based supervision (Abilio Ramos et al., 2019; Man et al., 2014) and onboard execution (Cordon et al., 2017; Lazarowska, 2019; Montewka et al., 2017; Porathe, 2019; Smolarek, 2008; Yousefi & Seyedjavadin, 2012) of safety-critical tasks. The impact of human factors was being evaluated and analysed (Ahvenjärvi, 2016; Lutzhoft et al., 2019; Rødseth & Burmeister, 2015).

There are two approaches to safety modelling: quantitative (QN) and qualitative (QL). In general, the former focuses on studying the magnitude of relevant relationships between the components of the system. Meanwhile, the latter consists in studying their causes and consequences. These are often of a complex nature that cannot be quantified easily or described using numbers. Sometimes, a more complex characterization is needed. Reportedly, both types of research methods can be successfully applied to studying any given system or phenomenon. They will however differ in the way that the actual research is performed and how

their outcome can be summarized (Filho et al., 2019; Montewka et al., 2018; Torkildson et al., 2018).

In order to study potential differences between the outcome of both QN and QL research, we investigate the results that were obtained in the course of the safety analysis of an on-board decision support and evaluation system being under development. The objective of the system is to provide assistance to the Officer in Charge of a Navigational Watch (OOW) in his/her efforts to avoid collisions at sea. The system will also evaluate the decisions taken and executed by an OOW.

There exist a number of decision support systems (DSS) for collision avoidance (Lazarowska, 2012; Pietrzykowski et al., 2017; Szłapczyński & Śmierzchalski, 2009; Zhang et al., 2015). These have been developed to various Technology Readiness Levels (TRLs). None of them includes a tool to verify whether the action taken by an OOW complied with the indications of a DSS. This gap is intended to be bridged by the developed and herein-described system, code-named 'Evaluation of Navigation' (EN).

Furthermore, the actual impact of DSSs on the safety of navigation has not been investigated, neither quantitatively nor qualitatively. Summary of such analysis will be presented within the paper together with a comparison of results obtained by an application of both types of methods.

DOI: 10.1201/9781003216582-18

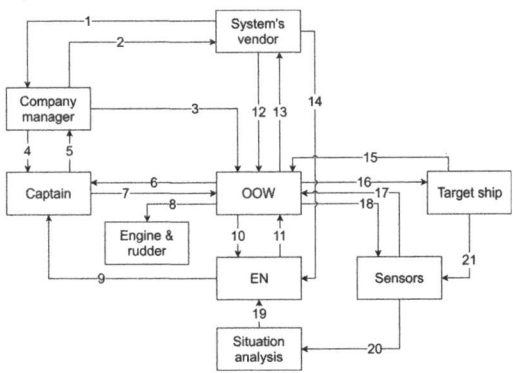

Figure 1. An outline of the EN system, (Montewka et al., 2020).

2 MATERIALS AND METHODS

The envisaged EN system is expected to assist navigator on board in twofold. First in assessing the navigational situation in multiple ship encounters. To this end the EN delivers collision evasive plan accounting for the collision regulation at sea (IMO, 2003), and updates this information as the encounter develops. Second, the EN evaluates the performance of a navigator on board, by collecting and comparing the information on the planned and executed collision evasive actions. Based on that a score is assigned to a given maneuver, which further makes up a safety score for a navigator. Finally, this information is forwarded to the Master and eventually shipping company for consideration. The information on the planned and actually performed maneuvers is expected to be further used for training purposes for bridge personnel, eventually improving their skills.

The general flowchart of the EN system is depicted in Figure 1, where the anticipated relations between the various actors and elements of the system are shown. The arrows point the direction of information flow and data feed while the number denumerates the interaction for further analysis. Note that although simplified, the model is concentrated around the OOW who plays a crucial role in ensuring the safety of navigation.

The major component of the EN system is a situation analysis module based on NavDec® (Pietrzykowski et al., 2017). It is operated by the navigator, collecting relevant data from the own ship's navigation systems such as AIS, ARPA, GPS, log, gyro-compass. Its basic function is to analyse and evaluate the navigational situation within up to 8 nautical miles according to the COLREG. If the given encounter is classified as a developing close-quarters situation, the system suggests a solution by informing a navigator on the required course or speed alteration and time horizon when the action should be taken.

2.1 *Quantitative approach*

The EN is expected to be a decision support tool for an OOW and an instrument for a master and shipping company to assess the performance of the navigator in close-quarter situations. Therefore, it can be expected that the EN will affect the performance of the OOW, thus the safety of navigation, at various levels, from motivation and skills through execution.

It is found that behaviour of the OOW is shaped by the context of the tasks that (s)he performs, and by human nature itself, therefore, it cannot be considered random (Hollnagel, 1998). To evaluate the human performance in a given context we utilize a method for second-generation human reliability analysis (HRA) called CREAM. The method has been developed in the 1990s and is continuously being improved. This method emphasizes the influence of context on human performance and advocates a deeper insight into the characteristics of human performance. By this, a better understanding of the nature of errors arising in the cognitive process of humans can be achieved (He et al., 2008; Sun et al., 2012; Zhang et al., 2019).

For the purpose of the performed research, CREAM was applied to calculate the probability of an accident at sea that stems from the human error in two contextual settings. First case pertains to the situation where the ship is equipped with standard navigational technology. The second case is the the ship equipped with the EN. These are then compared and the relative change in the probability of accident for a ship resulting from the introduction of the EN is obtained. The EN intends to do the following: 1) provide the navigator with the solution for collision evasive action; 2) compare the action taken by the navigator (course alteration/speed change) with the generated solution; 3) score the actions taken and notify navigators' supervisor(s).

To calculate the human error probability (HEP) from CREAM, a task that is analysed here, which is collision avoidance, needs to be decomposed, and the CFP for each sub-task (detection, assessment, action) calculated, see Figure 2.

Then the logic relation between the sub-tasks is to be determined, which translates into the way how the sub-tasks are linked, either through parallel or serial connection. Subsequently, the level of dependency between the sub-tasks shall be evaluated and attributed to one of the two categories: high or low dependence. This results in the HEP for the analysed task.

Subsequently, the areas potentially affected by the EN, called common performance conditions CPC, as well as the direction and strength of the influence need to be determined. To this end, experts' judgment is adopted. A panel of seven experts was consulted, including persons involved in the following industrial and academic fields: ship operation and management, marine IT systems design, risk and safety assessment of sociotechnical systems.

Finally, the effect of the EN on safety is measured through the anticipated changes on the probability of an accident, before and after the prospective implementation of EN onboard the analyzed ship.

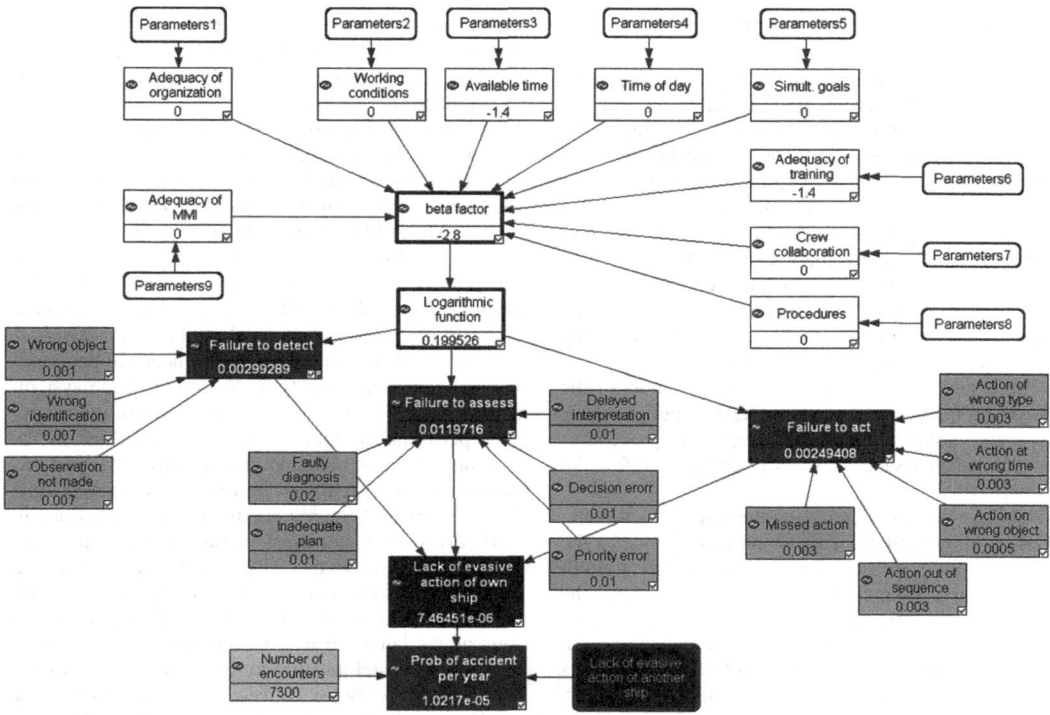

Figure 2. A structure of framework estimating the probability of accident, (Montewka et al., 2020).

2.2 Qualitative approach

System-Theoretic Process Analysis (STPA) is a tool to analyze complex socio-technical systems with focus on ensuring their safety in all circumstances. It is rooted in System-Theoretic Accident Model and Process (STAMP), an approach to safety in which the safety is controlled through proper interactions between system's components (Leveson & Thomas, 2018). Within STAMP/STPA, it is postulated that the solution to better model safety and help avoid accidents may lie within a holistic, systemic approach to safety. Moreover, systems or processes are viewed as nets of interactions between components and not as nets of components themselves. As a result, it is postulated that enforcing that interactions (also referred to as 'control actions' and 'feedbacks') remain as intended throughout the system lifetime is crucial and sufficient for maintaining the overall safety of a system (Leveson, 2011).

Such enforcements may be based on different foundations and consist of different solutions, such as organizational or technical. These solutions are sometimes referred to as 'mitigation measures' as their purpose is to mitigate interactions' shift towards improper interactions (Wróbel et al., 2018). It is acknowledged that systems have a natural tendency to shift towards unsafe conditions, for instance due to wear and tear, corrosion, personnel loss of skill etc. It can be noted that especially for technical solutions, they may consist in ensuring that given component maintains its reliability – an approach to safety in its reliability-oriented form.

As a first step of the analysis, a hierarchical control structure (HCS) is elaborated to depict the functional properties of the system in question. Its simplified model can be seen in Figure 1. Herein, relevant components of the system are connected by interactions to depict how they can affect others in virtually any circumstances. Each of these control actions is then analyzed with focus on how it can be incorrect, why and how to ensure its correctness. Within hereby report, seven potential categories of interactions' incorrectness are applied as below:

1. Interaction not provided;
2. Interaction provided when not expected;
3. Interaction provided too early;
4. Interaction provided too late;
5. Interaction stopped too soon;
6. Interaction provided too long;
7. Interaction provided incorrectly.

Upon determining whether a given category of incorrectness is applicable to an interaction in question and how such incorrectness can occur, potential ways of mitigating it are elaborated. This is normally achieved through literature review or expert elicitation. Herein, the same panel of seven experts has been elicited through a dedicated workshop.

3 RESULTS AND DISCUSSION

The results of quantitative analysis (QN) are presented in Figure 3, while the outcome of qualitative analysis (QL) are presented in Figure 4. The QN indicates the reduction of the human error probability can be expected on board ship as a result of EN installation. However, its extent remains unknow, as well as the number of anticipated error producing factors and their combinations. Therefore the results shall be presented in a form of interval, where the updated probability of accident is expected to fall, rather than a single tone value, which does not reflect the associated uncertainty (Aven et al., 2014; Bjerga et al., 2016).

The results of QL analysis reveal, that most of the mitigation measures focus on OOWs' performance, skills and ability to follow operational procedures issued by his/her supervisors. Supporting therefore the QN analysis.

In both methods of analysing the safety of a system, OOW is in a spotlight. This is most likely due to the fact that (s)he is normally responsible for the safety of navigation including collision avoidance. Although the process is a complex one (Gil et al., 2019) and multiple actors are involved, the OOWs are in charge of a direct decision-making and execution. With OOWs being the most important line of defense, their performance highly influences this of the entire system.

Within CREAM, only the performance of a person is calculated while in STPA it is embedded in the overall performance of the system. Nevertheless, OOW remains its central component. However, the approaches applied must have differed as can be seen through comparing Figure 1 and Figure 2. These differences make the results of the QN and QL approaches not easily comparable one to another at the first sight. As it can be noted that white background events (Figure 2) such as *Working conditions* or *Procedures* relate to *Company manager* or *Captain* influencing OOW's daily routine (Figure 1, 3). Meanwhile, black-background events mark components which an OOW can have influence on. One of the reasons for this incomparability is that QN approach analyses *events* (Wrobel et al., 2016) while

QL focuses on *interactions*. Respective models of the systems must have been developed accordingly. One of potential solutions of this issue may lay in the application of fuzzy grey mapping (Navas de Maya et al., 2019; Xue et al., 2019). However, these two approaches can be seen complementary, as the QL yields the detailed structure of the analysed subject, which can be then inputted to QN model, (Utne et al., 2020). Often some reduction or combination of elements of the structure as obtained in the course of QL may be necessary, but then it provides the wide picture of the subject, allowing definition of relevant contributing factors, that may be omitted otherwise.

While CREAM calculates the influence of particular factors on the total safety performance of the system, STPA aims at identifying potential ways of improving such performance. Due to the rationale behind STPA making it focused on actual interactions between components of the system, it is impossible to include vague factors such as 'Working conditions' as it was done in CREAM. However, a consideration of such factors can be introduced in further steps of the analysis where mitigation measures are elaborated. Herein, 'Working conditions' take a form of 'Operational procedures', as long as the latter are well thought of, implemented to a good effect and adhered to. Such procedures do not need to cover only the collision avoidance itself. They might include requirements not directly related to watchkeeping, just to name regulations on a rest time (International Labour Convention, 2006) so that OOWs working conditions are satisfactory.

As a result, QN methods can help determine whether a developed system can improve the safety of navigation. Meanwhile, QL methods deliver potential solutions on how to achieve such improvements.

However, it is difficult to predict how effective the solutions developed through QL will be in achieving the safety increase predicted through QN. Since both approaches have their own methodologies, assumptions and rationale, the results are not directly comparable. Therefore, the impact of elaborated mitigation measures should rather be measured *a posteriori*. Moreover, some additional mitigation measures not previously elaborated can be implemented into the system during its operations (Weber et al., 2018). These can be based on user experience and tacit knowledge. By this, QN calculations can serve as a method of preliminarily assessing whether the system being developed could succeed in reducing the operational risks.

The uncertainties can result from:

- inaccurate estimation of risk influencing factors - QN;
- system model inconsistency with its actual layout (due to the fact the model is created in an initial stage of system development) - QN, QL;
- incompleteness of the list of mitigation measures (for same reasons as above) - QL;
- experts bias, framing or lack of sufficient insight - QN, QL.

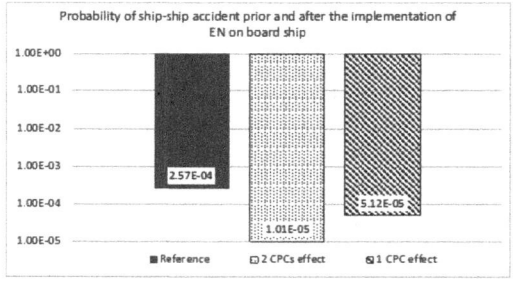

Figure 3. Expected impact of EN on operational risk, (Montewka et al., 2020).

Interaction number	Interaction description	Incorrectness category applicable							Mitigation measures
		1	2	3	4	5	6	7	
1	System documentation								Documentation to be delivered upon installation and request
2	Operational requirements								Must be clearly stated
3	Training on the use of system								Complete training to be provided to all OOWs; operational procedures to be followed
4	Operational procedures								Must be comprehensive and issued in advance
5	Operational feedback								Crew's participation in system to be welcome; procedures to be implemented
6	Operational reports								Operational procedures to be implemented and followed
7	Operational orders								Must be issued in accordance with operational procedures and industry standards; must be followed at all times unless causing a hazard
8	Set-points								OOW must be properly trained in both maritime operations and engineering; operational procedures to be implemented and followed
9	Evaluation-related data								OOW performance limits to be elaborated during system development and fine-tuned; software testing
10	Manually input data								Proper training to be provided to OOW on the operations of the system
11	Anti-collision solution recommendation								Sensor data feed to be secured; operational procedures to be followed;
12	Training on the use of system								Complete training to be provided to all OOWs; operational procedures to be followed
13	Suggestions for improvement								OOWs' participation in the development of system to be welcome; user-experience meetings to be held
14	Updates								Code integrity to be verified; software to be tested before update dissemination
15	Visual look-out								Operational procedures to implemented and followed
16	Communication								COLREG and operational procedures to be followed; OOW to keep in mind potential for misunderstandings
17	Data read and verification								Proper training to be provided to OOWs; operational procedures to be followed
18	Settings								OOW to familiarise himself/herself with the operations of installed equipment
19	Data feed								Redundant wiring; software compatibility testing
20	Data								Implementation of highly-reliable sensors; redundant wiring
21	Sensing								Sensors to be properly set-up; sensors to be installed in accordance with industry standards/legislation
Total occurrences		21	8	3	18	15	1	20	

Figure 4. Results of QL analysis.

4 CONCLUSIONS

The performed research formed a part of the development process of a decision support and evaluation system. The purpose of the research was to estimate the potential gains the system can produce in terms of operational safety and elaborate hazard mitigation measures helpful in doing so. In this respect, both methods produced valuable results that could easily be implemented into the process of system development. It may therefore be raised that using both qualitative and quantitative approaches can help gain a comprehensive insight into the safety features of a system.

However, at the present stage it remains debatable whether the results could be compared directly. Even if hazard mitigation measures as elaborated are implemented into the system it does not necessarily mean that the operational risks are reduced to the extent quantified. The reasons for this can be abundant.

While quantitative research in safety usually focuses on identifying risks of potential losses occurrence, qualitative research is focused on finding ways of preventing such losses. It may therefore be scientifically interesting to validate the results of quantitative part of the study and see if the mitigation measures identified and implemented actually led to the increase of safety to an extent predicted. This may however be difficult to achieve as some hazard mitigation techniques are introduced into the system during its operation. These are often unpredictable and so their impact cannot be estimated.

ACKNOWLEDGEMENTS

The research presented was supported by the National Centre for Research and Development, Contract No POIR.01.01.01-00-0860/17.

REFERENCES

Abilio Ramos, M., Utne, I.B., Mosleh, A., 2019. Collision avoidance on maritime autonomous surface ships: Operators' tasks and human failure events. Saf. Sci. 116, 33–44. https://doi.org/10.1016/J.SSCI.2019.02.038

Ahvenjärvi, S., 2016. The Human Element and Autonomous Ships. TransNav, Int. J. Mar. Navig. Saf. Sea Transp. 10, 517–521. https://doi.org/10.12716/1001.10.03.18

Aven, T., Zio, E., Baraldi, P., Flage, R., 2014. Uncertainty in Risk Assessment: The Representation and Treatment of Uncertainties by Probabilistic and Non-Probabilistic Methods. John Wiley & Sons, Ltd.

Bjerga, T., Aven, T., Zio, E., 2016. Uncertainty treatment in risk analysis of complex systems: The cases of STAMP and FRAM. Reliab. Eng. Syst. Saf. 156, 203–209. https://doi.org/10.1016/j.ress.2016.08.004

Cordon, J.R., Mestre, J.M., Walliser, J., 2017. Human factors in seafaring: The role of situation awareness. Saf. Sci. 93, 256–265. https://doi.org/10.1016/j.ssci.2016.12.018

Filho, A., Jun, G., Waterson, P., 2019. Four studies, two methods, one accident—another look at the reliability and validity of Accimap and STAMP for systemic accident analysis. Saf. Sci. 113, 310–317. https://doi.org/10.1016/j.ssci.2018.12.002

Gil, M., Wróbel, K., Montewka, J., 2019. Toward a Method Evaluating Control Actions in STPA-Based Model of Ship-Ship Collision Avoidance Process. J. Offshore Mech. Arct. Eng. 141. https://doi.org/10.1115/1.4042387

He, X., Wang, Y., Shen, Z., Huang, X., 2008. A simplified CREAM prospective quantification process and its application. Reliab. Eng. Syst. Saf. 93, 298–306. https://doi.org/10.1016/j.ress.2006.10.026

Hollnagel, E., 1998. Cognitive Reliability and Error Analysis Method (CREAM). Elsevier Science.

IMO, 2003. COLREG: Convention on the International Regulations for Preventing Collisions at Sea, 1972. International Maritime Organization.

International Labour Convention, 2006. Maritime Labour Convention. International Labour Organization, Geneva.

Lazarowska, A., 2019. Research on algorithms for autonomous navigation of ships. WMU J. Marit. Aff. 81–87. https://doi.org/10.1007/s13437-019-00172-0

Lazarowska, A., 2012. Decision support system for collision avoidance at sea. Polish Marit. Res. 19, 19–24. https://doi.org/10.2478/v10012-012-0018-2

Leveson, N.G., 2011. Engineering a Safer World - Systems Thinking Applied to Safety. MIT Press, Cambridge, MA.

Leveson, N.G., Thomas, J.P., 2018. STPA Handbook. https://doi.org/10.2143/JECS.64.3.2961411

Lutzhoft, M., Hynnekleiv, A., Earthy, J. V, Petersen, E.S., 2019. Human-centred maritime autonomy - An ethnography of the future, in: Proceeding of MTEC/ICMASS 2019. Trondheim. https://doi.org/10.1088/1742-6596/1357/1/012032

Man, Y., Lundh, M., Porathe, T., 2014. Seeking Harmony in Shore-based Unmanned Ship Handling-From the Perspective of Human Factors, What Is the Difference We Need to Focus on from Being Onboard to Onshore? Adv. Hum. Asp. Transp. Part I 7, 231.

Montewka, J., Goerlandt, F., Innes-Jones, G., Owen, D., Hifi, Y., Puisa, R., 2017. Enhancing human performance in ship operations by modifying global design factors at the design stage. Reliab. Eng. Syst. Saf. 159, 283–300. https://doi.org/10.1016/j.ress.2016.11.009

Montewka, J., Wrobel, K., Heikkilä, E., Valdez Banda, O. A., Goerlandt, F., Haugen, S., 2018. Challenges, solution proposals and research directions in safety and risk assessment of autonomous shipping, in: PSAM 14th Probabilistic Safety Assessment and Management Conference. Los Angeles.

Montewka, J., Wróbel, K., Mąka, M., Nozdrzykowski, Ł., Banaś, P., 2020. Quantitative model evaluating the effect of novel decision support tool on the probability of ship-ship accident, in: Paper Submitted to ESREL&PSAM Conference in Venice, 21-26.06.2020.

Navas de Maya, B., Kurt, R.E., Turan, O., 2019. Marine Accident Learning with Fuzzy Cognitive Maps (MALFCMs): A Case Study on Fishing Vessels, in: 29th European Safety and Reliability Conference. Hannover, pp. 331–337. https://doi.org/10.3850/978-981-11-2724-3

Pietrzykowski, Z., Wołejsza, P., Borkowski, P., 2017. Decision Support in Collision Situations at Sea. J. Navig. 70, 447–464. https://doi.org/10.1017/S0373463316000746

Porathe, T., 2019. Maritime Autonomous Surface Ships (MASS) and the COLREGS: Do We Need Quantified Rules Or Is "the Ordinary Practice of Seamen" Specific Enough? TransNav, Int. J. Mar. Navig. Saf. Sea Transp. 13, 511–517. https://doi.org/10.12716/1001.13.03.04

Rødseth, Ø.J., Burmeister, H.-C., 2015. Risk Assessment for an Unmanned Merchant Ship. TransNav, Int. J. Mar. Navig. Saf. Sea Transp. 9, 357–364. https://doi.org/10.12716/1001.09.03.08

Smolarek, L., 2008. Human Reliability at Ship Safety Consideration. J. Konbin 5, 191–206. https://doi.org/10.2478/v10040-008-0048-0

Sun, Z., Li, Z., Gong, E., Xie, H., 2012. Estimating Human Error Probability using a modified CREAM. Reliab. Eng. Syst. Saf. 100, 28–32. https://doi.org/10.1016/j.ress.2011.12.017

Szłapczyński, R., Śmierzchalski, R., 2009. Supporting navigator's decisions by visualizing ship collision risk. Polish Marit. Res. https://doi.org/10.2478/v10012-008-0015-7

Torkildson, E.N., Li, J., Johnsen, S.O., Glomsrud, J.A., 2018. Empirical studies of methods for safety and security co-analysis of autonomous boat, in: Safety and Reliability - Safe Societies in a Changing World - Proceedings of the 28th International European Safety and Reliability Conference, ESREL 2018. pp. 2949–2958. https://doi.org/10.1201/9781351174664-369

Utne, I.B., Rokseth, B., Sørensen, A.J., Vinnem, J.E., 2020. Towards supervisory risk control of autonomous ships. Reliab. Eng. Syst. Saf. 196, 106757. https://doi.org/10.1016/j.ress.2019.106757

Weber, D.E., MacGregor, S.C., Provan, D.J., Rae, A., 2018. "We can stop work, but then nothing gets done." Factors that support and hinder a workforce to discontinue work for safety. Saf. Sci. 108, 149–160. https://doi.org/10.1016/j.ssci.2018.04.032

Wrobel, K., Krata, P., Montewka, J., Hinz, T., 2016. Towards the Development of a Risk Model for Unmanned Vessels Design and Operations. TransNav,

Int. J. Mar. Navig. Saf. Sea Transp. 10, 267–274. https://doi.org/10.12716/1001.10.02.09

Wróbel, K., Montewka, J., Kujala, P., 2018. Towards the development of a system-theoretic model for safety assessment of autonomous merchant vessels. Reliab. Eng. Syst. Saf. 178, 209–224. https://doi.org/10.1016/J.RESS.2018.05.019

Xue, J., Wu, C., Chen, Z., Van Gelder, P.H.A.J.M., Yan, X., 2019. Modeling human-like decision-making for inbound smart ships based on fuzzy decision trees. Expert Syst. Appl. 115, 172–188. https://doi.org/10.1016/j.eswa.2018.07.044

Yousefi, H., Seyedjavadin, R., 2012. Crew Resource Management : The Role of Human Factors and Bridge Resource Management in Reducing Maritime Casualties. TransNav, Int. J. Mar. Navig. Saf. Sea Transp. 6, 391–396.

Zhang, J., Zhang, D., Yan, X., Haugen, S., Soares, C.G., 2015. A distributed anti-collision decision support formulation in multi-ship encounter situations under COLREGs. Ocean Eng. 105, 336–348. https://doi.org/https://doi.org/10.1016/j.oceaneng.2015.06.054

Zhang, S., He, W., Chen, D., Chu, J., Fan, H., 2019. A dynamic human reliability assessment approach for manned submersibles using PMV-CREAM. Int. J. Nav. Archit. Ocean Eng. 11, 782–795. https://doi.org/10.1016/j.ijnaoe.2019.03.002

Developments in Maritime Technology and Engineering – Guedes Soares & Santos (eds)
© 2021 Copyright the Author(s), ISBN 978-0-367-77376-2

Spatial distribution of ship near collisions clusters off the coast of Portugal using AIS data

H. Rong, A.P. Teixeira & C. Guedes Soares
Centre for Marine Technology and Ocean Engineering (CENTEC), Instituto Superior Técnico, Universidade de Lisboa, Lisbon, Portugal

ABSTRACT: This paper assesses the spatial distribution of near collision clusters off the continental coast of Portugal. The Moran's I and Getis-Ord Gi* spatial autocorrelation methods are used to determine whether ship near collisions show spatial clustering from global and local perspectives. The application of the developed approach to Automatic Identification System data of the maritime traffic off the coast of Portugal shows that there are several hotspot areas where the density of ship near collisions is relatively high. The identification of near collisions clusters and their characterization provide the basis for stablishing geographical maps of collision risk for improving maritime safety and reducing the occurrence of ship-ship collisions.

1 INTRODUCTION

Maritime transportation has been the world's foremost means for domestic and international trade. According to United Nations Conference on Trade and Development (UNCTAD, 2019) projections, international maritime trade will continue rising at a compound annual growth rate of 3.4 per cent over the 2019–2024 period. Safety and more recently security issues have been highlighted in the last decade due to the challenges imposed by the increasing demand for more ships with larger capacities and higher travelling speeds (Teixeira & Guedes Soares, 2018). As suggested by Guedes Soares and Teixeira (2001), groundings, ship-ship collisions and fires are the most common accident types globally. More specifically, ship-ship collision is one of the most frequently accident types occurring on local waterways with high traffic density.

Several studies indicate that the distribution of ship conflicts is location dependent (Goerlandt et al., 2012; Wu et al., 2016; Yoo, 2018; Rong et al., 2016, 2019) and often is larger at port approaches, where there is higher traffic concentration (Zhang et al. 2016). In terms of maritime traffic, the analysis of motion patterns is essential as it provides ship's manoeuvre information and the corresponding ship behaviour. AIS data-based ship traffic analysis provides information on the ship navigation activities and on the maritime traffic patterns. In this context, Rong et al (2020) presented a data mining approach for probabilistic characterization of maritime traffic in terms of ship types, sizes, and final destinations off the continental coast of Portugal.

As the occurrence of maritime accidents is relatively rare, the analysis of near collisions and collision candidates have been applied to assess the collision risk of the maritime transportation system (Goerlandt et al., 2012; Weng et al., 2012 Perera & Guedes Soares, 2015), as recently reviewed by Chen et al. (2019). Silveira et al. (2013) proposed a method to determine the number of collision candidates based on available Automatic Identification System (AIS) data with the concept of collision diameter defined by Pedersen (1995). The concept of "ship domain" has been used to identify near collision scenarios by using simulation models of ship navigation in restricted waters and AIS data (Rong et al. 2015), which allowed the characterization of near ship collision scenarios off the coast of Portugal in terms of ship types and dimensions, crossing angle, relative velocity, among others (Rong et al. 2016).

Several indicators have been adopted to gain further insight into the safety performance of the maritime traffic. Qu et al. (2011) introduced collision risk indices including speed dispersion, degree of acceleration and deceleration, and number of fuzzy ship domain overlaps. Zhang et al. (2015) proposed the Vessel Conflict Ranking Operator (VCRO) model to identify potential near collisions from AIS data, and those near collisions were further analysed for collision assessment of maritime transportation. Yoo (2018) has estimated the near collisions density in coastal areas and has visualized the corresponding high collision risk locations. These studies show that the analysis of near collisions can be used for ship-ship collision assessment.

DOI: 10.1201/9781003216582-19

Rong et al. (2019) have adopted collision risk indicators, including the Distance at the Closest Point of Approach (DCPA), Time to Closest Point of Approach (TCPA) and the relative distance between ships, to assess the collision risk of near collision scenarios. Spatial density maps of the collision risk were then derived using the Kernel Density Estimation (KDE) method to identify hotspots of collision risk off the continental coast of Portugal. Among the existing spatial statistical methods, spatial autocorrelation analysis has been employed to identify spatial clusters in traffic safety research. Zhang et al. (2019) analysed ship traffic demand, speed and traffic density distribution of ship traffic in the Singapore Strait. Several hotspot areas were identified by applying the spatial autocorrelation method, which coincided with the spatial distribution of ship accidents.

In this study, the maritime traffic off the coast of Portugal is analysed and the Moran's I spatial autocorrelation method is adopted to determine the spatial pattern of near collision locations from a global perspective. Then, the Getis-Ord Gi^* method is applied to identify near collision hotspots. Finally, the near collision hotspots are characterised in terms of the relative distance, relative bearing and relative speed of the vessels involved in the near collisions. In particular, the spatial pattern of near collisions involving vessels carrying hazardous materials are investigated.

2 SHIP NEAR COLLISIONS

Near collisions are identified using the concept of "ship safety domain" as collision criterion. The ship safety domain is a generalization of a safe space around a ship for collision avoidance with respect to other ships or stationary obstacles. The concept of ship safety domain first introduced by Fujii & Tanaka (1971) has been widely applied in navigational safety studies. In this paper the ship domain proposed by Fujii & Tanaka (1971) is adopted, which consists of an ellipse centred at the position of the ship with semi-major a and semi-minor b equal to 4 times and 1.6 times the ship length L, respectively.

For detecting the near collisions, ship domains are defined for all ships based on position and Course Over Ground (COG) data at all-time steps.

Let (x_i, y_i, φ_i) and (x_j, y_j, φ_j) denote the position and Course Over Ground (COG) of vessel i of interest and vessel j, respectively, shown in Figure 1. The relative distance and the relative bearing between vessel i and vessel j represented by D_{ij} and θ can be calculated by:

$$D_{ij} = \sqrt{\left(x_i - x_j\right)^2 + \left(y_i - y_j\right)^2} \tag{1}$$

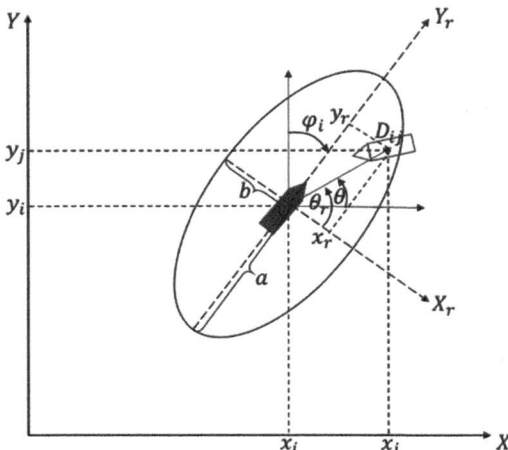

Figure 1. Ship domain violation.

$$\theta = \arctan\left(\frac{y_j - y_i}{x_j - x_i}\right) \tag{2}$$

Hence, vessel j violates the ship domain of vessel i if the following condition holds:

$$\frac{\left(D_{ij} * cos(\theta - \varphi_i)\right)^2}{a^2} + \frac{\left(D_{ij} * sin(\theta - \varphi_i)\right)^2}{b^2} < 1 \tag{3}$$

If the ship domain is violated by another ship, the encounter is defined as a near collision and the location of this event as well as relevant details are recorded.

In this study, three types of encounters identified in the International Regulations for Preventing Collisions at Sea (COLREGSs) are considered, namely head-on, overtaking and crossing situations, as shown in Figure 2:

- Overtaking: a situation in which two vessels are proceeding on the same route, lying on almost parallel courses, with own vessel's relative bearing from 112.5° to 247.5°.
- Head-on: a situation in which vessels are lying on almost reciprocal courses and the course difference falls in the range ±5°.
- Crossing: covers the area of own vessel's relative bearing from 5° to 112.5° or from 247.5° to 355°.

3 SPATIAL AUTOCORRELATION ANALYSIS

Spatial autocorrelation describes the correlation between the values of a single variable that is due to

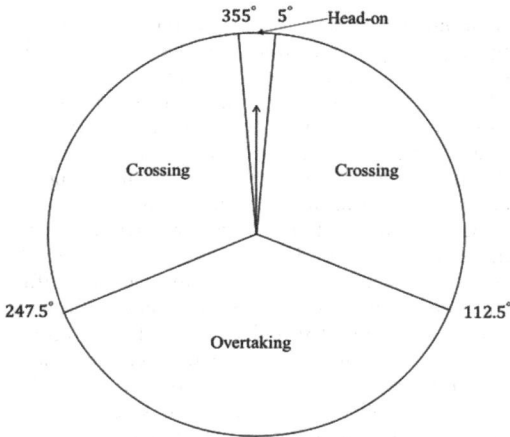

Figure 2. Encounter types.

the proximity of these values in the geographical space. The aim of spatial autocorrelation is to study and understand the spatial structure and spatial dependence in the study area. In this study, the spatial autocorrelation is performed over a complete study area using the Moran's I index and at specific locations with the Getis-Ord Gi* statistic.

3.1 Global spatial autocorrelation

In statistics, global Moran's I is a correlation coefficient that measures the overall spatial autocorrelation given a set of attribute values and the associate positions (Getis and Ord, 1992). It evaluates whether the pattern expressed is clustered, dispersed, or random.

Specifically, the global Moran's I statistic detects general spatial patterns using the near collision data by measuring attribute similarity and proximity of such near collisions with Eq. 4.

$$I = \frac{N}{W} \frac{\sum_i^N \sum_j^N w_{i,j}(x_i - \bar{x})(x_j - \bar{x})}{\sum_i (x_i - \bar{x})^2} \quad (4)$$

where N is the number of spatial units; x is the variable of interest, in this study, it refers the number of near collisions; $\bar{x}$ is the mean of x; $w_{i,j}$ is the element of spatial matrix with weights representing connectivity relationship between location i and neighboring location j, and W is the sum of all element $w_{i,j}$ given by:

$$W = \sum_i^N \sum_j^N w_{i,j} \quad (5)$$

The global Moran's I is within $[-1, 1]$. When estimating data spatial distribution pattern if $I > 0$ the near collisions are positively correlation in spatial

space, and the values approaching 1 indicate a strong clustering. If $I < 0$, the near collisions have negative correlation in spatial space, which means the near collisions are distributed dispersedly. If $I = 0$ the near collisions are truly randomly dispersed (perfect randomness).

The expected value of Moran's index under the null hypothesis of no spatial autocorrelation is:

$$E[I] = \frac{-1}{N-1} \quad (6)$$

where N is the sample size. At large sample size, the expected value approaches zero.

The variance of Moran's I is given by:

$$Var[I] = E[I^2] - E[I]^2 \quad (7)$$

Details on the calculation of the variance of Moran's index can be found in Appendix.

For statistical hypothesis testing, Moran's I values can be transformed to z-scores that represent the relationship to the mean of a group of values, measured in terms of standard deviations from the mean by

$$Z(I) = \frac{(I - E[I])}{\sqrt{Var[I]}} \quad (8)$$

3.2 Local spatial autocorrelation

The Getis-Ord G_i^* local statistic is frequently used in Hot Spot Analyses. The approach consists of looking at each feature of a data set within the context of neighboring features. A feature with a high value is interesting but may not be a statistically significant hot spot. To be a statistically significant hot spot, a feature will have a high value and be surrounded by other features with high values as well.

Given a set of weighted features, a positive value of the Getis-Ord G_i^* statistic indicates clustering of features with high attribute values, and a negative value denotes clustering of features with low attribute values. The Getis-Ord G_i^* statistic is expressed by (Getis & Ord, 1992):

$$G_i^* = \frac{\sum_{j=1}^n w_{i,j}x_j - \bar{x}\sum_{j=1}^n w_{i,j}}{S\sqrt{\frac{n\sum_{j=1}^n w_{i,j}^2 - \left(\sum_{j=1}^n w_{i,j}\right)^2}{n-1}}} \quad (9)$$

where x_j indicates the number of near collisions at location j, $w_{i,j}$ is the element of spatial matrix with weights representing connectivity relationship between location i and neighboring location j, $\bar{x}$ and S are the sample mean and the sample variance of variable x defined by Eqs. 10 and 11, respectively.

$$\bar{x} = \frac{\sum_{j=1}^{n} x_j}{n} \qquad (10)$$

$$s^2 = \frac{\sum_{j=1}^{n} x_j^2}{n} - \bar{x}^2 \qquad (11)$$

The standardized G_i^* is essentially a Z-score and therefore can be used to assess the statistical significance of the hot spots. A close to zero G_i^* value implies random distribution of the observed near collisions. For statistically significant positive z-scores, the larger the z-score is, the more intense the clustering of high values (hot spot). Conversely, for negative z-scores, the smaller the z-score is, the more intense the clustering of low values (cold spot). In summary, if the calculated index values are greater than a threshold associated with a statistical significance, the location of a cluster is identified as a hot spot. Therefore, any near collision locations that are nearby or that encompass such a cluster are identified as hot spots.

4 CASE STUDY

Portugal has jurisdiction over a considerable maritime area, of around 1.720.560 square kilometres. This maritime area under jurisdiction of Portugal is located at the crossroad of very important maritime routes, including traffic from Mediterranean Sea to North America, from Africa to northern Europe and from northern Europe to South America. In this study, the AIS data from Oct. to Dec. 2015 are used and a ship domain-based method is applied to identify near collisions off the continental coast of Portugal based on collected historical AIS data.

Most of the messages received from AIS (Automatic Identification System) transponders installed onboard ships are correct. However, there are inaccurate ship positions and speed information errors among the AIS messages. Therefore, the AIS data cleaning procedure proposed by Rong et al., (2016) and Qu et al. (2011) are adopted to filter the AIS data position errors and to update those inaccurate records, respectively. In addition, ship trajectories derived from AIS data are of different length and are defined from different number points mainly due to speed variations of the ships. In order to capture the snapshot of the maritime transportation situation, ship trajectories' data are interpolated with a predefined time interval regarding ship position, ship speed and heading based on the procedure proposed by Rong et al. (2019). This synchronizing procedure is essential for identifying ship domain violations, as the two ships do not report AIS data at the same time.

Figure 3a illustrates the maritime traffic in the study area. The maritime areas off the continental coast of Portugal are crossed by a complex network of routes where routes connecting northern Europe and the Mediterranean Sea meet vessels bound to and leaving from national ports. Two Traffic Separation Schemes (TSSs) located off Cape Roca and off Cape San Vicente organize the Northbound Traffic (blue polygon in Figure 1b), Southbound Traffic (red polygon in Figure 1b) and two-way Traffic (green polygon in Figure 1b) of dangerous and non-dangerous ships off the coast of Portugal and influence significantly the routes of the ships entering and leaving the national ports. The four lanes in each TSS are:

- Northbound non-dangerous cargo traffic lane;
- Northbound dangerous cargo traffic lane;
- Southbound non-dangerous cargo traffic lane;
- Southbound dangerous cargo traffic lane.

There are 57 ship motion patterns identified from the historical AIS data (Figure 3b) consisting of 25 northbound and 25 southbound traffic lanes and 7 two-way traffic lanes. The ship routes are estimated based on the analysis of the corresponding motion patterns. As shown in Figure 3b, the traffic-groups consist of maritime traffic in the Traffic Separation Scheme lanes, connecting TSSs lanes to the main ports and from port to port, namely: Southbound traffic from TSS off Cape Roca to TSS off Cape S. Vicente and vice versa; from TSSs lanes to the ports of Viana do Castelo, Leixões, Aveiro, Figueira da Foz, Lisbon, Setubal and Sines. The number of ship trajectories in each traffic-group reveals that most of ship traffic off the continental coast of Portugal follows the main traffic route in North/South direction. The remaining maritime traffic is those entering and leaving the ports off the continental coast of Portugal.

Figure 4 shows a total number of 1671 near collision scenarios identified off the continental coast of

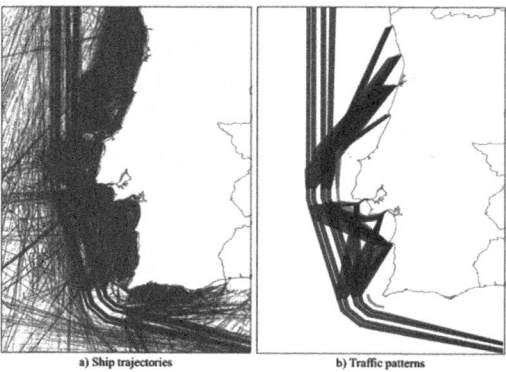

a) Ship trajectories   b) Traffic patterns

Figure 3. Maritime traffic in the study area.

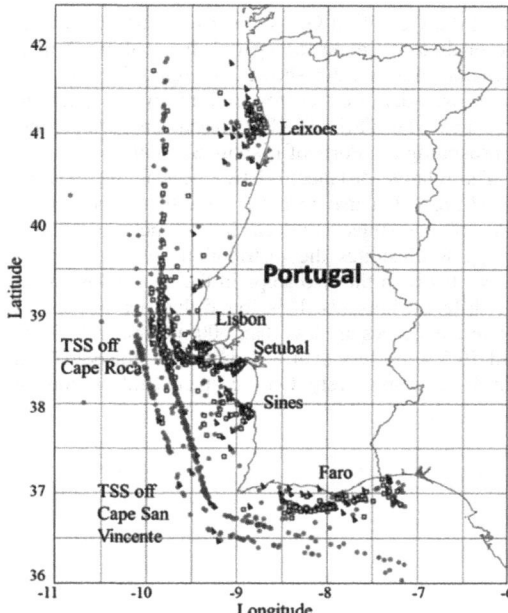

Figure 4. Near collision scenarios detected off the continental coast of Portugal.

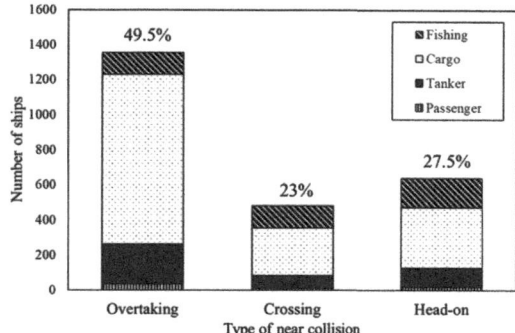

Figure 5. Ship type distribution by near collision type.

Table 1. Global spatial autocorrelation of near collisions by encounters.

	Moran's I	Z-score	Spatial pattern
All encounters	0.285	2.60 (*)	Clustered
Overtaking	0.315	3.24 (**)	Clustered
Crossing	0.424	4.57 (**)	Clustered
Head-on	0.322	3.55 (**)	Clustered

* : p-value < 0.01; **: p-value < 0.001

Portugal. There are 827, 384 and 460 near collisions identified for overtaking, crossing and head-on near encounters, respectively. It should be mentioned that ship domain violations involving tugs and pilot vessels as well as in port areas have been excluded in this study. Overtaking near collision locations are represented by red circles, crossing cases by blue triangles and head-on cases by yellow squares. It can be seen that near collisions occur mainly on major shipping lanes and on the approaches to the main ports of Portugal. More specifically, most overtaking cases are located on the ship traffic lane formed by the Traffic Separation Schemes (TSS) and crossings and head-on cases are correlated to ship traffic approaching and leaving the main ports of Portugal.

Figure 5 shows the ship type distribution for different near collision scenarios. According to Figure 5, overtaking account for the biggest proportion (49.5%), followed by head-on (27.5%), and then crossing (23.0%). It can be seen that cargo ship is the main ship type involved in near collisions (accounting for 71.3%, 55.8% and 54.2% in overtaking, crossing and head-on near collisions, respectively). The next highest ratio is observed for fishing vessels, followed by tankers. It should be mention that the near collisions involving only fishing vessels are excluded from the following study.

The spatial correlation analysis determines whether the spatial pattern of the different near collision types off the coast of Portugal presented cluster, dispersion and random patterns. Table 1 shows the global spatial autocorrelation results in terms of Moran's I values.

The z-scores in parentheses for the level of statistical significance (p-value) is indicated by asterisks for each near collision type. As revealed by Rong et al., (2019), the sea area of the TSS off Cape Roca and the approaches to Lisbon, Setubal and Sines ports are high collision risk locations, and the collision risk of the different near collision types is also strongly location-dependent. The results quantitatively confirmed that the spatial distribution of near collisions is significantly clustered, as indicated by small p-values and high Moran's I values. In addition, the locations of crossing near collisions are highly clustered with a Moran's I value of 0.424, followed by head-on near collisions (0.322) and overtaking near collisions (0.315). Regarding the spatial correlation of overtaking near collisions, due to high volume of traffic flow guided and organized by the TSS off Cape Roca and TSS off Cape San Vincent, most overtake cases are located in the ship traffic lane between the two TSSs and the spatial distribution of overtaking near collisions is almost evenly distributed.

Local spatial autocorrelation analysis using the Getis-Ord Gi* statistic is performed for different near collisions types. Since the number and the spatial distribution of each near collision type is different, different distance bands are used in the analysis of the local spatial autocorrelation. Additionally, Z score values equal or greater than 1.96 (p-value less than 0.05) are used to assess the significance of the spatial pattern (clustered, random and dispersed). Figure 6 shows the local spatial autocorrelation of near collisions for different near collision types off

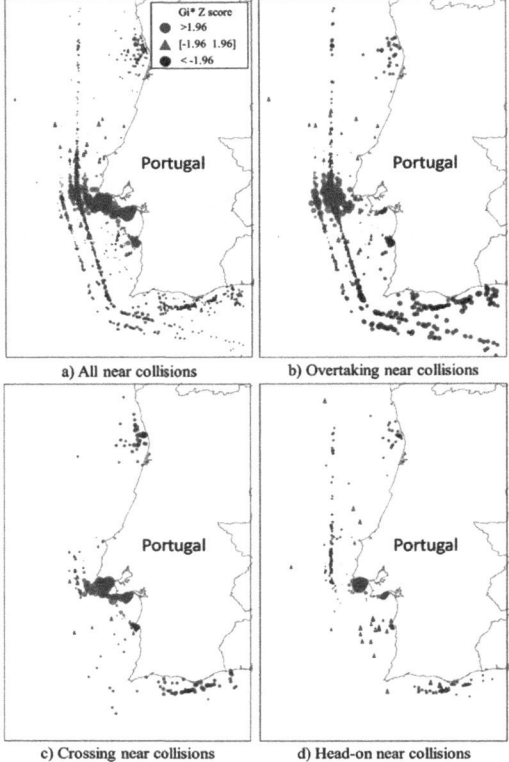

a) All near collisions   b) Overtaking near collisions

c) Crossing near collisions   d) Head-on near collisions

Figure 6. Local spatial autocorrelation of near collisions for different near collision types.

the coast of Portugal. In the figure, near collisions with Gi* Z score values of 1.96 or greater at the 5% significant level are marked by red circle, near collisions with Gi* Z score values of -1.96 or lower are marked by grey circle and near collisions with Gi* Z score values within -1.96 and 1.96 are marked by green triangle. It is clear that the high positive Gi* Z score of different near collision types is strongly location dependent. According to the local spatial autocorrelation of all near collisions (Figure 6a), the sea area at the TSS off Cape Roca and the approaches to Lisbon, Setubal and Sines ports are all high positive Gi* Z score locations. Although the local spatial autocorrelation of overtaking near collisions (Figure 6b) presents similar spatial pattern to that of all near collisions, this type of near collision presents slightly more clustering intensity on TSS off Cape Roca. Regarding crossing and head-on near collisions (Figures 6c&d), the waterways approaching the ports of Lisbon and Setubal and the junction area near TSS off Cape Roca have higher positive Gi* Z scores.

To identify the hotspots of near collisions off the coast of Portugal, the study area is discretized into a 150 × 100 grid and the near collisions are grouped into each grid. Local spatial autocorrelation analysis

using the Getis-Ord Gi* statistics is performed over the grids to estimate the spatial distribution of near collision clusters. As shown in Figure 7, four hotspots are identified corresponding to the junction area near the TSS off Cape Roca, and waterways approaching the Ports of Lisbon, Setubal and Sines.

The relative distance, relative bearing and relative speed are calculated to better investigate the behaviour of the ships involved in the near collisions. Figure 8 illustrates the distributions of relative distances between the two ships in the near collisions in the different hotspots. It is found that nearly 50% of relative distances are less than 500 meters, and 83.4% and 76.5% are less than 700 meters in hotspot 2 and hotspot 3, respectively. On the other hand, in hotspot

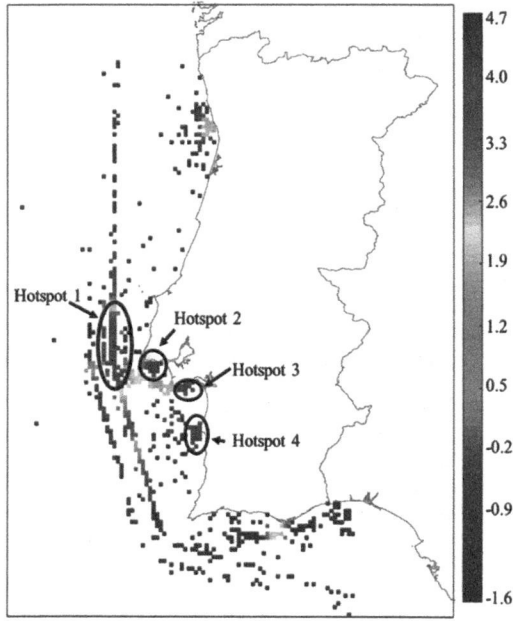

Figure 7. Hotspots of near collisions identified off the coast of Portugal.

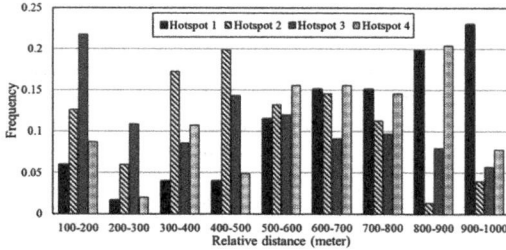

Figure 8. Relative distance distributions of near collision in the four hotspots.

1 and hotspot 4, the relative distances beyond 500 meters account for 84.5% and 73.8%, respectively.

Figure 9 shows the relative bearing distribution between two near collision involved ships. It is seen that for all hotspots, the range between 120- and 240-degrees accounts for nearly 40%-50%. Such distribution of the relative bearing well explains overtaking near collisions that account for the highest proportion among all near collisions. The relative bearing distributed evenly from 30 degree to 150 degree and from 210 degree to 330 degree in all hotspots. However, in hotspot 2 and hotspot 3, relative bearings between 330 degree and 30 degree, and between 150 degree and 210 degree take higher proportion than that in hotspot 1 and hotspot 4 (68.5% and 69.1% vs. 51.4% and 48.8%). The reason is that several ship routes converge in a limited geographical area on the waterway approach to Lisbon and Setubal ports, which results in more overtaking and head-on encounters between ships.

The relative speed is also investigated as it is an important factor affecting the collision risk. The relative speed between vessel i and vessel j can be calculated by:

$$\left|V_{ij}\right| = \sqrt{\left(v_i \cos\varphi_i - v_j \cos\varphi_j\right)^2 + \left(v_i \sin\varphi_i - v_j \sin\varphi_j\right)^2} \tag{12}$$

Figure 10 shows the relative speed distribution of near collision involved ships in the four hotspots. In terms of hotspot 1, it is found that in almost 66.3% of the cases the relative speeds are smaller than 4 knots and more than 80% are smaller than 8 knots. By combining the relative distance and relative speed distributions, it seems that hotspot 1 is relatively safe compared with the other hotspots as the average relative distance is higher and the average relative speed is lower.

Tankers and cargo ships carrying hazardous materials are highly concerned as they pose a high environmental risk. In this study, the local spatial autocorrelation of near collision involving tankers or

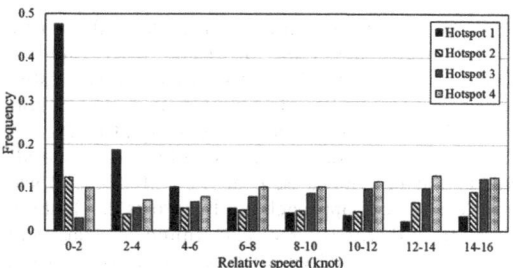

Figure 10. Relative speed distributions of near collision in the four hotspots.

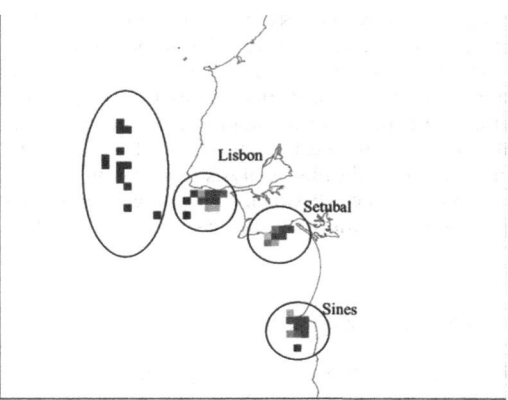

Figure 11. Local spatial autocorrelation of near collisions involving vessels carrying hazardous materials.

cargo-hazardous is presented. As shown in Figure 11, high levels of spatial correlation are observed at the entrance of Ports of Lisbon, Setubal and Sines. In addition, the spatial pattern of near collisions involving tankers/cargo-hazardous ships is not clustered in hotspot 1, which means that hotspot 1 is not a hotspot for this type of vessels.

5 CONCLUSIONS

This paper analyses near collision hotspots off the continental coast of Portugal. By detecting ship domain violations from historical AIS data, a total number of 1671 near collisions are identified, consisting of 827, 384 and 460 overtaking, crossing and head-on near collision scenarios, respectively. The global Moran's I statistic is used to investigate whether near collisions present a global positive spatial autocorrelation. The results of global Moran's indices suggest that the spatial distribution of near collisions is significantly clustered. In particular, the smallest value of the global Moran's index is obtained for overtaking near collisions, reflecting the fact that

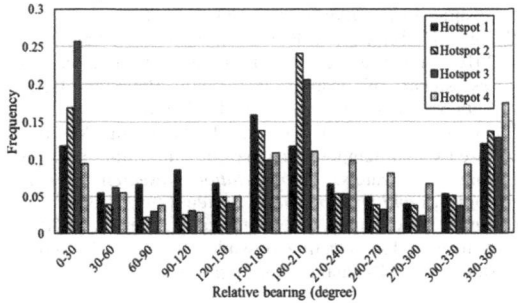

Figure 9. Relative bearing distribution of near collision in the four hotspots.

these cases are distributed evenly along the traffic lane between the TSSs off Cape Roca and off Cape San Vicente. The Getis-Ord Gi* local statistic is then applied to detect of near collision hotspots. By adopting Z-score values equal or greater than 1.96, four hotspots are identified corresponding to the junction area near TSS off Cape Roca, and waterways approaching the Ports of Lisbon, Setubal and Sines. By performing a statistical study on the relative distance, relative bearing and relative speed of the ships involved in the near collisions, it is found that hotspot 1 is relatively safe compared with the other hotspots, due to the higher average relative distance and lower average relative speed of the ships. In addition, hotspot 1 presents less near collision clustering intensity for vessels carrying hazardous materials.

If consistent patterns are observed on near collision hotspots due to distinct attributes of the maritime traffic, then an association exists between these attributes and the near collisions on these hotspots. In the future, a detailed analysis focusing on the spatial co-occurrences between ship near collision types and relevant maritime traffic characteristics will be performed.

ACKNOWLEDGEMENTS

The paper contributes to the project "Integrated System for Traffic Monitoring and Maritime Risk Assessment (MoniRisk)", which has been co-funded by the European Regional Development Fund (Fundo Europeu de Desenvolvimento Regional (FEDER) and by the Portuguese Foundation for Science and Technology (Fundação para a Ciência e a Tecnologia – FCT) under contract no. 028746. This work also contributes to the Strategic Research Plan of the Centre for Marine Technology and Ocean Engineering (CENTEC), which is financed by the Portuguese Foundation for Science and Technology under contract UIDB/UIDP/00134/2020.

REFERENCES

Chen P, Huang Y, Mou J, van Gelder PHAJM., 2019. Probabilistic risk analysis for ship-ship collision: State-of-the-art. *Safety Science*, 117:108–22.

Fujii, Y., Tanaka, K., 1971, Traffic capacity. *Journal of Navigation*, 24 (4), 543–552.

Getis, A. and Ord, J.K. (1992), The Analysis of Spatial Association by Use of Distance Statistics. *Geographical Analysis*, 24: 189–206.

Goerlandt, F., Montewka, J., Lammi, H., Kujala, P., 2012, Analysis of near collisions in the Gulf of Finland, *Advances in Safety, Reliability and Risk Management*, Berenguer, Grall & Guedes Soares (Eds), Taylor & Francis Group. London. 2880–2886.

Guedes Soares, C., Teixeira, AP., 2001, Risk assessment in maritime transportation. *Reliability Engineering and System Safety* 74(3): 299–309.

Pedersen, P.T. 1995. Collision and grounding Mechanics, *The Danish Society of Naval Architects and Marine Engineers*: 125–57.

Qu, X., Meng, Q., Li, S. 2011. Ship collision risk assessment for the Singapore Strait. *Accident Analysis and Prevention* 43(6): 2030–2036.

Perera, L. P. and Guedes Soares, C. 2015. Collision risk detection and quantification in ship navigation with integrated bridge systems. *Ocean Engineering*. 109: 344–354.

Rong, H., Teixeira, A.P., Guedes Soares, C. 2015. Simulation and analysis of maritime traffic in the Tagus River Estuary using AIS data. *Maritime Technology and Engineering*, Guedes Soares & Santos (Eds), Taylor & Francis Group. London: 185–193.

Rong, H., Teixeira, A.P., Guedes Soares, C. 2016. Assessment and characterization of near ship collision scenarios off the coast of Portugal. *Maritime Technology and Engineering 3*, Guedes Soares, C. & Santos T. A., (Eds.), Taylor & Francis Group. London: 871–878.

Rong, H., Teixeira, A.P., Guedes Soares, C. 2019. Risk of ship near collision scenarios off the coast of Portugal, M. Beer and E. Zio (eds). *29th European Safety and Reliability Conference (ESREL 2019)*. Hannover, Germany.

Rong, H., Teixeira, A. P., & Guedes Soares, C. 2020. Data mining approach to shipping route characterization and anomaly detection based on AIS data. *Ocean Engineering*, 198, 106936.

Silveira, P.A.M., Teixeira, A.P., Guedes Soares, C. 2013. Use of AIS Data to Characterise Marine Traffic Patterns and Ship Collision Risk off the Coast of Portugal. *The Journal of Navigation* 66: 879–898.

Teixeira, A.P. & Guedes Soares, C., 2018. Risk of maritime traffic in coastal waters. *Proceedings of the ASME 37th International Conference on Ocean, Offshore and Arctic Engineering, OMAE2018*, paper OMAE2018-77312. VOL 11A Article Number: UNSP V11AT12A025

UNCTAD, 2019. *Review of Maritime Transport*. United Nations Conference on Trade and Development.

Weng, JX., Meng, Q., Qu, XB., 2012, Vessel Collision Frequency Estimation in the Singapore Strait. *Journal of Navigation* 65, 207–221.

Wu, X., Mehta, A.L., Zaloom, V.A., Craig, B.N. 2016. Analysis of waterway transportation in Southeast Texas waterway based on AIS data. *Ocean Engineering* 121: 196–209.

Yoo, S. L. 2018. Near-miss density map for safe navigation of ships. *Ocean Engineering* 163: 15–21.

Zhang, J. F.; Teixeira, A. P.; Guedes Soares, C.; Yan, X. P., 2016; and Liu, K. H. Maritime transportation risk assessment of Tianjin Port with Bayesian Belief Networks. *Risk Analysis*. 36(6):1171–1187.

Zhang, L, Meng, Q, Fang Fwa, T. 2019. Big AIS data based spatial-temporal analyses of ship traffic in Singapore port waters. *Transportation Research Part E: Logistics and Transportation Review* 129:287–304.

Zhang, W., Goerlandt, F., Montewka, J., Kujala, P. 2015. A method for detecting possible near miss ship collisions from AIS data. *Ocean Engineering* 107: 60–69.

Appendix

$$Var[I] = \mathrm{E}\left[I^2\right] - E[I]^2 = \frac{NS_4 - S_3 S_5}{W^2 \prod_{i=1}^{3}(N-i)}$$

where:

$$S_1 = \frac{1}{2}\sum_i \sum_j \left(w_{ij} + w_{ji}\right)^2;$$

$$S_2 = \sum_i \left(\sum_j w_{ij} + \sum_j w_{ji}\right)^2;$$

$$S_3 = \frac{N^{-1}\sum_i \left(x_i - \bar{x}\right)^4}{\left(N^{-1}\sum_i \left(x_i - \bar{x}\right)^2\right)^2};$$

$$S_4 = \left(N^2 - 3N + 3\right)S_1 - NS_2 + 3W^2;$$

$$S_5 = \left(N^2 - N\right)S_1 - 2NS_2 + 6W^2.$$

Developments in Maritime Technology and Engineering – Guedes Soares & Santos (eds)
© 2021 Copyright the Author(s), ISBN 978-0-367-77376-2

Spatial-temporal analysis of ship traffic in Azores based on AIS data

H. Rong, A.P. Teixeira & C. Guedes Soares
Centre for Marine Technology and Ocean Engineering (CENTEC), Instituto Superior Técnico, Universidade de Lisboa, Lisbon, Portugal

ABSTRACT: A method is proposed to automatically produce synthetic maritime traffic representations from historical Automatic Identification System data. The method consists of a two-phase analysis of maritime traffic data, which includes ship traffic motion pattern identification as well as ship route characterization. Given the location of a port, ship trajectories and ship characteristics are extracted enabling to estimate ship traffic volume, composition and origin-destination topology. In addition, each ship route is individually analysed in terms of ship type distribution and daily and hourly patterns, which allows the characterization of the typical behaviour of a group of similar ships along a particular route. Furthermore, the speed distribution at the study area in terms of time of day is analysed to capture the traffic state changes over time. The method is applied to obtain a light and structured representation of the maritime traffic off the Azores Archipelago.

1 INTRODUCTION

Maritime transportation is responsible for approximately 90% of international trade and provides the most efficient means of transporting large quantities of goods for large distances. Safety and more recently security issues have been highlighted in the last decade due to the challenges imposed by the increasing demand for more ships with larger capacities and higher travelling speeds (Teixeira & Guedes Soares, 2018).

The need to ensure the safety of ship navigation has led to the implementation of Automatic Identification System (AIS), which is imposed by International Maritime Organization (IMO). The AIS autonomously broadcasting kinematic information (including ship position, Speed Over Ground, Course Over Ground, heading, rate of turn and estimated arrival time) and static information (including ship name, Maritime Mobile Service Identity, ship type and ship size). Ship trajectory information provided by the AIS data is one of the most important data sources for collision risk assessment (Mou et al., 2010; Qu et al., 2012; Rong et al. 2016; Rong at al., 2019a).

To maintain the operational efficiency and safety of maritime waterways, a deep insight into the ship traffic is of great importance. Zhang et al. (2017) developed a tangible analytical approach to analyse ship traffic demand and the spatial–temporal dynamics of ship traffic in Singapore port. Silveira et al. (2013) proposed a method to characterise marine traffic patterns and ship collision risk off the Coast of Portugal based on the available AIS data. Rong et al. (2018) proposed a data mining method to automatically identify maritime traffic junctions and applied a multinomial logistic regression model for predicting the ships destination based on a set of characteristics of the ships' behaviour at the junction.

The information provided by large amounts of AIS data can be transformed into useful information for intelligent maritime traffic applications, e.g. ship path prediction, anomaly detection and, thus, plays a central role in future autonomous maritime operations. Rong et al. (2019b) have adopted Gaussian Process models for probabilistic ship trajectory prediction within particular motion patterns derived from historical AIS data and for collision probability evaluation (Rong et al. 2020a). Rong et al (2020b) presented a data mining approach for probabilistic modelling of ship routes, which are represented by mean route and the corresponding 95% confidence boundary, and the off-route behaviour of ships can be automatically detected in real-time. Xu et al. (2019) developed an autonomous vessels guidance and control system by using ship route obtained from AIS data.

Furthermore, as the intelligent maritime systems develop rapidly, many research works have proposed for anomaly detection and motion prediction algorithms and, therefore, it is quite important to have a database that could serve as a benchmark for comparing the performance of different methods and algorithms.

In this study, a data mining approach is proposed for characterization of the maritime traffic off the Azores Archipelago based on AIS data in terms of traffic composition, speed distribution, temporal pattern and spatial distribution of average speed.

DOI: 10.1201/9781003216582-20

2 DATA PREPROCESSING

The Automatic Identification System (AIS) is a ship self-reporting system of messages broadcasted by ships carrying an AIS transponder. The recent increase of terrestrial networks and satellite constellations of receivers make AIS one of the main sources of information for maritime traffic characterization.

In this study, the reception range of terrestrial AIS-stations causes data missing, which may affect the performance of spatial-temporal analysis. Besides, the raw AIS data also contain erroneous speed and position data occurring during data collection, transmission and reception procedure. A speed error of AIS data refers to a ship speed larger than a threshold (30 knot), which may be caused by data noise. Ship position error refers to locations that may be within the normal range of the study area, but the values of latitude and/or longitude significantly deviate from a normal trajectory with an abnormal change of speed. A data cleaning procedure is therefore proposed to eliminate these errors and to correct the inaccurate records, and then trajectory interpolation is adopted for restoring the ship trajectory.

The pre-processing procedure consists of:
Step 1: Correct the erroneous position data.

With the longitude, latitude and interval time information between two successive locations, the value of speed can be calculated, then the erroneous position data is detected as:

$$\frac{\|p_{j+1} - p_j\|}{t_{j+1} - t_j} > 30knot \tag{1}$$

where $\|p_j + 1 - p_j\|$ represents the distance between points p_j and $p_j + 1$. In this study, the Haversine formula is applied to calculate the distance between latitude (λ) and longitude (φ) coordinates:

$$\|p_{j+1} - p_j\| = 2R_e sin^{-1}$$

$$\left(\sqrt{sin^2\left(\frac{\lambda_{j+1} - \lambda_j}{2}\right) + cos(\lambda_{j+1})cos(\lambda_j)sin^2\left(\frac{\varphi_{j+1} - \varphi_j}{2}\right)} \right) \tag{2}$$

where, R_e is the Earth radius.

Figure 1a shows a ship trajectory observed from AIS data with erroneous position points. Black circles represent ship locations, and the black line represents the ship trajectory. The incorrect AIS information greatly affects the representation of ship trajectory. By applying the described erroneous position detection method, there are two erroneous positions are identified and represented as red triangle in Figure 1b.

Step 2: Update the erroneous speed data.

Errors in speed data can be identified and corrected directly from the data sequence by detecting whether the speed data exceed a threshold. Suppose $SOG^i_{t_j}$ is

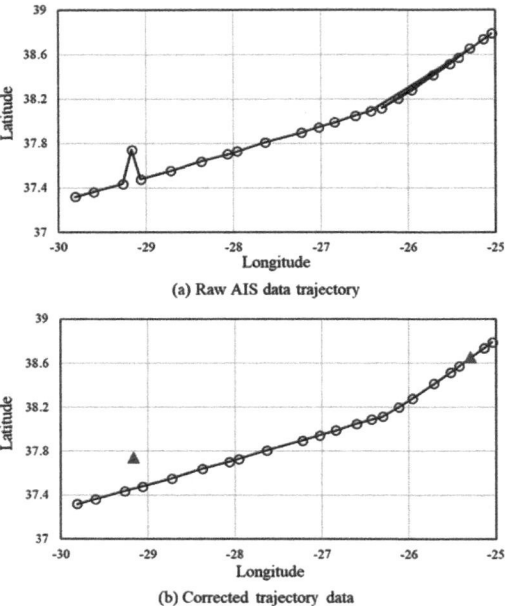

(a) Raw AIS data trajectory

(b) Corrected trajectory data

Figure 1. Erroneous AIS position detection.

the *SOG* of ship i at time t_j, then the $SOG^i_{t_j}$ is corrected according to the below equation:

$$SOG'_{t_j}i = \begin{cases} SOG^i_{t_j} & if \ SOG^i_{t_j} < 30 \\ \frac{\|p_{j+1} - p_j\|}{t_{j+1} - t_j} & otherwise \end{cases} \tag{3}$$

Based on the pre-processing procedure described above, erroneous position and speed data are removed and corrected. Limited by the coverage range of terrestrial AIS-stations, data missing can occur even though the AIS transponder correctly broadcasts the messages. In order to solve this problem, a linear interpolation with 10s-interval is performed for ship trajectory regarding ship position, ship speed and heading. It should be mentioned that one minute is set as the threshold interval for detecting AIS data missing between two successive points. The principle of trajectory linear interpolation assumes that the ship is in a uniform linear motion during the data missing time.

3 STUDY AREA

This research work studies the marine traffic off the Azores Archipelago based on Automatic Identification System (AIS) data. Computer programs for decoding, visualization and analysis of the AIS data are developed. From the analysis of the AIS data available, the maritime traffic off the Azores Archipelago is characterized and a statistical analysis of the marine traffic is presented.

The AIS data employed in this study are from 1st October to 31th December 2015, and the study area is bounded by parallels 35°N and 41°N, and by meridians 23°W and 33°W. The AIS data are received and stored by two terrestrial AIS-stations located at Faial Island and Sao Miguel Islands, as shown in Figure 2. The detail information of terrestrial AIS-stations can be found in Table 1.

Figure 3 shows the histogram of ship speed in the study area. The ship speed is constrained by the manoeuvrability of the ship. The abnormal speed refers to the low-frequency "very high" measurement errors. It can be seen that erroneous ship speed values are above 30 knots, some of which are even over 50 knots, which are caused by the data noise. In this study, the threshold speed is set to 30 knots for filtering abnormal speed data. After removing the ship position errors and updating ship speed, ship trajectories are interpolated in the time domain.

To be able to visualize traffic patterns using the corrected ship trajectories data, the study area is discretized into a 1200 × 2000 grids and trajectory points are grouped into each grid cell. The side of each grid cell is equivalent to 0.005 degree of latitude and longitude. Each value of the grid cell is incremented when the trajectory points is located in, and the trajectory points are plotted on the map with colors estimated according to the value of grid cells. Figure 4 shows the density map of ship trajectories on the study area. It can be seen that high traffic density occurs at well-defined traffic patterns, which indicates the main routes connecting the ports.

The ship movements data provided by AIS can be grouped based on similarity behavior, which can help to get an overview of the general motion patterns. In order to deduce the main ship routes, the ship trajectories that follow the same itinerary are extracted from the AIS data and then grouped together. The concept of itinerary can be defined by a pair of ports or spatial zones. The locations and the polygons that define the main ports of Azores are taken from the GeoNames ontology (http://www.geonames.org/). Once the polygons of ports are defined, the ship trajectories connecting a pair of ports are extracted and grouped together.

Figure 5 shows the motion patterns and the traffic topology at the study area. A total number of 527 ship trajectories are extracted, from which 509 ship

Table 1. Detail information of terrestrial AIS-stations in Azores Archipelago.

AIS station		#1	#2
Location		Faial Island	Sao Miguel Island
Position		(38.52°N, 28.69°W)	(37.79°N, 25.52°W)
Reception distance	Average (NM)	46.1	52.3
	Max (NM)	101.2	97.4

trajectories are kept after filtering out erroneous trajectories, which is enough to perform statistical analysis over this set of ship trajectories. The 509 ship trajectories grouped into 14 motion patterns are shown in Figure 5a. It is found that ships with the same motion pattern tend to follow main routes. The traffic topology at the study area is depicted in Figure 5b, where labeled white circles stand for the ports. Besides ship traffic between the ports, Eastbound/Westbound traffic accounted for a larger proportion (544 ship trajectories versus 509 ship trajectories). Eastbound/Westbound traffic refer to ship trajectories that do not visit a port and the travel distance is larger than 200 nautical miles. Eastbound and Westbound ship trajectories are

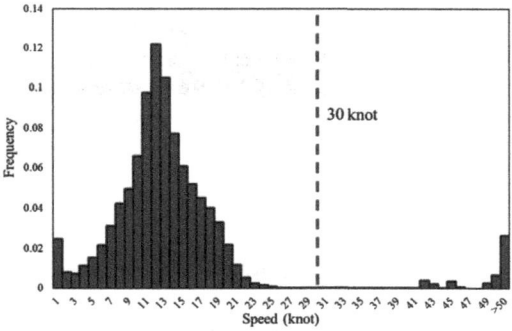

Figure 3. Ship speed distribution.

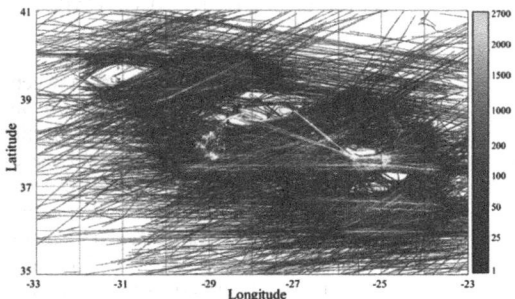

Figure 4. Density plot of AIS data in study area.

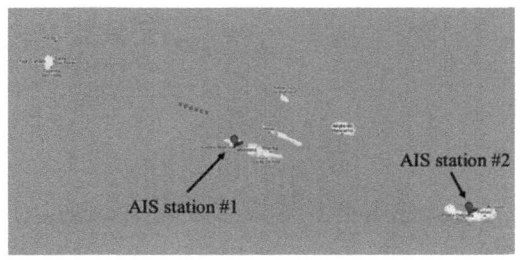

Figure 2. Terrestrial AIS-stations in Azores Archipelago.

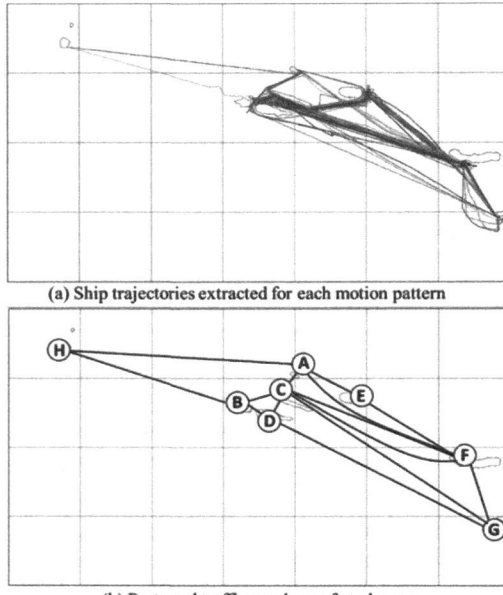

(a) Ship trajectories extracted for each motion pattern

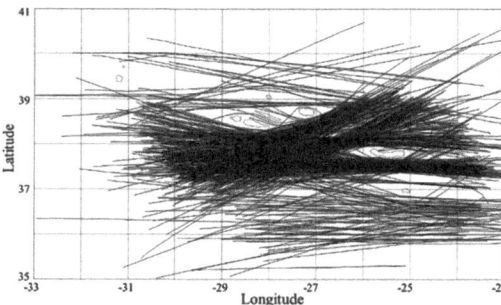

(b) Ports and traffic topology of study area

Figure 5. Motion patterns at the study area.

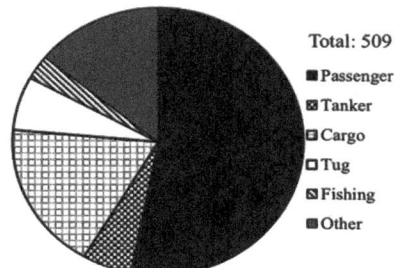

Total: 509

■ Passenger
▨ Tanker
□ Cargo
□ Tug
◪ Fishing
■ Other

(a) Ship type distribution of traffic between ports

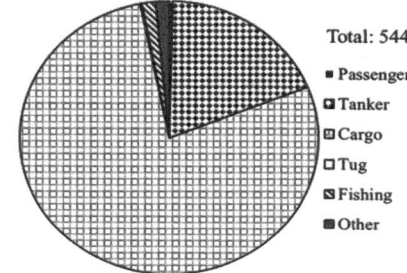

Total: 544

■ Passenger
□ Tanker
▨ Cargo
□ Tug
◪ Fishing
■ Other

(b) Ship type distribution of Eastbound and Westbound traffic

Figure 7. Ship type distribution.

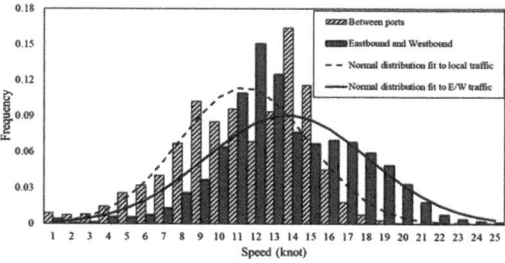

Figure 6. Eastbound and Westbound traffic at the study area.

shown in Figure 6, and the corresponding number of ship trajectories is illustrated in Table 2.

Table 2. Number of trajectories in each traffic-group.

Motion pattern	No. of trajectories
Eastbound	259
Westbound	285

Figure 8 shows the ship speed distributions for ship traffic between ports and Eastbound/Westbound traffic. Average speeds are calculated for ships with speed over 1 knot to discard vessels at anchor or drifting. It is seen that the average ship speed of Eastbound/Westbound traffic is slightly higher than speed of local traffic (13.68 knot versus 11.36 knot), which is limited due to the geophysical characteristics of restricted waterways. Ships sail at their design speed in the open sea when environmental conditions permit. According to the ship speed distributions in Figure 8, most of the ships sail at 11–19 knots, which account for 77.1%.

4 CHARACTERISTICS OF THE MARITIME TRAFFIC

The ship type distributions of trajectories between ports and Eastbound/Westbound traffic are shown in Figure 7a and Figure 7b, respectively. Ship traffic between ports consists mainly of passenger and cargo ships, which account for 53.0% and 17.9%, respectively. Among Eastbound and Westbound traffic, cargo ships and tankers dominate the traffic accounting for 96.8% in total (cargo ships take 78.2% and tankers take 18.6%). The statistical analysis shows that the ship traffic within Azores Archipelago and Eastbound/Westbound traffic are used by different ship types.

Figure 8. Ship speed distribution.

Table 3 illustrates statistical summary of the speed distributions for ship traffic between ports and Eastbound/Westbound traffic. A *t*-test shows that the speed distributions of both traffics are well approximated by a normal distribution (*p*-value equal to 0.49 and 0.38, respectively). The dash and solid lines in Figure 8 represent the normal distribution fit for ship traffic between ports and Eastbound/Westbound traffic, respectively. It is seen that the standard deviation of the ship traffic between ports is smaller than that of Eastbound/Westbound traffic.

The ship length distribution of Westbound and Eastbound ship traffic is illustrated in Figure 9. It is seen that ships with the length beyond 100 meters account for 98.2% and 97.4% for Eastbound traffic and Westbound traffic, respectively. Moreover, more than half of the ships' lengths are between 160 and 240.

The extracted ship trajectories also enable the characterization of leaving port times. As an example, the ship trajectory analysis of Port Horta is shown in Figure 10. Port Horta is located on the southeast coast of Faial island, the ship trajectories leaving and entering the port are shown in Figure 10a. An analysis of ship distribution (Figure 10b) indicates that Port Horta is a ferry port as the ship traffic consist mainly of passenger ships which account for 85%. In addition, the daily and hourly patterns of outbound traffic can be obtained based on the analysis of ship trajectories (Figure 10c).

By analysing all ship trajectories of the traffic between ports, an overview of daily/hourly patterns of ship traffic can be obtained. Figure 11 shows the daily and hourly patterns of ship trajectories of traffic between ports and Eastbound/Westbound traffic. It should be mentioned that the time

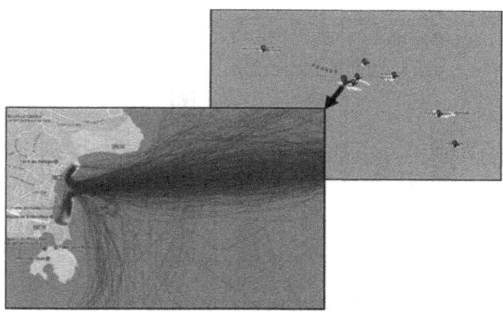

(a) Inbound and outbound of ship trajectories of Port Horta

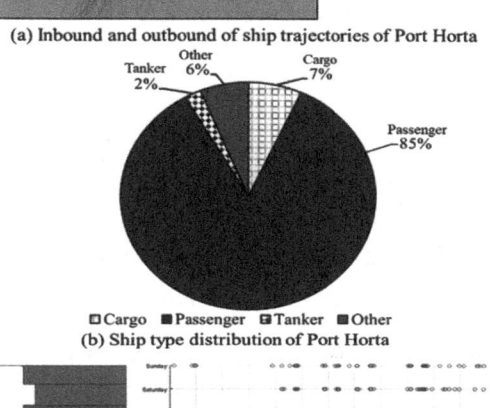

(b) Ship type distribution of Port Horta

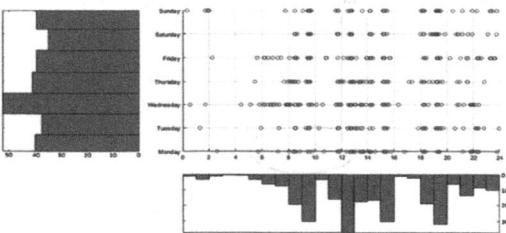

(c) Time histogram of outbound traffic of Port Horta

Figure 10. Analysis of ship trajectories of Port Horta.

of Eastbound/Westbound traffic refers to ship trajectories passing an observation line at longitude: 28°W. According to the daily pattern of ship trajectories (Figure 11a), both the marine traffic between ports and Eastbound/Westbound traffic are quite stable over weekdays. As the ship traffic between ports consists mainly of passenger ship. The multiple peaks of hourly pattern shown in Figure 11b are mainly due to the passenger ships' time schedule. However, the hourly pattern of Eastbound/Westbound traffic is evenly distributed.

The spatial-temporal analysis aims at investigating the ship traffic in space over time. In this study, an analysis of the evolution of sailing speed over time is performed. Specifically, the study area is divided into grids with the side length equivalent to 0.5 degree and a day is divided into daytime and nighttime. Then, the average speed within each grid is estimated based on the analysis of ship trajectories. Night-time is taken as 18:00pm-6:00am. The average speed distribution at the study area as function of the time of day is shown in Figure 12. It can be

Table 3. Summary of ship speed distributions.

	Mean	Std.	p-value
Traffic between ports	11.36	3.50	0.49
E/W traffic	13.68	4.41	0.38

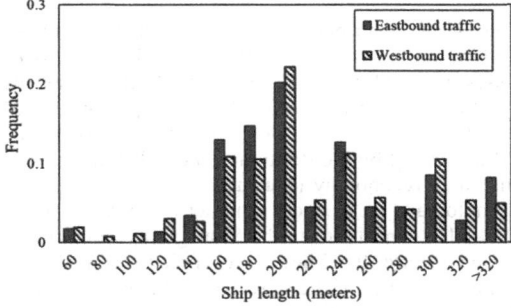

Figure 9. Ship length distribution for Eastbound and Westbound ship traffic.

189

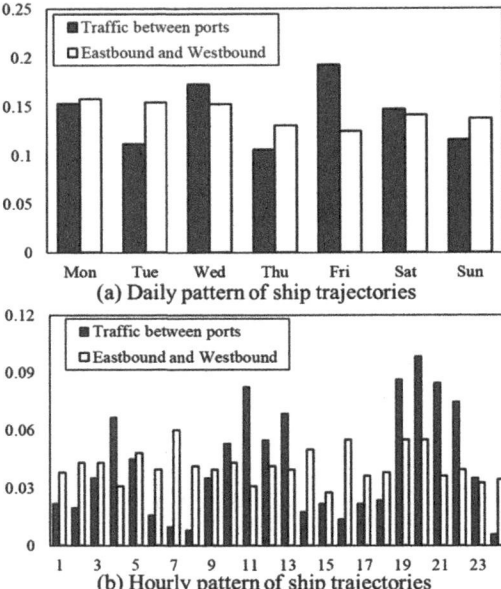

(a) Daily pattern of ship trajectories

(b) Hourly pattern of ship trajectories

Figure 11. Daily and hourly patterns of ship trajectories.

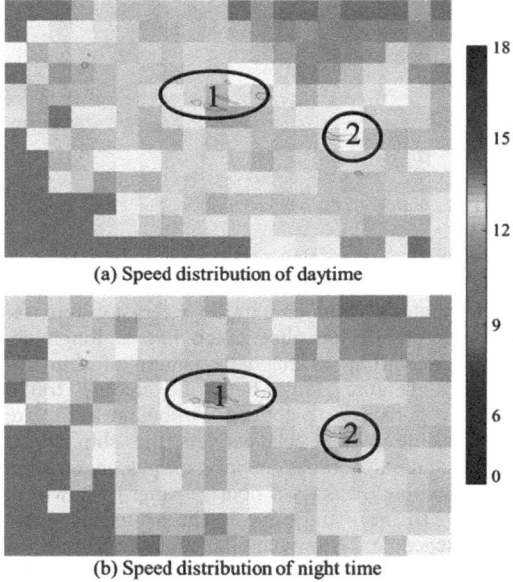

(a) Speed distribution of daytime

(b) Speed distribution of night time

Figure 12. Speed distribution at the study area divided by time of day.

seen that the overall speed distribution on study area do not change significantly between daytime and night, and the average speed near Azores Archipelago is typically lower than that in open sea. These areas are marked with number 1 and 2 in Figure 12. In addition, the speeds during the day are higher than that during the night in area 1 and 2.

5 CONCLUSIONS

This paper studies the maritime traffic off the Azores Archipelago based on collected historical AIS data. First, a data cleaning method is applied to filter erroneous position data and update speed data. A linear interpolation with a 10s-interval is performed for all ship trajectories regarding ship position, ship speed and heading in order to overcome the data missing problem.

After data pre-processing, a visualization of traffic density exhibits several well-defined traffic patterns within the study area. The maritime traffic on the study area consists mainly of traffic between ports and Eastbound/Westbound traffic. By integrating the main ports information of Azores Archipelago, a total number of 509 ship trajectories connecting the main ports are extracted resulting in 14 ship main routes at the study area.

In addition, a statistical analysis of the traffic between ports and Eastbound/Westbound traffic is performed. It is found that the traffic between ports consists mainly passenger and cargo ships (account for 53.0% and 17.9%, respectively). Regarding ship type distributions of Eastbound and Westbound traffic, ships with length larger than 100 meters accounted for 98.2% and 97.4%, respectively. The speed distributions of ship traffic between ports and Eastbound/Westbound traffic reveal that ships sail at higher speed in Eastbound/Westbound traffic than between ports, and most ships sail at 11–19 knots. Moreover, normal distributions are fitted to the ship traffic velocities, and it is found that the standard deviation of the ship traffic velocity between ports is smaller than that of Eastbound/Westbound traffic. The daily and hourly patterns of ship traffic are also analysed which provide an overview of ship traffic temporal pattern.

To better capture ship traffic in space and over time, the study area was divided into grids and average speed of each grid are estimated for daytime and night time. It is fount that average speed near Azores Archipelago is typically higher than that in open sea.

ACKNOWLEDGEMENTS

The paper contributes to the project "Integrated System for Traffic Monitoring and Maritime Risk Assessment (MoniRisk)", which has been co-funded by the European Regional Development Fund (Fundo Europeu de Desenvolvimento Regional (FEDER) and by the Portuguese Foundation for Science and Technology (Fundação para a Ciência e a Tecnologia – FCT) under contract no. 028746. This work also contributes to the Strategic Research Plan of the Centre for Marine Technology and Ocean Engineering (CENTEC), which is financed by the Portuguese Foundation for Science and Technology (Fundação para a Ciência e Tecnologia - FCT) under contract UIDB/UIDP/00134/2020.

REFERENCES

Mou, J.M. Tak, C.V.D. Ligteringen, H. 2010. Study on collision avoidance in busy waterways by using AIS data. *Ocean Engineering* 37(5–6): 483–490.

Qu, X., Meng, Q., Li, S. 2011. Ship collision risk assessment for the Singapore Strait. *Accident Analysis and Prevention* 43(6): 2030–2036.

Rong, H., Teixeira, A.P., Guedes Soares, C. 2016. Assessment and characterization of near ship collision scenarios off the coast of Portugal. *Maritime Technology and Engineering 3*, Guedes Soares, C. & Santos T. A., (Eds.), Taylor & Francis Group. London: 871–878.

Rong, H., Teixeira, A.P., Guedes Soares, C. 2018. A model for predicting ship destination routes based on AIS data. *Maritime Transportation and Harvesting of Sea Resources*, Guedes Soares & Teixeira (eds), Taylor & Francis Group, London: 257–264.

Rong, H., Teixeira, A.P., Guedes Soares, C. 2019a. Risk of ship near collision scenarios off the coast of Portugal, M. Beer and E. Zio (eds). *29th European Safety and Reliability Conference (ESREL 2019)*. Hannover, Germany.

Rong, H., Teixeira, A.P., Guedes Soares, C. 2019b. Ship trajectory uncertainty prediction based on a Gaussian Process model. *Ocean Engineering* 182: 499–511.

Rong, H., Teixeira, A.P., Guedes Soares, C. 2020a. Collision probability assessment based on uncertainty prediction of ship trajectories. *Developments in the Collision and Grounding of Ships and Offshore Structures*, Guedes Soares (ed.), Taylor & Francis Group, London: 283–290.

Rong, H., Teixeira, A. P., & Guedes Soares, C. 2020b. Data mining approach to shipping route characterization and anomaly detection based on AIS data. *Ocean Engineering*, 198, 106936.

Silveira, P.A.M., Teixeira, A.P., Guedes Soares, C. 2013. Use of AIS Data to Characterise Marine Traffic Patterns and Ship Collision Risk off the Coast of Portugal. *The Journal of Navigation* 66: 879–898.

Teixeira, A.P. & Guedes Soares, C., 2018. Risk of maritime traffic in coastal waters. *Proc.37th International Conference on Ocean, Offshore and Arctic Engineering*, Volume 11A: Honoring Symposium for Prof. Carlos Guedes Soares on Marine Technology and Ocean Engineering V11AT12A025, https://doi.org/10.1115/OMAE2018-77312

Xu H, Rong H, Guedes Soares C. 2019. Use of AIS data for guidance and control of path-following autonomous vessels. *Ocean Engineering* 194: 106635.

Zhang, L., Meng, Q., & Fang Fwa, T. 2017. Big AIS data based spatial-temporal analyses of ship traffic in Singapore port waters. *Transportation Research Part E: Logistics and Transportation Review* 129: 287–304.

Developments in Maritime Technology and Engineering – Guedes Soares & Santos (eds)
© 2021 Copyright the Author(s), ISBN 978-0-367-77376-2

Ship traffic risk complexity modelling based on complex network theory

A. Xuri Xin, C. Qing Yu & D. Xiaoli Wu
School of Navigation, Wuhan University of Technology, Wuhan, China

B. Kezhong Liu
School of Navigation, Wuhan University of Technology, Hubei Key Laboratory of Inland Shipping Technology,
National Engineering Research Center for Water Transport Safety, Wuhan, China

ABSTRACT: With the rapid increase in ship traffic volume, the traffic situation in some busy waters becomes more complex, bringing great challenges to the development of shipping industry. To fully understand the ship traffic situations and provide decision-making basis for maritime safety management, we propose a ship traffic risk complexity model based on complex network theory. First, a real-time ship-ship collision risk model that combines a dynamic ship domain model is developed to describe the between-ship proximity degree spatially and temporally. Then four complex indicators is selected and the Projection Pursuit Evaluation model is adopted to quantitatively evaluate the real-time traffic situations by synthesizing the multi-dimension problems to the projection value with one dimension. A case study is finally conducted using the AIS data in the outside waters of Ningbo-Zhoushan Port. Results show that the proposed model can help to identify the structural characteristics of ship traffic situations as well as detect the evolutionary characteristics of traffic situations.

1 INTRODUCTION

Maritime transport has contributed significantly to the world economy and around 10.7 billion tons of goods was transported by ships in 2018. With the increase of demand for cargo delivery, it is expected that the marine traffic will continue to increase over the next few decades. However, the significant increase in ship traffic will result in high traffic density and the increase in the traffic collision risk, especially in the busy waters. Moreover, the ship collision accidents usually pose threats to individuals and societies in terms of loss of life, damage to property, and environmental pollution. Therefore, the navigational safety of vessels is a matter of great concern to the crews, Maritime and Port Authority, and other relevant stakeholders.

Although many researchers have developed various models for collision risk assessment(Chen et al, 2018.; Zhen et al. 2017; Huang & van Gelder, 2019; Weng & Shan, 2015), the majority of these models emphasized on the ship-ship collision risk estimation. Obviously, it is inadequate to comprehensively assess the navigational risk from the perspective of the whole ship traffic network. (Wen et al., 2015) developed a marine traffic complexity model to evaluate the status of traffic situation, aiming at using the complexity to investigate the degree of crowding and risk of collision. (Zhang et al., 2019) proposed a New Vessel Conflict Ranking Operator (NVCRO) that combines density complexity and a multi-vessel collision risk operator for assessing

regional vessel collision risk. These model provided some references for ship traffic risk complexity modelling in the waters with high traffic intensities.

In recent years, the complex network theory has been increasingly applied to describe air traffic situation (Zekun et al, 2019; Wen et al., 2018). Complex network is an abstraction of abundant real complex systems and reflects various interactions and relations inside the complex systems. Similar to air traffic, a ship traffic situation is also a time-variable complex system and thus can be abstracted and described from the perspective of complex networks. As a result, this study will develop a ship traffic risk complexity model based on complex network theory to evaluate the real-time ship traffic conflict relationships and the ship traffic situation.

The contribution of this study is twofold. First, we develop a ship-ship collision risk measurement model that both considers the spatial and temporal relationship of encountering ships for the real-time collision conflict estimation. Second, we screen the complex network indicators for the ship traffic situation assessment and further process the multi-dimension data into one-dimension assessment problem by adopting the Projection Pursuit Evaluation model and the Moth-Flame Optimization. The proposed model is expected to quantitatively describe the real-time traffic situation and can be used as the foundation for maritime safety management.

DOI: 10.1201/9781003216582-21

2 METHODOLOGY

The methodology of the paper mainly includes three parts, which are ship-ship collision risk assessment model, selection of ship traffic complex network indicators, index weight determination method. The ship-ship collision risk model is used to unveil how dangerous of the ship-ship encounters, the complex network indicators are selected for reflecting the complexity of the ship traffic network, while the index weight determination method is used to transform multi-indicator evaluation problems into one-indicator problems.

2.1 Collision risk measurement for real-time ship-ship encounter

Lots of research (Zhen et al., 2017; Mou et al., 2010) applied both the Closest Point of Approach (DCPA) and Time to the Closest Point of Approach (TCPA) to measure the collision risk between ships. According to these research, we construct a real-time ship-ship collision risk measurement model based on a dynamic ship domain model, in which the between-ship proximity degree and the between-ship proximity degree changing rate are used to describe the spatial and temporal relationship of encountering ships, respectively.

(1) between-ship proximity degree

Ship domain has been widely concerned by scholars since it is an important concept for ship-ship collision risk. We can identify the scenarios with collision potentials according to the intrusion or overlap of ship domains. To reflect the effects of some factors such as ship size, ship speed, bearing etc. on the safety distance between ships, we adopted a dynamic ship domain model (see Figure 1) to identify the between-ship proximity degree. Moreover, it is the fourth type of ship domain-based safety criteria according to (Szlapczynski & Szlapczynska,

2017), so it can take into account both ships' attributes when measuring the collision risk of two approaching ships. The expression of the model is constructed as follows:

$$SD_i^x = r_x \cdot L_i \cdot g_x(v_i) \quad x = f, a, p, s \quad (1)$$

$$g_x(v_i) = \mu_x \cdot v_i^2 + \lambda_x \cdot v_i + 1 \quad (2)$$

where SD_i^x represents the distance from ship center to the domain boundaries in four axial directions; L_i and v_i are the ship i's length and speed; f, a, s and p represent the fore, aft, starboard and port side; r_x, λ_x and μ_x are the model coefficients.

According to the definition of the forth type domain, a ship collision scenario occurs when the ship domains overlap (see Figure 2). Thus, the proposed between-ship proximity degree is expressed as follows:

$$E_{ij}^A(t) = \frac{\left\| \vec{D}_{ij}^t \right\|}{d_i^\beta(t) + d_j^\beta(t)} \quad (3)$$

$$\left\| \vec{D}_{ij}^t \right\| = \sqrt{(x_i(t) - x_j(t))^2 + (y_i(t) - y_j(t))^2} \quad (4)$$

where $\left\| \vec{D}_{ij}^t \right\|$ is the distance of ship i and j at time t. $d_i^\beta(t)$ is the distance between ship i's center and the domain boundaries in the given bearings. Clearly, a smaller $E_{ij}^A(t)$ means that the two ships are spatially

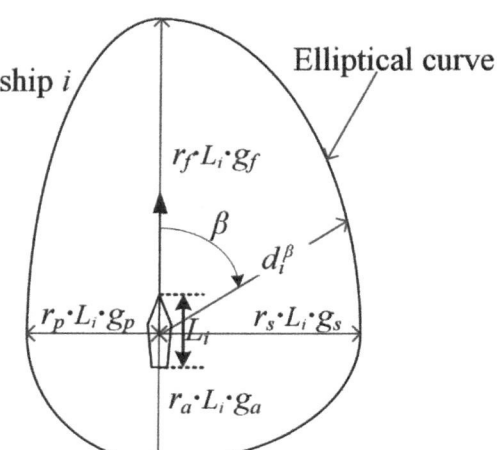

Figure 1. Illustration of dynamic ship domain.

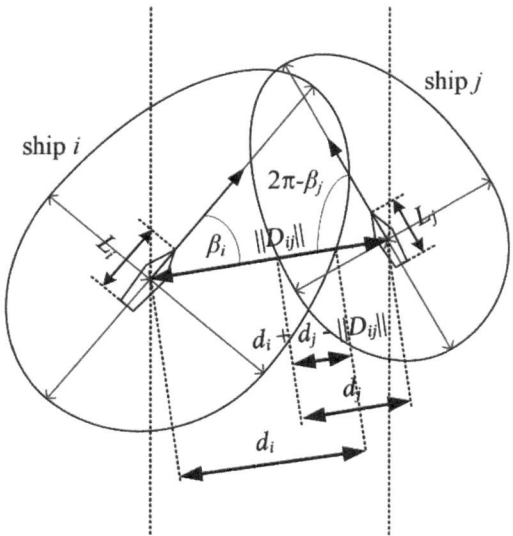

Figure 2. Illustration of domain overlap.

closer and corresponding collision risk is larger. Moreover, $E_{ij}^A(t) < 1$ indicates that the two ships' domains are overlapped.

(2) between-ship proximity degree changing rate

In real situation, the time factors like TCPA have an important influence on measuring the collision risk of encountering ships. For example, the time duration available for ship captains determines the difficulty of collision avoidance when two ships are approaching. Therefore, the between-ship proximity degree changing rate or called the spatial approaching rate is designed to represent the influence of time factors on ship collision risk, as follows:

$$\Delta V_{ij}^A(t) = E_{ij}^A(t) - E_{ij}^A(t-1) \tag{5}$$

From the formula, $\Delta V_{ij}^A(t) < 0$ means that the two ships are approaching. Correspondingly, the collision risk of the two ships will increase. In contrast, $\Delta V_{ij}^A(t) > 0$ indicates the two ships are diverging and their collision risk will decrease.

(3) Real-time ship-ship collision risk assessment model

According to the above analysis, the between-ship proximity degree and the spatial approaching rate are combined together to describe the between-ship proximity spatially and temporally, as follows:

$$C_{ij}^A(t) = \frac{1}{E_{ij}^A(t)} \cdot (1 - \alpha \cdot \Delta V_{ij}^A(t)) \tag{6}$$

where $C_{ij}^A(t)$ is the collision risk between ship i and j at time t, α is the adjustment coefficient for between-ship spatial proximity. From the formula, it can be found that $C_{ij}^A(t)$ will increase with the reduction of $E_{ij}^A(t)$. In addition, under the same between-ship proximity degree, a larger spatial approaching rate (larger divergence) implies a lower collision risk. Thus, the model can effectively compute the collision risk between ships.

It should be noted only when the distance between two encountering ships is less than the distance threshold (e.g. 6 n miles), they enter each other's early-warning zone and their collision risk need to be measured. Moreover, for two diverging ships, when their distance is larger than a relative small distance threshold (e.g. 2 n miles), their collision risk would not need to be paid attention to.

2.2 Ship traffic complex network indicators

Generally, a network is defined as a graph structure G composed of a node set and an edge set. In ship traffic complex network, the ships stand for the nodes $V = \{v_i \mid i \in N\}$ and the collision risk relationship are used to represent the edge between nodes. To evaluate the multi-ship traffic situation, the chosen indicators must have excellent representativeness for ship traffic network performance. Hence, Number of nodes, number of edges, node strength and clustering coefficient are selected for analysis.

(1) Number of nodes
Number of nodes N is the total number of nodes in the network. In general, a larger number of ships in a given waters means a higher density and complicity of the traffic situation.

(2) Number of edges
Number of edges is the total number of edges in the network, as follows:

$$D = \sum_{i=1}^N d_i / 2 \tag{7}$$

where d_i (called degree) is the number of connections between node v_i and the rest of nodes in the network.

(3) Strength
Node strength is defined as the sum of weights of the edges connected to one node v_i, as follows:

$$s_i = \sum_{j=1}^{d_i} w_{ij} \tag{8}$$

where w_{ij} is the weight (the real-time collision risk in this study) of edge between v_i and v_j. The strength of whole network is expressed as follows:

$$S = \sum_{i=1}^N s_i / 2 \tag{9}$$

(4) Clustering coefficient
Clustering coefficient describes the proximity of one node to other nearby nodes. If one node v_i is connected with d_i nodes via d_i edges, up to $d_i(d_i - 1)/2$ edges can be formed by these d_i nodes. The ratio of the real number of edges $N_\Delta(i)$ between these d_i nodes to the theoretical maximum number of edges is defined as the clustering coefficient C_i of node v_i:

$$C_i = \frac{2 \cdot N_\Delta(i)}{d_i(d_i - 1)} \tag{10}$$

The whole network's clustering coefficient C is defined as the sum of C_i of all nodes:

$$C = \sum_{i=1}^N C_i \tag{11}$$

2.3 Index weight determination method

According to the above analysis, the ship traffic risk complex assessment is a high-dimensional, complex, and variable problem. For these type of problems, lots of methods can be chosen, including Analytic hierarchy process (AHP), entropy weight method, the Technique for Order Preference by Similarity to Ideal Solution (TOPSIS), fuzzy comprehensive evaluation, gray relational analysis, and support vector machine (SVM). Each of these methods has their own disadvantages. For example, the gray relational analysis has a relative low resolution. The fuzzy comprehensive assessment method have some deficiencies in selecting assessment factors, determining weights and choosing operations.

Projection Pursuit Evaluation (PPE) model was first proposed by (Friedman & Tukey, 1974). It is a data-driven method which can avoid the interference of subjective factors, so the results are more objective and reasonable compared with the above traditional methods. The principle of the method is to project high-dimensional data to a low-dimensional space and explore these characteristics according to projection index values that reflect the structures and characteristics of high-dimensional data. Therefore, this method opens a path to solve high-dimensional problems using a one-dimensional method and it is chosen for the multi-indicator assessment of ship traffic network complicity.

The basic modeling steps of PPE model are as follows:

(1) Data normalization

Different evaluation indexes have different magnitude and different ranges, so the evaluation indexes should be normalized before decision-making. In this study, as a larger value of each chosen indicator means a higher complexity (called benefit indicators), they should be normalized as follows:

$$x(i,j) = \frac{x'(i,j) - x_{\min}(j)}{x_{\max}(j) - x_{\min}(j)} \quad (12)$$
$$i = 1, 2, \cdots, n; j = 1, 2, \cdots, m.$$

where $x'(i, j)$ is the ith sample of the jth indicator, $x_{\max}(j)$ and $x_{\min}(j)$ are the maximum and minimum values of the jth index, n and m are the number of samples and the number of evaluation indicators, and $x(i, j)$ is the index value after normalization.

(2) Construction of the projection index function

After obtaining the projection direction $p = \{p(1), p(2), p(3), \ldots, p(m)\}$, the n dimension data can be synthesized to the projection value $h(a)$ with one dimension, as follows:

$$h(i) = \sum_{j=1}^{m} p(j)x(i,j) \quad i = 1, 2, \cdots, n \quad (13)$$

where p is a unit vector.

The projection index function is constructed as follows:

$$Q(p) = S_h \cdot D_h \quad (14)$$

where S_h and D_h are the standard deviation and local kernel density of $h(i)$, respectively.

$$S_h = \sqrt{\frac{\sum_{i=1}^{n} (h(i) - E(h))^2}{n-1}} \quad (15)$$

$$D_h = \sum_{i=1}^{n} \sum_{j=1}^{n} (R - r(i,j)) \cdot u(R - r(i,j)) \quad (16)$$

where $E(h)$ is the expectation value of $h(i)$, $r(i, j)$ is the distance between $h(i)$ and $h(j)$, R is the window radius and is set to $1/4 \times r_{\max}(i, j)$, $u(i)$ is a unit step function as follows:

$$u(R - r(i,j)) = \begin{cases} 1, R - r(i,j) > 0 \\ 0, R - r(i,j) < 0 \end{cases} \quad (17)$$

(3) Optimization of the projection index function

When the projection index function takes the maximum value, the corresponding projection direction is the best projection direction which can best reflect the data characteristics. Therefore, the problem of finding the best projection direction is transformed into solving the following constrained optimization problem.

$$\begin{cases} Max \ Q(p) = S_h \cdot D_h \\ s.t. \sum_{j=1}^{m} p^2(j) = 1 \end{cases} \quad (18)$$

The Moth-Flame Optimization proposed by (Mirjalili) is a great optimization algorithm by simulating the flight mode of moths adjusting their flight direction according to the light at night. Its feasibility and reliability has been proved by lots of scholars and has broad application prospects, so in this study this algorithm is used to determine the projection direction.

3 CASE STUDY

3.1 Data collection

In order to evaluate the ship traffic risk complexity model, a case study is conducted by using the AIS data in the outside waters of Ningbo-Zhoushan Port

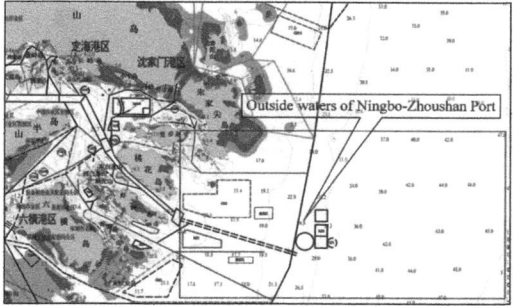

Figure 3. The Ningbo-Zhoushan Port and its outside waters.

since the AIS data is extremely promising to obtain deep insights into the safety of maritime transportation. The Ningbo-Zhoushan Port (see Figure 3) is one of the largest ports in the world in terms of cargo throughput and is of great importance to the Local economic development. The outside waters is located between longitudes 122°18E-122°50E and between latitudes 29°35N-29°52N, and link with two importance routes, namely the Shrimp main gate waterway and the Broom gate waterway. Moreover, lots of ships are anchorage at the area to wait for the scheduling optimization by maritime and port administrations before entering the port. Thus, it is a main area for ship-ship encounters and the ships' navigational safety is of utmost concern to the Maritime and Port Authorities.

In view of the importance of ship traffic safety in the area, we collected one month's AIS data from October. 21, 2018 to November. 31, 2018. The information in AIS data including MMSI (Maritime Mobile Service Identity) number, time, position (longitude and latitude), speed over ground (SOG), course over ground (COG), heading, ship type and ship size, is extracted for carrying out the experiments. As there are many errors in the original dataset, the data

processing method was employed to eliminate the inaccurate data records before testing the model.

3.2 Experiments and result analysis

At first, we selected a scenario to validate the real-time ship-ship collision risk assessment model. The scenario involved two encountering ships and their trajectories in geometric space are shown in Figure 4 (a). In the figure, 'x' marks the ships' starting locations and 'o' their final locations. Clearly, the two ships are converging and ship 2 has changed its direction for collision avoidance before reaching their minimal distance. Figure 4 (b) provides the corresponding evolution process of between-ship collision risk, in which Risk 1 and Risk 2 represent the collision risk without consider time factors and the proposed model. Both curves increase before two ships reach the minimal distance and then decline, which are in line with the characteristics of risk changes in the course of ship encounters. However, compared with Risk 1, Risk 2 has considered the spatial approaching rate, which can reflect the influence of the change rate of space proximity. Correspondingly, Risk 2′ collision risks are higher when two ships are converging and lower when two ships are diverging compared to Risk 1. Thus, the proposed model can provide us with more satisfactory results for the real-time collision risk assessment.

To obtain the ship traffic risk complexity, the weight of different complex indicators need to be determined by using PPE model. After executing the MFO optimization algorithm, the optimal projection direction $p = [0.76, 0.21, 0.56, 0.28]$ were obtained. As the projection direction value represents the importance of the indicators, the importance degree is ranked as follows: $N > S > C > D$.

Based on the above projection direction results, a ship traffic risk network complexity diagram is illustrated (see Figure 5) and a case of ship traffic risk complexity evolution on November 1st, 2018 is

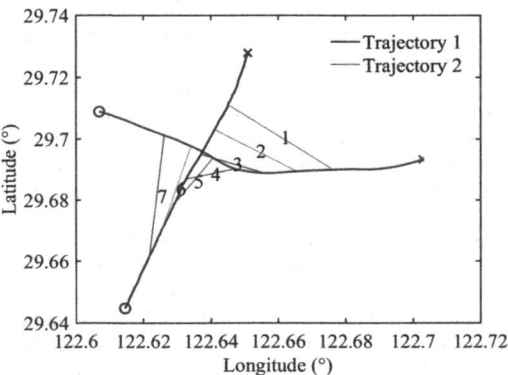

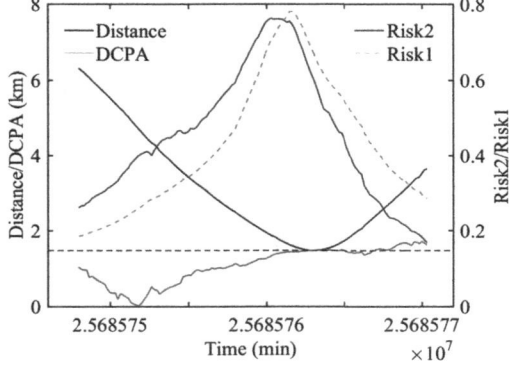

Figure 4. (a) Ship trajectories in geometric space; (b) evolution process of between-ship collision risk.

197

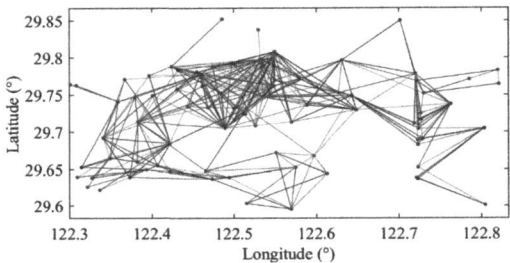

Figure 5. Illustration of ship traffic network at 02:00:00 on November 1st.

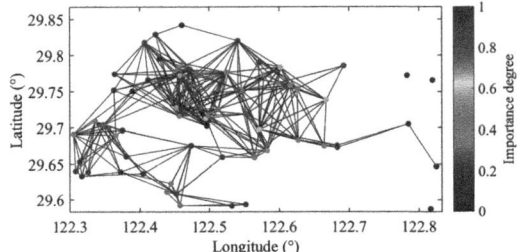

Figure 7. Illustration of node importance identification based on node deletion method.

Table 1. Case study of complexity index evolution.

Time	Number of nodes	Number of edges	Strength	Clustering coefficient	Projection value
02:00:00	89	458	72.9	44.0	0.81
02:10:00	93	522	98.1	41.4	0.90
02:20:00	92	520	87.6	46.3	0.88
02:30:00	92	522	76.2	42.9	0.84
02:40:00	90	460	63.5	44.8	0.80
02:50:00	86	349	48.4	40.3	0.71

further presented (see Table 1). At 02:00:00, there are 89 ships and the whole traffic situation is medium complexity. Next, the number of ships in the area gradually decreases and the complexity indicators change accordingly, the complexity of the traffic situation grows. After passing the peak, the ship successively leave the area and the complexity of the traffic situation decreases. Last, the number of ships further declines and the ship traffic situation become low complexity. In terms of this kind of analysis, we can further calculate the life cycle and transition probability of different complexity patterns in the future.

Figure 6 provides the statistical probabilities of ship traffic risk situation complexity patterns for different time periods. The ship traffic complexity was

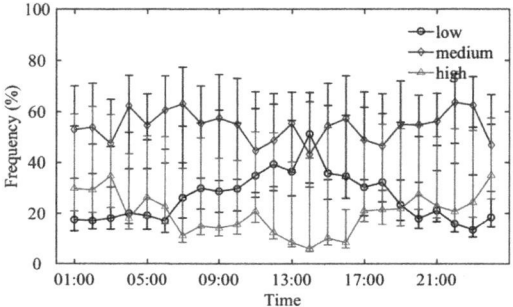

Figure 6. Frequency of occurrence of different patterns.

divided into three patterns according to the projection values, which are low, medium and high-complexity patterns. As the time in AIS data is reported in Universal Time Coordinates (UTC), it can be see that the traffic risk situation complexity patterns are higher at 06:00 am-14:00 pm local time (22:00 pm-06:00 am in Figure 6). As a result, the navigation safety of ships during these periods should be paid more attention to by the Maritime and Port Authorities.

In addition, each ship's risk level in the ship traffic network can be identified by using node deletion method. We can delete a node (represent a ship) and then calculate the network performance. By compare it with the original network, the change degree of network performance can be determined. A greater change of the network performance means that the node is more important. Figure 7 provides an illustration about real-time ship collision risk degree identification based on node deletion method. From the figure, we can easily find which ship has the most importance degree. This is expected to facilitate the Maritime and Port Authorities to identify the ships that need to be monitored in real time.

4 CONCLUSIONS

In this study, we develop a ship traffic risk complexity model based on complex network theory to evaluate the interactions among ship and the overall behavioral characteristics of the ship traffic system. We first establish a real-time ship-ship collision risk measurement model by considering between-ship proximity degree and between-ship proximity degree changing rate to describe the spatial and temporal relationship of encountering ships. After that, we select four complex network indicators to reflect the complexity of the ship traffic network, and further adopt PPE model to process the multi-dimension data into one-dimension assessment problem.

Finally, the AIS data in the outside waters of Ningbo-Zhoushan Port is collected to conduct experiments to test the performance of the model. Results show that the model can exhibit the real-time traffic situation directly, and help crews and traffic controllers monitor the traffic situation.

It should be noted that the proposed model still has a room for improvement. For example, the value of coefficient α in the real-time ship-ship collision risk model need to be further analyzed in the future to make a trade-off between the influence of space factor and time factor on the collision risk of encountering ships. In addition, it would be interesting to determine the weight of complex network indicators by combining the subjective knowledge and objective methods.

REFERENCES

Chen, Pengfei, et al. "Ship Collision Candidate Detection Method : A Velocity Obstacle Approach." *Ocean Engineering*, vol. 170, 2018, pp. 186–98, doi:10.1016/j.oceaneng.2018.10.023.

Friedman, Jerome H., and John W. Tukey. "A Projection Pursuit Algorithm for Exploratory Data Analysis." *IEEE Transactions on Computers*, vol. 100, no. 9, IEEE, 1974, pp. 881–90.

Huang, Yamin, and PHAJM van Gelder. "Time-Varying Risk Measurement for Ship Collision Prevention." *Risk Analysis*, Wiley Online Library, 2019.

Mirjalili, Seyedali. "Moth-Flame Optimization Algorithm: A Novel Nature-Inspired Heuristic Paradigm." *Knowledge-Based Systems*, vol. 89, 2015, pp. 228–49.

Mou, Jun Min, et al. "Study on Collision Avoidance in Busy Waterways by Using AIS Data." *Ocean Engineering*, vol. 37, no. 5-6, 2010, pp. 483–90.

Szlapczynski, Rafal, and Joanna Szlapczynska. "Review of Ship Safety Domains: Models and Applications." *Ocean Engineering*, vol. 145C, 2017, pp. 277–89.

Wen, Xiangxi, et al. "Node Importance Evaluation in Aviation Network Based on 'No Return' Node Deletion Method." *Physica A: Statistical Mechanics and Its Applications*, vol. 503, 2018, pp. 546–59.

Wen, Yuanqiao, et al. "Modelling of Marine Traffic Flow Complexity." *Ocean Engineering*, vol. 104, 2015, pp. 500–10.

Weng, Jinxian, and Xue Shan. "Ship Collision Frequency Estimation in Port Fairways: A Case Study." *Journal of Navigation*, vol. 68, no. 3, 2015, pp. 602–18.

Zekun, Wang, et al. "Identification of Key Nodes in Aircraft State Network Based on Complex Network Theory." *IEEE Access*, vol. 7, IEEE, 2019, pp. 60957–67.

Zhang, Weibin, et al. "Towards a Model of Regional Vessel Near-Miss Collision Risk Assessment for Open Waters Based on AIS Data." *The Journal of Navigation*, vol. 72, no. 6, 2019, pp. 1449–68.

Zhen, Rong, et al. "A Novel Analytic Framework of Real-Time Multi-Vessel Collision Risk Assessment for Maritime Traffic Surveillance." *Ocean Engineering*, vol. 145, 2017, pp. 492–501.

Developments in Maritime Technology and Engineering – Guedes Soares & Santos (eds)

Research on the evaluation method of a navigation plan based on the fuzzy comprehensive evaluation

Y.F. Zhang & J. Qiao

Shanghai Ship and Shipping Research Institute, Shanghai, China

ABSTRACT: The evaluation of navigation plan can help the crew to choose the safest and most economical route, but there are many factors involved in the evaluation of navigation plan, and it is a very complex problem to choose a suitable method to consider all factors comprehensively. In this paper, the fuzzy comprehensive evaluation method is used to calculate the weight of each factor, the ship factors, the environment factors and the economic cost factors. Two actual navigation plans are used to verify the feasibility of this method, and the calculated results can meet the expectation of the crew.

1 INTRODUCTION

The safety of ship navigation has always been the focus of the shipping industry. In the process of designing the route, analyzing the meteorological information that the ship may encounter and predicting the movement of the ship in the course of navigation can effectively evaluate the safety of the ship route.

Since the beginning of the 21st century, maritime accidents are still frequent. In June 2008, the Philippine "Princess of the Stars" ferry sank due to the impact of typhoon, and only 42 passengers survived ("Headed for Disaster: The Last Voyage of M/V Princess of the Stars."). In April 2014, the South Korean ferry "Sewol" sank in the sea area near South Korea, and only 172 people survived on board ("In the Absence: South Korea's Sewol Ferry Disaster."). In June 2015, China's ferry "Dongfang zhi Xing" encountered a rare severe convective weather in the Yangtze River The midstream sank, with 454 people on board and 442 killed ("China Rights Capsized Yangtze Ship."). In order to improve the safety management of ship transportation, it is necessary to evaluate the risk of ship route, so as to take accurate safety measures to ensure the safety of navigation.

2 THE STUDY OF FACTORS

When making the navigation plan, the crew will comprehensively design the important waypoints of the route according to the ship's conditions, weather, hydrological conditions and the data of the main navigation areas.

Navigation safety is the key factor that the pilot first considers, and the navigation safety of the ship mainly involves the safety condition of the ship itself, the sea condition of the planned route and the sea condition in the navigation process. However, the fuel cost also play an important role on select better route, the economic factors of the ship operation should be included as well. Therefore, the evaluation of the planned route of the target ship mainly involves the factors of the ship itself, the environmental factors in the route and the economic factors.

Ship's own factors:

The ship's own factors will also affect the navigation safety, and for new ships, ship age and tonnage can best reflect the ship's own factors.

1) If the draft, trim, loading state and stability of the ship are not up to the standard requirements, it leads to dangerous navigation and even capsizing, which will affect the safety of the ship;
2) The structural strength of the hull also has a greater impact on the navigation safety of the ship. The structural condition of the hull has a greater problem, and the deck and important structures have a potential safety hazard, which will affect the navigation safety of the ship;
3) The ship's equipment also affects the ship's navigation safety. If the navigation aid or navigation equipment is abnormal, the ship's navigation will lack necessary auxiliary instruments.

Route geographical environment factors:

There may be all kinds of obstacles in the navigation area of the planned route, such as open reef, reef, shallow water area, intertidal zone, sunken ship and no navigation area, etc. the existence of obstacles will seriously affect the navigation safety, and encounter these obstacles will lead to the ship grounding or even sinking.

DOI: 10.1201/9781003216582-22

Meteorological factors:

1) Wind, if the wind exceeds the safe wind level of the ship, the navigation of the ship will be greatly affected;
2) Wave, the wave will make the ship body move violently. If the ship is in the state of no-load or half load, the ship will be empty when sailing in the huge waves. At the same time, the increase of the frequency of the waves on the deck will also lead to the damage of the ship structure;
3) Visibility, the visibility reduction caused by fog, haze, rain, snow and other reasons will seriously affect the navigation safety of the ship;
4) The size of the water area and the depth or width of the water area passing by the navigation do not meet the minimum navigation standard of the target ship, which will also affect the navigation safety of the ship.

The environmental factors of the route are the key factors to determine the safety of the route. Before sailing, it is the most important task to study the safety of the route. If the safety factor is not up to the standard, it will directly determine that the route cannot be the planned route of the target ship. Therefore, the target ship needs to re-select the route that meets the safety standard as the main route. Generally speaking, whether the route can be safely navigable depends on the environmental factors of the route. Wind speed, wave height, visibility, ice area, storm, water depth and reef will affect the safety of the route. Once encountering the serious environment, the maritime directors of many countries strictly will prohibit the navigation of ships. Therefore, once there is bad weather on the planned route, the level of the route should be directly unsafe and the target ship is prohibited from sailing along the planned route.

In the case of ensuring the safety of navigation, economic factors should also be taken into account to evaluate the level of the route, the most direct embodiment of which is the cost control. Ship as a large fixed asset will produce tangible or intangible loss of equipment in the process of use. Assuming that the service life of the target ship is a fixed value, using the average service life method to calculate the depreciation cost, we can find that reducing the service life of each voyage can increase the utilization rate of the ship, so as to obtain more profits within the fixed service life, indirectly reduce the depreciation cost of the ship, increase the total number of voyage and improve the operating profit of the ship. In addition, in a certain voyage, the reduction of fuel consumption cost can directly reduce the operating cost of the voyage. Therefore, the economic factors to evaluate the level of navigation plan can consider the reduction rate of ship voyage time and fuel consumption cost.

In short, according to the data provided by the planned route of the target ship, the feasibility evaluation basis is as follows: the loading condition of the ship, the meteorological condition in the navigation area, the fuel consumption used for navigation, the navigation time, the service years of the ship and the tonnage of the ship. The above factors provide a reliable basis for the evaluation of navigation plan level in this chapter.

3 CONSTRUCTION OF EVALUATION METHOD

3.1 *Principle of fuzzy comprehensive evaluation method*

The fuzzy comprehensive evaluation method can transform the qualitative evaluation of the problem into the quantitative evaluation, and use the fuzzy mathematics method to make an overall and objective evaluation of the events restricted by multiple factors. Fuzzy comprehensive evaluation can clearly divide multiple factors into factor set or comment set (Kahraman, 2003).

In the course of route evaluation, the environmental factors, economic factors and ship's own factors in the course of navigation are taken into account, and then the tiny factors of various factors on route evaluation are considered respectively.

The fuzzy comprehensive evaluation method based on AHP uses the normative thinking in AHP to reduce the subjective randomness. According to the characteristics of each factor, this method divides factors into different categories of factor sets, which are arranged from low to high (Forman, 1998). This method needs to carry out primary comprehensive evaluation for each level, and then carry out higher-level comprehensive evaluation for each major category of factor set on this basis comprehensive evaluation can get a quantitative comprehensive evaluation result without too much subjective intention (Vaidya, 2006).

3.2 *Evaluation model of ship navigation plan level*

The evaluation system of navigation plan grade is composed of top layer, middle layer and bottom layer. The top level is the result of comprehensive evaluation of navigation plan level, and the middle level includes navigation safety factor, navigation economy factor and ship factor. The basic steps are shown in Figure 1.

The main steps of using AHP based fuzzy comprehensive evaluation to evaluate the navigation plan are as follows:

(1) Build factor set
Establish factor set $U=\{U_1, U_2,...,U_m\}$, and the evaluation object factor set u is a set of M evaluation indexes. According to the evaluation index system established in the previous section, the main factor set of route registration evaluation can be obtained as $U=\{U_1, U_2\}$; the second level subset is: $U_i = \{U_{i1}, U_{i2},..., U_{im}\}$
(2) Define the domain of hierarchy
Determine the comment set of each level, $V=\{V_1, V_2,..., V_s\}$. Where, $Vi(i=1,2,...,s)$ indicates the ith level in S-level.

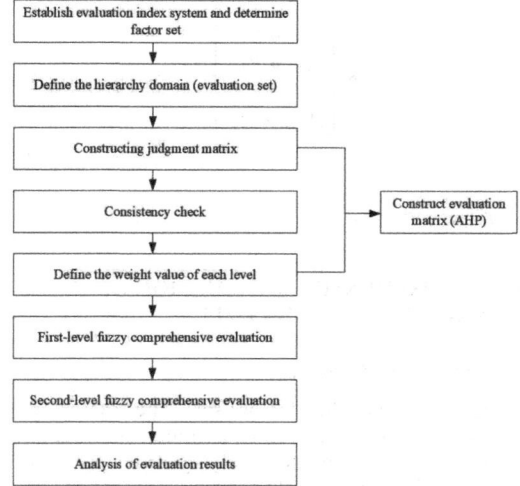

Figure 1. Navigation plan evaluation process.

(3) Determine weight matrix

According to the importance of the comprehensive evaluation index of route evaluation in each level, the weight number of each level factor set and single factor in each factor set is determined by AHP. The first level weight set is $A = (a_1, a_2, ..., a_m)$; the second layer weight set is $A_i = (a_{i1}, a_{i2}, ..., a_{im})$, where $i = 1, 2, ..., n$.

1) Constructing a judgment matrix

The judgment matrix is determined according to the importance of each influencing factor at the same level. The value of each element in the judgment matrix is determined by the 1-9 scale method, and it is carried out in combination with the opinions of experts.

$$A = (a_{ij})_{n*n} = \begin{bmatrix} a_{11} & a_{12} & \cdots & a_{1n} \\ a_{21} & a_{22} & \cdots & a_{2n} \\ \cdots & \cdots & \cdots & \cdots \\ a_{n1} & a_{n2} & \cdots & a_{nn} \end{bmatrix} \quad (1)$$

In the equation 1, a_{ij} is the value of i index compared with j index, using 1-9 scale method, as shown in Table 1.

2) Consistency check

To ensure the accuracy of the judgment matrix, it is necessary to check the consistency of the judgment matrix, and the test formula is shown in equation 2.

$$CI = \frac{\lambda_{\max} - n}{n - 1} \quad (2)$$

The closer the CI value is to 0, the less consistent the subjective judgment is, the more reasonable the matrix is. To achieve the consistency of evaluation criteria for planning processing, select RI value indicators, for $n = 1, 2, ..., 9$, and the average random consistency index is shown in Table 2. Therefore, the consistency index of the judgment matrix is equation 3.

$$CR = \frac{CI}{RI} < 0.10 \quad (3)$$

When $CR < 0.10$, the construction of the judgment matrix is reasonable.

3) Define the weight loss of each level

The method of product square root is mainly used to determine the weight of the layered index, as shown in equation 4.

$$b_i = \sqrt[n]{\prod_{j=1}^{n} a_{ij}} (i = 1, 2, ..., n) \quad (4)$$

And then normalize b_i, shown as equation 5.

$$\omega_j = \frac{b_i}{\sum_{k=1}^{n} b_k} (j = 1, 2, ..., n) \quad (5)$$

Finally, the maximum eigenvalue is calculated as equation 6.

Table 2. Consistency index of average random.

n	1	2	3	4	5	6	7	8	9
RI	0.0	0.0	0.58	0.90	1.12	1.24	1.32	1.41	1.45

Table 1. Analytic hierarchy process of the scaling method.

	Equally important	Slightly important	Important	Very important	Extremely important
Value	1	3	5	7	9
Note	Take 2, 4, 6 and 8 as the middle value of the above evaluation values				

$$\lambda_{\max} = \frac{1}{n}\sum_{i=1}^{n}\frac{\sum_{j=1}^{n}\alpha_{ij}\omega_j}{\omega_i} \qquad (6)$$

4) First-level fuzzy comprehensive evaluation

Each factor of the first level is determined by the sub-factors of the next level, and the single factor of this level is based on the comprehensive evaluation of multiple factors of the next level. The second level single factor evaluation matrix is set as R_i, and the number of rows of the matrix is determined by the number of factors in the second level of U. the matrix is shown as equation 7.

$$R_i = \begin{bmatrix} r_{i11} & r_{i12} & \cdots & r_{i1p} \\ r_{i21} & r_{i22} & \cdots & r_{i2p} \\ \cdots & \cdots & \cdots & \cdots \\ r_{in1} & r_{in2} & \cdots & r_{inp} \end{bmatrix} \qquad (7)$$

After weighting, the first level fuzzy comprehensive evaluation vector Bi is obtained as equation 8.

$$Bi = Ai \bullet Ri = [bi1, bi2, \ldots, bip] \qquad (8)$$

In the equation 8, $\bullet$ is a fuzzy operator, which can use "and", "or" and matrix multiplication. In this paper, the standard matrix multiplication operator is used.

5) Second-level fuzzy comprehensive evaluation

Each U_i is regarded as an element, and B_i is a single factor evaluation factor, shown as equation 9, which can form the evaluation matrix Ri again.

$$B_i = \begin{bmatrix} B1 \\ B2 \\ \cdots \\ Bn \end{bmatrix} = \begin{bmatrix} A_1 \bullet R_1 \\ A_2 \bullet R_2 \\ \cdots \\ A_m \bullet R_m \end{bmatrix} \qquad (9)$$

The second-level fuzzy comprehensive evaluation is shown as equation 10.

$$B = A \bullet R_i = A \bullet \begin{bmatrix} A_1 \bullet R_1 \\ A_2 \bullet R_2 \\ \cdots \\ A_m m \bullet R_m \end{bmatrix} = (b_1, b_2, \ldots, b_p) \qquad (10)$$

4 COMPREHENSIVE EVALUATION OF NAVIGATION PLAN LEVEL

4.1 Define the factor set

First, determine the factor set. The factor set needs to consider the use of as much information as possible to comprehensively consider the level of the navigation plan. Determining the factor set can make a more scientific judgment standard. Therefore, this paper makes a necessary selection of the factors that affect the level of a navigation plan. The first level indicators are divided into two categories: navigation environment factor U1, route economic factor U2 and ship factor U3; the second level indicators are wind speed U11, relative wind direction U12, wave height U13, relative wave direction U14, visibility U15, time-saving cost U21, energy-saving cost U22, ship age U31 and shipping tonnage U32, as shown in Table 3.

4.2 Determine the domain of hierarchy

Generally, the evaluation level should be differentiated. If the evaluation level is too simple, it will lead to a significant error in the navigation plan level. If the evaluation level is too detailed, the evaluation process is too complicated. Therefore, the classification level is divided into five sections, which is more in line with the commonly used criteria at home and abroad, $V=\{V_1, V_2, V_3, V4, V_5\}$, which can be expressed as $V=\{poor, fail, pass, good, excellent\}$, the corresponding score in Table 4 is a fuzzy number so that it can be quantified.

Table 3. Factor set for route evaluation of target ship.

	Factor								
First level index	Environment factor U1					Economic factor U2		Ship factor U3	
Second level index	Wind velocity U11	Relative Wind direction U12	Wave height U13	Relative Wave direction U14	Visibility U15	Time cost U21	Fuel cost U22	Age of ship U31	Shipping ton U32

Table 4. Grade evaluation score in this paper.

Value	Comment sets
1	Poor
2	Fail
3	Pass
4	Good
5	Excellent

4.3 Define weight value

In this step, we need to combine the expert scoring method to compare the two evaluation indexes to obtain statistical data to form a judgment matrix.

Through analysis and calculation, the absolute weight and relative weight of each factor are obtained, as shown in Table 5, Table 6 and Table 7. As there are only two secondary influencing factors of economy factor U2 and ship factor U3, consistency inspection is not suitable.

4.4 Comprehensive evaluation

In order to further calculate the level of navigation plan, each influencing factor is evaluated according to Table 4 and the expert's suggestions. Table 8 is the judgment basis of wind speed U11, relative wind direction U12, wave height U13, relative wave

Table 5. Evaluation result of the comparison of the importance of the influence factors in the target layer.

Target layer index	Weight value	Consistency check
Environment factor U1	0.640	
Economic factor U2	0.160	$\lambda_{max} = 3.009$ $CR = 0.0077$
Ship factor U3	0.200	

Table 6. Evaluation result of the comparison of importance in factors influencing the criterion layer.

Criterion layer index	Weight value	Consistency check
Wind velocity U11	0.116	
Relative wind direction U12	0.052	
Wave height U13	0.404	$\lambda_{max} = 5.07$ $CR = 0.0156$
Relative Wave direction U14	0.214	
Visibility U15	0.214	
Time cost U21	0.333	Un-consistency check
Fuel cost U22	0.667	
Age of ship U31	0.5	Un-consistency check
Shipping ton U32	0.5	

direction U14 and visibility U15 in environmental factors; Table 9 is the judgment basis of time-saving cost U21 and energy-saving cost U22 in economic factors; Table 10 is the judgment basis of ship age U31 and ship tonnage U32 in ship factors.

Wind, wave and visibility rating criteria are mainly based on meteorological forecast standard (Meaden et al., 2007). In order to classify the wind direction and wave direction, the absolute wind direction and wave direction should be converted into relative wind direction and wave direction according to the heading. The crosswind and wave are relative wind direction angles, and the relative wave direction angles are in the range of 30 ° - 60 °.

In the comprehensive evaluation, the most important thing is to consider the safety of the ship. Therefore, the evaluation standard of the navigation plan

Table 7. Grade evaluation score in this paper.

First level index	Weight	Second level index	Relative weight	Relative weight
Environment factor U1	0.640	Wind velocity U11	0.116	0.074
		Relative wind direction U12	0.052	0.033
		Wave height U13	0.404	0.259
		Relative Wave direction U14	0.214	0.137
		Visibility U15	0.214	0.137
Economic factor U2	0.160	Time cost U21	0.333	0.053
		Fuel cost U22	0.667	0.107
Ship factor U3	0.200	Age of ship U31	0.5	0.10
		Shipping ton U32	0.5	0.10

Table 8.   Judgment basis of environmental factor comments.

Value	Comment sets	Wind speed	Wind direction	Wave height	Wave direction	Visibility
1	Poor	Wind level > 11	Over 80% of the segments are crosswind	Wave height of some segments >11m	Over 80% of the segments are crosswave	Visibility of some segments are 1
2	Fail	Wind level > 7 or Wind level < 11	60%-80% of the segments are crosswind	Wave height of some segments 6m~11m	60%-80% of the segments are crosswave	Visibility of some segments are 2-3
3	Pass	The number of way-points of wind level 5-7 ≥ 50%	40%-60% of the segments are crosswind	More than 50% of route wave height 2.4m-6m	40%-60% of the segments are crosswavecrosswave	Visibility of some segments are 3-4
4	Good	The number of way-points of wind level 3-5 ≥ 50%	20%~40% of the segments are crosswind	More than 50% of route wave height 1.25m-2.4m	20%~40% of the segments are crosswave	Visibility of some segments are 4-6
5	Excellent	The number of way-points of wind level less 3 ≥ 50%	Less 20% of the segments are crosswind	Less than 50% of route wave height <1.25m	Less 20% of the segments are crosswave	Visibility of the whole segment > 6

Table 9.   Judgment basis of economic factor comments.

Value	Comment sets	Energy-saving ratio	Time-saving ratio
1	Poor	≤ 0.5%	≤0.5%
2	Fail	0.5%~1.0%	0.5%~1.0%
3	Pass	1%~1.5%	1%~1.5%
4	Good	1.5%~2.5%	1.5%~2.5%
5	Excellent	≥2.5%	≥1.5%

Table 10.   Judgment basis of ship factor comments.

Value	Comment sets	Age of ship	Shipping ton
1	Poor	≥ 21 years	≤ 30000 ton
2	Fail	15-20 years	30000-50000 ton
3	Pass	10-14 years	50000-90000 ton
4	Good	5-9 years	90000-150000 ton
5	Excellent	0-4 years	≥ 150000 ton

determines a premise: once the evaluation set of wind speed and wave height is "poor", the navigation plan level is directly determined as "poor".

5   CASE STUDY

In the process of speed optimization, the relevant data of navigation plan can be obtained: new route after dividing the designed route, wind speed, wind direction, wave height, wave direction, visibility, total navigation time and total fuel consumption of each turning point. This paper selects two navigation plans of two target ships for evaluation. The target ship is a 400000-ton ultra large ore ship newly built-in 2019. Since the routes of the target ship are all fixed routes, two independent navigation plans are selected for classification.

1)   Navigation plan I
The sailing plan is that the target ship will set out from Singapore to Malacca Strait, cross the Indian Ocean, and finally reach the Cape of good hope. According to the speed optimization, there are 43 turning points, 27 turning points with wind speed between 3 and 5 levels, 32 turning points for cross wind navigation, 25 turning points with wave height between 1.25m and 2.4m, 31 turning points for cross wave navigation, all turning points with visibility higher than level 6; according to the navigation plan, the total navigation time is about 480 hours, and the total navigation after speed optimization The running time is about 460 hours; according to the design speed given by the crew, the total fuel consumption is estimated to be 572.25 tons, and the total fuel consumption after speed optimization is 501 tons. Therefore, the comment collection is summarized in Table 11.
The evaluation result is 4.327, so the navigation plan is right according to the classification standard.

Table 11.   Grades of navigation plan I.

Criterion layer index	Value
Wind velocity U11	4
Relative wind direction U12	3
Wave height U13	4
Relative Wave direction U14	3
Visibility U15	5
Time cost U21	5
Fuel cost U22	5
Age of ship U31	5
Shipping ton U32	5

Table 12. Grades of navigation plan II.

Criterion layer index	Value
Wind velocity U11	4
Relative wind direction U12	4
Wave height U13	4
Relative wave direction U14	4
Visibility U15	5
Time cost U21	4
Fuel cost U22	5
Age of ship U31	5
Shipping ton U32	5

2) Navigation Plan II

The voyage plan is for target ship II to start from Cape of Good Hope, cross the Atlantic Ocean, and finally arrive in Brazil. According to the speed optimization, the divided routes have 25 turning points, 14 turning points with wind speed between 3 and 5 levels, six turning points for crosswind navigation, 16 turning points with wave height between 1.25m and 2.4m, eight turning points for cross wave navigation, all turning points with visibility higher than level 6; according to the navigation plan, the total navigation time is about 360 hours, and the total navigation time after the speed optimization About 352 hours; according to the design speed given by the crew, the total fuel consumption is estimated to be 345 tons, and the total fuel consumption after speed optimization is 260 tons. Therefore, according to the comment set, it is summarized into Table 12.

The evaluation result is 4.42, so the navigation plan is right according to the classification standard.

6 CONCLUSIONS

Through the analysis and research of this paper, the method of navigation plan level evaluation is established. In this paper, the fuzzy comprehensive evaluation method is used to evaluate the navigation plan grade. Based on the principle of the fuzzy comprehensive evaluation method, the navigation plan level is divided into poor, fail, pass, good and excellent. Applying the algorithm principle established in this paper, the route evaluation of target ships is carried out to evaluate the navigation plan I and II. The main factors of the planning level include meteorological factors, economic factors and ship's factors. According to the algorithm requirements, the three factors are further distinguished into more subtle factors. Finally, this paper evaluates some routes and gets the corresponding grades.

To further provide trustworthy evaluation results for crews, more factors should be considered in further research, for example, the damage level of cargo, proficiency of crew and pirate area. In short, the paper provide the foundation for next stage, and the fuzzy evaluation method has been proved to be applicable to the evaluation of navigation plan.

ACKNOWLEDGEMENTS

This paper was supported by the project from Ministry of Industry and Technology of PRC:Smart Ship 1.0 - Intelligent Management System of Ship Integrated Energy Efficiency(719C-6166A).

REFERENCES

"China Rights Capsized Yangtze Ship." BBC News. Last modified June 5, 2015. https://www.bbc.com/news/world-asia-china-33011557.

Forman, E. and Peniwati, K. 1998. Aggregating individual judgments and priorities with the analytic hierarchy process. European journal of operational research, 108 (1), pp.165–169.

"Headed for Disaster: The Last Voyage of M/V Princess of the Stars." Rappler. Last modified July 2, 2015. https://www.rappler.com/move-ph/issues/disasters/97429-disaster-mv-princess-stars-sulpicio-lines.

"In the Absence: South Korea's Sewol Ferry Disaster." Breaking News, World News and Video from Al Jazeera. Last modified December 1, 2019. https://www.aljazeera.com/programmes/witness/2019/11/absence-south-korea-sewol-ferry-disaster-191125082444128.html.

Kahraman, C., Cebeci, U. and Ulukan, Z. 2003. Multi-criteria supplier selection using fuzzy AHP. Logistics information management.

Meaden, G.T., Kochev, S., Kolendowicz, L., Kosa-Kiss, A., Marcinoniene, I., Sioutas, M., Tooming, H. and Tyrrell, J. 2007. Comparing the theoretical versions of the Beaufort scale, the T-Scale and the Fujita scale. Atmospheric research, 83(2-4), pp.446–449.

Vaidya, O.S. and Kumar, S. 2006. Analytic hierarchy process: An overview of applications. European Journal of operational research, 169(1), pp.1–29.

Maritime safety and reliability

Developments in Maritime Technology and Engineering – Guedes Soares & Santos (eds)
© 2021 Copyright the Author(s), ISBN 978-0-367-77376-2

Performance-based leading risk indicators of safety barriers on liquefied natural gas carriers

M. Abdelmalek & C. Guedes Soares

Centre for Marine Technology and Ocean Engineering (CENTEC), Instituto Superior Tecnico, Universidade de Lisboa, Lisbon, Portugal

ABSTRACT: This paper presents an approach for building leading risk indicators of the various Safety Barriers of the Liquefied Natural Gas (LNG) carriers. This target is achieved by introducing a literature review on the concept and definitions of safety barriers and safety barrier management. The second step is the identification of the critical variables to each safety barrier and the methods of determining the critical ones that need to be included in the continuous monitoring process. This is followed by providing the method of quantifying the leading indicators' values to provide real-time information on the risk level of safety barriers. The method is built on the related literature to the oil and gas industry, however, the case study is implemented on the containment safety barriers of the LNG carriers during the cargo unloading operation. Finally, discussions of the benefits of the method and the possibility of utilising it with other risk studies are presented.

1 INTRODUCTION

The International Maritime Organisation (IMO) has presented guidelines on Formal Risk Assessment (FSA) for the maritime units (IMO, 2018). The guidelines provide information on risk identification, as well as, on the cost-effectiveness of the various safety measures. Furthermore, it mentions the need for including human errors in the FSA process. However, so far, the maritime industry is lacking formal guidelines for Safety Barrier Management (SBM) in the various lifecycle phases, in particular in the operations phase. A good example of the availability of guidelines on SBM is the oil and gas industry. Moreover, Haugen, et al. (2018) identified the need for the maritime industry to migrate from a reactive to a proactive nature in safety.

This is crucial for types of ships with the potential of severe accidental consequences to human, environment and assets such as LNG carriers. Therefore, this paper contributes in utilising the SBM approach that is used in the oil and gas industry (e.g. PSA, 2013; Hauge & Oien, 2016; and DNV GL, 2014) to establish a method for monitoring the performance of SBs dynamically through real-time performance-based leading indicators.

The following section presents the background knowledge on Safety Barrier (SB) concepts, definitions, SBM and Dynamic Barrier Management (DBM). Afterwards, section 3 introduces information on the importance of utilising leading indicators to SBs during the operations phase to manage them dynamically. In addition, it presents the proposed

method for building the real-time leading indicators of the SB performance. Section 4 provides the application of the proposed methodology on the cargo containment system of the LNG carriers. In addition, a discussion is given to illustrate the benefit of the provided approach, as well as, the possibility of utilising it in integration the Quantitative Risk Analysis (QRA) with SBM in the operations phase. Lastly, section 5 summarises the presented information in the research and the potential areas for improvements.

2 BARRIERS AND BARRIER MANAGEMENT

2.1 *Safety barrier concept and definitions*

Sklet (2006) indicated that the concept of safety barriers has been initially introduced by Gibson (1961) under the name of the *energy-flow* model. According to this model, accidents occur when a vulnerable object(s) are exposed to harmful energies. This scenario resulted from the absence of the proper barrier between the impacted target(s) and the source of harmful energy.

Subsequently, the *Swiss cheese* model has been introduced by Reason (1997). The Swiss cheese model represents the weakness points of all barriers as Swiss cheese holes. Accordingly, accidents occur when these holes are lined up allowing the harmful energy to affect the vulnerable object(s). The more the safety barrier degrades, the more the size of holes increase and consequently increasing the possibility of accident occurrence. Therefore, it is mandatory to

DOI: 10.1201/9781003216582-23

establish a consistent SBM in the operations phase in order to minimise the size of these holes. Also, it is necessary to establish the barriers in a manner that avoids the alignment of the holes, regardless of their sizes, in two or more consecutive Swiss cheese layers.

So far there is no consensus in the literature on a unified definition of the term barrier. However, in PSA (2013), a barrier is defined as a measure or set of measures that are implemented to prevent the occurrence or reduce the impact of the hazardous event. Other than that, Sklet (2006) stated that *"a barrier is a physical and/or non-physical means planned to prevent, control, or mitigate undesired events or accidents"*. The definitions above indicate abridged explanations of the term barrier.

Moreover, barriers are arranged hierarchically to simplify their application in the process of SBM of the facility of interest. The purpose of this categorisation is to clarify between the definition of barriers and what barriers do (Sobral & Guedes Soares, 2019). According to NORSOK (2010) and Sklet (2006), the term barrier can be split into barrier functions, barrier systems, and barrier elements.

Based on DNV GL (2014), a barrier function is defined as *"The purpose or the role of the barrier"*, while barrier elements are *"Technical, operational, or organisational measures which alone or together realises one or more barrier functions"*. Other than that, NORSOK (2010) presented the following definitions of the barrier functions and elements in addition to barrier systems:

- **Barrier function**: *Function planned to prevent, control, or mitigate undesired or accidental events.*
- **Barrier system**: *a system designed and implemented to perform one or more barrier functions.*
- **Barrier element**: *Technical, operational or organisational component in a barrier system.*

However, barrier sub-functions can be used to assist in the identification of barrier elements during the SBM process (Hauge & Oien, 2016). Moreover, the following definitions introduce the meaning of the technical, operational, and organisational (TO&O) barrier elements as provided in DNV GL (2014):

- **Technical barrier element:** *"Engineered systems, structures, or other design features which realise one or several functions."*
- **Operational barrier element:** *"A task performed by an operator, or team of operators, which realises one or several barrier functions."*
- **Organisational barrier element:** *"Personnel responsible for, or directly involved in, realising one or different barrier function."*

Furthermore, DNV GL (2014) defined performance influencing factors (PIFs) as *" conditions which are significant for the ability of barrier function or element to perform as intended"*. In other words, PIFs has no direct role in the accidental scenarios,

but, it has an indirect effect on the performance of safety barriers (Johansen & Rausand, 2015).

A clear example of PIFs can be given on the competence and workload in relation to the organisational barrier elements. And maintenance and testing with respect to the performance of technical barrier elements. However, Johansen & Rausand (2015), clarified between the meaning of maintenance and testing as barrier elements and as PIFs. In the PIFs context, testing refers to the activities performed to verify the functional requirements and condition of the technical barrier element. On the other hand, testing is accounted as an operational barrier element when needed to detect deviations from the safe operating conditions, for instance, visual leak detection. However, lists of the various generic PIFs of the TO&O barrier elements are provided in Aven et al., (2006) and Rahimi & Rausand (2013).

2.2 *Safety barriers classification*

Barriers can be classified based on their function, system, and nature. Based on Johansen & Rausand (2015), barrier functions classified as proactive (prevention) and reactive (mitigation) barriers. The proactive barriers are established to prevent and reduce the occurrence of the hazardous events, whereas, the mitigation barriers are established to reduce the associated consequences of those events as illustrated in Figure 1.

Other than that, Rausand (2011) classified barrier systems into active and passive systems. An active system is defined as *"A system that is dependent on the actions of an operator, a control system, and/or some energy sources to perform its functions, e.g. active fire systems"*. On the other hand, a passive system is described as *" A system that is integrated into the design of the workplace and does not require any human actions, energy sources, or information sources to perform its function, e.g. passive fire system"*.

Barriers can also be classified based on their nature as introduced by Hollnagel (2004). This method classifies barriers as material barriers,

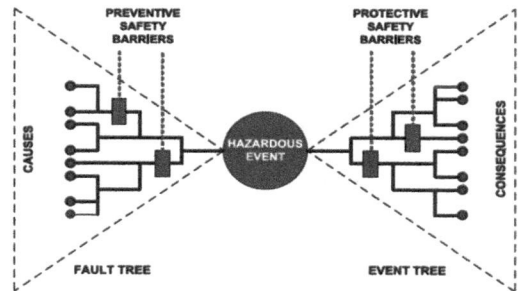

Figure 1. Bow tie diagram of safety barriers (Sobral & Guedes Soares, 2019).

functional barriers, symbolic barriers, and incorporeal barriers. However, further information on barrier classification can be found in Hollnagel, (2004) and Sobral & Guedes Soares (2019).

2.3 Safety barrier management

According to PSA (2013), barrier management is defined as "*Coordinated activities to establish and maintain barriers so that they maintain their function at all times*". The process of SBM in the planning phase has been represented by Hauge & Oien (2016) as illustrated in Figure 2.

The first step of the process is the identification of the facility and dividing the facility into areas to identify the protection requirements for each of them. The following step is to determine the Defined Situations of Hazard and Accident (DSHA) and identifying the required barrier functions. This step is carried out for each area to ensure the sufficient prevention and mitigation of the various DSHAs. The following step is concerned with the establishment of the required TO&O barrier elements to formulate the predefined barrier functions of each area.

Subsequently, step 4 refers to the identification of the performance requirements of the established TO&O barrier elements. Such requirements state the operational specifications of the TO&O elements such as experience, competence, response time, system capacity, reliability, robustness, and etc. (Sobral & Guedes Soares, 2019; Rausand, 2011; Sklet, 2006; and Hauge & Oien, 2006). Step 5 consists of the identification of the PIFs that influence the performance of the various TO&O barrier elements.

The last step in the process of SBM in the planning phase is the identification of the verification activities of the performance requirements during the operations phase. Those activities involve direct activities to barrier elements such as inspection and testing, as well as, indirect activities through monitoring the PIFs (Hauge & Oein, 2016). The verification step has been presented under a different name in PSA (2013) which is called monitoring and review. As per Johansen & Rausand (2015), this step establishes the rules to ensure that barrier elements are being operated within the acceptable operating envelope as described on the barrier strategy and performance standards.

2.4 Dynamic barrier management

The purpose of the Dynamic Barrier Management (DBM), is the provision of the real-time status of safety barriers' performance to reflect its impact on the risk level (Pitblado et al., 2016). In detail, DBM is using various sources of information (e.g. maintenance and inspection findings, training and competence records, incidents and near-miss reports, leading and lagging indicators, etc.) to produce a nearly real-time status of the barrier performance. Consequently, the information obtained from the DBM regarding the status of safety barriers can be used to support the strategic decision making of the daily activities, plus assisting in the migration from the static maintenance and inspection intervals of safety barriers to a real-time risk-based planning approaches. Figure 3 indicates the process of DBM.

Moreover, Pitblado et al., (2016), introduced a three-loop framework for DBM. The purpose of the first loop is to provide real-time information on the status of safety barriers. The second loop is the performance loop which is used to support the planning of the maintenance and inspection activities. Lastly, the safety risk mitigation loop is meant to provide supportive information to the decisions related to operational safety.

However, Hosseinnia et al., (2019), identified the difficulty to implement this framework with complex problems. In addition, it has been mentioned that the proposed DBM framework does not provide clear steps on the requirements, as well as, the method for establishing these loops. Accordingly, Hosseinnia et al., (2019) proposed the use of systems engineering tool called SPADE to facilitate the implementation of the DBM. However, further details on systems engineering and SPADE model can be found on (Zhang et al., 2018; and Haskins, 2008).

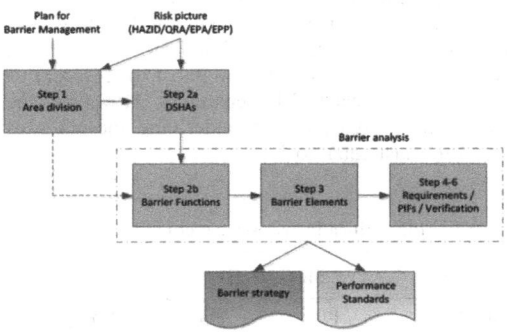

Figure 2. Barrier management process (Hauge & Oien, 2016).

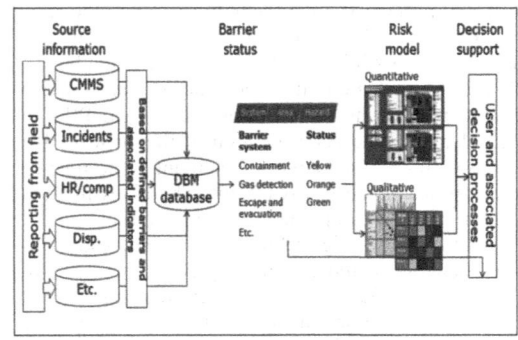

Figure 3. Concept of dynamic barrier management (Pitblado et al., 2016).

3 ESTABLISHING LEADING INDICATORS

3.1 *What is an indicator?*

Oien (2001) defined an indicator as "*a measurable or operational variable that can be used to describe the condition of a broader phenomenon or aspect of reality*". Furthermore, indicators are an important input to the process of monitoring the performance of safety barriers (PSA, 2013). Also, indicators perform a key role in providing decision makers with the required information while managing safety barriers during the operation phase. Other than that, indicators are one of the main input sources of information to the DBM process (Pitblado et al., 2016).

Indicators can be leading or lagging. The purpose of the leading indicators is the provision of early warning regarding the degradation of the system of interest. On the other hand, lagging indicators are utilised to provide information regarding the performance of the concerned systems based on specific time intervals in the past. According to Rambo (2011), the first type of indicators is known as a Key Risk Indicator (KRI), while the latter is known as a Key Performance Indicator (KPI).

The current study aims to develop leading indicators to provide real-time information regarding the performance of the various TO&O barrier elements at the operation phase. The proposed indicators are representing the status of the PIFs that influencing the various TO&O barriers. However, such indicators are not the PIF itself, but, they are measurable representations of the PIF (Oien et al., 2011). In other words, the indicators provide information on the condition of PIFs by using relative risk scores corresponding to the quantitative or qualitative values of the underlying measures of each PIF. The relative score is important because each measure has its unique values for each risk level. The proposed method of building the performance-based leading indicators is built on the provided work by Okstad et al., (2014) and Oien (2001).

3.2 *Methodology of building the leading indicators*

In step (1), the process starts with screening the relevant PIFs to each TO&O barrier element. However, due to the possibility of having a large number of PIFs for each element, it is necessary to establish criteria to restrict the selection on the critical PIFs to each element. The determination of the critical PIFs to the various TO&O barrier element is performed based on a criticality assessment. The assessment is carried out by implementing a combination of strength of knowledge (SoK) assessment, and combined sensitivity and dependency (CSD) analysis of the PIFs that influencing the barrier elements.

The SoK assessment is used to elaborate on the strength of the available information regarding each PIF during the process of barrier analysis. The selected criteria are provided by Aven (2014). The SoK criteria have been proposed initially to assess the strength of the background knowledge of the various elements of risk assessment. For example, to check the amount of the available data, the strength of the assumptions made, and uncertainty in a particular phenomenon.

The SoK criteria are utilised to assess the strength of the supportive information regarding the ability of a particular PIF to influence the performance of a barrier element. For example, in the situation of a PIF of a barrier element of a new system, which is lacking specific operational knowledge the SoK will be assigned as poor, and vice versa. Other than that, it can be used to determine the amount of the available information of a particular PIF during the design phase regardless of the novelty of the SB of interest. This is applicable for PIFs with information mostly collected from the operations phase. The SoK criteria of the PIFs are classified as poor (P), medium (M) and strong (S) knowledge.

Also, the CSD analysis is used to determine the effect of each PIF on the relevant barrier element as well as the corresponding barrier system and function. In other words, it is identifying how likely the performance of a barrier element, system and/or function is willing to change with respect to change in the condition of one of the underlying PIFs. The CSD assessment of the various PIFs can be performed quantitatively or qualitatively. In the first scenario, several approaches can be used to reflect the impact of PIFs quantitatively on the performance of safety barriers [see, e.g. Aven et al., 2006; Vinnem *et al.*, 2012; Rahimi & Rausand, 2013]. However, in this study, the CSD analysis of PIFs is performed qualitatively by classifying the level of CSD to low (L), medium (M) and high (H). Finally, according to the criticality matrix, as shown in Table 1, all PIFs fall in the red region will be included in the process of building the leading indicators.

The following step (2) consists of the assignment of the underlying quantitative measures of each critical PIFs. For example, the identification of the measures that will indicate the value of the competency PIF of a particular organisational barrier element (e.g. years of experience, training compliance percentage, supervisor assessment, etc.). Afterwards, in step (3) each of the underlying measures will be assigned a relative weight from 0 to 1 to represent its importance to each PIF separately. However, the sum of all weights of the underlying measures of each PIF must equal to 1.

Table 1. PIFs criticality assessment criteria.

Degree Of	CSD (L/M/H)		
SoK (P/M/S)	P/L	P/M	P/H
	M/L	M/M	M/H
	S/L	S/M	S/H

The following step (4) consists of identifying a common relative risk score to the numerical and qualitative input values for each underlying measure. The scoring process is created to relate the various intervals of the input data to a particular risk level. For example, what is the number of the relative risk score to a particular number of overdue maintenance tasks of the maintenance and testing PIF of barrier element X. Table 2 indicates the classification of the relative score risk intervals. Besides, Table 3 indicates an example of the scoring and weighting of the underlying measures of a particular PIF's indicator.

The calculation of the indicator's value of the corresponding PIF (i) to a particular barrier element X at a specific time (t) can be performed by using equation (1).

$$S_{x_i}(t) = \sum\nolimits_{x,i,j} v_{x,i,j} . w_{x,i,j}(t) \qquad (1)$$

$S_{x_i}(t)$ Indicates the risk level of the PIF (i) w.r.t the relevant barrier element X at time (t). Also, $v_{x,i,j}$ denotes the relative risk score of a particular underlying measure (j) to its corresponding PIF (x, i) at time (t), while $w_{x,i,j}$, refers to the relative importance weight of the measure (j) w.r.t the PIF (x, i).

The values obtained from the formula above are providing insight on the condition of the various critical PIFs to a particular barrier element based on the inputs from the underlying measures. Similarly, the formula can be repeated on a higher hierarchal level to provide information on the total risk level of a particular barrier element. This can be obtained by summing the product of the various risk values of the PIFs and the relative importance weight of each of them w.r.t to the barrier element of interest as shown in equation (2).

Table 2. Relative risk score intervals.

Risk Level	Low	Medium	High
Intervals	0 - 2	>2 - 4	- 6

Table 3. Example of scoring and weighing the underlying measures of a particular PIF.

PIF (x,i)	Measure (x,i,j)	Weight ($w_{x,i,j}$)	Relative Score of $v_{x,i,j}$					
			1	2	3	4	5	6
Technical Condition	Inspection Findings	0.4						
	Overdue Maint. activities	0.6						

$$TS_X(t) = \sum\nolimits_{x,i} S_{x,i} . w_{x,i}(t) \qquad (2)$$

TS_X Denotes the total risk indicator's value of the barrier element X at the time (t). Whereas, $w_{x,i}$ refers to the relative importance weight of each PIF indicator w.r.t the relevant barrier element.

4 CASE STUDY

4.1 Case description

This case study is implemented on the Prevent Loss of Primary Containment (LOPC) barrier function during the cargo unloading operation of a generic LNG carrier. In particular, the case will present a general overview of the relevant ship and the linked ship/shore barrier elements and systems to the PLOPC barrier function. Afterwards, an example is given on implementing the proposed monitoring method on one barrier element from each type of the presented TO&O barrier elements. The selected elements are cargo primary containment, cargo engineer's performance and the related tasks to the Emergency Shutdown (ESD) 1.

The cargo transfer system of the LNG carriers is generally consisting of cryogenic pumps for each tank connected to a discharge line fitted with ESD valve and isolating valve. The tanks' discharge lines are connected to the main cargo manifold, which is also fitted with isolating valve and ESD valve. The main manifold is ended up with a number of pipe branches that should be connected to the shore facility loading arms (ISGINTT, 2010). However, the case presents the barrier elements that exposed to the external environment, i.e. from the main deck to the unloading arms, since the production interruption risk is out of the scope of the current study.

4.2 Safety barriers of the PLOPC barrier function

The on board PLOPC barrier function of the activity of interest is realized with three sub-functions. The first one represents the cargo containment, which involves the pipework, valves, strainers, expansion bellows and etc. The other one is the process securing which represents the ESD 1 barrier function that is responsible for stopping the transfer and returning the process to the static condition (SIGTTO, 2009). The ESD 1 can function as a preventive and/or mitigation measure. So, it has been considered in this study as a part of the PLOPC function to secure the process segment in case of the realization of internal and/or external hazards that can cause in process leak. The ESD 2 which is responsible for activating the Powered Emergency Released Coupling (PERC) is not considered as it is managed completely from the production facility (SIGTTO, 2009). The last sub-function is the maintain ship position and stability which is significant to keep the ship within the safe operating envelope during the cargo unloading operation (DESFA, 2018). Accordingly, the third sub-function can be classified as indirect as it does not act as a direct containment measure. However, Table 4 presents the common barrier sub-functions, systems and TO&O elements of the PLOPC barrier function (DNV GL, 2015; ISGINTT, 2010; El Sayed et al., 2013; SIGTTO, 2009; and DESFA, 2018).

4.3 Application of the performance-based leading indicator's approach

The most relevant PIFs of the selected TO&O barrier elements are presented in Table 5 (DNV GL, 2014; Aven et al., 2006; and Rhimi & Rausand, 2013). The selection of the critical PIFs is determined qualitatively based on the presented criteria in section 3.

For the cargo primary containment element, the technical condition and maintenance and testing PIFs are considered to be the PIFs with a poor SoK as they are depending on the knowledge and feedback from the operations phase. On the other hand, the degree of CSD is assigned to be high for those PIFs. The high sensitivity has been determined for those PIFs due to their high relative importance weight to the technical systems when compared with the other PIFs. Also, those PIFs are dynamic as they are changing over time with the changes in the operational conditions. In addition, such PIFs are involving human and organisational factors (HOFs) in the execution of the maintenance and testing activities, as well as the identification of the unsafe conditions of the technical elements by all crewmembers.

Table 4. TO&O barrier elements/systems of the PLOPC barrier function.

Function	PLOPC while unloading cargo		
Group	Direct sub-functions		Indirect sub-functions
Sub-function	Contain the cargo	Process securing (ESD 1)	Maintain ship position and stability
Technical elements/systems	• Pipework. • Expansion bellows. • Cargo Isolation valves (including ESD valves as a cargo containing element) • Cargo pipe's strainer. • Process monitoring systems.	• ESD valves (as a cargo isolating element) • Ship/Shore Link (SSL) systems. • ESD push buttons. • Process monitoring systems. • Communications radios. • CCTV System.	• Ballast pump. • Ballast system's pipes and valves. • Ballast control & monitoring system. • Draft monitoring system. • Wind speed monitoring system. • Mooring lines. • Mooring winches. • Ballast monitoring system. • CCTV System.
Organizational elements	• Master performance. • Cargo engineer performance. • Duty AB (watcher) performance.	• Master performance. • Duty AB (watcher) performance. • Port Engineer performance. • Cargo Engineer performance • Duty officer performance.	• Duty officer performance • Duty AB (watcher) performance. • Master performance.
Operational elements	• Cargo handling related tasks.	• ESD related tasks.	• Marine operations related tasks.

Table 5. PIFs of the selected TO&O barrier elements.

Element	Cargo primary containment	Cargo Engineer Performance	ESD 1 related tasks
PIFs	• Technical Condition • Maintenance and Testing • Cargo characteristics • Operating parameters • Quality of CMMS • System geometry • Materials properties	• Competence • HMI • Supervision • Workload • Communications	• Complexity • Task related knowledge • CoW System

On the other hand, the high dependency of the technical condition and maintenance and testing PIFs is determined based on the existence of these PIFs in all technical barrier elements. For example, such PIFs are existing in the mechanical, electrical and control barrier elements and systems. Therefore, the variation in one of those factors is deemed to be critical to the performance of the cargo primary containment elements.

With regards to the cargo engineer which is in charge of monitoring and control of the cargo unloading operation. The corresponding critical PIFs that influences his performance during cargo unloading are determined to be the competence and workload. The SoK level of both factors is assigned a poor score. The assignment of this score to the competence PIF is driven from the lack of knowledge on the real performance of the cargo engineer during the cargo unloading operation. In this case, the dependence on the formal training and years of experience can camouflage the degree of confidence in the performance of the cargo engineer. Therefore, these factors have been strictly assigned this level of SoK to overcome the simplification when monitoring the performance of the cargo engineer. Moreover, the workload PIF is assigned the same level of SoK, as it is completely managed during the performance of the concerned task.

Other than that, both PIFs are assigned a high degree of CSD. Both PIFs are highly sensitive due to their considerable influence on the occurrence of human errors. Also, they are highly dependence as they are influencing the performance of the cargo engineer and the other PIFs influencing the same barrier element.

The critical PIFs to the ESD 1 related tasks barrier element are the Control of Work (CoW) system and the task-related knowledge. The minimum requirements of the CoW system are work permit, work instruction and the Job Risk Assessment (JRA). The lack of adherence to the CoW system can result in the occurrence of LOPC as a result of e.g. lack of knowledge of the performed task and the lack of awareness of the required safety measures. In addition, the task-related knowledge PIF indicates the task-specific knowledge and the necessary training to complete the task safely.

From the SoK criteria perspective, the CoW PIF is assigned a poor score, while the task-related knowledge is considered to be based on a medium level of knowledge. The first is assigned such score as its relevant information are obtained during the operation phase, while the latter assigned medium score due to the availability of some information during the design phase. For example, the pre-employment training of the involved employees in the task. Furthermore, both factors are appointed high degree of CSD due to their significant importance to the barrier element of interest.

Task complexity is an important PIF to the ESD 1 related tasks barrier element. The task complexity is

sensitive to some degree to the performance of the task. However, from the SoK perspective, the tasks' steps and sequence is considered to be pre-defined based on strong knowledge. However, the impact of the task complexity PIF can be indirectly tested by the measures assigned to the task-related knowledge PIF.

The proposed underlying measures of the critical PIFs of the selected TO&O barrier elements are indicated in Table 6 with their relative importance weights. Afterwards, the methods of assigning the quantitative and qualitative values of each measure and the real-time risk level calculation are demonstrated on the cargo primary containment element. However, more measures or a different set of measures can be used for the formal application of the method.

Cargo pipelines and its attachments are visually inspected and pressure tested periodically by the ship crew. However, thorough NDT examination of the cargo pipelines and valves is performed on each dry dock entry which, approximately, occurs each 5 years. Besides, it is assumed that after each dry dock the asset failure rates return to its initial condition (Rausand & Hoyland, 2004). Therefore, the assigned numerical interval of the life time measure is from 0 to 5 years. Furthermore, the management of change (MOC) measure, represents the number of deviations from and/or inhibitions of safe operating procedures and corresponding safety measures of the containment system, respectively. The assigned numerical risk interval of this measure is from 0 to 3. Lastly, the number of the open unsafe observation denotes the relevant unsafe conditions to the cargo containments elements. These observations do not include the findings of the regular maintenance and testing activities. For this measure the risk interval is from 0 to 4.

Other than that, for the maintenance and testing PIF, the risk interval value of the overdue maintenance and testing activities and the number of findings is assigned from 0 to 3.

The scoring and weighting of the PIFs and their underlying measures are provided in Table 7.

By assuming the relative risk score of the highlighted values of each measure in Table 7, with using equations number 1 and 2, we obtain the subsequent results. The risk score of the technical condition, and maintenance and testing PIFs are 2.4 and 2.8, respectively. Consequently, the risk score of the cargo primary containment barrier element is 2.64. In other words, this barrier element falls at the onset of the medium risk region. However, similar calculations can be used for the remaining barrier elements to determine the real-time risk level of the main barrier function.

4.4 Discussion

The proposed performance-based leading indicators method provides real-time information regarding the risk level of safety barriers. The methodology is acting on gathering real-time information from various input sources to produce barrier based indicators. The risk scores of safety barriers are providing early warnings

Table 6. Underlying measures of the PIFs of the selected barrier elements.

Barrier Element (X)	Cargo primary containment (1)					
PIFs (i)	Technical Condition (1)			Maintenance and testing (2)		
PIFs Relative Score	0.4			0.6		
Measure (j)	Time in operation in years (1)	No. of MOCs (2)	No. of open unsafe observations (3)	No. of overdue maint. tasks (1)	No. of overdue testing tasks (2)	No. of findings need intervention (3)
Importance score	0.2	0.4	0.4	0.33	0.33	0.34

Barrier Element (X)	Cargo engineer performance (2)					
PIFs (i)	Competence (1)			Workload and Fatigue (2)		
PIFs Relative Score	0.7			0.3		
Measure (j)	Experience in years (1)	Mandatory training compliance % (2)	Supervisor qualitative feedback (3)	No. of days on board (1)	No. of working hours on shift (2)	No. of reported medical conditions (3)
Importance score	0.3	0.3	0.4	0.35	0.4	0.25

Barrier Element (X)	ESD 1 related tasks (3)					
PIFs (i)	CoW system (1)			Task related knowledge (2)		
PIFs Relative Score	0.4			0.6		
Measure (j)	Average monthly frequency of the reported deficiencies in CoW sys. (1)	Number of SIMOPS during the unloading operation. (2)	Completion of the CoW training % (3)	Completion of the task related training %. (1)	Frequency of task related reported unsafe acts/conditions. (2)	Feedback from supervisor. (3)
Importance score	0.3	0.4	0.3	0.3	0.3	0.4

Table 7. Risk scoring and weighting of the measures of the cargo containment element PIFs.

Barrier Element (X)	PIF (x, i)	PIF's Weight ($w_{x,i}$)	Measure (x,i,j)	Weight ($w_{x,i,j}$)	Relative Score of $x_{i,j}$					
					1	2	3	4	5	6
Cargo Primary Containment (1)	Technical Condition (1,1)	0.4	Years in operation (1,1,1)	0.2	NA	1	2	3	4	5
			No. of open MOCs (1,1,2)	0.4	NA	NA	NA	1	2	3
			No. of open unsafe observations (1,1,3)	0.4	NA	NA	1	2	3	4
	Maintenance and testing (1,2)	0.6	No. of overdue maintenance activities (1,2,1)	0.3	NA	NA	NA	1	2	3
			No. of overdue testing activities (1,2,2)	0.3	0	NA	NA	1	2	3
			No. of findings need intervention (1,2,3)	0.4	NA	NA	NA	1	2	3

about the degradation of the performance of safety barriers. Accordingly, corrective actions can be taken to maintain safety barriers on LNG carriers within the safe level.

Moreover, the proposed monitoring approach provides a practical solution for implementing the presented concept of DBM by Pitblado et al., (2016) within the maritime risk management domain. Besides, the aim of the research was achieved by utilising various sources of the related literature to the oil and gas industry. For example, the process of defining, classifying and conceptualization of SB and SBM was built on the provided information on (Sklet, 2006; PSA, 2013; DNV GL, 2014; and Hauge & Oien, 2016). Furthermore, the definition and building of the barrier based indicators were driven from the literature provided by (Oien, 2001; Oien et al., 2011 and Okstad et al., 2014). However, the literature above was used in an integrated manner to produce a barrier oriented real-time information on board the LNG ships. This also contributes to the efforts that are spent to improve the proactivity of managing risk in the maritime industry.

Other than that, the provided methodology can act as an efficient solution to integrate the ship's QRA with SBM during the operations phase. This can be achieved by updating the failure rates of SBs in QRA continuously based on the values provided by the leading indicators. The deviation in failure rates of each safety barrier can be calculated by using one of the available methodologies for quantifying the effect of PIFs on safety measures. The Risk_OMT and BORA-methodologies can be used to reflect the impact of the underlying PIFs on each safety barrier (Vinnem et al., 2012; and Aven et al., 2006). On the other hand, a computerized Bayesian-Belief Network (BBN) tool can be used to reflect the changes of failure rates continuously on the numbers included in the QRA study (Meng et al., 2019). Accordingly, risk picture can be updated dynamically to assure the continuous compliance with the regulatory requirements of the safety level of the asset.

Furthermore, this approach is beneficial in providing early warning signals based on the instantaneous changes in the operational variables. In addition, it is ensuring the involvement of the various departments and expertise in identifying the status of the various TO&O safety barriers and consequently the final risk picture. However, the sensitivity to the ongoing operations and the involvement of the various expertise in SBM provides significant improvements in preventing the occurrence of surprising and extreme events (Aven, 2014).

5 CONCLUSION

This paper provided an approach for building performance-based leading indicators of the SBs on board the LNG carriers. The method has been presented to contribute to the efforts that are being made for improving the risk analysis methods in the maritime industry. In particular, for the ships that are transferring hazardous cargoes such as LNG.

The target of the paper has been achieved by presenting a scientific background knowledge on SBs, SBM and DBM. In addition to an introduction on the definitions and types of indicators that are used to monitor the performance of the TO&O elements of the asset. Moreover, the methodology of building the leading indicators has been presented and applied on the cargo primary containment barrier element of the PLOPC barrier function of the LNG ship during the cargo unloading operation.

Finally, a discussion has been given to highlight the benefits of using the proposed method. Furthermore, the discussion has also introduced the possibility of utilising the presented approach for integrating the QRA study of the LNG carrier with SBM during the operations phase of the ship.

ACKNOWLEDGEMENTS

This work contributes to the Strategic Research Plan of the Centre for Marine Technology and Ocean Engineering (CENTEC), which is financed by the Portuguese Foundation for Science and Technology (Fundação para a Ciência e Tecnologia - FCT) under contract UIDB/UIDP/00134/2020.

REFERENCES

Aven, T. (2014) *Risk, Surprises and Black Swans: Fundamental Ideas and Concepts in Risk Assessment and Risk Management*. New York: Routledge.

Aven, T., Sklet, S. and Vinnem, J. E. (2006) 'Barrier and operational risk analysis of hydrocarbon releases (BORA-Release). Part I. Method description', *Journal of Hazardous Materials*. 137(2): 681–691.

DESFA, (2018) 'LNG Terminal Marine Procedures Manual, Revithoussa, Greece.

DNV GL (2014) 'Barrier Management in Operation for the Rig Industry – A Joint Industry Project', *Offshore Technology Conference Asia*.

DNV GL (2015) 'Part 5 Ship Types, Section 7 Liquified gas tankers - Rules for Classification.

Elsayed, T., Marghany, K., & Abdulkader, S. (2014). Risk assessment of liquefied natural gas carriers using fuzzy TOPSIS. *Ships and Offshore Structures*, 9(4): 355–364.

Gibson, J. (1961). "The contribution of experimental psychology to the formulation of the problem of safety– a brief for basic research." *Behavioral approaches to accident research*: 77–89.

Haskins C., 2008, Systems Engineering analyzed, synthesized, and applied to sustainable industrial park development. PhD Thesis, NTNU, Trondheim, Norway.

Hauge, S. and Øien, K. (2016) 'Guidance for barrier management in the petroleum industry, *SINTEF Report A27623'* Trondheim, Norway.

Haugen, S., Ventikos, N.P., Teixeira, A.P., Montekwa, J. (2018) 'Trends and needs for research in maritime risk’. *Maritime Transportation and Harvesting of Sea Resources*, Guedes Soares & Teixeira (Eds). Taylor & Francis Group, London. 313–321.

Hollnagel, E. (2004) *Barriers and accident prevention*. Aldershot, UK: Ashgate Publishing Limited.

Hosseinnia, B., Haskins, C. Reniersk, G., Paltrinieri, N., (2019) 'A Guideline for the Dynamic Barrier Management Framework Based on System Thinking'. *Chemical Engineering Transaction*, AIDIC, 103–108.

IMO (2018) 'Revised guidelines for formal safety assessment (FSA) for use in IMO rule-making process.

ISGINTT. (2010). International Safety Guide for Inland Navigation Tanks Barges and Terminals. First Edition, Strasbourg, France.

Johansen, I. L. and Rausand, M. (2015) 'Barrier management in the offshore oil and gas industry', *Journal of Loss Prevention in the Process Industries* (34): 49–55.

Meng, X. Chen, G. Zhu, G. Zhu, Y. (2019) 'Dynamic quantitative risk assessment of accidents induced by leakage on offshore platforms using DEMATEC-BN',

International Journal of Naval Archeticture & Ocean Engineering (11): 22–23.

NORSOK (2010) 'Risk and emergency preparedness assessment - Norsok Standard Z-013', Norway.

Øien, K., Utne, I. B. and Herrera, I. A. (2011) 'Building Safety indicators: Part 1 - Theoretical foundation', *Safety Science*, 49(2):148–161.

Øien, K., (2001) 'Risk Indicators as a tool for risk control'. *Reliability Engineering & Sytem Safety*, (74): 129–145.

Okstad, E. . *et al.* (2014) 'Monitoring the risk picture by using QRA and barrier based indicators', in *Safety, Reliability and Risk Analysis: Beyond the Horizon - Proceedings of the European Safety and Reliability Conference, ESREL 2013.*

Sobral, J. Guedes Soares, C., (2019) 'Assessment of the adequacy of safety barriers to hazards'. *Safety Science*, (114): 40–48.

Pitblado, R., Fisher, M., Nelson, B., Flotakes, H., Molazemi, K., Stokke, A., (2016) 'Dynamic Barrier Management - *Managing safety barrier degradations'. Hazard 26* (IchemE).

PSA (2013) 'Principles for barrier management in the petroleum industry' (Technical report), Petroleum Safety Authority, Norway.

Rambo, F. (2011) 'Set Up Risk Indicators as an Early Warning System and Leverage Actionable Reports for Risk Monitoring, Retrieved 20/11/2019 https://sapexperts.wispubs.com

Rahimi, M. & Rausand, M. (2013) 'Prediction of failure rates for new subsea systems: a practical approach and an illustrative example. " Proceedings of the Institution of Mechanical Engineers, Part O', *Risk and Reliability: 1–12.*

Rausand, M. (2011) *Risk Assessment - Theory, Methods and Applications. Statistics in Practice*, John Wiley & Sons.

Rausand, M. & Hoyland, A. (2004). *System reliability theory: models, statistica methods and applications.* John Wiley & Sons.

Reason, J. T. (1997). *Managing the risks of organizational accidents*, Ashgate Aldershot.

SIGTTO. (2009) 'ESD Arrangement & Linked Ship/Shore Systems for Liquefied Gas Carriers'. (Technical Report). London, UK.

Sklet, S. (2006) 'Safety barriers: Definition, classification, and performance', *Loss Prevention in the Process Industries*.19(5):494–506.

Vinnem, J. E. *et al.* (2012) 'Risk modelling of maintenance work on major process equipment on offshore petroleum installations', *Journal of Loss Prevention in the Process Industries*. 25(2): 274–292.

Zhang J., Haskins C., Liu Y., & Lundteigen M.A., 2018, A systems engineering–based approach for framing reliability, availability, and maintainability: A case study for subsea design. *Systems Engineering*: 576–592.

Developments in Maritime Technology and Engineering – Guedes Soares & Santos (eds)
© 2021 Copyright the Author(s), ISBN 978-0-367-77376-2

Statistical characterization of risk influencing factors in ship collision accidents

P. Antão, A.P. Teixeira & C. Guedes Soares
Centre for Marine Technology and Ocean Engineering (CENTEC), Instituto Superior Técnico, Universidade de Lisboa, Lisbon, Portugal

ABSTRACT: This paper presents an analysis of ship collision accidents worldwide based on the available data. Data from 746 collision accidents from the period between 2006-2016 are analysed. The data are retrieved from the Global Integrated Shipping Information System of the International Maritime Organization consisting of 1276 ships involved in the collisions. The period of the study covers 5 years, prior and after the introduction of the New Inspection Regime in 2011, in order to identify potential impacts of its introduction into the collision accident rate. Additionally, information from the European Quality Information System on the world fleet statistics is also retrieved in order to estimate the risk levels. The analysis performed in the present paper provides the basis for a later identification of collision Risk Influencing Factors, fundamental for the development of risk models and the estimation of the risk levels of a given ship in a given scenario.

1 INTRODUCTION

The world seaborne trade is constantly growing, although it lost momentum in 2018, with volumes increasing by only 2.7 per cent (UNCTAD, 2019). Nevertheless, and according to the United Nations statistics, from the 11 billion tons shipped internationally, 7.8 billion tons are classified as dry cargo, with Asia being by far the largest trading region, with almost half of this value.

This trade growth aligned with the increase of the number of vessels navigating translates into more congested waters, resulting in higher collision probabilities. Despite the efforts in decreasing the number of ship collisions, they represent, still today, a significant amount of the maritime accidents worldwide, with percentages near 25% (EMSA, 2019).

Given this record is not surprising that significant efforts have been made in developing models for the estimation of collision risk (Teixeira & Guedes Soares, 2018). Goerlandt et al. (2012) have developed a collision probability model based on a time-domain micro simulation of maritime traffic, for which the input was obtained through a detailed analysis of data from the Automatic Identification System (AIS). The use of AIS, as data source for the collision modelling, has been applied in several other studies (e.g. Zhang et al., 2015; Silveira et al., 2013, Chen et al. 2019). Zhang et al., (2015) proposed a method for detecting possible near ship-ship collisions using AIS data. The developed Vessel Conflict Ranking Operator (VCRO) considers factors that affect the complexity of an encounter between two ships, as the distance between the two ships, the relative speed of the ships, and the difference between the headings of the ships. Similar approaches have been used by other authors, either by applying other techniques or by applying in different geographies and traffic characteristics (Goerlandt & Kujala, 2011; Li et al, 2018; Montewka et al, 2010; Rong et al., 2016; Zhang et al., 2018). In order to develop risk models, it is necessary to identify key parameters, or Risk Influencing Factors, that can increase or decrease the probability of occurrence of a given accident. In fact, the introduction of the New Inspection Regime (NIR) of the Paris Memorandum of Understanding (PMoU), which is the main element of the Port State Control Directive 2009/16/EC (EC, 2009), takes into consideration a set of variables in order to estimate the risk level of a given vessel and to establish the inspection prioritization.

The main aim is the identification of 'High Risk Vessels' in a given scenario. This has been addressed in several studies over the years (e.g. Sage, 2005; Stróżyna, & Abramowicz, 2015). The present work aims at providing the basis for the identification of collision Risk Influencing Factors, fundamental for the development of risk models and the estimation of the risk levels of a given ship in a given scenario.

2 DATA ON SHIP COLLISIONS

For the analysis of ship collisions, several data sources are available that could be the basis to the

DOI: 10.1201/9781003216582-24

present study. However, few have free access or have a wide geographical coverage. For these reasons (but not only) the information of the Global Integrated Shipping Information System (*GISIS*) of the International Maritime Organization is used in this study. The data retrieved from the GISIS consists of 1276 ships involved in 746 collisions in the period 2006-2016. The period of analysis covers 5 years, prior and after the introduction of the New Inspection Regime in 2011, in order to identify potential impacts of its introduction into the collision accident rate. A detailed analysis of the accident data set considered in the present study is presented in Chapter 3.

For the calculation of the collision rate over the period of analysis, it is necessary to know the fleet at risk. Similarly, to the accident data, information concerning the fleet at risk is also identified and included into the current analysis. Data is retrieved from Equasis statistics for the period between 2006-2016 (Equasis, 2019). The European Commission and the French Maritime Administration decided to cooperate in developing the Equasis information system in order to collect existing safety-related information on ships from both public and private sources and make it available on the Internet. The main aims of this system are to reduce substandard shipping and promoting the exchange of unbiased information and transparency in maritime transport. The mentioned reports collect a significant amount of information, ranging from ship characteristics (size, Gross Tonnage, ship type, age), class, flag, Protection and Indemnity (P&I), and Port State Control performance indicators (detention rate, number of inspections, ship performance).

Within these annual analyses, the previous variables are combined in order to provide distributions per ship type. These combinations are fundamental for the determination of conditional probabilities in Bayesian Networks modelling to be addressed in further studies. The focus of the present study is mainly on the statistical analysis of the different variables relevant to collision events. Figure 1 shows an example of the information retrieved for the current analysis. The figure presents the evolution of the number of ships

during the period in question. In average there was an increase of nearly 29% in the overall number of ships, from 69,572 in 2006 to 89,804 in 2016. This increase is more significant in particular for ship types such as *Other Tankers* (+250%), *Offshore Vessels* (+98%), *Bulk Carriers* and *Tugs* (+60%).

However, there are cases in which there was a drop on their number. The most evident case is of *General Cargo* vessel, where a decrease of nearly 10% may be found. It is worth noticing that this drop is more relevant after 2010, which means after the introduction of the NIR. In fact, and according to official reports from the Paris MoU (Paris MoU, 2019a), more than 50% of the detained ships within port inspections are of *General Cargo* or *multipurpose* vessels.

Regarding vessels size evolution (GT), it can be observed in Figure 2 that the previous trend for *General Cargo* vessels is also found there. This means that a decrease on the number of vessels has not been translated into an increase on their size. In opposition, one may see that *Bulk Carriers, Oil and Chemical Tankers* and *Container Ships* have increased substantially their sizes during the decade under analysis. In fact, this increase is, in average of 100%, when comparing 2016 with 2006. The increase of global trading during this period, although coinciding with a world financial crisis, explains this significant increase on these types of cargo vessels. Although not completely explicit in the figure due to the scale, is the *Other Tankers* category that presents a higher increase (+150%). This category contains ship types as Fruit Juice Tanker, Water Tanker and Wine Tanker.

A significant aspect is the ship size evolution during the decade under analysis. Figure 3 presents this evolution with particular interest in the Very Large vessels, which, in overall, present a grow of nearly 120% in relation to 2006. This trend can be easily observed in Figure 4, which presents the ratio between the number of *Very Large* vessels over *Small* ones, per year. Clearly there is a growing trend for larger vessels in the latest years, given the increase of the gross tonnage associated to each of the vessels. This is particularly relevant when

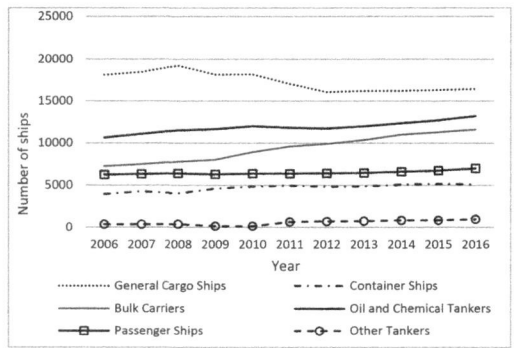

Figure 1. Number of ships per year.

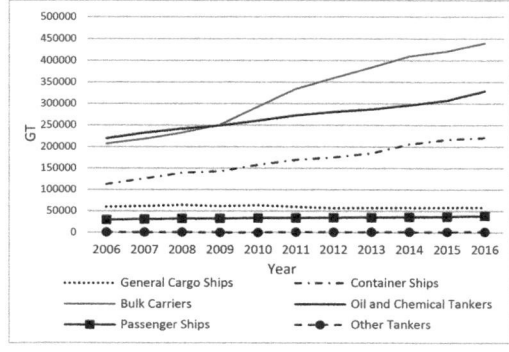

Figure 2. Gross Tonnage of ships per year.

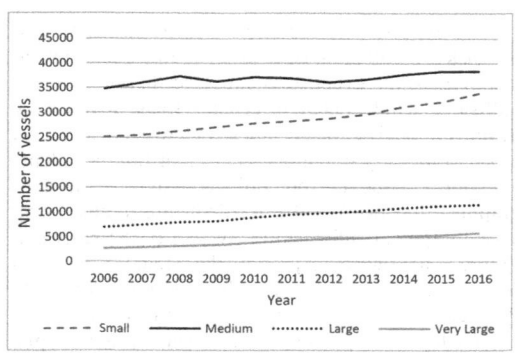

Figure 3. Evolution of the number of ships per year and size.

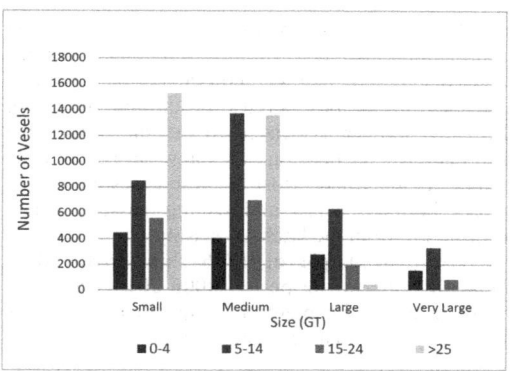

Figure 5. Distribution of the number of vessels per age and size, Year=2016.

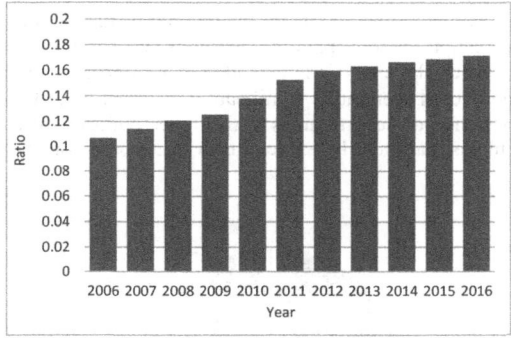

Figure 4. Ratio between the number of very large vessels over small ones, per year.

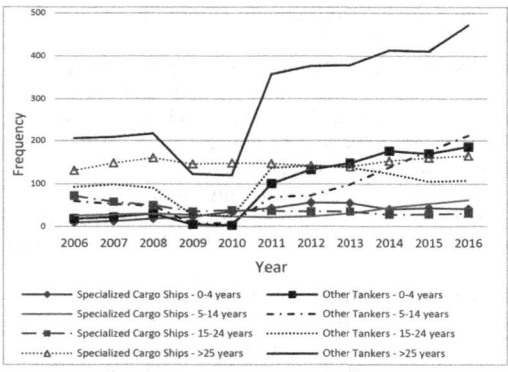

Figure 6. Evolution of the number of ships per age and per year.

analysing collision events and their consequences, since the potential impacts are substantially different.

The trend of cargo vessels with larger transport capacities can also be observed in Figure 5 and Figure 6, which present, respectively, the distribution of the number of vessels per age and size (for the more recent year of the sample) and its evolution for two specific ship types. Regarding the overall evolution, it can be observed that most of the *Small* and *Medium* size vessels within the fleet are older vessels, largely with ages above 25 years. The opposite trend can be found in large size vessels, as the case of the Very Large ones, where up to 57% of the fleet is composed of vessels with less than 14 years. This shows the trend on the investment on newer and larger vessels. This can also be observed in Figure 6, where the evolution of the number of ships per age and per year is presented. One may see a significant drop in the number of older vessels prior to the introduction of the New Inspection Regime. In the cases of *Bulk Carriers* and *Containers* vessels, there was a nearly 50% drop on the number of vessels with ages >25 between 2011 and 2013.

3 SAMPLE OF COLLISION ACCIDENTS

As mentioned in Chapter 2, the accident data was taken from the GISIS system. The data set consists of 3,914 casualty events, between 2006-2016, from which 746 are collisions, or 19% of the considered sample. This number is not very different from the one determined by Guedes Soares & Teixeira (2001) for a different time period. The system provides a wide range of information regarding the *incident summary* and some ship details. In the initial information one can find 31 variables as for example: IMO number of the involved ships, flag state, date of the incident, coordinates of the incident, type of casualty, initial event, ship type, consequences, Gross Tonnage (GT), Classification Society, fatalities, among others.

However, in the case of collisions, the information provided is only related to one of the ships involved in the incident. Thus, at least half of the information concerning collisions is not included in this initial sample. In order to overcome this pitfall, the missing information for each of the remaining vessels, namely, ship type, flag, GT, consequences, classification society, fatalities, is retrieved through the use of the IMO number associated to the ship which the information was missing. In many cases, no IMO number is identified. Most of these cases are occurrences between cargo and fishing vessels, where information from the latest is missing.

Overall, a sample of 1276 ships involved in 746 collisions in the period 2006-2016 was compiled. It should be mentioned that there are cases in which more than 2 vessels are involved in the incident.

Figure 7 presents the results in terms of the severity for the overall sample of 3914 accidents. IMO classifies incidents as *less serious*, *serious* and *very serious,* depending on the outcome of the maritime casualty. *Very serious* are marine casualties involving the total loss of the ship, a fatality or severe damage to the environment. From the abovementioned figure it can be concluded that nearly half of the incidents are classified as *Very Serious* (49%). This is a significantly high number and shows that most of the recorded maritime accidents within the IMO system have severe consequences to the ship, human life or to the environment.

However, this can, in certain way, be explained by the fact that there is a possible underreporting of less serious casualties, as already mentioned in previous studies (e.g. Psarros et al., 2010). Furthermore, maritime investigation branches are required to analyse serious and very serious accidents, and report the findings to IMO, which leads to a possible unbalance between the reported and unreported ones.

It is interesting to notice that this trend is not followed when the sample is just composed by collision accidents. In this specific case, as presented in Figure 8, the percentage of *Very Serious* accidents decreases to 39%. This lower frequency can be explained by a lower energy involved in some accident/contacts with low impact loads. This will be partly corroborated when analysing the consequences of the collisions.

One important variable towards the calculation of the risk of collision is related to the geographical locations of the event. It is expected that most of the collisions would occur in coastal areas, with heavy traffic in specific traffic lanes, or crossing areas. In order to identify hot spots where maritime accidents are more frequent – not only collisions - the geographical locations of the accidents are plotted using a Geographical Information System (GIS) software. Figure 9 to Figure 11 present the accident locations of all accident types and collisions, respectively in Portuguese and European waters and worldwide.

In what concerns Portuguese waters, and despite the heavy traffic that coastal lanes present on the daily basis, the occurrence of maritime incidents during the period under analysis is seldom. In fact, observing Figure 9, it can be concluded that collisions in particular, are almost inexistent (less than 10 events when extended until Gibraltar). This is a surprising outcome given the amount of fishing and recreation vessels operating near these coastal areas.

When the analysis is extended to European waters the overall scenario is significantly different, with the existence of several areas with higher collision risk. This is particularly evident near the coasts of the

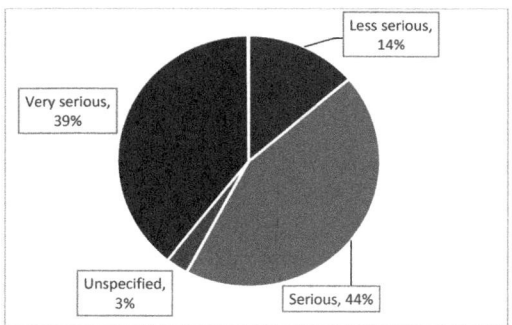

Figure 8. Type of casualty – Collision events, N=746.

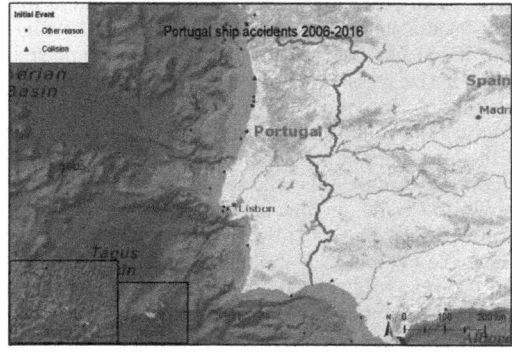

Figure 9. Geographical location of accidents in Portuguese waters (2006-2016).

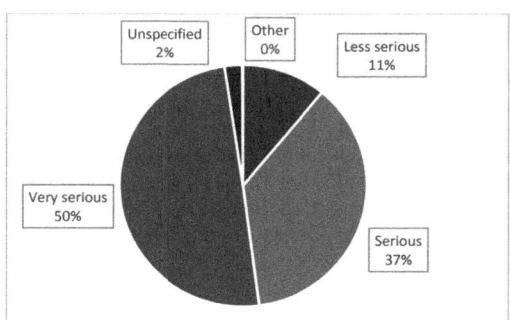

Figure 7. Type of casualty - All accidents, N=3914.

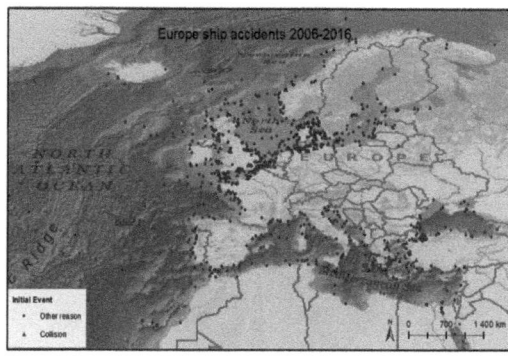

Figure 10. Geographical location of accidents in European waters (2006-2016).

North of France, Netherlands and Germany. These are areas that correspond to the accesses to large European ports, as for example, Le Havre, Antwerp, Rotterdam and Bremen. This means areas where there are frequent crossings of the navigation traffic lanes, thus with a higher risk exposure to the occurrence of collisions. However, given the array of accident occurrences than one can observe in Figure 10, it can be concluded that collisions are still an event with low frequently in these geographical areas.

When compared with the European waters, the Southeast Asia, in particular the Chinese, Vietnamese and Japanese coasts are particularly prone to high accident rates. These are neither unknown nor less expected results given the fact that these are one of the world areas with higher maritime traffic. The economic growth of China over the last decades lead to a significant increase of the maritime transportation from and towards this area. Furthermore, these are areas subjected to severe weather conditions,

which combined with a lower regulation enforcement and the frequent presence of small size fishing vessels leads to a higher frequency of accidents. In fact, this is one of the findings of the consequence analysis addressed in the next chapter of the present study.

4 ANALYSIS OF COLLISION EVENTS

Based on the data set described in the previous chapter, an analysis is conducted on the circumstances of the collisions events. This analysis, from one end, allows establishing a "framework" of these circumstances and, from the other, establishes the basis for the identification of the relevant variables to be considered while modelling and estimating the collision risk. Thus, variables possible to be compiled and having a large sample in order to draw consubstantiated conclusions are considered. In this regard, the following variables are considered: type of vessel; age; length; classification society; GT; hour; month and year distributions; flag; and consequences. Although the initial GISIS sample provides additional variables (for example, number of fatalities, etc.) one could find that the percentage of missing information is significant which would not allow to draw reliable conclusions.

Figure 12 shows the distribution of the ship types involved in the collisions. It can be observed that *General Cargo* (17%), *Container ships* (16%) and *Bulk Carriers* (14%), represent nearly 50% of the involved ships. These are ship types with larger number of port calls or doing coastal navigation, thus with more exposure to high traffic. Furthermore, these are ships types with larger fleets, thus with larger number of vessels at risk. Worth of notice is the low frequency of collisions involving oil tankers and related vessels. These are typically vessels navigating port-to-port,

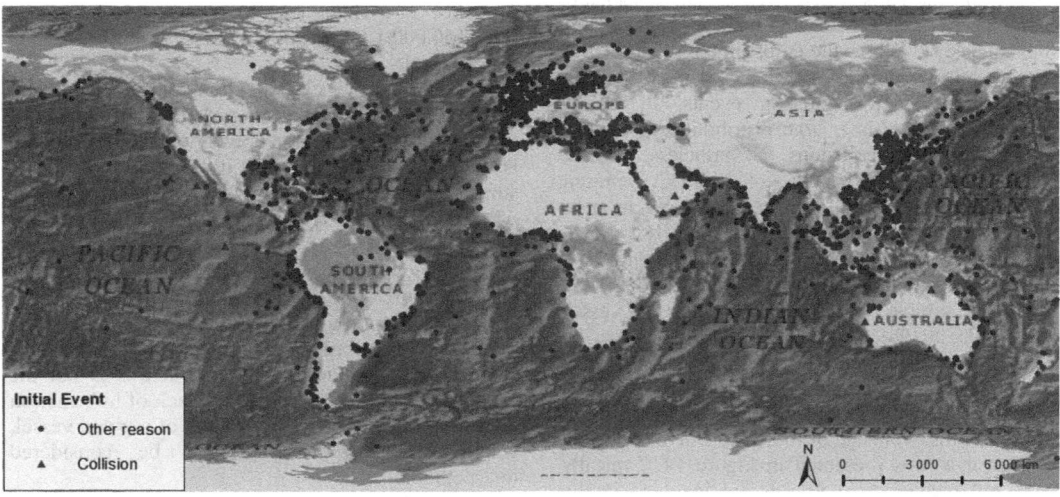

Figure 11. Geographical location of accidents worldwide (2006-2016).

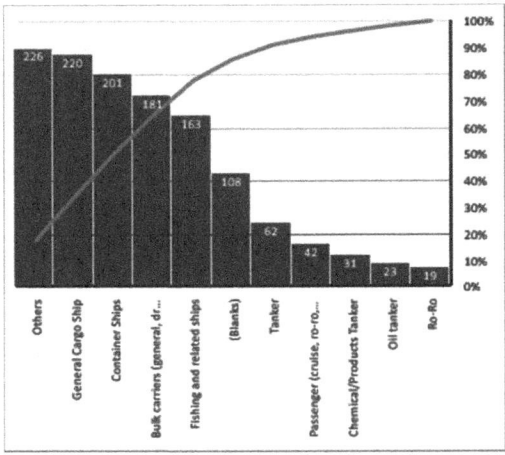

Figure 12. Distribution of the ship types involved in collisions, N=1276.

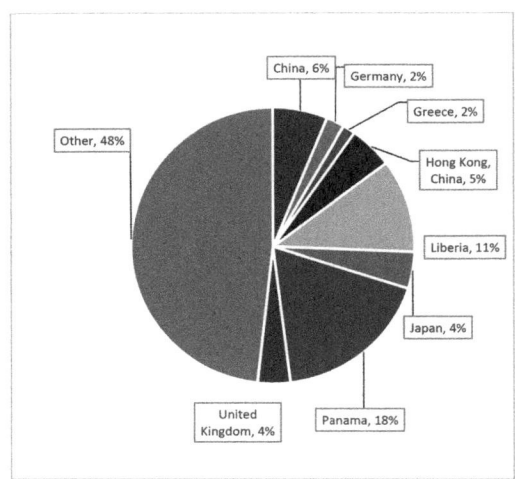

Figure 13. Ship flags distribution - Collision events, N=1186.

therefore less likely to be found in congestion waters or high traffic. Additionally, also fishing vessels present a low collision rate (~13%). Unlikely to the previous case, this low rate can be eventually justified by underreporting of several cases involving this ship type. Finally, it should be mentioned that in 8% of the cases there is no information regarding the vessel involved. Many of these cases are related to small vessels without the IMO number being properly identified in the database. Cases are, in which the IMO number is provided, but no further information is available in GISIS concerning this and other details.

The flag has always been seen as a key factor in any ship risk profile estimation. In fact, the ship flag is one of the variables considered presently in the NIR for the identification and prioritization of port inspections within the Paris MoU. Lack of safety standards are for years' issues associated with the use Flags of Convenience (FOC), as the enforcement of safety policies are softer. Figure 13 shows the results of the ship flags distribution for collision events. It can be observed that there is a large distribution of flags of ships involved in collisions. Nevertheless, Panama (18%) and Liberia (11%), two of the FOC, represent nearly 30% of the overall sample. This is not an unexpected result given the type of the vessels typically involved in collision events, in which these flags have a wide representation. It should be mentioned that most of the sample is constituted by white flag vessels (Paris MoU, 2019b)

One important variable in collision accidents is the size of the vessels since the consequences are highly related to the energy involved in the event. The distribution of Gross Tonnage of the vessels involved in the collision events is presented in Figure 14. Typically, small and medium size vessels are involved in this type of event. In fact, 70% of the vessels, with GT provided in the sample, have value below 25000t. Worth

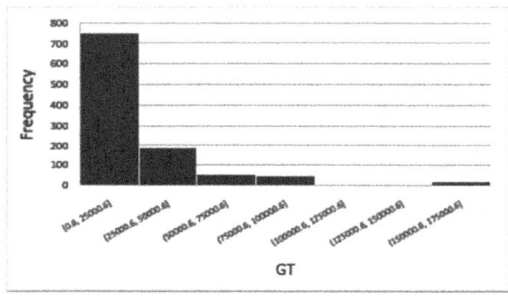

Figure 14. Distribution of gross Tonnage of the vessels involved in collision events, N=1072.

of notice is a slight increase on very large vessels (GT>150.000t) but these only represent 1.6% of the overall results.

Even though the classification society of the vessels involved in collisions is not provided in most of the records of the data set (nearly 60% of the sample is with missing information) it is important to make some considerations about it. From those cases where information is present, Figure 15 shows that more than 75% are from vessels with classification society belonging to the International Association of Classification (IACS), i.e., high-standard ones. It could be expected that a significant percentage of the vessels involved in these events would have substandard classification societies, but this is not the case. It should be highlighted the lack of information regarding the classification society of a given vessel, which, to some extent, can be considered unacceptable.

Figure 16 and Figure 17 present the frequency of collision events per year and per month, respectively.

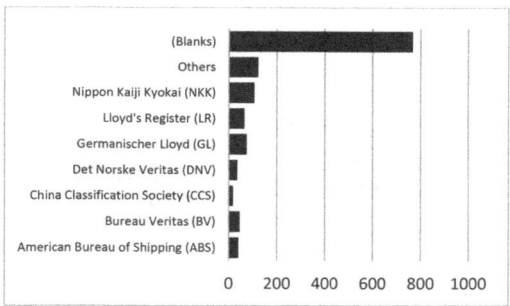

Figure 15. Distribution of classification society of ships involved in collision events, N=1276.

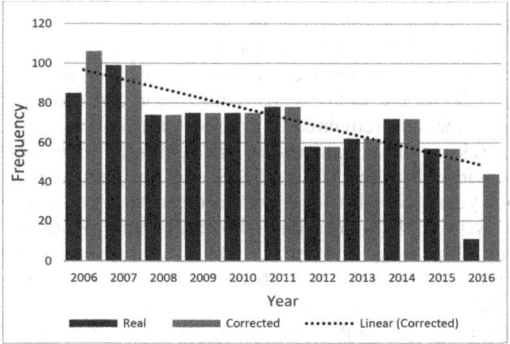

Figure 16. Frequency of collision events per year, N=746.

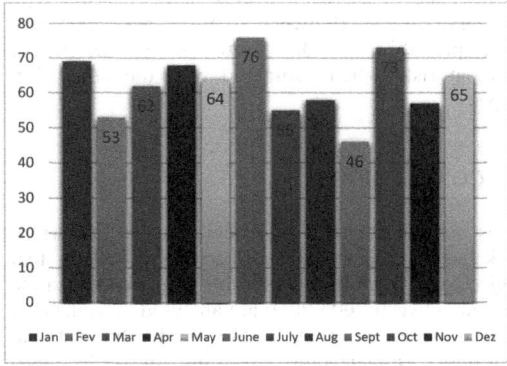

Figure 17. Frequency of collision events per month, N=746.

Regarding the former, it should be mentioned, that the retrieved sample covers the period between May 2011 and April 2016, and thus with the necessity of extrapolating the expected number of collisions in these specific two years (identified as "corrected" in the figure). The results clearly show

that there is a downward trend in the occurrence of collisions in the considered period. It can be observed two significant drops in the number of occurrences. One in 2008, that coincides with the economic crises, in which the overall world trade was affected, and the second, after 2011, which corresponds to the introduction of the NIR in the Paris MoU area.

Regarding the monthly distribution, it can be observed that there is an almost evenly distribution in the number of occurrences throughout all the months. However, it is worth notice that in the Summer months there is a slightly decrease of the number of occurrences. If from one side there is better weather conditions, from another, there are more recreation vessels navigating, thus increasing the risk of a collision. It should be mentioned that similar distribution is made for the time of day when collision events have occurred. The day was divided in 4 periods of 6 hours each in order to obtain more aggregated outcomes. The results show that nearly 60% of the collisions have occurred between 6 p.m. and 6 a.m.

This result clearly shows that *visibility* is a critical factor in the occurrence of collisions, despite all the technological advances, with the introduction of several bridge aids over the last decades. This result can also explain the previous one of less collisions in summer months – months which present more hours of daylight.

Concerning, the distribution of the building date of ships involved in collisions (Figure 18), it can be seen that almost 50% of the vessels are relatively new, since they were built in the period between 1995-2010, and additional 5% of vessels built after 2010. Nevertheless, a significant part of the occurrences (45%) involve vessels with more than 12 years at the time of the accident (best case scenario). It should be mentioned that the sample presents 3 cases of vessels with building dates earlier than 1950.

Figure 19 shows the distribution of the length of ships involved in the collisions. The results are

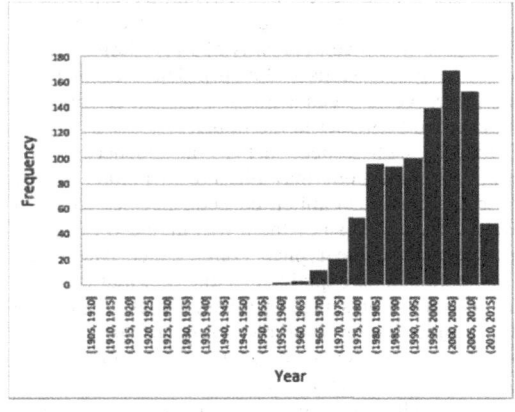

Figure 18. Distribution of building date of ships involved in collisions, N=894.

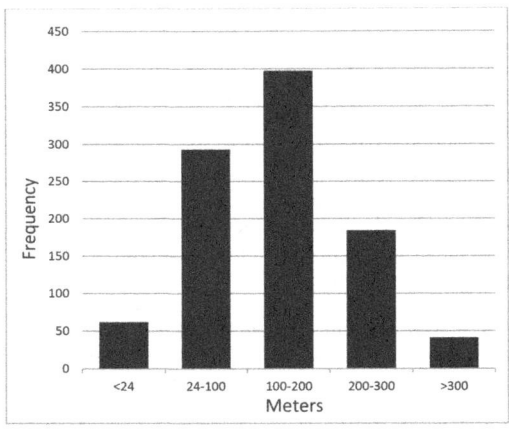

Figure 19. Distribution of the length of ships involved in collision events, N=977.

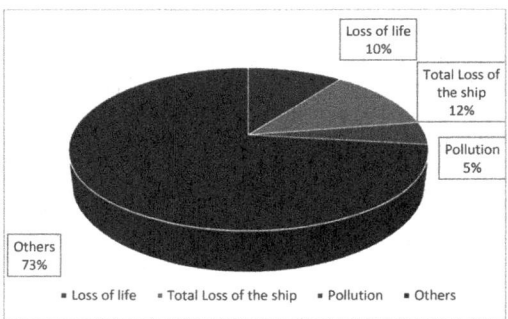

Figure 20. Explicit consequences of collision events, N=1276.

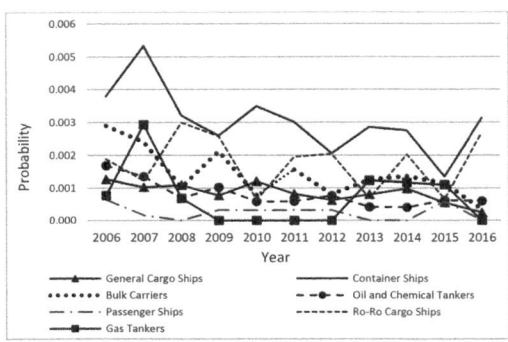

Figure 21. Collision probability per ship type and year.

aligned with earlier ones (in particular with the ones obtained for the GT). Most of the vessels involved present lengths below 200m (77%). These are vessels which have better manoeuvrability capabilities to avoid a collision course, if detected earlier. It is interesting to notice the nearly 5% very large vessels involved in collisions (>300 m). The current trend for larger vessels can explain this value, which is expected to be higher in the future.

Finally, with respect to the consequences of the collision events, one may observe in Figure 20, the distribution of the explicit consequences of collision events. "Explicit" since there was the need to aggregate them into the typical consequence categories. This need can be explained by the fact that the consequences are presented in a very widely manner in the sample, as for example, "*Loss of life, Serious injuries, Total loss of the ship*", "*Total loss of the ship; Pollution; Loss of life*".

The results show that only in 10% of the cases there is loss of human lives but in 12% there is the total loss of the ship. Typically, these are cases where

there is a large difference in the ship sizes, leading to the loss of the smaller one (in many of the cases it involves a fishing vessel). It is worth mentioning that in 34% of the cases there is no major consequence since the ship remained fit to proceed its course.

Taking in consideration the collision data set provided by the IMO's GISIS and the fleet at risk per year given by the Equasis information system, it is possible to estimate the collision probability per ship type and year, which is presented in Figure 21. The figure shows that *Bulk Carriers, Container ships* and *Ro-Ro Cargo vessels* are three of the ship types with higher probabilities of collision. While the two first ship types are in line with the results already presented, the case of Ro-Ro Cargo vessels is more surprising, with probabilities of collision of 0.002. Although the occurrence of collision events with this ship type is very seldom (between 1 to 5 per year) this high probability is due to the small size of the fleet at risk, when compared with the other ship types.

5 CONCLUSIONS

This paper presents a statistical analysis of various risk factors of ship collision accidents. Towards that objective, 746 collision accidents that occurred in the period between 2006-2016 are analysed. The data set was retrieved from the GISIS of the International Maritime Organization consisting of 1276 ships involved in the collisions. Two of the main objectives of the present study are to identify the circumstances in which collision events developed and to identify collision Risk Influencing Factors, fundamental for the development of risk models and the estimation of the risk levels of a given ship in a given scenario.

In the present study several variables are analysed, namely, ship type, flag, GT, consequences, classification society, ship length, etc. towards the abovementioned objectives. Findings show that *General Cargo* (17%), *Container ships* (16%) and

Bulk Carriers (14%), represent nearly 50% of the ships involved in collisions. Typically, small size ships are involved in these occurrences, although this trend is changing. It is worth notice that visibility is still a paramount factor to the probability of collision. Furthermore, and in order to estimate the collision probability per ship type and year, information concerning the fleet at risk is retrieved from the Equasis information system for the period in consideration.

Results show that *Bulk Carriers, Container ships* and *Ro-Ro Cargo vessels* are the three ship types presenting higher probabilities of collision.

ACKNOWLEDGMENTS

The paper has been conducted through the project "Integrated System for Traffic Monitoring and Maritime Risk Assessment (MoniRisk)", which has been co-funded by the European Regional Development Fund (Fundo Europeu de Desenvolvimento Regional (FEDER) and by the Portuguese Foundation for Science and Technology (Fundação para a Ciência e a Tecnologia – FCT) under contract no. 028746. This work contributes to the Strategic Research Plan of the Centre for Marine Technology and Ocean Engineering (CENTEC), which is financed by the Portuguese Foundation for Science and Technology (Fundação para a Ciência e Tecnologia - FCT) under contract UIDB/UIDP/00134/2020.

REFERENCES

Chen P, Huang Y, Mou J, van Gelder PHAJM., 2019. Probabilistic risk analysis for ship-ship collision: State-of-the-art. *Safety Science*, 117, pp. 108–22.

EC, 2009, Directive 2009/16/EC of the European Parliament and of the Council of 23 April 2009 on port State control, European Commission.

EMSA, 2018, *Annual Overview of Marine Casualties and Incidents 2018*. European Maritime Safety Agency.

EQUASIS, 2019, https://www.equasis.org/EquasisWeb/public/PublicStatistic?fs=HomePage (accessed in December 2019)

Goerlandt, F., and Kujala, P., 2011, Traffic Simulation Based Ship Collision Probability Modeling, *Reliab. Eng. Syst. Saf.*, 96(1), pp. 91–107.

Goerlandt, F., Hanninen, M., Stahlberg, K., Montewka, J., & Kujala, P., 2012. Simplified Risk Analysis of Tanker Collisions in the Gulf of Finland, *Int. Journal on Marine Navigation and Safety of Sea Transportation*, 6 (3), pp 381–387.

Guedes Soares, C. and Teixeira, A. P. 2001. Risk Assessment in Maritime Transportation. *Reliability Engineering and System Safety*. 74(3):299–309. 10.1115/OMAE2018-77312

Li, C., Li, W. and Ning, J. 2018, Calculation of Ship Collision Risk Index Based on Adaptive Fuzzy Neural Network, Advances in Intelligent Systems Research (AISR), *3rd International Conference on Modelling, Simulation and Applied Mathematics (MSAM 2018)*, Vol.160, pp. 223–227.

Montewka, J., Hinz, Kujala, P., Matusiak, J. 2010. Probability modelling of vessel collisions, *Reliability Engineering and System Safety*, 95, pp. 573–589.

Paris MoU, 2019a, https://www.parismou.org/system/files/2012-12-DetentionLists.pdf; https://www.parismou.org/system/files/2019-11-DetentionLists.pdf (accessed in December 2019)

Paris MoU, 2019b, https://www.parismou.org/detentions-banning/white-grey-and-black-list (accessed in December 2019)

Psarros, G., Skjong, R., Eide, MS. 2010, Under-reporting of maritime accidents, *Accident Analysis and Prevention*, 42(2), pp. 619–625

Rong, H., Teixeira, A. P., and Guedes Soares, C., 2016, Assessment and Characterization of near Ship Collision Scenarios off the Coast of Portugal, *Maritime Technology and Engineering 3*, Guedes Soares C. & Santos, TA: eds., Taylor & Francis Group, London, pp. 871–878.

Sage, B., 2005, Identification of 'High Risk Vessels' in coastal waters, *Marine Policy*, 29, pp. 349–355.

Silveira, P.A.M., Teixeira, A.P., Guedes Soares, C. 2013. Use of AIS Data to Characterise Marine Traffic Patterns and Ship Collision Risk off the Coast of Portugal. *The Journal of Navigation*, 66, pp. 879–898.

Stróżyna, M. and Abramowicz, W. 2015, A Dynamic Risk Assessment for Decision Support Systems in the Maritime Domain, *Studia Ekonomiczne. Zeszyty Naukowe Uniwersytetu Ekonomicznego w Katowicach*, Nr 243.

Teixeira, A.P., Guedes Soares, C. 2018, Risk of Maritime Traffic in Coastal Waters, *Proc.37th International Conference on Ocean, Offshore and Arctic Engineering*, Volume 11A: Honoring Symposium for Prof. Carlos Guedes Soares on Marine Technology and Ocean Engineering V11AT12A025, https://doi.org/10.1115/OMAE2018-77312

UNCTAD, 2019. Handbook of Statistics 2019. United Nations. (statistics available in http://unctadstat.unctad.org/).

Zhang, J., Teixeira, A.P., Guedes Soares, C., Yan, X., 2018. Quantitative assessment of collision risk influence factors in the Tianjin port, *Safety Science*, 110, pp. 363–371.

Zhang, W., Goerlandt, F., Montewka, J., Kujala, P. 2015, A method for detecting possible near miss ship collisions from AIS data, *Ocean Engineering*, 107, pp. 60–69.

Developments in Maritime Technology and Engineering – Guedes Soares & Santos (eds)
© 2021 Copyright the Author(s), ISBN 978-0-367-77376-2

Virtual model of FiFi system for autonomous detection and response to failure

S. Ferreno-Gonzalez, V. Diaz-Casas & M. Miguez-Gonzalez
University of A Coruna, Spain

C. Garcia-Sangabino
Navantia S.A. SME, Spain

ABSTRACT: This paper presents the results of the implementation of a virtualized model of the behavior of the firefighting (FiFi) system that allows autonomous detection and response of the system to failures. The developed system which is fed by this virtual model is based on Artificial Intellligence (AI) and will be implemented in the ship's digital twin, providing it with intelligence. For the elaboration of this virtualized model, we have started with data obtained in the tests carried out on board, as well as the results of simulations of the FiFi system. In these on-board tests/simulations, a double characterization of the system has been carried out: obtaining, on the one hand, data that represents normal operation, and on the other, the rupture and the collapse (total or partial) of the pipes. The combination of both has allowed to obtain a sufficiently broad and representative data collection to carry out the training of an AI - based system.

1 INTRODUCTION

Today, most ships have some type of automation in the failure control doctrine (especially in the case of warships, which for security reasons try to incorporate the latest technological advances), but there are very few developments based on Artificial Intelligence.

Artificial Intelligence (AI) is a facilitating technology that is implemented in many sectors of our society, while in others it is in an incipient phase of its development, such as the naval and maritime industry.

The implementation of AI-based systems in the control of system's failures in ships will allow, from the readings and analysis of the data obtained from the sensors installed in the ship's systems, the automatic (or almost automatic) detection of failures in those systems, allowing "real" time to achieve the following:

- Detection of the failure. Identification, by anomalous/out of range data readings in sensors present in the different systems of the vessel, whether: pressure sensors, flow meters, vibration or temperature sensors, etc.
- Identification of false alarms. Early decision for differentiating a real failure, for example: a sensor fail, signal loss, etc.
- Location in the space of the detected failure. By comparing the sensor data, we can locate with relative accuracy the location of the fault within the system.

- Identification of actions to take. Once the fault has been identified and located, the system will be able to provide assistance in decision - making, through messages of actions to be taken or even performing corrective actions itself automatically (for example, in the case of a break in a section pipeline, the system may advise the person in charge of system control to close certain valves).

In this work, the development of a virtual model of the FiFi system of a ship is presented. This model has been born to feed our AI - based system for detection and location of failures and the localization thereof within the system based on pressure sensors of the system.

This involves the following activities:

- Development of a simulation model of real system behavior.
- Validation of the simulation model.
- Characterization of system behavior. Obtaining data under different operating conditions.
- IA system training.

2 BACKGROUND

2.1 *Failure detection in pipelines*

The detection of failures in pipes is a problem that has been raised in many fields of engineering for

DOI: 10.1201/9781003216582-25

many years. Achieving rapid detection and the location of breakage or strangulation in pipe systems has been the subject of numerous developments. The methods for detecting leaks in pipes vary from manual inspection by trained technicians to advanced satellite images.

Traditional techniques for the detection of problems in pipes, used in water distribution systems, but also in the case of hydrocarbons are normally classified into 2 large groups: those based on software and those based on hardware: Procedures Software-based, employ different computational techniques, while hardware-based methods are classified according to the sensors and equipment used for fault detection.

Datta & Sarkar (2016) describe in depth the main methodologies and technologies and are summarized in Table 1.

Many of these methods are focused on the use of sensors and systems, which in some cases imply an extra cost of the installation, so the priority in this case is to detect the fault with the available means, thus, in his case, we are limited to the use of pressure measurements in the system.

Moreover, and if we focus on the application to the ship's FiFi system, we see that, in addition, many of them are not recommended for use in marine environments, due to the problems generated in systems that work with seawater.

Therefore, in this work it has been decided to take into consideration software-based detection methods.

There are applications based on logs and pressure measurements in piping systems for fault detection. Abdulshaheed et. al (2017), Sousa et. al (2015), Giustolisia et al (2015) and Berardi et al (2015) have been used different methods and almost all aimed at detecting faults in potable water pipelines.

Table 1. Pipeline failures detection methods.

Blockage detection techniques	Leakage detection techniques
Vibration analysis	Negative pressure wave based leak detection system
Pulse echo methodology leak detection	Fiber sensor based
Acoustic reflectometry	Support vector machine (SVM) based pipeline leakage detection
Transient wave blockage interaction and blockage detection	Piezoelectric acoustic emission (AE) sensor
Stochastic successive linear estimator (SLE)	Filter diagonalization method (FDM)
	Harmonic wavelet analysis
	Genetic algorithm (GA) in combination with the inverse transient method
	Computational fluid dynamic (CFD) simulation

Some of these developments, however, are based on AI applications such as the development of Expert Systems, Laurentys et al (2011), and others even employ neural networks, similar to our development. For example, Zhao et al (2014) employs Neural Networks algorithms that are used to detect problems in underwater pipe systems. However, no reference has been found in which NNs are applied to ship systems.

2.2 Artificial intelligence in shipbuilding

We can define Artificial Intelligence as a Scientific discipline that deals with creating computer programs that execute operations comparable to those performed by the human mind, such as learning or logical reasoning.

Within Artificial Intelligence we find different sub-disciplines, being those based on Computational Intelligence which are currently occupying the center of developments based on this concept. Computational Intelligence is a term that encompasses numerous disciplines of Artificial Intelligence, mostly of biological inspiration, and presented in contrast to those based on classical symbolic reasoning, which are used in order to solve complex problems difficult to solve with more traditional computational techniques. Artificial neural networks, evolutionary computing, diffuse logic, swarm intelligence and artificial immune systems are usually included in Computational Intelligence.

Some developments based on AI application in shipbuilding are described below.

Indra & Spanish Navy:

The SOPRENE Project aims to investigate the application of artificial intelligence techniques to improve ship maintenance.

The data available at the Center for Monitoring and Analysis of Monitored Data of the Spanish Navy (CESADAR) will be used. These data come from the sensorized equipment of the vessels and are recorded while the vessels are navigating to study the advantages that their analysis can provide in order to reinforce the predictive maintenance of the vessels, avoiding unforeseen breakdowns, increasing their availability and saving costs.

Fujitsu & Maritime and Port Authority of Singapore:

This solution aims to detect ship collision risks and predict areas where risks are concentrated as 'critical points of dynamic risk'. This technology has the potential to be implemented in a Vessel Traffic Service system to help maritime controllers proactively in the traffic management, with the aim of improving navigation safety.

Stena Line & Hitachi:

Stena Line, in collaboration with the technological company, Hitachi, has developed an artificial intelligence assistance system that will help determine what is the most efficient way in terms of fuel to operate a ship on a specific route. It takes into account the currents, weather conditions, water depth and speed, combining data in a way that would be impossible in a manual system.

Although it is an advance that is still under development, a first pilot is already being carried out on Stena Scandinavica on its route between the Swedish port of Gothenburg and the German port of Kiel.

Rolls-Royce:

The company that has advanced the most in the implementation of AI in ships is Rolls Royce. At this moment it has already carried out the first tests with an autonomous ship, Falco, a ferry of about 54 meters in length, which has the Rolls-Royce Ship Intelligence system and includes a series of cameras and sensors that are placed along the ship, and are responsible for scanning the waters looking for other ships.

The Falco was being able to navigate the waters safely, taking into account the presence of other vessels, and make the usual trip between Parainen and Nauvo. In addition, it is capable of performing all the necessary procedures for docking alone.

As we can see, among the applications of AI applied to ships and the marine world in general, we do not find any related to the integrity of the ship's systems.

3 FIRE FIGHTING SYSTEM

3.1 *System description*

The saltwater firefighting system supplies pressurized saltwater to the ship's fire nozzles. It also supplies the pressurized saltwater for other specialized firefighting services (sprinklers, foam station…), as well as other saltwater systems and consumers on the ship, such as:

- Wastewater Treatment Plant
- Cannon Cooling
- Decontamination Station
- Garbage Cooling
- Stabilizing Fins
- Propellers
- Transversal thruster
- etc

The firefighting system will be a pressurized system that will be fed with salt water by means of four centrifugal pumps of 200 m³/h and 12 bar, designed for continuous operation, fitted with a speed regulator and that will suck directly from the respective collectors of saltwater located in the same chamber in which they are located.

The ship is divided transversely by a bulkhead in frame 66, which divides the ship into two main fire zones.

FiFi pumps are located under the waterline, with the following location:

- Two bombs in the Auxiliary Room.
- A pump in the Bow Engine Room.
- A pump from the Stern Engine Room.

Each fire zone has 2 pumps with a capacity of: 200m³/h and 12 Bar (400 m³/h in each zone). In addition, FiFi pumps must have sufficient capacity to supply the maximum flow required by the FiFi of the ship with a pump out of order. The maximum flow in this condition is 600 m³/h. In addition, the system will be constantly pressurized, the operation of a pump being sufficient to maintain the pressure of 7 kg cm² (6.87 bar) at the highest point of the circuit.

The system will be arranged forming a vertical ring system, which will be subdivided into two main horizontal collectors. The top manifold will be located on the main deck and the bottom manifold under the second deck.

Each of the two horizontal longitudinal collectors that make up the system are able (in an emergency condition and with a speed of less than 6.4 m/s) to supply the total flow required by the FiFi system.

3.2 *System selection*

The main reason for selecting this system, and not others (like the fuel system, the cooling system or the lub oil system), was based on the fact that the firefighting system is one of the ship's vital systems, and therefore it is of great importance, that in case of impact by missile or collision, the system is capable of reconfiguring and maintaining its operation: both the one that performs continuously (cooling of elements) but especially the tasks of extinguishing possible fires (more, if we take into account that the impact of a missile can carry a high risk of fire).

4 FIFI SYSTEM MODEL DEVELOPMENT

4.1 *Model description*

A model for simulating the hydraulic behavior of a FiFi system based on a real base ship has been developed, with the objective of faithfully reflecting its behavior, because this model will be the main source of data to train the autonomous system. A section of this model is shown in Figure 1.

For the development of this model it has been used a commercial software to simulate hydraulic circuits in the stationary and transient regimes.

It has replicated the system (pipes, valves, pumps) and the different consumers of the system (FiFi Monitors, bilge ejectors, firefighting stations, etc.)

The software allows incorporating (although not done graphically) the characteristics of the pipes: materials, dimensions, and even accessories (elbows, tees …).

With the model, the identification of the system failure modes (possible sections of pipeline affected and failure in FiFi pumps), shown in Figure 2, was carried out, and operating simulations were performed:

Figure 1. Example of FiFi model (Section).

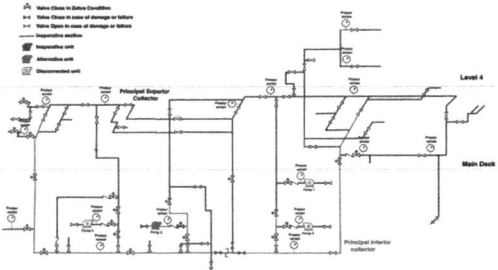

Figure 2. Example FAILURE MODE. Damaged pipe identification + inoperative section.

- normal operation of the system (opening of saltwater consumers of the FiFi system) and
- possible breaks and/or strangulation of pipes (by impact) located in different sections of the main collectors or branches of the system (according to the failure modes identified).

The results obtained in these simulations are the pressure values in the nodes defined in the positions where the vessel's pressure sensors are, since we start from the point at which the only source of information we will have on the ship real is what the existing pressure sensors on the ship offer us.

4.2 *Model validation*

With the simulation model of the FiFi system at work, tests were carried out onboard the real base ship. These tests were proposed with a double objective:

On the one hand, the performance of these tests was considered with the objective of obtaining real values of system operation beyond the theoretical values obtained in the simulations, and that present (or can present) alterations produced by the ship's own operation, which will imply the presence of noise in the pressure signal recorded by the system sensors.

On the other hand, it was considered necessary to carry out tests to validate the developed model, which, although it was estimated in view of the

simulations performed and the base to the experience that the model was valid, said statement can only be corroborated by real tests and comparison with the results obtained under the same conditions in the model.

The tests were divided into two parts: A first test in which we tried to characterize the main seawater consumers of the FiFi system (FiFi water cannons, Bilge Ejectors, FiFi Stations). For this, repeated openings/closures were carried out. In addition, it was also used to start/stop the pumps of the system, to have data on the pressure curves that are generated at the start and stop of the pumps.

In the second part of the tests, the breakage of a pipe section and the partial and total collapse of the pipe were simulated. Circumstances in which we could find ourselves if, for example, there is an impact in which this system is affected.

Next, in Figure 3 we see, as an example, one of the graphs obtained in the performance of the consumer characterization tests.

In it, an opening was carried out (sequentially of the bilge ejectors) with the pump 3 running.

As we see in the graph, the pressure drop caused by the opening of the ejectors is quickly corrected by the pump (always staying above 8 bar) and it is with the opening of the last ejector when a significant pressure drop occurs, since the pump is not able to supply the required flow/pressure. Pump 4 is started, whereby the pressure is recovered. Finally, the ejectors are closed in the reverse order of starting and the pump 4 is stopped.

Various tests were carried out, and we are waiting for the availability of the ship (this or another of the base ships that are being considered) to perform new tests and have a larger real database.

To validate the developed model, the events that were carried out in the tests on board the ship were replicated.

The results obtained in the steady-state simulations were compared, as in the examples in Figure 4. In Test 1, the opening/closing of the FiFi water cannon was carried out and subsequently the sequential opening of the bilge ejectors (and the corresponding closure).

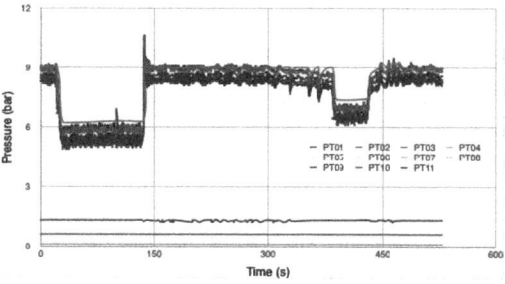

Figure 3. Results of FiFi System characterization under normal operation. Example.

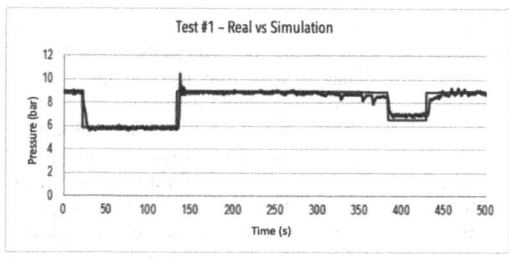

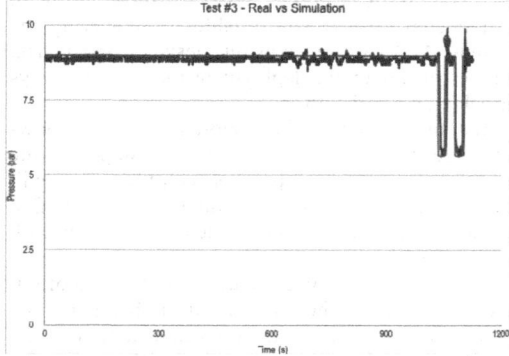

Figure 4. Comparative test results vs. simulations stationary model.

In test 3 was carried out the opening/closing of different consumers, those who have low flow show little variation any (or none) in the pressure of the system is therefore that in the graphic seems to be nothing until opened the water cannon.

It is evident that these simulations do not realistically reflect the system, but they serve so that, with few computational requirements, we can validate the developed model, for example, it can be verified that the pressure drops correspond to what we find in the tests.

These analyses were completed with the study of phenomena in transitory state (opening/closing of consumers or valves) or also the start/stop of the pump.

In them we can observe the pressure peaks that can occur in the different simulated events, as well as verify that the slopes of the curves correspond to what was observed in reality.

4.3 Characterization of pressure failures

To complete the characterization of the system, complementary analyzes have been carried out to identify the pressure drops corresponding to a normal operating state of the system against a pipe breakage. This study was based on the detailed analysis of the slopes generated in the pressure curves collected in the actual data, which will allow us to identify which falls correspond to a normal operation of the system. Those that move away from these, will be susceptible of being identified as a failure in the system.

Having data corresponding to the variation in pressure produced with respect to time will allow us to refine better in the differentiation of normal operating states of the systems against an anomaly produced in them.

The analyzes carried out in the characterization of FiFi monitors and bilge ejectors are shown as an example in Figure 5.

The slopes of the curve were analyzed, by stretches equi- spaced in time, which are shown in Figure 6.

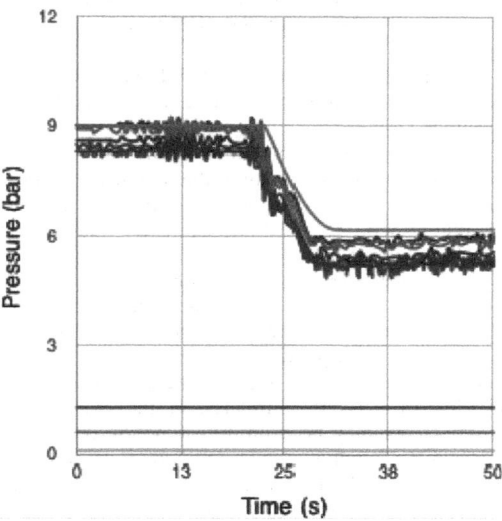

Figure 5. Pressure drop detail for FiFi Monitor Opening.

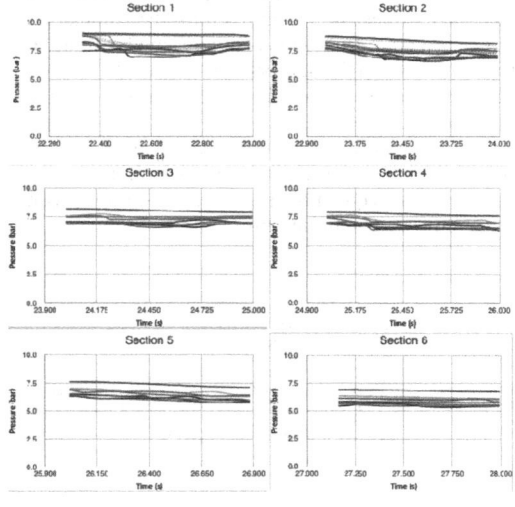

Figure 6. Pressure drop analysis for FiFi Monitor Opening.

5 AI SYSTEM FOR FAILURE DETECTION

Neural Networks (NN) are a computational model that tries to simulate the functioning of neurons in the human brain, with the objective of solving complex computational problems that are not easily programmable. (Martín & Sanz 2006). That is why that what is done is to "teach" or rather "help learn" this network artificial neural.

The general architecture of a NN consists of an input layer, hidden layer (or layers) and an output layer. (Lippmann 1987)

There are several types of layers: Convolutional, Activation, Pooling, Dropout, Dense and SoftMax layer.

The developed algorithm, based on neural networks, has a general scheme like the one displayed in Figure 7.

The inputs of our neural network will be the series of data corresponding to the pressure measurements in each of the sensors (time, pressure), while the output will be a 0/1 value, that is, failure or normal operation of the system.

Each node represents a neuron, and the arrows represent a connection between a neuron of one layer and the next. In these links the output value of a neuron is multiplied by a weight value. These weights can increase or decrease the activation status of adjacent neurons. Similarly, at the exit of the neuron, there may be a limiting function or threshold, which modifies the result value or imposes a limit to be able to spread to another neuron. It is known as an activation function.

Systems based on neural networks requires an important source of input data information to feed and train the NN and make it as accurate as possible. It is for this reason that it was decided to develop this virtual system model that would allow the system to be fed with data corresponding to simulations in a stationary/transient regime. From the simulations we can generate (in theory) infinite combinations of pressure results in the sensors that we decide

and with the sampling frequency that we consider appropriate.

The advantages of using these values, is the volume of data that we can have, as well as the versatility in its acquisition (number of sensors, sampling frequency, sampling time).

The main drawback is that the reliability of the data can never be total, since that can only be achieved with real data. That is why that feeding the system should be completed using actual data from tests.

On the other hand, the main advantage obtained by using real data as a source of information, is the guarantee that the data we are obtaining fits perfectly to what happens in the real system, but has the great disadvantage that it is not simple achieve a volume of data at a sampling frequency that serves to power the system. Currently, data is recorded with a frequency that is not suitable for powering the system. And carrying concrete tests to obtain data implies an important deployment that is not always possible on a real ship.

Without forgetting the importance of the knowledge generated by the study of the slopes in the pressure curves that, in parallel with the above, will allow the system to determine when we find a normal operation of the system, and when with a failure in it.

In this first stage of generation and training of our network, we have chosen to use data obtained in the simulation, because to date (and until new tests are carried out on board the ship) the available data are scarce and also do not have high precision, so they have not been considered optimal.

A total of 24 simulations (100-second simulations with 50 Hz data collection) have been selected: 12 of normal system operation and 12 of failure simulation.

A total of 120,000 measurements distributed in 2400 samples of 50 elements have been obtained.

With this data, 2 neural networks have been used: A fitting neural network and a pattern recognition method. In order to compare the results obtained in each one:

5.1 *Fitting neural network*

In this case we train a neural network on a set of inputs in order to produce an associated set of target outputs.

The network scheme that has been used is shown in Figure 8.

Next, we divide our data into data training, data test and data validation. And then, we can train our new network. In this case, Bayesian regularization algorithm have been used.

After the training process of the neural network, we can analyze the regression curves obtained,

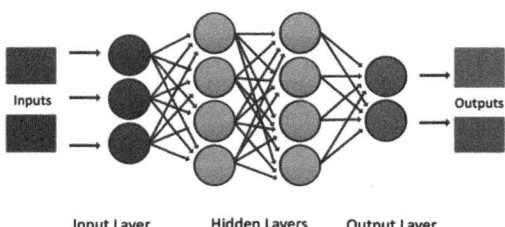

Figure 7. Neural Network General scheme.

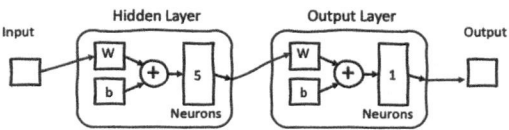

Figure 8. Neural Network scheme.

displayed in Figure 9, where we can see relationship between the targets and the outputs of the system.

The scheme used is similar to of Fitting Neural Network.

As we can see, the network seems to work correctly. Outputs and targets are near. Outputs > 0.8 indicates we have a failure.

5.2 *Pattern recognition network*

Pattern recognition networks are feedforward networks that can be trained to classify inputs according to target classes.

The scheme used is similar to of Fitting Neural Network (Figure 8).

In this case, we analyze the receiver operating characteristic curve (ROC) graphic that illustrates the diagnostic ability of a binary classifier system as its discrimination threshold is varied.

The ROC curve is created by plotting the true positive rate against the false positive. The true-positive rate is also known as sensitivity, recall or probability of detection. The false-positive rate is also known as probability of false alarm.

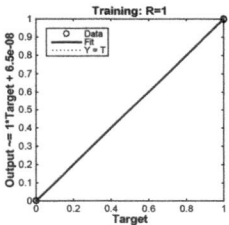 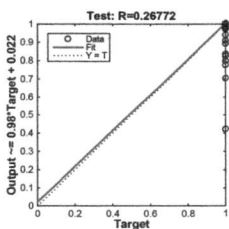

Figure 9. Regression analysis.

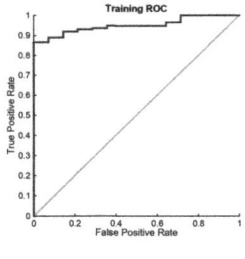 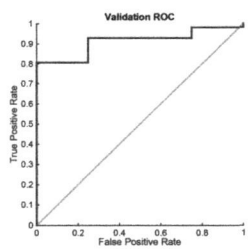

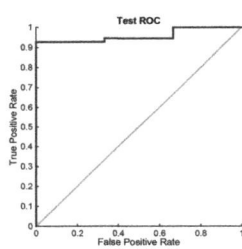 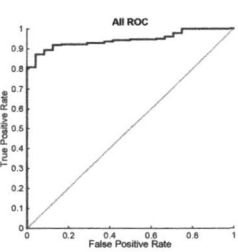

Figure 10. ROC curve.

As we can see in Figure 10, the graphs indicate that the results are remarkably good, we are always in the true positive zone.

6 CONCLUSIONS

The development of a virtual model to simulate the operation of a ship's FiFi system has been presented in this work. In order to simulate different operating conditions and system failures that allow obtaining an important database (pressure values in the system sensors), which allow feeding and training an AI-based system for fault detection in ship systems, in this case applied to the firefighting system.

For this, a system simulation model was developed that allowed generating different states of the system, which was complemented and validated with the performance of tests.

An analysis of the slopes of the pressure curves that occur with the operation of the system was carried out, in order to characterize what conditions, correspond to the normal operation of the system and from there be able to detect failures in the system.

Currently, work is being done on the learning phase of the system based on neural networks.

REFERENCES

Datta, S & Sarkar, S. 2016. A review on different pipeline fault detection methods. *Journal of Loss Prevention in the Process Industries* 41: 97–106.

Abdulshaheed, A. Mustapha, F. & Ghavamian, A. 2017. A pressure-based method for monitoring leaks in a pipe distribution system: A Review. *Renewable and Sustainable Energy Reviews* 69: 902–911.

Sousa, J. Ribeiro, L. Muranho, J. & Marques, A.S. 2015. Locating leaks in water distribution networks with simulated annealing and graph theory. *Procedia Engineering* 119: 63–71.

Giustolisia, O. Campisano, A. Ugarelli, R., Laucelli, D. & Berardi, L. 2015. Leakage management: WDNetXL Pressure Control Module. *Procedia Engineering* 119: 82–90.

Berardi, L. Laucelli, D. Ugarelli, R. & Giustolisi, O. 2015. Leakage management: planning remote real time controlled pressure reduction in Oppegård municipality. *Procedia Engineering* 119: 72–81.

Laurentys, C.A. Bomfim, C.H.M. Menezes, B.R. Caminhas, W.M. 2011. Design of a pipeline leakage detection using expert system: A novel approach. *Applied Soft Computing* 11: 1057–1066.

Lippmann, R.P. 1987. An introduction to the computing with neural nets, IEEE ASSP Magazine April 1987: 4–22.

Martín del Brío, B. and Sanz Molina, A. 2006. Redes Neuronales y Sistemas Borrosos. RA-MA Editorial. ISBN: 84-7897-743-0.

Zhao, X. Ba, Q. Zhoub, L. Li, W. & Ou, J. 2014. BP neural network recognition algorithm for scour monitoring ofsubsea pipelines based on active thermometry. *Optik* 125: 5426–5431.

Developments in Maritime Technology and Engineering – Guedes Soares & Santos (eds)
© 2021 Copyright the Author(s), ISBN 978-0-367-77376-2

A FMEA for a floating offshore wind turbine considering costs of failures

He Li

Centre for Marine Technology and Ocean Engineering (CENTEC), Instituto Superior Técnico, Universidade de Lisboa, Lisbon, Portugal
Center for System Reliability and Safety, University of Electronic Science and Technology of China, Chengdu, P.R. China

C. Guedes Soares

Centre for Marine Technology and Ocean Engineering (CENTEC), Instituto Superior Técnico, Universidade de Lisboa, Lisbon, Portugal

ABSTRACT: The paper extends the conventional failure mode and effects analysis methodology by considering costs of failures into generating risk priority numbers as a basis to finalize the failure analysis of a floating offshore wind turbine at both components and at systems levels. Accordingly, critical systems as well as components of the floating offshore wind turbine are ascertained. Recommendations on corrective designs, e.g. implementation of robust design to the pitch assembly, and preventive actions on safe operations, e.g. monitoring vibration of the gearbox and temperature of the converter, are concluded. The results of this analysis are validated by a comparison with other studies published related to failure analysis of offshore wind turbines.

1 INTRODUCTION

The economic efficiency of fixed-foundation offshore wind turbines decreases sharply with the increase of water depth (Uzunoglu et al. 2016). As a consequence, floating offshore wind turbines have been developed in various designs which can explore the offshore wind resources in deeper waters, located away from the coast (Diaz & Guedes Soares, 2020). Floating offshore wind turbines are promising facilities that have been reducing the electricity generation cost until a level comparable to fossil energy installations (Sinha & Steel 2015, Castro-Santos et al. 2020a,b). While the reliability of these platforms is of concern (Li et al. 2019, 2020a), the operation and maintenance (O & M) cost of floating offshore wind turbines would be higher than fixed-foundation ones due mainly to the structural complexity, harsher environmental conditions, larger whether windows of maintenance, and costly maintenance (Castro-Santos et al. 2016). Moreover, the O & M cost is reported to consume the largest proportion (about 25%-35%) of the economic benefit of floating offshore wind turbines, which can be compared with 20%-30% that of similar fixed-foundation structures (Sinha & Steel 2015).

To this end, reducing the O & M cost of floating offshore wind turbines turns out to be a primary issue that need to be addressed by a comprehensive failure analysis (Sinha & Steel 2015, Kang et al. 2017, Bhardwaj et al. 2019, Scheu et al. 2019), which

involves identifying the critical systems as well as components that give more rise to malfunctions of floating offshore wind turbines, and the components that are more critical and need to be considered particularly in both system design and operation stages. Accordingly, it is necessary to suggest recommendations on corrections and preventive actions against disastrous failures occurrences.

Failure analysis is designed to seek weak links, e.g. systems, components, failure modes, and failure causes of floating offshore wind turbines as a basis to recommend corrective and preventive actions for achieving economic maintenance strategies (Kang et al. 2019a, 2020a,b; Kang & Guedes Soares 2019a, Santos et al. 2018, Sinha & Steel 2015). Failure analysis can be conducted by logical deduction methodologies (qualitative, e.g. checklist, what-if analysis, etc.), numerical methodologies (quantitative, e.g. Monte Carlo simulation and structural reliability methods), and logical-numerical methodologies like semi-quantitative, e.g. fault tree analysis (FTA) and failure mode and effects analysis (FMEA) (Bouti & Kadi 1994; Kang et al. 2017, 2019b; Li et al. 2020b).

Among others, FMEA is selected and proved to be a sufficient tool for failure analysis of complex systems in various of scenarios and fields, for instance, wind turbines (Arabian-Hoseynabadi et al. 2010, Shafiee & Dinmohammadi 2014, Bharatbhai 2015, Du et al. 2017, Kang et al. 2017, Tazi et al. 2017, Cevasco et al. 2018, Scheu et al. 2019), and

DOI: 10.1201/9781003216582-26

floating offshore facilities (Kang et al. 2017, Scheu et al. 2019).

Generally, FMEA evaluates and rank items, e.g. systems, components, failure modes, and failure causes, of floating offshore wind turbines by calculating their Risk Priority Numbers (RPNs, see Equation (1)), that are in general products of severity (S), occurrence (O), and detection (D). Values of S, O, and D reflect local or global effect, likelihoods, as well as difficulties in identification of failures occurrence. In practice, the values of S, O, and D should be evaluated subjectively either by engineers in the operation stage or by designers in the design stage.

$$RPN = S \times O \times D \qquad (1)$$

However, the RPN generated by the aforementioned methodology is uninformative and without any physical meaning, which makes the results computed not to be comparable with what been concluded from other floating wind turbines, wind farms, or projects. On the other hand, costs of failures need to be modeled into the failure analysis of economy-sensitive-productions, e.g. floating offshore wind turbines.

To this end, Cost Priority Numbers (CPNs) as alternatives of RPNs were developed for conducting the failure analysis of offshore wind turbines, see Shafiee & Dinmohammadi (2014), Du et al. (2017), Tazi et al. (2017), and Cevasco et al. (2018). The idea of CPNs is to construct a full objective index for failure analysis of wind turbines. Specifically, in CPN-based technique, detection probability (DP) and failure probability (FP, or failure rate (FR)) of components of wind turbines (both onshore as well as offshore) are usually introduced as representations of values of detection and occurrence in RPN-based methodologies.

Moreover, the severity of CPNs has various representations according to the data available, but mostly are economic indices, e.g. Loss of Electricity Generation (LEG), Cost of Labor for Maintenance (CL), Cost of Replaced Materials (CRM), Downtime (DT), and Cost of Transportation (CT). Indices considered in CPN-based FMEAs of wind turbines are available in Table 1.

Overall, RPN-based FMEAs are better for use in the design stage of floating offshore wind turbines to analyze their critical failure modes together with their root causes, while, CPN-based techniques can be implemented in the operation stage owing to the operational data such as costs of failures of components may available. On the other hand, RPN-based FMEAs neglects operational data such as FR, CT, DP, DT, FP, CRM, CL, and LEG when analyzing failures or risk of floating offshore wind turbines, while, CPN-based FMEAs abandon valuable objective information from experts. Aims to remove the aforementioned restrictions, the paper proposed a new way of generating RPNs to finalize the failure

Table 1. The indices considered in CPN-based FMEAs of wind turbines.

| Indices | Literature | | | |
	[A]	[B]	[C]	[D]
FR	√	√		√
CT	√		√	
DP	√		√	
DT			√	√
FP			√	
CRM		√		
CL	√	√		
LEG	√	√	√	√
Type of Turbines	Off	Off	Off and On	On

Off: Offshore; On: Onshore; [A]: Du et al. (2017); [B]: Cevasco et al. (2018); [C]: Shafiee & Dinmohammadi (2014); [D]: Tazi et al. (2017)

analysis of a general floating offshore wind turbine at components and systems levels, which takes both risk factors, e.g. S, O, and D, as well as economic factors, e.g. costs of failures of components into calculation.

The rest of this paper is organized as follows: The proposed FMEA methodology is introduced in Section 2. The results are demonstrated in Section 3. The comparison with other studies conducted by conventional FMEAs is implemented in Section 4. Conclusions are provided in Section 5.

2 THE PROPOSED FMEA

A new FMEA methodology that considers both objective as well as subjective indices, is proposed as a basis to complete the failure analysis of a floating offshore wind turbine at component and system levels in this section. Specifically, four indices are modeled for producing the final RPNs of components/systems of the floating offshore wind turbine, that are, S. O, D, and the costs of failures. The severity represents the consequence of a failure from the risk point of view. The occurrence is the likelihood of a failure to happen. The detection is an indicator for numerically measuring the difficulty in detecting a failure before it takes place. The failure cost is the total cost that is paid for a failure repair, which generally including loss of electricity generation, cost of labor for maintenance, cost of replaced materials, cost of transportation, and others. The schedule of carrying out the developed FMEA methodology is summarized as:

(i) Recognize the floating offshore wind turbine to be analyzed and identify its components.

(ii) Require the values of S, O, and D of the components of the floating offshore wind turbine from specialists in the field.

(iii) Compute the risk-based RPN of each component of the floating offshore wind turbine by Eq. (1).

(iv) Collect the failure cost of each component of the floating offshore wind turbine according to the operation records (if data was available) or refer to that of other offshore wind turbines with a similar configuration (if data was insufficient).

(v) Upgrade the RPN calculated in step (iii) by the cost of failure of each component determined in step (iv) according to

$$RPN_i^{Comp} = C_i \times RPN_i = C_i \times S_i \times O_i \times D_i \quad (2)$$

(vi) Calculate the RPN of each system of the floating offshore wind turbine as summations of RPNs of components belonging to the system obeys

$$RPN_j^{Syst} = \sum_{i=1}^{n} RPN_i^{Comp} \quad (3)$$

In Eqs. 2-3, RPN_j^{Syst} and RPN_i^{Comp} are the RPNs of system j as well as component i of the floating offshore wind turbine, respectively. C_i represents the cost of failure of component i of the floating offshore wind turbine. S_i, O_i, and D_i are values of S, O, and D of component i of the floating offshore wind turbine.

3 RESULTS

3.1 The floating offshore wind turbine considered

A general floating offshore wind turbine is created in this paper, which is comprised of three systems, that are, support structure, blade system, and the wind turbine system, with 13 components namely blade, hub, main bearing, main shaft, generator, gearbox, converter, transformer, mooring and floating foundation facilities, tower, pitch devises, yaw devises, as well as control and electrical devices. The components considered in the floating offshore wind turbine are demonstrated in Figure 1.

Values of S, O, D, and costs of failures of components of the blade and the wind turbine systems are collected from publications available, including Arabian-Hoseynabadi et al. (2010), Shafiee & Dinmohammadi (2014), Bharatbhai (2015), Sinha & Steel (2015), Du et al. (2017), Kang et al. (2017), Tazi et al. (2017), Cevasco et al. (2018), Scheu et al. (2019). Values of S, O, and D assigned in the aforementioned publications vary somewhat because those offshore wind turbines be analyzed are installed in various offshore wind farms around the world.

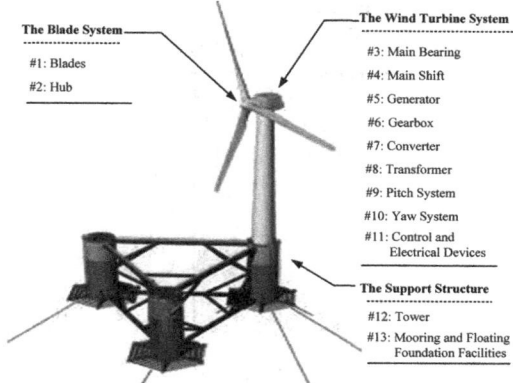

Figure 1. The components of the floating offshore wind turbine.

Hence, average values of S, O, and D of each component (with the rating scale from 1 to 10) were applied in this study aiming to provide an overall perspective of floating offshore wind turbines failures. Moreover, the floating offshore wind turbine is relatively new and values of S, O, and D of their support structures are unavailable from publications. Hence, that information was collected from an engineer in Goldwind company (China) from the designer's point of view.

3.2 The results of the proposed FMEA

The RPNs of 13 components of the floating offshore wind turbine are computed by the proposed FMEA methodology, the results of which are demonstrated in Figure 2. According to the results calculated, main bearing, gearbox, pitch devices, converter, and tower are identified as critical assemblies, for those components together give rise to more than 80% of RPN to the floating offshore wind turbine. Specifically, the main bearing is ascertained to be the riskiest component of the floating offshore wind turbine by contributing more than a quarter RPN to the whole facility. The highest RPN of the main bearing calculated due mainly to the considerable high failure cost resulted from the complex maintenance and expensive replacement. The failure frequency of the main bearings of offshore wind turbines, however, are quite low according to Shafiee & Dinmohammadi (2014) and Cevasco et al. (2018), hence, periodical inspections may appropriate to the components from the operation and maintenance point of view.

The gearbox is identified as the second important assembly of the floating offshore wind turbine, which leads to over 18% RPN of the facility. The outcome is the same as what been concluded in Sinha & Steel (2015) and Cevasco et al. (2018), in which the gearbox is ascertained to be the most important assembly of the offshore wind turbine

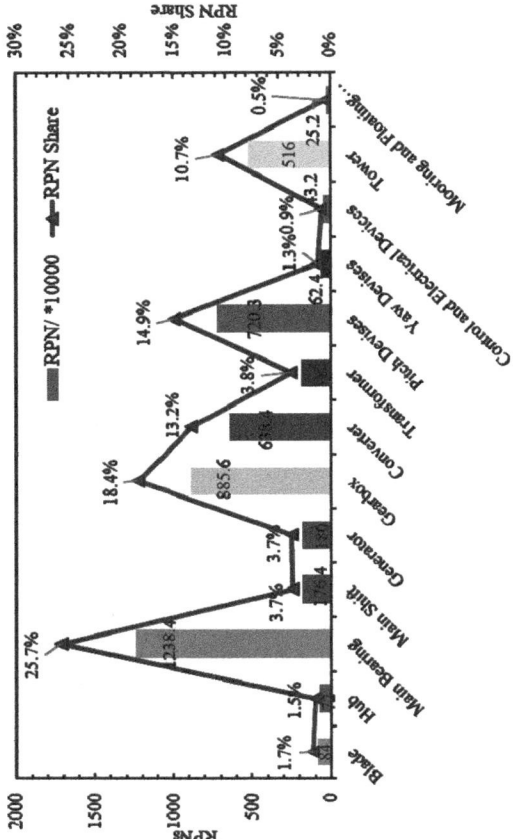

Figure 2. The RPNs of components of the floating offshore wind turbine.

frequency than the latter, while, the cost of which is lower. The conclusions are confirmed by Shafiee & Dinmohammadi (2014) and Du et al. (2017), in which the converter and the tower are identified as one of the top critical parts of the offshore wind turbines. Knowing that converter open and tower crack resulted from overheat and defective welding, respectively, are common failure modes of the converter as well as the tower of the floating offshore wind turbine. Accordingly, temperature monitoring to the converter and strict quality check of the tower welding are recommended for ensuring these assemblies free from unexpected failures.

From the system point of view, the wind turbine system is undoubtedly the most critical system, which includes quantitatively more components than the other two systems. Specifically, the wind turbine system with 9 components gives rise to 84.6% of RPN to the floating offshore wind turbine followed by the support structure (6.1%) and the bade system (1.6%), the latter each consist of only two components. The conclusions achieved are in line with the practice: (i) it is the wind turbine that realizes the function of electricity generation with multiple assemblies which would be critical than any other parts of the floating offshore wind turbine, and (ii) the support structure is an critical system of the floating offshore wind turbine owing mostly to the harsh sea conditions it experienced.

4 COMPARISON

A comparison between results of this study and that of publications associated with the failure/risk analysis of offshore wind turbines was carried out to clarify the similarities and differences of studies, see Table 2. The comparison, however, takes only 13 assemblies of three systems into consideration. For instance, sensors in Du et al. (2017) and Scheu et al. (2019), hydraulic system in Du et al. (2017) and Tazi et al. (2017), brake in Arabian-Hoseynabadi et al. (2010), Shafiee & Dinmohammadi (2014), and Tazi et al. (2017) are neglected in the comparison.

On the other hand, components of wind turbines in various publications are not constantly the same physically according to the differences in system configuration and the form of decomposition of wind turbines when conducting FMEAs. For instance, tower and substructure in Scheu et al. (2019), tower/ foundation in Arabian-Hoseynabadi et al. (2010) and Cevasco et al. (2018), are regarded as equal as the tower in this study.

A big agreement between the analytic results of this study and that concluded in Shafiee & Dinmohammadi (2014) is reached in the comparison except for the ranks of blades, main bearing, and tower. In detail, Shafiee & Dinmohammadi (2014) produce the CPN of components by considering the probability of occurrence, cost of failure, and probability of detection, which is similar to this analysis.

from either risk or economy point of view. The failure of the gearbox could result in catastrophic damage e.g. shutdown and lead to vast economic loss, which call for special attention to this assembly during the operation stage of the floating offshore wind turbine, for instance, observation of vibration signal of the gearbox, to avoid a slight failure transforms into catastrophic consequences.

Pitch devices are auxiliary components of the floating offshore wind turbine for which were designed to adjust the attack angle of blades to seek an optimized efficiency of energy generation. However, the consequences of malfunctions of the pitch devices are not as severe as gearbox, generator and other assemblies that always result in the shutdown of the floating offshore wind turbine. Instead, the failures of pitch devices could reduce the electricity production efficiency. The suggestions like the robust design to pitch devices (both electric-driven and hydraulic-driven assemblies) are recommended.

The converter and tower are critical components of the floating offshore wind turbine according to the results of the analysis. The former fails more in

Table 2. The results (criticality ranks of components) comparison between the proposed and conventional FMEAs.

Components	[*]	[A]	[B]	[C]	[D]	[E]	[F]
Blade	9	4	4	2	2	1	7
Hub	10	7	5	9	—	—	—
Main Bearing	1	—	—	11	—	—	—
Main Shift	8	—	—	8	3	8	3
Generator	7	5	2	6	4	2	4
Gearbox	2	3	1	3	1	4	5
Converter	4	—	6	4	6	—	6
Transformer	6	—	10	5	—	—	2
Pitch Devises	3	—	3	7	—	6	8
Yaw Devises	11	6	8	10	5	5	6
CED	12	2	7	—	7	3	—
Tower	5	1	9	1	—	7	1
MFFF	13	—	—	—	—	—	—

—: The component not considered; CED: Control and Electrical Devices; MFFF: Mooring and Floating Foundation Facilities; [*]: This Paper; [A]: Du et al. (2017); [B]: Cevasco et al. (2018); [C]: Shafiee & Dinmohammadi (2014); [D]: Tazi et al. (2017); [E]: Arabian-Hoseynabadi et al. (2010); [F]: Scheu et al. (2019).

A similar conclusion can be drawn from the comparison with Du et al. (2017), which considers the same indices as those in Shafiee & Dinmohammadi (2014). A slight difference, however, between the results of Du et al. (2017), Shafiee & Dinmohammadi (2014), and this paper is the criticality ranks of blades, main bearing, and tower, which resulted from the different ways of deriving the economic loss of failures.

However, disagreements between this research and that achieved by Arabian-Hoseynabadi et al. (2010) and Scheu et al. (2019) are distinguishing as those analyses neglect failures costs of components in their failure analysis. Overall, the disagreements proved that the costs of failures of components of the floating offshore wind turbine are of great need to be modeled into the failure analysis to gain a deeper understanding of failures of floating offshore wind turbines and to furtherly seek an economic maintenance strategy. Moreover, the different criticalities of components ranked in Tazi et al. (2017), Cevasco et al. (2018), and this study due mainly to indices selected. For example, the mentioned studies consider both failure probability as well as profit loss as indices rather than S, O, D, and costs of failures that been used in this study.

Unlike existing techniques, the proposed FMEA generates a higher RPN for a component with higher failure cost. Hence, the RPNs produced by the proposed FMEA methodology is more decided by failure costs of the components of the floating offshore wind turbine due mainly to the differences in value scales of risk factors (the scale of risk-based RPN is [1, 1000]) as well as the economic factor (the costs of failures are larger than zero and theoretically without any limitation).

5 CONCLUSIONS

This paper extended the conventional FMEA methodology by considering an additional economic index namely costs of failures into the process of the failure analysis of a floating offshore wind turbine at both components and at systems levels. The wind turbine system was recognized as the most critical system. Moreover, the main bearing, gearbox, pitch devices, converter, and tower among others are identified as critical components. Recommendations on corrective design, e.g. robust design to pitch assembly, and preventive actions on safe operations, e.g. monitoring vibration of gearbox and temperature of the converter, are concluded. The comparison between this study and published studies validated the results of this research and which also demonstrated the similarities and differences among results of FMEAs of offshore wind turbines.

ACKNOWLEDGMENTS

This study was completed within the project ARC-WIND - Adaptation and implementation of floating wind energy conversion technology for the Atlantic region, which is co-financed by the European Regional Development Fund through the Interreg Atlantic Area Programme under contract EAPA 344/2016. The first author has been supported by a scholarship from China Scholarship Council (CSC) under Grant No. 201806070048. This work contributes to the Strategic Research Plan of the Centre for Marine Technology and Ocean Engineering (CENTEC), which is financed by the Portuguese Foundation for Science and Technology (Fundação para a Ciência e Tecnologia - FCT) under contract UIDB/UIDP/00134/2020.

REFERENCES

Arabian-Hoseynabadi, H., Oraee, H. & Tavner, P.J. 2010. Failure modes and effects analysis (FMEA) for wind turbines. *International Journal of Electrical Power & Energy Systems* 32: 817–824.

Bharatbhai, M.G. 2015. Failure mode and effect analysis of repower 5M wind turbine. *International Journal of Advance Research in Engineering, Science & Technology* 2: 2394–2444.

Bhardwaj, U., Teixeira, A.P. & Guedes Soares, C. 2019. Reliability prediction of an offshore wind turbine gearbox. *Renewable Energy* 141: 693–706.

Bouti, A. & Kadi, D.A. 1994. A state-of-the-art review of FMEA/FMECA. *International Journal of reliability, quality and safety engineering* 1(4): 515–543.

Carroll, J., McDonald, A. & McMillan, D. 2016. Failure rate, repair time and unscheduled O&M cost analysis of offshore wind turbines. *Wind Energy* 19(6): 1107–1119.

Castro-Santos, L., Bento, A.R., Silva, D., Salvacao, N. & Guedes Soares, C. 2020a. Economic feasibility of floating offshore wind farms in the north of Spain. *Journal of Marine Science and Engineering* 8(1): 58–76.

Castro-Santos, L., Martins, E. & Guedes Soares, C. 2016. Cost assessment methodology for combined wind and wave floating offshore renewable energy systems. *Renewable Energy* 97: 866–880.

Castro-Santos, L., Silva, D., Bento, A.R., Salvacao, N. & Guedes Soares, C. 2020b. Economic feasibility of floating offshore wind farms in Portugal. *Ocean Engineering* 207: 107393.

Cevasco, D., Collu, M. & Lin, Z. 2018. O&M cost-Based FMECA: Identification and ranking of the most critical components for 2-4 MW geared offshore wind turbines. *Journal of Physics: Conference Series* 1102(1): 1–11.

Diaz, H.M. & Guedes Soares, C. 2020. Review of the current status, technology and future trends of offshore wind farms. *Ocean Engineering*. 209: 107381.

Du, M., Yi, J., Guo, J., Cheng, L., Ma, S. & He, Q. 2017. An improved FMECA method for wind turbines health management. *Energy and Power Engineering* 9: 36–45.

Kang, J. & Guedes Soares, C. 2020a. An opportunistic maintenance policy for the offshore wind farms. *Ocean Engineering* 216: 108075.

Kang, J., Sun, L., Sun H. & Wu, C. 2017. Risk assessment of floating offshore wind turbine based on correlation-FMEA. *Ocean Engineering* 129: 382–388.

Kang, J., Sobral, J. & Guedes Soares, C. 2019a. Review of condition-based maintenance strategies for offshore wind energy. *Journal of Marine Science and Application* 18(1): 1–16.

Kang, J., Sun, L. & Guedes Soares, C. 2019b. Fault Tree Analysis of floating offshore wind turbines. *Renewable Energy* 133: 1455–1467.

Kang, J., Wang, Z. & Guedes Soares, C. 2020b Condition-based maintenance for offshore wind turbines based on support vector machine. *Energies* 13: 3518.

Li, H. & Guedes Soares, C. 2019. Reliability Analysis of Floating Offshore Wind Turbines Support Structure using Hierarchical Bayesian Network, *In Proceedings of the 29th European Safety and Reliability Conference (ESREL 2019)*. Hannover, 22-26 September 2019. pp. 2489–2495.

Li, H., Guedes Soares, C. & Huang, H.Z. 2020a. Reliability analysis of floating offshore wind turbine using Bayesian Networks. *Ocean Engineering* 217: 107827.

Li, H., Teixeira, A.P. & Guedes Soares, C. 2020b. A Two-Stage Failure Mode and Effect Analysis of an Offshore Wind Turbines. *Renewable Energy* 162: 1438–1461.

Santos, F.P., Teixeira, A.P. & Guedes Soares, C. 2018. Maintenance planning of an offshore wind turbine using stochastic Petri Nets with predicates. *Journal of Offshore Mechanics and Arctic Engineering* 140(2): 021904.

Scheu, M.N., Tremps, L., Smolka, U., Kolios, A. & Brennan, F. 2019. A systematic Failure Mode Effects and Criticality Analysis for offshore wind turbine systems towards integrated condition based maintenance strategies. *Ocean Engineering* 176: 118–133.

Shafiee, M. & Dinmohammadi, F. 2014. An FMEA-based risk assessment approach for wind turbine systems: a comparative study of onshore and offshore. *Energies* 7: 619–642.

Sinha, Y. & Steel, J.A. 2015. A progressive study into offshore wind farm maintenance optimisation using risk based failure analysis. *Renewable and Sustainable Energy Reviews* 42: 735–742.

Tazi, N., Châtelet, E. & Bouzidi, Y. 2017. Using a hybrid cost-FMEA analysis for wind turbine reliability analysis. *Energies* 10: 276.

Uzunoglu, E., Karmakar, D. & Guedes Soares, C. 2016. Floating offshore wind platforms. L. Castro-Santos & V. Diaz-Casas (Eds.). *Floating Offshore Wind Farms*. Springer International Publishing Switzerland, pp. 53–76.

Appendix A. The values of indices and computed RPNs.

Systems	Components	S	O	D	Risk-Based RPN	Failures Costs	RPN/*10000
BS	Blade	6	7	5	210	4000	84
	Hub	6	5	4	120	6000	72
	Main Bearing	6	4	6	144	86000	1238.4
	Main Shift	3	4	7	84	21000	176.4
	Generator	4	6	5	120	15000	180
	Gearbox	6	3	6	108	82000	885.6
WT	Converter	4	7	6	168	38000	638.4
	Transformer	4	7	5	140	13000	182
	Pitch Devises	3	7	7	147	49000	720.3
	Yaw Devises	2	3	4	24	26000	62.4
	CED	6	6	6	216	2000	43.2
SS	Tower	8	5	3	120	43000	516
	MFFF	7	6	3	126	2000	25.2

BS: Blade System; WT: Wind Turbine; SS: Support Structure

Developments in Maritime Technology and Engineering – Guedes Soares & Santos (eds)
© *2021 Copyright the Author(s), ISBN 978-0-367-77376-2*

2004 post tsunami reconnaissance in Southern Thailand: The resilience and impact after 15 years

T. Ornthammarath
Department of Civil and Environmental Engineering, Mahidol University, Thailand

A. Raby
School of Engineering, Computing and Mathematics, University of Plymouth, Plymouth, UK

P. Latcharote
Department of Civil and Environmental Engineering, Mahidol University, Thailand

W. Mortimer
School of Engineering, Computing and Mathematics, University of Plymouth, Plymouth, UK

ABSTRACT: In order to investigate the vulnerability of buildings and population in Southern Thailand after the 2004 Tsunami, a site-specific survey of the affected areas of Phuket and Phang Nga has been performed. Eight local towns, situated in the inundation zone, were found to be highly populated following increasing tourism activities, but with inadequate evacuation signage. Currently, the Department of Disaster Prevention and Mitigation (DDPM) in Thailand aims to respond to different disaster situations, but a lack of understanding of potential geohazard sources for major tourist area could hinder existing emergency plans. Furthermore, understanding the level of hazard will provide capacity in the DDPM to organize the required performance level, which could lead to more appropriate disaster mitigation, providing more rapid response after major events. The present study will act as reference for future study, and it could provide insight for other developed areas exposed to tsunami hazard.

1 INTRODUCTION

In this paper we report on a study which aims to assess existing vulnerabilities and develop preliminary tsunami-risk assessments for the highly touristic regions of Phuket, and adjacent Phang Nga province, where a mosaic of vulnerable coastline communities and dense infrastructure is at risk from a range of natural hazards. Understanding of the 2004 tsunami hazard and impact can greatly help with mitigating the physical, environmental, and socio-economic impacts of future cascading scenarios. The results of this study could be used to facilitate communities and authorities to develop comprehensive and integrated approaches to disaster risk reduction (DRR) from cascading natural hazards, ultimately with the aim of reducing loss of life and earnings from future events and developing more resilient communities. The development of comprehensive hazard and risk plans for earthquake and tsunami hazards will guide the formulation of disaster management practices and procedures including evacuation strategies and directing appropriate information for different population groups.

Buildings and infrastructure within the inundation zone of the 2004 Tsunami received wide-spread and significant damage. However, findings from this recent visit to eight heavily affected local towns shows the area has largely returned to normal in the 15 years since the event.

In this report, a brief overview of the vulnerabilities that was observed over the course of a 5-day reconnaissance mission that took place 15 years after the earthquake is presented. After providing a general summary of the regional tectonic settings and tsunami impacts in southern Thailand, the post 2004 earthquake and tsunami warning system in Thailand is discussed, and the resilience and shortcomings of existing tsunami evacuation plans for both local residential and government buildings is evaluated.

2 THE 26 DECEMBER 2004 EARTHQUAKE AND COASTAL DAMAGE IN SOUTHERN THAILAND

Southern Thailand is situated on the seismically quiescent Sunda plate, which is located on the boundary

DOI: 10.1201/9781003216582-27

between two converging plates: The Indian-Australian plates in the east and the NW-drifting Philippine Sea plate in the west. Within the boundary of this broad 'Stable Sunda' zone, only around 20 well-located earthquakes with magnitude greater than 5 have occurred during the years 1964 to 2007. Geodetic data also indicate that strains measured within the Stable Sunda zone are low (Simons et al., 2007). However, this region is situated about 600 km from the Sumatra subduction zone, where the 26 December 2004 earthquake occurred (Figure 1). A convergence rate of 65–70 mm/year as a result of Australia moving toward South East Asia is reported by McCaffrey (1996).

The first observed tsunami wave reached southern Thailand about 2 hours after the main shock, and tsunami inundation heights were reported by several studies including EEFIT (2006), Warnitchai (2005), Tsuji et al. (2006) and Rossetto et al. (2007). Large tsunami inundation heights of 5.7 m with inundation distances of 300 m were reported in several tourist areas in Phuket including Patong and Kamala beaches (Figure 2). However, larger tsunami inundation heights of 11 m and inundation distances of approximately 2 km, were recorded in Khaolak, Phang Nga province causing the highest casualties in Thailand (about 4,000 total deaths). The largest measured tsunami inundation height was 19.6 m at Ko Phra Thong Island (Table 1).

The death toll in Southern Thailand associated with the 2004 northern Sumatra earthquake and tsunami was over 8,000 people. With at least half of these casualties estimated to be foreigners. Despite this, in 2017, Phuket ranked 11th in the Euro monitor

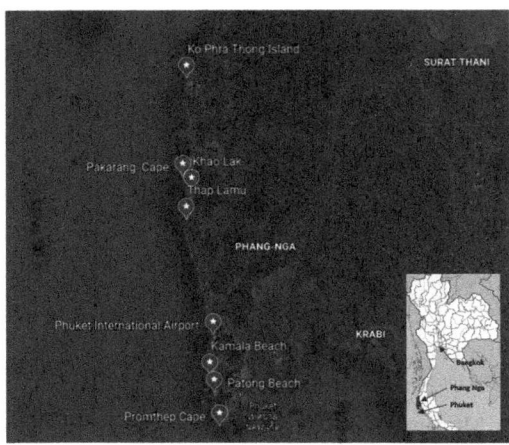

Figure 2. The locations of tsunami reconnaissance survey in December 2019 as reported in Table 1.

Table 1. Tsunami inundation heights (m) in southern Thailand reported by EEFIT (2006), Warnitchai (2005); Tsuji (2006), Rosetto et al. (2007).

Location	Inundation height (m)	Inundation distance (km)
Khao Lak	11.0	2.0
Kamala Beach	5.7	0.6
Patong Beach	5.0	0.3
Ko Phra Thong Is.	19.6	1.0

International list of "Top City Destinations" with 12 million visitors annually.

There is increasing recognition that one event (such as an earthquake) can subsequently cause a chain of adverse events (e.g., landslides, tsunamis, health epidemics etc.) that cumulatively have a much larger impact than the trigger hazard alone. Furthermore, evacuation plans and mitigation measures may need to be modified or completely redesigned in the event of interlinked disasters.

3 THE POST-2004 TSUNAMI WARNING SYSTEM IN THAILAND

Prior to the 2004 northern Sumatra earthquake, there was no tsunami warning system in the Indian Ocean, including the coastal region of Thailand. The lack of an active early warning system contributed to the high death tolls in southern Thailand even though there was a delay of up to 120 minutes between the earthquake original time and the first impact of the tsunami in Thailand. The Thai Meteorological Department (TMD) is a governmental organization responsible for monitoring,

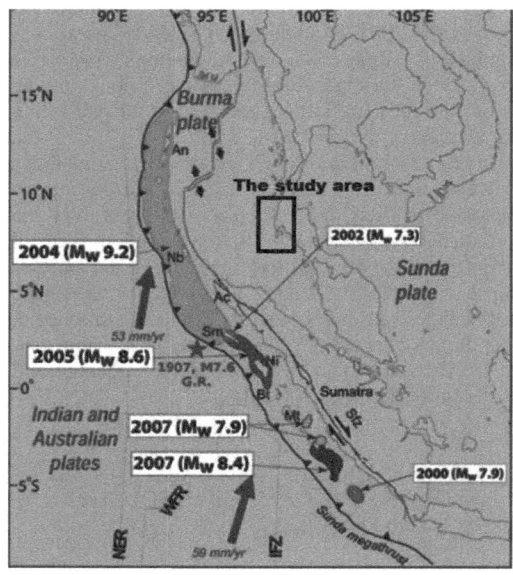

Figure 1. Rupture zone of 2004 Great Sumatra earthquake and other large events along Andaman-Sumatra Trench. Black rectangles represent current study area, Modified from Kanamori et al. (2010).

analyzing, and warning for natural hazards including earthquake information and tsunami warning. Currently, TMD maintains a real-time seismic activity monitoring network, which has been considerably expanded following the 2004 event. As of January 2018, around 80 digital seismic stations (Figure 3) are operational. These will be very valuable since they will be able to provide real-time tsunami warning system to the National Disaster Warning Center (NDWC) within 15 minutes of any earthquake. NDWC acts as the center coordinating with other governmental agencies and has responsibility to make decisions, announce all warnings, and evacuate people in high risk areas.

Warning messages will be disseminated by NDWC via television, radio, e-mail, SMS and warning towers. There are now 62 tsunami warning towers (Figure 4) installed in popular tourist and high-risk areas along the coastline of the Andaman Sea. These towers are maintained by the Department of Disaster Prevention and Mitigation (DDPM). Routine weekly tsunami warning tower checks are undertaken to confirm there are no operational problems. Moreover, DDPM are responsible for organising annual evacuation drills for local people in tsunami hazard zones.

In order to forecast tsunami arrival time from various sources, TMD also perform tsunami source inversion with time reverse imaging. Tsunami arrival time maps (Figure 5) from different tsunamigenic sources, which could have direct impact on the Thai coastlines both in the Gulf of Thailand and Andaman Sea, have been pre-determined for warning purposes.

Figure 4. Observed Tsunami warning towers in (Left) Khao Lak, Phang Nga and (Right) Kamala beach in Phuket. (Photo taken in December 2019).

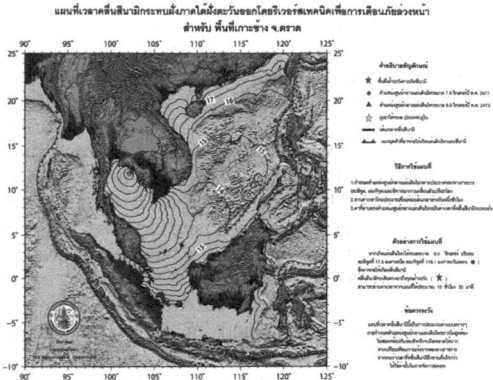

Figure 5. Tsunami arrival times which could reach the Thai coastline in the Gulf of Thailand from simulated earthquake occurred along the Manila Trench.

4 OBSERVED TSUNAMI RECOVERY IN PHUKET

4.1 Patong beach

Patong beach is located on the west coast of Phuket Island, approximately 15 km west of Phuket city center and 40 km south of Phuket International airport. The beautiful beaches, landscapes and nightlife of Patong, attracts over a million tourists annually, with many hotels, restaurants and amusements situated along the 3-km coastline. The most severe damage in the 2004 tsunami was recorded on buildings of 1 to 3 stories. Reinforced concrete (RC) structures, which are a popular building design in Patong, largely survived the wave impacts. One positive outcome of the rebuilding of Patong in response to the growing tourism industry, is that many more RC buildings have been constructed in the last 15 years. The increase in RC constructions in Patong could improve infrastructure resilience to future tsunami events.

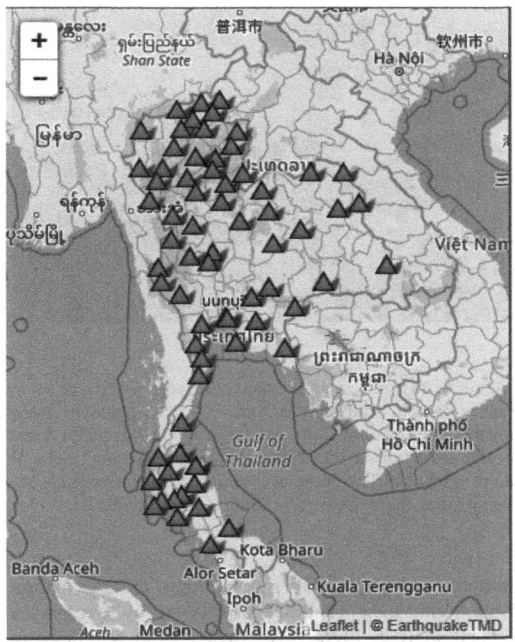

Figure 3. The current real-time TMD seismic station for tsunami warning system in Thailand.

During the recent visit, tsunami warning and evacuation signs were observed along the coastline every 100 m. Unfortunately, these signs are not easily distinguished against the chaotic backdrop of advertising and shop signage, which lines the crowded tourist areas (e.g. Figure 6 a). Moreover, the tsunami evacuation signage was generally too small to be effectively noticed and there was no clear demarcation of a tsunami hazard zone, where wave inundation could be expected. Relying on the current signage would be confusing in a real evacuation scenario. Fortunately, there is designated tsunami evacuation routes leading from Patong beach through the town to recognised shelters or higher ground. However, as the signage is small and disorganized, we found the routes to be confusing when following them in a relaxed non-emergency scenario. Should there be a real evacuation, this would be extremely difficult for beach goers to follow safely.

The tsunami evacuation routes indicated by these signs seem to be unclear for visitors. In addition, a tourist information board had not been designed to give possible tsunami evacuation routes or marked inundation zones (Figure 6 b). Furthermore, during the 2012 Mw 8.6 undersea earthquake in the Indian Ocean, which led to a tsunami warning across the region, heavy traffic jams in Patong were the main issues, since many people opted to use cars as a mean for evacuation through the narrow streets.

Signs for tsunami evacuation building were observed during the fieldtrip, but we found examples that were poorly maintained, with their visibility affected by other objects or large trees growing in the way. We also found examples of ambiguous signage where it was not clear which exact building was an evacuation structure (Figure 6c). Major improvement could be made by putting a clear tsunami evacuation building sign on the top of these buildings. Several mid-rise buildings (> 4-storey) could be seen along the coastline, but there was no clear indication that they were suitable tsunami evacuation buildings.

4.2 Kamala beach

Another west coast town in Phuket heavily damaged by the 2004 tsunami was Kamala beach, about 8.5 km north of Patong beach. Most of the buildings in this area were of 1-3 storey constructed with RC or timber. Most timber buildings were washed away, but many examples show that the RC buildings largely survived the tsunami wave impact with varying degrees of damage. As can be expected, the most severely damaged RC buildings were generally located along the coastline, while RC buildings situated on higher ground suffered less damage. This area is less populated compared to Patong beach, but many low-rise (< 2 storey) tourist resorts have since been built in the intervening 15 years.

Tsunami evacuation signs were observed at about 100m spacing along the shoreline, but there are fewer of them away from the beach. Again, there were

(a)

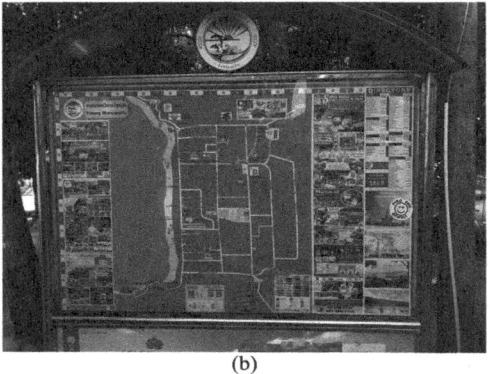

(b)

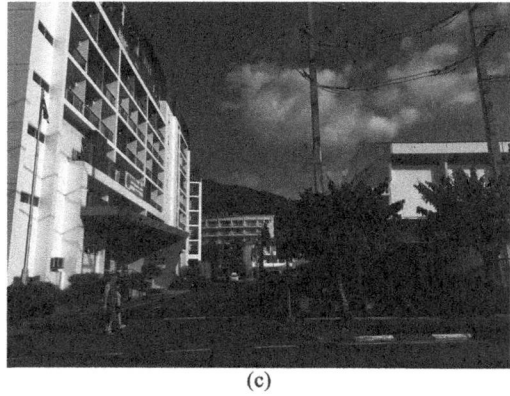

(c)

Figure 6. (a) A tsunami evacuation sign (red circle) in Patong Beach. The sign is too small compared to other advertisement boards. (b) An information board for local tourist attractions could be improved with tsunami evacuation routes. (c) The tsunami evacuation shelter sign is confusing since there are two buildings in the same area.

numerous examples of signs, along the evacuation route, with their visibility obstructed by overgrown vegetation (e.g. Figure 7). Most of the buildings in Kamala beach are of 1-2 storey RC construction, including a police station and a community hospital, but there is no evacuation shelter within this area.

Figure 7. A poorly maintained tsunami evacuation sign in Kamala beach.

When speaking to Kamala beach residents, there was report of a tsunami warning tower being out-of-service for a period of a few years, due to local authority maintenance issues, although this issue has now been resolved and the warning tower is fully functional, after complaints by concerned residents.

5 OBSERVED TSUNAMI RECOVERY IN PHANG NGA

5.1 Khao Lak

Located up the west coast of Phang Nga province, 60 km North-West of Phang Nga city center and 100 km North of Phuket, Khao Lak was the worst tsunami affected region in Thailand, during 2004. An estimated 71% of all Thailand's tsunami related deaths occurred here (EEFIT, 2006). Many luxurious hotel complexes line the coastline, comprising bamboo outhouses and cottages with little to no sea defences standing between the beach and complex. In addition, Khao Lak sustained the greatest tsunami height on the Thai coastline (> 10 m). Most wooden and bamboo cottages near the shoreline were washed away, and a key transport bridge collapsed. The observed inundation zone was as far as the main road running parallel to the coastline at a 2 km distance inland.

Figure 8. A tsunami evacuation shelter along the beach in Khao Lak.

Among all the visited tsunami affected area, Khao Lak has by far the highest tsunami awareness among the local residents. Numerous 4-storey RC constructed Tsunami evacuation shelters were observed along the coastline (e.g. Figure 8) as well as 2 km inland. These evacuation shelter structures appear to a standard design, regardless of location. The first two storeys are open without shelter space to reduce drag and impact from an in-coming tsunami wave. The third storey is an enclosed room, with a toilet and open floor plan accessed by two entrances on either side leading from two external stairways. When inspecting the shelter in Figure 8, we found one of the entrances to be padlocked and it was unclear who held the key, and whether the key could be accessed in an emergency. The fourth storey, roughly 15 m above ground is an open planned roof top with railings around the perimeter, which could be used by evacuees to brace against incoming water.

5.2 Ko Phra Thong Island

Ko Phra Thong Island is located 50 km North of Khao Lak with an area of 88 km^2 and separated from the mainland by sea-channels roughly 10 km wide. The island has a very low population with about 300 full-time inhabitants, including indigenous Moken people. About 80 fatalities were associated with the 2004 tsunami, due to the island's flat terrain.

The majority of survivors mostly lived in the eastern part of Island and many have since decided to move to the mainland as a result of the devastation. Most structures on the Island were constructed from traditional wooden or bamboo materials which were not resilient to the tsunami wave and lead to high death tolls. Figure 9 shows a typical coastal community on Koh Phra Thong located on an East coast sea-channel.

Figure 9. Wooden buildings along the beach in Ko Phra Thong Island.

6 CONCLUSIONS

On a visit to the affected areas of Phang Nga province in Southern Thailand, many changes have been observed in the 15 years since the 2004 tsunami event. Major improvements to the tsunami detection and warning system in Thailand, and other neighboring countries in Indian Ocean have been made since 2004. However, this is not a simple success story, as there are several weaknesses in the evacuation system, as identified in this study. The processes by which the local people and authorities need to maintain tsunami awareness including properly designed tsunami evacuation routes, clear and visible signage and well-maintained evacuation shelters are yet to be effectively implemented. Perhaps, during the past 15 years, perceived tsunami risk has been downplayed by locals, leading to a relaxation of the maintenance and upkeep of suitable evacuation infrastructure and procedures.

Furthermore, evacuation plans and mitigation measures may need to be modified or completely redesigned in accordance to different cascading disaster events. These preliminary findings discussed here could be used to form the basis of future disaster risk reduction (DRR) strategies which will require engineers, local government agencies, policy makers and the public (residents and tourists) to cooperate in the future.

ACKNOWLEDGEMENTS

This project is partially supported by the European Union's Horizon 2020 Research and Innovation Programme RISE under grant agreement no. 730888 (RESET).

REFERENCES

EEFIT (2006) The Indian Ocean tsunami of 26 December 2004: mission findings in Sri Lanka & Thailand. Pomonis A, Rossetto T, Peiris N, Wilkinson SM, Del Re D, Koo R, Gallocher S, Earthquake Engineering Field Investigation Team (EEFIT) Report, The Institution of Structural Engineers, London, UK. http://www.eefit.org.uk

McCaffrey R. 1996., Slip partitioning at convergent plate boundaries of SE Asia. In: Hall R, Blundell DJ (eds) Tectonic evolution of Southeast Asia. Geol Soc Spec Publ 106:3–18

Rossetto T, Peiris N, Pomonis A, Wilkinsin SM, Del Re D, Koo R, Gallocher S (2007) The Indian Ocean tsunami of December 26, 2004: Observations in Sri Lanka and Thailand. Nat Hazards 42:105–124

Simons, W., A. Socquet, Vigny, C., Ambrosius, B., Haji Abu, S., Promthong, C., Subarya, C., Sarsito, D.A., Matheussen, S., Morgan, P., Spakman, W. 2007., A decade of GPS in SE Asia: Resolving Sundaland Motion and Boundaries J. Geophys. Res., 112, B06420, doi:10.1029/2005JB003868R, 2007.

Tsuji, Y, Y Namegaya, H Matsumoto, SI Iwasaki W. Kanbua, M. Sriwichai, V. Meesuk, 2006. The 2004 Indian tsunami in Thailand: Surveyed runup heights and tide gauge records Earth, Planets and Space, Volume 58, Issue 2, pp 223–232

Warnitchai P (2005) Lessons learned from the 26 December 2004 tsunami disaster in Thailand. Scientific forum on the tsunami: its impact and recovery. A regional symposium. Asian Institute Technology, Thailand, 6–7th June 2005

Developments in Maritime Technology and Engineering – Guedes Soares & Santos (eds)
© 2021 Copyright the Author(s), ISBN 978-0-367-77376-2

Quantitative ecological risk assessment of oil spills near an island in the Atlantic

P.G.S.C. Siqueira, H.O. Duarte, M.J.C. Moura, S.Q.P. Silva, L.F. Lara, M.A. Silva & M.C. Araújo
Federal University of Pernambuco, Recife, Brazil

E.A.L. Droguett
University of Chile, Santiago, Chile

ABSTRACT: Industrial accidents, such as toxic spills, have caused catastrophic damage to ecological environments (animals and plants), so that an effective method to assess ecological risks has been demanded. The most recent case: the oil spill that is affecting the Northeast coast of Brazil's environment. Fernando de Noronha Archipelago (FNA) is a marine protected area off the coast of Brazil. A recent study performed a Quantitative Risk Analysis (QRA) in FNA and the results obtained were that the industrial activity under consideration causes the marine ecosystem of FNA to be endangered, according to the IUCN risk categories. However, this study was conservative and did not assess the fate and transport of the oil in the ocean. In this new study we intend to perform an improved QRA in FNA, including the fate and transport model of the oil, by coupling the Ocean Modeling System – ROMS Regional Oceanic Circulation model with the MEDSLIK-II hydrocarbon dispersion model to simulate the evolution of the oil plume.

1   INTRODUCTION

Oceanic islands are considered hotspots of biodiversity and host a large amount of endemic species. Additionally, their isolation makes them a repository of threatened species as priorities regions for legal conservation acts (Gillespie, 2001; Gove et al., 2016; Whittaker & Fernández-Palacios, 2007). Among Brazilian oceanic islands, Fernando de Noronha Archipelago (FNA) is the best studied and has the largest number of species, both marine and terrestrial, a fact attributed to its extension and the heterogeneity of its habitats. FNA has the status of Conservation Unit allowing the protection of endemic species and the maintenance of a healthy island ecosystem (Serafini & França, 2010). Two thirds of FNA consist of the Marine National Park of FN (PARNAMAR-FN), a marine protected area (MPA) that reaches to the 50-meter isobathic line (ICMBio, 2013). The basic objective of the creation of PARNAMAR-FN is to preserve natural ecosystems with great ecological significance and scenic beauty, enabling scientific research, activities of environmental education, recreation and ecotourism (BRASIL, 2000).

Most projects devised and developed for preserving PARNAMAR-FN focus on the conservation of a single representative species (e.g., spinner dolphin, turtles, sharks, coral reefs) (Amaral et al., 2009; Silva, 2010; Ferreira & Maida, 2006; Garla, 2004; Lira et al., 2009; Maida & Ferreira, 1995; TAMAR, 2006).

Managers need to assess and manage ecological risks (hereafter ecorisks) caused by routine (i.e., high frequency/low consequence) human activities within PARNAMAR-FN. These assessments should also contemplate improbable, large events (i.e., low frequency/high consequence). Taleb (Taleb, 2007) argues that surprises shape the world's history more than average events, mainly because humans restrict their thinking primarily to the "average/usual/common/probable" and are always surprised by the "improbable". Recent research efforts have shown the importance and feasibility of including improbable large events in model-based ecological risk assessments (Burgman et al., 2012; Duarte et al., 2012).

In average, 75 ships navigate daily on routes near FN (Medeiros, 2009), using landmarks to determine the ship's position at sea more precisely and consistently. Many of these ships are oil tankers. A first round of a Quantitative Ecological Risk Assessment (QERA) of industrial accidents was conducted (Duarte & Droguett, 2016) focusing on whether the risks of catastrophic oil spills are tolerable, or whether they need management (i.e., recovery or control measures that may reduce risks). However, this study was conservative and did not assess the fate and transport of the oil in the ocean, so the exposure of the ecosystem to the pollutants was roughly estimated. In this new study, we intend to perform an improved QERA in FN, including the fate and transport model of the oil, by

DOI: 10.1201/9781003216582-28

using computational routines which numerically solve the mathematical equations that describe this phenomenon. It is assessed the accidental event of an oil spill from tankers navigating near FN. Thus, this work aims to quantify the ecorisks to PARNAMAR-FN caused by potential oil spills nearby.

The wind – shear and stokes drift as a function of the wind speed – and surface currents are the major forcing that contribute for the oil transport in the aquatic environment (Spaulding, 2017). Beyond transport, oil spills impact depends mainly on the environmental conditions that controls the weathering processes at the site of the spill (e.g., currents, climate, waves) and the time required to engage mitigation operations (Lee et al., 2015; Marta-Almeida et al., 2013; NRC, 2003). Weathering is a general definition for changes in oil properties due to physical, chemical and biological processes when the spill is exposed to environmental conditions (e.g., in aquatic systems). The main weathering processes occurs on the surface (spreading, evaporation, photodegradation and emulsification), in the water column (dissolution, dispersion, adsorption to particulate matter, biodegradation) and the sediment (biodegradation).

The rest of this chapter is structured as follows. In the section 2, the methodology used for systematically conducting this QERA is presented, alongside with some partial results. In the section 3, the methodology is applied to the study region, using *S.stellata* as key-species, the results are presented, the risks, categorized and their implications are discussed. In section 4, we conclude remarks about the QERA results and its advantages and limitations are presented.

2 METHODOLOGY

The QERA methodology used (Duarte et al., 2019) is based on population modeling (Forbes et al., 2011) and considers risk as a measure of frequency of occurrence of undesirable events and the magnitude of adverse ecological effects over time, and it can be used to both assess ecological and microbial risks. The methodology can also consider infrequent events (IEs), such as industrial accidents that leads to oil spills in the ocean. It is capable of quantifying risks of a population becoming extinct (or exploding) over time in scenarios (SCNs). For more details on the methodology and other case studies, see Duarte et al. (2019). The steps of the methodology are as follows:

2.1 *Characterization of the problem:*

The Fernando de Noronha Archipelago (FNA) is an island off the coast of Brazil (03°51'S, 03°30'W) in the Western Equatorial South Atlantic. It is composed of 21 islands and islets, with a total area of 26 km² (IBAMA, 2005). Two Federal Conservation Units (UCs) are located in the archipelago: the Fernando de Noronha Marine National Park – PARNAMAR/FN (BRASIL, 1988), and the Fernando de Noronha

Environmental Protection Area – APA/FN (BRASIL, 1986). It is known that a number of oil tankers pass off FN using remarkable points on land to better locate their vessel at sea, so accidental oil spills are possible. Additionally, the lack of infrastructure and mitigation plans at FNA could increase the impact when facing an oil spill scenario. The instantaneous volume of oil spilled was defined as 60,000 tonnes, which corresponds to half the capacity of Transpetro's Suezmax ship, of 26°API (intermediary oil type) which is the most used ship in the Brazilian oil and gas exploratory activities (ANP, 2017).

The bioindicator species chosen to represent the ecosystem was *Siderastrea stellata*, an endemic, common coral species in Brazil. It is the main reef building organism in FN (Castro & Pires, 2001). Corals are sessile, if the region is affected by a pollutant, they will be exposed and suffer the effects of pollution(Yender et al., 2010). Corals serve as food and shelter to many types of animals, such as worms, crustaceans, sponges, sea urchins, and many species of fish (Yender et al., 2010). The loss of coral will affect both humans and terrestrial organisms, because it protects the shore, supports tourism and facilitates fisheries.

2.2 *Identification of hazards and consolidation of scenarios (SCNs)*

The monthly averaged currents and winds for March and July served as forcing for the MEDSLIK-II simulations. The choice of these periods allows analyzing the highest risk scenario since March and July correspond, respectively, to the start of the central branch of the South Equatorial Current (cSEC) intensification (with a lower wind intensity) and the period of higher intensity of the cSEC (with the increase of the southeast trade winds intensity) (Lumpkin & Garzoli, 2005; Molinari, 1982). For the MEDSLIK-II run, three initial spill releasing points were determined: T1 (3° 41.9'S; 31° 41.9'W), T2 (4° 00.0'S; 32° 08.6'W) and T3 (3°50.0'S; 31° 46.0'W) on the east side of the FNA (Figure 1). This results in a total of six scenarios evaluated (two seasons and three releasing points), with hourly departures and figures every 6 hours, from the start of the spill to 48 hours. These releasing points were chosen based on the minimum distance positions from the FNA of the main vessel tracks registered in Marine Traffic website (MARINETRAFFIC, 2017). Only spill positions on the eastern side of FNA were accounted, considering the dominant westward direction of the wind (SE trades) and the surface current (cSEC). Three accidental SCNs were consolidated on each releasing point:

- SCN-1: fire/explosion; very low frequency, even lower than collision SCNs, according to historical records (IMO, 2004, 2007a, 2007b, 2008b, 2008a, 2008c).
- SCN-2: head-on collision of two ships on the same route; very low frequency; the large lateral

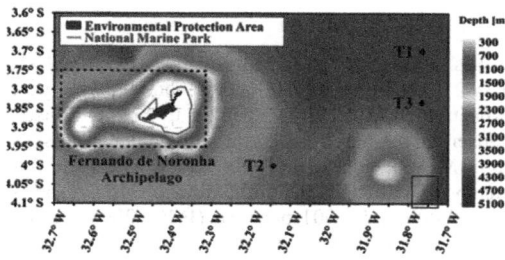

Figure 1. Location of the Fernando de Noronha Archipelago (FNA) (black dashed line). The area delimited in red corresponds to the Marine National Park (PARNAMAR) and in blue, to the environmental protection area (APA) of Fernando de Noronha. T1, T2 and T3 show the coordinates of three release oil spills scenarios considered in simulations (Adapted from: Queiroz et al. (2019)).

sea room that permits a hard-over turn to avoid collision, but still possible due to human error (MacElrevey & MacElrerey, 2004).

- SCN-3: overtaking parallel collision on the same route; very low frequency, but still higher than SCN-2; the lower is the relative velocity of two ships, the more likely is that problems in ship control occur (MacElrevey & MacElrerey, 2004).

Thus, from the less to the most frequent: SCN-1, SCN-2 and SCN-3. All were considered to have the same critical severity class III (i.e., considerable ecological damage, reaching the coast of FN and short recovery time). These SCNs were selected to the next step.

2.3 Exposure assessment

Some meteorological conditions (e.g., topography, tide conditions, and the distribution of temperature and salinity of the water) influence the dispersion of oil in the ocean, and thus the chemical concentration to which the coral population is exposed after an accident.

The dominant winds are the southeast trade winds, reaching greater intensity between July and August (Tchamabi et al., 2017). The highest sea surface temperatures (SST) occurs between March and June, with values surpassing 28°C, due to the occurrence of the southwestern tropical Atlantic warm pool (Cintra et al., 2015), and the lowest between August and November, (SST $\sim$ 26.5°C) (Hounsougbo et al., 2015; Silva, 2009; Tchamabi et al., 2017). In surface, the central branch of the South Equatorial Current (cSEC) flows westward until it reaches the North Brazil Current (NBC) near the coast (Lumpkin & Garzoli, 2005; Stramma & Schott, 1999). The cSEC is stronger between March and July, and weaker between August and February (Lumpkin & Garzoli, 2005; Tchamabi et al., 2017).

Dose-response studies between oil and S. stellata were not available. For that, it was considered

a dose-response study for a related species of coral (*P.damicornis*) (Shafir et al., 2007). It was obtained, by interpolation from data in the paper, that an exposure to 0.1124% concentration of the crude oil WSP caused an average of 52.7% mortality rate among *P. damicornis* (Shafir et al., 2007). Since all three SCNs occur in the same points, it was assumed they all have the same consequence, i.e., that an accident that causes the FNA coast to be exposed to 0.1124% concentration of oil will kill 52.7% of *S. stellata* populations exposed to this concentration.

2.3.1 Fate and transport modeling

ROMS is a three-dimensional, free-surface, numerical oceanic model that solves Navier-Stokes' primitive equations on a rotating Earth, considering Boussinesq's approach and hydrostatic equilibrium. In this model, the pressure gradients are solved in a barotropic-baroclinic coupled system that recovers processes lost or altered by rigid lid assumption (Shchepetkin & McWilliams, 2003, 2005). For the coupling, it was used the version ROMS_AGRIF/ROMSTOOLS developed by the French Institute of Research for Development – IRD (*French Institut de Recherche pour le Développment*) (Penven et al., 2008).

The configuration of the ROMS model for the simulation of circulation was based on the one applied and validated by Tchamabi et al. (2017). The computational domain of the model consisted of the limits of the mesh between 3°S-5°10'S and 35°W-31°W with horizontal and vertical resolution of 1/70° (~1.5 km) and 40 sigma levels, respectively. Stretching parameters were chosen in order to increase the resolution at the surface and bottom as $\theta_s = 6$ and $\theta_b = 0.6$ (Song & Haidvogel, 1994). Bathymetry was obtained from the Navy Hydrographic Center (CHM) merged with data from the GEBCO (Global Earth Bathymetric Chart of the Oceans, site) database with a resolution of 0.5′, with smoothing after interpolation.

The advective scheme corresponded to RSUP3 (Lemarié et al. 2012; Marchesiello et al.2009). We adopted the non-local scheme of planetary layer K-profile-parametrization for vertical mixing to for unresolved processes of scale below the grid (Large et al. 1994). The surface forcing were the input of heat and freshwater flow from the Comprehensive Ocean-Atmosphere Data Set (COADS) with monthly average fields with grid intervals at each 0.5° (da Silva et al. 1994). Wind shear was derived from the monthly NASA Quick Scatterometer (QuiksCAT) climatology, known as SCOW (Scatterometer Climatology of Ocean Winds) with grid resolution ranges of 0.25° (Risien & Chelton, 2008). Horizontal boundary conditions have been opened at all 4 borders. To improve the numerical solution, high resolution (9.28 km) TSM data from Advanced Very High Resolution Radiometer (AVHRR) – Pathfinder were used to calculate the heat flow (Casey & Cornillon, 1999). The model was initiated and forced with the temperature and salinity distributions extracted from the monthly climatology of the World Ocean Atlas

2013 (WOA), and the initial speed determined by geostrophy. The simulations were integrated for 10 years spin-up with a minimum time step of 4 minutes and the average output every 2 days.

MEDSLIK-II is a Lagrangean oil model that can be coupled with ocean circulation models to solve the problems of transport, diffusion and transformation of oil on the sea surface. In MEDSLIK-II, the oil spill is represented by the state variables called "spill", "particle" and "structural". The spill type state variables are used to solve the variation in oil concentration due to weathering processes (evaporation, spreading, dispersion, emulsification and coastal adhesion) acting on the total volume of oil. These processes are presented as empirical relationships between the volume of the spill, the wind at 10 meters (W) and the SST. Next, the spill is divided in N particles of defined volume and the state variables "particle" are defined to solve advective-diffusive processes and for the subsequent calculation of the concentration at the surface, water column and coast (structural type variables) (De Dominicis et al., 2013a,b).

For coupling the ROMS to MEDSLIK-II (Figure 2), subroutines have been created in the source code and some secondary MEDSLIK-II codes (medslik_II.sh, Extract_II.for and medslik_II.for). These new routines are necessary to recognize the ROMS output files (in netCDF format) as forcing input (SST, currents and winds) for MEDSLIK-II. Thus, we use the results from ROMS simulation as forcing for MEDSLIK-II.

For each accidental scenario, the fate and transport model predicted the oil concentration (in tonnes/km²) on the surface of the water, in each cell of the grid. It was assumed a shallow depth (the corals live in shallow waters (Barros & Pires, 2006)) in order to determine the volume of water and then, the weight of oil per km³, and then obtaining the concentration of oil in the water.

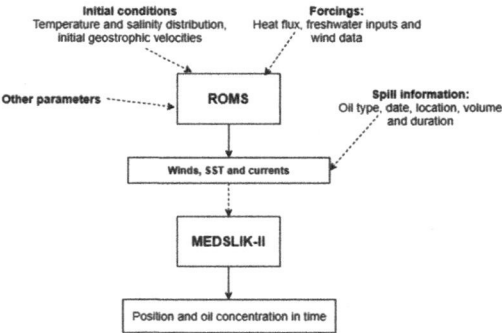

Figure 2. Simplified scheme of the coupled models ROMS and MEDSLIK-II with the inputs (dashed line) and outputs (solid line) (Adapted from: Queiroz et al. (2019)).

2.4 Frequency estimate

It was assumed that the frequencies for SCN-1, SCN-2 and SCN-3 are 0.001, 0.005 and 0.01 per year respectively, which are expected to be pessimistic (conservative) based on historical accident statistics from IMO for double hull oil tankers, where the frequency of collision and fire/explosion are, respectively, 0.009 and 0.001 per year (IMO, 2004, 2007a, 2007b, 2008b, 2008c, 2008a).

2.5 Population modeling

A metapopulation was built to describe the population without disturbance and call it the benchmark scenario (SCN-0). It consists of 15 colonies distributed alongside the coast of Rata Islet, the location where the oil reaches. The assessment endpoint of the model is the area (cm^2) of a colony i at a time step t, $A_i(t)$. When in equilibrium (undisturbed reef), the area in the next time-step depends on the area and growth rate at the present time step, respectively $A_i(t)$ and $R_i(t)$, as follows:

$$A_i(t+1) = R_i(t)A_i(t) \qquad (1)$$

The growth parameters is the ratio of the new area to the previous one, i.e., $R_i(t) = A_i(t+1)/A_i(t)$, where $A_i(t) = \pi \times r_i^2(t)$ is the radius of colony i at time t, $r_i(t) = r_i(0) + y(t)$, and y is the radial growth given from the literature data (M.M.L Barros & Pires, 2006). The initial area of the metapopulation, i.e., the sum of the area of each colony, was 150 cm².

The location and size of each colony were defined by expert opinion. The uncertainty was considered by calculating its standard deviation and considering it is normally distributed (Ku, 1966). A carrying capacity parameter, K (i.e., 255 cm²) was included to limit the maximum area of a colony (M.M.L Barros & Pires, 2006), i.e., $A_i(t) \leq K$.

SCN-1, SCN-2 and SCN-3 were integrated in the metapopulation model. For each model replication (total of 10,000), an oil spill was selected to occur (or not) every time-step according to its frequency of occurrence (from step 4: frequency estimates). If it occurs, it is assumed to kill 55% of each population in the metapopulation (from step 3: exposure assessment). All SCNs were simulated in RAMAS software (Akçakaya & Root, 2013).

2.6 Risk quantification and categorization

The final step consists of simulating SCNs separately (for isolated risks or together (for evaluating integrated strategies or cumulated risks). The risks may be represented as a probability-consequence curves derived from multiple runs (Monte Carlo method) (Kalos & Whitlock, 2008) and may be expressed in many ways.

There is a variety of software available for population model construction and probabilistic simulation via Monte Carlo method. In this paper it was used the software RAMAS Metapop v.6.0 (Akçakaya & Root, 2013) which allows for fast stochastic simulation of a SCN, so one can simulate hundreds of SCNs by varying the most uncertain parameters in one's model within a range of plausibility, compare them, and pick the most relevant ones (e.g., the ones that maximize/minimize the risks).

For this QERA, the main result is a cumulative density function (CDF) for the time to half loss (i.e., the time required by a population to reach below half its initial size). Although RAMAS software builds the CDF automatically, it is important to understand that the resulting CDF is built by using a Monte Carlo simulation. For every Monte Carlo run, a "single-point estimate" for the discretized time to half loss (e.g. in years), T, is calculated. After many Monte Carlo runs (e.g. 10,000), one will have a set of "single-point estimates" for the time to half loss and the number of occurrences of a "single-point estimate". Thus, one can calculate the probability of occurrence of each "single-point estimate" (e.g. (T) = number of occurrences of T/10,000). Then, for each time t, it is possible to cumulate the probabilities of all T lower than t, which results in the CDF for the time to half loss, i.e.: $F_T(t) = P(T<t)$: In summary, $F_T(t)$ means the probability that half loss will occur at or before a time t. This function can be plotted in a graph. The great advantage of a Monte Carlo simulation over deterministic analysis is that results show not only what could happen, but how likely each outcome is.

The methodology proposing the following categorization of risks concerning the probability of the 50% population size decline:

* CRITICALLY ENDANGERED (CE): >50% of probability within 10 years or 3 generations, whichever is longer (up to a maximum of 100 years);
* ENDANGERED (EN): >20% of probability within 20 years or 5 generations, whichever is longer (up to a maximum of 100 years);
* VULNERABLE (VU): >10% of probability within 100 years;
* NEGLIGIBLE (NE): <10% of probability within 100 years.

3 RESULTS

This QERA was conducted in 2015 during field research in the island of Fernando de Noronha (FN). Although large oil tankers do not dock on FN, they pass nearby to locate themselves at sea more accurately. The main motivation was to evaluate if such passages cause considerable risks to FN ecosystem. For more details on the first round QERA, see

Duarte & Droguett (2016); and for the fate and transport modeling of oil spills nearby FNA, see Queiroz et al. (2019).

3.1 Ocean circulation results

The releasing point T3 presented a critical situation, in which the oil plume reaches the FNA coast in March 45 hours after the release. Due to weathering processes, only 2.2% of the oil spill adhered to the shore, which corresponds to approximately 1350 tonnes of oil, mostly on Rata Islet (Figure 3). The highest concentration was about 1000 tonnes/km², and considering a depth of about 1m, the concentration is then 1.124 milliliters of oil per liter of water (0.1124 %). Regarding the month of July, the spill reached near to FNA coast 48 hours after the release.

The wind velocity used as forcing in ROMS was 4.66 m/s (±0.14 m/s) and 7.08 m/s (±0.14 m/s) in March (Figure 4a, b) and July, respectively. SST was higher in March than in July, with mean values of 28.26°C (±0.10°C) and 26.85°C (±0.09°C), respectively (Figure 4c, d). The mean surface current intensity varied from 0.39 m/s (±0.06 m/s) with predominance of the zonal component in March to 0.34 (±0.04 m/s) in July, with a small deviation to the northeast (Figure 4c, d). The oil transport was predominantly westward mainly due to the surface current direction. The influence of the wind seasonality increases the deviation to southwest in July, when the wind is stronger.

3.2 Risk quantification and categorization

The CDF for the time to half loss is shown in Figure 5. The level of risk to S. stellata is as follows: the cumulative risk which is caused by potential oil spills in FNA is deemed to make *S. stellata* NEGLIGIBLE due to SCN-1 (explosion) or SCN-2 (head-on collision) or SCN-3 (parallel collision). Considering scenarios in isolation, risk categories are also NEGLIGIBLE to all three SCNs.

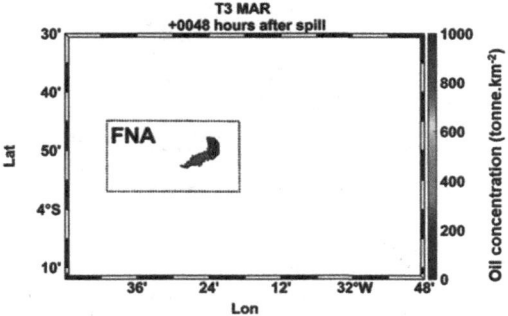

Figure 3. Oil concentration in the water (tonne/km²) for the scenario T3-March 48h after the spill (Adapted from: Queiroz et al (2019)).

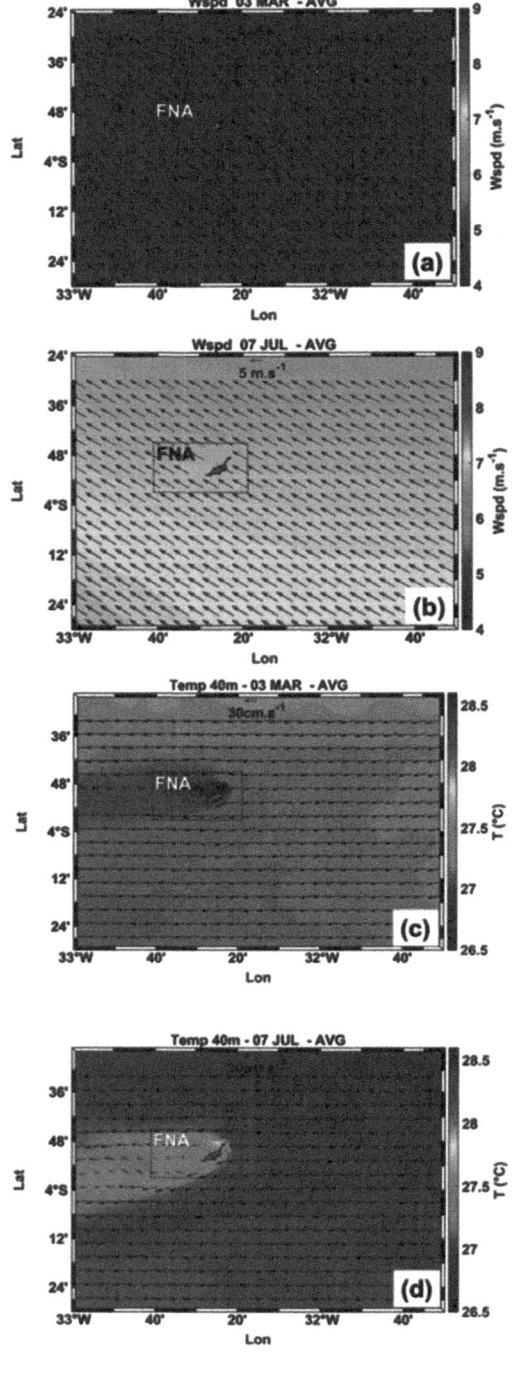

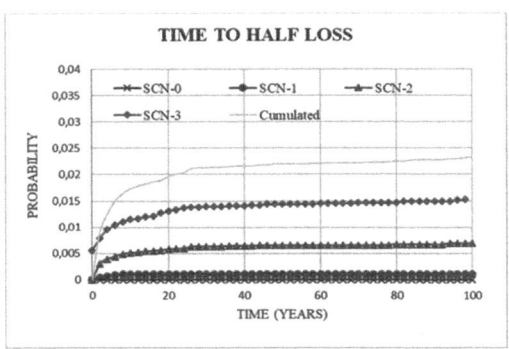

Figure 5 . Probability of half loss for cumulated risks (sum of risks of all scenarios), SCN-3 (overtaking parallel collision), SCN-2 (head-on collision), SCN-1 (fire/explosion) and SCN-0 (population without disturbance).

More results can be seen in Table 1, where a comparison is made between IUCN and our proposed risk categories. Note that the chosen probabilities of accidents were conservative, and thus they are likely to overestimate occurrences of oil spills on this route.

Table 1. ASs, their parameters and risk results to S.stellata metapopulation. The parameter Fj(t) is the frequency per year of each SCN j; Mj is the mortality rate to all colonies in case of SCN j. Values with a + or − symbol mean that they are being compared to the benchmark SCN-0.

Accidental scenario parameters

Scenario	SCN-0	SCN-1	SCN-2	SCN-3
$F_j(t)$	0	0.001	0.005	0.01
M_j	Not applicable	52.7%	52.7%	52.7%

Results

Scenario	SCN-0	SCN-1	SCN-2	SCN-3
Risk of half loss	0	+0.0011	+0.0069	+0.0153
Extinction risk	0	0	+0.0001	+0.0001
Median time (years) to half loss	>100	>100	>100	>100
Expected minimum size (cm²)	198.2	-34.7	-36.9	-39.5
IUCN risk category	NE	NE	NE	NE
Proposed risk category	NE	NE	NE	NE

Figure 4. Wind intensity (colors, m/s) and direction (arrows) for (a) March and (b) July; Sea Surface Temperature (colors, C) and surface current (arrows) for (c) March and (d) July (Adapted from: Queiroz et al. (2019)).

3.3 Discussion

The no-accident scenario (SCN-0) shows significant stability in population behavior, with zero risk of extinction and half loss. This indicates that any extinction risk estimated while analyzing the other scenarios is due solely to the added risk of potential oil spills.

All three AS caused the *S. stellata* metapopulation a very low risk of extinction, with the maximum risk being 0.04%, which would have been enough to classify an AS as negligible according to IUCN risk criteria (IUCN, 2001). This indicates that, according to IUCN criteria, FNA population of *S. stellata* is not exposed to significant risk due to potential oil spills on nearby ship routes.

SCN-1, SCN-2 and SCN-3 cause, respectively, 0.11%, 0.69% and 1.53% risk of half loss. According to the present criteria, these scenarios are categorized as NEGLIGIBLE. This indicates that nearby ship routes cause tolerable risks to FNA. Note, however, that the chosen probabilities of accidents were conservative, thus likely overestimating occurrences of oil spills on this route.

4 CONCLUSION

A second round QERA for accidental oil spills nearby the FNA has been performed following a methodology for QERA of industrial accidents (Duarte et al., 2019). In this paper, it was included the fate and transport model of the oil plume in the ocean, providing results of the concentration of oil to which the *S. stellata* is being exposed.

The weathering processes that would tend to reduce the risk in the tropical region of FNA happen in a slower pace than the advective forcing. In turn, the high speed of oil transport reduces the time window for possible mitigation actions, which makes preventive actions the most viable options in the face of possible spills. According to the results of the simulation, the place where the oil reached the coast is part of PARNAMAR-FN, highlighting the need for a delimitation of a critical restrictive region for the traffic of large ships. In addition, new steps consist of running new simulations for FNA and other sensitive regions of the Southwest Atlantic, considering different types and volumes of oil ranging from transoceanic cargos fuel capacity to small and large tankers crude oil capacity. These new simulations should be integrated with the QERA, in order to quantify the risks for this variety of accidental scenarios, in such a way that can be used as tools for mitigation plans.

According to both IUCN categories and the conservative categories here presented, the risk are negligible for all accidental scenarios, in the spill scenario considered (including the sum of risks of all scenarios), where the undesirable consequence is not the extinction, but half loss of the initial population size. The conclusions drawn are that the industrial activity under consideration causes the marine ecosystem of FNA to be in NEGLIGIBLE risk.

The results here presented contribute to support the process of decision making by the governors and any interested parts. Additionally, the work done with this analysis contributes to the development of scientific advance in general, since there is no other methodology that integrates the fate and transport modeling in such a way that can be transformed in a risk measure, along with the other integrated steps.

ACKNOWLEDGMENTS

We acknowledge the Fundação de Amparo à Ciência e Tecnologia do Estado de Pernambuco (FACEPE) for the financial support, under the grant number BIC-0588-3.08/19. We also acknowledge the Programa de Recursos Humanos-47 of the Agência Nacional de Petróleo (ANP/PRH-47) for providing student scholarship to S.Q. and M.A. thanks the INCT AmbTropic - National Institute on Science and Technology for Tropical Marine Environments, CNPq/FABESB (565054/2010-4 and 8936/2011). This work is a contribution to the LMI TAPIOCA.

REFERENCES

Akçakaya, H. R. & Root, W. T. 2013. *RAMAS GIS: Linking Spatial Data with Population Viability Analysis (version 6)*. Setauket, New York: Applied Biomathematics.

Amaral, F. M. D., Ramos, C. A. C., Leão, Z., Kikuchi, R., Lima, K. K. M., Longo, L. & Vasconcelos, S. L. 2009. Checklist and morphometry of benthic cnidarians from the Fernando de Noronha archipelago, Pernambuco, Brazil. *Cah. Biol. Mar.*, *50*(3), 227–290.

ANP. 2017. *Boletim Mensal da Produção de Petróleo e Gás Natural*. Retrieved from http://www.anp.gov.br/publicacoes/boletins-anp/2395-boletimmensalda-producao-de-petroleo-e-gas-natural

Barros, M.M.L & Pires, D. O. 2006. Aspects of the life history of Siderastrea stellata in the tropical Western Atlantic, Brazil. *Invertebrate Reproduction and Development*, 49(4), 237–244.

BRASIL. 1986. *Decreto n. 92.775 de 05 de junho de 1986: Declara Área de Proteção Ambiental o território federal de Fernando de Noronha, o Atol das Rocas e os Penedos de São Pedro e São Paulo, e dá outras providências*. Brasília - DF, Brasil: Diário Oficial da União, Poder Executivo. Retrieved from https://www.planalto.gov.br/ccivil_03/decreto/1980-%0A1989/1985-1987/d92755.htm

BRASIL. 1988. *Decreto n. 96.693 de 14 de setembro de 1988: Cria o Parque Nacional Marinho de Fernando de Noronha e dá outras providências*. Brasília - DF, Brasil: Diário Oficial da União, Poder Executivo. Retrieved from https://www.planalto.gov.br/ccivil_03/decreto/1980-1989/d96693.htm

BRASIL. 2000. Lei N°9.985, de 18 de Julho de 2000.

Burgman, M. A., Franklin, J., Hayes, K. R., Hosack, G. R., Peters, G. W. & Sisson, S. A. 2012. Modeling Extreme Risks in Ecology. *Risk Analysis*, *32*(11), 1956–1966.

Casey, K. S. & Cornillon, P. 1999. A Comparison of Satellite and In Situ–Based Sea Surface Temperature Climatologies. *Journal of Climate*, *12*(6), 1848–1863. https://doi.org/10.1175/1520-0442(1999)012<1848:ACOSAI>2.0.CO;2

Castro, C. B. & Pires, D. O. 2001. Brazilian coral reefs: what we already know and what is still missing. *Bull. Mar. Sci.*, *69*(2), 357–371.

Cintra, M. M., Lentini, C. A. D., Servain, J., Araujo, M. & Marone, E. 2015. Physical processes that drive the seasonal evolution of the Southwestern Tropical Atlantic Warm Pool. *Dynamics of Atmospheres and Oceans, 72*, 1–11. https://doi.org/10.1016/j.dynatmoce.2015.08.001

da Silva, J. M. 2010. *The Dolphins of Noronha*. São Paulo: Bambu.

da Silva, A. M., Young, C. C. & Levitus, S. 1994. *Atlas of Surface Marine Data, Volume 1: Algorithms and Procedures* (NOAA Atlas). U.S. Department of Commerce, NOAA, NESDIS.

de Barros, Monica Moraes Lins & Pires, D. O. 2006. Aspects of the life history of Siderastrea stellata in the tropical Western Atlantic, Brazil. *Invertebr. Reprod. Dev.*, *49*(4), 237–244.

De Dominicis, M., Pinardi, N., Zodiatis, G. & Archetti, R. 2013. MEDSLIK-II, a Lagrangian marine surface oil spill model for short-term forecasting – Part 2: Numerical simulations and validations. *Geoscientific Model Development*, *6*(6), 1871–1888. https://doi.org/10.5194/gmd-6-1871-2013

De Dominicis, M., Pinardi, N., Zodiatis, G. & Lardner, R. 2013. MEDSLIK-II, a Lagrangian marine surface oil spill model for short-term forecasting – Part 1: Theory. *Geoscientific Model Development*, *6*(6), 1851–1869. https://doi.org/10.5194/gmd-6-1851-2013

Duarte, H. D. O., Droguett, E. L., Araújo, M. & Teixeira, S. F. 2012. Quantitative Ecological Risk Assessment of Industrial Accidents: The Case of Oil Ship Transportation in the Coastal Tropical Area of Northeastern Brazil. *Human and Ecological Risk Assessment: An International Journal*, *7039* (September2013), 120912081853008. https://doi.org/10.1080/10807039.2012.723187

Duarte, H. O. & Droguett, E. L. 2016. Quantitative ecological risk assessment of accidental oil spills on ship routes nearby a marine national park in Brazil. *Human and Ecological Risk Assessment*, *22*(2), 350–368. https://doi.org/10.1080/10807039.2015.1067760

Duarte, H. O., Droguett, E. L., Moura, M. das C., Siqueira, P. G. S. C. & Lira, J. C. 2019. A novel quantitative ecological and microbial risk assessment methodology: theory and practice. *Human and Ecological Risk Assessment*, 1–24. https://doi.org/10.1080/10807039.2019.1596736

Ferreira, B. P. & Maida, M. 2006. *Monitoramento dos Recifes de Coral do Brasil: Situação Atual e Perspectivas* (Vol. 1). Brasília: MMA.

Forbes, V. E., Calow, P., Grimm, V., Hayashi, T. I., Jager, T., Katholm, A. & Stillman, R. A. 2011. Adding Value to Ecological Risk Assessment with Population Modeling. *Hum Ecol Risk Assess*, *17*(2), 287–299.

Garla, R. C. 2004. Ecologia e conservação dos tubarões do Arquipélago de Fernando de Noronha, com ênfase no tubarão-cabeça-de-cesto Carcharhinus perezi (Poey, 1876) (Carcharhiniformes, Carcharhinidae). Rio Claro, SP, Brazil: Univ. Est. Paulista, Inst. Biocienc., Publ. Avulsa.

Gillespie, R. G. 2001. Oceanid Islands: Models of Diversity. *Enciclopedia of Biodiversity*, 1–13.

Gove, J. M., McManus, M. A., Neuheimer, A. B., Polovina, J. J., Drazen, J. C., Smith, C. R. & Williams, G. J. 2016. Near-island biological hotspots in barren ocean basins. *Nature Communications*, *7*(1), 10581. https://doi.org/10.1038/ncomms10581

Hounsou-gbo, G. A., Araujo, M., Bourlès, B., Veleda, D. & Servain, J. 2015. Tropical Atlantic Contributions to Strong Rainfall Variability Along the Northeast Brazilian Coast. *Advances in Meteorology, 2015*, 1–13. https://doi.org/10.1155/2015/902084

IBAMA. 2005. *Plano de Manejo da APA Fernando de Noronha - Rocas - São Pedro e São Paulo: Resumo Executivo*. Retrieved from http://www.icmbio.gov.br/portal/images/stories/imgs-unidadescoservacao/Resumo Executivo_f.pdf

ICMBio. 2013. PARNAMAR - Parque Nacional Marinho de Fernando de Noronha. Retrieved from http://www.parnanoronha.com.br/paginas/91-o-parque.aspx

IMO. 2004. International Convention for the Control and Management of Ships Ballast Water & Sediments. *Diplomatic Conference*. London: International Maritime Organisation (IMO).

IMO. 2007a. *Formal Safety Assessment FSA—Container Vessels*. London: International Maritime Organization.

IMO. 2007b. *Formal Safety Assessment FSA—Liquefied Natural Gas(LNG) Carriers*. London: International Maritime Organization.

IMO. 2008a. *Formal Safety Assessment FSA—Cruise Ships*. London: International Maritime Organization.

IMO. 2008b. *Formal Safety Assessment FSA-Crude Oil Tankers*. London: International Maritime Organization.

IMO. 2008c. *Formal Safety Assessment FSA—RoPax Ships*. London: International Maritime Organization.

IUCN. 2001. *IUCN Red List Categories: Version 3.1*. Gland, Switzerland and Cambridge, UK: IUCN Species Survival Comission.

Kalos, M. H. & Whitlock, P. A. 2008. *Monte Carlo Methods*. Weinheim: Wiley-VCH.

Ku, H. (1966). Notes on the use of propagation of error formulas. *J. Res. Natl. Bur. Stand., Sect. C, 70C*(4), 263–273.

Large, W. G., McWilliams, J. C. & Doney, S. C. 1994. Oceanic vertical mixing: A review and a model with a nonlocal boundary layer parameterization. *Reviews of Geophysics*, *32*(4), 363. https://doi.org/10.1029/94RG01872

Lee, K., Boufadel, M., Chen, B., Foght, J., Hodson, P., Swanson, S., & Venosa, A. (2015). *The Behaviour and Environmental Impacts of Crude Oil Released into Aqueous Environments* (Ottawa). The Royal Society of Canada.

Lemarié, F., Debreu, L., Shchepetkin, A. F. & McWilliams, J. C. 2012. On the stability and accuracy of the harmonic and biharmonic isoneutral mixing operators in ocean models. *Ocean Modelling*, *52–53*, 9–35. https://doi.org/10.1016/j.ocemod.2012.04.007

Lira, S. M. A., Amaral, F. M. D. & Farrapeira, C. M. R. 2009. Population growth by the white sea-urchin Tripneustes ventricosus (Lamarck, 1816) (Echinodermata) at the Fernando de Noronha Archipelago Brazil. *Pan-American Journal of Aquatic Sciences*, *4*, 1–2.

Lumpkin, R. & Garzoli, S. L. 2005. Near-surface circulation in the Tropical Atlantic Ocean. *Deep Sea Research*

Part I: Oceanographic Research Papers, 52(3), 495–518. https://doi.org/10.1016/j.dsr.2004.09.001

MacElrevey, D. H. & MacElrerey, D. E. 2004. Shiphandling for the mariner. Atglen, PA: Schiffer Publishing.

Maida, M. & Ferreira, B. P. 1995. Preliminary evaluation of Suest Bay reef, Fernando de Noronha, with emphasis on the scleractinian corals. Boletim Técnico Científico Do CEPENE, 3(1), 37–47.

Marchesiello, P., Debreu, L. & Couvelard, X. 2009. Spurious diapycnal mixing in terrain-following coordinate models: The problem and a solution. Ocean Modelling, 26(3–4), 156–169. https://doi.org/10.1016/j.ocemod.2008.09.004

MARINETRAFFIC. 2017. MarineTraffic: Global Ship Tracking Intelligence | AIS Marine Traffic. Retrieved from http://www.marinetraffic.com

Marta-Almeida, M., Ruiz-Villarreal, M., Pereira, J., Otero, P., Cirano, M., Zhang, X. & Hetland, R. D. 2013. Efficient tools for marine operational forecast and oil spill tracking. Marine Pollution Bulletin, 71(1–2), 139–151. https://doi.org/10.1016/j.marpolbul.2013.03.022

Medeiros, R. C. 2009. O Arquipélago de Fernando de Noronha e a presença militarnaval: uma condicionante Estratégia (I). Sagres. Retrieved from http://www.sagres.org.br/artigos/marinha_afn.pdf

Molinari, R. L. 1982. Observations of eastward currents in the tropical South Atlantic Ocean: 1978–1980. Journal of Geophysical Research, 87(C12), 9707. https://doi.org/10.1029/JC087iC12p09707

NRC. 2003. Oil in the Sea III: Inputs, Fates and Effects</i>. (T. N. A. Press, Ed.). Washing, DC.

Penven, P., Marchesiello, P., Debreu, L. & Lefèvre, J. 2008. Software tools for pre- and post-processing of oceanic regional simulations. Environmental Modelling & Software, 23(5), 660–662. https://doi.org/10.1016/j.envsoft.2007.07.004

Queiroz, S., Fazekas, L., Silva, M. A. & Araújo, M. 2019. Simulation of Oil Spills Near a Tropical Island in the Equatorial Southwest Atlantic. Tropical Oceanography, 47(1), 17–37. https://doi.org/10.5914/tropocean.v47i1.243115

Risien, C. M. & Chelton, D. B. 2008. A Global Climatology of Surface Wind and Wind Stress Fields from Eight Years of QuikSCAT Scatterometer Data. Journal of Physical Oceanography, 38(11), 2379–2413. https://doi.org/10.1175/2008JPO3881.1

Serafini, T. Z. & França, G. 2010. Ilhas oceânicas brasileiras: biodiversidade conhecida e sua relação com o histórico de uso e ocupação humana. Revista de Gestão Costeira Integrada/Journal of Integrated Coastal Zone Management, 10(3), 281–301.

Shafir, A., Van-Rijn, J. & Rinkevich, B. 2007. Short and long term toxicity of crude oil and oil dispersants to two representative coral species. Environ. Sci. Technol., 41, 5571–5574.

Shchepetkin, A. F. & McWilliams, J. C. 2003. A method for computing horizontal pressure-gradient force in an oceanic model with a nonaligned vertical coordinate. Journal of Geophysical Research, 108(C3), 3090.

Shchepetkin, A. F. & McWilliams, J. C. 2005. The regional oceanic modeling system (ROMS): A split-explicit, free-surface, topography-following-coordinate oceanic model. Ocean Modelling, 9(4), 347–404.

Silva, M., Araújo, M., Servain, J. & Pierrick, P. (2009). Circulation and heat budget in a regional climatological simulation of the Southwestern Tropical Atlantic. Tropical Oceanography, 37(1–2). https://doi.org/10.5914/tropocean.v37i1-2.5156

Song, Y. & Haidvogel, D. 1994. A Semi-implicit Ocean Circulation Model Using a Generalized Topography-Following Coordinate System. Journal of Computational Physics, 115(1), 228–244. https://doi.org/10.1006/jcph.1994.1189

Spaulding, M. L. 2017. State of the art review and future directions in oil spill modeling. Marine Pollution Bulletin, 115(1–2), 7–19. https://doi.org/10.1016/j.marpolbul.2017.01.001

Stramma, L. & Schott, F. 1999. The mean flow field of the tropical Atlantic Ocean. Deep Sea Research Part II: Topical Studies in Oceanography, 46(1–2), 279–303. https://doi.org/10.1016/S0967-0645(98)00109-X

Taleb, N. N. 2007. The Black Swan: The Impact of the Highly Improbable. Paris: Random House & Penguin.

TAMAR. 2006. Life in the deep blue (1st ed.). São Paulo: Bambu.

Tchamabi, C. C., Araujo, M., Silva, M. & Bourlès, B. 2017. A study of the Brazilian Fernando de Noronha island and Rocas atoll wakes in the tropical Atlantic. Ocean Modelling, 111, 9–18. https://doi.org/10.1016/j.ocemod.2016.12.009

Whittaker, R. J. & Fernández-Palacios, J. M. 2007. Island Biogeography: Ecology, Evolution and Conservation (2nd ed.). Great Britain: Oxford University Press.

Yender, R. A., Michel, J., Shigenaka, G., Hoff, R. Z., Mearns, A. & Hunter, C. L. 2010. Oil Spills in Coral Reef: Planning and Response Considerations. Florida: National Oceanic and Atmospheric Administration - NOAA.

Developments in Maritime Technology and Engineering – Guedes Soares & Santos (eds)
© 2021 Copyright the Author(s), ISBN 978-0-367-77376-2

Reliability analysis of critical systems installed in ships based on degradation mechanisms

J. Sobral
Centre for Marine Technology and Ocean Engineering (CENTEC), Instituto Superior Técnico, Universidade de Lisboa, Lisbon, Portugal
Mechanical Engineering Department, ISEL – Instituto Superior de Engenharia de Lisboa, Lisbon, Portugal

C. Guedes Soares
Centre for Marine Technology and Ocean Engineering (CENTEC), Instituto Superior Técnico, Universidade de Lisboa, Lisbon, Portugal

ABSTRACT: This paper presents a methodology that can be applied to acquire information about the reliability of items subjected to failures associated with degradation, using mathematical models for extrapolation (like the Linear, the Exponential, the Power, the Logarithmic, the Gompertz or the Lloyd-Lipow) followed by a life data analysis based on the estimated times to failure. The work refers the main systems installed on ship engine rooms and presents a demonstrative example of application of the methodology for oil filters belonging to the lubrication system. Following the methodology proposed it is possible to observe the risk incurred and most of time correct manufacturers' specifications with real field data.

1 INTRODUCTION

According to Brocken (2016) in the last years the amount of tonnage moved by ships has been growing significantly. For example, in the year 2000 a seaborne trade of 30,823 billion of ton-miles was registered while in 2013 this number raised to 50,374 billion of ton-miles transported corresponding to an increase of 63,4% in 13 years. This increasing tendency is ongoing, demanding more ships, bigger ships and higher performance for the existing ones.

The type of ship will determine the level of auxiliary and support systems that are required. They differ in complexity and quantity. For example, a bulker or a merchant ship do not require much support systems compared to a passenger ship type, while offshore vessel and service ships have more complexity and specific equipment.

There are several systems installed in ships that can be considered as critical. A failure on a critical system of a ship may correspond to huge costs or even put in risk the crew and the ship or cause a severe environmental impact. Thus, it is important to assure high reliability for the items that belong to the referred critical systems. For failures that occur upon degradation processes and where it is possible to monitor some parameter related to that degradation mechanism, it allows having data that can be converted in crucial information regarding the

decision making process towards the best decisions on maintenance and thus avoiding the occurrence of failures on those critical systems.

The engine room is one of the most important areas of a ship, whatever the ship type. Thus, it must include reliable assets, assuring a high performance of the ship itself. In fact, there are some systems that play an important role such as the main engine, the steering gear, the fuel system, the electrical system, the cooling water system and the diesel generator.

The maintenance of these systems is of outmost importance because their failures will in most of the cases cause serious unwanted situations. However, the best results are achieved if a good maintenance is done to high reliable systems. Inspections can be done to make sure that critical systems are operating properly and, if not, maintenance activities are performed. Activities include the check of lubrication oil levels, the cleaning of filters and turbochargers, pumping fuel from one tank to another and other tasks.

Some systems are more error-prone than others and it is important to identify failure causes and promote the best solutions.

Some failures are more difficult to anticipate once they are sudden or the physics of failure is unknown. In these cases, classical reliability studies can be applied based on a life data analysis, regarding historic data (times to failure) or manufacturers information (standard preventive maintenance planning).

DOI: 10.1201/9781003216582-29

However, there are other types of failures that can be easily predicted because they occur in accordance with a degradation mechanism over time that can be monitored. It is about this type of failures that a methodology is proposed in this paper, using degradation analysis models and reliability analysis.

The paper is structured into 6 sections. Section 2 refers to critical ship systems and the most common failures. Section 3 describes what is underneath a degradation analysis in the scope of the reliability acting as the basis for the proposed methodology and in section 4 a demonstrative example is given. Section 5 presents some conclusions and an overall perspective for the applicability of the proposed methodology to distinct items.

2 SYSTEMS AND FAILURES

The Safety Investigation Authority of Finland produced a report in 2016 (SIA, 2016) with a safety study about power failures on ships. This study was done using the Bow-Tie method.

The diagram developed describes the risks included in an incident as well as the relations between hazards, top event, threats and consequences. In addition to these four factors the model also includes preventive barriers (decrease the probability of occurrence of the incident causes), protective barriers (mitigate the consequences) as well as escalation factors, which can prevent barriers from working. Figure 1 illustrates the diagram developed in the study.

The main or central event was a power failure, corresponding to a failure that disables the continuity of operations and the electricity supply to critical systems. This type of failure is one that is included in the designations of ships not under control, which can led to large accidents and ship losses (Wu et al, 2017, 2018).

Rubio et al. (2018) stated that the diesel engine is a widely used machine in naval sector both as a propeller and auxiliary generator sets, being the most critical equipment of vessel platform. Therefore, diesel engine reliability optimization has a transcendental impact on vessel availability, safety and life cycle costs. The authors describe the development of a 4-stroke high speed marine diesel engine failure simulator to reproduce the engine response to failures without having to provoke them in real engine and also it is possible to know the symptoms

of one failure before this failure becomes dangerous for the correct behavior of the engine.

Regarding the engine room and observing the annual reports of the German Ship Safety Division as referred by Brocken (2016), it can be seen by system group the amount of technical failures as presented in Table 1 (Summary from 2001 to 2004, 2010 and 2011).

As can be seen the main engine and steering gear are the most reported failures in terms of quantity. Only catastrophic failures are considered. For example, if one of two engines fails in operation it is not registered as a technical failure. It means that the real number of failures is by far bigger than the described. Figure 2 shows the evolution of the amount of technical failures reported.

The cost of a failure is not only related to the cost of the machinery damage itself, but most of time also due to the substantial financial claims when towing is required. Some of the major engine incidents are related to (Codan, 2015):

- Thermal overload;
- Fuel oil problems;
- Cooling water problems;
- Lubricating oil problems;
- Turbocharger problems;
- Alignment problems;
- Torsion vibration problems;
- Reduction gear problems;
- Failure in alarms and safety systems;

Table 1. Failures per system group (adapted from Brocken (2016)).

System Group	Amount	Percentage
Main Engine	139	38,72%
Steering Gear	75	20,89%
Fuel System	40	11,14%
Electrical System	32	8,91%
Cooling Water System	30	8,36%
Shafting	24	6,69%
Diesel Generator	13	3,62%
Other	6	1,67%
Total	359	100.0

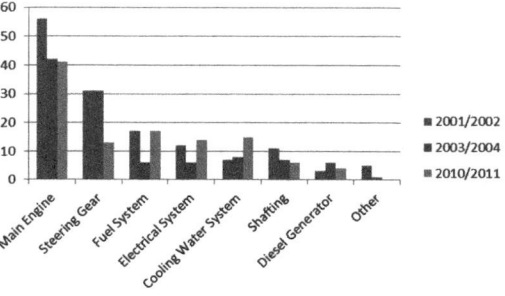

Figure 2. Evolution of the amount of technical failures (adapted from Brocken (2016)).

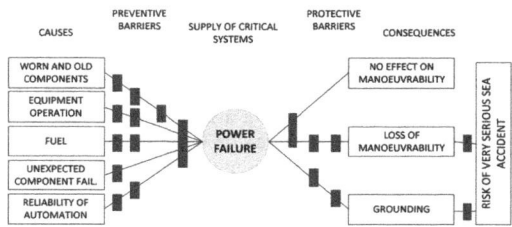

Figure 1. Bow tie diagram (adapted from SIA (2016)).

- Outflow of oil;
- Water in engine room.

Mishra et al. (2017) studied the causes of engine failures in ships and the maintenance required. The authors point out several factors that contribute to the failure of the engine, as poor design, contaminated fuel, unskilled manpower, lack of proper maintenance, long running hours, poor quality of fuel, poor fuel management technique, lack of vigilance and modification of the part. They indicate that preventive maintenance should be applied. They also refer that operation reliability of the diesel engine is mostly affected by wear and corrosion.

The maintenance of ship diesel engines can be grouped on the following activities:

- Lubrication – checking levels; changing oil, oil filters; performing oil sampling for trending analysis to optimize oil change intervals and to detect engine wear;
- Fuel system – changing fuel filters, fuel injectors; checking water separators; and doing fuel quality analysis to make sure fuel contains proper lubricants and additives;
- Cooling system – fluid level checks; coolant sampling for trending analysis; draining, flushing and refilling the system when required;
- Air intake system – inspecting and changing air filters; inspecting the turbocharger to make sure there is no fouling of the compressor blades from crankcase gases;
- Exhaust system – inspecting for leaks, corrosion, wet stacking;
- Valves and heads – inspecting, adjusting and recording of valve train wear for trending analysis; inspecting and recording of cylinder head wear for trending analysis;
- Emissions systems – inspecting crankcase ventilation systems, and diesel particulate filters;
- Mechanical systems – inspecting resilient engine mounts and torsion couplings; general inspecting for leaks, wear or deterioration;
- Operating systems – downloading data from digital engine management system to note and review alarm conditions.

The inadequate monitoring and maintenance of the condition of lubricating oil is one of the main causes for failures of most of the systems referred.

A limit situation occurred in March 2019 when a cruise ship was rescued from a storm off Norway's frigid North Sea coast because its engines didn't have enough lubricating oil. Sensors detected the oil shortage and automatically shut down the engines to prevent a breakdown. This triggered an alarm indicating a low level of lubrication oil.

In most of situations concerning engine failures due to oil, it is the quality of oil that is not the desired one. This aspect is very important and will be used in the demonstrative example of section 5.

3 DEGRADATION ANALYSIS METHODOLOGY

Degradation analysis can be seen as a prior step of a reliability study once it gathers precious information regarding the estimation of the time to failure of a certain failure mechanism. It involves the measurement of some degradation parameter over the time and based on the evolution of such measures and the definition of a threshold or limit value for that degradation it is possible to extrapolate for the potential time to failure using some known and adjustable mathematical models. Thus, the measurement of the parameter must be related to the degradation mechanism. The idea is to obtain degradation data and convert it to time to failure data in order to apply classical reliability theory. A degradation analysis is very important because one doesn't need to wait for a failure to have data concerning items life (Oliveira & Colosimo, 2004).

Barata et al (2002) developed a generic model for continuously-monitored deteriorating systems by using Monte Carlo Simulation. A non-repairable single component subjected to stochastic degradation was first considered and then the modelling was generalized to multi-component repairable systems. The effects of a preventive 'on-condition' maintenance strategy were investigated through the study of some desired reliability and availability characteristics.

Ali et al. (2012) give an example of a pressurized vessel, describing that the eventual failure of the vessel is directly related to the degree of metal loss, and then it can be used to predict or estimate the time to failure for the vessel.

Najjar et al. (2020) propose a theoretical model to simulate the performance and determine the degradation percentages of a Heat Recovery Steam Generator (HRSG) components at different operating loads and conditions. The results show that the degradation rate increases with time and with the operational load. On the other hand, there was no significant effect observed on the degradation rate by ambient conditions.

Cheng et al. (2019) present a wear model developed to describe the wear behavior considering the time-varying conditions of contact forces and sliding distances between the balls and the screw/nut raceways for ball screw mechanism (BSM). It allowed having a model for prediction of the wear of BSM, namely the positioning accuracy error/wear depth from the time-varying loading and rotational speed for BSM.

Fang et al. (2020) developed a methodology to observe the degradation processes where of the performance characteristics are positively correlated, investigating a bivariate degradation model of such a system. To analyze the accelerated degradation data, a flexible class of bivariate stochastic processes is proposed to incorporate the effects of environmental stress variables and the dependency between two degradation processes is modeled by a copula function. The authors present two real-world examples to demonstrate the applicability of the proposed modeling framework of system reliability on correlated degradation processes.

Sobral & Guedes Soares (2016) present a paper about the degradation of a HVAC filter where the data gathered along the time allows observing tendencies and determining time to failure of such filters, assuming a pre-defined value as the critical one, corresponding to asset failure.

Basically, for the parameter to be monitored the best condition monitoring technique must be selected among the available and technically feasible ones (thermography, vibration values, oil sample, pressure reading, temperature reading).

Then, the threshold value, limit value or alarm value must be defined. This definition may be established in accordance to regulations or manufacturer recommendations, or supported on field historic data.

The data related to the degradation is continuously or periodically recorded, manually or automatically registered and this represents an important step because it is the basis for the extrapolation and prediction of time to failure.

The estimation of the time to failure using the data gathered over time is obtained using an extrapolation method. This method is selected using one of the models presented in Table 2.

In Table 2 "a" and "b" are the model parameters determined in function of several observations and extrapolated through regression analysis to estimate the critical value.

Attention should be paid to the extrapolation model. The fitness to a specific model can be observed in correlation graphics "degradation x time" or using specific software which ranks the models previously mentioned in accordance to a data goodness of fit test.

Thus, based on the data gathered and on the extrapolation model selected it is possible to estimate the time to failure for similar assets working in identical conditions.

If dealing with data related to a single item that is in operation the estimation is direct. When recording data from several identical items it will result in a set of estimations of times to failure that allows using the classical life data analysis for reliability prediction with a statistical distribution that best fits the several times to failure estimated.

This distribution can be represented by its probability density function (pdf). Once defined the pdf the reliability function, the cumulative probability of failure function, the failure rate function or the mean time to failure (MTTF) can easily be determined.

If using a Weibull distribution (triparametric function), the pdf is expressed by:

$$f(t) = \frac{\beta}{\eta}\left(\frac{t-\gamma}{\eta}\right)^{\beta-1}.e^{-\left(\frac{t-\gamma}{\eta}\right)^{\beta}} \tag{1}$$

where:
γ = Location parameter
β = Shape parameter
η = Scale parameter or characteristic life parameter
t = time (number of cycles, etc...)
Thus, the reliability function is determined by:

$$R(t) = 1 - \int_0^t f(t).dt \tag{2}$$

$$R(t) = e^{-\left(\frac{t-\gamma}{\eta}\right)^{\beta}} \tag{3}$$

and the failure rate by:

$$\lambda(t) = \frac{\beta}{\eta}.\left(\frac{t-\gamma}{\eta}\right)^{\beta-1} \tag{4}$$

Consequently, the mean time to failure can be expressed by:

$$\overline{T} = \int_0^\infty t.f(t)dt \tag{5}$$

or

$$MTTF = T = \gamma + \eta.\Gamma.\left(\frac{1}{\beta}+1\right) \tag{6}$$

For other distributions the method is similar. Thus, the proposed methodology for time to failure estimation can be represented as shown in Figure 3.

Next section presents a demonstrative case study showing the applicability of the proposed methodology.

Table 2. Extrapolation models.

Model	Degradation parameter	Time
Linear	$y = at + b$	$t = \frac{y-b}{a}$
Exponential	$y = b.e^{at}$	$t = \frac{\ln y - \ln b}{a}$
Power	$y = b.t^a$	$t = \left(\frac{y}{b}\right)^{\frac{1}{a}}$
Logarithmic	$y = a.\ln(t) + b$	$t = e^{\frac{y-b}{a}}$
Gompertz	$y = a.b^{(c^t)}$	$t = \frac{\ln y}{\ln(a.b^c)}$
Lloyd-Lipow	$y = a - \frac{b}{t}$	$t = \frac{b}{a-y}$

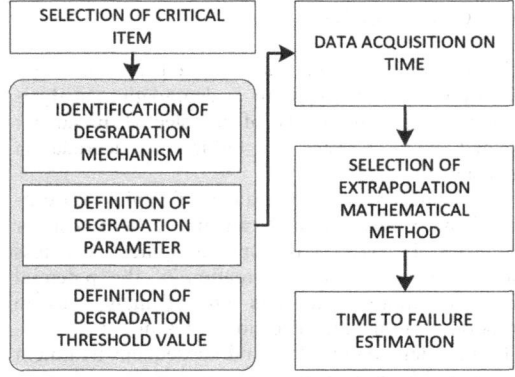

Figure 3. Degradation analysis methodology.

Table 3. Data related to filter #A.

Reading Number	Time Between Measures [h]	Cumulative Op. Time [h]	DOP [bar]	Filter ID
1	-	165	0.0500	A
2	180	345	0.0650	A
3	150	495	0.0720	A
4	200	695	0.0770	A
5	195	890	0.0910	A
6	225	1115	0.1120	A
7	260	1375	0.1200	A
8	185	1560	0.1380	A
9	160	1720	0.1630	A
10	180	165	0.0500	A

4 DEMONSTRATIVE EXAMPLE

Based on the referred importance of lubrication systems on critical equipment installed in the ships engine room, the demonstrative example gives an idea on how to apply the degradation analysis methodology proposed in the previous section to lubricating oil filters. The filters are assumed to be the critical items in the lubrication system because their clog can reduce drastically the efficiency of all lubrication system.

If the maintenance crew has information about the estimated time for failure of the filters they have time to plan an intervention (filter cleaning or replacement) in order to avoid such situation.

The engine internal lubricating oil system include the engine driven lubricating oil pump, the electrically driven pre-lubricating oil pump, thermostatic valve, filters and lubricating oil cooler. The lubricating oil pumps are located in the free end of the engine, while the automatic filter, cooler and thermostatic valve are integrated into one module.

Assume that data of lubricating system is given by three identical filters working in identical conditions (environment, speed, time). The drop of pressure (DOP) registered in those filters is directly linked to the clog and the filter clog can be referred in time, being expected to increase as time goes by. Based on this, the parameter to be monitored is the DOP in each filter.

Till the moment the maintenance crew follows filter manufacturer indications, performing an overhaul (replacement of internal elements) after the first operating 250 horas and then at each 2000 hours. The threshold is defined when the DOP (differential pressure) reaches 0.5 to 1 bar. Let's assume 0.5 bar as the limit value for the DOP.

Table 3 shows the data recorded on periodic inspections for one of these filters, as an example of the data acquired.

Based on the data collected on the three filters the mathematical extrapolation model that best fits the data was chosen. It was concluded that the model that better fits to data is the linear one. Table 4 outlines the values for linear extrapolation model for the three monitored filters.

The individual time to failure for each Filter corresponding to the time when each one achieves the threshold value is presented in Figure 4, showing the extrapolated values as well as the threshold value of clogging of 0.5 bar.

The estimated time to failure for Filter #A will be at about 2971 operating hours, for Filter #B will be at about 3419 operating hours and the Filter #C will be at about 3573 operating hours.

Table 4. Parameters of the linear extrapolation mathematical model.

Filter ID	Parameter "a"	Parameter "b"
#A	0.0100	0.0002
#B	0.0159	0.0001
#C	-0.0056	0.0001

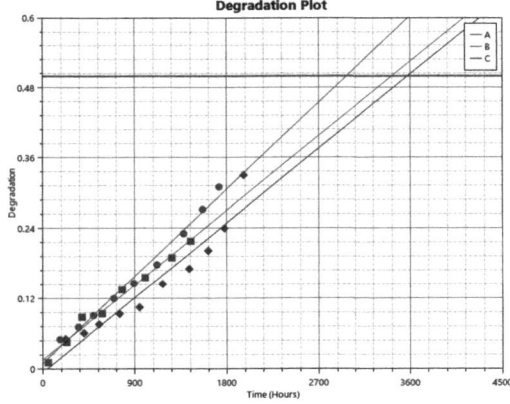

Figure 4. Extrapolated values for individual times to failure (Compass, 2019).

With these three times to failure it is possible to apply a reliability analysis. The distribution that best fits these times to failure is the Weibull biparametric distribution (rho = 0.9780) with a shape parameter (β) of 10.235 and a scale parameter or characteristic life (η) of about 3462 operating hours. The scale parameter being higher than one means that these filters are on their degradation phase (increasing failure rate). Figure 5 shows the instantaneous probability of failure corresponding to the probability density function for the referred filters operating on the known specific conditions.

Figure 6 shows the reliability curve for this type of filters operating on the known specific conditions.

Establishing a risk of 5% for the probability of failure, it means that we can perform a maintenance overhaul at each 2590 hours of operation of the filters.

Although this decision is based in reliability theory, algorithms, mathematical models and some assumptions the decision making process must include other factors as for example the economic ones.

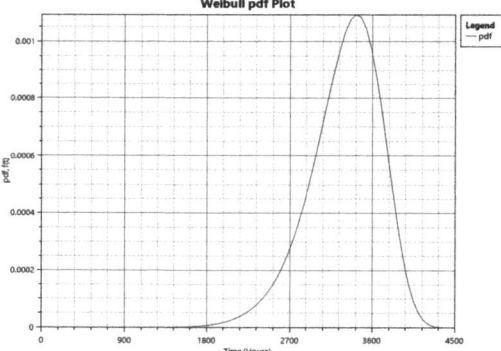

Figure 5. Probability density function curve (Compass, 2019).

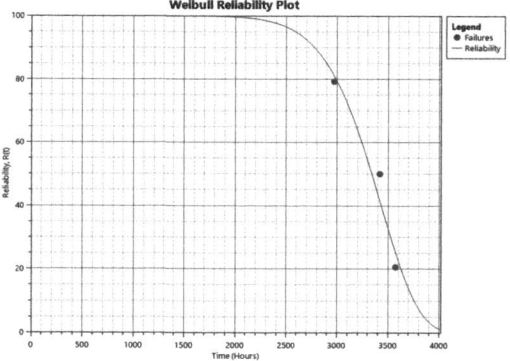

Figure 6. Reliability function (Compass, 2019).

5 CONCLUSIONS

The methodology that is proposed in this work shows how useful can be a degradation analysis based on measurements of a selected parameter related to a physics of failure or degradation mechanism.

It assumes special importance if applied on critical items where a failure can induce high costs or increases the risk for persons, equipment, business or environment. The information of the potential times to failure makes this type of analysis as an important tool for decision making processes. Through degradation data and subsequent treatment it is possible to have information about items' mean life, failure rate, probability of failure or reliability and decide when is the right moment to intervene.

It is possible to observe from the demonstrative example that for the time suggested by the filter manufacturer the filters present a reliability of 0.9964, which means that the probability of reaching a DOP of 0.5 bar is less than 1% (0.0036). This is a sign that if following manufacturer's indications is most of time incorrect. In this case the filters can stay working longer. For example if we want to have a risk of 10% of probability of failure (B10 life) the filters can operate for about 2779 hours (more 39% of time than manufacturer's indication). But in some cases it is the contrary, especially for example when the items are operating under worse conditions than the ones predicted by the manufacturer.

In this demonstrative example for a the probability of failure of 10% but with a level of confidence of 90% (bilateral), the maintenance of the filters can be done between 2272 and 3400 hours of operation of the filters. For lubrication systems operating 24/24 an operation of 2779 hours may signify maintenance each 117 days and not each 83 days.

The methodology presented fits well and is suitable to be applied in situations where the items are subjected to different conditions once the data collected and recorded on periodic inspections include the influence of those real conditions. This will be reflected on the extrapolation model that will point out the potential time to failure (time corresponding to the established threshold value for the degradation parameter).

Finally, this methodology can be applied to several other systems and items installed on ships where degradation occurs and is possible to monitor the evolution of a parameter that is characteristic of such degradation.

ACKNOWLEDGEMENTS

This work contributes to the Strategic Research Plan of the Centre for Marine Technology and Ocean Engineering (CENTEC), which is financed by the Portuguese Foundation for Science and Technology (Fundação para a Ciência e Tecnologia - FCT) under contract UIDB/UIDP/00134/2020.

REFERENCES

Ali, A., Majid, M.A.A. & Muhammad, M. (2012). Comparative study between degradation analysis and API 510 Remaining Life Evaluation Method for feed gas filter vessel reliability analysis. *Journal of Applied Sciences*, 12 (23), pp. 2448–2453

Barata, J.; Guedes Soares, C.; Marseguerra, M., and Zio, E. 2002. Simulation Modelling of Repairable Multi-Component Deteriorating Systems for "On Condition" Maintenance Optimization. *Reliability Engineering and System Safety*. 76(3):255–264.

Brocken, E.M. (2016). Improving the Reliability Of Ship Machinery. A Step Towards Unmanned Shipping, *Master of Science Thesis*, Delft University of Technology, Delft, The Netherlands

Cheng, Q., Qi, B., Liu Z., Zhang, C. & Deyi X. (2019). An accuracy degradation analysis of ball screw mechanism considering time-varying motion and loading working conditions. *Mechanism and Machine Theory*, Vol. 134, pp. 1–23

Codan (2015). Machinery Damage due to Engine Failure, http://www.codanmarine.com/start/claims/general_advice_and_instruction/machinery_damage_due_to_engine_failure

Compass (2019). Reliability4All (software). Degradation Analysis (DA) module. Compass. Sao Paulo. Brazil

Fang, G., Pan, R. & Hong, Y. (2020). Copula-based reliability analysis of degrading systems with dependent failures. *Reliability Engineering & System Safety*, Vol. 193, Article 106618

Mishra, C.S., Ali, F., Adam, S., Tesfariam, A., Yemane, A., Amanuel, H., Teklemariam, U. & Alem, Y. (2017). Study of the causes of engine failure in ship and maintenance required to prevent engine failure. *International Journal of Contemporary Research and Review*, Vol. 8, Issue 7, pp. 20218–20241

Najjar, Y.S.H., Alalul, O.F.A. & Abu-Shamleh, A. (2020). Degradation analysis of a combined cycle heat recovery steam generator under full and part load conditions. *Sustainable Energy Technologies and Assessments*, Vol. 37, 100587

Oliveira, V. & Colosimo, E. (2004). Comparison of Methods to Estimate the time-to-failure Distribution in Degradation Tests. *Quality and Reliability Engineering International*, Vol. 20, pp. 363–373

Rubio J.A.P., Vera-García, F., Grau, J.H., Cámara, J.M. & Hernandez, D.A. (2018). Marine diesel engine failure simulator based on thermodynamic model. *Applied Thermal Engineering*, Vol. 144, pp. 982–995

SIA (2016). Power failures on ships. Investigation number: M2016-S1, *Investigation report: S2/2017*, ISBN: 978-951-836-507-8 (PDF)

Sobral, J. & Guedes Soares, C. (2016). Preventive Maintenance of Critical Assets based on Degradation Mechanisms and Failure Forecast. *IFAC-PapersOnLine*, 49-28, pp. 097–102

Wu, B.; Yan, X. P.; Wang, Y.; Zhang, D., & Guedes Soares, C. 2017. Three-Stage Decision-Making Model under Restricted Conditions for Emergency Response to Ships Not under Control. *Risk Analysis*. 37 (12):2455–2474.

Wu, B.; Zong, L. K.; Yan, X. P., & Guedes Soares, C. 2018. Incorporating evidential reasoning and TOPSIS into group decision-making under uncertainty for handling ships without command. *Ocean Engineering*. 164:590–603.

Developments in Maritime Technology and Engineering – Guedes Soares & Santos (eds)
© 2021 Copyright the Author(s), ISBN 978-0-367-77376-2

Data-driven cognitive modeling and semantic reasoning of ship behavior

Rongxin Song, Yuanqiao Wen, Liang Huang, Fan Zhang & Chunhui Zhou
Wuhan University of Technology, Wuhan, China

ABSTRACT: Aiming at the cognition problem of systematic ship behavior in harbor, a behavior recognition model based on semantic reasoning based on discrete event system modeling theory was proposed. Firstly, the hierarchical modeling of the behavior of ships in port waters is divided into data layer, event layer, activity layer and process layer. Based on a certain theoretical understanding of ship behavior on different time scales and space scales, build a ship behavior cognitive model; Secondly, in the data layer, the trajectory key point detection and segmentation extraction of the motion trajectory of port ship AIS data are performed by integrating port navigation rules to realize the labeling of ship trajectories. Finally, based on the labeling results of the data layer, the ontology is used to make inferences to discover the implicit ship behavior, and realize the trajectory of the ship from the data layer to the semantic layer. Experiments were performed using Xiamen Port data. The experimental results show that the behavioral cognitive ontology based on discrete system modeling can realize the cognitive and semantic reasoning of ship behavior at different time and space scales.

1 INTRODUCTION

Facing the huge number of ships in the port waters, it is very difficult for the ships to navigate safely. At the same time, the recognition and expression of the maritime traffic situation in the port have an important role in assessing the safety situation of maritime traffic. The recognition of ship navigational situation, the identification and modelling of ship behavior is the basis of regional navigation situational awareness. Use ship automatic identification system data to model ship behavior, use discrete event system simulation modelling to model ship behavior in port waters, hierarchically model ship behavior at different time scales, and establish corresponding Ship behavior ontology. By identifying the key points of the ship's trajectory, the semantic labeling of the ship's trajectory at the data layer is realized, and the ontological reasoning function is used to realize the recognition of the ship's behavior.

At present, the identification of ship behavior mainly includes the following aspects to work:1) semantic trajectory: (Yan et al., 2013, de Vries et al., 2010) considered the motion characteristics of the ship itself, the ontology was used to enrich the semantics and behavior of the trajectory; used the ontology to enrich the ship's trajectory and fill the gap of the ship's movement data and semantic knowledge including ship's events. 2) Behavior identification and prediction: (Gao et al., 2018) developed an online real-time ship behavior prediction model by constructing (BI-LSTM-RNN) and which can predict the navigational behaviors of ship greatly;

(Jiang et al., 2020) explored animal's hidden behaviors pattern from time-series and further predict the behavior efficiently; (Gómez-Romero et al., 2015, Noor et al., 2018, Wu et al., 2019) construct the ontology model of basic data by fusing a variety of background information and the modelling of intent and the establishment of an ontology to achieve formal expression and intention reasoning of data.

In terms of knowledge reasoning, (R and Uma, 2018, Ding et al., 2019) expounded knowledge expression and reasoning using an ontology,(Salguero and Espinilla, 2018) used data-driven and knowledge-driven Activities are modelled and expressed.

In terms of behavior cognition, the cognition of ship behavior is relatively little, especially in activities with large time and space scales. (Giatrakos et al., 2019) defined and explored complex events, including detects and recognizes abnormal behaviors in the maritime regulatory field; (Zhang et al., 2019) analyzes the ship activity chain, aims to investigate the variation of ship traffic state in spatial space over time. (Ding et al., 2019) designed a malware knowledge base for individuals and families is constructed, and the behavior of the malware is mined through the Apriori algorithm.

In the human activity cognition of smart home, related research is relatively mature. (Liu et al., 2016, Noor et al., 2018, Jalal et al., 2018) designed algorithm to identify and activity recognition using sensor set in a smart room; (Gayathri et al., 2017) recognized the activities in the home, and the problem of uncertainty was solved with probability, which has a good recognition rate. (Gayathri et al., 2020) presented an unsupervised model combining

DOI: 10.1201/9781003216582-30

temporal reasoning and fuzzy ontology to represent the activity in a smart room.

(Gayathri et al., 2017) designed an activity recognition system that augments ontology-based activity recognition with probabilistic reasoning through Markov Logic Network to model various activities. (Roy et al., 2017) using uncertain observations to recognition activities in smart environments.

There is relatively little research on the behavior cognition of ships in the maritime transportation system of the port. The article models ship behaviors at different time scales based on the AIS time series data of the ship. The key-points detection algorithm realizes inference and cognition of ship behaviors at different levels.

We proposed related work in the third part, and the fourth part carried out hierarchical modelling of the ship's behavior in the port and marked the data process layer of the ship's behavior to complete the extraction of the ship's trajectory data to the trajectory key point sequence. The fifth part establishes the ontology layer of ship behaviors, expounds the recognition process of ship behaviors at different scales, and realizes the semantic reasoning of ship behaviors. Section VI is used as an experimental part to verify the rationality of the ontology by selecting typical ship behaviors for identification reasoning. Section VII presents the conclusions and looks forward to future work.

2 PORT SHIP BEHAVIOR MODELING

2.1 Modelling of discrete event systems

Discrete event systems are the system in which the state of the system changes discretely only at certain random points in time (Carapezza and Roy, 2010). In a port water traffic management system, changes in the state of interest of traffic managers often occur randomly, so, the water traffic management system can be regarded as a discrete event system. Based on the various elements of the discrete event system and the characterization of ship behavior in port waters at different time scales, the ship behavior is divided into three layers, as shown in Figure 1, including the event layer, the activity layer, and the process layer. Ships and port rules are used as entities in the system, Cause the state of the system to change through the interaction between entities.

Event layer: An event is an action that causes a change in the state of the water transportation system. In the maritime transportation system, an event refers to an action or a group of operations that changes the state of the system. For example, the action of a ship entering an anchorage can be regarded as an event. The ship sails outside the anchorage until it enters the anchorage. Interested change. As the key node of the entire ship behavior chain, the time and space scale of the event is the smallest in the entire behavior chain. Events are used

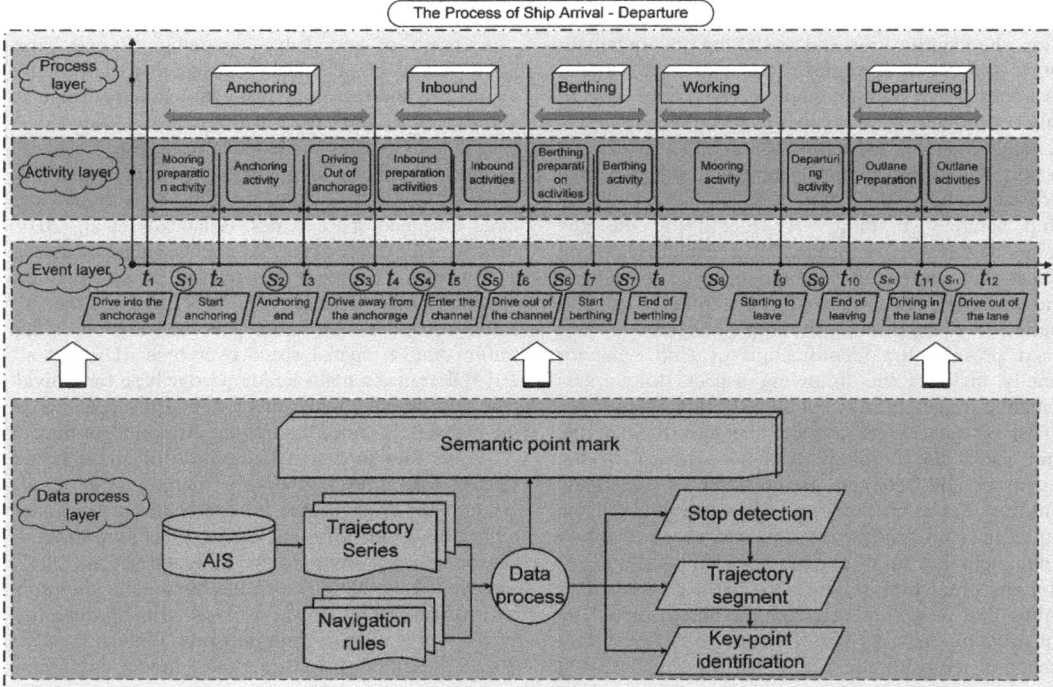

Figure 1. Semantic modeling process of ship entry and exit.

as trigger conditions for different behavior activities. For example, the event that a ship sails into an anchorage separates navigational activities outside the anchorage from preparation for anchoring, which plays a role in triggering the end of the previous activity and the start of the next activity.

Ship events can be expressed as:

$$Event = <t_i, e_i>, t_i, t_{i+1} \in [t_1, t_2, \cdots, t_n]$$

t can be instant or interval, most of the t is considered to be instant.

Activity layer: An activity represents the ship's behavior performed by an entity to complete a task. Ship activities are usually carried out within a large time and space scale, and their behaviors represent the actions that the ship is engaged in a task, such as anchoring activities, port activities, etc.

Ship activity can be expressed as:

$$Activity = <t_i, e_i, A, t_{i+1}, e_{i+1}>, t_i, t_{i+1} \in [t_1, t_2, \cdots, t_n]$$

Process layer: A process is a long-term behavior of a ship in the port waters. A process consists of events and activities. Events also act as the start and end of different processes. The process in the water transportation system usually includes four processes: anchoring process, inbound fairway process, in-port operation and out-port process.

The process can be expressed as:

$$process = <t_i, e_i, A_i, e_{i+1}, A_{i+1}, \cdots, A_{i+k}, e_{i+k+1}, t_{i+k+1}>,$$
$$t_i, t_{i+k+1} \in [t_1, t_2, \cdots, t_n]$$

2.2 Trajectory key points recognition

The ship's trajectory contains a series of key points, including the ship's stay points, the entry and exit anchorage points, and the entry and exit fairway points. The key points play an important role in a series of ship activities and events. Therefore, the identification of key points in the trajectory is of great significance for identifying ship behavior, Figure 2 shows the flow chart of trajectory point processing.

1. Stop points identification

The stopping point represents the behavior of the ship when it is stopped or at rest, usually including two states of anchoring and leaving. In order to explore the different motion behaviors of the ship, the ship motion trajectory is first segmented according to the stopping points. According to the statistics of the stay data of the ship, speed at the stay point is generally around 0.5kn, so the ship's speed can be filtered to obtain the stay point of the ship.

2. Identification of key points of ship accessing fairway

The fairway is an open area, consisting of straight or broken lines. Therefore, it is necessary to calculate the intersection of the ship trajectory and the fairway first, considering that the calculated intersection is not necessarily in the ship AIS data sequence, so it is necessary to replace the intersection with the original trajectory points. By comparing the intersection points of each ship with the coordinate points in the original AIS trajectory sequence, the intersection points of the ship and the fairway are approximately replaced to realize the data annotation of the ship entering and leaving the fairway.

3. Identification of key points of a ship arriving in and leaving out of the anchorage

Anchorage is usually a polygonal area or a circular area. When identifying the entry and exit of the anchorage, it is judged by whether the ship is located on different sides of the anchorage before and after. For example, if the previous moment is outside the anchorage and the latter moment is in the anchorage, it means that the vessel has entered the anchorage during this period, and the behavior of the vessel leaving the anchorage can be obtained in the same way.

3 SHIP BEHAVIOR RECOGNITION BASED ON ONTOLOGY REASONING

3.1 Ship behavior recognition based on ontology reasoning

Ontology originates from the field of philosophy and has now been extended to the field of artificial intelligence. It can abstract knowledge in the real world into concepts, relationships, and attributes in the ontology, and form a computer knowledge base. It can further perform operations such as knowledge reasoning and querying. In order to realize the ship behavior cognition at different time scales, the ontology modeling method is used to separate the ship's semantic behavior recognition model into a data layer and an ontology layer, as shown in Figure 3. Figure 4 illustrates the ontology graph constructed by protégé. The key points of ship trajectory are marked at the data processing layer to obtain the ship Time series of trajectory keypoints. Furthermore, by calling the data interface in the ship, events at different levels of the ship are identified, and activities and processes are reasoned based on the logical relationship between ship behaviors at different scales.

3.2 Event recognition

Based on the key point identification, the ship events are labelled semantically according to the trajectory

271

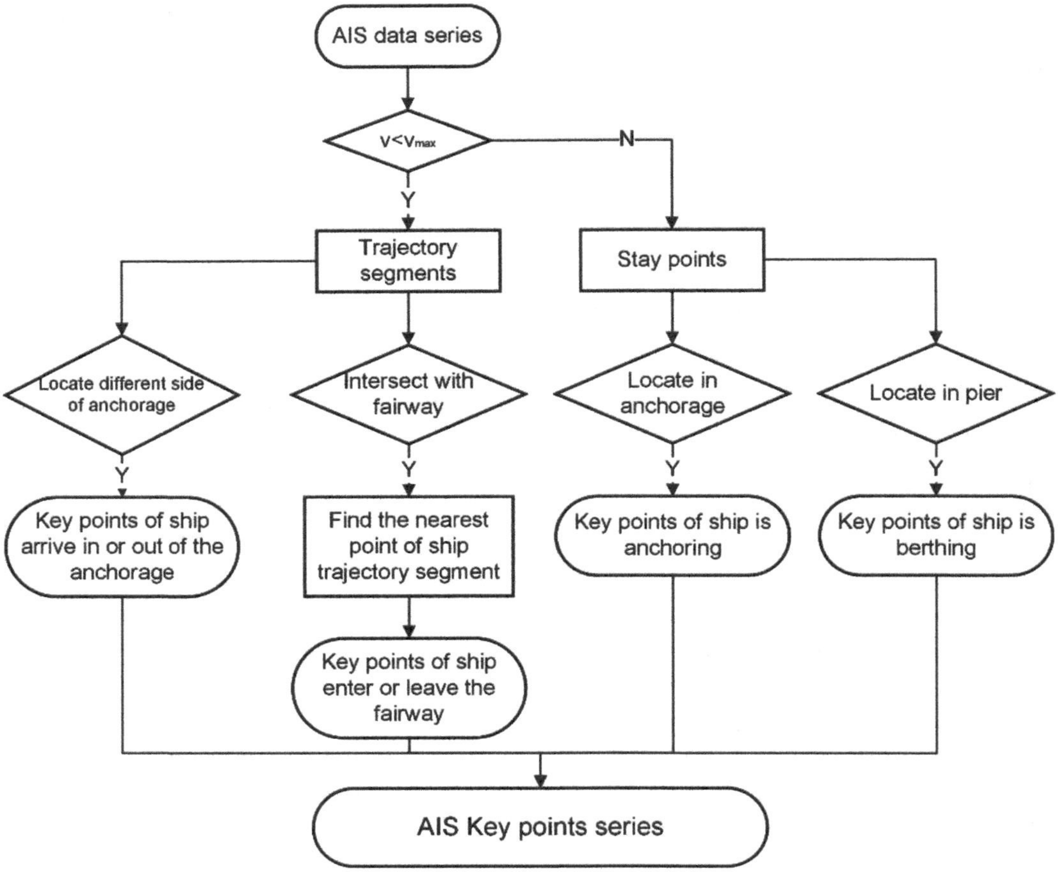

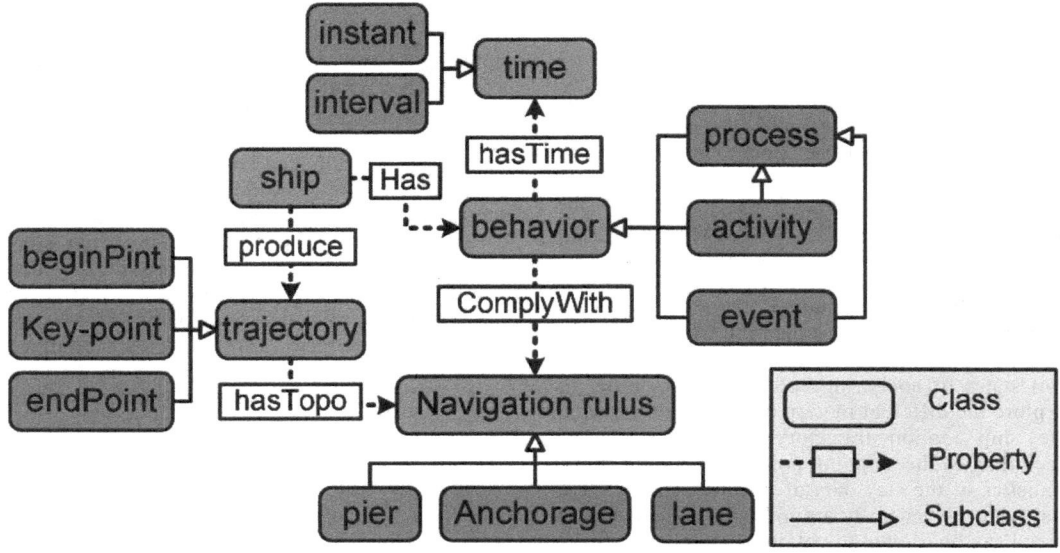

Figure 2. Key-points identification from AIS data flowchart.

Figure 3. Ship behavior ontology model.

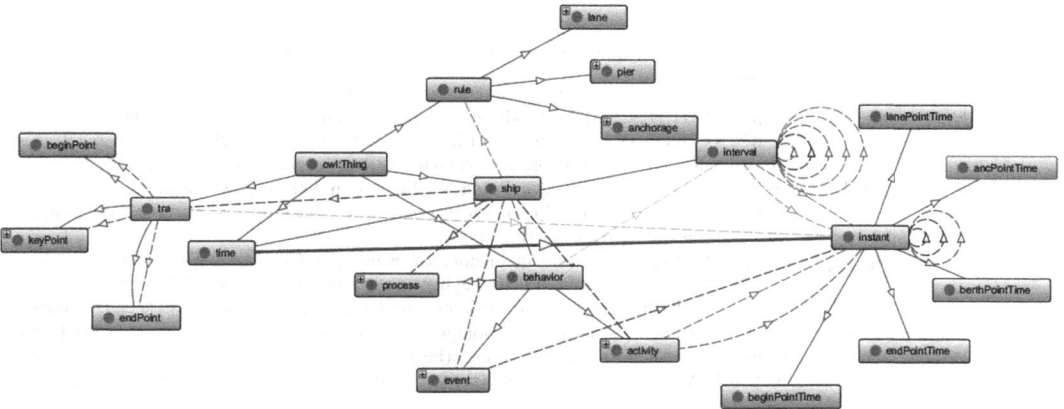

Figure 4. Rule-based port ship behavior ontology.

key point sequence. Furthermore, the use of SWRL rules can be used to recognize the implicit semantic behavior of ships. For example, the MMSI of the ship called "DONGHAIJIU131" is 412046080. It enters the 1# anchorage at 2016-04-05 21:42:19, stays at 1# anchorage and starts anchoring at 2016-04-05 21:59:32 and located at (24.352°N, 345.0°E), etc., including the identification of semantic behavior events such as when entering the main fairway and leaving the main fairway.

3.3 Activity and process identification

Vessel activities and processes are used to characterize the behavior of ships in a large time range. Because events are the demarcation points of ship activities and processes, ship activities can be characterized by identifying ship events, including preparing for anchoring activities and preparing to enter the fairway activities, ship sailing activities, etc. Eventually, the ship's progress is formed based on the order of events and activities, and reasoning is performed. This part mainly defines the activities and processes according to the Ontology SWRL (Semantic Web Rule Language) rules, as shown in Table 1. Based on the existing ship behavior events, the inferred activities and processes are derived by reasoning.

4 EXPERIMENT

4.1 Experimental platform

The experiment uses Python ontology library: owlready2, the library shapely for calculating spatial topological relationships, and the ontology software protégé for design and inference.

4.2 Data processing

The data of Xiamen Port from March to April 2016 were selected for the experiment. By processing the data of abnormal ship course and speed, and then matching the static data of the ship, the ship's AIS data was obtained, a total of 173,187 items.

4.3 Data layer processing

Stop point detection and labelling
According to empirical data, it can be obtained that the speed of the ship during the stay is maintained at about 0.5 kn, so the speed threshold of the stop point is set to $v_{max} = 0.5kn$. The ship's stopping point data is obtained by screening the ship' speed, and the ship trajectory is segmented.

Key points for ship navigation
Based on the segmentation of the ship's trajectory, first of all, based on the data of anchorage and berth in the port to determine whether the ship is located at the anchorage or berth, it can be obtained whether the ship is anchored or moored. Secondly, analyze the topological relationship between each trajectory segment of the ship and the navigation rules in the port, find the entry point and exit point of the ship through the traffic rules, and obtain the entering and exiting fairway points of the ship. By judging whether the ship is located on the opposite side of the anchorage, it is judged that the ship has entered or exited the anchorage.

4.4 Ontology layer behavior recognition and semantic reasoning

As the behavior recognition process of ships in anchorages, fairway and docks is similar, the behaviors of

Table 1. SWRL rules for ship events or activities.

Event or Activity	SWRL rules
Event of ship entering anchorage	ship(?s)^hasTra(?s,?tra)^hasKeyPoint(?tra,?p)^ancValue(?an,?y)^swrlb:lessThan(?x,"0"^^xsd:integer)^swrlb:abs(?y,?x)^hasAncValue(?p,?x)^atTime(?p,?t)^anchorage(?an)->hasEvent(?s,ArriveInAnchorage)^eventOccurTime(ArriveInAnchorage,?t)
Event of ship leaving anchorage incident	ship(?s)^hasTra(?s,?tra)^hasKeyPoint(?tra,?p)^hasAncValue(?p,?x)^anchorage(?an)^ancValue(?an,?y)^atTime(?p,?t)^swrlb:equal(?x,?y)->hasEvent(?s,ArriveOutAnchorage)^eventOccurTime(ArriveOutAnchorage,?t)
Event of ship end anchoring	ship(?s)^hasEvent(?s,ArriveOutAnchorage)^eventOccurTime(ArriveOutAnchorage,?T)^hasTra(?s,?tra)^hasBegin(?tra,?p)^beginPoint(?p)^speed(?p,?x)^swrlb:lessThan(?x,3)^atTime(?p,?t)^before(?t,?T)->eventOccurTime(endAnchor,?t)^hasEvent(?s,endAnchor)
Activities of ship preparing for anchoring	ship(?s)^hasEvent(?s,ArriveInAnchorage)^eventOccurTime(ArriveInAnchorage,?t)^hasEvent(?s,beginAnchor)^eventOccurTime(beginAnchor,?T)^before(?t,?T)->hasActivity(?s,prepareToAnchor)^occursEndTime(prepareToAnchor,?T)^occursBeginTime(prepareToAnchor,?t)
Anchoring activities	ship(?s)^hasEvent(?s,beginAnchor)^eventOccurTime(beginAnchor,?T)^hasEvent(?s,endAnchor)^eventOccurTime(endAnchor,?t)^before(?T,?t)->occursEndTime(anchoring,?T)^occursBeginTime(anchoring,?t)^hasActivity(?s,anchoring)
Activities of leaving anchorage	ship(?s)^hasEvent(?s,ArriveOutAnchorage)^hasEvent(?s,endAnchor)^eventOccurTime(ArriveOutAnchorage,?t)^eventOccurTime(endAnchor,?T)^before(?T,?t)->hasActivity(?s,leavingAnchor)^occursBeginTime(leavingAnchor,?T)^occursEndTime(leavingAnchor,?t)
Process of Anchoring	ship(?s)^hasEvent(?s,?ArriveInAnchorage)^hasKeyTime(ArriveInAnchorage,?t1)^hasEvent(?s,?ArriveOutAnchorage)^hasKeyTime(ArriveOutAnchorage,?t2)->hasActivity(?s,Anchoring)^hasInterval(t1,t2)

ships in anchorages were selected as experiments to conduct behavior recognition and reasoning.

According to the results of behavior modelling, the behavior of the ship at the anchorage is divided into six normal behaviors, including three events, two activities, and a process. The SWRL rules for events and activities in protégé are shown in Table 1.

4.5 Experimental results

By semantically labelling the sequence of key points of the AIS trajectory in the data layer, add the ship trajectory instance to the ontology, as shown in Figure 5, The track of donghaijiu131 is divided into 15 track segments, and the track "Donghaijiu_131_558" represents the track segment number 558 of donghaijiu131. Each track segment contains the key points, start point and endpoint of the track. As shown in Figure 6, "Donghaijiu_131_540_1anc_point" means that the tracking number 540 of donghaijiu131 and the 1# anchor point intersect. In the same way, "Donghai Jiu_131_540_1lane_point" indicates that the track segment numbered 540 of Donghaijiu131 intersects with channel 1. As shown in Figure 7, the time at which the trajectory key point occurs is related to the time attribute "atTime" to obtain the time of occurrence of the key point and the data attribute of the key point. At this moment, the value of the speed, course and trajectory key point of the ship, the value of the trajectory

Figure 5. Segmentation of ship trajectory.

Figure 6. Keypoint sequence labelling result of ship trajectory segment.

key point indicates that the key point is located at the rule number. For example, "hasAncValue: 5" indicates that the key point is located at the anchor 5#.

After running the pellet inference engine, events that occur at the anchorage of the ship can be detected, Such as entering the anchorage, exiting the anchorage and starting anchoring, etc., as well as ship activities, departure activities and anchoring activities in a longer time scale, such as Arriving the anchorage, arriving out of the anchorage and starting anchoring, etc., Besides, there are ship activities, departure activities and anchoring activities in a longer time scale, as shown in Figure 8.

The data layer in this model plays an important role in identifying the behavior of ships and determines the accuracy of ship behavior recognition. The data layer marks the sequence of the ship's key points, and the key points of the ship are identified by calculating the spatial topological relationship between the ship's trajectory segment and the traffic rules, In the ontology layer, the built ship behavior ontology is used to add ship trajectory segments and trajectory key points to the ontology. By judging the speed of the start and endpoints of the trajectory segment, identify whether the start and end of the trajectory segment are moored, thereby judging ship incidents and activities, thereby realizing the recognition of the ship's behavior.

Figure 7. The object properties and data properties of a key point.

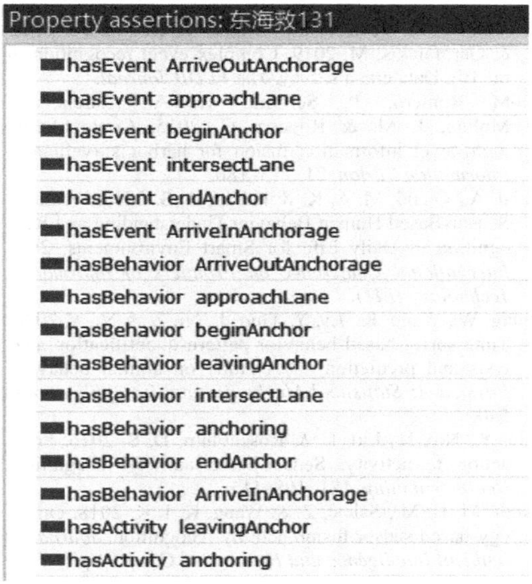

Figure 8. Ship behavior recognition based on SWRL rule reasoning.

5 CONCLUSIONS

By hierarchically modelling the behavior of ships in port waters at a different time and space scales, using the ontology tools to model the behavior, integrating SWRL rule language to reason about the implicit semantic behavior of ships, and using ship AIS data to realize the ship's behavior. Intelligent cognition. It lays the foundation for realizing a ship's autonomous behavior cognition and behavior intent reasoning.

REFERENCES

De Vries, G. K. D., Van Hage, W. R. & Van Someren, M. 2010. Comparing Vessel Trajectories Using Geographical Domain Knowledge and Alignments. *2010 IEEE International Conference on Data Mining Workshops.*

Ding, Y., Wu, R. & Zhang, X. 2019. Ontology-based knowledge representation for malware individuals and families. *Computers & Security*, 87.

Gao, M., Shi, G. & Li, S. 2018. Online Prediction of Ship Behavior with Automatic Identification System Sensor Data Using Bidirectional Long Short-Term Memory Recurrent Neural Network. *Sensors (Basel)*, 18.

Gayathri, K. S., Easwarakumar, K. S. & Elias, S. 2017. Probabilistic ontology based activity recognition in smart homes using Markov Logic Network. *Knowledge-Based Systems*, 121, 173–184.

Gayathri, K. S., Easwarakumar, K. S. & Elias, S. 2020. Fuzzy Ontology Based Activity Recognition for Assistive Health Care Using Smart Home. *International Journal of Intelligent Information Technologies*, 16, 17–31.

Giatrakos, N., Alevizos, E., Artikis, A., Deligiannakis, A. & Garofalakis, M. 2019. Complex event recognition in the Big Data era: a survey. *The VLDB Journal*.

GóMez-Romero, J., Serrano, M. A., García, J., Molina, J. M. & Rogova, G. 2015. Context-based multi-level information fusion for harbor surveillance. *Information Fusion*, 21, 173–186.

Jalal, A., Quaid, M. A. K. & Hasan, A. S. 2018. Wearable Sensor-Based Human Behavior Understanding and Recognition in Daily Life for Smart Environments. *2018 International Conference on Frontiers of Information Technology (FIT)*.

Jiang, W., Wang, K., Lv, Y., Guo, J., Ni, Z. & Ni, Y. 2020. Time series based behavior pattern quantification analysis and prediction — A study on animal behavior. *Physica A: Statistical Mechanics and its Applications*, 540.

Liu, Y., Nie, L., Liu, L. & Rosenblum, D. S. 2016. From action to activity: Sensor-based activity recognition. *Neurocomputing*, 181, 108–115.

Noor, M. H. M., Salcic, Z. & Wang, K. I. K. 2018. Ontology-based sensor fusion activity recognition. *Journal of Ambient Intelligence and Humanized Computing*.

R, G. & Uma, V. 2018. Ontology based knowledge representation technique, domain modeling languages and planners for robotic path planning: A survey. *ICT Express*, 4, 69–74.

Roy, P. C., Abidi, S. R. & Abidi, S. S. R. 2017. Possibilistic activity recognition with uncertain observations to support medication adherence in an assisted ambient living setting. *Knowledge-Based Systems*, 133, 156–173.

Salguero, A. G. & Espinilla, M. 2018. Ontology-based feature generation to improve accuracy of activity recognition in smart environments. *Computers & Electrical Engineering*, 68, 1–13.

Wu, L., Yang, L., Huang, Z., Wang, Y., Chai, Y., Peng, X. & Liu, Y. 2019. Inferring demographics from human trajectories and geographical context. *Computers, Environment and Urban Systems*, 77.

Yan, Z., Chakrabotry, D., Parent, C., Spaccapietra, S. & ABERER, K. 2013. Semantic trajectories. *ACM Transactions on Intelligent Systems and Technology*, 4.

Zhang, L., Meng, Q. & Fang Fwa, T. 2019. Big AIS data based spatial-temporal analyses of ship traffic in Singapore port waters. *Transportation Research Part E: Logistics and Transportation Review*, 129, 287–304.

Developments in Maritime Technology and Engineering – Guedes Soares & Santos (eds)
© *2021 Copyright the Author(s), ISBN 978-0-367-77376-2*

GIL breakdown fault location based on sound source recognition technology

J. Tang, Q.Q. Wang, Z. Wang & Y.N. Zhang
Naval Architecture and Marine Engineering, Tianjin University, Tianjin, China

Y.L. Li
China Electric Power Research Institute, Beijing, China

ABSTRACT: Existing Gas-Insulated Metal-enclosed Transmission Line(GIL) breakdown point positioning methods have poor anti-interference ability. In order to improve the accuracy of GIL breakdown point location, based on the characteristics of sound waves generated by GIL breakdown discharge, this paper proposes a precise location and simple method for locating acoustic faults. The arc breakdown site was regarded as the sound source, and the sound wave signals in the pipe gallery were collected. We use Time Difference of Arrival (TDOA) method to locate the arc. Using the linear microphone array and the TDOA positioning algorithm, a digital positioning system for the GIL breakdown point was built, and a simulation experiment of the location of the explosion sound source was performed. The experimental results show that the positioning error is less than 5%, which provides a new idea for the on-site positioning of GIL breakdown points. It has a good application prospect in mid-to-long distance transmission systems such as offshore nuclear power plants.

1 INTRODUCTION

Gas-insulated Metal-enclosed Transmission Line (GIL) is a new type of power transmission equipment with significant advantages such as large transmission capacity, small loss, high operational reliability, long life, and low environmental impact. It is especially suitable for power transmission in medium and long distance situations, such as offshore nuclear power transmission systems (Chen, et al 2009, Chen, et al 2014, Fan 2008, Koch & Hillers 2002, Koch 2012, Zhou, et al 2003). However, because the environment in which GIL is used in an underground gallery or subsea tunnel generally. Once an insulation defect occurs in the GIL, which causes an arc breakdown failure during GIL operation, it will be very difficult to troubleshoot and locate the fault point in an ultra-long transmission system (Li et al 2018). In order to improve the efficiency of UHV GIL troubleshooting, reduce maintenance costs, and ensure its long-term safe and stable operation, it is urgent to carry out research on UHV GIL arc breakdown fault location technology.

According to the characteristics of the vibration signal generated on the surface of the busbar shell when the GIL has a breakdown discharge, the vibration sensor is used to collect the vibration signal to achieve the location of the breakdown position (Hou, P.Y 2015). This method requires the sensor to

contact the GIL to resist interference difference and has poor anti-interference ability. Due to the short transmission distance of vibration signals, for GIL positioning, this method requires a large number of sensors which are expensive. Tang, B (2017) of the Mechanical and Electrical Engineering Bureau of China Yangtze River Three Gorges Group Corporation judges the breakdown position by detecting the radio frequency signal emitted by the discharge pulse. This method requires a monitoring system with a large dynamic range and an online automatic positioning function, and the detection cost is high. Sun, et al (2018). of Xi'an Jiaotong University proposed a two-dimensional positioning method of GIS (Gas-insulated combination switch) breakdown point based on spatial audible sound waves, which is simple to implement and accurate.

For GIL, when an arc breakdown occurs inside, with the breakdown of the insulating gas, a large amount of energy is released through the generation of photons and molecular vibrations, and the vibrations of gas molecules appear as sound waves on a macro scale. Based on the sound source localization system, this paper collects the sound wave signals in the pipe gallery by linear microphone array, and uses appropriate analysis algorithms to achieve the accurate location of fault points, which provides a new idea for GIL breakdown fault location. The method is simple and accurate.

DOI: 10.1201/9781003216582-31

2 SOUND SOURCE LOCALIZATION

2.1 Principle of fault sound localization

The sonic fault location method is based on the time difference location method at both ends. The schematic diagram is shown in Figure 1. A standard sound source is used to simulate the discharge sound of GIL. The wave velocity v of a sound wave signal when it propagates in a medium such as aluminum alloy and air is usually constant. In Figure 1, two acoustic signal microphones are placed at L1 and L2 where have some distance from the sound source point, and the time when the acoustic signal wave heads reach L1 and L2 is T1 and T2, then the time difference between the wave heads reaching the two microphones:

$$\Delta T = |T_1 - T_2| = \frac{|L_1 - L_2|}{v} = \frac{\Delta L}{v} \qquad (1)$$

Then we can get the distance between the discharge point and the two microphones:

$$\Delta L = \Delta T \cdot v \qquad (2)$$

If the sound wave signal reaches the NO.1-microphone first and then the NO.2-microphone, in the case of known L, we can get:

$$L_1 = \frac{L - \Delta T \cdot v}{2} \qquad (3)$$

$$L_2 = \frac{L + \Delta T \cdot v}{2} \qquad (4)$$

Similarly, if the sound wave signal reaches the NO.2-microphone microphone first and then the NO.1-microphone microphone, we can get:

$$L_1 = \frac{L + \Delta T \cdot v}{2} \qquad (5)$$

$$L_2 = \frac{L - \Delta T \cdot v}{2} \qquad (6)$$

The main working principle of the sound wave positioning system: first, the sound wave signal head is received by two sound wave microphone microphones, and the time when the sound wave signal wave head arrives is recorded. The exact position of the discharge point is calculated based on the position of the microphone microphone and the average speed of sound wave propagation.

2.2 Multi-microphone localization algorithm

For UHV GIL, because the length of the pipe gallery is too long, the reflection and attenuation of sound waves will occur due to obstacles during the transmission in the pipe gallery. Multi-microphone method can be used to locate the discharge point. The generalized cross-correlation Time Difference of Arrival (TDOA) method seeks the cross-power spectrum between two signals and gives a certain weight in the frequency domain, highlighting the relevant signal part and suppressing the part affected by noise(Bechler D, Kroschel K 2005). In order to make the peak of the correlation function at the time delay more prominent, and then inversely transform to the time domain to obtain the cross-correlation function between the two signals, and finally estimate the time delay between the two signals. This method has a relatively small amount of calculation, good real-time performance, low hardware cost, and is suitable for single sound source localization. Assuming that the two microphones are m_1 and m_2, and the received signals are $x_1(t)$ and $x_2(t)$, the cross-correlation function $R_{x_1x_2}(\tau)$ can be expressed as:

$$R_{x_1x_2}(\tau) = \int_0^{\pi} \psi_{12}(\omega)G_{x_1x_2}(\omega)e^{-j\omega t}d\omega = \int_0^{\pi} \phi_{x_1x_2}(\omega)e^{-j\omega t}d\omega \qquad (7)$$

In the above formula: $G_{x_1x_2}(\omega)$ is the cross power spectrum of the signals picked up by the microphones; m_1 and m_2; $\psi_{12}(\omega)$ is the weighting function; and $\phi_{x_1x_2}(\omega)$ is the generalized cross-correlation spectrum. We can get the result of delay:

$$\hat{\tau}_\phi = argmax(R_{x_1x_2}(\tau)) \qquad (8)$$

The data processing system receives a plurality of pulse signals at the front end of the acoustic wave, and marks the time when each pulse signal arrives

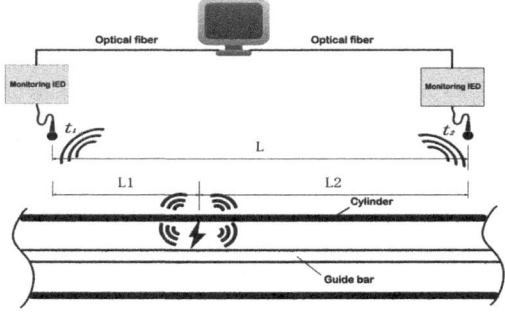

Figure 1. Schematic diagram of the time difference positioning method at both ends.

with absolute time. Analyze the signal waveform characteristics, find the two microphone front ends where the acoustic signal first arrives, calculate the time difference of signal arrival, and then further locate the exact location of the sound source point according to the improved TDOA method.

3 EXPERIMENT AND ANALYSIS

3.1 GIL fault acoustic localization test system

Each time a partial discharge occurs in the GIL, there will be a neutralization of positive and negative charges. At the same time, there will be a current pulse with a very short duration. The wave head time will generally not exceed a few nanosecond. At the same time, due to the SF6 gas insulation strength and breakdown field in the GIL The strength is very high. When a partial discharge occurs in a small area, the effect of the current pulse makes the local area where the partial discharge occurs instantaneously heat and expand, forming an effect similar to "explosion". The area returns to its original volume. This expansion and contraction volume change due to partial discharge causes an instantaneous change in the density of the medium, resulting in the Acoustic Emission (AE) phenomenon, which is transmitted from the partial discharge point to the surrounding air in a spherical wave manner, which can be Microphone detection (Wang Chengjiang 2007). Since the GIL breakdown discharge fault sound has the characteristics of a typical explosion pulse point sound source, this article uses a burst sound source to emit a pulse sound wave to simulate the GIL breakdown discharge fault sound for testing system experiments.

The experimental site is located in the traffic corridor of the Shisanling Pumped Storage Power Station, which is the passageway for the external traffic and transportation of the underground powerhouse of the power station. The length of the section is about 400m, and the environment of this traffic corridor is similar to that of Sutong GIL comprehensive corridor. In this paper, the straight section of the traffic corridor is used as the test section for the sound localization test of GIL discharge breakdown faults. The schematic diagram of the positioning experiment system is shown in Figure 2. The data acquisition and processing system and the experimental console are placed in the middle of the straight section of the traffic corridor. The solid circles indicate the microphone positions, and eight microphones form a linear array.

In the experiment, we used the NI-cDAQ9174 chassis and two NI-9234 acquisition cards to build a data acquisition system. Eight 130E22 microphones were fixed on the support, and the distance from the ground was 1.45m. After being connected to the data transmission cable, an acoustic signal was formed The sensor array and linear microphone array field layout is shown in Figure 3.

3.2 GIL fault sound localization experiment

In this paper, the explosion pulse sound source is used to simulate the breakdown sound of GIL discharge breakdown. We placed the microphone array at the center axis position of the traffic corridor to perform the sound source localization experiment. The system designed two sound source location experiments. The working system is shown in Figure 4. The solid circle indicates the microphone position, and the triangle indicates Sound source location.

Eight microphones are arranged at equal intervals of 20m to form a linear array with a length of 140m. Then we place it on the central axis of the pipe gallery and place the sound source at the midpoint between the central axis microphones as shown in Figure 4. After the explosion pulse sound is emitted, the sound signal is collected by the microphone array and transmitted to the data acquisition and processing system for data acquisition and analysis.

Figure 3. Site layout of linear microphone array.

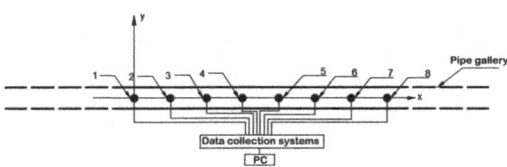

Figure 2. Schematic diagram of the sound field propagation characteristics test system of the experimental site.

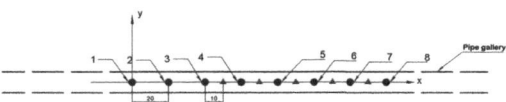

Figure 4. Schematic diagram of the experimental system of case 1.

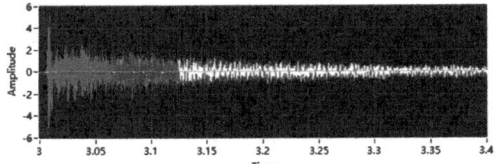

Figure 5. Explosion pulse sound waveform.

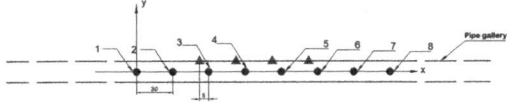

Figure 6. Schematic diagram of the experimental system of case 2.

A total of 5 experiments were performed in the case 1 test. The coordinates of the explosion sound source are shown in Figure 4. The explosion pulse sound waves collected by the microphone microphones in the microphone microphone array during the experiment are shown in Figure 5. As can be seen from Figure 5, the signal waveform has a typical shock response and high amplitude pulses, and the delay of each microphone sigal can be easily obtained.

The channel delay is calculated based on the data collected by each microphone. A set of data is recorded for each of the five experiments, and the calculation is performed separately. The sound source localization results calculated based on the data collected by each microphone are shown in Table 1.

In the case 1 experiment, each sound source and the microphone are located on the same central axis, which is far from both sides of the pipe gallery and the top wall surface, and more sound energy is transmitted along the central axis of the pipe gallery. In the high-pressure Sutong GIL integrated pipe gallery, GIL pipelines are arranged on both sides of the pipe gallery, close to the wall surface of the pipe gallery, and the wall surface of the pipe gallery is rough, and there are various interference sources such as various pipelines. In order to verify the effect of wall reflection on the sound field propagation and sound source localization system, the sound source localization experiment of the case 2 was performed. The case 2 experimental system is shown in Figure 6.

Eight microphones form a linear array with a length of 210m, and the microphone spacing is 30m, which is placed on the central axis of the pipe gallery. In Case 2 experiment, the sound source is located between the two microphones and is close to the wall surface of the pipe gallery on one side. The positional relationship between the sound source and the microphone is: the

distance from the right microphone to the X axis is 5m, and the left microphone. The distance in the X-axis direction is 25m. The distance between the sound source and the center axis of the microphone in the Y-axis direction is 3m, and it is close to the wall surface of the pipe gallery. A total of 4 experiments were performed in the test of case 2. The channel delay was calculated based on the data collected by each microphone. Each experiment recorded one set of data and performed calculations. Since the Y coordinate of the GIL pipeline is known in practical applications, Therefore, the main purpose of the experiment is to locate the X-axis coordinate position of the sound source, so the Y-axis coordinate of the sound source is not output during calculation. The sound source localization results calculated according to the data collected by each microphone are shown in Table 2.

3.3 Experimental result and analysis

It can be seen from the experimental results of case 1 and 2 that the system can locate the sound source more accurately. Even when the sound source is close to the wall, the sound source coordinates can be accurately located under the influence of wall reflection and scattering.

According to Table 1, when the sound source is located in the middle of microphones NO.3-7, the location of the sound source is still very accurate in the case 1. It proves that the method adopted by the system can effectively carry out the sound source. Positioning. However, when the sound source is located between the No. 7 and No. 8 microphones, a large error occurs in the measurement results. The same experimental phenomenon also exists in the case 2 experiment. In order to verify whether the error appears to be an episodic phenomenon, in the experiment of case 1, four repeated tests were

Table 1. Sound source localization results in Case 1.

Serial number	Actual coordinates (m)	Measured coordinates (m)	Error (m)
1	(50, 0)	(51, 0)	1
2	(70, 0)	(70, 0)	0
3	(90, 0)	(89, 0)	1
4	(110, 0)	(109, 0)	1
5	(130, 0)	(126, 0)	4

Table 2. Sound source localization results in Case 2.

Serial number	Actual coordinates (m)	Measured coordinates (m)	Error (m)
1	(55, 3)	(60, 3)	5
2	(85, 3)	(85, 3)	0
3	(115, 3)	(115, 3)	0
4	(145, 3)	(141, 3)	4

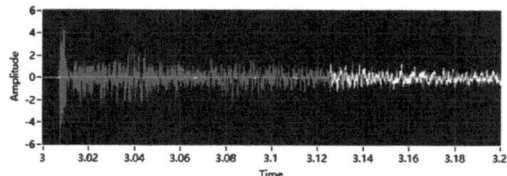

Figure 7. Comparison of microphone acquisition results.

performed between microphone NO.7 and microphone NO.8. The measured sound source coordinates were (126,0), (123, 0), (125,0), (127,0). It proves that this error phenomenon is not a sporadic error in the system, and then we analyse its cause.

Analyse the test data of the 5th group of the case 1 experiment. We select the time-domain waveform collected by the microphone No. 8 closest to the sound source and the microphone No.1 farthest from the sound source, as shown in Figure 7. The purple signal in the figure is the signal collected by the microphone No. 8 and the white signal is collected by microphone 1.

As can be seen in Figure 7, when the blue sound is transmitted to the microphone No. 1, the signal amplitude has been greatly attenuated. The signal envelope energy method is used to calculate the signal strength. The obtained results show that the microphone No. 8 has a sound pressure attenuation compared to the microphone No. 1. More than 20dB.

4 CONCLUSION

In this paper, a linear microphone array is used, according to the characteristics of the source of the GIL breakdown discharge fault. We use the TDOA (Time Delay Estimation) sound source localization algorithm to locate the specific location of the arc breakdown sound, and then determine the location of the arc breakdown failure.

At the same time, verification experiments carried out in the traffic gallery of the Shisanling Storage Power Plant show that the system can accurately locate the sound source location with an average positioning error of less than 5%. The poor positioning effect of the linear array edge area is due to the lack of sufficient microphones for calibration. Therefore, in practical applications, it is necessary to calibrate the effective range of the microphone array.

In the past 20 years, nuclear power plants have emerged with their huge advantages, and offshore floating nuclear power plants have also received extensive attention from the industry. For example, the world's first offshore nuclear power plant "Lomonosov" built by Russia has been put into use. Various countries have also carried out research on offshore nuclear power plants, and it can be foreseen that GIL technology will be well applied in offshore power transmission systems in the future. In this paper, the sound source localization system designed for GIL technology is compared with the traditional GIL breakdown fault location system. It has a simple layout, high applicability, and good economy. At the same time, there is no need to contact the transmission line, which avoids the strong electrical interference to the sensor system during GIL breakdown. It has considerable application prospects in the maintenance and inspection of long-distance GIL transmission systems.

REFERENCES

Bechler, D & Kroschel, K (2005). Reliability criteria evaluation for TDOA estimates in a variety of real environments[C]//IEEE International Conference on Acoustics. IEEE.

Chen, X.S, et al (2009). Prospect of high voltage long distance compressed-air insulated transmission lines[J]. High Voltage Engineering, 35(12):3137–3142.

Chen Min, et al (2014). Technology Research and Field Applications of GIS Equipment Breakdown Discharge Location[J]. High Voltage Apparatus, 50(06):81–90.

Fan, J.B(2008). Gas insulated metal enclosed transmission line (GIL) and its application[J]. Electric Power, 41 (8):38–43.

Hou, P.Y, et al (2015). Research on Location Method of GIL Breakdown Discharge Based on Vibration Method [J]. Shanghai Electric Technology, (04): 33–37 + 55.

Koch, H & Hillers, T (2002). Second generation gas-insulated line[J]. Power Engineering Journal, (3):111–116.

Koch, H (2012). Gas insulated transmission lines (GIL) [M]. Chichester, United Kingdom: Wiley-IEEE Press, 33–38.

Li, Y.L, et al (2018). Simulation study of acoustic propagation characteristics in Suzhou-Nantong GIL multi-utility tunnel[J]. Technical Acoustics, 37(04):297–302.

Sun, Z.M, et al (2018). Location Method of Breakdown Points in GIS Based on Audible Sound Waves[J]. Journal of Xi'an Jiaotong University, 52(10):88–94+109.

Tang, B (2017). Online Monitoring System for Partial Discharge in 500 kV GIS/GIL in Xiluodu Hydropower Station[J]. Hydropower and New Energy, (02): 14–16 +37.

Wang, C.J, et al (2007). Experiment of Acoustic Emission in Discharge of Transmission Lines[J]. Automation of Electric Power Systems, (19): 80–84.

Zhou, H, et al (2003). Insulation Characteristics of N_2/ SF_6 Gas Mixtures[J]. High Voltage Apparatus, (05): 13–15+19

A risk analysis of autonomous vessels in complex urban waterways

O.A. Valdez Banda & P. Kujala
Research Group on Maritime Risk and Safety, Department of Mechanical Engineering (Marine Technology), Aalto University, Aalto, Finland

Y. Sapsathiarn, O. Mokkhavas & W. Punurai
Faculty of Engineering, Department of Civil and Environmental Engineering, Mahidol University, Phuttamonthon Nakhon Pathom, Thailand

C. Suvanjumrat & J. Priyadumkol
Faculty of Engineering, Department of Mechanical Engineering, Mahidol University, Phuttamonthon Nakhon Pathom, Thailand

ABSTRACT: Autonomous vessels have become one potential alternative for the concept of urban mobility. In complex urban waterways, autonomous vessels could assist in the transporting of goods and people. The recent development of technologies enabling autonomous systems has supported the initial design of new autonomous vessels concepts. However, before these concepts are further designed and operated, it is necessary to analyze their risks to develop strategies for ensuring the safety of people and the protection of the natural environment. In this study, an initial analysis is elaborated to assess the application of autonomous vessels in Chao Phraya River in Bangkok and Larn island in eastern Thailand. The aim is to provide an initial analysis of the risks of autonomous vessels by developing an initial overview of the application of the autonomous vessels in the mentioned operational context. The capabilities of the proposed process to analyze these risks are also evaluated.

1 INTRODUCTION

The concepts of autonomous shipping and smart mobility are of high level of importance in the development of a more sustainable urban transport (Levander 2016). The recent progress in the development of technologies for developing autonomous systems has supported the planning and initial design of new concepts of autonomous vessels for urban mobility (Tannum and Ulvensoen 2019). In addition, significant efforts are currently made by ship manufacturers and technology developers for making an efficient integration of the components needed to have an autonomous vessel ready for operation (Yara 2019 and DNV GL 2019).

One critical aspect to further develop the design of new concepts and their potential sub sequential implementation is the analysis and management of the risks of the vessel operation (Valdez Banda et al. 2019). This type of analysis must support the design and operation of the autonomous vessel and the entire ecosystem where it operates. The prioritization on the analysis of the vessel and its operational ecosystem is critical because the efficiency of a smart vessel depends on efficiency design of the smart environment (operational context) (Renn 2016).

This study presents an preliminary analysis to assess the potential application of autonomous vessels in Chao Phraya River in Bangkok and Larn Island in eastern Thailand. The process for analysis is based on the process presented in Valdez Banda et al. (2019). The process aims at providing an initial coherent, transparent, and traceable safety input information for analyzing the potential implementation of autonomous vessels in complex urban waterways.

The initial results of this analysis provide an informative but yet realistic evaluation of the application of the autonomous vessels in the mentioned context. Moreover, the results provide a representation of the process capabilities to initialize the analysis and management of the risks of autonomous vessels in complex urban waterways.

2 RESEARCH BACKGROUND AND DATA

2.1 Operational context (Chao Phraya River)

The Chao Phraya River begins at the confluence of the Ping and Nan rivers at Nakhon Sawan in Nakhon Sawan Province. It flows south for 372 kilometres (231 mi) from the central plains to Bangkok and the Gulf of

DOI: 10.1201/9781003216582-32

Thailand. In Chai Nat, the river then splits into the main course and the Tha Chin River, which then flows parallel to the main river and exits in the Gulf of Thailand about 35 kilometres west of Bangkok in Samut Sakhon. In the low alluvial plain which begins below the Chainat Dam, there are many small canals which split off from the main river (McCarthy 2005). These canals interact among both rivers and allow water, and in some cases, urban traffic in Bangkok area.

In Bangkok area, several ferry lines transport people every day along 31 kilometers of the Chao Phraya River. The river is one of the existing alternatives to the oversaturated urban transport system of the city. Urban water transportation in the river has had some issues with finding subsidy and heavy government regulations that have delayed the modernization of the system (Tanko and Burke 2016). Current efforts are been allocated in modernizing the ferry piers (smart piers) similar to the existing train (Skytrain) system by BTS. The modernization of the system is essential as every day more than 100 000 passengers move along the river (Statistics by MD, 2019a).

In the context of the Chao Phraya river, this study focuses on the analysis of the river crossing between Siriraj Piyamaharajkarun Hospital and Thammasat University, Tha Phra Chan Campus (Tha Prachan - Wang Lang Pier). Figure 1 presents the geographical location of the area under analysis. In this area, more than 5000 people cross the river everyday (Statistics by MD, 2019a). Table 1 presents traffic statistics between Tha Prachan - Wang Lang Pier.

The pier at Thammasart University is composed of two steel docks (size = 7.00*11.00 m and

Table 1. Passenger and service statistics, ship dimensions on the operation between Tha Prachan - Wang Lang Pier in 2019 (MD, 2019a).

Context element	Description
Total number of passengers	1,947,523
Total number of boat trips	31,334
Average daily passengers	5,321
-Week	2,542
-Weekend	6,263
Average daily trips	86
-Week	88
-Weekend	81
Total ships in service	5
-Week	2
-Weekend	2
Fares	
-General	3.5 Bahts
-Elder	2
-Children (under 90cm)	Free
-Bicycle	6.5
Ship (ref. No SP.69)	
-Width	4.5 m
-Length	16 m
-Depth	1.2 m

4.00*11.00 m) with two steel bridges (size = 1.32*9.00 m and 1.54*9.00 m). The pier at Siriraj hospital has two steel docks (size = 7.00*11.00 m both) with two steel bridges (size = 1.59*7.40 m and 1.51*7.40 m) (MD, 2019a).

2.2 Operational context (Larn island)

Larn Island (Koh Larn) is a small island in the eastern part of Thailand. There are 7.5 kilometers from Pattaya shore that is a famous place for sea-activities such as swimming, diving, sailing and other recreational activities. It is approximately 120 kilometers away from Bangkok. Figure 2 presents the Larn Island location. Passengers have to transit to ferry or speed boat at Balihai pier from Pattaya. There are two major piers; Naban pier and Tawaen beach pier with

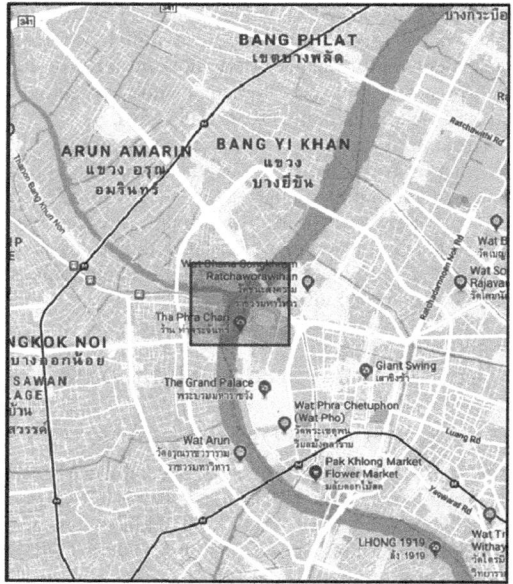

Figure 1. Graphical location of the are under analysis in Chao Phraya River (Tha Prachan - Wang Lang Pier).

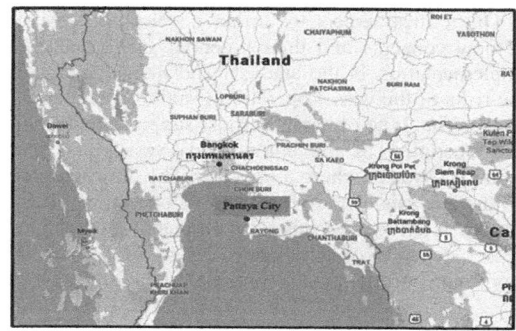

Figure 2. Larn Island geographical location.

another one floating pier at Samae beach as shown in Figure 3. Naban pier is a traditional pier for passenger and multi-purpose cargo which is focusing on this research. There are 5.84 million passengers in 2018 at Larn Island while 2.36 million passengers are transferred to/from Naban beach pier or estimate 40 percent of the whole passenger there (MD, 2019b).

The shortest distance for sea transportation to Larn island is Naban pier – Balihai pier route which approximately 8.4 kilometers. The peak period is from January to May with around 27,400 persons per day moving in this route. The average passenger in 2018 was 16,000 passenger per day, approximately 378 voyages (MD, 2019b).

The overall length of Naban pier is 180 meters including with 40 meters of general-purpose yard. There is an open passenger waiting area and small parking at the hinterland. The pier has laid into the sea as "I" platform. The sea depth at the end of the pier is about 3.5 meters. The water depth and tourists demand are critical aspects for the coordination of traffic.

Two main types of vessels (ferry and speed boat) are utilized for passenger transport from and to the island. Ferries are operated for daily transportation with a fixed schedule (see Table 2). The service time from Balihai pier to Larn pier is 40-45 minutes depending on weather condition. Mostly, the ferries length are between 15-30 meters, width 4-10 meters and the draft is not over 3 meter. The passenger capacity is between 100-250 passengers.

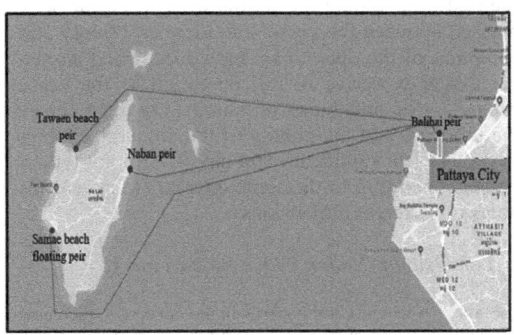

Figure 3. Naban pier and Tawaen beach piers location.

Table 2. Ferry service timetable at Larn Island.

Route	Service time AM	Service Time PM
Balihai pier (Pattaya)	7, 10:00	12, 14, 15:30, 17, 18:00
– Naban pier (Larn island) Naban pier (Larn island)		
– Balihai pier (Pattaya)	6:30, 7:30, 9:30	12, 14, 15:30, 17, 18:00

2.3 Hazards and accident information

In the Context of Chao Phraya River, information about accidents and previous safety analysis are not available and therefore not included in this analysis. However, as the context of analysis is narrowed down to the potential autonomous operation of the vessel crossing from Siriraj Piyamaharajkarun Hospital to Thammasat University, Tha Phra Chan Campus, the information utilized consists of available traffic statistics in (MD, 2019b), site visits for the analysis of the operational context, interviews with maritime authorities and the analysis presented in Valdez Banda and Kannos (2018).

In the Context of Larn Island, The Nation 2019 the most relevant risks reported are:

– The falling passengers' belongings into the sea.
– Injury onboard ferry from sea condition.
– Injury onboard speed boat from the heavy waves attached.
– Injury from the ferry transfer.
– Sinking passenger ferry from flooding
– Fire onboard the ferry
– Collision with fixed objects (rocks and buoys)

This information is complemented with the analysis to the concept of an autonomous ferry as presented in Valdez Banda and Kannos 2018. The environmental context described in Section 2.2 (DASTA 2015) is considered to adapt the described scenarios to the mentioned context.

3 METHODOLOGY

3.1 The process of analysis

The process is adapted from the original process presented in Valdez Banda et al. 2019. The process follows four steps:

– Step one (Hazard identification). It focuses on the collection of information about aspects and contextual scenarios (incl. the causes and sources of risk) which can negatively influence the safety of the system(s).
– Step two (Detailed hazard description). This step focuses on the elaboration of the detailed descriptions of the effects of the hazards, providing a comprehensive argumentation about the relevancy of the hazards and an estimation of their potential severity and type of consequences.
– Step three (Risk Control Options). The step focuses on defining of risk mitigation actions. These actions represent the initial specifications of the safety controls. Controls are set to eliminate the risk source (preventive safety controls) or mitigate the possible consequences (reactive safety controls).
– Step 4 (Identification and analysis of Unsafe control actions "UCAs") to each safety control UCAs are identified. The UCAs represent

actions that could lead to a hazardous state in the system. Hazardous states result from inadequate controls or enforcement of the safety control (Leveson, 2011). The identification of UCAs continues with a definition of how these can occur in the design and implementation of those actions. With this definition, safety specifications are provided to manage and prevent the UCAs.

As the vessels and system operations in both study cases are currently entirely man depended, the method considers the process for transferring the human knowledge towards automation and the potential interaction between new autonomous vessels concepts and the vessels that are entirely human operated. This is done by analyzing the human role in the operational context with site visits and video recording.

4 RESULTS

4.1 Accidents and hazards

Chao Phraya River (Tha Prachan - Wang Lang Pier)

Table 3. Accidents and hazards in the context of the vessel operating between Tha Prachan - Wang Lang Pier in 2019.

Accident	Hazard
Collision with pier	Object detection sensor error
	Software incorrect design
	Mechanical failure
	Heavy weather/river/ conditions
	Strong currents
	Position reference equipment failure
Collision with moving object (e.g. another vessel)	Object detection sensor error
	Software incorrect design
	Technical failure
Collision with a fixed object (e.g. buoys, beacons, etc.)	Object detection sensor error
	Software incorrect design
	Mechanical failure
	Heavy weather/river/ conditions
	Strong currents
	Position reference equipment failure

(Continued)

Table 3. (Cont.)

Accident	Hazard
Capsizing	Overloading of the vessel
	Shifting of weights
	Flooding
Fire onboard	Ignition of electrical equipment
	Generated by a passenger
Passenger over board	Unintended falling
	Intended jumping
Emergency on board and/or the pier	Person being injured
	Medical emergency

During the analysis of the context in site visits and video recording fundamental issues are detected. High traffic complexity, the context has not specific "rules of the road", so the operation of boats on this area is entirely dependent on the judgment and expertise of the vessel operators. Operational time, vessel speed operations vary during the entire journey. The average of trips per day is an important indicator of the expected performance of the vessel and crew on-board.

A docking operation to get on/off the vessel takes around 3-4 minutes (for a vessel with capacity of 90 passengers max.). Two persons on board and 1-3 persons on the pier control the docking, embarking and disembarking. There are not physical safety barriers focused on avoiding the falling of a person or being pressed between the vessel and the pier. Energy consumption of the operations, high speeds and passenger demands derive in full operation of the vessel capacity. Air and noise pollution is evident during the journeys and the quality of the air. Traffic diversity, there is a significant amount of different type of vessels operating in the same context (ferry and fast boats of different dimensions).

Larn Island (Naban peir – Balihai pier)

Table 4. Accidents and hazards in the context of the vessel operating between Naban peir – Balihai pier at (Pattaya-Larn Island).

Accident	Hazard
All those listed in Table 3	
Grounding	Software incorrect design
	Mechanical failure
	Position reference equipment failure
	Heavy weather/sea condition
	Strong currents

As mentioned in the report by The nation (2019), injuries happen as a combination of sea conditions and operational speeds. The process for docking in Chao Phraya river lacks of safety barriers to avoid that a person fall into the water or gets pressed between the vessel and the piers.

4.2 Detailed hazard description

This step provides information about the potential causal factors of the hazards, the definition of initial mitigation actions, an initial estimation of the difficulty and cost for their implementation, and the definition of the initial mitigation actions. This detailed hazard is taken from the analysis done in Valdez Banda and Kannos 2018. Table 5 presents an example of these detailed descriptions.

4.3 Risk control options

Table 6 presents the Risk Control Options for the Hazard (Object detection sensor error). The entire list of risk control options of the hazards listed in Section 4.1.

4.4 Unsafe Control Actions (UCAs)

Table 7 presents the detected UCAs for the Risk Control Option (RCO) Sensor system redundancy and diversity. The UCAs of the listed RCOs for all hazards is presented in Valdez Banda et al. 2018.

Table 5. Detailed description and initial mitigation actions for hazard H1 (Object detection sensor error (Valdez and Kannos 2018).

Hazard	Object detection sensor error

Hazard effect/description
In case of object detection sensor error, the information about objects around the vessel is not reliable and thus the vessel may not be able to navigate safely and avoid collisions with moving objects according to the rules of the roador collisions with fixed objects. This hazard may not affect the ship operation significantly in most cases, but in amore severe scenario, the hazard can have a negative impact on people, property, and environment. It can result ininjuries, the loss of human life, severe damage or loss of property (own and others property) and environmentaleffects such as oil spills or other damage of a sensitive waterway or sea area.

Causal factors
-Loss of power
-Equipment malfunction
-Dirt
-Heavy rain
-Overheating
-Equipment interference
-Inappropriate maintenance
-Incorrect sensor set and/or positioning of the sensors
-Targets impossible to detect
-Corrupted readings
-Complete equipment failure

Table 6. The detailed description and initial mitigation actions for hazard H1 (Object detection sensor error (Valdez and Kannos 2018).

Hazard	Object detection sensor error

Risk Control Options (RCOs)
-Sensor system redundancy and diversity
-UPS (Uninterrupted Power Source)
-Appropriate cooling and cleaning systems
-Thorough commissioning of equipment set
-Appropriate and continuous maintenance program
-Continuing system diagnosis and proof testing
-Autonomous Integrity monitoring

Table 7. Unsafe Control Actions (UCAs) of the RCO "Sensor system redundancy and diversity" (Valdez and Kannos 2018).

RCO	Sensor system redundancy and diversity

Unsafe Control Actions (UCAs)
UCA 1. Sensor does not function properly and there is no other sensor available
Potential causes
-Lack of economic resources
UCA 2. Equipment chosen to provide the redundancy are not suitable
Potential causes
-Lack of economic resources
-Lack of knowledge of sensors characteristics when planning the equipment set needed
UCA 3. Sensor failure is not detected
Potential causes
-Not enough coverage with the diagnosis
UCA 4. External or common cause failures takes several equipment down at the same time
Potential causes
-Inappropriate system design
-Incorrect installation
-Incorrect usage
-Environmental conditions

5 DISCUSSION

5.1 The process application and results

The implementation of the process focuses on the generation of initial risk management information that can guide the design and planning of the operation of the autonomous vessel concepts.

Step one defines the main accidents that may result in damages and injuries during the operations of the autonomous vessel and its entire operational system. The hazards represent the obvious initial states of the system which endanger the mission and operation of the vessels.

Step two incorporates a justification of why the hazard analysis is relevant and the initial estimation of its severity and its consequences. Moreover, potential causal factors are also identified and analyzed in this step. These provide a systemic representation of the causes linked to different components included in the design and operation of the vessels and the entire operational ecosystem.

Step three provides specific alternatives with valuable information for further development of safety controls and the analysis of the operational context and functionality of the vessels and other components in the operational environment.

Step four goes deeper into understanding how the vessels and the ecosystem can still drift into a risky status and potentially trigger an accident.

The site visits and recorded videos provide critical information and realistic perception of the level of complexity of the current operation and the evident challenge in the inclusion of autonomous vessels in the two selected operational contexts. The complexity linked to the functionality of a muddled water traffic system in Chao Phraya River demands a strategic plan to incorporate the autonomous vessel and the other components of its smart operational ecosystem.

Clear strategic specifications for planning the design of this smart ecosystem is essential. This includes the autonomous vessels and their interaction with other manned vessels. In discussions with experts of urban waterway and sea operation in Thailand, the experts mentioned the need for a gradual inclusion of the smart vessels and systems into the operational context. Due to the system complexity, the planning and making an entire disruption the operational context seems to be a non-viable alternative. As mentioned by the experts, there is a need for (gradually) evidence of the improvement of waterway traffic and the benefits obtained by the incursion of these modern systems. This evidence allows confirming the success of the planning, design and implementation of autonomous vessels and understanding the process to successfully updated waterway operations and urban waterway mobility.

The gradual transformation of urban waterway mobility in Thailand has started with the planning of smart piers in Chao Phraya River. These aim to support the process for embarking and disembarking of the vessels and provide specific safety controllers that prevent accidents in the pier. The information on the analysis of the accident "Emergency on the pier" provides useful information to support and guide the design of the pier.

Finally, the potential incursion of autonomous vessels in the specific operational context introduced in this study can significantly support and boost the development of a more sustainable ecosystem for urban waterways. In the next years, the strategic tasks for developing such ecosystem will entirely influence the development of the maritime traffic and maritime industry in Thailand and the rest of the world (The Nation, 2019). Every country will develop a specific approach based on traffic demands and the operational characteristics linked to it (ferry and fast boats of different dimensions).

The overall and general analysis presented in this study represent only an initial reference for the development of this analysis. The study is limited to the scarce information of accidents and incidents, the initial discussions with experts, video recording and site visits of the operational context and the information obtained in the study by Valdez Banda and Kannos (2018). This represents the need for continuing/completing the risk analysis in the initial planning phase of the specific vessels and service concepts.

6 CONCLUSIONS

This study presents an initial risk analysis process for the planning of new concepts of autonomous vessels for urban mobility in two specific operational contexts of urban mobility. The process supports the generation of specific information to guide the design of the new concepts and evaluate the critical aspects for the inclusion of smart vessels in complex urban waterways. The provided controls and critical aspects detected in the analysis can represent the basis of the initial risk management strategy of the autonomous vessels and its operating ecosystem.

The implementation of the process represents an initial useful alternative for analyzing hazards and proposing safety controls with a systemic approach that covers the operational context of the autonomous vessel. The process should continue with the extraction of relevant information to plan, design and construct the autonomous vessel and its entire operational system.

This study and the potential extension of the application of the presented process must be complemented with the incursion of new data such as accident data and statistics of the selected operational context and the discussion of the concepts with the stakeholders that are responsible for the management of safety of the vessels and their operational ecosystem.

ACKNOWLEDGMENTS

The work presented in this study was carried out as part of the "Reliability and Safety Engineering and Technology for large maritime systems" (RESET). RESET is partially supported by the European Union's Horizon 2020 Research and Innovation Programme RISE under grant agreement no. 730888 (RESET).

REFERENCES

Designated Areas for Sustainable Tourism Administratio (DASTA), 2015. The Study and design of port development project for Koh Larn tourism (Final report).

DNV GL, 2019. The Revolt concept. A new inspirational ship concept. Available at. https://www.dnvgl.com/technology-innovation/revolt/index.html.

Levander O., 2016. Ship intelligence – a new era in shipping. Smart Sh. Technol., London. RINA; 2016, p. 25–32.

Leveson, N. 2011. Engineering a Safer World: Systems Thinking Applied to Safety. MIT Press.

McCarthy J., 2005. From Bangkok to Korat – Elephants". Surveying and exploring in Siam.

MD (Marine Department of Thailand, Bangkok office). 2019a. Summary report of ferry motorbike passanger in Chao Phraya River. Report No. 2562.

MD (Marine Department of Thailand, Pattaya office). 2019b. The statistics report of Pattaya passenger to Larn Island.

Renn O., 2016. Keynote lecture in the 4th European STAMP Workshop and Conference 2016 in Zurich.

Tanko M. and Burke M.I., 2016. Transport innovations and their effect on cities: the emergence ofurban linear ferries worldwide. World Conference on Transport Research - WCTR 2016 Shanghai. 10-15 July 2016.

Tannum M.S. and Ulvensoen J.H., 2019. Urban mobility at sea and on waterways in Norway. Journal of Physics: Conference Series Vol 1357 P. 012018, October 2019.

The Nation 2019. Ferries charge up smog battle. Article from 13.02.2019. Available at https://www.nationthailand.com/Corporate/30364085.

Valdez Banda, O.A. and Kannos, S. 2018. Hazards analysis process for autonomous vessels. ÄLYVESI (Samart City Ferries) project report. July 2018 (https://www.aboamare.fi/Results).

Valdez Banda O.A., Kannos S., Goerlandt, F., Van Gelder P.H.A.J.M., Bergström M., Kujala P. 2019. A systemic hazard analysis and management process for the concept design phase of an autonomous vessel. Reliability Engineering & System Safety, Volume 191, November 2019, 106584.

YARA, 2019. YARA Birkeland (unmanned vessel concept) project. Available at https://www.ship-technology.com/projects/yara-birkeland-autonomous-container-vessel/.

Ship design

Developments in Maritime Technology and Engineering – Guedes Soares & Santos (eds)
© 2021 Copyright the Author(s), ISBN 978-0-367-77376-2

RIM driven propellers design using a simulation based design optimization approach

S. Gaggero
Department of Electrical, Electronic, Telecommunications Engineering and Naval Architecture University of Genoa, Genoa, Italy

ABSTRACT: A Simulation Based Design Optimization approach is proposed for the design of RIM driven propellers operating in an accelerating duct. The tool relies on RANS analyses of parametrically described geometries driven by an automatic, multi-objective optimization loop for the design of propellers with improved performances simultaneously in terms of both propulsive efficiency and cavitation inception. Pareto convergence is achieved by using a mix of fully resolved RANS analyses and surrogate models aimed at significantly improving the computational efficiency of the procedure. The effectiveness of the design approach is verified by comparing the devised RIM propellers with the performance of a reference ducted propeller at the same functioning point while design trends and guidelines are extracted from the analysis of the amount of data collected during the optimization process.

1 INTRODUCTION

RIM driven thrusters represent one of the most innovative propulsive concepts of the last years, even if the idea of driving a propeller via its blade tips dates back to the work of Saunders (1965). Only recently, however, thanks to the availability of materials and innovative manufacturing technologies (such as brushless permanent magnet electric motors (Hughes et al. 2000)) the functional implementation of this kind of propulsors was possible. The attention to the underwater radiated noise without sacrificing high level of efficiency encourages the use of unconventional propulsor systems. Increasing the cavitation-free speed most of the times results in a reduction of the propulsive efficiency and a consequent increase in fuel burn, which is completely against current ambitions to reduce emissions of GHG, NOx, SOx and particulate matter. Solutions capable of ensuring simultaneous improvements of efficiency and radiated noise levels, like ducted propellers and the employment of Energy Saving Devices are, then, preferred and, among the others, RIM driven propellers try to comply with these needs. A RIM driven thrust, in fact, is a propeller inside a nozzle, driven from the tip of the blades through a rotating ring (namely the rim) embedded into the nozzle. From the hydrodynamic point of view, then, a RIM propeller is very similar to a ducted propulsor, but contrary to usual ducted configurations, there is also no gap between the propeller tip and the inner surface of the nozzle. This means that local energy losses can be minimized by the almost vanishing of both the characteristic tip vortex form blade tip and the tip leakage vortex generated in the gap of ducted configurations (Baltazar et al. 2012). The possible increase of efficiency (which however could be affected by the additional torque associated to the frictional forces acting on the tip ring), then, can simultaneously supports the reduction of the radiated noise and the resulting vibrations normally associated to tip vortexes.

However, their hydrodynamic design, also due to the scarce literature concerning this type of propulsive solution, is challenging with respect to largely studied conventional propellers. The most important hydrodynamic issues related to RIM propellers regards the blade loading and the rim effect associated to the flow between the rotor and the stator of the electric motor embedded directly into the rim (i.e. the rotor) and into the nozzle (i.e. the stator).

Without any clearance between the inner surface of the nozzle and the blade tip, the radial circulation, which is zero at the tip for conventional propellers, has a finite value and, under certain conditions of pitch, chord and camber distributions, could reach its maximum (Cao et al. 2012). Zero circulation, on the contrary, characterizes the blade root, increasing the risk of stronger root vortexes. The possibility to increase (or to maintain sufficiently high) the load in the outer part of the blade is beneficial from the point of view of the propulsive efficiency. Moreover, with the gap between the blade and the inner surface of the nozzle sealed, the risk of tip vortexes is significantly

DOI: 10.1201/9781003216582-33

reduced (and the leakage avoided) also in case of high loads, preventing the development of noisy (cavitating) tip vortexes. The hydrodynamic design of the blade, however, has to consider that the higher load could lead to excessive values of suction pressure and a higher risk of bubble cavitation, which from the noise point of view is as harmful as the cavitating tip vortex, and that traditional potential flow based design approaches may suffer from too crude approximations especially in the case of unusual configurations and severe working conditions.

The first attempts to design RIM driven propellers having in mind these opportunities, which easily can turn into new design constraints, were those of Sparenberg (1969) and Mishkevich (2000), which used a modified lightly and moderately loaded lifting line theory to characterize both the blade and the nozzle loading distributions. Yakovlev et al. (2011) proposed an even simplified method using an hydrodynamic model consisting in 2D calculations for a lattice of hydrofoils and considering the pitch distribution as the only design parameter of the approach.

Only recently RANS calculations were directly employed to predict the hydrodynamic performance of whole propulsor system (Cao et al. 2012). Their use for design purposes, on the other hand, would address in a more consistent way the peculiarities of this kind of propulsors and would allow to exploit their potentialities at their best for performance maximization and noise reduction. Few examples were proposed in literature concerning with the use of RANS in a design procedure, which for the nature of RANS calculations themselves, obviously refer to the optimization paradigm. Gaggero (2019) and Gaggero (2020) developed a Simulation Based Design Optimization (SBDO) approach entirely based on RANS calculations for the design of RIM driven propellers operating in accelerating and in decelerating ducts. In both cases the devised optimal RIM configurations shown superior performance with respect to the reference conventional ducted propellers thanks to comparable values of efficiency and significantly improved cavitation inception speeds. The SBDO for RIM propeller design was similar to those already developed for conventional and unconventional propulsive solutions (Gaggero 2018, Gaggero et al. 2017, Gaggero et al. 2016) but suffered from the intrinsic low computational efficiency of RANS analyses which allowed for accurate and effective characterization of the propeller performance but made the design process hardly exploitable in the preliminary design stage.

Having this in mind, in present work an updated SBDO for the design of RIM propellers is proposed. The design process is still based on the paradigm of optimization, i.e. on the systematic analysis of hundreds (thousands) of different configurations, iteratively modified based on how their performance compare with the design objectives and constraints. Rather than by using fully resolved RANS analyses, the process is carried out taking advantage of surrogate models, which can be easily seen as response functions fitted to basic data obtained by evaluating the objectives and the constraints selected for the design of a certain number of configurations, adequately sampled over the design space. The surrogate model replaces the RANS solver in the design loop, allowing for very fast and efficient evaluations of the quantities of interest throughout the process. RANS calculations are then needed only for the characterization of the initial sampled configurations and for the final verification of the optimized candidates, which in turn can be used to re-train the surrogate models and iteratively enhance the design process.

2 SIMULATION BASED DESIGN OPTIMIZATION

In the last decades, the design of conventional and non-conventional marine propellers has significantly changed since secondary effects, like pressure pulses and radiated noise, are becoming the most important constraints in order to comply with comfort requirements and even stricter environmental regulations.

Traditional design tools, in this context, show all their limitations but simultaneously, the propeller design process, similarly to what happened for many complex engineering systems, has been substantially modified also by the availability of new simulations tools which can comply with at least some of these recent design requirements. The increased robustness and accuracy of numerical algorithms together with the exponential development of hardware resources foster the application of simulation methodologies capable of reducing or even replacing the experimental load. In fact they provide, already in an early design stage, most of the data necessary for the assessment of the additional design criteria and objectives to address the growth of technical and commercial requirements.

Simulation-Based Design (SBD) frameworks using medium- and high-fidelity tools have, consequently, become a standard also for propellers design. Obviously, the highest fidelity at the lowest possible cost is essential but not trivial. Accurate design analysis methods are computationally expensive but necessary to provide the flexibility and the validity needed in case of novel concepts and unconventional configurations which, instead, represent the application limits of computationally efficient, low-order, approaches. Establishing the performance of a design, or the relative merits of trade-off alternatives, when a decision about a design path has to be faced, is crucial in the constant search of improvements imposed by an increased market competition. Then, larger design spaces and more alternatives have to be explored and compared, each time by using higher fidelity simulations and complex geometrical representations, to allow for potentially bigger improvements and to avoid as much as possible designs optima falsified by the insufficient accuracy or by the inherent limitations of the low-fidelity flow solvers. In this context, simulation-based design frameworks easily turn into Simulation-Based Design Optimization frameworks (SBDO), where simulation tools, optimization algorithms and parametric descriptions of the geometry are fully and automatically integrated to deal with large numbers of decision parameters and stringent and mutually conflicting design objectives and constraints.

These are the reasons which foster, on one side, the use of RANS calculations (Section 2.1) in place of the more computationally efficient BEM analyses, previously adopted in similar SBDO (Bertetta et al. 2012) for this specific design of RIM driven propellers and, on the other, the application of meta-models, like Kriging approximation (Forrester et al. 2008, Nielsen et al. 2002, Couckuyt et al. 2012, Couckuyt et al. 2014) (Section 2.3). RANS are used for their superior reliability in dealing with unconventional configurations like RIMs, while meta-models are used to support the computational efficiency of a design process which most of the times has to comply with limited resources of both computational power and time.

On the contrary, the parametric description of the RIM blade can be handled using, as usual, B-Spline curves approximating the classical design table (Section 2.2). Also in the case of RIM driven propellers, the design table can be considered as the inherent parametric description of the blade and the relatively low number of control points needed for the parametrization provided by this type of curves allows for not excessively large design spaces maintaining, at the same time, a design variability consistent with the requirements of fairness generally expected for propeller blade shapes.

The entire process, finally, from pre-processing of geometries to post-processing of RANS results and monitoring of constraints and objectives, has been arranged within the modeFRONTIER environment (Esteco 2018) using a genetic type optimization algorithm (Section 2.3).

2.1 RANS Analyses for RIM driven propellers

Calculations were carried out using RANS for a single-phase, incompressible fluid. StarCCM+ (SIEMENS 2017) was used as solver on an unstructured mesh made of polyhedra using the Finite Volume method and second order discretization schemes for all the physical quantities. The computational grid was arranged following well-established guidelines (Gaggero and Villa 2018) cavitating and consists, in average, of 1.3 Million cells. By using a Moving Reference Frame and by exploiting the periodicity of the problem, only one blade passage is considered in the analyses while local mesh refinements account for the propeller and the duct wake (Figure 1).

The computational domain has a cross area at the propeller plane one hundred times higher than the area of the propeller disk ($D_{domain}/D_{prop} = 10$), the inlet is placed four propeller diameters in front of the propeller, the outlet four propeller diameters aft, and

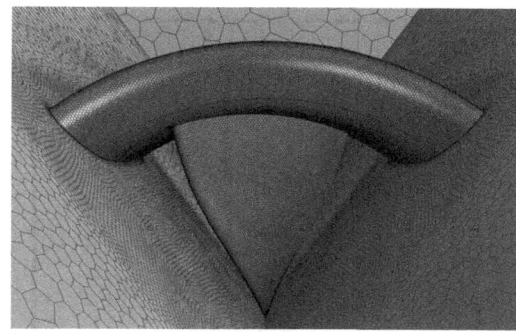

Figure 1. Polyhedral mesh around an illustrative RIM driven propeller.

uniform inflow is assumed as the initial condition The *Realizable* $k - \varepsilon$ turbulence model is used to compute the turbulent viscosity in a simulation where the turbulence intensity is set equal to 1% and the turbulent to molecular viscosity ratio is equal to 10.

2.2 Parametric description of the RIM blade

The parametric description of the geometry has been realized using the approach already devised for conventional propellers (Bertetta et al. 2012). B-Spline curves are used to describe the radial distributions of the canonical quantities adopted in the blade design table. Their control points turn into the free parameters of the optimization process. In this particular case, chord, pitch and the maximum sectional camber of conventional NACA hydrofoils (NACA66 thickness, a=0.8 camber line) were considered in the design.

None of the usual geometrical constraints used in conventional propeller designs is forced, with the exception of those related to the size of the computational domain (i.e the blade passage used for RANS analyses), since a widening of the design space can have only a beneficial influence in searching for an optimal, geometrically non-traditional, configuration. An example of the parametrization (e.g. the chord) is given in Figure 2. The chord distribution is described by a five-point control polygon for a total of 6 free parameters because some of the coordinates of the control points (i.e the radial position of the innermost and outermost points) cannot change to avoid cusp-shaped distribution. Pitch (four-point control polygon) and camber (four-point control polygon) have similar constraints, for a total of 18 free parameters.

2.3 Optimization strategy, constraints and objectives

The design of RIM driven propellers prevents the use of simplified computational models and, due to the peculiarities of the functioning, requires the monitoring of a certain number of quantities for a consistent design. In particular, by identifying any candidate

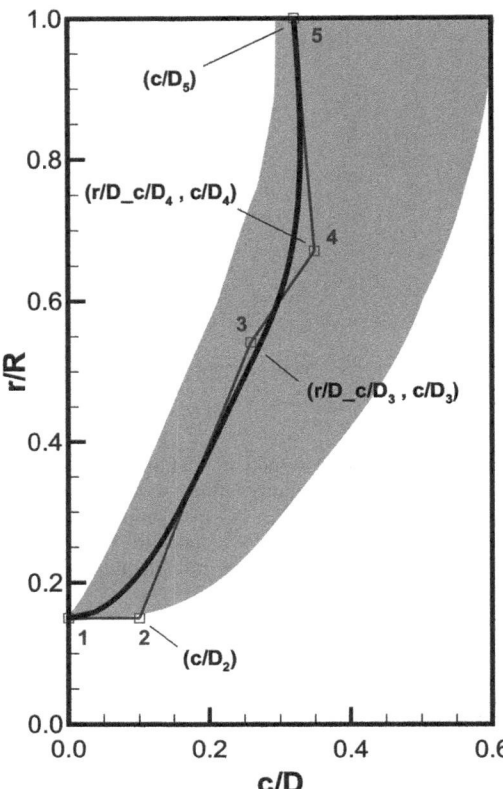

Figure 2. B-Spline parametric description of the chord distribution along the radial coordinate.

$\mathbf{p} = [p_1, p_2, ..., p_{18}]$ of the optimization process by its free parameters p_i ranging between assigned minimum and maximum values compatible with the choice of the computational domain, the design turns into a bound-constrained optimization problem:

$$\begin{cases} maximize(\eta_o(\mathbf{p})) \\ minimize(cavitation(\mathbf{p})) \\ \min(p_i) \leq p_i \leq \max(p_i) \\ 0.98K_{Ttref.} \leq K_{Tt}(p_i) \leq 1.02K_{Ttref.} \end{cases} \quad (1)$$

where the propulsive efficiency is maximized together with minimization of the cavitation risk for a given thrust, which namely is the thrust of the reference ducted propeller. Since providing the same thrust could result excessively demanding, a tolerance of 2% was accepted as reasonable threshold used to filter out unfeasible designs.

These requirements foster the development of efficient optimization work-flows since the computational effort of an multi-objective optimization activity entirely based on RANS calculation could be

prohibitive. An example of efficient work-flow is given in Gaggero (2020), where a sort of ``dual-step'' process was arranged to save computational time: the performance of the geometries devised by the optimization were monitored throughout their RANS calculation (i.e. before reaching convergence) and rejected (saving a certain amount of resources) if a preliminary, less stringent, constraint on the delivered thrust was not satisfied.

Another way to speed up the search of the optimal configuration is to make use of surrogates in the optimization process. A surrogate model is, essentially, a response function fitted to basic data obtained by evaluating objectives and constraints of the design at few points, through which build a meta-model of the phenomena under investigation. Among available surrogates, the meta-models used in this SBDO are of Kriging type. The basic idea of a Kriging model is to predict the value of a function at a given point by computing a weighted average of the known values of the function in the neighbourhood of the point itself. Kriging can be then seen as a method of interpolation (eventually of approximation) where the interpolated values are modelled by a Gaussian process governed by prior covariances. This interpolation can used at no computational cost in place of the expensive RANS calculations to gather data, correlations, tendencies from the model and, in the end, to build an entire design by the optimization process.

A robust surrogate definition presupposes a sampling of the design space to gather all the necessary data to train the meta-model with respect to design objectives and constraints. While predicting the propulsive efficiency and the propeller total thrust to accomplish the optimization problem of Eq. (1) and to build the relative meta-models is straightforward, the minimization of cavitation poses some issues, especially from the computational point of view, when using RANS calculations. In principle, cavitation can be reliably predicted, especially for what regards leading edge sheet phenomena, using multiphase approaches with the homogeneous mixture assumption but computational times for truly cavitating analyses are prohibitive and their inclusion in the optimization loop hardly feasible. Then, rather than minimizing cavitation intended as the portion of the propeller blade covered by vapour, as usually done in the case of optimization based design approaches using BEM, the objective is fulfilled in terms of risk of cavitation inception, which means maximization of the pressure (that is reduction of the suction peaks) on the propeller blade. Cavitation inception is monitored by the suction peaks at the blade leading edge, representative of the risk of sheet cavitation, on both suction and pressure side, but also by the value of the pressure at midchord, in order to account for the risk of bubble cavitation. Only computationally more efficient steady, non-cavitating calculations are, then, needed for this type of analysis. The minimization of cavitation, then, can be seen as the maximization of the inception speed using only pressure data sampled on the blade surface.

To this aim, the blade surface is "divided" in seven zones where pressure (the pressure coefficient non-dimensionalized with respect to the propeller rate of revolution) is collected and processed at the end of any case which fulfils the design constraint on the delivered thrust. Leading edge phenomena (within the 10% of the chord from leading edge) are monitored close to tip ($C_{PN1back}$ and $C_{PN6face}$ in the tip zone, $r/R > 0.9$) and in correspondence of intermediate radial positions ($C_{PN3back}$ and $C_{PN7face}$ in the midspan zone, $0.5 < r/R < 0.9$) on both suction and pressure side. Midchord inception (between 30 and 70% of the chord), instead, is considered only on the suction side ($C_{PN2back}$ and $C_{PN4back}$ at the tip and at midspan) while at the blade root suction peaks are collected regardless their chordwise position ($C_{PN5back}$) as shown in Figure 3. From a mere cavitation free design, the cavitation inception has to be avoided anywhere. Then, the minimization of the highest suction peak on the entire propeller blade could be considered sufficient as a design objective. Collecting data at different locations, however, provides a more detailed characterization of the phenomena occurring on the propeller, which could be useful in the post-processing of data for the selection of the "optimal" design among the Pareto configurations and for a better characterization of the propeller functioning. This choice turns Eq. (1) into a eight objective and one constraint optimization problem.

For the training of the surrogate models, data through high-fidelity RANS calculations are collected for each of the initial geometries which were used to fill the 18-dimensions (i.e 18 free parameters) design space. The Sobol sequencing was used to this aim, selecting 360 pseudo-random geometries within the range of variations allowed for the design parameters. Preliminary Kriging models were build for each quantity of interest (the design objectives of

efficiency and cavitation and the constraint, i.e the total delivered thrust) using the data collected for this initial sampling and were used for a preliminary optimization. In order to improve the accuracy of the surrogate models for the final design, indeed, Kriging meta-models are "re-trained" using in addition to the initial sampling of the design space, the information gathered from the RANS analyses of some of the most promising candidates identified during the preliminary optimization. In this way a certain "clustering" of the sampling of the design space around reasonable design optima is realized with the aim of improving the predictive capabilities of the meta-models. This re-training consisted in adding to the initial sampling of the design space additional 36 geometries (10% of the initial), for a total of 396 RANS calculations needed to build the final Kriging meta-models.

Besides the re-training, the final calculations were carried out making use of the approximating properties of Kriging models developed in Staum (2009), Couckuyt et al. (2012) and Couckuyt et al. (2014). The approximating property of Kriging models (better known as stochastic or regression Kriging) was developed to deal with stochastic data but results particularly useful to deal with deterministic, but noisy, information. While for integral quantities like thrust or efficiency it is not expected to have any problem of this kind, dealing with the pressure peaks at the leading edge of the blade could result in "dirty" data arising from unstructured, warped, skewed and, more in general, low-quality cells more likely observable at the blade leading edge. Even if the highest values of suction collected for the initial samples are averaged zone per zone, the risk of having non-smooth variations of the quantities of interest, in particular as a result of the clustering of candidates adopted for the re-training of the model, can not be ignored. Exploiting the approximating property of the model can mitigate this problem.

The final SBDO process based on Kriging is then built with these choices (re-trained and approximating meta-models). The design by optimization consists in the evolution of the initial population which samples the design space for one hundred generations using the NSGA-2 multi-objective genetic algorithm, for a total of 36000 configurations.

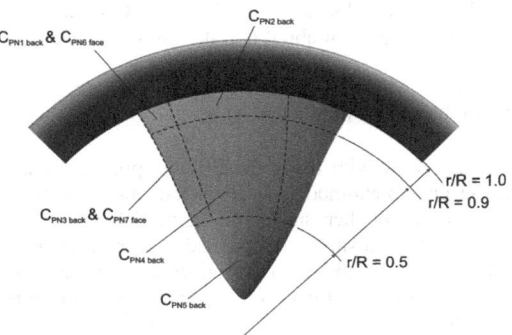

Figure 3. Identification of blade zones for cavitation inception estimation: tip ($r/R > 0.9$), midspan ($0.5 < r/R < 0.9$) and root ($r/R < 0.5$) zone. Leading edge within 10% of the chord.

3 THE REFERENCE DUCTED PROPELLER

The design of the RIM propellers was carried out starting from the analysis of a conventional ducted propeller which was considered as the performance reference. The objectives and the constraints of the design process were defined based on its performance (Table 1). The reference geometry was designed in the framework of the BESST (Breakthrough in European Ship and Shipbuilding Technologies) European Community's Seventh Framework Programme (FP7/2007-2013) funded project. It is a four-balded propeller with an expanded area ratio of 0.69 and a pitch over diameter ratio equal to 1.56 at r/R = 0.7, designed to operate in the well-established accelerating Nozzle 19A.

A preliminary assessment of the reference propeller performance is given in Figure 4. Calculations were carried out using the same global mesh arrangement devised for the analysis of RIM driven propellers, with the addition of a local refinement in the gap at the tip of the blade, from which significant tip leakage vortexes are supposed to originate (Baltazar et al. 2012, Gaggero et al. 2014). With these choices, the computational grid for the reference ducted propeller consists of about 2.3 Million cells.

The comparison with experiments is satisfactorily. At the advance coefficient selected for the design ($J = 0.7$), the thrust provided by the propeller blade is underestimated of less than 1.5% while the total thrust (blades and hub) is very close to the measured value, with a difference of about 0.6% mainly due to the slight overestimation of the thrust provided by the nozzle. The agreement is reasonable also in off design conditions, with difference, for what regards the thrust provided by the blades, always lower than 2% and discrepancy in the total thrust within 1%. Only very unloaded functioning conditions are predicted with a substantial overestimation of the nozzle thrust, which leads to an overprediction of the total delivered thrust of about 3%. Also the torque, which generally poses the major issues for this kind of analyses, differs from the measurements less than 1% over almost all the range of advance coefficient considered in the analysis.

Using this reference mesh, in addition, the original ducted propeller has been analyzed to collect the pressure values over the blade which are used to assess the compliance with the design objective of the optimized RIM geometries. Also for the original ducted propeller, pressures were collected over the zones identified for the RIM blades. Results, which serve as reference during the optimization, are summarized in Table 1.

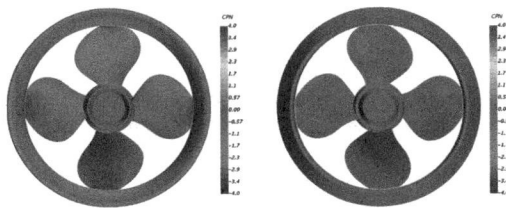

Figure 5. Pressure coefficient distributions (suction and pressure side) computed for the reference ducted propeller at the design advance coefficient of $J = 0.7$.

Table 1. Reference value for the original ducted propeller.

Efficiency	η_o	0.579
Total delivered thrust	K_{Tt}	0.427
Min. LE press. at tip (back)	$C_{PN1back}$	-8.41
Min. midchord press. at tip (back)	$C_{PN2back}$	-3.71
Min. LE press. at midspan (back)	$C_{PN3back}$	-2.07
Min. midchord press. at midspan (back)	$C_{PN4back}$	-2.68
Min. press. at root (back)	$C_{PN5back}$	-2.24
Min. LE press. at tip (face)	$C_{PN6face}$	-5.89
Min. LE press. at midspan (face)	$C_{PN7face}$	-6.63

The original ducted propeller (Figure 5) has a non-negligible risk of cavitation at the tip, where the cross-flow in the gap between the blade and the nozzle determines significantly lower values of suction on the back of the blade. Cavitation inception is predicted (in any case over a small portion of the blade suction side) for a cavitation index σ_N higher than 8, leading, on the basis of the position of occurrence, to leading edge sheet cavitation and cavitation of the tip leakage vortex. For considerable zone at midchord, the risk of bubble cavitation is for σ_N of about 3.7 while on the pressure side the leading edge sheet cavitation is supposed to start in correspondence of a cavitation index higher than 6 for positions corresponding to intermediate radial locations ($0.5 < r/R < 0.9$) an for a cavitation index of about 6 at the tip of the blade.

4 OPTIMIZATION RESULTS

The design activity using the SBDO approach based on Kriging meta-models was carried out for a five-blade RIM propeller, since with the available parametric description a value of expanded area ratio comparable to that of the reference ducted propeller was possible only with 5 blades. Four blades, with the current limits in the parametrization, would have resulted in a significantly lower expanded area, with obvious difficulties for what regards the compliance with the design objectives dealing with cavitation inception.

Criteria for the selection of the optimal geometries were based on the cavitation postponement and the

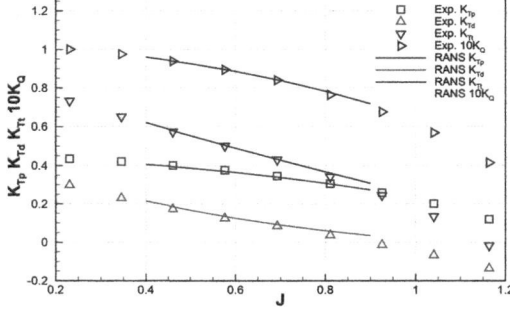

Figure 4. Open water performance of the reference ducted propeller. Comparison between owing tank tests and RANS calculations.

maximization of the propulsive efficiency. The overall best geometry (KR-2622) is the one that realizes the largest margins versus any kind of cavitation (sheet from leading edge, bubble at midchord, both suction and pressure side), then the one which simultaneously (i.e. everywhere on the blade) realizes the lowest values of suction (the lowest $-C_{PN}$). A second best (KR-2335) is selected among the geometries which are characterized everywhere by an inception lower than σ_N equal to 3.6. This threshold is a critical value since it determines in the case of the ducted propeller (Table 1) the inception of bubble cavitation at midchord in proximity of the tip. Among all the configurations satisfying $-C_{PN} < 3.6$ the one with the highest efficiency is preferred. A third choice (KR-12878) is selected by loosening the constraint of cavitation inception on the pressure side based on the (poor) performance of the reference ducted propeller. The inception is fixed on the pressure to a value equal to 6: among the feasible configurations, the one with the highest margins against back side cavitation is preferred.

RANS calculations were carried out for these three geometries. Moreover, in order to assess the reliability of the meta-models, all the geometries filtered out during the selection (i.e those with inception below 3.6 regardless their efficiency, those identified by loosening the threshold on the pressure side) plus a certain number of configurations having progressively higher values of inception were considered for detailed RANS analyses useful for the validation of the Kriging surrogates. Figures 6 and 7 summarize the correlation between predicted (with the Kriging models) and computed (with RANS) values, while Table 2 collects some metrics of the meta-models. Classical error measures are the percentage error $err\%$, its mean $\overline{err\%}$, its standard deviation $\sigma(err\%)$ and the R^2 coefficient of determination. In addition, the monotonicity G of the surrogate $(\phi_i^{RANS} \leq \phi_j^{RANS} \Rightarrow \phi_i^{surr.} \leq \phi_j^{surr.})$ is monitored: zero value indicates global monotonicity while the higher the magnitude, the more significant is the monotonicity loss.

The performance of the meta-models are conflicting. The meta-models for efficiency, midchord pressure coefficient at tip on the back $C_{PN2back}$ and

midchord pressure coefficient at midspan $C_{PN4back}$ are very well correlated to the direct RANS calculations ($R^2 > 0.9$). Also the pressure coefficient at root $C_{PN5back}$ and the pressure coefficient at the leading edge for intermediate radial positions on the back and on the face sides ($C_{PN3back}$ and $C_{PN7face}$ respectively) show a decent correlation ($R^2 > 0.7$) which, instead, is completely lost in the case of the pressure coefficient at the leading edge in correspondence of the tip region on both the blade sides ($C_{PN1back}$ and $C_{PN6face}$). In these two cases, data are spread, with relatively high values for the average error. This confirms the issues with leading edge phenomena, strongly influenced by the local arrangement and the quality of the computational mesh, which suggested the use of a Regression Kriging with approximation rather than with interpolation capabilities. Also the performance of the surrogate devoted to modelling the thrust are not positive, since data are vaguely correlated. In any case, most of the samples, also when computed with the RANS, fall within the constraint of $\pm 2\%$. All are within a threshold of $\pm 3\%$.

Based on these results, three additional geometries were added to the discussion. The criteria for their selection are the same of above: the simultaneous lowest value of inception, the highest efficiency for an inception fixed equal everywhere to 3.6 and the highest margin against suction side cavitation when the pressure side cavitation index is fixed equal to 6. In this case, however, the selection is based directly on RANS performance and then is limited to the geometries analysed for validation purposes at the end of the optimization process. Those geometries are, respectively, RANSE-1380, RANSE-1869 and RANSE-2136 which characteristics are discussed and compared with the corresponding configurations selected using the Kriging data based. Kriging models identified KR-2622 as the overall best configuration, able to postpone inception up to $\sigma_N = 3.45$ in correspondence of which the surrogate predicted a risk of back side cavitation at midchord (bubble cavitation) on the tip region ($C_{PN2back}$). Once analysed with the RANS, this geometry still provides good performance. On the back side it is still the best geometry, with the worst inception again at midchord for an even lower (with respect to Kriging predictions) value of cavitation index equal to 3.40. Instead, on the pressure side, in particular at the tip, its performance are degraded. The inception at 3.22, predicted by the Kriging, increases to 3.58 when truly RANS analyses are considered. RANSE-1380, instead, ensures an inception on the pressure side for a cavitation index equal 3.50 when analysed with the RANS. Its worst inception, based on RANS analyses, is for $\sigma_N = 3.53$ at midchord on the tip suction side making this configuration the overall best if truly RANS analyses are considered. The general behaviour of these two geometries, observable in Figure 8 and 11 is, in any case, very similar. The shape of the

Table 2. Reference value for the original ducted propeller.

	R^2	$\overline{err\%}$	$\sigma(err\%)$	G
η_o	0.907	0.331	0.278	363
K_{Ti}	0.173	0.968	0.729	497
$C_{PN1back}$	-0.871	10.67	13.03	3948
$C_{PN2back}$	0.988	1.574	1.123	283
$C_{PN3back}$	0.624	5.508	6.458	1698
$C_{PN4back}$	0.991	1.296	1.111	213
$C_{PN5back}$	0.784	8.877	7.164	737
$C_{PN6face}$	-0.377	30.01	37.40	1247
$C_{PN7face}$	0.765	27.153	35.49	540

blades, which are very similar themselves, reflects this similarity in performance.

The second best from the Kriging based SBDO is KR-2335. Meta-models predict for this geometry an inception for $\sigma_N = 3.52$ on the midchord at the tip suction side ($C_{PN2back}$) with a maximum efficiency of 0.564, the highest among the geometries with threshold on cavitation equal to 3.60. RANS analyses confirm the efficiency (0.565) and the good performance on the suction side (the inception is for $\sigma_N = 3.44$ again on the midchord at the tip) of this propeller but highlight some issues on the pressure side, where predictions of surrogates have shown their limits. On the pressure side ($C_{PN6face}$), indeed, RANS calculations indicate an inception for a cavitation index of 3.78 (while for Kriging the threshold was 3.45) which definitely prevents the acceptability of this geometry. Looking instead to the set of geometries analysed with RANS calculations, RANSE-1869 can be identified as the second best candidate. It generally has worse performance (truly RANS analyses) with respect to KR-2335: on the suction side the inception is predicted for $\sigma_N = 3.55$ ($C_{PN2back}$) and only at the leading edge this geometry has a postponed inception (3.40 at the tip, 3.00 for intermediate radial positions) with respect to KR-2335. On the face side, instead, cavitation inception is predicted for $\sigma_N = 2.90$ and then perfectly in line with the selection criteria adopted for this second best geometry. The side effect of this choice is a slightly lower value of efficiency, which is calculated by RANS equal to 0.560. Also for these two selections, the blades (Figure 9 and 12) look very similar.

Finally the third best was selected by loosening the constraint on the face side cavitation. KR-12878 and RANSE-2136 are the geometries which, respectively for the Kriging and for the RANS, provide the highest margins against suction side cavitation. For these two geometries the highest discrepancies between the

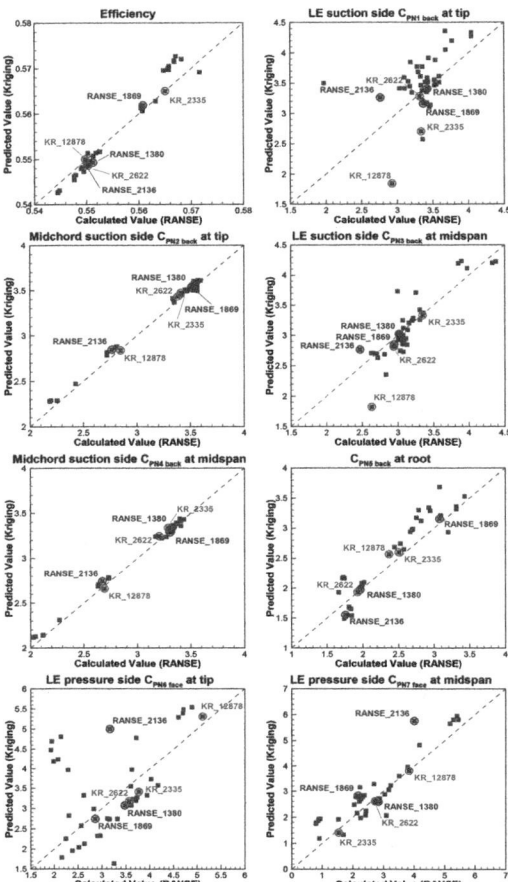

Figure 7. Correlation between predicted (Kriging) and calculated (RANSE) values. Propeller efficiency.

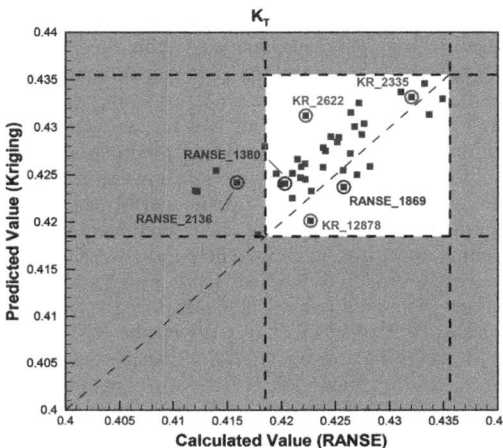

Figure 6. Correlation between predicted (Kriging) and calculated (RANSE) values. Propeller delivered thrust.

predicted and the calculated values concern the suction side pressure distribution at tip ($C_{PN1back}$) on the leading edge for KR-12878, which is substantially underestimated by the Kriging with respect to truly RANS analyses, and the leading edge pressure at tip on the face side for RANSE-2136, which is instead substantially overestimated by the Kriging model. Based on the RANS analyses, in any case, both these two configurations provide very good performance on the suction side, where the inception is postponed to 2.94 ($C_{PN1back}$) for KR-12878 and to 2.80 for RANSE-2136 (again $C_{PN1back}$). On the pressure side RANSE-2136 shows inception for $\sigma_N = 4.00$ in correspondence of the leading edge for intermediate radial positions ($C_{PN7face}$). KR-12878 has inception for $\sigma_N = 5.00$ at the tip of the blade ($C_{PN6face}$). Again, blade shapes (Figures 10 and 13) are very similar and present the highest blade area ratio.

None of the selected geometries has an efficiency higher than that of the reference propeller. Actually, for some of the analysed geometries (see, for

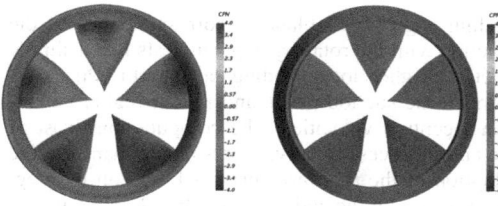

Figure 8. Pressure coefficient distribution for KR-2622.

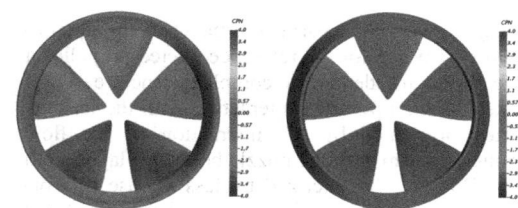

Figure 13. Pressure coefficient distribution for RANSE-2136.

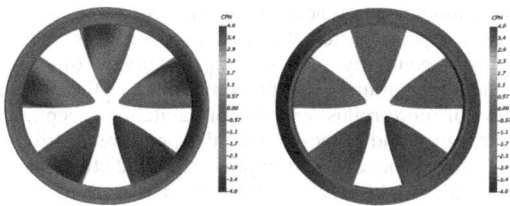

Figure 9. Pressure coefficient distribution for KR-2335.

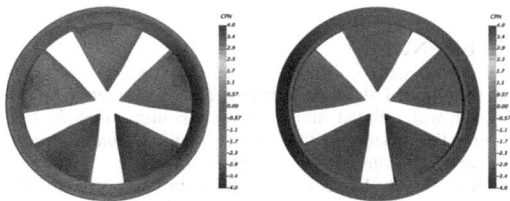

Figure 10. Pressure coefficient distribution for KR-12878.

instance, Figure 7) the Kriging meta-model predicted efficiencies higher than 0.57 (then closer to the baseline) which were mainly confirmed by the RANS analyses. The criteria adopted to identify the optimal geometries, however, preferred the reduction of the risk of cavitation, using efficiency only as a secondary objective. Since efficiency and cavitation are contrasting objectives, the lower values of efficiency for the selected geometries are not unexpected. In addition, also the role of the ring through which blades are connected at the tip and which rotates together with the propeller has to be considered. It definitely adds a substantial portion of wetted surface (i.e. friction) to the propeller, contributing in average to an increment of the total absorbed torque close to 5%.

Figure 14, finally, compares the vortex structures developed at the design advance coefficient by the reference ducted propeller and by the two overall best geometries, KR-2622 and RANSE-1380, selected using the Kriging and the RANS calculations. The comparison points out the effectiveness of the RIM configuration for the mitigation of the vortical structures developed by the blade, evidenced in

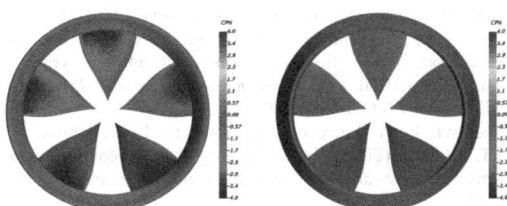

Figure 11. Pressure coefficient distribution for RANSE-1380.

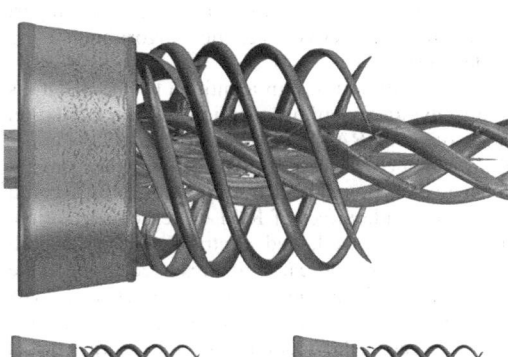

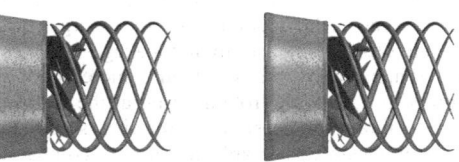

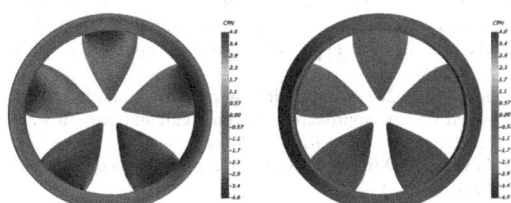

Figure 12. Pressure coefficient distribution for RANSE-1869.

Figure 14. Development of tip, leakage and hub vortexes. Reference ducted propeller (top), KR-2622 (bottom left) and RANSE-1380 (bottom right).

the figure with an isosurface of the Q-factor equal to $6500\ s^{-2}$. While for the reference ducted propeller it is possible to identify a complex structure represented by the tip vortex interacting with the leakage vortex originated by the interaction of the flow through the gap with the nozzle boundary layer, both the RIMs are characterized by less intense tip vortexes. Tip vortex as such is almost disappeared and only a sort of leakage vortex with a very reduced pitch is observable. It intensity, measured by its extension in the propeller wake, is significantly reduced as well as also the hub vortex, although the absence of the hub can favour the cross-flow, is substantially negligible.

5 CONCLUSIONS

In this paper a Simulation-Based Design Optimization approach for the design of RIM driven propellers inside accelerating nozzles has been developed. The process consists of the automatic handling of thousands of geometries, described using parametric B-Spline curves, from pre-processing and meshing to their numerical solution. Kriging based meta-models are employed to speed up the convergence to the optimum since the performance of each candidate, and their compliance with design objectives and constraints, are assessed using the surrogates iteratively trained, for each quantity of interest, on the initial sampling of the design space and on intermediate optimal geometries. For the specific case under investigation, a five-bladed RIM driven propeller in an accelerating duct able to minimize the risk of any type of cavitation while delivering a given thrust at the design advance coefficient has been designed. This problem was formalized as a 9 objective, 18 free parameters optimization, subjected to one constraint, which was solved with the support of a multi-objective genetic algorithm using modeFRONTIER.

The overall best design identified by the SBDO is able to avoid cavitation inception up to a cavitation index of 3.53 (RANS-1380), which is a substantial improvement with respect to the reference conventional ducted propeller. Root vortexes, which could represent a side effect of RIM driven configuration, are reduced as well, and contribute to the overall reduction of the vortical structures shed in the wake by the RIM driven configuration. The cost of these improvements is a slightly lower propulsive efficiency because the reduction of the energy losses associated to less intense tip and leakage vortexes is counterbalanced by the substantial increase of the absorbed torque of the rotating ring at the tip of the blades. Among the feasible geometries, some (KR-2355, for instance) are able to provide comparable values of efficiency. A sensibly higher risk of pressure side cavitation at the tip of the blade, however, characterizes these geometries. The location of the minimum of the pressure, on the pressure side at the

leading edge of the blade in correspondence of the junction with the rotating ring, suggests that a simple fillet or higher local leading edge radii can easily dampen the suction peaks and favour higher cavitation inception velocities. The computational cost of the entire process, finally, is reasonable thanks to the adoption of the response surfaces based on Kriging. The process, in the end, required 396 RANS calculations to characterize the geometries initially selected in the design space (360) and those (36) used for the re-training of the models. The metrics of the devised meta-models show some limitations of these surrogates, especially for the quantities strongly influenced even by small imperfections of the computational mesh. Multi-step training or the definition of more robust criteria for cavitation inception (averaging and smoothing of the pressure distribution, for instance) could improve the predictive characteristics and the reliability of the surrogate models which, in any case, can be considered a powerful and efficient tool at least for the very preliminary design phase.

REFERENCES

Baltazar, J., J. Falcão de Campos, & J. Bosschers (2012). Open-water thrust and torque predictions of a ducted propeller system with a panel method. *International Journal of Rotating Machinery 2012*.

Bertetta, D., S. Brizzolara, S. Gaggero, M. Viviani, & L. Savio (2012). Cpp propeller cavitation and noise optimization at different pitches with panel code and validation by cavitation tunnel measurements. *Ocean engineering 53*, 177–195.

Cao, Q.-M., F.-W. Hong, D.-H. Tang, F.-L. Hu, & L.-Z. Lu (2012). Prediction of loading distribution and hydrodynamic measurements for propeller blades in a rim driven thruster. *Journal of Hydrodynamics 24*(1), 50–57.

Couckuyt, I., T. Dhaene, & P. Demeester (2014). oodace toolbox: a flexible object-oriented kriging implementation. *The Journal of Machine Learning Research 15*(1), 3183–3186.

Couckuyt, I., A. Forrester, D. Gorissen, F. De Turck, & T. Dhaene (2012). Blind kriging: Implementation and performance analysis. *Advances in Engineering Software 49*, 1–13.

Esteco (2018). *modeFrontier, Release 2018*. Trieste, Italy: Esteco.

Forrester, A., A. Sobester, & A. Keane (2008). *Engineering design via surrogate modelling: a practical guide*. John Wiley & Sons.

Gaggero, S. (2018). Design of pbcf energy saving devices using optimization strategies: A step towards a complete viscous design approach. *Ocean Engineering 159*, 517–538.

Gaggero, S. (2019). Rim driven propellers: optimization based design approach using rans calculations. In *Proceedings of the RINA Propellers and Impellers: Research, Design, Construction and Application Conference*, pp. 1–12.

Gaggero, S. (2020). Numerical design of a rim-driven thruster using a rans-based optimization approach. *Applied Ocean Research 94*, 101941.

Gaggero, S., J. Gonzalez-Adalid, & M. P. Sobrino (2016). Design and analysis of a new generation of clt propellers. *Applied Ocean Research 59*, 424–450.

Gaggero, S., G. Tani, D. Villa, M. Viviani, P. Ausonio, P. Travi, G. Bizzarri, & F. Serra (2017). Efficient and multi-objective cavitating propeller optimization: An application to a high-speed craft. *Applied Ocean Research 64*, 31–57.

Gaggero, S., G. Tani, M. Viviani, & F. Conti (2014). A study on the numerical prediction of propellers cavitating tip vortex. *Ocean engineering 92*, 137–161.

Gaggero, S. & D. Villa (2018). Cavitating propeller performance in inclined shaft conditions with openfoam: Pptc 2015 test case. *Journal of Marine Science and Application 17*(1), 1–20.

Hughes, A., S. A. Sharkh, S. Turnock, et al. (2000). Design and testing of a novel electromagnetic tip-driven thruster. In *The Tenth International Offshore and Polar Engineering Conference*. International Society of Offshore and Polar Engineers.

Mishkevich, V. (2000). Theoretical modeling of ring propulsor in steady flow. In *Proceedings of Propellers Shafting Symposium. Society of Naval Architects and Marine Engineers, Virginia Beach, VA, USA*.

Nielsen, H. B., S. N. Lophaven, & J. Søndergaard (2002). DACE - a matlab kriging toolbox.

Saunders, H. E. (1965). Hydrodynamics in ship design. *Soc Nav Archit Mar Eng*.

SIEMENS (2017). *StarCCM+, version 12.06.011*. Siemens PLM Software.

Sparenberg, J. (1969). On optimum propellers with a duct of finite length. *Journal of Ship Research 13*(02), 129–136.

Staum, J. (2009). Better simulation metamodeling: The why, what, and how of stochastic kriging. In *Proceedings of the 2009 Winter Simulation Conference (WSC)*, pp. 119–133. IEEE.

Yakovlev, A. Y., M. A. Sokolov, & N. V. Marinich (2011). Numerical design and experimental verification of a rim-driven thruster. In *Proceedings of Second International Symposium on Marine Propulsors*.

A damage prediction model of oil tankers for design applications based on the regulations

H. Jafaryeganeh & C. Guedes Soares
Centre for Marine Technology and Ocean Engineering (CENTEC), Instituto Superior Técnico, Universidade de Lisboa, Lisbon, Portugal

C.L. Siow
Department of Aeronautics, Automotive and Ocean Engineering, Faculty of Mechanical Engineering, Universiti Teknologi Malaysia

ABSTRACT: This work deals with the development of the existing probabilistic approach of oil outflow calculation in the MARPOL regulation to be used for predicting the other consequences of the damaged oil tankers. For this purpose, the probabilistic model is provided for a parametric model of the internal layout of the oil tankers. The model is applied for evaluation of the probability of a series of damage cases for the feasible designed layout according the regulation requirements. These series of damaged scenarios are defined according to the requirement of damage stability calculations. For each designed layout, the stability of the ship is affected due to applying each damage scenario. Consequently, the imposed load is changed by damaged scenarios. In this regard, still water bending moment is investigated as one of the major indications of the imposed load to the ship hull girder. Then, the probability of the damage cases is incorporated into the magnitude and location of the calculated value of the maximum hogging and sagging still water bending moment. The provided prediction model can be used for estimation of the imposed load to the ship hull structure in the preliminary step of ship design in the framework of risk-based design.

1 INTRODUCTION

The probabilistic models of damage stability regulations intend to predict the survivability of the ships due to the damage consequences such as loss of bouncy because of flooding, weakness of stability that may result in capsize. During the past decades, the probabilistic approaches have been developed from the initial implementations in the regulations of SOLAS (IMO 1960) for the passenger vessels (Vassalos 2009). Later in SOLAS Chapter II-1 (IMO 2006), the probabilistic methods were harmonized with the other regulatory criteria and became into force for assessing the damage stability of dry cargo and passenger ships, though, similar approaches need to be developed for oil tankers (Papanikolaou and Eliopoulou 2008).

For the oil tankers, MARPOL (IMO 2003) recommended a probabilistic model for assessing the damage location and size, though, the regulation applies the model for prediction of oil outflow in the event of probable damage cases. The approach is based on probability density functions (PDFs) due to collect statistics for damage accidents from the classification societies. Those accidents are consisting of fifty-two cases of collisions and sixty-three cases of groundings for the

tankers with *30,000* tonnes deadweight and above, in the period between 1980 to 1990. The statistics can be updated by availability of new data and consequently the suggested PDFs may have changes in future developments of the regulatory recommendations.

A prediction model of damage consequences can be applied in the framework of risk-based design of oil tankers. For this purpose, the parametric model can be integrated with the prediction models to provide the opportunity for the designers to measure the risk of the design and improve the designs based on the estimated risk (Jung et al. 2018). In this regard, (Papanikolaou et al. 2010) used a model for internal spaces of oil tankers to predict the oil outflow parameter by MARPOL probabilistic approach. Then, the tanker design is defined as the optimization problem to reduce the oil outflow as one the major harmful effect of oil tanker damage to the environment.

On the other hand, one of the major consequences of damage to oil tankers is variation of the induced still water bending moment (SWBM) to the ship structure, which should be checked whether the section modulus is still acceptable after damage (Hussein & Guedes Soares 2009). The prediction model of SWBM in the intact condition has been investigated

DOI: 10.1201/9781003216582-34

in (Guedes Soares & Moan 1988) for different types of ships. For oil tankers, (Hørte et al. 2007) proposed a model for perdition of Max. SWBM of intact condition in sagging, which was focused on the full load condition. Later, their model is developed by (Teixeira & Guedes Soares 2011) for hogging condition in the full ballast conditions. Due to application in the risked-based design, the prediction models need to be provided for the damage conditions.

Regarding SWBM for tankers in damage conditions, the probabilistic approach of MARPOL has been used to provide the perdition models. However, the damage cases and loading cases were mostly assumed to be limited for specific applications. For instance, (Downes et al. 2007) analyzed the SWBM considering the uncertainty due to few possible damage scenarios, which were assumed to induced in the midship positions. (Bužančić Primorac et al. 2015) provide a statistical model of still water bending moment of Suezmax double hull oil in full load conditions. (Rodrigues et al. 2015) applied the probabilistic approach of MARPOL to predict the SWBM in the progressive flooding, they used full load condition as the only loading case for analysis of the damaged tankers. However, the hogging bending moment mostly required to be investigated in the full ballast conditions (Teixeira et al. 2013). Moreover, the provided models were not intended to be used in the design applications that usually require an iterative process.

On the other hand, a few design models are provided to incorporate the SWBM in the design of tankers, such as (Chen et al. 2010), (Yu & Lin 2013) and (Yu et al. 2015), majority of them are focused on the imposed load due to intact condition of the oil tankers (Jafaryeganeh et al. 2019). The lack of adoption of prediction methods with a parametric model is observed to integrate the consequences of the damage of oil tankers in the context of risk-based design.

In this work, initially, the probabilistic approach of obligatory regulation is elaborated, then, the equations are reformulated to be compatible with the predefined damage scenario. The provided method is adopted with a parametric model of the internal layout of tankers. The main applications are described to predict the major consequence due to damage to oil tankers. Finally, the model is examined for a case study and the predicted results are discussed. The probabilistic model has the ability to apply for other consequences of damage such as variation in the stability criteria of the ships due to flooding.

2 MODELING THE UNCERTAINTY OF DAMAGE CONSEQUENCES

The PDFs of MARPOL (IMO 2003) regulations are used to predict the side and bottom damage probabilities separately. In this method, a generic damage case is assumed for the oil tanker. Then the probability of happening of the damage case is evaluated by the integration of the mentioned probability distribution functions.

The properties of the generic damage case are in below:

- Longitudinal location of damage (x_l)
- Longitudinal extent of damage (l)
- Vertical location of damage (z_l)
- Vertical extent of damage (h)
- Transverse location of damage (y_l)
- Transverse extent of damage (b)

For side damage, the joint probability distribution of side damage extends (g_{ls}) can be calculated as below:

$$g_{ls}(x_l, y_l, z_l) = \int_0^l \int_0^b \int_0^h f_{s2} f_{s3} f_{s4} \, dz \, dy \, dx \qquad (1)$$

where f_{s2}, f_{s3} and f_{s4} are functions for the longitudinal, transverse and vertical extent of side damage respectively. So, the cumulative probability of breaching from the side (P_{Bs}) at the location of (x_l, y_l, z_l) is calculated as below:

$$p_{Bs} = \int_0^{x_l} \int_0^{z_l} g_{ls} \, dz \, dx \qquad (2)$$

Similarly, the joint probability distribution of bottom damage (g_{lb}) and cumulative probability of breaching form the bottom (P_{Bb}) are calculated at the location of (x_l, y_l, z_l) with the dimensions of l, b and h, as below:

$$g_{lb}(x_l, y_l, z_l) = \int_0^l \int_0^h \int_0^b f_{b2} f_{b3} f_{b4} \, dy \, dz \, dx \qquad (3)$$

$$p_{Bb} = \int_0^{x_l} \int_0^{y_l} g_{lb} \, dy \, dx \qquad (4)$$

where f_{sb2}, f_{b3} and f_{b4} are functions for the longitudinal, transverse and vertical extent of bottom damage respectively.

2.1 Simplified approach

To ease the application of the PDFs, (IMO 2004) assumed the damage region as an equivalent rectilinear block described by six boundaries.

The probability of breaching form side can be calculated by the following equation:

$$P_{Bs} = (1 - P_{Sf} - P_{Sa})(1 - P_{Su} - P_{Sl})(1 - P_{Sy}) \tag{5}$$

where P_{Sf}, P_{Sa}, P_{Su}, P_{Sl} are the probability the breaching from the side, where the damages lie entirely forward, aft, above and below of the location of the boundary planes, respectively.

The probability of breaching form bottom can be calculated by the following equation:

$$P_{Bb} = (1 - P_{Bf} - P_{Ba})(1 - P_{Bp} - P_{Bs})(1 - P_{Bz}) \tag{6}$$

where P_{Bf}, P_{Ba}, P_{Bp}, P_{Bz} are probability the breaching from the bottom, where the damages lie entirely forward, aft, above and below of the location of the boundary planes, respectively.

These probabilities of damage from side and bottom are functions of damage length and depth, location of damage and the main dimensions of the oil tanker. These functions values are calculated by linear interpolation from the presented tables of probabilities in the regulation 23. Annex I, MARPOL.

2.2 Adoption of the prediction model with the damage scenarios

In general, a series of damage scenarios are applied to evaluate the damage stability of the vessels. The scenarios are defined by intersection of hypothetical damage box with the model of ship including the internal compartments. For the oil tankers, the size and location of the hypothetical damage boxes are determined in the MARPOL regulations. A model is presented in (Jafaryeganeh et al. 2018) for identification all the possible damage scenarios based on the obligatory regulation for a parametric model of internal layout. The number of all possible damage cases is depended on the main dimensions of the ship, configuration of the transverse and longitudinal bulkhead and dimensions of the internal layout. However, due to simplification of the probabilistic predictions, this study assumes a series of predefined damage scenarios for evaluations of the damage consequences. The critical cases of damaged compartments are used for definition of the scenarios. These cases are defined by positioning the damage box at the intersection point of the longitudinal, transversal and vertical watertight members. Figure 1 shows a sample for a critical scenario of damage from bottom, the illustrated case includes four damaged ballast and cargo tanks, which have common watertight bulkheads in between. Consequently, a load case is defined for each damage scenario by the combination of the damaged compartments and intact spaces.

A compartment can be damaged from the side or bottom. So, the probability of breaching of

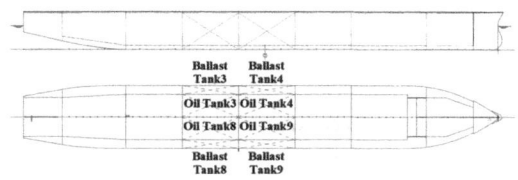

Figure 1. A sample of a critical damage scenario.

a compartment can be calculated by the accumulation of the probability of the breaching from side or bottom, as below:

$$p_{B_j} = p_{Bb_j} \cup p_{Bs_j} \tag{7}$$

$$p_{Bb_j} \cup p_{Bs_j} = p_{Bb_j} + p_{Bs_j} - p_{Bb_j} \cap p_{Bs_j} \tag{8}$$

where P_{Bi}, P_{Bsi}, and P_{Bbi} are the cumulative, side and the bottom probability of breaching of jth space respectively, in the considered damage scenario. These probabilities are assumed to be independent, thus, the probability of breaching can be calculated as below:

$$p_{B_j} = p_{Bb_j} + p_{Bs_j} - p_{Bb_j} \times p_{Bs_j} \tag{9}$$

Each damage scenario includes a series of intact and damage spaces. The spaces can be considered compartments or tanks. The probability of happening a damage scenario is calculated by the accumulation of the probability of the damage spaces, as below:

$$P^C_{DC_i} = \sum_1^m P_{B_j} \tag{10}$$

where $P^C_{DСi}$ accumulated probability for i^{th} damage case, P_{Bj} is the cumulative probability of breaching form side or bottom for j^{th} space and m is the number of damaged spaces in the ith damage case.

It should be noted that a damaged space can participate in different damage scenarios. For instance, in the damage case 10 and 12, the ballast and oil tank number 9 are the common damaged space in both scenarios. These two different damage cases are shown in Figure 2 and Figure 3 with the participation of the ballast and oil tank number 9. Thus, the accumulation of those damage probabilities for a defined series of scenarios may exceed one. On the other hand, a damage scenario needs to be compared with the other possible cases to quantify the frequency of happening the considered case. Thus, the probability of a damage scenario measured relative to the other damage cases. The relative probability

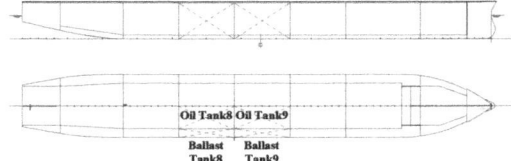

Figure 2. Presentation of a damage scenario number 10.

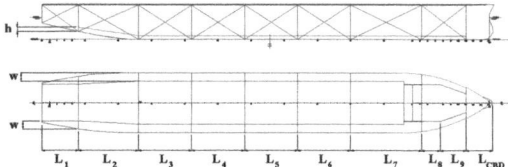

Figure 4. Presentation of the design variable for the parametric model of the internal layout of a typical oil tanker.

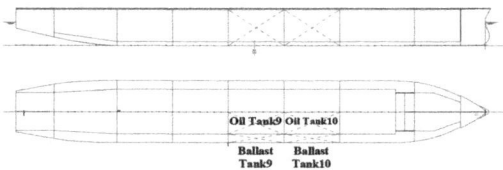

Figure 3. Presentation of a damage scenario number 12.

of happening each damage case (P_{DCi}) is defined below for a set of defined damage scenarios:

$$P_{DC_i} = \frac{P^C_{DC_i}}{\sum_1^n P^C_{DC_i}} \quad (11)$$

where n is the number of damage cases for the assumed set of damage scenarios in each design of internal layout.

2.3 Adoption of the prediction model with the parametric model of the internal layout

The probabilistic approaches for prediction of damage consequences are adopted with the parametric model, which was developed in (Jafaryeganeh et al. 2018). The parametric model was provided for generating of variety of possible designs of oil tankers within the regulatory constraints. The model was applied for the design of internal tanks and compartments. The assumed constant parameters are the main dimension of ship hull, the form of outer shell and the configuration of the internal layouts. The parameters that are considered as the design variables are the main dimensions of the internal layouts of the oil tankers.

Figure 4 shows the main parameter for the design of the internal layout for a typical oil tanker.

By considering the fixed dimensions of the hull, the introduced probability of breaching becomes the functions of the dimension of the watertight internal members. Thus, probability of breaching from side and bottom depends on a set of variables that presents the relevant dimensions of internal layout of hull. As matter of fact, variation in the dimensions of internal layout results in the change of the probability of

damage and consequence of the breaching. So, the provided model can be applied in the risk-based design at the early stages of design, where the dimension of internal layout needs to be specified.

The parametric model is provided by combination of CAESES (FRIENDSHIP-SYSTEMS 2019) and Maxsurf Stability (Bentley Systems 2018) for generating of different hull compartment layout. The CAESES is used as the programming based that control the stability solver by the COM interface. It also checks the criteria that are related to measuring the dimensions with outer shell hull by introduced technique in (Jafaryeganeh et al. 2016). Maxsurf is applied for definition of the series of damage scenarios and evaluating the hydrostatic behavior of the damaged ship. In brief, the tools facilitate the automatic procedure for generation of verity design, identification of the feasible solutions, prediction of probability of damage for each design and predicting the behavior of the designs for the defined damage scenarios.

3 APPLICATIONS

The consequence of the damage can be predicted based on the introduced probabilities of the damage cases. Those damage effects include the oil outflow, imposed load to the ship structure due to damage and stability indications such as metacentric height (GM) and properties of righting moment curve (GZ-curve).

Since this study is concentrated on the oil tankers, the impact of oil outflow and still water bending moment are investigated as the prominent consequences of damage for oil tankers. Though, the application can be extended to the prediction of damage stability. For the oil outflow, (Jafaryeganeh et al. 2020a,b,c) presented a comprehensive description for the uncertainty modeling and design applications. They used the prediction model for optimization of internal hull layout in the framework of robust-based optimization. Thus, here the focus is on the detail formulation of uncertainty modeling for Still Water Bending Moment (SWBM) due to damage cases.

3.1 Application on the SWBM prediction

SWBM is one of those consequences that can be significantly affected by the damage accidents. The Maximum SWBM is an indication for measuring the effect of damage on the ship hull structure.

The maximum SWBM can be calculated for each damage case, then, the probability of inducing the predicted value can be calculated by dedicating the probability of happening of the damage case. Since no difference indication is considered between the side and bottom damage for the calculated value of SWBM, their probabilities can be aggregated and incorporate into the damage consequence. Finally, the expected value of maximum hogging and sagging SWBM (μ_{SWBM}^{Hog} and μ_{SWBM}^{Sag}) due to the set of damage scenarios can be obtained by following equation:

$$\mu_{SWBM}^{Hog} = \sum_{1}^{n} (P_{DC_i} \times SWBM_{DC_i}^{Hog}) \qquad (12)$$

$$\mu_{SWBM}^{Sag} = \sum_{1}^{n} (P_{DC_i} \times SWBM_{DC_i}^{Sag}) \qquad (13)$$

where $SWBM_{DC_i}^{Hog}$ and $SWBM_{DC_i}^{Sag}$ are the maximum hogging and sagging SWBM in the ith damage case, respectively.

By the same way, the expected value of the longitudinal position of the maximum hogging and sagging SWBM ($\mu_{X_{SWBM}}^{Hog}$ and $\mu_{X_{SWBM}}^{Sag}$) are calculated in below equations:

$$\mu_{X_{SWBM}}^{Hog} = \sum_{1}^{n} (P_{DC_i} \times X_{SWBM_{DC_i}}^{Hog}) \qquad (14)$$

$$\mu_{X_{SWBM}}^{Sag} = \sum_{1}^{n} (P_{DC_i} \times X_{SWBM_{DC_i}}^{Sag}) \qquad (15)$$

where $X_{SWBM_{DC_i}}^{Hog}$ and $X_{SWBM_{DC_i}}^{Sag}$ are the longitudinal position of the maximum hogging and sagging SWBM alongside the ship length for the ith damage case, respectively.

Also, the standard deviation of the maximum hogging and sagging SWBM due to the defined set of damage scenarios for each design (σ_{SWBM}^{Hog} and σ_{SWBM}^{Sag}) are calculated as below:

$$\sigma_{SWBM}^{Hog} = \sqrt{\sum_{1}^{n} P_{DC_i} \times (SWBM_{DC_i}^{Hog} - \mu_{SWBM}^{Hog})^2} \qquad (16)$$

$$\sigma_{SWBM}^{Sag} = \sqrt{\sum_{1}^{n} P_{DC_i} \times (SWBM_{DC_i}^{Sag} - \mu_{SWBM}^{Sag})^2} \qquad (17)$$

Accordingly, the standard deviation of the longitudinal position of the maximum hogging and

sagging SWBM ($\sigma_{X_{SWBM}}^{Hog}$ and $\sigma_{X_{SWBM}}^{Sag}$) are calculated in below equations:

$$\sigma_{X_{SWBM}}^{Hog} = \sqrt{\sum_{1}^{n} P_{DC_i} \times (X_{SWBM_{DC_i}}^{Hog} - \mu_{X_{SWBM}}^{Hog})^2} \qquad (18)$$

$$\sigma_{X_{SWBM}}^{Sag} = \sqrt{\sum_{1}^{n} P_{DC_i} \times (X_{SWBM_{DC_i}}^{Sag} - \mu_{X_{SWBM}}^{Sag})^2} \qquad (19)$$

3.2 The procedure of prediction of damage consequences for the design evaluation

Figure 5 presents the procedure for predictions of the prediction of the main consequences of the damage of the oil tanker. Once, the feasible designs are provided for internal hull layout of the oil tanker, the parametric model is modified according to the properties of the design variable of each design. The probabilities of side and bottom damages are calculated for the provided dimensions of each oil tank. Then, the expected value of the oil outflow can be predicted by aggregation and non-denationalization of the predicted oil outflow for the oil tanks.

In the same way, the probability of the side and bottom damages is calculated for the other compartments to be used in the evaluation of the probability of the damage scenarios.

In this work, a series of predefined damage scenarios are considered as the input of the program. Also, the two load cases are only assumed to be applied in the oil tanker during the operation. Because, the oil tankers usually operate in two modes of fully loaded to deliver the oil cargo, and then fully ballast to return in the location of the oil reservoirs for recharging again. The combination of each damage scenario and load case result in verities of possible cases for predictions.

The stability is calculated for each case, accordingly, the SWBM is evaluated for all the sections alongside the ship. Then, the following properties are recoded for each case:

- The maximum value of the imposed SWBM
- The position of the maximum imposed moment alongside the ship
- The direction of the imposed moment that could be hogging or sagging

The obtained values are multiplied by the probability of the relevant case of damage and finally aggregated to evaluate the expected value and standard deviation of the maximum imposed moment due to probable damage cases.

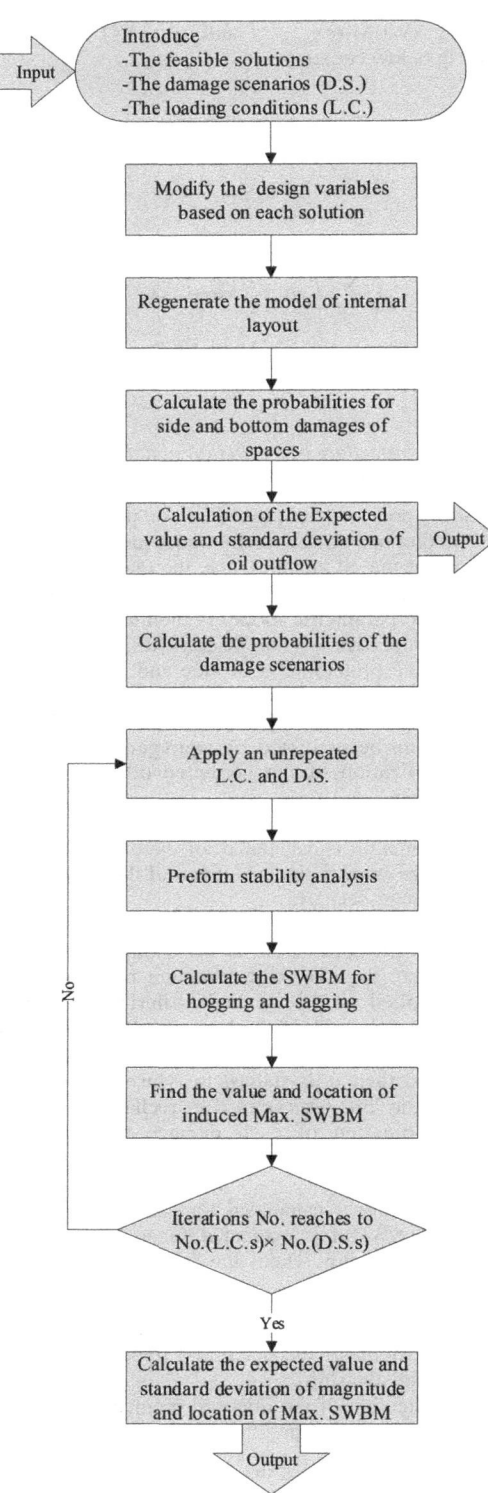

Figure 5. The procedure of prediction of damage consequences with the indication of input and output of the produced program.

4 PREDICTION FOR A CASE STUDY

A shuttle tanker is selected as a case study. In the reference design, the main subdivision consists of eight transverse bulkheads and a central longitudinal bulkhead.

Figure 4 shows the general arrangement of the taker. The verities of feasible design can be provided based on the regulations of MARPOL, SOLAS (IMO 2012), LLC (IMO 1966). For instance, (Jafaryeganeh et al. 2019) introduced *264* feasible solutions during a deterministic optimization procedure. However, in this study, only the reference general arrangement is used to demonstrate the application of the provided probabilistic model.

5 RESULTS AND DISCUSSION

Initially, the results are presented for the combination cases due to applying each damage scenario and load case. Then the results of the prediction model are presented for the consequences of the probable damage cases.

5.1 *Maximum SWBM for the damage cases*

The maximum still water bending moment (Max. SWBM) is measured due to each damage case for both loading conditions. Then, these measured quantities are categorized based on the hogging or sagging directions. Figure 6 and Figure 7 present the maximum bending movement relative to the damage cases for hogging and sagging direction, respectively. In general, the maximum bending moment in hogging have higher values than the sagging direction.

The majority of Max. SWBMs in hogging are due to full ballast conditions. However, in a few damage cases, both loading conditions result in Max. SWBM in hogging direction. These damage cases are indicated with I.D. numbers of *1, 8, 9, 17, 24* and *25*. Even in these damage cases, the Max. SWBM in full ballast case is higher than the Max. SWBM in fully loaded conditions.

All the measured Max. SWBMs in sagging direction are due to full loaded conditions. The damage cases that result in higher values in Max. SWBM in sagging direction have also higher values of the Max. SWBM in hogging direction relative to the other damage cases. However, vise verse relation is not observed i.e. the damage cases that result in higher values in the Max. SWBM in hogging direction do not have necessarily higher values of the Max. SWBM in sagging direction relative to the other damage cases.

The position of the induced Mx. SWBM alongside the ship length is also presented in Figure 8 and Figure 9 for hogging and sagging, respectively.

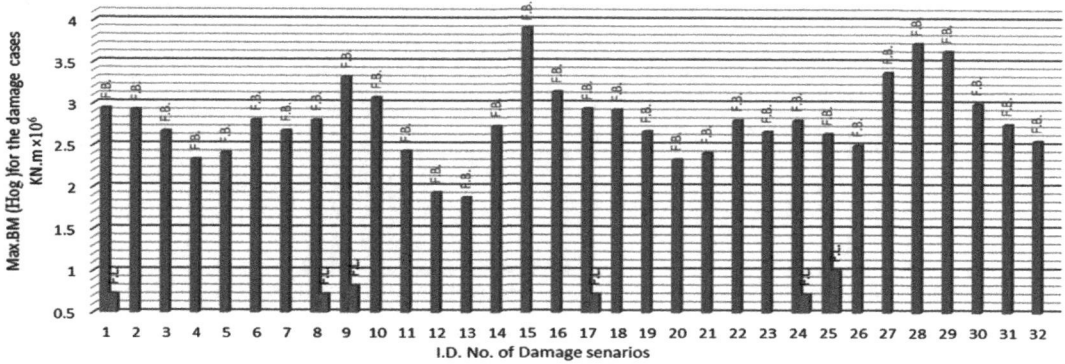

Figure 6. Max. SWBM that result in hogging, measured for each damage case in all the loading conditions.

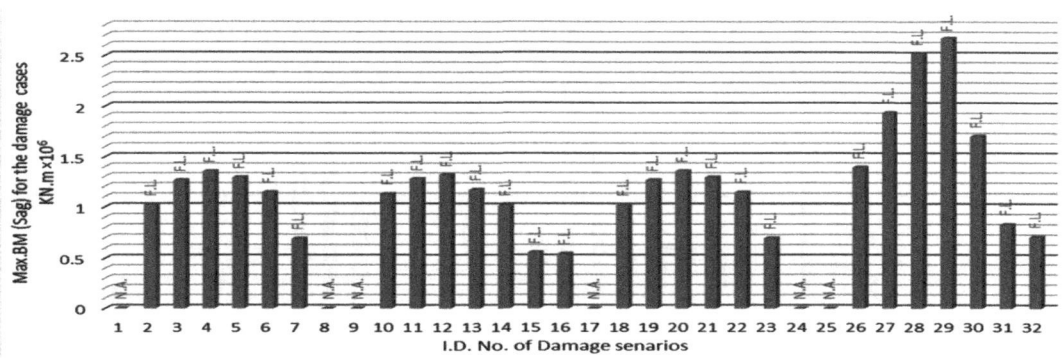

Figure 7. Max. SWBM that result in sagging, measured for each damage case in all the loading conditions.

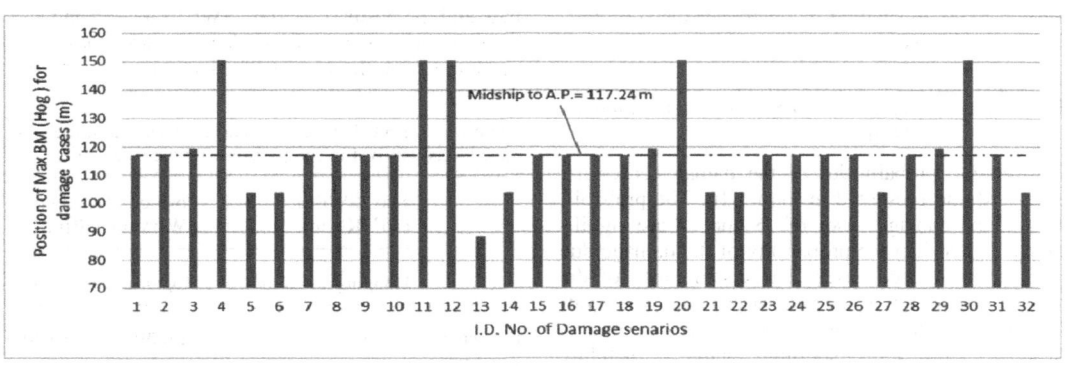

Figure 8. The position of Max. SWBM that result in hogging for all damage cases and loading conditions, measured from A.P.

Max. SWBMs in hogging are majorly induced in the position of midship section (Figure 8). Those damage cases that result in induced moments at the position out of midship section do not have the highest values of the Max. SWBM in hogging among the defined damage cases.

Max. SWBMs in sagging are majorly induced in the positions out of midship section (Figure 9).

311

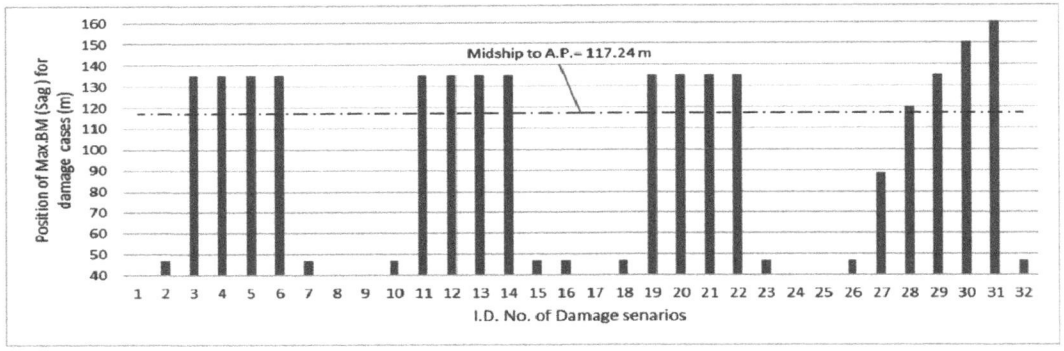

Figure 9. The position of Max. SWBM that result in Sagging for all damage cases and loading conditions, measured from A.P.

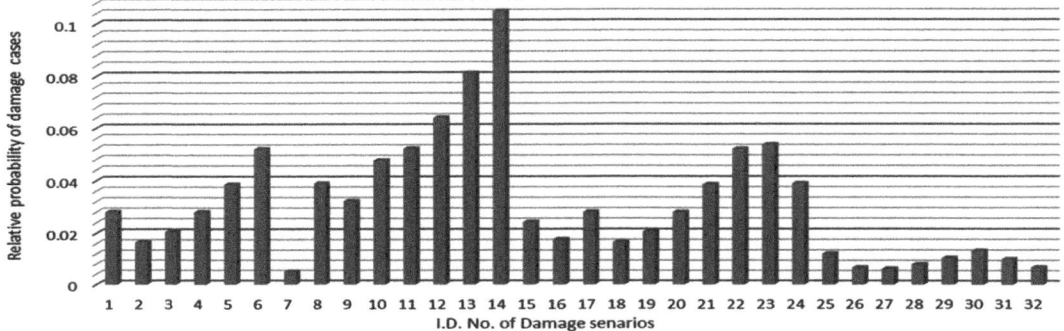

Figure 10. Relative probability of happening of each damage case.

However, those damage cases that result in larger Max. SWBM in sagging have smaller distance relative to midship section.

5.2 Prediction of maximum SWBM due to damage by a probability model

The relative probabilities of the damage cases are evaluated and presented in Figure 10. The probabilities of the damage cases of *13* and *14* are significantly higher than the probability of the damage due to other introduced cases. The dissimilarity of the relative probability of damage cases needs to be considered in the consequence of the damage scenarios, especially for the maximum induced bending moment. Thus, the relative probabilities of the damage cases are multiplied by the obtained value for magnitude and the positions of induced of the Max. SWBM. These results need to be aggregated for simplification of applying in the design process and interpretations. Thus, the expected values and the relevant divergences are discussed for the prediction of the quantities after the considered damage scenarios.

Table 1. presents the expected value of the magnitude and the position of Max. SWBM for sagging and hogging due to considered damages scenarios. The magnitudes of the expected values are close to the

Table 1. Expected value and standard deviation of the predicted Max. SWBM in comparison with the intact case.

	Max. SWBM ($10^6 \times$KN.m)		Position of the Max. SWBM to A.P.(m)	
	Hogging	Sagging	Hogging	Sagging
Expected value for damage cases	2.724	0.920	117.098	89.407
Standard deviation for damage cases	0.499	0.515	17.111	54.790
Intact condition	2.913	1.153	117.001	135.093

magnitude of Max. SWBM for the intact condition. The standard deviations of the magnitudes of Max. SWBM is almost in the same range for the sagging and hogging bending moments. Although, the predicted value due to hogging in damage cases is quite larger than the sagging direction, as well as the intact ship follows a similar trend.

Figure 11 shows the expected value and standard deviation of the predicted values alongside the ship length in comparison with the permissible still water bending moment according to common structural rules (IACS 2014). The predicted moment due to damages for hogging is induced in the same position of the intact ship, while, the moment due to sagging is not expected to induce in the same position of intact ship. Moreover, the position of hogging moments (intact and damage) are close to the midship section, while, the position of sagging moments (intact and damage) have significant distance to the midship section.

The standard deviation for the predicted position of sagging is higher than the standard deviation of the predicted position of hogging.

5.3 Uncertainty prediction of the oil outflow

Table 2 shows the non-dimensional expected value and standard deviation for the oil outflow prediction in comparison with the allowable value according to the MARPOL (IMO 2006). The marginal distance relative to the allowable value is also measured.

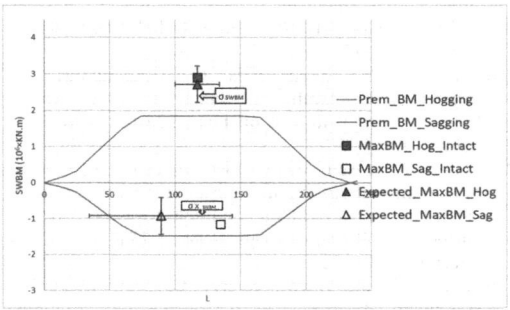

Figure 11. Expected value and standard deviation of the magnitude and the position of Max. SWBM alongside the ship length for hogging and sagging.

Table 2. Expected value and standard deviation of the predicted oil outflow in comparison with the allowable value.

Particular	Value
The expected value of oil outflow	8.232×10^{-3}
The standard deviation of oil outflow	1.893×10^{-3}
Allowable value	21.000×10^{-3}
Distance to constraint	19.107×10^{-3}

The obtained expected value for oil outflow indicates that the design passes the criteria. Even if the standard deviation of oil outflow is considered as uncertain domain for such a prediction, the design has considerable distance to the upper limit of oil outflow.

6 CONCLUSIONS

A model is developed for prediction of the damage consequences of oil tankers based on the recommended probabilistic approaches in the regulations. The model is compatible with the design requirement for examining the variety of the design layout, not only to check the obligatory criteria but also to improve the designs in the framework of risk-based design. The applied probabilistic method is adopted to the predefined damage scenarios to simplify the verification of the internal layout designs for oil tankers.

Max. SWBM in hogging has larger magnitude than Max. SWBM sagging, though the sagging is usually happening in the full load conditions. So, the sagging moments may result in more dangerous consequences due to probable damage accidents for oil tankers. Thus, separate investigation of the sagging and hogging needs to be considered in the process of robust design that implies the maximum imposed load to the ship structure as an objective to be minimized.

The expected position of the imposed sagging moment has a considerable distance to the midship section, while the expected position of the imposed hogging moment is close to the midship section. Moreover, the uncertainty of the predicted position of the imposed sagging moment is relatively higher than the predicted position of the imposed hogging moment. Thus, reduction of such uncertainties can be considered as an effective objective in the design of tankers.

By applying the provided model, the designer will be able to consider the risk of accidents in the initial stage of design. So, future works will implement the provided model to improve the hull layout design in the framework of optimization under uncertainty. The marginal distance to the upper limit of regulatory criteria can be considered as an objective to be maximized in future design frameworks. Though, the generating of the robust design will face a changeling task of runtime increment for evaluation of large number of damage scenarios for each design case.

ACKNOWLEDGEMENTS

This work was performed within the project RESET (REliability and Safety Engineering and Technology for large maritime engineering systems), which is partially financed by the European Union Horizon 2020 research and innovation program, under the Marie Skodowska-Curie grant agreement No. 73088 RESET. This work was performed within the scope of the Strategic Research Plan of the Centre for Marine Technology and Ocean Engineering (CENTEC), which is financed by the Portuguese Foundation for Science and Technology (Fundação para a Ciência e Tecnologia - FCT) under contract UIDB/UIDP/00134/2020.

REFERENCES

Bentley Systems. 2018. MAXSURF,Stability. Pennsylvania, (accessed August 2016): www.bentley.com/maxsurf.

Bužančić Primorac B, Slapničar V, Munić I, Grubišić V, Ćorak M, Parunov J. 2015. Statistics of still water bending moment of damaged suezmax oil tanker. In: Altosole M, Francescutto A, editors. 18th Int Conf Ships Shipp Res NAV 2015. Lecco, Italy; p. 580–589.

Chen J, Lin Y, Huo JZ, Zhang MX, Ji ZS. 2010. Optimization of ship's subdivision arrangement for offshore sequential ballast water exchange using a nondominated sorting genetic algorithm. Ocean Eng. 37 (11–12):978–988.

Downes J, Moore C, Incecik A, Stumpf E, McGregor J. 2007. A Method for the Quantitative Assessment of Performance of Alternative Designs in the Accidental Condition. 10th Int Symp Pract Des Ships Other Float Struct. 2(October).

CAESES. 2019. FRIENDSHIP-SYSTEMS. Potsdam, Germany. (accessed August 2016): www.CAESES.com.

Guedes Soares C, Moan T. 1988. Statistical Analysis of Still Water Load Effects in Ship Structures. Trans Soc Nav Archit Mar Eng New York. 96(4):129–156.

Hørte T, Wang G, White N. 2007. Calibration of the Hull Girder Ultimate Capacity Criterion for Double Hull Tankers. 10th Int Symp Pract Des Ships Other Float Struct. 1:553, 564.

Hussein AW, Guedes Soares C. 2009. Reliability and residual strength of double hull tankers designed according to the new IACS common structural rules. Ocean Eng. 36(17–18):1446–1459.

IACS. 2014. Common Structural Rules for Bulk Carriers and Oil Tankers. TB report no. Pt 1 – Residual strength, Ch 5, Sec 3. International Association of Classification Societies.

IMO. 1960. SOLAS. International Convention for the Safety of Life at Sea. In: International Maritime Organization, London.

IMO. 1966. ICLL. International Convention on Load Lines. In: International Maritime Organization, London.

IMO. 2003. Revised interim guidelines for the approval of alternative methods of design and construction of oil tankers under regulation 13F (5) of ANNEX I of MARPOL 73/78.

IMO. 2004. MEPC. 122 (52) Explanatory notes on matters related to the accidental oil outflow performance under regulation 23 of the revised MARPOL Annex I. In: International Maritime Organization, London; Vol. 122. p. 1–49.

IMO. 2006. MARPOL. International Convention for the Prevention of Pollution from Ships, Annex I. In: International Maritime Organization, London.

IMO. 2012. SOLAS. International Convention for the Safety of Life at Sea, Chapter II. In: International Maritime Organization, London.

Jafaryeganeh H, Ventura M, Guedes Soares C. 2016. Parametric modelling for adaptive internal compartment design of container ships. In: Guedes Soares C, Santos TA, editors. Marit Technol Eng III: London, UK: Taylor & Francis Group; p. 655–661.

Jafaryeganeh H, Ventura M, Guedes Soares C. 2018. Parametric modelling of tanker internal compartment layout for survivability improvement within the framework of regulations. In: Guedes Soares C, Teixeira A, editors. Marit Transp Harvest Sea Resour: London, UK: Taylor & Francis Group; p. 23–30.

Jafaryeganeh H, Ventura M, Guedes Soares C. 2019. Multi-Objective Optimization of Internal Compartment Layout of Oil Tankers. J Sh Prod Des. 35(4):374–385.

Jafaryeganeh, H.; Ventura, M., and Guedes Soares, C. 2020a Application of multi-criteria decision making methods for selection of ship internal layout design from a Pareto optimal set. Ocean Engineering. 202:107151-1 - 107151–14.

Jafaryeganeh, H.; Ventura, M., and Guedes Soares, C. 2020b Effect of normalization techniques in multi-criteria decision making methods for the design of ship internal layout from a Pareto optimal set. Structural and Multidisciplinary Optimization. 62:1849–1863

Jafaryeganeh, H.; Ventura, M., and Guedes Soares, C. 2020c. Robust-based optimization of the hull internal layout of oil tanker. Ocean Engineering. 216:107846

Jung SK, Roh M Il, Kim KS. 2018. Arrangement method of a naval surface ship considering stability, operability, and survivability. Ocean Eng. 152:316–333.

Papanikolaou A, Eliopoulou E. 2008. On the development of the new harmonised damage stability regulations for dry cargo and passenger ships. Reliab Eng Syst Saf. 93(9):1305–1316.

Papanikolaou A, Zaraphonitis G, Boulougouris E, Langbecker U, Matho S, Sames P. 2010. Multi-objective optimization of oil tanker design. J Mar Sci Technol. 15(4):359–373.

Rodrigues JM, Teixeira AP, Guedes Soares C. 2015. Probabilistic analysis of the hull-girder still water loads on a shuttle tanker in full load condition, for parametrically distributed collision damage spaces. Mar Struct. 44 (December):101–124.

Teixeira, A. P. and Guedes Soares, C. 2010. Reliability assessment of intact and damaged ship structures. Guedes Soares, C. & Parunov J., (Eds.). Advanced Ship Design for Pollution Prevention. London, U.K.: Taylor & Francis Group; pp. 79–93.

Teixeira A, Guedes Soares C, Chen NZ, Wang G. 2013. Uncertainty analysis of load combination factors for global longitudinal bending moments of double-hull tankers. J Sh Res. 57(1):1–17.

Vassalos D. 2009. Designing for Damage Stability and Survivability – Contemporary Developments and Implementation. 1(July):59–72.

Yu YY, Lin Y. 2013. Optimization of ship inner shell to improve the safety of seagoing transport ship. Int J Nav Archit Ocean Eng. 5(3):454–467.

Yu YY, Lin Y, Chen M, Li K. 2015. A new method for ship inner shell optimization based on parametric technique. Int J Nav Archit Ocean Eng. 7(1):142–156.

Developments in Maritime Technology and Engineering – Guedes Soares & Santos (eds)
© 2021 Copyright the Author(s), ISBN 978-0-367-77376-2

An application of a multi-objective evolutionary strategy to the ship hull form optimization

Linqiang Lan, Yafeng Sun & Weilin Luo
School of Mechanical Engineering and Automation, Fuzhou University, Fuzhou, China

ABSTRACT: A multi-objective evolutionary strategy is proposed for the optimization design of ship hull form. Parametric hull form represented by NURBS is optimized for the wave-making resistance based on multi-objective evolutionary algorithm. Rankine-source panel method is used to evaluate the hydrodynamic performance. The optimization strategy applied on the full parametric modeling of the hull, hydrodynamic performance calculation, design evaluation and shape variation is implemented, which is built on a fully automatic integration optimization framework. A forebody of a ro-ro ship is taken as a verification model. The optimal results indicate the effectiveness of the optimization strategy proposed.

1   INTRODUCTION

In the elements of ship hydrodynamic performance, the most fundamental and critical problem is resistance performance. The resistance performance of a ship strongly depends on the shape of ship. However, the optimization of hull form is a complicated problem, because it is a bound, multi-variable, multi-objective problem with many nonlinear constraints imposed by practical demands. In the recent year, a number of scholars have tried to optimize the ship hull form for the minimization of resistance using different optimization strategy. Some researchers have taken several objective functions into consideration, and others only a single objective function. For example, Peril et al. (2001) used sequential quadratic programming(SQP) algorithm, CG algorithm and SD algorithm to achieve the total resistance optimization of a tanker ship hull. Saha et al. (2004) optimized the wave resistance of the Wigley hull and the Series 60 hull by using the SQP algorithm. Peri et al. (2005) applied a global optimization(GO) algorithm to the optimization of hull. Tahara Y et al. (2011) presented a simulation-based design(SBD) framework based on global optimization(GO) algorithm for the optimization of ship hull. Mahmood,S et al. (2012) addressed the genetic algorithm(GA) to optimize Series60 for the total resistance. Zhang et al. (2009; 2012; 2015) performed hull form optimization using NGA, SGA, NLP algorithm to achieve the optimization of the wave-making resistance or total resistance.

In this paper, a multi-objective evolutionary strategy is adopted for the optimization of the hull form with respect to the wave-making resistance performance. The optimization of hull form focus on three main points: geometric modeling, an accurate hydrodynamic performance calculation and an optimization strategy. The fully parametric model generates and modifies the entire hull surface using NURBS under the Friendship-Modeler. The wave-making resistance is calculated using Rankine-source panel method. A fully automatic integration optimization framework is built in which a multi-objective evolutionary strategy is adopted to improve the efficiency at the scheme design process. A forebody of a ro-ro ship is taken as a verification model. The effectiveness of the proposed optimization strategy is indicated by optimal results.

2   PARAMETRIC MODELING

A major focus of the any optimization procedure of hull form is to link a set of hull form parameters identifying the variants to a faired hull form. Parametric Modeling provides the best way to establish the relationship and make sure that the design variation among the entire optimization process is effective and feasible. In the traditional process of geometric modeling of hull, the hull geometry is constructed from points via lines to surfaces. The modification of geometry is prompted by manipulating compatibly some points in design process, which is a labour-intensive and time-consuming work. Under the Friendship-Modeler, the parametric model has been representation. The process of curves generation starts with a set of given data elements that are used to define the complete parametric curve. In the Friendship-Modeler, the classic ship-building technology is adopted by defining a set of longitudinal lines, expressed as basic curves, including differential, integral and topological information to

DOI: 10.1201/9781003216582-35

describe the ship hull. As usual, transversal basic curves are able to be formed as soon as the longitudinal basic curves are defined by fair B-Spline parametric curves. A series of surface can be set up by corresponding transversal sections that stem from the parametric hull model. Therefore, the design of the fully parametric hull surfaces makes it possible to transform hull form efficiently and effectively (Harries, 1998; Nowacki et al., 1977; Nowacki et al., 1995; Nowacki and Kaklis, 1998). All of the hull topologies, the hull sections and whole hull surface can be generate using the basic curves specifically and precisely. Generally speaking, there are three stages in the formation process of hull in Figure 1:

(1) Parametric design of an appropriate series of basic curves like DWL, FOS, CPC, FOB, DEC etc. These basic curves are generated in accordance with a set of prominent transversal distinct transversal curves such as the main frame section, the transom and additional sections in the fore or aft body.
(2) Parametric modeling of design sections originated from these basic curves.
(3) Surface generation by interpolating or highly approximating.

The general expression of the fairness parametric curve with constraints, the starting point is regarded as a free form curve vector $r(t)$ parameterized by t:

$$r(t) = (x(t), y(t), z(t)) \tag{1}$$

To satisfy the m order fairness criterion, the equation L_m is represented as:

$$L_m = \int_0^1 (D^m r(t))^2 \tag{2}$$

where $D^m = d^m / dt^m$
A set of constraints can be embedded:
(1) Distance constraints: For $n + 1$ given data points P_i, the Euclidean distance is adopted between the points and the $r(t)$ associated with the parameter knot t_i and weighted by w_i, and finally squared to limit the maximum positive error tolerance ε:

$$A = \sum_{i=0}^{n} [w_i(r(t_i) - P_i)]^2 \leq \varepsilon, \ \varepsilon \leq 0 \tag{3}$$

Parametric design of basic curves → Parametric modeling of design sections → Generation of hull surface

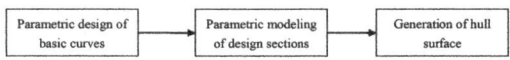

Figure 1. Hull surface generation process by complete parametric design.

(2) End constraints: As for tangent vectors Q_i and curvature vectors K_i, the first point for $i = 0$ and the last point for $i = n$ on the curve, expressed as following:

$$T_1 = D^1 r(t_i) - Q_i = 0 \tag{4}$$

$$T_2 = D^5 r(t_i) - K_i = 0 \tag{5}$$

(3) Area constraints: Considering the actual area S under a curve, the value should match a given area S_0:

$$F = S - S_0 = 0 \tag{6}$$

(4) Other constraints: Some other constraints should be taken into consideration. To deal with the constraints optimization problem, the equations can be formulated to minimize the objective functional I:

$$I = L_m + \lambda A' + \mu_1 E_1 + \mu_2 E_2 + vF \tag{7}$$

where $A' = A + d^2$, λ, μ_i and v are the Lagrange multipliers and d^2 is a slack variable.
Basic curves are fully expressed by NURBS by the parametric method of accumulation chord length:

$$Q(t) = \begin{pmatrix} x(t) \\ y(t) \end{pmatrix} = \frac{\sum_{i=0}^{n} w_i N_{i,k}(t) d_i}{\sum_{i=0}^{n} w_i N_{i,k}(t)} \tag{8}$$

where w_i is the weight factors, d_i is the control vertexes, $N_{i,k}(t)$ is the B-Spline basis function of order k, $t(0 \leq t \leq 1)$ is the nodes.
(8) is determined by minimizing the order fairness criterion:

$$E_n = \int_{t_B}^{t_e} \left(\left(\frac{d^m x}{dt^n} \right)^2 + \left(\frac{d^m y}{dt^n} \right)^2 \right) dt \tag{9}$$

where E_n is the strain energy of the curves, subscripts B and E are the beginning and ending points of the curve. (9) is subject to some equality constraints with respect to the given hull form parameters. The free variables of a nonlinear constrained optimization problem are the control points of the curves, which are calculated by solving a nonlinear equation group.
In this paper, the forebody hull form is built and distorted based on NURBS, but at the same time the aftbody is not changed. There are some parameters including global parameters and basic curves to determine the parametric modeling of the forebody

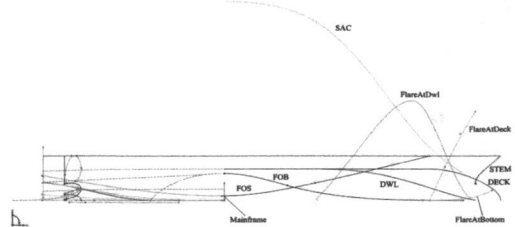

Figure 2. Basic curves of full parametric design of hull form.

hull form. The Basic curves of full parametric hull form are presented in Figure 2.Table 1 shows the basic curves to describe hull form of the whole ship. And, the design parameters of the forebody are shown in Table 2.

In the Friendship-Modeler, the whole hull is fully represented by B-Spline curves and surfaces that decide from the characteristic parameter based on a variational formulation.

A large number of highly quality hull shapes can be obtained quickly through this kind of means. Because of whole parametric curves is fairness in the procedure, the hull surface that is need for manual process has high quality and is smooth.

3 HYDRODYNAMIC NUMERICAL CALCULATION

In this paper, the wave-making resistance coefficient is calculated using Rankine-source panel method. For a ship traveling with a steady forward speed U in the calm water of infinite depth, the wave-making velocity potential $\varphi(x,y,z)$ satisfies the following equations:
in the fluid domain:

Table 1. Basic curves describing hull form.

	Curve	Symbol
Position	Design waterline	DWL
	Flat of side curve	FOS
	Center plane curve	SPC
	Flat of bottom curve	FOB
	Deck	DEC
Area	Sectional area curve	SAC
Curvature	Curvature at Beginning and End	CAB CAE
Tangent angle	Tangent angle at Beginning and End	TAB TAE
Centroid of area	Vertical moments of sectional area	VMS
	Lateral moments of sectional area	LMS

Table 2. The design parameters of the ro-ro ship forebody.

Parameters	Initial value	Value range
MainparallelOffset	15.877	[8.6,27.8]
SACareaCoeffAtFosEmerge	0.889	[0.82,0.91]
SACtanAtFwdFrame	-59.358	[-68,-52]
SACtanAtFp	-30.803	[-65,-24]
SACmzAtFosEmerge	-9.043	[-13,-6]
DwltanAtFp	-11.257	[-18,-5]
DwltanAtFos	-0.913	[-1,0]
DwlAreaCoeffFor	0.604	[0.58,0.62]
FOSxEmerge	1.108	[1.1,1.18]
FOBtanForEnd	-8.924	[-18,0]
FOBareaCoeffFor	0.507	[0.5,0.56]
FwdFrameAreaCoeff	0.876	[0.8,0.9]
FlareAtDwlAtFwdFrame	4.121	[0,14]
DeadriseAtFwdFrame	14.702	[0,18]
BowLength	5.81	[5,7]
BowBeamAtFp	4.602	[4.2,5.16]
BowBeamAtElevAtFp	6.816	[5.5,7.5]
BowZTopFp	-0.628	[-0.68.-0.56]
BowxAtBase	0.624	[0,1]

$$\nabla^2 \phi = 0 \qquad (10)$$

on the free surface:

$$\xi = \frac{1}{g}\left(-U_s \cdot \nabla\varphi + \frac{1}{2}\nabla\varphi \cdot \nabla\varphi\right) \qquad (11)$$

on the free surface:

$$\left(\nabla\varphi - U_s \cdot \vec{n}\right) \cdot \nabla\xi = \varphi_z \qquad (12)$$

on the hull surface:

$$\frac{\partial \varphi}{\partial n} = U_s \cdot \vec{n} \qquad (13)$$

where $U_s = (U,0,0)$, g is the acceleration of gravity, ξ is the free surface elevation.

Rankine method is used to solve velocity potential:

$$\varphi(P) = -\frac{1}{4\pi}\iint_{SB}\sigma(Q)\left(\frac{1}{r(P,Q)} + \frac{1}{r'(P,Q')}\right)ds$$
$$-\frac{1}{4\pi}\iint_{SF}\sigma(Q)\left(\frac{1}{r(P,Q)} + \frac{1}{r'(P,Q')}\right)ds \qquad (14)$$

where $P(x, y, z)$ is the field point, $Q(x_0, y_0, z_0)$ and $Q'(x_0, -y_0, z_0)$ are source points, $\sigma(Q)$ is the intensity of source point at the surface ds, r, r' are the distance between the field point and source points.

The nonlinear wave-making resistance coefficient can be calculated as follows:

$$p = \rho \left(U\varphi_x - \frac{1}{2}\nabla\varphi \cdot \nabla\varphi + gz \right) \quad (15)$$

$$C_w = \frac{R_w}{(0.5\rho U^2 S_0)} \quad (16)$$

where ρ is the mass density of the fluid, R_w is the wave-making resistance.

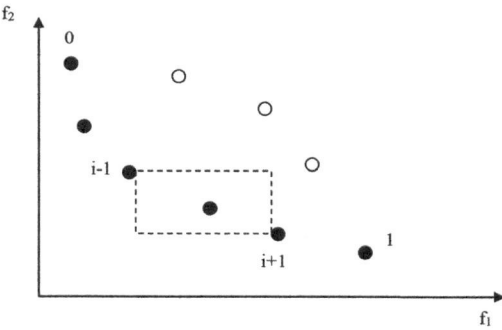

Figure 3. Crowding-distance calculation (Points marked in filled circles are solutions of the same non-dominated front).

4 OPTIMIZATION STRATEGY

The optimization algorithm is very important for the optimization process. In this paper, the multi-objective evolutionary algorithm NSGA-II is adopted as optimization strategy to search for the optimal hull form for the minimal total resistance. Contrary to classic optimization methods, the multi-objective evolutionary algorithm NSGA-II can maintain a number of candidate solutions and determine their relative merit (Deb et al., 2002). Comparing to the multi-objective evolutionary algorithms (EAs), the NSGA-II alleviates three difficulties including: $O(MN^2)$ computational complexity, non-elitism approach, and the need for specifying a sharing parameter.

4.1 Fast non-dominated sorting approach

In the fast non-dominated sorting approach, there are two entities domination count n_p that solutions dominate the solution p and S_p that solution p dominates. Domination count of all solutions in the first non-dominated front is 0. Domination count of each member (q) of S_p is reduces and if for individual q the domination count becomes 0, the individual q is put into a separate list Q that belongs to the second non-dominated front. The above process is continued with every member of Q until all fronts are defined.

4.2 Diversity Preservation

In order to the shortcoming of NSGA, the NSGA-II replaces the sharing function method with a crowded-comparison method. The crowded-comparison method is auto-definition parameters for sustaining diversity among population and has an excellent computational complexity.

(1) Density estimation: In order to obtain the density of solution in the population, the average distance of two points on either side of the point with every objective. The quantity regards as an estimation using

the nearest neighbors as the crowding distance. Figure 3 shows the crowding distance and the cuboid.

(2) Crowded-Comparison Operator: In the search process, the crowded-comparison operator is used to make sure that the algorithm can converge to a uniform distribution of Pareto-optimal. Each individual i belongs to two attributes including 1) non-domination rank(i_{rank}); 2) crowding-distance($i_{distance}$). If $i_{rank} < j_{rank}$, or $i_{rank} = j_{rank}$ and $i_{distance} > j_{distance}$, the lower(better) solution is selected. On the contrary, both solutions is the same front, the solution in a lesser crowded region is selected.

4.3 Main Loop

In the NSGA-II, the excellent individuals in all population can be got by selection, crossover and mutation using genetic operation after fast non-dominated sorting. Repeating the above procedure, the optimal solution isn't obtained until it meets the end condition. The flow chart for the flow of NSGA-II is presented in Figure 4.

5 OPTIMIZATION PROCEDURE OF THE HULL FORM

In this paper, the forebody of a ro-ro ship is optimizing for minimum the wave-making resistance coefficient. The optimal forebody form is generated by variation operators and the wave-making resistance coefficient is evaluated using Rankine-source panel method. The principal parameters of the ro-ro ship are shown in Table 3.

5.1 Optimization configuration

The wave-making resistance coefficient is evaluated in the design speed. Corresponding to the design speed, the Froude number is $F_n = 0.28$. The optimization function is defined as following:

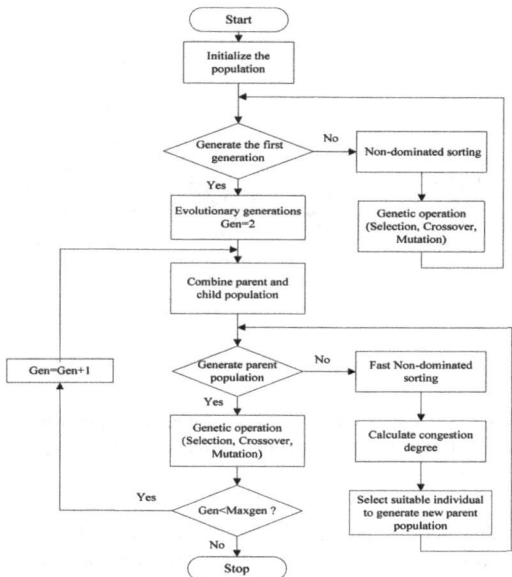

Figure 4. The flow chart of the NSGA-II.

Table 3. Principal parameters of a ro-ro ship.

Parameter	Value
L/m	165
B/m	24.8
T/m	8.7
C_B	0.65
Design F_n	0.28

$$Min\{C_w\}$$

Design variables:

The design variables defining the forebody are shown in Table 2.

Constraints:

The displacement: $|\Delta_o - \Delta_p|/\Delta_o \leq 1\%$

where Δ_o is the initial displacement of hull, and Δ_p is the displacement after optimization.

The centre of buoyancy longitudinal position: $|L_o - L_p|/L_o \leq 1\%$

where L_o is the initial centre of buoyancy longitudinal position of hull, and L_p is the displacement after optimization.

5.2 Optimization flow

In the optimization procedure, the Sobol algorithm is used to construct the design of experiment (DOE).

The Sobol sequence belongs to the quasi-random sequence that uses a base of two to establish sequentially better uniform partitions of the unit interval, then recorder the coordinates. The Sobol sequence is widely used because their common is regarded as a replacement of uniformly distributed random numbers. In order to obtain a Sobol sequence, a series of direction numbers are chosen. The initial direction numbers is selected at random, therefore the selected dimensions can get different realizations of the Sobol sequence. In order to a Sobol sequence, a set of direction numbers $v_{i,j}$ will be defined (Taghavifar et al., 2016). The design of experiment (DOE) is applied for the global search, which has high coverage rate and is better than the general global method. Then the NSGA-II is used to find the optimal forebody of hull that meets the objective and constraints in a small Pareto range. The fully automatic optimization integration framework is achieved by integrating the FRIENDSHIP with the CFD software SHIP-FLOW. The process of optimization is illustrated in Figure 5.

5.3 Optimization results

The form parameters are imported to the curve generator to obtain a set of feature curves. Then fairing surfaces can be derived automatically through surface generator. The parameterized hull surfaces are shown as Figure 6. from margin; Lock anchor).

The optimization results are depicted on Table 4 and Table 5. As can be seen, the wave-making resistance has obviously decreased while the constraints are satisfied after optimization. In the Figure 7, the comparison of the wave contours on free surface before and after

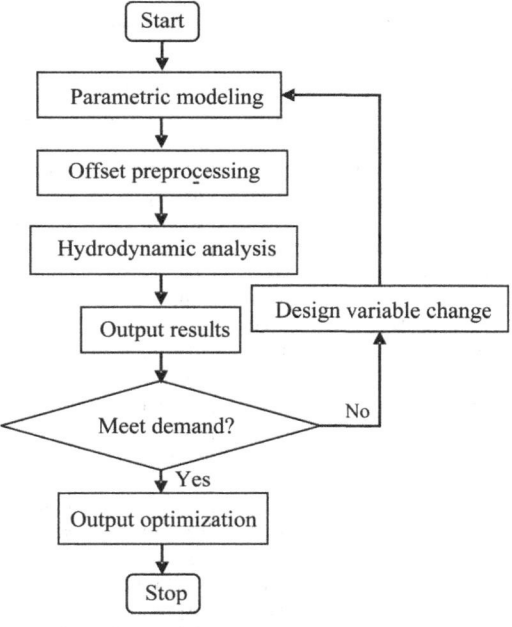

Figure 5. The flow chart of automation optimization.

319

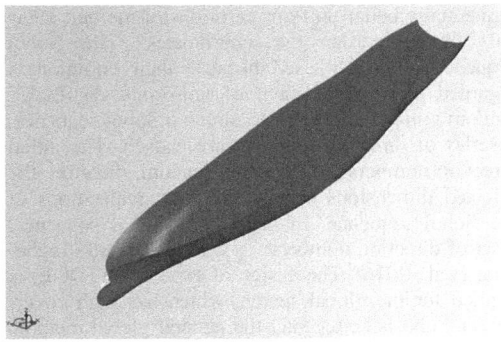

Figure 6. Hull surfaces of the Ro-Ro ship.

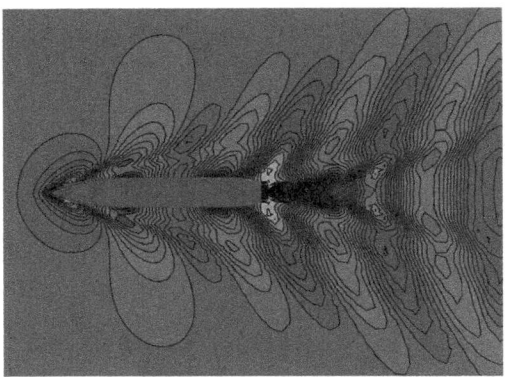

Figure 7. Comparison of the wave contours on free surface before and after optimization, lower part-before; upper part-after.

Table 4. Optimization results of the objective and constraints.

Description	Initial value	Optimized value	Variation
Nondimensional wave-making resistance C_w	0.0020368	0.0019177	5.85% ($\downarrow$)
Displacement $\Delta(t)$	22357.695	22408.439	0.227% ($\uparrow$)
Longitudal coordiante of boyance center L_{cb} (m)	0.515798	0.514674	0.218% ($\downarrow$)

optimization is presented. Comparison of the longitudinal wave cuts before and after optimization is shown in Figure 8. As can be seen, the wave numbers and the wave amplitude reduce obviously around the forebody after the lines of the forebody are optimized, which implies a decrease of wave-making resistance. In Figure 9, the dashed red lines represent the optimized results while black lines are with the original form.

Table 5. Optimization results of design variables.

Parameters	Initial value	Value range	Optimal value
MainparallelOffset	15.877	[8.6,27.8]	16.074
SACareaCoeffAtFosEmerge	0.889	[0.82,0.91]	0.861
SACtanAtFwdFrame	-59.358	[-68,-52]	-60.528
SACtanAtFp	-30.803	[-65,-24]	-32.034
SACmzAtFosEmerge	-9.043	[-13,-6]	-9.739
DwltanAtFp	-11.257	[-18,-5]	-10.535
DwltanAtFos	-0.913	[-1,0]	-0.945
DwlAreaCoeffFor	0.604	[0.58,0.62]	0.595
FOSxEmerge	1.108	[1.1,1.18]	1.121
FOBtanForEnd	-8.924	[-18,0]	-9.018
FOBareaCoeffFor	0.507	[0.5,0.56]	0.516
FwdFrameAreaCoeff	0.876	[0.8,0.9]	0.864
FlareAtDwlAtFwdFrame	4.121	[0,14]	5.283
DeadriseAtFwdFrame	14.702	[0,18]	14.031
BowLength	5.810	[5,7]	6.102
BowBeamAtFp	4.602	[4.2,5.16]	4.752
BowBeamAtElevAtFp	6.816	[5.5,7.5]	7.205
BowZTopFp	-0.628	[-0.68.-0.56]	-0.633
BowxAtBase	0.624	[0,1]	0.269

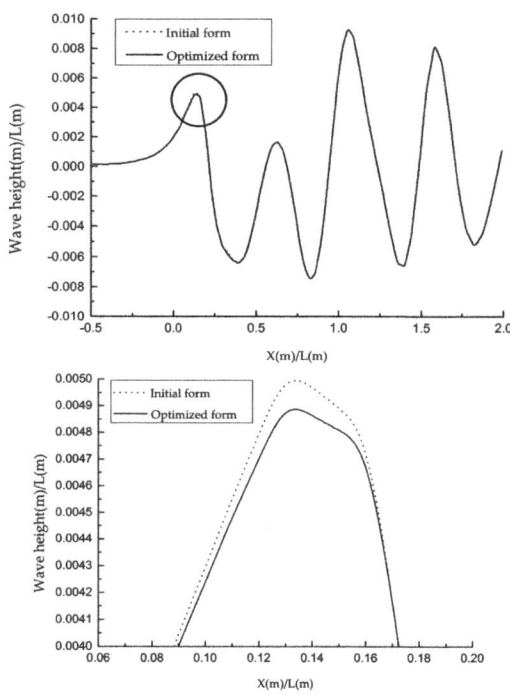

Figure 8. Comparison of the longitudinal wave cuts before and after optimization ($y/L = 0.1$).

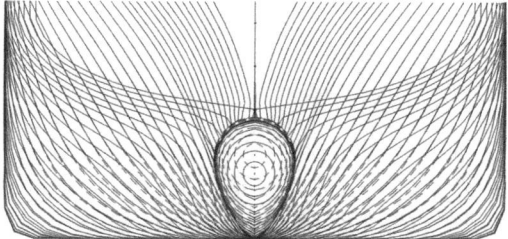

Figure 9. Comparison of the lines before and after optimization.

6 CONCLUSIONS

In this paper, a multi-objective evolutionary strategy is proposed for the optimization design of ship hull form. A fully automatic optimization integration framework is achieved. Through the framework, the full parametric modeling of the hull, hydrodynamic performance calculation, design evaluation and shape variation are implemented. A forebody of a ro-ro ship is taken as a verification model. The effectiveness of the proposed optimization strategy is demonstrated by optimal results. On the basis of the results, some recommendations are drawn: Parametric Modeling is the beginning to the hull optimization. And NURBS offers a simple, flexible and suited method to generate and modify ship hull quickly and makes sure that the design variation among the entire optimization process is effective and feasible. Multi-objective evolutionary strategies provide an effective and efficient tool to accomplish a multi-objective optimization. In this study, only hull is considered. In the next work, appendages including rudder and propeller will be taken into consideration in the optimization design of a ship.

ACKNOWLEDGMENTS

This work was partially supported by the Special Item supported by the Fujian Provincial Department of Ocean and Fisheries (no. MHGX-16).

REFERENCES

Deb, K., Pratap, A., Agarwal, S., & Meyarivan, T. (2002). A fast and elitist multiobjective genetic algorithm: nsga-ii. *IEEE Transactions on Evolutionary Computation*, 6(2), 182–197.

Harries S (1998). *Parametric design and hydrodynamic optimization of ship hull forms*. Ph.D. Thesis, Institut fü¨r Schiffs-und Meerestechnik, Technische Universita¨t Berlin, Germany; Mensch & Buch Verlag, Berlin.

Mahmood, S and D. B. Huang (2012). Computational Fluid Dynamics Based Bulbous Bow Optimization Using a Genetic Algorithm. *Journal of Marine Science and Technology*,11:286–294.

Nowacki H, Creutz G, Munchmeyer FC (1977) Ship lines creation by computer—objectives, methods, and results, symposium on computer-aided hull surface definition. *Annapolis*, MD.

Nowacki H, Boor MIG, Oleksiewicz B (1995) *Computational geometry for ships*. World Scientific, Singapore.

Nowacki H, Kaklis PD (eds) (1998) Creating FAIR and SHAPEpreserving curves and surfaces. *B.G. Teubner*, Stuttgart

Peri D, Campana EF (2005) High-fidelity models and multiobjective global optimization algorithms in simulation based design. *Journal of Ship Research*, 49(3):159–175.

Peri, D., Rossetti, M., & Campana, E. F. (2001). Design optimization of ship hulls via CFD techniques. *Journal of Ship Research*, 45(2), 140–149.

Saha GK, Suzuki K, Kai H (2004). Hydrodynamic optimization of ship hull forms in shallow water. *Journal of Marine Scienceand Technology*, 9(2), 51–62.

Tahara,Y.,D.Peri, E. F. Campana, and F.Stern.(2011). Single- and Multiobjective Design Optimization of a Fast Multihull Ship: Numerical and Experimental Results. *Journal of Marine Science and Technology*, 16: 412–433.

Taghavifar H, Jafarmadar S, Taghavifar H, A Navid (2016). Application of DoE evaluation to introduce the optimum injection strategy-chamber geometry of diesel engine using surrogate epsilon-SVR. *Applied Thermal Engineering*, 106,56–66

Zhang, B. J. (2009). The optimization of the hull form with the minimum wave making resistance based on Rankine source method. *Journal of Hydrodynamics*, 21(2), 277–284.

Zhang, B. J. (2012). Research on optimization of hull lines for minimum resistance based on Rankine source method. *Journal of Marine Science and Technology*, 20 (1), 89–94.

Zhang, B. J. (2015). The design of a hull form with the minimum total resistance. *Journal of Marine Science and Technology* 23(5), 591–597.

Developments in Maritime Technology and Engineering – Guedes Soares & Santos (eds)
© *2021 Copyright the Author(s), ISBN 978-0-367-77376-2*

Remarks about trends in fast ferry design

M.J. Legaz
Department of Sciences and Techniques of Navigation and Naval Construction University of Cádiz, Spain

C. Guedes Soares
Centre for Marine Technology and Ocean Engineering (CENTEC), Superior Technical Institute, Universidade de Lisboa, Lisbon, Portugal

ABSTRACT: Several maritime routes that operate with fast ferries can be found around the world. These maritime routes are usually regular routes established between island communities, coastal zones and inland waterways. The maritime transport on these routes has to be efficient to fulfil the modern requirements of this industry. The design of fast ferries plays an important role in the effectiveness and efficiency of this mode of transport. Ship design has to meet several requirements: fast ships need to be competitive with land-based transport schedules or even be able to reduce the journey times. The ships have to be efficient and eco-friendly, using sustainable propulsion systems and hydrodynamically optimized hulls. Likewise, the ships dynamic behavior is also important – passengers have to feel comfortable onboard during the journey. In this paper ship dynamic damping is commented upon along with some remarks about the methods and bibliography available to meet the requirements of modern fast ferry design.

1 INTRODUCTION

Worldwide, ferries play a very important role in global transportation. This role is almost comparable with the airline industry. The majority of fast ferry maritime routes are established between islands or in areas in which maritime transportation seems to have some advantages. For example, areas with particular geography where the use of land-based transportation would involve excessive distances.

From mainland United Kingdom to its islands and Ireland, France, Spain and Holland more than 39 million passenger journeys are made by ferries. This number of journeys is allowing ferry operators to invest in new ships and new routes (Batchelor, 2018). UK maritime routes can be seen in Figure 2.

Around the Mediterranean, several ferry lines and smaller operators are established. Some of the Mediterranean routes are shown in Figure 1.

In the next four years, the ferry industry will grow. The investment for new vessels and facilities will be more than £1bn in the UK alone. Because of this market growth, many ferry operators have made an investment in new sustainable vessels (Batchelor, 2018). The design of these new vessels plays an important role in the quality of maritime routes.

Ship design has to face several challenges to accomplish qualitative improvement. These challenges are: the ships have to directly compete with land transport or be fast enough to be an acceptable alternative, they also need to be efficient and eco-friendly and the passengers have to feel comfortable even in the worst weather conditions.

A review of the methods available to meet these challenges will be made in the next sections. In section 2, a review of the types of fast ships for passenger transport is made. In section 3, some eco-friendly European projects and eco-friendly catamarans are shown. Some remarks are made about the hull optimization process in section 4. The ways of obtaining more comfortable ships is explained in section 5.

2 REQUIREMENTS FOR FAST SHIPS

In order to compete with land transportation or to reduce journey time, ships have to be relatively fast. The ships that operate between islands or in coastal lines used to be classified in the category of "fast ferries", among them can be found: hydrofoils, hovercrafts, trimarans, catamarans and mono-hulls. Nowadays, the trend in fast ferries design is to have catamarans and mono-hulls.

Catamarans are quite popular, as short sea-route passenger ferries. The operators choose them for their high speed and low construction and maintenance costs. On the high-speed ferry market, the percentage of catamarans is about 60% (Yun & Bliault, 2012). About speed, several catamarans in service are capable of speeds in the region of 35-40 knots.

DOI: 10.1201/9781003216582-36

Figure 1. Mediterranean maritime routes. Source: (Ferrie-sonline, 2017).

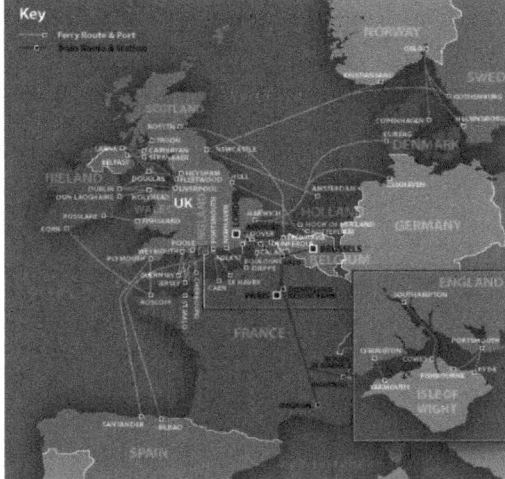

Figure 2. UK maritime routes. Source: (FerryUK24, 2019).

These vessels have displacements ranging from a few hundred tons to around 3,850 tons. In sheltered waters, small catamarans have reached speeds up to 50 knots (Lamb, 2003).

Other advantages of catamaran design are:

– Low wave-making because of high length/beam ratio. The previous one allows the designer to minimise the propulsion engine power rating for a given service speed.
– High transverse stability due to the distance between hulls. The transverse metacentric height (GM) will be around ten times higher than a mono-hulls.
– The centre of the bridge between two hulls provides large deck area. These areas can be used for spacious passenger cabins and other working cabins. The catamaran deck area is about 40-50% greater than that of the equivalent mono-hull.
– The distance in the twin propulsors allows high manoeuvrability, course stability and larger turning moment.

– Because of the hull slenderness compared with mono-hulls, catamarans have a reduced speed loss and low impact and slamming loads.
– Catamarans have a high safety level against flooding. They have many bulkheads in both side hulls, so the individual compartment volumes are small. Therefore, in the case of damage or a hull breach, the volume of flooding is less than in the mono-hulls. This fact allows us to have a high safety level against hull damage (Yun & Bliault, 2012).

3 ECO-FRIENDLY SHIPS

Global Warming, or Climate Change, due to the effect of increasing greenhouse gases in the atmosphere has received much attention from many authorities during the last few of years. The incremental increase in the average global temperature will have many negative implications for the world's ecosystems and on the life of human beings.

The governments of many nations have called for the global reduction of carbon dioxide (CO_2) emissions. These calls to reduce greenhouse gas emissions will also affect the maritime transportation industry. The International Maritime Organisation (IMO) established an index to measure the CO_2 emissions in new vessels, the Energy Efficiency Design Index (EEDI). The EEDI determines how clean a ship is by the amount of CO_2 it emits. It can be expressed as the ratio of its environmental impact divided by the benefits it brings to society, i.e. EEDI = environmental impact/benefit to society. It means, EEDI = CO_2 emissions/transport work (IRS, 2011). Use of the EEDI was made mandatory for new ships, and the Ship Energy Efficiency Management Plan (SEEMP) for all ships, at the Marine Environment Protection Committee (MEPC) 62 (July 2011) with the adoption of amendments to MARPOL Annex VI (resolution MEPC.203(62)) by parties to MARPOL Annex VI (MARPOL, 2011).

The previous index is directly related to the type and size of the ship. For passenger catamarans, some values of this index can be found in (Walsh & Bows, 2012). In the context of this document, when a reference to Eco-friendly propulsion system is made, this means a system of propulsion that can reduce the emissions of CO_2 and get good values of the EEDI.

Numerous European projects and research work have been undertaken in order to design eco-friendly ships. These lines of investigation have looked into obtaining less polluting marine engines, the use of batteries on short-shipping routes, the use of combined liquid natural gas (LNG) and diesel engines and the usage of renewable energy systems combined with more traditional, but cleaner, engines. Information on utilising other clean technology, such as hydrogen fuel cells, can also be found. Some European projects are discussed hereafter.

The E-Ferry project, with the objective of making a considerable reduction in the CO_2 emissions and air pollution in sea transportation, develops the E-Ferry design. The E-Ferry is a medium-size ferry for passengers, cars trucks and cargo which aims to validate the possibility of 100% electrically powered. The medium size ferries normally sail among island communities, coastal zones and inland waterways. One important characteristic of the batteries is their capacity of charging of up to 4 MWh, letting to short ports stops. Figure 4 shows a picture of the batteries. The E-Ferry battery pack is one of the largest ever put on a ferry. Other aspects of the design and construction are being taken into account in order to make the E-Ferry efficient. Some of them also possess an optimized hull-shape, lightweight equipment and widespread use of carbon composite materials (E-Ferry, 2015). A picture of the E-Ferry can be seen in Figure 3.

In constrained depths and in confined waters, the E-Ferry could sail to speed of 13-15 knots. This speed is higher than its normal speed of 10-12 knots (E-ferry, 2015).

There have been other European projects similar to the E-Ferry: SEABUBBLE, BB Green and GFF with the aim of developing electric ferries and small boats for the transportation of passengers.

In order to show more European projects, the Green Fast Ferry (GFF) should be mentioned. It has

the aim of designing a fast ferry, which is able to reach a speed of 30 knots with zero emissions. It is a high-speed passenger, full-electric craft. The autonomy of the ferry is 26 km in 30 minutes. The time for recharge is less than 20 minutes. The technology used by GFF is of the Air Supported Vessel (ASV). Because of this technology, a reduction of energy at least 40% is obtained. The technology consists of a pressurized air cushion underneath the hull. The ASV is used only in sheltered or inland waters (GFF, 2016).

Balearia and Buquebus are two companies that are investing in eco-friendly fast ferries:

The LNG powered high-speed catamaran of the Spanish shipping company, Balearia. The Balearia catamaran has a length of 125 meters and a beam of 28 meters: this being one of the longest, and highest capacity fast ferry catamarans. The capacity will be 1,200 passengers and 500 cars. The propulsion system will be formed from four Wärtsilä dual LNG/diesel engines delivering 8800 kW each. The catamaran will have a service speed of 35 knots, and a top speed of over 40 knots, with a range of 400 nautical miles. It's equipped with two tanks to store the chilled LNG fuel. Balearia aims to extend the use of LNG fuelled ferries across their routes in the Mediterranean. The first eco-route will be established operating in the Balearic Islands. The project has received a rating of excellent from the European Union (Baleraria, 2018). The Balearia catamaran can be seen in Figure 5.

The Francisco, a high-speed Ro-Ro ferry operated by Buquebus, operates with LNG as the primary fuel. The Francisco (formerly the Lopez Mena) is the first dual-fuelled high-speed Ro-Ro ferry to operate with LNG as its primary fuel. The high-speed ferry was built by Incat for its operator Buquebus. It commenced operation in 2013 on the River Plate between Buenos Aires, Argentina, and Montevideo, Uruguay.

The catamaran Francisco has a length of 99 m, a beam of 26.94m and a load capacity of 1000 passengers and 150 cars. The propulsion system consists of two Wärtsilä LJX 1720 SR waterjets driven by two GE Energy LM2500 gas turbines. A waterjet and gas turbine is located in each hull. Four Caterpillar C18 generators, rated at 340kW each, and two Caterpillar C9 generators, rated at 200kW each, are also used to power auxiliary electrical propulsion systems.

Figure 3. E-Ferry. Source: (E-ferry, 2015).

Figure 4. Batteries of the E-Ferry project. Source: (E-ferry, 2015).

Figure 5. Balearia catamaran. Source: (Balearia, 2018).

Figure 6. Catamaran Francisco. Source: (Francisco, 2013).

Francisco can reach a speed of 51.8 knots and has a deadweight capacity of 450 tonnes. Francisco has obtained a classification of 1A1 HSLC R4 Car Ferry B Gas Fuelled E0 (Francisco, 2013). Figure 6 shows an image of Francisco.

For eco-friendly fast ferries hydrogen fuel cell technology is also used (Hydrogenessis, 2013) as well as the hybrid, diesel-electric, propulsion system (MV Hallaign Ferry, 2012). A combination of a conventional system of propulsion with sources of renewable energy is another possibility for environment-friendly systems. Renewable power applications can be used in ships of all sizes for primary, hybrid and auxiliary propulsion.

Renewable energy available sources for ships are: solar/photo-voltaic power, wave energy and biofuels.

Wind, the energy of the wind can be captured using soft-sails, fixed-sails, Flettner rotors, kite sails and/or wind turbines.

- The so-called soft-sails are conventional sails attached to yards and masts. This is an ancient technology, brought up to date, that directly harnesses the propulsive force of the wind.
- Fixed-sails can be defined as rigid wings with a rotary mast. Projects that use this technology are UT Wind Challenger (UT wind challenger, 2009) and Effship's (Effship's, 2013).
- Flettner Rotors use the Magnus effect, that produces an aerodynamic force as the wind passes over an already revolving cylinder, to obtain propulsion.
- Kite sails, attached to the bow of the ship, can be operated at the appropriate altitude to maximise the force obtained from the wind for propulsion.
- For ships, wind turbines have been used as a propulsion system over the years. Until now, they have not been particularly successful.

Another clean energy source can be obtained using solar energy; the electricity is generated by photovoltaic cells. These cells can be placed in different ways on board; in fixed wings sails or horizontally on deck (IRENA, 2015).

Currently, various environment-friendly possibilities can be found for new fast ferry designs. The designer can choose one or a combination of them, as the best option for a specific design project.

4 EFFICIENCY IN HULL FORMS

The word optimization can refer to different aspects. In this section, an efficient ship means a ship in which its hull form offers the least resistance to moving through the water.

The resistance, or drag, the ship experiences by sailing through the seawater, is very much dependent on its hull form. For this reason, the process followed to design and define the hull form is really important. For many years, a technique to check the resistance of a ships hull have been tank tests using scale models. In these tests, the resistance of the hull form can be relatively straightforwardly measured. The theory of dimensional analysis can be utilized in order to get the resistance of the ship. With the advances in computers and associated calculating power, computational fluid mechanics (CFD) are applied to study the characteristics of the fluid flowing around the hull of a ship (Min et al. 1999). Even so, tank tests are still necessary to confirm, or otherwise, the calculated prediction.

Nowadays, the tendency followed by researchers in different countries is the combination of computer-aided design (CAD), CFD and methods of optimisation. (Zhang & Zhang, 2019). Figure 7 shows the cyclic nature of the process.

In the field of CAD, different methods to insert a hull into CAD software can be found, depending on the type of CAD software that is being used. The hull can be introduced in CAD software in a parametric or non-parametric way (conventional design). In a non-parametric way, the hull is introduced by a set of data points, depending on the type of software used (NURBS, B-Spline, etc), as an interpolation technique to adjust the hull form. The software has usually programmed some fairness criteria. In a parametric way, the hull is defined by a set of variables and parameters. In the parametric design, any changes can be propagated throughout the model when updating any related parts, therefore only a few modifications are required in order to achieve a new fair hull form.

With respect to optimisation, different decisions have to be made as to whether the problem is a single-

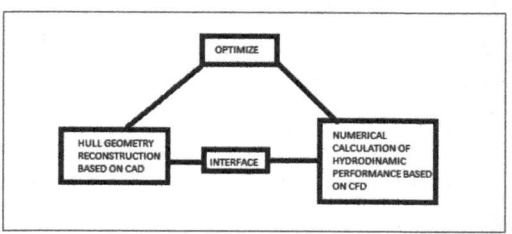

Figure 7. The optimization process of hydrodynamic performance based on CFD technology. Source: Own elaboration.

object or multi-object problem. In a single-object problem, only one component of the problem has to be minimised. In the multi-object problem, several components of the problem have to be minimised.

When we are faced with a multi-object problem, a suitable process of making decisions has to be adopted since an improvement in one component might cause a deterioration of another within the problem. This model is not able to combine the best performance for every objective, so the decision-making process plays a critical role in optimisation. The aim of the decision-making process is to find a solution that gives the best compromise solution. This solution can be found in the Pareto front, thus obtaining a set of feasible outcomes. Another important point is that various numerical algorithms can be utilized to solve the optimisation problem.

An example of research work carried out in this subject is the labour of Zaraphonitis and others (Zaraphonitis et al. 2019), in which a hull form optimisation procedure is carried out in order to minimise the wash and total resistance of high-speed ships. In this research the combination of three software programmes is made: firstly, NAPA a ship design software package used for the generation of several hull forms, secondly, the CFD software "Shipflow" to perform the hydrodynamic evaluation of each hull form and, thirdly, the optimization of the hull was made with mode Frontier software, using the method of Genetic algorithms. This research studied two class of ships, a semi-displacement mono-hull and a high-speed catamaran.

5   COMFORT ONBOARD

Another significant design aspect that a new fast ferry design has to achieve is the comfort of the passengers. Passengers can suffer due to the vertical and horizontal accelerations caused by the motion of the ship, which can result in discomfort, dizziness and nausea. The catamarans present very similar profiles of rolling and pitching in natural periods. This can cause substantial discomfort to the passengers and crew, especially in the case of shipping in quartering seas (Yun & Bliault, 2012).

Some works can be found in which a modification of hull form is performed in order to reduce the vertical accelerations combined with roll and pitch movement. In this respect, the work of Piscopo & Scamardella (2015) can be relevant. In the research, the Motion Sickness Incidence (MSI) index is used to optimising the hull form of a wave-piercing high-speed catamaran. The hull is modelled with the parametric method in order to generate two families of derived hull forms, the former varying the prismatic coefficient and the position of the longitudinal centre of buoyancy. After some time, the optimal hull shape is obtained, a comparison between the vertical accelerations at some critical points on the main deck and the parent ones are made. The MSI index can be expressed as the percentage of passengers who vomit after two hours of exposure to certain motion.

The problem of modifying the hull form to optimise passenger comfort is that the new hull form is probably not as efficient as before at cutting through the water. It is extremely difficult to modify the pitch and roll natural period without generating extra inefficiency in the hull form (Yun & Bliault, 2012). For this regard, it is necessary to pay attention to the modification of the damping coefficient of the hulls in roll and pitch.

With the aim of increasing damping, on modern high-speed catamarans have extensively used automated pitch control hydrofoils and transverse anti-rolling which improve their seaworthiness significantly. Other technologies utilised to minimise the movements of the ship are anti-rolling tanks and bilge keels.

The work of De la Cruz and others can be read as an example about improving the comfort of a fast ferry. The work is based on a ship sailing in heavy seas. A ship with two flaps at the stern and a T-foil near the bow is studied. Control inputs were applied to them in order to improve overall comfort (De la Cruz et al. 2004).

6   CONCLUSIONS

The main conclusions obtained in this study are:

- In ship design, as in most engineering design, a compromise solution must be taken based on the different objectives and requirements of the project.
- In this study, it is made a revision of the methods and systems currently available that can help to meet the requirement of a new fast ferry design.
- The study is based on fast catamaran ferries because they are a normally selected option. A revision of European projects in hull optimisation methods and damping systems that can be used in this type of ships has been carried out.
- Ultimately, the responsibility for deciding which are the most appropriate technology development and methodologies to use is that of the design team.

ACKNOWLEDGEMENTS

Research supported in part by programa de fomento e impulso de la investigación y la transferencia en la UCA 2019/20 and programa de apoyo a la investigación de la fundación Seneca-Agencia de Ciencia y Tecnología de la Región de Murcia 20928/PI/18, which supported the visit of the first author to CENTEC.

REFERENCES

BaleariaCatamaran.2018 https://www.balearia.com/es/noti cias/fast-ferry-gas-en [accessed 2019 January 1].

Batchelor, E. 2018. What 's driving the global ferry indus-try? Ship Technology, 2018:1–6 https://www.ship-tech nology.com/features/whats-driving-global-ferry-indus try/[accessed 2019 January 1].

De la Cruz J.M., Aranda J., Girón-Sierra J.M. & Velasco F. J. 2004. Improving the comfort of a fast ferry. IEEE Control Systems Magazine,24: 47–60

E-ferryProject.2015 https://cordis.europa.eu/project/rcn/ 193367/factsheet/en [accessed 2019 January 1].

Effship's Project..2013 http://www.effship.com/PublicPre sentations/Final_Seminar_2013-03-21/06_EffShip-Wind_propulsion-Bjorn_Allenstrom_SSPA.pdf [accessed 2019 January 1]

Francisco Catamaran. 2013 https://www.ship-technology. com/projects/francisco-high-speed-ferry/[accessed 2019 January 1].

Ferriesonline.2017 http://www.cemar.it/principale_eng. html [accessed 2019 January 1].

FerryUK24. 2019 http://ferryuk24.co.uk/ [accessed 2019 January 1].

Hydrogenessis Ferry Project. 2013 https://www.ship-tech nology.com/projects/hydrogenesis-passenger-ferry [accessed 2019 January 1].

GFF Project.2016 https://cordis.europa.eu/project/rcn/ 206526/factsheet/en [accessed 2019 January 1].

MV Hallaig Hybrid Ferry Project. 2012 https://www.ship-technology.com/projects/mv-hallaig-hybrid-ferry [accessed 2019 January 1].

Indian Register of Shipping. 2011. Implementing Energy Efficiency Design Index (EEDI). Powai

IRENA. 2015. Renewable energy options for shipping. Technology brief, 2015:1–60

Lamb R. 2003. High-speed, small naval vessel technology development plan. NSWCCD-20-TR-2003/09:1-130. West Bethesda.

MARPOL.2011 https://www.marpol-annex-vi.com/eedi-seemp/[accessed 2019 January 1].

Min K.S., Choi J.E., Yum D.J., Shon J.H., Chung S.H. & Park, D.W. 1999. Study on the CFD application for VLLC Hull-Form design. Twenty-Second Symposium on Naval Hydrodynamics.

Piscopo V., Scamardela A. 2015. The overall motion sick-ness incidence applied to catamarans. International Jour-nal of Naval Architecture and Ocean Engineering, 7: 665–669

UT Wind Challenger Project. 2009 http://wind.k.u-tokyo. ac.jp/index_en.html [accessed 2019 January 1].

Walsh C., Bows A. 2012. Size matters: Exploring the importance of vessel characteristics to inform estimates of shipping emissions. Applied Energy, 98:1–10

Yun L. & Bliault, A. Springer 2012. High Performance Marine Vessels. New York.

Zaraphonitis G., Papanikolaou A., Mourkayannis D. 2014. Hull-form optimization of high speed vessels with respect to wash and powering. IMDC 2003: The 8th International Marine Design Conference. Athens.

Zhang B.J. & Zhang S.L.2019. Research on Ship Design and Optimization Based on Simulation-Based Design (SBD) Technique. Springer, Shanghai, pp -1–233.

Developments in Maritime Technology and Engineering – Guedes Soares & Santos (eds)
© 2021 Copyright the Author(s), ISBN 978-0-367-77376-2

Cultural image for yacht appearance personalized design

L.S. Ma & Y. Wu

School of Art and Design, Wuhan University of Technology, Wuhan City, China

ABSTRACT: In view of the emotional intention of Chinese consumers, how to design a yacht with Chinese style is the current problem encountered in yacht design. The key technology is how to configure the personalized component module so that its modeling image can truly meet the emotional requirements of consumers. From the perspective of cultural design, Firstly, this paper establish a transformation relationship model from cultural elements to design elements, and build Chinese cultural symbol library to guide the direction of cultural design. Secondly, the classic brand yacht design cases are selected and analyze the aesthetic characteristics. The personalized genes and inheritance genes are refined. Shape grammar and Bezier curve are used to derive the personalized design genes to generate the yacht personalized design components. Thirdly, using the ant colony optimization thought as the guiding ideology of personalized design component configuration, eye tracking is used to select the personalized component that meets the consumer's preference from the personalized component module library for plan configuration. Finally, from the perspective of physiology and psychology, the validity of this method was verified by eye movement experiment and scale experiment.

1 INTRODUCTION

Yacht is a representative luxury, and its customers are the top consumer groups. Therefore, yacht appearance design is different from general industrial product design In addition to considering functional requirements, yacht appearance design should pay more attention to the interpretation of emotional images by appearance. From the beginning to the development of China's yacht manufacturing industry, the technology is gradually becoming mature, but the yacht aesthetic needs to be improved. At present, the yacht design in China rarely considers the personalized emotional needs of consumers, so it fails to create a yacht with Chinese characteristics and lacks the design of constantly innovating.

Yacht design is a very important link in the yacht industry chain. Scholars have studied the innovative design of yacht appearance from different aspects，an innovative method for an automated process planning and quality control of the coating process of luxury yachts is put forward by Volker B (Volker et al. 2018). It uses mathematical characteristics and limitations to describe yacht aesthetic criteria, and shows the current manufacturing and quality control process. In the preliminary ship design, McDonald T P puts forward a Library based approach for Exploring Style, which is used in the ship appearance design (Mcdonald, 2010). Shahroz K proposed a new design framework for the generation and parametric modification of yacht hull surface, which was applied to the innovative design of yacht hull (Shahrozk et al. 2017).Zhang L studied the

yacht design method guided by consumption psychology and behavior, and discussed the relationship between consumption psychology, behavior characteristics and yacht appearance preference(Zhang,2011). Dai G M established a mathematical statistics model of the relationship between the design elements of the yacht appearance and perceptive imagines, and gave corresponding yacht design suggestions(DaiI, 2011). To sum up, the yacht creative design based on people's emotional requirements is rarely involved at present.

Nowadays, the consumption pattern has changed from the previous function-oriented to the consumer-oriented, and consumers' personalized demand for products is becoming more and more urgent. It is difficult for the homogenized products to bring more sense of quality to consumers. Yacht is no exception. The personalized customization of yacht will be the mainstream of the future business model. Customers can design the appearance of yacht according to their own preferences, so as to meet their personalized requirements.

The style of yacht is deeply influenced by different cultures. According to the regional style, yacht can be divided into Italian yacht, British yacht, French yacht, American yacht, Taiwan yacht, etc. Yacht design needs culture. Culture is an important factor to arouse consumers' emotion, and reasonable integration of cultural elements is an effective means to attract consumers to buy. Therefore, the development of yacht design requires the integration of culture and other design factors that enhance emotions on the basis of respecting consumers 'individualized wishes, to arouse consumers' "sensibility" resonance.

DOI: 10.1201/9781003216582-37

2 YACHT DESIGN POSITIONING

(1) Crowd positioning: Chinese consumers are chosen as the target group of this study.

With the rapid development of China's economy and the increasing purchasing power of Chinese consumers, China has become a powerful luxury sales market in the world. Therefore, Chinese consumers are chosen as the target group of this study.

(2) Style positioning Design objective: Chinese style yacht.

Different national cultural backgrounds do affect the modeling style of yachts, which essentially reflects the influence of traditional ideology and culture on art. As far as Chinese yacht design is concerned, to enter the international market, it is necessary to integrate the national complex into our own yacht design. Only the most national things can obtain the international community's pass. Therefore, the "Chinese style" of yacht design is not only a problem faced by designers but also a problem to be solved by the Chinese yacht industry.

(3) Demand orientation: emotional requirement

Maslow's hierarchy of needs theory divides human requirements from low to high: physiological requirements, security requirements, emotional and belonging requirements, respect requirements and self-realization requirements. At present, people's physiological requirements and safety requirements have been basically met. On this basis, people have begun to pursue higher-level requirements. The personalized customization of the yacht is a concrete manifestation of Maslow's hierarchy of needs theory. Customization can allow users to obtain satisfaction on the emotional and belonging level. The personalized design of yachts enables users to participate in the process of yacht design and obtain emotional and self-realization satisfaction through yacht design. In addition to personalized design to meet consumers 'emotional needs, consumers' emotions about local culture or country will also be transferred to corresponding products. In the Chinese yacht market, for Chinese consumers, in the process of personalized yacht design, the cultural connotation of Chinese elements can be integrated to stimulate the emotional resonance of Chinese consumers, to realize the localization strategy of yacht design, and design a Chinese style Yacht.

3 METHOD

3.1 Research ideas

This study proposes a personalized design method of yacht appearance based on cultural image. The research idea of this study is shown in Figure 1.

Figure 1. Research ideas.

3.2 Cultural design theory

3.2.1 Establish the target cultural image

Beyond the material core of luxury, there must be a very important cultural connotation or emotional attribute that supports its high-level value. Cultural identity positively affects consumers' attitude towards domestic products. National identity positively affects consumers' willingness to purchase domestic products. Local identity positively affects consumers' preference for local brands. If we can control some cultural images that can arouse consumers before the yacht design and development, it will greatly improve the design efficiency and better meet the hidden needs of consumers. Therefore, this study first needs to establish cultural reference images. In terms of yacht design, designers express their yacht design concepts through cultural semantics, and consumers choose from the morphological and linguistic information of the yacht.

"Kirin, Phoenix, Turtle, and Dragon are four spiritual creatures." As shown in Figure 2. They are the creatures worshiped by Chinese ancestors, symbolizing auspicious signs and metaphors for rare and precious things. Therefore, this paper takes the traditional Chinese classic totem culture as an example to establish an image set of Chinese cultural elements to guide the design of the yacht.

3.2.2 Conversion from cultural elements to design elements

In the process of interaction, people use symbols to express behaviors, customs, values and other recessive gene attributes, so as to form a specific culture.

Figure 2. Establishment of target cultural image.

In semiotics, symbols have dominant gene attributes such as semantic, pragmatic, syntactic and context. Therefore, from the perspective of symbol interaction, we can construct the mapping relationship between cultural elements and design elements, and realize the transformation from cultural elements to design elements, such as "semantic-value", "syntactic-symbol", "context-behavior", "pragmatic-custom".

The knowledge of cultural design is often multidimensional. According to the idea of "one element and many features" of extension basis element theory, the design information of yacht can be described from two dimensions of "semantic-value" and "syntactic-symbol". Some cultural characteristics are regarded as knowledge and fed back to the new design and development. Phoenix is the auspicious bird in people's mind, symbolizing the peace of the world, and also a metaphor for people with talent and virtue. In this study, the phoenix totem is taken as an example, and the transformation mechanism of cultural elements and design elements is represented by the fusion of extension basis element theory and graphic semantic method, as shown in Figure 3.

3.3.3 The personalized design process of yacht appearance

This practical research is based on the yacht industrial design project of China Sunbird Yacht Co., Ltd. the vision of Sunbird company is to "move the world", hoping to be a yacht affordable by Chinese people, providing all-round personalized

services from scheme design, product manufacturing to maintenance service.

3.3.4 Design gene analysis of Sunbird Yacht

The case of this project is the personalized design of "phoenix" series of Sunbird brand yachts. "Phoenix" is a type of yacht independently designed and developed by Sunbird company. "Phoenix" series yachts are based on the ancient god bird "phoenix" in Chinese culture. The aesthetic style of yacht is mostly reflected in side view, such as aesthetic characteristics, linear relationship, scale value, etc. Therefore, this study obtains a large number of side view pictures of yacht appearance design through the official website of Sunbird, carries out comparative analysis from the perspective of design gene (as shown in Figure 4), determines inheritance gene (red analysis line in Figure 4) and personalized design gene (blue analysis line in Figure 4), provides reference for personalized design, and makes the yacht modeling design of Sunbird brand innovative while maintaining stability and continuity.

The inheritance genes and personalized design genes of Sunbird yachts are obtained, as shown in Table 1. For inherited genes, they cannot be changed in the innovative design of yachts, and the design genes of the original brands should be preserved. For personalized design genes, they can be derived from shape grammar and Bezier curve into a variety of creative design with cultural image

3.3.5 Genetic variation of personalized design gene

Yacht personalized design gene is not only the characteristic line of yacht modeling, but also the basic unit of yacht modeling evolution operation. Therefore, the shape grammar can be used to carry out genetic and mutation operations on personalized design genes. The variation rule of shape grammar focuses on "heredity", which is adjusted, crossed and transformed on the basis of the original shape (Wang et al. 2016), and its rules include: $r_1 \rightarrow$ Replacement, $r_2 \rightarrow$ Additions and deletions, $r_3 \rightarrow$ Scaling, $r_4 \rightarrow$ Duplicate, $r_5 \rightarrow$ Rotating, $r_6 \rightarrow$ Shear, $r_7 \rightarrow$ Coordinatechange. Assuming the shape A, the process TA of performing the geometric transformation is expressed as:

$$TA = [x, y, 1] \cdot \begin{bmatrix} 1 & 0 & 0 \\ 0 & 1 & 0 \\ \Delta_x & \Delta_y & 1 \end{bmatrix}, \text{ where } \Delta_x \text{ and } \Delta_y$$

represent the coordinate momentum of the shape A.

Since the yacht is a geometric feature space composed of multiple information layers of points, lines and planes, the characteristic surface of the yacht model contains characteristic lines, and the characteristic lines contain characteristic points. Therefore, characteristic points are the important representation of yacht modeling features. Bezier curve divides the characteristic line into several anchor points

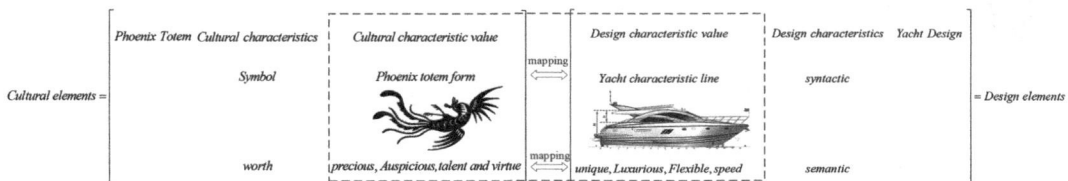

Figure 3. Conversion mechanism between cultural elements and design elements.

Figure 4. Design gene analysis of Sunbird Yacht.

Table 1. Inherited genes and personalized design genes localization design.

Inherited genes	Personalized design genes
Top outline of ship room	porthole
Cabin height (h)/Height above waterline(H)	sheer curve
Posture of yacht Center	Cabin porthole
Angle between bow	Radar frame shape

$p_i(i = 1, 2, \cdots, n)$. In this study, a cubic Bezier curve is adopted, and its expression is as follows: $B(t) = P_0(1 - t)^3 + 3P_1t(1 - t)^2 + 3P_2t^2(1 - t)^2 + P_3t^3, t \in [0, 1]$ When applying shape grammar reasoning to yacht modeling image design cases, adjusting the control points of Bezier curve anchor points can realize the inheritance and variation of yacht personalized design genes, and can better continue the modeling style. This research project takes yacht porthole design as an example, and its process is shown in Figure 5.

As the saying goes, the eyes are the window of the soul. In this study, the eye symbol features of the phoenix bird totem are extracted to form the initial shape of the inheritance and variation of the yacht porthole morphology. By using Bezier curve to describe the outline of the appearance, using shape grammar rules and adjusting the control line of

Bezier curve anchor point, the modification and innovation of yacht porthole are carried out. The deduction process is a recursive process until it satisfies the user's perceptual image.

This method can realize the derivative design of the personalized design gene of sunbird yacht and form a variety of design schemes, as shown in Figure 6. Figure a is a derivative design of the sheer curve of a yacht. Figure b is the derivative design of the porthole of yacht. Figure c is a derivative design of the cabin porthole. Figure d is the derivative design of the radar frame shape.

3.3.6 *Scheme configuration of personalized design component based on eye tracking*

In the process of yacht personalized design, the key technology is how to configure the personalized component module so that its modeling image can truly meet the emotional requirements of consumers. Scholars study this issue from different aspects. Norman believe that the personalized customization process is the interaction process between users and products. The emotional state of user interaction process can best reflect the emotional state of users and the experience state (Dlana et al. 2012). Zhang points out that personalized customization is a process for customers to interactively select product design elements, and the key lies in how to conduct virtual display of products(Zhang et al. 2014). Hunt puts forward individual unique needs, the pursuit of optimal products and the degree of attention to product aesthetics will affect consumers' perceived value of personalized customization(Hunt et al. 2014). Dong uses shape grammar to match knowledge from module division, module deduction and module reorganization(Dong et al. 2017). Through the above research, it is found that in the configuration process of the personalized design components of the yacht appearance, the fuzziness and uncertainty brought by the interfering factors to

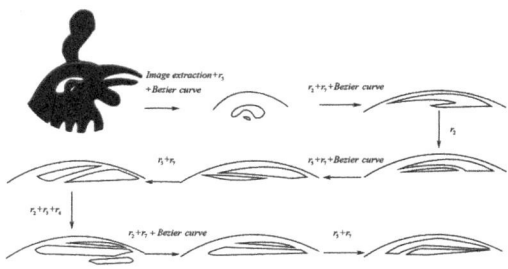

Figure 5. Genetic and variant design process of porthole.

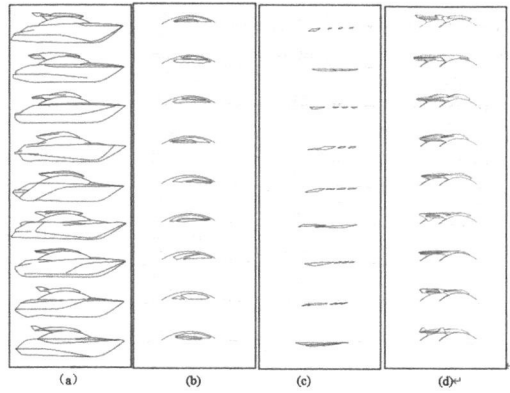

Figure 6. Derivative design scheme of personalized design gene of Sunbird Yacht.

the consumer's will cannot be effectively eliminated, and it is impossible to determine whether the configuration image truly conforms to the emotional intention of the consumer.

As an effective means to acquire users' tacit knowledge and design object knowledge, eye movement tracking technology makes the tacit knowledge that is difficult to express explicit and forms the knowledge system needed for design(Wang et al. 2017). Through the data analysis and transformation methods such as hot spot map, scanning path, fixation time, fixation points and areas of interest, etc., the attractiveness of personalized design modules to consumers' vision can be obtained, and the explicit knowledge of consumers can be tapped .From the physiological point of view, eye tracking technology is used to monitor the personalized design component configuration process of yacht appearance. By using objective data instead of subjective intention, we can truly reflect the design intention of the subjects, so as to quickly design the yacht appearance design to meet the consumers' perceptual image.

3.3.7 Selection of personalized design components

The configuration process of yacht appearance design component is essentially a dynamic one-way connected graph. There is a source point representing the input of personalized design component information and an end point representing the output of yacht appearance design configuration scheme. There are 4 personalized design genes for yacht appearance design, which need to be derivative design. The part of personalized design is recorded as $\{PDG_1, PDG_2, PDG_3, PDG_4\}$. Each personalized design part (PDG) has 9 personalized design component schemes, which are recorded as $\{X_{n,1}, X_{n,2}, X_{n,e}, \cdots, X_{n,9}\}, 1 \leq e \leq 9$, where n represents the number of personalized design part (PDG), $1 \leq n \leq 4$. Each personalized design component corresponds to an area of interest, which is recorded as $\{AOI_{n,1}, AOI_{n,2}, AOI_{n,e} \cdots, AOI_{n,9}\}, 1 \leq e \leq 9)$. Path represents the scanning path of personalized design component$X_{n,i}$ to $X_{n+1,j}$, where $1 \leq n \leq 3, 1 \leq i \leq 9, 1 \leq j \leq 9$. Each path has an $X_{n,e}(ph_1, ph_1, \cdots)$attribute constraint, and the constraint value contains multiple pheromones. There are multiple paths from the starting point to the end point, indicating that there are multiple configuration schemes of personalized design components to meet the needs of consumers, as shown in Figure 7.

In the ant colony algorithm, ants will probabilistically choose a path with more pheromones, indicating that this is the better path . In the personalized design of yacht appearance, pheromone is affected by the multi-dimensional information such as the consumer's own heterogeneity, the appearance of the design scheme, cultural semantics and color, ect., which brings the problems of fuzziness and uncertainty to consumers when they choose the

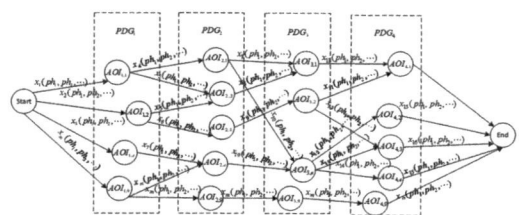

Figure 7. Component configuration process based on Ant Colony Optimization.

personalized design components. In order to further quantify the significance weight of the personalized design component $X_{n,e}, 1 \leq n \leq 4, 1 \leq e \leq 9$ in the personalized design part (PDG$_n$), the number of fixation points corresponding to the area of interest (AOI$_{n,e}, 1 \leq n \leq 4, 1 \leq e \leq 9$) of the personalized design component is counted as the measurement index. Finally, the personalized design component with more attention points is selected as the optimal solution of the personalized design part (PDG$_n$), thus forming the personalized design configuration scheme of yacht appearance.

3.3.8 Configuration experiment of personalized design component.

Eye movement experiment 1: configuration experiment of personalized design component

The purpose of this experiment is to conduct research on the configuration method of personalized design components of the yacht's appearance. The main tester introduces the experimental methods and tasks to the subjects, and performs visual calibration on each subject separately. The eyes of the subjects and the display remain flat, about one arm away. After the calibration, an instruction was issued to ask the subjects to view the area of interest (AOI) in the first 4 sample images and to look at the design components in accordance with their personal preferences for 20,000 ms.

40 subjects successively viewed 4 visual stimulation images of the personalized design components of the yacht's appearance. Taking the distribution of the number of fixation points formed by the 20th subject as an example, as shown in Figure 8. The specific experimental data is shown in Table 2.

The personalized design component corresponding to the interest area with the highest number of fixation points among the four personalized design parts was selected as the program configuration component of the 20th subject. The configured design program is shown in Figure 9.

4 METHOD VERIFICATION

In order to verify the effectiveness of the personalized design method of yacht appearance based on

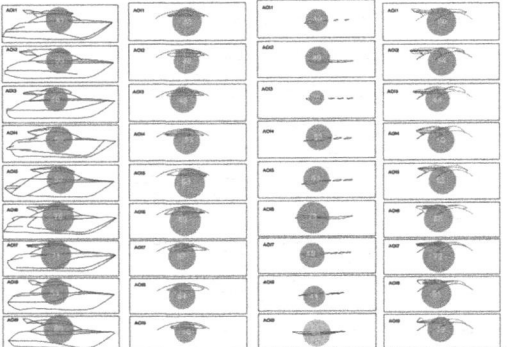

Figure 8. The distribution of the number of fixation points of the 20th subject.

Table 2. The 20th subject-data of fixation points.

AOI	sheer porthole	Cabin curve	Radar frame porthole	shape
AOI1	57	58	40	60
AOI2	55	48	49	54
AOI3	46	46	26	48
AOI4	50	53	53	50
AOI5	52	63	53	52
AOI6	78	69	78	68
AOI7	71	57	45	72
AOI8	53	54	41	55
AOI9	60	40	65	49

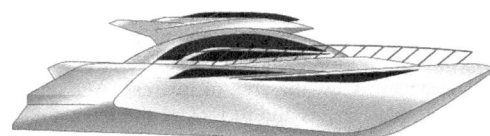

Figure 9. The 20th subject-configuration plan.

cultural imagery, this study used eye movement experiments and 7-level LIKERT scale (-3 ~ + 3) experiments for comparison and verification.

Eye movement experiment 2: effectiveness of configuration scheme

Task: Prepare three experimental samples, one of which is the optimal configuration scheme after the configuration experiment of personalized design components, and the other two are nonoptimal solutions. Subjects were required to watch the area of interest (AOI) in the three samples, and the time was 20000ms.

In this experiment, the heat map distribution of the collected subjects is used as a measurement index, and the gray scale histogram is used to quantify and compare the heat map of the area of interest

(AOI), and the high frequency part of the image is reflected by the variance index. The magnitude of the variance of the grayscale histogram indicates the contrast of the picture. The larger the variance, the higher the contrast, indicating that the subject pays more attention to the area of interest (AOI). Taking three AOI_1, AOI_2, AOI_3 heat maps of interest area of the 20th subject's personalized design configuration scheme for yacht appearance as an example, using Python programming of ANACONDA software to convert (AOI) heat map to gray histogram and calculate its variance. The result is shown in Figure 10, and the variance of heat map is shown in Table 3.

From the above data, the variance of Figure 3 is the largest, indicating that this subject has the highest attention to the third configuration scheme, which is also the optimal configuration scheme obtained by the subject after the configuration experiment of personalized design components.

Control experiment 3: 7-level LIKERT scale experiment

Likert scale was first proposed by American social psychologist R.A. Likert in 1932. Likert scale is one of the most commonly used forms of attitude scale in social survey and psychological test. It can be used for individual or group's attitude, evaluation or intention to something. In order to further explain whether the design scheme selected through eye movement mode is the configuration scheme that consumers really like and can meet the emotional needs of consumers. In this study, a 7-level LIKERT scale experiment was set as a control Each subject was asked to evaluate the seven level Likert scale (- 3 ~ + 3) on three kinds of yacht appearance personalized design configuration schemes. Among them, - 3 represents "extremely dislike", + 3 represents "very

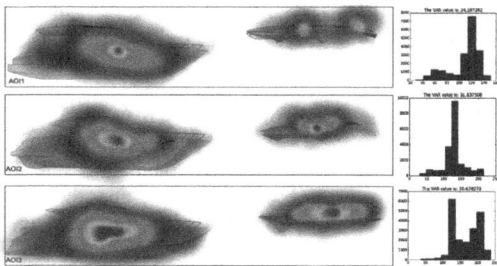

Figure 10. Configuration scheme AOI- heat map and gray histogram.

Table 3. Hot-spot graph variance of the configuration scheme AOI.

AOI	AOI_1	AOI_2	AOI_3
δAOI^2	24.197262	31.637508	38.678273

Table 4. Yacht personalized design component configuration scheme 7 level LIKERT scale.

Pair difference number

Mean value	Standard deviation	Standard error mean value	Confidence interval of 95% difference number		T	df	Sig.
			Lower limit	Upper limit			
29.65000	2.10504	.56260	28.43459	30.86541	52.702	13	269

Table 5. Paired sample verification.

Configuration scheme	LIKERT scale	Score
	-3 -2 -1 0 +1 +2 +3	0
	-3 -2 -1 0 +1 +2 +3	+1
	-3 -2 -1 0 +1 +2 +3	+3

like", and the intermediate value indicates that the degree of progress gradually increases, as shown in Table 4. There are three personalized configuration schemes on each table, one of which is the optimal scheme obtained by the subject through eye movement experiments of individual design components, and the other two are non-optimal solutions.

SPSS22 software was used to perform paired T test on the eye movement variance data and 7-level LIKERT scale data of each subject to verify whether the method was feasible. The results are shown in Table 5.

Table 5 shows the t-test analysis results of paired samples of eye movement experiment data and questionnaire survey data, and lists the mean value, standard deviation, standard error mean value and confidence interval of 95% difference number of the difference between the experimental data obtained from the eye movement experiment and the scale survey. Due to the significance level $\alpha = 0.05$, the results show $^{***}p=0.269>0.05$, indicating that there is no significant difference between the two results, and this method is feasible.

5 CONCLUSIONS AND FUTURE RESEARCH

The personalized design of Chinese-style yachts involves the conversion of cultural elements to design elements. This study takes Chinese cultural imagery as the design guide, and integrates totem cultural symbols into the personalized genetic design of the yacht. The shape grammar and Bezier curve are used to carry out the derivative design of the personalized design gene to form the personalized component module library. Then use eye tracking system combined with ant colony optimization ideas to conduct individual component module matching research, thereby forming a Chinese style yacht personalized customization program. In order to verify the effectiveness of this method, the eye movement experiment and the 7-level LIKERT scale (-3~+3) were used in this study for comparison and verification. The results prove that this method is feasible and the personalized configuration scheme designed can effectively meet the emotional needs of consumers. The summary of the research is as follows:

(1) The optimization idea of the ant algorithm can be used as the guiding idea of the configuration process of the personalized design components of the yacht to optimize the configuration of the design components.
(2) In the process of cultural design, the key point is to transform cultural elements into design elements, which can be achieved by establishing a multi-dimensional mapping relationship of "semantic-value", "syntactic-symbol", "context-behavior", and "pragmatic-custom".
(3) In the process of yacht personalized design, the eye movement mode can be used to help consumers select and configure personalized components, and the result can effectively avoid the ambiguity and uncertainty caused by subjective factors.

In the research of this article, the final personalized yacht design scheme best meets the consumer's emotional image, but it does not indicate whether it is the design scheme with the most Chinese style image, and its solution is only consistent with the Chinese cultural image. In future research, we will focus on the Chinese style yacht design, quantify the Chinese style yacht design indicators, and lay the foundation for the Chinese style yacht design.

ACKNOWLEDGEMENTS

This work is Supported by "the Fundamental Research Funds for the Central Universities (WUT:2020IVA073;WUT:2019■186;WUT:2020■016GX)".

REFERENCES

Dai, I G M. 2011.Research of small and medium-sized motor yacht form design based on imagines acknowledge theory. *South China University of Technology.*

Dlana P S, Ahmad R, Huang LQ .2012 .Models of Joint Economic Lot-sizing Problem with Time-based Temporary Price Discounts. *International Journal of Production Economics*, 139(1) 145–154.

Dong H H, Wang W W, Lu M M, et al. 2017.Modular Design of Personalized Customization Model. *Packaging Engineering*, 03:129–133.

Hunt D M, Radford S T, Evans K R. 2013. Individual differences in consumer value for mass customized products. *Journal of consumer behavior*, 12 (4):327–336.

Mcdonald T P.2010.A Library Based Approach for Exploring Style in Preliminary Ship Design. *UCL.*

Shahrozk, Erkan G, Kemal M D. 2017.A novel design framework for generation and parametric modification of yacht hull surfaces. *Ocean Engineering,* 136:243–259.

Volker B, Berend D, Marca D, Robert K.2018. Intelligent Computation in Manufacturing Engineering. *In 12th CIRP:Mathematical description of aesthetic criteria for process planning and quality control of luxury yachts,* Italy, July 18–20.

Wang Z Y,Li H W . 2016.Automotive Styling Feature Extraction and Cognition Based on the Eye Tracking Technology. *Packaging Engineering*, 20:54–58.

Zhang H L.2011. The research on yacht design based on consumer Psychology and consumer behavior.*Hunan university.*

Zhang X F,Huang R Q .2014.Virtual display design and evaluation of clothing: a design process support system. *International Journal of Technology and Design Education*, 24 (2): 223:240.

Design and application of propeller boss cap fins to ducted propellers

M. Martinelli, D. Villa & S. Gaggero

Department of Electrical, Electronic, Telecommunications Engineering and Naval Architecture, University of Genoa, Genoa, Italy

ABSTRACT: The interest in Energy Saving Devices (ESD) has grown significantly during the last years. Recent regulations, like the IMO EEOI (Energy Efficiency Operational Indicator) and EEDI (Energy Efficiency Design Index), impose further restrictions on emissions for both new designs and existing vessels. Moreover, these regulations encourage the development and the application of efficient, unconventional, propulsive solutions to fulfil these new requirements. The paper presents a design approach, based on the paradigm of optimization applied to Propeller Boss Cap Fin (PBCF) energy saving devices for ducted propellers. PBCF, indeed, were extensively studied in the case of conventional propellers, but nowadays there are only very few researches regarding the combination of these devices with unconventional propulsors, such as ducted propellers. Since the particular functioning of propellers inside nozzles can influence significantly the entire flow field aft the propeller, normally exploited by the PBCF for energy losses recovery, ad-hoc design procedures are required. By using a combination of RANSE analyses, a parametric description of the PBCF blade and an optimization algorithm, the paper illustrates the results of the design of PBCFs for propellers operating both in accelerating and decelerating nozzles, discussing the effectiveness of such kind of devices in case of ducted propellers.

1 INTRODUCTION

Nowadays, PBCFs have been applied to several types of vessel, particularly to cargo vessels, and both with Fixed Pitch (FPP) and Controllable Pitch (CPP) propellers. Consequently, a lot of samples of propellers combined with PBCFs are available in the literature, both in the model and ship scale. The installation of a PBCF generally improves the propulsive global efficiency, and this is due to two factors. The former is the generation of a torque, which contrasts that absorbed by the propeller, and the generation of and additional thrust, which can be concordant or not (in that case it is a resistance) to that delivered by the propeller. In both cases, the reduction of torque is more effective on propulsive efficiency than the increase of thrust, yielding to an efficiency improvement. The latter is the reduction, or sometimes the elimination, of the hub vortex developed from the propeller hub. This vortex causes several problems, such as vibrations, cavitation and efficiency losses caused by the drag of the hub associated with the presence of this vortex. Therefore, a reduction or an elimination of this implicates various benefits to the propulsive system. These aspects have been known since when these ESDs were introduced (Ouchi et al. 1988). The benefits in terms of efficiency, and consequently in fuel-saving or higher design speed, are well known and documented by tests in both model and ship scale. Nevertheless, most of them regard only conventional propellers and the effect of the PBCFs

applied to them. Nowadays, non-conventional solutions have not been investigated deeply yet.

Actually, in literature, there are very few in-depth studies about the combined effect of a non-conventional propeller and the PBCFs. One of these is the ducted propeller, that will be analysed in this paper in both configuration, accelerating and decelerating. Theoretically, a collateral effect of ducted propellers is the generation of a non-negligible hub vortex, which causes loss of efficiency, since ducted propeller are usually highly loaded propulsors, especially in the accelerating configuration. Since the main goal of PBCFs installation is to strongly reduce this hub vortex, or sometimes to delete it completely, there are good reasons to suppose that this solution could bring improvements in the propulsive global efficiency of the system.

Since there are not established design procedures for PBCF, one of the most important goals of this paper is to build a design tool based on the paradigm of shape optimization, which iteratively and automatically changes the geometry of the ESD based on the results of high-fidelity RANS analyses to maximize the propulsive efficiency of the system at a specific advance coefficient. The tool takes as input data the geometric features of the propeller, the hub, the PBCF and the duct and determines the performances of a huge number (thousands) of cases by systematically varying only PBCF geometrical parameters, in a way similar to that already exploited for propellers (Gaggero 2020, Gaggero

DOI: 10.1201/9781003216582-38

et al. 2017) or PBCF applied to conventional propulsors (Gaggero 2018, Mizzi et al. 2017).

Moreover, to give a global perspective of the behaviour of the combination ducted propeller – optimum PBCF, a range of advance coefficients will be taken for the computation of the entire propeller open water diagrams. Consequently, it will be possible to understand how the PBCF affects propulsive global efficiency even when the propeller does not work at the design functioning condition. Summarizing, this paper wants to investigate the possibility of combining a ducted propeller, both with an accelerating and decelerating duct, with custom designed PBCFs by using an iterative algorithm based on optimization process, for a given range of advance coefficients.

2 THE REFERENCE DUCTED PROPELLERS

To investigate the possibility of improving ducted propellers efficiency, two different cases have been considered in the analyses. The first one presents an accelerating duct, while the second one a decelerating duct. Both propulsors were developed in the framework of the EU FP7/2007-2013 research project BESST - Breakthrough in European Ship and Shipbuilding Technologies and geometries have been kindly provided by Fincantieri SpA. Since towing tank and cavitation tunnel measurements are available in model scale, all the analyses and the design activities have been carried out for the model propellers, the most important features of which are summarized in Table 1.

In both cases, some preliminary computations using RANS (StarCCM+, SIEMENS (2017)) have been performed to support the research. These analyses helped to determine the thrust and torque coefficients used as reference in the optimization process. According to the relevant literature (Ouchi et al. 1988), the PBCF is effective if it reduces the absorbed torque of the propulsor at almost constant delivered thrust. A reduction of the hub vortex and of the associated hub resistance is another possible positive consequence of the presence of a PBCF. On the other hand, nothing can be said about the thrust coefficient, since the variation of the PBCF geometrical parameters, such as pitch, camber or thickness distributions, might change the PBCF force into a thrust or a resistance. In the first

case, the total thrust will be increased, and theoretically, a PBCF that reduces torque and increases thrust is the best solution. In the second case, the influence on the PBCF on global propulsive efficiency will depend both on the variations of thrust and torque coefficients. These preliminary calculations, hence, are useful to obtain the thrust and torque values of propellers blades, propellers hub and ducts for both the accelerating and the decelerating cases to be used as reference for the assessment of the role of the PBCF not only at the design functioning point but for several values of advance coefficient.

In Figure 1 a global view of this reference grid used in the optimization is shown. It consist of about one million cells used to discretize a computational domain which consists only in one blade passage since periodic boundary conditions and a moving reference frame allow for this simplification and for steady calculations. It is possible to notice three different grid densities, corresponding to three different volumes of refinement. These volumes are more refined close to the most interesting regions, e.g. the propeller, the hub, the duct and the vorticity expansion area. Moreover, the mesh presents a global refinement to ensure polyhedral cells as regular as possible in the entire domain. A second refinement surrounding the hub, the propeller, the duct and the vorticity expansion area is used to capture the features of the propeller wake. Finally, the third refinement is the smallest and includes the hub, the propeller, the duct and the PBCF and it is aimed at modelling with a sufficient accuracy the most relevant flow features of ducted propellers, like the tip leakage vortex from the blade/duct gap (Figure 2 and 3) and the vortexes from the hub/PBCF. The aim of subdividing the domain into many volumes is to catch in the best computationally efficient way the computation outputs in the most important regions. Consequently, there will be a loss of accuracy in the areas far from the smallest refinement. Anyway, this can be accepted for the comparative purposes of the analyses in an optimization approach.

For what concerns the modelling of the boundary layer, dedicated prism cells have been employed at the walls, as in Figure 2. Their total thickness has been computed taking advantage of boundary layer

Table 1. Ducted propellers main characteristics.

	Propeller A	Propeller B
Nozzle type	Accelerating	Decelerating
Propeller type	CPP	CPP
$D[m]$ (model scale)	0.230	0.230
$\frac{A_E}{A_0}$	0.689	0.725
$\left(\frac{P}{D}\right)_{0.7R}$	1.566	1.354
Number of blades	4	4
$n[rps]$ (model scale)	20	20

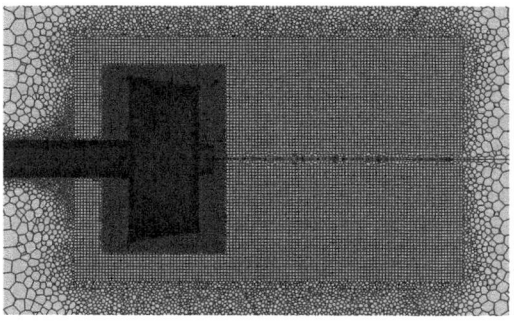

Figure 1. Longitudinal view of the computational grid.

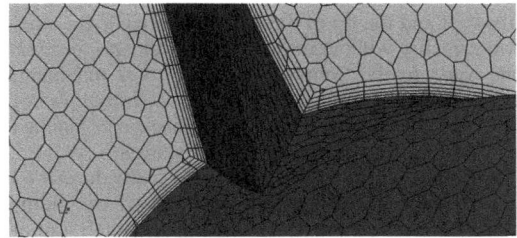

Figure 2. Details of the prism layers at the blade root.

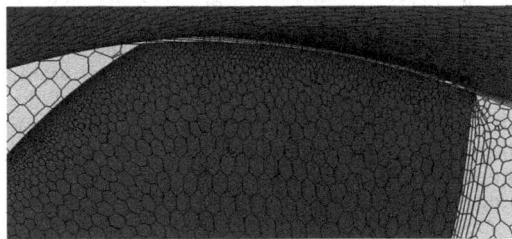

Figure 3. Particulars of the grid in the blade/nozzle gap.

empirical formulae for fully developed turbulent flow. The turbulent model chosen to solve the RANS equations is the $k - \varepsilon$.

By using RANS equations and this computational grid, the open water diagrams of both propellers have been calculated and compared to experimental data. Besides, to understand if the optimum PBCF brings some improvements even to the same propellers but operating without the duct, the open water diagrams related to propellers without ducts have been computed too and then compared to the available experimental data. Results are summarized in Figures 4 to 7.

Data have been normalized to comply with company internal information security policies, dividing outputs by some reference values. Specifically, J, the advance coefficient, has been divided by its value at the design condition; K_T, the thrust coefficient, by

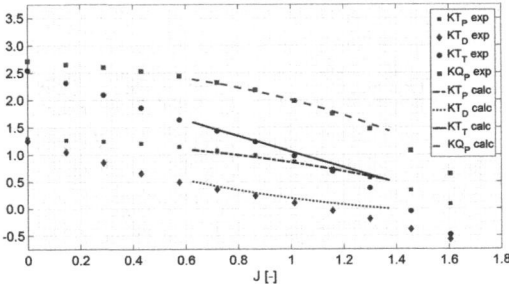

Figure 4. Open water propeller performance. Comparison between experiments and calculations. Accelerating ducted propeller (with duct).

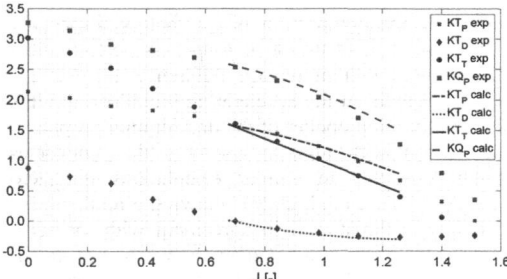

Figure 5. Open water propeller performance. Comparison between experiments and calculations. Decelerating ducted propeller (with duct).

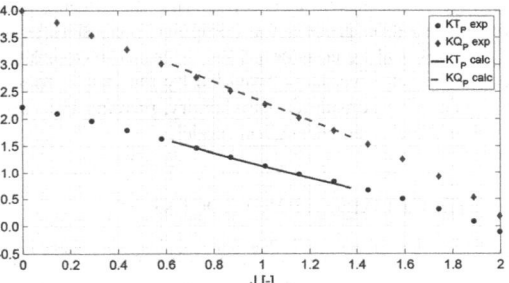

Figure 6. Open water propeller performance. Comparison between experiments and calculations. Accelerating ducted propeller (without duct).

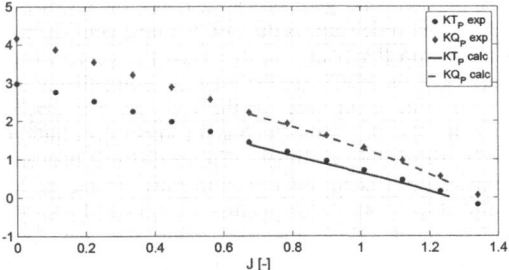

Figure 7. Open water propeller performance. Comparison between experiments and calculations. Decelerating ducted propeller (without duct).

K_T^{REF}, which is the value of total thrust coefficient at the design condition, comprehensive of both propeller and duct contributions. A similar procedure has been applied to K_Q, the torque coefficient, which has been divided by K_Q^{REF} and then doubled. Even though in the plots the values are shown as J, K_T and K_Q, they stand for the normalized ones. Finally, as usual in literature, the values of K_Q are shown multiplied by ten.

Overall, the comparison is satisfactory. Especially in the case without the duct, propellers performances are computed with an average difference less than 2% for the propeller of the accelerating propulsor and less than 4% for the propeller of the decelerating propulsor. Calculations with nozzles resemble the tendencies already observed in similar calculations (Gaggero et al. 2014, Villa et al. 2020), showing a total (blades plus nozzle) thrust well in agreement with measurements as the result of a balancing between an overestimation of the blade thrust and of the nozzle resistance (decelerating duct). In the case of the accelerating ducted propeller the agreement is particularly good at very loaded functioning conditions (i.e. the conditions this type of propulsors is designed for) while a certain overestimation of the thrust can be appreciated in correspondence of very high advance coefficients. For the purpose of the design, also considering that throughout the optimization process the computed propellers performances will be considered the reference baseline to assess the improvements provided by the PBCF, these results can be considered satisfactory, proving the validity of the chosen simulation model.

3 THE OPTIMIZATION PROCESS

The optimization process takes advantages of three different softwares, each with a specific task: an optimizer (in this case ModeFRONTIER Esteco (2018)), a RANS solver (StarCCM+, SIEMENS (2017)) and a coding language. To find the optimum PBCF suitable for both ducted propellers, a tool which generates parametrically a PBCF is required. The first step of the optimization is the writing of a code which takes as input data the geometrical features of a generic PBCF. The code realizes the PBCF model considering it as a propeller blade. In this way, the geometrical features of the PBCF are the same as a propeller. Particularly, the input data for the code are the PBCF diameter, the chord, the pitch and camber distribution along the radial coordinate. These distributions are obtained by using control polygons which build B-Spline curves of each geometrical feature by varying the points of the polygons, which then turn into the free variables of the optimization process. The abscissae of each polygon are the radial stations, in which the first and the last, sorted by the minor to the greater, are fixed. In particular, the first radial station should be equal to the radius of the propeller hub, the last to the radius of the PBCF. The other stations are free to vary, but the ranges have been chosen to avoid intersections between each other. For instance, the upper limit of the variation range of the second control point (in radial direction) must be smaller than the lower limit of the third control point.

Regarding the pitch, the thickness and the camber distributions, which are the geometrical features together with the chord of the PBCF allowed to change in the design process, the control polygons consist of a four-points figure, in which two abscissae

(the root and the tip of the PBCF) are fixed, the other two can vary without intersections. Ordinates, instead, can vary freely from a minimum to a maximum value, as will be discussed forward. Two examples of pitch and camber distributions obtained and handled by control polygons are shown in Figure 8.

Also the chord distribution has been described in the same way, but since the chord on diameter ratio is maintained constant for each radial station, the corresponding control polygon collapses into a straight line, adding only one parameter (i.e the chord over diameter ratio) to the design process. Summarizing, there are sixteen independent variables which generate a PBCF: six values each define the pitch and the camber distributions (two as abscissae and four as ordinates), d, the diameter (in the percentage of the propeller diameter D), $\frac{c}{d}$, the chord on PBCF diameter, Δx, the translation with respect to the propeller plane and θ, the rotation around the propeller axis.

Once the geometry parametrization is defined, it is necessary to establish the ranges in which these sixteen free variables can vary. The limits of these ranges can be fixed by observing the following fact:

- The translation along the propeller axis cannot be greater than S, the maximum available space on the hub. This distance can be measured and is a physical limit because the translation will be a fraction of it. To avoid infeasible configurations, e.g. PBCF fins intersecting the trailing edge of the blades or protruding out the aft of the hub, only 90% of this distance will be available, leaving a clearance of 5% forward and aft . Knowing the diameter, the chord and the PBCF pitch distribution, the encumbrance of the PBCF over the hub can be computed. The cases in which this encumbrance is greater than 90% of the available space can be discarded;
- The rotation depends on the propeller blades number. Then, since both the examined propellers are four bladed propellers, the rotation range ranges from 0° to 89°, because a rotation of 90° means the same situation of 0°;

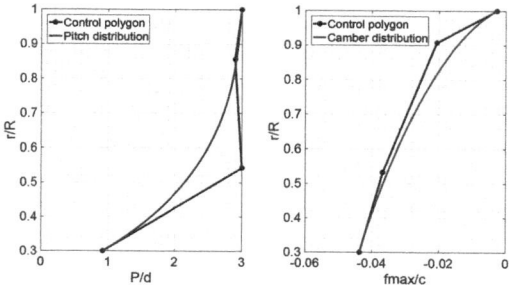

Figure 8. Pitch and camber distributions examples.

- The PBCF diameter cannot be greater than 50% of propeller diameter, because the increase of the (frictional) resistance (i.e. wetted area) will nullify the torque reduction, and cannot be smaller than 32.5%, because on the contrary it will easily produce a negligible contribution to the propulsive performance of the system;
- The chord cannot be greater than 60% of the PBCF diameter, because of encumbrance, and smaller than 20%, to avoid the generation of a too stocky profile due to minimum strength requirements.

The ranges of variations of the control points of pitch and camber have been selected to enlarge as much as possible the design space, allowing a large variety of shapes which are necessary to explore unconventional and potentially favourable geometries. Except for the radial coordinates of the control polygons, which should avoid overlapping, values of pitch and camber cover the ranges summarized in Table 2 allowing, in the case of camber, any profile shape, both symmetrical and not.

Hence, the optimizer fills the design space, which is an hypercube having a number of dimension equal to the free variables number, with an initial SOBOL spacing to have an idea of how the PBCF performances behave generally (i.e how performances vary depending on different PBCF shapes). The optimizer chooses the values of these variables and transfers them to the parametric description of the ESD, which generates the correspondent PBCF in the form of an *.stl* file which is then transferred to the RANS solver. The process runs the RANS solver using the mesh and the turbulent model previously specified, extracting at the end of each simulations the trends of torque and thrust of the propeller, PBCF, hub and duct and finally computing the total propulsive efficiency. Based on these results the choice of the free parameters is updated and the process is repeated until convergence to the maximum achievable efficiency is obtained. Testing about one thousands configurations was necessary for both the accelerating and the decelerating cases to have converged trends. Ten days of computational time on a mid-end workstation were necessary to complete each optimization process.

4 OPTIMIZATION RESULTS - THE ACCELERATING DUCTED PROPELLER

The first design activity regards the accelerating ducted propeller. Results of the optimization process are collected in terms of trend of efficiency with respect to the configuration case (Figure 9) but also with respect to the delivered thrust (Figure 10) and the absorbed torque (Figure 11). The optimum PBCF configuration is specified with a black circle, the reference η_o and the references K_T and K_Q of the propeller operating at the same advance coefficient but without the PBCF respectively with a solid and a dashed line.

From these plots, some considerations can be deduced. First of all, in most of the feasible cases (those that satisfy the geometrical constraints), the reference efficiency value is exceeded. Secondly, the total thrust coefficient value is smaller than the reference in most of the cases; this implies that the PBCF fins produce resistance instead of thrust and that the reduction of the hub drag is not sufficient to balance this additional resistance. Finally, the total torque coefficient is frequently reduced: this means that in only a few cases that of the PBCF is an absorbed torque while, generally, it contributes to a reduction of the total absorbed torque of the propulsive system. The parameters generating the optimum PBCF are reported in Table 3. Despite the optimization process, the best PBCF produces an almost negligible improvement of efficiency, in the measure of an increase of about 0.5% at the design

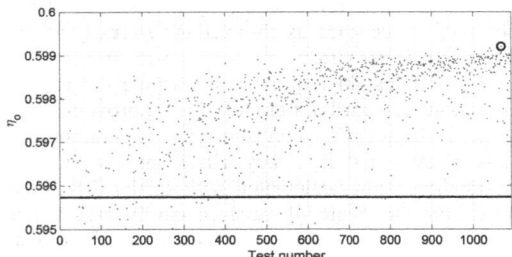

Figure 9. Evolution of the total propulsive efficiency during the optimization process. Accelerating ducted propeller.

Table 2. Ranges of variations of the free parameters describing the PBCF geometry.

Variable	Maximum	Minimum
d	$0.325D$	$0.5D$
$\frac{P}{d}$	0.7	4
$\frac{c}{d}$	0.2	0.6
$\frac{f_{max}}{c}$	-0.1	0
$\left(\frac{r}{R}\right)_2$	0.6	0.75
$\left(\frac{r}{R}\right)_3$	0.8	0.95
Δx	$0.05S[m]$	$0.95S[m]$
θ	$0°$	$89°$

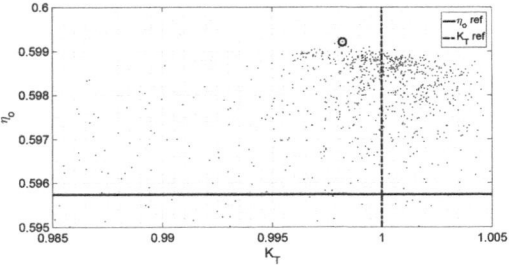

Figure 10. Thrust - efficiency relationship for the Accelerating ducted propeller.

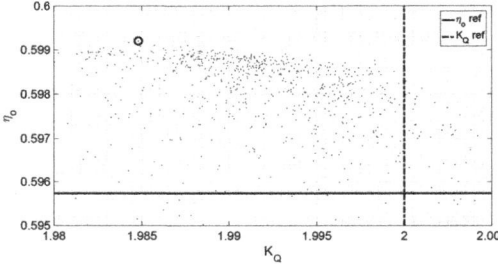

Figure 11. Torque - efficiency relationship for the Accelerating ducted propeller.

Table 3. Geometrical characteristics of the optimum PBCF for the accelerating ducted propeller.

Parameter	Value	Parameter	Value
$\frac{d}{D}$	0.27	Δx	0.54S
$\frac{c}{d}$	0.313	θ	71
$\left(\frac{r}{R}\right)_2 \left(\frac{P}{d}\right)$	0.7	$\left(\frac{r}{R}\right)_2 \left(\frac{f_{max}}{c}\right)$	0.511
$\left(\frac{r}{R}\right)_3 \left(\frac{P}{d}\right)$	0.756	$\left(\frac{r}{R}\right)_3 \left(\frac{f_{max}}{c}\right)$	0.774
$\left(\frac{P}{d}\right)_1$	0.50	$\left(\frac{f_{max}}{c}\right)_1$	−0.0502
$\left(\frac{P}{d}\right)_2$	1.61	$\left(\frac{f_{max}}{c}\right)_2$	−0.0356
$\left(\frac{P}{d}\right)_3$	2.615	$\left(\frac{f_{max}}{c}\right)_3$	−0.0238
$\left(\frac{P}{d}\right)_4$	3.2	$\left(\frac{f_{max}}{c}\right)_4$	−0.0039

point. This is achieved totally by a reduction of the absorbed torque since, as observed in Figures 10 all the cases deliver less thrust.

A better insight into the reasons (and the role of the relative contributions) of this improvement is given in Figure 12 where the relative variations of thrust, torque and efficiency with respect to the corresponding values calculated without the PBCF are given for the range of advance coefficients under investigation. For instance, in the case of the thrust, the variations is calculated as:

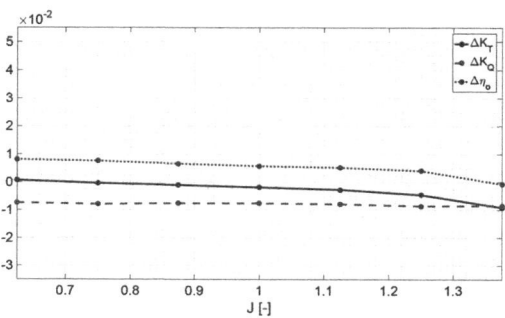

Figure 12. Variations of total thrust, torque and efficiency with respect to the calculations without the PBCF. Accelerating ducted propeller.

$$\Delta K_T = \frac{K_T^{PBCF} - K_T}{K_T} \qquad (1)$$

where the superscript *PBCF* means that the coefficient has been computed considering the presence of the PBCF.

The optimum PBCF maintains everywhere a negative ΔK_Q, which means that the fins torque contrasts that absorbed by the propeller for any advance coefficient considered in the analysis. For what concerns ΔK_T, if the propeller works in a loaded condition, the PBCF can produce a little and negligible thrust, that is the best condition. On less loaded working points this thrust changes to resistance, but this contribution is negligible almost everywhere, except for working points very far from design condition. At the design point the total thrust is almost zero and the improvement in efficiency is entirely due to the reduction of the absorbed torque. An even more detailed explanation of the influence of the PBCF on the propulsive characteristics of the system is given in Figure 13, where each component of the propulsor (i.e propeller blades, subscript P, PBCF fins, subscript F hub, subscript H and duct, subscript D) is considered separately. For the propeller blades (subscript P), for instance, the variation is computed according to:

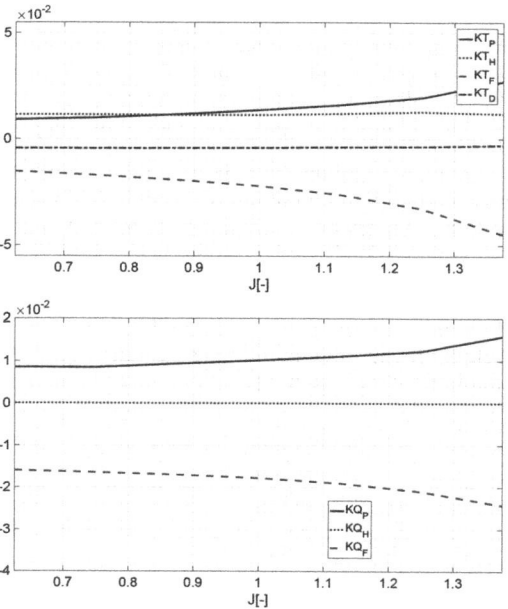

Figure 13. Influence of PBCF on thrust (top) and torque (bottom) coefficients. Accelerating ducted propeller.

$$\Delta K_{TP} = \frac{K_{TP}^{PBCF} - K_{TP}}{K_T} \qquad (2)$$

The presence of the PBCF alters the blades thrust. This is due to a sort of blockage effect at the root of the propeller blades induced by the presence of the PBCF fins. The increase of the blades delivered thrust is, however, more than balanced by the additional resistance produced by the fins. Also the duct is partially affected by the presence of the PBCF, since its thrust is slightly reduced. The nullification of the hub vortex is finally responsible of the reduction (i.e positive value of ΔK_{TH}) of the resistance of the hub. Similarly to the thrust, also the torque of the blades increases as a consequence of the blockage. Anyhow, in this case the fins have a positive influence since the torque delivered to the shaft is significantly higher.

The PBCF designed for the ducted propeller results particularly effective when the propeller operated without the duct. The same analyses of the ducted propeller (relative variations and influence of the PBCF on each component of the propulsive system) have been carried out also in this case. Results are summarized in Figures 14 and 15. The presence of the PBCF designed for the ducted propeller in the absence of the duct itself results in an outstanding improvement of the propulsive efficiency of the propulsor, up to 5% in off-design conditions (low advance coefficients). At the design functioning, a 3% increase is achieved thanks to an almost constant reduction of both the absorbed torque (regardless the advance coefficient) and of the hub resistance. With respect to the case with the duct, the most evident difference is exactly the influence of ΔK_{QF} while ΔK_{TF} does not vary substantially in the two cases.

These differences clearly reflect on the pressure distributions and on the vortical structures characterizing the propellers. The comparison of the pressure distributions with and without the PBCF, including or not the duct (respectively Figures 16 and 17) clearly evidences the effect of the PBCF on the hub surface: in the

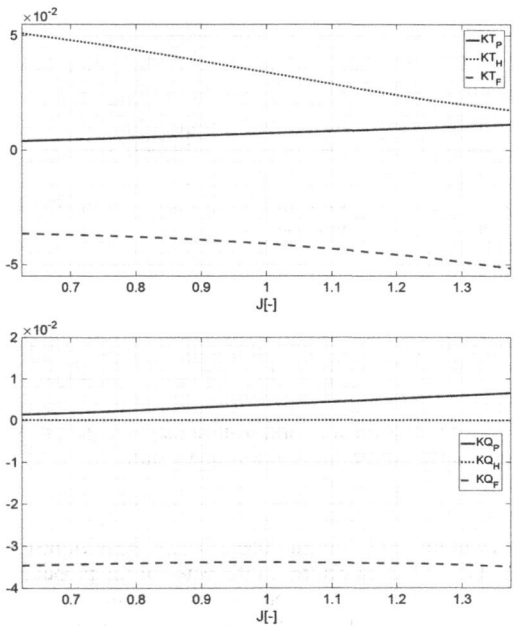

Figure 15. Influence of PBCF on thrust (top) and torque (bottom) coefficients. Accelerating ducted propeller without duct.

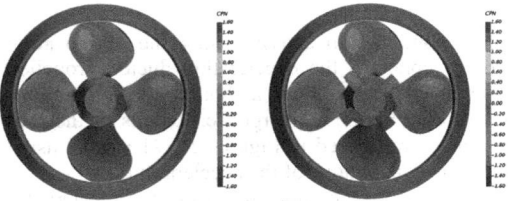

Figure 16. Non-dimensional pressure distribution (C_{PN}) on the blade pressure side and on the hub. Accelerating ducted propeller without and with the optimal PBCF.

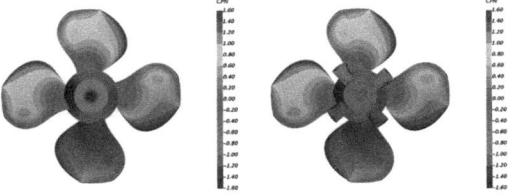

Figure 17. Non-dimensional pressure distribution (C_{PN}) on the blade pressure side and on the hub. Accelerating ducted propeller without the duct.

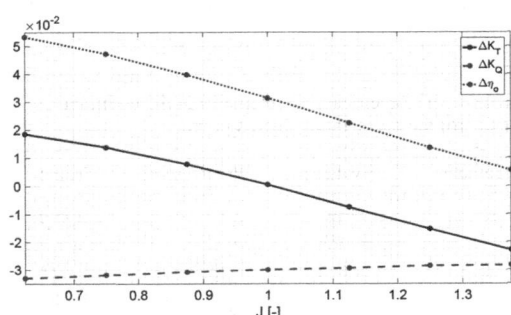

Figure 14. Variations of total thrust, torque and efficiency with respect to the calculations without the PBCF. Accelerating ducted propeller without the duct.

reference case, the hub face presents a remarkable low-pressure area, responsible of the hub drag, completely nullified by the PBCF especially in the case

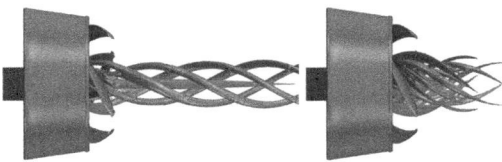

Figure 18. Vortical structures without and with the PBCF. Accelerating ducted propeller.

Figure 19. Vortical structures without and with the PBCF. Accelerating ducted propeller without the duct.

without the duct. Similar considerations arise comparing the vortical structures in the wake of the propeller compared in Figures 18 and 19. Also in terms of vorticity the role of the PBCF, which completely cancels the the hub vortex, is clear, showing the benefits of the PBCF installation.

5 OPTIMIZATION RESULTS - THE DECELERATING DUCTED PROPELLER

The second design activity using the optimization process concerns the decelerating ducted propeller. Also in this case, about a thousand configurations were tested before convergence. Results of the process are summarized in Figures 20, 21 and 22 using the same conventions of the accelerating case.

Compared to the same analysis of the accelerating case, the decelerating propeller presents some important differences. First of all, both in the $\eta o(K_T)$ and $\eta o(K_Q)$ trends, it is evident that the reference values of both thrust and torque coefficients are rarely exceeded. This fact yields to a negative ΔKT and ΔKQ almost in all the tested cases, while in the

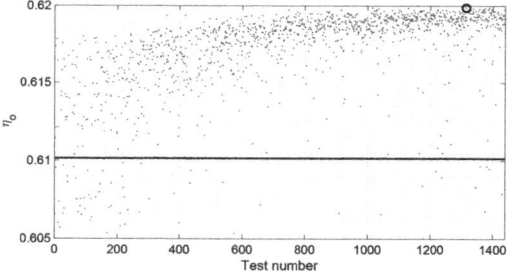

Figure 20. Evolution of the total propulsive efficiency during the optimization process. Decelerating ducted propeller.

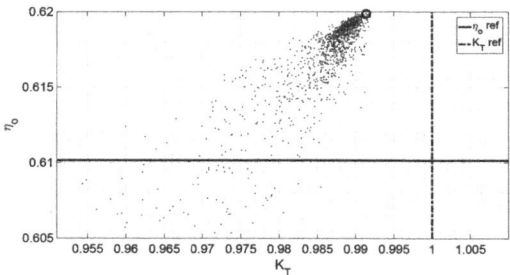

Figure 21. Thrust - efficiency relationship for the Decelerating ducted propeller.

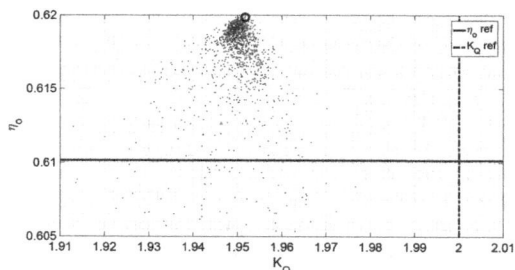

Figure 22. Torque - efficiency relationship for the Decelerating ducted propeller.

accelerating cases at least the thrust coefficient is often greater than the reference value. This means that in the decelerating ducted propeller all the tested PBCFs in the design process produce a resistance. Nevertheless, the torque coefficient difference is greater than that of thrust, and this explains why in this case the improvement of efficiency is so relevant. The parameters defining the geometry of the optimal PBCF for the decelerating ducted propeller, which ensures an increase of efficiency of about 1.2%, are summarized in Table 4. For this propulsor too, the analyses of the influence of the optimal PBCF are proposed also in the case without the duct. Results, in terms of variations and influence of the

Table 4. Geometrical characteristics of the optimum PBCF for the decelerating ducted propeller.

Parameter	Value	Parameter	Value
$\frac{d}{D}$	0.38	Δx	0.57S
$\frac{c}{d}$	0.365	θ	60
$\left(\frac{r}{R}\right)2\left(\frac{P}{d}\right)$	0.645	$\left(\frac{r}{R}\right)2\left(\frac{fmax}{c}\right)$	0.421
$\left(\frac{r}{R}\right)3\left(\frac{P}{d}\right)$	0.8	$\left(\frac{r}{R}\right)3\left(\frac{fmax}{c}\right)$	0.75
$\left(\frac{P}{d}\right)1$	0.70	$\left(\frac{fmax}{c}\right)1$	−0.08
$\left(\frac{P}{d}\right)2$	1.225	$\left(\frac{fmax}{c}\right)2$	0
$\left(\frac{P}{d}\right)3$	3.6	$\left(\frac{fmax}{c}\right)3$	−0.0693
$\left(\frac{P}{d}\right)4$	3.24	$\left(\frac{fmax}{c}\right)4$	−0.0015

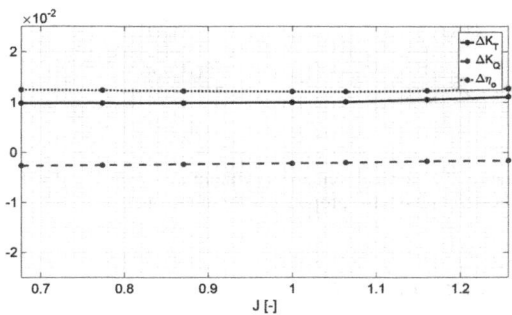

Figure 23. Variations of total thrust, torque and efficiency with respect to the calculations without the PBCF. Decelerating ducted propeller.

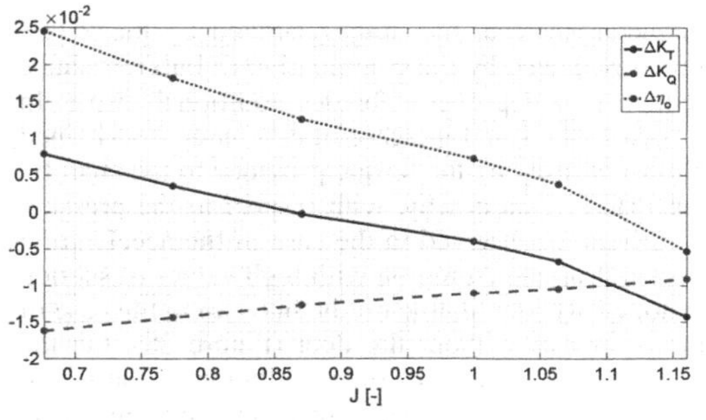

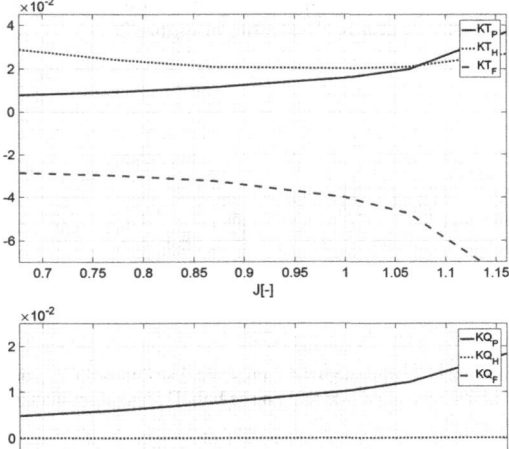

Figure 25. Variations of total thrust, torque and efficiency with respect to the calculations without the PBCF. Decelerating ducted propeller without the duct.

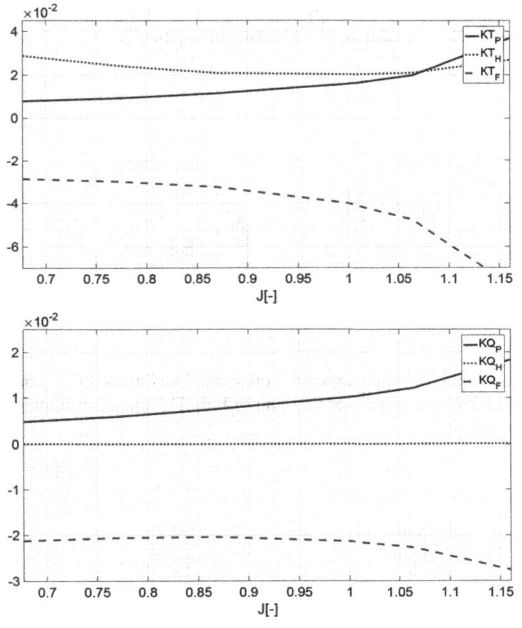

PBCF on the performances of the components of the propulsors, are given in Figures 23 to 26.

Trends, compared to the accelerating duct, are quite different. The installation of a PBCF improves significantly the performance of the decelerating ducted propeller, since the increase of efficiency is almost constant for a relatively wide range of advance coefficients.

Without the decelerating duct, the same PBCF still improves the efficiency of the propeller but the outstanding performances observed for the accelerating case are not replicated. Without the decelerating

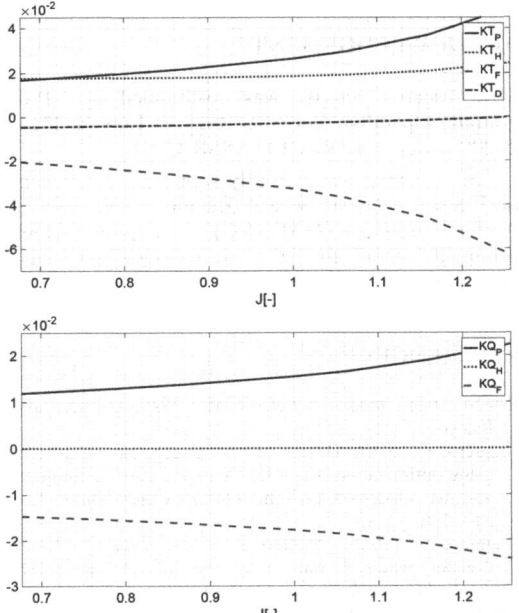

Figure 24. Influence of PBCF on thrust (top) and torque (bottom) coefficients. Decelerating ducted propeller.

Figure 26. Influence of PBCF on thrust (top) and torque (bottom) coefficients. Decelerating ducted propeller without duct.

duct, the PBCF increase up to a 2.5% the efficiency but only in very off design functioning: at the design point, the increase is lower than that achieved with the presence of the duct. Reasons of these improvements can be found in the analyses of Figures 24 and 26. When the duct is included, the reduction of the hub resistance is fundamental for the increase of the efficiency since the additional thrust by the propeller blades is almost perfectly balanced by the resistance of the fins and the reduction in torque provided by the PBCF is nullified by the increase of the required torque by the propeller blades. When the duct is not included, instead, the drag generated by the PBCF fins is balanced by the combination of hub resistance reduction and propeller blades additional thrust. In this case, the small increase of efficiency is sustained almost entirely by the torque provided to the shaft by the PBCF. This is why, with respect to the pressure distributions observed in the case of the accelerating ducted propeller, a region with high values of suction is not easily identifiable on the hub cap in the case of the propulsor without the duct (Figure 28). On the contrary, the benefits of the installation of the PBCF in presence of the nozzle are clearly evidenced by the vortical structures developed by the propeller without and with the PBCF shown in Figure 29 and 30.

Even if the reduction of suction on the hub is hardly visible also for this propulsive configuration (Figure 27) despite a significant reduction of the hub drag, the avoidance of a strong hub vortex thanks to the PBCF action is observable in Figure 29.

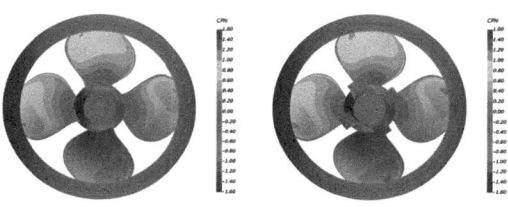

Figure 27. Non-dimensional pressure distribution (C_{PN}) on the blade pressure side and on the hub. Decelerating ducted propeller.

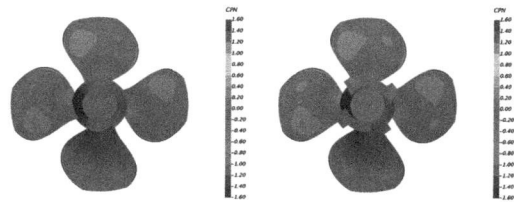

Figure 28. Non-dimensional pressure distribution (C_{PN}) on the blade pressure side and on the hub. Decelerating ducted propeller without the duct.

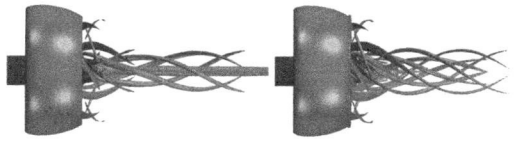

Figure 29. Vortical structures without and with the PBCF. Decelerating ducted propeller.

Figure 30. Vortical structures without and with the PBCF. Decelerating ducted propeller without the duct.

6 CONCLUSIONS

A framework for the hydrodynamic optimization of PBCF energy saving devices applied to ducted propellers, relying on a parametric description of the fins, high-fidelity RANS calculations and a genetic type optimization algorithm, has been presented. The effectiveness of the process has been proven considering both accelerating and decelerating type nozzles, for each of which optimal PBCFs have been developed. Observed improvements of the propulsive efficiency varied between a 0.5% for the accelerating ducted propeller up to 1.2% in the case of the decelerating configuration. An opposite trend (higher efficiencies for the propeller of the accelerating ducted propulsor) was highlighted when the designed PBCFs were applied to the propellers without their reference ducts. Off-design analyses, as well as the study of the influence of the ESD fins on the propulsive characteristics of the systems finally shed a light on the prevalent functioning conditions of these devices.

ACKNOWLEDGEMENTS

This research activity was co-founded by ``Programma Operativo Por FSE Regione Liguria 2014-2020'' under grant RLOF18ASSRIC/4/1.

This research was partially supported by Fincantieri S.p.A., Naval Vessel Business Unit, Naval Architecture Dept. (MM-ARC), Genoa, Italy, which provided data for the reference test case geometries.

REFERENCES

Esteco (2018). *modeFrontier, Release 2018*. Trieste, Italy: Esteco.

Gaggero, S. (2018). Design of pbcf energy saving devices using optimization strategies: A step towards a complete viscous design approach. *Ocean Engineering 159*, 517–538.

Gaggero, S. (2020). Numerical design of a rim-driven thruster using a rans-based optimization approach. *Applied Ocean Research 94*, 101941.

Gaggero, S., G. Tani, D. Villa, M. Viviani, P. Ausonio, P. Travi, G. Bizzarri, & F. Serra (2017). Efficient and multi-objective cavitating propeller optimization: an application to a high-speed craft. *Applied Ocean Research 64*, 31–57.

Gaggero, S., G. Tani, M. Viviani, & F. Conti (2014). A study on the numerical prediction of propellers cavitating tip vortex. *Ocean engineering 92*, 137–161.

Mizzi, K., Y. K. Demirel, C. Banks, O. Turan, P. Kaklis, & M. Atlar (2017). Design optimisation of propeller boss cap fins for enhanced propeller performance. *Applied Ocean Research 62*, 210–222.

Ouchi, K., M. Ogura, Y. Kono, H. Orito, T. Shiotsu, M. Tamashima, & H. Koizuka (1988). A research and development of pbcf (propeller boss cap fins). *Journal of the Society of Naval Architects of Japan 1988(163)*, 66–78.

SIEMENS (2017). *StarCCM+, version 12.06.011*. Siemens PLM Software.

Villa, D., S. Gaggero, G. Tani, & M. Viviani (2020). Numerical and experimental comparison of ducted and non-ducted propellers. *Journal of Marine Science and Engineering 8(4)*, 257.

Developments in Maritime Technology and Engineering – Guedes Soares & Santos (eds)
© 2021 Copyright the Author(s), ISBN 978-0-367-77376-2

Airborne noise emissions from marine vessels: An analysis based on measurements in port

L. Mocerino, F. Quaranta & M. Viscardi
University of Naples "Federico II", Italy

E. Rizzuto
University of Genoa, Italy

ABSTRACT: The problem of the airborne noise transmitted outside the ship represents an acoustic pollution affecting the surrounding environment and third parties. Despite this type of emission can be quite annoying and has raised serious complaints from the inhabitants of areas close to ports, the acoustic impact of vessels, to date, has not been subjected to systematic control. In this context, recent is the new class notation issued by a classification society focusing on the characterization and evaluation of the ship source. This paper aims at analyzing the provisions of this new class notation from its effectiveness in characterizing the ship as a source of airborne noise radiated outside the vessel. Monitoring campaigns and results of numerical models will be exploited to identify a new method for verifying the notation in future applications.

1 INTRODUCTION

The sustainability of anthropogenic activities is a key topic of technical discussion, research and development in recent years. In a broad sense, sustainability implies controlling the effects on the ecosystem of our planet, including also those directly affecting humans. As known, one of the actions by which anthropogenic activities affect environment is trough emissions in form of pollutants (solid, liquid or gases) or energy (thermal, acoustical, electro-magnetical or other).

The impact of ships on the environment has been considered in the last decades under the aspects of discharge/release of pollutants directly from board (oil and solid), release of toxic compounds from hull paints, importation of exotic biological species through ballast water, emission in the atmosphere of pollutants having local as well as global effects. The latter aspect is being taken into particular consideration as far as greenhouse gases are concerned, due to implications in terms of climate changes (Mocerino et al., 2018, 2020).

Noise has been only recently categorized among other forms of energy pollution emissions, but has been since long time considered for the impact on human health. The effects on health and on the quality of life and work conditions have caused worries in the public opinion and have generated through years an international trend towards the development of thematic development projects all over the world. Noise has negative impacts on human health and psychophysical well-being: this has prompted increasing efforts in the containment of noise emitted by plants, buildings, infrastructures and transport means. In 2108, the World Health Organization (WHO) published updated guidelines aimed at tackling the noise problem associated with most anthropogenic activities (WHO 2018). The Guidelines provide a set of source-specific public health recommendations on long-term exposure to environmental noise, covering, among other sources, road traffic, railway, aircraft, wind turbines, and leisure noise.

All anthropogenic activities in general generate noise emissions. This applies to shipping, as well as to the other transportation industries. In the maritime sector, three types of acoustic impacts are recognized: inside the ship, in water, and in air (Borelli 2015).

The scientific community, for years, has approached the problem of the acoustic impact of ships only in terms of noise propagation on board (ibidem). Noise control inside the ship was at first focused mainly at ensuring proper working conditions on board and at preventing communication problems in key areas on board. Health and safety were the main targets.

Later, the subject of comfort for both passengers and crew has acquired increasing importance (Kurt 2016). Nowadays comfort is one of the strategic aspects of the success of cruise vessels and pleasure crafts, reaching a considerable importance within the design process of these types of vessels.

Among various factors affecting the perception of comfort, noise and vibrations are certainly the main ones. In a comfort-oriented project, stationary vibrations

DOI: 10.1201/9781003216582-39

and noise, originating in machinery normally in continuous operation during navigation (propulsion, ventilation, conditioning), are considered the primary source of disturbance in accommodation areas. The dimensions of the ship usually play a positive role in containing noise levels on board, because of the possibility of keeping a distance between source and receivers and of interposing countermeasures (which generally imply a toll in terms of weight, space (in addition to cost).

The noise radiated by ships in water normally increases with the dimensions and the speed of the ship. Propellers, engines and water flow around the hull are certainly the main causes of the radiated noise in the water. Effects are shown on the marine fauna, ranging from behavioral changes to abandonment of breeding or feeding areas to masking of long distance intra-species communication (particularly for large cetaceans). The International Maritime Organisation has launched investigations on the subject since about a decade and, even though global requirements have not yet been issued, local regulatory bodies have issued specific requirements in various sensitive zones around the globe (International Ocean Noise Coalition, Hoyt 2018). Strange enough, regulators have so far neglected the third aspect of acoustic impact from ships, the one regarding emissions outboard in air.

This may be due to the fact, as recognized in Badino (2011), that the impact of airborne noise radiated from ships is strongly influenced by local effects (distribution of population in the areas interested by the propagation, orography and topography of the area: see Di Bella (2013), port lay-out and meteorological effects) (Curcuruto et al., 2015). In addition, from a regulatory viewpoint, many Institutions are potentially involved in the control activity (Cost Guard, Port Authority, Class Societies, Municipalities, and Health Agencies). This makes more complicate a coordination and an effective controlling action.

As a matter of facts, to date, noise pollution due to ships is not regulated by general requirements: only some indications are present at national and/or local level regulating noise emissions by ship traffic.

On the other hand, the noise emitted in navigation by ships approaching ports or sailing in channels/ straits close to the coast or during maneuvering phases or during loading/unloading operations at quay may actually result in a serious deterioration of the living conditions of the inhabitants of close areas. This applies in particular to ports placed in the proximity of urban areas (Murphy 2014), as it happens quite frequently in the Mediterranean sea.

It is not a coincidence that the noise associated with port operations ranks in the top ten priorities of European ports (ESPO 2018): from 2016, high noise levels occupy always the third place among concerns, preceded only by bad air quality and energy consumption; (ESPO 2018).

In this context, some pre-regulatory activities have been developed worldwide.

From 2020, the World Port Sustainability Program (WPSP) will include the airborne noise of the ship in the Environmental Ship Index formula to calculate the environmental impact of ships with possible reductions in port fees for less impacting units (and, accordingly, for quieter ships (Mocerino 2019, DNV 2019).

The general goal of the project NEPTUNES-Noise Exploration Program To Understand Noise Emitted by Seagoing Ships is to increase awareness and gain support to reduce noise from seagoing ships (Witte 2019). The ports that have joined the project are among the most important in the world: Amsterdam, Vancouver, Rotterdam, Hamburg, Copenhagen Malmo being just a few examples.

The SILENV project (Ships oriented Innovative soLutions to rEduce Noise & Vibrations) was developed in the context of the transport theme and focuses on the reduction of the acoustical impact of ships (SILENV 2012). Regarding airborne noise emitted in port areas, two cases were identified: noise generated by sailing close to the coast, and noise emitted by ships in the harbor. After an analysis and summary of the present regulations and previous projects, one of the results of the project was a proposal for limit values of noise emissions from ships and recommendations on measurement methods. This can be considered the first attempt to set limits to the specific component of the noise emitted from ships, that, on the other hand is one of the sources acting in a port area, together with loading/unloading gears, ground transportation means: cars, trucks, trains, conveyors and other activities connected with maintenance/repairing. The paper is structured as follows: after a first introduction on the topic, the regulatory framework has been exploited. The second part of the paper aims to analyze the new class notation. Finally, the results of a numerical model and experimental campaign has been used to verify the potential and the difficult linked to the verification of the new class notation.

2 REGULATORY FRAMEWORK FOR THE VARIOUS FORMS OF NOISE IMPACT

2.1 Class rules and statutory regulations on noise

The first requirements concerning ship noise have traditionally regarded noise on board. To protect personnel from noise, SOLAS regulation requires ships to be constructed to reduce onboard noise in accordance with the IMO Noise Code MSC.337(91): "Code on noise levels on-board ships". These ships must be constructed to reduce onboard noise and to protect personnel from noise, ensuring better working and living conditions on board (IMO 2012).

The so-called Comfort Classes are voluntary class notations setting comfort requirements on vibration and noise in different onboard spaces. The Classification Societies issue a comfort class notation after full-scale measurements carried out according to the notation requirements. Noise limits are fixed for passengers based on the time spent in the space, the use of the space and the noise expectation. For the crew, the limits are based on this subdivision of the spaces: machinery spaces, navigation spaces, accommodation spaces and service spaces.

Both IMO provisions and Comfort Classes requirements are based on human sensitivity to noise. The two groups of requirement reflect different perspectives and aims: the former is focused on the preservation of an acceptable level for the working environment on board, the latter is aimed at achieving comfort (for both passengers and crew, at different levels). The requirements reflect the different targets in different levels.

The IMO MEPC in 2008 included in the agenda of the 58th session the "Noise from commercial shipping and its adverse impact on marine life" (IMO 2009). Since then, measurement standards have been developed for commercial ships, including limits for the underwater noise emitted. Among them, the DNV Silent Class Notation (DNV 2010, DNV 2019) and the BV Rule Note 614 (Bureau Veritas 2014). As noted in Badino et al. (2012) all these requirements are basically technology based, i.e. the limits reflect the state of the art of modern shipbuilding: they represent a reasonable level of underwater noise containment, based on what is technically achievable nowadays. The actual impact of these (or other) limits on the marine fauna is not evaluated and cannot be evaluated at the present state of knowledge. A proper consideration of such impact in a given area would imply to have a full characterization of the perception of the underwater noise by the receiver (marine fauna) i.e. audiograms for the numberless species of animals populating the area and a precise information of the geographical distribution in water (in three dimensions) of the animals themselves.

To control the airborne noise the European Commission introduced the Environmental Noise Directive 2002/49/EC (European Commission 2011, 2016); this directive introduces tools for planning and measuring noise pollution in sensitive areas such as roads, airports and, of course, ports; the directive allows the Member States to set noise limits, but this was rarely done for ports specifically.

2.2 LR CLASS requirement on ABN

To meet the demand for a control of the airborne noise radiated by a ship, in January 2019 the Lloyd's Register issued a class notation on airborne noise emission of ships (ABN) setting a procedure for awarding this notation (Lloyd's Register, 2019). The notation applies to new or existing self-propelled ships of length greater or equal of 24m and may be awarded when the measured values fulfill the defined criteria.

The notation identifies two principal operating conditions: the harbor moored and the free sailing conditions. In the former one, the ship is moored at pier with all equipment normally operating at harbor (including possibly a situation with power provided

from shore) and the main propulsion system turned off. For vessels with garages such as passenger ferries and ro-ro ships, the car deck ventilation shall be included in the noise emissions while shall not be included the noise directly emitted from cars/trucks and from passengers during loading and unloading operations. The free sailing condition may correspond to the sailing of the vessels inside port areas or along the coast in channels or in straits with all equipment normally operating during the sailing condition (main engines, ventilation systems, auxiliary power in operation). The standard sailing speed is 5 knots; the assessment, at a minimum, shall be performed in the frequency range 31.5–8000 Hz in 1/1-octave bands.

In the following, formulas and definitions are recalled of the noise levels to be calculated, and monitored, as well as all the definitions necessary for the application of the class notation.

- $L_{WA, ship}$ is the "Ship Sound Power Level", the energy sum of all single-source sound power levels for given operating (eq.1);

$$L_{WA,i} = 10 \log \left(\frac{P_i}{P_0} \right), dB$$

$$L_{Wa,ship} = 10 \log \left(\sum_{i=1}^{n} 10^{\frac{L_{WA,i}}{10}} \right), dB \quad (1)$$

where i refers to the single source number, n is the total number of single sources on a ship, P_i is the sound power of source number i in W and P_o is the reference sound power (1 pW).

- $L_{Aeq,T}$ is the "Equivalent Continuous A-Weighted Sound Pressure Level", the sound pressure levels (eq. 2);

$$L_{Aeq,T} = 10 \log \left(\frac{1}{Tp_0} \int_0^T p_A(t)^2 dt \right), dB \quad (2)$$

where p_A is the instantaneous A-weighted sound pressure in Pa and T is the specified time interval in s.

- $L_{pAS,max}$ is the "Maximum Sound Pressure Level", the maximum A-Weighted root-mean-square sound pressure level measured with constant SLOW during passage of the ship or during the defined operating condition, according to IEC 61672-1;
- Distance to shipside is the distance to the hull side in horizontal direction as shown in Figure 1;

Table 1 shows the limit values for the assessment criteria:

The acronyms in the table correspond respectively to Super Quiet, Quiet, Standard, Inland Waterways

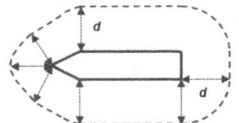 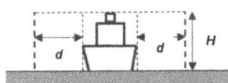 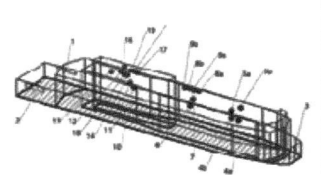 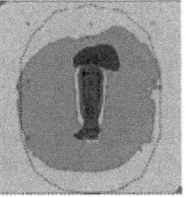

Figure 1. Distance to shipside.

Figure 2. Example of 3D calculation model and noise contour map.

Table 1. Assessment criteria values Lloyd's Register, 2019.

Sound power	Harbor moored		Free sailing		Distance to ship side
	dB(A)		dB(A)		m
	$L_{WA,ship}$	$L_{Aeq,T}$	$L_{WA,ship}$	$L_{pAS,}$ max	d
SQ	82	40	92	50	50
Q	88	40	98	50	100
S	96	40	106	50	250
IW	101	65	111	75	25
C	108	40	-	-	1000

and Commercial notation levels. To obtain the ABN notation the ship shall meet all four requirements at the same level. The ABN(*) notation at a given level may be awarded only if the airborne noise levels measured are less than the corresponding limit. The classification of a vessel requires therefore the correct evaluation of two different quantities (Lloyd's Register, 2019). On one hand there are the ship sound power levels, $L_{WA,ship}$, intended as the energy-based sum of all single-source sound power levels. The application of the rule requires the correct evaluation of this quantity for the different acoustic sources: these values are to be then added to each other. To determine the single source power levels of each source onboard the standards to be followed are ISO3744 or ISO9614.

The second set of limits refers to the sound pressure level (in terms of $L_{Aeq,T}$ and $L_{pAS,max}$) measured at specific distances from the vessel.

It appears evident that, because of the formulation of the criteria, it is necessary to verify at which distance a given "limit SPL" is verified. Generally speaking, a huge number of measurements should be required during the test campaign.

For new constructions, it is therefore recommended to realize a 3D calculation model of the ship according to ISO 9613-2 (Figure 2). This model shall include the main geometry and individual noise sources of the vessels, screening, reflection and absorption by the ship structure, as to realize a model of the ship through which identifying the specific SPL values at given target points.

Following this recommendation, each source shall be characterized by its sound power level and, if relevant, a directivity index. In the predictive model main sources are considered the exhaust stacks and funnels of main and auxiliary engines, the ventilation air intakes and exhaust, all the external fans, any special equipment in operation (as cranes and/or cargo pumps) and, if relevant, the hull radiated noise. The calculated noise emitted from the exhaust stacks shall also include, if present: silencers, scrubbers and filters (Lloyd's Register, 2019). The requirement prescribes also the way for modeling the sources on board: small ventilation openings and exhaust stack openings may be in general be modeled as point sources, larger ventilation grilles shall be modeled as surface sources. The design stage calculation report shall include primarily: the selected assessment criteria in each operating condition, all user input according to ISO9613, the single source sound power levels, the determined ship sound power levels, any deviation from calculation method and a calculated sound pressure levels as color-coded noise contour maps.

On-site measured levels for the verification of compliance to the class notation should be carried out. For both the two preset operating conditions of "free sailing" and "harbor moored", the notation mentions on-site measurements following preferably a "near-field" method, defined according to the following definition: "The near field is limited to a distance from the source equal to about a wavelength of sound or equal to three times the largest dimension of the sound source (whichever is the larger)".

The limit weather conditions in which it would be preferable to carry out the measurements should not exceed 3 on the Beaufort scale and a sea state 2, according to WMO sea state code. The ship under test should be ballasted to the design draught with all equipment, normally in operation, actually running. The sound pressure levels $L_{Aeq,T}$ and the $L_{pAS,max}$ shall be determined by updating the 3D calculation model with the single-source sound power levels and verifying at a relevant distance and at least two heights (3.5m and at ship height above sea level) the compliance to the assessment criteria (Lloyd's Register, 2019).

3 CLASSIFICATION CRITERIA: DISCUSSION ON EXPERIMENTAL AND NUMERICAL APPROACH LIMITS AND POTENTIALITY

A main difficulty in modeling the acoustic emission of ship in air consists in the complexity of the ship source in terms of size and distribution of the onboard sources. The correct modeling of the surrounding environment, too, can require attention when the spatial conformation can strongly influence the propagation patterns.

The numerical tools available for the characterization of the phenomenon are well assessed, but the effectiveness of results is strongly dependent on the correct definition of the aforementioned parameters.

From a theoretical point of view, noise source can be correctly modeled through their specific spectral power and directivity patterns (spherical, linear, plane or defined by a directivity solid) but these pieces of information may not be available with the correct approximation.

In addition, a proper modelling of the near field diffraction effect, due to the short distance of the noise source to the ship boundary surfaces, presents difficulties because of the simplification in the vessel geometry unavoidably present in the numerical models.

As suggested by the ABN notation, experimental measurements aboard the ship, both in maneuvering and mooring conditions, can be useful for a better characterization of the noise sources while ground measurements in the port area help in assessing their on-shore effects. In synthesis, a good experimental campaign is needed both for a proper source characterization (source power) and for an effective assessment of the sound pressure at receiver (for updating the model).

Unfortunately, ships radiate noise from different sources on board, located at very different heights, including at low levels above water. Obstacles, reflections, refractions, absorption and shielding from remote sensors, will affect the acoustic field. In addition to that, various vessels may be located close to each other, not only in transient but also in stationary conditions and can present a similar "acoustic signature". All these aspects make the recognition of the contribution of every single ship source quite complicate. When the monitoring activity is performed for checking compliance with legislative requirements, this problem becomes critical.

Instruments mostly used during this type of measurement are certainly sound level meters but also sound intensity probes as well as noise camera (beam forming systems) can be useful when dealing with not accessible/high-placed noise sources.

To isolate the specific noise contribution emitted from a single source, excluding other components, simultaneous records from arrays of microphones arranged in specific geometrical configurations can be analyzed. By post-processing the time signals, accounting for amplitude and phases, it is possible to identify levels that can be attributed to sources in a given direction and distance i.e. to perform spatial filtering (f.i. with beam forming techniques) (Figure 3). The computed noise levels from all sources within a specified range of angles are accumulated, giving an estimation of the total L_{Aeq} noise level arriving to the receiver from that direction.

Preliminary applications (see Coppola et al. 2017, 2018) of the concept have demonstrated good performance in source location and inherent noise filtering, with a correct representation of the true noise value to be assigned to the specific source.

Sound level meter stations represent, instead, the standard device for noise monitoring at the receivers. These transducers, being aimed at the characterization of the noise impact in a receiving position, measure the sound level in a given location but they do not provide information about the direction and distance of the source, which needs to be derived with additional information.

With the aim of a preliminary analysis of the new class notation proposed by Lloyd Register, in this work reference is made to a part of the data acquired during the experimental campaign of May 2017 carried out in the Port of Naples.

3.1 Comparisons between class notation and the results of an experimental campaign.

The port of Naples is one of the largest of the Mediterranean in term of quantity and diversity of traffic (Figure 4). It consist of a wide area, with different sub areas, specifically dedicated to: local connection fast boat, Cruise ship area, long and short distances ferries, commercial activities (ADSP 2020). The harbor area is adjacent to the city center without any discontinuity as evident in the next picture.

Figure 5, highlights the measuring points of the experimental campaigns.

An application of the ABN notation in this specific context by means of numerical models can be carried out by means of ray tracing-based methods.

Figure 3. Example sound camera result.

Figure 4. The port of Naples.

Figure 5. The cruise ship and long-range ferries area.

These methods, however, need to be applied with specific attention to the anthropomorphic characteristics of the area, because the specific geography of the port can produce discrepancies of results.

Figure 6, shows an example of the terrain modeling when evaluating the acoustic footprint of a long-range ferry at quay.

The adopted software, Terrain by Olive Tree Lab-Suite, is based on a ray-tracing approach implementing International Standard ISO 9613-2 methodology.

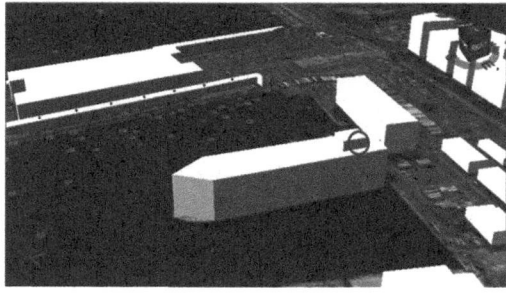

Figure 6. Example of 3D of a ship for a ray-tracing model.

The model is based on high frequency resolution calculations, but results can also be represented in 1/1 and 1/3 octave, both as sounds spectra and cartographic maps. In this simulation, only the incinerator was considered (as a spherical source) being identified as the main noise source.

Nevertheless, it appears evident that the noise transmission paths to the target receivers are strongly affected from the presence of buildings in the surroundings (Figure 7). This fact, when verifying the ABN, produces different noise levels at similar distances from the source.

The next table, Table 2, shows a direct comparison of noise level estimated at the target point and relative experimental measurements.

It is recalled that points D, I and L are located along the balcony on the right side of the ship, S,T and U are located on the opposite left side on the quay area and V is located on the back of the ship (see Figure 5).

Numerical and experimental values are quite similar in some locations but differ significantly in other ones.

This may be due to model uncertainties both in the characteristics of the source and of the environment. On one hand, the source (in lack of more detailed information) was modelled as omnidirectional and this may turn out to be imprecise. On the other hand, the surrounding environment is definitely asymmetric and different effects affect the propagation in different

Figure 7. Numerical model transmission paths.

Table 2. Numerical and experimental values comparison with a sound source of 100 dB.

Point	Experimental value dB	Numerical value dB	Difference dB
D	63	63	0
I	63	59	-4
L	60	55	-5
S	61	58	-3
T	64	64	0
U	65	67	+2
V	71	65	-6

directions (different interaction of direct, diffracted and reflected paths).This may imply a different degree of realism of predicted values along the various transmission paths.

4 CONCLUSIONS

The problem of noise pollution generated by ships in port is becoming more and more of interest for the legislative bodies. The publication of this additional class notation confirms this.

Preliminary study, the notation has been explained and briefly compared to results obtained during experimental campaigns and simulations carried out on a particular case study. The results of these analysis shows how the verification of the class notation will be not very simple and fast; on the other hand all the actual state of the art of the software should allow the possibility of approach the problem in the best way.

In future studies the same notation will be compared to simulations made ad-hoc and to a wider database of experimental measurements in order to reach a more complete verification of the same.

For examples, the future simulations will take into account the directivity of the sources of noise and of the effect of buildings near the mooring zones (aspects of which the notation does not take care).

REFERENCES

ADSP 2020, https://adsptirrenocentrale.it/
Badino, A., Borelli, D., Gaggero, T., Rizzuto, E., & Schenone, C. 2011. Normative framework for noise emissions from ships: present situation and future trends. *Advances in Marine Structures*. Leiden: CRC Press/Balkema.
Badino, A., Borelli, D., Gaggero, T., Rizzuto, E., & Schenone, C. 2012. Noise emitted from ships: impact inside and outside the vessels. *Procedia-Social and Behavioral Sciences*, 48, 868–879.
Borelli, D., Gaggero, T., Rizzuto, E., & Schenone, C. 2015. Analysis of noise on board a ship during navigation and manoeuvres. *Ocean Engineering*, 105, 256–269.
Bureau Veritas 2014, Rule Note NR614 DT R00E, Underwater Radiated Noise (URN). *Bureau Veritas*.
Coppola, T., Mocerino, L., Rizzuto, E., Viscardi, M., & Siano, D. 2018. Airborne Noise Prediction of a Ro/Ro Pax Ferry in the Port of Naples. In *Technology and Science for the Ships of the Future: Proceedings of NAV 2018: 19th International Conference on Ship & Maritime Research* (p. 157). IOS Press.
Curcuruto, S., Marsico, G., Atzori, D., Mazzocchi, E., Betti, R., Fabozzi, T., … & Bisegna, F. 2015. Environmental impact of noise sources in port areas: a case study. In *Proceedings of the 22nd International Congress on Sound and Vibration*, Florence, Italy.
Di Bella, A., & Remigi, F. 2013. Prediction of noise of moored ships. In *Proceedings of Meetings on Acoustics ICA2013* (Vol. 19, No. 1, p. 010053). Acoustical Society of America.
DNV 2010. 6, Chapter 24-*Silent Class Notation*. Rules for Classification of Ships. sl: DNV.
DNV 2019, https://www.dnvgl.com/expert-story/maritime-impact/QUIET-A-DNV-GL-class-notation-for-airborne-external-noise-from-ships.html.
ESPO 2018, European Sea Ports Organisation, https://www.ecoports.com/publications/top-10-environmental-priorities-of-eu-ports-2018.
European Commission 2011, Report from the Commission to the European Parliament and the Council on the implementation of the Environmental Noise Directive in accordance with Article 11 of the Directive 2002/49/EC, COM, Brussels.
European Commission 2016, Directive (EU) 2016/1629 of the European Parliament and of the Council of 14 September 2016 laying down technical requirements for inland waterway vessels, amending Directive 2009/100/EC and repealing Directive 2006/87/EC.
Hoyt, E. 2018. Marine protected areas. In Encyclopedia of marine mammals (pp. 569–580). Academic Press.
IMO 2009, MEPC 59 Noise from Commercial Shipping and its Adverse Impacts on Marine Life – Report of the Correspondence Group International Ocean Noise Coalition, Oceannoisecoalition.Org.
IMO 2012, RESOLUTION MSC.337(91), ADOPTION OF THE CODE ON NOISE LEVELS ON BOARD SHIPS MSC 91/22/Add.1, Annex 1.
Witte, R. 2019. NEPTUNES Measurement protocol and Noise Label. In INTER-NOISE and NOISE-CON. *Congress and Conference Proceedings* (Vol. 259, No. 2, pp. 7389–7392). Institute of Noise Control Engineering.
Kurt, R. E., Khalid, H., Turan, O., Houben, M., Bos, J., & Helvacioglu, I. H. 2016. Towards human-oriented norms: Considering the effects of noise exposure on board ships. *Ocean Engineering*, 120, 101–107.
Mocerino, L., Quaranta, F., Rizzuto, E. 2018. Climate changes and maritime transportation: A state of the art. In *NAV International Conference on Ship and Shipping Research*, (221499). https://doi. org/10.3233/978-1-61499-870-9-1005.
Mocerino, L., & Rizzuto, E. 2019. Preliminary approach to the application of the Environmental Ship Index. In *Sustainable Development and Innovations in Marine Technologies: Proceedings of the 18th International Congress of the Maritime Association of the Mediterranean* (IMAM 2019), September 9-11, 2019, Varna, Bulgaria (p. 285). CRC Press.
Mocerino, L., Murena, F., Quaranta, F., & Toscano, D. 2020. A methodology for the design of an effective air quality monitoring network in port areas. *Scientific Reports*, 10(1),1–10.
Murphy, E., & King, E. A. 2014. An assessment of residential exposure to environmental noise at a shipping port. *Environment international*, 63, 207–215.
Lloyd's Register, 2019, Design and Construction Additional Design and Construction, Procedure for the Determination of Airborne Noise Emissions from Marine Vessels, Lloyd's Register Group Limited 2019 Published by Lloyd's Register Group Limited.
SILENV 2012.Ships oriented Innovative soLutions to rEduce Noise and Vibrations, FP7. ⟨http://www.silenv. eu/⟩.
Viscardi, M., Coppola, T., Quaranta, F., Rizzuto, E., & Siano, D. 2017. On field experimental characterisation of the ship sources of acoustic pollution within a commercial harbour. In *Proceedings of 24th International Congress on Sound and Vibration*, ICVS24, London, UK.
WHO, World Health Organization 2018. Environmental noise guidelines for the European region.

Developments in Maritime Technology and Engineering – Guedes Soares & Santos (eds)
© 2021 Copyright the Author(s), ISBN 978-0-367-77376-2

Second generation intact stability criteria: Application of operational limitations and guidance to a megayacht unit

N. Petacco, P. Gualeni & G. Stio
DITEN - Department. of Electric, Electronic and Telecommunication Engineering and Naval Architecture
University of Genoa, Italy

ABSTRACT: Second Generation Intact Stability criteria (SGISc) are under finalization at IMO aiming to provide a set of not mandatory criteria covering phenomena related to intact stability in waves. SGISc introduce also operational guidance and/or operational limitations to the navigation. The former entails the analysis of operational data (e.g. speed and headings) as complementary aspects to the ship design strategies. Operational limitations consist of restrictions to the environmental conditions or geographical area where the vessel may sail. This paper focuses on the operational guidance and limitations as proposed at the 7th session of IMO Ship Design and Construction subcommittee. An overview of the current rules concerning the operational aspects of the criteria is given. A comprehensive assessment of a mega yacht unit is carried out for the excessive accelerations phenomena, except for the direct stability assessment. Results are going to be presented with special attention to the operational aspects.

1 INTRODUCTION

In the latest ten years, the development of the so-called Second Generation Intact Stability Criteria (SGISc) has been addressed by the Sub-Committee on Safety Design and Construction (SDC) at the International Maritime Organization (IMO). As stated in the introductory parts of Intact Stability code (IMO 2008), the perceived need of new criteria able to assess the ship stability waves is at the origin of this activity. Therefore, the SDC sub-committee decided to tackle this issue with an innovative framework and considering the physically based approach as the leading principle. Four different phenomena have then been identified:

- Righting arm variation problems due to waves (parametric roll and pure loss of stability);
- Stability under dead ship condition;
- Maneuvering-related failures in waves (surf-riding/broaching-to);
- Excessive lateral acceleration.

The physics at the basis of this phenomena has been widely studied in literature. A comprehensive description of all the stability failure modes is given in Belenky et al. (2011) and in U.S. Coast Guard (2019).

An innovative method to tackle the SGISc, namely the multi-layered approach, has been proposed in 53/3/5 (2010). In the latest years the proposal has been modified and improved, the last version of the multi-layered approach is given in SDC 7/5/1 (2019). For each stability failure mode, three different assessment levels have been formulated, each one with an increasing level of accuracy. The assessment of Level 1 (LV 1) is very simple and fast to be carried out, but it give conservative results; on the other hand, assessment of Level 2 (LV 2) requires more information and calculations effort, even if the outcomes are more accurate. Finally, the Direct Stability Assessment (DSA) represents the third assessment level and it is the most accurate but also the most computational time consuming. This tool, in principle, should be based on non-linear time domain seakeeping numerical simulations, and it should be able to consider directly the coupling factors among selected ship motions involved in each stability failure addressed. As a supplementary level, some operational restrictions could be introduced in the assessment. These measures are divided in two categories: restrictions acting only on the environmental condition where the ship is supposed to sail, namely Operational Limitations (OL), and restrictions acting on the vessel operability such as headings and speeds in relation to the environmental condition encountered, i.e. Operational Guidance (OG).

At the beginning of criteria development, levels are intended to be carried out sequentially from LV1 to DSA passing trough LV2, but during the 5th session of SDC it has been clarified that no hierarchy among levels exists. It means that, even if the sequential application of vulnerability levels appears the most rational path to tackle the SGISc, there is no need to approach them with a hierarchical procedure. Therefore, ship designers may carry out the assessment by applying any level regardless the conservativeness adopted. The same applies also for OG and OL, which can be

DOI: 10.1201/9781003216582-40

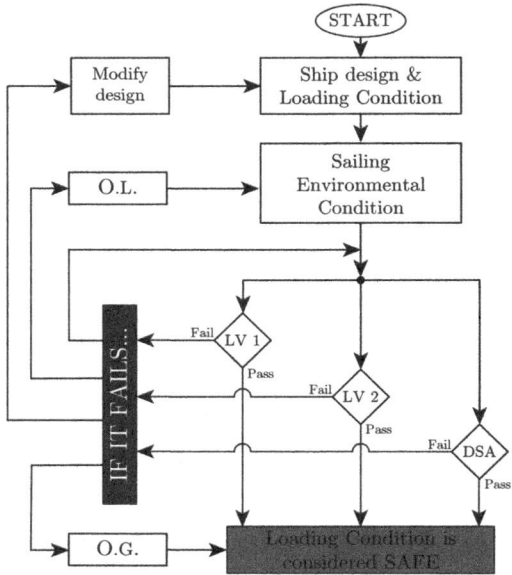

Figure 1. Simplified scheme of the application logic of second generation intact stability criteria.

carried out even before LV1 or LV2. A simplified schematization of the multilayered approach and its application logic is shown in Figure 1.

The finalized draft of Interim Guidelines for the application of all vulnerability levels, OG and OL has been presented at the SDC 6th session (SDC 6/WP.6 2019) and it is expected that SGISc will enter into force during 2020 for a trial period. For the time being, these new criteria should be treated as a supplementary assessment tool, in order to give a further guide to ship designers to assess ship stability under different aspects. An exhaustive overview about the steps that lead to the development of SGISc has been made by francescutto(2019).

In this paper, a focus on the OG and OL for the excessive acceleration phenomenon, as proposed at the 7th session of SDC by the intersessional correspondence group, is given (SDC 7/5 2019). An overview of the current rules concerning the operational aspects of the criteria will be given. Besides, an assessment of a mega yacht unit with the SGISc will be carried out, except for the third level. Results are going to be presented, when applicable by means of polar diagrams, and analyzed with special attention to the operational aspect that such set of rules can imply.

2 OPERATIONAL MEASURES IN THE SGISC FRAMEWORK

Although a more and more accurate ship design may increase notably the safety level of vessels, it is recognized that operational measures are required as a complementary tool to fully address the safety

performance (Liw°ang 2019, Baˇckalov et al. 2015). In particular, this aspect assumes more relevance when complex phenomena related to navigation are addressed, such as those tackled by the SGISc. It is obvious that phenomena which are not controllable by human active role are excluded, for example the dead ship condition. Moreover, when too many restrictions to a loading condition are introduced by the operational measures, it means that this loading condition should not be deemed as acceptable in principle. Operational measures may be divided in two categories, depending on which sailing factor they treat: the environment where the vessel is sailing or the operational ship features.

2.1 Operational guidance

Operational Guidance may be considered as an additional vulnerability level able to identify which situations should be avoided. The assumed situation is defined as the combination of the sailing condition (V_S and μ) and the sea state characteristics, i.e significant wave height H_S, zero-crossing wave period T_Z, wind direction and gust characteristics. For this reason, OG could be considered as an independent tool able to evaluate ship stability in a seaway condition. A schematic representation of the operational guidance structure is proposed in Figure 2.

OG provides a set of information about the ship handling in specified sea state condition. With reference to vessel speed and heading, the guidance determines which actions should be carried out by the master in order to reduce the stability failure probability. Since the OG are drawn up during the design phase, it should embrace all the possible environmental conditions that might be encountered during the ship life. For this reason, detailed sea state and wind forecast information are required on board during the navigation, to allow the master to plan the best sailing condition (i.e. combination of ship forward speed V_S and mean wave encounter direction μ) according to the OG. Within the SGISc three equivalent approaches to draw up OG are provided:

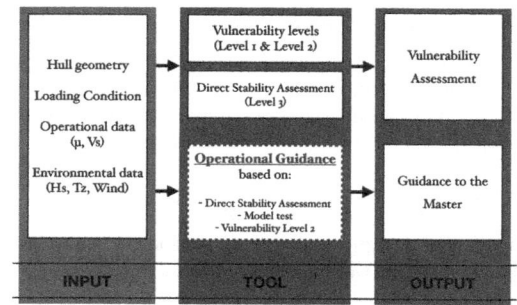

Figure 2. Scheme of operational guidance structure. It is considered as an independent assessment tool based on numerical simulation, model test or vulnerability assessment level.

- Probabilistic operational guidance;
- Deterministic operational guidance;
- Simplified operational guidance.

The simplified approach is based on the same methodology of LV1 and LV2, while the probabilistic and the deterministic approaches may share the same numerical calculation method of DSA. The numerical method should reproduce ship motions in irregular seas and should detect the failure event, defined as the exceedance of a specific roll angle or a lateral acceleration. The roll angle threshold is set equal to the minimum angle among 40 [deg], the vanishing angle or the downfloading angle evaluated in calm water. The threshold for lateral acceleration is set equal to gravity acceleration g = 9.81 [m/sec^2]. Moreover, lateral acceleration should evaluated at the highest position where crew or passengers may be present. Depending on the stability failure investigated, the numerical tool should replicate at least three degrees of freedom in the time domain, considering coupling factors among motions. Detailed technical requirements of the numerical tool are given in (SDC 7/5 2019). The development of a comprehensive numerical tool able to comply with the SGISc requirements is a challenge among experts and researchers, therefore great efforts have been made in this field (Shigunov et al. 2019, Kuroda et al. 2019)in the last years.

2.1.1 Probabilistic operational guidance
This kind of approach takes into consideration a probabilistic criteria, such as the probability of stability failure over a time period or the stability failure rate. For each loading condition, the allowed sailing condition are those for which the criterion does not exceed a corresponding probabilistic standard threshold. The relation between criterion and the standard is given in ((1)).

$$r < 10^{-6} \qquad (1)$$

where r [1/sec] is the upper boundary of 95% confidence interval of the stability failure rate.

The probabilistic approach requires that the numerical simulation proceeds until a failure event occurs, i.e. the exceedance of roll angle or lateral acceleration threshold. The duration of the simulation is used to calculate the failure rate. The criterion r is evaluated as the average over all the numerical simulations that have been carried out.

2.1.2 Deterministic operational guidance
This approach results to be faster than the previous one. It takes into account a deterministic criteria, such as maximum roll amplitude evaluated in a given exposure time. This simplification leads to a lower level of accuracy compared to the previous approach, therefore the standard threshold is conservatively selected, pursuing an equivalent level of safety. For a given loading condition, a sailing condition is deemed safe by this approach if ((2)) is observed.

$$\alpha \cdot X_{3h} < X_{lim} \qquad (2)$$

where $\alpha = 2$ is a scaling factor to provide an equivalent safety level, X_{3h} is the mean-three hour maximum amplitude of roll angle or lateral acceleration while X_{lim} is the stability failure event threshold, i.e. roll angle or lateral acceleration.

The deterministic criterion may be calculated either using numerical simulation, models test or their combination. It is recommended that tests or numerical simulations have at least an overall duration of 15 hours for each combination of V_S, μ and environmental condition. There is the possibility to split the test or simulation into several part, not shorter than three hours.

2.1.3 Simplified operational guidance
Finally, simplified operational guidance represents the simplest but also the less accurate approach to draw up OG. In principle, simplified operation guidance may consist in any approximated conservative estimation method that has been proven to provide a superior conservative level compared to design assessment requirements, for example first and second vulnerability levels. Design assessments correspond to the application of LV2 or DSA.

Within the guidelines on SGISc, examples of recommended approach based on LV1 or LV2 are suggested. Briefly, these examples are summarized below:

- For pure loss of stability failure mode, it is suggested to avoid forward speed greater than $0.752 \cdot \sqrt{L_{PP}}$ [m/sec] in following to beam wave directions for those sea states not complying with LV2.
- For parametric rolling failure mode, forward speeds not compliant with second check of LV2 should be avoided in each wave directions for the considered sea states.
- For surf-riding/broaching failure mode two examples are provided. The first one suggests to avoid forward speed greater to $0.94 \cdot \sqrt{L_{PP}}$ [m/sec] in quartering seas for selected wavelength and wave height. The second approach is based on a modified version of LV2 taking into account diffraction wave forces; the critical speed obtained by the modified criterion should be avoided in following to beam seas for the considered sea state.
- For excessive acceleration failure mode, navigation is not recommended for those sea states not compliant with the modified LV2. The criterion modification introduces the waves encounter direction within the lateral acceleration Response Amplitude Operator (RAO) as well as the encounter wave frequency is considered.

2.2 Operational limitations

Operational Limitations set restrictions to the ship operability in the considered loading condition in terms of specific sea state or geographical area. Therefore, the OL may be considered as a tool acting directly on the environmental input data of another methodology in the framework of SGISc, such as vulnerability levels, DSA or OG. This is possible thanks to the modularity of the SGISc structure, mainly based on the physics behind the phenomena. In fact, criteria have been thought with the possibility to easily introduce modifications in the boundary conditions or in specific formulations. A schematic representation of the logical application of operational limitations is proposed in Figure 3.

According to how OL affects the environmental conditions, two typology of restrictions may be identified: related to the navigation areas or related to wave characteristics.

The first limitation defines specific operational areas, either geographical or typology (i.e. sheltered water), routes or specific year periods where the navigation is forbidden because of the high risk of stability failure. The environmental data can be modified acting on the selection of the appropriate wave scatter table, where the joint probability of the range of average zero-crossing wave period and significant wave height is reported, together with wind statistics. Since the OL related to areas or routes and season refer to fixed environmental condition, it is not required that detailed weather forecasts are present on board.

The restriction related to wave characteristics defines for which wave heights the risk to experience a failure event during the navigation is deemed acceptable. The OL related to maximum significant wave height modifies the wave scatter table used for the assessment. Sea states having a wave height higher than the selected one are cut off, obtaining

a limited scatter table. These limitations are strictly related to the encounter sea conditions, thus it is necessary to have on board detailed weather forecast, inclusive of significant wave height. An interesting application of OL have been carried out by rudakovic:opLim.

3 APPLICATION CASE

In this paper, the vulnerability assessment of the excessive acceleration failure mode of a representative mega-yacht unit has been carried out. Main dimensions of the hull and the design loading condition are given in Table 1.

Excessive accelerations criterion requires the definition of the vertical and longitudinal coordinates of the highest location where crew or passengers may be present. In this analysis, the point has been located on the Sun Deck at 15.70 [m] from the keel line and its longitudinal relative position (with reference to the ship length) is 0.40 [-], as shown in Figure 4.

Initially, KG limiting curves have been evaluated for both LV1 and LV2. This first analysis has been carried out by means of an *in-house* computational codes, developed at the University of Genoa by Coraddu et al. (2011) and Petacco (2019). Thereafter, OL have been applied to the second vulnerability level and KG limiting curves have been computed again. Restrictions take into account both OL related to maximum significant wave height and OL related to geographical area. In the first application of limitations, four different limited scatter tables have been made starting from the full scatter table of North Atlantic ocean (IACS 2001). Maximum significant wave heights from 1.5 [m] to 15.5 [m], with steps of 2.0 [m], have been chosen to define the limited scatter tables. The second restriction consists in replacing the North Atlantic wave scatter table, defined by the criterion, with the Mediterranean wave scatter table. Wind statistics have been kept the same as those defined by the criterion.

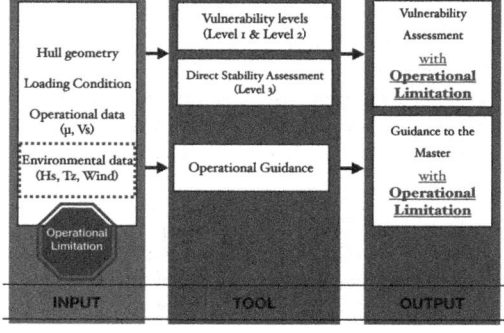

Figure 3. Scheme of logical application of the operational limitations. It acts on the environmental input data of the assessment tool in the SGISc framework, such as vulnerability levels, direct stability assessment or operational guidance.

Table 1. Hull dimensions and loading condition of the investigated mega-yacht unit.

Main dimension and Loading conditions		
L_{PP}	64.94	[m]
B	13.20	[m]
D	7.50	[m]
d	3.30	[m]
V_S	17.4	[kn]
KG	5.50	[m]
GM	1.91	[m]
T_{roll}	8.35	[sec]
C_B	0.562	[-]
C_m	0.918	[-]

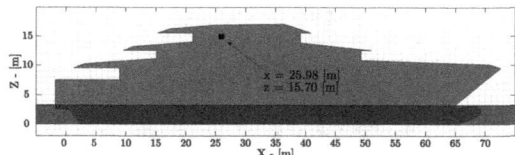

Figure 4 . Identification of the highest location where crew or passengers may be present on the longitudinal view of megayacht unit.

To complete the analysis, OG has been drawn up. In particular, simplified operational guidances have been evaluated for all the sea states contained within the North Atlantic scatter table. For sake of convenience, only a selection of representative polar diagrams have been reported in the following section.

In the next section, results will be shown in terms of KG limiting curves and polar diagrams and commented paying attention also to the operational aspect that such measures can imply.

4 RESULTS

Results of the excessive acceleration criteria application are shown in Figure 4. On the horizontal axis the assessed draught range is reported, while the limiting KGs are represented on the vertical axis. Because of the physics behind the excessive acceleration failure mode, the plotted curves should be considered as minimum limiting curves, i.e. the allowed domain is above the curve. This means that a loading condition is considered not vulnerable to the stability failure mode if the KG is located above the line. In the figure the limiting curves are represented, referring to the first vulnerability criterion and the second vulnerability criteria with and without OL. In the analysis, both typologies of limitations have been applied, i.e relating to the navigation

areas and relating to the wave characteristics. For the first limitation, the North Atlantic wave scatter table has been replaced by the wave scatter table of Mediterranean sea (dash-dot line). The second restriction consists of cutting the North Atlantic scatter table up to a maximum significant wave height (dashed line). An allowed significant wave height equal to 7.5 [m] (the same of the maximum wave height of Mediterranean sea) has been set in order to consistently compare the two typologies of limitation. The Intact Stability code has also been applied and its maximum KG limiting curve is represented by a black line. This upper limit has been calculated according to the analysis of the righting lever curve properties (paragraph 2.2 of the Intact stability code) and it represents the maximum value of KG after which the unit fails to comply with the Intact Stability code.

In the analysed draught ranges, the limitation on the maximum wave height is the least conservative allowing the maximum design domain. On the contrary, as expected, the most conservative assessment is represented by the LV1 criterion, located in the upper part of the graph. Moreover, it appears that no design domain between Intact Stability code and LV1 curves exists for draught lower than 3.2 [m]. On the other hand, LV2 and OL limiting curves ensure a sufficient region where the unit may be designed.

With reference to the design loading condition (black cross), it appears that the vessel does not meet the requirements of the two vulnerability levels. In fact, the loading condition marker is located far below the LV1 and LV2 limiting curves. Therefore, to comply with the SGISc, the OL should be applied: the loading condition is still not compliant even if the restriction on the sailing area is applied; the introduction of a limitation on the maximum wave height permits to made the loading condition compliant with SGISc.

In Figure 6, the relationship between the minimum KG and the limiting significant wave height

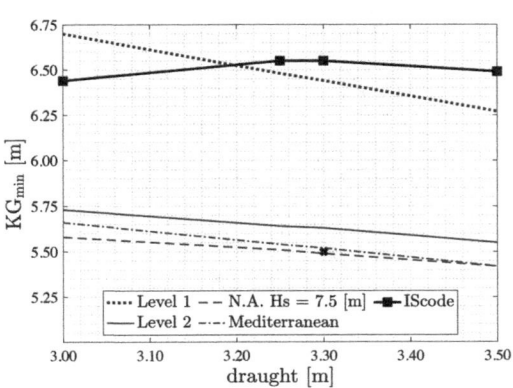

Figure 5. KG minimum limit curves for EA, both typologies of OL have been applied.

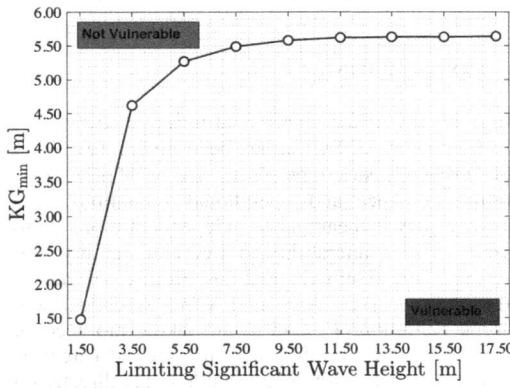

Figure 6. Relationship between the selected limiting significative wave height and the minimum allowed KG for EA.

for the excessive accelerations failure mode is shown. On the vertical axis, the minimum KG allowed by the application of OL is reported . The selected limiting wave height is reported on the horizontal axis: Hs equal to 17.5 [m] represents the North Atlantic full wave scatter table. Figure 6 shows how the minimum KGs are affected by the selection of the maximum significant wave height in OL. It seems that choosing limiting wave heights from 7.50 [m] up to 17.5 [m], i.e. the full wave scatter diagram of North Atlantic, affects slightly the limiting KGs. Below the limiting wave height equal to 7.5 [m], the minimum KG curve falls down quickly enlarging the design domain.

As concern the OG, the simplified methodology proposed SDC 6/WP.6 (2019), has been applied. Results have been presented by means polar plots. In the plots, heading seas direction is equal to 0 [deg] while following seas have a direction of 180 [deg]. Ship speeds are reported in knots along the radius of the graphs. Conditions that should be avoided, according to the simplified OG for the excessive accelerations failure mode, are indicated by red areas. To carry out a complete analysis and draw up the OG to the master, all the sea states within the wave scatter diagram should be assessed. The North Atlantic scatter table consists of 272 different sea states, 197 of which have a weighting factor different from zero. Due to the large amount of combinations between significant wave height and zero-crossing wave period, only polar diagrams for a limited selection of sea states are presented in this paper. Sea states having a zero-crossing wave period T_z close to the ship natural roll period are shown in Figure 7. Significant wave heights ranges from 3.5 [m] to 9.5 [m] with a step of 2.0 [m]. As expected, the higher the wave height, the larger the red area to be avoided become. It is interesting to see how forward ship speed affects vessel vulnerability to the failure mode: even if beam seas are deemed the worst sailing condition when referring to the excessive acceleration failure mode, Figure 5 shows that at higher speeds the vessel seems not to be vulnerable for the considered sea state.

5 CONCLUSIONS

An overview of the operational measures defined in the framework of SGISc has been given. Two main categories of restrictions are defined by SGISc: operational guidance and operational limitations. The first category is meant as an independent tool able to assess vessel vulnerability to a certain stability failure mode for assumed situations. Both environmental condition (sea state and wind statistics) and sailing condition (V_S and μ) are taken into account in the OG. On the other side, the second category, i.e. Operational Limitations, is able to act directly on the environmental conditions; thus OL should be always associated to another methodology in order

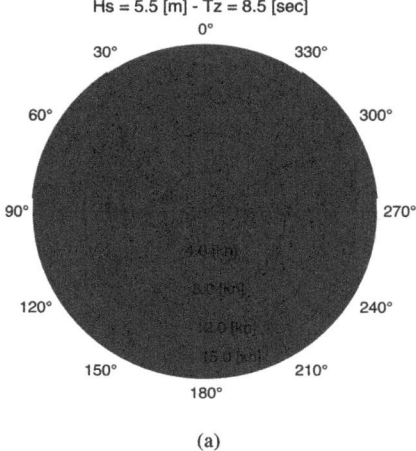

(a)

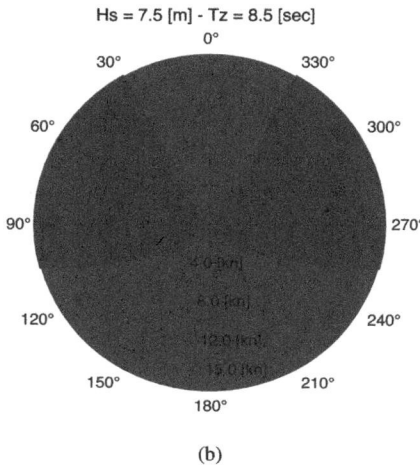

(b)

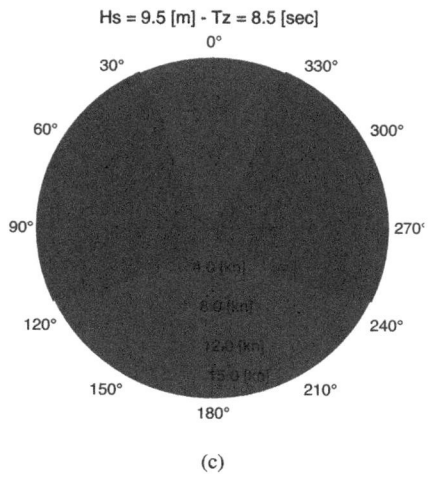

(c)

Figure 7. Representation of polar diagrams for a selection of sea states.

to assess the vulnerability of the vessel. According to the typology of the restriction made, OL can be classified as a limitation on the maximum wave height or a limitation to the operative geographical area.

An application of both typologies of OL and simplified OG for the excessive acceleration failure mode has been carried out. A representative megayacht unit has been investigated. Since the design loading condition is considered vulnerable by LV1 and LV2, it has been decided to apply both OL typologies of restriction. The design loading condition has been considered not vulnerable by changing the North Atlantic scatter table with the Mediterranean sea table. Moreover, outcomes show the high strictness of LV1, in fact taking into account the Intact Stability code limiting curve, it appears that for low values of draught no design domain exists. As concerns LV2, its limiting curve is located far below the Intact Stability curve allowing a sufficient design area for each draught investigated.

Then, an application of the simplified OG is presented. OG should be performed for each sea state listed in the wave scatter table, in this investigation for 197 cases. As expected, polar plots point out that the higher the wave height, the larger the forbidden operational area. Moreover, it is interesting to appreciate the effect of the sailing condition on the vulnerability judgement, e.g. for some wave heights in beam seas, even if this is the worst environmental condition for the excessive acceleration failure mode, the vessel is not considered vulnerable when sailing at high speed. This might be due to the influence of roll damping, in particular of the lift component. Nevertheless, it is worth to note that sailing at the maximum speed in certain sea states is not always practicable. Therefore, although a sailing condition improves the safety level in terms of stability, this may lead to a weakening of safety of other ship design aspects, such as seakeeping, longitudinal or local strength of structures.

REFERENCES

Bakalov, I., G. Bulian, A. Rosén, V. Shigunov, & N. Themelis (2015). Ship and safety in intact condition through operational measures. In *Proceedings of 12th International Conference on the Stability of Ships and Ocean Vehicles*, Glasgow, UK.

Belenky, V., C. Bassler, & K. Spyrou (2011, December). Development of second generation intact stability criteria.

Hydromechanics Department Report, Naval Warfare Center Carderock Division, Carderock, USA.

Coraddu, A., P. Gualeni, & D. Villa (2011). Investigations about wave profile effects on ship stability. In *XVI Congress of International Maritime Association of the Mediterranean*, Genova, Italy.

Francescutto, A. (2019). *Sustainable development and innovations in marine technology*, Volume 3 of *Marine Technology and Ocean Engineering series*, Chapter The development of second generation intact stability criteria, pp. 3–8. Taylor and Francis group.

IACS (2001, November). Standard Wave Data. Recommendation n.34 - Rev.1 34, International Association of Classification Society.

IMO (2008, December). Adoption of the international code on Intact Stability. Resolution MSC.267(85), International Maritime Organization, London, UK.

Kuroda, T., S. Hara, H. Houtani, & D. Ota (2019). Direct stability assessment for excessive acceleration failure mode and validation by model test. *Ocean Engineering* 187.

Liwång, H. (2019). Exposure, vulnerability and recoverability in relation to a ship's intact stability. *Ocean Engineering* 187.

Petacco, N. (2019, May). *Second Generation Intact Stability Criteria: Analysis, Implementation and Applications to significant ship typologies*. Doctoral thesis, University of Genova, Italy.

Rudakovi´c, S. & I. Bakalov (2019). Operational limitations of a river-sea container vessel in the framework of the second generation intact stability criteria. *Ocean Engineering 183*, 409–418.

SDC 6/WP.6 (2019). Finalization of Second Generation Intact Stability criteria. Report of the experts' group on intact stability, International Maritime Organization, London, UK.

SDC 7/5 (2019, October). Finalization of Second Generation Intact Stability criteria. Report of the correspondence group (part 1), International Maritime Organization, London, UK.

SDC 7/5/1 (2019, December). Finalization of Second Generation Intact Stability criteria. Report of the correspondence group (part 2), International Maritime Organization, London, UK.

Shigunov, V., N. Themelis, & K. Spyrou (2019). *Contemporary Ideas on Ship Stability*, Volume 119 of *Fluid Mechanics and Its Applications*, Chapter Critical Wave Groups Versus Direct Monte-Carlo Simulations for Typical Stability Failure Modes of a Container Ship, pp. 407–421. Springer, Cham.

SLF 53/3/5 (2010). Comments on the structure of new generation intact stability criteria. Submitted by Poland, International Maritime Organization, London, UK.

U.S. Coast Guard (2019). Continued development of Second Generation Intact Stability criteria. Naval architecture division report, U.S. Coast Guard Office of Design and Engineering Standards, USA.

Developments in Maritime Technology and Engineering – Guedes Soares & Santos (eds)
© *2021 Copyright the Author(s), ISBN 978-0-367-77376-2*

The preliminary design of rubber mounted pillars for pleasure and passenger yachts

G. Vergassola, D. Boote & L. Falcinelli
DITEN, University of Genoa, Genoa, Italy

ABSTRACT: In this research, the Naval Section of the Department of Electrical, Electronic, Telecommunications Engineering and Naval Architecture of the University of Genoa has developed a preliminary study for the use of pillars mounted on rubber supports that should be able to accomplish their structural role and concurrently reduce the vibration transmission. In particular, in this paper, the results of static and natural mode analyses has been carried out on FE mock up models of a simplified two-deck ship structure in order to verify the impact of rubber mountings on pillars, both in the transmission of loads and vibrations.

1 INTRODUCTION

The recent trend in pleasure and passenger yacht design is to increase the comfort level of owners, guests and crews. This approach is backed up both by Classification Societies that have developed new stricter rules and additional notations in terms of noise and vibration level onboard (Registro Italiano Navale, 2017a, 2017b), and by owners themselves, who often require level even lower than CS rules (Pais et al., 2018).

In this perspective, it is necessary to design proper insulation plans and new structural solutions (Vergassola, et al., 2018; Vergassola, 2019) that are able to limit the noise and vibration propagation from main sources (main engines, gearboxes, shafts, propellers, exhausts) to receivers.

The role of pillars commonly installed on engine rooms is crucial, since they are necessary from a pure structural point of view (in order to reduce the longitudinal beam spans) (Boote et al., 2017a) but they transmit high level of structure borne noise and vibrations (Biot et al., 2014; Pais et al., 2018).

In this research, the Naval Section of the Department of Electrical, Electronic, Telecommunications Engineering and Naval Architecture of the University of Genoa has developed a preliminary study for the use of pillars mounted on rubber supports that should be able to accomplish their structural role and concurrently reduce the vibration transmission.

In particular, in this paper, the results of static and natural mode analyses has been carried out on FE mock up models of a simplified two-deck ship structure in order to verify the impact of rubber mountings on pillars, both in the transmission of loads and vibrations.

2 RUBBER MOUNTED PILLAR DESIGN

The idea of the research is to create a pillar that is connected to the ship structures by a rubber compound. In Figure 1, a render representation of the rubber mounted pillar is presented; it can be easily seen that, between the metallic tube and the rubber compound, a flange has been installed, in order to distribute, as uniformly as possible, the axial load from the pillar to the support.

The dimensions of the rubber mounted pillar have been selected in compliance with the standard structural requirements of a superyacht, having an external diameter of 140 mm, a thickness of 6 mm with consequent internal diameter of 128 mm. The transversal flange has a diameter of 260 mm

With this configuration, the transmission of sound is completely shifted to the bolts, since are rigidly connected to the metallic structures (Fragasso et al. 2019). For this reason, it has been added an acoustic washer (Figure 2) between the flange and the bolts' heads; this approach has also the positive aspect of permitting to couple different metallic materials without creating galvanic currents.

During a compression load, the pillar system works together with the rubber, while, in traction condition, the stiffness is guaranteed only by bolts and in minimal part by the washer.

Two different rubber rigidities have been tested, corresponding to 1/3 and 1/2 of the pillar rigidity respectively; higher rubber rigidity decreases the static deflection of the decks, but, in the meanwhile, amplifies the transmission of vibrations and noises (Hecquet et al., 2017).

DOI: 10.1201/9781003216582-41

a)

b)

Figure 1. Rubber connection between the two part of the pillars: a) global view and b) close up of the connection.

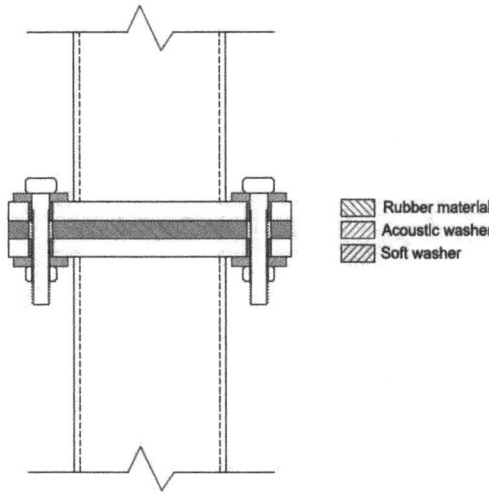

Figure 2. Section of the experimental pillar.

Figure 3. Mock up model.

3 THE NUMERICAL MODEL

In order to numerically test the static and dynamic behavior of rubber mounted pillar, a mock up of a ship structure has been created in a commercial FE software.

It corresponds to a portion of a two deck yacht, with a length of 8250 mm and an half-width of 5200 mm, as reported in Figure 3.

All the plating and the primary stiffeners have been modeled by using 2D linear shell elements, while the secondary stiffeners have been modeled by using 1D linear beam elements, in order to reduce the number of degrees of freedom and, consequently, the computational times (Boote et al. 2017b; Vergassola & Boote, 2019).

The upper deck has been loaded with an uniform pressure of 10 kN/m^2 and it has been constrained as it follows:

- Symmetry condition in correspondence of the symmetry plane of the yacht;
- Vertical support condition on the lower deck edges;
- Fixities in correspondence to the transversal bulkheads;

The load intensity has been chosen in order to simulate all the outfitting present on closed decks. This value is commonly used by Classification Society rules.

Two different schematizations of the rubber mounted pillar have been investigated; in the first case, a full 3D model (Figure 4) has been created in order to ensure the highest possible accuracy for the abovementioned Study Case 3 (see Table 1).

Nevertheless, this detail cannot be used in large numerical models, as a superyacht is, since it should require long discretization time and it increases the

a)

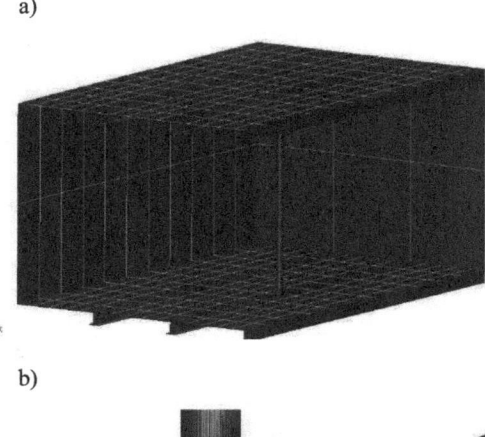

b)

Figure 4. Detail of the 3D rubber mounted pillar: a) global view and b) close up.

Table 1. Structural layouts analyzed.

Study Case #	Type of pillar	r
1	No pillar	0
2	Standard pillar	1
3	Pillar on rubber mounting on lower deck	0.33
4	Pillar on rubber mounting on lower deck	0.5
5	Pillar on rubber mounting on upper deck	0.33
6	Pillar on rubber mounting on lower and upper deck	0.33

complexity of the global model; for this reasons, a simplified model has been tested as well.

It is represented by a non-linear spring instead of the full 3D description of the rubber mounting (Figure 5) connected to 1D beam elements representing the pillar; the spring coefficient has been calibrated as the rubber rigidity if the pillar is under compressive stresses, while it has been imposed as the bolt stiffness during tensile ones.

With this schematization, different rubber behaviors could be tested by modifying the spring constant; a parameter 'r' is so introduced, representing the ratio between the rubber (and so the spring constant) and pillar rigidity.

In order to assess different possible layouts, the structural schemes reported in Table 1 have been studied.

Figure 5. Detail of the 1D rubber mounted pillar (in yellow).

4 RESUTS OF THE STATIC ANALYSES

4.1 Study Case 1

In Figure 6, the displacement plot of the Study Case 0 is reported.

As it can be seen, the vertical pressure deforms the upper deck structure up to 31 mm at the center of the deck, where there is no primary longitudinal and transversal stiffeners. Naturally, lower deck is unloaded, and side and bulkhead with very low loads, in fact the displacements are concentrated only on the upper deck.

a)

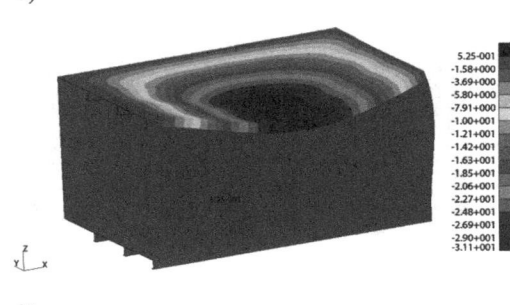

b)

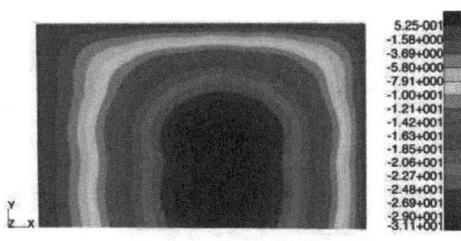

Figure 6. Displacement plot, Study Case 0: a) isometric view and b) top view.

367

4.2 Study case 2

To the previous model, a standard pillar in the mid of central stringer is added. As it was easily predictable, the introduction of pillar leads to a reduction of a deflection of stringer and to a new distribution of load in the system (Figure 7).

Pillar reduces the maximum displacement on the upper deck by halving the span of the central reinforced beam, reduces the tension on the extremity of the stringer but it generates a displacement field on the lower deck. Moreover, this is a direct connection between the two decks which can transmit vibration and noise.

4.3 Study case 3 with 3D pillar

The first study case in which the rubber mounted pillar must be accurately verified is the Study Case 3. The rubber material has been modelled by its stress strain graph, obtained from a local manufacturer.

As it can be seen by comparing Figure 7 and 8, even if the rubber is relatively soft, there is no significant difference (2.22 mm vs. 2.29 mm) in the static response between a standard pillar and a rubber mounted one.

Also the buckling verification is fully verified since a maximum stress of 24.48 MPa is acting on the pillar that has a critical stress of 110.49 MPa.

So, it can be deduced that the rubber mounted pillar achieves its static structural goal, and no remarkable difference with standard bolted/welded one can be noted.

a)

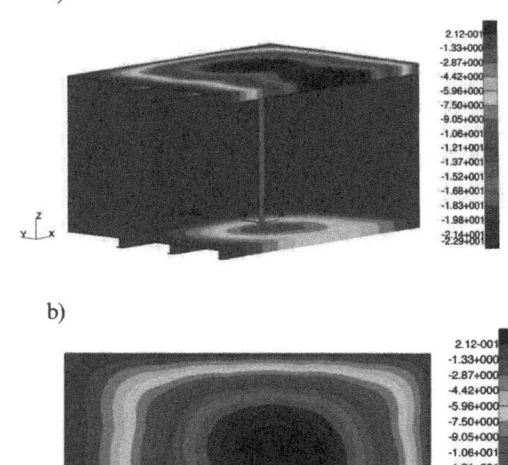

b)

Figure 8. Displacement plot, Study Case 3 with 3D pillar: a) isometric view and b) top view.

4.4 Study case 3 with 1D pillar

Since the modelling of a 3D pillar is very time consuming, the aforementioned simplification has been carried out, and in Figure 9, the displacement results

a)

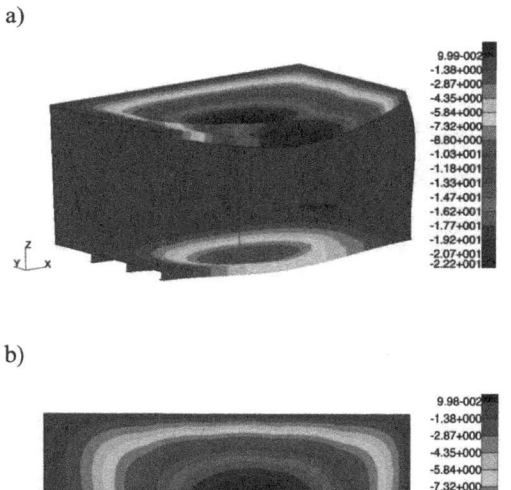

b)

Figure 7. Displacement plot, Study Case 1: a) isometric view and b) top view.

a)

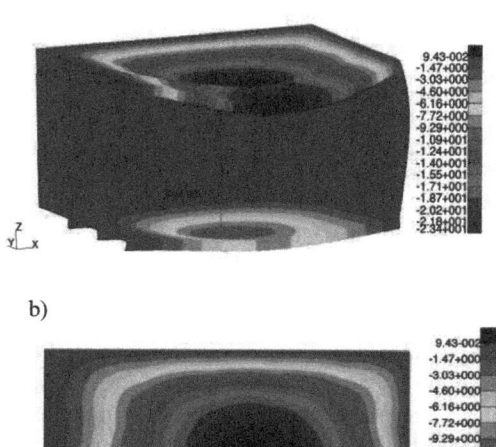

b)

Figure 9. Displacement plot, Study Case 3 with 1D pillar: a) isometric view and b) top view.

of Study Case 3 with 1D pillar connected to a non linear spring are reported.

The difference between the complete modelling of pillar and the simplified approach is about 2% and it can be considered as an optimum compromise between accuracy and modelling time.

The principal difference is due to the fact that the 3D version and the 1D one are linked to the main structure with a different connection and so the stress field, and consequently the displacement one, is different.

Nevertheless, the simplified version is on safety side, since it tends to amplify the displacements. For this reason, the following study cases have been analyzed only be using the simplified 1D approach.

4.5 Study case 4

This study case presents an harder rubber compound and in Figure 10 the displacement plot is reported.

The harder rubber compound and so a more rigid spring does not lead to a remarkable difference in deck displacements, both from a qualitative and quantitative point of view.

Nevertheless, it must be noted that a softer connection is more able to uncouple the pillar from the hull structure, with a consequence on its vibroacustic behavior.

4.6 Study case 5

It has also been tested the structural behavior of the mock up in the case of a pillar on rubber mounting on the upper deck. This solution could be useful to uncouple the pillar itself from the upper deck and it can be used in case of critical vibration propagation from a lower deck (e.g. in engine room) to an upper one, as often happens in superyacht and passenger yachts.

In Figure 11, the displacement plots are reported.

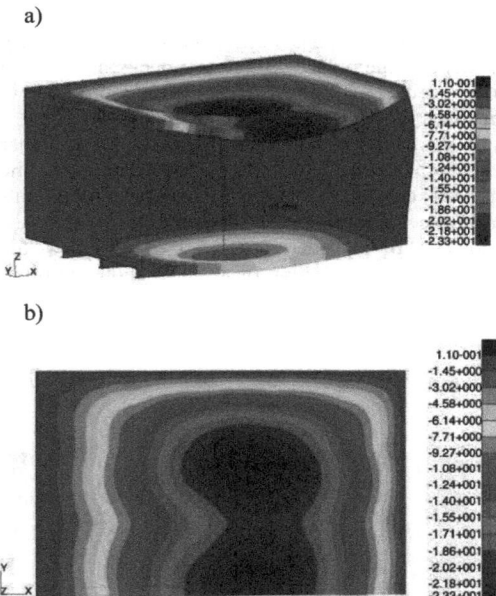

Figure 11. Displacement plot, Study Case 5: a) isometric view and b) top view.

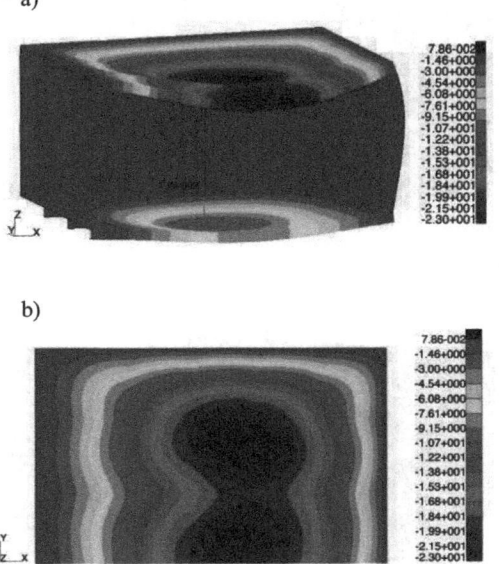

Figure 10. Displacement plot, Study Case 4: a) isometric view and b) top view.

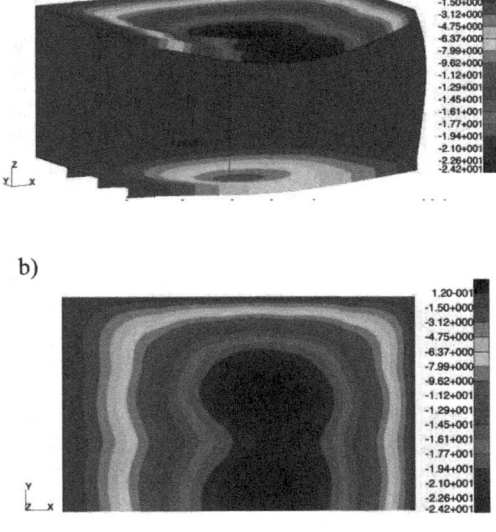

Figure 12. Displacement plot, Study Case 6: a) isometric view and b) top view.

From a static point of view, this version is fully equivalent with Study Case 3, so it can be deduced that the positioning of the rubber mounting is not affecting the structural behavior of the pillar itself.

4.7 Study case 6

The last scenario has been studied by completely uncoupling the pillar with the two decks; it has been achieved by positioning the rubber mounting both in the lower and upper deck connection (Figure 12).

In this case, the maximum displacement is higher with respect to all the other structural layout; the reason is due to the fact that a double rubber mounting reduces the stiffness of the pillar.

In fact, by considering a series of springs as the structural schematization of a rubber mounted pillar it can be obtained that;

$$\frac{1}{k_{tot}} = \frac{1}{k_{pillar}} + n \cdot \frac{1}{k_{rubber}} \qquad (1)$$

where k is the stiffness of the structural element and n is the number of the rubber mounting. From Eq. (1) appears clear that doubling the number of rubber mounting the total stiffness decreases and so this solution increases the maximum displacement of 4%.

5 RESULTS OF THE NATURAL FREQUENT ANALYSES

In order to fully comprehend the role of rubber mounted pillars on the dynamic response of super-yachts and passenger yachts, eigenmodal analyses

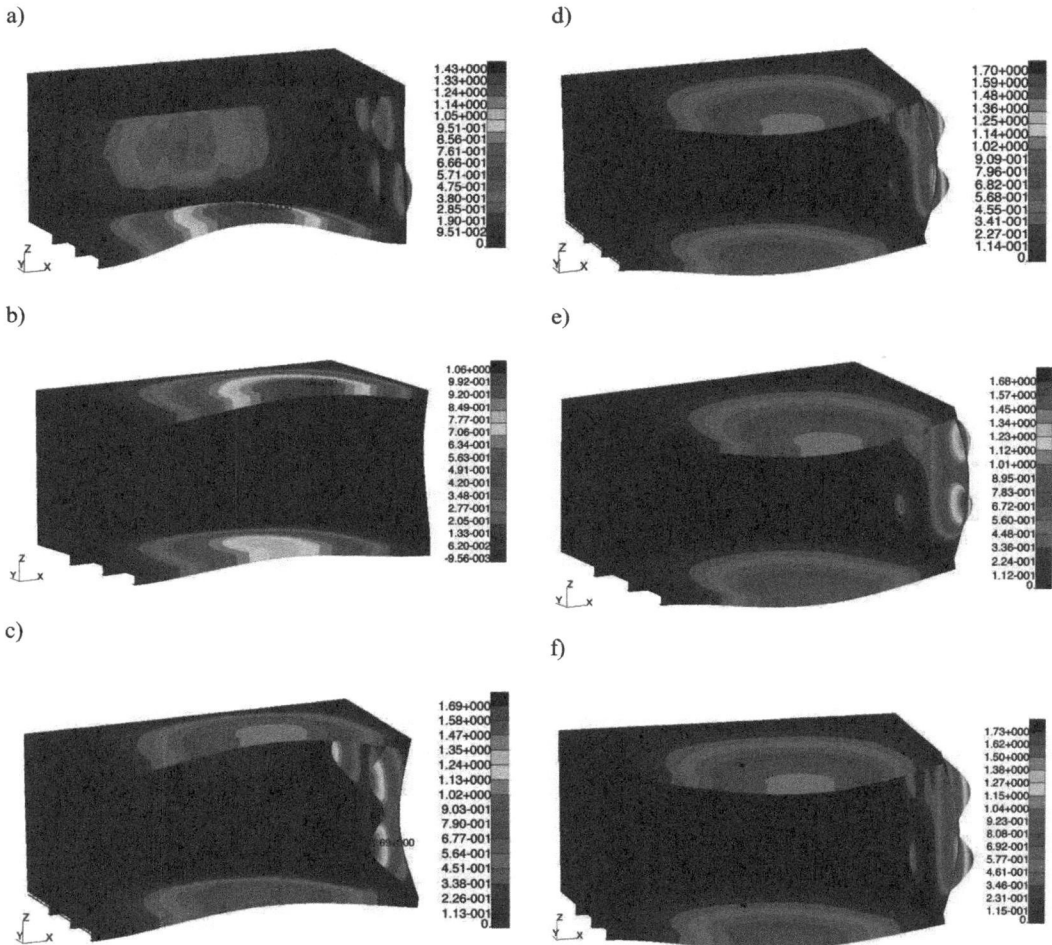

a) b) c) d) e) f)

Figure 13. First mode shape: a) Study Case 1, b) Study Case 2, c) Study Case 3, d) Study Case 4, e) Study Case 5, f) Study Case 6.

Table 2. First eigenfrequency of each Study Case.

Study Case #	Eigenfrequency [Hz]
1	23.04
2	16.95
3	15.74
4	15.36
5	15.82
6	15.08

Table 3. Maximum vertical displacement.

Study Case #	Maximum Displacement [mm]
1	31.1
2	22.9
3 with 3D pillar	23.4
3 with 1D pillar	23.4
4	23.0
5	23.3
6	24.2

have been carried out for the aforementioned Study Cases.

Since it is a linearized analysis, only the eigenmodes in which the pillar results under compression have been considered, since a nonlinear spring cannot be implemented in the FE software. So, it has been substituted by a linear spring having a spring constant equal to the rubber compound stiffness.

In Figure 13, the first mode shape of each Study Case has been reported; the eigenfrequency of each model are reported in the following Table 2.

As it can be noted, the mode shape is not affected by the different pillar design (positive or negative curvature of decks is negligible since the solution is sinusoidal), and the eigenfrequencies are in the same range (15-16 Hz).

It can be deduced that the only difference between the proposed solutions are in the motion amplitude of the structural response, that cannot be evaluated by an eigenmodal analysis, but requires a frequency response one.

6 CONCLUSIONS

The present research is the starting point of a comprehensive evaluation of different pillar design for superyachts and passenger yachts. In particular, it is devoted to the implementation of rubber mounted pillars in structural design of pleasure crafts, that could resolve the problem of critical transmission of vibrations and consequently noise, that commonly affect the comfort standard of such vessels.

As a first step, the static structural response of 6 different solutions (Table 1) has been analyzed by using a commercial FE code and the results in terms of maximum displacements are reported in Table 3.

It has been deduced that the different design solutions have a negligible impact (with the only exception of double rubber mounting) on the structural response to vertical loads, i.e. the only one pillars have to withstands, both due to outfit weight and from global bending.

From the dynamic point of view, an eigenmodal analysis has been carried out and the results, reported in Table 2 and Figure 12, highlight that, also for the first eigenfrequency and mode shape, there are no appreciable differences due to the different pillar designs.

This research is undergoing in view of the frequency response analyses, which will give the frequency response function of the herein presented designs. This is the first requirement to better understand if the rubber mounted pillar is able to reduce the vibration transmission.

REFERENCES

Biot, M., Moro, L., & Mendoza Vassallo, P. N. (2014). Prediction of the structure-borne noise due to marine diesel engines on board cruise ships. In *21st International Congress on Sound and Vibration 2014, ICSV 2014.*

Boote, D., Vergassola, G., & Di Matteo, V. (2017). Strength analysis of superyacht superstructures with large openings. *International Review of Mechanical Engineering, 11*(1), 1–9. https://doi.org/10.15866/ireme.v11i1.9289

Boote, D., Vergassola, G., Pais, T., & Kramer, M. (2017). Finite element structural analysis of big yacht superstructures. *International Review of Mechanical Engineering, 11*(4). https://doi.org/10.15866/ireme.v11i4.9231

Fragasso, J., Moro, L., Lye, L. M., & Quinton, B. W. T. (2019). Characterization of resilient mounts for marine diesel engines: Prediction of static response via nonlinear analysis and response surface methodology. *Ocean Engineering.* https://doi.org/10.1016/j.oceaneng.2018.10.051

Hecquet, A., De'Vidovich, B., Brocco, E., Biot, M., Licciulli, F., Fabro, G., … Moro, L. (2017). On the experimental characterization of resilient mounting elements. In *Progress in the Analysis and Design of Marine Structures - Proceedings of the 6th International Conference on Marine Structures, MARSTRUCT 2017.* https://doi.org/10.1201/9781315157368-13

Pais, T., Boote, D., & Vergassola, G. (2018). Vibration analysis for the comfort assessment of a superyacht under hydrodynamic loads due to mechanical propulsion. *Ocean Engineering, 155*(February), 310–323. https://doi.org/10.1016/j.oceaneng.2018.02.058

Pais, T., Boote, D., Vergassola, G., & Di Iorio, M. (2018). Engine foundation re-design due to modification of the shaft line arrangements. *Transactions of the Royal Institution of Naval Architects Part B: International Journal of Small Craft Technology, 160*(B1), 17–30. https://doi.org/10.3940/rina.2018.ijsct.b1.208

Registro Italiano Navale. (2017a). COM Yacht Additional Class Notation.

Registro Italiano Navale. (2017b). Rules for the classification of pleasure craft.

Vergassola, G. (2019). The prediction of noise propagation onboard pleasure crafts in the early design stage. *Journal of Ocean Engineering and Marine Energy*. https://doi.org/10.1007/s40722-019-00149-4

Vergassola, G., & Boote, D. (2019). Numerical and experimental comparison of the dynamic behaviour of superyacht structure. *Ships and Offshore Structures, 14* (sup1), 1–8. https://doi.org/10.1080/17445302.2018.1546451

Vergassola, G., Pais, T., & Boote, D. (2018). Numerical tools and experimental procedures for the prediction of noise propagation on board superyachts. *Transactions of the Royal Institution of Naval Architects Part B: International Journal of Small Craft Technology, 160* (B1), B9–B16. https://doi.org/10.3940/rina.ijsct.2018.b1.207

Developments in Maritime Technology and Engineering – Guedes Soares & Santos (eds)
© 2021 Copyright the Author(s), ISBN 978-0-367-77376-2

Transforming Chinese ancient ship art and cultural features into modern yacht: A design DNA model of yacht localization design

D.T. Zhou, X.F. Yuan, Y. Wu & C.X. Pan
School of Art and Design, Wuhan University of Technology, Wuhan City, Hubei Province, China

ABSTRACT: There is a broad consensus that excellent yacht art design can stimulate consumers' desire to purchase, and evidence has shown that local culture is the main driving force and inspiration source of yacht art design. Chinese ancient ships have produced very rich art forms, but these artistic features are rarely seen in yacht design, although these yacht products are designed to attract Chinese consumers. This paper works towards studying how to analyze the artistic and cultural characteristics of ancient Chinese ships, and consciously turn them into elements and inspiration of yacht localization design to solve this problem. Finally, a design DNA model of yacht localization design is established. To demonstrate the feasibility and validity of the model, two sets of comparative experiments were conducted with the ancient Chinese "Fu-boat" as the design object. The experimental results show that the cultural value of yacht design elements produced by using design DNA model at the same time has greater scope and depth and better design effect. The results prevented a kind of local cultural expression model and cross-cultural design communication tool for designers to establish and enhance the artistic and cultural value in yacht design.

1 INTRODUCTION

With the improvement of living standards, yachts are no longer just "state ship" showing the power of the state, king or president, but can be considered as "the Toy" of a private owner (Ruggiero, 2016). As the development of yacht technology and function tends to similar, yacht design demand has changed and increased, except for achieving stable structure, excellent performance and harmonious human-machine relationship, yacht design also needs to show clear symbolic meanings, such as fashion, style, art and value to satisfy consumers' emotional demands for yachts and the needs of aesthetic modeling.

It can be said that a new kind of Renaissance going on in the yacht design field under the trend of increasing homogenization of yacht technology at present (Ruggiero, 2016). Although yachts are not works of art, excellent yacht design can also create artistic aesthetics to increase visibility and competitiveness. Studies have shown that the yacht art design is a process of aesthetic creation and emotional expression, while cultural background is the main influencing factor of art design, such as nationality, region, historical and cultural tradition, philosophical concept, folk custom, aesthetic emotion, value, etc. Facing the consumers of different cultural backgrounds, the artistic and cultural characteristics of yacht design need to show aesthetic and emotional resonance in line with their requests in order to attract attention. Therefore, the yacht artistic

aesthetic design becomes the main factor to stimulate consumer spending.

Nowadays, besides the yacht material, structure, safety and other objective parameters, yacht modeling style, symbolic meaning, cultural values and other subjective factors become an important medium of user recognition, and is a key means to attract user preferences and consumption (Eliasson et al. 2014). With the market competition changes, designers and researchers are looking for ways to create this unique and attractive artistic image. There are usually two ways to establish and enhance the uniqueness and value of yacht design:

(1) The first method is to satisfy the personalized needs of users for the purpose of yacht design, such as yacht bionic aesthetics (Li et al. 2017), Kansei design (Li et al. 2017; Dogan et al. 2018), computational generation design of yacht schemes (Khan et al, 2017; Khan et al. 2019), modular design (Zheng et al. 2016), custom yacht design (Vallicelli et al. 2019), yacht interior design (Campolongo, 2017; Jin et al. 2017) and so on. This kind of method is to conduct personalized and customized yacht design based on the needs of users, and its obtaining yacht design features can improve the satisfaction of some users. However, in most cases, the technical difficulty of the design process is higher, design cost is high, which is not suitable for popular design and application, and the obtained yacht products is not suitable for extensive marketing.

DOI: 10.1201/9781003216582-42

(2) The second method is to embed local cultural attributes in the yacht art design. *"products are more likely to be exported in global markets and that designs fail if the culture of the target country is not taken into account"*, Rose said (2005). Hsu (2015) argues that *"the global market has gradually moved its focus towards local styles, and creative design applications have become a global competitive advantage of an industry"*. According to Piardi (2017), *"a typical cruise ship in port doesn't interact with the city, but it's still a transit guest"*. Therefore, cultural characteristics are considered to be unique features that can be embedded in yacht design, which can not only enhance the identity of yacht brands in the global market, but also enhance the emotional experience and aesthetic styling of individual consumers. Such as Ruggiero (2018) argues that *"the use of colour for the outer form of ships has been taking on multiple valence: identification, symbolic, apotropaic aspects"*. The findings are small, but designers have noted the importance of linking yacht design to cultural identity to enhance product value.

In summary, embedding local cultural characteristics as a design strategy to enhance the brand, uniqueness and attractiveness of yachts, for design, cultural value added and created the value core of yacht products, the same design drives cultural innovation and development. At this point, the second method has more extensive research significance and application value than the first method. However, there is still a research gap in the design of yacht with local characteristics, that is, the lack of standards, specifications and effective method reference in the process of cultural embedded yacht design. Research on practical design shows that a clear, comprehensive and effective understanding of the local characteristics of cultural objects is the basis of yacht art design (Li et al. 2018). Therefore, helping designers understand the artistic features and values of cultural objects, so that consciously apply cultural elements to yacht design becomes critical.

In the thousands of years of historical development, Chinese ancient ships produced very rich localized art forms, but the yacht design failed to apply these cultural characteristics (unclear, inaccurate, and superficial), although these yachts are aimed at Chinese consumers. Therefore, how to transform the art and culture of Chinese ancient ships into modern yacht design is the research objective of this paper. This paper argues that providing yacht design elements through an analytical model of Chinese ancient ships in terms of form, color, rules, symbols, and other art and culture aspects can enhance the cultural value of yacht design, this model is called a design DNA model of yacht localization design (Yacht Design DNA for short). To demonstrate the feasibility and validity of the model, two sets of comparative experiments were conducted with the

ancient Chinese "Fu-boat" (a wooden sailing boat manufactured in ancient China's Fujian Province) as the design object. The aim of this paper is to bridge the gap between culture (ancient ship cultural) and design (yacht design), a more practical motivation is to provide designers with a clear model of local culture expression and an effective cross-cultural design communication tool to support yacht localization design.

2 CULTURAL LEVELS AND ARTISTIC FEATURES OF CHINESE ANCIENT SHIPS

2.1 *Characteristics of four cultural levels*

Hofstede's cultural dimensions theory defines culture as: a mental process shared by many people with the same education and life experience in the same environment. Culture describes the evolution of human civilization as a result of the process, which involves customs, art, thought, behavior and religion (Lin, 2007). In reviewing the conceptualization and measurement of culture, there are a wide range of conflicting academic views on which values, specifications and beliefs should be measured to represent the concept of culture. Hoft (1996) proposes a descriptive model (a cultural iceberg model) to better understand cultural identity, which holds that only 10% of cultural features, such as color and texture, are easy to identify, while the remaining 90%, such as value and faith, are difficult to identify and easy to ignore. Leong (2003) proposed three levels of cultural themes: the outer physical level of culture, the middle level of behavior, and the inner level of invisibility. Hofstede (2010) constructed a hierarchical model (onion model) composed of symbols, heroes, rituals and values to study cultural differences. Based on the Hofstede onion model, Hung (2018) considers that culture can be described at different levels of cultural elements, including object, symbol, behavior, rite and value, and constructs a cultural DNA model referring to the structure of human DNA.

Analysis indicated that culture is composed of different layers (levels), just like onions, each layer is not independent, but interactive. Based on the above theories, this paper constructs an onion model to formulate culture to help designers better understand the symbols, rules, behaviors, thoughts and beliefs in local or cross-cultural, so as to better carry out creative design or cross-cultural design. As categorized in Figure 1, there are four main levels:

(1) "Outer level" is visible and tangible, and it is a symbol with recognition significance, manifested as obvious physical form, such as color, shape, texture, lines, surface lines, decorative patterns, etc. Outer cultural elements are tangible communication carriers of other cultural levels.
(2) "Middle level" is a visible and behavioral process of creating tangible culture, it is the application

process of certain technology and knowledge, such as craft, skill, operation, technology and craftsmanship. Middle behavior realizes the externalization of outer symbol or other cultural level.

(3) "Inner level" is non-material and invisible, a customary pattern of behaviors, ideas or aesthetics that people of all ages in a particular cultural region have followed together, and manifested as conscious behaviors that can be accepted by the public, such as ceremonies, ritual, rules, convention, etc. Inner tradition stipulates the content of middle-level behavior.

(4) "Nuclear level" is implicit and non-material, and is the value that is assumed, pursued, and implemented by the public in the local area, as the performance of human ideology, such as emotion, story, attitude, anticipate and belief. Nuclear value directly affects the emotional expression regulated by inner traditions.

2.2 Characteristics of the artistic level of Fu-boat

Chinese ancient ships have various forms and complete functions, from the invention of canoes to the manufacture of giant sea ships. It shows that Chinese ancient ships had independence and creativity in the course of water activities. As a typical representative of Chinese ancient seagoing ships, Fu-boat (a wooden sailing ship manufactured in ancient China's Fujian Province, also known as Foochow junk) is the most important trading boat on the maritime Silk Road (Lu, 2014). The famous "Keying" (Sailing ship of the Qing Dynasty in China) is a typical style of the Fu-boat, see Figure 2a.

According to the cultural elements of ancient ships in different levels (see Figure 1), this paper analyzes the artistic and cultural characteristics of the Fu-boat, as follows:

(1) Outer symbol (see Figure 2b). The main features of Fu-boat are V-shape bottoms, point bow, small

Figure 2a. Keying: representative style of Fu-boat.

square head, wide and aft-extending stern, with both ends curved upward (Hull with two ends upturned), and horseshoe-shaped stern in vertical section, with wide and flat deck, and boat handrail protruding outward. A lion's head (fierce and deterrent as an auspicious and apotropaic beast) or "Yin-Yang" Mirror (symbolizing round Sun) held between two waves is painted on Fu-boat's bow. The "Dragon-eye" (eyes carved out of camphor wood) are decorated on the bow, which are flat hemispherical in shape and black and white in color, like bulging eyes. "Liu-Yu" (a fish abstracted from the shape of Shuttles hoppfish) is painted on both sides of the stern, with a red fish body, a white neck and tail up, as if swimming. Colorful patterns are painted on the horseshoe-shaped stern, commonly known as the "Colorful But", such as a "Yi-Niao" (a type of water bird with great flying power) is painted on the stern's upper panel, also decorated with Phoenix Peony (auspicious patterns), Eight Immortals (ancient Chinese mythological figures), Eight Treasures (eight auspicious objects of Tibetan Buddhism) and other auspicious patterns. There are many colorful flowers painted in the stern near the ship's edge, which is called "Jiao-Lian-Hua" (Mazu's dress pattern, MAZU is the Chinese belief in the sea god). Fu-boat's bottom was white, its sides usually black, its windows white, and the railings around its face red, and the red and yellow Tai Chi patterns are painted on the "Narcissus Gate" (the gate through which people go up and down or carry goods) in the middle of the shipboard.

(2) Middle behavior. Water-tight compartments (bulkheads that separate compartments into impermeable compartments) and fish-scale overlap (the longitudinal zigzag hull increases the stability of the ship's lateral sway) is the most unique Fu-boat Construction technology. The Chinese Zodiac (Rat, cattle, tiger, rabbit, dragon, snake, horse, sheep, monkey, chicken, dog, pig) is used to refer to different parts and components on ships (see Figure 2c), such as the "water snake" is a longitudinal thickened member on both sides of the shipboard, which acts as

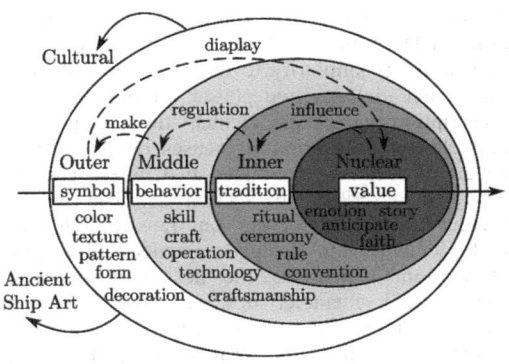

Figure 1. Onion model: 4 levels of cultural characteristics and ancient ship art.

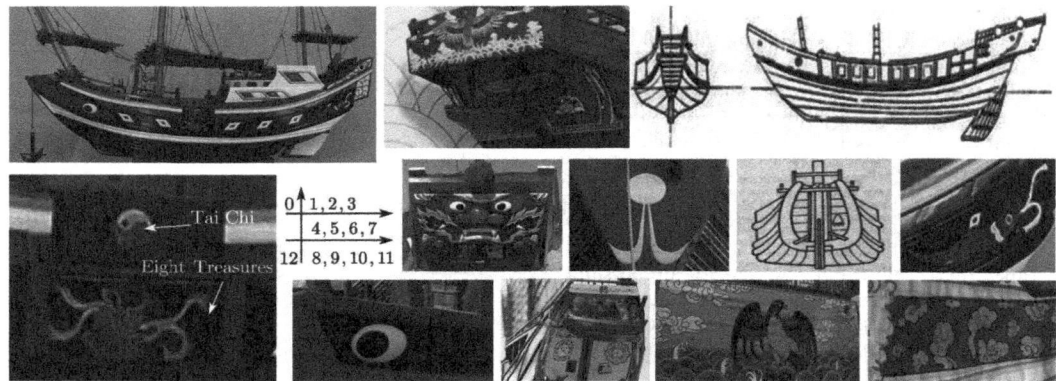

Figure 2b. Characteristics of the artistic level of Fu-boat.

a standard line for measuring the draft depth and head-tail balance of the ship, and the size of the "Dragon-eye" can be used to measure the size of the ship (proportional relationship between "eye" and keel). This kind of structure characteristic with humanization, arranged in different positions on the ship, can make the sailor enter the clear position quickly in the operation, each has its own duty, reduce mistake and guarantee safety. Judging from the shape and structure of Fu-boat, the oblique trend is the main aesthetic characteristics of its technical performance, such as the window frames on the boat are rarely vertical and basically inclined.

(3) Inner tradition. Proportions and traditional rules are displayed in the design and construction of Fu-boat, such as the cabin depth is 40% of the deck width, and the keel length to deck width is approximately 3:1. The size and direction of the "Dragon-eye" on the bow have specific rules, such as the keel length should be 110:4 in proportion to the eyes, the eyes of the fishing vessel should look down to signal the search for catch, and the eyes of the merchant vessel should look forward. Legend has it that "Yi-Niao" flew backward (symbolizing fearlessness in storms and extreme speed), so Yi-Niao is painted on the stern as a symbol of being able to push the boat forward. The shrines of Mazu (the patron saint of navigation in ancient Chinese legends) are often placed on all kinds of Seagoing ships, so that praying and offering sacrifices to the gods at sea in order to obtain the timely blessing of the Gods. Parts of the twelve zodiac animals in Fu-boat are mascots of traditional Chinese culture, which are believed to protect the safety of maritime navigation.

(4) Nuclear value. Behind each of Fu-boat's symbols are legends, some magical, some tragic, some real and fantasy, that have become a distinctive part of ancient ship culture. These devout beliefs, complex God system and bizarre folklore provide rich themes and content for ship design, and also reflect the unique aesthetic values of local culture. In Fu-boat culture, praying culture often runs through the creation process, and all cultural values tend to be consistent in survival, profit and harm avoidance. It can be said that satisfying people's own needs is the emotional foundation of ancient ship design, which has a specific utilitarian significance and pursues spiritual satisfaction and pleasure.

3 A DESIGN DNA MODEL OF YACHT LOCALIZATION DESIGN

The above research has determined how to describe the level characteristics of cultural objects, but to realize the cultural value expressed in yacht design, designers must also consider how to transform and encode cultural elements into design features, as well as studying how to consciously and in detail analyze and integrating the application of cultural elements in the design process. Referring to the new concept of genetic engineering model applied to product design (Li et al, 2018; Hung et al, 2018), a model of embedding culture level into yacht design is constructed, as shown in Figure 3, which is called a design DNA model of yacht localization design because of its similarity to DNA ladder structure. The model includes two phases: cultural element transcription and design element translation. Cultural element transcription focuses on identifying the artistic features contained in Chinese ancient ships (as described in Section 2). The design element translation focuses on transforming cultural elements into design elements and design information, which are the design forms with core cultural characteristics extracted from the ancient ship art, and is the basic characteristic element and expression form to solve the yacht localization design. Corresponding to the four cultural levels described in the onion model (see Figure 1), the process of translating design

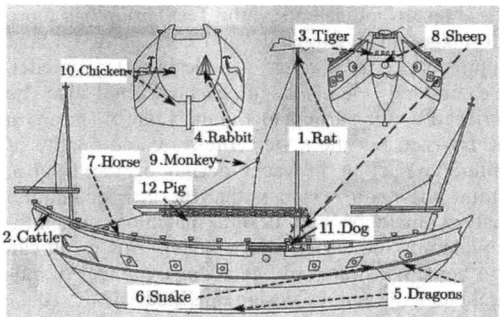

1:bolt for fixing mast   2:balustrade 3:anchor exit   4:bathroom
5:dragon eye and boat keel 6:measuring the draft depth 7:deck
8:parts of fixing rope, the guide channel of the anchors 9:pulley
10:parts for fixing rudder 11:anchor winch 12:sail rack

Figure 2c. The position of Chinese Zodiac parts on Fu-boat.

elements is also divided into four parts: Appearance, Function, Situation and Emotion, as follows (see Figure 3):

(1) Outer Symbol-Appearance: "Appearance" mainly includes the shape, form, color and decoration of the yacht, such as the cultural elements of ship painting. "Appearance" element matches the "Symbolic" elements of the outer-level culture, mainly through visual perception to meet the cultural demand directly. At this level, the principle of form isomorphism can be used for intuitively transformation between them.

(2) Middle Behavior-Function: "Function" mainly includes the function and technical features of the yacht, such as yacht's inclined porthole design (reference Fu-boat's Oblique Aesthetics), protruding waterline design (reference Fu-boat's water snake function). "Function" element matches the "Behavior" element of the middle-level culture, mainly through the "behavior similarity" of function and technology to create cultural features, so that consumers can experience the derivation of cultural functions. At this level, the principle of functional imitation can be used to transform both.

(3) Inner Tradition-Situation: "Situation" mainly includes the decoration scene, structural proportion or area division of yacht. "Situation" element and the culture inner-level "Tradition" element to match, mainly through the situation perception way to create the cultural characteristic. At this level, the principle of situation consistency can be used to transform both.

(4) Nuclear Value-Emotion: "Emotion" mainly includes the design concept and conceptual connotation of yacht, such as the awe of natural law. "Emotion" element matches with the "Value" element of the nuclear-level culture, and

resonates with consumers thoughts and psychology by reflecting or expressing the stories, emotions and values represented by cultural objects. At this level, the principle of emotional connection can be used to transform both.

4 EXPERIMENTAL VERIFICATION

4.1 Participants

Twelve graduate students of design institute were recruited online as participants, consisting of 6 women (Chinese students) and 6 men (2 Chinese students and 4 International Students), ranging in age from 25 to 28, and the experience and capabilities of yacht design and cultural creative product design are similar. The 12 participants were randomly divided into two equal groups, each consisting of 4 Chinese students (3 women and 1 man) and 2 international students (men). In order to reduce the communication barriers of cross-cultural design teams and simulate the mature design team as much as possible, the design practice simulations within the group was carried out twice before the experiment. All participants were informed of the purpose of the experiment and signed the experimental consent form, promise rewards to participants.

4.2 Materials and task

The necessary design tools were provided to the participants, including laptops, hand drawing tools, paper, whiteboards, In addition, a design DNA model of yacht localization design was randomly provided to a group of participants, called the Test Group, the team was told to design according to four levels of "Fu-boat" cultural and artistic information provided by the design DNA model (developed jointly by designers and ancient ship experts) and four levels of matching rules between cultural and design elements. The Control Group, who was not provided with design DNA model, was told they could do as they pleased. The specific tasks are as follows:

(1) Design practice. Test Group and the Control Group were asked to design a yacht concept that reflected the "Fu-boat" culture. Each participant was required to submit more than one design proposal within 180 minutes, and the display method was not restricted, members of the group can communicate and discuss with each other.

(2) Organize the reply. Participants in both groups had to explain their designs, and participants in the same group can reply together about how ship design reflects Fu-boat's cultural and artistic identity.

(3) Data statistics. According to the integrity of the yacht design, whether the cultural characteristics reflected in the design are acceptable and which

level of culture reflected, etc., experts evaluated the yacht design scheme submitted by the participants. The reply content and judgment data were recorded and counted by the researchers.

(4) Experimental interview. After the experiment, all participants were interviewed and asked how they felt during the experiment, such as whether there are communication obstacles, whether there are problems in the design method and suggestions for the yacht localization design.

4.3 Results and discussion

The experimental results are shown in Tables 1 (Text Group) and Tables 2 (Control Group): The statistical data of "Number of per cultural level" shows that the symbolic features of the Outer-level culture are the most commonly used yacht design (T: 45.7% and C: 68.0%), which shows that the form, color and pattern of Fu-boat are the most easily perceived cultural creative elements, such as "Dragon-eye" and "Colorful Butt", which are used most in both sets of projects. Compared with other levels, the value characteristic of nuclear-level culture is the least attention and application, in which the Control Group's Nuclear-level element accounts for 0%, while the Test Group paid much more attention to it than the Control Group, reaching 11.4%. In addition, compared with the Test Group (22.9% and 20.0%), the Control Group applied fewer middle-level elements (20.0%) and Inner-level elements (12.0%) of Fu-boat culture.

These results indicated that the Test Group had more advantages in the quantity and content of Fu-boat culture under the guidance of design DNA model. The cultural elements of Fu-boat can also be retrieved by the participants in the Control Group on the Internet, but the lower efficiency and quality of culture extraction in yacht design of the Control Group, the main reason is that the lack of cultural analysis methods and design matching principles provided by the design DNA model. This result can also be demonstrated by the average number of cultural elements and the design qualified rate (T: 100% and C: 61.1%) in Tables 1 and Tables 2.

The "Number of qualified designs" in Tables 1 and 2 show that the Test Group submitted 9 qualified yacht designs (judged by experts to be in line with the design theme), which were less than the Control Group's 11 yacht designs, the result shows that the design DNA model limited the creativity of participants in the Test Group. Further analysis found that the participants in the Test Group under the guidance of design DNA model directly copy cultural symbols applied to yacht design, lack of change and innovation of cultural symbols. The Control Group developed and innovated the symbol freely, derived many kinds of symbol features, and applied them to different yacht design schemes. However, the qualified rate of the Control Group is 61.1%, which was significantly lower than that of the Test Group (100%), and main reason was that the innovative symbols applied in the yacht design of the Control Group did not conform to the Fu-boat's cultural characteristics,

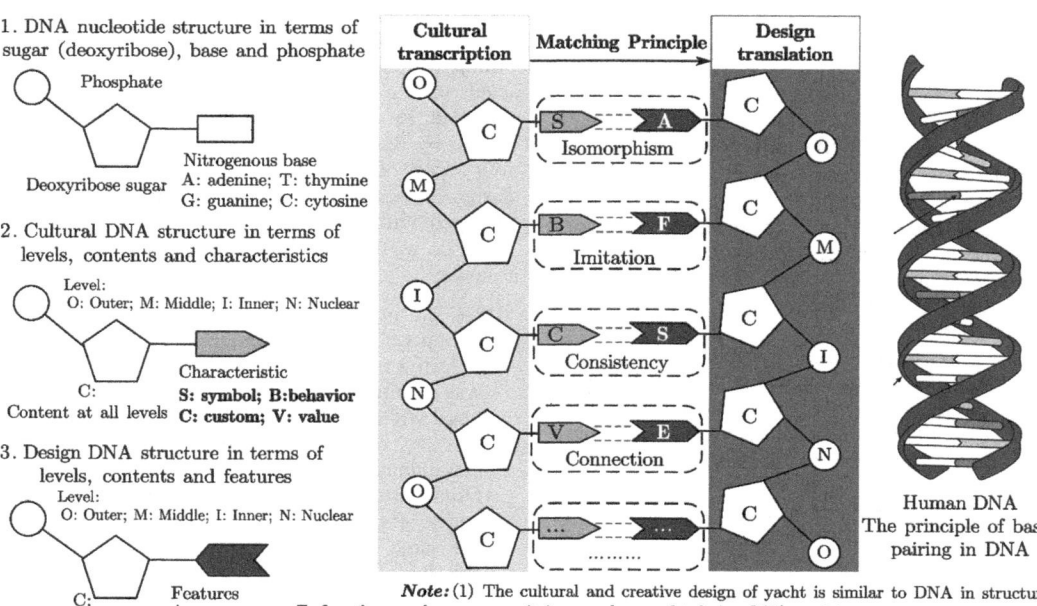

Figure 3. A design DNA model of yacht localization design.

which indicated that the participants in the Control Group (who were not provided with a design DNA model) had a deviation in their understanding of Fu-boat culture. In summary, the design DNA model constrains the design innovation capabilities of participants, but it avoids the illogical (wrong) cultural interpretation of the design scheme.

The design results of 4 international students in Table 1 and Table 2 were compared: the number, scope and level of the cultural characteristics of Fu-boat applied by 2 international students in the Test Group are higher than that of 2 international students in the Control Group, which indicated that design DNA model facilitated the exchange (communication) of Fu-boat cultures among designers from different cultural backgrounds. The analysis of the experimental interview also proves that the design DNA model has a positive effect on the cross-cultural exchange of yacht design. Although the existing experimental sample size cannot explain the application value of design DNA model in cross-cultural design communication, from the current work results, it is reasonable to assume that design DNA model can provide smooth design communication and cooperation for designers from different cultural backgrounds, which provides a way for cross-cultural design research.

5 CONCLUSION AND SUGGESTIONS

Yacht art design has become the main factor to stimulate consumption, and how to embed the unique local cultural characteristics in the yacht art design is the key design method to enhance the brand and attraction of yacht design. In yacht design, embedding cultural attributes is a process of comprehensively thinking and identifying cultural characteristics, and then consciously defining and expressing them as design features. To this process, a design DNA model of yacht localization design is proposed, which includes two phases: Cultural Transcription and Design Translation. The cultural transcription part established an onion model to express the

Table 1   Test (T) group: Statistics on the application of cultural characteristics of Fu-boat.

| Participant | Number of per cultural level* | | | | Total of cultural characteristics | Total of designs | Number of qualified designs | Average cultural characteristics of per design |
	O	M	I	N				
T1	2	1	1	0	4	1	1	4.0
T2	3	2	2	1	8	2	2	4.0
T3	4	1	1	1	7	2	2	3.5
T4	3	1	1	1	6	1	1	6.0
T5**	3	1	1	1	6	2	2	3.0
T6**	1	2	1	0	4	1	1	4.0
Proportion of cultural characteristics	45.7%	22.9%	20.0%	11.4%	35	9	9	Qualified rate: 100%

* O = Outer level, M = Middle level. I = Inner level, N = Nuclear level; ** T5 and T6 are international students.

Table 2   Control (C) group: Statistics on the application of cultural characteristics of Fu-boat.

| Participant | Number of per cultural level* | | | | Total of cultural characteristics | Total of designs | Number of qualified designs | Average cultural characteristics of per design |
	O	M	I	N				
C1	2	1	0	0	3	3	1	3.0
C2	4	1	1	0	6	2	2	3.5
C3	4	1	1	0	6	2	2	3.0
C4	4	0	1	0	5	4	3	1.0
C5**	2	1	0	0	3	3	2	1.5
C6**	1	1	0	0	2	4	1	2.0
Proportion of cultural characteristics	68.0%	20.0%	12.0%	00.0%	25	18	11	Qualified rate: 61.1%

* O = Outer level, M = Middle level. I = Inner level, N = Nuclear level;** C5 and C6 are international students.

content and relationship of the outer-level, middle-level, inner-level and nuclear-level of culture, helping the designer to obtain the knowledge content and value range of the cultural objects. The design translation section instructs designers to combine cultural elements and design consciously to realize the application of design features of different cultural levels in yacht design. The results of the comparative experiment have proved that the Fu-boat cultural characteristics can be transformed into the yacht design characteristics more effectively than the free-design, and the work of this paper provides some methods for designers to clearly analyze the content and value of culture, and avoid the illogical interpretation in the process of cultural creativity, and it provides the future development direction for the cross-cultural design.

Considering the constraint of design DNA model on designer's creativity in the process of yacht creative design. Future studies can refer to the characteristics of DNA mutation and variation in genetic engineering to perfect the matching principle between cultural levels and design features, so as to inspiring more design creativity and cultural derivation. Furthermore, in the application of design DNA models, considering the efficiency and accuracy of designers in retrieving ancient ship culture. A detailed online database of Chinese ancient ship culture and design features needs to be developed, provides fast and accurate creative knowledge for yacht and cruise ship art design.

ACKNOWLEDGEMENTS

This work is supported by "National Key R&D Program of China (2018YFB1308500)" and "the Fundamental Research Funds for the Central Universities (WUT: 2020VI077; WUT: 2020III046)".

REFERENCES

Campolongo, M. 2017. House and Yacht: the Aesthetics of the Interior as a Link between Different Sectors. *The Design Journal* 20(1): 209–218.

Dogan, K. M., Suzuki, H., & Gunpinar, E. 2018. Eye tracking for screening design parameters in adjective-based design of yacht hull. *Ocean Engineering* 166: 262–277.

Eliasson, R., Larsson, L., & Orych, M. 2014. *Principles of yacht design*. A&C Black.

Hofstede, G., Hofstede, G.J., & Minkov, M. 2010. *Organizations and cultures: Software of the mind*. New York: Mc Graw Hill.

Hoft, N. Developing a cultural model. 1996. In: Del Galdo, E.M., Nielsen, J. (eds.), *International User Interfaces*: 41–73. Wiley, New York.

Hsu, C. H., & Tsai, W. C. 2015. A Design Strategy of Cultural and Creative Products on the Global Market. In: Rau P. (eds), *Cross-Cultural Design Methods, Practice and Impact. CCD 2015.* Lecture Notes in Computer Science 9180. Springer, Cham.

Hung, Y. H., & Lee, W. T. 2018. The Need for a Cultural Representation Tool in Cultural Product Design. *In Computational Studies on Cultural Variation and Heredity*: 107–114. Springer, Singapore.

Jin, Z., & Kim, C. S. 2017. Influence of Interior Color and Material Matching Design of the Yacht to the Consumer Psychology. *In 3rd International Conference on Arts, Design and Contemporary Education (ICADCE 2017)*. Atlantis Press.

Khan, S., Gunpinar, E., & Dogan, K. M. 2017. A novel design framework for generation and parametric modification of yacht hull surfaces. *Ocean Engineering* 136: 243–259.

Khan, S., Gunpinar, E., & Sener, B. 2019. Gen Yacht: An interactive generative design system for computer-aided yacht hull design. *Ocean Engineering* 191: 106462.

Leong, B. D., & Clark, H. 2003. Culture-based knowledge towards new design thinking and practice—A dialogue. *Design Issues* 19(3): 48–58.

Li, S. J., Li, H. Y., & Yang, W. M. 2017. The study of yacht shape feature based on eye tracking experiments. *In Mechatronics and Manufacturing Technologies-Proceedings of The International Conference (MMT 2016)*: 253. World Scientific.

Li, Y., & Cai, W. 2017. A Study on the Bionic Design Method for Yacht Shape. *In The 27th International Ocean and Polar Engineering Conference*. International Society of Offshore and Polar Engineers.

Li, Y., Li, J., & Yan, Q. 2018. Design method and application of DNA in the design of cultural creative products. *In International Conference on Cross-Cultural Design*: 172–185. Springer, Cham.

Lin, R. T. 2007. Transforming Taiwan aboriginal cultural features into modern product design: A case study of a cross-cultural product design model. *International Journal of Design*: 1(2).

Lu, X. 2014. The Naval Architecture of Ancient Fujian Style Sea Going Sailing Junks: A Manuscript. *In Proceedings of the 2nd Asia-Pacific Regional Conference on Underwater Cultural Heritage*, Honolulu, Hawai'I.

Piardi, S., Mori, L., Attardo, M., Cornetto, D., & Vicari, S. 2017. New Concept Design for an Event Cruise Ship. *Journal of Shipping and Ocean Engineering* 4: 174–179.

Röse,K. 2005.Cultural issues and their representation in cultural variables for human-machine- systems. *In Proceedings of the 11th International Conference on Human-Computer Interaction*. Las Vegas, NV, USA.

Ruggiero M. E. 2018. The color of ships: communication and identity. *Cultura e Scienza del Colore - Color Culture and Science* 09: 7–15.

Ruggiero, V. 2016. Changes in design approach for large yachts. *III International Multidisciplinary scientific conference on social sciences & arts-SGEM* 4(2): 155–164.

Vallicelli A., Di Nicolantonio M., Lagatta J., & Biagi A. 2019. "Just in Time" Product Design: Case Study of a High-Customizable Chase Boat. In: Stanton N. (eds), *Advances in Human Aspects of Transportation. AHFE 2018.* Advances in Intelligent Systems and Computing 786: Springer, Cham.

Zheng, G., Deng, Y., & Du, J. 2016. The Study on the Modularity Design Principles of the Packaging and the Styling of Yacht. *Art and Design Review* 4(03): 113–118.

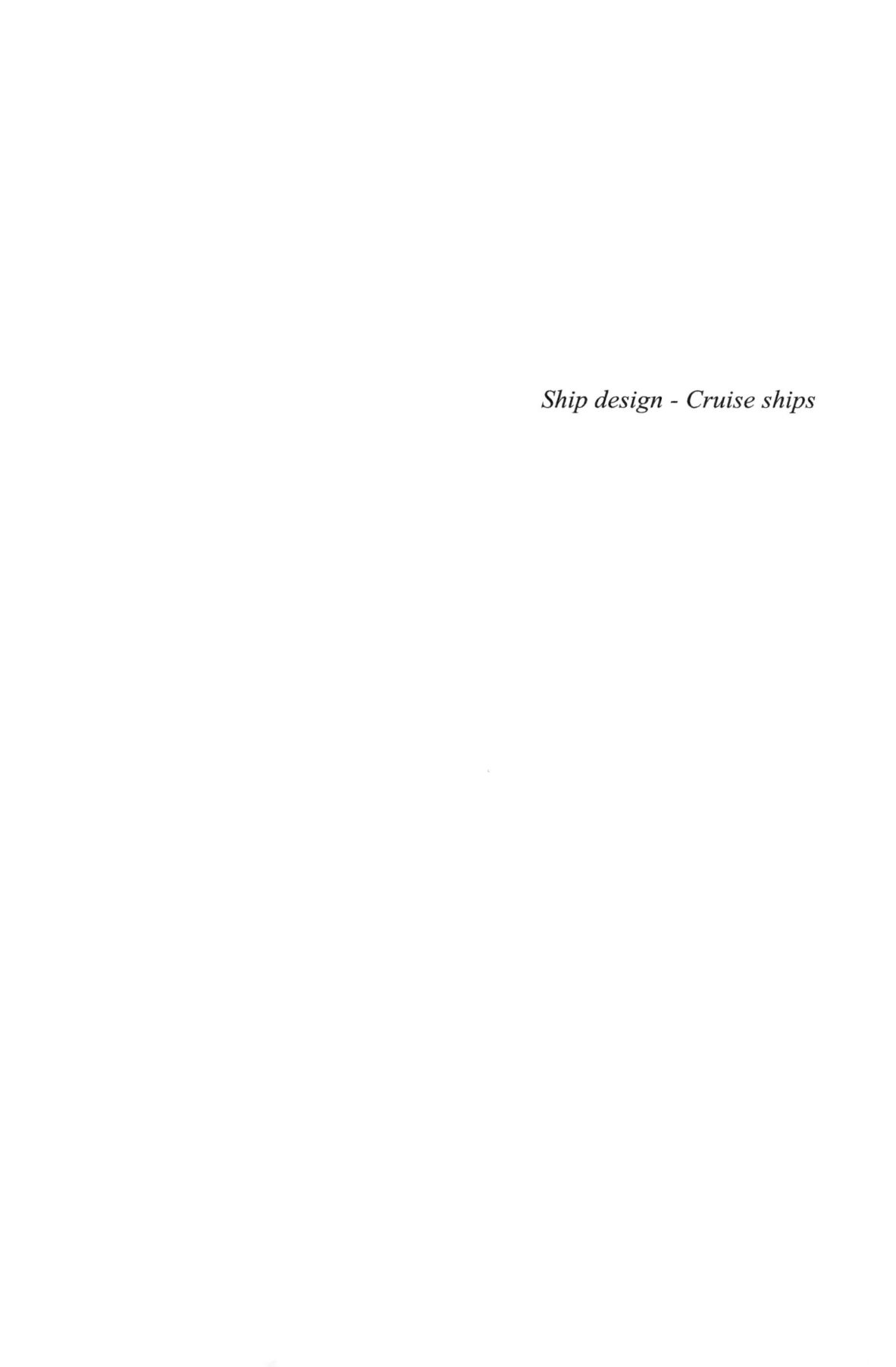

Ship design - Cruise ships

Developments in Maritime Technology and Engineering – Guedes Soares & Santos (eds)
© *2021 Copyright the Author(s), ISBN 978-0-367-77376-2*

Cruise color analysis system for interior color scheme under complicated ocean lighting conditions

Jingguang He, Qingnan Li & Jinjin Wang
School of Art and Design, Wuhan University of Technology, China

ABSTRACT: Color scheme is an important factor in cruise interior design. By considering materials, mechanisms and styles, a variety of cruise interior environmental design works are presented for a visual and aesthetic impression. Various color models, based on specific requirements and configuration of colors, have been developed to follow the line with the artistic theory and principle of creation to achieve practicality and aesthetics. Although success has been made, its performance still suffers from complicated ocean lighting conditions on cruise interior color scheme task. In this paper, we suggest a cruise interior color analysis system to learn the connectivities of global color scheme interaction graph based on artistic styles, consumer characteristics and typical functional spaces under complicated ocean lighting conditions, in which an Interactive-Scan-Flood Fill algorithm is proposed to scan the whole image for potential seeds by human-machine interactive faceted search and exploration. The result shows that our color suggestion provides designers with the prior knowledge so as to point out their shortcomings for cruise creative arts and design.

1 INTRODUCTION

Due to the large difference in artistic styles around the world, introspection is a poor guide to understand how much of our knowledge of an artistic style comes from color scheme. Recently, rapid development of deep-learning technology has motivated researchers to focus on irrelevant details in color scheme that human vision often ignored. Deep-learning based approaches are proposed to extract effective representations for capturing the statistics of universal artistic styles, and have been applied in a variety of applications such as neural image colorization and style transfer for cruise interior design.

The pioneering works (Gatys et al. 2015a, Gatys et al. 2015b, Gatys et al. 2016) are proposed to extract deep features using convolutional neural networks (CNNs), which have explored the relationship between statistical color properties and artistic styles for creative arts and design. Inspired by prior works, the current literature on stylization task using deep learning networks can be divided into two groups: (i) methods that directly extract deep features for stylization. (ii) methods that propose diversity hyperparameters for users to control the stylization strength.

The first group of methods demonstrates that deep image representations can be generated on basis of convolutional neural networks, in which the high-level content is preserved for content reconstruction, while the correlations between different features in different layers are computed for style reconstruction. Such

approaches pave a way to suggest future insights for designers, and can be further subdivided in three groups: (i) methods developed for efficiency (Li and Wand 2016b, Johnson et al. 2016, Ulyanov et al. 2016). (ii) methods developed for visual quality (Li and Wand 2016a,Wang et al. 2017, Sanakoyeu et al. 2018). (iii) methods developed for style diversities (Dumoulin et al. 2016, Chen et al. 2017, Li et al. 2017a, Huang and Belongie 2017, Li et al. 2017b). These approaches investigate various aspects of deep architectures and have amassed a large number of research studies.

However, since creative arts and design are required to satisfy the preferences of different users, the human-machine interaction is important and serves as the key factor affecting the stylization in deep feature space. The second group of methods has been developed to provide extensive user control in deep features generated from summary statistics on CNN responses. During the stylization, these methods introduce intuitive ways to provide various parameters for adjusting spatial variation in styles, color transformation, and diversity optimization (Champandard 2016, Gatys et al. 2017, Liao et al. 2017, Risser et al. 2017, Wang et al. 2019). Therefore, the representations of an image style within parametric neural models allow more intuitive control over the stylization outcome than deep net architectures without human-machine interaction.

Deep-learning based methods can generate deep features, and have achieved perceptual effect on the stylization result, however, environment variations such as

DOI: 10.1201/9781003216582-43

Figure 1. Sun glare scenarios perturb the generated deep features, which may reduce the quality of stylization.

Figure 2. The realistic colors in the yellow bounding box are distorted by the complicated ocean lighting conditions. The ceiling and carpets in the red bounding box appear blue at a higher degree.

lighting and weather conditions (e.g. sun glare scenarios) limit the use of these methods to acquire the most realistic colors that the network should have generated, as shown in Figure 1. Therefore, the generated deep features are not discriminant or very noisy under complicated ocean lighting conditions, which lead to a decrease in the color scheme for cruise interior design.

Facing the aforementioned challenges, a cruise interior color analysis system, affectionately dubbed "Color Analysis System" (CAS), is specifically designed to overcome the above artistic, technical, and creative challenges. The proposed CAS starts by formulating a set of scenarios with typical artistic styles, while the Interactive-Scan-Flood Fill (ISFF) algorithm is proposed to learn the connectivities of global color scheme interaction graph based on artistic styles, consumer characteristics and typical functional spaces, guiding cruise interior color scheme by the way of human-machine interaction.

The key contributions of our work are as follows:

- An Interactive-Scan-Flood Fill algorithm is proposed to select the most realistic colors under complicated ocean lighting conditions by the way of human-machine interaction.
- In view of the complicated ocean lighting conditions, the proposed Interactive-Scan-Flood Fill algorithm scans the whole image to set potential starting seeds for flood filling with interactive threshold.

In the remainder of this paper, we first statement that complicated ocean lighting conditions perturb the deep feature generation, and then detail the Interactive-Scan-Flood Fill algorithm. Finally, we present our quantitative results under complicated ocean lighting conditions, which show that our proposed CAS can provide designers the prior knowledge and point out their shortcomings for cruise creative arts and design.

2 FORMULATION OF THE PROBLEM

Recent success in visual recognition has inspired a family of data-driven methods that use deep

networks for stylization. However, due to the complicated ocean lighting conditions, the generated deep features suffer from understanding of the presented realistic colors. Let us consider an image of cruise ocean view stateroom under complicated ocean lighting conditions, in which local image patterns are ambiguous for objects with appearance variation, as shown in Figure 2.

The complicated ocean lighting conditions have significant influence on the materials within the yellow bounding box. Specifically, compared with the diffuse reflective materials such as carpets, the ceiling appears blue at a higher degree due to its specular reflection. Therefore, a grand challenge in cruise interior color scheme is the task of understanding the most realistic colors under complicated ocean lighting conditions.

Color analysis system is proposed to find the most realistic and accurate color of objects under complicated ocean lighting conditions by the way of human-machine interaction. The most experienced designers in the world are invited to work with the system and provide analysis towards understanding of the connectivities of global color scheme interaction graph based on artistic styles, color schemes and typical functional spaces under complicated ocean lighting conditions.

3 SYSTEM DESIGN

As the prior work suggested (Csikszentmihalyi 1997, Benedetti et al. 2014), several factors are considered in our system design: F1) Clear goals; F2) Immediate feedback; F3) Balanced challenges; F4) Merged action and awareness; F5) No distractions; F6) No fear; F7) No self-consciousness; F8) Ignorance of time; F9) Autolectic activity. We draw from several of the aforementioned theories but also add novel goals that are specific to complicated ocean lighting conditions:

Algorithm 1 Interactive-Scan-Flood Fill(img)

Input:	img
Output:	resultImg

1:      mousePressEvent event;
2:      **for** each $i \in [0, num(perceived\ colors)]$ **do**
3:          // The realistic colors perceived by artists
4:          x ← event.x();
5:          y ← event.y();
6:          c_i ← img[x][y];
7:          **for** h in $range(img.height())$ **do**
8:             // Main Filling Process
9:             **for** w in $range(img.width())$ **do**
10:                **if** img[h][w] == -255 or |img[h][w]-c_i| > preset threshold α **then**
11:                  break;
12:                **end if**
13:                seed← (h,w);
14:                region r_{c_i} ← floodfill(img, seed) with filling color c_i and interactive threshold β;
15:                resultImg m_{c_i} ← region r_{c_i};
16:                img r_{c_i} ← -255;
17:             **end for**
18:          **end for**
19:          resultImg.append(resultImg m_{c_i})
20:      **end for**

Kickstart Users often do not know how to import an image. The CAS should offer a simple and obvious starting point.

Easy to learn Little or no training for users. Reducing the time investment necessary to engage in cruise interior color analysis.

Easy to use The CAS UI should be simple and intuitive. The proposed tools should be easily found and predictably called in response to a request from users immediately.

WYSIWYG Accurate visual feedback about the results of their interactions with the system.

Accurate results The proposed Interactive-Scan-Flood Fill should give different users accurate results since the presented color corresponds to an object material no matter how the lighting condition changes.

The above design goals are considered to offer the user enough supports for color analysis under complicated ocean lighting conditions, as illustrated in Figure 3. In the following, the Interactive-Scan-Flood Fill algorithm is detailed.

Considering the distorted colors introduced by complicated ocean lighting conditions, we propose an Interactive-Scan-Flood Fill algorithm illustrated below. Given a cruise interior image with a set of K realistic color labels $\{c_i\}_{i\in\{1,...,K\}}$ verified by the most experienced designers in the world, we first use flood filling algorithm to fill the connected regions with label color c_i, and then we search through all other pixels in the image, setting all the pixels that are similar to label color c_i as the new seeds for flood filling recursively within an interactive threshold β. Finally, both the connected and unconnected regions are filled with the ith realistic color and transformed to a result image labeled with m_{c_i}. For the orgin image, these selected regions are transformed into "black" areas since the next iteration should avoid them to ensure that the sum of proportions of colors is 100%. The pseudocode is given in algorithm 1.

Taking complicated ocean lighting conditions into account, our proposed Interactive-Scan-Flood Fill algorithm scans through the whole image to set potential starting seeds for flood filling with interactive threshold, which can successfully deal with complicated regions filled with distorted colors.

4 IMPLEMENTATION

The Interactive-Scan-Flood Fill algorithm is designed to learn the connectivities of global color scheme interaction graph, as illustrated in Figure 4.

World famous artists and designers are invited to provide comparative analysis of color scheme of typical functional spaces (eg. ocean view stateroom, balcony stateroom, main dining room and cruise atrium) using our proposed CAS. The CAS can handle complicated ocean lighting conditions in ocean view or balcony staterooms. Moreover, the Interactive-Scan-Flood Fill algorithm is specially designed to handle

   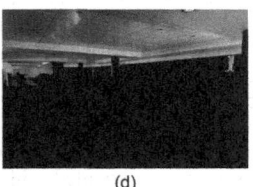

       (a)   (b)   (c)   (d)

Figure 3. Evolution of an image during the implementation of Interactive-Scan-Flood Fill (ISFF) algorithm. We illustrate how ISFF acts on an original image (a): first select the realistic color as color label, and then fills the connected regions with interactive threshold. However, in view of complicated ocean lighting conditions, some regions are not selected (b) or even over selected (c). In the following, ISFF scans the whole image for unconnected regions with carefully chosen threshold by the way of interaction-machine, as shown in (d).

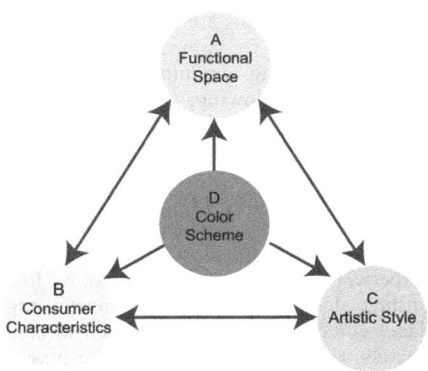

A. Functional Space: cruise stateroom, cruise dining space, cruise entertainment space, cruise health and beauty space, cruise business space, cruise children's space, etc.

B. Consumer Characteristics.

C. Artistic Style: new England-style, Mediterranean-style, modern style, etc.

D. Color Scheme: holistic view regarding functional space, consumer characteristics and artistic style.

Figure 4. The CAS is proposed to learn the connectivities of global color scheme interaction graph based on artistic styles, consumer characteristics and typical functional spaces.

annuli, 'holes' or even more complicated structures in regions of cruise main dining room and atrium.

4.1 Ocean view stateroom

The ocean view stateroom serves as a rest space for tourists, where tourists can enjoy the elegant and casual life comfortably. Due to the complicated ocean lighting conditions, three major tones of furniture, floor and wall are generally considered to create a peaceful and harmonious environment. The overall color scheme design should be as uniform and harmonious as possible. We take 11 cruises for a detailed comparative analysis using our proposed CAS, as shown in Figure 5.

Warmer tones of medium or high brightness are usually used for ocean view staterooms, in which the yellow-gray tone is the most commonly used since the light and bright color tones lead to a sense of openness, making the room more spacious. While the beige is another commonly used color for ocean view stateroom, making tourists feel friendly and comfortable. Therefore, the beige with higher brightness and lower purity, coupled with brighter colors of specific artistic styles are applied to the ocean view staterooms in the above-mentioned international cruises, such as the Queen Mary 2, which make the entire stateroom elegant and vibrant in tranquil atmosphere.

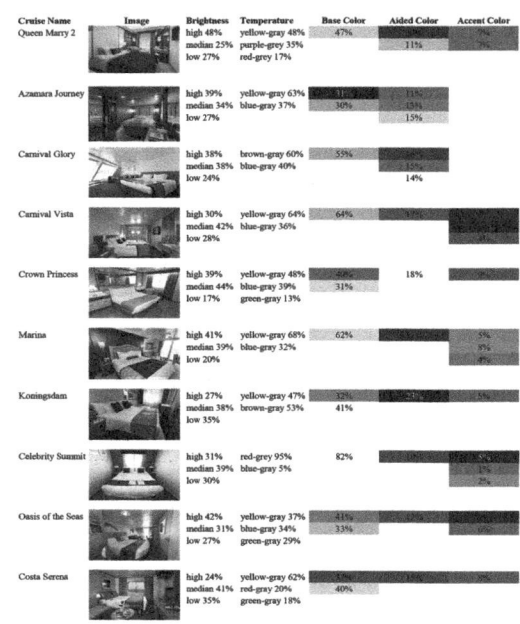

Figure 5. Color scheme of ocean view staterooms.

4.2 Balcony stateroom

Simplicity and sobriety are the basic requirements for tourists to color scheme of balcony staterooms. Basic objects such as beds, TVs, chairs, bedside tables, lamps, etc. have met the tourists' needs of rest and relaxation and constituted the main decorative elements of the space. How to reasonably color-match these objects is the primary consideration in the design of balcony staterooms. We take 13 cruises for a detailed comparative analysis using our proposed CAS, as shown in Figure 6.

The white and blue color scheme schemes are commonly adopted in the balcony staterooms of the above-mentioned large international cruises since bright colors can improve the brightness of the furniture and have an activating effect on human emotions. Particularly, the blue can be subdivided into several groups. The azure blue indicates calmness and introversion. The light blue is friendly, expanding, and creative. The dark blue represents solidity and calmness.

4.3 Main dining room

A moderate or high brightness warm tone is usually adopted in the color selection of the dining space, particularly, the warm yellow tone is the most commonly used color since it can not only stimulate the appetite well enough to create a pleasant and comfortable dining environment, but also give tourists a warm feeling and make the food look more delicious so as to enjoy it in the best state.

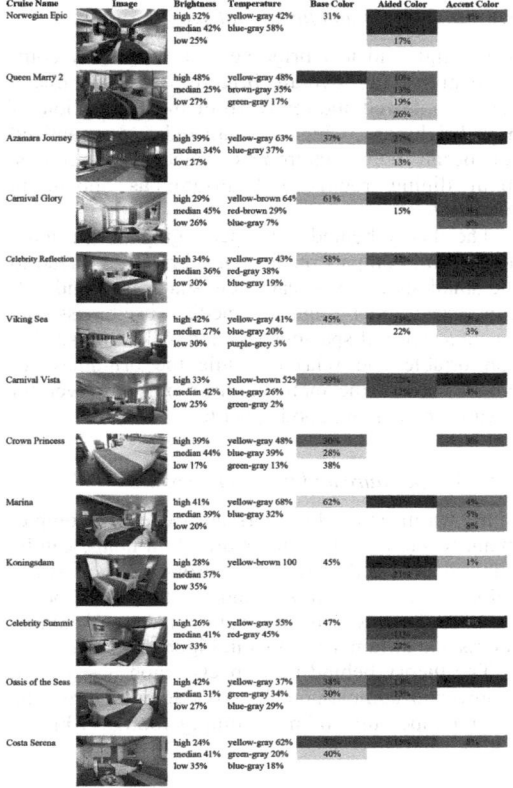

Figure 6. Color scheme of balcony staterooms.

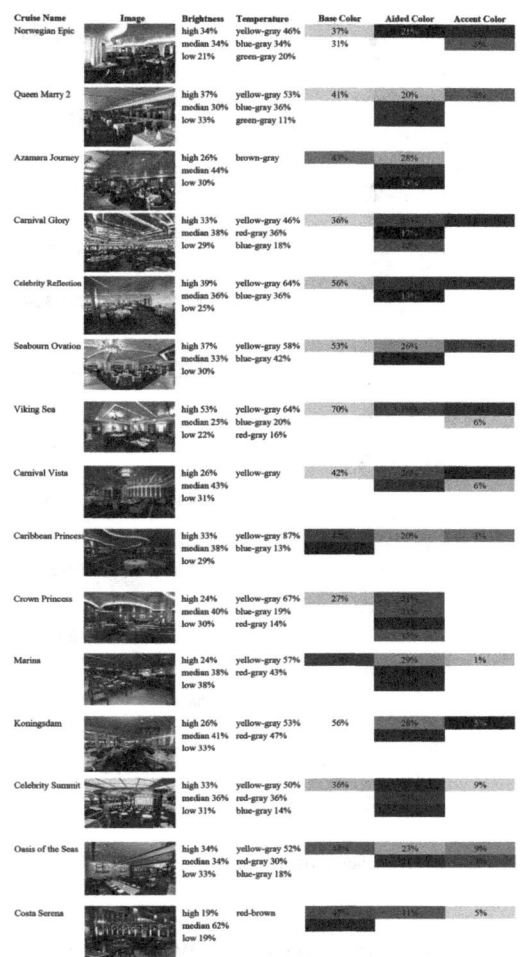

Figure 7. Color scheme of main dining room. Our proposed Interactive-Scan-Flood Fill is robust to annuli, 'holes' or even more complicated structures in regions of cruise main dining room.

In the color scheme of the dining space, the ceiling and wall are mostly decorated with warm and light colors with high brightness and low purity. The yellow or red-brown series with higher purity are often used in seats, floor, carpets, etc. Occasionally, bright yellow or blue is used to color some decorations in the entire interior space. In this paper, we take 15 cruises for a detailed comparative analysis using our proposed CAS, as shown in Figure 7.

In the dining space of large cruises, the flow of tourists is relatively high. In order to meet such functional requirements, firstly, warm colors with high brightness and saturation such as lemon yellow and vermilion are used for color scheme in main dining room, which belong to inflated colors that can widen the space psychologically and reduce the crowded feeling if there are many guests. Secondly, warm colors such as red, yellow, and orange make tourists feel that the time is longer than it actually is, and therefore eat much faster than usual, which will increase the flow of tourists in main dining room. Furthermore, bright colored spaces and interior decorations can leave a clean and efficient impression on visitors. Most of the cruises listed above use the color scheme with red, orange and yellow as the main tone.

4.4 · Atrium

The atrium square is the hub of cruise that serves as the public activity center. It reflects the luxury characteristics of a large international cruise. Since yellow symbolizes noble and gorgeous that can not only enhance the entire space but also make tourists feel like being in a palace, golden yellow based spatial layout shows magnificent and resplendent in the extreme, and becomes the main choice in color selection of the atrium.

In the color scheme of the atrium, light yellow or blue are often used for the ceiling, meanwhile, the colors in low lightness are often used for carpets and tiles. Green and purple are occasionally used to color some decorations in the atrium. The atrium color scheme, with distinct feature in each cruise, has

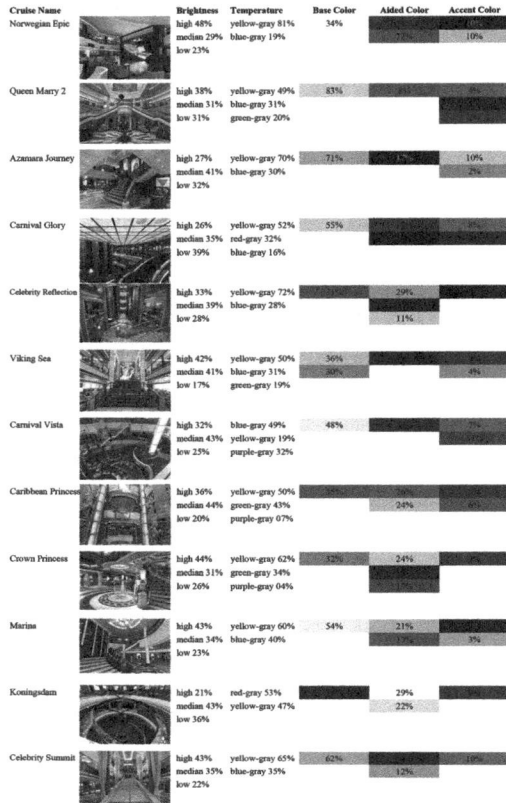

Figure 8. Color scheme of Atrium.

a strong visual impact on tourists. We take 15 cruises for a detailed comparative analysis using our proposed CAS, as shown in Figure 8.

Significant investments and achievements have been made for cruise atrium that paves the way to multiple functional spaces. The atrium in the cruise not only solves the functional requirements of the upper and lower floors, but also carries the cruise spiritual culture and information representations. Therefore, the stylization of the whole cruise is considered in the color scheme to highlight the luxury and elegance.

5 COLOR THEORY FOR DESIGNERS

We have deeply explored the connectivities of global interaction graph based on artistic styles, color schemes and typical functional spaces under complicated ocean lighting conditions. In this section, we focus on the theoretical research on color brightness and temperature, which verifies the feasibility of our color analysis system and provides a guidance on cruise creative arts and design.

5.1 Brightness and functional spaces

Since high and low brightness are relatively complementary, the overall brightness of an image can be inferred indirectly from the proportion of high brightness, which is analyzed and visualized for ocean view staterooms, balcony staterooms, main dining rooms and atrium, as shown in Figure 9.

The theory behind the aspects graph has offered a powerful comparison for color brightness. Due to the small spaces of ocean view staterooms and balcony staterooms, high or median brightness can create a sense of spaciousness and make tourists feel comfortable and relaxed, while the brightness of colors used in the main dining rooms is lower for a warm feeling and good appetite.

5.2 Temperature and functional spaces

Since warm and cold colors are relatively complementary, the overall temperature of an image can be inferred indirectly from the proportion of warm colors, which is analyzed and visualized for ocean view staterooms, balcony staterooms, main dining rooms and atrium, as shown in Figure 10.

The theory behind the aspects graph has offered a powerful comparison for color temperature. The color temperature of main dining rooms is higher

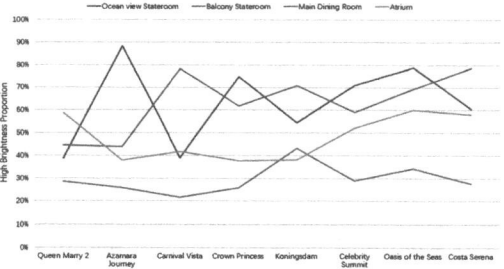

Figure 9. Results on high brightness variation in different cruise functional spaces.

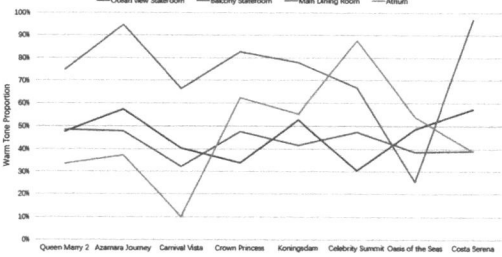

Figure 10. Results on warm tone variation in different cruise functional spaces.

than that of interior staterooms, which is suitable for the dining space with enthusiasm and passion. The atrium color temperature is more uniform in longitudinal variation due to a wide range of different kinds of consumer characteristics and artistic styles.

6 CONCLUSIONS

In this paper, we propose a cruise interior color analysis system to learn the connectivities of global color scheme interaction graph based on artistic styles, consumer characteristics and typical functional spaces under complicated ocean lighting conditions, in which an Interactive-Scan-Flood Fill algorithm is proposed to scan the whole image for potential seeds by the way of human-machine interaction. We believe the efficiency and simplicity of our proposed color analysis system will benefit future research and open up various directions for related creative arts and design.

REFERENCES

Benedetti, L., H. Winnemöller, M. Corsini, & R. Scopigno (2014). Painting with bob: assisted creativity for novices. In *Proceedings of the 27th annual ACM symposium on User interface software and technology*, pp. 419–428.

Champandard, A. J. (2016). Semantic style transfer and turning two-bit doodles into fine artworks. *arXiv preprint arXiv:1603.01768*.

Chen, D., L. Yuan, J. Liao, N. Yu, & G. Hua (2017). Stylebank: An explicit representation for neural image style transfer. In *Proceedings of the IEEE conference on computer vision and pattern recognition*, pp. 1897–1906.

Csikszentmihalyi, M. (1997). Flow and the psychology of discovery and invention. *HarperPerennial, New York 39*.

Dumoulin, V., J. Shlens, & M. Kudlur (2016). A learned representation for artistic style. *arXiv preprint arXiv:1610.07629*.

Gatys, L., A. S. Ecker, & M. Bethge (2015a). Texture synthesis using convolutional neural networks. In *Advances in neural information processing systems*, pp. 262–270.

Gatys, L. A., A. S. Ecker, & M. Bethge (2015b). A neural algorithm of artistic style. *arXiv preprint arXiv:1508.06576*.

Gatys, L. A., A. S. Ecker, & M. Bethge (2016). Image style transfer using convolutional neural networks. In *Proceedings of the IEEE conference on computer vision and pattern recognition*, pp. 2414–2423.

Gatys, L. A., A. S. Ecker, M. Bethge, A. Hertzmann, & E. Shechtman (2017). Controlling perceptual factors in neural style transfer. In *Proceedings of the IEEE Conference on Computer Vision and Pattern Recognition*, pp. 3985–3993.

Huang, X. & S. Belongie (2017). Arbitrary style transfer in real-time with adaptive instance normalization. In *Proceedings of the IEEE International Conference on Computer Vision*, pp. 1501–1510.

Johnson, J., A. Alahi, & L. Fei-Fei (2016). Perceptual losses for real-time style transfer and super-resolution. In *European conference on computer vision*, pp. 694–711. Springer.

Li, C. & M. Wand (2016a). Combining markov random fields and convolutional neural networks for image synthesis. In *Proceedings of the IEEE Conference on Computer Vision and Pattern Recognition*, pp. 2479–2486.

Li, C. & M. Wand (2016b). Precomputed real-time texture synthesis with markovian generative adversarial networks. In *European conference on computer vision*, pp. 702–716. Springer.

Li, Y., C. Fang, J. Yang, Z. Wang, X. Lu, & M.-H. Yang (2017a). Diversified texture synthesis with feed-forward networks. In *Proceedings of the IEEE Conference on Computer Vision and Pattern Recognition*, pp. 3920–3928.

Li, Y., C. Fang, J. Yang, Z. Wang, X. Lu, & M.-H. Yang (2017b). Universal style transfer via feature transforms. In *Advances in neural information processing systems*, pp. 386–396.

Liao, J., Y. Yao, L. Yuan, G. Hua, & S. B. Kang (2017). Visual attribute transfer through deep image analogy. *arXiv preprint arXiv:1705.01088*.

Risser, E., P. Wilmot, & C. Barnes (2017). Stable and controllable neural texture synthesis and style transfer using histogram losses. *arXiv preprint arXiv:1701.08893*.

Sanakoyeu, A., D. Kotovenko, S. Lang, & B. Ommer (2018). A style-aware content loss for real-time hd style transfer. In *Proceedings of the European Conference on Computer Vision (ECCV)*, pp. 698–714.

Ulyanov, D., V. Lebedev, A. Vedaldi, & V. S. Lempitsky (2016). Texture networks: Feed-forward synthesis of textures and stylized images. In *ICML*, Volume 1, pp. 4.

Wang, X., G. Oxholm, D. Zhang, & Y.-F. Wang (2017). Multimodal transfer: A hierarchical deep convolutional neural network for fast artistic style transfer. In *Proceedings of the IEEE Conference on Computer Vision and Pattern Recognition*, pp. 5239–5247.

Wang, Z., L. Zhao, L. Qiu, H. Chen, Q. Mo, S. Lin, W. Xing, & D. Lu (2019). Diversified arbitrary style transfer via deep feature perturbation. *arXiv preprint arXiv:1909.08223*.

Developments in Maritime Technology and Engineering – Guedes Soares & Santos (eds)
© 2021 Copyright the Author(s), ISBN 978-0-367-77376-2

Research on cruise space planning and design based on spatial syntax and correlation analysis

Z. Li & W.B. Shao
School of Art and Design, Wuhan University of Technology, Wuhan, China

ABSTRACT: The space reconstruction and function accumulation brought by the cruise entertainment and large-scale cause the space logic to be discrete and disordered. This paper takes Symphony of The Seas as the research object, using Depthmap software and combining with space syntax theory and correlation analysis to elaborate the quantitative analysis of cruise space planning and design. In short, the spatial characteristics represented by the integration and control of the cruise functional space should be consistent with the functional positioning. The focus of the study is to analyze the spatial quantitative parameters from the perspective of feasibility and visibility, to obtain the cruise space design guidance with data basis, and on this basis, to inspire the specific design matching the spatial function at the macro, meso and micro planning levels.

1 INTRODUCTION

1.1 Cruise overview

According to the "2019 China Cruise Development Report" released by the China Communications and Transportation Association, China's cruise tourism market has been in a period of optimization and adjustment in the past three years. Domestic and foreign cruise companies have continuously adjusted their market strategies to adapt to the new development situation. At the same time, the cruise ship manufacturing side is also accelerating the exploration of new manufacturing technologies and design methods, and gradually entered the construction stage of the entire industry chain.

Modern sightseeing cruise ships are born out of ocean liners. Their specifications and structures mostly originate from the old system. It is easy to conflict with the space reconstruction and function accumulation brought by entertainment and large-scale, resulting in discrete and disordered space logic. In order to comprehensively study the comprehensive planning and design problems of modern cruises in terms of function, aesthetics, services, etc., it is necessary and scientific to carry out in-depth coordination and integration of mathematical tools. These methods will provide principles and technical support for the spatial planning and design of cruise ships.

1.2 Space syntax theory

Space syntax theory was proposed by Bill Hillile and his team at UCL Bartlett Faculty of the Built Environment in the 1970s. As a new method to describe the form of space, space syntax theory believes that space

itself does not have importance. What is important is the organizational relationship between spaces. Therefore, the basic method of its application is to divide the space. The syntactic uses the concept of linguistics, which is extended to the rule of the combination relationship of spatial units. Space syntax specializes the combination relationship between spatial units into a spatial composition diagram, and then performs topological analysis on spatial accessibility through convex space segmentation and axis translation, and uses mathematical tools to calculate analysis variables. Through a series of quantitative parameters to quantitatively describe the spatial structure, it can intuitively express the difficult to perceive spatial characteristics and its relationship with human social activities.

The theory of space syntax points out that the fabric relationship is the medium of the interaction between space form and social function. It has coordinated the contradiction between form and function in space design to a certain extent, and is widely used in urban area, block building group and comprehensive building planning research. Liu Z. analyzes the planning and application of space syntax in commercial blocks from both qualitative and quantitative aspects, and guides the design of self-functional formats and traffic nodes in commercial blocks. Fu W. introduced space syntax theory to quantitatively analyze the deck layout of Quantum of the Seas, and established evaluation criteria to judge the streamline planning scheme of tourist activities.

1.3 Ideas for cruise space planning

Le Corbusier once proposed the concept of functional zoning of "work, residence, entertainment,

DOI: 10.1201/9781003216582-44

transportation" in urban planning. The spatial form of buildings in each zoning is closely combined with the use function. Its basic theory has guided the spatial function design of cruise ships. The characteristics of the cruise ship are prominently manifested in the islanding of resources and the concentration of functions. During the construction process, convenient modular design and assembly methods are also used, so the functional areas are clearly divided. However, this mechanized model mostly comes from paradigm-based experience, and lacks in-depth study of spatial organization forms.

The current large-scale design trend of cruise ships is manifested in a more spacious event space, a substantial increase in passenger capacity and a form of upgrading consumption, in order to maximize economic efficiency and brand reputation. However, after all, the internal space is limited. The characteristics of high density and three-dimensional functional structure are obvious. The blind addition of functional nodes by ignoring the main body of the space will cause contradictions in spatial planning. Therefore, the cruise ship space design solution should replace expansionary growth with structural adjustment. Space syntax theory emphasizes the exploration of the relationship between space and society through human activities, and advocates the use of space organization to restrict user behavior, and user behavior can also reversely shape the social function of space(Tao 2015).

2 TECHNICAL ROUTE AND ANALYSIS METHOD OF SPACE PLANNING AND DESIGN OF CRUISE

2.1 Research path

The core of the research on cruise spatial planning is to analyze the relationship between spatial organization and social functions. As a collection of multiple independent spaces, the flow of tourists on the cruise line reflects the spatial organization. This article takes Symphony of the Seas as the research object and Depthmap as the research tool to explain the quantitative analysis of the cruise spatial planning and design.

Symphony of the Seas has a tonnage of up to 230,000 tons. It is currently the largest and most expensive cruise ship among existing cruises. It is like a sea city, with complex space, complete functions, and great research value (Figure 1). One of its spatial characteristics is the vertical distribution, the living room is set on the lower level, and the functional space that can promote consumer activities is located on the upper level with better lighting and better vision. Specifically, the deck1,2 are crew decks, the deck7, 9, 10, 11, 12, and 18 are cabin decks, the deck4, 5, 15, and 16 are other functional decks, and the deck3, 6, 8, 14, and 17 are composite decks. Among them, the functional space of the deck8 is more complete and contains a patio garden, the deck15 is a semi-open-air entertainment space, and the deck17 is a senior suite

Figure 1. Symphony of the seas central park (https://www.rcclchina.com.cn).

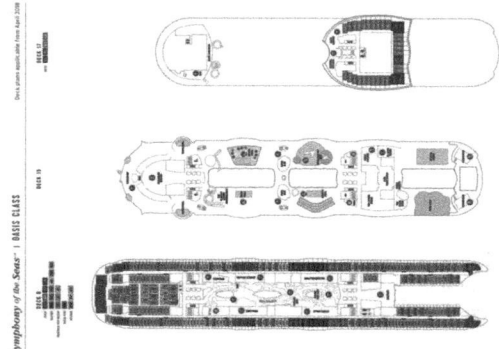

Figure 2. Symphony of the seas case analysis deck plan (https://www.rcclchina.com.cn).

and functional space. The three representative decks above are selected as analysis objects (Figure 2).

Depthmap software is a space syntax analysis software applied to the environment of architectural space and urban space. By importing the relationship diagram after the target space system is segmented into the software, the method of convex space segmentation, axis analysis, viewing zone analysis and other methods are used Parameter calculation and interpretation, the calculation results are expressed in a color cloud, and the values such as connectivity and integration are expressed from red (warm tones) to blue (cold tones) from large to small, and the space can be intuitively determined according to the color feature.

2.2 Technical methods

2.2.1 Spatial syntax analysis
In order to simplify the model, the curves of the cruise ship were straightened, the closed space at the stairwell is treated as a closed space, and the open space stairwell is treated as a feasible deck (Figure 3 left), and then imported into the Depthmap environment for axial map calculation to obtain Deck8, Deck15, Deck17 visual layer and feasible layer axis

schematic (Figure 3 right). The color of the axis in the figure corresponds to the integration value, which represents the theoretical probability of the straight line passing through the space.

Then the axis translation analysis is carried out, and the spatial cognition is deepened by reading the structure relationship. It can be seen from the figure that the integration level of the Deck17 feasible layer and the visual layer is consistent, because there is no obvious visual occlusion on this layer, the duplex suite on three sides, and the lounge, lounge and church on the other side region. The most integrated area is located in the lounge in front of the elevator, with good accessibility, which is conducive to guiding users to consume.

There is a big difference between the Deck15 feasible layer and the visible layer. The reason is that the two surrounding walls of the hull of the hull cause 2-3 spaces in the upper layer to be visible and not feasible. According to the axis diagram, it is found that the areas with high integration of feasible layers are concentrated in the horizontal passage at the bow elevator and the pools on both sides. The two decks are wide and suitable for viewing, which is consistent with the actual situation. The areas with low integration are the bow sunbathing area and the bow deck surfing area, both of which are chargeable project areas, which can promote secondary consumption and increase cruise income.

Deck8 integrates cabin space and functional space, indoor space and open space, the most complex form. The area with the highest integration value of the feasible layer is between Central Park Cafe and Western Restaurant, and the integration value of the nearby areas is higher. The highest level of integration in the visible layer is the longitudinal road on the side of the Italian restaurant in the open air area, along the road is a high-end shopping space, which often attracts tourists to stop; the dining space is arranged on the lower level of the integration, that is, relatively quiet, in line with its commercial positioning. The cabin aisle presents a relatively uniform medium integration degree, which is connected to the central patio, avoiding the ring-enclosed enclosed structure of the ordinary residential layer, and the spatial characteristics reflected by the integration degree of the entire internal and external spaces are obviously different. Combined with the schematic diagram of the axis and viewing the actual photos, it is found that the difference between the visible layer of Deck 8 and the feasible layer is that there are some loose low-lying bush landscapes in the park in the open area, which can't be passed but the view is not blocked.

2.2.2 *Correlation analysis*

Correlation analysis is used to investigate the degree of correlation between two variables, which is more objective and authentic than traditional empirical design. Select the connection degree of the local variables and the integration degree of the overall variables from the space syntax quantization parameters, and calculate the fitting degree of the two as the intelligibility degree-greater than 0.5 is considered to have a basic fitting relationship, and greater than 0.7 indicates a better linear distribution relationship.

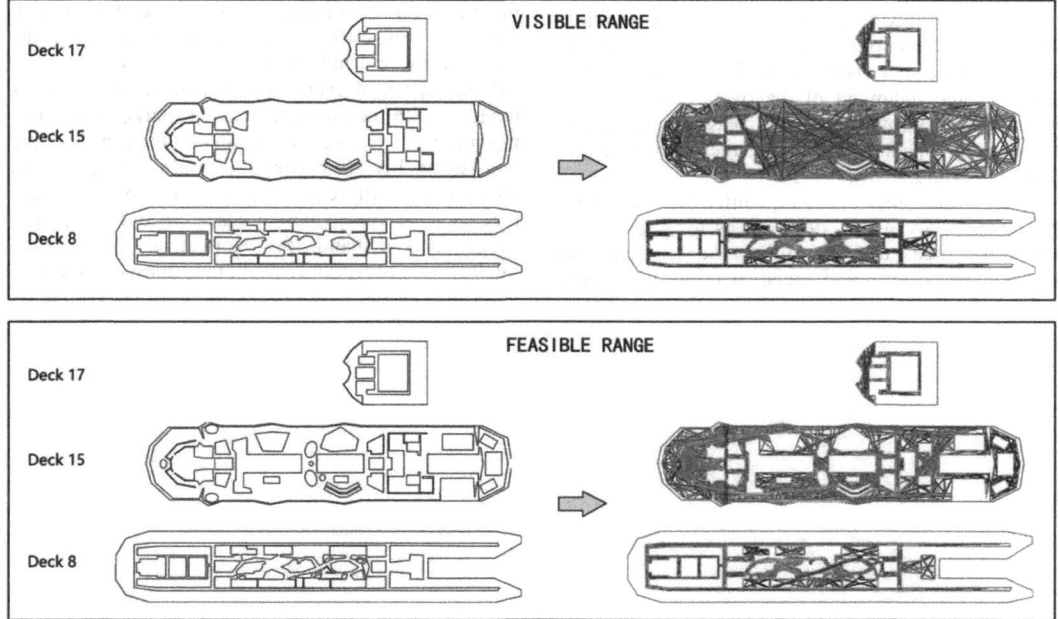

Figure 3. Axis diagram.

Analysis and processing were performed on the connectivity and integration of the feasible layers of Deck8 and Deck15, respectively, and the fitting degrees of the two were 0.621 and 0.677, respectively (Figure 4). It can be found that the two levels have a good fit, that is, tourists can intuitively establish an understanding of the overall spatial structure based on their real-time position, and the degree of understanding is high. This is because the layout of Deck8 is a simple hollow shape, the difference between the inside and outside spaces is obvious, and each forms a small independent system, which is particularly easy to allocate functions and be easily perceived; while Deck15 is an open-air deck, the commercial enclosure structure divides the front and rear spaces into noisy- Two types of static distribution, showing extremely strong spatial planning potential.

2.3 *Results discussion*

Space syntax and correlation analysis is helpful to find out the impact of the large-scale introduction of entertainment and commercial facilities on the spatial relationship, and improve the perception and experience of tourists. Symphony of the Seas' Deck15 has high visual penetration, but the surrounding space structure causes the tourists to move complicatedly. Therefore, the installation of guide signs can be used to promote the consumption of the stern project. In view of the conflict between Deck8's visual and feasible, the height of the landscape in the open park can be increased appropriately, so that the line of sight is interrupted to reduce the integration of the visual layer, and then the surrounding commercial atmosphere can be highlighted. It can be seen that the visual accessibility often needs to undergo multiple spatial transformations, and the spatial accessibility can't rely on direct vision alone, but depends on the cognition and judgment of the overall spatial planning-the reality of visual and feasible regulation. Above is the spatial movement direction and behavior pattern(Tang 2018).

Space syntax analysis can quantify the integration and connectivity of the deck, so as to quickly find high-value areas and coordinate the contradiction between spatial form and social function. Inevitably, the accuracy loss caused by simplified drawings in the theoretical analysis, the definition and division of obstructions may cause errors from the actual, and subsequent research can further improve the relevant technical methods to achieve the best.

3 THEORETICAL MODEL CONSTRUCTION OF CRUISE SHIP SPATIAL PLANNING AND DESIGN

3.1 *Level planning*

With the trend of large-scale construction of modern cruise ships and the core of activities is gradually multi-polarized. The formation of the cruise ship's spatial structure is spontaneous, and the system elements penetrate each other in the process of continuation, conflict, and fusion. This can be discussed from three planning levels in conjunction with the space syntax theory.

Macro level: Due to the spatial divestiture of cruise ships, the macro level planning design mainly considers higher-level planning such as brand image, market preference, product positioning, design style, and consumption form to establish the overall spatial form planning direction and improve the commercial layout(Shao 2018). The essence of cruise space design is to better serve commercialization activities, plan engineering requirements and function settings, and graft or develop secondary consumption behaviors through space design as much as possible within the allowable range. At the same time, due to the influence of laws and regulations, structural safety, fire safety, orderly evacuation and life-saving equipment also limit the overall layout of the ship. In addition, the cruise design also needs to describe the spatial form and influencing factors required for the external space of cruise ship benchmarking in combination with the characteristics of market port, driving route and sea area. For example, the cabins of the Norwegian Joy are narrow, the routes are fixed, and the service is differentiated. Coupled with market factors such as fuzzy market research and disordered supply and demand, the Norwegian Cruise Line finally chose to withdraw from the Chinese market.

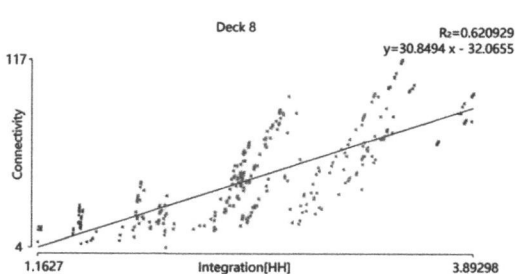
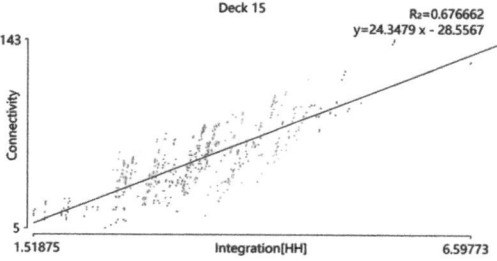

Figure 4. Intelligibility diagram.

Meso level: The overall frame design of the cruise liner is planned in the vertical direction, with emphasis on spatial cognition and spatial layout. The deck is used as the basic skeleton of the cruise space, and the functional space, cabin area, crew area, etc. are set up and down to build a basic space layout with a cognitive path. The spatial planning at this level mainly considers the maximization of functions. For example, recreational facilities such as water parks and sports fields require open spaces and good views. Therefore, they are arranged on the top deck, rather than simply on the circular streamlines connecting the vertical layers. Similarly, constructing an intelligibility model with the elevator as the path channel on the vertical section will help optimize the layout when planning and designing a new type of functional space.

Micro level: The horizontal functional space of cruise ships focuses on the construction of visual and feasible space systems. Staff streamline design are the rules of the mobile experience of tourists in the cruise space. Based on the design of streamlines, they calculate the spatial integration and connection values that change with the layout of the plan, and analyze the relationship between spatial organization attributes and flow behavior and consumption habits. Comparing visibility and feasibility and choosing the best option, the integration will eventually become the basic component of a complex space system. At this time, the implementation of space syntax theory connects the spatial structure and social relations, which fully takes into account the needs of the cruise company and the user to achieve a win-win situation.

3.2 Design path

Functional space planning: The cruise ship is an offshore building with the compound characteristics of an urban commercial complex. The basic requirements for the planning of cruise ship functional space include: the safety of diversified design, the comfort of reducing the high-density constraints of space, and the coordination of form and function imply aesthetic artistry. Planning is embodied in form, and form serves function. In the space planning of cruise ships based on the space syntax theory, the public space should be located in the area with the highest integration and connection value, while maintaining a high degree of openness. Areas with a high degree of integration and control have great commercial value and are suitable for layout of shopping and consumption spaces, while areas with low privacy are good and suitable for high-end areas such as lounges. In short, the spatial characteristics represented by the degree of integration and control of the cruise ship's functional space should be consistent with the functional positioning.

Service model planning: The core of service design is not the product itself, but focuses on creating value through the service itself. According to the characteristics of cruise ships, the activity streamline is derived from dynamic programming, so its service mode is a function-oriented service based on streamline design. The function-oriented design includes information platform design and service activity design, etc., which jointly construct the user's abstract experience of the cruise space. The active streamline is dynamic, the spatial structure is static, and the dynamic and static comprehensive design is spatial planning. Service design can optimize the space planning of cruise ships, such as arranging functional spaces with high commercial value in areas with high integration values, and at the same time conducting secondary guidance on its activity streamline to get people sharing (Pan 2018).

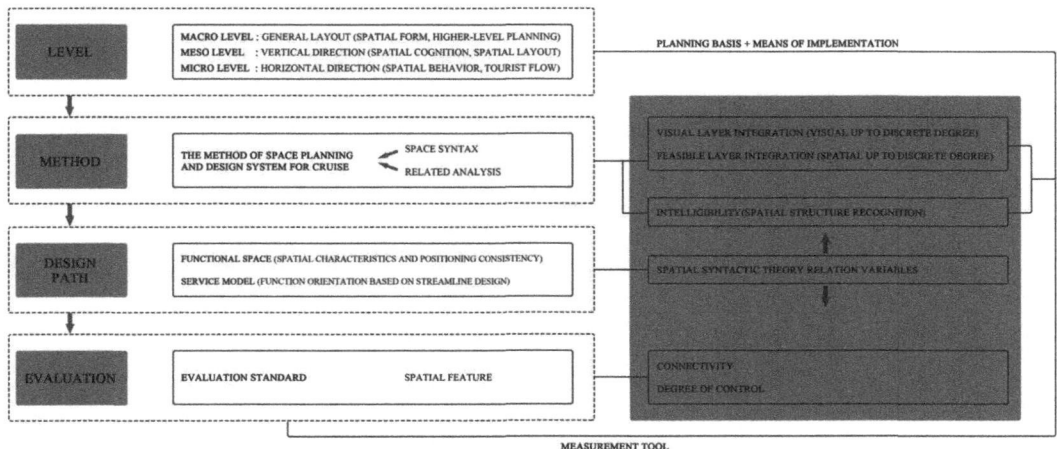

Figure 5. Theoretical model.

3.3 Theoretical model framework

The significance of constructing the theoretical model is to transform spatial elements into data information and realize the relationship mapping of "mathematics-physical space"(Li 2018). This relationship will facilitate the use of creativity and the advantages of computer computing to monitor and complement each module. This is a thinking order based on quantitative data that can be dynamically tracked and evaluated (Figure 5).

The left side of the theoretical model is the logic of the space planning design of the cruise ship. After the execution layer is refined, the mechanism within which the space syntax theory functions can be seen. In the figure, the black arrow indicates the thinking process, the solid black line indicates the decomposition. The focus of the model is to promote specific spatial planning and design that matches the spatial positioning at multiple planning levels. Secondly, the spatial parameters are analyzed from both visual and feasible perspectives, and the scientificity of the design is quantitatively verified. Finally, this model can also be derived when it is applied to the spatial planning of local sites, and its in-depth research is expected to result in multiple innovative solutions.

4 CONCLUSIONS

At present, the design research and engineering research in cruise theory are relatively independent. The method proposed in this paper can verify and modify the planning and design based on the engineering layout, until the preset optimal state. Although the space syntax used in this paper has been frequently used in architectural design, it needs a lot of practice to verify whether the visualization means such as axis diagram and comprehensibility have practical significance in cruise design. The theoretical model obtained in the process of theoretical summary only extends to the planning aspect in the design level. It is necessary to add modular theoretical units to the system design by combining ship engineering and tourism services in the industrial chain.

The cruise ship itself is an important destination for tourism, and spatial planning and design are crucial. As an island-oriented miniature city, cruise ships should follow the formal laws of space and society in spatial planning and design, and use multidisciplinary theory to promote the construction of relevant theoretical systems to guide practical activities. Aiming at the multi-level planning of cruise ship design, using spatial syntactic theory and other research methods to carry out data analysis of spatial organization and relationship, and then combined with the actual situation to construct a specific planning and design model, has important reference value for modern cruise ship design.

REFERENCES

Li, Z. & Li. X. 2018. Service Model of Open Design in Digital Form. *Packaging Engineering* 39(18):191–195.

Pan,C.X.&Wang,X.Y.2018. Building Cruise Guidance Service System Based on Tourist Streamline Behavior-Taking Ocean Quantum Cruise as an Example. *Decoration* 9:85–87.

Shao, J.W. & Pan, C.X. 2018. Personalized Design Path of Chinese Luxury Cruises Based on Tourist Behavior. *Packaging Engineering* 39(10):140–145.

Tang, Z.& Yuan,S. 2018. Analysis of the Space Structure and Touring Behavior of Fangta Garden Based on Big Data and Spatial Syntax. *Decoration* 12:70–73.

Tao, W.& Ding,C.B. 2015. The Application of Space Syntax Theory in the Spatial Planning of Urban Recreation System. *Planner* 31(08), 26–31.

Developments in Maritime Technology and Engineering – Guedes Soares & Santos (eds)
© 2021 Copyright the Author(s), ISBN 978-0-367-77376-2

Research on ventilation and particle flow pattern in cruise cabin design based on FLUENT

Zhuo Li
School of Art and Design, Wuhan University of Technology, Wuhan, China

Yafeng Yang
School of Automobile Engineering, Wuhan University of Technology, Wuhan, China

ABSTRACT: Affected by the maritime season, the unstable changes of bio-aerosols on cruise routes have an important impact on human health. Especially particles smaller than 10 microns are easily attached by bacteria and viruses. In this paper, solid particles sized 2.5 microns are selected as the research object. Based on the Euler multiphase flow model, the particulate matter propagation characteristics of the cruise cabin under natural ventilation are simulated and studied. Two typical cruise ship cabins are numerically simulated. The flow field characteristics and the spatial distribution of particles in the cruise ship cabins are analyzed quantitatively and qualitatively. The multi-index evaluation method was used to focus on the influence of the location and size of the air outlet on the airflow characteristics and the transmission process of particulate matter. The results of the study show that the air quality in the cabin can be improved by optimizing the air inlet and adjusting the layout of the indoor furniture. It provides a theoretical basis for reducing the risk of passenger infection. The conclusion would directly guide the layout and optimization of cruise cabin design.

1 INTRODUCTION

In modern life, people spend about 80% of their time in the indoor environment. Indoor air quality is directly related to people's health. Microbial particles carrying viruses can spread in the air. Indoor environments are more likely to be exposed and infected with bacteria. During the stay of cruise ships, diversified ethnic composition, dense population contact, enclosed cabin environment and long-term isolation life are all likely to cause the accumulation of microbial particles, which further increases the risk of disease transmission. Therefore, to build a healthy, safe, and comfortable cruise environment is an important goal of cruise ship design.

Following the outbreak of SARS (Severe Acute Respiratory Syndrome) that swept the world in 2003, Bird Flu in 2005, Swine Influenza in 2009, and Covid-19 bring serious public health problems and economic losses to the world. In these tragic incidents, serious infections broke out in Japan's Diamond Princess cruise ship and many US warships, which posed huge challenges to cruise ship design and infectious disease prevention. Faced with the existing building and space environment, effectively preventing biological pollution and the spread of infectious diseases to build a healthy and safe building indoor environment is currently an urgent task and problem to be solved. However, to deal

with the existing problems, we must first understand the causes, hazards and transmission methods of the bio-contaminants. Improving indoor air quality is a complex issue involving multiple disciplines and several fields, which requires comprehensive multidisciplinary knowledge for exploration and practice. From the perspective of design, this paper discusses how to optimize spatial flow at the minimum cost under the constraints of economic costs and engineering field.

2 MICROBIAL AEROSOL TRANSMISSION PROPERTIES

2.1 *Causes and propagation laws of pollutants*

Biologically-derived pollutants mainly exist in two forms in the air. One type is the spread of droplets, microorganisms attach to people by sneezing, coughing, singing, talking, they are sprayed from the mouth and nose. The second is to attach to aerosol particles to form microbial aerosol. Most of the particles with aerodynamic diameter greater than $10\mu m$ will land with the microorganisms, while the particles less than $10\mu m$ (PM_{10}) mostly carry microorganisms floating in the air for a long time.

Microbial aerosol refers to a colloidal system formed by microorganisms suspended in the air. It

DOI: 10.1201/9781003216582-45

includes dispersed phase microbial particles and continuous phase air medium, so it is biphasic. The distribution range is very wide, from 0.001 to 100 μm. The particle spectrum is also very wide, mostly from about 0.002 to 30 μm depends on the physical properties of the microbial population (Yu, 2002).

In general, bacteria, fungus, etc. exist in the form of aerosol and spread in the air. Most particles with a diameter of less than 1μm are exhaled after inhalation, aerosol particles between 1~5μm will enter and stay in the lungs, particles with a diameter of 5~15μm will adhere to the nasal cavity, trachea or in the bronchi. The 15~20μm particles have a strong deposition effect and tend to settle on the ground (Liu, 2007). Among them, the most harmful to the human body is 1 to 5μm airborne particles. Once inhaled, they can directly invade the alveoli and cause the human body to suffer from various infectious diseases. Therefore, the transmission law and mechanism of infectious disease aerosols in cruise cabin environment need to be studied urgently.

In the cruise environment, the spread of disease is affected by various factors. For example, the rate of microbial reproduction is directly related to the fluctuation of the temperature difference between day and night, the seasonal change of the route, the invasion of plants and biological allergens in the area of the voyage, etc. Dozens of biological aerosols, including viruses, bacteria, actinomyces, fungus, microbial composition, plant fragments, protozoan and insect fragments and excreta, cell products and proteins, are very likely to cause nose, suffocation, throat and upper respiratory tract, bronchus reactions, or measles, rubella and deliberately reactive dermatitis. An adult would breathe 10m^3 air per24 hours, weighing 16kg. According to the second level of China's air quality standard, each cubic meter contains 300 μg of various microbial biomass to calculate. Each person inhales 3000μg complex particles (50 μg microbial particles involved), which is a huge and heavy load for a lung which only worked by 300 million alveoli and 100m^2 capillaries. These substances are quickly absorbed through the lungs and transported to various parts of the body causing harm, second only to intravenous injection in speed (Ding, 2009).

There may be multiple infection sources and routes of transmission on cruise ship. Including contact transmission and media transmission. Contact transmission includes direct contact transmission and indirect contact transmission, and media transmission mainly transmitted through air, soil, and water. Among them, airborne transmission is the most difficult to prevent. The most typical airborne microorganisms are Mycobacterium Tuberculosis, Rubella Viruses and Chicken Pox (Tu, 2004). SARS can also be transmitted only by air. After the droplets with bacteria or viruses ejected from the pathogen source propel in the air has traveled a certain distance. The formed aerosol can stay in the air for a very long time and may settle in the conjunctiva, nasal mucosa and mouth of the host. It is always high risks of infection by transmission in crowded enclosed space such as cruise cabins, elevators, movie theaters.

2.2 *Research on computer simulation flow field*

CFD tools can be used to simulate and study the movement and distribution of indoor pollutants. Chen has conducted a large number of simulation studies on microbial particulates in ventilated rooms and cabin environment (Chen, 2006), and his research results show that the computer-simulated pollutant propagation laws are in good agreement with the experimental data. Kato conducted a simulation study on the propagation of droplets generated by coughing in indoor airless and air-conditioned ventilation environment (Zhu, 2009), and conducted experimental studies on the propagation distance of droplets. The cough airflow was visually analyzed by visual means. At the same time, the Lagrange calculation simulation method was used to analyze and study the propagation of droplets generated by cough.

The influence of ventilation on microorganisms in indoor air is mainly reflected in the concentration distribution, diffusion, spread of microorganisms and the control effect. Zhang (2009) simulated the problem of particulate matter diffusion in the ventilation room with three types of air flow organization, upper air supply, side air supply and lower air supply. And confirmed the simulation results with Murakami's (2012) experimental data of the top-to-back airflow and side-to-back airflow as well as floor airflow experiment. Indicating that the air distribution structure of the lower air supply has better particle removal ability, but there is a problem of secondary dust. Pereira (2008) studied the effects of four airflow patterns on the indoor particulate concentration in the office environment, such as top-to-top airflow, top-to-back airflow, under-floor air supply and split air-conditioning air supply, and experimentally compared in detail the removal effects of various airflow forms on pollutants.

3 ANALYSIS OF AEROSOL INDOOR MOTION CHARACTERISTICS

3.1 *Aerosol microbial characteristics*

For the control of microbial pollution, understanding the characteristics of microbes is an essential prerequisite for conducting research and finding control methods. The general and specific characteristics of microorganisms are: ① small individuals, ② rapid reproduction, ③ wide distribution and variety, ④ high variability and strong adaptability to the environment. Most microorganisms in nature are beneficial. A few microorganisms can cause biological pollution and do harm to mankind, namely viruses, bacteria, fungus. Viruses cannot exist alone in the environment, most of them use bacteria as their hosts to replicate and reproduce. Bacteria are pathogenic after being eroded by viruses, and it is easy to form aerosol particles or droplets to spread in the air and cause infectious diseases. Therefore, as long as the concentration of bacteria in the air is controlled, the possibility of infection could be reduced.

3.2 Analysis of aerosol indoor motion characteristics

Several microbial aerosol particles that are harmful to humans generally have a particle size of about 1~2.5 μm, which have good airflow following ability. The indoor propagation is mainly affected by the initial speed provided by the germ source and the indoor flow field distribution, which moves in accordance with the airflow. Therefore, different airflow organization forms, outlet forms and positions have a decisive effect on the distribution of microbial particles. It is worth researching to find ways and means to eliminate and control indoor biological pollution quickly from the perspective of airflow organization and air supply form.

Due to the impact of the Covid-19 epidemic, the paper mainly uses the theoretical analysis method to analyze indoor biological pollution. Firstly, the stress analysis is performed on microorganisms of 1~2.5 μm with a threat to human health. By analyzing the aerodynamic characteristics and forces of biological particles, the distribution of indoor bio aerosol diffusion is mathematically calculated, the diffusion distance of bacterial particles in the room and the law of airflow following is obtained. the indoor biological aerosol is controlled by adjusting the position of outlet and room layout.

4 RESEARCH METHODS

In this study, Fluent is used to simulate the flow field, and the control equation for the calculation of air flow is shown in Equation (1).

$$\frac{\partial}{\partial t}(\varphi) + \nabla \cdot (\mathbf{u}\varphi) = \nabla \cdot (\Gamma_\varphi \nabla \varphi) + S_\varphi \quad (1)$$

where φ is an arbitrary variable, u is the velocity vector. Γ_φ is the effective diffusion coefficient for each dependent variable j. S_φ is the source term of the general equation. When φ is mass, the equation changes into the continuity equation. With φ, Γ_φ taking corresponding expressions, Equation (1) can be expressed as continuous equations, momentum equations, energy equations, and turbulent flow energy and turbulent flow energy dissipation rate.

4.1 Turbulence model

Since the 1990s, CFD theory has been initially applied to the study of indoor flow fields, adding objects to the room is populated such as furniture, appliances and people. Related research results show that the k-ε model gets good adaptability to indoor airflow simulation. In the paper, Reynolds average Navier Stokes method and k-ε turbulence model are used to predict the airflow organization.

4.2 Particle model

In CFD, the model of the propagation mechanism of particulate matter in the room can be divided into the Euler model and the Lagrange model. The Euler model regards the particles as the continuous phase. The conservation equation of aerosol particles in the finite volume method has the same form as the fluid phase. The Lagrange method treats particles as discrete phases, tracks the trajectory of each particle, and on the trajectory the particles exchange material and energy with the adjacent fluid phase, calculate the characteristics of the particles in each control body, and finally obtain the particle concentration Spatial distribution and other relevant parameters. In this paper, the slip flux model in the Euler model is adopted, considering the sliding and gravity effects between the particle phase and the air phase. Compared with the Lagrangian method, the calculation time is shorter, which is convenient for application in engineering practice. As shown in equation (2).

$$\frac{\partial\left[(V_j + V_{sj})C\right]}{\partial x_j} = \frac{\partial}{\partial x_j}\left(\frac{v_t}{\sigma_C}\frac{\partial C}{\partial x_j}\right) + S_C \quad (2)$$

Where V_j and V_{sj} are the averaged fluid (air) velocity and the gravitational settling velocity of particles in j direction, respectively, C is the averaged concentration of particles (mass or number per unit volume), σ_C is the turbulent Schmidt number of C, which is usually equal to 1.0.

5 RESEARCH ON FLUENT-BASED DISCRETE PHASE MODEL AND TURBULENCE MODEL

5.1 Establish geometric model of cruise cabin

Choose a typical balcony cabin with a length, width and height of 6x2.5x2.7 meter as the experimental cabin. The window sill is about 1 meter above the ground and the sized 1.5x1 meters. The state is set to natural ventilation, the balcony door is used as the air supply port, while the cabin door is close to the negative pressure area of the guest room corridor, so it is the outlet. A stream field model of the ordinary cabin of a cruise ship is constructed, shown in Figure 1.

5.2 Meshing and numerical calculation settings

Meshed by ANSA tool, the entire fluid domain is divided into 1 million grids totally and skewness is less than 0.9. Simulate the natural state of a clear sea at sea (approximately wind scale is force 2 to 3), the

Figure 1. Construction of geometric model of typical balcony room.

doors and windows are fully opened, the simulation conditions are set as follows. The particle size is 2.5μm, surface released from window, speeding (wind speed) with 3 m/s, concentration is 15 to 45μg/m^3, the particle density is 1.5 g/cm^3, and the particle size is 2.5μm.

5.3 Microbial concentration analysis

Using the k-ε turbulence model and the slip flux model, the SIMPLE algorithm is selected as the velocity-pressure coupling method, discretized the equation with first-order upwind style finite volume method. The simulation calculation is carried out in the Fluent environment, following results are obtained.

Figure 2 shows the microbial concentration cloud in X direction. Observe the particle concentration in 4 sections of abcd at the same distance from the head of the bed to the end of the bed. The overall pattern is as follows. The concentration distribution of the indoor flow field is uneven, big difference between the left and right along the axis. The concentration gradually increases toward both sides, higher top than bottom, highest in corner. The cavum between bed and bathroom is most likely to produce eddy currents that allow bacteria to stay or distribute for a long time, while the air velocity is faster in the aisle space (near the air inlet), so the microbial particles are not easy to attach.

Known by the floor plan, there is a sofa place between bed and bathroom, located in a medium-concentration area. It is not a good place to relax and stay for a long time, while the desk area oppositely is much safe. Since bed

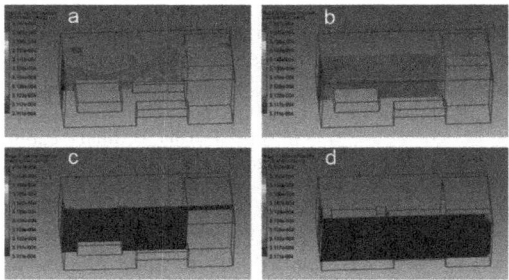

Figure 2. X direction microbial concentration cloud.

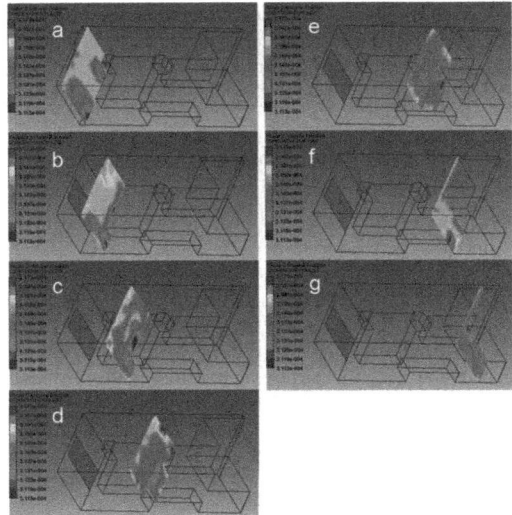

Figure 3. Y direction microbial concentration cloud.

head is immersed in a higher concentration area, switching sleep direction is more rational, it would effectively improve the breathing environment at night.

Seven equidistant sectional particle concentration cloud images from a to g in the Y direction are shown in the Figure 3 above. The microorganisms concentration expands and strengthens in a fan shape radially from air inlet. The concentration in the bed area is higher than that in the open position, the turbulence is easily formed at the air inlet due to the blocking of the bed, easily bacteria bred. To improve the situation, optimizing the size, orientation and position of the air inlet will be helpful.

According to the discipline of human activity, take three positions of the standing, sitting, and lying in the Z direction according to the height of the mouth and nose, h=1.6, 1.1, 0.6 meters. The concentrated cloud images is shown in the figure 4. The standing concentration in the cabin is the lowest and the lying position is higher, so the indoor flow field at night should be especially improved. Considering that the doors and windows are usually closed at night, it can be optimized by the form of indoor air conditioning.

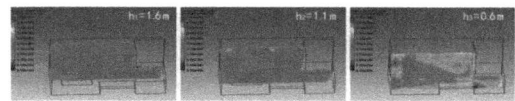

Figure 4. Z direction microbial concentration cloud.

6 OPTIMIZATION OF CABIN VENTILATION MODEL BASED ON FLUENT

The previous is a qualitative discussion on the force of microbial aerosol and the followability of airflow. The distribution of indoor microbial aerosol concentration was discussed through calculation and testing. However, the performance evaluation of the airflow system pattern itself and the actual control effect of different airflow system patterns on indoor microbial aerosols also need to be discussed and verified by the airflow performance experiments.

6.1 Brief introduction of airflow

A good airflow form is to rationally arrange the location of the air inlet and outlet, distribute the air volume and select the air outlet form, induce the airflow through the overall organization of the airflow and the air outlet, and achieve the best ventilation effect with the minimum amount of ventilation. Among them, the airflow organization is to arrange the supply and exhaust vents reasonably, so that fresh air can form a relatively uniform and stable temperature and humidity, air velocity and cleanliness in the working area, reduce the dead angle of the air supply, and improve the indoor ventilation effect to meet the human comfort. This is also an effective measure to improve indoor air quality to meet human health requirements.

Unidirectional flow ventilation means that air is sent out from one side and is exhausted through the opposite air outlet to provide low turbulence "piston flow" across the entire room. This system is mainly used for ventilation of clean room, and its main task is to remove pollutant particles from the room.

There are also many kinds of spatial airflow distribution, which mainly depend on the location and form of the air inlet and outlet. The air inlet plays a leading role, and its layout determines the form of airflow distribution. After a preliminary demonstration, this subject adopted the addition of a lower air supply outlet to conduct the experiment.

6.2 Establish geometric model of cruise cabin

Select a larger suite in the cruise ship to build a cabin model, as shown in the Figure 5, the purpose is to investigate whether there is a large change in the concentration of microbial particles in a large non-box space. This room type is similar to the height of the building room, so it can also be used as a flow field analysis of civil building space.

The air inlet is set as a balcony, enter the room through the floor sliding door, and exhaust the air through the opposite window through the living room and dining room. According to the architectural experience, an additional flat strip-shaped air outlet with a distance of 100mm from the ground is added at the air inlet and the air outlet, and the area is 1000x250mm. In order to investigate the change

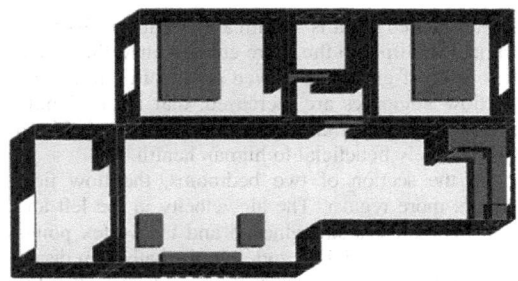

Figure 5. Geometric model building of suite.

of the flow field before and after the addition of the opening.

The conditions of microbial particle release are the same as above.

6.3 Comparison of microbial propagation speed

Figure 6 shows the air velocity streamlines of the three slices in the X direction, the changes is obvious before and after the opening. After the bottom opening is added, the sectional view a of the air outlet axis reflects the north-south living space of the living room, and the streamline becomes significantly smoother, the air outlet speed is slightly reduced but the air inlet speed is significantly enhanced; the mid-to-high layer flow field speed becomes faster, and the vortex point on the upper right is destroyed and greatly weakened, indicating that the microbial aggregation strength is reduced and the suspension time is shortened. The middle vortex point is lower, not only is it closer to the ground and is easily taken away by wind currents, but the reduction in Z-direction height is beneficial to human activity. It is farther from the nose and mouth.

The sectional view b at the center of the width of the room reflects that the streamline is more uniform,

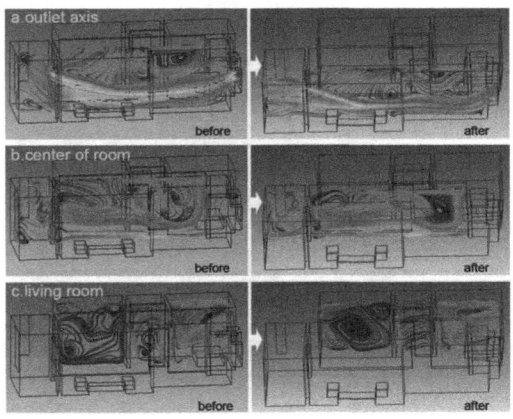

Figure 6. X velocity streamline diagram.

the turbulence point is significantly reduced from the original multiple to the more concentrated three, and the center of gravity is moved downward, almost all the flow velocities are increased, that is, the length of stay in space per unit time becomes shorter, which is objectively beneficial to human health.

On the section of two bedrooms, the flow field appears more regular. The air velocity in the left and right bedrooms is strengthened and the vortex points are reduced, the original vortex in the bathroom disappears and becomes more circulated, indicating that the air has passed straight here and the improvement is the best.

6.4 *Comparison of microbial concentration*

Figure 7 reflects the X-direction microbial concentration comparison before and after the addition of the bottom opening. The increase in blue is most intuitive in view b and view c, indicating that the concentration has been significantly reduced, and the area of low concentration has increased, which is conducive to human activities indoors. The concentration at the top of the bathroom in view d has increased. It should be that the air velocity at the corners of the two walls is small, causing particulates to accumulate.

Figure 8 reflects the comparison of the microbial concentration in the Y direction before and after adding the bottom opening. In view abc, the increase in blue is obvious, that is, the concentration in these

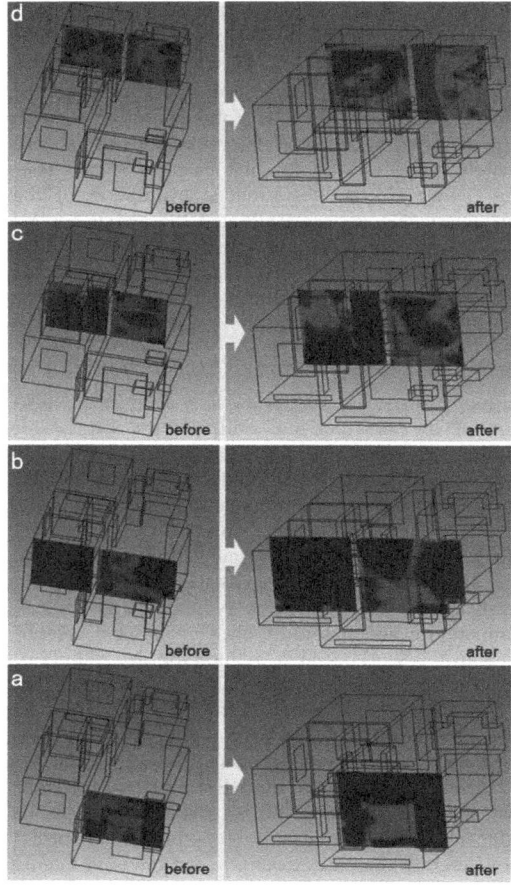

Figure 8. Y direction microbial concentration cloud.

spaces has been reduced. View c is the bathroom and view d is the bedroom, the red area is slightly raised, and these positions are concentrated on the walls and corners, indicating that the bottom opening cannot be the effective transformation of these places. It is still the key part of daily manual sterilization. And red continues to the wall in the Y direction. It is not recommended to install wardrobes, tea cabinets and other furniture on this wall to avoid the human body staying in this area and avoid long-term storage of clothing, towels, etc.

7 CONCLUSION

The paper uses a combination of theoretical analysis and simulation experiments to analyze the movement laws of microbial particulates in cruise cabins, and proposes a more feasible method to improve the airflow form of the air supply, which can play a positive role in indoor microbial pollution distribution control, the following are the conclusions.

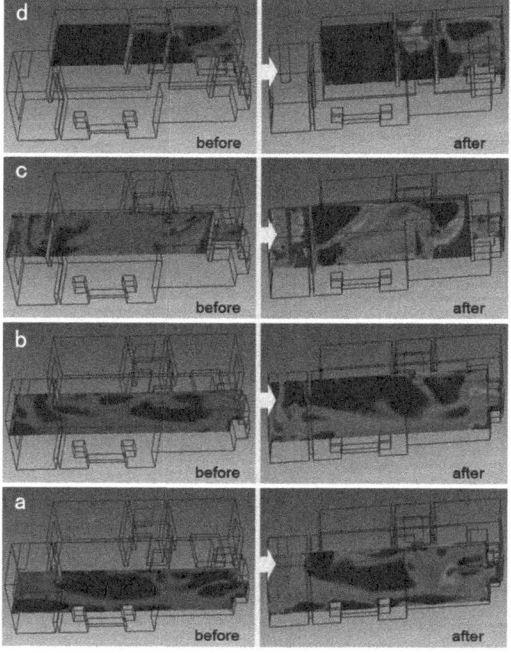

Figure 7. X direction microbial concentration cloud.

(1) In this paper, the theoretical calculation and analysis of the stress of microbial particles with a particle size of about 1 to 2.5μm in the indoor airflow field are carried out. The pathogenic microbial aerosol of this particle size has good airflow followability.

(2) Under the condition of natural ventilation, a CFD model for predicting the distribution of indoor biological concentration was established with the help of FLUENT.

(3) Based on the original research, the prediction model was modified according to the basic characteristics of the airflow structure, and it was proposed that the addition of the airflow structure could improve the original flow field and was confirmed through simulation experiments.

(4) Under the condition that it is not possible to actively modify the house type and the tuyere, a method based on the perspective of art & design is proposed to interfere with the flow field, that is, to optimize the indoor microbial flow rate, flow direction and concentration by appropriately modifying the interior and arranging the furniture.

The air health of the cruise cabin is affected by various factors. This paper is limited to the airflow movement under natural ventilation. It should also consider the factors of airflow organization and tuyere form, tuyere position, and the number of air changes. It is also necessary to carry out in-depth analysis and comparison of the temperature field, velocity field and concentration field of microorganisms in combination with the geographical and seasonal climate of the route.

In addition, the diversity and accuracy of biological pollution prediction and evaluation methods need to be strengthened, especially in combination with the settlement characteristics of different bacterial species along the route to predict and control the distribution of biological pollution. Strive to create a better quality cruise ship health environment.

REFERENCES

Che Fengxiang. 1983. Atmospheric microbiota and its air pollution. Beijing: *Fascicule of Oversea military information (Fifth volume)*.

Ding Yan. 2009. Study on the influence of the form of supply air flow on the spread of indoor biological pollutants. *Tianjin University Master Thesis*.

Liu Shusheng. 2007. Study on transport mechanisms of bio-aerosol produced by mouth in the indoor environment. *Tianjin University Master Thesis*.

Murakami, S. Diffusion characteristics of airborne particles with gravitational settling in a convection-dominant indoor flow field. *ASHRAE Transactions* 98\(Part 1): 82–97.

Pereira, M.L. et al. 2008. Determination of particle concentration in the breathing zone for four different types of office ventilation systems. *Building and Environment*: 1–8.

Q. Chen & Z. Zhang. 2006. Experimental measurements and numerical simulations of particle transport and distribution in ventilated rooms. Building and Environment (40): 3396–3408.

Tu Guangbei. 2004. The control of hospital infections and airborne. Beijing: *China Biosafety Committee, Biosafety Special Lecture Expert Collection*.

Yu Xihua. 2002. Modern air microbiology. Beijing: *People's Military Medical Press*.

Zhang Zhao. 2009. Experimental and numerical investigation of airflow and contaminant transport in an airline cabin mockup. *Building and Environment* (44): 85–94.

Zhu Shengwei, et al. 2006. Study on transport characteristics of saliva droplets produced by coughing in a calm indoor environment. *Building and Environment* (12): 1691–1702.

Developments in Maritime Technology and Engineering – Guedes Soares & Santos (eds)
© 2021 Copyright the Author(s), ISBN 978-0-367-77376-2

Study of the design and styling of large cruise ships over the last 200 years

Jiefeng Lv, Jiankun Sun & Jing Chen
School of Art and Design, Wuhan University of Technology, Wuhan, China

ABSTRACT: This research was based on the appearance and modeling cases of 655 large cruise ships which range from the birth of cruise ships to now and 171 ships were selected as samples. First, a research on holistic image perception of the obfuscated cruise ship appearance by using fuzzy cluster analysis method is carry out. According to statistical calculations, its style can be divided into three historical stages and eight styles. Then its evolutionary context and preliminarily inferring trends. The influencing factors of the appearance style of the cruise ship can be identified by the characteristics of each style by observation method.

1 INTRODUCTION

The history of design plays an important role in design practice and research (Li 2009). With time goes by, the appearance of cruise has gradually evolved into a modern cruise. There have been studies on the style of cruise ships, but most of them are interiors (Andrew 2013) and cabins (Peter 2007)There are relatively few studies on the appearance design of cruise ships. It can help designers to master the basic forms and laws of the appearance of cruise ships by combing the development of the appearance style of cruise ships and analyzing their style characteristics. Then understand the internal mechanism and trend of the evolution of cruise aesthetic style, and put forward a contextual and forward-looking appearance molding design plan.

At present, scholars have used fuzzy modeling to quantify qualitative aesthetic preferences to study the standards that affect people's aesthetic preferences for cruise ship design. Similarly, perceptual quantification can provide positive assistance to style division. For this purpose, the fuzzy set theory (Zadeh 1965) provides a theoretical basis and an analytical tool for the classification of such fuzzy boundaries, and the later development of fuzzy clustering analysis method also provides a new idea for the classification of cruise ship appearance modeling style. Based on the perceptual image cognition of the appearance of the cruise ship (Li 2003), the image scale diagram (Zhang & Zhao 2002) was used to obtain the key features of the cruise ship shape.

The perceptual cognition of the style is quantified through fuzzy calculation (Zhou & Gu 2010). The observation analysis method, principal component analysis, cluster analysis and other methods (Zhu et al. 2000) can be used to study the style type, development vein, modeling characteristics and influencing factors of the cruise ship modeling.

2 COLLECTION AND SAMPLING OF INFORMATION ON CRUISE SHIP APPEARANCE

2.1 Information collection

By using Cruise Ship as keywords to search, 655 cruise ships which contains almost all cruise ships from 1818 to 2018 were finally sorted out after screening, supplementing and checking. Collecting information includes images from different perspectives of the cruise ship and textual information such as the name, the company, first voyage, size, historical background. (Figure 1).

2.2 Sample selection

According to the calculation formula of statistical minimum sampling quantity:

$$n \approx \frac{(Z_\alpha/2)^2 \sigma^2}{E^2} \qquad (1)$$

And n is the number of samples taken. δ^2 is the variance,which represents the deviation between the values mean of sample and whole.E is the sampling error, which can be set according to the percentage of the sampling value mean. Zα/2 expressed confidence or reliability coefficient. The larger the sample size, the higher the confidence.

As can be seen from the calculation, selecting 171 out of 655 cruise ships can reach 95% confidence degree and the error can be within 8%.

By using the sampling procedure, the 171 cruise ships (Table 1) were randomly sampled on an equal scale of every 25 years with serial numbers 001 to 171 according to the time of their first voyage.

DOI: 10.1201/9781003216582-46

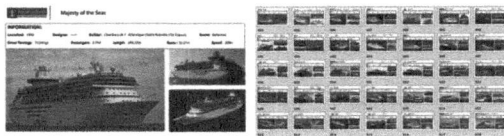

Figure 1. Cruise information sample and some thumbnails.

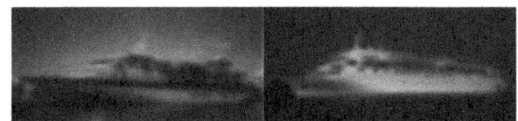

Figure 2. Sample card.

3 CRUISE SHIP APPEARANCE STYLING STYLE HISTORICAL DIVISION

3.1 *The first round of survey on holistic perceptual image of cruise ship*

To eliminate or reduce the influence of painting on the judgment of modeling style and details that interfere with overall style perception, the images should de-colored and blurred. Through three feasibility tests, the optimal size of the card was determined to be 148.5mm*70mm, and the Gaussian blur radius was 14 pixels at the resolution of 150dpi.Then make sample cards (Figure 2) and mark the number, name of ship and time of first voyage on the back.

Since the identification of aesthetic style requires professional knowledge and experience in design, 32 testers with professional backgrounds were selected in this survey. After informing them of the purpose of the experiment and the follow-up arrangements, the subjects were asked to freely classify the images of the 171 sample cruise ships according to the overall appearance style.

The first round of the survey took about 15 hours to collect 32 pieces of valid survey data, and the categories of respondents ranged from 3 to 12. After the completion of the division, the tester shall feedback the observation path, main focus points, classification reasons and description adjectives of each category of cruise ships.

3.2 *The division in three historical stages*

Convert the classification results of each tester into a gradient ribbon, Categories containing no.1 (the earliest cruise sample) are shown in the lightest color, while categories containing no.171 (the latest cruise sample) are shown in the darkest color. The other categories are colored from light to dark according to the minimum number contained.

Finally, 32 ribbons were formed (Figure 3), which more intuitively showed the correlation between

Figure 3. A ribbon diagram of the initial survey results and after merging and assigning values.

individual samples in each classification scheme and the correlation between schemes.

Although the color distribution of each ribbon is different, it can be roughly classified into several intervals from the shallow to the deep. There are some deep and light colors without crossing, and the boundary is clear. Some dark and light colors cross each other, showing the state of interpenetration, indicating that the boundaries of interval division are fuzzy.

The cross-dense colors were merged into the same color (Figure 3), and the whole showed obvious tertiary differentiation. Assign 1 to the lightest color, 2 to the middle color, 3 to the darkest color. The color bands were quantified in the unified standard by assignment, and the accurate classification results were obtained by SPSS calculation and analysis (Figure 4).

The results showed that 171 cruise ship samples were divided into three categories. No.1-9 and 12 belong to the first category. No.10, 11, 12-95, 97, 100 and 102 belong to the second category. No. 96, 98, 99, 101, 103-171 belong to the third category.

In order to ensure that there is no intersection between the first class and the third class, all intersection items are classified into the second class, categorize No.1-9 as the first class, No.10-102 as the second class, and No.103-171 as the third class finally.

At the same time, the results with concomitant probability less than 0.05 indicate that there is a significant correlation between the classification of cruise ships and the serial number.

Table 1. Sample list of cruise ships.

The time interval	1818-1842	1843-1867	1868-1892	1893-1917	1918-1942	1943-1967	1968-1992	1993-2018	Total
The total number	5	12	37	107	125	84	59	226	655
Sampling number	1	3	10	28	33	22	15	59	171

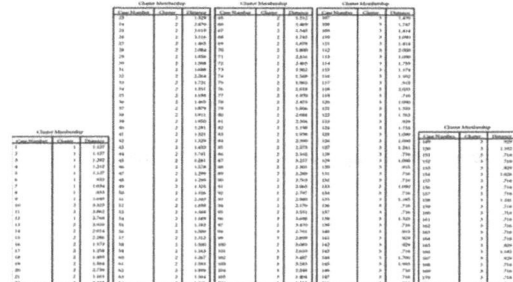

Figure 4. Cruise age division results.

Therefore, according to the chronological order, the three stages of the overall style of the cruise ship appearance can be named as "the early age", "the middle age" and "the late age" respectively. At the same time, the results with concomitant probability less than 0.05 indicate that there is a significant correlation between the classification of cruise ships and the serial number.

Therefore, according to the chronological order, the three stages of the overall style of the cruise ship appearance can be named as "the early age", "the middle age" and "the late age" respectively.

There are four main purposes for dividing the overall style of cruise ship appearance into historical stages: Firstly, reducing the number of single questionnaires and researching fatigue to improve accuracy. Secondly, some cruise ships are significantly different from others in overall feeling, so the same set of image scale diagram can be improper to measure. So corresponding image scale diagram should be made again according to their specific characteristics. Thirdly, it is the first time to verify whether the overall appearance style of the cruise ship is related to time directly. The fourth purpose is to provide objective basis for the next stage of the survey, including style representation points and image vocabulary, which can provide objective materials for the image scale map.

4 CLASSIFICATION OF THE APPEARANCE STYLE TYPES OF LARGE CRUISE SHIPS IN DIFFERENT HISTORICAL STAGES

4.1 Correlation theory of multidimensional image scale graph

The second round of research was carried out mainly by using multi-dimensional image scale graph to quantify the perceptual evaluation of the researchers on the feature cluster.

Based on the fuzzy cluster analysis method, the survey data was analyzed by SPSS software, and the style division in each era was obtained.

The multi-dimensional image scale can be divided by using the form of matrix:

$$
F = \begin{bmatrix} a_1 a_2 \ldots a_j \end{bmatrix} = \begin{bmatrix} x_{11}\, x_{12}\, x_{13} \ldots x_{14} \\ x_{21}\, x_{22}\, x_{23} \ldots x_{24} \\ \ldots\ldots\ldots\ldots\ldots \\ x_{i1}\, x_{i2}\, x_{i3} \ldots x_{ij} \end{bmatrix} \quad ((2))
$$

F is the product style description matrix with i×j, which describes the j dimension image scale divided into i discrete point quantities.

Matrix F exists:1) $X_{ij} \in \{0,1\}$; 2) $\sum_{i=1}^{m} x_{ij} = 1$. And the image cognition is quantified by the description matrix.

4.2 The second round of survey on the classification of cruise ship styles at various historical stages

According to the first round of respondents' feedback on the image vocabulary of cruise styling style and considering literature and expert opinions, more than 200 image words were collected.

Corresponding to the first round of survey, the main attention points and observation paths of the respondents on the appearance and styling of the cruise ship, the descriptive words which are suitable for cruise characteristics in each age can be selected according to the weight.

The style in "Early age" involves 6 stylistic representation points and corresponding words pairs. The style in "Middle age" involves 23 and "Late age" involves 24 (It can be showed at Figure 5). Combined with the sample images and words pairs of 171 cruise ships, three types of image scale aiming at the characteristics of different times are formed (Figure 6). The images can be divided into 9, 93 and 69, respectively. In order to facilitate the use of SPSS for data analysis, Fi= {-2, -1,0,1,2} is used here as a fifth-order psychological measurement scale. Such as the feature points: the overall sense of order (orderly – messy),"- 2" means "Most orderly", "1", means "more orderly" and "0" means "neutral", "1" means "a little bit messy", "2", means "very messy".

In the second round of research, 32 copies of research forms were issued, and 27 copies of effective research forms were recovered. Since the

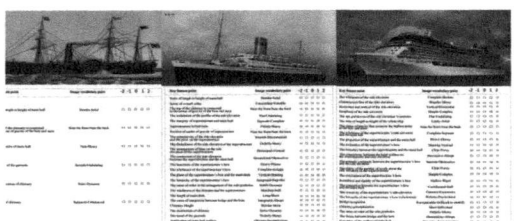

Figure 5. Sample of image scale diagram.

evaluation level of the scale does not have actual mathematical significance and only serves as an indicator. So neither the average nor the percentile is appropriate. Finally, the most objective image scale of each cruise ship sample was obtained by the highest percentage of the evaluation grade.

Due to the large number of sample data, the style subdivision in the "Middle age" was taken as an example to illustrate the division process. In the middle period, a total of 93 cruise ship samples were processed by SPSS software. The results are shown in table 2, where the analysis results of KMO and Bartlett are given, showing that the value of KMO is 0.756, which is very close to 1. It shows that this group of data is suitable for factor analysis. Bartlett's analysis results show that the associated probability is 0.00, indicating that the data is from a normally distributed population, which is suitable for further analysis.

The results also showed that only the eigenvalues of the first seven factors were greater than 1, and the cumulative contribution of the sum of the eigenvalues of the first seven factors accounted for 65.184% of the total eigenvalues. Therefore, the first 7 factors are extracted as the main factors to replace the original variables.

Through the clustering results of the system and considering the factors, it may be impossible to carry out the generic analysis due to over-classification, it can be roughly divided into 2 to 5 categories. The clustering results are visualized as a tree (Figure 6) according to the data obtained. After comprehensive consideration, it is concluded that the most reasonable way is to divide it into three styles.

According to the style representation, the three styles of the "Middle age" are named as "industrial style", "decorative style" and "tower style". And get the scatter diagram of style classification (Figure 7) to provide reference for the following stylistic venation chart.

Using the same method, the styles of "early times" can be divided into 2 kinds and named as "paddle wheel sailing style" and "machete style". The style

Table 2. Analysis result of KMO & Bartlett test.

KMO and Bartlett's Test (KMO 和 Bartlett 的检验)		
Kaiser-Meyer-Olkin Measure of Sampling Adequacy. (取样足够度的 Kaiser-Meyer-Olkin 度量。)		.756
Bartlett's Test of Sphericity (Bartlett 的球形度检验)	Approx. Chi-Square (近似卡方)	725.244
	df	253
	Sig.	.000

Total Variance Explained (解释的总方差)									
Component (成份)	Initial Eigenvalues (初始特征值)			Extraction Sums of Squared Loadings (提取平方和载入)			Rotation Sums of Squared Loadings (旋转平方和载入)		
	Total (合计)	% of Variance (方差的 %)	Cumulative % (累积 %)	Total (合计)	% of Variance (方差)	Cumulative % (累积)	Total (合计)	% of Variance (方差)	Cumulative % (累积)
1	5.480	23.886	23.886	5.480	23.886	23.886	3.315	14.415	14.415
2	2.674	11.627	35.485	2.674	11.627	35.485	3.134	13.663	27.998
3	1.680	7.218	42.712	1.660	7.218	42.712	2.338	10.164	38.162
4	1.486	6.343	48.956	1.456	6.343	48.955	1.690	7.345	45.304
5	1.360	5.949	54.905	1.360	5.949	54.905	1.584	6.758	52.261
6	1.329	5.559	60.242	1.229	5.339	60.242	1.547	6.725	58.989
7	1.137	4.942	65.184	1.137	4.942	65.184	1.655	6.195	65.184
8	.980	4.174	69.358						
9	.847	3.682	73.040						
10	.838	3.602	76.642						
11	.769	3.358	79.000						
12	.623	2.711	82.911						
13	.580	2.595	85.106						
14	.505	2.194	87.300						
15	.465	2.022	89.392						
16	.437	1.869	91.281						
17	.378	1.644	92.925						
18	.352	1.530	94.457						
19	.297	1.293	95.750						
20	.283	1.231	96.981						
21	.273	1.189	98.170						
22	.200	1.059	99.230						
23	.182	.790	100.000						

提取方法：主成份分析。(提取方法：主成分分析。) Extraction Method: Principal Component Analysis.

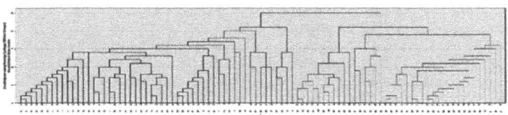

Figure 6. "Middle age" cruise style cluster tree.

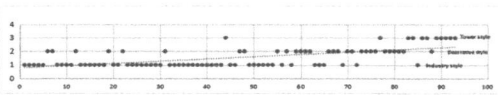

Figure 7. Scatterplot of the second historical stage.

of "the late age" can be divided into 3 kinds and named as "melt style", "mecha style" and "enclosing style".

5 EXPERIMENTAL RESULTS AND FEATURE ANALYSIS

5.1 Context of the appearance style of large cruise ships

Through the above fuzzy clustering analysis, the development of the appearance modeling style of large cruise ships is clearly divided into three ages and eight styles .The style of "Early age" included the "paddle wheel sailing style" (late 1810s-late 1850s) and the "machete style" (1860s-1880s). The two styles appeared in turn without long periods of transition or blending.

"Middle age" includes three styles: "industrial style", "decorative style" and "tower style". "Industrial style" began to became the mainstream in the 1880s. Followed by "decorative style" which emerged in the 1930s while "industrial style" gradually retreated. In the 1960s, "decorative style" retreated and "tower style" prevailed.

"Late age" is an era in which styles coexist. The styles which contain "melt style","mecha style" and "enclosing style" blend with each other. Among them,"melt style" and"enclosing style" play a leading role. The two have lived side by side throughout the ages. However, the "mecha style" appeared in the middle of the period and gradually subsided in the later period.

5.2 Analysis of the appearance style characteristics of large cruise ships

5.2.1 Early age

This time cruise ship modelling is full of historical sense obviously. The whole is emaciated and lightsome. The main characteristics in this era are:

➢ The model is composed of the main hull and the mast sail. Visually separated by gunwale into upper and lower parts. The contrast between reality and fiction is strong.

➢ The ship is slender and straight, with simple and compact structure.

➢ The sailing area is vast, the number is large, the layout is complicated, and it gradually becomes dense. But the outline is a stable "mountain" shape, which is interspersed with 3~6 long vertical mast and many horizontal short girder, oblique cable. Besides, the lines are complicated. The number of chimneys is small (mostly 1-2), the shape is slender, the visual effect is weak.

In addition to the above common features, the different styles of cruise ships in the early times are mainly reflected in the modeling of the main hull and superstructure and overall posture.

(1) Paddle wheel sailing style
  • Main hull: The profile of the side elevation appears as a long and narrow rectangle or inverted trapezoid; The paddle wheel with two outboard sides slightly to the rear is the unique external feature of this style. The shape is round and protruding from the side, forming the visual center of the main hull.
  • Superstructure: Basically no superstructure; Most of them are equipped with fore-mast sails, so the shape is more complex; The chimney is located in the middle of the deck near the front, basically upright.
  • Overall posture: Visual stability and equilibrium . The sense of dynamic is not strong.

(2) Machete style
  • Main hull: Shaped like a machete. Most of the bows appear as the axe head, occasionally flying shear. There is no paddle wheel on the side. The lifeboats were arranged in horizontal order, The horizontal direction line is more fluent and rhythmic.
  • Superstructure: The superstructure appeared gradually, but the height proportion was very small and the recognition was low. The chimney moves back and leans back a little.
  • Overall posture: More powerful and speedy.

5.2.2 *Middle age*
The cruise ship modeling in this era is full of more industrial sense. The whole is strong and rough.The main characteristics in this era are:

➢ The model is composed of the main hull, superstructure and chimney. It can be divided into three or two long, narrow sections in horizontal direction visually. The volume is full of a strong sense of piling up.

➢ The shape of the main hull is simple, with a narrow and inverted trapezoid side elevation. Lateral dotted portholes are mostly distributed on the side. The sails were gradually reduced and removed, and the unrigged masts towered over the fore and aft decks.

➢ The superstructure is located in the middle and rear position above the main hull (so the bow deck is long and the stern deck is short). Most of them have 4-5 decks with larger volume. Show "convex" shape with solid thick and sharp edges. Pipes and machinery are exposed. Upper side windows are clear and orderly. The lifeboats which are located at the back of the top floor of the upper building and exposed on side aisles form horizontal parallel line and correspond with window line in the head of superstructure.

➢ The chimney is inserted into the top of the superstructure, making the shape more robust and strong and becomes the most recognizable part.

In addition to the above common features, the different styles of cruise ships in the middle age are mainly reflected in the modeling of the main hull, the superstructure and the chimney, the combination of the superstructure and the main hull, and the overall style.

(1) Industrial style
  • Main hull: The side elevation is close to the machete style. Small openings, few in number, and loose arrangement of portholes appeared on both sides. The porthole is more conspicuous on the light-colored painted ship body, which enriches the details of the ship.
  • Superstructure: The overall height of the superstructure is equivalent to the height of the main hull above the waterline. Its modelling is basic element with square form block, resemble building blocks to pile up. The contour line is more zigzag, and the tail shows a ladder shape. The side elevation is divided into pieces and turns into sharp corners.
  • Chimney: The number of chimneys has increased (1~4), most of which are cylindrical of equal size. The chimneys are regularly arranged and lean back slightly. Due to the specificity of height and direction, it becomes the visual center of the whole ship.
  • Combination of the superstructure and the main hull: The width of superstructure is roughly as same as the main hull, but visually distinct and rigidly jointed.
  • Overall style: Integral modelling style is hale and disorderly. The sense of industry is strongest.

(2) Decorative style
  • Main hull: The head column gradually leans forward, and the profile of the side elevation changes from rectangle to inverted trapezoid, so the rising trend of the bow is gradually strong. The main hull is reinforced with decorative ribbons that surround the hull. The number of portholes which were

arranged more closely and develops a line in the horizontal direction increased.

- Superstructure: The volume of the superstructure has increased. Its modelling still is given priority to with square form piece which is more simplified. It is erected vertically in the front elevation, but bent in the horizontal direction and slightly raised in the forward direction. The intersection of the front and side elevation is transited with rounded corners. The upper part of the bow seems to suspend over the bridge extending beyond the two facades.
- Chimney: The number of chimneys is reduced (1~3 mainly). They are mostly longitudinal flat cylindrical, and more thick and short. It starts to be decorative. For example, the top of the diagonal line is high in front and low in back, and the outer wall has a circular decorative ribbon.
- Combination of the superstructure and the main hull: The width of superstructure is as same as the main hull. The two sides of the hull are smoothly joined or slightly creased with the side elevation of the superstructure.
- Overall style: Style expression is whole square with local fluidity. Decorative details have increased.

(3) Tower style
- Main hull: The front column is further forward and slightly reversed, which is more impact-like with the rising bow
- Superstructure: The volume of the superstructure further increases. It is still dominated by square blocks, and the sense of accumulation is enhanced. And the accumulation of the head and tail is inclined like a ladder, making the superstructure become as a whole a flat platform. The front face is more rounded in the forward arc, which can be smoothly transferred to the side elevation, or rounded with the side elevation. Front elevation portholes are scattered, but side elevation portholes are dense and orderly. The response between the two is weak.
- Chimney: The number further reduces (mostly 1, minority 2). The shape gradually changes into a flat round platform, which is consistent with the superstructure. And part of the top of the cruise ship appeared wing-like decoration.
- Combination of the superstructure and the main hull: The superstructure is more closely and naturally connected to the main hull. The middle of the bulwarks naturally rise vertically to the side elevation of the superstructure.

- Overall style: The overall style is sharper and full of a strong sense of forward speed.

5.2.3 *Late age*

Because of the appearance of a variety of decorative modeling techniques, this era of yacht modeling more artistic. The overall shape is thick and full. Use more surface, curve to make it more round and smooth. The main characteristics of this era are:

➢ The model is composed of the main hull and superstructure. Except for the sharp angle formed by the front elevation of bow and superstructure, the other parts are closely integrated into one. The visual boundary between them is blurred.

➢ The main hull inherits the characteristics from the late middle period, with the stem column remaining tilted. Part of the curved surface between the bow and the side is clear in the turning point, forming a forward angular line consistent with the posture of the stem. Most ships are straight stern.

➢ The proportion of the volume of the superstructure further increases (the bow deck became shorter and there was basically no aft deck). The lateral elevation is a flat trapezoid. The shape is more overall and coordinated, and the details are more abundant and exquisite. The visual texture formed by the horizontal close connection of the side facade balconies and windows has become the main visual feature of the whole ship. The position of the lifeboat gradually moved from the top of the superstructure to the main deck line in a horizontal order, separating the ship from the superstructure. Superstructure top modelling is relatively rich and changeable, because head and tail are much taller and show "cave in waist" form. The height of the bow mast has been shortened and the visual effect has been greatly reduced. Most of the cruise ship tail has a chimney, and the proportion of the volume reduced. The specific enhancement of chimney shape (mostly related to corporate brand image) is still one of the main visual features of the whole ship.

In addition to the above common features, the different styles of cruise ships in the later age are mainly reflected in the superstructure and the overall style.

(1) Enclosing style
- Front of superstructure: The surface is flat and backward, usually steep. The slope is flat or slightly forward arcs. A small square or round window is scattered on a large surface and is not connected with a side window. The bridge window is located at the top of the inclined plane, and three sides of the bridge are pointed out. The front portion wraps over the side portion.

- Side of superstructure: The side is a flat trapezoid, and the contour line is basically straight. The changes of pattern contour and texture which are developed from windows, balconies, verandas and lifeboats on the outer walls are various. The order sense in horizontal direction is formed by the side windows and balconies, disturbed by the verandas and lifeboats frequently.
- Back of superstructure: The tail of the superstructure is usually slightly forward or vertical. More with the inclined or straight stern smooth connection. Sometimes it has big corners.
- Overall style: The lines and faces are straight, the shape is neat and smooth, and the whole surface (side, front of superstructure) is covered with fine texture (side of superstructure).

(2) Mecha style
- Front of superstructure: The front of the superstructure is in a stepped shape with a backward tendency and is generally relatively flat. Most have several rows of horizontal windows extending to the side window. The projecting bridge piled up the steps along the trend. The front turns to the side with sharp edges.
- Side of superstructure: The contour line is flat with variation. The patterns of the side walls composed of balconies, verandas and lifeboats are regular and neat (mostly consistent with the profile of the side elevation). The texture is uniform and compact, with a strong sense of horizontal order.
- Back of superstructure: The rear part of the superstructure is vertical or slightly inclined. To form a vertical or large angle with the stern.
- Overall style: The overall lines are sharp and angular. Modeling layers are rich and orderly, with a sense of mecha.

(3) Melt style
- Front of superstructure: The Angle of the front is various, the shape is rich and varied. The points, lines, planes and bodies are interwoven, so that the bridge is not prominent.
- Side of superstructure: The outline is very zigzag. There is irregularity in the regular pattern of the side walls. The elevation is also uneven, through more convex, concave, hollow out to create the vertical direction of the strewn at random changes.
- Back of superstructure: More vertical, with straight stern into a straight line. There is also a forward dip, forming a large angle with a straight or slanted stern.
- Overall style: The overall lines are smooth and the modeling elements are rich. Modeling methods are various, with mixed feeling.

6 ANALYSIS OF INFLUENCING FACTORS AND TRENDS OF APPEARANCE STYLE OF LARGE CRUISE SHIPS

According to the results of previous experiments, the division of the eight styles of the appearance of large cruise ships has a strong stamp of The Times.

Through the analysis of its characteristics, it can be seen that the main factors affecting the style of cruise ship are the volume, the proportion and shape between different parts the shape situation, and the modeling techniques .The meaning of the era is not limited to time, but also implies the background of the era, functional needs, scientific and technological level, social factors and other contents. (Figure 16).

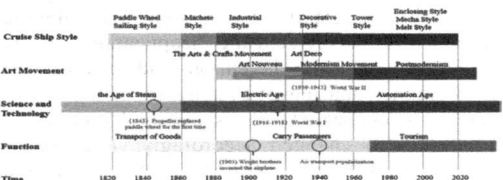

Figure 8. Context of cruise ship style.

6.1 *Functional requirements*

After nearly two hundred years of development, the function of the cruise ship changes strongly, which has a huge impact on the appearance and shape. In the early 19th century, passenger liners were specialized for carrying mail, goods, or settlers. Early cruise ships used the lower part of the main deck as a storage site for goods, and there was no excessive demand for light. Therefore, in the appearance of the cruise ship, the main hull accounted for a large proportion and the performance is relatively stable.

Later, it evolved into a means of transportation across oceans and continents. Especially after the 1880s, the immigration boom from Europe to the United States and Canada led to a huge increase in passenger demand. And the long journey, which can be as short as a week, or as long as several months, not only increases the space for meals and accommodation, but also the leisure places such as bars gradually appear on the cruise ship. Comprehensive market and passenger demand, cruise ship volume is increasing, mainly in the superstructure volume is expanding.

With the improvement of air transport, cruise ship has lost its original shipping advantage and transformed into a mobile leisure tourism site. In order to meet the needs of leisure and entertainment, the volume of the cruise ship, especially the superstructure, is increased. At the same time, there are more windows in the appearance of the cruise ship, more facilities on the top deck and more changes in the skyline.

Much of the change in the function of the cruise ship is due to drastic and long-term social changes. At

411

present, the cruise market is steadily growing. In terms of the types of ships and tourism, there are no similar competitive products in the market in a short period of time. Therefore, it is expected that in the future, the main development direction of cruise will be to improve the leisure and entertainment experience.

6.2 *Science and technology level*

Cruise ships' appearance was greatly affected by the level of technology at the time. The invention of the steam engine gradually replaced the original natural power such as manpower and wind power. It directly led to the emergence of paddle wheels and chimneys, while the sails and masts remained as auxiliary. Then the huge advantage of the propeller as a transmission device made it directly replace the paddle wheel, and the invention of the double propeller also accelerated the disappearance of the sail. Afterwards, the invention and maturity of diesel engines and power systems made the power plant more concealed. The shape and number of chimneys also changed accordingly. The use of advanced navigation systems such as radar has led to the transformation of the mast's function, which in turn has affected the shape of the mast from tower to tower, and the number has been reduced. Technology fundamentally affects the types, shapes, and arrangement of appearance components, and is a decisive factor for related shapes. It can be seen that the innovation of the exterior design of cruise ships not only comes from aesthetics, but also requires continuous innovation and improvement with the help of technology.

Mastering the development trend of technology can help designers to subvert the totality of the appearance of cruise ships. It is bold to assume that if new energy sources such as solar energy, batteries or fuel cells are subsequently used, it may not be necessary to install a chimney. If there is a more compact communication device, the mast can also be changed. At the same time, combined with the development demands of its functions, emerging technologies will be used in the updating of cruise power, entertainment, communications and other content. And the current technology has entered the intelligent stage, and corresponding feedback should be provided for the newly added functions in the design.

6.3 *Other social factors*

Other social factors refer to a combination of trends, market aesthetics, and competitors under current social conditions. Style is essentially a reflection of social conditions at the time.

During World War II many cruise ships were transformed and used for war. Although the design and construction of cruise ships suffered considerable damage in wartime, the changes in the shape of cruise ships have not stopped. From industrial style to decorative style, there are many people's longing for a better life.

The mainstream design style is also an important factor that affects the appearance and shape. The change in cruise style is closely integrated with the development of mainstream design trends.

Designer William James designed the "RMS Ocean" by drawing on the design principles of hotels on land, thus opening up a new era of cruise design (Geoff 2018). In the end of 19 century, The designers were mainly influenced by the Arts & Crafts movement and Art Nouveau style, especially on interior. It emphasized the overall harmony and mostly used linear patterns as design elements. In the middle age, the characteristics of its design style range from the chaotic industrial style to the simple and orderly tower style. It is the result of exploring the combination of technology and art through the arts & crafts movement, decorative art movement and other art movements. The designers of "the late age" were mainly influenced by modernism and postmodernism. The shape is simple and atmospheric. Encourage the integration of multiple design elements, combined with the presentation of modern high-tech.

The current integration of design styles is diverse and there is no phenomenon like one-size-fits-all. Therefore, the choice of style is also more free, which also gives the designer a richer design inspiration.

7 CONCLUSIONS

Firstly, this study verifies the possibility of applying fuzzy clustering methods to aesthetic research. Appropriate and innovative use of fuzzy clustering methods, quantify the cognition of the appearance design style of large cruise ships in perceptual way. The above can make a rational analysis of art problems, so as to obtain a relatively objective style division result, and can provide a path for subsequent similar research.

Secondly, by using the strategy of step-by-step research through image blurring and the method of digitizing parameter images for perceptual evaluation, the appearance and style of large cruise ships is clearly divided into eight major styles. It also shows the development of the appearance style of large cruise ships in various eras and types. And make analogous descriptions of the characteristics of different eras and different styles to clarify the modeling differences between the times and styles. This strategy can also provide guidance and reference for the appearance design of modern large cruise ships. At the same time, combined with the background of time, explore the factors that affect the appearance of the style such as technology, function, social factors. It also speculates on the development trend of the appearance of cruise ships, and can help the forward-looking design of cruise ships.

REFERENCES

Lunn, G. 2018. 200 years of world luxury cruise. Shanghai. Shanghai Jiao Tong University Press.

Li,L.X. 2009. Research methods of art design. Jiangsu: Jiangsu Fine Art Press, pp161.

Li,Y.Z .2003. New design concept: Kansei Engineering. New art magazine. pp20–25.

Peters, A. 2013. Ship Decoration 1630-1780. Barnsley: Seaforth Publishing

Plowman, P. 2007. Australian Cruise Ships. New South Wales(Australia): Rosenberg Publishing.

Zadeh, L.A. 1965. Fuzzy sets. Information and Control, pp 338–353.

Zhang, J& Zhao, J. H. 2002. Image scale method and cognitive research. Art & Design. pp 21.

Zhou,M.Y.& Gu, M. 2010. esearch on product perceptual design model based on key modeling features.Packaging engineering.pp 1-3+24。

Zhu, S.S., Luo,S.J. &Zhao, J. H. 2000. Research on the Scale of NC Machine Modeling Image Based on Ergonomics. Journal of computer-aided design and graphics, pp 873–875.

Developments in Maritime Technology and Engineering – Guedes Soares & Santos (eds)
© 2021 Copyright the Author(s), ISBN 978-0-367-77376-2

Comparative life cycle assessment of battery- and diesel engine-powered river cruise ship

M. Perčić, I. Ančić & N. Vladimir
Faculty of Mechanical Engineering and Naval Architecture, University of Zagreb, Zagreb, Croatia

A. Fan & Y. He
School of Energy and Power Engineering, Wuhan University of Technology, Wuhan, China

ABSTRACT: Exhaust gases from marine engines are one of the major causes of marine environmental pollution. Although the researches into ship emissions are more focused on ocean-going vessels, emissions from short-sea and inland navigation also contribute to the global amount of pollutant emissions produced by combustion of fuel oil. While this share is small, short sea and inland ships usually operate in highly populated areas and consequently affect both human health and environment. In order to evaluate the environmental impact of river cruise ship that occasionally operates in Croatian inland waterways, its life cycle assessment (LCA) has been performed. Two different power system designs were investigated, i.e. lithium-ion battery-powered ship and diesel engine-powered ship. The analyses were performed by means of general LCA software GREET 2018, The analysis showed that diesel engine-powered ship emits 46.63 kg CO_2-eq/nm, versus battery-powered ship with 20.39 kg CO_2-eq/nm.

1 INTRODUCTION

Anthropogenic greenhouse gases (GHGs) are causing the greenhouse effect, and therefore the global warming. These GHGs refer to emissions of carbon dioxide (CO_2), methane (CH_4), nitrous oxide (N_2O) and fluorinated gases (UNFCCC 2001). This air pollution nowadays represents probably one of the most important environmental problems that needs to be resolved. Along with industry and road traffic, shipping sector contributes to this problem. Exhaust gases released from combustion of fuel in marine engines are considered to be one of the major causes of marine environmental pollution. The most pernicious emissions released from the engine are CO_2, carbon monoxide (CO), nitrogen oxides (NO_X), sulphur oxides (SO_X) and particulate matter (PM) (IMO 2014). The presence of these gases has negative effect both on the environment and on the human health causing respiratory diseases.

It is fair to say that research into emissions from shipping and the wider impact on air quality as well as climate changes has been mainly directed to ocean-going vessels, and less on the inland ships and their emissions. The reason for this is mainly the general opinion that these emissions have a small contribution to total transport emissions. However, it is important to mention that inland waterway transportation is regularly realized within highly populated areas, and therefore its effect should not be ignored (Keuken et al. 2014). The inland waterway transport is, together with road and rail transport, one of the main three land transport modes. Goods are transported by ships via inland waterways, such as canals, rivers and lakes, between inland ports and wharfs (ECA 2015). Beside transportations of passengers and cargo, inland waterways are nowadays highly used for tourism purposes (Wiegmans et al. 2015).

Quantification of CO_2 emissions can be achieved by Carbon Footprint (CF) calculation. The CF term represents a measure of the total amount of CO_2 emissions that is directly and indirectly caused by an activity or is accumulated over the life stages of a product (Wiedmann & Minx 2008). CF calculation serves as a tool to assess the negative impact of the CO_2 emission and it can be expressed in tons of CO_2 or in tons of CO_2 equivalent (CO_2-eq).

With the aim to increase the energy efficiency of ships, conventional power systems (diesel-mechanical propulsion) can be replaced by alternative hybrid power systems (HPS) or integrated power system (IPS), that result in reduced pollutant emissions. The HPS are characterized by the use of different types of power sources, while the main characteristic of IPS is the centralized electric power generation and the application of electric propulsion. For example, Ančić et al. (2018a) proved that ro-ro passenger ships with IPS or HPS are more energy efficient compared to the fleet average which is

DOI: 10.1201/9781003216582-47

using mechanical propulsion. Kalikatzarakis et al. (2018) analyzed a tugboat powered by a hybrid propulsion plant with power supply that can be recharged with renewable shore power. That hybrid configuration has the additional challenge to determine the optimal power-split between three or more different power sources in real-time, and to optimally deplete the battery packs over the mission profile. Motivated by the extensive exploitation of electric power in ships, Kanellos et al. (2017) proposed an optimal power management method for ship electric power systems comprising integrated full electric propulsion, energy storage and shore power supply facility. Gagatsi et al. (2016) have presented a fully electrified ferry (E-ferry concept) as a new paradigm in short-sea shipping. So far, typical electric ro-ro passenger ship could use batteries as the main power source on short trips and they could be charged whilst connected to the shore power. As batteries continue to develop, the electric propulsion would replace conventional one on longer distance trips. Battery-powered ferries seem to be the most environmentally friendly, but there are limitations that are connected to high speed of a ship, long distance trips, increased time in ports due to charging the battery and capacity limitations of the electricity grid (Kullmann 2016). However, electrification of a ship results in releasing zero emissions during the operation. In order to assess the environmental impact of that ship, the emissions released from processes of electricity generation and battery manufacturing need to be considered. That can be achieved by performing Life cycle assessment (LCA). LCA provides quantification of emissions through life cycle of a specific product. This technique is evaluating the environmental impact of a product from its production, through its use and up to eventual reuse, recycling or disposal (Ling-Chin et al. 2016). The results of LCA can be presented in amount of different emissions, which are released from processes during its life cycle. This assessment also represents useful tool for comparison of different power system configurations. Such kind of research was performed by Jwa & Lim (2018). Comparative LCAs of lithium-ion (Li-ion) battery electric bus and diesel bus were completed, from extraction of fuel and generation of energy to vehicle operation. Results showed that vehicle powered by diesel engine has higher emissions than the one powered by Li-ion battery.

The aim of this paper is to perform comparative LCA of battery- and diesel engine-powered river cruise ship operating in Croatian inland waterways. The paper is structured into six sections. In the next section the methodology of LCA of the river cruise ship is presented. The third section is dedicated to LCA of diesel engine-powered ship and LCA of battery-powered ship in fourth section. The fifth section contains the results of performed LCA comparison of different power system designs and discussion. Finally, concluding remarks are drawn in sixth section.

2 METHODOLOGY

2.1 LCA

According to International Organization for Standardization (ISO 1997), LCA is a technique for assessing environmental impact of a product throughout its life cycle (i.e. cradle-to-grave) which includes:

- Raw material,
- Production or manufacturing,
- Use of product,
- End of life treatment,
- Recycling and final disposal.

In this paper, two LCAs are performed by means of GREET 2018 software. Processes of raw material recovery, power source production and its supply to the vessel are referred as "Well to Pump" (WTP), while WTP processes and use of power source in vessel operations as "Well to Wheel" (WTW), Figure 1. Even though the classical LCA includes disposal process of a product as a final phase of life cycle, in this paper the LCA is performed from the WTW point of view and disposal is not included into the assessment.

In this paper, the system boundary of the LCA is defined by GREET 2018, which takes into account emissions related to the main product and other products that are related to it somehow, but do not considers, for example, emissions released by building the infrastructure or vehicle manufacturing. For the river cruise ship powered by diesel engines, the LCA begins with an extraction of crude oil. After that, the crude oil is transported to refinery where it is processed into the diesel fuel. Diesel is then transported by tank trucks to the gas stations, and ultimately ends up in the vessel. Electricity generation followed by electricity transmission, distribution, battery manufacturing and ship operation constitute the life cycle of electricity as power source for battery-powered ship. The comparison of those two different power systems configurations is based on the results that represent total emissions of harmful gases throughout the entire power system configuration life cycles, Figure 2. During their life cycles, other pollutants are also released, such as NOx, particulate matter (PM), hydrocarbons (HC) and CO, but this paper follows ongoing research trends in the field of CF and is focused on the anthropogenic GHG emissions only. In order to evaluate the contribution to greenhouse effect from each of GHGs, the global warming potential (GWP) term has been developed. It represents a measure of how much energy the emission

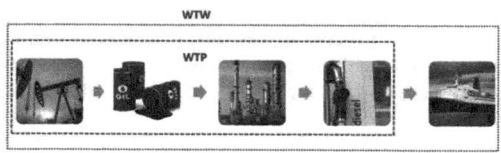

Figure 1. WTW and WTP display of diesel-powered ship.

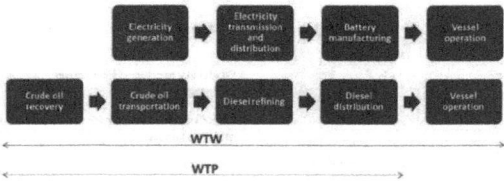

Figure 2. Life cycles of power system configurations.

of one ton of a gas will absorb over a given period, relative to emission of one ton of CO_2. The time range usually used is 100 years and typically, GHGs are expressed in CO_2-eq (EPA 2019). GHG emissions are converted into CO_2-eq by multiplying the GHG emissions with their GWP values, as prescribed in (EPA 2018). Therefore, the LCA results are expressed in unit of CO_2-eq.

Total amount of GHG emissions from battery- and diesel engine-powered river cruise ship are calculated by using default data from GREET 2018 as well as by adapting the software with the data typical for some processes in Croatia.

2.2 Ship particulars

The Croatian inland waterway network consists of natural streams of the Danube River in length of 137.5 km, Sava River 446 km, Drava River 198.6 km and Kupa River 5 km, Figure 3, (MSTIRC 2019).

The considered ship for the comparison of two different power system designs from a carbon footprint viewpoint is a diesel engine-powered river cruise ship, with main particulars given below:

Length overall: 109.9 m
Breadth: 11.4 m
Design speed: 12 kn

The vessel named MS Prinzessin Sisi, Figure 4, is equipped with two Caterpillar main engines with 783 kW each. More data on the vessel can be found in (CM 2019) and (Ship particulars 2019). The vessel can transport 156 passengers and it was built in 2000 and refurbished in 2015.

Even though MS Prinzessin Sisi does not formally belong to Croatian inland fleet, it navigates on

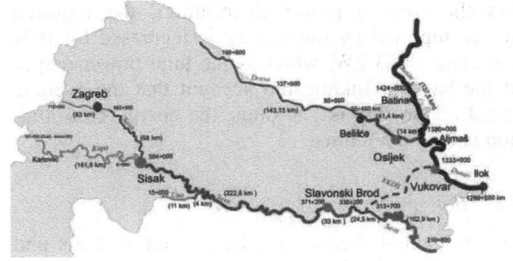

Figure 3. Inland waterway network of Croatia (MSTIRC 2019).

Figure 4. Analyzed river cruise ship in operation (CM 2019).

Danube River through Croatian area, on its way to the Black Sea. The design speed is 12 knots, but the average speed that ship achieves is 7.8 knots (MT 2019). Since the ship power is roughly proportional to the cube of its speed, average ship power on that route was calculated according to the following expression:

$$P_{average} = P_1 \cdot \left(\frac{v_{average}}{v_1} \right)^3 \qquad (1)$$

The calculated average power is 430 kW. Taking into account the average speed, the energy consumption is estimated at 55.1 kWh/nm. The fuel consumption of the ship has been calculated by multiplying energy consumption with specific fuel consumption (SFC). SFC is determined depending on the engine speed, as proposed by Ančić et al. (2018b), i.e. it is assumed that for high speed engines with engine load of 25%, the SFC yields 240 g/kWh, which is used in this assessment. The fuel consumption of this ship is then calculated and equals 13.2 kg/nm.

3 LCA OF DIESEL ENGINE-POWERED SHIP

3.1 Crude oil recovery

Production of domestic crude oil in Croatia is performed on exploitation fields in the continental part of the country. In addition to domestic production, Croatia also imports crude oil primarily from Azerbaijan, Iraq and Kazakhstan (CERA 2016). Due to the lack of data specific for Croatia on process of crude oil recovery, for this assessment, inputs, outputs and process parameters have been used from GREET 2018 database (process Conventional Crude Recovery).

3.2 Transportation of crude oil

It is assumed that the crude oil has been transported from Middle East via tankers and pipelines to Croatia. After tankers deliver crude oil to the offshore terminal in Omišalj on the island of Krk, it is then further transported through the oil pipeline system up to oil refineries in Rijeka and Sisak. For this

Table 1. Tailpipe emissions from inland passenger ship.

Emission	Emission factor	Tailpipe emission
	g emission/kg diesel	
CO_2	3206	43.32 kg CO_2/nm
CH_4	0.019	0.25 g CH_4/nm
N_2O	0.142	1.87 g N_2O/nm

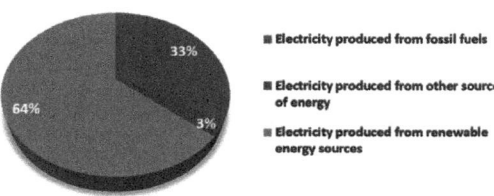

Figure 5. Shares of individual energy sources in total produced electricity in Croatia.

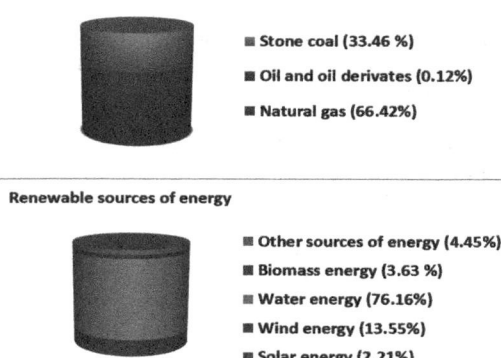

Figure 6. Energy sources for electricity generation in Croatia divided on fossil fuels and renewable energy sources.

assessment, due to reason of simplicity, it is assumed that diesel is produced only in refinery Rijeka. Length of oil pipeline from offshore terminal to this refinery equals 7 km (CERA 2016).

3.3 Production of diesel

After the transportation, crude oil is refined in the refinery in order to produce diesel fuel. It is assumed that diesel fuel used by MS Prinzessin Sisi corresponds to Conventional Diesel from GREET 2018 database. Therefore, for the process of diesel refining, the parameters are obtained from GREET 2018 default process of refining conventional diesel.

3.4 Transportation of diesel

After diesel is produced, it is mostly distributed by tank trucks to the gas stations. Mode parameters are obtained from default GREET 2018 mode for heavy-duty truck. Tank trucks transport diesel 450 km to the gas station.

3.5 Vessel operation

Previously determined ship energy need is 55.1 kWh/nm, while consumption of diesel is 13.2 kg/nm. Tailpipe emissions from diesel combustion in internal combustion engine have been calculated by multiplying ship fuel consumption by emission factors, as prescribed in (EPA 2019), Table 1.

4 LCA OF BATTERY-POWERED SHIP

4.1 Electricity generation, transmission and distribution

Electricity generation is the process of generating electric power from sources of primary energy. The main types of energy sources are shown in the Figure 5 with the exception of nuclear energy which production does not exist on the territory of Croatia (HEP 2019).

A more detailed breakdown of individual energy sources is provided in the Figure 6 (HEP 2019).

The electricity generation data are obtained from GREET 2018 database (Non distributed U.S. Mix). Processes of electricity generation by water, wind and solar energy are assumed to be emissions-free since these processes do not require other products

for the generation process and the emissions released by building up the generation facilities are not included into the system boundary. Shares of total electricity generation were adapted to the case study of Croatia. After its generation, electricity has been transmitted and distributed to consumers.

4.2 Vessel operation

Battery-powered ship is supplied with a power by the on-board battery only. It is assumed that the ship has two propellers powered by two electric motors and that the propulsion power system needs remain unchanged. Due to the losses in the electric motor and the electric power distribution, the required power supplied by the battery is increased by 10% and equals 473 kW, which is the total power output of the battery. Taking into account that the average speed of the ship is 7.8 knots, the energy consumption is 60.6 kWh/nm.

4.3 Battery

Environmental regulations, battery innovations and increase in fuel prices open the path to electrification of passenger ships in Europe. Leader in this area is Norway, with introduction of the first fully electric ferry using Li-ion batteries in 2014 (Gagatsi et al. 2016). Even though Li-ion batteries are quite expensive, they have by far the highest energy density compared to other types of batteries. Lead acid batteries appear to be more economical solution. However, the low material resistance in the marine environment and the short life period makes them more expensive in the life cycle of a ship (Dedes et al. 2012).

The considered ship, MS Prinzessin Sisi, navigates through Croatia on Danube river waterway that is 74.2 nautical miles long. The battery is charging on the state border of Croatia, i.e. the ship sails around 9.5h without recharging the battery. The minimum required capacity is around 4500 kWh. Due to safety component, this value is increased by one third of minimum required capacity and equals 6000 kWh. Typical power density of Li-ion battery is around 0.254 kWh per kg. Knowing this data, the weight of battery was easily calculated, and it is around 23.6 tons. The emissions from the process of Li-ion battery manufacturing are obtained from GREET 2018.

5 RESULTS AND DISCUSSION

In order to evaluate the environmental impact of two different ship power systems configurations for the same river cruise ship, LCAs are performed in which GHG emissions have been expressed in CO_2-eq per nautical mile.

The existing diesel engine-powered river cruise ship through its life cycle emits 46.63 kg CO_2-eq/nm. The main share in total GHG emissions has the ship operation with 42.88 kg CO_2-eq/nm, while the WTP GHG emissions are 3.75 kg CO_2-eq/nm, as presented in the Figure 7. WTP GHG emissions from diesel fuel are presented in Figure 8, where the process of diesel refining contributes the most to the release of GHG emissions.

Option of electrification of the existing river cruise ship that navigates through Croatian inland waterways has been explored by taking into account results from LCA of battery-powered ship. During the operation, the battery-powered ship has zero emission but during the production of battery, different emissions are released and considered for the total amount of GHG emissions during WTW

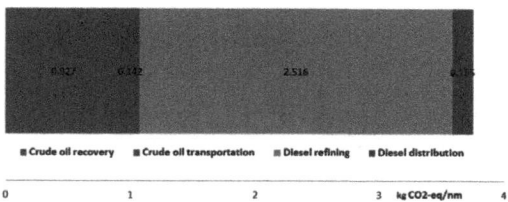

Figure 8. WTP GHG emissions from diesel fuel.

assessment. Results in the Figure 9 represent the WTP GHG emissions from electricity life cycle.

Processes that contribute the most to the GHG emissions are electricity generation from natural gas and coal.

The amount of WTP GHG emissions from electricity is 13.12 kg CO_2-eq/nm. During Li-ion battery manufacturing, certain emissions are released and they equal to 7.27 kg CO_2-eq/nm. WTW GHG emissions from battery-powered river cruise ship are presented in Figure 10, and they contain the emissions from WTP life cycle of electricity and emissions from battery manufacturing. Total WTW GHG emission of battery-powered ship amount 20.39 kg CO_2-eq/nm.

The comparison of LCA results of battery and diesel engine-powered river cruise ship is presented in the Figure 11. As can be seen, the diesel engine-powered ship release significantly higher amount of GHG emissions through its life cycle then the battery-powered option.

During its operation, diesel engine-powered river cruise ship emits 42.88 kg CO_2-eq/nm, while emissions from life cycle of diesel fuel, without its use in ship, amounts to 3.75 kg CO_2-eq/nm. Considering that battery-powered river cruise ship during its whole life cycle emits 20.39 kg CO_2-eq/nm, it can

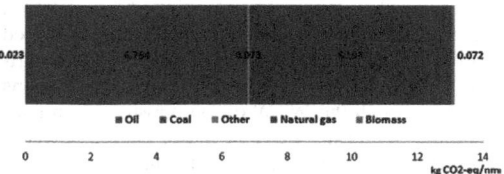

Figure 9. WTP GHG emissions from electricity generation.

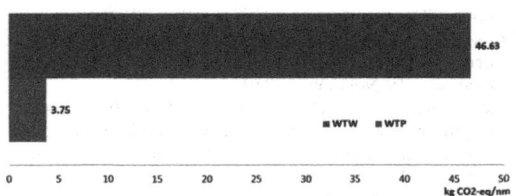

Figure 7. WTW and WTP GHG emissions of diesel engine-powered river cruise ship.

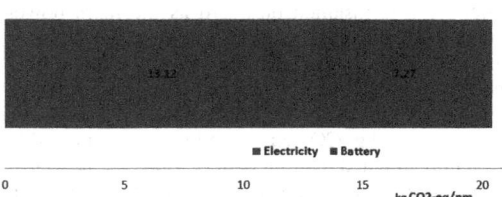

Figure 10. WTW GHG emissions of battery-powered ship.

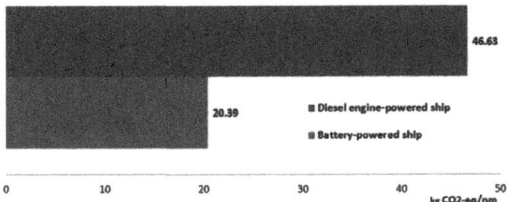

Figure 11. WTW GHG emissions from ship with different power system designs.

be concluded that electrification would significantly reduce environmental footprint of this ship.

Even though the ship is applying slow steaming with average speed of only 7.8 knots, further actions need to be taken in order to reduce GHGs release in the atmosphere. Electrification of the existing diesel engine-powered ship has its benefits due to lower GHG emissions, but it requires higher investment cost, but also the maintenance cost, which refers to the cost of battery replacement after approximately 10 years of its use. Even though the lifetime of a battery is lower than the lifetime of a diesel engine, leading to higher maintenance costs, due to the future more stringent regulation on inland waterways pollution and introduction of carbon pricing policy, electrification seems to be a viable solution that can achieve the decarbonization of the shipping industry.

6 CONCLUSIONS

In order to evaluate the electrification of the river cruise ship, which operates in Croatian inland waterways, from the environmental point of view, LCAs of diesel engine-powered ship and battery-powered ship were performed by means of GREET 2018 software. The analysis is focused on the emissions released through the WTW of the power system configuration. While the LCA of diesel engine-powered ship constitutes of the process of crude oil recovery and transportation to the refinery, diesel refining and its transportation to the pump and finally its use in the ship which results in tailpipe emissions, the LCA of the battery-powered ship comprises of the processes of electricity generation and its distribution, but also of the manufacturing process of the Li-ion battery that is installed on-board as the main power source. The obtained results show that WTW emissions of diesel engine-powered ship are much higher and amount 46.63 kg CO_2-eq/nm, than those released from WTW life cycle of battery-powered ship yielding to 20.39 kg CO_2-eq/nm. Since electrified ship with implemented Li-ion battery releases no gases during its operation and has more than twice lower amount of WTW GHG emissions than the existing ship powered by diesel engine, it can be concluded that electrification is great solution for decarbonization of the shipping industry and compliance with strict regulation on environmental protection. On the other hand, it is fair to say that complete insight into the feasibility of the electrification of river cruise ship will be achieved by comparing the existing and the battery power system also from the economic viewpoint, which will be subject of further studies.

ACKNOWLEDGEMENTS

This research was supported by the Croatian Science Foundation under the project "Green Modular Passenger Vessel for Mediterranean (GRiMM)", (Project No. UIP-2017-05-1253) as well as within the Croatian-Chinese bilateral project "Energy efficient and environmentally friendly power system options for inland green ships" between University of Zagreb, Faculty of Mechanical Engineering and Naval Architecture (Croatia) and Wuhan University of Technology (China). Also, the funding within the international collaborative project Global Core Research Center for Ships and Offshore Plants (GCRC-SOP, No. 2011-0030669), established by the Korean Government (MSIP) through the National Research Foundation of South Korea (NRF) is greatly acknowledged. Miss Maja Perčić, Ph.D. student is supported through the "Young researchers' career development project – training of doctoral students" of the Croatian Science Foundation, funded by the European Union from the European Social Fund.

In this paper, life cycle assessments were performed by GREET 18 software produced by UChicago Argonne, LLC under Contract No. DE-AC02-06CH11357 with the Department of Energy.

REFERENCES

Ančić, I., Vladimir, N. & Runko Luttenberger, L. 2018a. Energy efficiency of ro-ro passenger ships with integrated power system. *Ocean Engineering* 16: 350–357.

Ančić, I., Vladimir, N. & Cho, D.S. 2018b. Determining environmental pollution from ships using Index of Energy Efficiency and Environmental Eligibility (I4E). *Marine Policy* 95: 1–7.

Croatian Energy Regulatory Agency (CERA) 2016. Annual Report. https://www.hera.hr/en/docs/HERA_Annual_Report_2016.pdf [accessed 03 November 2019].

Cruise Mapper (CM) 2019. Ship particulars. https://www.cruisemapper.com/ships/MS-Prinzessin-Sisi-1367 [accessed 19 November 2019].

Dedes, E.K., Hudson, D.A. & Turnock, S.R. 2012. Assessing the potential of hybrid energy technology to reduce exhaust emissions from global shipping. *Energy Policy* 40: 204–218.

Environmental Protection Agency (EPA) 2018. Emission factors. https://www.epa.gov/sites/production/files/2018-03/documents/emission-factors_mar_2018_0.pdf [accessed 15 November 2019].

Environmental Protection Agency (EPA) 2019. Understanding the Global Warming Potentials. https://www.epa.gov/ghge-missions/understanding-global-warming-potentials [acce-ssed 19 November 2019].

European Court of Auditors (ECA) 2015. Inland Waterway-Transport in Europe: No significant improvements in modal share and navigability conditions since 2001. https://www.eca.europa.eu/Lists/ECADocuments/SR15_01/SR15_01_EN.pdf [accessed 19 November 2019].

Gagatsi, E., Estrup, T. & Halatsis, A. 2016. Exploring the potentials of electrical waterborne transport in Europe: The E-ferry concept. *Transportation Research Procedia* 14: 1571–1580.

HEP 2019. Structure of Croatian electric energy. http://www.hep.hr/elektra/trziste-elektricne-energije/izvori-elektricne-energije/1553# [accessed 05 November 2019].

International Maritime Organization (IMO) 2014. Third IMO GHG Study: Executive Summary and Final Report. http://www.imo.org/en/OurWork/Environment/PollutionPrevention/AirPollution/Documents/Third%20Greenhouse%20Gas%20Study/GHG3%20Executive%20Summary%20and%20Report.pdf [accessed 19 November 2019].

International Organization for Standardization (ISO) 1997. ISO 14040, Environmental Management-Life cycle assessment: Principles and framework. https://web.stanford.edu/class/cee214/Readings/ISOLCA.pdf. [accessed 10 November 2019].

Jwa, K. & Lim, O. 2018. Comparative life cycle assessment of lithium-ion battery electric bus and Diesel bus from well to wheel. *Energy Procedia* 145: 223–227.

Kalikatzarakis, M., Geertsma, R.D., Boonen, E.J., Visser, K. & Negenborn, R.R. 2018. Ship energy management for hybrid propulsion and power supply with shore charging. *Control Engineering Practice* 76: 133–154.

Kanellos, F., Anvari-Moghaddam, A. & Guerrero, J.M. 2017. A cost-effective and emission-aware power management systems for ships with integrated full electric propulsion. *Electric Power Systems Research* 150: 63–75.

Keuken, M.P., Moerman, M., Jonkers, J., Hulskotte, J., Denier van der Gon, H.A.C., Hoek, G. & Sokhi, R.S. 2014. Impact of inland shipping emissions on elemental carbon concentrations near waterways in The Netherlands. *Atmospheric Environment* 95: 1–9.

Kullmann, A.B. 2016. A Comparative Life Cycle Assessment of Conventional and All-Electric Car Ferries, Norwegian University of Science and Technology, Master thesis. https://brage.bibsys.no/xmlui/bitstream/handle/11250/2491 124/15570_FULLTEXT.pdf? sequence=1&isAllowed=y. [accessed 3 May 2019].

Ling-Chin, J., Heindrich, O. & Roskilly, A.P. 2016. Life cycle assessment (LCA)–from analysing methodology development to introducing an LCA framework for marine photovoltaic (PV) systems. *Renewable and Sustainable Energy Reviews* 59: 352–378.

Marine Traffic (MT) 2019. Ship particulars. https://www.marinetraffic.com/en/ais/details/ships/shipid:301692/mmsi:2564 17000/imo:0/vessel:PRINZESSIN_SISI [accessed 19 November 2019].

Ministry of the Sea, Transport and Infrastructure of the Republic of Croatia (MSTIRC) 2019. Inland navigation. https://mmpi.gov.hr/more-86/unutarnja-plovidba-110/110 [accessed 23 October 2019].

Ship particulars 2019. https://www.seereisenmagazin.de/ausgabe-4-2019/reportagen/flussreise-save-auf-dem-stillen-balkan-fluss.html [accessed 10 November 2019].

United Nations Framework Convention Climate Change (UNFCCC) 2001. Climate Change Information kit. https://unfccc.int/resource/iuckit/cckit2001en.pdf [accessed 10 November 2019].

Wiedmann, T. & Minx, J. 2008. A Definition of Carbon Footprint. In Pertsova C.C. (ed.), *Ecological Economics Research Trends*: 1–11. New York: Nova Science Publishers.

Wiegmans, B., Witte, P. & Spit, T.2015. Inland port performance: a statistical analysis of Dutch inland ports. *Transportation Research Procedia*8: 145–154.

Ship structures

Developments in Maritime Technology and Engineering – Guedes Soares & Santos (eds)
© 2021 Copyright the Author(s), ISBN 978-0-367-77376-2

A review on numerical approaches in the hydroelastic responses of very large floating elastic structures

I.B.S. Bispo, S.C. Mohapatra & C. Guedes Soares
Centre for Marine Technology and Ocean Engineering (CENTEC), Instituto Superior Técnico, Universidade de Lisboa, Lisbon, Portugal

ABSTRACT: This paper presents a review on the hydroelastic response of very large floating structures in near-shore and shallow water over sea-bottom with significant variability and complex bathymetry. A review is provided of the modelling approaches based on various numerical methods used in hydroelastic theory. At first, a basic general mathematical formulation of wave diffraction by a floating elastic structure over variable depth under linearized water wave theory and structural response is presented. Next, the hydroelastic analysis of various types of numerical models associated with different arrangements of floating elastic structures is reviewed. Finally, conclusions and useful recommendations in relation to future research developments and directions within the field of flexible floating structures are discussed.

1 INTRODUCTION

The study on the very large floating or submerged elastic structures have made significant progress in the development of various design techniques based on different methodologies for applications as floating platform, floating airports, floating offshore base, offshore storage facilities, energy islands, floating or submerged breakwaters, and floating city (Wang & Tay, 2011; Watanabe et al. 2004).

The very large floating structures (VLFSs) behaves as elastic in nature due to the unprecedent length compared to its thickness. Also, due to the large size of the structures, the computational burden becomes too cumbersome. To overcome these difficulties, often these types of structures are assumed to be infinitely long in comparison with the wavelength of the incident waves. The advantage of this type of elastic structure is that they are rapidly deployable, cost-efficient, and do not damage the marine ecosystem compared to rigid or fixed type structures.

Due to increase in marine and human activities in coastal regions, there is considerable significance in the application of horizontal flexible floating and submerged structures as breakwaters for protecting coastal infrastructures and other human activities (Guo et al., 2020c). When the floating or submerged flexible or elastic structures are constructed in a coastal region, it is complex to find a wide flat sea bottom area, which is relevant not only to make its installation easier, but also to ensure that its behaviour is somewhat more predictable. Therefore, the numerical modelling of hydroelastic problems over variable bottom topography near

the shoreline is of great importance for coastal development. Numerical methods play an important role on modelling and developing the understanding of the hydroelastic response of elastic or flexible floating structures to design breakwaters.

The hydroelastic analysis of the pontoon type VLFSs are based on the following basic assumptions:

- The VLFS is modelled as a thin elastic plate with free edges.
- The fluid is ideal, incompressible, inviscid, and the fluid motion is irrotational, so that the velocity potential exists.
- The amplitude of the incident wave and the motion of the VLFS are both small, and only the vertical motion of the structure is considered.
- There is no gap between the VLFS and the free surface of the fluid.

There are two major approaches followed to deal with the hydroelastic analysis of VLFSs, namely (i) frequency domain analysis and (ii) time domain analysis.

Commercial software such as WAMIT and AQWA use BIEM (panel method) and the output is always in frequency domain. On the other hand, the commonly used approaches for the time domain analysis of VLFS are the direct time integration method (Watanabe & Utsunomiya, 1996; Watanabe et al., 1998) and the method that uses the Fourier transform (Ohmatsu, 1998; Kashiwagi, 2000; and the papers cited therein). In the direct time integration method, the equations of motion are discretized for both the structure and the

DOI: 10.1201/9781003216582-48

fluid domain. In case of the Fourier transform method, initially, the frequency domain solutions for the fluid domain are obtained and then Fourier transform is applied to obtain the elastic motions of the VLFS. Then, by using the Finite Element Method (FEM) or Boundary Element Method (BEM) or other suitable computational methods, the equations are then solved directly in the time domain analysis. In the time domain analysis, due to the unbounded infinite or semi-infinite domain, either FEM or the BEM produces complicated large matrices that lead to large memory size and more computing time.

The first relevant papers on numerical methods for the analysis of elastic properties of structures were based on the use of FEM. To tackle the issue of representing the bottom of the sea when dealing with near shore applications, some approaches are available (Zhou et al, 2016). There are also studies focused on solving hydroelastic problems using FEM or BEM to analyse the hydroelastic response of floating structures with variable bathymetry (see Athanassoulis & Belibassakis, 2009; Belibassakis & Athanassoulis, 2005; Belibassakis et al., 2016; Song et al., 2005).

Another interesting aspect of BEM is the application of the integral form of Green's function in various boundary integral equation formulations associated with physical problems in hydroelastic problems in two or three-dimensions arising in ocean engineering and related branches. Recently, Green's functions associated with various model involving floating and submerged flexible/elastic structures of large dimensions were derived. For instance, Mohapatra & Sahoo (2014) derived the Green's function of the problem of oblique wave interaction with floating and submerged flexible structures in both finite and infinite water depths. Mohapatra & Guedes Soares (2016b) developed a three-dimensional hydroelastic model associated with wave interaction with elastic bottom (bottom is modelled as elastic beam equation) based on Green's function in three-dimensions. Mohapatra et al. (2018a) derived the Green's function associated with surface gravity wave interaction with a very large submerged elastic plate and demonstrated a physical problem of articulated submerged elastic plates in finite and infinite water depths. Further, Mohapatra et al. (2018b) derived the Green's function for wave interaction with a submerged flexible porous plate in two-dimensions. Mohapatra & Guedes Soares (2019) derived the three-dimensional Green's function for wave interaction with floating and submerged flexible or elastic structures. Guo et al. (2020b) derived the Green's function for wave interaction with submerged flexible porous membrane under oblique waves. Further, Mohapatra & Guedes Soares (2020) derived the Green's function associated with oblique wave interaction with a submerged flexible porous structure.

It is important to note that up to now there exist well documented review papers in the literature in the broad area of wave-structure interaction problems in the field of hydroelasticity. For instance, Chen et al. (2006) reviewed four different types of existing hydroelastic theories and their applications to the analysis of very large floating structures. Wang et al. (2010) reviewed the design and performance of anti-motion devices such as breakwaters, submerged plates, oscillating water column breakwaters, air-cushion, auxiliary attachments, and mechanical joints for mitigating the hydroelastic response of very large floating structures under wave action. Karmakar et al. (2011) reviewed the existing analytical hydroelasticity theories to analyse different types of floating and submerged structures applications for very large floating structures and large ice sheets. Dai et al. (2018) reviewed the recent research and developments on floating breakwaters for performance of different breakwaters and effectiveness of various wave attenuating devices. Recently, Guo et al. (2020a) presented a review development on the flexible structures for application to breakwaters based on analytical, numerical, and experimental methodologies.

From the above literature, it is confirmed that until now there is no review reported to the public related to the recent research and developments on models for the hydroelastic response of floating horizontal flexible structures of elastic structures based on numerical methods.

Therefore, the aim of this paper is to present a literature review on the hydroelastic analysis of floating elastic structures, particularly over variable bottom topography based on numerical methods. This review is organized as follows: Section 2 describes the modelling approaches and the analysis of their results based on numerical methods over complex bathymetry. Subsection 2.1 presents the basic mathematical formulation of wave diffraction by a generic floating elastic structure over variable bottom under the assumption of linearized water wave theory and structural response. Subsection 2.2 reviews on the hydroelastic response of floating elastic structures with different numerical models associated with FEM or BEM or hybrid approach for (i) ice sheets and generic floating plates over variable sea bottom, (ii) single body VLFS, and (iii) multibody VLFS. Section 3 summarizes the conclusion and future scope of the present review.

2 MODEL AND RESULT DESCRIPTION

2.1 *Mathematical description of the general physical problem*

The basic mathematical description modeling the problem of a long-narrow flexible structure under wave loads is described in the following, based on the formulation of diffraction of waves by a flexible, floating and mat-like structure which floats at a surface of an ideal incompressible and inviscid fluid of possible variable depth denoted by $h(x, y)$. This description is based on the one presented by Song et al. (2005), which is similar to other works referenced in this paper, although independently developed by them.

A diagram of the modeled system is presented in Figure 1 (extracted from Kyoung et al., 2005), at which, Ω is the (computational) fluid domain, S_F is the free-surface boundary, S_O is the floating structure boundary, S_B is the sea-bottom boundary and L, B and T define, respectively, the length, breadth and thickness of the floating flexible structure. In the same diagram is possible to see where the reference of the coordinate system is considered, at the center of the floating plate.

A similar approach was presented by (Song et al., 2005), in which only the inertial reference is adopted in the corner of the plate, instead of the center. The linear theory of waves is assumed, so the incoming waves (propagating to positive x-axis) are expected to have amplitudes much smaller than the wavelength and the fluid motion is supposed irrotational. The nomenclature of the regions and surfaces is equivalent in both works (see Figures 1 and 2). Within the scope of the linear theory of waves, assuming they propagate in still water and that the function $\Phi(x,y,z,t)$ represents the velocity potential of the fluid, satisfying the three-dimensional Laplace equation as:

$$\nabla^2\Phi(x,y,z,t) = 0. \tag{1}$$

The free surface boundary condition is given by (as in Mohapatra & Guedes Soares, 2016a)

$$g\frac{\partial\Phi}{\partial z} + \frac{\partial^2\Phi}{\partial t^2} = 0 \text{ on } z = \zeta(x,y) \ (in \ I) \tag{2}$$

The relevant boundary conditions are read as:

$$\frac{\partial\Phi}{\partial z} = \frac{\partial w}{\partial t} \text{ on } A \ (in \ II) \tag{3}$$

$$\frac{\partial\Phi}{\partial n} = 0 \text{ on } z = -h(x,y) \ (in \ I \ and \ II) \tag{4}$$

The elastic plate equation is given by

$$D\nabla^4 w + \rho g w + m\frac{\partial^2 w}{\partial^2 t} = -\rho\frac{\partial\Phi}{\partial t}. \tag{5}$$

The free edge conditions are given by

$$\frac{\partial^2 w}{\partial x^2} + \nu\frac{\partial^2 w}{\partial y^2} = 0, \tag{6a}$$

$$\frac{\partial^3 w}{\partial x^3} + (2-\nu)\frac{\partial^3 w}{\partial y^2\partial x} = 0 \ (x=0,L), \tag{6b}$$

$$\nu\frac{\partial^2 w}{\partial x^2} + \frac{\partial^2 w}{\partial y^2} = 0, \tag{7a}$$

$$\frac{\partial^3 w}{\partial y^3} + (2-\nu)\frac{\partial^3 w}{\partial y\partial x^2} = 0 \ (y=0,B), \tag{7b}$$

$$\frac{\partial^2 w}{\partial x\partial y} = 0 \text{ for } (x=0,L), \ (y=0,B), \tag{8}$$

where Φ is the velocity potential, w =plate deflection, ζ=wave height, A=fluid and structure interface, ν=Poisson's ratio, ρ= plate density, g is the gravitational acceleration, m is the mass per unit area, and n is the unit normal vector.

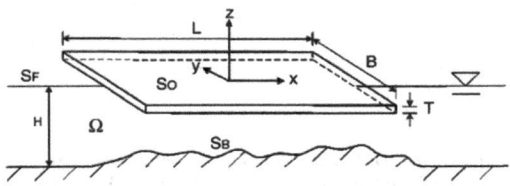

Figure 1 . Definition of domain, dimension, and origin of co-ordinate system (Kyoung et al., 2005).

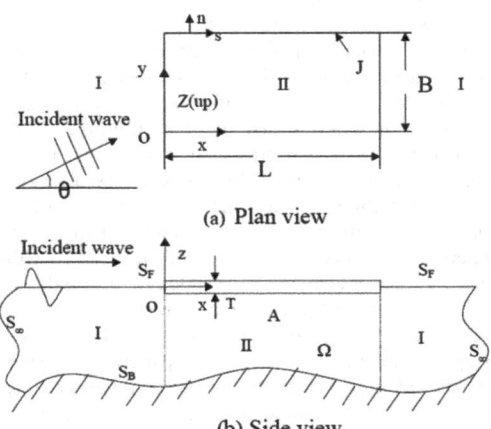

(a) Plan view

(b) Side view

Figure 2. Definition of domain, dimension, and origin of co-ordinate system (Song et al., 2005).

2.2 Models for the hydroelastic analysis of floating structure over variable bottom

In this Section, the hydroelastic response of floating ice sheets and generic floating plates, single body VLFS, and multi-body VLFS over variable bottom

427

topography will present based on numerical models in subsequent subsections.

2.2.1 *Waves propagating, ice sheets and generic floating plates over variable sea bottom*

Song et al. (2005) presented results of a 2-D comparison between a numerical method BEM for determination of hydroelastic responses of a VLFS and reduced scale trials. That work consists on the computation of the hydroelastic response of a long and thin plate subjected to a set of 4 distinct wavelengths, combined with 3 variations of the water depth, and also the evaluation of the flat and horizontal sea bottom versus 2 different breadth sizes of shoals. The linear wave theory is assumed and the deflection of a rectangular fixed size plate is computed by means of a BEM. Since linearity is assumed, as the wavelength increases or the slope caused by the shoal is steeper or abrupt, this assumption is no longer valid and, therefore, the problem becomes nonlinear and demands being solved by other means. Figure 2 displays a good agreement between numerical and experimental data with wavelength=200m, water depth=40m, bottom B refers to a shoal measuring 0.1m in height and 2.0m of breadth.

Kyoung et al. (2005) performed a similar study for four different bottom cases, using FEM to solve the equations in the fluid domain. The structure considered in the study is a pontoon modelled as a Kirchhoff plate (Figure 3) over variable bottom. The schematic diagram of different plate position over variable bottom is presented in Figure 4(a). A numerical method was developed by Kyoung et al. (2005) for the hydroelastic analysis of floating structure when it is located in coastal region. The FEM was used to solve the fluid region rigorously where the sea-bottom geometry is not flat. By comparison between the existing experimental data, the developed numerical method was validated.

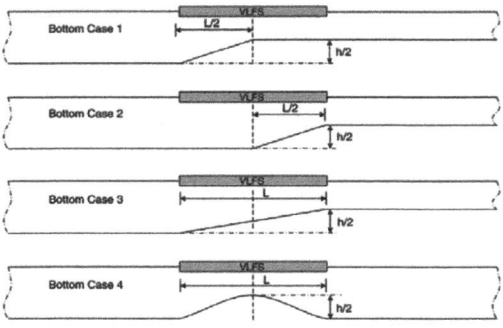

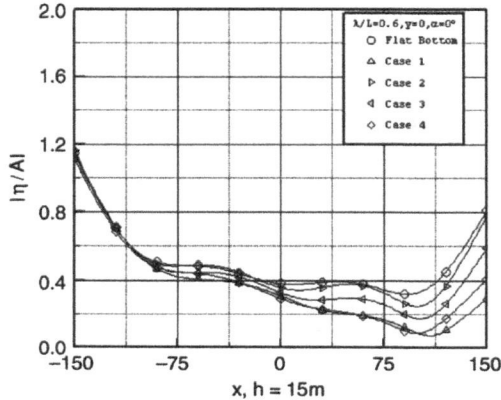

Figure 4. (a) Different position of plate over different bottom configurations (Kyoung et al. 2005). (b). Hydroelastic response of pontoon for L = 300 m, B = 60 m, h = 15 m, and $\lambda/L = 0.6$ (Kyoung et al., 2005).

In Figure 4(b), the hydroelastic response decreases at the rear end of the plate as the wavelength increases and the water depth decreases. There is an increase of the local deformation of the structure due to the effects of a variable sea bottom in comparison with the case of a flat bottom.

Buchner (2006) has presented a study where the sloping seabed is numerically modelled as a second body interacting with the floating structure. This second body in diffraction theory, when chosen properly with respect to its shape and size, can contribute to the correct calculation of added mass and damping of a floating object over a sloped sea. Figure 5 shows the propagation of simulated waves over a sloped seabed and the respective pattern of interference caused by it. The study shows that it was not possible to model a sloping seabed as a second body without special measures, since the refraction and interference effects are too strong and affect the wave exciting forces on the floating structure in an incorrect way.

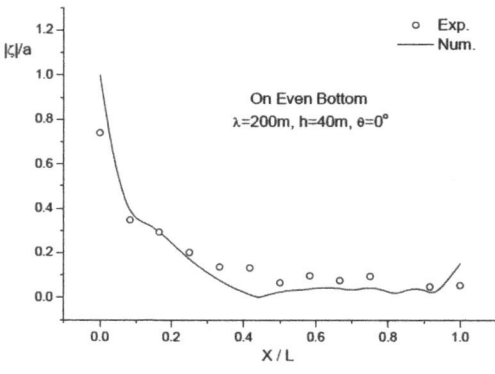

Figure 3. Hydroelastic response of a floating plate under regular waves (see Song et al., 2005).

Figure 5. Snapshots of a 31.4s/0.2 rad/s wave (top) and 10.47s/0.6 rad/s wave (below) showing the different interference patterns (Buchner, 2006).

A coupled-mode system (CMS) of horizontal equations was presented by Athanassoulis & Belibassakis (2009) for the hydroelastic analysis of large floating bodies or ice sheets of finite thickness over variable bathymetry. The method is based on the theory of shear deformable plates (or beams), and is derived by an enhanced representation of the elastic displacement field, containing additional elastic vertical modes and permitting shear strain and stress to vanish on both the upper and lower boundaries of the thick floating plate. This model extends 3rd order plate theories to plates and beams of general shape.

Figure 6 shows the elastic-plate deflection for three bottom profiles with (i) horizontal bottom (thin solid line), (ii) undulating bottom with 15% amplitude (dashed line), and (iii) undulating bottom with 30% amplitude (thick solid line). As the incident wave angle increases, the horizontal wavelength of the wave and of the plate deflection increases. In addition, refraction phenomena become more significant. The effects of bottom corrugations on the hydroelastic behaviour of the system are found to be of secondary importance.

The CMS model developed by Athanassoulis & Belibassakis (2009)) was extended by Belibassakis et al. (2016) and applied to the hydroelastic analysis of three-dimensional large floating bodies of shallow draft lying over variable bathymetry. This extended formulation finds useful applications to the study of interaction of water waves in coastal region with waters with ice floes or floating flexible structures. The effect of bathymetric variations on the calculated wave field and the elastic plate deflection, in the case of waves of period T=15s normally incident on the elastic structure over the smooth shoal in Figure 7.

More recently, numerical simulations were performed in Wei et al. (2017) based on the three-dimensional potential theory and FEM as in Song et al. (2005). The continuous floating structure is first discretized into rigid modules connected by elastic beams and the equations of motion of the entire floating structure. It was established according to the six degrees of freedom of each module by coupling the hydrodynamics of the modules with the structural stiffness matrix of the elastic beams in the frequency domain. The sketch of the numerical model (Wei et al., 2017) under uneven bottom is presented in Figure 8.

2.2.2 Single-body VLFS numerical models

Another work addressing the problem of a VLFS placed in water with variable depth, but comprising a 3D numerical simulation, was presented by Utsunomiya et al. (2008). The study was carried out by the application of FEM using beams and quadrilateral plate elements to model a hybrid-type VLFS as a single body, consisting of a floating runaway with 3120 m in total length, 524 m of width and 1.5 m of draft in the pontoon area, whilst having 11.5 m of draft in the semisubmersible region. Since the

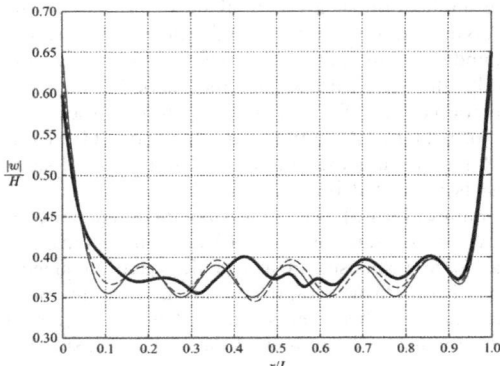

Figure 6. Effects of bottom corrugations on the modulus of the plate deflection (Athanassoulis & Belibassakis, 2009).

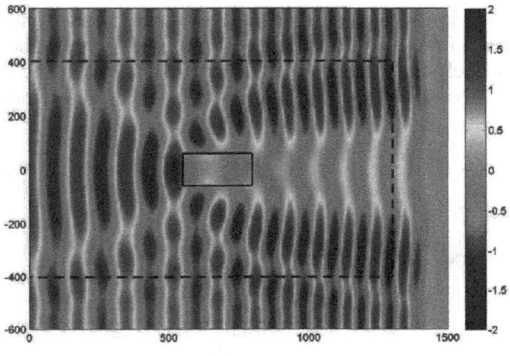

Figure 7. Effect of bathymetric variations on the wave field and the elastic plate deflection with T = 15s (Belibassakis et al., 2016).

Figure 8. Numerical model of a floating plate under an uneven bottom (Wei et al., 2017).

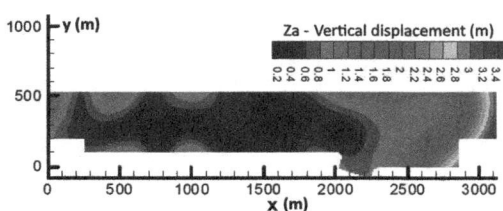

Figure 10. Vertical displacement $Z_{1/2}$(m) for irregular waves with $H_{1/2} = 4.15$m, $T_{1/2} = 8.3$s, and $\beta = 45°$ for $C_d = 0$ (Utsunomiya et al., 2008).

numerical mesh of the FEM was finely detailed, another one was developed to obtain faster results for the hydroelastic analysis, an equivalent plate model with same eigenfrequencies and eigenmodes of the first in air. The results of both were compared and good agreement between both has been observed. Both models were tested for three different configurations of the sea bottom: flat bottom at 20 m depth, a sloping bottom with 1/75 slope and a third case with variable bottom using bathymetry data.

In Figure 9, it was found that the effects of a variable sea bottom are significant with respect to the hydroelastic response of both models. Another important result is that the viscosity of the water must not be neglected when performing the simulations for the semi-submersible part of VLFS, since it can be verified that there is a significant discrepancy in the vertical displacement obtained in both cases of viscosity, which can be attained by applying a drag force coefficient C_d to the plate model in Figure 10.

Great progress has been made in the field of hydroelasticity theory and the corresponding numerical approaches to better meet the increasing requirements of analysing the global and local wave induced structural responses. Some theoretical creations and application results of the progress are roughly summarized by Wu et al. (2016), focusing on those made by the research group of the authors. More information about the hydroelasticity theories and applications may be found in the literature, such as the comprehensive analysis of the field performed by Wu & Cui (2009). The hydroelastic behaviour of nonlinear analysis associated boundary conditions and validation of numerical

methods was addressed by Ding et al. (2017) based on a numerical method dependent on a direct coupled method using Boussinesq equations and the Rankine panel method (Kring, 1994). Both accuracy and reliability of this method (THAFTS-BR) are validated by comparisons with the results of the software THAFTS, which was developed and continuously improved during the past three decades by China Ship Scientific Research Centre (CSSRC), being validated through model tests, full scale measurements and also compared to other similar software (BEM and other methods). In this case, it is emphasized that the existence of floating bodies in the flow field could not be considered in the current theoretical wave model and that the hydrodynamic analysis could not be directly used in the prediction of wave propagation near islands and reefs. From these characteristics rise the challenge of hydrodynamic analysis in these regions.

A division of the computational domain in two fictitious regions, inner and outer, as depicted in Figure 11(a), and the problem is then solved by applying a middle-scale model for the wave propagation in the outer region using Boussinesq equations.

On the fictitious boundary, the wave elevation and vertical distribution of fluid velocity are used to ensure continuity between the inner and outer regions, these values are then employed as initial for a boundary value problem to be solved by a local-scale model in the inner region, i.e. by the application of a potential flow solver with the Rankine sources distributed along the whole boundaries, including the wetted surface of the body, the free surface and the seabed. The computational grid for the inner region is shown in Figure 11(b). A comparison between THAFTS-BR results and measurements from a small-scale test is presented in Figure 12.

Figure 13(a-b) show that the inhomogeneity of waves may induce about 2 to 3 times the increase of the force responses in a specific wave frequency. The hydroelastic responses of a floating plate according to this method showed good agreement with published experimental data when considering an uneven sea bottom and the inhomogeneity of regular waves could induce an approximately 30% to 80% increase in the maximum bending moments/ shear forces compared to homogeneous waves.

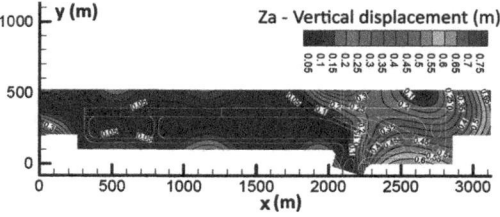

Figure 9. Vertical displacement amplitude for unit-amplitude waves at T = 8.3s and $\beta = 45°$ (Utsunomiya et al., 2008).

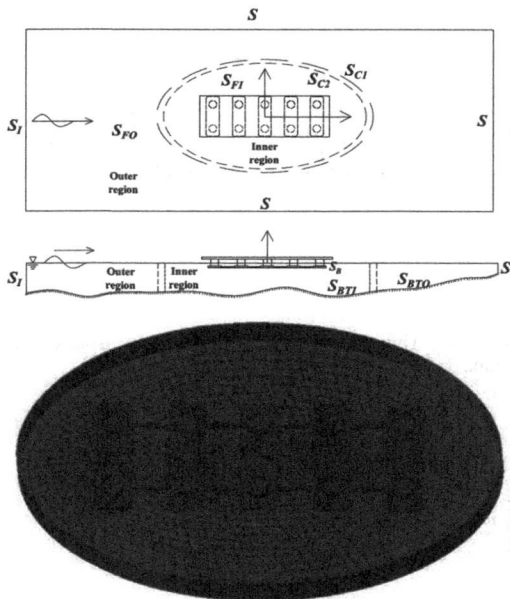

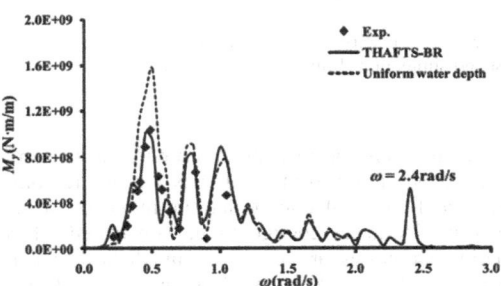

Figure 11. (a). Computational domain of direct coupled method (Ding et al., 2017). (b). Hydrodynamic grid of the inner region (Ding et al., 2017).

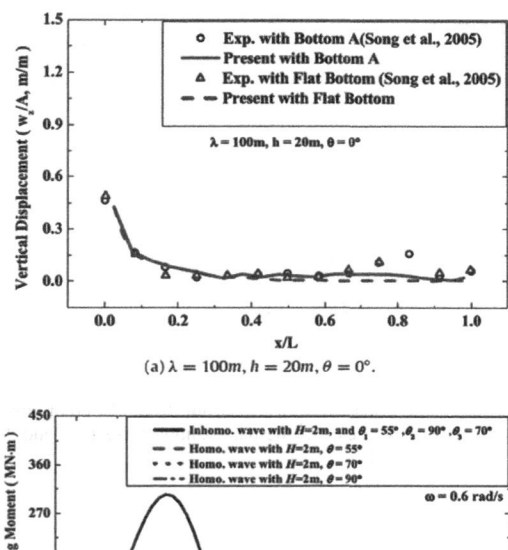

(a) $\lambda = 100m, h = 20m, \theta = 0°$.

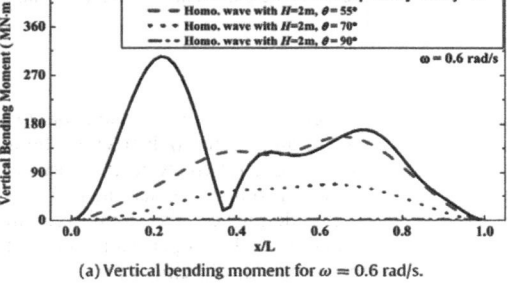

(a) Vertical bending moment for $\omega = 0.6$ rad/s.

Figure 12. Comparison of the wave loads between the model test and THAFTS-BR (Ding et al., 2017).

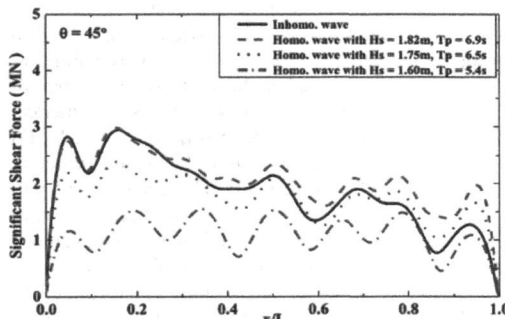

Figure 13. (a). Vertical displacement with $\lambda = 100m$, h= 20m, and $\theta = 0°$ along the centreline on an uneven bottom (Wei et al., 2017). (b). Vertical bending moment with $\omega = 0.6$ radm^{-1} of the structure (Wei et al., 2017). (c). The significant shear forces in homogenous and inhomogeneous waves at $\theta = 45°$ (Wei et al., 2018).

An extension of the later reference is presented by Wei et al. (2018), which shows also an application of irregular waves conditions at various wave incidence angles. According to their results, such as the one presented in Figure 13(c), it concludes that the vertical displacements, bending moments, shear forces, and torsional moments of the freely floating VLFS in inhomogeneous waves apparently differ from those in homogeneous waves in beam seas. It is also noted that a safe design may not be guaranteed for structures in inhomogeneous wave field even using the most severe homogeneous wave condition for the hydroelastic analysis of freely floating structures.

The work of Cheng et al. (2017) established a time domain fully nonlinear Numerical Wave Tank (NWT)

using a Higher Order BEM Method (HOBEM) and applied it to compute the hydroelastic responses of a floating elastic plate over sea-bottom of variable depth. This implementation was performed in two-dimensional potential theory and mixed Eulerian-Lagrangian technique, being validated through a series of comparisons with previous experimental references and also analytical models.

Figure 14 shows the different displacement components at the mid-position of the plate for 3-distinct

431

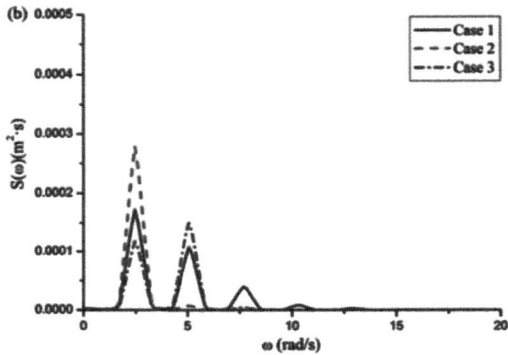

Figure 14. Comparisons of spectrum from different displacement components of the plate among Case 1, 2, and 3 (Cheng et al., 2017).

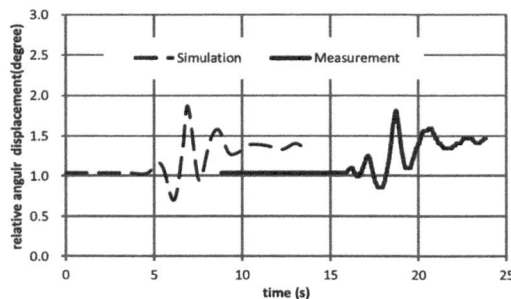

Figure 15. Time histories of relative angular displacement (Iijima & Fujikubo, 2018).

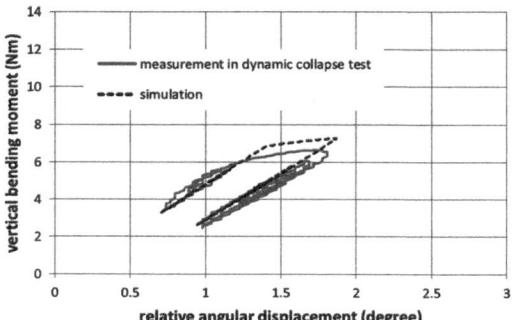

Figure 16. Capacity curves obtained in dynamic collapse test and simulation (Iijima & Fujikubo, 2018).

positioning cases, relatively to the obstacle on the sea-bottom. Cases 1, 2 and 3 refer to distinct positions of the floating elastic plate relatively to the obstacle on the sea-bottom. This result shows that the first harmonic displacement dominates the total vertical motion response and that the larger value of first order power of the vertical displacement for Case 2 is due to larger energy transfer from first or-der to the higher order modes for the other 2 cases.

A simulation method based on hydro-elastoplasticity approach was developed by (Iijima & Fujikubo, 2018) to predict the collapse behaviour of a VLFS. This implementation comprises a numerical wave tank of 16m long by 0.45m deep, with a VLFS of length 6m placed at its centre, being this structure modelled by 60 beam elements. This method was validated using the results of tests performed in a wave tank with a reduced scale model that was developed with the capability of reproducing the elastoplastic mechanism. A series of simulations is performed to reproduce the tank test results. In general, the developed method showed good agreement with the tank test results. The proposed method is effective in predicting the collapse behaviour of VLFS subjected to extreme vertical bending moment, as it can be observed in Figures 15-16.

The issue of a floating body near the island together with non-uniform submarine topography is also addressed by Lu et al. (2020), considering the three-dimensional linear hydroelasticity theory. It was noted by this work that commercial hydrodynamic numerical software, such as AQWA and SESAM, cannot analyse the behaviour of marine structures near reefs, including variable bathymetry. An alternative that has been used by many researchers is the 3D hydroelastic software THAFTS, which can deal with problems requiring the boundary conditions to be modified. Based on the simulations provided by this software, a three-dimensional hydroelastic analysis was carried out by the modal superposition method.

The motions and responses of a barge in uniform water depth and also in a complex terrain are compared and the results show that the influence of the later is significant enough to be considered when marine structures are deployed in shallow water, especially less than 50 m deep, close to islands. Figure 17 illustrates the effect of shallow water depth in the responses of the barge with different water depths.

2.2.3 *Multi-body VLFS numerical models*
A conceptual test of a floating mega island made out of large triangular pontoons is described by Waals et al. (2018). This first attempt has been carried out to calculate the motions response of the island using standard diffraction analysis of 87 triangular bodies in close proximity (see Figure 18). The results show a large difference between the numerical model, based on a BEM code and the reduced scale test performed in a towing tank (see Figure 19).

The diffraction calculations of that study were performed using an in-house linear diffraction code from MARIN, called DIFFRAC, based on potential flow and computes multi-body wave interaction using a BEM applying the zero-speed Green function as source on the mean wetted part of the island.

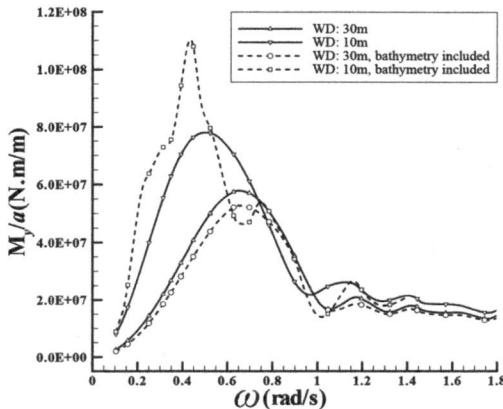

Figure 17. Comparison of RAOs of vertical bending moment at amidships with different water depths over complex bathymetry (Lu et al., 2020).

Figure 18. Model test of the island with 87 triangular pontoons (Waals et al., 2018).

Due to the large size of the island, the numerical simulations are somewhat challenging, since they require a large number of panels to have reliable and convergent results. Apart from that, for each degree of freedom of the island (6x87 in total) and for each direction of the incident wave, the water kinematics on all the panels must be computed, leading to a post-processing step that takes a considerable amount of time.

In order to ensure enough computational resources and RAM, the simulations were performed on a Linux computer cluster using nodes consisting of 20 CPU's and 512 GB RAM. Parallelization schemes must be applied to the computational expensive parts of the code to ensure that the results would be attained on reasonable time. The last but not least important result of the study was to indicate that available iterative solvers (like GMRES) did not converge due to the strong interaction between adjacent triangular elements of the islands.

A new study was carried out by Wu et al. (2018) by comprising a three-module VLFS of the same element studied by Ding et al. (2017) subjected to similar conditions (see Figure 20). The results are

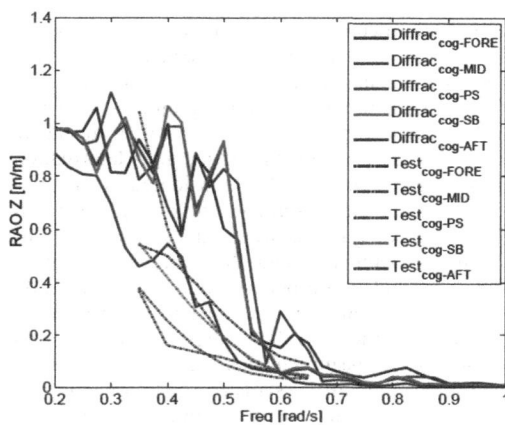

Figure 19. Heave RAO for the instrumented pontoons com-pared to results obtained from Waals et al. (2018).

also compared to experimental data and show good agreement, reassuring the effectiveness of this method and the robustness of THAFTS-BR software. The same conclusions from the previous study about the limitations of simplifications of the model are drawn as well. In Figure 21, one case of result showed a clear divergence is presented.

The method proposed by Lu et al. (2016) consisting of a discrete-module beam bending based

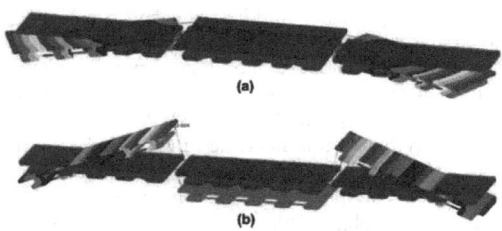

Figure 20. The typical dry modes for two distinct conditions (Wu et al., 2018).

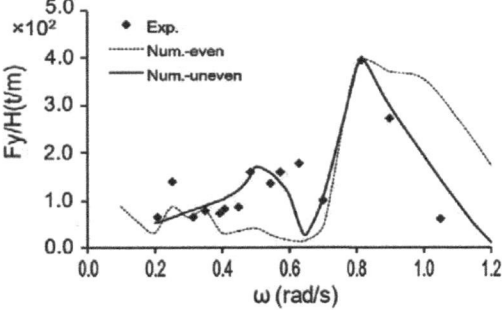

Figure 21. Comparisons of horizontal forces for the predictions by the direct coupled method with the test results (Wu et al., 2018).

hydroelastic method used to estimate the hydroelastic behaviour of a VLFS, was extended with the aid of FEM in Zhang & Lu (2018) to be applicable to a flexible structure with complex geometric features.

The central idea in that work is that an elastic beam should emulate the structural deformation effects and, for this, it will be considered as a macro-element, being then discretized following the standard FEM, leading to the structural stiffness matrix of this macro-element. Using the static condensation technique and rigid-body-motion transformation matrix, the equivalent stiffness matrix corresponding to the displacements at the centres of two end cross sections of this beam can then be derived from the macro-element's stiffness matrix. Subsequently, the hydroelastic response of a flexible structure with complex geometric features can be solved.

The extension of this hydroelastic response estimation method is validated by comparisons to an analytical determination of the stiffness matrix (Approach 1), to the hydroelasticity theory based on the mode-superposition method and to experimental results from previous works. The comparison of vertical response along the structure between experiment and mode decomposition (Zhang and Lu, 2018) is presented in Figure 22.

Yang et al. (2019) investigated the hydroelastic responses of a VLFS in the complicated geographic and wave environment near shore or islands based on three scales of approaches. The use of three different scale models for the numerical calculation, including mega-scale, middle-scale and local-scale, wave parameters can be obtained gradually from the three numerical models and then a more realistic simulation can be performed locally. Namely, these adopted numerical models are SWAN for mega-scale, WW3 for middle-scale and THAFTS for local-scale. Similar to the structure presented in previous works of the same research group, a three-module semi-submersible VLFS is investigated in

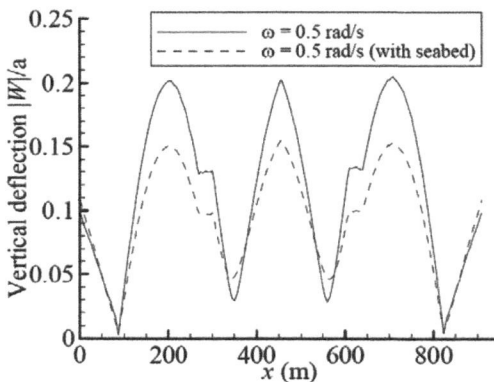

Figure 23. Comparison of vertical deflection along the centre line in x-direction on upper deck (Yang et al., 2019).

terms of its hydroelastic responses. Also, the numerical results were compared with on-site measurements near islands. Furthermore, the mathematical foundation of hydroelasticity is established considering the variable complicated seabed.

In Figure 23, both the numerical results and the data of model tests show that the variable seabed has a great effect on the response of floating structures, especially for larger wave periods. The numerical simulations also show that the method proposed in the paper can produce results consistent with the experiment. In addition, in Figure 23, results also provided evidences that a certain influence on the hydroelastic responses of the VLFS was due to the presence of a complicate seabed profile, including the vertical deflection responses.

3 CONCLUSIONS

This paper presents a brief review on the research and development of the models for the hydroelastic response of floating elastic structures based on numerical studies under effect of variable bottom topography. Most of the developed studies are based on codes of BEM using the panel method to discretize and compute the forces on the structures. Some of the references also used FEM to compute either the waves or the loads on the structures. More recent works could rely on more powerful computational infrastructure, showing that this feature is of great relevance when dealing with numerical models of wave propagation at complex bathymetry regions or intricate and detailed structures in which the stresses and loads are to be computed over a finer mesh.

On the numerical methods for the elastic structure models: FEM and BEM have been developed by researchers to investigate the dynamic behaviour of the structure under the action of waves and currents, while a focus on a FEM model associated with multi-mode motions based on coupling methodology will

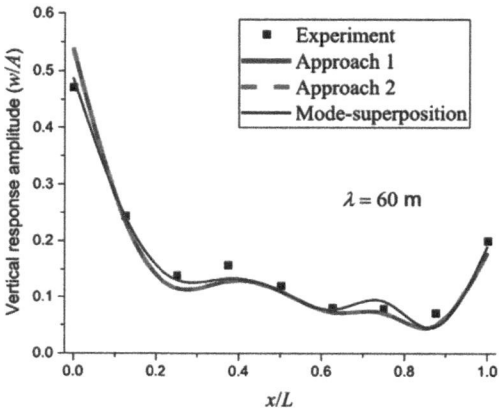

Figure 22. Vertical response along the structure. "Approach 2" denotes the presented method. Extracted from (Zhang & Lu, 2018).

allow for a better understanding of hydroelasticity in three-dimensions over variable bottom for ocean space utilization and wave energy conversion application.

The analytical model assumptions and structural characteristics can be used to simulate the structural deflections of the structure using numerical software like ANSYS or Simcenter Star-CCM+, but most commercial software at this point in time still lack some functionality for the hydroelastic analysis of VLFS deployed near shore or over complex bathymetry regions. The present endeavour is expected to be a good reference for future application studies to the researchers and engineers in the field of Marine Technology to study problems over variable bottom topography.

ACKNOWLEDGEMENTS

The work was performed within the project Hydroelastic behaviour of horizontal flexible floating structures for applications to Floating Breakwaters and Wave Energy Converters (HYDROELASTWEB), which is co-funded by the European Regional Development Fund (Fundo Europeu de Desenvolvimento Regional - FEDER) and by the Portuguese Foundation for Science and Technology (Fundação para a Ciência e a Tecnologia – FCT) under contract 031488_770 (PTDC/ECI-EGC/31488/2017). The second author has been contracted as a Researcher by the Portuguese Foundation for Science and Technology (Fundação para a Ciência e Tecnologia – FCT), through Scientific Employment Stimulus, Individual support under the Contract No. CEECIND/04879/2017. This work contributes to the Strategic Research Plan of the Centre for Marine Technology and Ocean Engineering (CENTEC), which is financed by the Portuguese Foundation for Science and Technology (Fundação para a Ciência e Tecnologia – FCT) under contract UIDB/UIDP/00134/2020.

REFERENCES

Athanassoulis, M. A. & Belibassakis, K. 2009. A novel coupled-mode theory with application to hydroelastic analysis of thick, non-uniform floating bodies over general bathymetry. *Proceedings of the Institution of Mechanical Engineers Part M: Journal of Engineering for the Maritime Environment* 223(3): 419–438.

Belibassakis, K. A. & Athanassoulis, G. A. 2005. A coupled-mode model for the hydroelastic analysis of large floating bodies over variable bathymetry regions. *Journal of Fluid Mechanics* 531: 221–249.

Gerostathis, Th. P., Belibassakis, K. A. & Athanassoulis, G. A. 2016. 3D hydroelastic analysis of very large floating bodies over variable bathymetry regions. *Journal of Ocean Engineering and Marine Energy* 2(2): 159–175.

Buchner, B. 2006. The motions of a ship on a sloped seabed. *Proceedings of the International Conference on Offshore Mechanics and Arctic Engineering*, Paper No. OMAE2006-92321, pp.339–347.

Chen, X. J., Wu, Y. S., Cui, W. C. & Jensen, J. J. (2006). Review of hydroelasticity theories for global response of marine structures. *Ocean Engineering* 33(3–4): 439–457.

Cheng, Y., Ji, C., Zhai, G. & Gaidai, O 2017. Fully nonlinear numerical investigation on hydroelastic responses of floating elastic plate over variable depth sea-bottom. *Marine Structures* 55: 37–61.

Dai, J., Wang, C. M., Utsunomiya, T. & Duan, W. 2018. Review of recent research and developments on floating breakwaters. *Ocean Engineering* 158: 132–151.

Ding, J., Tian, C., Wu, Y. Sheng, Li, Z. Wei, Ling, H. Jie & Ma, X. Zhou. 2017. Hydroelastic analysis and model tests of a single module VLFS deployed near islands and reefs. *Ocean Engineering* 144: 224–234.

Guo, Y.C., Mohapatra, S.C. & Guedes Soares, C. 2020a. Review of developments in porous membranes and net-type structures for breakwaters and fish cages. *Ocean Engineering* 200: 107027.

Guo, Y.C., Mohapatra, S.C. & Guedes Soares, C. 2020b. Wave energy dissipation of a submerged horizontal flexible porous membrane under oblique wave interaction. *Applied Ocean Research* 94: 101948.

Guo, Y.C., Mohapatra, S.C. & Guedes Soares, C. 2020c. Composite breakwater of a submerged horizontal flexible porous membrane with a lower rubble mound. *Applied Ocean Research* 104: 102371.

Iijima, K. & Fujikubo, M. 2018. Hydro-elastoplastic behaviour of VLFS under extreme vertical bending moment by segmented beam approach. *Marine Structures* 57: 1–17.

Karmakar, D., Bhattacharjee, J. & Sahoo, T. 2011. Contemporary approaches in the hydroelastic analysis of floating and submerged structures. *In Marine Technology and Engineering.* Guedes Soares, C. Garbatov Y. Fonseca N. & Teixeira A. P., (Eds.). London: Taylor & Francis, 1, 461–478.

Kashiwagi, M. 2000. A time-domain mode-expansion method for calculating transient elastic responses of a pontoon-type VLFS. *Journal of Marine Science and Technology* 5: 89–100.

Kring, D. C. 1994. Time-domain ship motions by a 3-D Rankine panel method. Massachusetts Institute of Technology (p. 134).

Kyoung, J. H., Hong, S. Y., Kim, B. W. & Cho, S. K. 2005. Hydroelastic response of a very large floating structure over a variable bottom topography. *Ocean Engineering* 32(17–18): 2040–2052.

Lu, Y., Temarel, P., Zhou, Y. & Tian, C. 2020. Hydroelasticity of the barge near island affected by bathymetry based on THAFTS software. *Proceedings of the Institution of Mechanical Engineers Part M: Journal of Engineering for the Maritime Environment* 234(1): 48–58.

Mohapatra, S.C. & Guedes Soares, C. 2019. Interaction of ocean waves with floating and submerged horizontal flexible structures in three-dimensions. *Applied Ocean Research* 83: 136–154.

Mohapatra, S.C. & Guedes Soares, C. 2020. Hydroelastic response of a flexible submerged porous plate for wave energy absorption. *Journal of Marine Science and Engineering* 8(9): 698.

Mohapatra, S.C. & Guedes Soares, C. 2016a. Effect of submerged horizontal flexible membrane on moored floating elastic plate. *In*: Guedes Soares C. and Santos, TA. (Eds), *Maritime Technology and Engineering* 3. London: Taylor & Francis Group, pp.1181–1188.

Mohapatra, S.C. & Guedes Soares, C. 2016b. Interaction of surface gravity wave motion with elastic bottom in three-dimensions. *Applied Ocean Research* 57: 125–139.

Mohapatra, S.C., Sahoo, T. & Guedes Soares, C. 2018a. Interaction between surface gravity wave and submerged horizontal flexible structures. *Journal of Hydrodynamics* 30(3): 481–498.

Mohapatra, S. C., Sahoo, T. & Guedes Soares, C. 2018b. Surface gravity wave interaction with a submerged horizontal flexible porous plate. *Applied Ocean Research* 78: 61–74.

Mohapatra, S.C. & Sahoo, T. 2014. Oblique wave diffraction by a flexible floating structure in the presence of a submerged flexible structure. *J Geophysical & Astrophysical Fluid Dynamics* 108(6): 615–638.

Ohmatsu, S. 1998. Numerical calculation of hydroelastic behaviour of VLFS in time domain. *Proceedings of the 2nd International Conference Hydroelasticity in Marine Technology*, Fukuoka, 89–97.

Song, H., Cui, W., Tao, L. & Liu, Y. 2005. Hydroelastic response of VLFS on uneven sea bottom. *Proceedings of the International Conference on Offshore Mechanics and Arctic Engineering - OMAE*, 433–443.

Utsunomiya, T., Noguchi, T., Kusaka, T., Watanabe, E., Yamamoto, S. & Ogamo, T. 2008. Hydroelastic analysis of a hybrid-type VLFS in water of variable depth. *Proceedings of the International Conference on Offshore Mechanics and Arctic Engineering - OMAE*, 3: 733–742.

Waals, O. J., Bunnik, T. H. J. & Otto, W. J. 2018. Model tests and numerical analysis for a floating mega island. *Proceedings of the International Conference on Offshore Mechanics and Arctic Engineering - OMAE*, 1: 1–11.

Watanabe, E., Utsunomiya, T. & Wang, C.M. 2004. Hydroelastic analysis of pontoon-type VLFS: a literature survey. *Engineering Structures* 26: 245–256.

Wei, W., Fu, S., Moan, T., Lu, Z. & Deng, S. 2017. A discrete-modules-based frequency domain hydroelasticity method for floating structures in inhomogeneous sea conditions. *Journal of Fluids and Structures* 74: 321–339.

Wei, W., Fu, S., Moan, T., Song, C. & Ren, T. 2018. A time-domain method for hydroelasticity of very large floating structures in inhomogeneous sea conditions. *Marine Structures* 57: 180–192.

Wu, Y. S., & Cui, W. C. 2009. Advances in the three-dimensional hydroelasticity of ships. *Proceedings of the Institution of Mechanical Engineers Part M: Journal of Engineering for the Maritime Environment* 223(3): 331–348.

Wu, You-sheng, Ding, J., Tian, C., Li, Z., Ling, H., Ma, X. & Gao, J. 2018. Numerical analysis and model tests of a three-module VLFS deployed near islands and reefs. *Journal of Ocean Engineering and Marine Energy* 4(2): 111–122.

Wu, You-sheng, Zou, M., Tian, C., Sima, C., Qi, L., Ding, J., Li, Z. & Lu, Y. 2016. Theory and applications of coupled fluid-structure interactions of ships in waves and ocean acoustic environment. *Journal of Hydrodynamics* 28(6): 923–936.

Yang, P., Liu, X., Wang, Z., Zong, Z. & Wu, Y. 2019. Hydroelastic responses of a 3-module VLFS in the waves influenced by complicated geographic environment. *Ocean Engineering* 184: 121–133.

Wang, C.M., Tay, Z.Y., Takagi, K. & Utsunomiya, T. 2010. Literature review of methods for mitigating hydroelastic response of VLFS under wave action. *Applied Mechanics Reviews* 63(3): 1-18-030802.

Wang, C.M. & Tay, Z.Y. 2011. Very large floating structures: applications, research, development. *Procedia Engineering* 4: 62–72.

Watanabe, E. & Utsunomiya, T. 1996. Transient response analysis of a VLFS at airplane landing. *Proceedings of. International Workshop on Very Large Floating Structure*. Y Watanabe (Ed.) Hayama, Kanagawa, Japan, November 25–28:243–247.

Watanabe, E., Utsunomiya, T. & Tanigaki, S. 1998. A transient response analysis of a very large floating structure by finite element method. *Structural Engineering/Earthquake Engineering*, JSCE, 15(2):155–163.

Zhang, X. & Lu, D. 2018. An extension of a discrete-module-beam-bending-based hydroelasticity method for a flexible structure with complex geometric features. *Ocean Engineering* 163(February): 22–28.

Zhou, XQ.; Sutulo, S., and Guedes Soares, C. 2016. A paving algorithm for dynamic generation of quadrilateral meshes for online numerical simulations of ship manoeuvring in shallow water. *Ocean Engineering*. 122:10–21

Developments in Maritime Technology and Engineering – Guedes Soares & Santos (eds)
© *2021 Copyright the Author(s), ISBN 978-0-367-77376-2*

Strength analysis of container ship subjected to torsional loading

K. Woloszyk, M. Jablonski & M. Bogdaniuk
Research and Development Department, Polish Register of Shipping, Gdansk, Poland
Faculty of Ocean Engineering and Ship Technology, Gdansk University of Technology, Gdansk, Poland

ABSTRACT: The objective of this work is to investigate the torsional response of a container ship, with particular consideration of the warping effect. Two different models are investigated. In the first case, the full length model of a ship is analyzed and a distributed load is applied with the use of a novel approach. The model is supported in the torsional centre of the cross-section, which is derived analytically. In the second case, the three cargo hold model is analyzed, where the influence of fore and aft parts of the ship is simulated by 'end-constraint beams'. A constant, concentrated torsional moment is applied in the shear centre of the fore and aft end cross-section. The structural responses of these two models are obtained and distributions of normal stress across the critical cross-sections are derived. The studies revealed that three cargo hold model subjected to torsional moment underestimates the level of normal stresses caused by the warping effect when compared to the full-length ship model subjected to a distributed load.

1 INTRODUCTION

In last decades, the worldwide capacity of container ships and the size of the vessels has been increasing rapidly (Tran and Haasis, 2015), with modern ships reaching up to 24 000 TEU and 400 m in length (Park and Suh, 2019). Due to this reason, the proper strength evaluation of container ships is of utmost importance. With the design striving to maximize container capacity and the need for vertical loading of cargo, the area of the strength deck has been reduced to a minimum. The combination of large length and wide cargo hatches makes the hull structure rather flexible, especially when subjected to torsional loading. Due to these factors, the classification societies' rules need prompt development to satisfy rising market demands.

The importance of torsional loading in the global analysis of container ships has been shown in several studies, both experimentally and numerically (Sun and Soares, 2003; Iijima, *et al.*, 2004; Senjanović, *et al.*, 2008; Villavicencio, *et al.*, 2015).

Nowadays, ship strength assessment is carried out with the application of FE analyses. The classification societies' rules support this application and provide guidelines accordingly, such as IACS Common Structural Rules (International Association of Classification Societies, 2018). This approach is particularly suitable for ships where hull girder response cannot be reliably determined with the use of beam theory, in which case the so-called 'three cargo hold' analysis is used.

Rörup, *et al.* (2017) has shown that the reduction of the container ship model to three cargo

holds, which is common for bulk carriers and tankers, can significantly underestimate warping stresses. The aft and fore parts of the ship restrain the cross-sections from deforming out of plane. Thus, the full-length model needs to be analysed in order to accurately reflect the influence of aft and fore parts.

The current study investigates the torsional response of a container ship with the use of two different models. First, with the full-length model where the load is applied according to newly developed methodology by the Polish Register of Shipping; second, with a model conforming to the requirements of Common Structural Rules for Oil Tankers and Bulk Carriers.

2 FINITE ELEMENT MODELING

2.1 *Analyzed ship*

In current study, the panamax container ship is analyzed with the dimensions presented in Table 1.

2.2 *Finite Element Model*

The FE model was created in a PRS in-house preprocessor. The FE model was made from QUAD4 elements for plating and primary supporting members and BAR elements for stiffeners. The calculations were performed in FEMAP with NX Nastran (Siemens PLM Software, 2019) software. The FE model is presented in Figure 1.

DOI: 10.1201/9781003216582-49

Table 1. Main ship dimensions.

Dimension	Value
Total Length L_{OA} [m]	220.00
Length between perpendiculars L_{PP}[m]	210.20
Breadth B [m]	32.24
Height H [m]	18.70
Draft T [m]	10.50

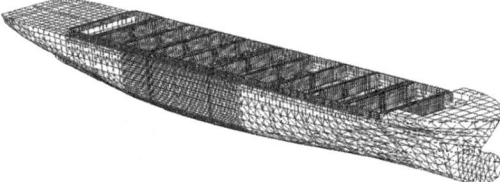

Figure 1. Finite Element model of the container ship.

The model was divided into two regions:

- Three cargo hold region in the midship area where the element size is equal to the distance between stiffeners;
- Aft and fore parts, where the element size is equal to the distance between primary supporting members.

Based on the reference model two separate models were derived, model A and B, respectively. The first one will be analyzed according to newly developed approach, which will be furtherly used in new PRS rules for container ships. The model is a full-length structural model of the ship. The second model is analyzed as prescribed in CSR for Bulk Carriers and Oil Tankers (International Association of Classification Societies, 2018). Thus, the range of the model is limited to three cargo holds and end-cross sections are additionally restrained with so-called 'end-constraint beams' to simulate the influence of fore and aft parts of the ship. The detailed loading and boundary conditions applied to both models are discussed subsequently.

2.3 Torsional moment

The loads formulas for new PRS Container Rules, which will be based on the long-term predictions of ship behaviour in different sea states are still in the development phase. Due to that, the torsional moment in still water was applied based on the BV Rules (Bureau Veritas, 2019):

$$M_{sw-t} = 0.04LB(S_hT_h + S_dT_d) \qquad (1)$$

where:
S_h - number of container stacks over the breadth in cargo hold amidships;
T_h - number of container tiers in cargo hold amidships;

S_d – number of container stacks over the breadth B on deck amidships;
T_d – number of container tiers on deck amidships.

In presented case study, the number of containers are as follows: $S_h = 11; T_h = 7; S_d = 13; T_d = 6$. The calculated M_{sw-t} is equal to 41517kNm. Based on that value, the still water torsional moment at any longitudinal position is calculated:

$$M_{sw-t-LC} = C_{WT}M_{sw-t}(f_{q-a} + f_{q-f}) \qquad (2)$$

where C_{WT} is loading factor taken as 1, and f_{q-a}, f_{q-f} are the distribution factors. The still water torsional moment along the ship length is presented in Figure 2.

The wave-induced torsional moment M_{wt} is calculated using BV rules accordingly, and it is based on the vertical shear forces values and distribution factors. The wave-induced torsional moment is presented in Figure 3, and total torsional moment (M_T) is presented in Figure 4. The maximum torsional moment is equal to 500750 kNm.

2.4 Boundary conditions – model A

The boundary conditions were realized with the use of rigid links (RBE2 elements) that connects the nodes of

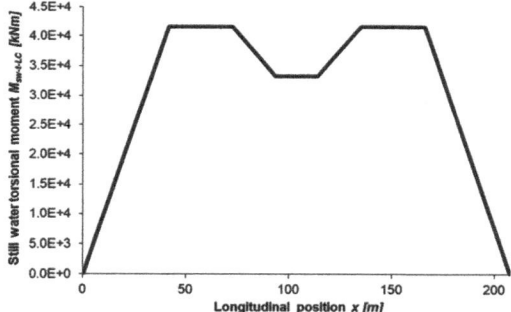

Figure 2. Still water torsional moment along the ship length.

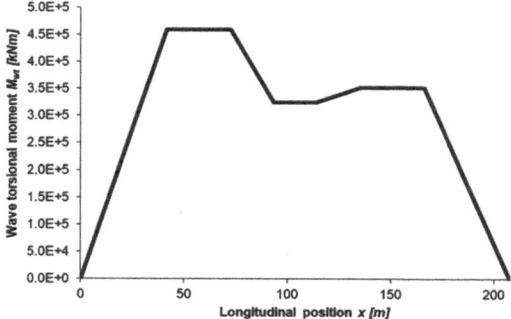

Figure 3. Wave-induced torsional moment along the ship length.

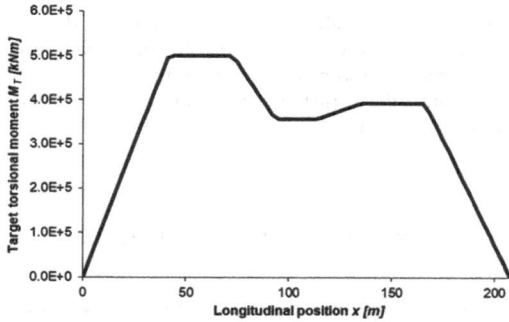

Figure 4. Total torsional moment along the ship length.

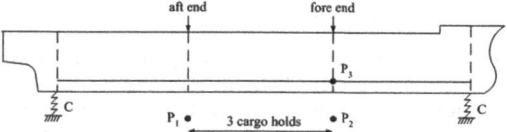

Figure 5. Position of independent points.

the selected cross-sections with the so-called 'master node'. Then, the displacement and rotations are restrained in the master node only. The master node in this case is the centre of torsion of the cross-section. The boundary conditions are presented in Table 2.

The position of points P_1, P_2 and P_3 is presented in Figure 5.

The centre of torsion position of aft and fore end cross-sections are calculated based on the simplified formula. The formula calculates the centre of torsion for an equivalent thin-walled channel section and follows the given equation:

$$z_{CT} = \tfrac{1}{2} h_{DB} - \frac{3H^2 t_H}{B_b t_b + 6H t_h} \qquad (3)$$

where:

z_{CT} – the vertical coordinate of the centre of torsion [m];

H – vertical distance between mid-height of the double bottom and top of continuous longitudinal hatch coaming or main deck level (in case the longitudinal hatch coamings are not continuous), in [m];

$B_b = B - B_s$, where B_s is the breadth of double side [m];

A_s – cross section area, in [cm^2], of all continuous hull structure members above the inner bottom level, computed for their as-built thickness decreased by $0.5t_c$;

A_b – cross section area, in [cm^2], of all continuous hull structure members below the inner bottom level, computed for their as-built thickness decreased by $0.5t_c$;

$t_H = 0.05A_s/H$ [mm];

$t_b = 0.1A_b/B_b$ [mm].

Additionally, in the extreme parts of the ship the four vertical springs are applied (see Figure 5) with the stiffness $C = 6 \cdot 10^5$ N/m. The springs are placed at bottom floors, symmetrically at the distance of b_c from C. L. (Figure 6). They are placed as far apart as possible. The application of elastic supports prevents excessive rotation of the ship cross-sections, especially in the case of a non-ideal equilibrium of torsional forces alongside the ship.

One can notice, that to obtain the torsional response of the ship, there is no need of supporting the model in the independent points. In this way, the cross-sections will rotate around the actual value of the torsional centre. However, the presented approach is suited to subsequent bending analysis. Thus, it will allow to obtain the target values of bending moments in the 'three cargo hold' region, similarly as it is required in CSR.

Table 2. Boundary conditions in Model A (RL – rigid link).

	Translation			Rotation		
Point	δ_x	δ_y	δ_z	θ_x	θ_y	θ_z
Independent points P_1 and P_2	-	Fix	Fix	-	-	-
Intersection of CL and inner bottom (point P_3)	Fix	-	-	-	-	-
Ship hull transverse cross-section at the ends of 3 cargo holds region	-	RL	RL	RL	-	-

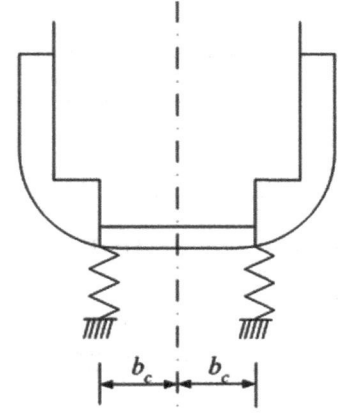

Figure 6. Springs position.

2.5 Boundary conditions – model B

Similarly, to model A, the boundary conditions were applied utilising rigid links in fore and aft ends of the three cargo hold model. In this case, the independent points were placed in the centre of gravity of the cross-sections. The summary of boundary conditions is presented in Table 3.

To simulate the stiffness of fore and aft parts of the ship end constraint beams between the cross-sectional nodes are applied. Their characteristics (moments of inertia and shear areas) are defined in CSR as a fraction of hull girder moment of inertia and cross-section.

2.6 Loading realization – model A

In model A, the load is applied in accordance to the novel PRS method. Target values of the torsional moment (see Figure 4) are achieved at web frames and transverse bulkheads with the application of a distributed vertical load. The loads are positioned at both sides of the ship, at the height of a deck closest to the neutral axis, as presented in Figure 7.

The values of distributed loads are calculated using following equation:

$$q = \frac{\left[M_{T-targ}(x_j) - M_{T-FEM}(x_j)\right] - \left[M_{T-targ}(x_{j-1}) - M_{T-FEM}(x_{j-1})\right]}{b_j l_j}$$

$$(4)$$

where:

x_j, x_{j-1} – the x-coordinate of considered subsequent web frames [m];

$M_{T-targ}(x)$ – target value of torsional moment, equal to $M_T(x)$;

$M_{T-FEM}(x)$ – value of torsional moment due to local loads (in presented work the local loads are omitted and the moment is equal to 0);

b_j – breadth, in [m], as shown in Figure 7, measured at $x = 0.5(x_j + x_{j-1})$;

$$l_j = \sqrt{(x_j - x_{j-1})^2 + (y_j - y_{j-1})^2};$$

y_j, y_{j-1} – the y-coordinates of point A, as shown in Figure 7, in [m].

2.7 Loading realization – model B

In model B, the load is applied in accordance to CSR. However, in this work the local loads were omitted (e.g. water pressure acting on bottom and sides). The target value of torsional moment, obtained at the longitudinal position of the aft bulkhead, was achieved through a concentrated load applied at the master node of fore cross section (see Table 3).

3 RESULTS AND DISCUSSION

3.1 Results - model A

The deformation of model A is presented in Figure 8. Its form is consistent with preliminary predictions. It can be noticed that in the three cargo hold region the cross-sections are not significantly out of plane.

The von Mises stresses in model A are presented in Figure 9. Only the three cargo hold region is shown. Maximum stresses can be observed in the region of transverse bulkheads, especially in the upper deck region, farthest from the centre of torsion.

3.2 Results - model B

The deformation of model B is presented in Figure 10. Similarly, to model A, the deformation is consistent

Table 3. Boundary conditions in Model B (RL – rigid link).

Point	Translation δ_x	δ_y	δ_z	Rotation θ_x	θ_y	θ_z
Aft end						
Independent point	-	Fix	Fix	Fix	-	-
Intersection of CL and inner bottom	Fix	-	-	-	-	-
Cross section	-	RL	RL	RL	-	-
End constraint beams						
Fore end						
Independent point	-	Fix	Fix	M_T	-	-
Cross section	-	RL	RL	RL	-	-
End constraint beams						

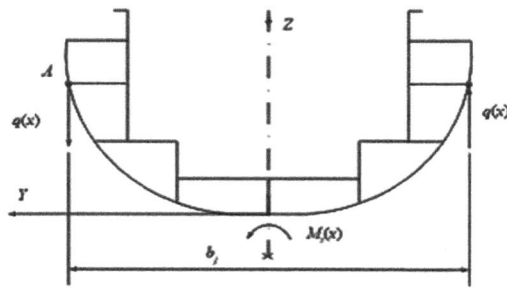

Figure 7. Realization of torsional loading.

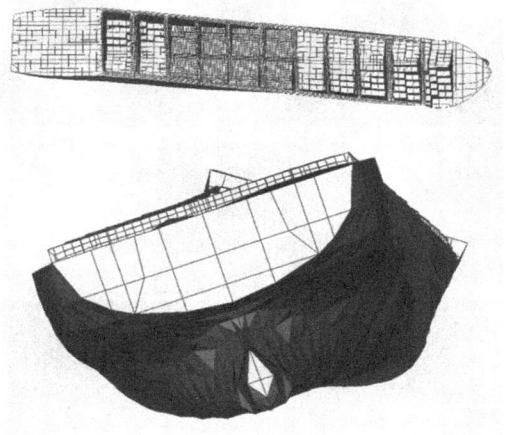

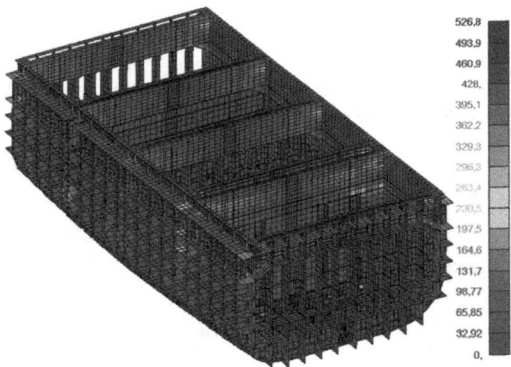

Figure 8. Deformation of model A, top view (top) and back view (bottom).

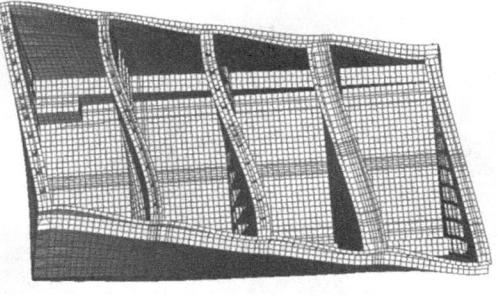

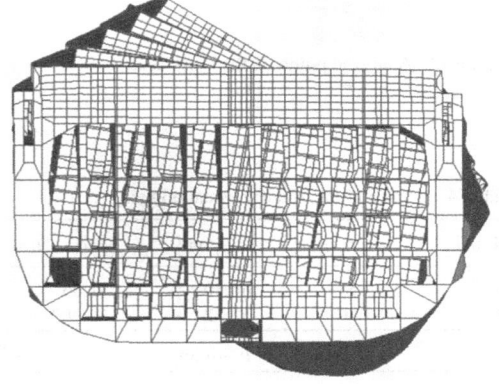

Figure 10. Deformation of model B, top view (top) and back view (bottom).

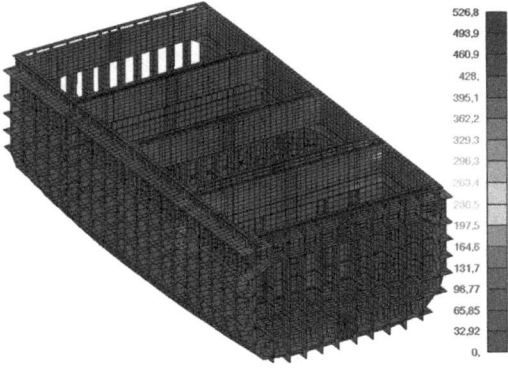

Figure 9. Von Mises stresses in the model A [MPa].

Figure 11. Von Mises stresses in the model B [MPa].

with applied loads. Additionally, compared to model A, the fore and end cross-sections are significantly out of plane.

In Figure 11, the von Mises stresses in model B are presented. Similarly, to model A, the highest stresses are in the region of transverse bulkhead and upper deck. However, the stress levels are significantly lower compared to model A.

3.3 *Detailed comparison between models A and B*

To present the differences in both models, deformations and stresses in selected regions are compared.

Firstly, the longitudinal translations of the points in the midship cross-section as presented in Figure 12 are compared. The results of the comparison are presented in Table 4.

Significant deviations between the longitudinal displacement in model A and model B can be noticed. The full-length model achieves a much

more accurate level of torsional rigidity on extreme cross sections.

The warping effect is mostly visible in upper deck regions near the transverse bulkhead. Thus, the normal stresses and von Mises stresses in two selected regions, as presented in Figure 13, are analyzed.

The distribution of normal stresses in region 1 for both models is presented in Figure 14, whereas the distribution of the shear stresses is presented in Figure 15. The summary of results, which contains

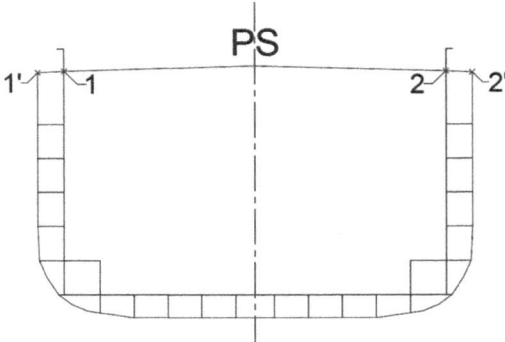

Figure 12. Analyzed points in the cross-section.

Table 4. Translations of the selected points.

Model	$u_{x,1}$ [mm]	$u_{x,2}$ [mm]	$\|u_{x,1} - u_{x,2}\|$
A	-40.53	40.50	81.03
B	-53.30	53.32	106.62
	$u_{x,1'}$ [mm]	$u_{x,2'}$ [mm]	$\|u_{x,1'} - u_{x,2'}\|$
A	-28.36	28.33	56.69
B	-50.75	50.72	101.47

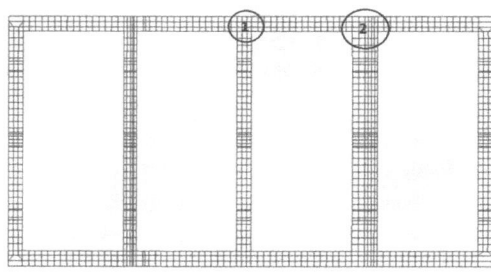

Figure 13. Analyzed deck regions.

maximum absolute values of normal, shear and von Mises stresses, is presented in Table 5.

The differences between stresses are in the range of 20-38%. It can be concluded that in the three cargo hold model, the warping stresses as well as shear stresses are significantly underestimated. The significant deviations are mainly originated from the differences in the deflection restraining of the fore and end cross-sections in both models. In model B, the end-constraint beams are letting the cross-sections easily to deflect (which is visible in translations results presented in Table 4), and thus the stress levels are significantly lower.

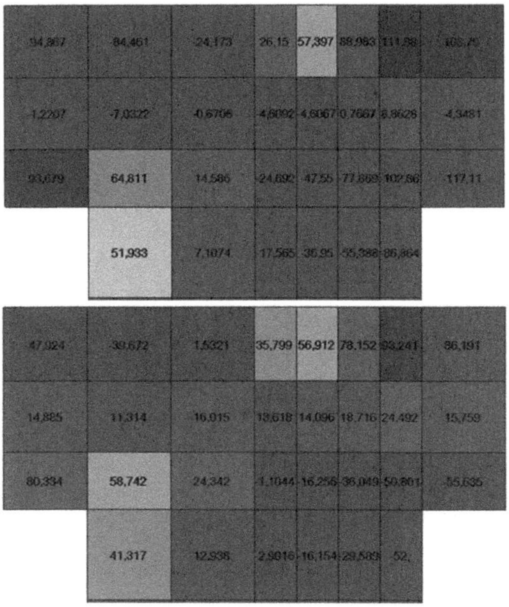

Figure 14. Normal stresses in longitudinal direction [MPa] in region 2 in model A (top) and model B (bottom).

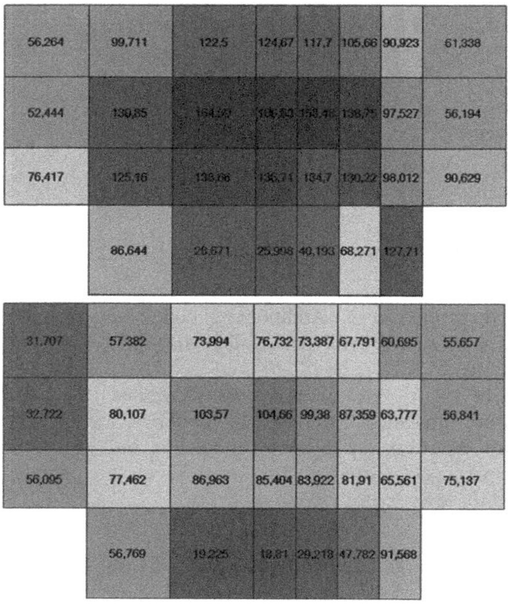

Figure 15. Shear stresses [MPa] in region 2 in model A (top) and model B (bottom).

3.4 Influence of centre of torsion position

In the case of model, A, the centre of torsion was derived using a simplified formula. As the actual position could slightly differ, its influence on the

Table 5. Maximum stresses in model A and B.

Region	Stresses [MPa]	Model A	Model B	Difference [%]
1	Normal	60.0	38.5	36
	Shear	120.2	79.7	38
	Von Mises	208.3	138.2	34
2	Normal	117.1	93.3	20
	Shear	166.4	103.5	38
	Von Mises	298.7	209.4	30

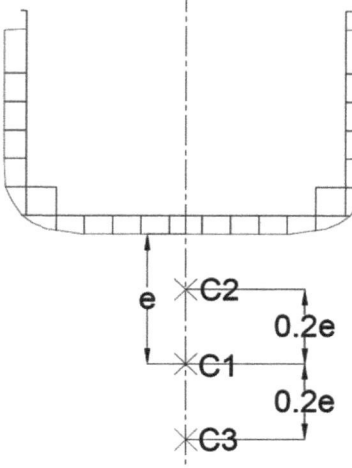

Figure 16. Positions of analyzed torsional centre.

Table 6. Translations for different positions of torsional centre.

Z_{CT}	$u_{x,1'}$ [mm]	$u_{x,2'}$ [mm]	$\lvert u_{x,1'} - u_{x,2'} \rvert$	Δ [%]
C1	-23.51	23.49	47.00	-
C2	-28.72	28.70	57.42	22.2
C3	-18.30	18.28	36.58	22.2

structural response is investigated. Three different positions are analyzed, as presented in Figure 16.

The initially calculated position of torsional centre (C1) is equal to e = 7.03m. The two other positions (C2 and C3) are analyzed. The results of longitudinal translations of point 1' and 2' (Figure 12) are presented in Table 6. As can be noticed, the differences in translation are directly proportional to the position change of the torsional centre.

The analysis of von Mises stresses shows that there are virtually no differences between the different positions of torsional centre. Based on the presented results it could be concluded that the simplified calculation of torsional centre will not have significant impact on the structural response.

3.5 Model A validity

To verify the validity of results obtained from the analysis of model A stress values were compared with those of a model similar to model A, but devoid of vertical supports at the end sections of the three cargo hold region. Instead, constraints prohibiting y-translation of the extreme nodes on the main deck of the model were applied. In elements with the highest stress values the difference between the two models was less than 2%. Thus, the obtained results are considered to accurately represent the distribution of stresses in the hull structure.

4 CONCLUSIONS

The work presented here proves, based on the FE analysis, that in order to provide reliable results of deformations and stresses of the container ship hull subjected to asymmetrical loads, it is necessary to use the full-length model of the ship. The presented work gives the basis to contest the validity of the methodology in which the torsional stiffness of the ship is modelled with the use of 'end-constraint beams', as it is required in IACS Common Structural Rules. The results obtained using this methodology underestimates the stresses even up to 50 % in some elements, when compared to the approach developed in this paper.

Despite the higher workload, structural analysis with the use of full-length models should be required in classification of container ships. Nowadays, the computational power of CPUs allows calculation of large and complex models in a few minutes. Additionally, the coarse mesh in fore and aft parts significantly decreases the computational time and modeling effort.

The methodology developed in PRS also shows little sensitivity of structural response with respect to position of torsional centre. This leads to a conclusion, that simplified formula for derivation of the centre of torsion could be used, as it has no influence on stress levels.

REFERENCES

Bureau Veritas (2019) Structural Rules for Container Ships NR625.

Iijima, K., Shigemi, T., Miyake, R. and Kumano, A. (2004) 'A practical method for torsional strength assessment of container ship structures', Marine Structures, 17(5), pp. 355–384.

International Association of Classification Societies (2018) Common Structural Rules (BC & OT).

Park, N. K. and Suh, S. C. (2019) 'Tendency toward Mega Containerships and the Constraints of Container Terminals', Journal of Marine Science and Engineering, 7(5), p. 131.

Rörup, J., Darie, I. and Maciolowski, B. (2017) 'Strength analysis of ship structures with open decks', Ships and Offshore Structures, 12(sup1), pp. S189–S199.

Senjanović, I., Tomašević, S., Rudan, S. and Senjanović, T. (2008) 'Role of transverse bulkheads in hull stiffness of large container ships', Engineering Structures, 30(9), pp. 2492–2509.

Siemens PLM Software (2019) 'FEMAP with NX Nastran 11.3'.

Sun, H.-H. and Soares, C. G. (2003) 'An experimental study of ultimate torsional strength of a ship-type hull girder with a large deck opening', Marine Structures, 16 (1), pp. 51–67.

Tran, N. K. and Haasis, H.-D. (2015) 'An empirical study of fleet expansion and growth of ship size in container liner shipping', International Journal of Production Economics, 159, pp. 241–253.

Villavicencio, R., Zhang, S. and Tong, J. (2015) 'A study of cross deck effects on warping stresses in large container ships', in Guedes Soares, C. and Shenoi, R. A. (eds) Analysis and Design of Marine Structures V. CRC Press, pp. 385–393.

Developments in Maritime Technology and Engineering – Guedes Soares & Santos (eds)
© 2021 Copyright the Author(s), ISBN 978-0-367-77376-2

Review of digital twin of ships and offshore structures

B.Q. Chen & C. Guedes Soares
Centre for Marine Technology and Ocean Engineering (CENTEC), Instituto Superior Técnico, Universidade de Lisboa, Lisbon, Portugal

P.M. Videiro
Federal University of Rio de Janeiro, Rio de Janeiro, Brazil
Centre for Marine Technology and Ocean Engineering (CENTEC), Instituto Superior Técnico, Universidade de Lisboa, Lisbon, Portugal

ABSTRACT: The digitalization of large structures as ships and offshore platforms supports the application of sophisticated virtual product models, which are referred to as digital twins, throughout all stages of construction and operation life. Particularly, more realistic virtual models of the structural systems are essential to bridge the gap between design and construction and to mirror the real and virtual worlds. In this paper, the vision of the digital twin and its evolution is reviewed, aiming at providing a coverage of the current applications and the challenges of the digital twins of the physical ships and offshore structures in design, construction and service life and enabling technologies along with recommendations and reflections.

1 INTRODUCTION

In today's highly competitive markets, the ambitions for shortening the time to market and for increasing the product development performance fuel the application of sophisticated virtual product models, which are frequently referred to as digital twins (DT) (Schleich et al. 2017).

The idea of using a twin model can be dated back to NASA's Apollo program where two identical space vehicles were built to allow mirroring the conditions of the space vehicle during the mission (Rosen et al. 2015). With the paired module, NASA engineers were able to test and refine potential recovery strategies on earth before giving instructions to the crew. In this sense, all types of prototypes used to mirror the real operating conditions for simulation of the real time behavior can be regarded as a twin.

In general, the digital twin is defined as an integrated multi-physics, multi-scale, probabilistic simulation of a complex product and uses the best available physical models, sensor updates, etc., to mirror the life of its corresponding twin (Glaessegen & Stargel 2012). Table 1 lists some key events during the evolutions of DT, including also the events in early ages before the introduction of the concept of DT. The model of digital twin was first introduced in 2002 as a concept for product lifecycle management (PLM). After its initial names of mirrored spaces model (MSM) and information mirroring model (IMM), the model was finally referred to as the Digital Twin in 2011 (Grieves 2019).

As shown in Figure 1, the digital twin is the digital or virtual version of the smart, connected product system (SCPS). The DT model continually adapts to changes in the environment or operation using real-time sensory data. Therefore, it can be used to forecast the future of the corresponding physical assets, to monitor and identify potential issues with the real physical twin (PT), and to allow the prediction of the remaining useful life (RUL) of PT by leveraging a combination of physics-based models and data-driven analytics (Kaur et al. 2020).

The DT concept consists of three distinct parts: the physical product, the digital or virtual product, and connections between the two products (Tao et al. 2018). These connections are the data that flow from the physical product to the digital or virtual product and the available information from the digital or virtual product to the physical environment.

Some of the applications of DT were in NASA's spacecraft (Caruso et al. 2010; Piascik et al. 2010, Glaessgen & Stargel 2012) and jet fighters by the US Air Force (Tuegel 2012). Later on, DT was used by many main PLM vendors like Dassault Systèmes, PTC, and Siemens. TESLA aims at developing a digital twin for every built car, hence enabling synchronous data transmission between the car and the factory. More recently, the digital twin model has been proposed for robust deployment of the Internet of Things (IoT) (Maher, 2018), and been applied to a wide range

DOI: 10.1201/9781003216582-50

Table 1. List of evolutions of digital twins.

Year	Event
1970	NASA pairing technology on Apollo 13 mission
1977	Flight simulators with computer simulation
1982	AutoCAD is born
1990s	AutoCAD used in nearly all engineering & design
2002	Dr. Grieves' concept of a digital twin emerges
2011	NASA & USAF papers on digital twin
2015	GE digital wind farm initiative
2017	Gartner lists digital twins as a top 10 tech trend
2018	DT used in major software & industrial companies

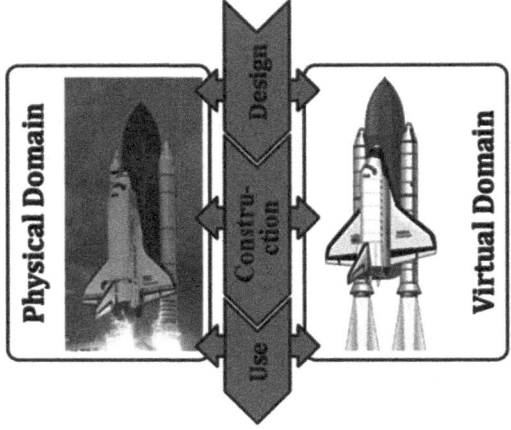

Figure 1. The vision of the digital twin throughout the product life cycle (Schleich et al. 2017).

of industries including aerospace (e.g. Li et al. 2017, Mandolla et al. 2019), automotive (e.g. Magargle et al. 2017, Damjanovic-Behrendt 2018), healthcare (e.g. Torkamani et al. 2017, Laaki et al. 2019, Liu et al. 2019, Jimenez et al. 2020), manufacturing (Post et al. 2017, Bilberg & Malik 2019, Howard 2019), smart city (e.g. Mohammadi & Taylor 2017, Ruohomäki et al. 2018), and many more (see in Figure 2).

The European H2020 project MAYA aims at developing a plant DT supporting activities in all factory lifecycle phases: from the design, through the optimization of the operational life, to the dismissal phase (Negri et al. 2017).

General Electric (GE) used DT to forecast the health and performance of their products over lifetime. They have also built a digital wind farm to redefine the future of wind power. Through collecting real-time data (e.g., weather, component messages, service reports) continuously, a digital twin can be built for each wind turbine to optimize equipment maintenance strategy, improve reliability, and increase annual energy production. A 20% gain in efficiency is expected to be achieved based on the application of DT (Tao et al. 2019).

Despite its fast development, the digital twin has a number of problems in practical use. A digital twin may fail to echo what is going on in the real world

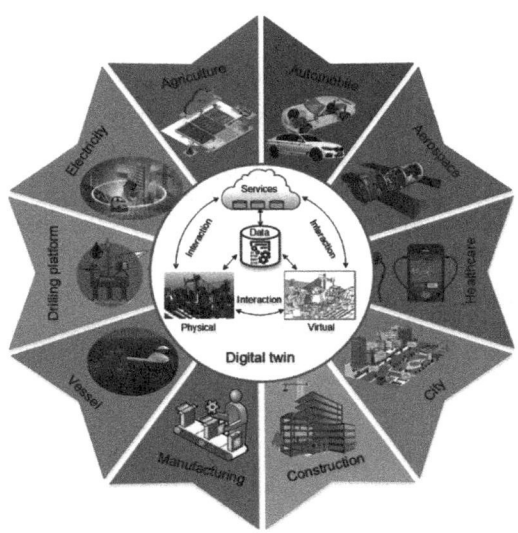

Figure 2. Digital twin applications in different fields (Qi et al. 2020).

and lead managers to make poor decisions (Tao & Qi 2019). There are no common methods, standards or norms. It can be difficult to aggregate data from thousands of sensors that track vibration, temperature, force, speed and power, for example. Furthermore, data can be spread among many owners and be held in various formats.

There are few connections between industry and academia, in part because of commercial secrecy. Most academic research focuses on improving modelling techniques rather than on optimizing data and implementing digital twins. The following steps are suggested to make research and development of digital twins more coherent: to unify data and model standards, to create a public database for sharing data and models, to develop products and services to help digital twins become easier to build and use, and to establish forums for practitioners and researchers.

The concept of DT has also been used in large structures as ships and offshore platforms throughout the stages of construction and operation life. In this paper, the vision of DT and its evolution, and the applications of DT in ships and offshore structures in design, construction and service life are reviewed. Further discussions and comments are also addressed.

2 DITIGAL TWIN APPLICATION IN SHIPBUILDING

A DT in shipbuilding means a digital copy of a real ship, and is used to help optimize the design, maintenance, production and sustainability of the physical ship (see Figure 3). The exact digital replicas will be updated throughout the lifecycle of the ships. Hence, the DT of the physical ships are helpful in the development of new vessels. With data gathered from the

Figure 3. Example of a digital twin (credit: Newport News Shipbuilding).

lifecycle of a vessel, one can generate complex operational simulations and, when needed, adjust the ship's design or layout using technologies like IoT, cloud computing, machine learning, augmented reality (AR) and virtual reality (VR).

It was reported that the Shipbuilding 4.0, at the principles of the Industry 4.0, will transform the design, manufacturing, operation, shipping, services, production systems, maintenance and value chains in all aspects of the shipbuilding industry (Stanic et al. 2018). The DT concept provides the ability to reproduce ships digitally, to reduce the effort involved in ship maintenance, and to mitigate safety risks. Today, there are estimates that the DT concept will be widely used in the near future. According to a prediction by Gartner, half of large industrial companies will use digital twins by 2021, achieving 10% improvement and effectiveness (Pettey 2017).

The US-based IT provider DXC Technology uses artificial intelligence (AI) on a DT setup to predict critical parts failures, simulating how a fire can spread during an emergency or an attack (see Figure 4), especially for navy and combat ships.

Through the use of DT, the subsequent spread of fire as happened on the HMAS Westralia in 1998 can be predicted. It has determined in the simulation that there are eight walls vulnerable to fire and they should be better insulated. It has also determined

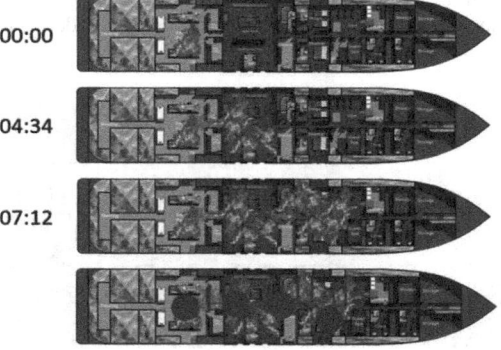

Figure 4. Simulation of fire spread and identification of weak points using DT and AI (Bernard 2019).

that four zones are not protected by fire suppression. Such simulation is not just based on the physical layout of the ship but also on a number of other factors such as compartment materials, flammable substances, ventilation, installed fire-suppression systems and so forth.

In this simulation of fire spread case, the physical and behavioral attributes are tagged to the DT, so the AI layer can determine factors such as how long it will take to burn through different materials (Bernard 2019).

The shipbuilding company Navantia specializes in DT-based simulations for tracking ships' life cycles, from the design stage to the delivery to the customer and beyond (Rivas 2018). The DT allows to link the physical with the virtual environment of the product by applying technologies such as cloud computing, machine learning or IoT. The data obtained during the whole life cycle of ships are analyzed in the virtual environment to generate simulations which identify corrective measures and recommend preventive actions. To digitalize the company's shipbuilding process, Navantia has announced most recently a joint agreement with Siemens.

The Bureau Veritas Group has partnered with Dassault Systèmes to develop a sophisticated asset integrity management platform. The Veristar AIM[3D] platform can provide a digital twin of marine or offshore asset for use throughout its life, allow significant cost savings in operational expenses, improve the safety, facilitate smooth data collection and reporting, and help decide on proper maintenance and eventual decommissioning of vessels at the end of their lifecycle. It was reported by Bureau Veritas. (2019) that seven generated digital twins for floating units (FPSOs, FSUs, FSRUs) for four different clients and a 25% operating cost reduction over a five-year period.

GE attempts to work with Military Sealift Command (MSC) to improve performance and mission readiness (Tao et al. 2019). High-speed data sampling for critical marine equipment is enabled to build a digital twin for the equipment. Enabled by the Predix platform, the difference between the real-time data from the physical twin and the simulated data from the DT can be detected. The difference may represent a performance degradation leading to a potential failure. Thus, the corresponding problem can be identified and solved, the reliability and availability of the equipment is increased, and the maintenance cost is reduced.

MPA Singapore has agreed with Keppel Offshore & Marine and the Technology Centre for Offshore and Marine (TCOMS) to jointly develop autonomous vessels for a variety of applications including undertaking harbor operations such as channeling, berthing, mooring, and towing operations. A DT of a 65-meter tugboat is used to simulate vessel behavior in multiple scenarios. Data analytics tools to improve the control and response of the tug will enhance situational awareness for safer shipping operations (Keppel Offshore & Marine 2019).

DNV GL attempts to establish a virtual sister ship for the physical twin to reduce costs, improve efficiency, and boost safety during the vessel's lifecycle through enabling collaboration among designers, builders, operators, and others. It has analyzed values brought by DT from the perspectives of different stakeholders in the maritime industry (DNV GL 2018):

- For ship owners: supporting visualization of ships and subsystems, data qualification and analytics, optimization of performance, internal and external communication, autonomous operations, and decommissioning;
- For equipment manufacturers: providing tools for system integration, performance demonstration, system quality assurance, and additional services.
- For authorities: offering a systematic framework to automatically produce information and high-quality reports;
- For universities and maritime academies: supporting related research and education, and training future and existing maritime personnel; and
- For consultancies: providing business opportunities.

Rolls-Royce Marine, in partnership with the Norwegian University of Technology Science (NTNU), DNV GL, SINTEF Ocean and Hyundai Heavy Industries, develops an open-source platform called Open Simulation Platform (OSP) to standardize the creation of digital twin models. They use cloud computing and machine learning technologies to simulate changes to and identify aspects that affect the ship's performance.

With digital twin, the traditional marine industry can benefit from advanced digital technologies, and predictive analytics for the critical marine equipment and PLM for the entire ship will be application spots in the future (see Figure 5).

Having an exact digital record of vessels is a key to keeping ships in-service and effective in their roles for longer. During a ship's lifespan with many possible changes, the digital model requires working with information that comes from different sources, time periods, and technologies.

3 DIGITAL TWIN APPLICATION IN OFFSHORE STRUCTURES

The oil industry is exploring the use of digital twin for ocean-based production platforms (Renzi et al. 2017). Digital twins could be of massive value to oil and gas operations and maintenance, since they can help operating teams better understand risks, help create an executable plan and work schedule, and help identify and manage changes in design (Menard 2017).

Figure 6 displays a visualization of the digital twin technology leveraged in the Ivar Aasen project off the Norwegian coast. Under the agreement, Aker BP and Siemens will develop digital lifecycle automation and performance analytics solutions for all future assets in the field.

In China, the ultra-deep water dual-rig semi-submersible drilling platform "Blue Whale I", from China International Marine Containers (CIMC), has been successfully used to exploit combustible ice in the South China Sea since 2017 (Zhang 2018). Advanced technologies such as AR/VR, IoT, virtualization, and data integration are used to form the DT of the platform, which has realized the visualization display, operation and maintenance management, research and development process display, and has provided an efficient remote means for state monitoring and data collection.

The British Petroleum (BP) has used DT to model physical projects, such as new oil fields and associated infrastructure (Zborowski 2018). They have a DT of a gas-collection facility in snowbound Alaska, and uses it to identify equipment for decommissioning, plan maintenance and perform planning for equipment installation.

Knezevic et al. (2018) combined two new technologies: reduced basis FEA (RB-FEA) and fast full load mapping (FFLM) for fast and powerful structural and hydrodynamic analyses (see Figure 7). The combined RB-FEA/FFLM analysis enables new capabilities for structural integrity management, and is viewed as a true enabler of DT of floating assets (e.g. semi-submersibles and FPSOs).

In the case study on a DT of a drilling semi-submersible considering a relatively coarse model with

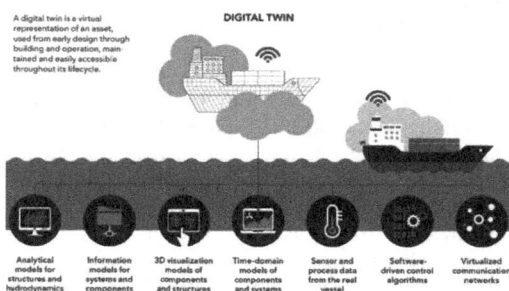

Figure 5. Elements of a digital twin ship (DNV GL 2018).

Figure 6. A visualization of the digital twin technology leveraged in the Ivar Aasen project.

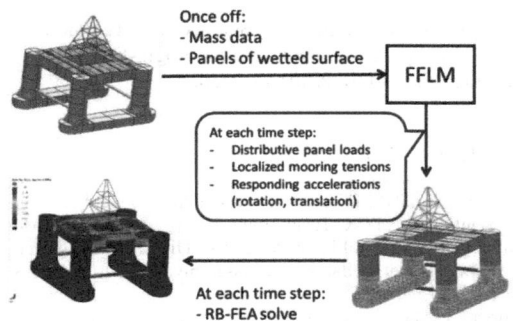

Figure 7. Overview of digital twin by combination of FFLM and RB-FEA (Knezevic et al. 2018).

400,000 FEA degrees of freedom, the RB-FEA solution requires 0.4 seconds, compared to 45 seconds for FEA. For each time step, wave pressures (generated by software such as WAMIT/WADAM) are mapped onto the hull, and the RB-FEA solve is performed in less than one second (Sharma et al. 2018). It was concluded that the approach makes it feasible to perform the entire hydrodynamic analysis in real-time.

Being implemented on one of Shell's platforms in North Sea (nearing the end of its design life of 50 years), the DT concept combined with operational modal analysis (OMA) and loading from real-time environmental loading significantly increases the predicted fatigue life of assets compared to current methodologies (Pedersen et al. 2019, Knezevic et al. 2019).

Sivalingam et al. (2018) proposed a novel methodology to predict the remaining useful life (RUL) of an offshore wind turbine in DT framework as a means of predictive maintenance strategy (see Figure 8). Virtual sensors are placed in the numerical turbine for predicting RUL for optimum operating and maintenance strategy comparison, from which is derived the risk of failure for each failure modes.

4 FURTHER DISCUSSION

Digitalization is a powerful tool for improving efficiency during construction works and for assurance of required safety levels during operation life of ships and offshore structures.

Qi et al. (2020) classified the tools for DT service applications into platform service tools, simulation service tools, optimization service tools, diagnostic and prognosis service tools, and listed the tools in each category (see Figure 9).

To add physical realism to DT, there are many open-source and commercial multi-physics simulation packages being used to solve the governing equations derived through physical modeling. For example, ANSYS Twin Builder launched in early 2019, containing extensive application-specific libraries and features third-party tool integration, is an appropriate software tool for DT modeling. The Twin Builder can enable engineers to quickly build, validate and deploy the digital models of physical assets. The built-in libraries provide rich components to create the desired system dynamics models at an appropriate level of detail. The GE has used a ANSYS DT to design megawatt-sized electric circuit breakers (Rasheed et al. 2019).

On the other hand, a challenge therefore also lays in properly using the new tools. As with all new concepts, there are both obstacles and possibilities in further application of the digital twin. In the marine application of DT, the following points could be considered to be elaborated:

- The DT construction shall start at design phase of ships and offshore structure and shall be updated with as built and as installed structure conditions.
- The DT should be calibrated during initial phase of operation life to reproduce structural response due to environmental conditions (i.e. waves, wind and current).
- The calibration of the twin model requires installation of sensors in the structure to read structural

Figure 8. 5MW Wind turbine configuration for fixed and floating application: Schematics of information mirroring model (Sivalingam et al. 2018).

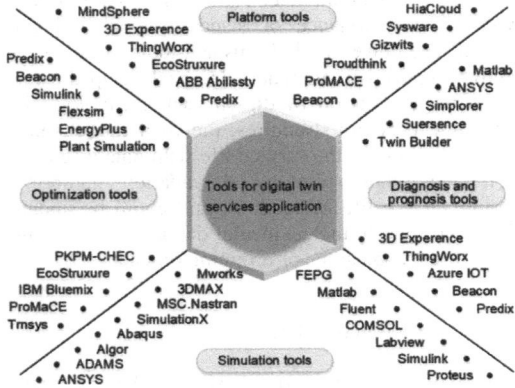

Figure 9. Tools for digital twin services applications (Qi et al. 2020).

response (tensions or strains) at key structural points and equipment for gathering environmental data.

- With the calibrated twin model, fatigue damage can be evaluated in real-time during the service life, considering actual environmental conditions. Thus, the owner and other authorities will be able to know remaining fatigue life and issue actions to optimize fatigue life or provide improvements or reinforcements in structural details to increase life.
- The twin model must be updated during the service life with data of corrosion and other structure degradation gathering during service life inspections.
- The twin model shall be constructed in such way that in case of accidents, as collisions, fires, explosions or dropping objects, which result in structural damages, the DT could be updatable in a fast way according to accident damage, permitting a quick evaluation of the asset safety and providing information for damage repair.
- The information provided by twin models regarding remaining fatigue life will be valuable for assessment of extension of asset service life.

5 CONCLUSIONS

With the recent wave of digitalization, the digital twin has been discussed as a very powerful technology in a variety of industries. The emergence of the digital twins provides an efficient way to realize remote monitoring and control, downtime prediction, and risk reduction for related infrastructure, equipment, and process in the shipbuilding and the oil and gas industry, which can greatly improve the management efficiency.

This paper provides a detailed coverage of the current applications and the challenges of the digital twins of the physical ships and offshore structures in design, construction and service life and enabling technologies along with recommendations and reflections.

More research is needed to improve traditional data collection and processing methods and to implement the communication interface between real and physical twins. The importance of standardization is highlighted. The universal platforms and tools for the digital twin applications are required to be developed.

In the application of digital twin technology in ships and offshore structures, the twin model must be updated during the service life with data of corrosion and other structural degradation gathering during service life inspections, and could be updatable according to the accidental damages from collisions, fires, explosions or dropping objects.

ACKNOWLEDGEMENTS

This work was performed within the Strategic Research Plan of the Centre for Marine Technology and Ocean Engineering, which is financed by Portuguese Foundation for Science and Technology (Fundação para a Ciência e Tecnologia-FCT) under contract UIDB/UIDP/00134/2020.

REFERENCES

Bernard, A. 2019. Digital twins revolutionize shipbuilding. White paper. DXC Technology.

Bilberg, A. & Malik, A.A. 2019. Digital twin driven human–robot collaborative assembly. *CIRP Annals* 68 (1): 499–502.

Bureau Veritas. 2019. Twintelligence: the changing face of asset management. *VERISTAR MAG*: 11.

Caruso, P., Dumbacher, D. & Grieves, M. 2010. Product lifecycle management and the quest for sustainable space exploration. *AIAA SPACE 2010 Conference & Exposition, Anaheim, California, 30 August - 02 September 2010*.

Damjanovic-Behrendt, V. 2018. A digital twin-based privacy enhancement mechanism for the automotive industry. *International Conference on Intelligent Systems (IS), Madeira, Portugal, 25-27 September 2018*, pp. 272–279.

DNV GL. 2018. Digital twins for blue Denmark. Digital twin report for DMA. Report No.: 2018–0006, Rev. A.

Howard, D. 2019. The digital twin: virtual validation in electronics development and design. 2019 Pan Pacific Microelectronics Symposium, 11–14 February 2019, Kauai, HI, USA. pp. 1–9.

Glaessgen, E. & Stargel, D. 2012. The digital twin paradigm for future NASA and US Air Force vehicles. *53rd AIAA/ASME/ASCE/AHS/ASC Structures, Structural Dynamics and Materials Conference, Hawaii, 23–26 April 2012*.

Grieves, M. 2019. Virtually intelligent product systems: digital and physical twins. In Flumerfelt, et al. (ed.), *Complex systems engineering: theory and practice*: 175–200. Reston: American Institute of Aeronautics and Astronautics.

Jimenez, J.I., Jahankhani, H. & Kendzierskyj, S. 2020. Health care in the cyberspace: medical cyber-physical system and digital twin challenges. In: *Digital Twin Technologies and Smart Cities*. Springer, Cham, pp. 79–92.

Kaur, M.J., Mishra, V.P. & Maheshwari, P. 2020. The convergence of digital twin, IoT, and machine learning: transforming data into action. In *Digital Twin Technologies and Smart Cities*: 3–17. Cham: Springer.

Keppel Offshore & Marine. 2019. Transforming business with digitalization. *Offshore Marine - A newsletter of Keppel Offshore & Marine* 4:12–13.

Knezevic, D., Kang, H., Sharma, P., Malinowski, G. & Nguyen, T.T. 2018. Structural integrity management of offshore structures via RB-FEA and fast full load mapping based digital twins. *28th International Ocean and Polar Engineering Conference, Sapporo, Japan, 10–15 June 2018*.

Knezevic, D., Fakas, E. & Riber, H.J. 2019. Predictive digital twins for structural integrity management and asset life extension - JIP concept and results. *Society of Petroleum Engineers Offshore Europe Conference and Exhibition, Aberdeen, UK, 3–6 September 2019*. SPE-195762-MS.

Laaki, H., Miche, Y. & Tammi, K. 2019. Prototyping a digital twin for real time remote control over mobile networks: application of remote surgery. *IEEE Access* 7: 20325–20336.

Li, C., Mahadevan, S., Ling, Y., Choze, S. & Wang, L. 2017. Dynamic Bayesian network for aircraft wing health monitoring digital twin. *AIAA Journal* 55(3): 930–941.

Liu, Y., Zhang, L., Yang, Y., Zhou, L., Ren, L., Wang, F., Liu, R., Pang, Z. & Deen, M.J., 2019. A novel cloud-based framework for the elderly healthcare services using digital twin. *IEEE Access* 7: 49088–49101.

Magargle, R., Johnson, L., Mandloi, P., Davoudabadi, P., Kesarkar, O., Krishnaswamy, S., Batteh, J. & Pitchaikani, A. 2017. A simulation based digital twin for model-driven health monitoring and predictive maintenance of an automotive braking system. *Proceedings of the 12th International Modelica Conference, Prague, Czech Republic, May 15-17 2017*, pp. 35–46.

Maher, D.P. 2018. On Software Standards and Solutions for a Trusted Internet of Things. *51st Hawaii International Conference on System Sciences (HICSS-51), Hawaii, 3–6 January 2018*.

Mandolla, C., Petruzzelli, A.M., Percoco, G. & Urbinati, A. 2019. Building a digital twin for additive manufacturing through the exploitation of blockchain: a case analysis of the aircraft industry. *Computers in Industry* 109: 134–152.

Menard, S. 2017. 3 ways digital twins are going to help improve oil and gas maintenance and operations. https://www.linkedin.com/pulse/3-ways-digitaltwins-going-help-improve-oil-gas-sophie-menard.

Mohammadi N. & Taylor J.E. 2017. Smart city digital twins. 2017 *IEEE Symposium Series on Computational Intelligence (SSCI), Honolulu, 27 November -1 December 2017*, pp. 1–5.

Negri, E., Fumagalli, L. & Macchi, M. 2017. A review of the roles of digital twin in cps-based production systems. *Procedia Manufacturing* 11: 939–948.

Pedersen, E.B., Jørgensen, D., Riber, H.J., Ballani, J., Vallaghé, S. & Paccaud, B. 2019. True fatigue life calculation using digital twin concept and operational modal analysis. *The 29th International Ocean and Polar Engineering Conference, Honolulu, Hawaii, USA, June - 16–21 2019*. ISOPE-I-19-354, pp. 3969–3977.

Pettey, C. 2017. Prepare for the impact of digital twins. Gartner report. available at https://go.nature.com/2krzbjd

Piascik, R., Vickers, J., Lowry, D., Scotti, S., Stewart, J. & Calomino, A. 2010. Technology area 12: Materials, structures, mechanical systems, and manufacturing road map. *NASA Office of Chief Technologist*.

Post, J., Groen, M. & Klaseboer, G. 2017. Physical model based digital twins in manufacturing processes. *10th Forming Technology Forum, Enschede, The Netherlands, 12–13 October 2017*.

Qi, Q., Tao, F., Hu, T., Anwer, N., Liu, A., Wei, Y., Wang, L. & Nee, A.Y.C. 2020. Enabling technologies and tools for digital twin. *Journal of Manufacturing Systems*. DOI: 10.1016/j.jmsy.2019.10.001.

Rasheed, A., San, O. & Kvamsdal, T. 2019. Digital twin: values, challenges and enablers. *Electrical Engineering and Systems Science*:1910.01719.

Renzi, D., Maniar, D., McNeill, S. & Del Vecchio, C., 2017, October. Developing a digital twin for floating production systems integrity management. *Offshore Technology Conference Brasil, Rio de Janeiro, 24–26 October 2017*.

Rivas, Á.R. 2018. Navantia's Shipyard 4.0 model overview. *Ship Science & Technology* 11(22): 77–85.

Rosen, R., von Wichert, G., Lo, G. & Bettenhausen, K.D. 2015. About the importance of autonomy and digital twins for the future of manufacturing. *IFAC-PapersOnLine* 48(3): 567–572.

Ruohomäki, T., Airaksinen, E., Huuska, P., Kesäniemi, O., Martikka, M. & Suomisto, J. 2018. Smart city platform enabling digital twin. *International Conference on Intelligent Systems (IS), Madeira, Portugal, 25-27 September 2018*, pp. 155–161.

Schleich, B., Anwer, N., Mathieu, L. & Wartzack, S. 2017. Shaping the digital twin for design and production engineering. *CIRP Annals* 66(1): 141–144.

Sharma, P., Knezevic, D., Huynh, P. & Malinowski, G. 2018. RB-FEA based digital twin for structural integrity assessment of offshore structures. *Offshore Technology Conference, Houston, Texas, USA, 30 April - 3 May 2018*. OTC-29005-MS.

Sivalingam, K., Sepulveda, M., Spring, M. & Davies, P. 2018. A review and methodology development for remaining useful life prediction of offshore fixed and floating wind turbine power converter with digital twin technology perspective. *2nd International Conference on Green Energy and Applications, Singapore, 24-26 March 2018*. pp. 197–204.

Stanic, V., Hadjina, M., Fafandjel, N. & Matulja, T. 2018. Toward Shipbuilding 4.0 - an Industry 4.0 changing the face of the shipbuilding industry. *Brodogradnja/Shipbuilding* 69(3): 111–128.

Tao, F., Cheng, J., Qi, Q., Zhang, M., Zhang, H. & Sui, F. 2018. Digital twin-driven product design, manufacturing and service with big data. *The International Journal of Advanced Manufacturing Technology* 94: 3563–3576.

Tao, F. & Qi, Q. 2019. Make more digital twins. *Nature* 573: 490–491.

Tao, F., Zhang, M. & Nee, A.Y.C. 2019. *Digital Twin Driven Smart Manufacturing*. London: Academic Press.

Torkamani, A., Andersen, K.G., Steinhubl, S.R. & Topol, E.J. 2017. High-definition medicine. *Cell* 170 (5):.828–843.

Tuegel, E. 2012. The airframe digital twin: some challenges to realization. *53rd AIAA/ASME/ASCE/AHS/ASC Structures, Structural Dynamics and Materials Conference, Honolulu, Hawaii, 23–26 April 2012*.

Zborowski, M. 2018. Finding meaning, application for the much-discussed "Digital Twin". *Journal of Petroleum Technology* 70(6): 26–32.

Zhang, G. 2018. Proposals on the Development of the Marine Economy, Deepwater Energy and Marine Engineering Equipment. *China Oil & Gas* 25(1): 3–8.

Developments in Maritime Technology and Engineering – Guedes Soares & Santos (eds)
© 2021 Copyright the Author(s), ISBN 978-0-367-77376-2

Study on the dynamical structural response under ice load with different ice conditions

Shifeng Ding, Li Zhou & Gu Yingjie
Jiangsu University of Science and Technology, Zhenjiang, Jiangsu, China

Chenkang Zhong & Jing Cao
Shanghai Rules and Research Institute, China Classification Society, Shanghai, China

ABSTRACT: The polar ships may encounter different ice conditions in polar waters, which may cause hull damage, including plate tearing, frame buckling, and stringer bending. This paper focuses on the dynamical structural response for the polar ships. The information of ice conditions in polar water is studied to distinguish the key factors, which effect the structural response greatly. And three common ice conditions consist of crushed ice, ice ridge and level ice are assumed to simulate the interaction between ship hull and sea ice. Moreover, the case study is carried out with an ice-strengthened vessel based on dynamical non-linear finite element method. The comparison of dynamical structural response is studied considering different ice conditions, and the related hull damage model and strength criteria is presented furtherly. The outcome is a reference for the structural design of ice-strengthened vessels.

1 INTRODUCTION

Ice navigation is a common condition for the polar shipping. While the sea ice has an impact on the hull structure, ice load (Hong, 2007) is induced, including dynamic local ice pressure and continuous ice resistance. The hull structure should be ice-strengthened appropriately to withstand the local pressure (Wisniewski 2003; Bo 2008), which refers to the ship velocity, ice condition, hull lines and other factors.

It is a hot academic question to simulate the interaction between ship and ice. The available methods consist test simulation in ice basin (Daley, 2007), analytic formula based on energy equilibrium (Zhenhui, 2010; Ji, 2018) and numerical simulation (Daley, 1998, 2002; Bond, 1998). In recent years, the finite element method (FEM) is widely used to analyze the structural response under the ice load, and a series of remarkable results are achieved. Jaehyun (2012) worked on the structural strength of an ice-strengthened LNG carrier under the ice impact load compared with experimental and numerical methods. Assuming the ship hull is large structural grillages, Hyunmin (2018) studied the strength under ice load by FEM and experimental methods. Kõrgesaar (2018) analyzed the bearing capacity of ship hull structure under the idealized ice load and the certain boundary conditions. Matsui (2018) simulated the structural response of ship's energy-saving propulsion device under ice load impact by FE method.

This paper focuses on the dynamical structural response under ice load with different ice conditions, including ice channel condition defined in FSICS (Finish transport safety agency, 2017), level ice condition defined in Polar Class Rule (IACS, 2016), and severe ice condition in which ice need to be repeated rammed, such as ridge condition defined in DNV's ice class (DNV,2013). The amplitude of ice load is assumed according to the rule requirement, while the shape of curve is based on ice basin model test. The dynamic structural response analysis is carried out on an ice-strengthened hull structure by nonlinear FE method. The results are compared to find out the difference caused by the ice load cases. Furthermore, the related hull damage model and strength criteria is studied to ensure the safety of hull structure under impact ice load. It provides a reference for the structural design of ice-strengthened vessels.

2 DYNAMIC ICE LOAD IDEALIZATION

Ice load is the key input data for the structural design of an ice-strengthened vessel, which is a random variable related to the impact process, ice shape, ice mass, ship velocity, hull lines angle and other factors. For the definite ice operation of a polar ship, the ice condition is the key to determinate the ice load. Due to various ice classes, the maximum value of ice load was presented differently in different ice rules,

DOI: 10.1201/9781003216582-51

considering the factors of ship velocity, hull lines angle, ice mass and so on. The time-domain tendency of ice load should be idealized furtherly for the simulation of the impact process between hull and ice.

2.1 Ice load idealization for ice channel condition

Ice channel condition is common in first-year ice waters where the light ice-strengthened vessel sails behind vessel with high ice class or icebreaker (See Figure 1), which is the application scenario for IC to IA ice class in FSICS.

The small ice blocks distributed in the ice channel, which have a continuous impact on the hull structure. The induced ice load could be idealized to be a constant value with time. The ice pressure could be calculated by formula (1) according to FSICS and the ice load pitch could be assumed to be the strip area around the hull shell near the water line.

$$p = c_d c_p c_a p_0 \tag{1}$$

Where, p is ice load, c_d is a factor which takes account of the influence of the size and engine output of the ship, c_p is a factor that reflects the magnitude of the load expected in the hull area in question relative to the bow area. c_a is a factor which takes account of the probability that the full length of the area under consideration will be under pressure at the same time.

2.2 Ice load idealization for level ice condition

Level ice condition is common in polar water where the ice thickness may reach 3m. As the level ice has an impact on the hull, the anti-force increases by time and won't reach the maximum value until the crushing of level ice as shown in Figure 2.

From Figure 2, the interaction between hull and ice is a dynamic process, which was usually assumed be to a quasi-static state in PC rule. In order to study on the dynamic character, an ice basin model test was carried out in former papers, and the time-domain ice pressure curve are shown in Figure 3 (Ding, 2019).

From the time-domain ice pressure curve, the interaction between hull and ice is random and nonlinear, which may last 6s. Furtherly, the ice load could be divided into a lot of independent impact

Figure 1. Ice operation in ice channel condition.

Figure 2. Ice operation in level ice condition.

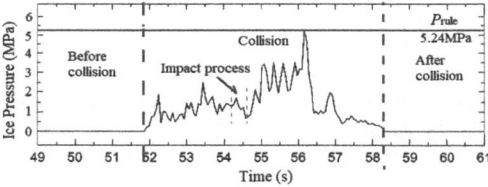

Figure 3. Total ice force and ice pressure by model test.

processes, and each process may last about 0.5s, including loading and unloading. Therefore, the dynamic tendency of ice load could be idealized with a series of loading and unloading processes, while the value and ice load pitch could be determined by ice load formula according to PC rule.

$$P_i = F_i^{0.22} \cdot CF_D^2 \cdot AR_i^{0.3} \tag{2}$$

Where: F_i is the basic nominal ice force, CF_D is the load patch dimensions class factor, AR is the load patch aspect ratio.

2.3 Ice load idealization for ice ridge condition

If the polar ship encounters severe ice condition, such as ice ridge, some special ice-operation methods could be used, taking ice ramming as shown in Figure 4 as the example. In that case, the vessel will sail ahead in the ice waters, until the ice load reaches the limitation. Then the vessel retreats and sails ahead again with a high velocity, subsequently, the ice ridge would be destroyed by repeated ramming.

Large ramming ice load would act on the hull during the ramming process, and the maximum ice pressure was presented in DNV's ice class rule as formula (3). The whole ramming action could be divided into loading process and unloading process.

$$p = F_B P_0 \tag{3}$$

Where, F_B is the correction factor for size of design contact area, P_0 is the basic ice pressure, taken large value for repeated ramming condition.

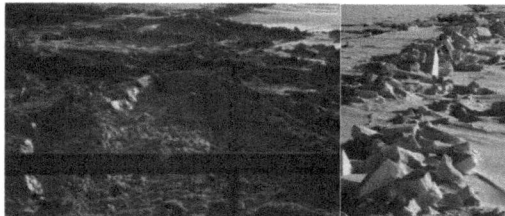

Figure 4. Ice operation in ice ridge condition.

2.4 Case study on ice load

An ice-strengthened vessel is used for studying the ice load acting on the hull structure with different ice conditions, and the main parameters are listed as Table 1.

Then the maximum value of ice pressure related to different ice conditions are calculated by formula (1) to (3) respectively, and the results are listed in Table 2.

Considering the dynamic character analyzed above, the time-domain ice pressure curves are idealized as Figure 5, lasting about 6s.

From Figure 5, the ice pressure of ice channel condition is assumed to be a constant value, which reflects the continuous action of many small ice blocks. The ice pressure of ice level condition is assumed to be consisted of several loading and unloading processes, which is used to simulate the continuous crushing process of the level ice. The ice pressure of ridge ice condition is assumed to be a single loading process with a single unloading process, which is used to simulate the load increasing as the ramming begin and the load decreasing when the ramming stop.

Table 1. The main parameters of an ice-strengthened vessel.

Parameter	Value
Ship Length	128.6 m
Ship Breadth	22.8 m
Draught	6.8 m
Angle of Water Line	20 degree
Angle of stem	21 degree
Displacement	16350 t

Table 2. The maximum value of ice pressure according to ice rule.

Ice condition	Assumed ice condition related to ice rule	Calculated ice pressure(N/mm^2)
Ice channel	Ice class B3	1.48
Level ice	Ice class PC 3	5.24
Ice ridge	Ice class icebreaker	5.76

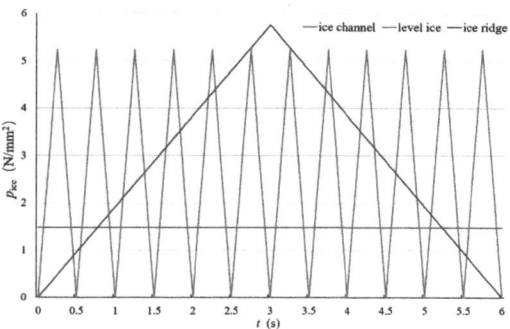

Figure 5. The idealized ice load with three ice conditions.

3 STRUCTURAL ANALYSIS UNDER ICE LOAD

3.1 Structural modeling

The stem of an ice-strengthened vessel is the key area of ice impact, and the FE model contains bow structure and bow intermediate structure as shown in Figure 6.

The plate and the web of frame are simulated by shell element and the flange of frame is simulated by beam element. Considering of the intermediate frame, the mesh size is taken as 0.5m, where s is the spacing of frame. The elastic-plastic material is used to construct the hull structure whose non-linear stress-strain relationship is shown in Figure 7. Young's modulus is 210GPa, yield stress is 355MPa. The nodes at the end of bow intermediate are constrained by displacement in three directions.

3.1.1 Application of ice load with different ice conditions

As the ice-strengthened vessel navigates in ice waters, the stem structure bears the ice load directly,

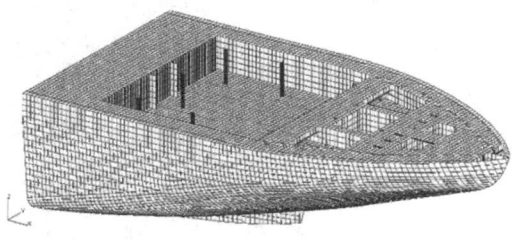

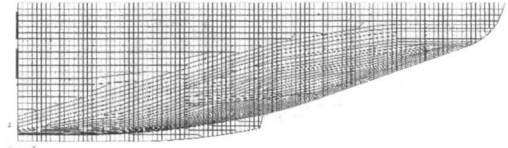

Figure 6. The FEM model of an ice-strengthened vessel.

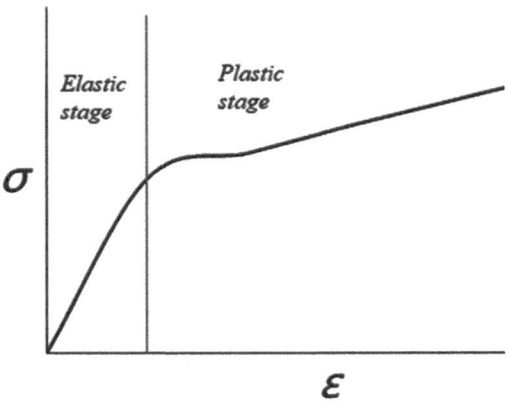

Figure 7. The stress-strain curve of material.

and an ice load pitch on the bow area is chosen for structural analysis, as shown in Figure 8.

From Figure 8, when the ship navigates in ice-channel, the sea ice blocks have impact on the bow firstly, and stem area is the dominant area for bearing ice-load. When the ship navigates in level-ice, the ice acts on bow structure and increases with the time until a piece of ice level destroyed, and the process is repeated subsequently. The ramming at ice ridge condition is the severest for ice navigation, the impact ice-load increases continually until the ship stops. Thus, the chosen load pitch is the position most likely to bear the large ice-load directly.

3.2 *Dynamic analysis on hull structure*

As the ice load acts on the hull structure, the dynamic structural stress changes with time and the diagrams at 1.2s, 3.0s and 4.8s are shown in

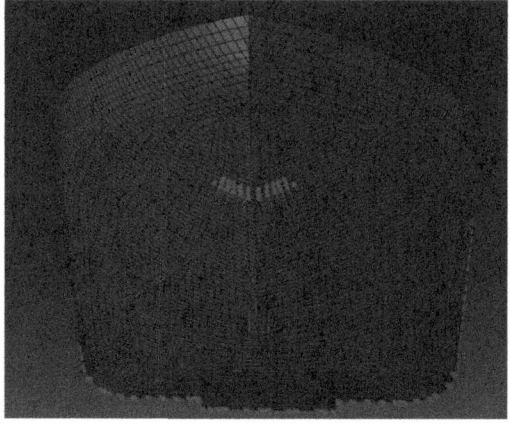

Figure 8. The idealized ice load acting on bow structure.

Figure 9. The time-domain dynamic stress curves of out shell and web plate in the ice load patch are shown in Figure10.

Although the outer shell interacts directly with sea ice, the largest stress appears at the web plate. The dynamic structural stress is concentrated at the connection location in the ice-loading area from Figure 9. It illustrates that the web frames are the main load bearing structural member under ice load. Moreover, the platform, inner deck and bulkhead is also under the impact of ice load by the time, because the impact energy induced by ice load would transmits along the hull structure with the time. It illustrates that the integral structural frame, not only the local structural member, undertakes the ice load.

The structural response is a dynamic process, the stress curves changes with the time under the idealized ice-load as shown in Figure 10. The stress curves differ due to the different ice conditions assumed by this paper.

The dynamic stress is under the elastic range in ice channel condition, because the ice load in this condition is much lower than others. The stress curve is almost very stable under the approximate constant ice load acting all the time. The maximum stress of shell is about 52.04MPa while that of web is about 94.34MPa, the difference is almost 44.3%.

The stress curve of level ice appears nonlinear and periodic tendency with time. It is obvious different from the channel condition, which is the characteristic of independent ice operation in arctic area. The stress increases with the beginning of a new ice-breaking process and decreases with the ending of the process. And the stress of shell is also much lower than that of web plate. The maximum stress of shell is about 93.31MPa, while that of web is about 343.13MPa, which is very close to the limit of yielding(355MPa), and the difference is almost 72.8%.

Ice ramming is used for the severe ice condition during the ice operation, such as thick first-year ice, ice ridge, multiyear ice. The induced ice load and structural response is both large, when the ship hits on the ice with a high velocity, which has the special requirements in ice class rule. We just focus on the process from the beginning to the ending of the impact action, which consists of a single loading process and a single unloading process. The maximum stress of shell is about 133.84MPa, while that of web is about 346.56MPa, which is very close to the limit of yielding, and the difference is almost 61.4%.

In conclusion, the structural response is distinct under different ice-conditions. The ice load and structural response is intense in the severe ice condition, vice versa. Hence, the ice-strengthen structure should be specially considered for the vessel with high ice-breaking capability.

(a) Ice Channel (1.8s) (b) Ice Channel (3.0s) (c) Ice Channel (4.8s)

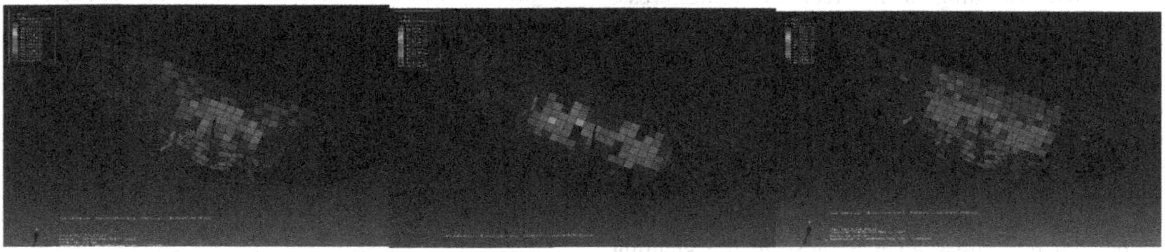

(a) Level Ice (1.8s) (b) Level Ice (3.0s) (c) Level Ice (4.8s)

(a) Ice Ridge (1.8s) (b) Ice Ridge (3.0s) (c) Ice Ridge (4.8s)

Figure 9. Dynamic structural stress diagram.

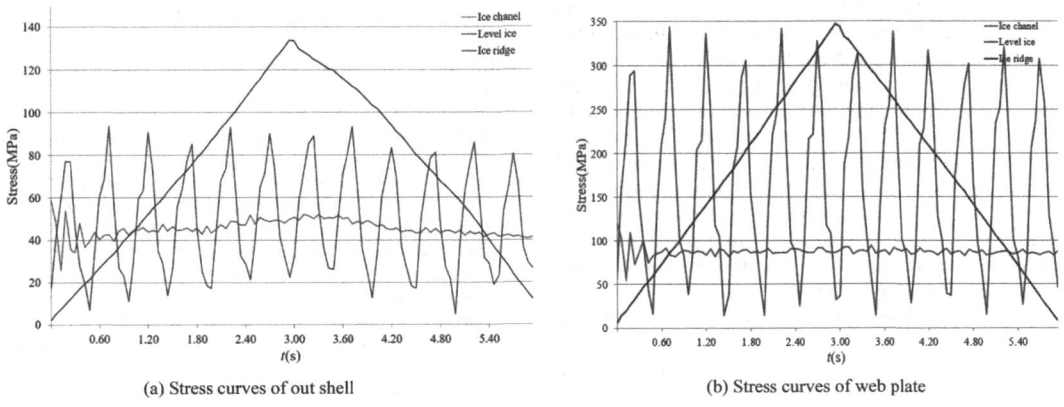

(a) Stress curves of out shell

(b) Stress curves of web plate

Figure 10. Time-domain dynamic stress. (a) Stress curves of out shell (b) Stress curves of web plate.

4 STUDY ON STRENGTH CRITERIA

The low strength criteria may decrease the safety level of ice operation, while the high strength criteria may reduce the economics of ice navigation. It is very important to work out the suitable strength criteria for ice structure of hull with ice-class notation. Two methods are presented referring to the PC rule (IACS, 2016).

One is the allowance stress for hull structure is determined according to the steel's yielding stress. For example, 235MPa for ordinary steel and 355MPa for AH36 steel. In this paper, the stress level in ice channel condition is much low than the yielding stress, and it illustrates the ice-strengthened hull structure is competent with the slight ice condition in that case study.

The other is the criteria of ultimate bearing capacity under ice pressure, which is used to the case of the dynamic stress level is close to the yielding stress. In this paper, the maximum dynamic stress in level ice condition is 343.13MPa, which is about 96.1% of the yielding stress, moreover, the maximum dynamic stress in ice-ridge condition is 343.13MPa, which is about 97.6% of the yielding stress. And the additional analysis on ultimate bearing capacity was carried out in other paper (Ding, 2019), and the result showed the ice structure could fulfill the requirements.

5 CONCLUSIONS

This paper focuses on the analysis of ice-strengthened structure of hull under ice load. The maximum values of local ice-pressure are calculated based on the ice class rules while the shape of the time-domain ice load is determined by referring to the experimental result of ice basin model test. And three idealized ice-loading curves are assumed to simulate the interaction of ship and ice in different ice conditions, including ice-channel condition, level ice condition and ice-ridge condition. Furtherly, the bow structure of an ice-strengthened ship is chosen to study the dynamic structural response under time-domain ice pressure. The results show that the stress is under the yielding stress level, and its tendency is stable in ice-channel condition. But it will change in severe ice conditions, such as level ice condition and ice ridge ice condition. The stress level reaches the yielding stress and causes stress redistribution, and the ultimate bearing capacity should be carried to study the limitation of ice operation. However, the strength criteria is presented for the ice-strengthened hull considering the different ice conditions, which is a good reference for structural design of polar ship.

ACKNOWLEDGEMENT

The first author greatly acknowledges the supports of State Key Laboratory of Ocean Engineering (Shanghai Jiao Tong University, Grant No. 1807). The second author greatly acknowledges the supports of the National Natural Science Foundation of China (Grant No. 51809124), Natural Science Foundation of Jiangsu Province of China (Grant No. BK20170576), Natural Science Foundation of the Higher Education Institutions of Jiangsu Province of China (Grant No. 17KJB580006) and State Key Laboratory of Ocean Engineering (Shanghai Jiao Tong University, Grant No. 1704).

REFERENCES

Bond, J., and Kennedy, S. 1998. Physical Testing and Finite Element Analysis of Icebreaking Ship Structures in the Post-yield Region. *8th Int Offshore and Polar Eng. Conf.*, Montreal, Canada, 577–585.

Daley, C. 2002. Derivation of Plastic Framing Requirements for Polar Ships. *Marine Structure*, 15, 543–559.

Daley, C, Hermanski, G, Pavic, M et al. 2007. Ultimate Strength of Frames and Grillages Subject to Lateral Loads-an Experimental Study. *10th Int Symp on Practical Design of Ships and Other Floating Structures*, Houston, Texas, USA.

Daley, C., Tuhkuri, J., and Riska, K. 1998. The Role of Discrete Failures in Local Ice Loads, *Cold Regions Science and Technology*, 27, 197–211.

Ding S. F., Zhou L., Zhong C. K., et al. 2019. A Structural Analysis Procedure Combining Linear and Nonlinear FE Methods for Polar Ship. *Proceedings of the Twenty-ninth (2019) International Ocean and Polar Engineering Conference Honolulu*, Hawaii, USA, June 16–21:815–821.

DNV. 2013. Ships for Navigation in Ice.

Finish Transport Safety Agency. 2017. Finnish-Swedish Ice Class Rules.

IACS. 2016. The Unified requirements for Polar Class.

Kim, H., Daley, C., Kim, H. 2018. Evaluation of Large Structural Grillages Subjected to Ice Loads in Experimental and Numerical Analysis. *Marine Structures*. 61, 467–502.

Kim, J., Kim, D., Song, H. 2012. Safety Assessment of Membrane Type Cargo Containment Systems in LNG Carrier Under the Ice-ship Repeated Impact. *Proc 22nd Int Offshore and Polar Eng. Conf.*, Rhodes, Greece, 1194–1201.

Kõrgesaar, M., Kujala, P., and Romanoff, J. 2018. Load Carrying Capacity of Ice-strengthened Frames under Idealized Ice Load and Boundary Conditions. *Marine Structures*, 58, 18–30.

Lin, H. and Amdahl, J. 2007. Plastic Design of Laterally Patch Loaded Plates for Ships. *Marine Structures*, 20, 124–142.

Liu, Z. and Amdahl, J. (2010). A New Formulation of the Impact Mechanics of Ship Collisions and Its Application to a Ship-iceberg Collision. *Marine Structures*, 2, 360–384.

RMRS. 2010. Rules for Building and Classing Steel Vessels.

Sadaoki, M., Uto, S., Yamada, Y. et al. 2018. Numerical Study on the Structural Response of Energy-saving Device of Ice-class Vessel due to Impact of Ice Block. *Int. J. of Naval Arch. and Ocean Eng.* 10, 367–375.

Wang, B., Yu, H., and Basu, R. 2008. Ship and Ice Collision Modeling and Strength of Evaluation of LNG Ship Structure. *Proc. 27th Int. Conf. on Offshore Mechanics and Arctic Eng.*, Estoril, Portugal, ASME, 1–8.

Wiśniewski, K., and Kołakowski, P. 2003. The Effect of Selected Parameters on Ship Collision Results by Dynamic FE Simulations. *Finite Elements in Analysis and Design*, 39, 985–1006.

Xu, J., Karr, D., and Oterkus, E. 2018. A Non-simultaneous Dynamic Ice-structure Interaction Model. *Ocean Engineering*, 16, 278–289.

Developments in Maritime Technology and Engineering – Guedes Soares & Santos (eds)
© 2021 Copyright the Author(s), ISBN 978-0-367-77376-2

Uncertainty analysis on the pseudo-shakedown phenomenon of rectangular plates subjected to dynamic pressure pulse

Xu He & C. Guedes Soares

Centre for Marine Technology and Ocean Engineering (CENTEC), Instituto Superior Técnico, Universidade de Lisboa, Lisboa, Lisbon, Portugal

ABSTRACT: An uncertainty analysis is presented on the pseudo-shakedown phenomenon of rigid, perfectly plastic rectangular plates subjected to repeated dynamic pressure pulses. The maximum transverse displacement for a structure subjected to a short duration pulse would be smaller than the corresponding static deflection associated with the pressure pulse having same peak value. Under this circumstance, the phenomenon of pseudo-shakedown would occur under the repeated dynamic pressure pulses. The uncertainties are considered for the variables involved in the plastic response of rectangular plate. Monte Carlo Simulations are used to generate the random variables. Based on the rigid, perfectly plastic analysis, the dynamic maximum displacement and the static maximum displacement of rectangular plate under pressure pulse can be calculated when inputting the uncertain variables. The histograms of these maximum displacements are obtained and their probability density functions are analyzed. The probability of failure is calculated. Finally, sensitivity analyses for random variables are performed by using the first-order second-moment method and linear regression method.

1 INTRODUCTION

Structural reliability methods have been widely used in marine structure design (for example, Mansour & Wirsching 1995, Guedes Soares & Teixeira 2000, Guedes Soares et al. 1996, 2010, Paik & Frieze 2001, Zhang et al. 2019), because the reliability-based design approaches can quantify the uncertain quantities during the design process, while the deterministic methods are not able to deal with the uncertainties. Thus, the reliability-based design methods are more flexible and rational. Typically, the main sources of uncertainties in structural analysis involves in the strength of various structural elements, the loads and load combinations, and modeling errors in analysis procedures. The emphasis of reliability analysis is on constructing realistic models and on quantifying the influence of these uncertainties for practical use.

In the last decades, in the field of marine structures, the uncertainty analysis on structural collapse ha been mainly focused on the compressive strength (Guedes Soares 1988, 1997, Gaspar et al. 2011, 2012, 2013, 2014), on the ultimate strength (Gordo et al. 1996, Teixeira & Guedes Soares 2005, 2007, Teixeira et al. 2013), on the fatigue strength (Guedes Soares & Garbatov 1996, 1999, Dong et al. 2018, 2020) of ship components. Very limited works are reported on the reliability analysis of ship structures under dynamic impact, such as subjected to the slamming loading, blast loading, ship collision.

However, many uncertainties also exist in the problem of structural impact.

A phenomenon known as pseudo-shakedown is an interesting characteristic of structures undergoing large deflection when subjected to repeated dynamic impacts. Pseudo-shakedown phenomenon was introduced by Jones (1973) for a rigid, perfectly plastic rectangular plate subjected to repeated impact loads having a triangular-shaped pressure-time history, based on the observation of repeated wave impact on ship bows. Shen and Jones (1992) proved the pseudo-shakedown phenomenon of rigid, perfectly plastic beams and plates subjected to repeated dynamic pressure pulses and comments were made on the number of repeated dynamic loadings necessary to reach a pseudo-shakedown state. Huang et al. (2000) performed repeated mass impact tests on fully clamped circular plates made from aluminum alloy and square plates made from mild steel. His motivation was to examine the existence of pseudo-shakedown phenomenon in the mass impact events. An energy criterion was postulated to judge the pseudo-shakedown state of structure. It showed that the structure will reach a pseudo-shakedown state when the maximal elastic strain energy to be absorbed in the deformed structure is equal to or larger than the external work of the dynamic load in magnitude. This hypothesis was confirmed by their experimental results. Jones (2014) extended the study of pseudo-shakedown phenomenon to circular and rectangular plates under identical

DOI: 10.1201/9781003216582-52

repeated mass impacts, allowing the accumulation of masses from earlier impacts on the plate surface, by using a rigid-plastic method of analysis. The result suggested that a pseudo-shakedown state is achieved when mass accumulation is retained.

For the design of a structure that works under repeated impacts, if the plastic deformation is allowed, the pseudo-shakedown state will be the limit state of the design, provided a fatigue failure does not occur. Therefore, the pseudo-shakedown phenomenon maybe beneficial to the structural safety. Thus, study on the occurrence of the pseudo-shakedown phenomenon should be concerned.

This paper carries out a study on the occurrence of pseudo-shakedown phenomenon when some uncertainties are considered. Section 2 introduces the condition that the pseudo-shakedown phenomenon occurs. Section 3 raises the reliability assessment problem. Section 4 presents a case study by using the Monte Carlo Simulation method. Section 5 performs sensitivity analyses for the uncertainty problem. Conclusions are drawn in Section 6.

2 PLASTIC BEHAVIOUR OF RECTANGULAR PLATE SUBJECTED TO PRESSURE PULSE

2.1 Static plastic behavior of rectangular plate subjected to a uniform pressure

The analysis methods for rectangular plates under pressure pulse load statically or dynamically have been developed by Jones (2012). Consider a fully clamped rectangular plate with length $2L$, width $2B$ and thickness H subjected to a uniformly distributed static transverse pressure, as shown in Figure 1, the variation of the maximum transverse displacement with the pressure is expressed as Equation (1), which is derived by a rigid, perfectly plastic analysis, in which the square yield condition (see Figure 2) is used and the finite displacement is taken into account.

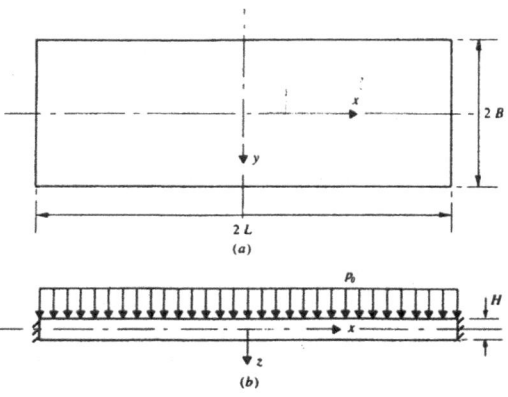

Figure 1. Fully clamped rectangular plate subjected to a uniformly distributed transverse pressure. (a) Plan view. (b) Side view.

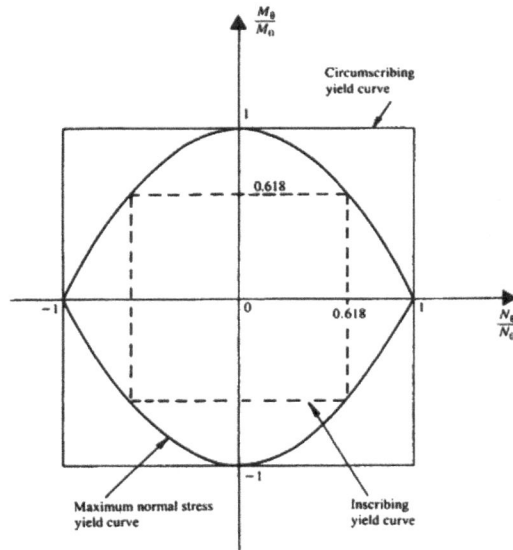

Figure 2. Square yield condition.

$$\left(W_f/H\right)_{static} = 2(\eta - 1)/a_2 H \qquad (1a)$$

where

$$\beta = B/L \qquad (1b)$$

$$\tan \phi = -\beta + \sqrt{3 + \beta^2} \qquad (1c)$$

$$a_2 = 2[1 + (1 - \beta \tan \phi)/(1 + \beta \cot \phi)]/H \qquad (1d)$$

and $\eta = P_0/P_c$ is the dimensionless pressure amplitude; $P_c = 12M_0/(B \tan \phi)^2$ is the static plastic collapse pressure for a fully clamped rectangular plate; $M_0 = \sigma_0 H^2/4$ is the fully plastic bending moment per unit length; σ_0 is the yield stress.

2.2 Dynamic plastic behavior of rectangular plate subjected to a uniform pressure

When the same rectangular plate is subjected to a dynamic pressure pulse with a rectangular pressure–time history, as shown in Figure 3, considering the influence of finite displacement and using the square yield condition in Figure 2, the maximum transverse displacement is expressed as Equation (2).

$$\left(W_f/H\right)_{dynamic} = \frac{(3 - \xi_0)\left[\{1 + 2\eta(\eta - 1)(1 - \cos a_3 \tau)\}^{\frac{1}{2}} - 1\right]}{2\{1 + (\xi_0 - 1)(\xi_0 - 2)\}}$$

$$(2a)$$

where

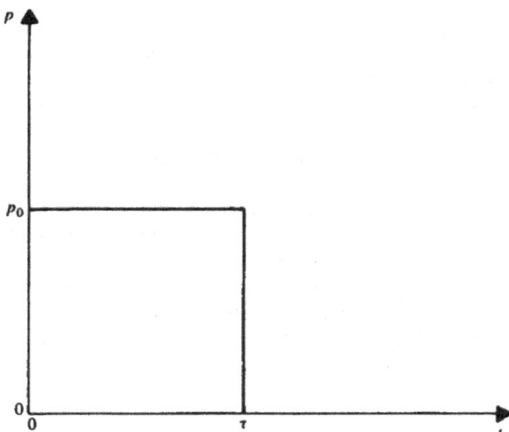

Figure 3. Rectangular pressure pulse.

$$a_3\tau = \frac{\sqrt{2}I'\{1 - \xi_0 + 1/(2 - \xi_0)\}^{1/2}}{\eta} \qquad (2b)$$

$$\xi_0 = \beta\tan\phi \qquad (2c)$$

$$I' = P_0\tau/(\mu HP_c)^{1/2} \qquad (2d)$$

and μ is the plate mass per unit area.

It was proved by Shen & Jones (1992) that the pseudo-shakedown phenomenon will occur for a rigid, perfectly plastic structure subjected to repeated identical pressure pulses when the maximum permanent displacement caused by the dynamic pressure pulse is lower than that which would occur for the same amplitude pressure applied statically, i.e. $(W_f)_{dynamic} < (W_f)_{static}$

3 RELIABILITY ASSESSMENT PROBLEM

3.1 Limit state function

The reliability problem is formulated by considering the occurrence of pseudo-shakedown phenomenon for rectangular plate structures subjected to repeated pressure pulses. As mentioned, for the marine structures that are very likely subjected to repeated pressure pulses during their lives, such as the slamming loads, if the plastic deformation is allowed, the pseudo-shakedown state will be the limit state of the design, provided a fatigue failure does not occur. Therefore, the displacement induced by the dynamic pressure pulse is expected to be limited to the displacement caused by the static pressure with same amplitude. Thus, the reliability analysis considers the limitation of maximum displacement, which can be described by a limit state function in the form of:

$$g(X) = (W_f)_{static} - (W_f)_{dynamic} \qquad (3)$$

where X is a vector of the variables that describe the uncertain quantities involved in the description of the dynamic maximum displacement in Equation (2) and the static maximum displacement in Equation (1).

3.2 Stochastic models

In the uncertainty analysis of the plastic behavior of rigid, perfectly plastic rectangular plates subjected to pressure pulse, the structural dimensions, the material properties and the pressure pulse loading, i.e. L, B, H, σ_0, ρ, η and τ are considered as basic random variables. The dynamic maximum displacement $(W_f/H)_{dynamic}$ in Eq. (2) and the static maximum displacement $(W_f/H)_{static}$ in Eq. (1) are implicit functions of these basic random variables. The random variables and the stochastic model adopted for each basic random variable are given in Table 1.

The uncertainty in structural dimensions (L, B and H) is typically represented by a normal distribution with the mean value equal to the deterministic design value and the standard deviation or variability defined by a prescribed Coefficient of Variation (COV). The material properties of mild steel are considered in the analysis, including the yield stress σ_0 and the material density ρ. The uncertainties in the yield stress and the material density can be represented by a lognormal distribution with the statistical parameters derived from a characteristic design value, respectively. The uncertainty on the impulsive loading (η and τ) is assumed to follow a normal distribution with the mean value equal to the prescribed design loading. Here, the magnitudes of the COV and the functions of probability distributions of the random variables are selected according to the recommendations in Guedes Soares (1988), Siddiqui et al. (2014) and Liu & Guedes Soares (2015).

Table 1. Random variables and statistical data used in the uncertainty analysis.

Random variable	Unit	Mean value	COV	Distribution
Half plate length L	m	1	0.01	Normal
Half plate width B	m	0.5	0.01	Normal
Plate thickness H	mm	9	0.04	Normal
Yield stress σ_0	MPa	286	0.08	Lognormal
Material density ρ	Kg/m³	7850	0.1	Lognormal
Pressure amplitude η	-	2.8	0.1	Normal
Loading duration τ	ms	1.5	0.1	Normal

On the other hand, a correlation may exist between two sets of parameters, i.e. the yield stress σ_0 and the material density ρ, the impulsive pressure η and the loading duration τ. However, no information about the magnitude of the correlation coefficients could be found in the literature. Here, both correlation coefficients are assumed to be 0.2. Although this assumption does not have applications for the realistic structures, it still can be a case study for preliminarily discussion of the uncertainty. As for other basic random variables related to the displacement functions, i.e. the structural dimensions, they are independent.

4 MONTE CARLO SIMULATIONS

4.1 Reliability assessment method

Monte Carlo simulation method can be used to assess the possibility of failure in structural reliability problem. This technique involves sampling a set of values of the basic variables at random from the probability density function to evaluate the model results, the possibility of failure, and moreover, the statistics computed on these results, such as mean value and standard deviation. Generally, as more trials are calculated, the statistical analysis can be more accurate. Compared with the FOSM method, the Monte Carlo method is capable to cope with any shape of possibility density function for the variables.

For the component reliability problem, the probability of failure p_f can be defined as:

$$p_f = P[g(X) \leq 0] \tag{4}$$

The empirical estimates of probability of failure can be calculated as

$$p_f = N_f(X)/N \tag{5}$$

where $N_f(X)$ is the number of $g(X) \leq 0$ in N times of sampling, which means the failure happens. When the number of sampling is large enough, p_f is approximately equal to failure probability.

The accuracy of simulations can be characterized by the COV of the estimated value for the probability of failure p_f, expressed as

$$COV(p_f) = \sqrt{(1 - p_f)p_f/N}/p_f \tag{6}$$

4.2 Generating samples of random variables

To generate sample values of these random variables, the inverse transformation technique is used. First N random numbers between 0 and 1 are generated. Then the random number x_i can be calculated from the inverse cumulative distribution function (CDF) of X.

The sample values of the independent random variables, i.e. L, B, H can be directly generated by using the inverse transformation technique. Some extra calculations are needed to generate samples of correlated random variables, i.e. σ_0 and ρ, η and τ. In the following, a method to generate correlated random variables is introduced.

According to the principle of the triangle decomposition of symmetrical matrix, correlated random variables X can be transformed to independent standard normal variables Y, letting that

$$X = [V]Y + \mu_X \tag{7}$$

where V is the transformation matrix, μ_X is the matrix composed of the mean values of random variables X.

Therefore, to generate the correlated random variables X, the key is to obtain the transformation matrix V.

For n random variables $X = [X_1, X_2, \cdots, X_n]^T$ which follow the normal distribution, the covariance matrix of X is

$$C_X = E\left[(X - \mu_X)(X - \mu_X)^T\right] = \begin{bmatrix} c_{11} & c_{12} & \cdots & c_{1n} \\ c_{21} & c_{22} & \cdots & c_{2n} \\ \vdots & \vdots & & \vdots \\ c_{n1} & c_{n2} & \cdots & c_{nn} \end{bmatrix}$$

$$= \begin{bmatrix} \sigma_{X_1}^2 & \mathrm{cov}(X_1, X_2) & \cdots & \mathrm{cov}(X_1, X_n) \\ \mathrm{cov}(X_2, X_1) & \sigma_{X_2}^2 & \cdots & \mathrm{cov}(X_2, X_n) \\ \vdots & \vdots & & \vdots \\ \mathrm{cov}(X_n, X_1) & \mathrm{cov}(X_n, X_2) & \cdots & \sigma_{X_n}^2 \end{bmatrix}$$

$$\tag{8}$$

where σ_{X_i} is the standard deviation of X_i. Because $\mathrm{cov}(X_i, X_j) = \mathrm{cov}(X_j, X_i)$, $[C_X]$ is a symmetrical matrix.

From Equation (7) and the definition of covariance matrix in Equation (8), leads to

$$C_X = E\left[(X - \mu_X)(X - \mu_X)^T\right] = E\left[(VY)(VY)^T\right]$$
$$= E\left[[V]YY^TV^T\right] = VE[YY^T]V^T = V[V^T]$$

$$\tag{9}$$

as Y is standard normal variables, hence $E[YY^T] = 1$.

It is apparent from Equation (9) that the transformation matrix V can be obtained by performing triangle decomposition for covariance matrix C_X. Specifically, the Cholesky decomposition can be used such that the transformation matrix V becomes a lower triangular matrix, whose element v_{ij} is

$$v_{ij} = \left(c_{ij} - \sum_{k=1}^{j-1} v_{jk}^2 \right)^{1/2}, \ j = 1, \ 2, \ \ldots, \ n \quad (10)$$

$$v_{ij} = \frac{1}{v_{jj}} \left(c_{ij} - \sum_{k=1}^{j-1} v_{ik} v_{jk} \right), j = 1, \ 2, \ldots, n, i = j+1, j+2, \ldots, n$$
$$(11)$$

where c_{ij} is the element of covariance matrix C_X.

Then the correlated random variables can be generated by calculation with Equation (7).

According to the above method, the samples of correlated random variables of η and τ can be generated as they follow the normal distribution.

With regards to the correlated random variables of σ_0 and ρ, as they follow the lognormal distribution, before using above mentioned method, a transformation calculation from the lognormal distribution to the normal distribution is needed. As the normal distributions of $\ln(\sigma_0)$ and $\ln(\rho)$ are known, then the samples for the correlated random variables $\ln(\sigma_0)$ and $\ln(\rho)$ can be obtained by using above method. Finally, the samples of σ_0 and ρ can be obtained by one Exponential operation. In addition, it is found that the correlation coefficient between independent random variables remains unchanged during these transformation calculations.

4.3 Case study

The Monte Carlo simulation is applied to estimate the probability of the nonoccurrence of pseudo-shakedown phenomenon of rigid, perfectly plastic rectangular plates subjected repeated dynamic pressure pulses. The sample size N of random variables used to calculate the model is 10002. The histogram of the dynamic maximum displacement and the static maximum displacement is shown in Figure 4 and Figure 5 respectively. In order to better represent the probability density function, the Gauss fit and Lognormal fit are used to fit the histograms, see Figures 4 and 5.

With respect to the relative frequency fit for the dynamic maximum displacement, the R-squares are 0.988 and 0.986 for the Gauss fit and the Lognormal fit respectively. It means that the Gauss fit is more accurate to fit the relative frequency of the dynamic maximum displacement. In this way, the normal distribution is suitable to be used as the probability density function of the dynamic maximum displacement. The mean and the standard deviation for this normal distribution are 11.25mm and 2.87mm respectively, while the mean and the standard deviation for the sample results of the Monte Carlo simulation are 11.51mm and 2.86mm respectively.

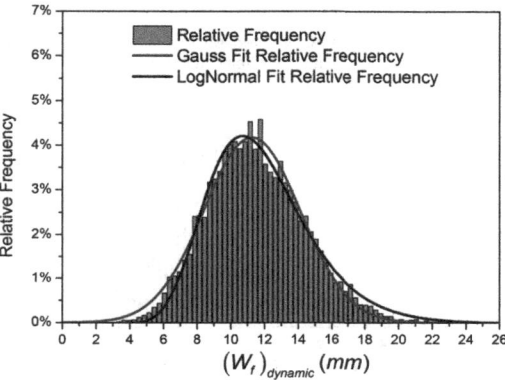

Figure 4. Relative frequency counts of the dynamic maximum displacement.

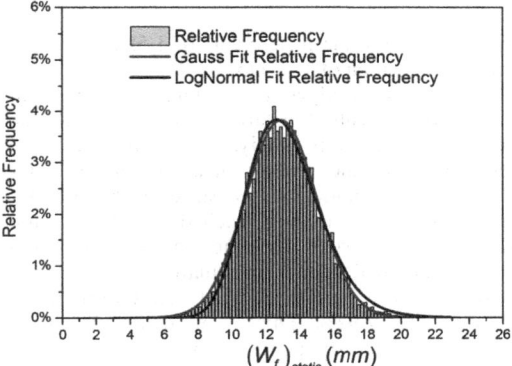

Figure 5. Relative frequency counts of of the static maximum displacement.

Concerning the relative frequency fit for the static maximum displacement, the R-squares are 0.993 and 0.987 for the Gauss fit and the Lognormal fit respectively. Similarly, the Gauss fit is more accurate to be used to fit the relative frequency of the static maximum displacement. The normal distribution is used as the probability density function of the static maximum displacement. The mean and the standard deviation for this normal distribution are 12.90mm and 2.10mm respectively, while the mean and the standard deviation for the sample results of the Monte Carlo simulation are 12.93mm and 2.09mm respectively.

It is apparent that the probability density functions for the dynamic maximum displacement and static maximum displacement are fitted satisfactorily by the normal distributions.

The histogram and the curve fit for the difference between the dynamic maximum displacement and the static maximum displacement is shown in Figure 6. The probability density function of the displacement difference can be expressed empirically by a normal

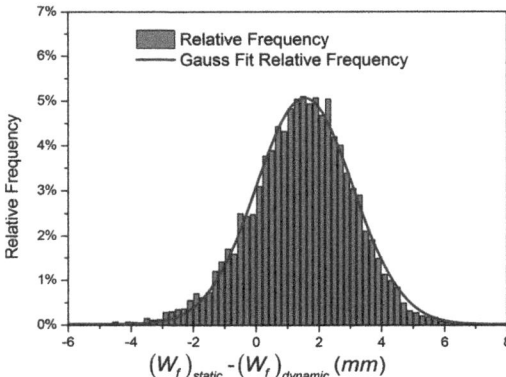

Figure 6. Relative frequency counts of the difference between the dynamic maximum displacement and the static maximum displacement.

distribution with the mean value of 1.55mm and the standard deviation of 1.55mm.

In order to analyze the failure probability of structural elements, the failure criterion should be determined. In this study, the failure can be regarded as the nonoccurrence of pseudo-shakedown phenomenon. Because if the pseudo-shakedown state is not reached for a structure subjected to repeated dynamic loadings, the structural failure will happen eventually.

In this case study, the analyzed failure probability p_f is 17.92%. This means that 1792 values of the simulated dynamic maximum displacement are greater than the static maximum displacements of the investigated plate (The sample size is 10002).

According to Equation (6), the COV for the estimated probability of failure is 0.021. As more trials are calculated, the COV becomes smaller and the estimated probability becomes more accurate. Here the relatively low COV value of 0.021 is acceptable, which means that the sample size of 10002 is enough.

5 SENSITIVITY ANALYSIS

Sensitivity analysis is used to identify the importance of the input variables to the model results. Therefore, sensitivity analysis can be helpful to guide to optimize the design parameters for engineering design. Sometimes a model may have many sources of uncertainty, then sensitivity analysis is useful to simplify the model by eliminating model inputs that have no effect on the output, or identifying and removing redundant parts of the model structure.

Sensitivity analyses are performed herein by using the FOSM method and the linear regression method, which belongs to local sensitivity methods

and global sensitivity methods respectively, and the sensitivity results obtained by two different ways are compared.

Based on the FOSM formulation, the first-order sensitivity factors for the model response with respect to the set of random input variables X_i ($i=1, 2, ..., n$) are defined as:

$$\alpha_{X_i} = \frac{\sigma_{X_i}\left(\frac{\partial g}{\partial X_i}\right)_{X=\mu_X}}{\sqrt{\sum_{i=1}^{n}\sigma_{X_i}^2\left(\frac{\partial g}{\partial X_i}\right)_{X=\mu_X}^2}} \quad (i = 1, 2, ..., n) \quad (12)$$

with $\rho_{ij} = 0 (i, j = 1, 2, ..., n)$.

The linear regression approach involves fitting the simple linear regression model to the Monte Carlo simulation data using a first-order polynomial of the form:

$$Y^{(i)} = b_0 + \sum_{j=1}^{n} b_{X_j} X_j^{(i)} \quad (13)$$

with $Y^{(i)}(i = 1, 2, ..., m)$ being the vectors of output Y, $X_j^{(i)}(j = 1, 2, ..., n\ i = 1, 2, ..., m)$ being the realizations of the input-output variables, $\{b_0, b_{X_j}\}$ being the regression coefficients determined by fitting the model to simulation data, typically using the least squares method.

The estimated regression coefficients of the first-order polynomial provide an approximation for the partial derivatives of the true model and can therefore be used to derive sensitivity measures for the input variables. However, sensitivity measures derived on the basis of the regression coefficients alone are not sufficient and therefore a normalization using the standard deviations of the input-output variables is in general adopted. The normalization adopted in the FOSM method can used for this purpose and the regression coefficients are then denoted as standardized regression coefficients (SRC):

$$\beta_{X_j} = b_{X_j}\frac{\sigma_{X_j}}{\sigma_Y} \quad (14)$$

with σ being the standard deviation.

The sensitivity factors of the random variables obtained by FOSM and SRC are shown and compared in Figure 7. In the model uncertainty factors, positive sensitivity indicates that the variable is benefit to the structural reliability and vice versa. It is observed that the sensitivity factors for each variable obtained by FOSM and SRC are similar. The basic random variables with positive contribution to the structural reliability are the half plate length L, half plate width B, the plate thickness H, and the density ρ. The yield

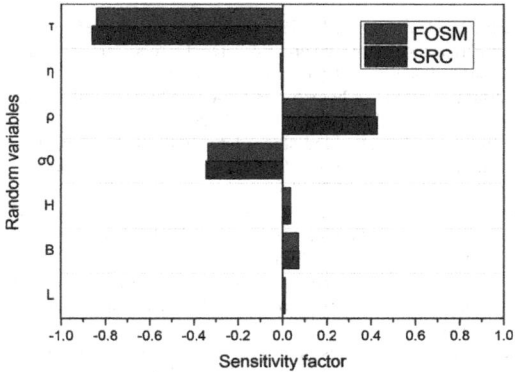

Figure 7. Sensitivity factors of random variables.

stress σ_0, dimensionless pressure amplitude η, the loading duration τ contribute negatively to the structural reliability.

It is noticed that the highest sensitivity factor is the loading duration, i.e. the occurrence of pseudo-shakedown phenomenon is most sensitive to the loading duration. The uncertainties of the density and the yield stress have moderate importance. On the other hand, the uncertainties of the dimension of the rectangular plate has small influence on the pseudo-shakedown phenomenon. The uncertainty of the dimensionless pressure amplitude has the least importance.

6 CONCLUSIONS

Monte Carlo simulation method are applied to analyze the uncertainty of the occurrence of pseudo-shakedown phenomenon of rigid, perfectly plastic rectangular plates subjected to pressure pulse. The limit function is determined by a criterion presented by Shen & Jones (1992).

The histograms of the dynamic maximum displacement and the static maximum displacement are drawn and analyzed. The probability density function of these displacements can be well presented by the normal probability distribution. The probability of failure is computed.

Finally, sensitivity analyses are conducted and compared by using FOSM method and linear regression method. In terms of the order of the importance of the random variables to the output, two methods provide the similar prediction. The pseudo-shakedown phenomenon is most sensitive to the loading duration. The second and the third most important factors are the material density and the yield stress respectively. Therefore, in order to guarantee the occurrence of the pseudo-shakedown phenomenon, the loading duration is the first element to look for.

The computation in this study is quite simple, because the limit function is an analytical function of the random variables considered, which can be calculated easily. But the method in this paper can still be a complete case study for uncertainty analysis, which can be extended to practical engineering design.

ACKNOWLEDGEMENTS

The first author is grateful to the support of the Scholarship from China Scholarship Council (CSC) under the Grant No. 201706950008.

REFERENCES

Dong, Y., Teixeira, A.P. & Guedes Soares, C. 2018. Time-variant fatigue reliability assessment of welded joints based on the PHI2 and response surface methods. *Reliability Engineering & System Safety* 177: 120–130.

Dong, Y., Teixeira, A.P. & Guedes Soares, C. 2020. Application of adaptive surrogate models in time-variant fatigue reliability assessment of welded joints with surface cracks. *Reliability Engineering & System Safety* 195: 106730.

Gaspar, B., Teixeira, A.P., Guedes Soares, C. & Wang, G. 2011. Assessment of IACS-CSR implicit safety levels for buckling strength of stiffened panels for double hull tankers. *Marine Structures* 24(4):478–502.

Gaspar, B., Naess, A., Leira, B.J. & Guedes Soares, C. 2012. System reliability analysis of a stiffened panel under combined uniaxial compression and lateral pressure loads. *Structural Safety* 39:30–43.

Gaspar, B. & Guedes Soares, C. 2013. Hull girder reliability using a Monte Carlo based simulation method. *Probabilistic Engineering Mechanics* 31: 65–75.

Gaspar, B., Bucher, C. & Guedes Soares, C. 2014. Reliability analysis of plate elements under uniaxial compression using an adaptive response surface approach. *Ships and offshore Structures* 10(2): 145–161.

Gordo, J.M., Guedes Soares, C. & Faulkner, D. 1996. Approximate assessment of the ultimate longitudinal strength of the hull girder. *Journal of Ship Research* 4(1):60–69.

Guedes Soares, C. 1988. Uncertainty modelling in plate buckling. *Structural Safety* 5(1): 17–34.

Guedes Soares, C. 1997. Probabilistic modelling of the strength of flat compression members. *In Probabilistic methods for structural design*: 113–140. Dordrecht: Springer.

Guedes Soares, C. & Garbatov, Y. 1996. Fatigue reliability of the ship hull girder. *Marine Structures* 9(3):495–516.

Guedes Soares, C. & Garbatov, Y. 1999. Reliability of corrosion protected and maintained ship hulls subjected to corrosion and fatigue. *Journal of ship research* 43(02): 65–78.

Guedes Soares, C. & Teixeira A.P. 2000. Structural reliability of two bulk carrier designs. *Marine Structures* 13(2):107–28.

Guedes Soares, C. Garbatov, Y. & Teixeira, A.P. 2010. Methods of structural reliability applied to design and maintenance planning of ship hulls and floating platforms. In C. Guedes Soares (eds), *Safety and reliability of industrial products, systems and structures*: 191–206. London: Taylor & Francis.

Guedes Soares, C., Dogliani M., Ostergaard, C., Parmentier, G. & Pedersen, P.T. 1996. Reliability based ship structural design. *Transactions of the Society of Naval Architects and Marine Engineers* (SNAME) 104:357–389.

Huang, Z.Q., Chen, Q.S. & Zhang, W.T. 2000. Pseudo-shakedown in the collision mechanics of ships. *International Journal of Impact Engineering* 24(1): 19–31.

Jones, N. 1973. Slamming damage. *Journal of Ship Research* 17(2): 80–86.

Jones, N. 2012. *Structural Impact*. 2nd ed. Cambridge: Cambridge University Press.

Jones, N. 2014. Pseudo-shakedown phenomenon for the mass impact loading of plating. *International Journal of Impact Engineering* 65: 33–39.

Liu, B. & Guedes Soares, C. 2015. Uncertainty analysis of the energy absorbed in beam and plate elements under impulsive loading. *Analysis and Design of Marine Structures:* 775–783. London: Taylor & Francis Group.

Mansour A.E. & Wirsching P.H. 1995. Sensitivity factors and their application to marine structures. *Marine Structures* 8: 229–255.

Paik J.K. & Frieze P.A. 2001. Ship structural safety and reliability. *Progress in Structural Engineering and Materials* 3:198–210.

Shen, W.Q. & Jones, N. 1992. The pseudo-shakedown of beams and plates when subjected to repeated dynamic loads. *Journal of Applied Mechanics* 59(1): 168–175.

Siddiqui, N.A., Khateeb, B.M.A., Almusallam, T.H., AlSalloum, Y.A., Iqbal, R.A. & Abbas, H. 2014. Reliability of RC shielded steel plates against the impact of sharp nose projectiles. *International Journal of Impact Engineering* 69: 122–135.

Teixeira A.P. & Guedes Soares, C. 2005. Assessment of partial safety factors for the longitudinal strength of tankers. In C. Guedes Soares, Y. Garbatov & N. Fonseca (eds), *Maritime transportation and exploitation of ocean and coastal resources:* 1601–1610. London: Francis and Taylor.

Teixeira A.P. & Guedes Soares, C. 2007. Probabilistic modelling of the ultimate strength of plates with random fields of corrosion. In: G. Deodatis & P.D. Spanos (eds), *Proc. of the 5th International Conference on Computational Stochastic Mechanics:* 653–661. Rotterdam: Millpress.

Teixeira, A.P., Guedes Soares, C., & Wang, G. 2013. Probabilistic modelling of the ultimate strength of ship plates with non-uniform corrosion. *Journal of Marine Science and Technology* 18: 115–132.

Zhang, S.M., Pedersen, P.T. & Villavicencio, R. 2019. *Probability and mechanics of ship collision and grounding*. Oxford: Butterworth-Heinemann.

Developments in Maritime Technology and Engineering – Guedes Soares & Santos (eds)
© 2021 Copyright the Author(s), ISBN 978-0-367-77376-2

Thermal load and residual strength of vessels under cabin fire

Chenfeng Li, Kun Zhang, Ziyang Wei, Xueqian Zhou, Huilong Ren & Weijun Xu
College of Shipbuilding Engineering, Harbin Engineering University, Harbin, China

ABSTRACT: In order to accurately assess the structure safety of ships undergone fires, a method, based on large eddy simulation of fires and a technique for structure thermodynamic coupling analysis, for analyzing the residual hull strength of ships subject to cabin fire is proposed in this paper. By taking the example of a surface ship, the fire accident in the engine room is simulated and the fire temperature is analyzed. The thermal load on the structure is obtained by using the heat flux at the interface between the two media and the heat conduction, and then the residual load carrying capacity is predicted for the ship hull undergone fire. The obtained results show that the ventilation has a significant influence on the development of fire and also on the temperature distribution; application of heat flux at the interface and the heat conduction is an effective way to obtain the temperature load on the structure; elevated temperature due to fire results in reduction of mechanic performance in the fire zone as well as the redistribution of stress distribution in the cross-section, which eventually lead to a decrease in the load carrying capacity of the hull. This study is of significance for anti-fire design of hull structures and for assessment of structure safety of ships undergone fire.

1 INTRODUCTION

Fire accidents are one of the main types of accidents that threaten the safety of ships, accounting for about 10% of ship accidents (Miao, 2000). It is typical that Ship cabins are arranged in a crowded manner, the spaces are relatively small, openings are usually located at the top, and there are many inflammables. Because of these factors, a fire, when it occurs, spreads out quickly and is difficult to extinguish, and it is likely to cause serious loss of life and property. In addition, the high temperature of the fire will also cause the degradation of the mechanical properties of the steel in the over-heated areas, and in turn reduce the strength of the hull. In order to improve the design level of fire resistance of ships, ensure the safety of personnel and reduce the property loss, it is necessary to investigate the structural safety of ships undergone fire.

Studies on the hull structure safety undergone fire are scarce, and there are only a few on related topics in the literature. Based on a unified probability model, Shetty et al (1998) evaluated the reliability of an offshore platform undergone fire accidents and performed an optimization study. It was shown that the thermal load and mechanical properties of steel at different temperatures have a significant impact on structural safety. Guedes Soares et al (2000) carried out a numerical study on the ultimate compressive strength of rectangular plates at elevated temperature, and analyzed the influence of thermal loads of different ranges and sizes on the ultimate strength; it has also been pointed out that, if the heating area is more

than 50% of the total area of the structure, the load bearing capacity of the structure will rapidly decrease. Fu & Wang (2018) simulated the temperature rise of the hull structure in a fire scenario using a standard fire temperature rise curve, and analyzed the ultimate strength of 10000TEU container ship undergone fire. In an attempt to reveal the cause of the accident of the Sanchi tanker sunk after fire, Li et al (2018) analyzed the influence of the fire on the mechanical properties of the material, and compared the loading carrying capacity of the hull before and after the fire. Liu et al (2018), based on the temperature load using fire scene simulations, carried out a study on the structural thermal response of deck slabs under cabin fire.

In this paper, based on a two-zone large eddy simulation method of fire scene simulation and structural thermal-mechanical coupling response analysis method, a structural temperature load analysis method and a hull residual strength analysis method are proposed for ships undergone cabin fires, which could be used as a reference for fire resistance design and structural safety assessment undergone fires.

2 BASIC THEORY AND METHOD

2.1 *Method for numerical simulation of fire scene*

The study is performed using *Fire Dynamic Simulation (FDS)*, a hydrodynamics-based fire analysis computer code developed by the National Institute of Standards and Technology (Mcgrattan et al, 2013). The code is

DOI: 10.1201/9781003216582-53

based on the two-zone large eddy theory and the combustion model. By establishing a fire scenario and directly solving the N-S equation of low Mach number flow driven by fire buoyancy. It is able to simulate the turbulent flow process of fires, especially the smoke and heat transfer process, and accurately simulate the distribution and spread of temperature and smoke in the fire scene.

The main governing equations of FDS are as follows (Wen et al, 2018):

Energy equation:

$$\frac{\partial(\rho h)}{\partial t} + \frac{\partial(\rho U_j h)}{\partial x_j} = \frac{\partial}{\partial x_j}(\Gamma_h \frac{\partial h}{\partial x_j}) + S_h \qquad (1)$$

N-S equation:

$$\frac{\partial(\rho U_j)}{\partial t} + \frac{\partial(U_j U_i)}{\partial x_j} = \frac{\partial}{\partial x_j}(\Gamma_U \frac{\partial U_J}{\partial x_j}) + S_{U_j} \qquad (2)$$

Continuity equation:

$$\frac{\partial \rho}{\partial t} + \Delta \rho \vec{u} = 0 \qquad (3)$$

Chemical equation:

$$frac\partial(\rho m)\partial t + \frac{\partial(\rho U_j m)}{\partial x_j} = \frac{\partial}{\partial x_j}\left(\Gamma_h \frac{\partial m}{\partial x_j}\right) + S \qquad (4)$$

where ρ is the density of the air, kg/m^3; h the enthalpy of the gas, KJ/mol; m the mass of air, kg; U_i and U_j the velocity components, m/s; Γ the exchange coefficient, which is related to model size; and S the source of fire.

2.2 Analysis method of structure temperature load based on heat conduction principle

Characteristics of the fire scene at any instant, such as temperature and gas pressure, flow velocity, CO_2 concentration, heat flux of the structure and environment, can be calculated with FDS. Heat flux, or heat flux density, refers to the heat flux that passes through a unit area per unit time. Because the FDS model and the structural thermal response analysis model nodes do not match, the heat flux is used instead, in order to accurately obtain the structural temperature load under fire, as the environmental boundary condition, and the temperature load for structural thermal response analysis is indirectly obtained by means of heat conduction analysis. The calculation principle is as follows (Kakac et al, 2018).

The temperature field is a function of space and time, which can be expressed as:

$$T = f(x, y, z, t) \qquad (5)$$

where the object isotherm is dense, the temperature change rate is large, and the temperature change rate is often different in different directions. Among the temperature change rates in different directions, the change rate along the isotherm normal direction is the largest, and its mathematical expression is:

$$\mathrm{grad}T = \frac{\partial T}{\partial n}\vec{n} = \frac{\partial T}{\partial x}i + \frac{\partial T}{\partial y}j + \frac{\partial T}{\partial z}k = \nabla T \qquad (6)$$

where grad T is the temperature gradient; $\partial T/\partial n$ is the rate of temperature change in the direction of the isotherm normal; $\vec{n}$ is a unit normal to the isotherm; and ∇ is the Hamilton operator.

The basic law of heat conduction, i.e., the Fourier's law takes the form:

$$q = -\lambda \mathrm{grad}T = -\lambda \nabla T \qquad (7)$$

where q is heat flux; λ is the thermal conductivity, a thermophysical parameter that characterizes the thermal conductivity of a substance.

The heat flux components in the three directions x, y, and z are:

$$q_x = -\lambda \frac{\partial T}{\partial x}; \ q_y = -\lambda \frac{\partial T}{\partial y}; \ q_z = -\lambda \frac{\partial T}{\partial z} \qquad (8)$$

Then the heat introduced and exported in a microelement can be obtained, taking the example of the x direction:

$$d\Phi_{in-x} = q_x dydz = -\lambda \frac{\partial T}{\partial x}dydz \qquad (9)$$

$$d\Phi_{out-x} = d\Phi_{in-x} + \frac{\partial}{\partial x}(-\lambda \frac{\partial T}{\partial x})dydzdx \qquad (10)$$

According to the law of energy conservation, i.e., the heat introduced into the micro-elements $d\Phi_{in}$ and the heat generated by the internal heat source dQ are equal to the heat exported out of the micro-element $d\Phi_{out}$ and the incremental conservation of the internal energy of the micro-element dU:

$$d\Phi_{in} + dQ = d\Phi_{out} + dU \qquad (11)$$

The increment of the internal energy of the micro-element can be written as:

$$dU = \rho c \frac{\partial t}{\partial \tau}dxdydz \qquad (12)$$

By substituting equations (9), (10) and (12) into equation (11), the heat transfer differential equation with heat flux as boundary and no internal heat source can be obtained:

$$\rho c \frac{\partial T}{\partial t} = \lambda \left(\frac{\partial^2 T}{\partial x^2} + \frac{\partial^2 T}{\partial y^2} + \frac{\partial^2 T}{\partial z^2} \right) = \lambda \nabla^2 T \qquad (13)$$

where c is specific heat capacity of object; and ∇^2 is the Laplace operator.

2.3 Theory of structural thermal-mechanical coupling response

When there is a temperature gradient in the structure, the coupling analysis of the displacement, stress and strain that occur during the interaction between the thermal stress generated within the structure and the plastic deformation generated by the structure under the thermal stress is referred to as thermo-mechanical coupling analysis (Melnik, 1998).

The governing equation of structural transient temperature field analysis:

$$\dot{u}_N^T(t)[K_u u_N(t) + M_T \dot{T}_N(t) - F(t)] = 0 \qquad (14)$$

where K_u is structural stiffness matrix; M_T is the thermodynamic stiffness matrix; and $F(t)$ the force vector.

The governing equation of structural thermal stress field analysis takes the form:

$$T_N^T(t)[C_u \dot{T}_N(t) + M_u \dot{u}_N(t) - D - R - K_T T_N(t)] = 0 \qquad (15)$$

where $T_N(t)$ is the node temperature vector; $u_N(t)$ the node displacement vector; C_u is hot melt matrix; M_u is the thermal coupling matrix; K_T is the heat transfer matrix; D is the dissipative vector; and R is the heat load vector.

By combining equations (14) and (15), the finite element method for structural thermal-mechanical coupling response analysis can be obtained:

$$\begin{bmatrix} K_u & M_T \\ M_u & C_u \end{bmatrix} \begin{bmatrix} \dot{u}_N(t) \\ \dot{T}_N(t) \end{bmatrix} = \begin{bmatrix} F(t) \\ Z(t) \end{bmatrix} \qquad (16)$$

$$Z(t) = D + R + K_T T_N(t) \qquad (17)$$

3 BASIC INFORMATION AND STRUCTURAL LAYOUT OF THE SHIP

According to the statistics of British Navy ship fire accidents, more than 50% of the fire accidents are

Figure 1. Finite element model of the ship.

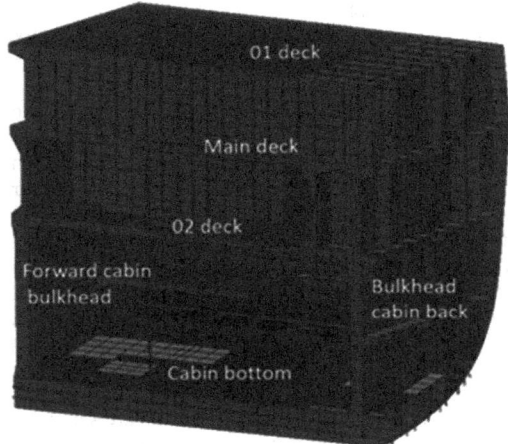

Figure 2. Profile view of engine room region.

engine room fires (Miao, 2000). For surface ships, the engine room is often located on the stern of the ship, and the hull beam has the largest bending moment. The fire in the engine room will seriously threaten the safety of the hull structure. For this reason, this study, taking the example of a surface ship, focuses on the simulation of engine room fire and analysis of the residual strength of the hull under fire.

The ship has a length of 120m, a width of 14.4m, and a draft of 3.87m. The finite element model for the entire ship is shown in Figure 1. The engine room is located in the middle of the ship. The engine room section is 24.15m long, 14.4m wide and 10.8m high (including 3 decks). The structural layout is shown in Figure 2. Table 1 shows the main thermo-physical parameters of the material (AISC, 2005).

4 PHOTOGRAPHS AND FIGURES

4.1 Construction of cabin fire scene

4.1.1 Selection of model range
The model range of the cabin fire scenario simulation is a three-cabin model with the engine room being the core. On the one hand, this selection can

Table 1. Physicochemical properties of heptane.

Temperature °C	Thermal Conductivity W/M °C	Specific heat capacity J/Kg °C	Heat transfer coefficient W/m² °C	Poisson's ratio	Linear expansion coefficient 10^{-5} / °C	Elastic Modulus GPa	Yield stress MPa
20	50	460	100	0.28	1.1	205	235
250	47	480	350	0.29	1.22	187	180
500	40	530	520	0.31	1.39	150	130
750	27	675	1000	0.35	1.48	70	40
1000	30	670	1500	0.4	1.34	20	25
1500	35	660	3000	0.45	1.33	19	2
1700	140	780	3100	0.48	1.32	18	1
2500	142	820	3500	0.5	1.31	12	0

help reduce the modeling workload and improve the calculation efficiency; on the other hand, it can achieve relatively continuous heat conduction through the connection between the front and rear cabins so as to ensure the calculation accuracy. In the fire scenario model, local structures such as stiffeners and brackets are ignored, and only the main structure members, such as vertical and horizontal bulkheads and decks, are taken into consideration, and their material properties are assigned.

Since ventilation has a significant impact on the development of the fire, external ventilation boundaries and internal ventilation conditions are set in the FDS model according to the cabin layout of the ship and the locations of the doors and vents. Figure 3 is the model of fire scene in the cabin with the outer panel hidden.

4.1.2 Simulation and setting of fire source
The setting of fire source is a key step for modeling fire scenarios, which mainly involves the choice of fire source type and the determination of heat release rate. Engine room fires are mostly diesel fires, and the fire types are Class B fires. The combustion characteristics of heptane are similar to diesel (Chen et al, 2015), and its physical and chemical properties are shown in Table 2. Therefore, heptane was selected as the fuel to simulate the oil fire in the engine room. The fire source is located in the engine room floor near the front bulkhead, with an area of 1m².

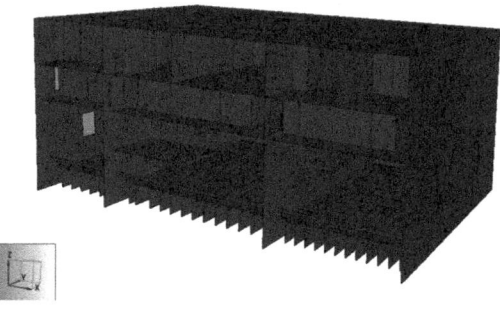

Figure 3. Three-cabin FDS model.

The heat release rate refers to the amount of heat released by the material in a unit time. The heat release rate models mainly include the t^2 stable fire source model, the segmented average method, the segmented linear method, and the method determined based on the mass loss rate (Hostikka & Keski-rahkonen, 2003). Among them, the t^2 stable fire source model uses a piecewise function to simulate the entire fire process. The heat release rate increases squared with time in the early stage of the fire, reaches a maximum heat release rate for a period of time, and then decays. The formula is as follows:

$$\bar{Q}(t) = \begin{cases} \alpha t^2 & t \leq t_g \\ \bar{Q}_{\max} & t_g \leq t \leq t_d \\ \bar{Q}_{\max} e^{-\frac{t-t_d}{\tau}t} & t \geq t_d \end{cases} \quad (18)$$

where $Q_{\max}$ is the maximum heat release rate achieved during combustion, for flammables such as gasoline and diesel, the maximum heat release rate is generally 1~3MW; t_g and t_d are the instants when the heat release rate reaches its maximum and begins to decay, respectively; τ is the decay time; α is the fire growth factor, engine oil fires are ultra-fast-growth fires, recommended to take the value of $0.1878KW \cdot s^{-2}$ (Hostikka & Keski-rahkonen, 2003).

The t^2 stable fire source model is more in line with the actual fire source, so this model is used to simulate the cabin fire and ignore the weakening stage. The high temperature load of the stable combustion stage is used to analyze the structural thermal response.

4.1.3 Setting of temperature measurement points and environmental slices
FDS provides two methods for obtaining the temperature of the structure and the temperature of the environment during the fire simulation, including temperature measurement points and environmental slices. Among them, the temperature measurement point simulation thermocouple can be used to measure the temperature change of the structure during the fire; the environmental slice can capture the fire dynamic

Table 2. Physicochemical properties of heptane.

Parameters	Units	Values	Parameters	Units	Values
Density	kg/m³	684	Thermal Conductivity	W/m°C	0.14
Specific heat capacity	J/kg°C	2246	Heat of combustion	kJ/g	44.6
Boiling point	°C	98.5			

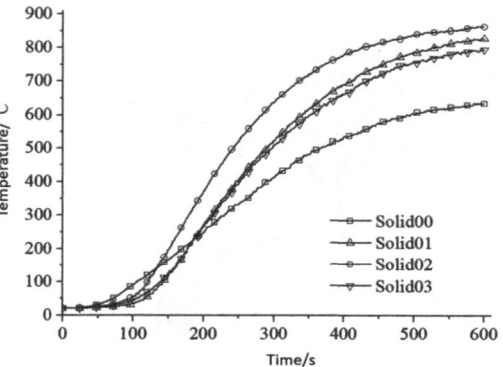

Figure 5. Measuring point temperature around heat source.

characteristics parameters including temperature, pressure, flow velocity, etc, depending on the section of the slice.

In order to track the changes of the fire source and the ambient temperature of the cabin under the fire in the engine room, four structural temperature measurement points were set around the fire source, and slices were set at the mid-longitudinal section, the inner floor and below Deck 2, as shown in Figure 4.

4.2 Engine room fire simulation results and analysis

4.2.1 Fire temperature

Figure 5 Shows the change of the temperature over time at the inner floor measurement point around the bottom fire source in the cabin. It can be found that: 1) the patterns of the temperature change at the structural measuring points are very similar, and the temperature of the measuring point near the opening of Deck 2 is relatively high; 2) the temperature of the measuring point facing away from the opening of the deck is relatively low; 3) in the initial stage of the fire, the temperature of the structure increased slowly, and 4) the temperature starts to increase rapidly at 150s and reaches a stable state at approximately 500s.

Figures 6 to 8 show the temperature distribution on different slices at different stages of the fire

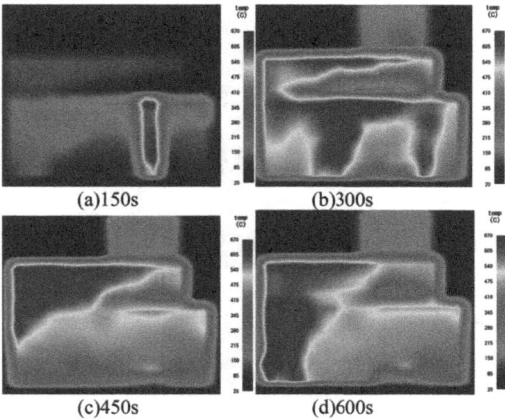

(a)150s (b)300s

(c)450s (d)600s

Figure 6. Environmental temperature distribution at longitudinal section at different instants.

development. Among them, Figure 6 is the ambient temperature distribution in the longitudinal section of the cabin. Figures 7 and 8 are the ambient temperature distributions in the cabin bottom and the cabin top (Deck 2), respectively. Further analysis in conjunction with Figure 5 reveals that:

1) At the beginning of the fire, the ambient temperature above the source of the fire rose sharply due to the upward spread of smoke, in contrast, the temperature of the structure around the source of fire near the ground increased slowly.

2) When the high-temperature smoke reaches the top of the cabin, it continues to move towards the oxygen-rich Deck 2 opening, and the ambient temperature of the cabin increases rapidly; the high-temperature smoke follows the hatch and enters the cabin between Deck 2 and the main deck, and continues to move to the hatch or vent.

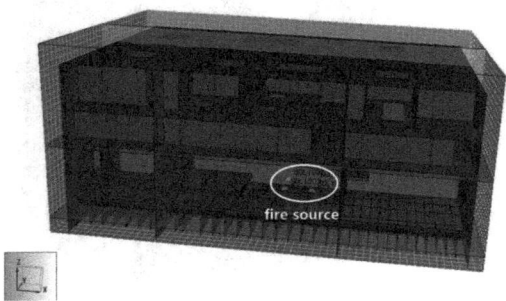

Figure 4. Slice and measuring point layout.

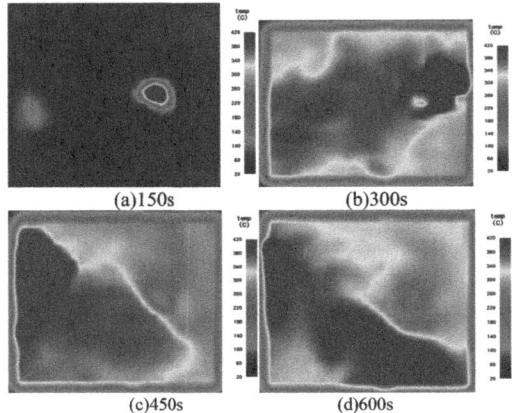

(a)150s   (b)300s

(c)450s   (d)600s

Figure 7. Environmental temperature distribution close to inner bottom at different instants.

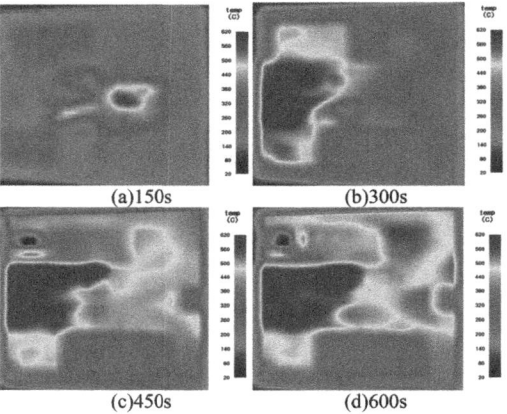

(a)150s   (b)300s

(c)450s   (d)600s

Figure 8. Environmental temperature distribution under Deck 2 at different instants.

3) When a steady state is reached, the ambient temperature at the top of the cabin is higher than the lower one, and the maximum ambient temperature reaches 700℃.

Therefore, the temperature distribution of cabin fires is very complicated. The ventilation conditions such as hatches and vents have a significant impact on the development of the fire and the temperature distribution. The structure temperature is related to the ambient temperature of the near structure, but it is not completely consistent.

4.2.2 Heat flux and structural temperature load

Figure 9 shows the heat flux extracted from the structural surface of the steady-state cabin segment. The heat flux distribution on the wall of the cabin of the

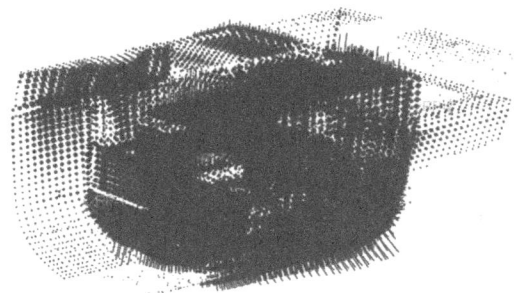

Figure 9. Heat flux distribution.

fired engine compartment is dense. The heat flux distribution in other cabins is relatively sparse due to the influence of ventilation conditions. This is also consistent with the heat flux distribution in an actual fire scenario.

Using the node coordinates as the index, the heat flux mapping of the structure surface is applied to the finite element mode of the structure (Figure 2), and the heat flux is used as the environmental boundary to obtain the cabin temperature load through thermal conduction analysis. As shown in Figure 10, the temperature of Deck 1 and the outer floor is close to the ambient temperature, the main deck temperature is around 400°C, and the local maximum temperature of the structure appears at the bulkhead adjacent to the opening of Deck 2, and the temperature is close to 700°C. Figure 11 shows the comparison between the structural temperature of the bulkhead and the ambient temperature. It can be found that the temperature distribution of the two is basically the same. Because of the proximity to the opening of Deck 2, as the high-temperature smoke flows, the ambient temperature and the heat flux on the surface of the structure are large. As a result, the structure temperature is also higher.

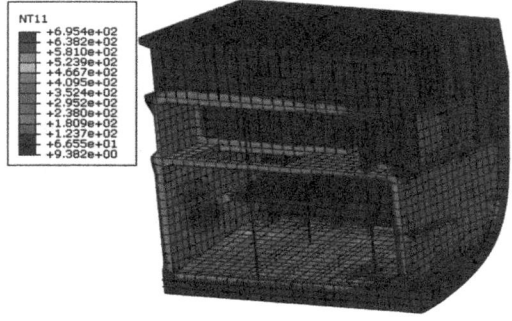

Figure 10. Temperature distribution of the engine room hold section.

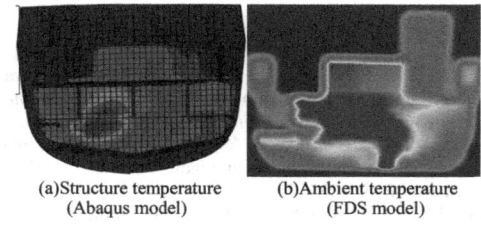

(a)Structure temperature (b)Ambient temperature
(Abaqus model) (FDS model)

Figure 11. Comparison of bulkhead temperature distribution.

5 ANALYSIS OF RESIDUAL STRENGTH OF HULL STRUCTURE UNDER FIRE

In order to analyze the degree of influence of high temperature on the hull load carrying capacity, the comparative analysis of the sag limit bearing capacity of the hull under normal conditions and cabin fire was carried out.

Figure 12 shows the bending moment-curvature curves of the hull girder under two conditions. The peak point of the curve is the ultimate bearing capacity of the hull corresponding to the condition. In the normal state, the sag limit bending moment of the hull is 7.04×108N·m, the residual load carrying capacity in the event of a fire is 6.56×108N·m, and the fire of 1m² lasting 600s in the engine room caused the hull load capacity to decrease by 6.8%.

Figures 13 and 14 show the structural stresses and deformations of the hull section under the two conditions. The maximum stress tends to the material yield limit at room temperature. Compared with the normal state without the influence of fire and high temperature, when the hull reaches the limit state in a fire accident, the high stress area of the deck is smaller but the structural deformation is greater, and the stress level on the bottom of the ship is lower. This is caused by the degradation of mechanical

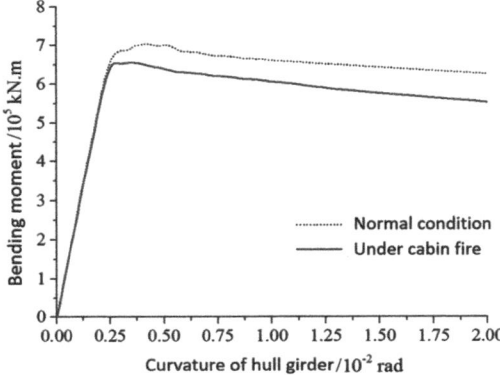

Figure 12. Comparison of moment versus curvature curves of hull girder undergone fire accident and normal condition.

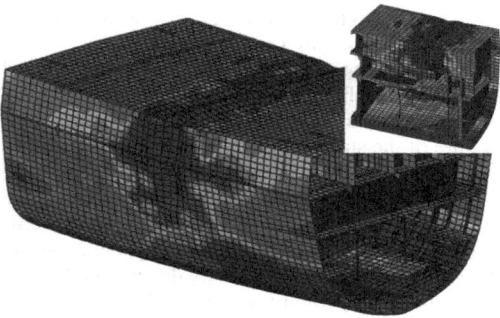

Figure 13. Stress and deformation of hull girder in ultimate limit state undergone cabin fire.

Figure 14. Stress and deformation of hull girder in ultimate limit state in normal condition.

properties such as the yield limit and elastic modulus of the steel material in the overheated area due to the high temperature of fire. Specifically, the decrease in the mechanical properties of the material results in a decrease in the stiffness and load-bearing capacity of the structure, and the decrease in the stress level of the structure in the fire zone further causes the cross-section and the axial downward shift, As a result, the plastic strain on the deck at the far end of the neutral axis increases, and the structural deformation becomes more significant. At the same time, the stress level at the bottom of the ship decreases, and the combined effect ultimately reduces the ultimate load carrying capacity of the section.

6 CONCLUSIONS

In this study, a two-zone large eddy simulation method is adopted to simulate the cabin fire scene, and a structural thermal-mechanical coupling response analysis method is applied to analyze the hull residual strength. Based on the results the following conclusions are drawn.

The two-zone large eddy simulation method implemented in FDS is suitable for the simulation of fire conditions in ship cabins. The measured temperature of the structure and the ambient temperature of the slice show that the ventilation conditions have a significant impact on the development of the fire and the temperature distribution. The structure temperature is related to the ambient temperature of the near structure, but does not solely depend on that.

Structural temperature load is the basis for carrying out structural thermal response analysis under fire. It is shown in this study that the use of heat flux as the environmental boundary, combined with heat conduction analysis, is an effective approach to obtain the temperature load on the hull.

The high temperature of the fire can cause degradation of the mechanical properties of the steel in the fire area and the redistribution of the structural stress, which in turn results in a reduction of the hull carrying capacity. The obtained results indicate that it is necessary to carry out systematic researches and analysis of typical cabin fires for fire resistant designs of ship structures and the assessment of structure safety of ship undergone fire.

REFERENCES

American Institute of Steel Construction (AISC), Inc, 2005. Specification for Structural Steel Buildings: AISC 36005.

Chen L.Y., Liu Y.H., Liu B.Y., Yi X.L. 2015. Numerical simulation study on fire characteristics of top open cabin. *Journal of Sichuan Soldiers*, 36(4): 139–143. (*in Chinese*)

Fu D.W., Wang D.Y. 2018. Research on residual ultimate strength of container ships in fire. *The Ocean Engineering*, 36(1): 83–90. (*in Chinese*).

Guedes Soares C, Teixeira A P. 2000. Strength of plates subjected to localized heat loads. *Constructional Steel Research*, 53: 335–358.

Hostikka S, Keski-rahkonen O. 2003. Probabilistic Simulation of fire scenarios. *Nuclear Engineering and Design*, 224(3):301–311.

Kakac S., Yener Y., Naveira-Cotta C.P. 2018. Heat Conduction (Fifth edition). Boca Raton, London, New York: CRC press Taylor & Francis Group.

Li C.F., Ren H.L., Zhou X.Q., Xu W.J. 2018. Deduction and analysis of the cause of the sinking of the Sanchi tanker. *Journal of Harbin Engineering University*, 39 (7): 1123–1131. (*in Chinese*)

Liu Y.S., Xue H.X., Zhou Jia, Tang W.Y. 2018. Dynamic Thermal Response Analysis of Ship Cabin Structure in Fire Scene. *China shipbuilding*, 59(4): 161–169.

Mcgrattan K, Hostikka S, Mcdermott R, Hostikka S. 2013. Fire Dynamics Simulator, User's Guide. Nist Special Publication.

Melnik N. 1998. Convergence of the operation-difference scheme to generalized solution of a coupled field theory problem. *Journal of Difference Equations and Applications*, 4(2):185–212.

Miao G.Q. 2000. Ship fire prevention and suppression. *Ship Fire and Safety*, 1–12. (*in Chinese*)

Shetty N. K., Guedes Soares C., Thoft-Christensen P, Jensen F.M. Fire safety assessment and optimal design of passive fire protection for offshore structures. *Reliability Engineering & System Safety*, 1998, 61(1–2): 139–149.

Wen J.X., Kang K., Donchev T., Karwatzki J. M. 2007. Validation of FDS for the prediction of medium-scale pool fires. *Fire Safety Journal*, 42(2): 127–138.

Response of steel stiffened plates under shock wave loadings

Kun Liu, Li Ke & Jiaxia Wang
School of Naval Architecture and Ocean Engineering, Jiangsu University of Science and Technology, Zhenjiang, Jiangsu, China

ABSTRACT: Experimental studies are performed to investigate the performance of stiffened plate under plane shock wave. The dynamic response is studied in an explosion-proof tank with a 400g TNT explosive, and the plastic deformation and acceleration of the specimen are investigated. The experimental results indicate that the deformation curves on structure surface have symmetry during the plastic deformation process. The plastic deformation which was mainly concentrated in non-stiffener area in the stiffened panel showed that stiffener is an efficient way to enhance the blast resistance of panel. The results can offer useful information for investigating the impact behavior of a stiffened panel during air blast.

1 INTRODUCTION

In the past few decades, some kinds of mechanical properties of stiffened panels have been extensively investigated. Moreover, scholars have conducted the studies on ship structures suffering the explosive loading to evaluate the safety and reliability of structures (Bulson & Ebrary, 1998). Therefore, it is worth noting that the antiknock performance of ships is a vital indicator of survivability with various conditions encountered in marine navigation, which means stiffened panels need to be designed against explosive loading. Previous scholars have studied the response of stiffened panels under the impact of explosion and analyzed the relevant dynamic behaviors (Rudrapatna 2000; Zong 2013), proposing the acoustic-structure coupling method and the phenomenological interactive failure criterion comprising bending, tension and transverse shear are useful to study the damage process and failure model of surface ships impacted by shock wave. Other scholars have also studied the results from experimental, theoretical and numerical studies on the deformation and failure model of thin square, circular and rectangular panels subjected to blast loading in order to predict the deformation energy (Nurick 1986; Teeling-Smith & Nurick 1991; Olson 1993; Nurick 1995). Then, the further researches of experimental and numerical work on the response of complex structures named built-in mild steel quadrangular panels with different stiffener configurations were carried out with uniform and localised blast loading (Yuen & Nurick 2005; Langdon 2005). The comparison results showed large inelastic deformation is well predicted for the various stiffener configurations and size for both maximum deflection and deformation shape by numerical

simulation. In addition, numerical simulations of air blast loading in the near-field acting on deformable steel plates have been performed and compared to experiments, and showed simplified models had a great effect on the calculation results (Zakrisson 2011). Current practice in designing structures against explosive loading usually considers only the near-field explosion generated by the explosive. However, there are more threats from far-field explosions for ships during navigation. Therefore, the study on the shock wave from far-field blast is extraordinarily necessary, which is closely related to dynamic response analysis and engineering practice.

Due to the constraints of cost and technology, the simulation of plane shock wave from far-field air blast is usually relied on some tools and instruments (Bernstein & Goettelman 1966). The typical configurations of plane wave generator with complicated structures and high cost include two-component explosives, multipoint initiation, mouse trap and inert materials (Yadav 1986; Chen 1993; Bucci 2012). Some scholars put forward the plane wave can be generated by a simple generator which is only composed of two identical cylindrical high explosive charges and an air-metal barrier (Zhang 2006; Xiong 2014). But it was designed just by means of theoretical analysis and numerical simulations, which means lack of experimental verification. Therefore, a comprehensive numerical study performed with the finite element software is proposed to simulate structural responses under blast pressures resulting from a far-field explosion (Ding 2015). This approach is proved to be a useful tool for obtaining the structural responses under far-field explosions. However, the far-field explosion test is usually performed in an open-air environment which meant the accuracy of the measurement results is difficult to control. As

DOI: 10.1201/9781003216582-54

reported by Kury (1965), cylinder test can be used to study explosion performance. It provides an idea that far-field air blast could be also simulated in a hermetic device. According to the approach, Zhang et al. (2015) tested the dynamic response of metallic structures under air blast loading in an explosion tank. And both the experimental phenomena and measured results are satisfactory. Therefore, it is feasible to convert the spherical wave generated by conventional explosion into the plane wave with a long duration of action when conducting far-field air blast tests.

In this paper, the experimental results of the stiffened panels under plane shock wave from far-field air blast are presented in a cylindrical device. Particular focus is placed on the deformation and the acceleration of the stiffened panel.

2 AIR BLAST EXPERIMENT

To guarantee the success of the experiment, this paper designs the stiffened panel system to ensure that it can constrain the test model as well as it can match the test device. The stiffened panel system including a stiffened panel specimen, a pressure plate, four square tubes and a support base can guarantee the effective explosion surface, as shown in Figure 1. During the process of constraint, the stiffened panel system is uniformly designed with circular holes with diameters of 26 mm on boundary, and the total number is 40, and each part of the system is fixed at the bolt-holes by M24 bolts (with the diameter of 24 mm). As shown in Figure 1(a), the pressure plate of the system can be used to increase the constrained area because of the bolt-constraints belongs to point-constraints which may cause panel tearing and affect measurement data during the blast test. The four square tubes are square in cross section with a side length of 80mm and a thickness of 6 mm. And the length and width of the stiffened panel as well as pressure plate are 1160 mm, and an inner width of the pressure plate is 1000 mm. In the design of the support base, it is necessary to ensure that it does not affect the measurement of the test data and the deformation is small during the explosion process. Therefore, the base should have a certain space and channel to facilitate the arrangement of related measurement data lines, in addition to providing a good support effect. Based on these requirements, the design height of the base is taken as 400 mm, and the circular holes with diameter of 50 mm are opened in the side wall surface. Table 1 summarizes the geometric information and material types of all the part of the stiffened panel system.

The plate of the stiffened panel specimen is manufactured from the mild steel and the stiffener is manufactured from the high-strength steel with the section profile of NO.12 flat-bulb steel. A summary of the properties of the material is given in Table 2.

2.1 *Experimental device*

The explosion which results in a spherical shock wave occurs in a very short time after the conventional explosive is detonated in the open-air environment. If the distance is far enough, the spherical wave can be approximated as a plane wave. However, the energy of the shock wave will be attenuated at this time, making it difficult to meet the energy and duration of action requirements. Moreover, it also brings great difficulty to experimental measurement. Therefore, it is necessary to select a device which can convert the spherical shock wave into a plane wave during the explosion, and maintain a sufficiently long action time. In this paper, the

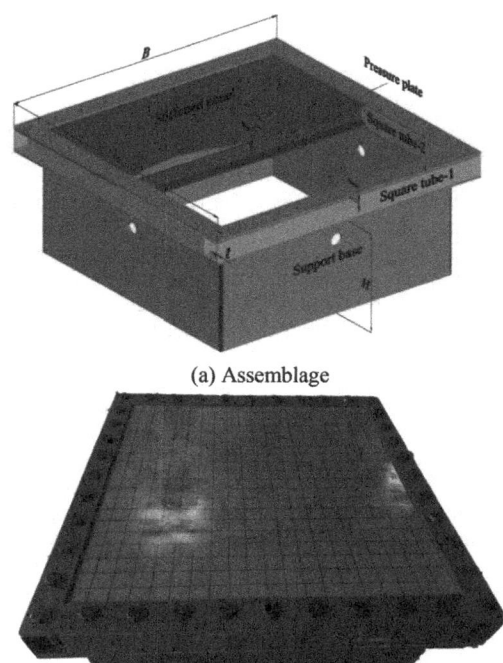

(a) Assemblage

(b) Field installation

Figure 1. Stiffened panel system in the blast text.

Table 1. Properties of Q235 and 945 steel used to produce the stiffened panel.

Category		L/mm	B/mm	H/mm	t/mm	Material
Stiffened panel	panel	1160	1160	-	5	mild steel
	stiffener	-	1000	120	-	high-strength steel
Square tube-1		1160	80	80	6	mild steel
Square tube-2		990	80	80	6	mild steel
Pressure plate		1160	1160	-	10	mild steel
Support base		1160	1160	400	10	mild steel

Table 2. Properties of mild and high-strength steel used to produce the stiffened panel.

| Symbol | Unit | Value | | Property |
		mild steel	high-strength steel	
ρ	kg/m³	7850	7850	Density
E	GPa	210	201	Young's modulus
$\sigma_{0.2\%}$	MPa	235	465	0.2% offset yield strength
σ_b	MPa	450	550	Tensile strength
ε_f		~0.32	~0.22	Failure strain

device is an explosion-proof tank with an inner diameter of 2 m and a height of 2.6 m. And the height above the ground is 1.6 m, and the depth below the ground is 1.0 m. The experimental device is depicted in Figure 2. It constitutes multiple systems including airtight system, explosion system and protection system, which can generate shock waves that occur during far-field air blast. The bottom of cavity is uniformly arranged with circular holes accounting for about 40% of the whole area and their diameter is 20 mm. The explosive is placed at a certain height from the bottom of the cavity, and the shock wave instantly fills the cavity after the explosive explodes. After reflection, it is filtered through circular holes, and the filtered shock wave uniformly acts on the surface of the specimen, converting a spherical wave into a plane wave.

2.2 Experiment procedure

As illustrated in Figure 3, it is 50 mm distance between lines on the surface of structure, and 3 measuring points are selected in the quarter area based on the symmetry of stiffened panel It is worth noting that the 3 points need to have a certain distance from the stiffener and cover as much as the

Figure 2. The explosion-proof tank used for blast tests.

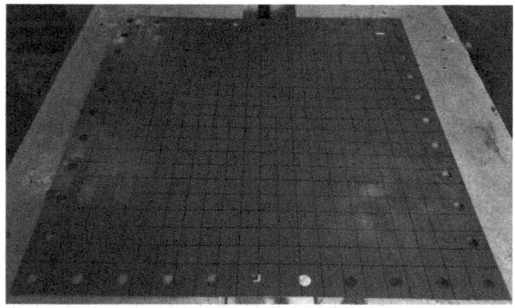

(a) Upper surface

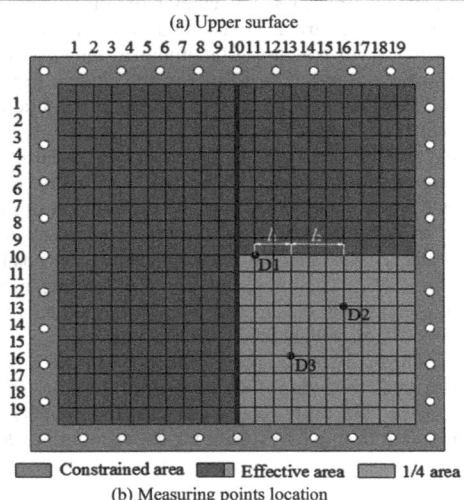

(b) Measuring points location

Figure 3. Stiffened panel structure and locations of measuring points.

area for acceleration measurement. Therefore, the 3 points are evenly distributed with 50 mm away from stiffener, where l_1 is 100 mm and l_2 is 150 mm. In addition, measuring points D1, D2, and D3 are used to set integrated electronics piezoelectric (IEPE) for the measurement.

The test procedure adopted in this study consists of four steps. Step one: The test stiffened panel was constrained respectively by the pressure plate and square tubes, and bolted onto the 400 mm high support base to form a stable stiffened system. Step two: the system is hoisted horizontally into the explosion-proof device by crane and pre-buried with fine sand. In this study, the blast loading is obtained from the shock wave which is developed in a 400 g cylindrical TNT explosive. Step three: The explosive is placed at a certain height directly above the cavity to ensure a distance of 1.2m from the surface of stiffened panel. Then, the shock wave goes through to atmosphere from the circular hole of the cavity and applies a plane blast load on the target specimen that is mounted at a distance from the circular hole of the cavity. Step four: the explosion-proof device is capped with the cover after everything is installed. It is difficult and inaccurate to measure the dynamic displacement of the stiffened panel during the air blast. In view of this situation, the plastic deformation of the stiffened panel structure is recorded by comparing the displacement of the measuring points before and after the air blast. A portable laser range finder (PLRF) is used to record the displacement, and the results of subtraction are the plastic deformation.

It should be explained that the experimental set-up used in the present study cannot directly determine the impulse imparted on stiffened panel. However, it is well known that the total impulse delivered onto the surface of structure results in the change in pressure of the panel. Due to that the panel is stationary initially and impacted plane shock wave after air blast, the impulse can be approximately equal to the momentum of the panel. Therefore, the impulse transmitted to the panel is replaced by the product of pressure and time which is recorded by a piezo resistive pressure sensor (PRPS). Figure 4 shows the position of the PRPS. As shown, the whole measuring instrument, including sensor and joint, is assumed to be symmetric on both sides of stiffened panel.

Additionally, some pre-explosion tests need to be performed to judge the sensitivity of the PRPS before to the truly blast test. In this paper, the equivalent of explosive used for experiment is 400 g.

2.3 Test results

The pressure time histories in each measuring point (i.e. P1, and P2) on the structure surface with truly blast test and on the same distance measured from the blast center with pre-explosion tests at the target points directly below the explosive as indicated in Figure 5 are recorded. It shows that the shock wave quickly reaches its peak under the air blast loading, and then gradually decays towards zero with the time of about 1s. The pressure amplitude curves measured by the PRPS at P1 and P2 are basically identical showing that the measuring instrument (PRPS) meets the requirement of blast tests. From the pressure measurement results presented in Figure 5, it can be

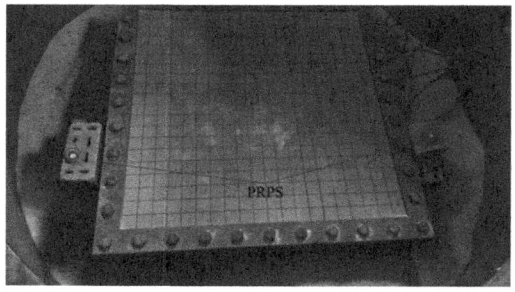

Figure 4. Installation location of PRPS.

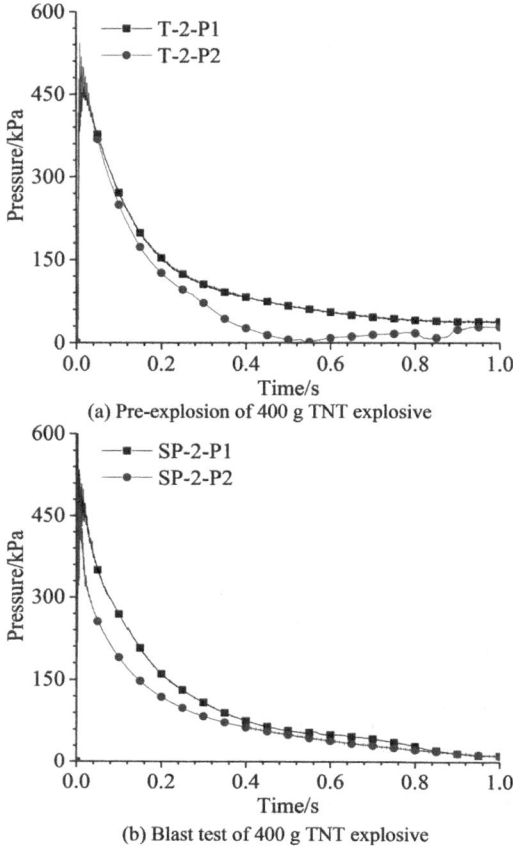

(a) Pre-explosion of 400 g TNT explosive

(b) Blast test of 400 g TNT explosive

Figure 5. Pressure amplitude curves of the pre-explosion and blast tests.

observed that the pressure on the structure surface reaches its peak in a very short time, and the amplitude curves of the two measurement points are obviously smooth, which showed that the test results were satisfactory. Therefore, it can be illustrated that the peak pressure of structure surface is 500 kPa impacted by the shock wave with 400 g TNT explosive.

In addition, it can be illustrated that the shock wave reaching on the structure surface is plane wave from the results that the pressure amplitude curves obtained at different positions on the structure surface under different TNT explosives.

The pressure time histories in each measuring point (i.e. P1, and P2) on the structure surface with truly blast test and on the same distance measured from the blast center with pre-explosion tests at the target points directly below the explosive as indicated in Figure 5 are recorded. It shows that the shock wave quickly reaches its peak under the air blast loading, and then gradually decays towards zero with the time of about 1s. The pressure amplitude curves measured by the PRPS at P1 and P2 are basically identical showing that the measuring instrument (PRPS) meets the requirement of blast tests. From the pressure measurement results presented in Figure 5, it can be observed that the pressure on the structure surface reaches its peak in a very short time, and the amplitude curves of the two measurement points are obviously smooth, which showed that the test results were satisfactory. Therefore, it can be illustrated that the peak pressure of structure surface is 500 kPa impacted by the shock wave with 400 g TNT explosive.

In addition, it can be illustrated that the shock wave reaching on the structure surface is plane wave from the results that the pressure amplitude curves obtained at different positions on the structure surface under different TNT explosives.

(1) Acceleration performance

When recording acceleration data for the stiffened panel, the sensitivity of the IEPE is 0.043 m/s², and the sampling frequency is 100 kHz. To clarify the influence of the shock wave on the accelerations of stiffened panel, the post-mortem analysis of accelerations with same surface but different position is conducted, as shown in Figure 6.

It is shown that the accelerations at the three measurement points increase rapidly under the 400g air blast loading, and then gradually decays towards zero with the time of 0.6s. Meanwhile, the acceleration peak relationship of the three measurement points is $a_{D3} > a_{D2} > a_{D1}$ and there is not much difference between the amplitude curves of measurement points D2 and D3, which indicates that the stiffened plate mainly depends on the stiffener to improve the impact resistance. There is not much difference in the mechanical property of the area far from the stiffener. After ignoring the individual sudden points, the maximum accelerations at the three measurement points are a_{D1}=9216.4 m/s², a_{D2}=11995.5 m/s² and a_{D3}= 10838.2m/s².

(2) Deformation performance

The geometric change of the specimen impacted by the shock wave arose from 400 g TNT explosive is illustrated in Figure 7. It can be clearly illustrated that the response of stiffened panel under vertical impact loading mainly concentrated in non-stiffener area, and the deformation of these areas has obvious symmetry. It is also evident that the maximum plastic deformation emerged on the center in non-stiffener area and grew along the direction toward the stiffener. Under the interaction of face panel and boundary, the corner of panel appears in a "plastic hinge", as shown in Figure 7a. Furthermore, there is a crack that looks like a "bulb" at the one connection between the stiffener and panel but another connection does not have crack, which may be caused by the welding technology rather than the pressure imbalance. Because of this, the overall deformation of the stiffened panel is not affected much and only if it is near the crack that the deformation will be slightly larger than the deformation near where crack does not occur, as shown in Figure 7b.

(a) Upper surface

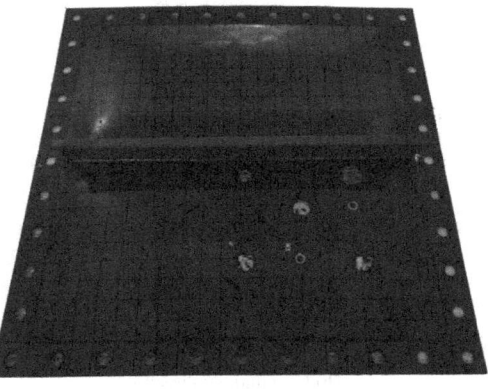

(b) Lower surface

Figure 7. Plastic deformation of stiffened panel.

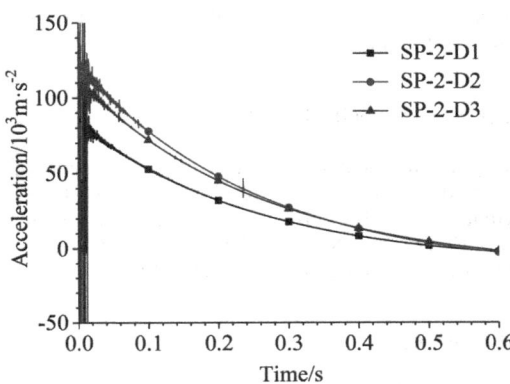

Figure 6. Pressure amplitude curves of the surface of stiffened panel.

It is worth mentioning that the structure surface has been uniformly painted locating points with the number of $19^2=361$. According to these points, it can effectively respond to plastic deformation at different positions of the stiffened panel. Among them, number L1~L19 indicate the direction of vertical stiffener (transverse direction), and number B1~B19 indicate the direction of parallel stiffener (longitudinal direction).

The deformations at different positions on the surface of the stiffened panel are recorded and the maximum plastic deformation is 38mm. As shown in Figure 3(b), 9 points with D1~D9 are covered by Locating points B4, B7, B10, L10, L13 and L16. In order to further study the deformation law of stiffened panel, the amount of deformation at typical locations such as B4, B7, B10, B13, B16, L4, L7, L10, L13 and L16 are used for analysis, as shown in Figure 8.

It is illustrated that the deformation of stiffened panel appears in a "double peaks" along the transverse direction, which means the stiffener can effectively resist the loading caused by air blast. It can be seen clearly that the peak curves of the deformation is smoother on the side near the stiffener, and changes linearly on the side near the boundary. Simultaneously, the peak appears at locating points L5 and L15. Comparing the deformation curves of the five locating points, it can be realized that the deformation of the stiffened panel has obvious symmetry in the transverse and longitudinal directions. And there is little difference in deformation near the center, such as the curves of B4, B7 and B10. Figure 8(b) shows the deformation trends which appear in a "arch shape" of the five locating points along the longitudinal direction. Except for the stiffener, the deformation at several points near the boundary is basically identical as well as does not change at locating points B8, B9, B10, B11 and B12. According to the deformation of the stiffened panel impact by plane shock wave, it can be considered its deformation mode is affected by the coupling of the stiffener and the boundary, which is stronger in the transverse direction and weaker in the longitudinal direction.

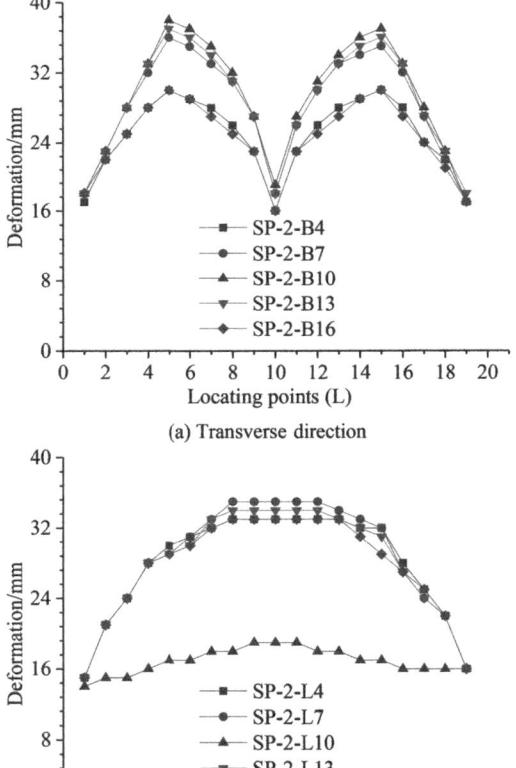

Figure 8. Plastic deformation of stiffened panel at typical locations.

3 CONCLUSIONS

The blast performances of the stiffened panel made of mild and high-strength steel was investigated by performing air blast experiment to simulate the plane shock wave from far-field air blast in an explosion-proof tank. Result of the plastic deformation which was mainly concentrated in non-stiffener area in the stiffened panel showed that stiffener is an efficient way to enhance the blast resistance of panel. The results offer useful information for investigating the impact behavior of a stiffened panel during air blast.

It can be considered the deformation mode of stiffened panel is affected by the coupling of the stiffener and the boundary, which is stronger in the transverse direction and weaker in the longitudinal direction. And the impact accelerations of the coupling area of the boundaries are generally larger than other area, which can be indicated the impact resistance of the coupling area is weaker.

ACKNOWLEDGEMENTS

We gratefully acknowledge the financial support of the National Natural Science Foundation of China (Grant No. 51609110; Grant 51779110; Grant 51809122) and Natural Science Foundation of Jiangsu Province (Grant No. BK20191461).

REFERENCES

Bernstein D, Goettelman R C. Generation of cylindrically symmetric implosions. Review of Scientific Instruments, 1966, 37(10): 1373–1375.

Bucci O M, Migliore M D, Panariello G, et al. On the synthesis of plane wave generators: performance limits, design paradigms and effective algorithms//2012 6th European Conference on Antennas and Propagation. Institute of Electrical and Electronics Engineers, 2012: 3500–3503.

Bulson P S, Ebrary I. Explosive loading of engineering structures: a history of research and a review of recent developments. International Journal of Impact Engineering, 1998, 21(5):415.

Chen C Y, Shiuan J H, Lan I F. Design of an inert material type plane wave generator. Propellants, explosives, pyrotechnics, 1993, 18(3): 139–145.

Ding C, Ngo T, Ghazlan A, et al. Numerical simulation of structural responses to a far-field explosion. Australian Journal of Structural Engineering, 2015, 16(3): 226–236.

Kury J W, Hornig H C, Lee E L, et al. Metal acceleration by chemical explosives//Fourth International Symposium on Detonation. White Oak, Maryland, 1965:3–13.

Langdon G S, Yuen S C K, Nurick G N. Experimental and numerical studies on the response of quadrangular stiffened plates. Part II: localised blast loading. International Journal of Impact Engineering, 2005, 31(1): 85–111.

Nurick G N, Olson M D, Fagnan J R, et al. Deformation and tearing of blast-loaded stiffened square plates. International Journal of Impact Engineering, 1995, 16(2): 273–291.

Nurick G N, Pearce H T, Martin J B. The deformation of thin plates subjected to impulsive loading. Inelastic behaviour of plates and shells. Springer, Berlin, Heidelberg, 1986:597–616.

Olson M D, Nurick G N, Fagnan J R. Deformation and rupture of blast loaded square plates-predictions and experiments. International Journal of Impact Engineering, 1993, 13(2): 279–291.

Rudrapatna N S, Vaziri R, Olson M D. Deformation and failure of blast-loaded stiffened plates. International Journal of Impact Engineering, 2000, 24(5): 457–474.

Teeling-Smith R G, Nurick G N. The deformation and tearing of thin circular plates subjected to impulsive loads. International Journal of Impact Engineering, 1991, 11 (1):77–91.

Xiong W, Zhang X, Guan Z, et al. Study of simple plane wave generator with an air-metal barrier. Defence Technology, 2014, 10(2):190–197.

Yadav H S, Kamat P V, Sundaram S G. Study of an Explosive-Driven Metal Plate. Propellants, Explosives, Pyrotechnics, 1986, 11(1): 16–22.

Yuen S C K, Nurick G N. Experimental and numerical studies on the response of quadrangular stiffened plates. Part I: subjected to uniform blast load. International Journal of Impact Engineering, 2005, 31(1):55–83.

Zakrisson B, Wikman B, Häggblad H Å. Numerical simulations of blast loads and structural deformation from near-field explosions in air. International Journal of Impact Engineering, 2011, 38(7): 597–612.

Zhang C S, Zou D H, Madenga V. Numerical simulation of wave propagation in grouted rock bolts and the effects of mesh density and wave frequency. International journal of rock mechanics and mining sciences, 2006, 4(43): 634–639.

Zhang P, Liu J, Cheng Y, et al. Dynamic response of metallic trapezoidal corrugated-core sandwich panels subjected to air blast loading-An experimental study. Materials & Design, 2015, 65: 221–230.

Zong Z, Zhao Y, Li H. A numerical study of whole ship structural damage resulting from close-in underwater explosion shock. Marine Structures, 2013, 31:24–43.

Developments in Maritime Technology and Engineering – Guedes Soares & Santos (eds)
© 2021 Copyright the Author(s), ISBN 978-0-367-77376-2

A case study on the impact of Lightship Weight (LWT) Distribution in the structural scantlings

A. Miranda
WestSEA Shipyard, Viana do Castelo, Portugal

ABSTRACT: The paper describes a case study during the design of a Split Hopper Dredger, which was shown to be highly depended on the assumed distribution of the Lightship Weight in Stillwater. Due to the referred case, a general overview of the possible methods for the LWT Distribution was analysed in this paper. The typical tasks taken nowadays by the Classification Societies, to review any newbuilding, have been also evaluated, exposing potential problems on their role to duly survey the LWT Distribution curve. It is then assumed that a reliable definition of a LWT Distribution Curve should be based on the geometric properties of a Trapezium for each weight element of the LWT List. Therefore, it is described all the best considerations that the Designer should follow, being aware of the common errors that must be totally avoided, in order to properly define a trustable shape of the LWT Distribution.

1 INTRODUCTION

WestSEA Shipyard has signed a contract at January 2017 to build a Split Hopper Dredger of about 70m length, and hopper capacity of 1000m3 (Figure 1, Table 1), to be delivered by July of 2018.

At the final stage of delivering, the original LWT Distribution was adjusted to the Inclining Test results. The entire structural design was completely not complying at that point. After a profound and detailed correction of the shape LWT Distribution, the same vessel was now complying with significant larger margins. This sudden change on the structural compliance confirmed that the shape of the LWT Distribution has an huge impact on the hull-girder stresses.

However, based on our experience, none of the Classification Societies involved in our previous newbuildings had ever questioned the shape of LWT Distribution adopted. Moreover, Class rules do not give any clear guidelines for a proper definition of the LWT Distribution.

Therefore, an important question rises regarding the reliability of the Structural Review: Are Classification Societies duly surveying it?

Considering the lack of updated guidelines, it is then suggested a common procedure that Designers should uniformly adopt to properly define a LWT Distribution, based on the Geometric properties of a Trapezium, highlighting the best practices to be followed, and identifying the common errors to be avoided.

2 CASE STUDY

2.1 Details of the case

At the Basic Design stage, there was an intense Class review with several Finite Elements (FE) analysis and discussions, between the original Designer, the Shipyard and the Classification Society. The building schedule was very tight, and the best structural solution of the Midship Section was agreed by all, proven by several FE calculations performed by the Class.

Considering the original Lightship Weight (LWT) Distribution and the Envelopes of Stillwater (SW) Bending Moments and Shear Forces (BM&SF), the Midship Section was finally approved with a margin gap of <5 MPa till the maximum allowable stresses.

During later stage of vessel's production, it was expected already an increase of LWT, as a result of several and additional structural reinforcements agreed and suggested by the Class review.

An internal Inclining Test has been conducted in end of March 2018, in order to better define the real LWT and its Centre of Gravity (LCG, TCG and VCG), as well as to determine the optimal Solid Ballast to be installed to minimize the natural ship's list (due to dredging deck equipment installed mainly at starboard).

The official Inclining Test has been performed at June 2018, at the final building stage of the ship.

2.2 Identification of the problems

With the results of that internal Inclining Test, it was confirmed a significant increase of LWT. The whole

DOI: 10.1201/9781003216582-55

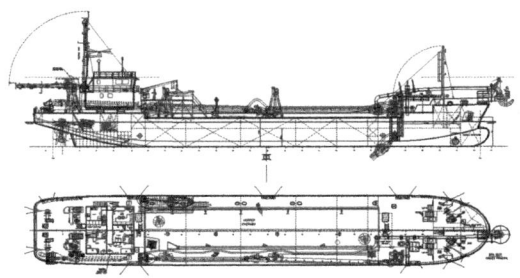

Figure 1. General arrangement of the ship.

Table 1. Main characteristics of the ship.

Length Overall	Loa	73.00	m
Breadth	B	11.40	m
Depth	D	4.75	m
Air Draught		8.50	m
Hopper capacity		1045	m³
Dredging Depth		-30	m
Crew		8	members
Navigation (not dredging)		Unrestricted	
Dredging (within 15/20 miles from shore/ port)		H.S.<2.5m	

ship needed to be re-evaluated for an increased Scantling Draught, necessary in order to increase Displacements. Due to this, it was already expected that some critical areas at Midship Section would have higher stresses than the allowable limits, considering the very small gap during Class review.

However, the worst situation to be solved was the significant higher SW BM above the envelop limits (sagging), after simulating the Intact Loading Conditions with cargo, and considering the LWT Distribution adjusted to the Inclining Test results. The few Ballast Tanks of the ship reduce any possibility to arrange another distribution of Deadweight that could lead to reduce the Sagging Bending Moments.

At this moment, the LWT Distribution was simply adjusted, adding a trapezoidal shape (with the length of the ship) to the original Designer's LWT Distribution, as shown in Figure 2, in order to reach the Inclining Test results.

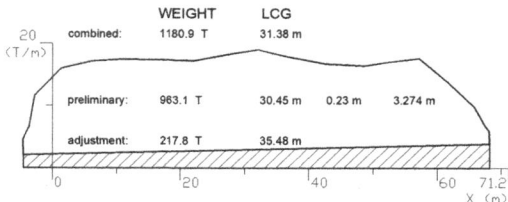

Figure 2. Original LWT distribution of the vessel, with the adjustment portion (in hatch) to achieve the inclining test results.

2.3 Main goals to be solved

Considering the very critical situation, which there was no possible solution to comply with the original SW BM&SF Envelops for the fully loaded cargo condition, it was decided to start re-analysing all the Class review input parameters.

The first and main target was to accurately check the shape of the original LWT Distribution. Based on the detailed LWT List, developed and monitored by the shipyard, it was introduced the longitudinal ends for each item, taking into account all the best guidelines as presented at chapter 4.

The resulted shape of the LWT Distribution (see Figure 3), given by our homemade software "Ficha-Gama" (see ch.4.6), was completely different from the original Designer's shape (Figure 2).

It is very interesting that the new and realistic LWT Distribution, perfectly match with the General Arrangement Ship's Profile, clearly identifying the major areas which the LWT items are concentrated: Aft Winches; Engine Rooms AFT and FWD; Superstructure; Split Hull Hydraulic Jacks AFT & FWD; Dredger Deck Equipment; Solid Ballast; FWD Winches and Chain Locker; Bow Thruster.

With the realistic LWT Distribution Shape, the SW BM&SF drastically reduced in fully Cargo loaded (Dredging scenario), showing after all a significant margin to the original SW BM&SF Envelopes.

This was indeed the key point to focus an entire new Scantling Class Review, with a complete set of new parameters. Considering that the ship delivery was only couple months ahead, the main goal was to re-approve the Hull Scantlings without any structural reinforcements.

2.3.1 Maximum Draught complying minimum summer freeboard

The first step was to re-calculate accurately the Minimum Summer International Freeboard, following the ILLC of 1966. All details have been exhaustively analysed, considering all possible deductions, having then the approval of Portuguese Flag Authorities (DGRM). It was defined the maximum Freeboard Summer Draught, which became to be our main target to subsequently verify the compliance in other subjects: 3.945m.

In the beginning of this Project, the Scantling Draught (Tscant) was pretty much lower, thus the Freeboard had a larger reserve. It is assumed, the original Designer had not calculated the Freeboard exhaustively, as there was not any necessity for it.

Based on the Guidelines of Reduced Freeboards for Dredgers (DR-68), as well as based on the Classification Society Rules for Dredgers Notation, it is possible to reduce up to 1/2 of the Summer Freeboard, only within the dredging operation range.

2.3.2 Max Draught complying stability criteria

All Stability Calculations have been then recalculated for these maximum Draughts, which proves that complies with all stability criteria for Unrestricted Navigation and for Dredging Operation (DR-68).

Fortunately, it was possible to ensure the contractual Cargo DWT as desired by the Owner, without compromising the delivery of the ship, and without penalties.

A complete analysis of Intact Stability Loading Conditions (more than 40) have been developed, for several Spoil densities, and for each spoil type (liquid or solid).

2.3.3 New SW BM & SF envelopes

The shapes of SW BM&SF, from all Loading Conditions (which were considering already the new and realistic LWT Distribution of Figure 3), have been collected, and later on analysed with the original SW BM&SF Envelopes. Due to this, it was possible to clearly determine the areas which the limits could be reduced or adjusted to obtain the new proposed Envelopes.

The Figures 4.1&4.2, as well the Figures 5.1&5.2, shows the original Envelopes in fine dashed lines. Over those original limits, are the relative key points bullets, on which the Class had initially reviewed at the Hopper's (Hold) ends. The new proposed Envelopes are then represented in coarse and thicker dashed lines, with the optimal shape that comfortably accommodate all BM&SF curves (in continuous lines) from all Loading Conditions.

At Dredging Conditions (Figure 5.1, 5.2), it is noted the most significant reduction on the new SW BM&SF limits. Therefore, we were confident that the actual Midship Section was able to withstand with the new Input parameters, because this scenario was the most demanding one at the Initial Structural Review by the Class.

At Navigation Condition (Figure 4.2), the SW SF limits had also reduced slightly. Regarding the SW BM, the Sagging Condition had also reduced

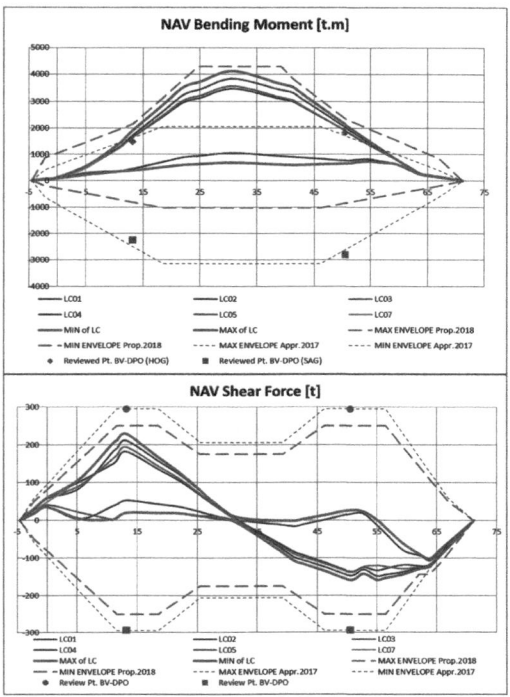

Figure 4.1 & 4.2. The BM & SF for unrestricted navigation operation.

significantly, however, by the other hand, the limits of Hogging Condition had increased (Figure 4.1). Therefore, optimal adjustments have been inputted in the shape of SW BM (hogging) Envelope in order to minimize the increase of the figure numbers at the Hopper's ends. The main goal was to keep similar the BM and SF values at those locations, in order to remain valid, as far as possible, the Structural class review already done at beginning of the project.

2.3.4 Maximum Draught complying steel structure review

After collecting all input from the Stability (SW Envelope Limits), it was time to check in detail all impacts at the Hull Structure, mainly higher pressures at bottom structures due to higher scantling/dredging Draughts, taking into consideration the new combined Vertical BM&SF.

As can be checked at the Basic Design Parameters differences (Table 2), the amount of the WV Hull Girder's Loads did not increased proportionally with the higher Draughts. It is assumed that the initial Designer, at a very preliminary project stage, had defined roughly a Class Rule Length (L_{Rule}), giving a pretty round figure number. At the later re-analysis Midship's carry on by the Shipyard, an accurate value for the L_{Rule} was obtained, being 94mm shorter.

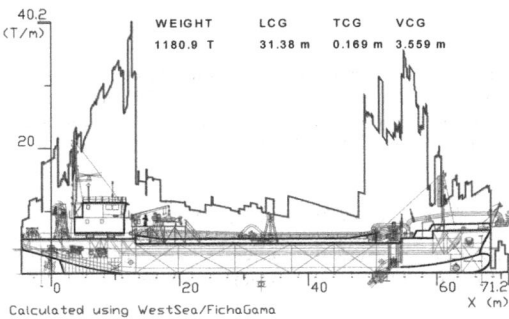

Figure 3. Realistic and detailed LWT distribution of the vessel, based on more than 320 items.

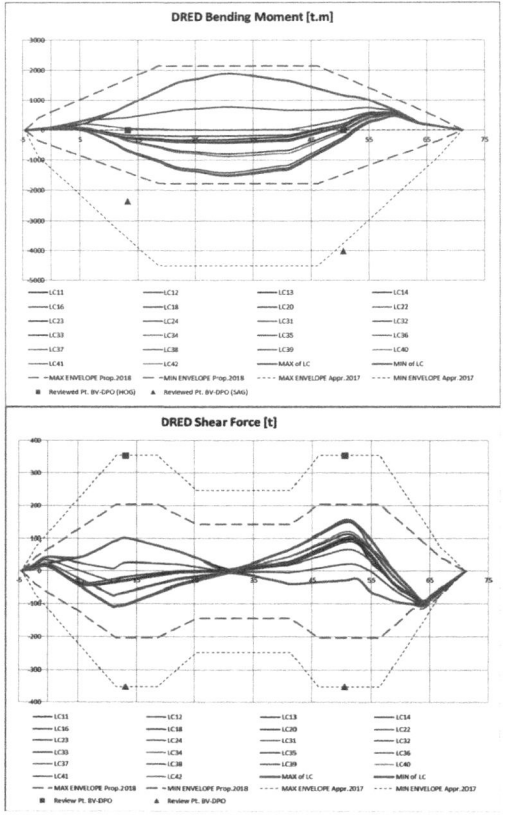

Figure 5.1& 5.2. The BM & SF for dredging operation.

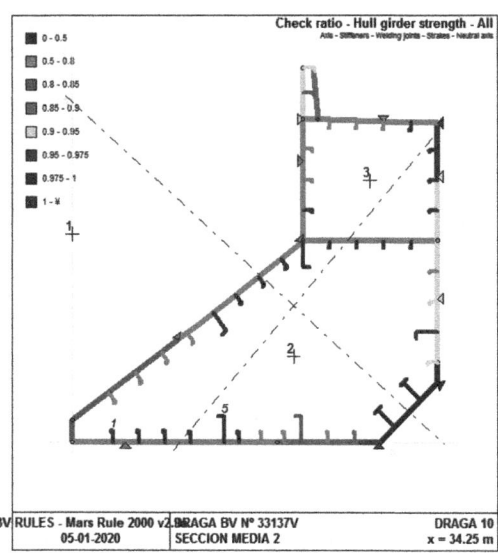

Figure 6. The overall stresses at structure considering the preliminary inputs, showing <5MPa of margin to maximum allowable stresses, at the critical scenario: Dredging condition.

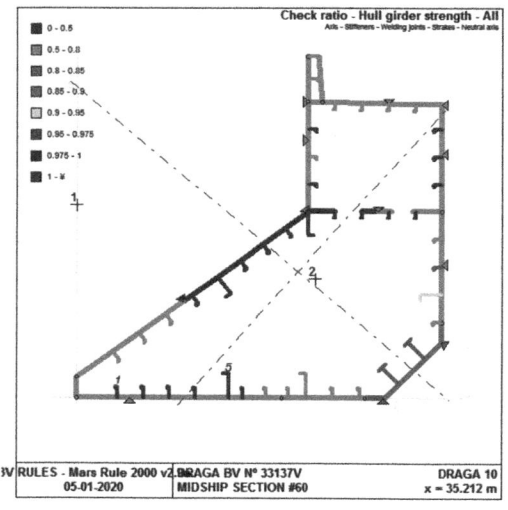

Figure 7. The overall stresses at structure considering the new proposed inputs, having now abt. 35MPa of margin up to maximum allowable stresses, in the most demanding scenario: Navigation condition.

Moreover, a detailed Block Coefficient (C_B) was determined according to the Class rules, based on the new L_{Rule}, which become slightly smaller. It is not sure how the original Designer had calculated the original C_B. Eventually based on a Length between Perpendiculars (Lpp) at 85% Depth.

Nevertheless, at the new Midship's Review, the accurate values of L_{Rule} and C_B presented to the Classification Society resulted in smaller Wave Bending Moments. The Class formulae to achieve this numbers are highly dependent to C_B and mainly to L_{Rule} rooted to square.

Therefore, in Dredging Scenario (initially the most demanding one), the combined Vertical WV + SW BM&SF had decreased with the new input parameters. Thus, the Structural Elements at the critical Bilge area are not highly stressed due the Hull Girder Loads as previously, leaving a larger gap to sustain the local pressures (due to higher Draughts).

The Navigation Scenario is now the most critical situation, but stress results are now pretty safer than originally, as shown in the Figures 6 and 7.

2.4 Difference of basic design parameters

The Table 2 gives the general change of input, for the requested new class structural review.

2.5 Case study conclusion

A detailed document, with ll above considerations, was submitted to the Classification Society

Table 2. Main characteristics and differences.

Design:		Original	Actual	
Length Class	L_{Rule}	69.20	69.11	m
Draught Scantling	T_{scant}	3.515	3.945	m
Draught Dredging	T_{dred}	3.958	4.353	m
Freeboard Scantling	FB_{scant}	1246	816	mm
Freeboard Dredging	FB_{dred}	803	408	mm
Block Coefficient	C_B	0.868	0.835	–
NAVIGATION				
Vertical SW BM	(hogging)	2055	4280	t.m
Vertical WV BM	(« «)	5910	5665	t.m
Combined Vertical BM HOG	7965	**9945**	(+25%)	7965
Vertical SW BM	(sagging)	-3145	-1000	t.m
Vertical WV BM	(« «)	-6180	-6030	t.m
Combined Vertical BM SAG	-9325	**-7030**	(-25%)	
Vertical SW SF		295	250	t
Vertical WV SF		170	205	t
Combined Vertical SF	465	**455**	(-2%)	
DREDGING				
Vertical SW BM	(hogging)	0	2140	t.m
Vertical WV BM	(« «)	3940	3775	t.m
Combined Vertical BM HOG	3940	**5915**	(+50%)	
Vertical SW BM	(sagging)	-4520	-1785	t.m
Vertical WV BM	(« «)	-4120	-4020	t.m
Combined Vertical BM SAG	**-8640**	**-5805**	(-33%)	
Vertical SW SF		355	205	t
Vertical WV SF		115	145	t
Combined Vertical SF	470	**350**	(-26%)	

in beginning of June, nearly after the Official Inclining Test when the Ship was practically concluded. Surprisingly, the entire re-analysis of the whole Ship Design took only 5 weeks, with total approval without remarks.

The only welding re-works done by the Shipyard was shifting the Plimsoll Mark (and Dredging Marks) to the new Scantlings/Dredging Draughts positions. All Hull structure remains the same, without any additional structure reinforcements. Moreover, the sea conditions for Dredging or Navigation operation remain the same.

And, after concluding all these intense work of structure reviews, the vessel has larger stress margins up to the maximum allowable stress limits, after all. It all depended on the shape difference of the LWT Distribution, only. And this raises an important question: are Classification Societies duly surveying the shape of LWT Distribution, to ensure ship's safety?

Considering the presented Case Study, it naturally leads us for the inconvenient doubt: what was the amount of DWT loss that was jeopardized by an increase of unnecessary structure reinforcements, based on the preliminary non-realistic LWT Distribution? There is actually no answer for this. Of course, the total responsibility of the project and given inputs are at Designer/ Shipyard side. But seems fair to think that Classification Societies should also have an higher attention to survey this vital information, in order to ensure higher reliability of their own Stress Hull Reviews.

3 ACTUAL ROLE OF CLASSIFICATION SOCIETIES

3.1 Shipyard's past experience

According to our personal past experience, which it has been involved in 16 different Projects, representing the delivery of more than 30 ships and its sister-ships, ranging between 70m to 190m length, none of the main Classification Societies had ever questioned the shape of the LWT Distribution adopted.

As a Shipyard, it is believed that this is a crucial subject that should be subject to review and verification by the Classification Societies.

3.2 Potential problems

The LWT Distribution is usually presented at the Intact Stability Manuals. The coordinates of that Distribution are presented, and the curve of it is plotted. However, there is few or none identification of the items that become parts of the LWT Distribution curve. Therefore, it is not possible to check the reliability of that shape.

The Inclining Tests of the ships are conducted by the Shipyards/Designers in very accurate procedures and at perfect conditions. The local Surveyors from the Classification Societies witness in detail all the data record. The final report of the most important test of the ship, which confirms the real LWT and the CoG (Centre of Gravity), is then reviewed and approved in detail by the Class Plan Approval Office (or/and also by the Flag).

These real results are then used in all the Final Stability Calculations, in the format of a LWT Curve (defined and adjusted by the Designer), together with the real figures of TCG and VCG.

Therefore, Classifications Societies can only confirm the correlation of the Inclining Test's results with the integral of LWT Distribution line (ton/m) and its centroid (LCG). But, there are several possible shapes of LWT Distribution curve, resulting in the same Weight and LCG. As it can possible be check in the Figure 8.

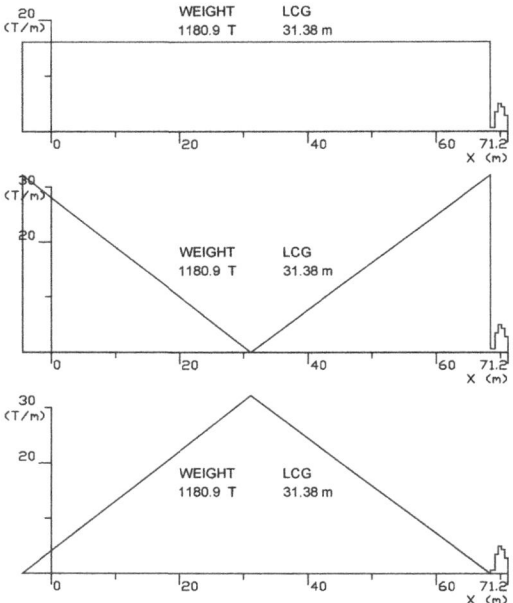

Figure 8. Some examples of an infinity of curve's shapes that have the same LWT and LCG. However, it results in significantly different values of BM & SF.

It is assumed that, inside the Classification Societies' Plan Approval Offices, the Department of Structural Review is completely distinct from the Stability Department. It is then suspicious if surveyors from Structural Department go through in detail to the shape of the SW BM&SF of all the Loading Conditions. Or even to the LWT Distribution shape.

There is a very small common area between these distinct departments (Figure 9), which they indeed survey, such as: scantling Draughts; SW BM&SF maximum values; minimum fwd draught in ballast condition.

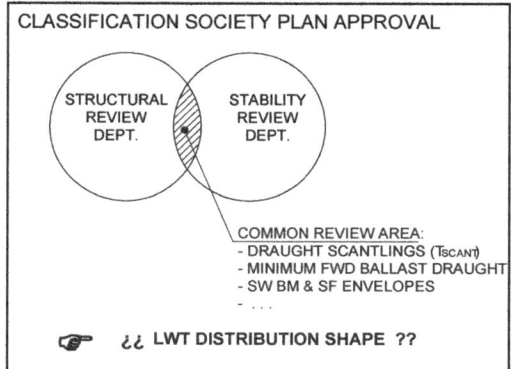

Figure 9. Common subjects that shall be review between departments.

Eventually, we could guess that the Structural Dept. informs the values of the Envelop Limits considered during Midship Section review, while the Stability Dept. goes through all Loading Conditions to check it.

Therefore, based on our experience and based on the Case Study presented, some easy solutions that could help Classification Societies to duly review the shape of the LWT Distribution, as closer to reality as it can be, was proposed by Miranda (2020).

4 DEFINING LWT DISTRIBUTION CURVE

Considering the highly importance of the LWT Distribution to determine the Global Hull Girder loads, seems to be very few updated literature or guidelines to define it properly. General overview will be given through the existing information.

4.1 Definition of the hull girder loads

As it is widely known among the Naval Architecture designers, the ship's stillwater global loads are based in the beam theory, which the bending moment, at elastic small-deflection condition, is given by:

$$\frac{d^2 BM}{dx^2} = f(x) \qquad (1)$$

Therefore, the Bending Moments can be given by the integration of the Shear Forces along the length of the ship. While, the Shear Forces is given by the integration of the resultant vertical forces, along the length of the ship. It is considered that BM and SF is zero at ship's ends, resembling to a simply supported beam.

The resultant vertical forces is determined by the Ship's WEIGHT Distribution subtracted to the Ship's BUOYANCY Distribution.

The nowadays Ship's Stability Software can easily compute all these steps, determining the Weight Distribution, Buoyancy Distribution, Shear Forces and Bending Moments.

Concerning the WEIGHT Distribution, it is composed by LWT Distribution and DWT Distribution.

The DWT Distribution is mainly composed by Cargo, Ballast, Fuel, Fresh Water, other fluids and supplies. The DWT Distribution of liquids, is also well determined by the software, as long as the shapes of the tanks are also modelled.

Other "dried" items that are part of the DWT are defined by the stability users, which necessarily need to introduce their weight, CoG, and longitudinal extents of each item, to properly calculate the Loading Conditions and its hull girder loads. Those items are, as example: Containers, Ro-Ro wheeled cargoes, other non-homogeneous cargoes, supplies or persons and luggage.

Only the LWT Distribution is an input totally from the user side, which the shape of the curve needs to be introduced in the software: ordinates in [t/m] along the ship's length [m].

4.2 Classification society rules

Nearly no information or guidelines can be found in the Classification Society Rules, that could recommend the best and reliable procedures to obtain the LWT Distribution.

At IACS (International Association of Classification Societies) publications, it was only found scattered information concerning LWT Distribution, but no information regarding guidelines to create the curve, neither to survey it.

There are, however, some guidelines to determine the preliminary maximum SW BM, mainly for preliminary design stages.

4.2.1 Preliminary stillwater bending moments equations

At preliminary design stages, which Intact Stability have not been yet exhaustively analysed, the Rules can give an empirical formula to determine the preliminary SW BM, having as parameters the Ship's Length, Breadth and Block Coefficient. The equation is, somehow similar to the equation that deter-mines the WV BM.

It is our opinion that these formulae can be duly reliable at any preliminary conceptual design stage. Later on, the preliminary figures of these SW BM equations are not relevant anymore, being only mandatory that all Loading Conditions complies with the maximum SW BM&SF Envelopes, defined at Midship Section Class review.

Even so, for the preliminary figures of SW BM, the same question can be raised again, regarding the reliability of it.

Surely the Classification Societies had reached a good approximation for those equations, based on a large database that Classification Societies can continuously collect from several projects.

However, it might be highly possible that this equation could have some uncertainty because of the LWT Distribution shape, from those database cases, which were not subjected to a detailed class review.

4.3 Specialized bibliography

At some Naval Architecture specialized bibliography, there are some approximation methods to determine the LWT Distribution.

An interesting review through all methods to determine the LWT Distribution has been exposed in Hansch (2008), mainly:

– Grouping Methods
– Approximation Methods
– Direct Distribution Methods

4.3.1 Grouping methods

Also known as "Bucket" Method (Figure 10), it literally assumes that several buckets are evenly distributed along the ship. Then, all items that compose the LWT are theoretically placed at the single bucket closer to the item's LCG, independently of their presence along the ship.

For instance, some significant items that spreads-on the entire ship (painting, designer margins, welding and mill tolerance weights, ...) will surely be concentrated at few central buckets, inducing a sagging effect at the LWT Distribution which is not correspondingly real.

Thus, it is not, by far, an accurate method. To make it a realistic method, it would necessarily take user action to identify each item that need to be subdivided through the buckets under the "projection" of the longitudinal "footprint" of that item.

4.3.2 Approximation methods

In the past, several approximation methods for distributing the Hull Weight have been defined, such as:

– Approximation by Comstock (1944).
– Approximation by Biles, frequently called the "coffin" diagram, (Munro-Smith, 1967).
– Approximation by Prohaska (Munro-Smith, 1967).
– Trapezoidal Approximation from PNA (1941).
– Parabolic Approximation by Cole (Munro-Smith, 1967).
– Approximation by Hughes (1988).

In general for the above methods, the Hull Weights is the LWT itself, excluding the items: anchor, chain, windlass, steering gear, main propulsion machinery. Those excluded items shall then be added in a trapezoidal or rectangular shape over the method's Hull Distribution shapes.

The referred paper Hansch (2008) collects all details and equations for these approximations.

Generally, the referred methods resembles to a "coffin" shaped distribution (Figure 11). Eventually, these methods were very useful in the past, especially at an initial design stage, and also due to the lack of generalized computer aids to assist in these kinds of calculations for LWT and also DWT.

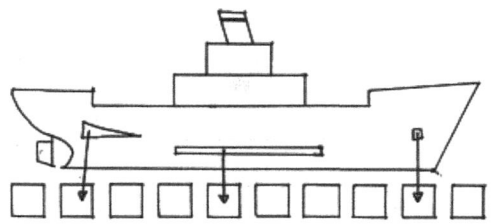

Figure 10. Bucket list illustration (Hansch, 2008).

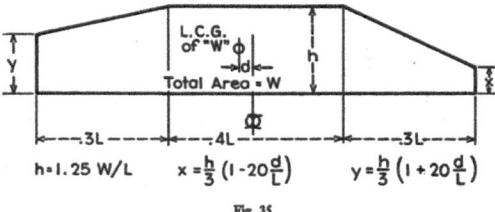

Fig. 35

Figure 11. Typical "coffin" shape distribution of the Hull Weights (Comstock, 1944).

However, based in our experience of defining the LWT Distribution, it is quite difficult to clearly identify a "coffin" shape.

This paper didn't go through all methods in detail, to evaluate the accuracy of them. Nevertheless, these methods might not be sufficiently accurate to be taken into consideration, for an advance detail stage of the project.

4.4 Direct distribution methods: Trapezoids

It is our understanding that this is the most reliable method. It is simply based on the properties of the geometric area of a Trapezoid. The Figure 12, and the correspondent equations, are stated at arithmetic references. It should be highlighted that the Centroid of this figure (Eq.3) is related to the largest base edge (i.e., "b").

$$Area_{TRAPEZIUM} = \frac{a+b}{2} \cdot h \qquad (2)$$

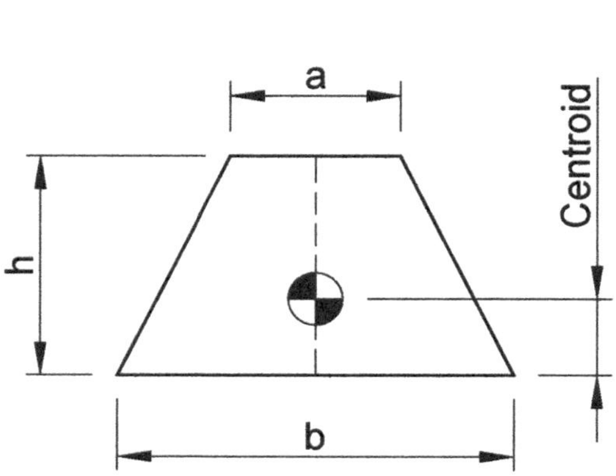

$$Centroid_{TRAPEZIUM} = \frac{h}{3} \cdot \left(\frac{2a+b}{a+b} \right) \qquad (3)$$

More details and Guidelines for a proper definition of the LWT Distribution is describe in the following chapter.

5 GUIDELINES FOR A PROPER DEFINITION OF A LWT DISTRIBUTION

For each item that composes the LWT List, the following prescriptions shall be adopted in order to determine a realistic LWT Distribution, based on the direct distribution method of a Trapezium.

5.1 Development of the weight distribution equations

Based on the following developed equations, it is possible to automatically determine the left (b_i) and right (a_i) bounds of the Trapezoid, which will become the ordinates at the limits of a trapezoidal item weight defined (Figure 13).

Assuming that it is known for each item i the following:

- Weight of item: Wi [ton]
- LCG of item: LCGi [m] (fr. #0)
- Footprint of item: Xini and Xfin [m] (fr. #0)

The above dimensions are related to the ship's global referential.

Therefore, it can be determined:

Footprint length: length$_i$ = Xfin – Xini , in [m]

Centroid relat. to left bound: lcg$_i$ = LCG$_i$ – Xini, [m]

Considering that:

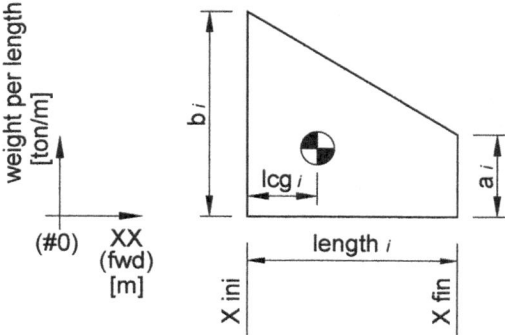

Figure 13. Geometric properties of a Trapezium, adjusted for the weight distribution approach.

Figure 12. Standard geometric properties of a Trapezium.

$$W_i = Area_{Trapz} = \frac{a+b}{2} \cdot length_i$$

$$lcg_i = Centroid = \frac{length_i}{3} \cdot \left(\frac{2a+b}{a+b}\right)$$

Then, it is possible to develop and achieve the equations for the determination of the left and right bounds, which are the ordinates of the Weight Distribution of that single item i:

$$a = \left[\left(\frac{3 \cdot lcg_i}{length_i}\right) - 1\right] \cdot \left(\frac{2 \cdot W_i}{length_i}\right), \text{ in[ton/m]} \quad (4)$$

$$b = \left[\left(\frac{3 \cdot lcg_i}{length_i}\right) - 1\right] \cdot \left(\frac{2 \cdot W_i}{length_i}\right), \text{ [ton/m]} \cdot \text{in[ton/m]} \quad (5)$$

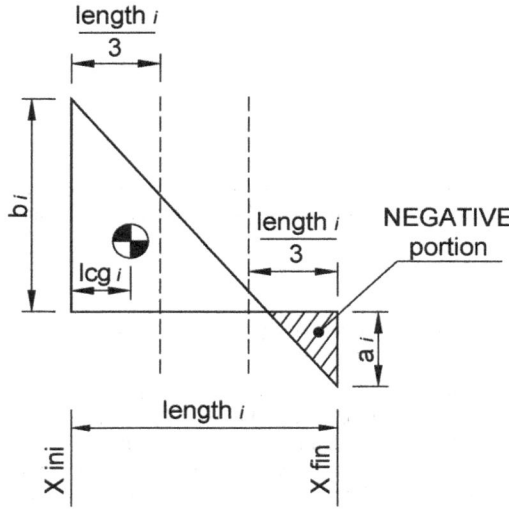

Figure 14. Weight distribution of an item which centroid does not lay in the middle third length.

5.2 Limitations of this approach: Lcg_i versus [Xini-Xfin] interval

The main limitation of this approach is that the LCG$_i$ shall lay in the middle third of the footprints interval.

$$\frac{length_i}{3} \leq lcg_i \leq 2 \cdot \frac{length_i}{3} \quad (6)$$

In case the above is not met, the weight trapezium shape will have some negative region (Figure 14), which is not a reasonable realistic approach.

A common reason for this could be a non-evenly distributed item, which comprises more than a few number of units.

Therefore, the Designer shall correct the parameters of this item. Eventually, the best option is to split this item into two different ones, at the wider space between them (Figure 15).

Nevertheless, a very simple condition code wherever the Designer is working on (Excel, Access, or other), can easily highlight those referred items that shall be re-checked: either verifying the Weight$_i$'s parameters (LCG or footprint locations), or splitting it into separated items.

5.3 Aggregation of different items in the same weight_i record

Considering that there are a large number of items that composes the LWT List, it is very

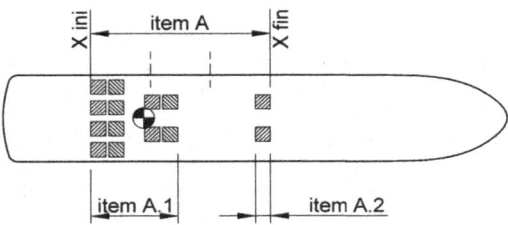

Figure 15. Example of a negative portion of a weight distribution "item A", which should be avoided. The designer is suggested to split the most distant units into a different weight group (A.1 and A.2).

typical that the Designer would aggregate similar units in a single Weight$_i$ record, for simplicity and time saving.

However, it is up to the sensitivity of the Designer to identify which items may or may not be aggregated in a single Weight$_i$ record.

As a guideline, items positioned in the Transverse direction of the ship, even if it is widely separated, can be aggregated in the same Weight$_i$ record, without jeopardizing the real LWT Distribution.

However, in the Longitudinal direction, may not be that trivial. As a guideline, if similar items are separated evenly in a longitudinal direction, they can also be aggregated to a single Weight$_i$ record, resulting in a trustable LWT Distribution shape results, as shown in the Figure 16.

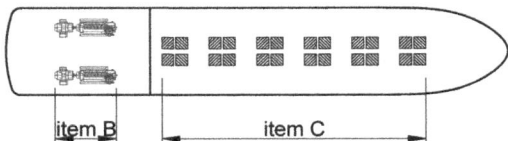

Figure 16. Example of aggregation of Transverse units ("B"), or closely distributed units in Longitudinal direction ("C").

5.4 *Limitations of this approach: Inducing flawy unrealistic sagging*

This subject may concern significantly more than the previous limitations describe at 4.2, because it is more difficult to identify it. Moreover, it could significantly induce sagging, or reduce hogging, compared to the real behaviour of the ship in stillwaters, if the Designer adopts this procedure for the majority of the LWT List.

Eventually, it might be possible that this had seriously contributed for the non-realistic LWT Distribution shape, originally defined in the presented Case Study.

A frequently inaccuracy action is to aggregate widely separated items, such as the various elements that constitute an Aft and Fwd Mooring Arrangement.

The Mooring weights items that are equally representative at Forward and Aft (Winches, Bollards, Fairleads, Rollers, Rope baskets, Cables, ...) can have different Weight Distributions as per Figure 17, resulting in the same Ship's LWT and CoG. However, the LWT Distribution shape would surely bring different figures.

5.5 *Verification and guidelines for LWT Distribution shape*

After completing the LWT List and creating the LWT Distribution, the Designer shall always analyse

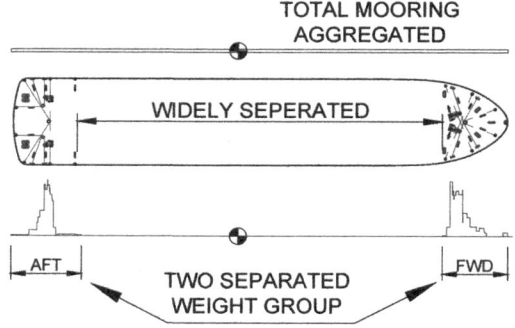

Figure 17. Example of an equivalent weight and LCG for two different weight distributions.

the existing Peaks and Valleys of the curve shape, in order to identify any other errors in the given input.

Placing the GA profile over the LWT Distribution can give a good support to check the reliability of the Curve Shape. It might be more effective than placing all CoG points over the GA Plan.

5.6 *Shipyard's software: FichaGama*

As it was presented by Rodrigues & Barbosa (2002), FichaGama was an homemade soft-ware developed in Office Access®. It was developed in our facilities to improve sequence production and logistic of Steel parts, receiving inputs from NUPAS® software (Steel Hull detail development).

Couple years after, an upgrade was developed in order to collect and combine the weights, CoG, and its longitudinal extend automatically from all items modelled in NUPAS®, and also CADMATIC® software (Outfit-ting detail development).

Alternatively, it can receive a list of LWT items in Excel format, if it contains, for each item: Weight, CoG, Longitudinal extension (Xini and Xfin).

Thus, FichaGama calculate the weight distribution for each item, based on the direct distribution theory from Trapezoids Centre of Area, as described in chapter 4.1. After, it combine all weights distribution elaborating the Global LWT Distribution of the ship, exporting it in AutoCAD® format, or even in AutoHYDRO® format (to be directly used in our Stability calculation software).

The software is able to highlight possible input errors as described at chapter 4.2. It is also able to highlight items which are extremely concentrated weights [t/m], or are defined outside the overall ship's boundaries.

There are also other known commercial software for calculating the ship's weight distribution. However, for the required needs, the homemade software FichaGama suits perfectly.

Eventually, depending on the Designer/shipyard's skills of code programming, an homemade tool can be developed (maybe in Access), based on the developed equations [4] and [5], in order to collect and elaborate the LWT Distribution Curve.

6 CONCLUSIONS

The presented Case Study clearly shows that LWT Distribution has a vital impact on the real hull girder loads. It is important to highlight that even if the Vessel is relatively short, with only 70m Length, the shape of the LWT Distribution revealed to still have a big impact on the Stresses results. Thus, it is our understanding that the Classification Societies shall give a major importance to the shape of it, at the same level as Inclining Tests are surveyed at the end of the construction.

Therefore, it is suggested that Guidelines shall be stated by Classification Societies, to clearly define the method and its best procedures, which Designers shall uniformly follow to define the LWT Distribution.

After a general overview of the existing methods, it seems to us that they are not sufficiently reliable, and should be disregarded.

It is our understanding that the direct distribution method of a Trapezium, is the most trustable one. Nevertheless, all potential problems that may lead to a flawy LWT Distribution, have been identified here, which the Designer shall take it into consideration.

REFERENCES

Comstock, John P. 1944. *Introduction to Naval Architect.* New York: Simmons-Boardman Publishing Corporation.

Hansch, David Laurence, 2008. Methods of Determining the Longitudinal Weight Distribution of a Ship. From Northrop Grumman Newport News; joint meeting of Society of Naval Architects and Marine Engineers (SNAME) and Society of Allied Weight Engineers (SAWE), 24 Jan 2008.

Hughes, Owen 1988. Ship Structural Design: A Rationally-Based, Computer-Aided Optimization Approach. Jersey City: The Society of Naval Architects and Marine Engineers.

Miranda, Andomarc 2020. Proposal for a reliable definition of a LWT Distribution Curve at an early stage of Design. In *Developments on Maritime Technology and Engineering*, C. Guedes Soares, T.A. Santos (ed.), Taylor and Francisci Group, London UK, (in press).

Munro-Smith, R. 1967. *Applied Naval Architecture.* New York: American Elsevier Publishing Company.

Rodrigues, Carlos & Barbosa, Edgar 2002. Evolução do Software de Projecto nos Estaleiros Navais de Viana do Castelo. In C. Guedes Soares (Ed.), *O Mar Fonte de Desenvolvimento Sustentado*; 8th Technical Conference of Marine Engineering, Viana do Castelo, 3-4 October 2002.

Rossel, Henry E. & Lawrance, B. Chapman (eds) 1941. *Principles of Naval Architecture Volume I.* New York: The Society of Naval Architects and Marine Engineers.

Developments in Maritime Technology and Engineering – Guedes Soares & Santos (eds)
© 2021 Copyright the Author(s), ISBN 978-0-367-77376-2

Proposal for a reliable definition of a lightship weight distribution curve at an early stage of design

A. Miranda

WestSEA Shipyard, Viana do Castelo, Portugal

ABSTRACT: The lightship weight (LWT) distribution shape has a major impact on the resulted Hull Girder Stresses results. However, only at a very advanced stage of the construction of a new ship, the Designer is able to conclude the List of LWT, with hundreds or thousands of items, and correspondently create the final LWT Distribution. In this paper, a simple approach is developed which could reach to a reliable shape of a LWT Distribution, sufficiently at an early stage of Design, based only on some dozens of items, and very approximate to what would be the final real curve shape. Thus, it is suggested a new role for the Classification Societies, which would easily review the LWT Distribution shape of all projects, improving the reliability of the structural Class Review.

1 INTRODUCTION

WestSEA Shipyard built in 2018 a Split Hopper Dredger of about 70m length, and hopper capacity of 1000m^3 (Figure 3). The delivery stage of this vessel was an interesting Case Study, which exposed the weakness of the Classification Societies on their Structural Class Review (Miranda, 2020).

During the Basic Design of that vessel, several Finite Elements analysis and discussions had been carried between the Designer, the Shipyard and the Classification Society. Final approval of the Midship Section was achieved, with a very small safety margin up to the maximum allowable stresses, and based on a Lightship Weight (LWT) Distribution defined preliminary by the Designer.

During the production of the vessel, the LWT had increased quite significantly. With the results of the final Inclining Test, the original LWT Distribution was adjusted, by simply adding a Trapezoidal shape (Figure 1).

The first check of the Intact Stability Loading Conditions, with the adjusted LWT Distribution (Figure 1), showed that the Still Water (SW) Bending Moments and Shear Forces (BM&SF) substantially exceeded the original SWBM Limit Envelopes defined at Basic Design. The whole panorama was, at the first sight, quite dramatic, especially because it was Shipyard's intention to also increase Scantling Draught, in order to minimize the loss of Deadweight (DWT) due to the increase of LWT.

The shipyard took the effort to find a quick solution, considering the tight schedule for delivery. Based on their own LWT List (with hundreds of

items), the resulted real LWT Distribution revealed to be significantly different from the original shape.

Simulating again the Loading Conditions based on the Real LWT Distribution, as shown in background line of Figure 5, the resulted SW BM&SF were clearly different, showing a significant margin till the Envelop Limits after all.

After a deep analysis through all project, it was possible to re-approve the Midship Section, keeping exactly the same steel structure already built, and assuming an higher Scantling Draught.

This major overturn on the Ship's Structure Class Review was simply based on the different shape of the LWT Distribution curve.

The referred case study revealed that the shape of the LWT Distribution has a major impact on the Hull Girder Stress results, without being seriously analysed by the Classification Societies.

According to our personal past experience, none of the main Classification Societies had ever questioned the shape of the LWT Distribution curve adopted in all of our recent projects/shipbuildings.

Therefore, the actual role of the Classification Societies may not be duly reliable nowadays.

2 PROPOSALS TO IMPROVE LWT DISTRIBUTION CLASS SURVEY

In this chapter, it will be presented a proposed new role for the Classification Societies, to duly review the LWT Distribution curve/shape.

It is our understanding that a New Document shall be submitted for review, as a satellite document of the Miship Section Class Review.

DOI: 10.1201/9781003216582-56

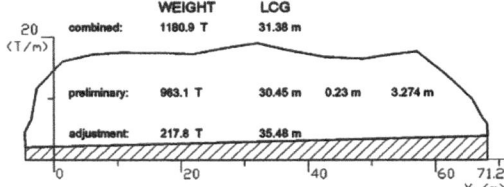

Figure 1. Original LWT Distribution of the vessel, with the adjustment portion (in hatch) to achieve the Inclining Test results.

Explanations are given to identify the most significant items that Class shall focus, saving time for review.

And a general view of the chronologic steps during the classification of a newbuilding, are presented as an example, in order to clearly indicate the key milestones due to the new Class role and Document.

2.1 Theory assumed for LWT distribution

The LWT Distributions presented in this paper were defined according to the direct distribution method, as suggested by Hansch (2008).

Therefore, the calculation of the LWT Distribution Curve, as typically is used in our Shipyard, is based on the Trapezium geometric properties, and considering the best suggested guidelines to properly define it (Miranda, 2020).

The obtain figures of those curves were obtained by our own software FichaGama created and presented by Rodrigues & Barbosa (2002).

2.2 Submission of new document to be reviewed

The LWT List, defined and controlled by the Designer and/or Shipyard, could reach an huge number of items: hundreds, or nearly thousands for larger vessels. Therefore, it is unbearable for the Class to survey exhaustively the entire list, even if it is provided for review.

Based on our experience, we had determined the major items that could give the best approximation of the real shape of the LWT Distribution.

The proposal is that the Designer/Shipyard should elaborate a new document, for each Ship's Project, to be subject to Class Review, with about 50 items only. Those items, named hereafter as <u>Major Items</u>, should be presented in the form of a list, containing the following figures:

– Item Description
– Weight
– LCG
– Xinitial and Xfinal (extension of the footprint)

The rest of the non-relevant Items shall then be combined to become one single item that will be uniformly distributed within the entire Ship's ends. For sake of clarification, it will be named hereafter as

<u>Miscellaneous Items</u>. The characteristics and its CoG of that combined portion should be determined, in order that the total Weight and CoG matches the considered Ship's LWT and its CoG, in the preliminary Stability Documents.

2.3 Selection of the most relevant major items

As guideline, the most expressive elements are those with heavier weights per item, which has a well-defined footprint or length.

As a thumb rule, for the following example cases, it was selected all the items which their Weight/unit is ≥ 0.1% of Ship's LWT.

Generally speaking, it is assumed that the following items are always eligible to be considered as being a Major Item:

– Steel Units Blocks
– Main Engines and Auxiliary Engines
– Steering Gears or Azimuthal Propulsion
– Sewage Treatment Plant
– Mooring Winches
– Chain Cables (at Chain Lockers)
– Anchors
– Solid Ballast
– Other single heavy Equipment

When applicable, the Dry and Liquid weights must always be determined and considered for each equipment.

2.4 Selection of the spread miscellaneous items

In this item, it should be combined all items which has lower weights. Or also, those items which have an average Weight per Length [ton/m] low.

For instance, if the LWT List has an item that is quite heavy, such as "Ship's Painting" or any Piping System which is actually spread through all ship, then it should be disregarded as being Major Items of the proposed New Document. It should be combined into the global Miscellaneous Items.

2.5 Characteristics to be subjected to class review

Proposed tasks of the Classification Society are divided into 3 areas, for different surveyors roles.

2.5.1 By plan approval offices
Classification Societies should then review the definition of the Trapezoids of each selected element. The Longitudinal Limits (Xini & Xfin) should also be roughly surveyed, based on the GA Plan. Consequently, the LCG of that item should lie within the middle 1/3 of the footprint's limits.

The combined Curve, aggregating all items from the proposed Document, should be calculated by Class in order to properly review the shape curve delivered by the Designer/Shipyard.

2.5.2 By local class surveyor at yard (SAY)

The Weight of each Steel Unit should be witnessed by the Local Class Surveyor. It is proposed that the Shipyards should measure the real Block's weight at the time that is being assembly. This kind of action is actually a procedure of our Shipyard, based in our philosophy in order to control and monitor the LWT evolution.

Eventually, if a considerable number of Unit Blocks that are measured, matches with the theoretic weights given by a reliable steel-preparation software, than it could be acceptable that the rest of the blocks need not be really measured, assuming valid the theoretic figures.

Solid Ballast shall also be witnessed by the Local Surveyors at Yard, to confirm the installed weights.

2.5.3 Class surveyors at equipment's certification

Regarding the other major equipment, the responsibility to duly measure the weights (dry and liquid service) should be on the Suppliers side, at the moment that each Class Certification or Type Approval is witnessed and issued by the Classification Society.

This action is not exclusively targeted to the Ship Project under review. It will be useful information valid for all equipment's family, thus for any other ship newbuilding.

Therefore, the procedure should be adopted by the Manufacturer to any kind of marine equipment that measure more than 1.0 ton (dry and liquids included), regardless the Lightship Weight of the vessel that will be indeed installed. It is only a matter of the Designer if that equipment shall be present on the referred document, depending on the specific Project.

These action surveyed by the Class, will be then useful to the majority and significant vessels, with LWT > 1000t.

2.6 Chronological tasks

2.6.1 Conceptual design stage

At this stage, the LWT of the new project is only estimation, based on the ship's reference and best Designer's experience.

In case that the LWT Distribution is not predictable at this stage, an empiric formula is already given by the Classification Societies' Rules for a preliminary maximum SW BM values.

2.6.2 Basic design stage

It is well known that a reliable and consolidated LWT list is typically defined at a very advanced stage of construction, due to unpredictable design changes, or ongoing project updates as a result of the design coordination.

Nevertheless, is not likely that the major items will change substantially their position related to the ship's general arrangement. Considering their over-

size, usually confined in its dedicated spaces, these major equipment may be already well established and positioned at the Basic Design stage (Figure 2).

However, regarding the Steel Unit Blocks, it might be still premature to have a reliable Steel Weight Distribution.

Thus, only for the stability documents, a preliminary LWT Distribution shall be predicted based on the established major equipment, and expected Unit Blocks' Weight.

The Midship Section should be reviewed and approved at this stage, which will be pending for the document containing the Approximate LWT Distribution.

2.6.3 Detail design stage

At this stage, starts the Hull Detail project. After modelling all the Ship's steel parts, it will be reliable to obtain the Weights, LCG and longitudinal extends of each Block units.

It shall be then the time to produce a reliable Approximate LWT Distribution, in order to finally concludes the Midship Section approval.

Therefore, looking generally to the schematic schedule taken on the construction of our NB.010

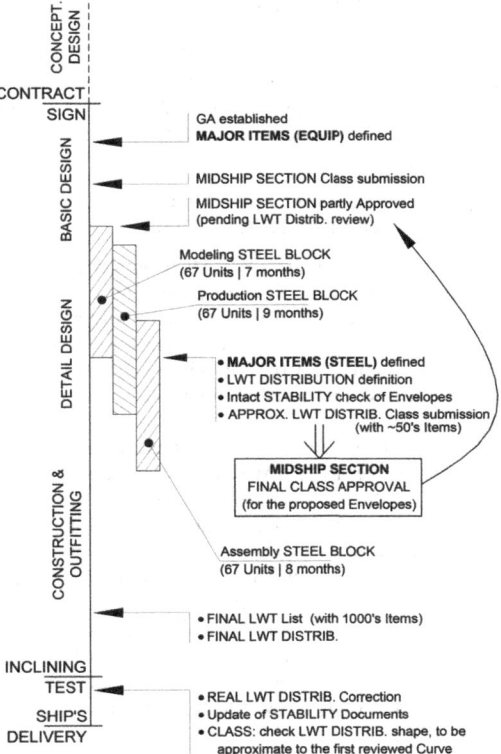

Figure 2. Schematic schedule of 29 Months of construction of NB.010 M/S "World Explorer" (Ocean Cruise Vessel, as example at Ch.3.3).

499

(Figure 2), it is possible to achieve a trustable LWT Distribution shape, sufficiently at an early stage of the project/building, confirming the proposed SW BM&SF Envelopes considered at the Midship Section Class review.

2.6.4 Ship's delivery stage

As nowadays is, it is expected that the Designer/Shipbuilder reach a consolidated final LWT List, close to the end of construction. Therefore, it should be then produced the correspondent final LWT Distribution.

Nearly before the delivery, the project is practically concluded with the conducting of the Inclining Test. At this event, the Final LWT Distribution curve should be adjusted to the Real LWT and CoG results, to be used in the final stability documents.

The real shape of this curve, meanwhile developed and monitored by the Designer, must be surveyed by the Class, in order to determine if it closely resembles to the presented curve made by surveyed Major items and spread out Miscellaneous, submitted in the Detail Steel Design Stage.

3 VERIFICATION OF THE RELIABILITY OF THE REVIEW PROPOSALS

Based on the previous chapter, it was checked the reliability of the proposed approach to get an Approximate LWT Distribution shape. A general check has been executed to **3** of our vessels:

– Cargo Vessel (Dredger abt.70m)
– River Passenger Vessel (abt. 80m)
– Ocean Passenger Vessel (abt. 120m)

Determination of the LWT Distribution was executed, step by step, in order to figure out the number and type of Major Items that could best represent the real and final LWT Distribution Shape. The fewer items, the best it will be for a straight forward review and reliable approval by the Classification Society.

In all shown figures of LWT Distributions, it is represented the Real LWT Distribution shape, in thinner grey line at background, based on hundreds or thousands of items that composes the final LWT List of the referred vessel.

The LWT of each presented ship has been grouped similarly, in order to clearly indicate the representative weight of each group inside the global LWT. At those same tables (Table 1, 3, 5), it shows the relative number of eligible Major Items that is worth to measured and/or survey.

To each ship example, it was determined the resulted maximum SW BM in equilibrium, without any addition of DWT, given by a Stability software. Therefore, it is possible to compare the real LWT Distribution with the approximation achieved with the proposed approach (Table 2, 4, 6).

3.1 Cargo vessel: Split hopper dredger (abt.70m)

This example is our NB.011 "José Duarte", a Split Hopper Dredger, of 70m Length, Breadth of 11m, and Draught(Scant.) 3.95m (Figure 3). This is the vessel from the presented Case Study (Miranda, 2020).

Special attention was taken to items weighing about ≥1.1 ton, per unit.

3.1.1 Type of LWT items
The Final LWT List has slightly more than 300 weight records, which resulted in the real LWT Distribution, as per Table 1.

3.1.2 Comparing LWT distribution approaches
In Figures 4.1, 4.2, 4.3, it is possible to check the representative of the Major items, compared to the global LWT. While at Figure 4.4, it is possible to see the Miscellaneous Items Distribution.

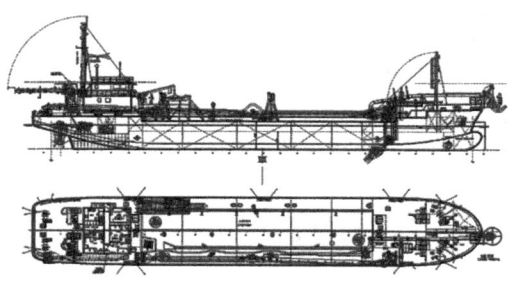

Figure 3. General Arrangement of the Ship.

Table 1. Main Groups of LWT Items, and the representative items to be surveyed for LWT Curve purposes. The items (*) are to be surveyed by SAY.

	SubTotal Weight [t]	Weight [t]	Major Items % of LWT	n° of Items
STEEL BLOCKS	813.5	813.5	(68.9%)	17*
STEEL Outfitting	53.6	0		
DECK EQUIP.	103.3	92.3	(7.8%)	25
CREW+PAX Outfit.	22.6	1.0	(0.1%)	1
MACHINERY Outft.	50.0	28.6	(2.4%)	6
OTHER Syst. (out ER)	102.2	27.8	(2.4%)	4
SOLID BALLAST	36.4	36.4	(3.1%)	1*
LWT List (322 items)	1180.9	999.7	(84.6%)	54

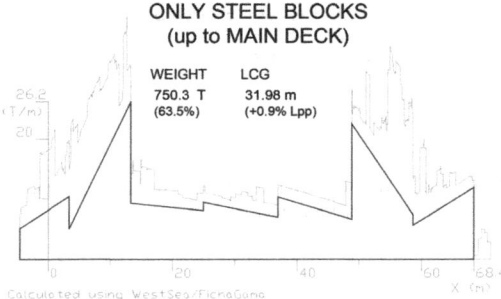

ONLY STEEL BLOCKS
(up to MAIN DECK)

WEIGHT	LCG
750.3 T (63.5%)	31.98 m (+0.9% Lpp)

Calculated using WestSea/FichaGama

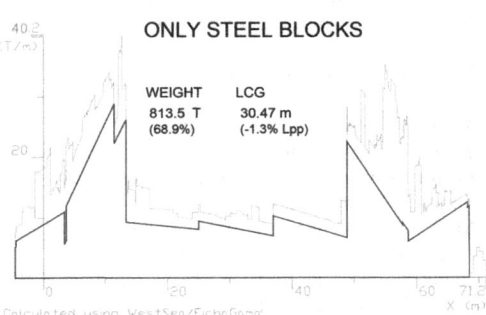

ONLY STEEL BLOCKS

WEIGHT	LCG
813.5 T (68.9%)	30.47 m (-1.3% Lpp)

Calculated using WestSea/FichaGama

ONLY STEEL BLOCKS
+ MAIN ITEMS

WEIGHT	LCG
999.7 T (84.6%)	31.90 m (+0.8% Lpp)

Calculated using WestSea/FichaGama

Figure 4.1, 4.2, 4.3. LWT Distribution, considering: only Steel Unit Blocks up to Freeboard Deck; entire Steel Unit Blocks of the Ship; all Steel Unit Blocks and Major Items.

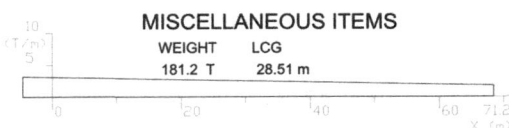

MISCELLANEOUS ITEMS

WEIGHT	LCG
181.2 T	28.51 m

Figure 4.4. LWT Distribution for Miscellaneous Items, laying completely within Steel Hull Length.

3.1.3 *Final reliable LWT distribution as per proposal*

The resulted shape of the LWT Distribution for Survey Purposes seems to be very approximate to the Real Curve (Figure 5).

FINAL APPROXIMATION

WEIGHT	LCG
1180.9 T	31.38 m

Calculated using WestSea/FichaGama

Figure 5. Final LWT Distribution, considering Major Items and Miscellaneous Items, as per proposal.

Table 2. Max SW Bending Moments of the Ship in equilibrium, only LWT Distribution (without DWT).

LWT Distribution type	Max SWBM [t.m]	Location [m] fr.#0	%Error REAL	Envelop. Limits
REAL	3013	30.70	—	72.8%
APROX 1 Misc	2826	31.90	-6.2%	69.8%

The resulted SW BM of the approach is also very accurate. Therefore, with this approach, it seems that the reliability of the Structural Class Reviews can be largely ensured.

3.2 *River passenger cruise vessel: Abt.80m*

This example is our NB.005 "Douro Elegance", a serie of 3 River Cruise Vessels, of 80m Length, Breadth of 11m, and Draught(Scant.) 1.8m (Figure 6).

In this example, special attention was taken to items weighing about ≥1.0 ton, per unit.

3.2.1 *Type of LWT Items*

The LWT List has nearly 1200 weight records, which resulted in the real LWT Distribution, as per Table 3.

3.2.2 *Comparing LWT distribution approaches*

In Figures 7.1, 7.2, 7.3, it is possible to check the representative of the Major items, compared to the

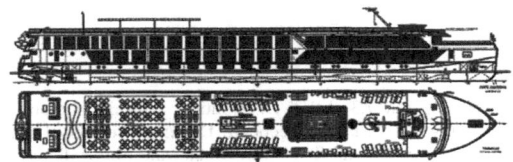

Figure 6. General Arrangement of the Ship.

501

Table 3. Main Groups of LWT Items, and the representative items to be surveyed for LWT Curve purposes. The items (*) are to be surveyed by SAY.

	SubTotal Weight [t]	Weight [t]	Major Items % of LWT	n° of Items
STEEL BLOCKS	547.5	547.5	(53.4%)	21*
STEEL Outfitting	26.4	0		
DECK EQUIP.	17.3	8.2	(0.8%)	7
CREW+PAX Outfit.	306.3	0		
MACHINERY Outft.	62.0	37.9	(3.7%)	9
OTHER Syst. (out ER)	66.6	6.4	(0.6%)	5
SOLID BALLAST	0	0		
LWT List (1163 items)	1026.0	600.0	(58.5%)	42

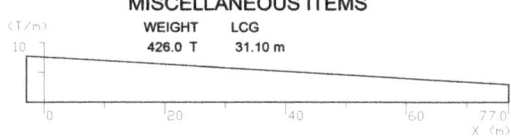

Figure 7.4. LWT Distribution for Miscellaneous Items, laying completely within Steel Hull Length.

global LWT. While at Figure 7.4, it is possible to see the Miscellaneous Items Distribution.

3.2.3 Final reliable LWT distribution as per proposal

Looking exclusively to the Figure 8, it seems that the resulted shape of the Approximate LWT Distribution for Survey Purposes seems to be already very approximate to the Real Curve.

However, looking to the resulted SW BM (Table 3), the approximation may not be that close.

Indeed, looking carefully at the extremities of the Figure 8, the suggested approach seems to be not approximate to the Real LWT Distribution curve indeed.

Thus, Passenger vessels, which have several Decks with different extension lengths, it might be possible to improve the similarity of both curves if the MISCELLANEOUS item is split by the number of Decks and their length (Figure 9).

For this new approach, it was considered only the length of enclosed superstructure. Bellow the Freeboard Deck, it was only considered the length of Decks, within collision bulkheads.

The resulted approximation of the LWT Distribution (Figure 10), considering the Miscellaneous Distribution per Deck (Figure 9), shows now a good approximation at the extremities. The correspondent max SW BM in equilibrium is closer to the Real LWT Distribution results, being deviated in 10% (if related to the SWBM Envelop Limits).

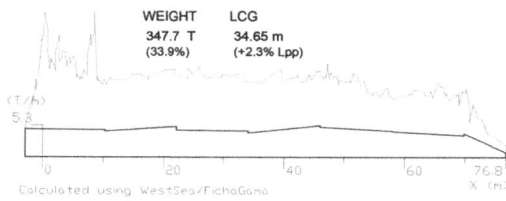

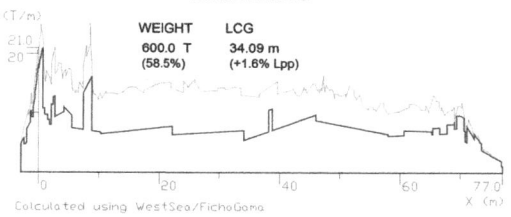

Figure 7.1, 7.2, 7.3. LWT Distribution, considering: only Steel Unit Blocks up to Freeboard Deck; entire Steel Unit Blocks of the Ship; all Steel Unit Blocks and Major Items.

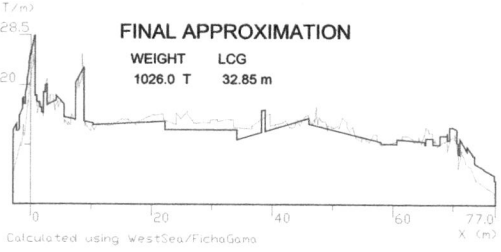

Figure 8. Final LWT Distribution, considering Major Items and one single Miscellaneous Item, as per proposal.

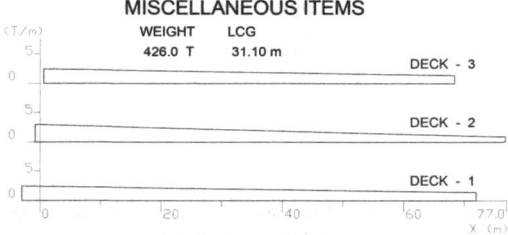

Figure 9. LWT Distribution for Miscellaneous Items, split per each Deck's length, keeping the same LCG for simplicity of calculations.

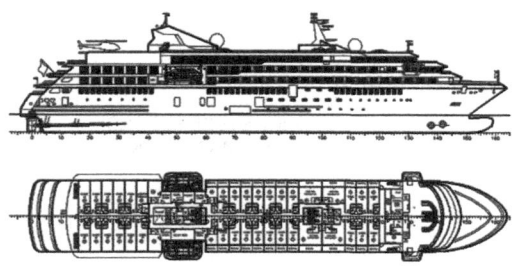

Figure 11. General Arrangement of the Ship.

Table 4. Max SW Bending Moments of the Ship in equilibrium, only LWT Distribution (without DWT).

LWT Distribution type	Max SWBM [t.m]	Location [m] fr.#0	%Error REAL	Envelop. Limits
REAL	1217	32.19	—	49.5%
APROX 1 Misc	1775	34.10	+46%	71.4%
APROX Misc. p/Dk	1480	34.10	+22%	59.5%

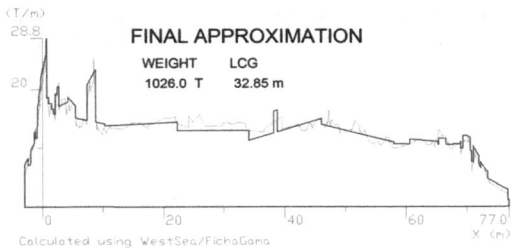

Figure 10. Improved approximation of the LWT Distribution approach.

Table 5. Main Groups of LWT Items, and the representative items to be surveyed for LWT Curve purposes. The items (*) are to be surveyed by SAY.

	SubTotal Weight [t]	Weight [t]	Major Items % of LWT	nº of Items
STEEL BLOCKS	2924	2924.4	(52.9%)	69*
STEEL Outfitting	204	0		
DECK EQUIP.	138	59.7	(1.1%)	4
CREW+PAX Outfit.	2271	40.6	(0.1%)	5
MACHINERY Outft.	262	146.8	(2.7%)	6
OTHER Syst. (out ER)	806	140.8	(2.5%)	7
LWT List (4098 items)	5532.1	3312.3	(59.9%)	91

3.3 Ocean passenger cruise vessel: Abt.120m

This example is our NB.010 "World Explorer", a serie of 7 Seagoing Passenger Cruise Vessels, especially for Polar Expeditions, of 120m Length, Breadth of 19m, and Draught(Scant.) 4.8m (Figure 11). In this example, special attention was taken to items weighing about ≥5.5 ton, per unit.

3.3.1 Type of LWT items
The LWT List has nearly 4100 weight records, which resulted in the real LWT Distribution, as per Table 5.

3.3.2 Comparing LWT distribution approaches
In Figures 12.1, 12.2, 12.3, it is possible to check the representative of the Major items, compared to the

global LWT. While at Figure 12.4, it is possible to see the Miscellaneous single Item Distribution.

3.3.3 Final reliable LWT distribution as per proposal
The resulted shape of the approximation LWT Distribution (Figure 13), considering one single item for MISCELLANEOUS, is already quite similar, however the uncertainty at the extremities is clearly more evident than the River Passenger vessel shown previously.

Therefore, only one single item for Miscellaneous shall not be followed for any Passenger Vessel. The % of error shown at Table 6 is quite significant.

So, it is deemed necessary for larger Passenger vessels to split the Miscellaneous group into the number of Decks, depending on the length of it: enclosed superstructure decks, or within collision bulkheads if bellow freeboard decks (Figure 14).

As it can be checked at Figure 15, the proposed approach shows now a very good approximation to the realistic LWT Distribution. The absolute error compared to the Real LWT Distribution is less than 4%. The accuracy of the LWT Distribution approach

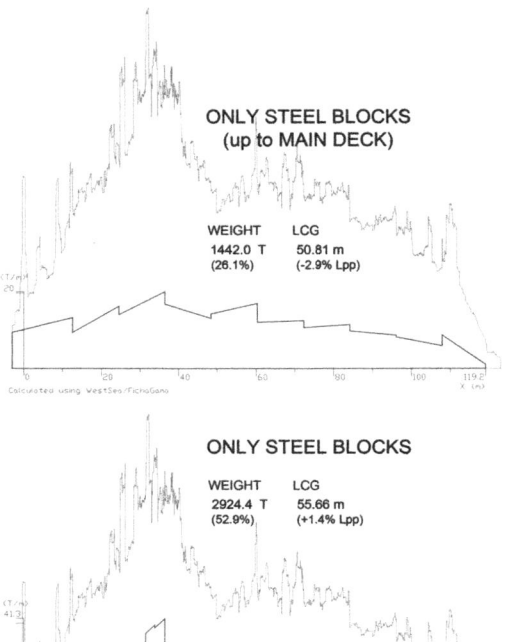

ONLY STEEL BLOCKS
(up to MAIN DECK)

WEIGHT	LCG
1442.0 T	50.81 m
(26.1%)	(-2.9% Lpp)

ONLY STEEL BLOCKS

WEIGHT	LCG
2924.4 T	55.66 m
(52.9%)	(+1.4% Lpp)

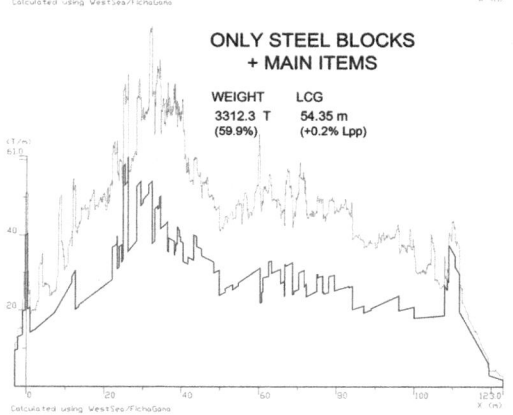

ONLY STEEL BLOCKS
+ MAIN ITEMS

WEIGHT	LCG
3312.3 T	54.35 m
(59.9%)	(+0.2% Lpp)

Figure 12.1, 12.2, 12.3. LWT Distribution, considering: only Steel Unit Blocks up to Freeboard Deck; entire Steel Unit Blocks of the Ship; all Steel Unit Blocks and Major Items.

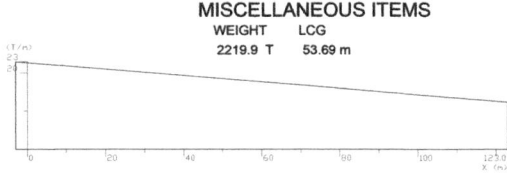

MISCELLANEOUS ITEMS

WEIGHT	LCG
2219.9 T	53.69 m

Figure 12.4. LWT Distribution for Miscellaneous Items (single item), lying completely within Steel Hull Length.

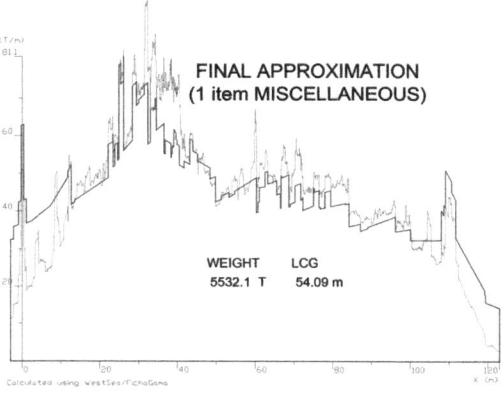

FINAL APPROXIMATION
(1 item MISCELLANEOUS)

WEIGHT	LCG
5532.1 T	54.09 m

Figure 13. Final LWT Distribution, considering Major Items and Miscellaneous Items (1 Trapezium), as per proposal. The presented curve has same weight and LCG of the Vessel.

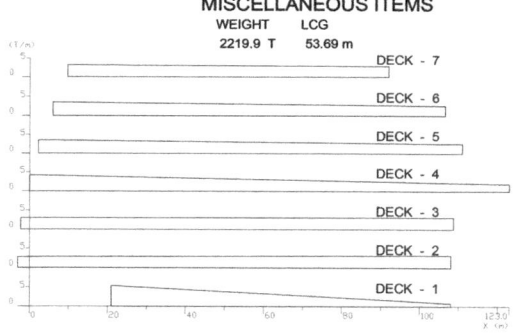

MISCELLANEOUS ITEMS

WEIGHT	LCG
2219.9 T	53.69 m

Figure 14. LWT Distribution for Miscellaneous Items, split per each Deck's length, keeping the same LCG for simplicity of calculations.

504

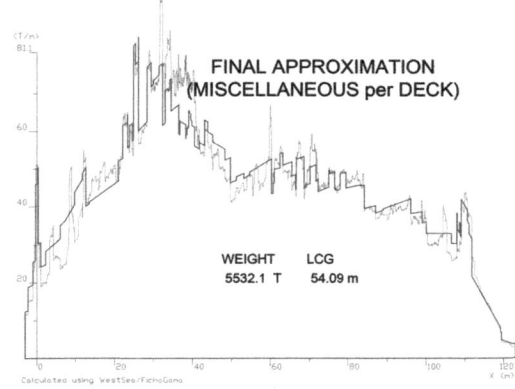

FINAL APPROXIMATION
(MISCELLANEOUS per DECK)

WEIGHT   LCG
5532.1 T   54.09 m

Calculated using WestSea/FichaGama

Figure 15. Improved approximation of the LWT Distribution approach.

with the real curve, compared to the SWBM Envelop limits, is very high, deviating only 3.5% (Table 6).

4 CONCLUSIONS

The Midship Section and the entire Hull Structure shall be duly reviewed based on a realistic and reliable LWT Distribution curve.

Therefore, it was proposed simple actions that Designers and Classification Societies could adopt, to duly create and survey the LWT Distribution, with a minimal increase in efforts by all actors, which can ensure the ship safety.

It is quite notorious that few dozens of Items and the Steel Block units can produce a very approximate LWT Distribution curve, compared to the real curve based on hundreds or thousands of items.

Table 6. Max SW Bending Moments of the Ship in equilibrium, only LWT Distribution (without DWT).

LWT Distribution type	Max SWBM [t.m]	Location [m] fr.#0	%Error REAL	Envelop. Limits
REAL	17636	55.88	—	91.1%
APROX 1 Misc	23866	55.88	+35%	123.3%
APROX Misc. p/Dk	18318	55.88	+3.9%	94.6%

Thus, the resulted SW Bending Moments obtain with this approach would also be very accurate, ensuring the reliability of the Midship Section Class Review, at a quite early stage of construction.

REFERENCES

Hansch, David Laurence, 2008. Methods of Determining the Longitudinal Weight Distribution of a Ship. From Northrop Grumman Newport News; joint meeting of Society of Naval Architects and Marine Engineers (SNAME) and Society of Allied Weight Engineers (SAWE), 24 Jan 2008.

Miranda, Andomarc 2020. A case study on the impact of Lightship Weight (LWT) Distribution in the structural scantlings. In *Developments on Maritime Technology and Engineering*, C. Guedes Soares, T. A. Santos (ed.), Taylor and Francis Group, London, UK (inpress)

Rodrigues, Carlos & Barbosa, Edgar 2002. Evolução do Software de Projecto nos Estaleiros Navais de Viana do Castelo. In C. Guedes Soares (Ed.), *O Mar Fonte de Desenvolvimento Sustentado*; 8th Technical Conference of Marine Engineering, Viana do Castelo, 3-4 October 2002.

Developments in Maritime Technology and Engineering – Guedes Soares & Santos (eds)
© 2021 Copyright the Author(s), ISBN 978-0-367-77376-2

Strength identification of ageing structures using shock pulse approach

A. Vitorino & Y. Garbatov

Centre of Marine Technology and Ocean Engineering (CENTEC), Instituto Superior Técnico, Universidade de Lisboa, Lisbon, Portugal

ABSTRACT: This work investigates the shock pulse approach in inspection of corroded steel plates in identifying the degree of degradation and ultimate capacity. A series of rectangular steel plates are artificially corroded using the random field approach. Employing the FEM, a shock-pulse load is subjected to the corroded plates and resulting transient structural response is analysed. Non-corroded steel plates of a wide range of thicknesses are firstly analyzed to estimate the ultimate strength and later they are subjected to a shock pulse load to identify their transient structural response. The inspection is performed by applying a shock-pulse load to the corroded plates and the resulting transient response is compared to the closest transient response of a non-corroded plate identifying the corrosion degradation level and the ultimate strength of the corroded plate. The developed approach can be used in a real conditions of inspection, presenting significant savings in both money and time.

1 INTRODUCTION

The tendency nowadays in sea transportation is building larger ships, carrying large and heavy cargoes around the world. This tendency, along with today's technology, leads to modern and efficient structural ship design. With a wide range of materials with different advantages available in the market, steel continues to be the most used material in shipbuilding, especially in larger ships.

Although steel has many advantages for this type of structures, having good mechanical properties and it is prone to corrosion deterioration over the years.

Over the years, the corrosion degradation in marine structures has been getting more attention regarding the corrosion control and prevention. The corrosion degradation phenomenon is complex, and the monitoring and inspection of corroded marine structures are an essential part of the structural maintenance planning and safety regulations.

Marine structures are subjected to periodic surveys, which includes thickness measurements of hull structures (IACS, 2015). The non-destructive testing and inspections are well documented as for an example in (Porter, 1992, Singh et al., 2019), where the capacity of different inspections techniques are examined, including visual inspection and more complex measurements methods such as dye penetrant, magnetic particle, eddy current, current potential drop, alternating current field measurement, radiography, ultrasonic, and many others.

Several practical difficulties are associated with the inspection process, including the associated costs, and the size of inspected structures. Additionally, the surveys require the vessel to be out of service for some time. It has to be recognised that complex non-destructive test, involving instrumentations and/or structure preparation is very rarely employed while visual inspections are the techniques widely used for ongoing ships structures.

The size of inspection tools need to be smaller than the standard manhole of ships (max 600x400mm), and the detector has to be able to access small areas, such as the corner of tanks. The weight of the tools should be light enough and easy to carry (Garbatov et al., 2006).

Wheel-tapper was a broaden inspection approach in the 20th century. Using a hammer to tap the train's wheels to check if they are cracked, damaged or good. Following this type of inspection, the shock pulse approach starts to be seen as a capable one, and it is investigated here to verify its applicability to perform an inspection of ship structures and identify the level of corrosion degradation.

A study performed here consists of analysing the structural response of corroded plates subjected to a shock pulse load, like a hammer tap, manual or automatic. The analysis of the shock pulse load can relate the transient structural response with an average corroded plate thickness and its ultimate strength, to determine the ship's overall strength and maintenance panning police (Garbatov & Guedes Soares, 2009).

The developed here approach includes two phases. In the first phase, an explicit analysis is performed to generate a database of shock-pulse structural responses and associated ultimate strength for a variety of structural configurations and corrosion severity as can be seen in Figure 1.

DOI: 10.1201/9781003216582-57

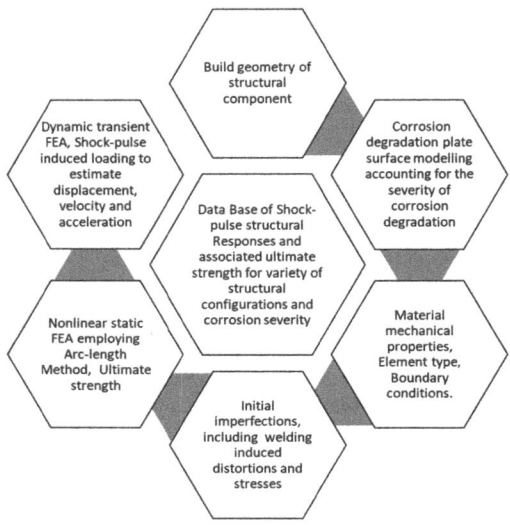

Figure 1. Explicit analysis.

The second phase covers the empiric-numerical analysis, which is based on the shock-pulse response measurement and numerically simulated here. A similarity analysis is performed, where an equivalent corroded plate thickness is identified based on the shock response similarity of the corroded and all stored in the database equivalent plates defining the equivalent thickness and corresponding ultimate strength.

2 SHOCK PULSE APPROACH

Several approaches are currently employed to identify structural degradation. The most common is the periodic measurement in determining the present structural component thickness. The corroded structural component thickness is compared with the acceptable level, and if it is below the acceptable level is replaced with a new one.

A more sophisticated approach in this respect is the use of a vibration analysis (ISO2631-1, 2010). This approach is employing "listening" for the roughness of the plate surface using a sophisticated spectrum analysis. In this case, sonic and ultrasonic noise and acoustic measurement techniques are also exploited. However, this approach is very much influenced by environmental conditions.

A change of mass, for an example due to corrosion degradation, changes the vibration spectrum. Accounting and measuring that is quite straightforward, but identifying the reasons and the degree of degradation are not so easy.

A different approach based on monitoring the shock pulse impacts caused by damage is also available and used nowadays (Sundström, 2010). This approach permits to identify different damage conditions over its entire life. The shock pulse approach is not influenced by the size of the structural components or background vibration noise.

The shock pulse approach, SPA is usually detecting the development of a mechanical shock wave generated by the impact between two colliding masses. At the moment of the impact, a compressive shock wave is developed in each mass, during a short time after the first contact of the colliding bodies time of the contact is very short and produces an increased acceleration at the impact point as can be seen from Figure 1.

Figure 2 shows as the mass is dropped with a velocity, v onto a structural component, which is initially at rest. At the moment of the impact, the structural component and the collapsed mass surface are colliding with a velocity equal to the dropping mass velocity. At the impact, a significant acceleration of the surfaces is generated. The magnitude of the acceleration is solely dependent upon the dropped mass and impact velocity, and it is not influenced by the relative sizes of the dropped mass and structural component or by the mechanical vibration.

The acceleration at the impact point generates a compressive wave, which propagates ultrasonically in all directions through the structural component and another wave also travels through the dropped mass. The magnitude of the wave front is an indirect measurement of the impact velocity.

In the second stage of the impact, the dropped mass and the structural component surfaces will deflect, and the energy of the motion will cause vibrations, which is typically detected by the vibration analysis. However, the SPA identifies and measures the magnitude of the mechanical impact by measuring the resultant compressive wave front as a shock pulse.

The SPA uses a piezo-electric accelerometer to measure the shock pulse, without being influenced by the background noise. The compressive wave front, defined as a shock pulse caused by a mechanical impact, sets up a dampened oscillation in the transducer at its resonant frequency.

The peak amplitude of the oscillation is proportional to the impact velocity, and the dampened transient is defined with a constant decay rate (ASTM-D-3332-94, 1994).

The measurement and analysis of the maximum value of the dampened transient is the base of the SPA for monitoring the condition of degrading structures.

A shock pulse response force, displacement, velocity and acceleration as a function of time, in the case of a simply supported noncorroded plate,

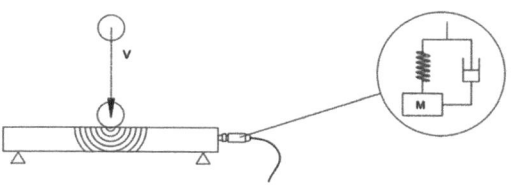

Figure 2. Shock pulse impact.

subjected to a shock pulse load defined as a dropped mass, $m = 50$ N, stiffness, $k=857$ kN/m, damping ratio, $x_i= 0.05$, $\Delta t=0.005$ s, initial displacement of $u_o =0$ m, initial velocity $v_o= 0$ m/s and the Newmark Integration parameters $\gamma= 0.55$ and $\beta= 0.27$ are showing in Figure 2 to Figure 6.

Marine structures are subjected to severe corrosion degradation, and corroded plate surfaces have a degree of roughness. When a mechanical impact is generated it will produce shock pulse, the magnitude of this pulse depends on the surface roughness condition, residual plate thickness and plate mass, boundary conditions and the initial velocity of the impacted mass.

Shock pulse generated by a dropped mass may increase up to several times when the plate is in

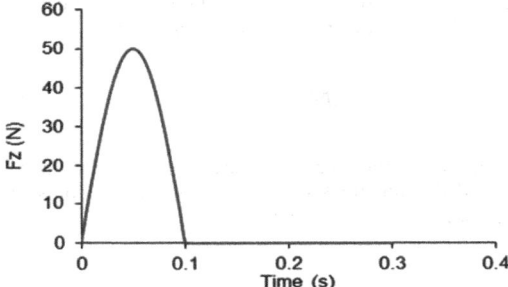

Figure 3. Shock pulse force.

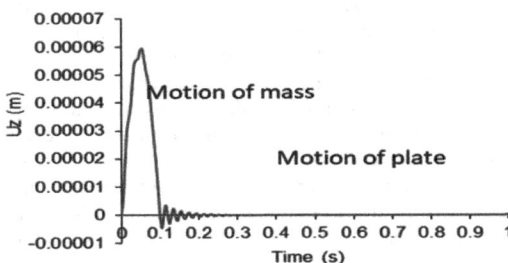

Figure 4. Shock pulse displacement.

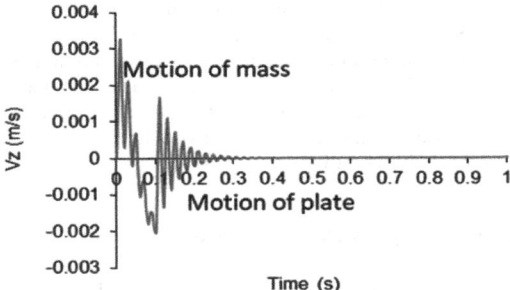

Figure 5. Shock pulse velocity.

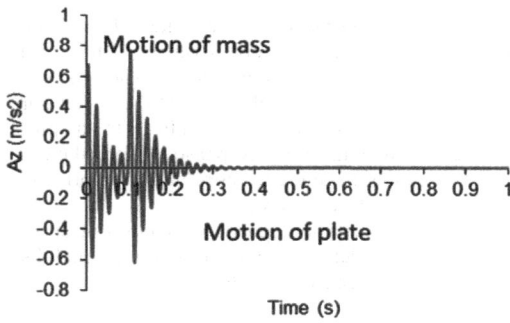

Figure 6. Shock pulse acceleration.

good condition. The SPA approach can accurately interpret the shock pulse through the service life of ageing structures from an initial stage of as-built to the final one when the replacement of the structural component is needed.

3 CORROSION DEGRADATION

Corrosion degradation is a process in which the material deteriorates when it is exposed to a corrosive environment. Corrosion is an electrochemical process, involving oxidation reactions where the metal ions and free electrons travel through a conducting solution and cathodic reactions where the cathode reduces the electrons. Moister can act as a reactive compound in the cathodic reactions (Melchers, 1999), and it allows the transfer of metal ions to occur more quickly. The cathodic reaction that is more associated with corrosion in maritime structures is the dominant reaction when pH is greater than 5, and the temperature is over ten degrees (Tomashov, 1996).

Regarding corrosion modelling, three fundamental approaches used nowadays. The first one is assuming that the corrosion rate is constant, leading to a linear approximation of the corrosion depth as a function of time that is an overestimation of the corrosion effects. The second approach is the experimental approach, which is based on the results of experiments for specific environmental conditions leading to a relationship of corrosion degradation as a function of specific governing parameters. For this, it would be necessary to know the environmental conditions where the structure will operate, and, there is a problem of generalisation of the experiments to full-scale conditions.

The approach used in the present study is based on the trend that is derived from the dominating corrosion degradation mechanism and then fitted to field data.

A model used here is the one developed as a nonlinear time-dependent function of corrosion wastage in (Guedes Soares & Garbatov, 1999). In this model, the corrosion degradation is separated into three phases. In the first phase, it is considered that there is no corrosion due to the effectiveness of the metal surface protection. The second one is when the corrosion degradation takes place, after the metal surface protection is damaged, leading to a plate thickness reduction. The third phase is when the corrosion degradation tends to stop, and the corrosion rate is reduced close to zero.

The non-linear time-dependent function of corrosion wastage model is assumed to be Lognormal distributed with a mean value defined as:

$$E[d(t)] = d_\infty\{1 - exp[-(t - \tau_c)/\tau_t]\}, t > \tau_c \quad (1)$$

where τ_c is the time that the metal surface protection is effective and τ_t is the transition time. The governing parameters of corrosion degradation used in the present study are d_∞=1.85 mm, τ_c=10.54 years and τ_t=17.54 years for ballast tanks of tankers as given in (Garbatov et al., 2007).

Another essential statistical parameter to model the corrosion degradation is the standard deviation, that was also used as found in (Garbatov et al., 2007) for deck plates of ballast tanks, by fitting the standard deviation as a function of time to a logarithmic function as:

$$StDev(t) = aLn(t - \tau_c) - b, t > \tau_c \quad (2)$$

where a and b are defined based on the regression analysis of the logarithmic function as 0.384 and 0.710 respectfully. The carrion depth is considered to follow the Lognormal distribution, as can be seen from Figure 7.

The corroded plate thickness is modelled by random non-correlated corrosion pits at the nodes of the artificially meshed plate resulting in an irregular corroded plate surface at the mesh nodes, equally spaced along the x and y directions.

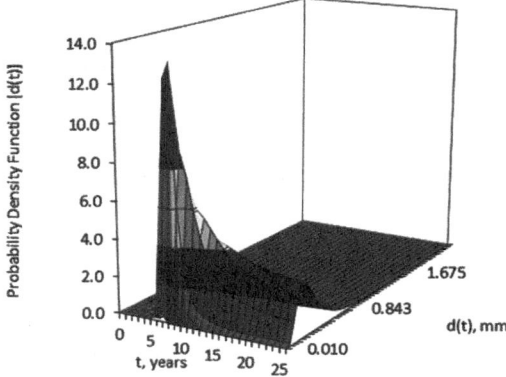

Figure 7. Time-variant Lognormal pdf of corrosion depth.

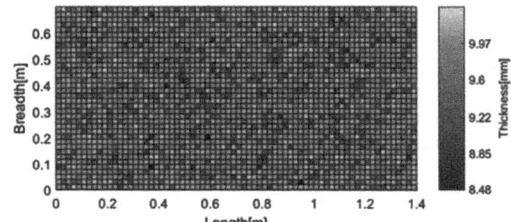

Figure 8. Plate nodal thicknesses for an as-built thickness of 12 mm at 20th year.

The corroded plate thickness, $d(i,j)^{cp}$, at any finite element nodal location (i,j) is defined by the random thicknesses of the non-corroded plate surface, $d(i,j)^{ip}$ reduced by the random corrosion depth, $d(i,j)^{cd}$ (Garbatov & Guedes Soares, 2019):

$$\mathbf{d}^{cp} = \mathbf{d}^{ip} - \mathbf{d}^{cd} \quad (3)$$

where **d** is the vector of the corroded and non-corroded plate thicknesses and corrosion depths at (i,j) locations are defined as:

$$d_{ij}^{cp}(t) = F_{ip}^{-1}[p_{ij}, E(d^{ip}), StDev(d^{ip})] \\ -F_{cd}^{-1}[p_{ij}, E(d^{cd}(t)), StDev(d^{cd}(t))] \quad (4)$$

where $F^{-1}[]$ is the inverse of the Log-normal cumulative distribution function of the probability, p_{ij}, which may take a value between 0 and 1.

The thickness of the non-linearly corroded plate is defined as a result of randomly distributed plate thicknesses for the randomly determined reference points at a specific time, during the service life. The corroded and non-corroded plate thicknesses are described by the Lognormal distribution. The non-corroded plate and the corresponding corrosion depths are considered as non-correlated.

The non-linear time-dependent function of corrosion wastage model is used to generate corroded plate surfaces with non-corroded thicknesses of 11, 12, 13, 14 and 15 mm respecting the ages of 15, 18, 20, 23 and 25 years. For each plate, 2,627 corrosion pits are created, one for each FE node that coincides with the plate's generated FE mesh. The corroded plate thickness at the nodal locations of the 20-year old corroded plate that has an as-built thickness of 12 mm is shown in Figure 8.

4 ULTIMATE STRENGTH ASSESSMENT

The plates analysed here are of a rectangular shape, with a length l =1.4, width b =0.7 and as-built thickness of t_p. The plate is built of steel with the

Table 1. Plate's boundary conditions.

	U_x	U_y	U_z	Rot_x
y=0	F	C	C	C
y=l	F	F	C	C
x=0	F	F	C	F
x=b	F	F	C	F
x=b/2, y=0	C	C	C	C

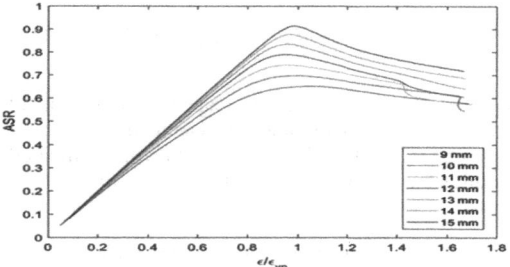

Figure 9. Normalized stress-strain relationship of non-corroded plates.

Young modulus, $E = 205.8 \; GPa$, Poisson coefficient, $v = 0.3$ and specific density of 7.8 t/m³. The boundary conditions used in the FE analysis as defined in (Silva et al., 2013) are shown in Table 1 (C for constrained and F for free).

These initial imperfections considered here is as proposed in (Smith et al., 1988):

$$w(x,y) = w_o sin(\pi x/b) sin(\pi y/l) \qquad (5)$$

$$w_o = 0.1 h_o \beta_{pl,o}^{2} \qquad (6)$$

$$\beta_{pl,o} = (b/t_p)\sqrt{\sigma_{yp}/E} \qquad (7)$$

where x, y, z are the plate coordinate system, $\beta_{pl,o}$ is the non-corroded plate slenderness as proposed in (Faulkner, 1975), w_o is the maximum out of the plane deflection, and E and σ_{yp} are, respectively, the elasticity modulus and the yield stress.

The finite elements used in the analysis is SEHLL181, having four nodes (one on each corner) and six degrees of freedom on each node. When using this type of elements, the thickness can be defined for each node of the element, resulting then on the thickness of the elements.

The uniaxial axial loads are applied on the edge $y = l$, and the reaction forces on the edge $y = 0$ are used to estimate the average stresses and the average stress ratio (ASR) is calculated as:

$$ASR = \sum \frac{R_{yi}}{\sigma_{yp} A_o} \qquad (8)$$

where R_{yi} is the reaction force at the ith node in the y-direction (these nodes are on the edge $y = 0$), $i = 1, \ldots k$ where k is the number of the nodes at $y = 0$, σ_{yp} is the yield stress point of the material and A_o is the sectional area of the plate at $y = 0$.

The ratio between the strain and the yield strain of the plate is also computed:

$$\frac{\varepsilon}{\varepsilon_y} = \frac{U_{yp}}{l \varepsilon_y} \qquad (9)$$

where ε is the plate strain, ε_y is the yield strain of the material, l is the plate length, and U_{yp} is the displacement of the plate in the y-direction at the nodal location $(b/2, l)$.

A total of 86 plates were analysed, including 25 corroded and 61 non-corroded ones with a thickness varying by 0.1 mm from 9 to 15 mm. The results of the FE analysis are plotted as ASR against $\varepsilon/\varepsilon_y$, in Figure 9.

5 TRANSIENT ANALYSIS

In the present study, a shock pulse impact on a corroded rectangular plate, induced in the centre area of the plate, is studied using the Finite Element Method, FEM by employing the commercial software ANSYS, 2009, Mechanical APDL 15.0, using the Newmark time integration method.

For dynamical problems of structural systems, the principle of virtual work in conjunction with the finite element method is most widely used by the semi-discrete equation of the motion:

$$\boldsymbol{m\ddot{u}}(t) + \boldsymbol{c\dot{u}}(t) + \boldsymbol{f^i}(t) = \boldsymbol{f^a}(t) \qquad (10)$$

where $\mathbf{m}$ and $\mathbf{c}$ are mass and damping matrices, $\ddot{u}(t)$ is the nodal acceleration vector, $\dot{u}(t)$ is the nodal velocity vector, $\mathbf{f^i}(t)$ is the internal load vector, and $\mathbf{f^a}(t)$ is the applied load vector.

As in the linear systems, the structural stiffness remains constant, and the internal load is linearly proportional to the nodal displacement leading to:

$$\boldsymbol{m\ddot{u}}(t) + \boldsymbol{c\dot{u}}(t) + \boldsymbol{ku}(t) = f^a(t) \qquad (11)$$

where $\mathbf{k}$ is the stiffness matrix, and $\mathbf{u}(t)$ is the nodal displacement vector. In the Newmark method, the semi-discrete motion equation for the single-step algorithm is employed, Hughes (1987):

$$m\ddot{u}_{n+1} + c\dot{u}_{n+1} + ku_{n+1} = f^a \quad (12)$$

where $\ddot{u}_{n+1}$, $\ddot{u}_{n+1}$, u_{n+1} are respectively, the nodal acceleration, velocity and displacement vectors at the time t_{n+1}. In addition, the Newmark method requires the following equations to be used:

$$\dot{u}_{n+1} = \dot{u}_n + [(1 - \delta)\ddot{u}_n + \delta\ddot{u}_{n+1}]\Delta t \quad (13)$$

$$u_{n+1} = u_n + \dot{u}_n\Delta t + [(0.5 - \alpha)\ddot{u}_n + \alpha\ddot{u}_{n+1}]\Delta t^2 \quad (14)$$

to solve the problem by an implicit algorithm.

The analysed rectangular plate is assumed to be simply supported. The shock pulse is generated as a half-sine pulse, and the shock force is modelled with the half-sine pulse with a 0.1-second duration, as:

$$F_z = F_o sin(k\pi t), \quad 0 < t < 0.1 \quad (15)$$

where F_z is the shock pulse force subjected perpendicularly to the plates' surface, F_o is the excitation amplitude, $\omega = k\pi$ is the excitation circular frequency, and t is the time.

To analyse the shock pulse plate response a transient analysis is performed, several computational parameters needed to be defined to achieve an adequate for this study and possible to perform with the available computational capacity. For the present study, a full analysis of the Newmark's method is assumed to be the most adequate, with a ramped stepped loading, which is defined according to ANSYS, 2009.

Several test-simulations are made for the non-corroded plate of a thickness of 10 mm. The analysis of these simulations has shown that after 5 seconds the displacement in the plate's centre would follow the same pattern and tend to zero and 5 seconds is chosen for the duration of the plate's response analysis. The dropped mass used in the study is $m = 50\,N$, $\Delta t = 0.1\,s$, initial displacement is $u_o = 0\,m$ and initial velocity $v_o = 0\,m/s$.

The response taken from the transient analysis is the displacement, U_z in the z-direction of a node near the centre of the plate.

Two different time stepping is explored to identify the most appropriate for the present study different sub-steps responses were analysed and the 100 sub-step response is chosen since it has more data regarding the transient process of the impacted plate as can be seen from Figure 10.

With all parameters defined, 86 transient analyses are completed, for different plate thicknesses and corrosion degradation levels.

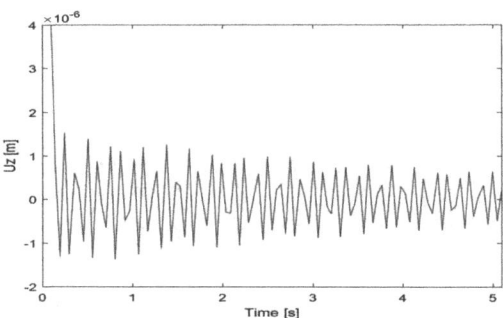

Figure 10. U_z as a function of time for a 10 mm non-corroded plate thickness, 100 sub-steps.

6 SIMILARITY

Measures of similarity of the shock pulse signal are required to identify the closest equivalent plate thickness and ultimate strength of a corrosion degraded steel plates. In this respect, many basic approaches such as matched filtering, cross-correlation, and beam formation are widely used to measures of similarity.

The matched filtering involves the projection of an unknown waveform onto a known template, the cross-correlation is used to determine the translation required to bring one waveform into the alignment with another; while beam formation brings the multiple signals into a phase so that they constructively interfere.

A measure of similarity is performed here to identify the level of corrosion degradation of the rectangular steel plate and corresponding stress-strain behaviour. Statistical treatment of a shock pulse signal is used to derive the measure of signal similarity. The approach is based on standard statistical relationships (Leemis, 1995).

A hypothesis test is performed where the null hypothesis is associated with the waveforms generated by shock pulse load subjected to the corroded plate, H_o shows no statistically significant difference with an equivalent non-corroded plate subjected to the same shock pulse load. Under the null hypothesis are compared two signals assumed to follow the Normal distribution.

The null hypothesis is defined as the signal means differ by Δ_o (which is expected to be zero) as a two-tailed test: $H_o : \mu_1 - \mu_2 = \Delta_o$. The alternative hypotheses are defined as a one-tailed, left tail test: $H_L : \mu_1 - \mu_2 < \Delta_o$ and one-tailed, right tail test: $H_R : \mu_1 - \mu_2 > \Delta_o$.

It is quite essential to note that this refers to the probability of the value under the assumption that H_o is true. To decide when H_o is to be rejected, a significance level is set at $p_{cr} < 0.05$, i.e. if the likelihood of a difference is less or equal to 5%, given that H_o is true.

It is considered that a signal of the shock pulse response of a corroded plate response is recorded, and series of replicas of simulated signals fixed to represent a variety of equivalent non-corroded steel plates of different thicknesses subjected to the same load and corresponding compressive stress-strain levels are compared.

However, the measured and the replica signals are representing similar but independent of each other conditions, and the two-sample are compared with each other.

A two-sample t-test is used to test the difference between two sample means μ_1 and μ_2. The errors in the two conditions are assumed to have identical independent distributions and Normally distributed. The difference between the two signals is also considered as a random variable. To verify the null hypothesis about the difference, the Welch (1947) generalised t-test can be employed:

$$t = [(\mu_1 - \mu_2) - \Delta_o]/\sqrt{\sigma_1^2/n_1 - \sigma_1^2/n_2} \qquad (16)$$

where σ_1^2 and σ_2^2 are the variances, and n_1 and n_2 are the numbers of reversals of the compared signal conditions.

To test the null hypothesis that the means are the same, the probability associated with one of the possible alternate hypotheses is checked. A t-test is used to test the alternative hypothesis that the means are different from (either higher than or less than) each other.

The degree of freedom is defined as:

$$df = \frac{\frac{\sigma_1^2}{n_1} - \frac{\sigma_2^2}{n_2}}{\left[\frac{\left(\frac{\sigma_1^2}{n_1}\right)^2}{(n_1-1)} + \frac{\left(\frac{\sigma_2^2}{n_2}\right)^2}{(n_2-1)}\right]} \qquad (17)$$

and the probability is defined as:

$$p = T(|t|, df, tails) \qquad (18)$$

where T is the Student distribution and the tails is the number tails=2.

If $p < p_{cr}$, H_o is to be rejected, otherwise H_o is to be accepted.

However, for this difference to be statistically significant, it must be large relative to the standard error of the difference. If there is very little precision in the estimates of the averages, then one still may not be able to say that that difference is statistically significant even if the observed difference is substantial (see Figure 11).

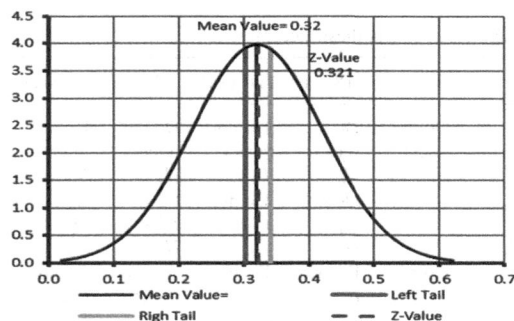

Figure 11. Similarity estimate.

7 RESULTS AND DISCUSSION

For every corroded plate, with different initial thicknesses and age, the similarity analysis is performed using the statistical descriptors of the shock pulse response of the corroded and equivalent non-corroded plates resulting in a total of 61 conditions to be checked.

To find the equivalent non-corroded plate that has a shock pulse response closest to each corroded plate, the t-value similarity index is defined, and the corresponding ultimate strength ratio is found. The t-value index, estimated for each corroded plate concerning the closest equivalent thickness non-corroded plate can be seen in Table 2.

To verify if the H_o-hypothesis is accepted or rejected for the t values presented in Table 2, $p = T(|t|, df, tails) < p_{cr}$ needs to be verified, where p_{cr}=0.05, tails=2 and df is defined using Eqn (15). The output from the verification is given in Table 3.

To evaluate the sensitivity and quality of the developed approach, the matching of the corrosion degradation level and resulting ultimate strength ratio, USR of the corroded plate and its equivalent non-corroded plate are analysed. The % differences in USR as a function of the thickness and age are presented in Table 4 to 8.

Table 4 to 8 present a difference between the corroded and equivalent non-corroded plate USR showing the most of them are less than 10% difference, with only a few exceptions above with a 20% difference. It can be seen that the approach

Table 2. t-value similarity index.

	15 year	18 year	20 year	23 year	25 year
11 mm	4.52E-04	3.73E-04	-4.44E-04	3.82E-02	9.28E-02
12 mm	-3.25E-02	-1.70E-02	-3.89E-03	7.38E-04	-2.20E-04
13 mm	-3.82E-02	1.53E-04	-2.49E-03	-2.24E-03	-1.08E-03
14 mm	-7.15E-02	-5.39E-02	-6.78E-02	-6.60E-02	-6.39E-02
15 mm	-1.08E-01	-9.98E-02	-9.25E-02	-8.21E-02	-7.88E-02

Table 3. H_o-hypothesis acceptance.

	15 year	18 year	20 year	23 year	25 year
11 mm	Do not reject H_o	Do not reject H_o	Do not reject H_o	Do not reject H_o	Do not reject H_o
12 mm	Do not reject H_o	Do not reject H_o	Do not reject H_o	Do not reject H_o	Do not reject H_o
13 mm	Do not reject H_o	Do not reject H_o	Do not reject H_o	Do not reject H_o	Do not reject H_o
14 mm	Do not reject H_o	Do not reject H_o	Do not reject H_o	Do not reject H_o	Do not reject H_o
15 mm	Do not reject H_o	Do not reject H_o	Do not reject H_o	Do not reject H_o	Do not reject H_o

Table 4. USR of corroded and equivalent non-corroded plates, h=11mm, USR=0.743.

	15 year	18 year	20 year	23 year	25 year
USR -C	0.615	0.605	0.535	0.450	0.391
USR-ENC	0.644	0.634	0.613	0.562	0.562
Difference	4.50%	4.55%	12.71%	19.94%	30.51%

Table 5. USR of corroded and equivalent non-corroded plates,, h=12mm, USR=0.789.

	15 year	18 year	20 year	23 year	25 year
USR -C	0.615	0.605	0.535	0.450	0.391
USR-ENC	0.644	0.634	0.613	0.562	0.562
Difference	4.50%	4.55%	12.71%	19.94%	30.51%

Table 6. USR of corroded and equivalent non-corroded plates, h=13mm, USR=0.83.

	15 year	18 year	20 year	23 year	25 year
USR -C	0.747	0.706	0.651	0.528	0.517
USR-ENC	0.834	0.536	0.536	0.536	0.536
Difference	10.43%	-24.07%	-17.70%	1.54%	3.60%

Table 7. % USR of corroded and equivalent non-corroded plates, h=14mm, USR=0.876.

	15 year	18 year	20 year	23 year	25 year
USR -C	0.861	0.749	0.810	0.789	0.789
USR-ENC	0.876	0.876	0.876	0.876	0.876
Difference	1.71%	14.51%	7.54%	9.95%	10.02%

Table 8. % USR of corroded and equivalent non-corroded plates, h=15mm, USR=0.913.

	15 year	18 year	20 year	23 year	25 year
USR -C	0.896	0.874	0.853	0.820	0.812
USR-ENC	0.808	0.913	0.913	0.913	0.913
Difference	-9.78%	4.28%	6.49%	10.12%	11.01%

employed here is capable of identifying the corrosion degradation level of the corroded plate and its match to an equivalent non-corroded plate by comparing their shock pulse responses and agree that the corroded plate will have the same ultimate strength as the one of the associated equivalent non-corroded plate.

8 CONCLUSIONS

The overall objective of this study was to evaluate the applicability of the shock pulse approach in identifying the status of the degradation and strength level of corroded plates. The SPA showed an excellent acceptance regarding the estimated corrosion degradation and ultimate strength of a corroded plate since it is possible to associate a USR to each of the corroded plates studied, having most of the studied cases less than 10% difference concerning the associated equivalent non-corroded plate.

Some improvement of the applied approach may enhance the estimation as for an example, the measurements may be taken from several points of the plate instead of just one point. This way, the possibility of one corrosion pit thickness point causes an incorrect output may be reduced, and the overall results will be more accurate.

It can be concluded that besides the fact that the SPA can be further improved, it is capable of identifying the corrosion degradation level of a corroded plate and its match to an equivalent non-corroded plate and associated ultimate strength.

The analysed shock pulse approach may complement or even replace the commonly used thickness measurements by ultra-sonic and spare time in the marine structural periodic surveys. Besides, it may provide information about the ultimate strength of inspected plates.

REFERENCES

ANSYS 2009. Advanced analysis techniques guide, South pointe, 275 Technology Drive, Canonsburg, PA 15317 ANSYS, Inc.

ASTM-D-3332-94 1994. Standard test methods for mechanical shock fragility of products using shock machines. Philadelphia, PA: ASTM.

Faulkner, D., 1975. A review of effective plating for use in the analysis of stiffened plating in bending and compression. *Journal of Ship Research*, 19, 1–17.

Garbatov, Y. & Guedes Soares, C., 2009. Structural maintenance planning based on historical data of corroded deck plates of tankers. *Reliability Engineering & System Safety*, 94, 1806–1817.

Garbatov, Y. & Guedes Soares, C., 2019. Spatial corrosion wastage modelling of steel plates exposed to marine environments. *Journal of Offshore Mechanics and Arctic Engineering*, 141.

Garbatov, Y., Guedes Soares, C., Ok, D., Pu, Y., Rizzo, C. M., Rizzuto, E., Rouhan, A. & Parmentier, G., 2006, Modelling strength degradation phenomena and inspections used for reliability assessment based on maintenance planning, *Proceedings of the 25th International Conference on Offshore Mechanics and Arctic Engineering* (OMAE2006), 4-9 June, Hamburg, Germany, ASME, New York, Paper OMAE2006-92090.

Garbatov, Y., Guedes Soares, C. & Wang, G. 2007. Nonlinear time-dependent corrosion wastage of deck plates of ballast and cargo tanks of tankers. *Journal of Offshore Mechanics and Arctic Engineering*, 129, 48–55.

Guedes Soares, C. & Garbatov, Y. 1999. Reliability of maintained, corrosion protected plates subjected to non-linear corrosion and compressive loads. *Marine Structures*, 12, 425–445.

IACS 2015. Common structural rules for bulk carriers and oil tankers. London: International Association of Classification Societies.

ISO2631-1 2010. Mechanical vibration and Shock-Evaluation of human exposure to whole-body vibration —Part 1: General requirements. *AMD 1*. International Standards Organization.

Leemis, L., 1995. *Reliability, probabilistic models and statistical methods*, Prentice-Hall.

Melchers, M. E., 1999. Corrosion uncertainty modelling for steel structures. *Journal of Constructional Steel Research*, 52, 3–19.

Porter, R., 1992. Non-destructive examination in shipbuilding. *Welding Review*, 11, 9–12.

Silva, J. E., Garbatov, Y. & Guedes Soares, C. 2013. Ultimate strength assessment of rectangular steel plates subjected to a random localised corrosion degradation. *Engineering Structures*, 52, 295–305.

Singh, R., Raj, B., Mudali, K. & Singh, P. 2019. *Non-destructive evaluation of corrosion and corrosion-assisted cracking*, Wiley-American Ceramic Society.

Smith, C. S., Davidson, P. C., Chapman, J. C. & Dowling, J. P. 1988. Strength and stiffness of ships' plating under in-plane compression and tension. *Transactions RINA*, 130, 277–296.

Sundström, T. 2010. An introduction to the SPM HD method. R&D, SPM Instrument AB.

Tomashov, N. D. 1996. *Theory of corrosion and protection of metals*, The Mac Millan Co.

Welch, B. L. 1947. The generalisation of Student's problem when several different population variances are involved. *Biometrika*, 34, 28–35.

Developments in Maritime Technology and Engineering – Guedes Soares & Santos (eds)
© 2021 Copyright the Author(s), ISBN 978-0-367-77376-2

Accelerated large scale test set-up design in natural corrosion marine environment

K. Woloszyk
Faculty of Ocean Engineering and Ship Technology, Gdansk Univeristy of Technology, Gdansk, Poland

Y. Garbatov
Centre for Marine Technology and Ocean Engineering (CENTEC), Instituto Superior Técnico, University of Lisbon, Lisboa, Portugal

ABSTRACT: The standards for conducting small-scale specimen tests are well developed, but there is a lack of direct guidelines for conducting corrosion tests for large-scale specimens. The objective here is to develop a methodology which may be used in designing an accelerated corrosion test of large-scale structural components subjected to a natural corrosion marine environment. Different factors influencing corrosion degradation of steel structures are analysed, including salinity and pH of the water, water velocity, oxygen content etc. The most important governing factors of corrosion degradation are identified, and the process of corrosion testing is defined in order to produce accelerated corrosion degradation in relatively short interval of time, developing a natural corrosion degradation mechanism, leading to irregular random corroded surfaces and material properties changes. Existing commercial testing equipment, that can be employed in creating accelerated corrosion conditions for testing large scale specimens, are investigated and the most efficient salt corrosion testing tank is identified. Several features of the corrosion test set-up are discussed.

1 INTRODUCTION

Ship and offshore structures are subjected to a severe marine environment. In order to evaluate the influence of corrosion degradation into the strength of various structural elements, the experimental testing is needed (Saad-Eldeen, *et al.*, 2013; Garbatov, *et al.*, 2017, 2018). Based on the experimental results, numerical models for predicting the strength behaviour of corroded structures could be developed (Garbatov, *et al.*, 2015; Woloszyk, *et al.*, 2018; Woloszyk and Garbatov, 2019a, 2019b). One can test members taken from real structures after some time of exploitation, or create artificial corroding environment in the laboratory. In the first case, the initial properties of structural members, as well as corrosion conditions, are usually missing. In the second case, the variables can be controlled.

In the case of small-scale specimens, the corrosion testing procedures are already well-developed (Baboian, 2005). In the case of large-scale specimens, there are no strict rules and existing guidelines are rather modest (ASTM Norma G 52, 2006). In this way, one needs to choose the proper environmental conditions to be simulated in the experimental domain.

The most common type of corrosion degradation for constructional steel is the so-called 'uniform' corrosion. In this case, the loss of material is relatively uniform, where the surface is rough and

irregular. The corrosion then is measured via material weight-loss after some exposure time.

In the presented study, the main factors influencing corrosion degradation of steel structures are discussed and corrosion test set-up for large-scale specimens is presented. The simplified methodology for calculating the corrosion rate and acceleration factor compared to natural marine conditions is evaluated. The corrosion measuring techniques are discussed and an optimum number of thickness measurements, depending on the exposure time, is obtained.

2 FACTORS INFLUENCING CORROSION DEGRADATION

The corrosion degradation is influenced by many factors, mainly: biological, chemical and physical (Melchers, 1999). In the case of biological factors, their importance seems to be not significant, and thus there will be not discussed furtherly.

2.1 *Chemical factors*

The main influencing chemical factors in terms of corrosion degradation are content of the oxygen and sodium chloride in the seawater conditions.

DOI: 10.1201/9781003216582-58

The dissolved oxygen concentration undoubtedly influences the corrosion rate of carbon steel. The maximum saturation of oxygen in water is a function of the temperature and salinity (Weiss, 1970). Near the ocean surface, the content of dissolved oxygen usually reaches the maximum level of saturation due to contact with the atmosphere (Pedeferri, 2018). Two sets of conditions, however, could lead to super-saturated conditions. First one is the production of oxygen due to the photosynthesis of the microscopic marine plants. During intense periods, the seawater could be even supersaturated reaching 200% of dissolved oxygen comparing to the equilibrium conditions. The second case that the saturation is above the maximum level is the increased level of air bubbles due to the wave actions. In this case, the super-saturation does not usually exceed 110 %.

The linear relation between the corrosion rate and content of the dissolved oxygen for mean ocean temperature and salinity could be found in (Guedes Soares, et al., 2011):

$$r(O_2) = 0.0286 \cdot [O_2] + 0.086 \quad (1)$$

The corrosivity of the water increases proportionally with the increase of the salinity. This phenomenon is caused due to the increase of water conductivity. However, if the salinity exceeds 3 %, the water corrosivity decreases (Kirk and Pikul, 2009). With the increase of salinity, the solubility of the oxygen is reduced, and thus 3.5 % of salinity results in the highest corrosion rate. Thus, the 3.5 % salinity is to be set up in testing conditions.

The pH of the seawater is usually very constant and is between 8.1 and 8.3 (Melchers, 2003), thus its effect on the corrosion rate is negligible. Even for higher ranges of pH, the corrosion rate seems to be independent.

2.2 Physical factors

The main physical factors that influence the corrosion rate are temperature, water velocity and water pressure. Water pressure is not essential since the testing tank will be rather shallow.

Based on the measurement data (Laque, 1975), the influence of temperature (T) into the corrosion rate in seawater could be modeled by linear relationship:

$$r(T) = 0.00356 \cdot T + 0.0392 \left[\frac{mm}{y}\right] \quad (2)$$

However, the linear relationship is valid up to some moment. This is related to a decrease in the maximum oxygen saturation with the temperature increase.

In the case of the water velocity, it has a big impact on the corrosion rate. In high water velocities, oxygen more easily reaches the metal surface

and protective films are significantly reduced. Additionally, the corrosion products are more frequently removed. The relationship between the flow velocity (v) may be modelled with the use of an exponential function (Laque, 1975):

$$r(v) = 0.934(1 - Exp(-0.446(v + 0.282))) \quad (3)$$

Due to that, even a minimal water velocity may significantly speed-up the corrosion process compared to the stagnation conditions.

3 THE CORROSION TEST SET-UP

Based on the analysis presented in the previous section, the leading natural factors that influence the corrosion rate are temperature, content of dissolved oxygen in water, water velocity and salinity. The control of these three factors may result in accelerated corrosion degradation of the specimens.

A very high rate of corrosion degradation, reaching even couple of mm/year may be as a result of electrochemical corrosion. In this case, the water and specimens are subjected to a DC power input source. However, the resulting corrosion morphology is very different from those obtained in the case of natural conditions. Xiao, et al., 2020 presented a comparative analysis between electrochemically and naturally corroded specimens. Due to the different corrosion mechanisms, the surfaces of the corroded specimens diverged significantly. Additionally, the mechanical properties were also different depending on the type of corrosion for the same degradation level.

Since the aim of the corrosion test is to simulate the marine environmental conditions, one should avoid the electrochemically stimulated corrosion degradation process. Thus, only natural parameters could be controlled in order to magnify the corrosion rate.

3.1 Specimens dimensions

The specimens that will be subjected to the corrosion degradation process are stiffened plates of a 1.2m length and 0.4 m width with a stiffener of a -0.1 m height. There are made from normal strength steel of S235. Three different thicknesses are investigated: 5, 6 and 8 mm. The sample specimen during the welding process is presented in Figure 1. The total number of nine specimens will be tested, three specimens for each thickness.

Additionally, the standard coupon specimens will be tested in order to provide information about mechanical properties for different values of degree of corrosion degradation. The total number of 27 specimens will be tested. Both big and small-scale specimens are presented in Figure 2.

Figure 1. Specimens to be tested.

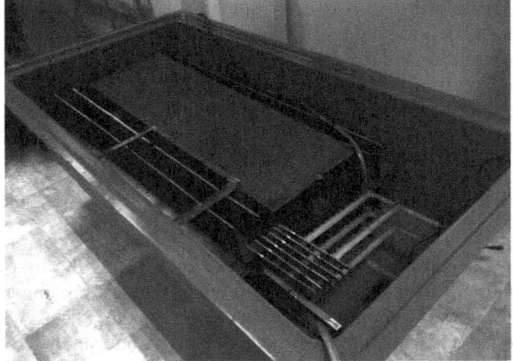

Figure 2. Corrosion tank with placed specimens.

3.2   Corrosion tank

Due to the scale of specimens, the corrosion tank needs to provide sufficient space for water circulation. A 900 l tank with the external dimensions of 2000 x 1100 x 750 mm is chosen for that purpose of the experiment. The big-scale specimens were placed alongside the tank edge as presented in Figure 2. The small-scale tensile specimens were placed on the specially designed grillage structure. The tank is made from the GRP laminate. The advantages of this solution are the excellent corrosion resistance and lower weight compared to metal tanks, with a similar price.

3.3   Oxygen content control

As discussed in the previous section, the content of the dissolved oxygen in the surface of the seawater usually reaches the maximum value. However, in testing conditions, one can easily obtain the supersaturated conditions with the use of aerator (Colt and Westers, 1982). Such aerators are commonly used in fish ponds (Clay and Kovari, 1984). Due to that, the corrosion rate could be significantly magnified. The aeration pump with an efficiency of 70 l/min is

chosen to aerate the water circulating in the tank. It is assumed that it should create conditions with a constant supersaturation of about 300 %. The actual value of the dissolved oxygen content will be analyzed in laboratory conditions during testing. It could be done with the use of a special oxygen meter.

3.4   Temperature control

The tank will be placed in a closed room when temperature oscillates about 20 °C. However, to speed up the corrosion rate, water heating may be applied. Excessive heating (above 40 °C) of such amount of water requires heaters with enormous power, leading to high energy consumption. Due to that, the heaters with moderate energy consumption of a total power of 600W are used. The maximum temperature of 34 °C could be obtained and the temperature is continuously controlled in order to keep the same level during the whole corrosion test.

3.5   Water velocity control

As noticed previously, even minimal water velocity may produce faster corrosion degradation. To control the circulation water velocity, a circulation pump as presented in Figure 3, is used. The maximum efficiency of the pump is about 20,000 l/h. With these dimensions of the tank, the water circulates with a velocity of approximately 0.03 m/s.

3.6   Salinity control

In order to provide the proper salinity conditions, natural saltwater should be applied. Such salt comes from the process of the water evaporation from the real seawater. The salinity level should be controlled during the corrosion degradation process. To measure the salinity, equipment which measures the salinity based on the water density is used. With the salinity increase, the water density increases as well. However, the density will be temperature-dependent

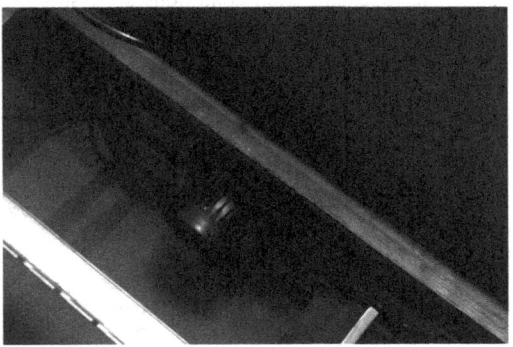

Figure 3. Circulation pump used during corrosion testing.

and a proper calibration table needs to be used in order to provide the assumed value of salt level.

4 ESTIMATION OF CORROSION ACCELERATION FACTOR

The corrosion rate in any conditions may be calculated based on the Eqns 1 to 3. However, the influence of all discussed factors could be combined into one equation (Pedeferri, 2018):

$$r = 0.012 \cdot 2^{\frac{T-25}{T}} \cdot \{[0_2] + 2.72\} \cdot (1 + \sqrt{v}) \quad (4)$$

In both cases, the results are somewhat similar, and due to simplicity, Eqn 4 will be used to estimate the corrosion rate.

The mean values of environmental factors in real ocean conditions (in the surface layer) are presented in Table 1 (Venkatesan, et al., 2002). The coupon not subjected to any flow velocity is assumed as a reference. In the surface layer, approximately fully aerated conditions are existing. Additionally, in the testing tank, both surfaces of the specimens will be corroded. In ships, usually, only one surface is corroded significantly. This will double the corrosion rate in testing conditions. The simulated testing conditions and corrosion rates calculated using Eqn 4 are presented in Table 1.

Based on the corrosion rates in real and testing conditions, the acceleration factor can be calculated:

$$a = \frac{r_{test}}{r_{real}} = 9.35[-] \quad (5)$$

In this way, the same corrosion degradation, which in normal conditions is obtained after 20 years, will be obtained after approximately 2 years.

One can notice that there is a possibility to increase the corrosion rate even more significantly. In the case of the water salinity, in both real and testing conditions, there is already an optimum value. The supersaturation conditions are also hard to be increased. The parameters that could be

Table 1. Environmental conditions.

Factor	Real conditions	Testing conditions
Salinity [‰]	34.3	35
Temperature [°C]	15.5	34
Oxygen saturation [%]	100	300
Water velocity [m/s]	0	0.03
Corroding surfaces	1	2
Corrosion rate [mm/y]	0.0787	0.736

increased are the temperature and water velocity. However, this will require more financial and energy resources. The proposed methodology is somehow a compromise between costs and time of the corrosion test.

Additionally, one needs to be aware, that corrosion degradation is a very complicated process and estimated corrosion rates are subjected to high uncertainties (Zakowski, et al., 2014). Thus, the corrosion rate needs to be monitored continuously during the whole corrosion process.

5 THE CORROSION MEASURING TECHNIQUES

During corrosion test, the specific degradation level is to be obtained. To monitor the corrosion rate, measurements need to be performed. There are various methods to measure corrosion degradation (Pedeferri, 2018), such as:

- Weight measurements of corrosion coupons;
- Electric Resistance;
- Electrochemical Impedance Spectroscopy (EIS);
- Acoustic Emission methods;

The simplest, and actually very precise, is the first method, which is based on the weight measurements. The small coupons made from the same sheet of metal as big-scale specimens are taken and analyzed after some period of time. The corrosion degradation is obtained as a reduction of the current mass compared to its initial one. However, the specimens need to be cleaned from corrosion products, which could disturb the measurements. The detailed analysis of the mass measurements, conducted for the need of the present study, of coupons with and without the corrosion products shown, that the difference could reach the level of 0.5 %. To reduce the uncertainty related to the measuring equipment, one can choose a very precise laboratory weight with an accuracy of 0.1 g.

The Electric Resistance method is very widely used, especially for measuring the material loss in the interior of plants and pipelines. The electric resistance probe is made from the same material and calculates the corrosion rate trough the measurement of the electrical resistance. With the corrosion development, the electrical resistance increases due to the reduction of the cross-section area. The main disadvantage of this method is for a low corrosion rate, and the resistance variations are small. Thus, the instrumentation with considerable accuracy needs to be used.

The electrochemical impedance spectroscopy has been proven to be an excellent method for measuring corrosion rates. The probe sends the electrical signal with a wide range of frequencies. Then, the impedance characteristics are measured creating a full spectrum. Depending on the shape of EIS, the parameters of the corrosion can be obtained. Nevertheless, this methodology requires knowledge of the operator and proper equipment.

Figure 4. Typical ultrasonic thickness gauge.

The acoustic emission methods can be used in thickness measurements and for defects detection. In the case of thickness measurements, the probe sends an acoustic signal with a known velocity and measure the time that is required to capture the signal again in the probe. Based on that, the thickness of the element can be determined. The typical ultrasonic thickness gauge consisting of the probe and measuring device is presented in Figure 4. The accuracy of such equipment is usually around ±0.05mm in case of thin steel plate elements. However, when the surface of the analyzed element is irregular, like in severely corroded plates, the measurements can be significantly disturbed (Cegla and Gajdacsi, 2016). Additionally, such equipment requires proper calibration prior to the measurements.

In the present study, two types of measurements will be carried out. The first one is based on the weight measurements of coupon specimens, which is the most accurate one. To monitor the actual corrosion degradation, the specimens are taken out of the tank and measured from time to time. However, the mass measurements are giving information about the mean corrosion depth. Due to that, to see the thickness deviations around the specimen, the ultrasonic thickness measurements will be carried out. To minimize the possible measuring error, the optimum number of thickness measurements can be defined, which is discussed in the next chapter.

6 OPTIMUM NUMBER OF THICKNESS MEASUREMENTS

All measuring techniques are subjected to some uncertainty level. With the corrosion growth, the surface becomes more and more irregular. By measuring the thickness in a couple points only, one cannot obtain the correct value of the degree of degradation. However, the use of proper statistical tools could be justified in order to provide the optimum number of thickness measurement points, which should lead to a proper estimation of the degradation level.

To estimate the corrosion degradation propagation, the nonlinear time-dependent model of Guedes Soares and Garbatov (1998) is used. The corrosion depth in the function of time can be estimated as follows:

$$d(t) = \begin{cases} d_\infty\left[1 - \exp\left(-\frac{t-\tau_c}{\tau_t}\right)\right], & t > \tau_c \\ 0, & t \le \tau_c \end{cases} \quad (6)$$

where d_∞ is the long-term corrosion depth, τ_c is the coating life and τ_t is the transition time.

The following values can be fitted based on real measurements, and the actual values can be found in (Garbatov, et al., 2006). Since the tested plates are exposed to seawater, their corrosion will be similar to the corrosion of deck plates of ballast tanks. The transition time is equal to 17.5 years in this case and the long-term corrosion depth is calibrated in order to obtain the same mean corrosion rate, as calculated via Eqn 4 for real conditions (0.0787 mm/year). The long-term corrosion depth is estimated equal to 2.36 mm. The corrosion depth in a function of time in the real conditions as presented in Figure 5.

For testing conditions, the transition time will be the one for real conditions divided by the acceleration factor. For the calculated acceleration factor (9.35), it will be equal to 1.87 years. For all specimens, 25% of corrosion degradation needs to be obtained. In the case of 5 mm specimens, the target corrosion depth is 1.25 mm. This corrosion depth will be obtained after 13.2 years in real conditions and 514 days in testing conditions. In the case of 8 mm specimen, the target is a 2 mm corrosion depth and it will be achieved after 32.8 years in real conditions and 1,282 days in testing conditions.

The corrosion depth as given in Figure 4 represents the mean value. According to Garbatov, et al., 2006, the standard deviation value of corrosion depth will be time-dependent and can be determined as (for coating life equal to 0 year):

$$StDev(t) = 0.384Ln(t + 10.54) - 0.71 \quad (7)$$

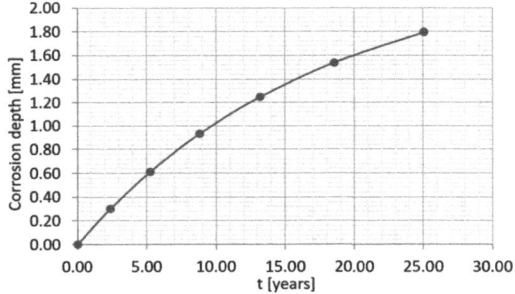

Figure 5. Corrosion depth as a function of time.

It can be noticed, that with the time increment, the standard deviation increase as well.

Based on the mean value and standard deviation of corrosion depth in the time domain, assuming that the Normal distribution of the corrosion depth is known, the needed number of thickness measurements conditional to the confidence level and assumed margin of error can be obtained (Dowdy, et al., 2004). Considering n number of measurements, the measured value m should lie within the following margin:

$$\bar{d}(t) - \frac{u_a \sigma(t)}{\sqrt{n}} \leq m \leq \bar{d}(t) + \frac{u_a \sigma(t)}{\sqrt{n}} \qquad (8)$$

where u_a is the percentile of the N(0,1) distribution with the assumed confidence level α, $\bar{d}(t)$ and $\sigma(t)$ are mean value and standard deviation of corrosion depth, respectively.

In this case, the $(u_a \sigma(t))/\sqrt{n}$ is the standard error $(\Delta d(t))$ of the estimation, when divided by $\bar{d}(t)$, it is expressed as $Err = \Delta d(t)/\bar{d}(t)$. For the assumed standard error and confidence level, the required number of measurements is estimated to:

$$n = \left(\frac{u_a \sigma(t)}{Err} \right)^2 \qquad (9)$$

To ensure the proper estimation of the degradation level, the thickness measurements need to be taken a couple of times during testing. The number of measurements will vary with time since the standard deviation is growing. To find the number of the optimum measurements, three different confidence levels (90%, 95% and 97.5%) and standard errors (5%, 10%, 15%) are investigated. The number of measurements as a function of the testing time with different values of decisive factors is presented in Figure 6. The example of a 5 mm plate is presented here.

One can notice that the relationship between the number of measurements and testing time is almost linear. The last day of measurements (516) is related to the time obtaining the target value of corrosion degradation for a 5 mm plate thickness (1.25 mm). To see the influence of the decisive factors in the number of measurements, two additional plots are presented. Figure 7 shows the influence of the assumed error in the number of measurements. Figure 8 presents the influence of the confidence level.

As can be seen, both relationships are nonlinear. With the decrease of the standard error, the measuring effort significantly increases. In this case, the relationship is highly nonlinear. As for the confidence level, the measuring effort increase for higher values of this variable. However, the relationship is only slightly nonlinear.

Based on this analysis, the confidence level of 95% and an error value of 10% were chosen as an optimum value that will maximize measurement precision with the possible minimum measuring effort. The measuring program with an estimated corrosion depth range is presented in Table 2.

After each measuring period, the corrosion depth needs to be recorded. Based on that information, the measuring program needs to be updated accordingly, since the corrosion could be higher or lower from the assessed one.

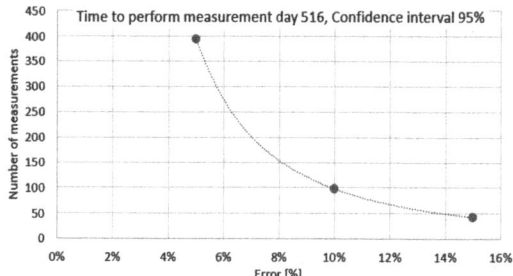

Figure 7. Number of measurements as a function of standard error.

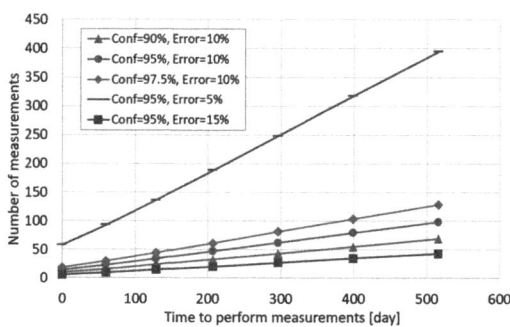

Figure 6. Number of measurements as a function of testing time.

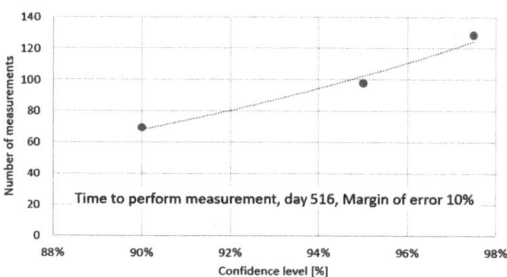

Figure 8. Number of measurements as a function of confidence level.

Table 2. Measuring program.

Measurement time [days]	$\bar{d}$ [mm]	$\bar{d} - \Delta d$ [mm]	$\bar{d} + \Delta d$ [mm]	n [-]
0	0.00	0.00	0.00	14
60	0.20	0.18	0.22	23
128	0.40	0.36	0.44	34
207	0.61	0.55	0.68	47
296	0.83	0.75	0.91	62
399	1.04	0.94	1.15	79
516	1.25	1.12	1.37	98

7 CONCLUSIONS

The presented study analyses various issues related to large-scale accelerated corrosion tests. The study shows, that the water temperature, oxygen content and water velocity are the main factors that can accelerate the corrosion quickly. The proposed testing set-up shows that the test could be performed without the use of very specialized equipment, which makes the proposal economically efficient. Nevertheless, the corrosion rate can be significantly magnified (up to 10 times). The main advantage of the proposed methodology is the control of only natural factors, avoiding the usage of the electric current, resulting in a corrosion mechanism similar to the real seawater conditions.

The optimum number of measuring points for thickness measurements of specimens is derived with the use of the simple statistical methodology.

ACKNOWLEDGEMENTS

This work has been supported by the National Science Centre, Poland (grant No. 2018/31/N/ST8/02380).

REFERENCES

ASTM Norma G 52 (2006) 'Standard Practice for Exposing and Evaluating Metals and Alloys in Surface Seawater', *Annual Book of ASTM Standards*.

Baboian, R. (2005) *Corrosion Tests and Standards: Application and Interpretation-Second Edition, Corrosion Tests and Standards: Application and Interpretation-Second Edition*. ASTM International.

Cegla, F. and Gajdacsi, A. (2016) 'Mitigating the effects of surface morphology changes during ultrasonic wall thickness monitoring', in *AIP Conference Proceedings*, p. 170001.

Clay, C. H. and Kovari, J. (eds) (1984) *Inland Aquaculture Engineering: Lectures Presented at the ADCP Interregional Training Course in Inland Aquaculture Engineering, Budapest, 6 June-3 September 1983*. Rome: Food & Agriculture, Org.

Colt, J. and Westers, H. (1982) 'Production of Gas Supersaturation by Aeration', *Transactions of the American Fisheries Society*. Wiley, 111(3), pp. 342–360.

Dowdy, S. M., Wearden, S. and Chilko, D. M. (2004) *Statistics for research*. Wiley-Interscience.

Garbatov, Y., Guedes Soares, C. and Wang, G. (2006) 'Nonlinear Time-Dependent Corrosion Wastage of Deck Plates of Ballast and Cargo Tanks of Tankers', *Journal of Offshore Mechanics and Arctic Engineering*, 129(1), pp. 48–55.

Garbatov, Y., Saad-Eldeen, S. and Guedes Soares, C. (2015) 'Hull girder ultimate strength assessment based on experimental results and the dimensional theory', *Engineering Structures*, 100, pp. 742–750.

Garbatov, Y., Saad-Eldeen, S., Guedes Soares, C., Parunov, J. and Kodvanj, J. (2018) 'Tensile test analysis of corroded cleaned aged steel specimens', *Corrosion Engineering, Science and Technology*. Taylor & Francis, pp. 1–9.

Garbatov, Y., Tekgoz, M. and Guedes Soares, C. (2017) 'Experimental and numerical strength assessment of stiffened plates subjected to severe non-uniform corrosion degradation and compressive load', *Ships and Offshore Structures*, 12(4), pp. 461–473.

Guedes Soares, C. and Garbatov, Y. (1998) 'Non-linear time-dependent model of corrosion for the reliability assessment of maintained structural components', *A. A. Balkema, Safety and Reliability*, 2, pp. 929–936.

Guedes Soares, C., Garbatov, Y. and Zayed, A. (2011) 'Effect of environmental factors on steel plate corrosion under marine immersion conditions', *Corrosion Engineering, Science and Technology*. Taylor & Francis, 46 (4), pp. 524–541.

Kirk, W. and Pikul, S. (2009) 'Seawater Corrosivity Around the World: Results from Three Years of Testing', in Baloun, C. H. (ed.) *Corrosion in Natural Waters*. 100 Barr Harbor Drive, PO Box C700, West Conshohocken, PA 19428–2959: ASTM International, pp. 2–36.

Laque, F. L. (1975) *Marine corrosion causes and prevention*. Wiley.

Melchers, R. E. (1999) 'Corrosion uncertainty modelling for steel structures', *Journal of Constructional Steel Research*, 52(1), pp. 3–19.

Melchers, R. E. (2003) 'Probabilistic Models for Corrosion in Structural Reliability Assessment—Part 2: Models Based on Mechanics', *Journal of Offshore Mechanics and Arctic Engineering*, 125(4), pp. 272–280.

Pedeferri, P. (2018) *Corrosion Science and Engineering*. Cham: Springer International Publishing (Engineering Materials).

Saad-Eldeen, S., Garbatov, Y. and Guedes Soares, C. (2013) 'Experimental assessment of corroded steel box-girders subjected to uniform bending', *Ships and Offshore Structures*, 8(6), pp. 653–662.

Venkatesan, R., Venkatasamy, M. A., Bhaskaran, T. A., Dwarakadasa, E. S. and Ravindran, M. (2002) 'Corrosion of ferrous alloys in deep-sea environments', *British Corrosion Journal*, 37(4), pp. 257–266.

Weiss, R. F. (1970) 'The solubility of nitrogen, oxygen and argon in water and seawater', *Deep-Sea Research and Oceanographic Abstracts*, 17(4), pp. 721–735.

Woloszyk, K. and Garbatov, Y. (2019a) 'Structural Reliability Assessment of Corroded Tanker Ship Based on Experimentally Estimated Ultimate Strength', *Polish Maritime Research*, 26(2), pp. 47–54.

Woloszyk, K. and Garbatov, Y. (2019b) 'Uncertainty assessment of ultimate strength of corroded stiffened plates subjected to different maintenance actions', in Georgiev, P. and Guedes Soares, C. (eds) *Sustainable Development and Innovations in Marine Technologies*. CRC Press, pp. 429–436.

Woloszyk, K., Kahsin, M. and Garbatov, Y. (2018) 'Numerical assessment of ultimate strength of severe corroded stiffened plates', *Engineering Structures*, 168, pp. 346–354.

Xiao, L., Peng, J., Zhang, J., Ma, Y. and Cai, C. S. (2020) 'Comparative assessment of mechanical properties of HPS between electrochemical corrosion and spray corrosion', *Construction and Building Materials*. Elsevier Ltd, 237, p. 117735.

Zakowski, K., Narozny, M., Szocinski, M. and Darowicki, K. (2014) 'Influence of water salinity on corrosion risk—the case of the southern Baltic Sea coast', *Environmental Monitoring and Assessment*, 186 (8), pp. 4871–4879.

Developments in Maritime Technology and Engineering – Guedes Soares & Santos (eds)
© 2021 Copyright the Author(s), ISBN 978-0-367-77376-2

An investigation on the overall stability of ring-stiffened cylindrical shell

Xinyao Lv & Junjie Ruan
College of Shipbuilding Engineering, Harbin Engineering University, Harbin, China

Weijun Xu, Chenfeng Li & Xueqian Zhou
College of Shipbuilding Engineering, Harbin Engineering University, Harbin, China
International Joint Laboratory of Naval Architecture and Offshore Technology between Harbin Engineering University and University of Lisbon

ABSTRACT: Existing methods for improving the overall stability of ring-stiffened cylindrical shell are use of extra-large stiffeners and reduction of the length of compartment, etc. However, the overall stability of the ring-stiffened cylindrical shell of some dimensions cannot be effectively improved using these methods. This phenomenon is usually referred to as the abnormal characteristics. Therefore, it is significant to study the abnormal characteristics of pressure hull when the overall stability of the ring-stiffened cylindrical shell is being evaluated. In this paper, the instability equation of a ring-stiffened cylindrical shell under uniform pressure is presented based on the Ritz method. Then, the effect of improving the overall stability by adding intermediate ring-stiffener, deceasing stiffener-spacing and increasing plate thickness are discussed with the theoretical method. Finally, the feasibility of adding longitudinal and transverse stiffeners to improve the overall stability of cylindrical shell structure with abnormal characteristics and the influence of some related structural parameters on the overall stability are discussed.

1 INTRODUCTION

The dimensions of ring-stiffened cylindrical shell have kept increasing so as to install large equipment inside the compartment of submarines (Graham, 1991; Xie & Xu, 1994). Compared with ordinary ring-stiffened cylindrical shells, the ratio of radius to length of ring stiffened cylindrical shell with long compartment is relatively small, and thus the stability of long compartment shell structures has become more and more serious (Wang & Chen, 2010).

The stability of long cylindrical compartments has drawn the attention of many researches. Park et al (Park & Yim, 1993) analyzed the stability and optimized the mechanical performance of a ring stiffened cylindrical shell with one or two large stiffeners. It has been found that the difference between the numerical simulation value and the theoretical value gradually increases with increasing of the compartment length. When the length of compartment exceeds a certain value, the discrepancies of critical buckling pressure between the results obtained by the traditional stability theory and that by the numerical simulation are significant, which indicates the traditional stability theoretical formula is not suitable for evaluating the stability of a cylindrical shell with long compartment. Xie & Wu (2007) analyzed the abnormal characteristics of

long compartment and concluded that the half wave number of axial instability along the length direction of cylindrical shell is not equal to 1, i.e. m≠1 is the basis to determine whether the compartment is in abnormal characteristics condition. Jaunky et al (1998) proposed an optimal design strategy for composite grid stiffened columns subjected to global and local buckling constraints and strength constraints using discrete optimization algorithm. It was concluded that the results for grid-stiffened cylindrical shell subjected to axial compression indicated that, for simply supported cylinders that buckle globally, there is no significant difference in weight between optimal designs obtained with and without strain constraints. By adopting Donnell's thin shell theory and Vlasov's thin-walled beam theory, Wang & Lin (1973) investigated the stability characteristics of simply supported thin shells with stringers under axial compression, where the shell and stiffener were treated as discrete components respectively. Andrianov et al (2006) used the theory of structural orthotropic with momentless pre-buckling state to obtain an engineering approach to the computation of critical buckling pressure for a stringer-stiffened cylindrical shells, and some simple analytical expressions that governing non-axially symmetric pre-buckling state components for the cylindrical shell with stringer-stiffened were proposed. Seleim & Roorda

DOI: 10.1201/9781003216582-59

(1987) studied, theoretically and experimentally, the post-buckling behavior of ring-stiffened cylindrical shell under lateral pressure in which, the energy method and terms including fourth order of the buckling displacements were taken into account. Comparison of the numerical results with the experiment showed a fair accuracy. It was concluded that the post-buckling behavior are stable and symmetric for the shell and general instability modes. Das et al (2003) evaluated the buckling and ultimate strength of ring-stiffened shells, ring-stiffened shells and truss-stiffened shells under various buckling modes such as axial compression, radial pressure and combined load. Sobhaniaragh et al (2017) studied the mechanical buckling behavior of the continuously graded Carbon Nano-Tube (CNT)-reinforced shells stiffened by stringer and rings. By using a variation method based on a Third-order Shear Deformation Theory (TSDT), the equilibrium equations of the shells stiffened by stringer and rings were obtained. It has been found that the aggregation tendency of the CNTs has a negative effect on the mechanical buckling behavior of the continuously graded stiffened cylindrical shell. Hotała et al (2014) experimentally investigated the stability of stiffened cylindrical shell of steel silos. Models with short ribs topped with intermediate ring and without an intermediate ring had been studied. The authors believed that the use of longitudinal ribs of considerable length topped with intermediate ring has little influence on the local instability of cylindrical shell under compression, and the use of vertical ribs in the support zone of shells increases the global load capacity. Tian et al (1999) proposed a Ritz method for elastic buckling analysis of ring stiffened shells under general pressure. The general buckling solutions for shells under various end conditions, stiffener rib distributions and pressure distributions were presented. It was concluded that the appropriate distribution of ring stiffeners can lead to a significant increase of the critical buckling capacity of a ring-stiffened cylindrical shell. Radha & Rajagopalan (2006) investigated inelastic buckling failure of submarine pressure hull structures with the Johnson-Osterfield inelastic correction method, an imperfection approach and a finite element method, respectively. It was concluded that, for the preliminary examination of submarine pressure hull analysis, the imperfection approach is suitable for inelastic ring buckling failure; finite element analysis is for the rigorous analysis of the submarine pressure hull, and the Johnson–Ostenfeld correction for plasticity for the inelastic overall buckling failure. Do et al (2018) examined the effect of damages on the load carrying capacity of stringer-stiffened cylinders under external hydrostatic pressure, and intact and damaged models' hydrostatic pressure tests were performed to characterize the ultimate strength of stringer-stiffened cylinders. It was concluded that the effects of damage on the ultimate strength of stringer-stiffened cylinders were extremely low. Cho et al

(2018) experimentally investigated the failure modes of ring-stiffened cylinder models subjected to external hydrostatic pressure. The investigation identified that the failure modes of a ring-stiffened cylinder including shell yielding, local shell buckling between ring stiffeners, overall buckling of the shell together with the stiffeners, and interactive buckling mode combining local and overall buckling.

Although many researches are dedicated to cylindrical shell, the studies on the overall stability of a ring-stiffened cylindrical shell with long compartment is rare, especially the overall stability of the cylindrical shell with abnormal characteristics. This paper investigates the overall stability of ring-stiffened cylindrical shell with long compartment using the Ritz method and finite element method. The feasibility of adding longitudinal and transverse stiffeners to improve the overall stability of cylindrical shell structure with abnormal characteristics is discussed and the influence of various parameters on the critical pressure is explored.

2 ABNORMAL CHARACTERISTICS OF RING-STIFFENED CYLINDRICAL SHELL

According to the energy method for overall potential energy equation for ring-stiffened cylindrical shell (Shi & Wang, 1997), the overall instability equation of ring-stiffened cylindrical shell with uniform external pressure is

$$
\begin{aligned}
T_1 m^2 \alpha^2 & + T_2(n^2 - 1) = \frac{D}{R^2}(m^2\alpha^2 + n^2 - 1) \\
& + \frac{Etm^4\alpha^4}{(m^2\alpha^2 + n^2)^2} + \frac{EI}{R^2 l}(n^2 - 1)^2
\end{aligned}
\tag{1}
$$

where T_1 and T_2 are axial pressure along the length of cylindrical shell and circumferential pressure produced by the uniform external pressure; R, l and t are the radius, length and thickness of the ring-stiffened cylindrical shell respectively; D and E the flexural rigidity and Young modulus; m the half wave number of axial instability along the length direction of cylindrical shell, n the wave numbers of circumferential instability along the circumferential direction; I the moment of inertia of stiffener taking into account the strip plate; and α is a geometrical parameter:

$$
\alpha = \frac{\pi R}{L}
\tag{2}
$$

In order to analyze the influence of the geometrical properties of cylindrical shell with abnormal characteristics on the critical pressure, two kinds of external pressure are considered: $P^{(1)}$ and $P^{(2)}$, which correspond to the axial pressure and circumferential pressure. Then, it holds that

$$\begin{cases} T_1 = 0.5P^{(1)}R \\ T_2 = 0.5P^{(2)}R \end{cases} \qquad (3)$$

According to the equations above, the overall stability equation of ring-stiffened cylindrical shell with uniform external pressure can be obtained

$$0.5P^{(1)}m^2\alpha^2 + P^{(2)}(n^2 - 1) = \frac{D}{R^3}(m^2\alpha^2 + n^2 - 1)$$
$$+ \frac{Etm^4\alpha^4}{R(m^2\alpha^2 + n^2)^2} + \frac{EI}{R^3 l}(n^2 - 1)^2$$
$$(4)$$

For the ease of calculation and discussion, the following parameters are introduced,

$$\beta = \frac{10^6 I}{R^3 l} \qquad (5)$$

$$\gamma = \frac{100t}{R} \qquad (6)$$

$$D = \frac{Et^3}{12(1 - \mu^2)} \qquad (7)$$

where β denotes the influence of stiffness of ring stiffener on the overall critical buckling pressure; γ the influence of shell thickness on the overall critical buckling pressure; D the flexural rigidity of strip plate, and μ the Poisson's ratio of material.

The overall stability equation of ring-stiffened cylindrical shell under theoretical critical pressure P_E, under theoretical axial pressure $P_E^{(1)}$ and theoretical circumferential pressure $P_E^{(2)}$ can be written in the form of α, β and γ,

$$P_E = \frac{E \times 10^{-6}}{n^2 - 1 + 0.5m^2\alpha^2} \left[\frac{\gamma^3}{12(1 - \mu^2)}(m^2\alpha^2 + n^2 - 1)^2 \right.$$
$$\left. + \frac{\gamma m^4\alpha^4 \times 10^4}{(m^2\alpha^2 + n^2)^2} + \beta(n^2 - 1)^2 \right]$$
$$(8)$$

$$P_E^{(1)} = \frac{2E \times 10^{-6}}{m^2\alpha^2} \left[\frac{\gamma^3}{12(1 - \mu^2)}(m^2\alpha^2 + n^2 - 1)^2 \right.$$
$$\left. + \frac{\gamma m^4\alpha^4 \times 10^4}{(m^2\alpha^2 + n^2)^2} + \beta(n^2 - 1)^2 \right]$$
$$(9)$$

$$P_E^{(2)} = \frac{E \times 10^{-6}}{n^2 - 1} \left[\frac{\gamma^3}{12(1 - \mu^2)}(m^2\alpha^2 + n^2 - 1)^2 \right.$$
$$\left. + \frac{\gamma m^4\alpha^4 \times 10^4}{(m^2\alpha^2 + n^2)^2} + \beta(n^2 - 1)^2 \right]$$
$$(10)$$

From Eqs.(8)-(10), it can be found that the theoretical critical buckling pressures P_E, $P_E^{(1)}$ and $P_E^{(2)}$ are determined by the parameters α, β and γ, that is, the critical buckling pressures are related to the dimension parameters, such as length of compartment L, radius of cylindrical shell R, thickness of shell t, and moment of inertia of stiffener I. In order to explore the influence of R and L of cylindrical shell on its overall stability, the parameter of a compartment model γ is assumed to be 0.7 according to the study of Wang. et.al (1997), and β is assumed to as small as possible so as to neglect the influence of ring-stiffener on the overall stability.

For the case of β=0.4 and γ =0.7, α is varied, and the corresponding critical buckling pressure of the compartment structure is obtained and given in Table 1, and relationship between the critical buckling pressure of the compartment structure and α is plotted in Figure 1.

In order to explore the influence of circumferential stiffness of the long compartment structure with abnormal characteristics on its overall stability, the length of compartment structure is assumed to be very long so that the value of α is very small. For this reason, α is set to be 0.2, and the parameter β is varied for discussion.

For the case of α=0.2, γ=0.7, the parameter β is varied from 0.5 to 5.0 with an interval of 0.5, then the corresponding critical buckling pressures of the compartment are obtained and given in Table 2, and the relationship between the critical buckling

Table 1. Critical buckling pressures of the compartment structure for a variety of α.

α	P_E/(MPa)	$P_E^{(1)}$/(MPa)	$P_E^{(2)}$/(MPa)
0.5	4.07	12.46	4.24
1.0	8.12	12.48	8.62
1.5	12.23	12.48	14.22
2.0	12.26	12.47	16.67
2.5	12.32	12.48	22.86
3.0	12.23	12.48	25.55
3.5	12.23	12.48	27.14
4.0	12.29	12.48	29.92
4.5	12.32	12.48	32.96

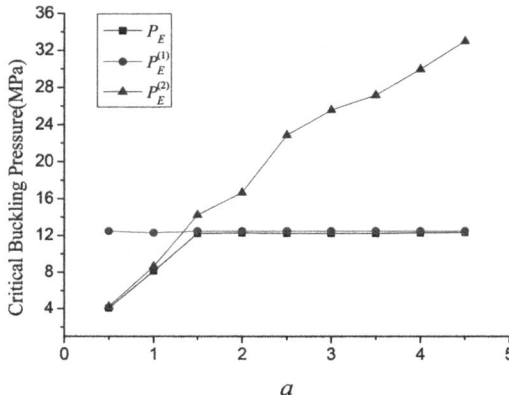

Figure 1. Relationship between the critical buckling pressure and a.

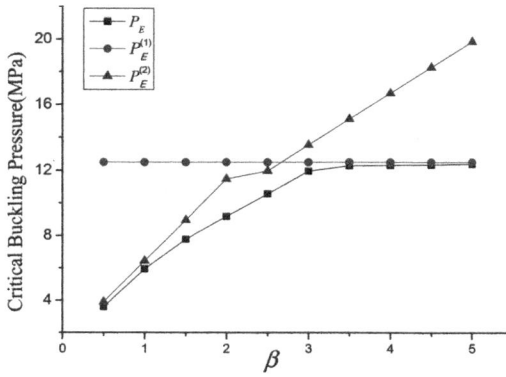

Figure 2. Relationship between the critical buckling pressures and β.

Table 2. Critical buckling pressures of the compartment for a variety of β.

β	P_E/(MPa)	$P_E^{(1)}$/(Mpa)	$P_E^{(2)}$/(Mpa)
0.5	3.60	12.48	3.90
1.0	5.93	12.48	6.42
1.5	7.76	12.48	8.94
2.0	9.16	12.48	11.46
2.5	10.55	12.48	11.95
3.0	11.94	12.48	13.53
3.5	12.26	12.48	15.10
4.0	12.29	12.48	16.68
4.5	12.32	12.48	18.25
5.0	12.36	12.48	19.83

pressures of the compartment and β is plotted in Figure 2.

According to Figure 1 and Figure 2, it can be found that the patterns of P_E, $P_E^{(1)}$, $P_E^{(2)}$ are very different characteristics.

The theoretical buckling pressure of the cylindrical shell $P_E^{(1)}$ barely changes with α or β, while $P_E^{(2)}$ increases monotonically with α or β. P_E increases with α and β until a certain value, named as α_i (for $i=1.5$) and β_j (for $j=4$), then it approaches to the value of $P_E^{(1)}$.

P_E is smaller than $P_E^{(1)}$ and $P_E^{(2)}$, thus the requirement for the overall stability for the ring-stiffened cylindrical shell is satisfied because the overall stability of the cylindrical shell is mainly determined by $P_E^{(2)}$ when $\alpha<\alpha_i$, $\beta<\beta_i$, while it is mainly restricted by $P_E^{(1)}$ when $\alpha>\alpha_i$, $\beta>\beta_i$ and P_E will no longer increase. The reason is that the overall critical buckling pressure is not only determined by the circumferential pressure or axial pressure. For the case of ring-stiffened cylindrical shell is subjected only to axial external pressure, decreasing the length of the compartment or increasing the stiffness of ring stiffener will not increase the longitudinal stiffness of the cylindrical shell structure, thus the $P_E^{(1)}$ will not vary with α and β; when the ring-stiffened cylindrical shell is subjected only to circumferential external pressure, decreasing the length of the compartment or increasing the stiffness of ring stiffener will increase the circumferential stiffness of the cylindrical shell structure, i.e. $P_E^{(2)}$ increases with α and β.

In general, the overall stability of pressured shell of submarine is improved by minimizing the space between ring stiffeners or increasing the stiffness of ring stiffener. This measure is effective only when $\alpha<\alpha_i$, $\beta<\beta_i$; that is to say, if $\alpha>\alpha_i$ or $\beta>\beta_i$, the overall stability of the ring-stiffened cylindrical shell cannot be improved by increasing the circumferential stiffness, but by enhancing the longitudinal stiffness. This phenomenon is regarded as abnormal characteristics of overall stability of the ring-stiffened cylindrical shell structures.

3 ABNORMAL CHARACTERISTICS OF THE CYLNDRICAL SHELL WITH INTERMEDIATE RING-STIFFENER

The overall stability formula of the cylindrical shell with intermediate ring-stiffener can be obtained by slightly modifying the general stability formula of ring-stiffened cylindrical shell (Eq.(6)). By replacing L and I/l with l and i/λ in Eq.(11)

$$P_E = \frac{1}{n^2 - 1 + 0.5m^2\alpha^2}\left[\frac{D}{R^3}(m^2\alpha^2 + n^2 - 1)\right.$$
$$\left. + \frac{Etm^4\alpha^4}{R(m^2\alpha^2 + n^2)^2} + \frac{E(I+i)}{R^3 l}(n^2 - 1)^2\right] \quad (11)$$

the following theoretical critical bucking pressure formula of the cylindrical shell with intermediate ring-stiffener can be obtained.

$$P_E = \frac{1}{n^2 - 1 + 0.5m^2\alpha_1^2}\left[\frac{D}{R^3}(m^2\alpha_1^2 + n^2 - 1)\right.$$
$$\left. + \frac{Etm^4\alpha_1^4}{R(m^2\alpha_1^2 + n^2)^2} + \frac{Ei}{R^3\lambda}(n^2 - 1)^2\right] \quad (12)$$

where $\alpha_1 = \pi R/l$ and $\lambda = 1/2$.

3.1 Influence of stiffener space on the theoretical critical pressure of ring-stiffened cylindrical shell

The structure parameters and the material properties of the calculation model are: length of compartment $L=13600$mm, radius $R=4100$mm, thickness of shell $t=35$mm, elastic modulus $E=2.0\times10^5$MPa, Possion's ratio $\mu=0.3$, yield strength $\sigma_s=800$MPa. The ring-stiffener is assumed to be T-shaped (T30C) and the intermediate ring-stiffener be the No.5 bulb-flat steel.

The stiffener space is varied from 500mm to 1000mm with an interval of 100mm, and the corresponding theoretical critical bucking pressure of the cylindrical shell and intermediate ring-stiffener are obtained and shown in Table 3 and Table 4.

The relationship between the overall critical buckling pressure of ring-stiffened cylindrical shell structure and stiffener space is plotted in Figure 3, and that between the critical buckling pressure of the intermediate ring-stiffener and stiffener space is plotted in Figure 4.

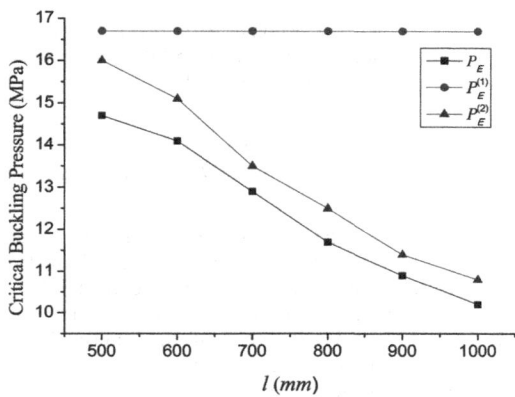

Figure 3. Relationship between the overall critical bucking pressures of the cylindrical shell and stiffener space l.

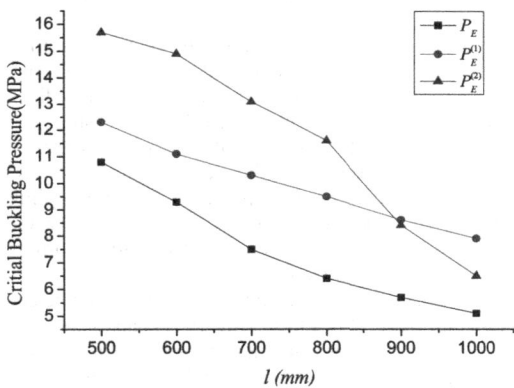

Figure 4. Relationship between critical buckling pressures of the intermediate ring-stiffener and stiffener space l.

It can be found from Figure 3 that the overall critical buckling pressure P_E decreases as the stiffener space increases, the changing trend always follows the relationship of $P_E<P_E^{(2)}<P_E^{(1)}$, and the overall critical buckling pressure P_E is mainly determined by $P_E^{(2)}$. This phenomenon is normal characteristics of ring-stiffened cylindrical shells. It is also found that the longitudinal stiffness of whole cylindrical shell is irrelevant to stiffener space, while the improvement of circumferential stiffness can be achieved by decreasing the stiffener space. It can be found from Figure 4 that the overall critical buckling pressure P_E is closer to $P_E^{(1)}$ when the length of stiffener space l is less than 900mm, and approaches $P_E^{(2)}$ when $l>900$mm. Abnormal characteristics of the intermediate ring-stiffener occurs when $l<900$mm. The phenomena indicate that, compared to the normal ring-stiffener in the same cylindrical shell structure, the intermediate ring-stiffener is prone to abnormal characteristics. The is

Table 3. Critical bucking pressures of the cylindrical shell for various stiffener spaces.

l/mm	500	600	700	800	900	1000
P_E/MPa	16.9	14.1	12.9	11.7	10.9	10.2
$P_E^{(1)}$/MPa	16.7	16.7	16.7	16.7	16.7	16.7
$P_E^{(2)}$/MPa	16.0	15.1	13.5	12.5	11.4	10.8

Table 4. Critical bucking pressures of the intermediate ring-stiffener for various stiffener spaces.

l/mm	500	600	700	800	900	1000
P_E/MPa	10.8	9.3	7.5	6.4	5.7	5.1
$P_E^{(1)}$/MPa	12.3	11.1	10.3	9.5	8.6	7.9
$P_E^{(2)}$/MPa	15.7	14.9	13.1	11.6	8.4	6.5

because L in Eq.(11) becomes l in Eq.(12), and as a result, α_1 is much larger than α, and it is closer and more prone to exceed the critical value.

3.2 Influence of shell thickness on the theoretical critical pressure of ring-stiffened cylindrical shell

For this analysis, the stiffener space l is set to be 600mm, and the thickness is varied from 35mm to 43mm with an interval of 2mm. The overall stability of the ring-stiffened cylindrical shell and the theoretical critical bucking pressure of the intermediate ring-stiffener are investigated, and the results are given in Table 5 and Table 6.

The relationship between the overall stability of ring-stiffened cylindrical shell structure and shell thickness is plotted in Figure 5, and the relationship between the critical buckling pressure of the intermediate ring-stiffener and shell thickness is plotted in Figure 6.

Figure 5 And Figure 6 show the overall stability of ring-stiffened cylindrical shell. It can be seen that the critical buckling pressure of the intermediate ring-stiffener can be improved by increasing the thickness of shell. Increase of thickness improves not only the circumferential stiffness but also the longitudinal stiffness. Compared to the same cylindrical shell structure but with normal ring-stiffener, the intermediate ring-stiffener is prone to abnormal characteristics because $P_E < P_E^{(1)} < P_E^{(2)}$.

The relationship of relevant structural parameters shown in Figure 7 (Park & Yim, 1993) can be used as a reference for the stability of intermediate ring-stiffener. For the ring-stiffened cylindrical shell with intermediate ring-stiffener, the parameters β and γ increase with the thickness of shell, which imposes an opposite effect on the critical value of structural

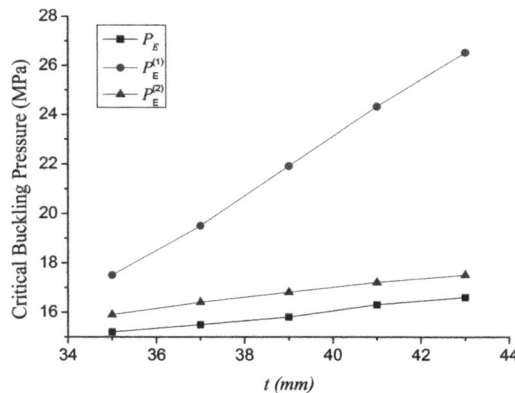

Figure 5. Relationship between the overall stability of ring-stiffened cylindrical shell structure and shell thickness t.

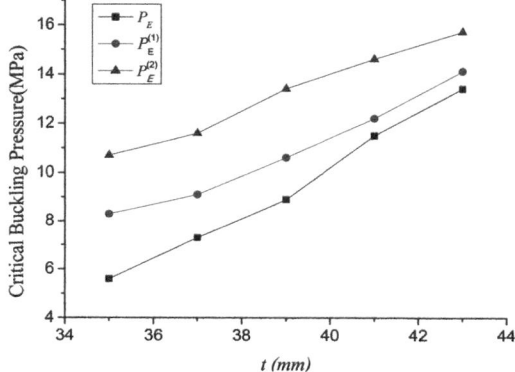

Figure 6. Relationship between the critical buckling pressure of the intermediate ring-stiffener and shell thickness t.

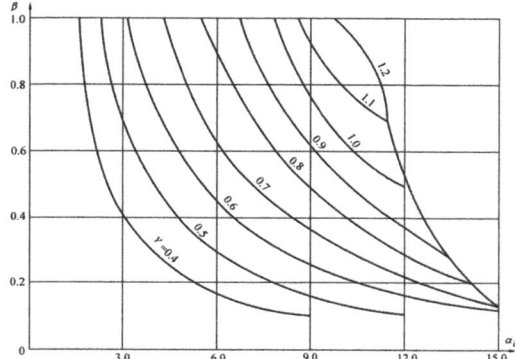

Figure 7. Relationship of structure parameters α_i, β and γ.

Table 5. Overall stability of the ring-stiffened cylindrical shell with various shell thickness t.

t/mm	35	37	39	41	43
P_E/MPa	15.2	15.5	15.8	16.3	16.6
$P_E^{(1)}$/MPa	17.5	19.5	21.9	24.3	26.5
$P_E^{(2)}$/MPa	15.9	16.4	16.8	17.2	17.5

Table 6. Critical bucking pressure of the intermediate ring-stiffener with various shell thickness t.

t/mm	35	37	39	41	43
P_E/MPa	5.6	7.3	8.9	11.5	13.4
$P_E^{(1)}$/MPa	8.3	9.1	10.6	12.2	14.1
$P_E^{(2)}$/MPa	10.7	11.6	13.4	14.6	15.7

parameter α_i. For the present model of cylindrical shell with a shell thickness of 35mm, the values obtained for α, β and γ of the intermediate ring-stiffener are 21.46,

0.337 and 0.853, respectively. As shown in Figure 7, when β=0.337 and γ =0.853, the corresponding critical value of structural parameter α_i can be found to be approximately 11.6, which is smaller than values obtained for α, thus abnormal characteristics of the intermediate ring-stiffener appears.

Ring-stiffened cylindrical shell structures with abnormal characteristics have a very serious problem of insufficient longitudinal stiffness, which can only be improved by thoroughly changing the form of the structure. Simply increasing the size of stiffener or adding extra-large stiffener will not improve the critical buckling pressure but will increase structure weight, it has nothing to do with increasing the theoretical critical buckling pressure. This means that whether the structure has abnormal characteristics must be investigated in the structure design of the intermediate ring-stiffener.

4 STABILITY ANALYSIS OF STRINGER AND RING-STIFFENED CYLINDRICAL SHELL

According to the study of Xie et al (2003), the formula for calculating the critical buckling pressure of the stringer and ring-stiffened cylindrical shell is

$$
P_E = \frac{1}{n^2 - 1 + 0.5m^2\alpha^2}\left[\frac{D}{R^3}(m^2\alpha^2 + n^2 - 1)\right.
$$
$$
\left. + \frac{Etm^4\alpha^4}{R(m^2\alpha^2 + n^2)^2} + \frac{EI}{R^3l}(n^2 - 1)^2 + \frac{EJ}{R^3b}m^4\alpha^4\right]
$$

(13)

where J is the moment of inertia of stringer, and it takes the form

$$
J = J_0 + \frac{b_e t^3}{12} + (y_{of} + \frac{t}{2})^2\frac{fb_e t}{f + b_e t}
$$

(14)

where J_0 is the moment of inertia of stringer itself; b_e is the width of attached strip plate, and b_e =b/6, where b is the space between two adjacent stringers; f is the cross section area of stringer; and y_{of} is the distance between the neutral axis of stringer and the surface of cylindrical shell.

4.1 *Determination of the calculated model for the cylindrical shell with abnormal characteristics*

In order to have an appropriate calculation model, the structural parameters of a large submarine are used to create the basic model, and to inspect whether the model has abnormal characteristics. The particulars of the model are as follows: length of compartment L=17000mm, radius of compartment R=4500mm, stiffener space l=600mm, thickness of shell t=33mm,

stiffness of ring-stiffener I=4.63×10^{4}cm^4, elasticity modulus E=2.0×10^{5}MPa, yield strength σ_s =785MPa, and Possion ratio μ=0.3.

A T-section ring-stiffener is adopted, and its panel width is 150mm, thickness of panel 24mm, web height 300mm, and web thickness 25mm.

Substituting α=0.83, β=8.473, γ=0.733 into the Eq. (13), then the critical buckling pressure P_E=13.49MPa (for m=25 and n=2). If α_1 is doubled while the other parameters remain unchanged, then the critical buckling pressure becomes P_E=13.61MPa (m=13, n=2); If β is doubled and the other parameters remain unchanged, then P_E=12.68MPa (m=25, n=2). It can be found from the above results that the change in the critical buckling pressure of the compartment structure is very small. Abnormal characteristics of the structure are present whether the spacing of stiffeners is reduced or the stiffness of stiffeners is increased. For this reason, a compartment length (L) of 13.2m is chosen for the model, while the other structure parameters remain unchanged. 2, 3 and 4 stringers of the same profile section are added to analyze their effects on the overall stability of the cylindrical shell structure.

4.2 *Influence of the number of stringers on the overall stability of stringer and ring-stiffened cylindrical shell with abnormal characteristics*

In order to study the effect of the number of stringers on the overall stability of ring-stiffened cylindrical shell, a parameters η=10^6J/R^3b is introduced, where J denotes the moment of inertia of longitudinal stringer, and b the arc length along the circumference between the stringers. Since the overall critical buckling pressure of ring-stiffened cylindrical shell with abnormal characteristics is determined by the critical buckling pressure of the axially compressed cylindrical shell, so the overall stability of the stringer and ring-stiffened cylindrical shell can be rewritten as

$$
P_E = \frac{E \times 10^{-6}}{0.5m^2\alpha^2}\left[\frac{\gamma^3}{12(1 - \mu^2)}(m^2\alpha^2 + n^2 - 1)^2\right.
$$
$$
\left. + \frac{\gamma m^4\alpha^4 \times 10^4}{(m^2\alpha^2 + n^2)^2} + \beta(n^2 - 1)^2 + \eta m^4\alpha^4\right]
$$

(15)

Then, the concept of equivalent thickness is introduced, which takes the form

$$
\bar{t} = t + \frac{F}{2\pi R}
$$

(16)

where F denotes cross section area of a single stringer, and m the number of stringers.

531

According to Eq. (15), the overall critical buckling pressure P_E of the cylindrical shell with two stringers is 20.32MPa (for m=2 and n=3), and the corresponding equivalent thickness due to two stringers is $\bar{t}$=33.8mm. If the improvement of overall critical buckling pressure of the cylindrical shell is only achieved by increasing the thickness of shell, then the corresponding thickness will reach t=39.8mm, which will result in an increase in the total weight of pressure hull. Thus the approach of adding stringers does not only effectively improve the overall stability of the pressure cylindrical shell, but also benefit, in comparison with increasing thickness of shell, the control of total weight of pressure hull.

Similarly, for the cylindrical shell with 3 and 4 stringers, the obtained critical buckling pressures of the cylindrical shell are P_E=20.35MPa (for m=2 and n=3) and P_E=20.38MPa (for m=2 and n=3), and the corresponding equivalent thickness are $\bar{t}$=34.2mm or $\bar{t}$=34.6mm respectively. It can be found that adding stringers can effectively increase the longitudinal stiffness of the cylindrical shell, and the overall critical buckling pressure is improved. However, the effects of different numbers of stringer on the overall stability of the cylindrical shell with abnormal characteristics seem to be almost the same. From the viewpoint of submarine design, for the pressured cylindrical shell structure, 4 stringers should be added. This is because the use of 4 stringers results in better stress distribution and symmetrical characteristics can be guaranteed. The geometric model and finite element models of the cylindrical shell with different numbers of stringers and transvers ring-stiffeners are shown in Figure 8.

The compartment finite element model is constructed using the finite element software package ABAQUS, and the shell element and quadrangular mesh type are adopted for the shell and stiffeners structures. The mesh is refined to eliminate the negative effect of grid on the calculation results. For the boundary conditions, the displacement for one end of the compartment is restricted in the x, y and z axis direction, and the other is restricted in x and y axis direction, while released in the z axis (along the length of the compartment). The rotations of two ends are not restricted.

In fact, whether the abnormal characteristics of a cylindrical shell exist is conditional. For a cylindrical shell with sufficient longitudinal stiffness, the circumferential stiffness may become insufficient as the diving depth increases. Thus, only increasing the longitudinal stiffness will not improve the overall stability of cylindrical shell, and the effective method is increasing the thickness of shell or dimension of ring-stiffeners. When the circumferential stiffness becomes more obvious instead of longitudinal stiffness, the abnormal characteristics of the cylindrical shell will disappear.

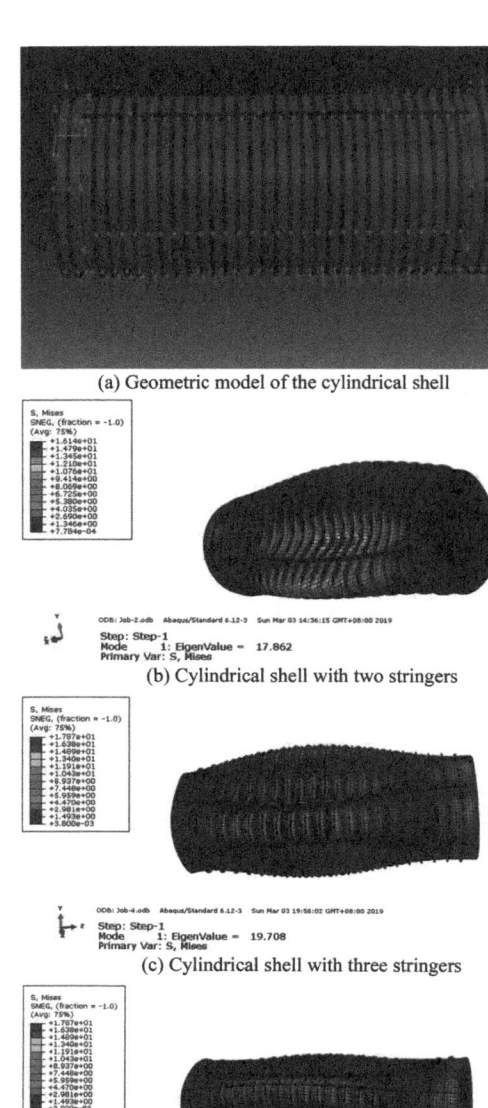

(a) Geometric model of the cylindrical shell

(b) Cylindrical shell with two stringers

(c) Cylindrical shell with three stringers

(d) Cylindrical shell with four stringers

Figure 8. Geometric model and finite element models of the cylindrical shell with different numbers of stringers and ring-stiffeners.

4.3 *Influence of structural parameters of the stringer and ring-stiffened cylindrical shell with abnormal characteristics on the overall stability*

In order to discuss the influence of structural parameters α, β, η on the instability pressure of a stringer and ring-stiffened cylindrical shell with abnormal characteristics, the critical buckling pressures calculated by using Eq.(15) are given in Table 7-Table 9. Corresponding half wave number of axial instability m and wave numbers of circumferential instability n are listed in the parentheses below the critical buckling pressure in tables.

In order to further discuss the influence of structural parameters on the critical buckling pressure of ring-stiffened cylindrical shell, the corresponding relationship curves are plotted in Figures 9 through 11.

From Table 7 and Figure 9, it can be found that the critical buckling pressure of the cylindrical shell with abnormal characteristics increase with the increasing of parameter η, which indicates that the critical buckling pressure can be improved by increasing the longitudinal stiffness.

When $\eta \leq 0.4$, the critical buckling pressure increases with η regardless the value of α, which indicates that stringers with relatively small stiffness make little contribution to the critical buckling pressure of the cylindrical shell. This is because, when a stringer with smaller longitudinal stiffness is attached to the cylindrical shell, the critical buckling pressure of overall compartment structure is mainly determined by the shell structure; however, when the longitudinal stiffness is relatively large, the critical buckling pressure of overall compartment structure is determined by the shell structure and stringer, and in this case the effect of stringer becomes more obvious. The effect of the adding stringers in improving the overall stability of cylindrical shell is better for short compartments than for long compartments. Thus it can be concluded that increase in longitudinal stiffness benefits the overall stability. It can also be found that the buckling mode of cylindrical shell changes with the longitudinal stiffness. The longitudinal buckling half wave number deceases with the increase of longitudinal stiffness.

It can be seen from Table 8 and Figure 10 that, for a cylindrical shell with abnormal characteristics, reducing the length of compartment structure or increasing the stiffener stiffness does not

Table 8. Critical buckling pressure of the stringer and ring-stiffened cylindrical shell with different values of a and β when $\eta=0.3$.

β \ α	2.0	2.5	3.0	3.5	4.0
2.5	37.897 (6,4)	37.96 (4,4)	37.897 (4,4)	37.616 (3,4)	37.897 (3,4)
3.0	38.175 (6,3)	38.41 (4,4)	38.175 (4,3)	38.024 (3,4)	38.175 (3,3)
3.5	38.264 (6,3)	38.602 (5,3)	38.264 (4,3)	38.432 (3,4)	38.264 (3,3)
4.0	38.353 (6,3)	38.684 (5,3)	38.353 (4,3)	38.733 (3,3)	38.353 (6,3)
5.0	38.531 (6,3)	38.848 (5,3)	38.351 (4,3)	38.965 (3,3)	38.531 (3,3)
6.0	38.709 (6,3)	39.011 (5,3)	38.709 (4,3)	39.197 (3,3)	38.709 (3,3)

Table 7. Critical buckling pressure of the stringer and ring-stiffened cylindrical shell with different values of α and η when $\beta=0.4$.

η \ α	2	3	4
0.1	23.358 (6,7)	23.372 (4,6)	23.372 (3,6)
0.2	28.063 (5,7)	28.058 (3,7)	28.414 (2,7)
0.3	30.974 (4,7)	31.298 (3,7)	30.974 (2,7)
0.4	33.447 (3,7)	34.887 (2,7)	33.534 (2,7)
0.5	34.887 (3,7)	34.887 (2,7)	36.094 (2,7)
0.6	35.795 (2,6)	36.327 (2,7)	38.654 (2,7)
0.7	36.435 (2,6)	37.767 (2,7)	41.214 (2,7)
0.8	37.075 (2,6)	39.207 (2,7)	43.774 (2,7)
0.9	37.715 (2,6)	40.647 (3,7)	46.334 (2,7)
1.0	38.355 (2,6)	42.087 (2,7)	48.894 (2,7)

Table 9. The critical buckling pressure of the stringer and ring-stiffened cylindrical shell with different values of β and η when $\alpha=0.2$.

η \ β	4	5	6
0.1	24.984 (74,2)	25.003 (75,2)	25.019 (75,2)
0.2	32.432 (63,3)	32.580 (64,3)	32.736 (64,3)
0.3	38.241 (57,3)	38.432 (58,3)	38.622 (58,3)
0.4	43.129 (53,3)	43.357 (53,3)	43.578 (54,3)
0.5	47.247 (48,4)	47.654 (50,3)	47.910 (50,3)
0.6	50.774 (46,3)	51.497 (48,3)	51.775 (48,3)
0.7	53.930 (43,3)	54.991 (46,3)	55.294 (46,3)
0.8	56.790 (41,3)	58.079 (42,4)	58.528 (44,3)
0.9	59.407 (40,3)	60.813 (40,4)	61.537 (43,3)
1.0	61.803 (38,3)	63.324 (39,4)	64.330 (41,3)

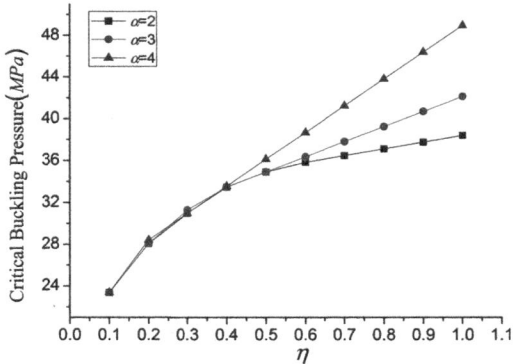

Figure 9. Relationship between the critical buckling pressure P_E and parameter η when β=0.4 for α=2.0, 3.0 and 4.0.

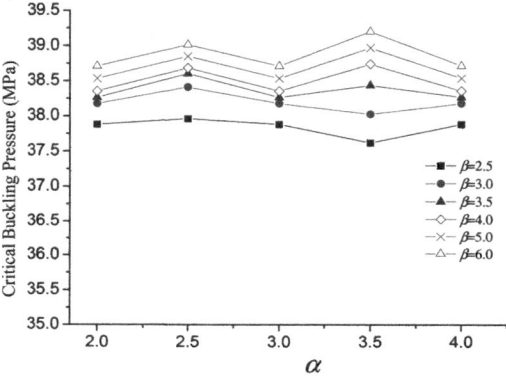

Figure 10. Relationship between the critical buckling pressure P_E and parameter α when η=0.3 for β=2.5, 3, 3.5, 4, 5 and 6.

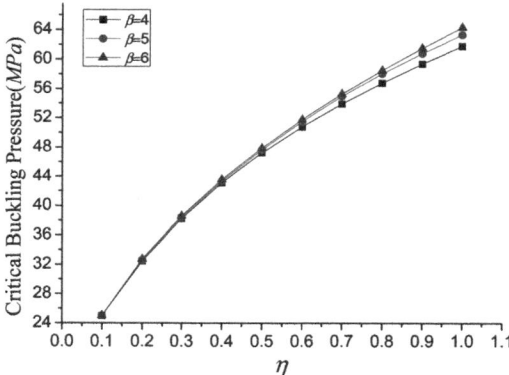

Figure 11. Relationship between the critical buckling pressure P_E and parameter η when α=0.2 for β=4.0, 5.0 and 6.0.

significantly improve the critical buckling pressure, i.e. these measures do not improve the overall stability of ring-stiffened cylindrical shell. It can be also found, however, by comparing Figure 10 with Figure 11, that increasing the longitudinal stiffness can effectively improve the overall stability.

5 CONCLUSIONS

The overall stability of ring-stiffened cylindrical shell with long compartment was investigated based on the Ritz method, then the effect of improving the overall stability by adding intermediate ring-stiffener, deceasing stiffener-spacing and increasing plate thickness were discussed in the theoretical method. Finally, the feasibility of adding stringer and ring stiffeners to improve the overall stability of cylindrical shell with abnormal characteristics was studied and discussed. According to the results and analysis, the following conclusions can be drawn:

(1) For a ring-stiffened cylindrical shell under uniform pressure, the overall stability of the cylindrical shell is generally improved with the increasing of circumferential stiffness, that is, the structural parameters that related to the circumferential stiffness. When the circumferential stiffness is enough to ensure the overall stability, while the longitudinal stiffness is insufficient, the cylindrical shell will still buckle, and the corresponding structure parameter is a critical value. Once the structural parameter exceeds the critical value, the buckling mode of the ring-stiffened cylindrical shell under uniform pressure is equivalent to that under pure axial external pressure, and abnormal characteristics phenomenon appears.

(2) Although the use of intermediate ring-stiffeners can improve the overall stability of the ring-stiffened cylindrical shell, however, the stability of the intermediate ring-stiffener itself cannot be neglected. In the structural de sign of the intermediate ring-stiffener, it is necessary to consider whether the structure has abnormal characteristics.

(3) The critical buckling pressure of compartment reinforced with a stringer of smaller stiffness is mainly determined by the shell structure; however, the critical buckling pressure is determined by both the shell structure and stringer if the stiffness of the stringer is large, and the effect of stringer becomes more obvious.

(4) Increasing of longitudinal stiffness benefits the overall stability of ring-stiffened cylindrical shell with abnormal characteristics; and the buckling mode of cylindrical shell changes with the longitudinal stiffness. Also, the longitudinal buckling half wave number deceases with the increase of longitudinal stiffness.

(5) According to the comparison between the analytical and numerical FEM results, it can be found that the error is about 13.77% when two stringers are attached, and the errors decease to 3.26% and 6.0% when 3 and 4 stringers are attached respectively. From the viewpoint of submarine design, for the pressure cylindrical shell structure with abnormal characteristics, stringers should be added to effectively increase the longitudinal stiffness of the cylindrical shell, the numbers should be controlled to reduce the weight, and the stringer should be symmetrical distributed to achieve stress homogenization.

ACKNOWLEDGEMENTS

This work is partially funded by National Natural Science Foundation of China (Grant Nos. 51679055 and 51779055).

REFERENCES

Andrianov I.V., Verbonol V.M., Awrejcewicz J. 2006. Buckling analysis of discretely stringer-stiffened cylindrical shells. *International Journal of Mechanical Sciences*, 48(12), pp.1505–1515.

Cho S. R., Muttaqie T., Do Q. T., Kim S., Kim S. M., Han D. H. 2018. Experimental investigations on the failure modes of ring-stiffened cylinders under external hydrostatic pressure. *International Journal of Naval Architecture and Ocean Engineering*, 10, pp. 711–729.

Das P.K., Thavalingam A., Bai Y. 2003. Buckling and ultimate strength criteria of stiffened shells under combined loading for reliability analysis. Thin-Walled Structures, 41(1), pp. 69–88.

Do Q. T., Muttaqiec T., Park S. H., Shin H.K., Cho S.R. 2018. Ultimate strength of intact and dented steel stringer-stiffened cylinders under hydrostatic pressure. *Thin-Walled Structures*, 132, pp. 442–460.

Graham D.D.1994. A structural research on submarines and submersibles. *Marine Structures*, 7, pp.213–256.

Hotała E.H., Skotny Ł. 2014. Experimental investigations on the stability of stiffened cylindrical shells of steel silos. *Journal of Constructional Steel Research*, 96, pp. 81–94.

Jaunky N., Knight Jr N.F., Amburb D. R. 1998. Optimal design of general stiffened composite circular cylinders for global buckling with strength constraints. *Composite Structures*, 41(3–4), pp.243–252.

Park C.M., Yim S.J. 1993.Ultimate strength analysis of ring-stiffened cylinders under hydrostatic pressure. *Proceedings of the 12th International Conference on Offshore Mechanics and Arctic Engineering*, Glasgow, June 20–24, pp.399–404.

Radha P., Rajagopalan K. 2006. Ultimate strength of submarine pressure hulls with failure governed by inelastic buckling. *Thin-Walled Structures*, 44(3), pp.309–313.

Seleim S.S., Roorda J. 1987. Theoretical and experimental results on the post-buckling of ring-stiffened cylinders. *Journal of Structural Mechanics*, 15(1), pp.69–87.

Shi D., Wang X. 1997. *Submarine Structures*. Harbin Engineering University Publishing. (in Chinese)

Sobhaniaragh B., Nejati M., Mansur W.J. 2017. Buckling modelling of ring and stringer stiffened cylindrical shells aggregated by graded CNTs. *Composites Part B*, 124, pp.120–133.

Tian J., Wang C.M., Swaddiwudhipong S. 1999. Elastic buckling analysis of ring-stiffened cylindrical shells under general pressure loading via the ritz method. *Thin-Walled Structures*, 35(1), pp.1–24.

Wang T.S., Lin Y.J. 1973. Stability of discretely stringer-stiffened cylindrical shells. *AIAA Journal*, 11 (6), pp.810–914.

Wang X., Chen J. 2010. A new method for stress calculation of ring stiffened cylindrical shells. *China Ship Research*, 5, pp.49–53.

Wang X., Deng J., Ning L. Xu J. 1997. A study of the stability features and buckling patterns of the ring–stiffened cylindrical shell. *Journal of Harbin Engineering University*, 18(3), pp. 13–21.(in Chinese)

Xie Z, Wu J. 2007. *Submarine Strength*. Harbin Engineering University Press. (in Chinese)

Xie Z., Wang Z., Wu J. 2003. *Submarine Structure Analysis*. Huazhong University of Science and Technology Press. (in Chinese)

Xie Z., Xu J. 1994. Overall stability of submarine thin-walled cylindrical shells with large radius. *China Shipbuilding*, 5, pp.82–88. (in Chinese)

Developments in Maritime Technology and Engineering – Guedes Soares & Santos (eds)
© 2021 Copyright the Author(s), ISBN 978-0-367-77376-2

Investigation on the performance of an OC3-Hywind Spar-type floating wind turbine impact by an offshore service vessel

Y. Zhang & Z. Hu
School of Engineering, Newcastle University, Newcastle upon Tyne, UK

ABSTRACT: This paper mainly focused on the global dynamic responses of an OC3-Hywind spar-type offshore floating wind turbine under the scenarios of ship impact. A theoretical collision model was developed and then it was combined with an in-house programme, DARwind, which was particularly used for fully coupled simulation of offshore floating wind turbines (OFWT), to predict the dynamic responses of OFWT after impact. With the newly developed hybrid approach, the time-domain results of OFWT in the ship-impact scenarios can be analyzed. Various cases with different impact velocities in still water were studied and it showed that responses of OFWT are sensitive to the impact velocity. To further investigate the dynamic responses of OFWT after ship collision in more complicated sea environment, wave-only condition and wave-wind combined condition were studied respectively and the ship collision in wave-wind combined condition was thought to be the most critical condition.

1 INTRODUCTION

1.1 *Potential ship-OWT collision hazard*

Offshore wind turbines have received increasing attention in the past decade as they are superior to the onshore wind turbines in many aspects, e.g. more consistent wind resources over the sea, less noise pollution and few space limitations. According to the global wind report (GWEC, 2019), the cumulative installed capacity of offshore wind farms has reached at 23GW in 2018.

At the same time, with the growing number of offshore wind farms, the potential ship collision hazard may rise accordingly. There were already some collision accidents that can be found in the Accident Statistics at Caithness Windfarm Information Forum (Caithness, 2019). On 23rd November 2012, five seamen got injured after their vessel impact the tower of a wind turbine at Sheringham shoal offshore wind farm, England. On 4th April 2014, a crew transfer vessel collided with a wind turbine at a high speed in Great Yarmouth offshore wind farm, England.

Additionally, some studies on the assessment of ship impact against OWT have been conducted to clarify the collision risk. Christensen et al. (2001) studied the collision frequencies for the wind farm and different factors, such as ship traffic, navigation routes, geometry of wind farms and bathymetry in the wind farm area were identified and analyzed. Biehl et al. (2006) combined the numerical simulations with statistical data and probabilities of occurrence of ship collision to evaluate the safety of wind farms. Dai et al. (2013) presented a comprehensive model on risk assessment. More scenarios and different types of ships were considered in their study. It was found that ship collisions against OWT even at low speed could result in structural damage OWT. Presencia et al. (2018) presented research of risk assessment particularly for maintenance vessel and OWT and different maintenance tasks, OWT components as well as ship types are analyzed. It was found that the collision risks are more likely to be accompanied by the maintenance of some certain components.

1.2 *Relevant studies on ship-OWT collision analysis*

The existing studies of ship collision against OWT mostly focused on the structural dynamic responses and non-linear finite element analysis (NLFEA) was the most popular method. A brief review regarding to this is presented as follow.

Biehl (2005) studied the collision between several common types of OWT and ships with LS-DYNA, a NLFEA tool. A series of detailed features were considered in the study, e.g. gravity and loads of turbine, soil interaction. Le Sourne & Barrera (2015) mainly analyzed the crashworthiness of a jacket-type OWT impact by a ship. The ship was modelled as both rigid and deformable body respectively to compare the results. Moulas et al. (2017) also adopted the NLFEA methods to study the collision between OWT and an offshore service vessel (OSV). The structural damage at the foundation of OWT in

DOI: 10.1201/9781003216582-60

different collision scenarios, including head-on collisions and side-way collisions, was evaluated. Bela et al. (2017) studied the structural responses of a monopile OWT impact by ships. The responses of rotor-nacelle assembly was studied as well to evaluate the safety of some equipment there. Pire et al. (2018) further derived a series of formulae to calculate the energy dissipation at the base of wind turbine jacket under ship collision.

More recently, Echeverry et al. (2019) studied the crashworthiness of a spar-type offshore wind turbine by ship impact with numerical methods. Hydrodynamic forces, ballast effects and mooring systems were all taken into consideration. Then the structural dynamic responses were more emphasized while the global motions were considered as well.

From the review, it is obvious that the ship collision against OFWT received less attention, especially the global responses. Unlike the bottom-fixed OWT, the floating platform brought more complicated loads. When impact by a ship, the OFWT would be more flexible and not response like a monopile or jacket OWT that only get serious local structural deformation. The impact energy would not only be dissipated due to plastic deformation, but also be transferred to the kinematic energy of OFWT, causing considerable translation and rotation motions. As a rigid-flexible coupled multi-body system, the global responses of various components are necessary to predict.

Currently, the global responses of OFWT under the scenarios of ship collision are difficult to predict due to the lack of dedicated analysis tool. In this study, a hybrid approach which combines a ship collision model for spar-type OFWT and a time-domain OFWT simulation tool together will be proposed. With the newly developed method, the global dynamic responses in the time domain of the whole OFWT system under ship-impact scenarios can be predicted.

2 THE WHOLE HYBRID APPROACH

The analysis approach mainly consists of two parts, a ship-OFWT impact dynamic model and a fully coupled aero-hydro-servo-elastic model for OFWT simulating after collision moment. In this section, both models will be described firstly and then the procedures that how to combine them together for analysis will be presented.

2.1 Ship-OFWT collision model

The ship-OFWT collision model presented in this sub-section is a further implementation of Stronge (2004) method and Liu (2010) method, so the basic assumptions in their studies are still applied here.

1) The collision duration is short. (Collision occurs and ends at the same moment and OFWT has no location change during collision)

2) The collision force is much larger than other forces. (All other forces are temporarily neglected during collision)

3) The deformations are limited to a small zone within the contact area. (The deformed element has negligible mass)

The striking ship has an initial velocity and impact the platform of OFWT at point C. Three coordinate systems are applied as shown in Figure 1. Two global coordinate systems are set at the COGs (centre of gravity) of the vessel and the OFWT when collision starts. A set of mutually perpendicular unit vectors n_1, n_2, n_3 are used and the vectors n_1, n_2 are in the common tangent plane at the collision point C while n_3 is normal to this plane. Then a so-called local coordinate system (n_1, n_2, n_3) can be obtained. The variables with a prime mark ' indicates the variables from ship, e.g. the direction vectors from COG of ship to collision point is represented by r' and r indicates the direction.

Let M and M' be the masses and I_{ij}, I_{ij}' be the inertia tensors for the second moments about the COGs of two bodies. We need note that the added mass and inertia factor are included in M, M', I_{ij}, I_{ij}' here to consider the effects of surrounding fluid. The added mass factor of the ship are estimated from empirical formulae (Popov et al., 1969) while the added mass factor of OC3 Spar-type OFWT are obtained from Jonkman (2010) report. V_i, V_i', ω_i, ω_i' denotes the velocities components of COG and angular velocities components. At the collision point C, the two bodies are undertaken mutual impact forces F and F_i' and the impulses to each body are written as

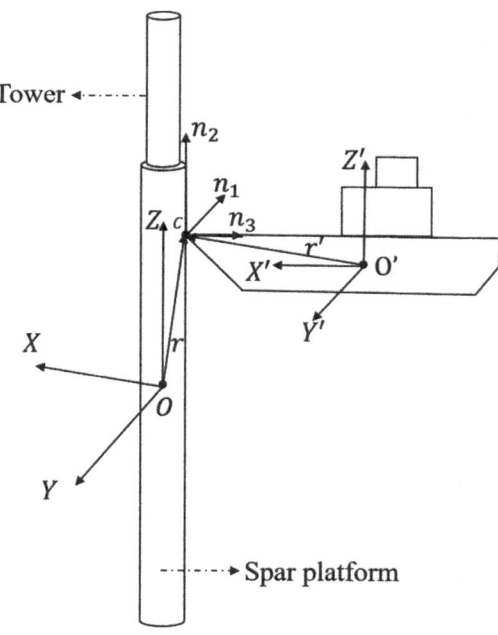

Figure 1. Illustration of coordinate system in the collision scenarios.

$$dP_i = F_i dt \tag{1}$$

$$dP_i' = F_i' dt \tag{2}$$

Then the equations for translation motion and rotational motion can be written as

$$M dV_i = F_i dt \tag{3}$$

$$I_{ij} d\omega_j = \varepsilon_{ijk} r_j dP_k \tag{4}$$

and

$$M' dV_i' = dP_i \tag{5}$$

$$I_{ij}' d\omega_j' = \varepsilon_{ijk} r_j' dP_k' \tag{6}$$

These equations are written in Einstein summation convention and the Levi-Civita symbol equals to 1 when the subscript indices are in cyclic order, or it equals to -1 for anticyclic indices or equals to 0 if there are any same indices

Next, the velocities of each body at the collision point C can be obtained as follows according to the COG velocity and the velocity relationship between two points on a rigid body

$$V_{ci} = V_i + \varepsilon_{ijk} \omega_j r_k \tag{7}$$

$$V_{ci}' = V_i' + \varepsilon_{ijk} \omega_j' r_k' \tag{8}$$

Then the relative velocity is expressed as

$$v_i = V_{ci} - V_{ci}' \tag{9}$$

As the deformation is assumed to be limited in a very small zone in the contact area, the deformed element has negligible mass. The incremental change in the reaction impulses is equal in magnitude but in converse direction. Thus, we have

$$dp_i = dP_i = -dP_i' \tag{10}$$

From all the above equations we get

$$dv_i = m_{ij}^{-1} dp_j \tag{11}$$

where

$$m_{ij}^{-1} = (M_{ij}^{-1} + M'_{ij}^{-1}) + \varepsilon_{ikm}\varepsilon_{jln}\left(I_{kl}^{-1} r_m r_n + I'_{kl}^{-1} r'_m r'_n\right) \tag{12}$$

The change of relative velocity can be expressed as

$$dv_i = v_i^t - v_i^0 \tag{13}$$

where v_i^t refers to relative velocity in $\mathbf{n_i}$ direction after collision and v_i^0 indicates relative velocity when collision occurs. If we ignore the possible repeated impacts and assume there's no sliding between OFWT and ship during the collision, then the relative velocity after collision can be expressed as

$$v_i^t = \begin{cases} 0 & i = 1 \\ 0 & i = 2 \\ -ev_3^0 & i = 3 \end{cases} \tag{14}$$

In the above equations, e is defined as coefficient of restitution and usually varies from 0 to 1. It takes the value of zero when the perfectly plastic collision occurs and equals to 1 in the pure elastic collision.

It is noticed that the above calculations are conducted in the local coordinate system ($\mathbf{n_1}$, $\mathbf{n_2}$, $\mathbf{n_3}$). Thus, the data in their global coordinate system must be transformed to the local system.

2.2 Full-coupled analysis tool for OFWT

DARwind is an integrated code based on coupled aero-hydro-servo-elastic methods for the simulation of OFWT developed by Chen and Hu. (2019). The main analysis procedure is presented as Figure 2.

In the code, hybrid coordinate dynamical analysis method (Likins, 1972) and cardan angle methods (Tupling & Pierrynowski, 1987) are adopted to describe the translation and rotation motions of OFWT. Bladed element momentum (BEM) theory (Hansen, 2015), as a very common and efficient method, is employed here to calculate the aerodynamic loads and some corrections are employed as well to provide more accurate results. As for the hydrodynamics module in the programme, potential-flow theory is mainly used and Morison's equation with the strip theory is applied at the same time to consider the viscous damping. Additionally, for some types of OFWT, such as semi-submersible floating wind turbine or TLP floating wind turbine, the second-order wave force are also considered. Then a quasi-static method (Dai et al, 2013) is utilized for the catenary mooring system

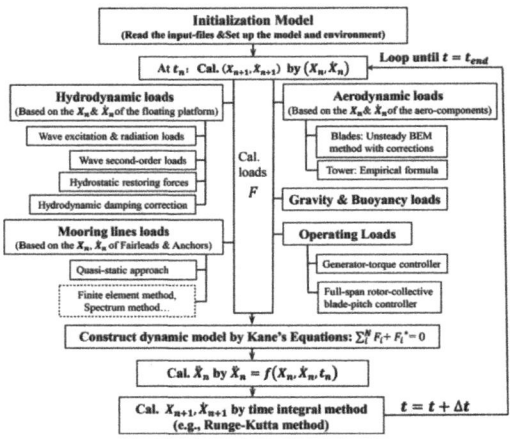

Figure 2. Analysis procedure of DARwind (Chen et al. 2019).

while a generator-torque controller as well as a full-span rotor-collective blade-pitch controller are included in control strategies. Kane's dynamic equation method (Kane & Levinson, 1983) is applied to calculate multi-body system and finally, with the Runge-kutta method, the time-domain results are obtained.

2.3 *Hybrid analysis approach*

An in-house programme is built based on the 3D collision model presented in section 2.1 and it is combined with the DARwind code to conduct the integrated analysis of OFWT under ship-collision scenarios. The procedure is illustrated as Figure 3.

Firstly, we need to use DARwind to conduct a certain period of simulation until collision occurs ($t=t_0$) and then we got the dynamic responses results from $0 - t_0$, shown as step 1 and 2.

At the moment of $t = t_0$, in order to finish the ship collision calculation, the collision analysis programme requires some input data, of which some can be extracted from the results at time t_0 (step 3) and some need to be prepared independently (step 4). Then the 6DOF platform velocities after collision can be output, shown as step 5.

Due to the assumptions of short collision duration, the collision is thought to occur and end at the same moment. If we distinct the moment before and after collision as t_0^- and t_0^+. Only the velocities of platform at t_0^+ are thought to be changed due to the ship impact while other characters of OFWT at t_0^+, such as platform location, aerodynamic loads, power, etc. all keep the same as those at t_0^-. Then, following step 6, a new input file for DARwind is formed. These input data are the same as those in the output file at t_0^- from step 2 except that the platform velocities are replaced with the results at $t = t_0^+$ output from collision analysis programme.

Finally, with the newly formed input file, another period of simulation, t_1, is conducted and the new results from t_0 to t_1 are calculated, shown as steps 7 and 8. The whole time-domain results from $0 - t_1$ are then obtained by integrating the results of $0 - t_0^-$ and $t_0^+ - t_1$, shown as step 9.

3 CASE STUDIES

A 5MW OC3 Hywind spar-type floating wind turbine (Jonkman et al. 2018, 2019) and an offshore

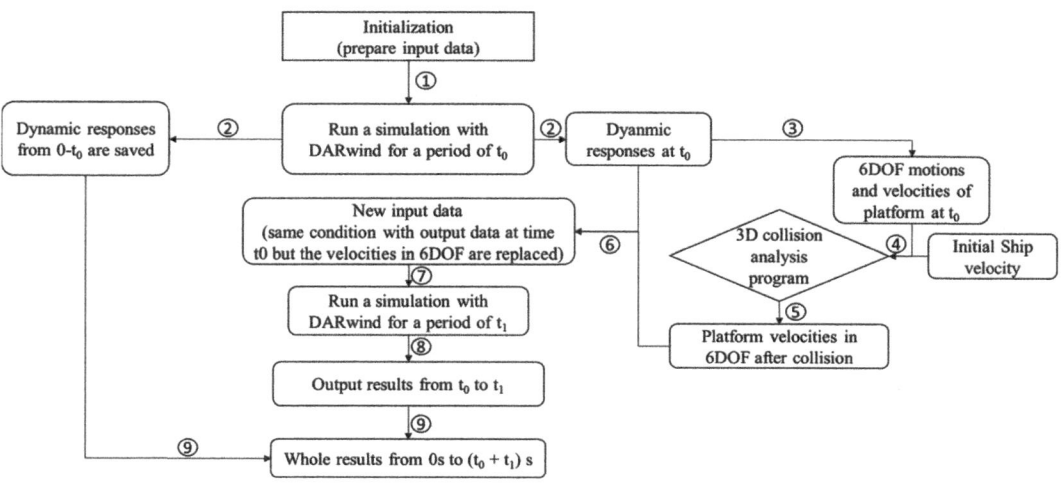

Figure 3. Hybrid analysis procedure.

service vessel are selected as an example, and the main properties of them are shown in Table 1 and Table 2. During all the collision cases, the coefficient of restitution e takes a value of 0.5, meaning the collision including both plastic and elastic deformation.

3.1 Influence of impact velocity

Different impact velocity implies varying impact energy, which are typically thought to be a key factor in the design stage. OFWT usually works in a wave-wind combined conditions and itself have uncertain initial velocities before ship collision. To investigate the influence of impact velocity, still water without wind were chosen to eliminate the initial motions of OFWT, meaning the OFWT keeps stable on the upright position in the still water when ship impacts it.

Head-on collisions are selected as they are thought to be the most dangerous scenarios in the ship collision accidents. The ship is given different impact velocities from 1m/s to 2.5m/s along the X axis of global coordinate system on OFWT. The most obvious motion responses should appear in surge and pitch directions, shown as Figure 4 and Figure 5.

The collision occurs and ends at the moment of $t = 0s$ and after ship impact, both motions in surge and pitch direction immediately have large increment and the whole trend shows a periodically varying pattern. This is due to the reason that after collision, the impact force will no longer act on the OFWT and it only stays at a non-wave and non-wind condition. Thus, it actually can be regarded as

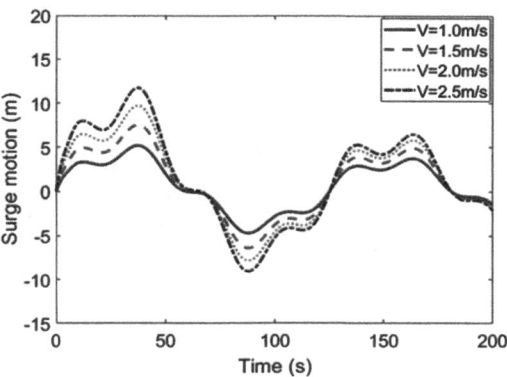

Figure 4. Surge motion after collision.

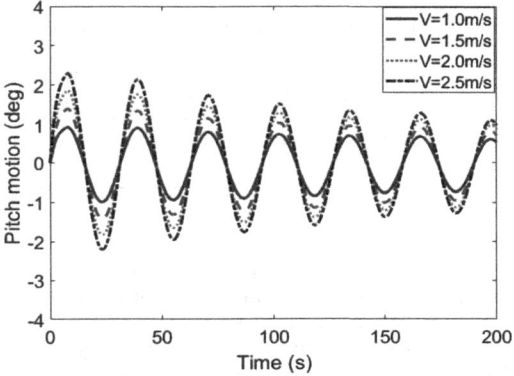

Figure 5. Pitch motion after collision.

a motion in still water with an initial velocity caused by the ship impact and finally released at the original upright position under the combination effects of mooring system and hydrostatic restoring force. Additionally, it is interesting to find that these motion responses of platform increase almost proportionally with the impact velocity rising and the maximum surge motion even reaches at 14m. As for the pitch motions, even in the scenario with the largest impact velocity, 3m/s, the maximum pitch motion after collision are still within 3 degrees, which is still in a normal range.

To further evaluate the safety of OFWT, the dynamic responses of nacelle (tower top) are investigated, the velocities and maximum accelerations are shown in Figure 6 and Table 3. It is quite interesting to find that the velocity of nacelle in surge direction has a lot of fluctuations and this is due to the elasticity of tower. As the OFWT is a typical rigid-flexible coupled multibody system, the vibrations of the flexible tower can be excited by the ship impact. Table 3 gave the maximum accelerations of nacelle in impact direction as theses data could help to assess the safety

Table 1. Main properties of OFWT.

Offshore floating wind turbine	
Total Draft	120 m
Total mass	8066048 kg
CM location below SWL	78 m
Water depth	320 m
Spar diameter (base)	9.4 m
Spar diameter (top)	6.5 m
Tower base above SWL	10 m
Tower height	90 m

Table 2. Main properties of OSV.

Offshore service vessel	
L.O.A	72.2 m
Beam	16 m
Depth	7.55/11 m
Draft	5.4 m
Displacement	4,000 ton

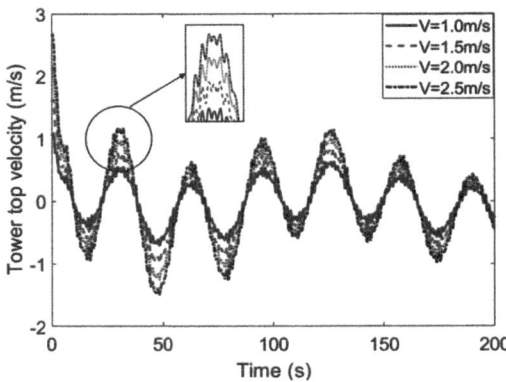

Figure 6. Velocity of nacelle in surge direction.

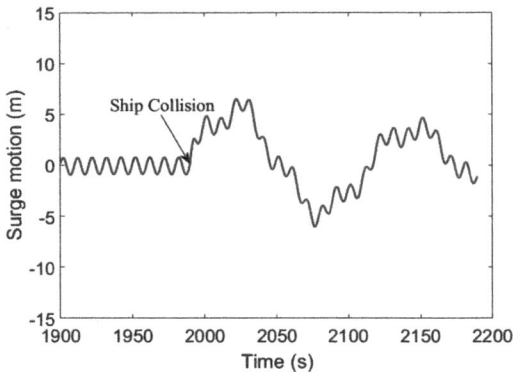

Figure 7. Surge motion responses in wave-only condition.

Table 3. Maximum acceleration of nacelle after ship collision.

Impact velocity (m/s)	1	1.5	2	2.5
Maximum acceleration (m/s^2)	0.584	0.721	0.864	1.011

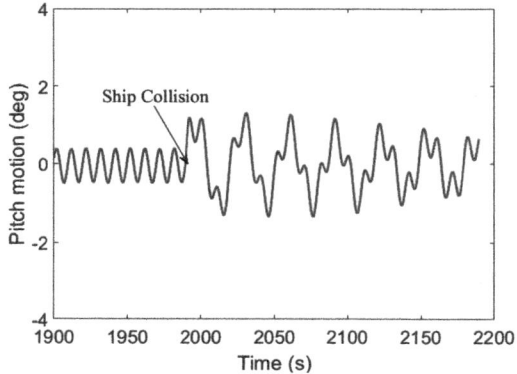

Figure 8. Pitch motion responses in wave-only condition.

of some equipment in the nacelle which is sensitive to the axis-direction acceleration. It is noticed that the maximum accelerations at tower top is less than 0.2g-0.3g, which is thought to be in a safe range (Bela 2017), where g is the gravitational acceleration.

3.2 Influence of environmental condition

Generally, OFWT works in a wave-wind combined condition, undertaking more complicated loads. How the OFWT responses after ship collision in different environmental conditions could have more practical implication.

Firstly, regular wave is applied, propagating along the positive direction of X axis in the global coordinate system of OFWT. The wave height as well as period are set as 4m, 10s and the initial velocity of ship is defined as 3m/s, along the wave propagation direction.

The whole time-series motion responses are shown in Figure 7 and Figure 8. It is noticed that OFWT are in the periodical oscillation under regular wave before collision and suddenly has obvious motion increment due to ship collision. Then the overall movement has a decay trend which is caused by the restoring effects of mooring system and water. It is noticed that the maximum value after collision are even smaller than that in still water. This can be explained by the wave effects and initial motions of OFWT. In wave condition, the OFWT already has initial velocity so the relative velocity at collision moment could be smaller, implying the collision extent may not be so serious.

Additionally, wave propagation could have offset effects on the motion as moving in the wave may consume more kinetic energy than that in still water.

Then, a typical rated wind speed 11.4m/s is chosen and other conditions keep the same. As additional load is introduced, the motions responses could be more complicated. To have a clear insight on how motions are influenced, both results with and without ship collision are compared in Figure 9.

Similar to the motions in wave-only condition, the motions in surge and pitch directions have some oscillations due to the wave effects. But the position that surge motion and pitch motion oscillate about are not at zero because of the wind load. After ship collision, the amplitude of surge motion and pitch motion have obvious growth but the amplitude of yaw motion decreases a little.

Indeed, the yaw motion of OFWT are mainly induced by the gyroscopic moments, which results from the coupling effects of rotational rotor and platform pitch motion. After collision, the obvious change of pitch motion and surge motion must result in the change of aerodynamic loads and change of

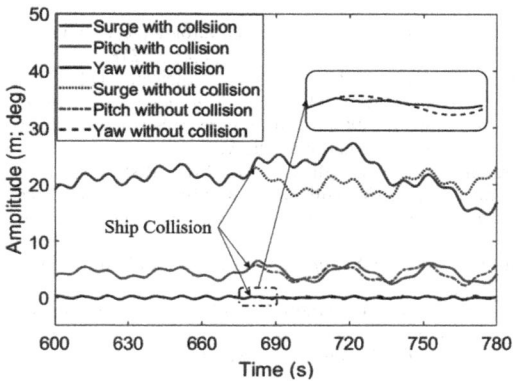

Figure 9. Comparison of motion responses in wave-wind conditions.

Table 4. Maximum accelerations of nacelle in different environmental conditions.

Impact scenarios	Wave only	Wave & wind
Maximum acceleration (m/s^2)	1.3065	14.0387

mooring loads as OFWT is a fully coupled system. Thus, the gyroscopic moment varies as well. Due to the combination of change of mooring responses and gyroscopic moment, the yaw motion changes accordingly.

Considering all the loads and structural coupled effects, the maximum accelerations of nacelle in these 2 cases are presented in Table 4. It is noticed that in the wind-wave combined condition plus ship impact, the maximum accelerations at tower top are extremely huge and much larger than the suggested limitation (Echeverry et al. 2019), 0.2-0.3g, where g is the gravitational acceleration. In this case, the equipment in the nacelle might fail to work but the detailed results cannot be exactly simulated in the current study. As for the wave-only condition, the maximum acceleration is thought to be acceptable.

4 CONCLUSIONS

This paper presented a theoretical model for ship-OFWT collision analysis and proposed an approach to combine it with an OWT simulation code, DARwind. The proposed hybrid approach is then used to conduct some case studies in different collision scenarios.

Impact velocity is found to have significant influence on the dynamic responses of spar-type OFWT. The maximum accelerations of nacelle in along ship-impact direction are all in safe range. The most dangerous collision scenarios should be in wave-wind conditions. The maximum surge motion is over 25m and the maximum accelerations at tower top are much larger (around 10 times larger) than that in still water and wave-only condition.

More cases and more detailed analysis on other components of OFWT such as tower, blade, mooring system, will be done in the future and these results may provide valuable information on the crashworthiness of OFWT in the design phase.

REFERENCES

Bela, A., Le Sourne, H., Buldgen, L. and Rigo, P., 2017. Ship collision analysis on offshore wind turbine monopile foundations. *Marine Structures, 51*, pp.220–241.

Biehl, F., 2005. Collision safety analysis of offshore wind turbines. In *4th LSDYNA European Conference*.

Caithness Windfarm Information Forum, 2019. available from http://www.caithnesswindfarms.co.uk/AccidentStatistics

Chen, J., Hu, Z., Liu, G. and Wan, D., 2019. Coupled aero-hydro-servo-elastic methods for floating wind turbines. *Renewable energy, 130*, pp.139–153.

Christensen, C.F., Andersen, L.W. and Pedersen, P.H., 2001. Ship collision risk for an offshore wind farm. In *Structural Safety and Reliability: Proceedings of the Eighth International Conference, ICOSSAR* (Vol. 1).

Dai, L., Ehlers, S., Rausand, M. and Utne, I.B., 2013. Risk of collision between service vessels and offshore wind turbines. *Reliability Engineering & System Safety, 109*, pp.18–31.

Echeverry, S., Márquez, L., Rigo, P. and Le Sourne, H., 2019, October. Numerical crashworthiness analysis of a spar floating offshore wind turbine impacted by a ship. In *Developments in the Collision and Grounding of Ships and Offshore Structures*, C. Guedes Saores (Eds), Taylor and Francis Group, pp. 85–95.

GWEC, G. W. E. C. 2019. Global wind report 2018. Brussels: GWEC.

Hansen, M.O., 2015. *Aerodynamics of wind turbines*. Routledge.

Jonkman, J., 2010. *Definition of the Floating System for Phase IV of OC3* (No. NREL/TP-500-47535). National Renewable Energy Lab. (NREL), Golden, CO (United States).

Jonkman, J., Butterfield, S., Musial, W. and Scott, G., 2009. *Definition of a 5-MW reference wind turbine for offshore system development* (No. NREL/TP-500-38060). National Renewable Energy Lab. (NREL), Golden, CO (United States).

Kane, T.R. and Levinson, D.A., 1983. The use of Kane's dynamical equations in robotics. *The International Journal of Robotics Research, 2*(3), pp.3–21.

Le Sourne, H., Barrera, A. and Maliakel, J.B., 2015. Numerical crashworthiness analysis of an offshore wind turbine jacket impacted by a ship. *Journal of Marine Science and Technology, 23*(5), pp.694–704.

Lehmann, E. and Biehl, F., 2004. Collision of ships and offshore wind turbines: calculation and risk evaluation. In *Third International Conference on Collision and Grounding of Ships (ICCGS)*. and Arctic Engineering (pp. 663–670). American Society of Mechanical Engineers.

Likins, P.W., 1972. Finite element appendage equations for hybrid coordinate dynamic analysis. *International Journal of Solids and Structures, 8*(5), pp.709–731.

Liu, Z. and Amdahl, J., 2010. A new formulation of the impact mechanics of ship collisions and its application to a ship–iceberg collision. *Marine Structures*, *23*(3), pp.360–384.

Masciola, M., Jonkman, J. and Robertson, A., 2013, June. Implementation of a multisegmented, quasi-static cable model. In *the Twenty-third International Offshore and Polar Engineering Conference*. International Society of Offshore and Polar Engineers.

Moulas, D., Shafiee, M. and Mehmanparast, A., 2017. Damage analysis of ship collisions with offshore wind turbine foundations. *Ocean Engineering*, *143*, pp.149–162.

Pire, T., Le Sourne, H., Echeverry, S. and Rigo, P., 2018. Analytical formulations to assess the energy dissipated at the base of an offshore wind turbine jacket impacted by a ship. *Marine Structures*, *59*, pp.192–218.

Popov, Y.N., Faddeev, O.V., Kheisin, D.E. and Yakovlev, A.A., 1969. *Strength of ships sailing in ice* (no. Fstc-ht-23-96-68). Army Foreign Science and Technology Center Charlottesville Va.

Presencia, C.E. and Shafiee, M., 2018. Risk analysis of maintenance ship collisions with offshore wind turbines. *International Journal of Sustainable Energy*, *37*(6), pp.576–596.

Stronge, W.J., 2018. *Impact mechanics*. Cambridge university press.

Tupling, S.J. and Pierrynowski, M.R., 1987. Use of cardan angles to locate rigid bodies in three-dimensional space. *Medical and Biological Engineering and computing*, *25* (5), pp.527–532.

Ship structures - Ultimate strength

Developments in Maritime Technology and Engineering – Guedes Soares & Santos (eds)
© 2021 Copyright the Author(s), ISBN 978-0-367-77376-2

Effect of pressure on collapse behavior of stiffened panel

L. Jiang
Lloyd's Register ATG, Halifax, Nova Scotia, Canada

S. Zhang
Lloyd's Register EMEA, Southampton, UK

ABSTRACT: During the operations of ships and offshore structures in the ocean environment, these structures are subjected to combined lateral pressure and in-plane stresses. However, in today's ship design and analysis procedures, the effects of the lateral pressure on the ultimate strength are often ignored. Previous studies have indicated that the lateral pressure could have a noticeable influence on the ultimate strength of stiffened panels subjected to combined longitudinal, transverse and shear stresses. The purpose of this paper is to present a systematic numerical study to quantify the lateral pressure effects on the ultimate strength of stiffened panels. The sensitivity of the panel's ultimate strength to lateral pressure is characterized as a function of the panel geometry, the pressure magnitude and the ratio of the in-plane stress components. The present numerical study is performed by using LR's in-house nonlinear finite element program VAST and by following the LR procedure for nonlinear structural mechanics analysis. The results and findings from this study are detailed in this paper.

1 INTRODUCTION

Stiffened panels are the basic building units and the primary load carrying components of ships and ship-shaped offshore structures. During the operations of these structures in the ocean environment, they are subjected to complex loading conditions, involving combined biaxial in-plane compression/tension, in-plane shear and lateral pressure. To ensure integrity of these structures under extreme loadings, ultimate limit states or ultimate strengths of the stiffened panels need to be evaluated.

It has been recognized that the ultimate limit states approach is superior to the traditional allowable stress approach as the latter is primarily based on linear elastic solutions (Paik & Seo, 2009a). In other words, the true margin of structural safety can only be determined after the ultimate strength of the structure is known. A comprehensive analytical treatment on nonlinear buckling behavior of plates and stiffened panels subjected to combined in-plane and pressure loads and various boundary conditions was recently published (Hughes & Paik, 2010).

In today's ship design practice, ultimate strength calculations are normally performed at two different levels, where the stiffeners are excluded and included, respectively. In analyses using both models, the ultimate strength assessments are normally performed by only considering the in-plane stresses, but the influence of the lateral pressure on the collapse is often ignored.

However, the validity of this approach has never been studied systematically.

To aid design of ships and ship-shaped structures, several semi-analytical methods have been proposed for quick ultimate strength assessments of stiffened panel structures (Zhang et al 2009, Byklum et al 2004, Paik & Lee, 2005). However, these methods normally contained approximations which lead to either over or under estimations of the ultimate strength (Paik et al, 2008), especially when lateral pressure is present. The nonlinear finite element method has been shown to be a powerful technique for predicting ultimate strength of ship structures under combined loads. The finite element method not only permits consideration of both large deflections/rotations and nonlinear material properties, but also allows the structural details to be modelled to the desired accuracy. Several finite element-based ultimate strength analyses of stiffened panels under combined in-plane stresses and lateral pressure have been published in the literature, such as Paik and Seo (2009b), Khedmati et al (2010) and Zhang (2013). However, more studies are considered necessary to draw meaningful conclusions on this topic.

This paper presents a systematic numerical study on the effects of the lateral pressure on nonlinear collapse characteristics of stiffened plates. The influences of various factors, such as the plate dimensions, the magnitude of pressure and the longitudinal to transverse stresses ratio were investigated by considering collapse of four representative ship panels. Reductions on the collapse stresses caused by lateral pressure were quantified.

DOI: 10.1201/9781003216582-61

2 STIFFENED PANEL DIMENSIONS

Four stiffened panel models were considered in the present study. These models were taken from the bottom of four different double hull oil tankers. The dimensions of these panels are summarized in Table 1 where a and b indicate the frame and stiffener spacings, respectively. The stiffener profiles are given in Figure 1, where all dimensions are in mm.

The material properties of steel were applied to all panel models. These included a Young's modulus of 205.800 GPa and a Poisson's ratio of 0.3. Panel L1 was fabricated using mild steel with an initial yield stress of 235 MPa, all other panels used high tensile steel with an initial yield stress of 315 MPa. A bilinear stress-strain curve was assumed with a constant hardening modulus of 1.0 GPa.

When ships and offshore structures operate in ocean waves, they are subjected to different levels of hydrostatic and hydrodynamic pressure. To investigate the sensitivity of stiffened panel's ultimate strength to the applied pressure level, a range of pressures were considered for each panel as indicated in Table 2, which represented different heights

Table 1. Panel dimensions and material yield stresses.

Model ID	Stiff. Type	a (mm)	b (mm)	σY (MPa)
L1	L	2630	715	235
L2	L	3264	841	315
T1	T	3700	830	315
T2	T	5700	850	315

Table 2. Lateral pressure levels (MPa).

Model ID	p0	p1	p2	p3	p4
L1	0.00	0.05	0.10	0.16	
L2	0.00	0.10	0.15	0.20	0.25
T1	0.00	0.07	0.12	0.20	
T2	0.00	0.10	0.20	0.30	

Table 3. In-plane load cases for collapse analyses.

Load Case ID	$\sigma_X : \sigma_Y : \tau_{XY}$
LC1	1.00: 0.00: 0.00
LC2	0.90: 0.10: 0.00
LC3	0.80: 0.20: 0.00
LC4	0.70: 0.30: 0.00
LC5	0.65: 0.35: 0.00
LC6	0.50: 0.50: 0.00
LC7	0.30: 0.70: 0.00
LC8	0.00: 1.00: 0.00

of the water head for parametric studies. The effects of pressure applied on either the plate side or the stiffener side were investigated. Please note that one meter of water head generates a pressure of 0.01 MPa.

To characterize the influence of the pressure load on nonlinear collapse behavior of ship panels under different combinations of in-plane stresses, eight load cases with a wide range of longitudinal to transverse stress ratios were considered in the present study. These load cases are summarized in Table 3.

3 FINITE ELEMENT MODEL AND ANALYSIS

The finite element models used in the present study contained four stiffeners, five stiffener spans and three frame spacings, including two complete and two half bays as indicated in Figure 2. The finite element meshes were constructed by following the procedure for nonlinear structural analyses developed by LR (Zhang & Jiang 2014, Jiang 2014), which recommended 8 elements between stiffeners, minimum 3 elements over web height and 1 or 2 elements in flange. The element size in the axial direction was determined to make the aspect ratio of the shell elements close to unity.

Boundary conditions on the longitudinal edge X=0 included constraints of the axial displacements (u_X=0) and the out-of-plane rotations (θ_Y=0). All the nodes on the other longitudinal edge X=3a was constrained to have zero out-of-plane rotation, but an identical displacement in the axial direction. The transverse and vertical displacements were not constrained on these edges. The boundary conditions were assumed

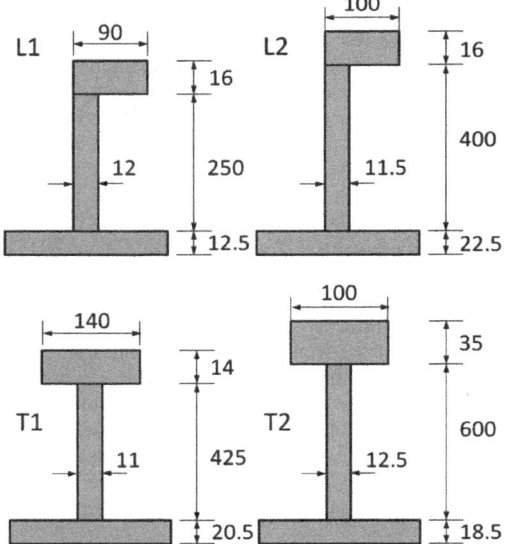

Figure 1. Stiffener profiles.

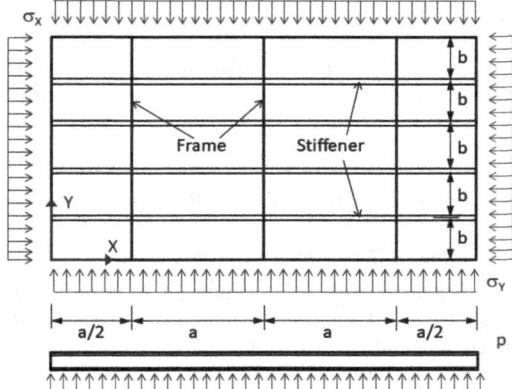

Figure 2. Panel model for finite element analysis.

at the transverse edges, Y=0 and Y=5b, for the out-of-plane deformations (u_Y=0, θ_X=0). In addition, multipoint constraint equations were applied to the in-plane displacements to enforce the transverse edges to remain straight and parallel to each other during large deformations. For more details on boundary conditions, please refer to Zhang and Jiang (2014).

In the finite element models, the transverse frames were not explicitly modelled, but their effects were represented approximately by applying constraints on vertical (Z) displacements along the frame locations as indicated in Figure 3. It should be noted that the effects of these approximations on the computed ultimate strength could be two-fold. On one hand, the application of the vertical (Z) constraints could result in an overrepresentation of the vertical bending stiffness of the frames. On the other hand, these simplifications on frames and stiffeners led to ignorance of the rotational restraints of the frames and stiffeners to the plates. So, its gross effect on the numerical solutions depended on the ratio of the bending and rotational stiffness of these structural members.

In the present nonlinear analyses, the lateral pressure was first applied incrementally to the desired level. This lateral pressure was then maintained while the combined in-plane stresses were applied up to and beyond the collapse point of the stiffened panels.

4 SENSITIVITY TO INITIAL IMPERFECTIONS

Proper definitions of initial geometric imperfections are crucial to nonlinear elastic-plastic collapse analyses. In the present study, the geometric imperfections were defined as the sum of a global imperfection and a local imperfection. The global imperfect-tion contained global bending and stiffener tripping modes with a maximum displacement of a/1000 as depicted in Figure 4.

The local imperfection was defined using the linear buckling mode shape obtained for the same set of load cases considered in the nonlinear collapse analysis. In these linear buckling analyses, additional boundary conditions were introduced to constrain vertical displacements along the longitudinal stiffeners to prevent global bending and stiffener tripping deformations. The magnitude of the local imperfection superimposed to the finite element model was b/200. Selected buckling modes for plate T1 are presented in Figure 5 where m denotes the number of half sine waves in the longitudinal direction between adjacent frames. These results indicated that under longitudinal stress (LC1), the m value is close to the aspect ratio (a/b) of the plate. However, when the transverse stress becomes more significant,

Figure 4. Global imperfection applied to the FE model of panel T1.

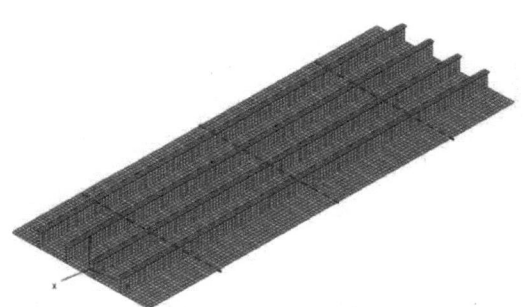

Figure 3. Finite element model for panel T1.

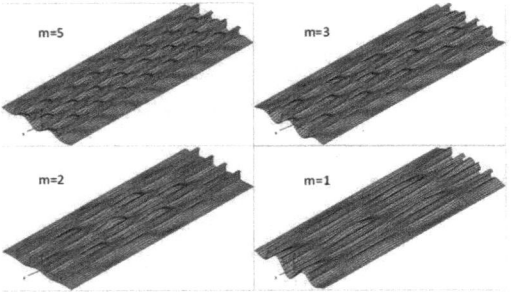

Figure 5. Local imperfections from buckling modes of panel T1 for load cases LC1, LC4, LC5 and LC8.

the m value continuously decreases and eventually becomes one for transverse stress dominated load cases (LC8).

To characterize the effect of the form of imperfections on the predicted nonlinear behavior, in this study, nonlinear analyses were performed for each stiffened panel model and each load case using three different linear buckling modes as the local imperfections. These included the modes for the uniaxial load cases (LC1 and LC8) and the present load case for which the nonlinear analysis was carried out (named 'current load case' in the paper).

Numerical solutions confirmed that the shape of the initial imperfections had a very significant effect on the nonlinear deformation and collapse behavior of the stiffened panels. Using the case of panel T1 with no pressure as an example, the ultimate stresses obtained for different in-plane load cases using various imperfections are given in Figure 6. These results indicated that the eight load cases considered in the present study can be divided into two groups, which are characterized by collapse under either longitudinal or transverse dominated deformations. For longitudinal dominated deformation cases, such as those for load cases LC1, LC2 and LC3, use of linear buckling mode for pure longitudinal stress LC1 resulted in the most conservative (lowest) ultimate stresses, whereas for panel collapses under transverse dominated deformations, such as those under load cases LC4 to LC8, the application of the buckling mode for pure transverse stress LC8 produced the minimum ultimate stresses. It is interesting to note that the linear buckling mode for the current load case does not normally produce the most conservative solution although it has been most commonly used to define geometric imperfections in nonlinear finite element collapse analyses of ship and offshore structures.

Similar trend on the sensitivity of imperfections on nonlinear collapse behavior were found for all plate and stiffened panel models under all lateral pressure levels. For each of these cases, the most

conservative interaction curve was constructed by taking the minimum ultimate stress values as indicated in the Figure 6. The effect of the pressure was then evaluated by comparing these minimum ultimate stresses.

5 RESULTS AND DISCUSSIONS

5.1 *Panel collapse under longitudinal stress*

The load-shortening curves obtained from stiffened panel model T2 subjected to pressure and collapsed under longitudinal stress (LC1) are presented in Figure 7 in terms of normalized axial stress and strain. These results indicated that the applications of lateral pressure on the plate side or the stiffener side reduced the ultimate load carrying capacity of all four stiffened panels. The influence of lateral pressure on the ultimate stresses of the four stiffened panels is summarized in Table 4. These levels of reduction on the ultimate strength predicted in the present work are consistent with those reported in the literature, such as Paik and Seo (2009b).

The results shown in Table 4 suggested that the lateral pressure applied to the plate side or stiffener side seemed to cause similar level of reductions on the ultimate stresses. However, this level of reduction was

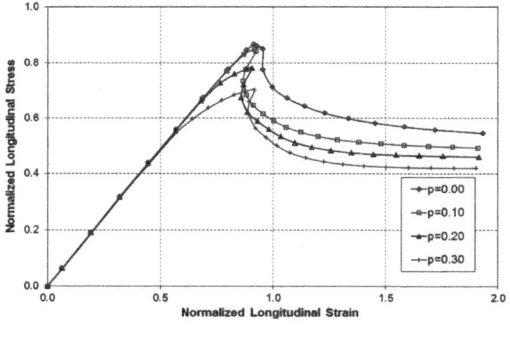

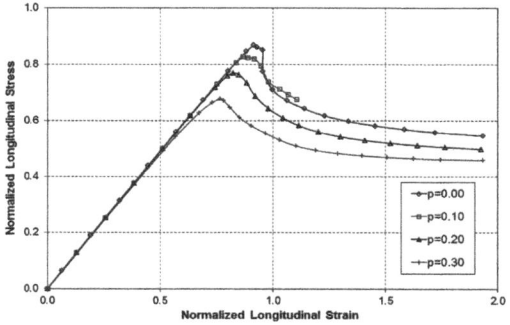

Figure 7. Normalized load-shortening curves of panel T2 under longitudinal stress with pressure applied on the plate (top) and stiffener (bottom) side.

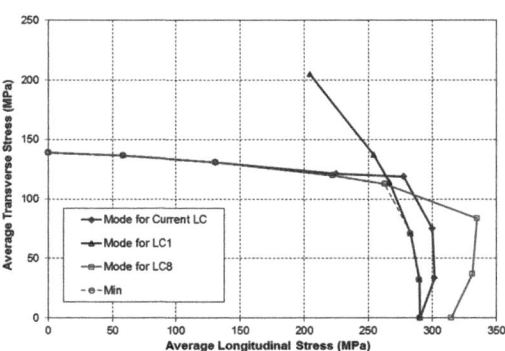

Figure 6. Interaction curves predicted using different local imperfections for panel T1.

550

Table 4. Comparison of ultimate stresses without and with lateral pressure.

		Plate Side		Stiffener Side	
Model ID	Pressure (MPa)	Ultimate Stress (MPa)	Reduction (%)	Ultimate Stress (MPa)	Reduction (%)
L1	0.00	205.28		205.28	
L1	0.05	199.15	-2.99	200.71	-2.23
L1	0.10	189.36	-7.76	193.24	-5.87
L1	0.16	176.46	-14.04	182.30	-11.19
L2	0.00	292.77		292.77	
L2	0.10	287.55	-1.78	285.90	-2.35
L2	0.15	281.77	-3.76	280.99	-4.02
L2	0.20	275.37	-5.94	274.75	-6.16
L2	0.25	268.43	-8.31	267.53	-8.62
T1	0.00	292.35		292.35	
T1	0.07	287.67	-1.60	287.74	-1.58
T1	0.12	280.93	-3.91	282.42	-3.40
T1	0.20	269.56	-7.80	271.23	-7.22
T2	0.00	273.37		273.37	
T2	0.10	267.76	-2.05	260.58	-4.68
T2	0.20	246.10	-9.98	242.23	-11.39
T2	0.30	220.80	-19.23	213.05	-22.07

heavily dependent upon the geometric and material properties of the stiffened panel. For instance, among the four panels considered in the present study, panels L1 and T4 were more sensitive to the lateral pressure.

This was because that panel L1 had weaker stiffeners and a lower yield strength, whereas panel T2 had a longer frame spacing compared to other stiffened panels.

The present numerical study also indicated that the lateral pressure had significant effects on the deformation patterns of the stiffened panels and the stress distributions in them. By examining the stress distributions, it was noted that the lateral pressure applied to the plate side caused the stiffeners to deform in a clamped-clamped mode. On the other hand, when the pressure was applied to the stiffener side, the stiffeners deformed in a simply-supported mode.

In addition to altering the stress patterns in stiffeners, the application of the lateral pressure also changed that shape of initial imperfection. If the lateral pressure was not applied, the deformation of the panel was controlled by the shape and magnitude of the initial imperfections. However, with the increase of the pressure level, the panel deformation and the final collapse mode were more and more dominated by the deformation pattern caused by the pressure.

Another interesting finding was that the direction of pressure application had a significant influence on the final collapse mode of the stiffeners. The pressure applied on the plate side seemed to cause severe local deformations at the stiffener-frame intersections. On the other hand, pressure applied on the stiffener side promoted

large stiffener deformations in the mid-span of the frame spacings. These final collapse patterns were consistent with the concentrated plastic deformations discussed before. In addition, it should be noted that the lateral pressure applied on both plate and stiffener side promoted collapse of stiffeners, but with different mechanisms.

The lateral pressure also had a significant influence on the final collapse modes of the plates. For all four stiffened panels considered in the present study, the plate always failed by formation of plastic hinges. However, the locations of the plastic hinges were heavily influenced by the pressure application. When the lateral pressure was applied on the plate side, the plastic hinge was consistently formed in the mid-span between transverse frames. However, application of the lateral pressure on the stiffener side promoted the plastic hinge(s) to be formed at the transverse frame locations instead.

5.2 Panel collapse under bi-axial stresses

In this section, we consider nonlinear collapse behavior of stiffened panels under bi-axial in-plane stresses and quantify the effect of lateral pressure on the ultimate strength of these panels. For each panel model and each in-plain stress combination, three nonlinear runs were carried out using different linear buckling modes as the local geometric imperfections and the minimum ultimate stress was then selected. These includes the buckling modes obtained from the purely longitudinal (LC1), purely transverse (LC8) stresses and the current load case for each the nonlinear analysis was performed. Due to the page limit, only representative results are presented here.

Figure 8 compares load-shortening curves obtained for panel model T1 under two bi-axial in-plane load cases, LC3 and LC4, with various levels of lateral pressure applied on the plate side. As indicated, the applied pressure had very little influence on the pre-collapse stiffness at the early stage of the loading process. However, increased pressure caused early appearance of the stress softening behavior, such as a lower peak stress and a softer post-collapse response. These sets of load-shortening curves were obtained using the linear buckling modes for pure longitudinal (LC1) and pure transverse (LC8) stresses as initial imperfections, respectively as these imperfections resulted in the lowest ultimate strengths among all nonlinear solutions using different local imperfections. From the load-shortening curves presented in Figure 8, we realized that the influence of the pressure was considerably less significant for panels that collapsed under transverse-dominated deformation mode. It is interesting to note that although LC4 was dominated by the longitudinal stress (σ_X:σ_Y=0.7:0.3), the stiffened panel

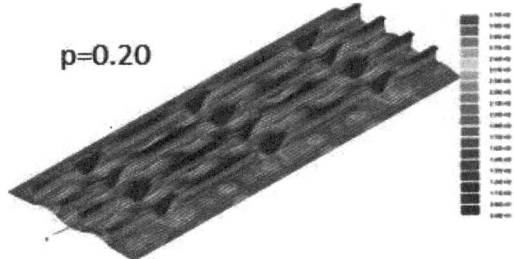

Figure 9. Deformed shape and VM stresses of panel T1 at ultimate state under LC3 with lateral pressure (p=0.20 MPa) applied on plate side.

collapsed into a transverse-dominated mode. Detailed reviews of the stress distributions indicated that for both collapse modes, the application of the lateral pressure on the plate side enhanced plastic deformations at the stiffener-frame intersections, and thus trigger stiffener failure in a clamped-clamped mode (Figure 9). This is consistent with the panels under purely longitudinal stresses as considered before. The influence of lateral pressure applied on the stiffener side on the load-shortening behavior of stiffened panels was also considered in the

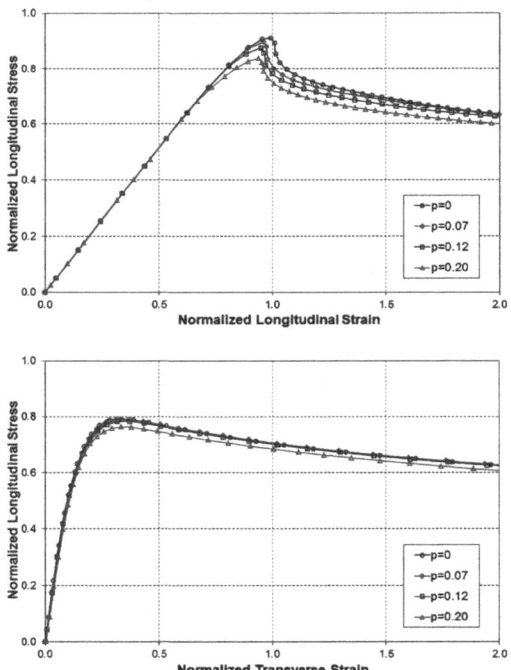

Figure 8. Load-shortening curves of panel T1 under LC3 (top) and LC4 (bottom) with lateral pressure applied on the plate side.

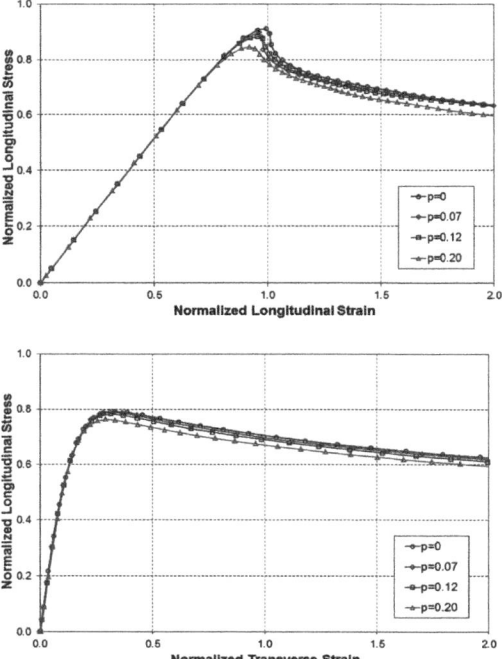

Figure 10. Load-shortening curves of panel T1 under LC3 (top) and LC4 (bottom) with lateral pressure applied on the stiffener side.

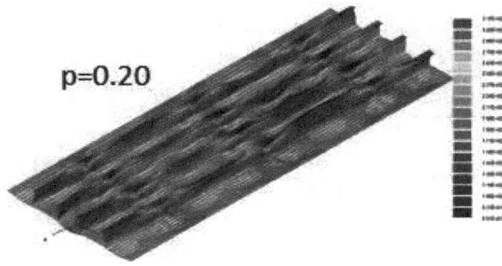

p=0.20

Figure 11. Deformed shape and VM stresses of panel T1 at ultimate state under LC3 with lateral pressure (p=0.20 MPa) applied on stiffener side.

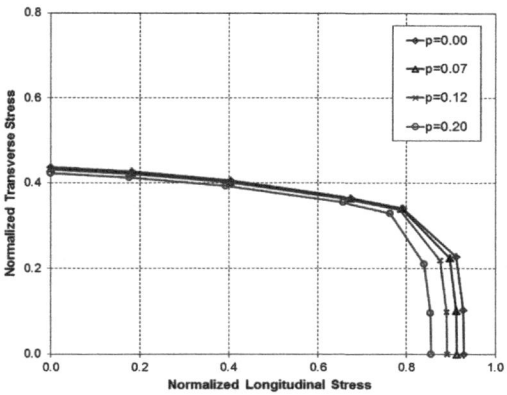

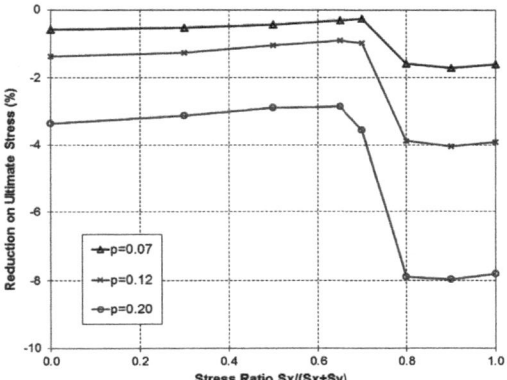

Figure 12. Normalized interaction curves and reduction on ultimate strength of panel T1 under different levels of lateral pressure on the plate side.

present study. The results presented in Figure 10 are for panel T1 subjected to two typical cases: LC3 and LC4. For these two load cases, use of buckling modes for load cases LC1 and LC8 were found to produce the lowest ultimate stresses, respectively. These results indicated pressure applications on both the plate and stiffener sides have similar influence on the collapse behavior of panels under combined in-plane stresses. However, the extent of the influence of the pressure on the stiffener side was slightly less than that on the plate side. A close examination of the deformed configurations and stress fields at the ultimate states (as shown in Figure 11) indicated that the application of pressure on the stiffener side consistently produced concentrated plastic deformations in the mid-span of stiffeners between frames, but less plastic strains at the stiffener-frame intersections. This is again consistent with the collapse behavior of stiffened panels subjected to pure longitudinal stress.

5.3 Interaction curves

Once the ultimate stresses were determined for each load case and pressure level, interaction curves could be generated as shown in Figure 12 and 13, where the influence of pressure in the ultimate strengths could be readily quantified. To better represent the reduction of ultimate stresses due to pressure, the percentage reductions are summarized in Tables 5 and 6.

From all four sets of numerical results presented in this section, we noticed that the percentage reductions of the ultimate stress always contain two distinct branches: one for longitudinal-dominated collapse modes and one transverse-dominated collapse modes. For the longitudinal-dominated branches, reductions of 4-8% on the ultimate stress were observed, whereas for the same panels, the highest reductions for the transverse-dominated branches were considerably less.

6 CONCLUSIONS

This paper presented a numerical study on the influence of lateral pressure on the collapse behavior of stiffened panels subjected to bi-axial in-plain compressive stresses using nonlinear finite element program VAST. Steel panel configurations from the bottoms of four different double-hull oil tankers were utilized in this investigation. From the numerical results, the following conclusions can be drawn:

- The present results indicated that the application of lateral pressure can significantly reduce the ultimate stress of a stiffened panel under axial compression. For instance, for a 30 m water head, the ultimate strength of the stiffened panel can be reduced by as much as 20%.
- The lateral pressures applied on the plate and stiffener sides cause similar levels of reductions on ultimate strength. However, the direction of pressure application has a noticeable effect on the post-collapse behavior indicated in the load-

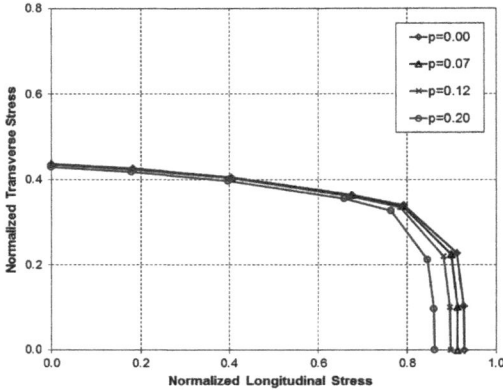

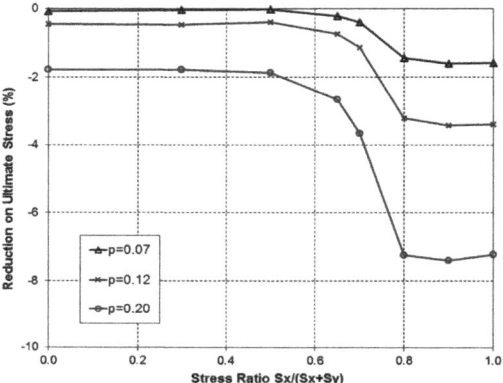

Figure 13. Normalized interaction curves and reduction on ultimate strength of panel T1 under different levels of lateral pressure on the stiffener side.

Table 5. Reduction (%) on ultimate stress of panel T1 due to lateral pressure application on the plate side.

Load Case	p=0.15	p=0.20	p=0.25
LC1	-1.60	-3.91	-7.80
LC2	-1.70	-4.04	-7.97
LC3	-1.58	-3.88	-7.89
LC4	-0.25	-0.97	-3.55
LC5	-0.31	-0.90	-2.86
LC6	-0.43	-1.05	-2.89
LC7	-0.52	-1.27	-3.13
LC8	-0.58	-1.36	-3.36

shortening curves. The pressure on the plate side results in an abrupt decrease on the load-carrying capacity right after the peak load is passed, however the pressure on the stiffener side leads to a more gradual reduction of the residual strength.
• The direction of pressure application has a significant influence on the deformation

Table 6. Reduction (%) on ultimate stress of panel T1 due to lateral pressure application on the stiffener side.

Load Case	p=0.15	p=0.20	p=0.25
LC1	-1.58	-3.40	-7.23
LC2	-1.59	-3.43	-7.41
LC3	-1.44	-3.21	-7.24
LC4	-0.40	-1.13	-3.66
LC5	-0.22	-0.75	-2.67
LC6	-0.03	-0.40	-1.87
LC7	-0.04	-0.47	-1.78
LC8	-0.08	-0.46	-1.78

patterns and stress fields in the stiffeners. The pressure on the plate side causes concentrated plastic deformations at the stiffener-frame intersections, but the pressure on the stiffener side results in large plastic deformations in the mid-span between transverse frames. These deformation patterns may be referred as the clamped-clamped mode and the simply-supported mode, respectively.
• The direction of pressure application also affects the location of the plastic hinges formed in the final collapse mode of the plate. The present results indicated that with pressure on the plate side, the plastic hinge was always formed in the mid-span of the transverse frame spacings, but with pressure on the stiffener side, the plastic hinges are normally formed along the transverse frames.
• The present results indicated that it is essential to consider the ultimate strength assessment of the grillages in practical ship design.

REFERENCES

Bathe, K.J. & Dvorkin, E.N. 1985. A four-node plate bending element based on Mindlin/Reissner plate theory and a mixed interpolation. *International Journal for Numerical Methods in Engineering* 21: 367–383.
Byklum, E., Steen, E. & Amdahl, J. 2004. A semi-analytical model for global buckling and post-buckling analysis of stiffened panels. *Thin-Walled Structures* 42: 701–717.
Hu, S.Z. & Jiang, L. 1998. Finite element simulation of the test procedure of stiffened panels. *Marine Structures* 11: 75–99.
Hughes, O.F. & Paik, J.K. 2010. *Ship Structural Analysis and Design*. Society of Naval Architects and Marine Engineers (SNAME).
Jiang, L. 2014. Some modelling considerations on nonlinear finite element collapse analyses of stiffened panels. Martec Technical Report TR-14-23, Martec Limited, Halifax, NS, Canada.
Khedmati, M.R., Bayatfar, A. & Rigo, P. 2010. Post-buckling behavior and strength of multi-stiffened aluminum panels under combined axial compression and lateral pressure. *Marine Structures* 23: 39–66.

MacKay, J.R., Jiang, L. & Glas, A.H. 2011. Accuracy of nonlinear finite element collapse predictions for submarine pressure hulls with and without artificial corrosion damage. *Marine Structures* 24: 292–317.

Paik, J.K. & Lee, M.S. 2005. A semi-analytical method for the elastic-plastic large deflection analysis of stiffened panels under combined biaxial compression/tension, biaxial in-plane bending, edge shear and lateral pressure loads. *Thin-Walled Structures* 43: 375–410.

Paik, J.K., Kim, B.J. & Seo, J.K. 2008. Methods for ultimate limit state assessment of ships and ship-shaped offshore structures: Part II: Stiffened panels. *Ocean Engineering* 35: 271–280.

Paik, J.K. & Seo, J.K. 2009a. Nonlinear finite element method models for ultimate strength analysis of steel stiffened-plate structures under combined biaxial compression and lateral pressure actions – Part I: Plate elements. *Thin-Walled Structures* 47: 1008–1017.

Paik, J.K. & Seo, J.K. 2009b. Nonlinear finite element method models for ultimate strength analysis of steel stiffened-plate structures under combined biaxial compression and lateral pressure actions – Part II: Stiffened panels. *Thin-Walled Structures* 47: 998–1007.

Rankin, C.C. & Brogan, F.A. 1986. An element independent corotational procedure for the treatment of large rotations. *ASME Journal of Pressure Vessel Technology* 108(2): 165–174.

Simo, J.C. & Hughes, T.J.R. 2006. *Computational Inelasticity.* Springer Science & Business Media.

VAST92, 2018, User's Manual, Version 9.1. Martec Limited, Halifax, Canada.

Zhang, S. & Khan, I. 2009. Buckling and ultimate strength of plates and stiffened panels. *Marine Structures* 22: 791–808.

Zhang, S. 2013. Buckling and ultimate strength assessments of ship structures. *ASME 32nd International Conference on Ocean, Offshore and Arctic Engineering, OMAE2013, June 9- 14,2013.* Nantes, France.

Zhang, S. & Jiang, L. 2014. A procedure for nonlinear structural collapse analysis, *ASME 33rd International Conference on Ocean, Offshore and Arctic Engineering, June 8-13, 2014.* San Francisco, CA, USA.

Developments in Maritime Technology and Engineering – Guedes Soares & Santos (eds)
© 2021 Copyright the Author(s), ISBN 978-0-367-77376-2

The effects of welding-induced residual stress on the buckling collapse behaviours of stiffened panels

Shen Li & Simon Benson

Marine, Offshore and Subsea Technology Group, School of Engineering, Newcastle University, UK

ABSTRACT: An assessment is conducted in this paper for the effects of welding-induced residual stress on the elastoplastic buckling collapse behaviour of stiffened panels subjected to in-plane compression. A series of nonlinear finite element analyses are completed which covers a range of plate slenderness ratios (β = 1.0~4.0) and column slenderness ratios (λ = 0.2~1.2) typical in marine application. The finite element models incorporate an idealised distribution of welding-induced residual stress with average severity. The numerical investigation indicates a reduced ultimate strength and increased ultimate strain due to the residual stress, as compared with the initial stress-free condition. Hence, a set of design formulae are proposed with the aid of regression analysis to predict the ultimate strength reduction and ultimate strain variation. The proposed design formulae are employed in combination with an empirical load-shortening curve formulation and simplified progressive collapse method to predict the ultimate bending capacity reduction of a box girder model. Validation is completed through equivalent finite element analysis.

1 INTRODUCTION

The assessment of ultimate ship hull strength is usually completed using the simplified progressive collapse method, as codified in the Common Structural Rule (CSR) issued by IACS (2019). One of the key steps in the progressive collapse analysis is the estimation the structural components' elastoplastic buckling behaviour under in-plane compression, which is normally represented by a load-shortening curve. A wide range of methodology can be employed to predict the compressive load-shortening curves of stiffened panels, including analytical approach (Dow and Smith, 1986; Yao and Nikolov, 1991 & 1992; Ueda and Rashed, 1984; Ueda et al., 1984), numerical simulation (Benson et al, 2013; Li et al., 2019) and empirical formulation (Li et al., 2020).

Among various methodology, the empirical formulation introduced by Li et al. (2020) is highly efficient, as only the basic dimensionless parameters are involved, in comparison to the analytical approach with rather complicated derivation and computationally expensive numerical simulation.

Following the previous development in Li et al. (2020), this paper aims to enhance the empirical formulation to accommodate the effects of welding-induced residual stress. A series of nonlinear finite element analyses (NLFEA) covering seven plate slenderness ratios (β = 1.0~4.0) and six column slenderness ratios (λ = 0.2~1.2) are completed to explore the influence of residual stress on load-shortening behaviours of stiffened panels. Based on the NLFEA results, a set of design equations are proposed to evaluate the ultimate strength reduction and ultimate strain variation due to welding-induced residual stress. The proposed design equations are employed in conjunction with empirical load-shortening curve formulation and simplified progressive collapse method to predict the ultimate bending capacity reduction of a box girder model. Validation is completed through equivalent finite element analysis.

2 BACKGROUND

During the welding process, the welding metal is melted with the mother plates until solidification (Yao and Fujikubo, 2016). Skrinkage of the stiffened panels would occur at the solidified part which is constrained by the surrounding unmelted part. As a result, the tensile stress is induced in the solidified part and the compressive stress is developed at the neighbouring part so as to achieve an equilibrium condition. In short, the residual stress caused by welding is induced as tension near the welding line, whereas a compressive stress field is developed in the adjacent part.

Smith et al. (1991) suggested a simplified welding-induced residual stress distribution for stiffened panels (Figure 1a), in which the residual stress field was idealised as tension and compression blocks. In addition, a triangular distribution shape was assumed for the compressive stress field of stiffener's web, while the tensile stress field was taken as a rectangular strip near the intersection with plating. Similar distribution was given by Yao and Fujikubo (2016) for fillet

DOI: 10.1201/9781003216582-62

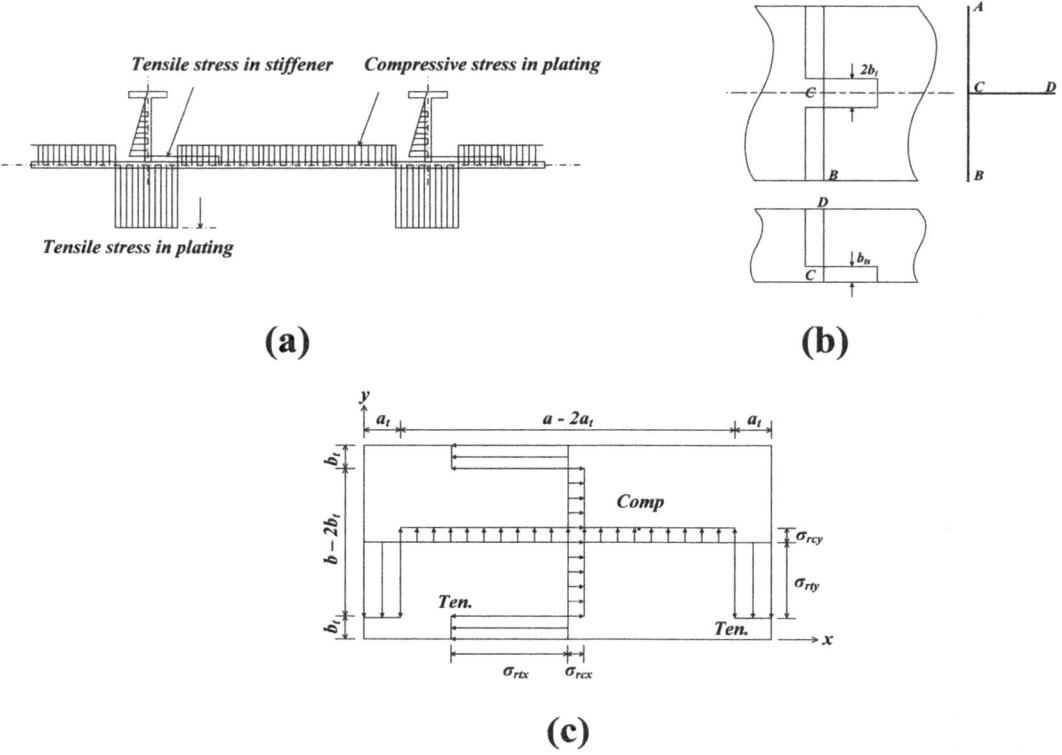

(a)

(b)

(c)

Figure 1. Idealised distribution of the welding-induced residual stress in stiffened panels.

welding (Figure 1b). However, an uniform distribution was assumed for the compressive stress field along the height of the stiffener's web. A more elaborated distribution pattern for the plating was given by Paik and Thayamballi (2003) considering the residual stress in both longitudianl and transverse directions (Figure 1c).

Since the residual stress is self-equilibrating, the equilibrium condition gives the relationship of Equation (1). To determine the magnitude of the residual stress, Yao (1980) suggested that the width of tensile block can be expressed as a function of plating thickness, web thickness and the weld heat input. Meanwhile, the tensile yield stress is equal to the material yield stress in the case of ordinary steel (Yao et al., 1998). An empirical formula was given by Smith et al. (1991) to estimate the compressive residual stress (Equation 2 to 4). Three different severities (slight, average and severe) were suggested.

$$2b_t\sigma_{rtx} = (b - 2b_t)\sigma_{rcx} \qquad (1)$$

$$\sigma_{rcx} = 0.05\sigma_{Yp} \; (slight) \qquad (2)$$

$$\sigma_{rcx} = 0.15\sigma_{Yp} \; (average) \qquad (3)$$

$$\sigma_{rcx} = 0.30\sigma_{Yp} \; (severe) \qquad (4)$$

Several case studies may be found in the litarature on the effects of welding-induced residual stress. Gannon et al. (2016) conducted a nonlinear collapse analysis on tee-bar and angle-bar stiffened plates under compression considering the welding-induced residual stress. A three-dimensional thermo-elasto-plastic finite element analysis was carried out to simulate the residual stress due to welding. The resutls suggested that the ultimate strength may be reduced by 12.5% because of the residual stress. Hansen (1996) concluded that residual stress may lead to decrease of ultimate strength by 25%, while Gordo and Guedes Soares (1993) found that the ultimate strength was reduced by 10% when the compressive residual stress in the plate was taken as 20% of the yield stress. Regarding to the hull girder strength, Gannon et al. (2012) reported that the hull girder strength can be reduced by 3.3% in the case that the ultimate strength of its component stiffened panels was decreased by 11%. As suggested by the literature survey, it is generally accepted that the welding-induced residual stress would lead to the reduction of ultimate compressive strength of stiffened panels. However, a practical solution to consider the welding-induced residual stress in the ultimate limit state design of stiffend plating structures is lacking.

3 NONLINEAR FINITE ELEMENT ANALYSIS

3.1 Scope of analysis

Nonlinear finite element analysis is carried out for a series of stiffened panels with and without residual stress under compressive load, in which the following parameters are varied systematically giving a total of 42 stiffened panels and 84 test cases:

- Plate slenderness ratio $\beta = b/t\sqrt{\sigma_{Yeq}/E} = 1.0$, 1.5, 2.0, 2.5, 3.0, 3.5, 4.0;
- Column slenderness ratio $\lambda = a/\pi r\sqrt{\sigma_{Yeq}/E} = 0.2, 0.4, 0.6, 0.8, 1.0, 1.2$;
- Stiffener area ratio $A_s/A = A_s/(A_s+bt) = 0.2$ where As is the stiffener cross-sectional area;
- Stiffener shape: all calculations refer to tee-bar stringers, which corresponds to the 114mm × 44.5mm Admiralty long-stalk tee bar section;
- Material property: the yield strength and Young's modulus in all calculations are 324MPa and 207000MPa respectively. An elastic-perfectly plastic behaviour is assumed;
- Initial distortion: local plate distortion w_{opl}, column-type distortion w_{oc} and stiffener sideway distortion w_{os} as defined by Equation (5) to (7) and schematically shown in Figure 2;
- Residual stress: it is only considered for the plating in the longitudinal direction following Figure 1(c). The width of tensile stress block b_t is taken assuming that the tensile residual stress equals the material yields stress and compressive residual stress corresponds an average-level magnitude as given by Equation (3) (More details in Appendix).

$$w_{opl} = A_o sin\left(\frac{m\pi x}{a}\right) sin\left(\frac{\pi y}{b}\right) \qquad (5)$$

$$m = \frac{a}{b} + 1$$

$$w_{oc} = B_o\ sin\left(\frac{\pi x}{a}\right) sin\left(\frac{\pi y}{B}\right) \qquad (6)$$

$$w_{os} = C_o\frac{z}{h_w}\left[0.8\ sin\left(\frac{\pi x}{a}\right) + 0.2\ sin\left(\frac{i\pi x}{a}\right)\right] \qquad (7)$$

$$i = a/h_w + 1$$

$$A_o = 0.1\beta^2 t$$

$$B_o = 0.0015a$$

$$C_o = 0.0015a$$

3.2 Finite element modelling

It was discussed by ISSC (2012) that the constrain of the end-rotation of stiffeners may over-estimate the ultimate compressive strength of stiffened panels. Meanwhile, Smith et al. (1991) indicated that the interaction between longitudinal adjacent structures should be considered. Thus, a two bays/two span model with eight identical stiffeners in each span is employed (Figure 3). The longitudinal girder and transverse frame are modelled with boundary condition constraining the out-of-plane movement. The present model extent allows for the end-rotation of stiffeners and the interactions between the adjacent structures in the longitudinal direction and may be a reasonable representation of typical continuous ship grillages.

The FE model is discretised with four-node shell element (S4R in ABAQUS) with reduced integration.

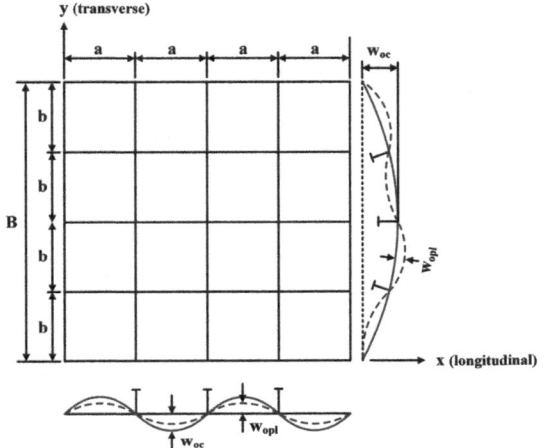
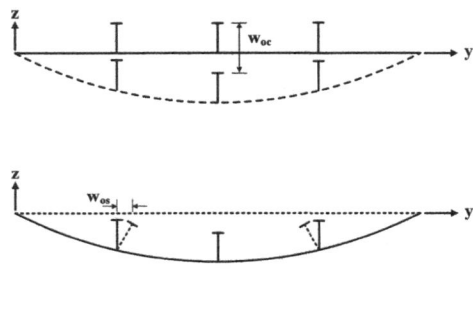

Figure 2. Schematic view of the initial distortions.

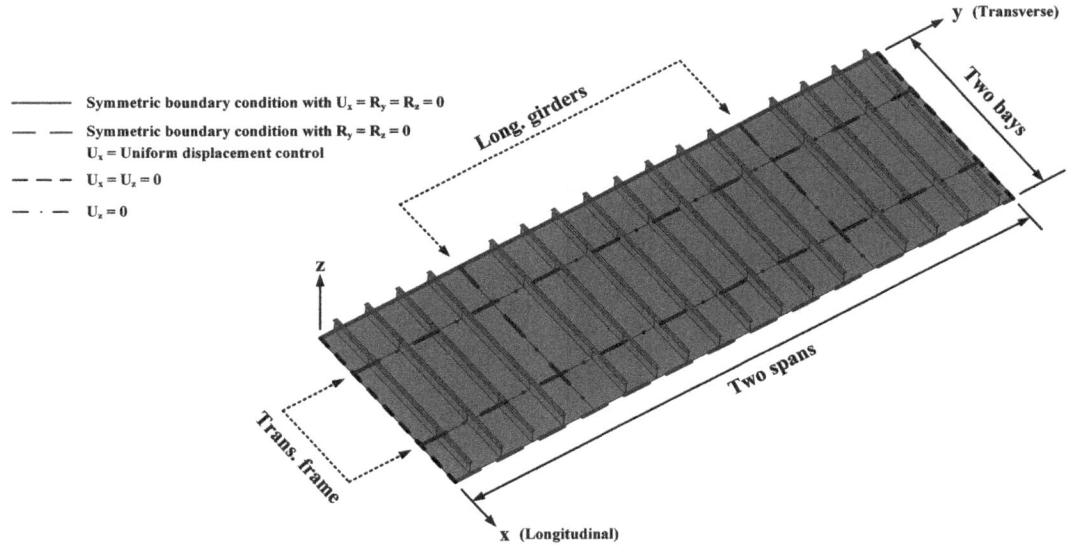

Figure 3. Boundary conditions of FE stiffened panel models.

For the local plating, the element number in the longitudinal direction is taken as 50λ dependent on the column slenderness ratio, while ten elements are employed in the transverse direction for all cases giving a characteristic plating mesh size of 50mm × 50mm. For the stiffener, six elements are used in both stiffener's web and flange.

The welding-induced residual stress are modelled by applying an initial stress field. An average-level magnitude is considered. A relaxation step is applied for the self-equilibrium of the initial stress field.

3.3 Results and discussions

A selection of the load-shortening curves of the tested stiffened panels are shown in Figure 4 to compare the predictions with and without welding

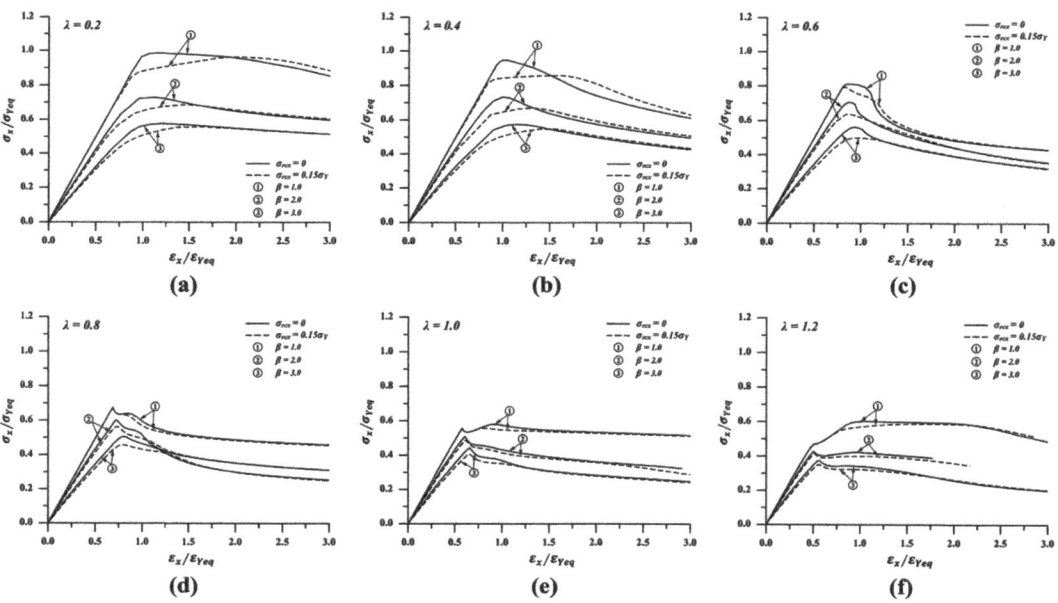

Figure 4. Selected load-shortening curves of stiffened panels with and without welding residual stress.

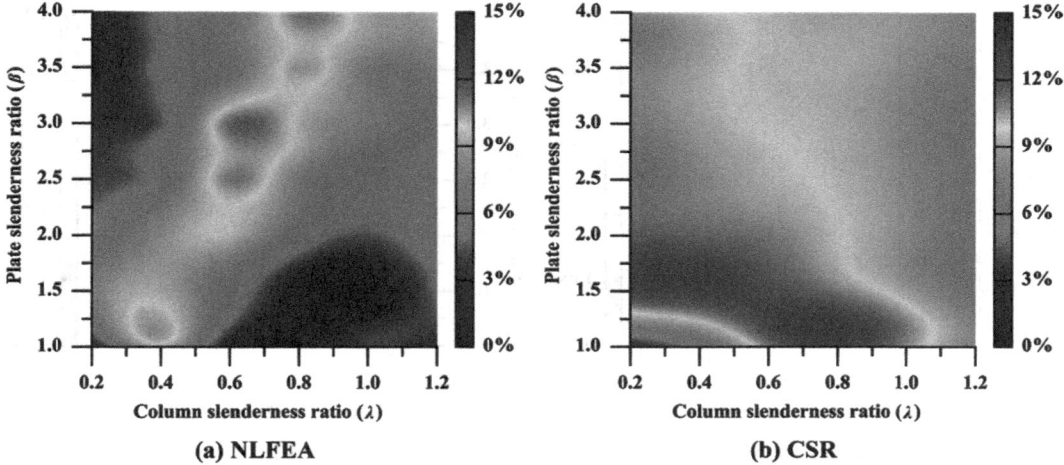

Figure 5. Ultimate strength reduction caused by welding residual stress (Cubic interpolation).

residual stress. From these comparisons, the following observations can be made:

- A reduced initial stiffness is observed for all tested cases as a result of the welding residual stress. However, the stiffness reduction is relatively insignificant, which may therefore be discarded in the ultimate limit state assessment;
- The ultimate compressive strength of stiffened panels is degraded due to the residual stress. A contour plot illustrating the strength reduction of different stiffened panel configurations is given in Figure 5 by a cubic interpolation of the obtained numerical results. Comparison with the CSR method is given where the effect of residual stress is accounted by modifying the edge function, as proposed by Gordo and Guedes Soares (1993). A maximum reduction of 11.2% is induced in the numerical simulation. The reduction is less considerable for stiffened panels with a combination of low λ and high β or combination of high λ and low β. However, the corresponding CSR prediction is generally overestimated;
- It appears that the collapse of stiffened panels generally takes place at a higher strain due to the welding residual stress, which is more substantial for low plate and column slenderness panels ($\beta \leq 1.5$ & $\lambda \leq 0.4$) as shown in Figure 6. This may be due to the presence of initial tensile stress field which induces certain hardening effect (Khan and Zhang, 2011).
- The post-collapse behaviour is unaffected by the residual stress with nearly identical response path.

4 DESIGN EQUATIONS

Design equations are proposed in the general form given by Equation (8) for predicting the ultimate strength reduction. Least-square regression is performed for different column slenderness ratios and thus six sets of coefficients are derived based on the present dataset, as summarised in Table 1 where the goodness-of-fit (R^2) is also indicated. For the prediciton of stiffened panels with a column slenderness ratio other than those tested in the present study, an interpolation between the results may be used:

In regard to ultimate strain variation, the same four-order polynomial form is adopted (Equation 9). However, regression analysis is only conducted for column slenderness ratio $\lambda = 0.2$ and 0.4, since only an insignificant ultimate strain variation is induced by the residual stress as indicated in Figure 7. Hence, two sets of coefficients are suggested and summarised in Table 2.

The statistical correlation of the prediction between the proposed design equations and the present NLFEA is shown in Figure 7. A good agreement is obtained with a mean value of 0.9804 and COV of 0.0930 respectively for strength reduction prediction and a mean value of 1.0021 and COV of 0.0282 for collapse strain variation prediction.

$$\frac{\sigma_{xu} - \sigma_{xu}^R}{\sigma_{xu}} = \sum_{i=0}^{4} C_i \beta^i \qquad (8)$$

$$\frac{\varepsilon_{xu}^R}{\varepsilon_{xu}} = \sum_{i=0}^{4} D_i \beta^i \qquad (9)$$

5 APPLICATION

The proposed design equations are applied in combination with the empirical load-shortening curve (LSC) formulation introduced by Li et al. (2020) and

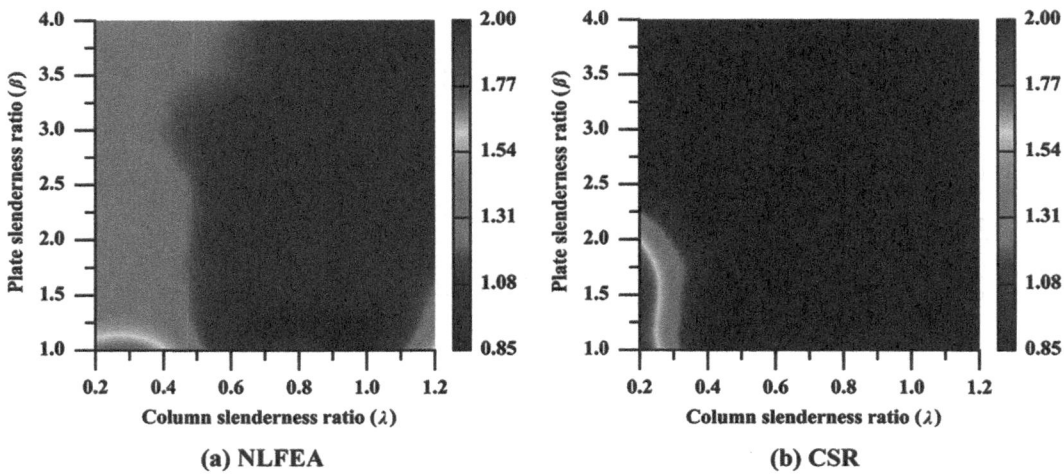

Figure 6. Ultimate strain variation caused by welding residual stress (Cubic interpolation).

Table 1. Coefficients of the proposed design equations to evaluate the ultimate strength reduction.

	C_0	C_1	C_2	C_3	C_4	R^2
$\lambda = 0.2$	-0.454	0.868	-0.497	0.1161	-0.00973	0.9404
$\lambda = 0.4$	-0.115	0.432	-0.292	0.0762	-0.00687	0.9655
$\lambda = 0.6$	0.117	-0.333	0.337	-0.1107	0.01163	0.9701
$\lambda = 0.8$	0.059	-0.190	0.178	-0.0518	0.00499	0.9944
$\lambda = 1.0$	0.534	-0.993	0.654	-0.1727	0.01612	0.9842
$\lambda = 1.2$	-0.256	0.511	-0.301	0.0752	-0.00661	0.9747

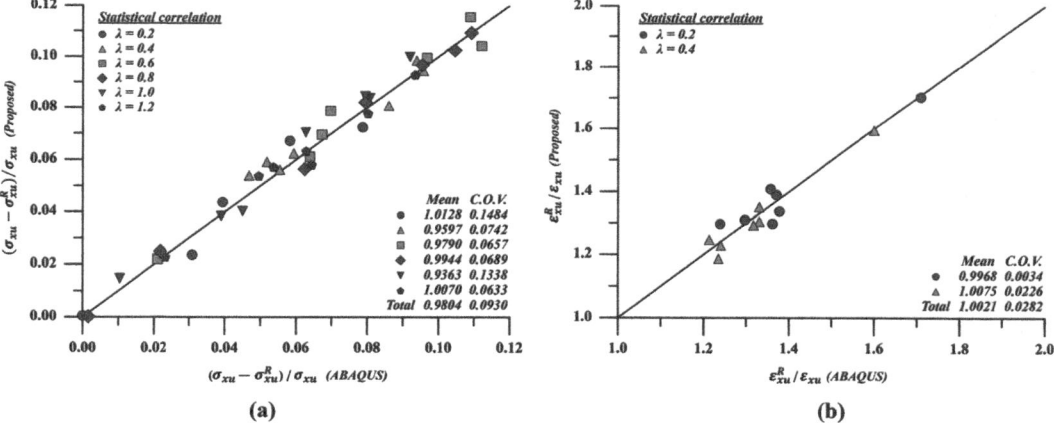

Figure 7. Correlation of the prediction by proposed design equations and NLFEA.

simplified progressive collapse method (ProColl) to predict the ulimate bending strength of a welded box girder model. Validation is performed through a comparison with equavilent NLFEA. The case study model was tested originally by Gordo and Guedes Soares (2013). The dimensions of the box girder model is illustrated in Figure 8. The model is made of mild steel with yield stress of 270 MPa and Young's modulus of 200000 MPa.

Detailed derivation of the empirical formulation can refer to Li et al. (2020). To accommodate the influence of welding residual stress on overall load-shortening behaviour, the formulation is modified for low slenderness stiffened panels ($\lambda \leq 0.4$) as

Table 2. Coefficients of the proposed design equations to evaluate the ultimate strain variation.

	D_0	D_1	D_2	D_3	D_4	R^2
$\lambda = 0.2$	5.740	-7.49	4.47	-1.117	0.0996	0.9104
$\lambda = 0.4$	3.777	-4.05	2.45	-0.644	0.0612	0.9573

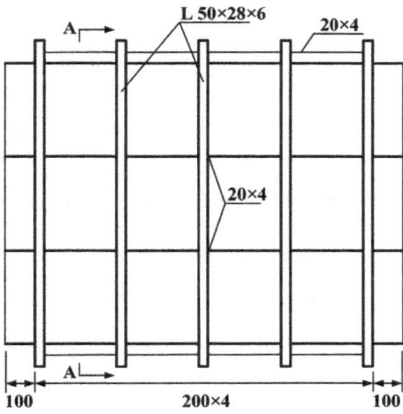

Figure 8. Schematic illustration of the case study model.

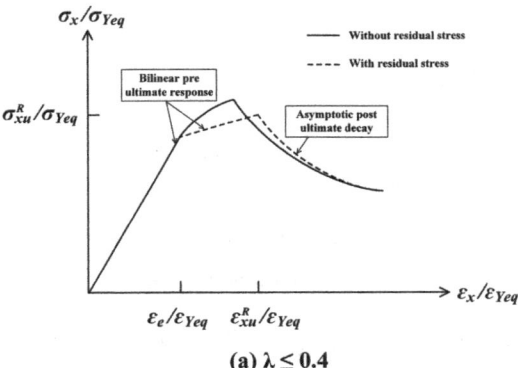

(a) $\lambda \le 0.4$

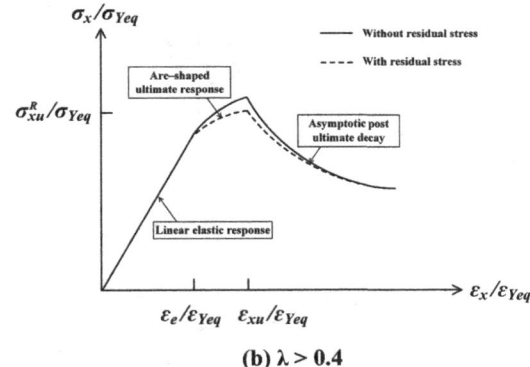

(b) $\lambda > 0.4$

Figure 9. Empirical load-shortening curve formulation.

schematically illustrated in Figure 9 and given by Equation (10) to (12). On the other hand, the LSC of high slenderness stiffened panels ($\lambda > 0.4$) follows the original formulation. The compressive ultimate strength of structural members are estimated by Zhang and Khan formula (2009).

The initial stress field of the FE model after relaxation is shown in Figure 10. Figure 11 compares the collapse mode of the box girder model with and without residual stress. In both cases, the collapse of the box girder is primarily induced by beam-column buckling. A comparison of the bending moment/curvature curves is given in Figure 12. Although the pre-collapse bending stiffness is underestimated by

the NLFEA, which may be attributed to multi-frame configuration of the box girder model, there is close correlation on the reduction of ultimate bending capacity of the case study model as given in Table 3. The accuracy of the proposed design equations may therefore be verified.

$$\frac{\sigma_x}{\sigma_{Yeq}} = \bar{E}_{To}\frac{\varepsilon_x}{\varepsilon_{Yeq}} \tag{10}$$

$$\frac{\sigma_x}{\sigma_{Yeq}} = \bar{E}_T\left(\frac{\varepsilon_x}{\varepsilon_{Yeq}} - \frac{\varepsilon_e}{\varepsilon_{Yeq}}\right) + \frac{\sigma_e}{\sigma_{Yeq}} \tag{11}$$

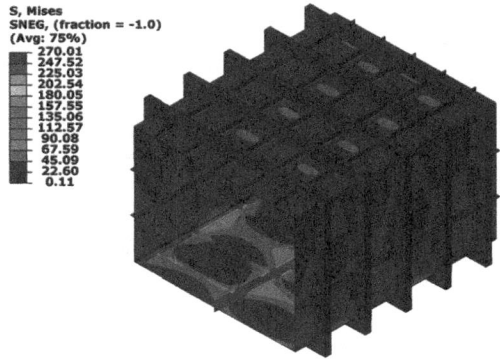

Figure 10. Initial stress field of box girder model.

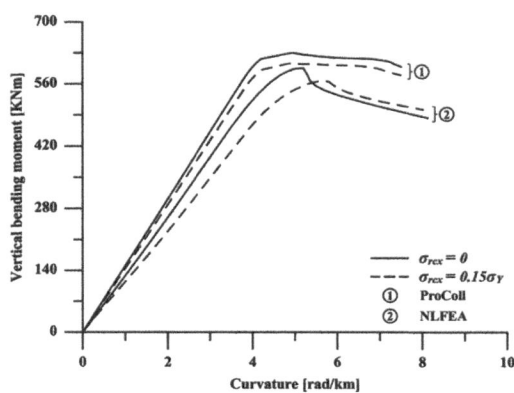

Figure 12. Comparison of bending moment/curvature curves.

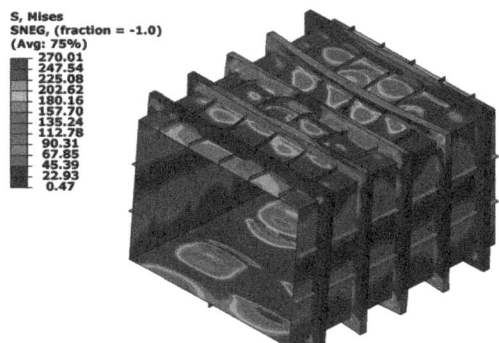

(a) Ultimate limit state of box girder with initial stress

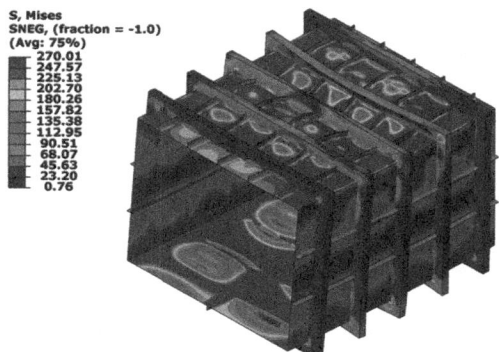

(b) Ultimate limit state of box girder with initial stress

Figure 11. Contour plots of box girder FE model.

$$\frac{\sigma_x}{\sigma_{Yeq}} = C \frac{\sigma_{xu}^R}{\sigma_{Yeq}} + (1 - C) \frac{\sigma_{xu}^R}{\sigma_{Yeq}} exp\left(\frac{\varepsilon_{xu}^R}{\varepsilon_{Yeq}} - \frac{\varepsilon_x}{\varepsilon_{Yeq}}\right)$$

(12)

Table 3. Summary of the computed ultimate bending strength.

	ProColl (KNm)	NLFEA (KNm)
Without residual stress	629.5752	595.1840
With residual stress	605.5636	566.9330
Reduction	-3.81%	-4.75%

where

$$\frac{\varepsilon_{xe}}{\varepsilon_{Yeq}} = \frac{\varepsilon_{xu}}{\varepsilon_{Yeq}} + Rsin\left[-tan^{-1}\left(\bar{E}_{To}\right)\right]$$

$$R = \frac{cos\left[tan^{-1}\left(\bar{E}_{To}\right)\right]\left(\bar{E}_T \frac{\varepsilon_{xu}}{\varepsilon_{Yeq}} - \frac{\sigma_{xu}}{\sigma_{Yeq}}\right)}{1 - cos\left[tan^{-1}\left(\bar{E}_{To}\right)\right]}$$

$$C = \left(0.7834 - 0.3174\sqrt{\lambda} - \frac{0.0060}{\beta^2}\right)\bigg/\left(1 - \frac{\sigma_{xu} - \sigma_{xu}^R}{\sigma_{xu}}\right)$$

$$\frac{\varepsilon_{xu}}{\varepsilon_{Yeq}} = -0.0004 + 1.005\,\lambda - 1.3126\lambda^2 + 1.7101/\sqrt{\beta}$$

$$\frac{\varepsilon_{xu}}{\varepsilon_{Yeq}} = -0.3752\,\lambda/\sqrt{\beta} - 0.7337/\beta$$

$$\frac{\varepsilon_{xu}}{\varepsilon_{Yeq}} = \frac{\varepsilon_{xu}}{\varepsilon_{Yeq}} \times \frac{\varepsilon_{xu}^R}{\varepsilon_{xu}}$$

$$\frac{\sigma_{xu}^R}{\sigma_{Yeq}} = \frac{\sigma_{xu}}{\varepsilon_{Yeq}} \times \left(1 - \frac{\sigma_{xu} - \sigma_{xu}^R}{\sigma_{xu}}\right)$$

6 CONCLUSIONS

A systematic nonlinear finite element analysis covering a range of plate slenderness ratio ($\beta = 1.0{\sim}4.0$) and column slenderness ratios ($\lambda = 0.2{\sim}1.2$) is conducted to assess the effect of welding residual stress. The finite element models incorporate an idealised distribution of welding-induced residual stress with average severity. Design formulae are proposed for prediction the collapse strength reduction and ultimate strain variation caused by residual stress. Application is presented to evaluate the ultimate bending strength of a welded box girder model. From this study, the following conclusion may be drawn:

- The welding-induced residual stress would deteriorate the ultimate compressive strength of stiffened panels and furthermore leads to a considerable variation of the ultimate strain for stock panels.
- The proposed design equations provide an efficient and accurate way to assess the effects of welding-induced residual stress on the progressive collapse behavior of stiffened panels;
- The design equations can be incorporated with the empirical formulation (Li et al., 2019b) to predict load-shortening curves of stiffened panels considering welding stress. The application in box girder ultimate bending strength prediction has demonstrated the accuracy of the proposed equations.

REFERENCES

18th ISSC, 2012. Ultimate Strength, Rostock, Germany.

Benson S, Downes J, Dow RS. 2013. Load shortening characteristics of marine grade aluminium alloy plates in longitudinal compression. *Thin-Walled Structures*, 70, 19–32.

Dow RS, Smith CS. 1986. FABSTRAN: A computer program for frame and beam static and transient response analysis (Nonlinear). ARE report TR86205.

Gannon L, Liu Y, Pegg N, Smith MJ, 2012. Effect of welding-induced residual stress and distortion on ship hull girder ultimate strength. *Marine Structures*, 28, 25–49.

Gannon L, Liu Y, Pegg N, Smith MJ, 2016. Nonlinear collapse analysis of stiffened plates considering welding-induced residual stress and distortion. *Ships and Offshore Structures*, 11, 228–244.

Gordo JM, Guedes Soares C, 1993. Approximate load shortening curves for stiffened plates under uniaxial compression. *Integrity of Offshore Structures*, 5, Proc 5th International Symposium on Integrity of Offshore Structures: EMAS, UK; Faulkner D., Cowling M. J. & Incecik A., (Eds.). pp. 189–211.

Gordo JM, Guedes Soares C, 2013. Experiments on three mild steel box girders of different spans under pure bending moment. *Analysis and Design of Marine Structures*. Guedes Soares, C. & Romanoff J., (Eds.), Taylor & Francis; pp. 337–346.

Hansen AM, 1996. Strength of midship sections. *Marine Structures*, 9, 471–494.

International Association of Classification Societies (IACS). 2019. *Common Structural Rules for Bulk Carriers and Oil Tankers*.

Khan I, Zhang SM, 2011. Effects of welding-induced residual stress on ultimate strength of plates and stiffened panels. *Ships and Offshore Structures*, 6, 297–309.

Li S, Hu ZQ, Benson S. 2019. An analytical method to predict the buckling and collapse behaviour of plates and stiffened panels under cyclic loading. *Engineering Structures*, 199, 109627.

Li S, Kim DK, Benson S. 2020. An adaptable algorithm to predict the load-shortening curves of stiffened panels in compression. In *Proceeding: 5th International Conference on Ships and Offshore Structures*, Glasgow, UK.

Paik JK, Thayamballi, AK, 2003. Ultimate limit state design of steel-plated structures, John Wiley & Sons, Chichester, UK.

Smith CS, Anderson N, Chapman JC, Davidson PJ, Dowling PJ. 1991. Strength of stiffened plating under combined compression and lateral pressure. *Transactions of Royal Institution of Naval Architects*, 131–147.

Ueda Y, Rashed SMH. 1984. The idealized structural unit method and its application to deep girder structures. *Computers & Structures*, 18, 277–293.

Ueda Y, Rashed S, Paik JK. 1984. Plate and stiffened panel units of the idealized structural unit method under in-plane loading. *Journal of the Society of Naval Architects of Japan*, 156, 366–376.

Yao T, Fujikubo M, 2016. Buckling and ultimate strength of ship and ship-like floating structures, Elsevier.

Yao T, 1980. Compressive ultimate strength of structural members in ship structure. PhD thesis, Osaka University, Japan (in Japanese).

Yao T, Fujikubo M, Yanagihara D, Varghese B, Niho O, 1998. Influences of welding imperfections on buckling/ultimate strength of ship bottom plating subjected to combined bi-axial thrust and lateral pressure. In *Proceeding: International symposium on thin-walled structures*, Singapore.

Yao T, Nikolov P. 1991. Progressive collapse analysis of a ship's hull under longitudinal bending (1st report). *Journal of the Society of Naval Architects of Japan*, 170, 449–461.

Yao T, Nikolov P. 1992. Progressive collapse analysis of a ship's hull under longitudinal bending (2nd report. *Journal of the Society of Naval Architects of Japan*, 172, 437–446.

Zhang S, Khan I. 2009. Buckling and ultimate capability of plates and stiffened panels in axial compression. *Marine Structures*, 22, 791–808.

APPENDIX

No.	a[mm]	b_p[mm]	t_p[mm]	h_w[mm]	t_w[mm]	b_f[mm]	t_f[mm]	σ_{Yeq}[MPa]	β	λ	A_s/A	b_t[mm]
1	593.7	311.1	12.3	104.8	5.1	44.5	9.5	324.0	1.0	0.2	0.2	20.3
2	1187.4	311.1	12.3	104.8	5.1	44.5	9.5	324.0	1.0	0.4	0.2	20.3
3	1781.1	311.1	12.3	104.8	5.1	44.5	9.5	324.0	1.0	0.6	0.2	20.3
4	2374.8	311.1	12.3	104.8	5.1	44.5	9.5	324.0	1.0	0.8	0.2	20.3
5	2968.5	311.1	12.3	104.8	5.1	44.5	9.5	324.0	1.0	1.0	0.2	20.3
6	3562.2	311.1	12.3	104.8	5.1	44.5	9.5	324.0	1.0	1.2	0.2	20.3
7	586.5	381.0	10.0	104.8	5.1	44.5	9.5	324.0	1.5	0.2	0.2	24.8
8	1173.1	381.0	10.0	104.8	5.1	44.5	9.5	324.0	1.5	0.4	0.2	24.8
9	1759.6	381.0	10.0	104.8	5.1	44.5	9.5	324.0	1.5	0.6	0.2	24.8
10	2346.2	381.0	10.0	104.8	5.1	44.5	9.5	324.0	1.5	0.8	0.2	24.8
11	2932.7	381.0	10.0	104.8	5.1	44.5	9.5	324.0	1.5	1.0	0.2	24.8
12	3519.3	381.0	10.0	104.8	5.1	44.5	9.5	324.0	1.5	1.2	0.2	24.8
13	582.4	440.0	8.7	104.8	5.1	44.5	9.5	324.0	2.0	0.2	0.2	28.7
14	1164.7	440.0	8.7	104.8	5.1	44.5	9.5	324.0	2.0	0.4	0.2	28.7
15	1747.1	440.0	8.7	104.8	5.1	44.5	9.5	324.0	2.0	0.6	0.2	28.7
16	2329.4	440.0	8.7	104.8	5.1	44.5	9.5	324.0	2.0	0.8	0.2	28.7
17	2911.8	440.0	8.7	104.8	5.1	44.5	9.5	324.0	2.0	1.0	0.2	28.7
18	3494.1	440.0	8.7	104.8	5.1	44.5	9.5	324.0	2.0	1.2	0.2	28.7
19	579.5	491.9	7.8	104.8	5.1	44.5	9.5	324.0	2.5	0.2	0.2	32.1
20	1159.1	491.9	7.8	104.8	5.1	44.5	9.5	324.0	2.5	0.4	0.2	32.1
21	1738.6	491.9	7.8	104.8	5.1	44.5	9.5	324.0	2.5	0.6	0.2	32.1
22	2318.1	491.9	7.8	104.8	5.1	44.5	9.5	324.0	2.5	0.8	0.2	32.1
23	2897.6	491.9	7.8	104.8	5.1	44.5	9.5	324.0	2.5	1.0	0.2	32.1
24	3477.2	491.9	7.8	104.8	5.1	44.5	9.5	324.0	2.5	1.2	0.2	32.1
25	594.1	538.8	7.1	104.8	5.1	44.5	9.5	324.0	3.0	0.2	0.2	35.1
26	1188.3	538.8	7.1	104.8	5.1	44.5	9.5	324.0	3.0	0.4	0.2	35.1
27	1782.4	538.8	7.1	104.8	5.1	44.5	9.5	324.0	3.0	0.6	0.2	35.1
28	2376.6	538.8	7.1	104.8	5.1	44.5	9.5	324.0	3.0	0.8	0.2	35.1
29	2970.7	538.8	7.1	104.8	5.1	44.5	9.5	324.0	3.0	1.0	0.2	35.1
30	3564.9	538.8	7.1	104.8	5.1	44.5	9.5	324.0	3.0	1.2	0.2	35.1

No.	a[mm]	b_p[mm]	t_p[mm]	h_w[mm]	t_w[mm]	b_f[mm]	t_f[mm]	σ_{Yeq}[MPa]	β	λ	A_s/A	b_t[mm]
31	575.9	582.0	6.6	104.8	5.1	44.5	9.5	324.0	3.5	0.2	0.2	38.0
32	1151.7	582.0	6.6	104.8	5.1	44.5	9.5	324.0	3.5	0.4	0.2	38.0
33	1727.6	582.0	6.6	104.8	5.1	44.5	9.5	324.0	3.5	0.6	0.2	38.0
34	2303.4	582.0	6.6	104.8	5.1	44.5	9.5	324.0	3.5	0.8	0.2	38.0
35	2879.3	582.0	6.6	104.8	5.1	44.5	9.5	324.0	3.5	1.0	0.2	38.0
36	3455.2	582.0	6.6	104.8	5.1	44.5	9.5	324.0	3.5	1.2	0.2	38.0
37	574.6	622.2	6.2	104.8	5.1	44.5	9.5	324.0	4.0	0.2	0.2	40.6
38	1149.2	622.2	6.2	104.8	5.1	44.5	9.5	324.0	4.0	0.4	0.2	40.6
39	1723.7	622.2	6.2	104.8	5.1	44.5	9.5	324.0	4.0	0.6	0.2	40.6
40	2298.3	622.2	6.2	104.8	5.1	44.5	9.5	324.0	4.0	0.8	0.2	40.6
41	2872.9	622.2	6.2	104.8	5.1	44.5	9.5	324.0	4.0	1.0	0.2	40.6
42	3447.5	622.2	6.2	104.8	5.1	44.5	9.5	324.0	4.0	1.2	0.2	40.6

Developments in Maritime Technology and Engineering – Guedes Soares & Santos (eds)
© 2021 Copyright the Author(s), ISBN 978-0-367-77376-2

Probabilistic evaluation of the computational uncertainty in ultimate ship hull strength prediction

Shen Li & S. Benson

Marine, Offshore and Subsea Technology Group, School of Engineering, Newcastle University, UK

ABSTRACT: Simplified progressive collapse method (Smith method) is codified in the Common Structural Rule (CSR) to calculate the ultimate bending capacity of ship hull girders. However, several benchmark studies have demonstrated a notable uncertainty in predicting the progressive collapse and ultimate limit state of ship hull girders, which is primarily attributed to load-shortening curve (LSC) of local structural components adopted by different participants. In this regard, this paper employs a probabilistic approach to assess the uncertainty of ultimate ship hull strength prediction caused by the critical features of structural component's LSC, e.g. ultimate compressive strength. Probability distribution of ultimate compressive strength estimation of stiffened panels is developed based on a dataset generated by different empirical formulae and nonlinear finite element method. An adaptable LSC formulation, with ability to accommodate different compressive strength of local components, is utilised in conjunction with the Monte-Carlo Simulation procedure where the simplified progressive collapse method is employed to complete the ultimate ship hull strength prediction in each sampling. Case study is conducted on a single hull VLCC and it is found that the CSR-based ultimate ship hull strength prediction follows a Weibull distribution when considering the computational uncertainty caused by different LSC models. The results and insights developed from this paper would be useful to improve the reliability-based design of marine structures.

1 INTRODUCTION

It is recommended by the Common Structural Rule (CSR) that the ultimate bending capacity of ship hull girders, including bulk carriers, oil tankers and container ships (IACS, 2015; IACS, 2019), can be calculated by the simplified progressive collapse method originally developed by Smith (1977). Progress in the development of Smith method includes differing formulations (Adamchak, 1982; Ueda and Rashed, 1984; Yao and Nikolov, 1991; Gordo and Guedes Soares, 1996) and extensions for biaxial bending (Smith and Dow, 1986), multi-frame collapse (Benson et al., 2013), torsion (Syrigou et al., 2018) and cyclic loading (Li et al., 2019; Li et al., 2020). The underpinning algorithms within these methods are established, relying on an evaluation of the load-shortening curve (LSC) of stiffened panel components under in-plane loading.

Various benchmark studies, such as ISSC (2000) and ISSC (2012), were performed to predict the ultimate bending strength of ship hull girders by applying the simplified progressive collapse method. These investigations demonstrated a notable uncertainty in the predictions by different participants. The sensitivity analysis completed by the Special Task Committee VI.2 of 14th ISSC (2000) concluded that the uncertainty is primarily attributed by the LSC of structural elements. An uncertainty of 10% on the compressive ultimate strength of a structural element can lead to an uncertainty of roughly 20% on the sagging ultimate bending moment. Additionally, variations of 25% in sagging and 12% in hogging may be induced as a result of the post-collapse characteristics.

Extending upon 14th ISSC (2000), Li et al. (2020) recently conducted a systematic investigation to analyse the influence of ultimate strength, ultimate strain, elastic stiffness and post-ultimate strength stiffness of structural components on hull girder strength calculation. It was indicated that the ultimate strength and post-collapse stiffness have the largest impact on the hull girder strength calculation among four considered parameters.

However, both of these studies were completed by a deterministic procedure. Whilst the most critical features can be identified from these deterministic evaluations, it is necessary to employ a probabilistic approach to quantify the realistic influence of each critical features of the LSC. Thus, this paper develops a probabilistic procedure to assess the uncertainty of ultimate ship hull strength prediction caused by the critical features of structural component's LSC. In this paper, the ultimate compressive strength of stiffened panel element is under consideration. Probability distribution of the ultimate compressive strength estimation of stiffened panels is developed based on

DOI: 10.1201/9781003216582-63

a dataset generated by empirical formulae and non-linear finite element method. An adaptable LSC formulation, with ability to accommodate different compressive strength of local components, is utilised in conjunction with the Monte-Carlo Simulation procedure where the simplified progressive collapse method is employed to complete the ship hull strength prediction at each sampling. Case study is conducted on a single hull VLCC under pure vertical bending.

2 METHODOLOGY

The overall methodology of the present study to investigate the computational uncertainty in ultimate ship hull strength prediction may be illustrated by the flow chart in Figure 1.

As identified by Li et al. (2020), the critical features of LSC of stiffened panel elements that have significant influences on the hull girder capacity prediction are the ultimate compressive strength and post-collapse characteristics. In the present paper, the probabilistic evaluation is confined to the former, while the influence of post-collapse characteristics could be examined by applying the same procedure.

A dataset containing the ultimate compressive strength prediction of a range of stiffened panels typical for ship structures are developed. Best-fit probability distribution is then selected for the histogram based on Anderson-Darling statistic, as employed by

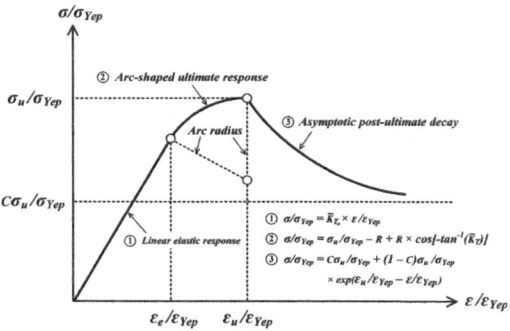

Figure 2. Adaptable LSC formulation (Li et al., 2020).

Kim et al. (2018). With the developed probability distribution of ultimate compressive strength of stiffened panels, an adaptable LSC formulation introduced by Li et al. (2020) is adopted which allows the derivation of a set of parametric LSCs with probabilistically varying features. The adaptable LSC formulation is schematically illustrated by Figure 2. Simplified progressive collapse analysis is conducted for predicting the ultimate bending capacity of ship hull girders where a Monte-Carlo Simulation procedure is combined to enable the sampling of ultimate compressive strength of each structural component. The principle of simplified progressive collapse method may refer to Smith (1977), Dow et al. (1981), Benson et al. (2013). All of the material property and geometric dimension are taken as their nominal values, since the aim is to evaluate the computational uncertainty due to the variability of the critical features of structural component load-shortening curves, which may also be considered as the epistemic uncertainty.

3 PREDICTION UNCERTAINTY OF ULTIMATE COMPRESSIVE STRENGTH OF STIFFENED PANELS

3.1 Dataset

The dataset is generated from the prediction by seven empirical formulae (Chalmers, 1993; Paik and Thayamballi, 1997; Zhang and Khan, 2009; Kim et al., 2017; Xu et al., 2018; Kim et al., 2019) and nonlinear finite element analysis with and without residual stress. All the adopted empirical formulae are proposed for ordinary design of stiffened panels and the present finite element modelling techniques are consistent with the ISSC recommendation. Meanwhile, all of the ultimate strength prediction is normalised by the corresponding CSR estimation. Hence, the dataset may be able to represent the variability of ultimate compressive strength prediction with reference to CSR.

Furthermore, in order to obtain an improved probabilistic characterization of the prediction uncertainty, the dataset is divided into four domains based on slenderness ratios, i.e. Domain 1 ($\beta > 1.9$ &

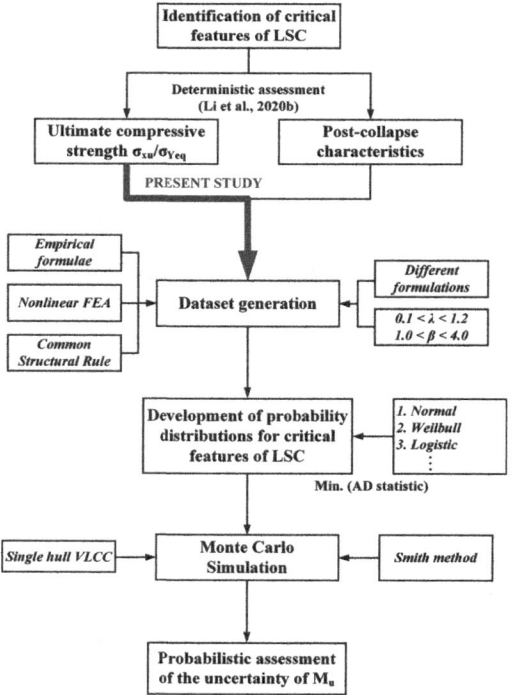

Figure 1. Flow chart of the methodology of the present study.

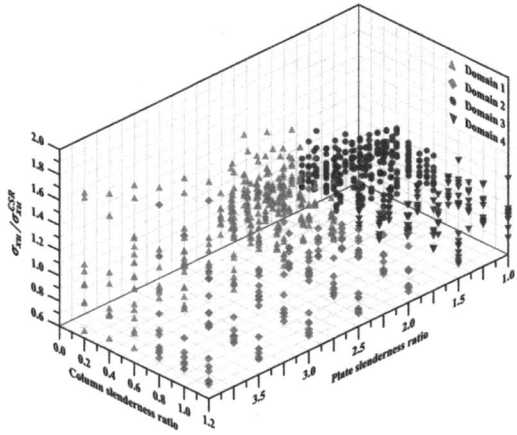

Figure 3. Dataset of the ultimate compressive strength normalised by CSR prediction.

$\lambda < 0.6$), Domain 2 ($\beta > 1.9$ & $\lambda > 0.6$), Domain 3 ($\beta < 1.9$ & $\lambda < 0.6$) and Domain 4 ($\beta < 1.9$ & $\lambda > 0.6$). The rationale of this sub-division follows the von-Karmon model for plate ultimate strength and Ozdemir et al. (2018) in which it was suggested that the primary failure mode of stiffened panels under compression may change from local plate buckling to beam–column buckling at the threshold of $\lambda = 0.6$.

The developed dataset with four domains is shown in Figure 3. Relevant descriptive statistics are summarised in Table 1. It can be seen that the dataset presents a fairly large variance, particularly in Domain 1 and Domain 2. The CSR prediction is generally underestimated for $\lambda < 0.6$ and overestimated for $\lambda > 0.6$ with respect to empirical formulae and NLFEA, as indicated by the mean values.

3.2 *Probability distribution of ultimate compressive strength of stiffened panels*

The histograms of the normalised ultimate compressive strength with best-fitted probability distribution are presented in Figure 4. The best-fit probability

Table 1. Descriptive statistics of the ultimate compressive strength normalised by CSR prediction.

	$\sigma_{xu}/\sigma_{xu}^{CSR}$			
	Domain 1	Domain 2	Domain 3	Domain 4
Max.	1.9078	1.9457	1.4292	1.2351
Min.	0.6627	0.6880	0.7944	0.6539
Mean	1.0707	0.9322	1.0065	0.8838
Median	1.0330	0.8404	0.9927	0.8768
COV	0.1895	0.2616	0.1077	0.1231

distribution is determined based on an examination of the Anderson-Darling statistic (AD statistic) on a series of probability distributions listed in Table 2 and Table 3. For Domain 1 and Domain 2, a 3-parameter Log-logistic distribution is selected (Equation 1). A 3-parameter Weibull and largest extreme value distributions are chosen for Domain 3 (Equation 2) and Domain 4 (Equation 3) respectively.

3-parameter Log-logistic distribution:

$$f(x) = \frac{1}{B} \frac{1}{(x-C)} \frac{exp\left[\frac{log(x-C)-A}{B}\right]}{\left\{1 + exp\left[\frac{log(x-C)-A}{B}\right]\right\}^2} \quad (1)$$

3-parameter Weibull distribution:

$$f(x) = \frac{B}{A}\left(\frac{x-C}{A}\right)^{B-1} exp\left[-\left(\frac{x-C}{A}\right)^B\right] \quad (2)$$

Largest extreme value distribution:

$$f(x) = \frac{1}{B} exp\left(\frac{x-A}{B}\right) exp\left[-exp\left(\frac{x-A}{B}\right)\right] \quad (3)$$

4 EVALUATION OF THE COMPUTATIONAL UNCERTAINTY IN SHIP HULL STRENGTH PREDICTION

4.1 *Case study model*

Single hull VLCC *Energy Concentration* is adopted for case study. The cross section of the hull girder model is shown in Figure 5. Detailed dimensions and material properties may refer to ISSC (2000). The slenderness ratios of stiffened panels of the present case study model is illustrated in Figure 6. It is apparent that most of the structural components below the neutral axis, which are under in-plane compressive in hogging, lie in Domain 3. Conversely, the components above the neutral axis, which undergo compressive load in sagging, lie in Domain 1 and Domain 3.

4.2 *Combined progressive collapse analysis and Monte-Carlo simulation*

The simplified progressive collapse analysis is performed with the aid of the adaptable LSC where sampling of the ultimate compressive strength follow Monte-Carlo procedure. Analysis is completed for both sagging and hogging, in which 2000 samplings are conducted for each case.

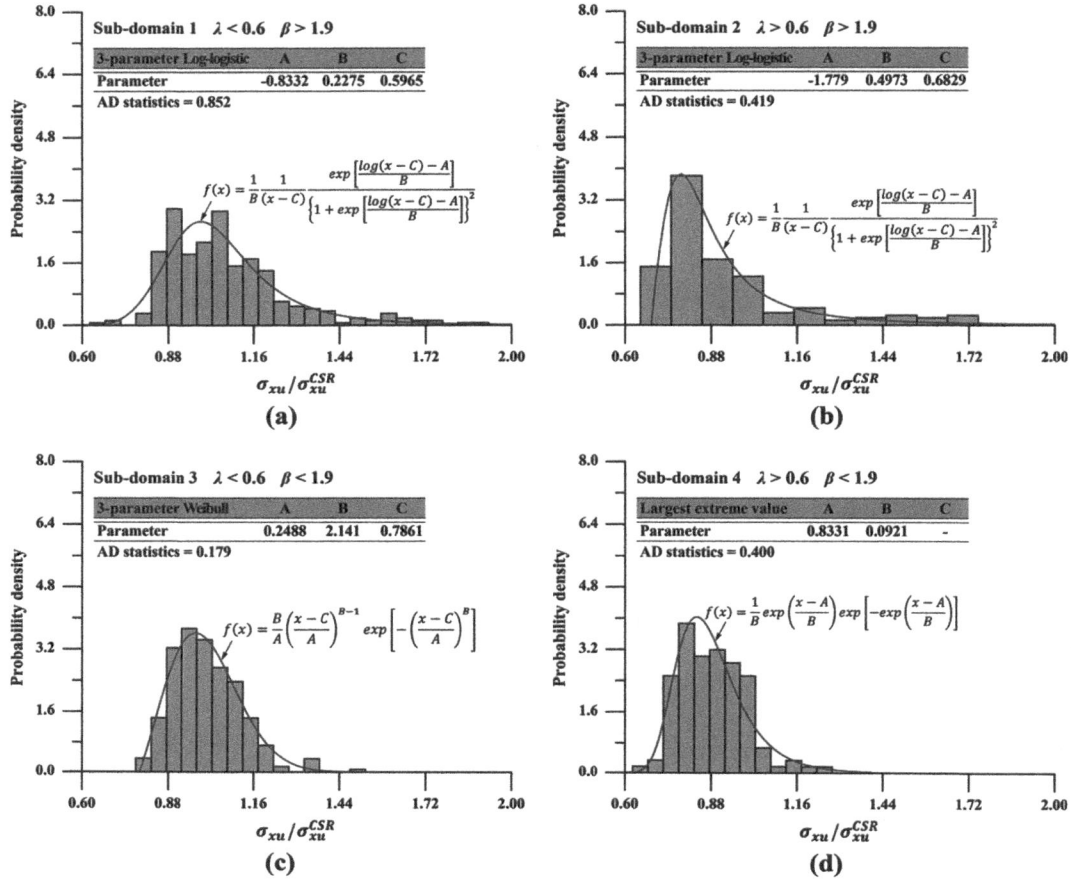

Figure 4. Histogram and best-fitted probability distribution of the ultimate compressive strength of stiffened panels.

Table 2. AD statistic of normalised ultimate compressive strength (Domain 1 and 2).

Distributions	AD statistic	
	Domain 1	Domain 2
Normal	8.258	13.185
Log–Normal	3.494	8.263
Exponential	104.567	47.486
2 para–exponential	45.916	3.567
Smallest extreme value	23.967	20.217
Weibull	14.352	14.514
3 para–Weibull	4.977	2.700
Largest extreme value	1.077	5.394
Logistic	4.327	8.515
Log–Logistic	2.144	5.484
3–Loglogistic	**0.852**	**0.419**

Table 3. AD statistic of normalised ultimate compressive strength (Domain 3 and 4).

Distributions	AD statistic	
	Domain 3	Domain 4
Normal	1.464	0.94
Log–Normal	0.594	0.491
Exponential	103.045	42.654
2 para–exponential	29.432	15.613
Smallest extreme value	8.932	4.361
Weibull	5.543	2.687
3 para–Weibull	**0.179**	0.495
Largest extreme value	0.508	**0.400**
Logistic	1.296	0.813
Log–Logistic	0.833	0.611
3–Loglogistic	0.633	0.522

In light of investigating the influence of ultimate compressive strength variation of structural components and being comparable with the corresponding CSR prediction, the initial stiffness, ultimate strain and post-collapse characteristics are consistent with the CSR formulation. Hence, the initial stiffness and

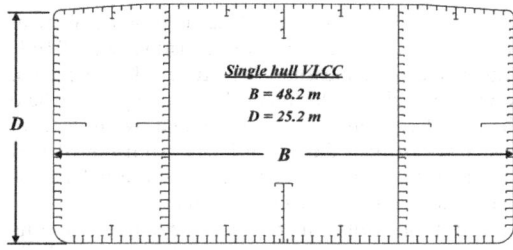

Figure 5. Cross section of the case study single hull VLCC.

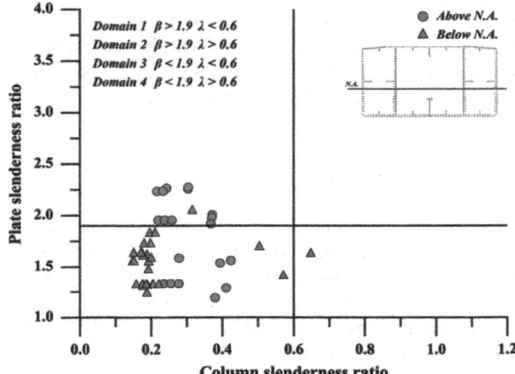

Figure 6. Stiffened panel element configurations of the case study model.

ultimate strain are both specified as unity. Additionally, consistency of the post-collapse strength decay rate is ensured by simplifying the post-collapse behaviour as bilinear with calibration on the stiffness from the original CSR (Figure 7). As verified in Figure 8, the baseline prediction using the adaptable LSC is equivalent to the original CSR.

Descriptive statistics of the predicted ultimate bending strength normalised by the CSR estimation are given in Table 4, while histograms are presented in Figure 9 with best-fitted probability distributions based on AD-statistic criterion (Table 5). It is clear that the

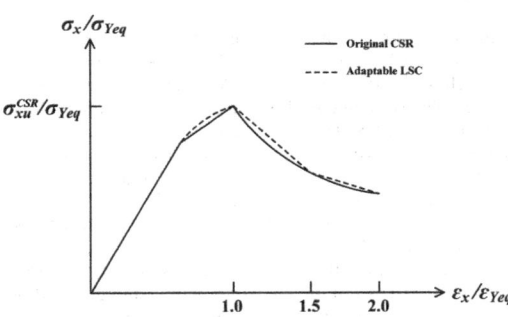

Figure 7. Calibration of the post-collapse decay.

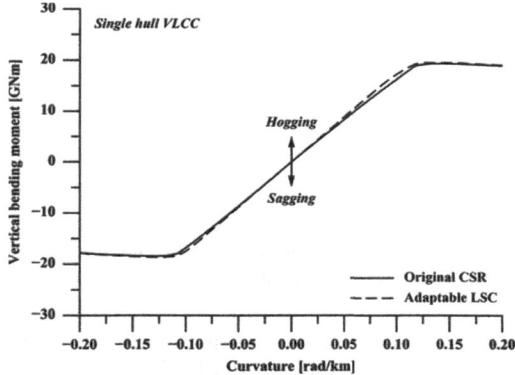

Figure 8. Verification of the bending moment/curvature curves.

Table 4. Descriptive statistics of the ultimate bending capacity of VLCC normalised by CSR prediction.

	M_u/M_u^{CSR}	
	Sagging	Hogging
Max.	1.3799	1.1835
Min.	0.9105	0.9016
Mean	1.0669	1.0220
Median	1.0570	1.0231
COV	0.0778	0.0481

Table 5. AD statistic of different fitted distribution density functions for normalised ultimate bending capacity.

	AD statistic	
Distributions	Sagging	Hogging
Normal	11.722	1.765
Log–Normal	5.739	2.139
Exponential	786.194	832.643
2 para–exponential	189.088	301.878
Smallest extreme value	67.005	19.931
Weibull	49.130	14.589
3 para–Weibull	1.040	1.052
Largest extreme value	2.147	17.425
Logistic	8.264	4.688
Log–Logistic	5.117	4.980
3–Loglogistic	2.228	4.766

ultimate sagging strength is subjected to a larger uncertainty than the hogging strength, as indicated by the COV. The mean value in both conditions are slightly higher than unity, indicated that the CSR prediction is relatively conservative for the present case study

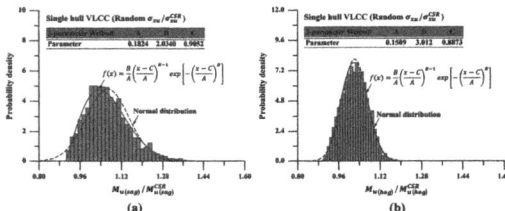

Figure 9. Histogram and best-fitted probability distribution of the ultimate bending capacity of the case study model.

model. A normal distribution with the predicted mean and variance are compared with the best-fit distributions in Figure 9. It can be seen that the normal distribution may not be appropriate for the sagging condition, as the histogram appears to be right-tail heavier.

5 DISCUSSIONS

The analysis results have demonstrated that a considerable computational uncertainty may exist in the prediction of ultimate ship hull strength due to the variability in the calculation ultimate compressive strength of structural segments. As indicated by Li et al. (2020a), the load-shedding in post-collapse regime may be another important factor. Hence, its effects and interaction with the ultimate compressive strength should be further investigated.

Compared with the existing deterministic procedure, the variation of the ultimate strength is estimated in the proposed procedure and can be used to assess the validity of the calculated mean value.

In terms of the further application, the present results (e.g. probability distribution and descriptive statistics) could be used in the reliability analysis of ship structures as an evaluation of the model uncertainty. In the ultimate limit state-based reliability analysis, the model uncertainty may be considered in a rather empirical way. Guedes Soares and Teixeira (2000) indicated that the material yield stress normally has a coefficient of variance of 7% to 8%. Combined with the additional uncertainty related to computation, a 15% COV was assumed for the ultimate ship hull strength. Similar application may refer to Gasper et al. (2014) and Frieze and Lin (1991), in which the uncertainty factor is assumed to follow the normal distribution with unity mean value and COV of 10% and 15% respectively. A more elaborated approach to account for the variability of material property and geometric dimension is the inverse application of the first order reliability method, as employed by Teixeira and Guedes Soares (2009). For sagging, it was proposed that the mean value between the ultimate bending moment and elastic bending moment was 1.014 and the standard deviation was 0.067. For hogging, it was proposed that the mean value between the

ultimate bending moment and elastic bending moment was 1.195 and the standard deviation was 0.081. Alternatively, as presented by Xu et al. (2015), the model correction factor method may be applied where the model correction factor was found by an iterative calculation so that the safety index estimations based on a realistic strength model and the corrected simplified model satisfies convergence criterion.

The probability distribution developed in this paper may be adopted in the reliability analysis where the strength calculation is completed in accordance with the CSR method. In this way, the uncertainty in strength computation using a standardised method is catered. Alternatively, the overall procedure can be employed for the ultimate strength calculation, which can further be combined with the uncertainty caused by material property and geometric dimension.

6 CONCLUSIONS

This paper examines the computational uncertainty in the calculation of ultimate ship hull strength using a probabilistic approach. Simplified progressive collapse method is adopted to predict the ultimate bending capacity of ship hull girder where an adaptable LSC formulation and Monte Carlo simulation are incorporated for sampling of the ultimate compressive strength of structural components. From this study, the following conclusions can be drawn:

- A considerable uncertainty exists in the prediction of ultimate bending capacity computation using simplified progressive collapse method of ship hull girders due to the LSC formulation;
- The computational uncertainty is more significant for sagging strength calculation of the present case study model;
- The results in the present study might be used to improve the reliability analysis of ship hull structures for evaluating the model uncertainty of ultimate strength calculation.

REFERENCES

14th ISSC, 2000. Ultimate Strength, Nagasaki, Japan.
18th ISSC, 2012. Ultimate Strength, Rostock, Germany.
Adamchak, J., 1982. ULTSTR: A program for estimating the collapse moment of a ship's hull under longitudinal bending, DTNSRDC report 82/076.
Benson, S., Downes, J., Dow, R.S., 2013. Compartment level progressive collapse analysis of lightweight ship structures. *Marine Structures* 31, 44–62.
Chalmers, D.W., 1993. Design of Ship's Structures. London: HMSO.
Dow, R.S., Hugill, R.C., Clark, J.D., Smith, C.S., 1981. "Evaluation of ultimate ship hull strength," Proc Extreme Load Response Symposium, Arlington, VA.

Frieze, P.A., Lin, Y.T., 1991. Ship longitudinal strength modelling for reliability modelling. *Proc Marine Structural Inspection, Maintenance and Monitoring Symposium*, Arlington, Virginia.

Gordo, J.M., Guedes Soares, C., 1996. Approximate method to evaluate the hull girder collapse strength. *Marine Structures*, 9, 449–470.

Guedes Soares, C. and Teixeira, A. P., 2000. Structural Reliability of Two Bulk Carrier Designs. *Marine Structures*, 13, 107–128.

Gaspar, B., Teixeira, A.P. and Guedes Soares, C., 2016. Effect of the Nonlinear Vertical Wave-Induced Bending Moments on the Ship Hull Girder Reliability. *Ocean Engineering*, 119, 193–207.

International Association of Classification Societies (IACS), 2015. Longitudinal Strength Standard for Container Ships.

International Association of Classification Societies (IACS), 2019. Common Structural Rules for Bulk Carriers and Oil Tankers.

Kim, D.K., Lim, H.L., Kim M.S., Hwang, O.J., Park, K.S., 2017. An empirical formulation for predicting the ultimate strength of stiffened panels subjected to longitudinal compression. *Ocean Engineering*, 140, 270–280.

Kim, D.K., Wong, E.W.C., Lee, E.B., Yu, S.Y. & Kim, Y.T., 2018. A method for the empirical formulation of current profile. *Ships and Offshore Structures*, 14, 176–192.

Kim, D.K., Lim, H.L., Yu, S.Y., 2019. Ultimate strength prediction of T-bar stiffened panel under longitudinal compression by data processing: A refined empirical formulation. *Ocean Engineering*, 192, 106522.

Li, S., Hu, Z., & Benson, S., 2019. Bending response of a damaged ship hull girder predicted by the cyclic progressive collapse method. *Developments in the Collision and Grounding of Ships and Offshore Structures*, 111–119.

Li, S., Hu, Z., & Benson, S., 2020. Progressive collapse analysis of ship hull girders subjected to extreme cyclic bending. *Marine Structures*, 73, 102803.

Li, S., Hu, Z., Benson, S., 2020. The sensitivity of ultimate ship hull strength to the structural component load-shortening curve. *Proc 30th Int Offshore and Polar Eng Conf*, Shanghai, China.

Li, S., Kim, D.K., Benson, S., 2020. An adaptable algorithm to predict the load–shortening curves of stiffened panels in compression. *Proc 5th Int Ships and Offshore Structures Conf*, Glasgow, UK.

Lin, Y.T., 1985. Ship longitudinal strength modelling. *PhD thesis*, University of Glasgow, Scotland.

Paik, J.K., Thayamballi, A.K., 1997. An empirical formulation for predicting the ultimate compressive strength of stiffened panels. *Proc 7th Int Offshore and Polar Eng Conf*, Honolulu, Hawaii, ISOPE, 328–338.

Smith, C.S., 1977. Influence of local compressive failure on ultimate longitudinal strength of a ship's hull. *Proc International Symposium on Practical Design of Ships and others Floating Structures (PRADS)*, Tokyo, Japan.

Smith, C.S., Dow, R.S., 1986. Ultimate strength of a ship's hull under biaxial bending. ARE TR 86204, Admiralty Research Establishment, Dunfermline UK

Syrigou, M., Benson, S.D., Dow, R.S., 2018. Progressive collapse assessment of intact box girders under combined bending and torsional loads. *Proc International Conference on Ships and Offshore Structures*, Gothenburg, Sweden.

Teixeira, A.P. and Guedes Soares, C., 2009. Reliability analysis of a tanker subjected to combined sea states. *Probabilistic Engineering Mechanics*, 24, 493–503.

Ueda, Y., Rashed, S.M.H., 1984. The idealized structural unit method and its application to deep girder structures. *Computers & Structures*, 18, 277–293.

Xu, M.C., Song, Z.J., Zhang, B.W., Pan, J., 2018. Empirical formula for predicting ultimate strength of stiffened panel of ship structure under combined longitudinal compression and lateral loads. *Ocean Engineering*, 162, 161–175.

Xu, M.C., Teixeira, A.P., and Guedes Soares, C., 2015. Reliability assessment of a tanker using the model correction factor method based on the IACS-CSR requirement for hull girder ultimate strength. *Probabilistic Engineering Mechanics*, 42, 42–53.

Yao, T., Nikolov, P., 1991. Progressive collapse analysis of a ship's hull under longitudinal bending. *Journal of the Society of Naval Architects of Japan*, 172, 437–471.

Zhang, S., Khan, I., 2009. Buckling and ultimate capability of plates and stiffened panels in axial compression. *Marine Structures*, 22, 791–808.

Developments in Maritime Technology and Engineering – Guedes Soares & Santos (eds)
© 2021 Copyright the Author(s), ISBN 978-0-367-77376-2

Influence of U-type stiffener on buckling strength of stiffened plate

Gui-jie Shi & De-yu Wang
State Key Laboratory of Ocean Engineering, Collaborative Innovation Center for Advanced Ship and Deep-Sea Exploration, School of Naval Architecture, Ocean & Civil Engineering, Shanghai Jiao Tong University, Shanghai, China

ABSTRACT: The U-type stiffener has two web plates welded to plate which will be result in a bigger constraint to resist buckling occurrence in its attached plate, which is widely used in hatch cover structures of ship cargo hold and deck panels of long bridge, due to its easier fabrication, maintenance and painting. However, there is still a gap between the existing evaluation criteria and actual buckling strength for U-type stiffened plate. In this paper, linear finite element method is used to study the elastic buckling strength of stiffened plate with and without U-type stiffener under axial compression. Based on numerical results, a simple equation for U-type stiffened plate buckling evaluation is proposed. Then, ultimate strength results of U-type stiffened plates prove that the proposed equation is reliable and reasonable. The conclusions in this paper can be applied to guide structural design and ship rule amendment.

1 INTRODUCTION

Common stiffeners, such as angle stiffener, bulb stiffener and T type stiffer, usually have one web plate connected with the attached plate. However, U-type stiffener has two web plates welded to the attached plate which will be result in bigger constraint to resist buckling occurrence in the attached plate between two U-type stiffeners. The U-type stiffener is widely used in hatch covers of ship cargo hold and deck panels of long bridge, due to its easier fabrication, maintenance and painting (Chen & Yang 2002; Shin et al. 2013).

The review of theoretical and design methods for stiffened panels did not identify either results for U stiffeners of the type used in the present panels (Gordo & Guedes Soares 2011).

Collapse model tests under axial compression were carried out on U-type stiffened panels of 200mm span (Gordo & Guedes Soares 2008), 300m span (Gordo & Guedes Soares 2011) and 400mm span (Gordo & Guedes Soares 2012). The test panels included one narrow model with one U-stiffener and another wide model with two U-stiffeners in width direction. These panels had made of very high tensile steel S690 on plating and mild steel on U-stiffener. The model tests showed that the panel collapses were induced by the flange plate buckling of U-stiffener and deformed almost without any ability to support a load after collapse. The ultimate strength of hybrid test panels reached about 2.0 time of minimum yield strength. And also the model tests on U-type stiffened panels found the plastic deformation that occurred during the first collapse of the wide panel did not affect much

the relationship between the structural modulus and the applied stress.

It should be noted in actual engineering the U-stiffener should be designed with a higher buckling strength than its attached plate in order to provide enough support to the whole panel. And also the scantlings of U-stiffener should satisfy the relevant industry standards for manufacturing.

The hatch covers should have enough capacity to withstand the design loads, including weather loads and cargo loads. The effect of ship motion, such as heave, pitch or roll should also been considered for increasing loads magnitude in static strength analysis. In addition, the hatch covers are loaded by forces due to elastic deformations of the ship's hull (IACS 2011). As the increase of applied loads, local buckling will takes place between stiffeners, and then the top panel will collapse as a stiffened plate under the action of combined lateral pressure and thrust loads (Yao et al. 2003). Therefore, the buckling collapse of the top panel will govern the ultimate strength and the overall collapse of the hatch cover.

In ship construction rules of classification societies, simply supported boundary conditions are always assumed in buckling evaluation of elementary panel plate. However, there are different types of stiffeners on plate boundary, which will give various degrees of constraint effect on the plate buckling strength. Therefore, stiffener effect on plate buckling can be represented by a kind of boundary correction factor.

One of boundary correction factors is F_{long}, which is referred to the longitudinal stiffener's effect on elementary panel plate under longitudinal compression. F_{long}

DOI: 10.1201/9781003216582-64

can be applicable to most part of ship structures including hatch covers, deck, upper tank, side tank, lower tank, outer bottom, inner bottom and longitudinal girder and so on. Although different types of stiffeners are used for building ship structures, this stiffener effect on plate buckling can be simplified into similar equations based on the analysis of mechanical principles.

However, the factor F_{long} of U-type stiffener hasn't regulated in IACS related documents, even in UR S21A (IACS 2011) unified requirements concerning scantlings of hatch covers. The buckling strength of U-type stiffened panel under axial compression can not use the same evaluation equations as that of the open profile stiffener. Apparently, UR S21A will greatly underestimate the ultimate strength of U-type stiffened panel under axial compression conditions, which will eventually unnecessarily increase the weight of hatch covers.

In this paper, a series of U-type stiffened plates are built based on the scantlings from actual ship structures and then the corresponding non-stiffened plates are assumed by removing U-type stiffener and at the same time applying simply boundary conditions on the connection U-type stiffener with panel plate. Elastic buckling strength of U-type stiffened plate and the corresponding non-stiffened plate is calculated by linear finite element method. The results show the effect of U-type stiffener on elastic buckling strength for stiffened

plate. Based on these numerical results, a simple equation for the boundary correction factor F_{long} is proposed. At last, the results by nonlinear finite element analysis prove that the proposed boundary correction factor can be applied to ultimate strength evaluation.

2 FE MODELING METHOD

As shown in Figure 1, U-type stiffener has two web plates, which are welded to the panel plate. The U-type stiffener and its attached plate compose a closed section, which has a bigger torsional constraint on the attached plate.

Three span panels allows for more realistic results by avoiding boundary conditions (Xu et al. 2013) problems for the middle plates related to eccentricity of load and for including the interference between adjacent panels (Gordo & Guedes Soares 2012).

Based on the Cartesian coordinate system, a FE model of U-type stiffened plate is built. X axis is assumed along the stiffener length direction, Y axis along panel width and Z axis perpendicular the XY plane. Considering the symmetry of the panel model and loading conditions, the FE model will include 3 U-type stiffeners and their attached plate, so the width of FE model is set to be $3b_d+3b_s$, in which b_d denotes the attached plate width under U-type stiffener and b_s denotes the attached plate width between

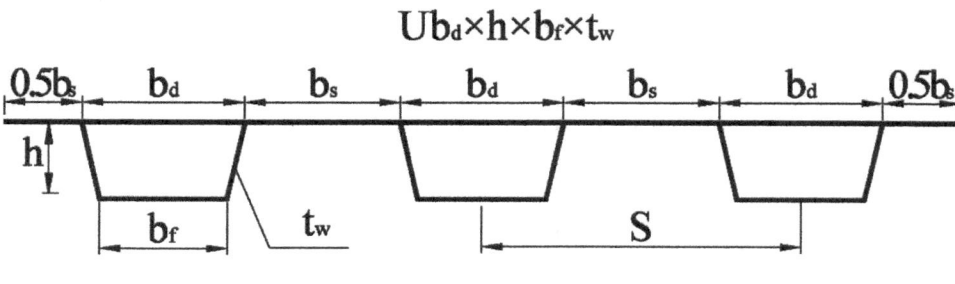

(a) Cross section

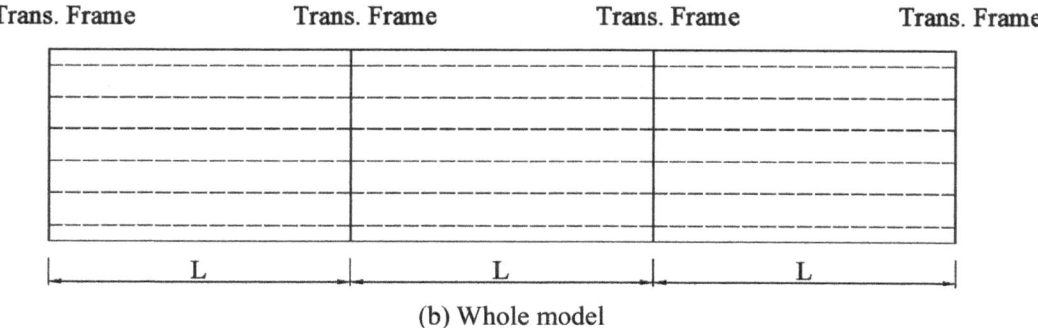

(b) Whole model

Figure 1. Schematic diagram of U-type stiffened model.

two U-type stiffeners, as shown in Figure 1. The length of FE model is set to be 3 frame spaces.

(1) The material properties of FE model are assumed to be linear elastic with elastic module E =206000 N/mm² and Poisson factor v=0.3.

Four-node shell element is applied in U-type stiffener web, flange and attached plate. A reasonable mesh density should be chosen to represent local deformations in structural members. The attached plate under U-type stiffener is divided into 10 elements, and U-type stiffener web is divided into 6 elements.

Simply supported constrains are supposed along the shorter edges of FE model, as shown in Figure 2. The detailed boundary conditions of FE model are as following:

(2) Simply supported conditions are applied at the two shorter edges including stiffeners to prevent side-way movement.

(3) One of shorter edges is constrained along X axis displacement and another shorter edge is applied uniform axial compression. After calculating the buckling strength of stiffened model, the U-type stiffener is removed to become un-stiffened model and at the same z=0 is applied at the connection between U-type stiffener and attached plate.

3 ANALYSIS OF BOUNDARY FACTOR F_{long}

IACS CSR rules offer an equation to calculate the elastic buckling strength of simply supported plates under uniform longitudinal compression, as shown:

$$\sigma_{\text{E}} = \frac{4\pi^2 E}{12(1-v^2)} \left(\frac{t_p}{b}\right)^2 \qquad (1)$$

where, b denotes the plate width, which is referred to bs and bd in this paper.

In order to study the effect of web thickness on the torsional constraint of longer edge, i.e. $F_{\text{long\_w}}$, the ratio between b_d and b_s is fixed, and a series of different web thickness of U-type stiffener is assigned in the FE model.

Then, the web thickness is fixed and a series of different width b_d and b_s is assigned in the FE model to study the effect of attached plate width on the constraint of longer edge, i.e. $F_{\text{long\_b}}$.

Owing to the two aspects, the U-type stiffener has an increase effect on the buckling strength of attached plate under axial compression, so the boundary correction factor $F_{\text{long}} = F_{\text{long\_w}} \cdot F_{\text{long\_b}}$

The detailed calculation procedure of the boundary correction factor F_{long} of U-type stiffened plate is as following:

(1) Based on hatch cover structures of actual ships, a U-type stiffened plate with representative scantlings is chosen to build FE model;

(2) Uniform axial compression is applied at the model end along the stiffener direction and then elastic buckling strength $\sigma_{\text{E\_with}}$ of the panel with U-type stiffeners is calculated by FEM;

(3) The U-type stiffeners are removed from FE model, Z=0 displacement constraint is applied at the connection between the stiffener and its attached plate, and at last elastic buckling strength $\sigma_{\text{E\_without}}$ of the panel without stiffener is calculated;

(4) For each panel, $F_{\text{long\_w}} = \sigma_{\text{E\_with}}/\sigma_{\text{E\_without}}$;

(5) In order to analyze the effect of U-type web on attached plate edge, the plate width b_d and b_s is fixed and U-type web thickness is assigned with different values. Then a series of results are obtained by repeating Step(1)~Step(4) to summarized a simplified equation for $F_{\text{long\_w}}$.

(6) Assuming the attached plate b_s in Figure 1 with simplified boundary conditions, i.e. $F_{\text{long\_w}} = 1.0$, the buckling strength of attached plate bs can be calculated by Eq(1), denoted as $\sigma_{\text{E\_Eq(1)}}$;

(7) For each panel, $F_{\text{long\_b}} = \sigma_{\text{E\_without}}/\sigma_{\text{E\_Eq(1)}}$;

(8) In order to analyze the effect of plate width on attached plate edge, the plate thickness is fixed and the plate width b_d and b_s is assigned with different values. Then a series of results are obtained by repeating Step(6)~Step(7) to summarize a simplified equation for $F_{\text{long\_b}}$.

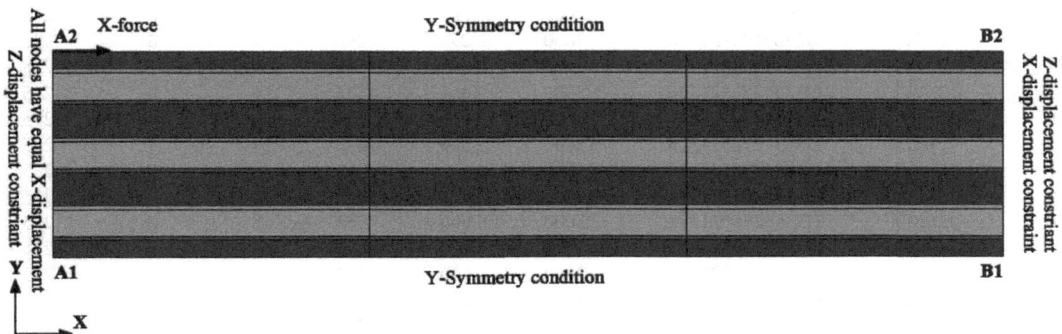

Figure 2. FE model of U-type stiffened plate.

Finally, the equation for stiffened plate is obtained by $F_{\text{long}} = F_{\text{long\_w}} \cdot F_{\text{long\_b}}$.

4 PROPOSAL EQUATION FOR BOUNDARY FACTOR F_{long}

A series of U-type stiffened models are chosen to carry out FE analysis and their scantlings are shown in Table 1. The results of elastic buckling strength are shown in Table 2 and Table 3. It is obvious that elastic buckling strength σ_{E_with} obtained by FE analysis is much bigger than $\sigma_{E\_\text{Eq}(1)}$ calculated by IACS CSR rules. The elastic buckling strength σ_{E_with} of stiffened plate obtained by FE analysis is 1.17~1.67 times bigger than $\sigma_{E\_\text{Eq}(1)}$. It is much conservative to evaluate the elastic buckling strength of U-type stiffened plate by only using an assumption of simply supported boundary condition.

After removing U-type stiffener from stiffened plate and adding displacement constrain z=0 at the connection of stiffener and attached plate, the elastic buckling strength $\sigma_{E\_\text{without}}$ is still bigger than $\sigma_{E\_\text{Eq}(1)}$ in some cases. This is because the attached plate b_s in Figure 1 is constrained by the shorter width of plate b_d. Furthermore, the shorter width the attached plate b_d is, the bigger constraint on attached plate b_s will

induce. Such boundary condition becomes more and more close to fixed constraint. So U-type stiffener will increase the buckling strength of stiffened plate owing to two aspects of torsional constraint on the longer edge, including the U-type stiffener web and different width of attached plate between U-type web.

Elemental panel plate between stiffeners is used as a basis to calculate the buckling strength. A correction factor is assumed as $F_{\text{long\_w}} = \sigma_{E\_\text{with}} / \sigma_{E\_\text{without}}$, which denotes the effect of U-type stiffener web on plate buckling strength. Based on the results in Table 2, the simplified equation for U-type stiffener is proposed as following:

$$F_{\text{long\_w}} = -0.07(tw/tp)^2 + 0.36(tw/tp) + 0.84 \quad (2)$$

The comparison of FE results and proposed formula is plotted in Figure 3, which shows that the proposed formula will give the consistent results as FE analysis of U1 and U2 stiffener.

Another correction factor is assumed as $F_{\text{long\_b}} = \sigma_{E\_\text{without}} / \sigma_{E\_\text{Eq}(1)}$, which denotes the effect of different attached plate width on buckling strength of U-type stiffener. In general, U-type stiffeners in cargo hatch cover of actual ships should satisfy $2.0b_d \geq b_s \geq b_d$. The comparison of FE analysis

Table 1. Scantlings of U-type stiffened plate.

			L	bs	bd	tw	tp	σ_Y
No.	Model	U stiffener	mm	mm	mm	mm	mm	MPa
1	U1S1T1	U290*170*213*6	3000	580	290	6	10	235
2	U1S1T2	U290*170*213*6	3000	580	290	6	9	235
3	U1S1T3	U290*170*213*6	3000	580	290	6	8	235
4	U1S1T4	U290*170*213*6	3000	580	290	6	7	235
5	U1S1T5	U290*170*213*6	3000	580	290	6	6	235
6	U1S1T6	U290*170*213*7	3000	580	290	7	6	235
7	U1S1T7	U290*170*213*8	3000	580	290	8	6	235
8	U1S1T8	U290*170*213*9	3000	580	290	9	6	235
9	U1S1T9	U290*170*213*10	3000	580	290	10	6	235
10	U1S2T5	U290*170*213*6	3000	508	290	6	6	235
11	U1S3T5	U290*170*213*6	3000	435	290	6	6	235
12	U1S4T5	U290*170*213*6	3000	363	290	6	6	235
13	U1S5T5	U290*170*213*6	3000	290	290	6	6	235
14	U2S1T1	U333*270*213*6	3000	666	333	6	10	235
15	U2S1T2	U333*270*213*6	3000	666	333	6	9	235
16	U2S1T3	U333*270*213*6	3000	666	333	6	8	235
17	U2S1T4	U333*270*213*6	3000	666	333	6	7	235
18	U2S1T5	U333*270*213*6	3000	666	333	6	6	235
19	U2S1T6	U333*270*213*7	3000	666	333	7	6	235
20	U2S1T7	U333*270*213*8	3000	666	333	8	6	235
21	U2S1T8	U333*270*213*9	3000	666	333	9	6	235
22	U2S1T9	U333*270*213*10	3000	666	333	10	6	235
23	U2S2T5	U333*270*213*6	3000	583	333	6	6	235
24	U2S3T5	U333*270*213*6	3000	500	333	6	6	235
25	U2S4T5	U333*270*213*6	3000	416	333	6	6	235
26	U2S5T5	U333*270*213*6	3000	333	333	6	6	235

Table 2. Elastic buckling strength of U-type stiffened plate when kept the same ratio between bs and bd (bs=2.0bd).

Model	σ_{E_with} MPa	$\sigma_{E\_without}$ MPa	$\sigma_{E\_with}/\sigma_{E\_without}$
U1S1T1	310	302	1.03
U1S1T2	255	245	1.04
U1S1T3	205	193	1.06
U1S1T4	161	148	1.09
U1S1T5	122	109	1.12
U1S1T6	126	109	1.15
U1S1T7	129	109	1.18
U1S1T8	131	109	1.21
U1S1T9	133	109	1.22
U2S1T1	234	228	1.02
U2S1T2	192	185	1.04
U2S1T3	154	146	1.05
U2S1T4	121	112	1.08
U2S1T5	91	82	1.11
U2S1T6	94	82	1.14
U2S1T7	97	82	1.17
U2S1T8	99	82	1.20
U2S1T9	100	82	1.22

Table 3. Elastic buckling strength of non-stiffened plate.

Model	$\sigma_{E\_without}$ MPa	$\sigma_{E\_Eq(1)}$ MPa	$\sigma_{E\_without}/\sigma_{E\_Eq(1)}$
U1S1T5	109	80	1.37
U1S2T5	139	104	1.34
U1S3T5	185	142	1.30
U1S4T5	251	204	1.23
U1S5T5	323	319	1.01
U2S1T5	82	60	1.36
U2S2T5	106	79	1.34
U2S3T5	140	107	1.30
U2S4T5	190	155	1.23
U2S5T5	244	242	1.01

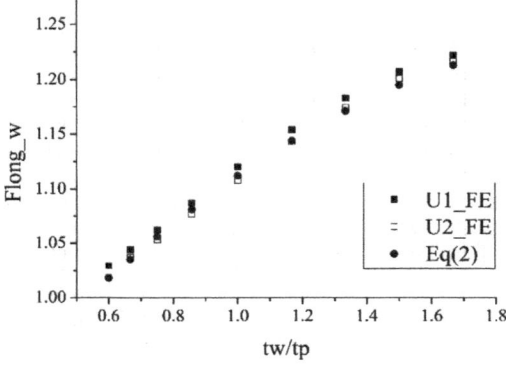

Figure 3. Boundary correction factor F_{long_w}.

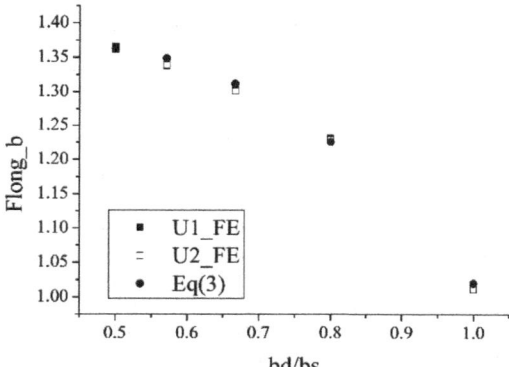

Figure 4. Boundary correction factor F_{long_b}.

results and Eq(1) for different width of attached plate are listed in Table 3. It is should be noted that U-type stiffeners are removed from stiffened plate and displacement constraints z=0 are added at the connection when analyzing the effect of different attached plate width.

Figure 4 Shows that the value of F_{long_b} will decrease as the increase of ratio b_d/b_s. Therefore, the simplified equation for U-type stiffener is proposed as:

$$F_{long\_b} = -0.74 * (b_d/b_s)^2 + 0.40 * (b_d/b_s) + 1.36 \tag{3}$$

where $2b_d \geq b_s \geq b_d$

In total, the calculation formula F_{long} of U-type stiffener is summarized as:

$$F_{long} = F_{long\_w} \bullet F_{long\_b} = [-0.07*(t_w/t_p)^2 + 0.36*(t_w/t_p) + 0.84]*[-0.74 * (b_d/b_s)^2 + 0.40 * (b_d/b_s) + 1.36] \tag{4}$$

The results in Table 4 show that the proposed Eq (4) has a good consistent with FE results.

5 EFFECT OF BOUNDARY FACTOR F_{long} ON ULTIMATE STRENGTH

Based on test data and numerical data of stiffened plates with open profile, such as flat bar, L-type and T-type, empirical expressions for predicting the ultimate strength under uniaxial compression can be derived. These empirical expression can be expressed in terms of the plate slenderness ratio β, which would be applicable in wide range of dimensions of stiffener and plate. One of these formulas

Table 4. Comparison of elastic buckling strength between Eq(4) and FE results.

Model	σ_{E_with} MPa	$\sigma_{E\_Eq(4)}$ MPa	$\sigma_{E\_Eq(4)}/\sigma_{E\_with}$
U1S1T1	310	307	0.99
U1S1T2	255	253	0.99
U1S1T3	205	204	0.99
U1S1T4	161	160	0.99
U1S1T5	122	121	0.99
U1S1T6	126	124	0.99
U1S1T7	129	127	0.99
U1S1T8	131	130	0.99
U1S1T9	133	132	0.99
U1S2T5	158	156	0.99
U1S3T5	212	207	0.98
U1S4T5	291	278	0.96
U1S5T5	406	362	0.89
U2S1T1	234	236	1.01
U2S1T2	192	194	1.01
U2S1T3	154	157	1.02
U2S1T4	121	123	1.02
U2S1T5	91	93	1.02
U2S1T6	94	96	1.02
U2S1T7	97	98	1.02
U2S1T8	99	101	1.02
U2S1T9	100	103	1.02
U2S2T5	118	120	1.02
U2S3T5	157	159	1.01
U2S4T5	218	214	0.98
U2S5T5	283	279	0.98

was proposed (Faulkner 1975) calculate the ultimate capacity accounting for the average initial imperfection, as following:

$$\sigma_U = \frac{2.25}{\beta} - \frac{1.25}{\beta^2} \qquad (5)$$

These existing formulas for ultimate strength evaluation of stiffened plate have not considered the closed-section effect on plate boundary, which may underestimate the ultimate strength of U-type stiffened plates.

In Sec4, it has been shown that the U-type stiffener increases the elastic buckling strength of attached plate under longitudinal compression. Such an increase of elastic buckling strength can lead to the increase of ultimate strength. The increase effect can by considered by adding the boundary factor into the plate slenderness β. The Eq.(5) should be changed as following for U-type stiffened plate:

$$\sigma_U = \frac{2.25}{\beta\sqrt{Flong}} - \frac{1.25}{\left(\beta\sqrt{Flong}\right)^2} \qquad (6)$$

The proposed Eq.(6) on ultimate strength of U-type stiffened can be validated the nonlinear finite element analysis(NFEA). The requirements of FE analysis procedure can be found in the reference (Shi & Gao 2019). Shell element type S4R is chosen in the analysis. The initial imperfection of the stiffened plate is imposed as the 1st elastic buckling mode. Based on IACS ship building standards (IACS 2017), the limit deformation of deck plating should be less than 6mm.

The results of ultimate strength by NFEA is listed in Table 5. The results show that the Eq(5) underestimate the ultimate strength of U-type stiffened plate by about 21%. The proposed Eq(6) by considering the boundary factor F_{long} will have a better estimation of ultimate strength. The mean value of the ratio between Eq(6) and NFEA is 0.91 with COV=0.08.

Table 5. Effect of proposed equation Flong on ultimate strength of stiffened plate.

Model	σ_{U_NFEA} MPa	$\sigma_{U\_Eq(5)}$ MPa	$\sigma_{U\_Eq(6)}$ MPa	$\sigma_{U\_Eq(5)}/\sigma_{U\_NFEA}$	$\sigma_{U\_Eq(6)}/\sigma_{U\_NFEA}$
U1S1T1	205	193	213	0.94	1.04
U1S1T2	201	181	202	0.90	1.01
U1S1T3	196	167	190	0.85	0.97
U1S1T4	191	151	175	0.79	0.92
U1S1T5	185	134	159	0.72	0.86
U1S1T6	185	134	161	0.72	0.87
U1S1T7	185	134	162	0.72	0.88
U1S1T8	185	134	164	0.72	0.88
U1S1T9	185	134	165	0.72	0.89
U1S2T5	190	149	174	0.78	0.91
U1S3T5	196	167	190	0.85	0.97
U1S4T5	204	189	206	0.93	1.01
U1S5T5	213	214	221	1.00	1.04
U2S1T1	200	177	198	0.89	0.99
U2S1T2	196	165	186	0.84	0.95
U2S1T3	191	151	173	0.79	0.91
U2S1T4	186	136	159	0.73	0.86
U2S1T5	180	120	143	0.67	0.79
U2S1T6	180	120	145	0.67	0.80
U2S1T7	180	120	146	0.67	0.81
U2S1T8	180	120	148	0.67	0.82
U2S1T9	180	120	149	0.67	0.82
U2S2T5	185	134	157	0.72	0.85
U2S3T5	191	151	173	0.79	0.91
U2S4T5	198	172	191	0.87	0.96
U2S5T5	207	198	206	0.96	1.00
		Mean		0.79	0.91
		COV		0.13	0.08

6 CONCLUSIONS

Common stiffener only has one web plate connected with attached plate. However, U-type stiffener has two web plates welded with plate, which forms a closed section on the attached plate. So U-type stiffener will greatly enhance the buckling capacity of stiffened plate.

The effect of U-type stiffener on plate buckling can be summarized in two aspects: U-type stiffener web plates will have a torsional constraint along the longer edge of attached plate and also different width of attached plate (i.e $b_d \neq b_s$) between two web plates will further increase torsional constraint.

The elastic buckling strength of stiffened plate and non-stiffened plate are calculated by linear finite element method. Based on the numerical results, a simplified equation F_{long} is proposed in this paper to estimate elastic buckling strength of U-type stiffened plate. At last, ultimate strength results by NFEA show that the proposed equation F_{long} is reliable and reasonable. Consequently, the proposed equation will fulfil the gap of UR S21A to estimate the buckling capacity of U-type stiffened plate.

ACKNOWLEDGEMENTS

This project is supported by National Natural Science Foundation of China (Grant No.51809168 and Grant No.51979163) and Young Teacher Initiation Program of Shanghai Jiao Tong University (17X100040060).

REFERENCES

Chen, S. & Yang, K. 2002. Inelastic behavior of orthotropic steel deck stiffened by U-shaped stiffeners. *Thin-Walled Structures*, 40 (6): 537–553.

Faulkner, D. 1975. A Review of Effective Plating for Use in the Analysis of Stiffened Plating in Bending and Compression. *Journal of Ship Research*, 19 (01): 1–17.

Gordo, J. M. & Guedes Soares, C. 2008. Compressive tests on short continuous panels. *Marine Structures*, 21 (2): 113–137.

Gordo, J. M. & Guedes Soares, C. 2011. Compressive tests on stiffened panels of intermediate slenderness. *Thin-Walled Structures*, 49 (6): 782–794.

Gordo, J. & Guedes Soares, C. 2012. Compressive Tests on Long Continuous Stiffened Panels. *Journal of Offshore Mechanics and Arctic Engineering*, 134: 021403.

IACS 2011. UR S21A: Evaluation of Scantlings of Hatch Covers and Hatch Coamings and Closing Arrangements of Cargo Holds of Ships. London.

IACS 2017. Rec 47 Shipbuilding and Repair Quality Standard. London.

Shi, G. & Gao, D. 2019. Analysis of hull girder ultimate strength for cruise ship with multi-layer superstructures. *Ships and Offshore Structures*, 14 (7): 698–708.

Shin, D. K., Le, V. A. & Kim, K. 2013. In-plane compressive strengths of HPS deck panel system stiffened with U-shaped ribs. *Thin-Walled Structures*, 63: 70–81.

Yao, T., Magaino, A. & Koiwa, T., et al. 2003. Collapse strength of hatch cover of bulk carrier subjected to lateral pressure load. *Marine Structures*, 16 (8): 687–709.

Developments in Maritime Technology and Engineering – Guedes Soares & Santos (eds)
© 2021 Copyright the Author(s), ISBN 978-0-367-77376-2

Uncertainty of ultimate strength of ship hull with pits

Xing Hua Shi, Haoran Shen & Jing Zhang
School of Naval Architecture and Ocean Engineering, Jiangsu University of Science and Technology, Zhenjiang, Jiangsu Province, P.R. China

C. Guedes Soares
Centre for Marine Technology and Engineering (CENTEC), Instituto Superior Técnico, Universidade de Lisboa, Lisbon, Portugal

ABSTRACT: Pitting corrosion may be a common progressive damage to ships, which may precipitate the collapse of the ship hull. In this paper, a method is introduced to study the ultimate strength of a ship hull with pits. A series of analysis will be performed to investigate the pit reduction on the ship hull ultimate strength, where different levels of corrosion damage will be included in different areas. As the pits happen randomly, the pit depths will change with time and area dramatically. Due to the high deviation of the pit size and locations, the ultimate strength of the ship hull with pits will be uncertain. By a series of Monte Carlo simulations, the uncertainty of the ultimate strength of a pitted ship hull will be assessed including the material, geometry randomness, as well as the sensitivity of the related factors.

1 INTRODUCTION

Failure of deck, bottom or side shell stiffened panels may lead to progressive collapse of the hull girder. During the recent decades, the reliability-based limit state design of ship or offshore structures has been the focus of researchers due to the importance of the safety of ships in their lifecycle (Guedes Soares et al, 1996). However, reliability-based design requires calculation of the ultimate state and the uncertainty, not only the elements as the structural panels and other members, especially the hull girder as the combination the elements.

Pitting corrosion is randomly distributed corrosion areas with local material reduction. In ships, pitting is usually observed to occur as a result of protection coating breakdown or ineffective catholic protection systems (Nakai et al., 2004). It is more likely to occur on the bottom plating of cargo oil tanks and some horizontal surfaces in ballast tanks (Magelssen, 2000, SRAJ, 2002). Although the bottom plates may be covered by a thin layer of oil behaving as a protective coat, severe localized corrosion will occur once this layer disappears.

Due to the increasing requirements by classification societies for the use of protective coatings, pitting corrosion is more frequently observed during ship surveys. As the ship loading is unbalanced, localized pits may be detrimental to ship structural strength. A large number of numerical studies have been carried out on simulated pitting corrosion patterns by varying the degree of pitting (DOP), pit depth, pit shape, pit size, pit distribution and location (Silva et al., 2013, Islam & Sumi, 2011, Khedmati et al., 2011, Huang et al., 2010, Ahmmad & Sumi, 2010).

The effects of pitting corrosion have mainly been studied on the plate elements under compressive, tensile, shear or bending loads. The localized corrosion pattern is either rectangular area or circular pits with various shapes (cylindrical, conical, semi-sphere or ellipsoidal). Structural properties such as aspect ratio and slenderness ratio have also been considered together with the corrosion.

However, the pitting corrosion may also be represented by a stochastic factor. Although the probabilistic characteristics of the corrosion process have been recognized, there are few studies of probabilistic modeling of the strength of pitting corroded structural elements.

The methods developed in the past decades for calculating the ultimate strength of the ship hull are mostly deterministic (Smith 1977, Gordo et al. 1996). However, due to the unavoidable presence of uncertainties in some input variables (such as geometry and material property parameters) and the simplification of modeling, the ultimate strength has to be evaluated using a probabilistic method. The statistical characteristics of the ultimate strength of structural members and hull girder are required for predicting the ultimate limit state based risk or reliability assessment for ship and offshore structures, in which the effect of structural uncertainty is taken account. The effect of the variability of the parameters and the uncertainty of the models have been analyzed in the assessment of the ultimate

DOI: 10.1201/9781003216582-65

strength of the structures (Guedes Soares 1988). A hull girder reliability assessment is performed including the ultimate collapse and local failure of midship cross section in vertical bending as possible failure modes (Guedes Soares & Ivanov, 1989).

The high-dimensional problems including the geometric and material properties of the midship cross section elements are considered as the basic random variables. The ultimate strength of a corroded plate with random initial distortions and random material and geometrical properties predicted by semi-empirical design equations or non-linear finite element analysis is adopted to demonstrate the accuracy of the proposed approach (Jiang and Guedes Soares, 2012). Gaspar & Guedes Soares (2013) determined the hull girder reliability assessment using a Monte Carlo based simulation method. The simulation method used was developed for component and system reliability problems and provides accurate estimates for the failure probability with reduced computational cost compared to crude Monte Carlo simulation. Campanile et al. (2014) studied the time-variant ultimate strength statistical properties of a bulk carrier in intact condition by Taylor series expansion method and Monte Carlo simulation, where the coefficient of variations is not only depend on time, but also on the assumed correlation among corrosion wastages of all structural elements.

The collapse of a ship hull with pits will be investigated, where the size of the pits will be different in different areas of the ship due to the varying corrosion rate. A series analysis is performed to study the ultimate strength of the ship hull with pits along the years. Due to the variability of the pit size, the uncertainty of the ship hull is analyzed including the randomness of the pits, geometry and material strength.

2 ULTIMATE STRENGTH OF SHIP HULL WITH PITTING CORROSION

2.1 Finite element model

The hull girder bending capacity evaluated by nonlinear finite element analysis (FEA). Figure 1 shows the FE model of the half of mid-ship cross section. The symmetric structural model includes deck, double bottom, side plating and longitudinal girders. The material of plates and stiffeners are AH steel with yield stress $\sigma_y = 315$MPa, elastic modulus $E = 2.058 \times 10^5$MPa, Poisson's ratio $\gamma = 0.3$. The shell 188 element, which is a four nodes element with six degrees of freedom at each node is adopted to the model the ship structure. Elastic-perfectly plastic constitutive model is adopted to take into account material nonlinearities.

2.2 Calculation method

The transverse frames are assumed to be strong and to serve as boundary conditions in their locations. The degrees, u_z and θ_x for the vertical members, u_y and θ_x for the horizontal structural components, are

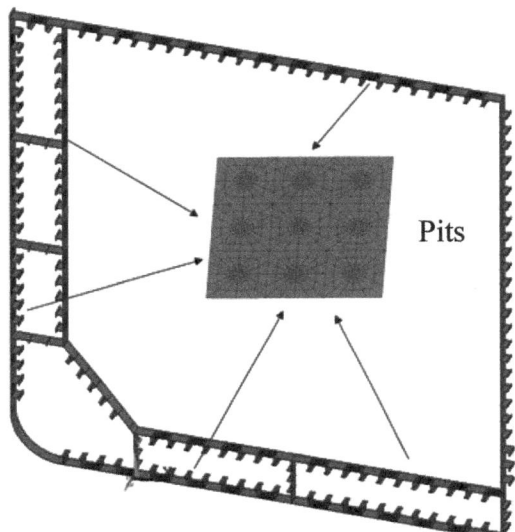

Figure 1. Model of ship hull with pits.

constrained at the transverse frame position. For the panels on deck and bottom, the symmetric boundary condition is applied at the end loading edge in the longitudinal direction, which means that the end loading edges have the same displacement an inverse rotation. The rotation degrees of freedom, θ_y and θ_z, are constrained at the edges (x = 0 and l) and l is defined as the length of the section. The displacements in the x direction, (u_x) at x = 0 and l are coupled to keep the hull cross-sections plane and constrained at x = 0 and vertical position of neutral axis. As shown in Figure 1, the coordinate of the model is defined as follows: x longitudinal direction of the ship hull, y transverse direction and z depth direction.

The displacement is applied on the nodes at the end of the hull cross section x = l. The moment M is calculated by the reaction force the end x = 0, is linear along the vertical direction and summed according to the distance between the nodes and the neutral axis.

The peak value of the vertical moment is defined as ultimate hull girder bending moment capacity M_u. The incremental-iterative method is adopted to simulate the progressive process damage of each structural element (stiffeners, stiffened panels and other elements) in the hull girder section and that represents the collapse behavior of the ship hull girder.

The M-R curve represents the relationship of the vertical moment and the applied rotation displacement, obtained by means of an incremental-iterative approach where rotation is applied incrementally and the adjustment for the instantaneous neutral axis is achieved iteratively. The bending moment acted on the hull girder section due to the imposed rotation R, is calculated for each step of the incremental procedure. Due to the incremental rotation of the hull girder section about its effective horizontal neutral axis, an increment in the axial strain ε is induced in each structural element in the section. In the hogging condition, the structural elements above the neutral axis are in tensile, while elements below the neutral axis are compressed and the contrast loading in the sagging bending condition.

The distance between the adjacent transverse frames, l is 2.205m as in (Xu et al., 2015). It will cost much time to perform enough number of analysis to obtain the uncertainty characteristics as the models need to have refined meshes. The value of l is decreased to reduce the calculation time. This is validated by comparing the results of the intact models calculated by different l length and Smith-based method (Smith 1977). The symmetric effect along the transverse direction is also validated. The intact models with different l, 0.6m and 2.205m are shown in Figure 2 without symmetric simplification along the transverse direction.

The ultimate strength results are shown in Figure 3 and Table 1. The curves of the short model with symmetric condition are close to the long models by FEA. The ultimate strength by FEA are larger than the results by Smith-based method obtained in (Xu et al., 2015). The model of l=0.6m, symmetric is adopted in this paper for the pitted ship hull analysis.

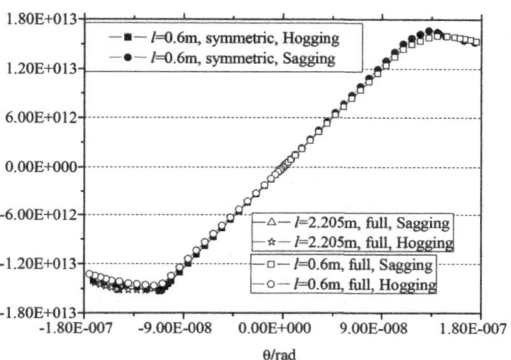

Figure 3. The results of intact model.

Table 1. Ultimate strength of different intact models.

Model	Hogging (Nmm)	Sagging (Nmm)
l=0.6m, symmetric	1.6673×10^{13}	-1.5314×10^{13}
l=0.6m, full	1.6559×10^{13}	-1.5197×10^{13}
l=2.205m, full	1.6022×10^{13}	-1.4656×10^{13}
Smith-based method	1.5824×10^{13}	-1.3242×10^{13}

2.3 Pit parameter

The ship structures suffer from corrosion induced by the sea water, salt, fog, cargos in the ship, among other effects (Guedes Soares et al, 2011; Zayed et al, 2018). Compared with the size of the hull, the corrosion damage of a single pit to the structure is very small. However, the accumulated damage will induce the collapse of the hull girder, as the number of pits on the hull structure is very large. In order to investigate the effect of pitting corrosion on the residual ultimate strength of ship hull structure, different pit corrosion is taken into account in deck plate, side outer plate, side inner plate, bottom plate and bottom inner plate.

The pit depth is calculated as the equivalent reduction of plate thickness caused by uniform corrosion with the same corrosion volume loss, where the time-varying corrosion loss model in (Paik et al., 2003) is used for different area. The pit diameter is assumed as 60mm for 10 years, 70mm for 15 years and 80mm for 20 years. Table 2 lists the size of pits on each structure in 10, 15 and 20 years of corrosion damage. The pitting corrosion damage of ship structure is simulated by sandwich shell element considering the thickness reduction.

2.4 Some results

Figure 4 to Figure 6 show the stress distribution of the ship hull with pitting corrosion damage. It can be clearly seen the stress distribution and deformation each element in sagging or hogging state. No matter in

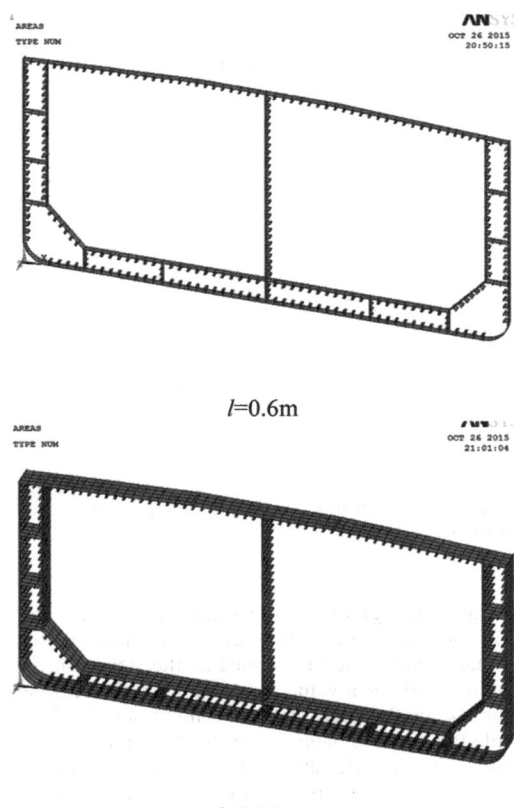

Figure 2. The intact models.

Table 2. Size of pitting corrosion in different area and periods.

Periods	Corrosion site	Side outer plate (mm)	Side inner plate (mm)	Bottom outer plate (mm)	Bottom inner plate (mm)	Deck plate (mm)
10 years	Pitting diameter	60	60	60	60	60
	Pitting depth	7.24	9.47	6.70	9.47	7.27
15 years	Pitting diameter	70	70	70	70	70
	Pitting depth	7.98	10.44	7.39	10.44	8.02
20 years	Pitting diameter	80	80	80	80	80
	Pitting depth	8.14	10.65	7.54	10.65	8.18

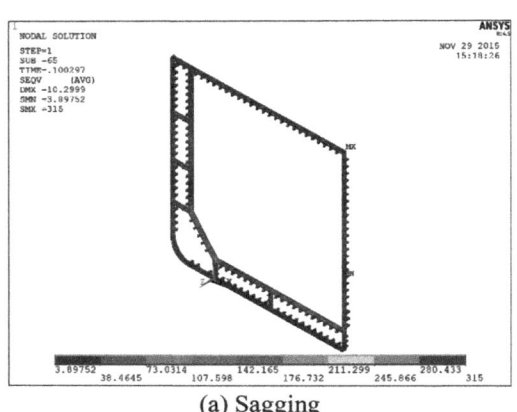

(a) Sagging

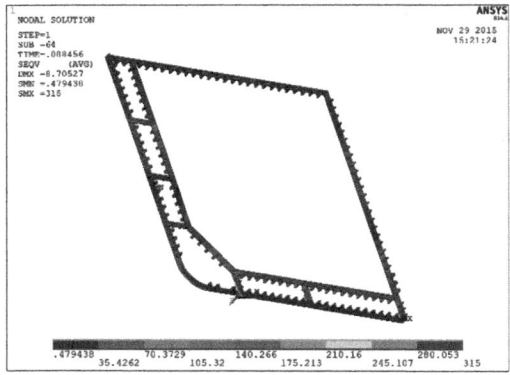

(a) Sagging

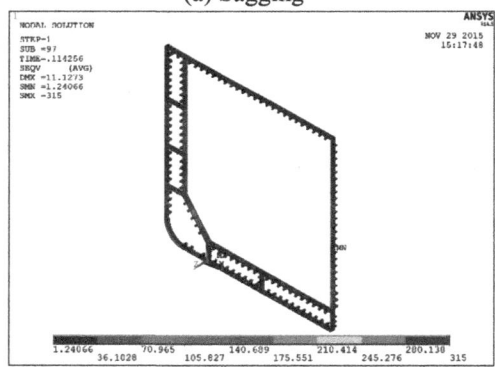

(b) Hogging

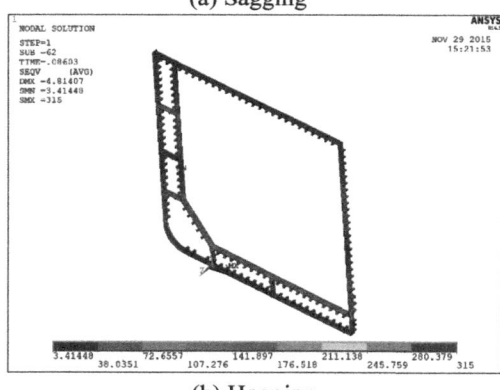

(b) Hogging

Figure 4. Stress distribution of hull subjected to pitting corrosion in 10 years.

Figure 5. Stress distribution of hull subjected to pitting corrosion in 15 years.

the condition of sagging or hogging, the maximum stress happens concentrated in the deck plate, bottom plate, side part near the deck or bottom. It is indicated that the main bearing elements with pit damage is located in the deck plate and bottom plate. These two locations reach the yield limit of materials firstly in tensile and also the most prone to buckling failure in compression. The stress of side plate, middle longitudinal bulkhead and longitudinal girder in the middle part along the depth direction of ship hull is relatively small.

Figure 4(a) gives the stress state of the ship hull with pits in 10 years in sagging condition at the ultimate state, which is defined as the step with the maximum moment value. It is found that the bucking happens for the deck plate in compression loading, while the yielding happens for the bottom plate. The bottom plate buckles and deck plate yields, as shown in Figure 4(b). The bilge structure has not reached the yield state, and its bearing capacity still has a certain reserve margin. It effectively connects the

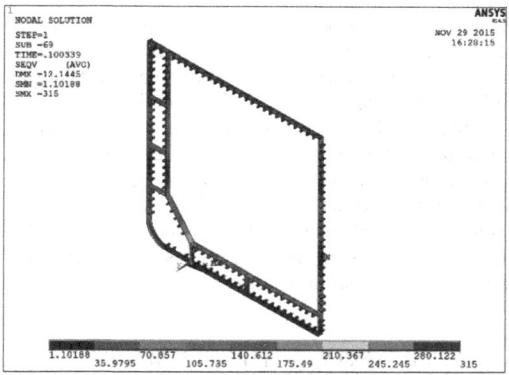

(a) Sagging

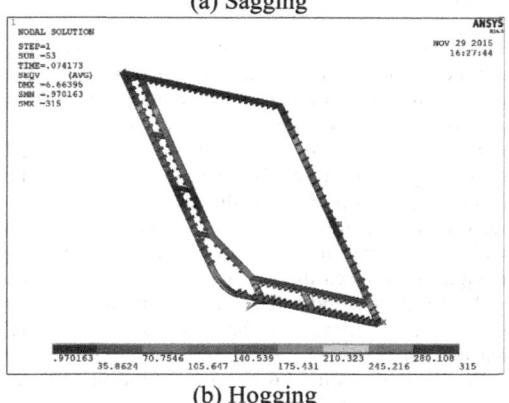

(b) Hogging

Figure 6. Stress distribution of hull subjected to pitting corrosion in 20 years.

Table 3. Ultimate moment of hull girder.

Condition	10-year corrosion wastage (Nmm)	15-year corrosion wastage (Nmm)	20-year corrosion wastage (Nmm)
Sagging	-1.4391×10^{13}	-1.4175×10^{13}	-1.3574×10^{13}
Hogging	1.5725×10^{13}	1.5336×10^{13}	1.4992×10^{13}

side inner plate and bottom inner plate structure. The plate and the bottom inner plate structure, the strong members between the side plates and the strong members between the bottom plates still have the ability to continue to carry, which provides strong support for the whole frame structure.

The Moment-Rotation curves are shown in Figure 7 (a), (b), (c) for the hull girder with the pit corrosion loss in 10 years, 15 years and 20 years, respectively. It is found that the bending moment of the hull girder increases with the increase of the applied rotation angle at the beginning linearly. The moment will decrease until a certain value, called ultimate bending moment.

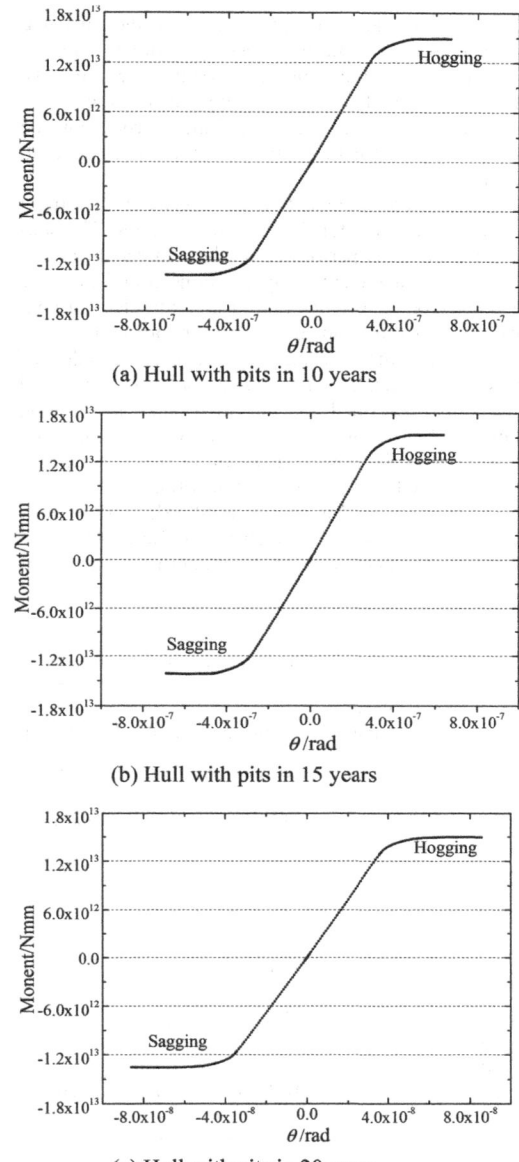

(a) Hull with pits in 10 years

(b) Hull with pits in 15 years

(c) Hull with pits in 20 years

Figure 7. M-R curves of ship's cross section.

3 STATISTICAL ANALYSIS OF THE ULTIMATE STRENGTH OF A SHIP HULL WITH RANDOM PITS

Stiffened plate is the basic component of ship hull structure. The failure of a single stiffened plate gradually extends to the failure of multiple stiffened plates, which eventually results in the collapse of the whole hull structure. The parameters of the structure as geometry, material strength and others may have significant influence of the failure of the stiffened plate. It is reasonable to assess the ultimate strength of the ship hull using probabilistic models. A statistical analysis is performed to investigate the ultimate strength of the ship hull considering the uncertainty of structure dimensions, material strength and pitting size.

Monte Carlo method is used to generate a sample including 100 data for the random variables. The statistical analysis is performed as the generated random samples are taken as the input parameters of the finite element model. A series nonlinear finite element analysis are performed including the random variables of structure thickness, yield stress, elastic modulus of material, pit depth. The elastic modulus and yield stress are described as the lognormal distribution (Guedes Soares & Garbatov, 1996), the structure dimension as the normal distribution. The pit depth is also described as the lognormal distribution as the statistical research in (Melchers, 2008). All random variables are assumed to be independent. Table 4 lists the statistical characteristics of the random parameters of

Table 4. Statistical characteristics of random parameters.

Name	Random variable	Mean value	COV	Distribution
Elastic modulus	E	205800MPa	0.06	Lognormal
Yield limit	σ_y	315MPa	0.06	Lognormal
Pitting depth	t_c	-	0.2	Lognormal
Deck thickness	t_d	20.5mm	0.03	Normal
Side thickness	t_{br}	16mm	0.03	Normal
Bottom thickness	t_b	17mm	0.03	Normal
Longitudinal bulkhead thickness	t_l	11mm	0.03	Normal
Bilge thickness	t_{bi}	16.5mm	0.03	Normal

the ship hull. The mean value of the pitting depth is taken as the value corresponding to 10 years, shown in Table 2. The different pit depth values are taken in different area, the same COV, as shown in Table 5.

Table 5 and Table 6 lists the statistical results of the ultimate bending moment of the hull structure with random pits in sagging and hogging, respectively. Compared with the hogging state, the mean value of the ultimate bending moment and the degree of dispersion are larger in sagging, and the coefficient of variation is almost the same.

4 UNCERTAINTY OF THE ULTIMATE STRENGTH

Based on the Monte Carlo simulation results using nonlinear finite element method, uncertainty of the ultimate bending moment of ship hull structure is analyzed. The histogram of the ultimate bending moment in sagging are shown in Figure 8. A normal probability distribution function is used to fit the histogram, according to the shape of the relative frequency of the ultimate moment. The correlation coefficient of the histogram of the ultimate moment in sagging between the normal distribution function is $R^2=0.855$. It is indicated that the ultimate bending moment in sagging can be described as normal distribution.

Figure 9 shows the histogram and the fitted function of the ultimate bending moment in hogging. The correlation coefficient of the ultimate bending moment in hogging is $R^2=0.947$, compared the histogram of ultimate bending moment in hogging and the normal distribution function.

The influence each random variable on the ultimate bending moment of ship hull structure can be obtained by sensitivity analysis. Figure 10 and Figure 11 give the sensitivity histogram of random variables in sagging and hogging, respectively. The first eight random variables including material parameters, structure dimensions and pitting depths are listed in the sensitivity curves for multiple linear regression analysis in order. It can be found that the

Table 5. Descriptive statistics of sagging.

Sagging	Mean value μ (N*mm)	Standard deviation σ	COV	Minimum value (N*mm)	Maximum value (N*mm)
Ultimate bending moment	-1.40×10^{13}	-2.34×10^{11}	0.0507	-1.23×10^{13}	-1.60×10^{13}

Table 6. Descriptive statistics of hogging.

Hogging	Mean value μ (N*mm)	Standard deviation σ	COV	Minimum value (N*mm)	Maximum value (N*mm)
Ultimate bending moment	1.56×10^{13}	6.66×10^{11}	0.0564	1.34×10^{13}	1.80×10^{13}

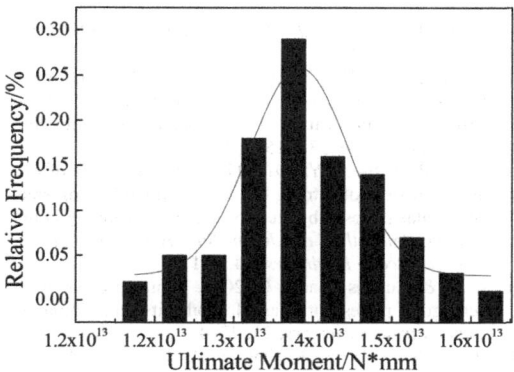

Figure 8. Histogram and fitting function of ultimate moment in sagging.

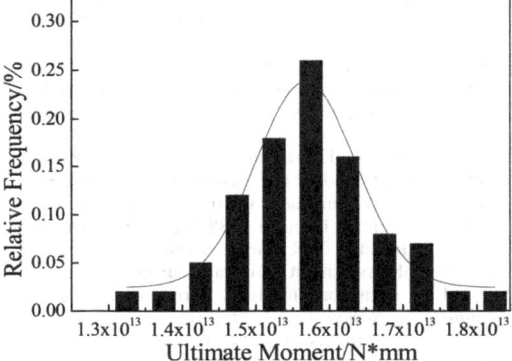

Figure 9. Histogram and fitting function of ultimate moment in hogging.

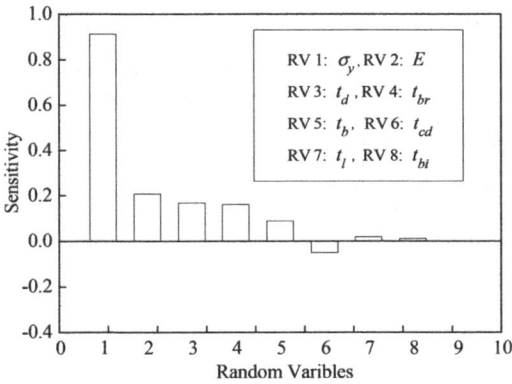

Figure 10. Sensitivity of random variables in sagging.

yield stress and elastic modulus are the dominant factors affecting the ultimate bending moment of ship hull structure with random pits in sagging. The deck thickness is the third influent factor for the ultimate bending moment in sagging. It is indicated

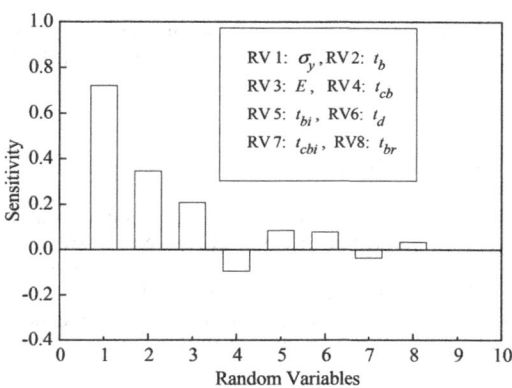

Figure 11. Sensitivity of random variables in hogging.

that the buckling of the deck plate in sagging may be dominant failure of the collapse of the ship hull with random pits in sagging. The pit depth in deck will affect the ultimate strength of the ship hull in sagging.

The dominant factors affecting the ultimate bending moment in hogging are different from that in sagging. As shown in Figure 11, the yield stress is the first factor that affect the ultimate bending moment. However, the bottom thickness is the second factor in hogging. The sensitivity value of pit depth in bottom plate in hogging is bigger than the pit depth in deck plate in sagging. The buckling failure of the bottom plate with pits may have significant influence on the collapse of the ship hull in hogging.

5 CONCLUSIONS

The ultimate strength of a ship hull with pits is investigated by a series calculations using nonlinear finite element method. Uncertainty analysis is performed to study the probability distribution of the ultimate bending moment. Some dominant factors that affect the ultimate bending moment are obtained in sagging or hogging. Some conclusions are as follows:

(1) The ultimate bending moment of ship hull decreases with the increase of pitting corrosion depth in sagging or hogging significantly, including the different pit depths in different area.
(2) Compared with the sagging, the point corrosion damage may have greater influence on the ultimate bending moment in hogging of the ship type presented in this paper.
(3) The buckling failure may be dominant the collapse of the ship hull with pits in sagging or hogging. The bottom plate thickness and the pit depth in bottom have obvious influence on the ultimate bending moment in hogging, while the

deck plate thickness and the pit depth in deck affect the ultimate bending moment of ship hull in sagging.

ACKNOWLEDGMENTS

The first author has been funded by National Natural Science Foundation of China (Grant No. 51809126, 51509113), Natural Science Foundation of Jiangsu Province, China (Grant No. BK20181468) and the Natural Science Foundation of the Higher Education Institutions of Jiangsu Province, China (Grant No. 16KJA580003).

REFERENCES

Ahmmad, M. M. & Sumi, Y. 2010. Strength and deformability of corroded steel plates under quasi-static tensile load. *Journal of Marine Science and Technology*, 15, 1–15.

Campanile, A., Piscopo, V. & Scamardella, A. 2014. Statistical properties of bulk carrier longitudinal strength. *Marine Structures*, 39, 438–462.

Gaspar, B. & Guedes Soares, C. 2013. Hull Girder Reliability using a Monte Carlo Based Simulation Method. *Probabilistic Engineering Mechanics*. 31:65–75.

Gordo, J. M.; Guedes Soares, C., and Faulkner, D. 1996. Approximate Assessment of the Ultimate Longitudinal Strength of the Hull Girder. *Journal of Ship Research*. 40(1):60–69.

Guedes Soares C. 1988. Uncertainty Modelling in Plate Buckling. *Structural Safety*, 5, 17–34.

Guedes Soares, C. 1992. Design equation for ship plate elements under uniaxial compression. *Journal of Constructional Steel Research*, 22, 99–114.

Guedes Soares, C. & Garbatov, Y. 1996. Reliability of Maintained Ship Hulls Subjected to Corrosion. *Journal of Ship Research*, 40, 235–243.

Guedes Soares, C. & Ivanov, A. D. 1989. Time dependent reliability of the primary ship structure. *Reliability Engineering and System Safety*, 26, 59–71.

Guedes Soares, C.; Garbatov, Y., and Zayed, A. 2011. Effect of environmental factors on steel plate corrosion under marine immersion conditions. *Corrosion Engineering Science and Technology*. 46(4):524–541.

Guedes Soares, C.; Dogliani, M.; Ostergaard, C.; Parmentier, G., and Pedersen, P. T. Reliability Based Ship Structural Design. *Transactions of the Society of Naval Architects and Marine Engineers* (SNAME). 1996; 104:357–389.

Huang, Y., Zhang, Y., Liu, G. & Zhang, Q. 2010. Ultimate strength assessment of hull structural plate with pitting corrosion damnification under biaxial compression. *Ocean Engineering*, 37, 1503–1512.

Islam, M. R. & Sumi, Y. 2011. Geometrical effects of pitting corrosion on strength and deformability of steel rectangular plates subjected to uniaxial tension and pure bending. *Journal of the Japan Society of Naval Architects and Ocean Engineers*, 14, 9–17.

Jiang, X. & Guedes Soares, C. 2012. Ultimate capacity of rectangular plates with partial depth pits under uniaxial loads. *Marine Structures*, 26, 27–41.

Khedmati, M.R., Mahdi Roshanali, M. & Nouri, Z.H.M.E. 2011. Strength of steel plates with both-sides randomly distributed with corrosion wastage under uniaxial compression. *Thin-Walled Structures*, 49, 325–342.

Magelssen, W. 2000. Corrosion/corrosion protection – what impact will this have on modern ship design. DNV.

Melchers, R. E. 2008. Extreme value statistics and long-term marine pitting corrosion of steel. *Probabilistic Engineering Mechanics*, 23, 482–488.

Nakai, T., Matsushita, H., Yamamoto, N. & Arai, H. 2004. Effect of pitting corrosion on local strength of hold frames of bulk carriers (1st report). *Marine Structures*, 17, 403–432.

Paik, J., Wang, G., Thayamballi, K., Lee, J. & Park, Y. 2003. Time-Dependent Risk Assessment of Aging Ships Accounting for General/Pit Corrosion, Fatigue Cracking and Local Denting Damage. ABS Technical Papers.

Silva, J. E., Garbatov, Y. & Guedes Soares, C. 2013. Ultimate strength assessment of rectangular steel plates subjected to a random localised corrosion degradation. *Engineering Structures*, 52, 295–305.

Smith, C., 1977, Influence of Local Compressive Failure on Ultimate Longitudinal Strength of a Ship Hull, *Proceedings of the International Symposium on Practical Design in Shipbuilding (PRADS)*, Japan, pp. 73–79.

SRAJ 2002. Study on cargo oil tank corrosion of oil tanker, in: Report of Ship Research Panel 242. The Shipbuilding Research Association of Japan, Tokyo.

Xu, M. C., Teixeira, A. P. & Guedes Soares, C. 2015. Reliability assessment of a tanker using the model correction factor method based on the IACS-CSR requirement for hull girder ultimate strength. *Probabilistic Engineering Mechanics*, 42, 42–53.

Zayed, A.; Garbatov, Y., and Guedes Soares, C. 2018. Corrosion degradation of ship hull steel plates accounting for local environmental conditions. *Ocean Engineering* 163: 299–306.

Developments in Maritime Technology and Engineering – Guedes Soares & Santos (eds)
© 2021 Copyright the Author(s), ISBN 978-0-367-77376-2

Study on residual strength of stiffened panels with dent damages

Kun Sun, Ling Zhu & Liang Xu
School of Transportation, Wuhan University of Technology, Wuhan, China, WUT-UL Joint International Laboratory of Extreme Load and Response, Wuhan, China

C. Guedes Soares
Centre for Marine Technology and Ocean Engineering (CENTEC), Instituto Superior Técnico, Universidade de Lisboa, Lisbon, Portugal, WUT-UL Joint International Laboratory of Extreme Load and Response, Wuhan, China

ABSTRACT: Experimental and numerical simulations are used to study the effects of dent damage on the residual strength of stiffened panels under uniaxial compression. Firstly, through the quasi-static lateral loading test, a dent with specified shape and location is imposed on the stiffened panels. Then, in order to analyze the effect of the dent damage on the residual strength of the stiffened panels, uniaxial compression tests of the intact stiffened panels and the stiffened panels with the dent damages are carried out under simply supported boundary conditions. The results show that the residual strength of the stiffened panels is reduced by the dent damage. The nonlinear finite element software ABAQUS is used to simulate the response of the panels, and the residual strength of the stiffened panels is obtained. The experimental results are in agreement with the numerical results, and have verified the numerical simulation. On this basis, further studies on the effects of parameters, such as the depth and area of the dent on the residual strength of the stiffened panels are carried out by means of numerical simulations.

1 INTRODUCTION

Ships may suffer from various kinds of damages in their life cycle. The actual examples are: when ore carrier is irregularly loaded, iron ore falls from a high place and impacts the inner bottom plate; the grab of the bulk carrier hits the inner bottom plate; the flight deck is dented by the impact of aircraft taking off and landing. The above conditions will cause local dents on the structure. Dent damage will reduce the strength and stiffness of the structure, affect the normal functioning ship structure and may even endanger the safety of ships.

Therefore, it is significant to evaluate whether a dented structure has enough residual strength to resist the load. For the design, in order to keep the integrity of the structure, it is necessary to consider the damage redundancy of the structure; for the maintenance and replacement, it is necessary to understand the integrity of the structure, and then corresponding decisions can be made. Therefore, based on the concept of residual strength, it is necessary to evaluate the residual strength of damaged structure.

Some experiments and FE studies have been performed to investigate the residual strength of plate and stiffened panels with dent damages.

Paik et al. (2003) analyzed the effect of dent shape, size (depth, diameter) and location on the residual

strength of steel plates under axial compression loads. Spherical and conical denting shape are considered, but the results are similar. The diameter of the dent can influence the ultimate strength significantly, while the depth has little effect as long as the diameter is small. The longitudinal location of the dent can affect the ultimate strength. When the dent location is close to the unloaded edge, ultimate strength is decreased by 20%. Further, a closed-form formula is empirically derived to predict ultimate strength of dented steel plates.

Luis et al. (2009) studied the effect of dimple imperfection on the collapse strength of plate assemblies. The position of dimple imperfection was analyzed as an important parameter. Longitudinal position of dimple imperfection could be important in some cases. The effect of dimple imperfection on the ultimate strength depended on the slenderness of the panels.

Witkowska et al. (2009) carried out finite element analysis about local damage on a stiffener. Different damage locations and numbers in one stiffener are considered. Local damage on a stiffener can reduce strength of the plate. When two damages are located on a stiffener, the second damage next to the first one decreases the strength of the plate, but does not change the general behavior.

Raviprakash et al. (2012) studied various dent parameters' effects on the ultimate strength of a square plate under uniaxial compression. These parameters

DOI: 10.1201/9781003216582-66

include dent length, dent width, dent depth and orientation angle of the dent. As the dent depth increases, the ultimate strength of plate decreases. Plates with a longitudinal dent have higher ultimate strength than plates with a transverse dent of same size.

Amante et al. (2014) analyzed the ultimate strength of dented FSPO side stiffened panels suffered from bow collisions. Various impact velocities and locations are considered. With the increase of the speed, the dent depth increases, and as a result, residual strength decreases. Dent on the stiffener can reduce strength by 9% at most, while dent on the plate can reduce 7.18% at most.

Saad-Eldeen et al. (2015) studied the ultimate strength of dented rectangular plates subjected to uniaxial compression loading. The effects of different boundary conditions, plate thicknesses and dent depths were analyzed by means of nonlinear finite element method.

Witkowska et al. (2015) studied ultimate strength of locally damaged plate panels. Three models were analyzed to evaluate the effect of adjacent intact plates and determine the minimum size of the model. The conclusion was that one span model was enough for analysis of ultimate strength but whole collapse behavior needed large panel.

Xu et al. (2015) carried out NLFEA on central dented narrow stiffened panels under axial compression, considering the residual strength caused by indentation. The residual stress caused by indentation can affect ultimate strength, but depends on the dimension of dent and panels and whether the tension residual strength can be neutralized be compression load or not.

Chujutalli et al. (2018) carried out experimental and FE analysis on the ultimate strength of smallscale stiffened panels with indented stiffeners. Seven specimens are tested, two of them are intact, the rest are dented. Panels with indentation at mid-span decreases ultimate strength more than indentation at a quarter of the span. Parametric study on dent depth and location is then carried out. A formula to predict residual strength of stiffened panels with dent is proposed.

So, a lot of numerical simulations have been carried out to determine the effect of local dent on the ultimate strength of plate and stiffened panels but relevant experiment studies are not as many as they could be. Chujutalli et al.□2018□studied the effect of dent damage on ultimate strength of stiffened panel under clamped boundary conditions. But the collapse behavior of stiffened panel under simply supported boundary condition haven't been studied. Although the ultimate strength of a specimen can be obtained by test, the stress distribution of the model can not be observed. Therefore, experimental and finite element methods were both necessary to study the ultimate state of model.

In this paper, experiments and simulations are conducted to study the effects of local dent on ultimate strength of stiffened panels under simply supported boundary condition.

2 DISCRIPTION OF MODEL

Stiffened panels are the most common structural components utilized in ship and offshore structure. Experiments are carried out on a series of stiffened panels which are longitudinal stiffened in Figure 2. The model is one bay under uniaxial compression loads. The cross-section layout can be observed in Figure 1.The span of the model is 900mm, width is 1000mm, thickness of plate is 3mm. Stiffener is 50 ×3mm. The slenderness of the plate β is 3.4, while the slenderness of the stiffener λ is 0.9. The material of the model is mild steel. There are 3 models in all, one of them is named SP0 as intact specimen, the other are named SP1 and SP2 as dented specimen.

3 EXPERIMENTAL PROCEDURE

3.1 *Tensile test*

To obtain the mechanical properties of material, quasi-static tensile tests are carried out. The material of the tensile test pieces is mild steel, which is the same as the specimens. The tensile rate is set as 2.00mm/min for all test pieces. Young's modulus, yield stress and Poisson's ratio are represented as E,σ_y and μ in Table 1. The stress-strain curve is presented in Figure 3. After the tensile tests, the engineering stress-strain relationship is converted to true stress-strain relationship.

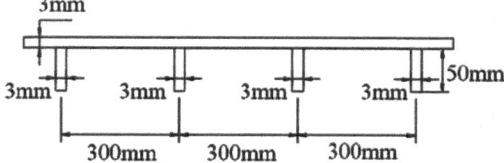

Figure 1. Detail of the model.

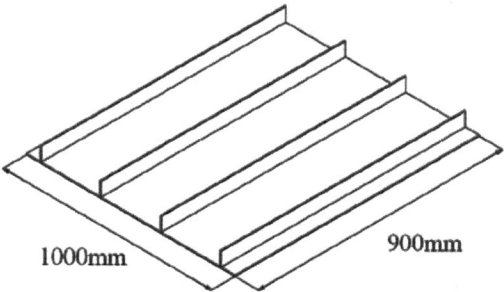

Figure 2. Geometry of the model.

Table 1. Mild steel material property.

E GPa	σy MPa	μ
210	193	0.3

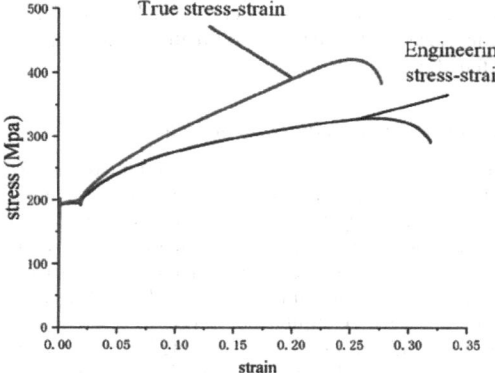

Figure 3. Stress-strain curves of material.

3.2 Quasi-static denting test

Due to difficulties of defining the dent dimension directly, the dimension of the indenter is used to define the dent dimension in Figure 4. Dent length is L_a, dent breadth is L_b, dent depth is L_d. Dent length and breadth are defined identical to the length and breadth of indenter so readers can refer to the dimension of dent easily. The dent depth is the distance between the bottom of the dent and plane of the plate. In experiment, a rectangular indenter of 170mm long and 60mm wide is applied.

To produce stiffened panels with dent damage, tests are implemented with jack and pressure transducer. An indenter is placed on the center of a panel, then upholders, jack and pressure transducer are put on one another in sequence, which is shown in Figure 5. Due to the restricted space and equipment of laboratory, stiffened panels can only be laid on the ground rather than supported at four edges. During the experiment,

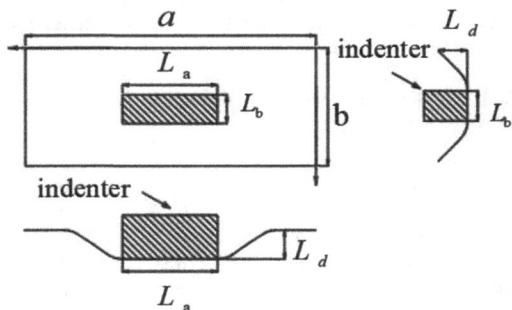

Figure 4. Geometry of dent.

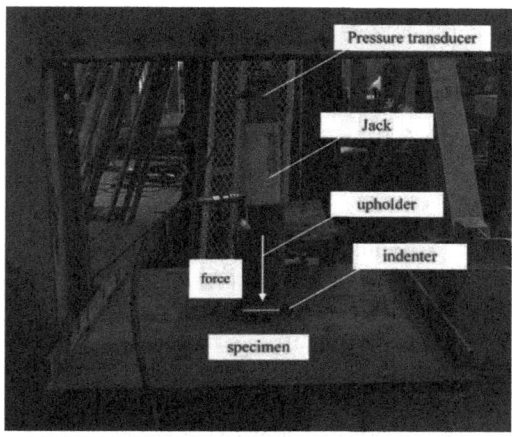

Figure 5. Quasi-static denting test.

load and displacement can be observed by pressure transducer and dial gauge respectively at the same time. Final dent depth is measured by means of Vernier caliper.

3.3 Uniaxial compression test

After the denting test, uniaxial compression tests are conducted with all 3 specimens in sequence. The set-up of uniaxial compression tests was shown in Figure 6. The boundary condition of 3 specimens is simply supported. Special tools are designed to

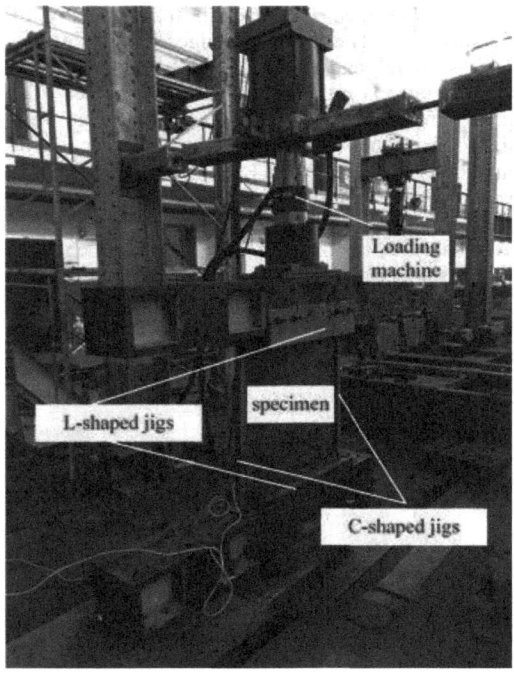

Figure 6. Uniaxial compression test.

593

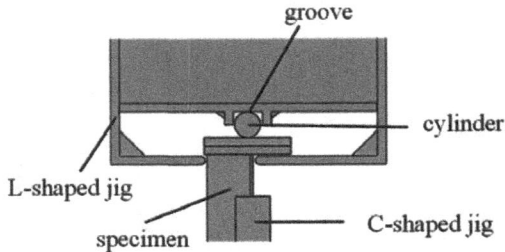

groove

cylinder

L-shaped jig

specimen

C-shaped jig

Figure 7. Side view of compression test.

simulate such boundary condition. Side view of compression tests was displayed in Figure 7. Loaded edges and unloaded edges are considered respectively because the tools mounted at the loading edges must be capable of withstanding and transferring loads. The plan is connecting loading edges of specimens with cylinders by bolts, then putting cylinder in the groove. At last, two L-shaped jigs are fixed on the loading equipment to constrain the boundary of the specimen.

With the help of cylinder and groove, rotational freedom of specimen is allowed. Lateral displacement of the edges is constrained by two jigs, so the edges of specimen can not move freely. Two C-shaped jigs are used to constrain unloaded edge, which is shown in Figure 8. They are sufficiently rigid to withstand the lateral deformation of specimen. The shape of the jigs is carefully designed to constrain the boundary of the specimen. The unloaded edges in the jigs are hold straight and tight, displacement is constrained but except for rotation. The contact zone of the unloaded edges and jigs is lubricated with oil, so freedom of rotation along the direction of unloaded edge is free. Load and displacement data can be acquired directly from the loading machine. The position of the specimen is calculated carefully so that the load is applied on neutral axis of specimen. The loading point is at the neutral axis of specimen, the location of the neutral axis is calculated as the static moment of specimen divided by the area of specimen.

4 NUMERICAL SIMULATIONS

FE software ABAQUS is utilized throughout the numerical simulation. The simulation work consists of two procedures: quasi-static denting simulation and uniaxial compression simulation. The stiffened panels are modeled as deformable shell element of type S4R. The indenter is modeled as discrete rigid body. Initial imperfection is considered in the simulation because it affects the ultimate strength of the stiffened panels.

4.1 Quasi-static denting simulation

Indenter and stiffened panels are modeled in this part. Static general analysis is adopted to simulate the whole denting process. Resilience of the panels and residual stress caused by indentation are considered in the simulation. The interaction properties between the indenter and panel are defined as frictionless, because Chujutalli et al (2018) find out that the final indentation is not affected by frictional coefficient. The boundary condition of stiffened panels is described in Table 2 and Figure 10. Constrains are applied on the edges of stiffeners. Linear displacements x, y and z were zero. Indenter's all six degrees of freedom are fixed except vertical displacement U_y. A reference point is established on the center of indenter to control the loading and unloading.

The first step is introducing initial imperfections in the stiffened panels. Superposition of the buckling mode method is used to simulate the initial imperfections of the models. The stiffened panels are analyzed for their buckling modes. Linear perturbation, buckle analysis step is adopted to acquire the buckling modes. The typical buckling eigenvalue which could reflects initial imperfection is imported to the model. First eigenvalue of buckling mode is adopted in the simulation as initial imperfection, which is shown in Figure 9. The second step is moving indenter towards the center of the panel by setting a predefined displacement of indenter until indenter reached the maximal displacement. Third step is removing the indenter. Forth step is simulating resilience process of the panel and acquire final plastic deformation. It can be observed from Figure 11 that stress decreases from center to edge, and stiffeners

Figure 8. C-shaped jig.

Table 2. Boundary condition of denting tests.

	X1X1'	X2X2'	X3X3'	X4X4'
U_x	0	0	0	0
U_y	0	0	0	0
U_z	0	0	0	0
UR_x	free	free	free	free
UR_y	free	free	free	free
UR_z	free	free	free	free

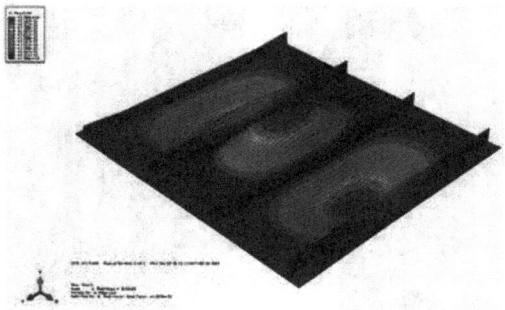

Figure 9. 1 eigenvalue of buckling mode.

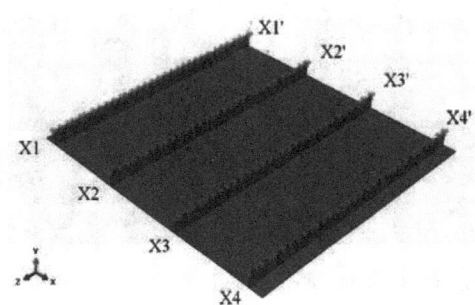

Figure 10. Boundary condition of the denting simulation.

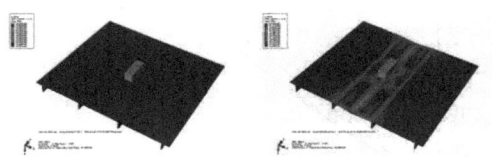

Figure 11. Indentation process.

stop stress from moving to the side of the panel. So, the effect of dent is mainly on the middle panel. As a result of deflection and residual stress caused by dent, the specimen is likely to collapse from the half span.

4.2 Uniaxial compression simulation

Static Riks method of ABAQUS is used to carry out uniaxial compression simulation. Material properties obtained from tensile tests are adopted in the simulation. Before uniaxial compression simulation, deformation and residual stress caused by indentation are imported as initial state of predefined field. Boundary condition of stiffened panels is simply supported at four edges. Detailed description can be

Table 3. Boundary condition of compression tests.

	AB	CD	AD	BC
U_x	0	0	0	0
U_y	0	0	0	0
U_z	free	free	free	free
UR_x	0	0	free	free
UR_y	0	0	0	0
UR_z	free	free	0	0

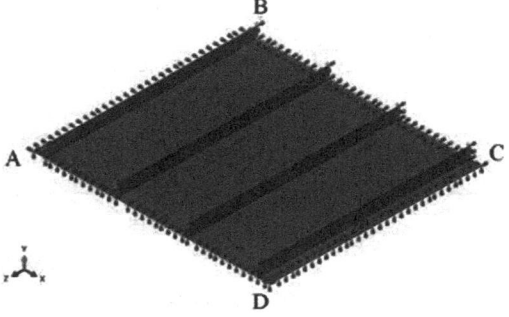

Figure 12. Boundary condition of axial compression simulation.

observed in Table 3 and Figure 12, which is the same as model tests. The boundary condition is defined according to the local coordinate system in Figure 12. Two loading edges are kinematically coupled with two reference points respectively to apply loads and obtain axial displacements.

5 RESULTS AND DISCUSSION

Quasi-static denting test can be seen in Figure 14, and data is recorded in Table 4. During the tests, two specimens are monitored by pressure transducer and dial gauge. Because the jack is controlled manually, the results can not be perfectly identical. But attempts are made to ensure results are close by observing lateral displacement and force through dial gauge and pressure transducer. According to the definition of dent in Figure 4, dent depth measured by a Vernier caliper is shown in Table 4. Dent depth of SP1 is 10.28mm, and force applied is 62.5KN. Dent depth of SP2 is 10.08mm, and the force applied is 61.3KN. Dent depth of SP1 differs 1.9% from SP2. The force applied on SP1 differs 1.9% from SP2.

Results of uniaxial compression tests are plotted in Figure 16-19, and the ultimate strength of each specimen is given in Table 5. By comparing SP0 with SP1 and SP2, it can be acknowledged that dent damage reduced ultimate strength of the stiffened panels. Residual strength of SP1 is the same as the strength of SP2, which is 100MPa. The ultimate

Table 4. Data of denting tests.

	Force(KN)	L_d(mm)
SP1	62.5	10.28
SP2	61.3	10.08

Table 5. Results of uniaxial compression tests.

	σ_u (MPa)	Reduction
SP0	111	0
SP1	100	-9.9%
SP2	100	-9.9%

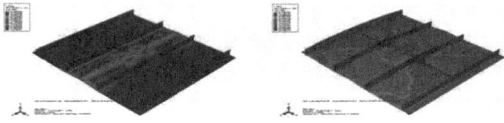

Figure 13. Uniaxial compression simulation.

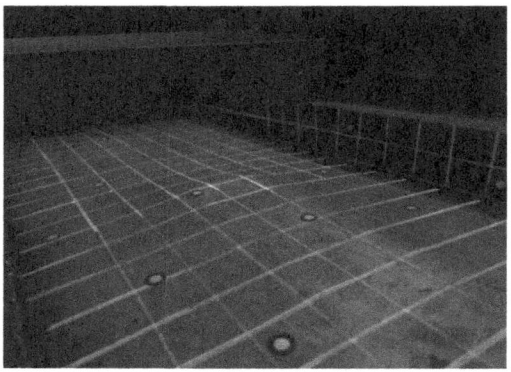

Figure 14. Local dent.

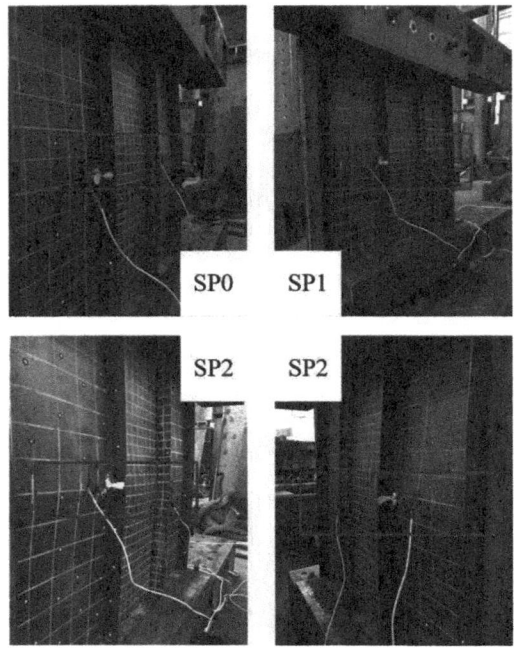

Figure 15. Buckling of specimen.

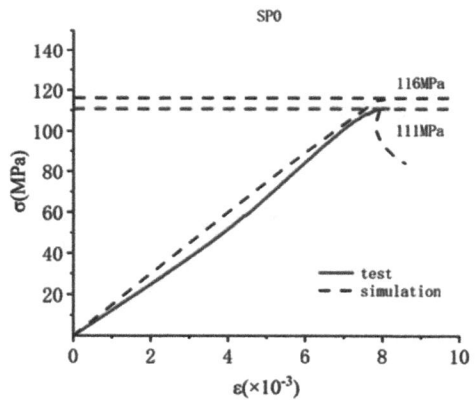

Figure 16. Stress-strain curve of SP0.

strength of the dented specimen is reduced by 9.9% when compared with the intact specimen.

As shown in Figure 15, the specimen begins to buckle from half span, which is similar to the deformation of numerical simulation in Figure 13. It can be seen in Figure 13 that the stiffeners begin to bend and stress concentrates mainly on half span of the specimen.

At last, the results of tests and simulations are compared for each specimen. As far as the intact stiffened panel SP0 is concerned, the simulation result is less than the test by 4.5% and the stiffness of the two curves is in good agreement. The simulation results of dented specimen SP1 and SP2 are both bigger than the tests. Simulation results of SP1 and SP2 are 4% bigger than the test results. The stiffness of the simulation and test curves in Figure 16 and Figure 18 are

almost in agreement. But the stiffness of simulation and test curves in Figure 17 are not in good agreement. The reason of divergence can be clearance of loading machine and specimen.

6 PARAMETRIC STUDY

The effects of dent depth, dent area as well as dent location on ultimate strength of stiffened panels are studied in this part.

The effect of dent depth on ultimate strength is analyzed firstly. The modelling process is the same as the

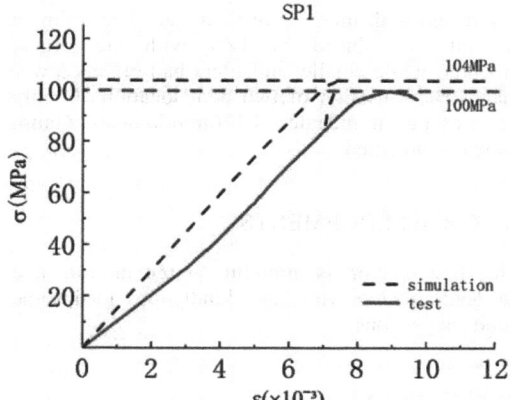

Figure 17. Stress-strain curve of SP1.

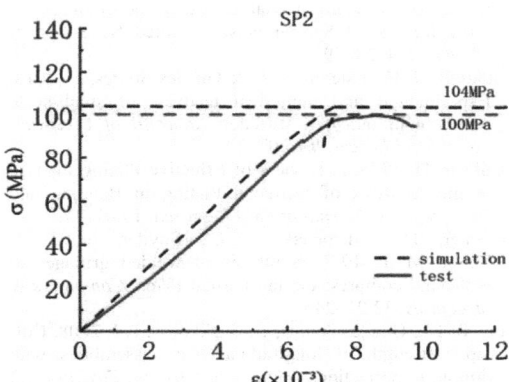

Figure 18. Stress-strain curve of SP2.

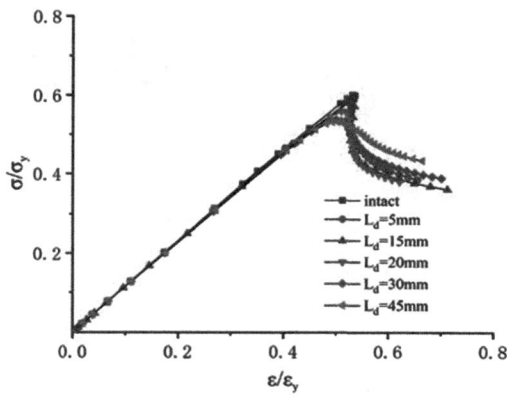

Figure 19. Dent depth analysis.

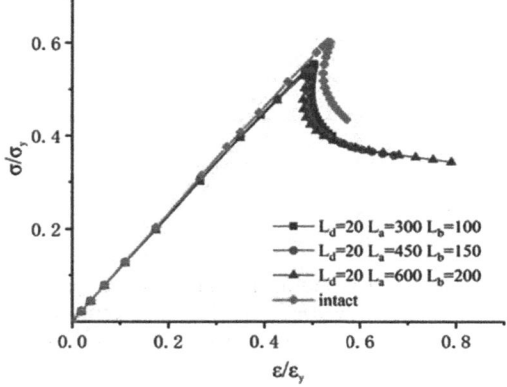

Figure 20. Dent size analysis.

method illustrated before. The same indenter and stiffened panels are modeled in this part. Therefore, damages have the same shape and size. By setting various pre-defined displacements of the indenter, local dents of different depths are acquired. The results of parametric study are normalized. Five different dent depths are computed, and the results were plotted in Figure 19.

It can be observed that with the increase of the dent depth the ultimate strength of the stiffened panels is reduced, but the variation is slightly small. Because the dent area is not large enough compared with area of a single panel. The biggest variation is 11% when dent depth is 45mm. When dent depth is smaller than 20mm, residual strength of the panel is reduced no more than 10%.

Secondly, the size of the dent is analyzed. Three different indenters are adopted in the simulation. Length and width of the indenter are represented with the method in Figure 4. The depth of the dent is controlled to be identical, which is 20mm. As shown in Figure 20, with the increase of the indenter's size, the

residual strength decreases slightly. The larger the area of the dent is, the less residual strength is. An indenter of 600mm long and 200mm wide can reduce the ultimate strength of the stiffened panel by 12%. But smaller indenter has little effect lower than 10%. The results of different indenter were displayed in Table 6.

At last, the effects of different dent locations on the ultimate strength are analyzed. Dents in the center and side of the panel are simulated. An indenter of 170mm long and 60mm wide is applied. A sample of dent location is shown in Figure 21. The results are plotted in Figure 22. The line of the side dent is lower than

Table 6. Results of dent size analysis.

L_a (mm)	L_b (mm)	$\sigma u/\sigma y$
300	100	0.56
450	150	0.54
600	200	0.53
0	0	0.60

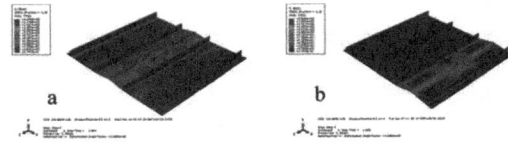

a   b

Figure 21. Dent location.

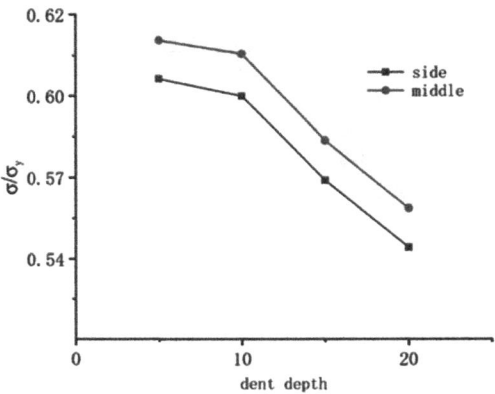

Figure 22. Dent location analysis.

the middle dent in the figure. But there are only 2% mean variation between them. The reason could be the indenter is relatively small compared with the panel.

7 CONCLUSIONS

In this paper, uniaxial compression tests are conducted on an intact stiffened panel and two stiffened panels with dent damage. As a result, stress-strain curves of each stiffened panel are obtained. Then relevant numerical simulations are carried out. By combining two methods, the ultimate capacity and collapse process can be both studied. Dent damage do reduce the ultimate strength of the stiffened panel. The test results show that the ultimate strength of the stiffened panel can be reduced by 9.9% when the dent is 10 mm deep, 170 mm long and 60 mm wide. The simulation results are larger than the test results, but the variations are all within 5%.

Results of the parametric study show that the residual strength is reduced with the increase of dent depth and 45mm dent depth can reduce the ultimate strength by 11%. Variation between the damaged and intact is lower than 10% when depth is smaller than 20mm. The result of dent size is similar to result of dent depth. The ultimate strength is reduced with increase of dent size. The ultimate strength is reduced by 12% with the largest indenter, while smaller indenters had effects lower than 10%. Variation of two dent locations is only 2%, as far as an indenter of 170mm long and 60mm wide is concerned.

ACKNOWLEDGEMENTS

The first author is grateful to Zexin Lin and Dr Kailing Guo who have kindly provided some good suggestions.

REFERENCES

Amante, D., Chujutalli, J. H. & Estefen, S. F. 2014 Residual Compressive Strength of Dented FPSO Side Shell Panel. In proceedings of 33rd International Conference on Offshore Mechanics and Arctic Engineering; 2014 June 8–13 San Francisco, United States. paper OMAE2014–24059

Chujutalli, J. H., Estefen, S. F. & Guedes Soares, C. 2018. Experimental and numerical analysis of small-scale panels with indented stiffeners. *Journal of Constructional Steel Research*.15:07–22.

Faulkner, D. 1975. A Review of Effective Plating for Use in the Analysis of Stiffened Plating in Bending and Compression, *Journal of Ship Research*. 19:01–17.

Faulkner, D., Adamchak, J. C., Snyder, G. J. & Vetter, M. F. 1973. Synthesis of welded grillages to withstand compression and normal loads. *Computers & Structures*. 32:21–246.

Luís, R. M., Guedes Soares, C. & Nikolov, P. I. 2008. Collapse strength of longitudinal plate assemblies with dimple imperfections. *Ships and Offshore Structures*. 3 (4):359–370.

Paik, J. K., Lee, J. M. & Lee, D. H. 2003. Ultimate strength of dented steel plates under axial compressive loads. *International Journal of Mechanical Sciences*. 45:433–488.

Raviprakash, A. V., Prabu, B. & Alagumurthi, N. 2012. Residual ultimate compressive strength of dented square plates. *Thin-Walled Structures*. 58:32–39.

Saad-Eldeen, S., Garbatov, Y. & Guedes Soares, C. 2015. Stress–strain analysis of dented rectangular plates subjected to uni-axial compressive loading. *Engineering Structures*. 99:78–91.

Witkowska, M. & Guedes Soares, C. 2009. Ultimate strength of stiffened plates with local damage on the stiffener. *Analysis and Design of Marine Structures*. In: Guedes Soares C., Das P.K. (Eds). Taylor & Francis, London, U.K. 145–154.

Witkowska, M. & Guedes Soares, C. 2015, Ultimate strength of locally damaged panels. *Thin-Walled Structures*. 97:225240.

Xu, M. & Guedes Soares, C. 2015. Effect of a central dent on the ultimate strength of narrow stiffened panels under axial compression. *International Journal of Mechanical Sciences*. 100:68–79.

Developments in Maritime Technology and Engineering – Guedes Soares & Santos (eds)
© 2021 Copyright the Author(s), ISBN 978-0-367-77376-2

Analysis of post-collapse behaviour of rectangular plate employing roof mode plastic solutions

M. Tekgoz & Y. Garbatov

Centre for Marine Technology and Ocean Engineering (CENTEC), Instituto Superior Técnico, Universidade de Lisboa, Lisbon, Portugal

ABSTRACT: The objective of this work is to develop a new approximate solution to study the large deflection behaviour of rectangular plates accounting for the geometric and material nonlinearities. The principle of the virtual work and rigid plastic solution are used to predict the plate pre-collapse and post-collapse responses, respectively. The newly developed model predicts the post-collapse response of rectangular plates using the roof mode plastic solution. Displacement and stresses as predicted by the new approximate solution are compared with the nonlinear finite element solution employing the commercial software ANSYS.

1 INTRODUCTION

The ship plating plays a significant role in the overall structural capacity of the ship structures. Therefore, it is crucial to evaluate their structural capacity under the external load that they are subjected.

The finite element solution has become a standard method to evaluate plate strength. Many authors, as for example Paik et al. (2008), Silva et al. (2013), Tekgoz et al. (2015), used the finite element solution to investigate the structural capacity of the unstiffened plate elements accounting for several aspects of the environmental phenomenon.

However, the approximate solutions to estimate the large non-linear response of plates enables to reduce the computational time with a reasonable accuracy given the fact that the ship plating constitutes a large part of the ship structure.

Here the large elastic and rigid plastic deflection methods have been implemented through the virtual work principle in predicting the large deflection response of plates, including their post-collapse regime. The coupling of these approaches is seen as a possible solution in the case of a large deflection plate behaviour (Fujita et al. 1977, Paik & Pedersen 1996 and Cui & Mansour 1999).

Okada et al. (1979) implemented the large elastic solution along with the rigid plastic solution to investigate the compressive strength of the long rectangular plates subjected to hydrostatic pressure.

Yao & Nikolov (1991) estimated the plate structural capacity using the large elastic solution with the rigid plastic mechanism based on the virtual work method under the longitudinal thrust to evaluate the ship structural capacity.

The large elastic deflection solutions have been implemented accounting for different aspects of problems, namely the structural capacity of the plates, in studying combined loading conditions as presented by Cui et al. (2002).

The objective of this work is to develop an approximate solution of the large deflection behaviour of rectangular plates accounting for the geometric and material nonlinearities, employing analytical and finite element methods.

The principle of the virtual work and rigid plastic solution are used to predict the plate pre-collapse and post-collapse regimes, respectively. The estimated displacement and stresses are compared with the analytical and finite element solution employing nonlinear analyses.

A simple model is presented in predicting the post-collapse response of rectangular plates accounting for the roof mode plastic solution demonstrating a reasonably good accuracy.

2 STRUCTURAL DESCRIPTION AND FEM MODELLING

2.1 Structural descriptions

Three different plates with different plate aspect ratios, a/b and plate thicknesses, t_p have been studied here, as shown in Table 1.

2.2 Finite element modelling

The large deflection structural response has been estimated by employing the FE method using commercial software, ANSYS (2009).

DOI: 10.1201/9781003216582-67

Table 1. Plate structural description.

Plates	Length, a	Breadth, b	t_p	a/b
	mm	*mm*	*mm*	-
1	400	500	5	0.8
2	3,270	880	10, 15, 20	4
3	4,950	830	10, 15, 20	6

Shell element, SHELL181 has been used to model the studied plates. The element type has four nodes with six degrees of freedom at each node, including translations and rotations about the x, y, and z-axes. The material is assumed to follow the elastic-perfectly plastic relationship. The material descriptors are defined as yield stress, $\sigma_y = 285$ *MPa*, Young's Modulus, $E=207,000$ *MPa* and Poisson coefficient, $v=0.3$. The boundary conditions of the plates are assumed to be simply supported for all FEM studies carried out here, keeping the unloaded edges straight, assuming that the plate fails before the stiffener as a part of the stiffened panel (see Figure 1 and 2). The applied boundary condition in the FE model is in compliance with the analytical solution presented here since the transverse membrane stresses are accounted for after the plate buckling in both cases occurs, leading to more optimistic results.

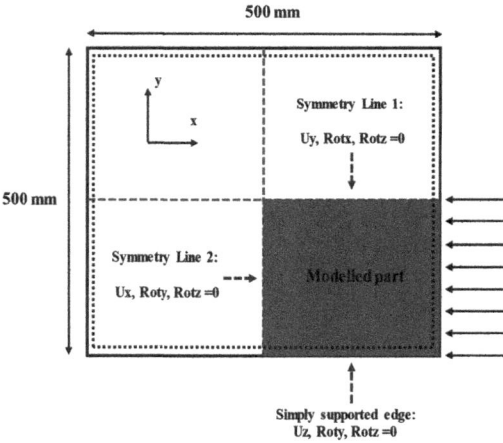

Figure 1. Boundary conditions and loading, Plate 1.

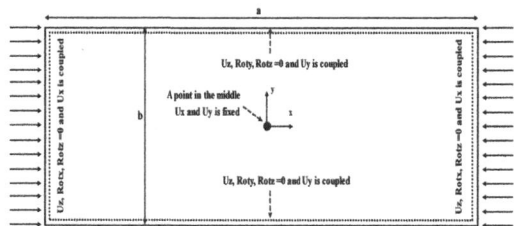

Figure 2. Boundary conditions and loading, Plate 2 and 3.

3 ANALYTICAL SOLUTION

3.1 *Elastic post-buckling solution for plate pre-collapse response*

The principle of the virtual work method has been implemented here to estimate the elastic post-buckling pre-collapse response of the rectangular plates studied here. Assuming that the system is already in equilibrium, any virtual work through the virtual displacements is zero leads to:

$$\delta\Delta U_e = \delta\Delta U_i \qquad (1)$$

where $\delta\Delta U_e$ is the incremental external virtual work and $\delta\Delta U_i$ is the incremental internal virtual work or internal virtual strain energy, where $\delta\Delta U_i$ follows the general form of:

$$\delta\Delta U_i = \iiint \left(\sigma_{xmp} + \sigma_{xb} + \Delta\sigma_{xmp} + \Delta\sigma_{xb}\right)\delta\left(\Delta\varepsilon_{xmp} + \Delta\varepsilon_{xb}\right) +$$
$$+ \left(\sigma_{ymp} + \sigma_{yb} + \Delta\sigma_{ymp} + \Delta\sigma_{yb}\right)\delta\left(\Delta\varepsilon_{ymp} + \Delta\varepsilon_{yb}\right) +$$
$$+ \left(\tau_{xymp} + \tau_{xyb} + \Delta\tau_{xymp} + \Delta\tau_{xyb}\right)\delta\left(\Delta\varepsilon_{xymp} + \Delta\varepsilon_{xyb}\right)\mathrm{d}x\mathrm{d}y\mathrm{d}z$$

$$(2a)$$

where σ, τ and $\Delta\sigma$, $\Delta\tau$ are the normal and shear stresses and their incremental form, respectively. $\delta\Delta\varepsilon$ is the incremental virtual form of the respective strains.

The incremental internal virtual energy can be transformed before the initial plate yielding, assuming that the total strain energy can be broken down into the mid-plane and bending strain energies as:

$$\delta\Delta U_i = \iint \left(N_{xmp} + \Delta N_{xmp}\right)\delta\left(\Delta\varepsilon_{xmp}\right) + \left(N_{ymp} + \Delta N_{ymp}\right)\delta\left(\Delta\varepsilon_{ymp}\right) +$$
$$+ \left(N_{xymp} + \Delta N_{xymp}\right)\delta\left(\Delta\varepsilon_{xymp}\right) + \left(M_x + \Delta M_x\right)\delta\left(\Delta\gamma_{xb}\right) +$$
$$+ \left(M_y + \Delta M_y\right)\delta\left(\Delta\gamma_{yb}\right) + \left(M_{xy} + \Delta M_{xy}\right)\delta\left(\Delta\gamma_{xyb}\right)\mathrm{d}x\mathrm{d}y$$

$$(2b)$$

where N and ΔN are the mid-plane forces and their incremental form, respectively, M and ΔM are the bending moments, and their incremental form $\delta\Delta\gamma$ is the incremental virtual form of the respective curvatures. At this stage, an initial plate deformed surface is assumed, defined as a uni-modal one as presented by Ueda & Yao (1985):

$$W_o = W_{o\,\max} \sin(m\pi x/a)\sin(n\pi y/b) \qquad (3)$$

$$W_z = W_{z\,\max} \sin(m\pi x/a)\sin(n\pi y/b) \qquad (4)$$

where a is the length of the plate, b is the breath of the plate, m and n are parameters defining the number of the half-waves considered. W_o and W_z are the initial

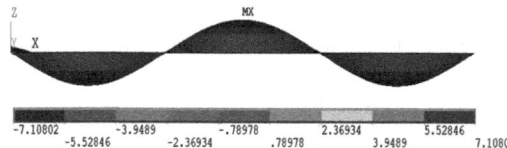

Figure 3. Initial imperfection.

and the final plate vertical displacements. For all FE analysis and the analytical studies here, n is set to 1 and m is calculated as the minimum integer of:

$$a/b \leq \sqrt{m(m+1)} \tag{5}$$

As for the mid-plate maximum initial imperfection, W_{omax} is given as the average initial imperfections unless stated otherwise in the respective study, as defined by Smith et al. (1988) (see Figure 3):

$$W_{o\,max} = 0.1\beta^2 t_p \tag{6}$$

where β represents plate slenderness and t_p is the plate thickness.

Assuming a large deflection structural response, the plane stress condition is defined as:

$$\varepsilon_{xmp} = \partial u/\partial x + 1/2(\partial(W_z - W_o)\partial x)^2 \tag{7}$$

$$\varepsilon_{ymp} = \partial v/\partial y + 1/2(\partial(W_z - W_o)\partial y)^2 \tag{8}$$

$$\varepsilon_{uymp} = \partial u/\partial x + \partial v/\partial y + \partial^2(W_z - W_o)/\partial x \partial y \tag{9}$$

where ε_{xmp}, ε_{ymp} and ε_{xymp} are the mid-plane membrane normal and shear strains in the longitudinal and transverse directions, u and v are the mid-plane displacements in the longitudinal and transverse directions respectively, and W_o and W_z are the initial and final plate displacements in the vertical direction.

The compatibility equation is defined by differentiating the mid-plane shear strains in x and y directions as:

$$\frac{\partial^2 \varepsilon_{xmp}}{\partial^2 y} + \frac{\partial^2 \varepsilon_{ymp}}{\partial^2 x} - \frac{\partial^2 \varepsilon_{xymp}}{\partial x \partial y} = \left(\frac{\partial^2 W_z}{\partial^2 xy}\right)^2 + \left(\frac{\partial^2 W_o}{\partial^2 xy}\right)^2$$
$$- \left(\frac{\partial^2 W_z}{\partial^2 x}\right)\left(\frac{\partial^2 W_z}{\partial^2 y}\right) + \left(\frac{\partial^2 W_o}{\partial^2 x}\right)\left(\frac{\partial^2 W_o}{\partial^2 y}\right) \tag{10}$$

where ε_{xmp}, ε_{ymp}, ε_{xymp} are defined as:

$$\varepsilon_{xmp} = \frac{1}{E}\left(\sigma_{xmp} - \nu\sigma_{ymp}\right) \tag{11}$$

$$\varepsilon_{ymp} = \frac{1}{E}\left(\sigma_{ymp} - \nu\sigma_{xmp}\right) \tag{12}$$

$$\varepsilon_{xymp} = \frac{2}{E}(1+\nu)\tau_{xymp} \tag{13}$$

According to the Airy's theory, the mid-plane stresses can be identified by the stress function, F satisfying the force equilibrium as:

$$\sigma_{xmp} = \frac{N_{xmp}}{t_p} = \frac{\partial^2 F}{\partial^2 y} \tag{14}$$

$$\sigma_{ymp} = \frac{N_{ymp}}{t_p} = \frac{\partial^2 F}{\partial^2 x} \tag{15}$$

$$\sigma_{xymp} = \frac{N_{xymp}}{t_p} = -\frac{\partial^2 F}{\partial^2 xy} \tag{16}$$

Substituting all terms in Eqn. (10), the stress function, F, is defined as:

$$\frac{\partial^4 F}{\partial^4 x} + \frac{\partial^4 F}{\partial^4 y} + 2\frac{\partial^4 F}{\partial^2 x \partial^2 y}$$
$$= \frac{E\left(W_{z\,max}^2 - W_{o\,max}^2\right)m^2\pi^4}{2a^2 b^2}\left(\cos\left(\frac{2m\pi x}{a}\right) + \cos\left(\frac{2\pi y}{b}\right)\right) \tag{17}$$

leading to:

$$F = \frac{E\left(W_{z\,max}^2 - W_{o\,max}^2\right)}{32}$$
$$\left(\frac{a^2}{m^2 b^2}\cos\left(\frac{2m\pi x}{a}\right) + \frac{m^2 b^2}{a^2}\cos\left(\frac{2\pi y}{b}\right)\right) - \frac{1}{2}\sigma_{xav}y^2 \tag{18}$$

where σ_{xav} is the average stress in the longitudinal direction.

At this stage, the mid-plane forces, N_{xmp}, N_{ymp}, N_{xymp} and their virtual incremental strains, $\delta\Delta\varepsilon_{xmp}$, $\delta\Delta\varepsilon_{ymp}$, $\delta\Delta\varepsilon_{xymp}$, can be defined using Eqn. (11) to Eqn. (16) respectively.

The bending moments, M_x, M_y and M_{xy} are defined as:

$$M_x = \int\limits_{-\frac{t_p}{2}}^{\frac{t_p}{2}} \sigma_{xb} z \, dz \qquad (19)$$

$$M_y = \int\limits_{-\frac{t_p}{2}}^{\frac{t_p}{2}} \sigma_{yb} z \, dz \qquad (20)$$

$$M_{xy} = \int\limits_{-\frac{t_p}{2}}^{\frac{t_p}{2}} \tau_{xyb} z \, dz \qquad (21)$$

where σ_{xb}, σ_{yb} and τ_{xyb} are the bending and shear stresses defined as:

$$\sigma_{xb} = \frac{E}{1-\nu^2}\left(\varepsilon_{xb} - \nu\varepsilon_{yb}\right)$$
$$= -\frac{Ez}{1-\nu^2}\left(\frac{\partial^2(W_z - W_o)}{\partial^2 x} + \nu\frac{\partial^2(W_z - W_o)}{\partial^2 y}\right) \qquad (22)$$

$$\sigma_{yb} = \frac{E}{1-\nu^2}\left(\varepsilon_{yb} - \nu\varepsilon_{xb}\right)$$
$$= -\frac{Ez}{1-\nu^2}\left(\frac{\partial^2(W_z - W_o)}{\partial^2 x} + \nu\frac{\partial^2(W_z - W_o)}{\partial^2 y}\right) \qquad (23)$$

$$\tau_{xyb} = \frac{E}{2(1+\nu)}\gamma_{xyb} = -\frac{Ez}{1+\nu^2}\left(\frac{\partial^2(W_z - W_o)}{\partial x \partial y}\right) \qquad (24)$$

and the curvature may be defined, following as:

$$\gamma_{xb} = -\frac{\partial^2(W_z - W_o)}{\partial^2 x} \qquad (25)$$

$$\gamma_{yb} = -\frac{\partial^2(W_z - W_o)}{\partial^2 y} \qquad (26)$$

$$\gamma_{xyb} = -2\frac{\partial^2(W_z - W_o)}{\partial x \partial y} \qquad (27)$$

Substituting all terms in from Eqn. (19) to Eqn. (27), the moments, M_x, M_y and M_{xy} and the incremental virtual curvatures, $\delta\Delta\gamma_{xmp}$, $\delta\Delta\gamma_{ymp}$, $\delta\Delta\gamma_{xymp}$ can be defined.

Substituting all the terms in Eqn. (2b), the internal incremental virtual work can be defined as:

$$\delta\Delta U_i = \frac{Eabt\pi^4\left(\left(W_{z\max} + \Delta W_{z\max}\right)^2 - W_{o\max}^2\right)\left(W_{z\max} + \Delta W_{z\max}\right)}{64}\left(\frac{m^4}{a^4} + \frac{1}{b^4}\right) +$$
$$+ \left(\begin{array}{c} \frac{Et^3\pi^4 m^2 b}{48(1-\nu^2)a}\left(\frac{m^2}{a^2} + \frac{\nu}{b^2}\right) + \\ + \frac{Et^3\pi^4 a}{48(1-\nu^2)b}\left(\frac{1}{b^2} + \frac{m^2\nu}{a^2}\right) + \\ + \frac{Et^3 m^2\pi^4}{24(1+\nu)ab} \end{array}\right)\left(W_{z\max} + \Delta W_{z\max} - W_{o\max}\right) \qquad (28)$$

and the incremental external virtual work, $\delta\Delta U_e$ is:

$$\delta\Delta U_e = (F_{ext} + \Delta F_{ext})\delta\Delta u = \left(\sigma_{xav}bt_p + \Delta\sigma_{xav}bt_p\right)\delta\Delta u \qquad (29)$$

where $\delta\Delta u$ is the incremental virtual displacements in the longitudinal direction, and it can be defined using Eqn. (7). Adding all terms in the Eqn. (29), the form of $\delta\Delta U_e$ is given as:

$$\delta\Delta U_e = (\sigma_{xav} + \Delta\sigma_{xav})\frac{bt_p m^2\pi^2(W_{z\max} + \Delta W_{z\max})}{4a} \qquad (30)$$

Finally, by equating the internal and external virtual work, one might define the force-displacement equilibrium of the system that follows the form of:

$$\frac{Ea^2\pi^2\left(\left(W_{z\max} + \Delta W_{z\max}\right)^2 - W_{o\max}^2\right)\left(W_{z\max} + \Delta W_{z\max}\right)}{16m^2}$$
$$\left(\frac{m^4}{a^4} + \frac{1}{b^4}\right) + \frac{Et_p^2\pi^2}{12(1-\nu^2)b^2}\left(\frac{bm}{a} + \frac{a}{mb}\right)^2$$
$$\left(W_{z\max} + \Delta W_{z\max} - W_{o\max}\right)$$
$$= (\sigma_{xav} + \Delta\sigma_{xav})(W_{z\max} + \Delta W_{z\max}) \qquad (31)$$

3.2 Rigid-plastic solution of plate post-collapse response

The rigid plastic solutions of the plate post-collapse response depend on the assumed plastic formation shape. In the present study, the roof mode type labelled as Model A and shown in Figure 4 is used, which can be applied to the plates that are in compliance with the rule as suggested by Yao &Nikolov (1991):

$$a/mb \leq 1 \qquad (32)$$

where a is the plate length, b is the plate breadth and m is the half wave number in the longitudinal direction. For this type of collapse, the solution is:

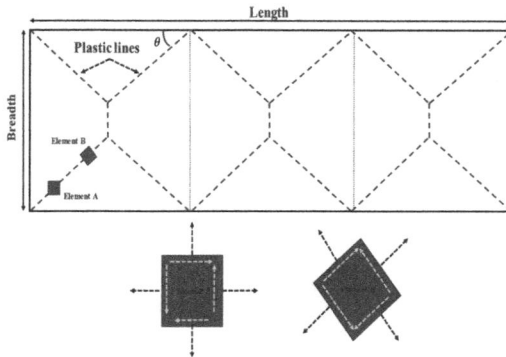

Figure 4. Rigid plastic solution, roof model, Model A, $a/mb<1$.

$$m_{45} + \left(\frac{1}{\alpha} - 1\right)\frac{m_{90}}{2} + \left(\frac{2}{\alpha} - 1\right)\overline{\sigma}\overline{W}_{z\max} = 0 \quad (33)$$

where

$$\overline{\sigma} = \frac{\sigma_{xav}}{\sigma_y} \quad (34)$$

$$\overline{W}_{z\max} = \frac{W_{z\max}}{t_p} \quad (35)$$

$$m_{90} = 1 - \overline{\sigma}^2 \quad (36)$$

$$m_{45} = \frac{4m_{90}}{\sqrt{1 + 15m_{90}}} \quad (37)$$

where σ_y is the material yield stress.

It is essential to point out here that there are particular locations on plates where the high shear stresses are developed on the mid-plane after the initial yielding, which might lead to the development of a plastic mechanism as shown in Figure 5.

However, if the developed shear stresses are not big enough, they cannot cause this type of plastic formations. It has been noticed that the post-collapse prediction based on this model, as shown in Eqn. (33), is weak and therefore, a new simple model, labelled as Model B, is introduced here to be used in predicting the post-collapse response of the plates assuming that the plastic failure may be modelled by a linear spring with an adjustable stiffness (see Figure 6).

Applying the principle of the virtual work, the internal, U_{int} and external, U_{ext} virtual work can be calculated as:

$$U_{\text{int}} = 2m(\varsigma)M_p b\delta\theta + K_s W_{z\max}^2 \quad (38)$$

$$U_{ext} = F_{ext}bW_{z\max}\delta\theta = \sigma_{xav}bt_p W_{z\max}\delta\theta \quad (39)$$

where

$$\delta\theta = 2mW_{z\max}/a \quad (40)$$

$$M_p = \frac{t_p^2\sigma_y}{4} \quad (41)$$

where $m(\varsigma)$ is taken as given in (Cui et al. (2002)):

$$m(\varsigma) = \frac{2\left(1 - \left(\frac{\sigma_{xav}}{\sigma_y}\right)^2\right)}{\sqrt{4 - 3\left(\frac{\sigma_{xav}}{\sigma_y}\right)^2}} \quad (42)$$

By equating $U_{ext} = U_{int}$, the average stress-vertical displacement, $W_{z\,max}$ relationship can be found as:

$$W_{z\max} = \frac{m(\varsigma)t_p^2 bm}{a\left(2Pbt_p\frac{m}{a} - \frac{K_s}{\sigma_y}\right)} \quad (43)$$

where $W_{z\,max}$ is the vertical plate displacement, as shown in Figure 6, K_s is the linear adjustable spring stiffness, $P = \sigma_{xav}/\sigma_y$ is the normalised strength and σ_y is the material yield stress.

4 RESULTS AND DISCUSSIONS

4.1 Average stress – vertical displacement relationships

Here, the large deflection response of plates with different plate thicknesses and aspect ratios ranging from 1 to 6 is analysed employing the FEM and analytical solutions.

Figure 7 shows the strength-displacement relationship of Plate 1 estimated by the FEM and analytical solutions, where Model A represents the rigid plastic solution as shown in Figure 4 and Model B represents the model as shown in Figure 6. The initial imperfection is set to 10 % of the plate thickness. It is important to point out here that the initial imperfection has been applied using the Eqn. (5) for Plate 2 and Plate 3.

Figure 8 to Figure 10 show the strength-displacement relationship of Plate 2. It has been

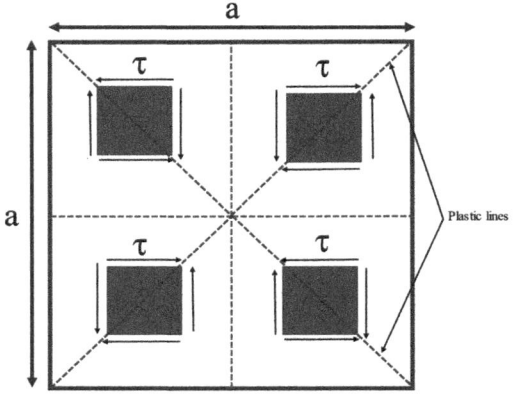

Figure 5. Progress in development of plastic mechanism.

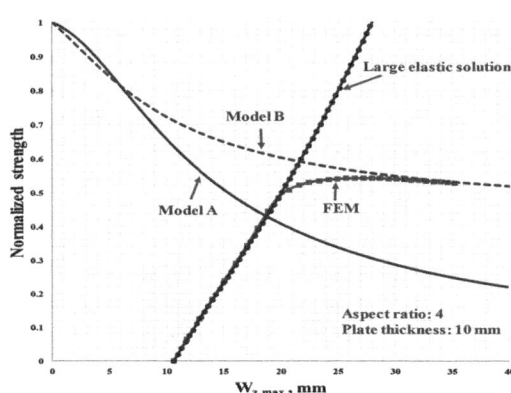

Figure 8. Strength-displacement relationship, $a/b=4$, t_p $=10$ mm.

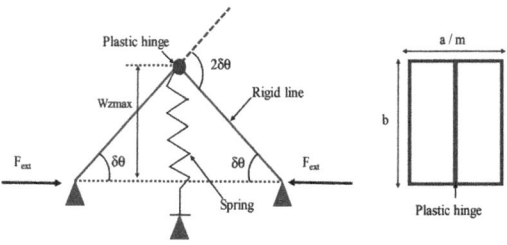

Figure 6. Simple model of plate post-collapse response, Model B.

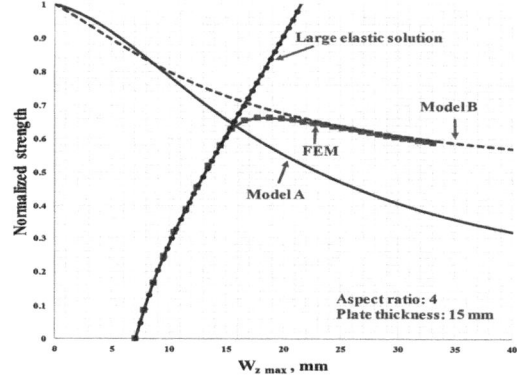

Figure 9. Strength-displacement relationship, $a/b=4$, t_p $=15$ mm.

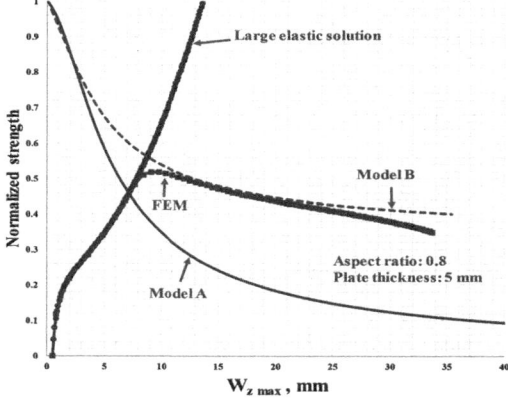

Figure 7. Strength-displacement relationship, $a/b=0.8$, $t_p=5$ mm.

observed that the rigid plastic solution, as shown in Figure 4, underestimates the plate post-collapse response. This might be attributed to the plastic form pattern assumed in this solution. The new developed simple solution, as shown in Figure 6, shows better

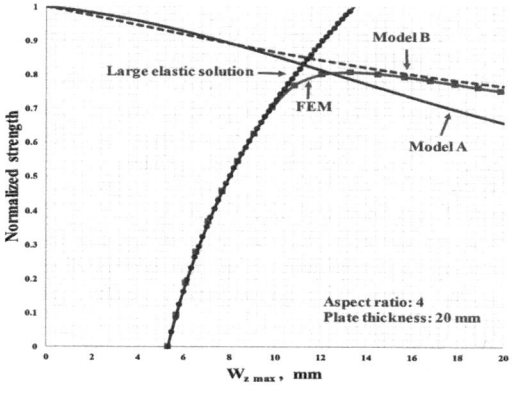

Figure 10. Strength- displacement relationship, $a/b=4$, $t_p=20$ mm.

agreement with the one predicted by the FEM solution.

Figure 11 To Figure 13 shows the strength-displacement relationship of Plate 3. Similar findings can be seen in the case of Plate 3.

For all studied cases, shown here, it turned out that the newly developed model predicts the post-collapse of the plates reasonably accurate.

According to all studied plate cases here, by changing the plate thickness and aspect ratios, the stiffness of the spring in the newly developed model has been defined as shown in Figure 14.

In the case of the simple model, Model B, the entire line through the plate breadth is assumed to be yielded and the spring is considered to be elastic, it is thought that the spring stiffness is to be based on the plate thickness rather than based on the ratio b/t_p where b is the plate breadth and t_p is the plate thickness (see Figure 6). The influence of the material properties in the simple model is included; however, its impact on the spring stiffness is still to be investigated in Eqn. (43).

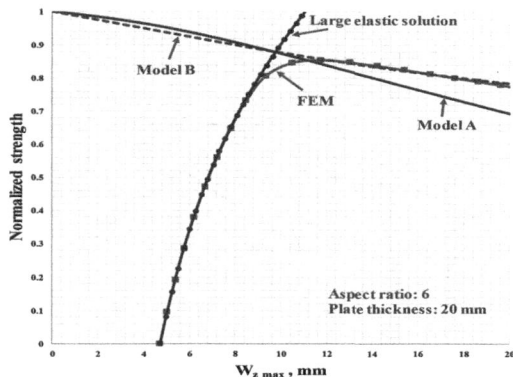

Figure 13. Strength-displacement relationship, $a/b=6$, $t_p =20$ *mm*.

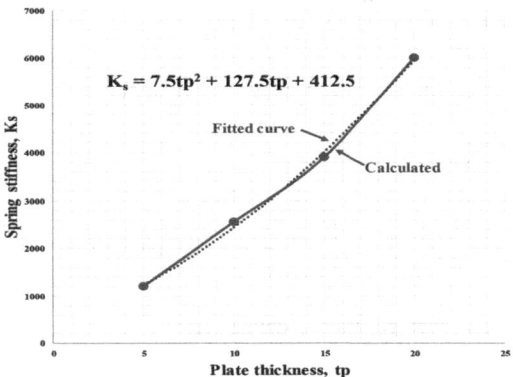

Figure 14. Spring stiffness - plate thickness relationship.

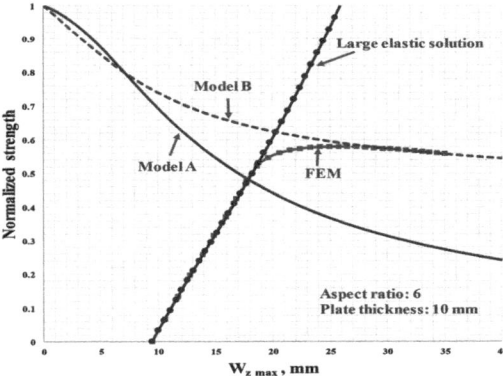

Figure 11. Strength-displacement relationship, $a/b=6$, $t_p =10$ *mm*.

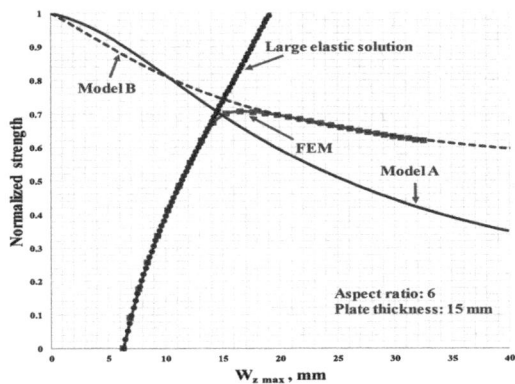

Figure 12. Strength-displacement relationship, $a/b=6$, $t_p =15$ *mm*.

It is essential to point out here that the analytical solution presented here does not fully represent the ones as predicted by the FEM solutions and only the response from A_0 to A_1 and from A_2 to A_3 has been estimated and shown in Figure 15. The dashed lines are still to be determined, which remains to be a future work to be carried out.

The reason for this is that the large elastic solution based on the virtual work method assumes that the plate is elastic when the plate initially yields, this form fails because of the material non-linearity, non-uniform tangent modulus. Thereby, the pre-assumed stress forms fail as given in Eqn. (14) to Eqn. (16).

4.2 *Average stress – average strain relationships*

The average stress and strain behaviours of the Plate 2 and 3 have been studied employing the new developed simple model, Model B along with the large elastic solution presented in Section 4.1, labelled as a present model here in this section.

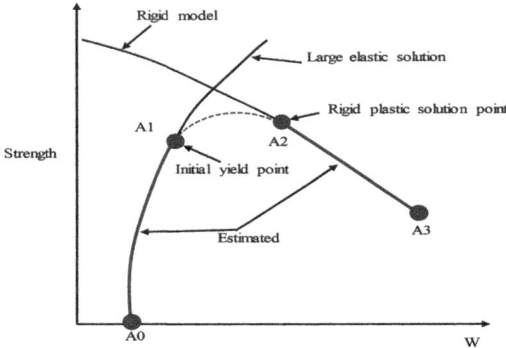

Figure 15. Estimated regions.

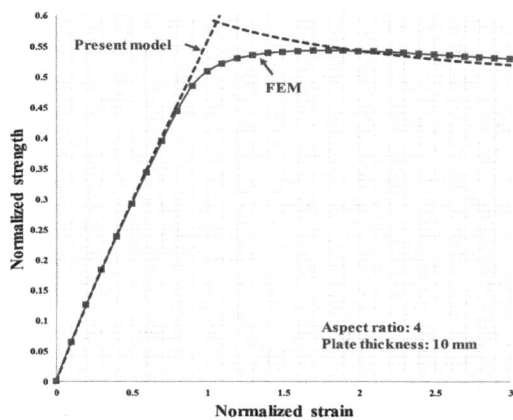

Figure 16. Strength-strain relationship, $a/b=4$, $t_p=10$ mm.

It is essential to point out that the average strains equal to the following form until the plate collapse, defined by using the Eqn. (7) as:

$$\partial u / \partial x = \varepsilon_{xmp} - 1/2(\partial(W_z - W_o)/\partial x)^2 \qquad (44)$$

leads to

$$u = \frac{1}{b}\int\int \varepsilon_{xp} - \frac{1}{b}\int 1/2(\partial(W_z - W_o)/\partial x)^2 dxdy \quad (45)$$

The average strain is defined as:

$$\varepsilon_{average} = \frac{\sigma_{xav}}{E} + \frac{m^2\pi^2\left(W_{z\max}^2 - W_{o\max}^2\right)}{8a^2} \qquad (46)$$

and after the plate collapse, it is assumed that the strain energy is stored in the spring model, and the average strain is defined as:

$$\varepsilon_{average} = \frac{K_s W_{z\max}^2}{(1 - \sigma_{xav})bt_p a} \qquad (47)$$

where K_s is the spring stiffness as defined in Section 4.1.

Figure 16 to Figure 18 show the strength-strain relationship of Plate 2. Figure 19 to Figure 21 show the strength-strain relationship of Plate 3. It has been observed that the presented model shows a reasonable accuracy with the one predicted by the FEM solution in terms of the average stress and strain behaviours. However, as the plate thickness gets smaller, the accuracy deteriorates from the ultimate capacity point of view.

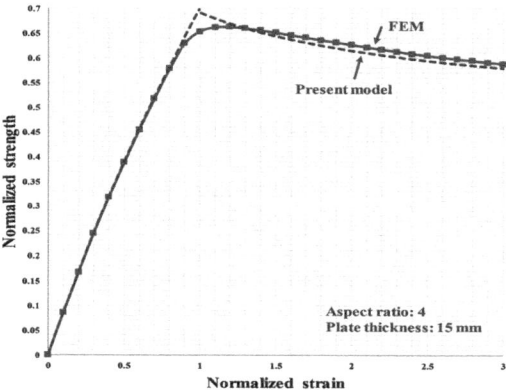

Figure 17. Strength-strain relationship, $a/b=4$, $t_p=15$ mm.

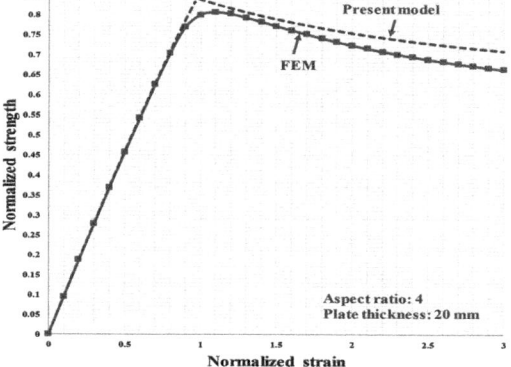

Figure 18. Strength-strain relationship, $a/b=4$, $t_p=20$ mm.

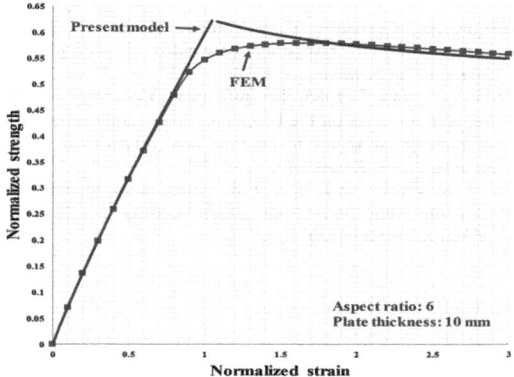

Figure 19. Strength-strain relationship, a/b=6, t_p=10 mm.

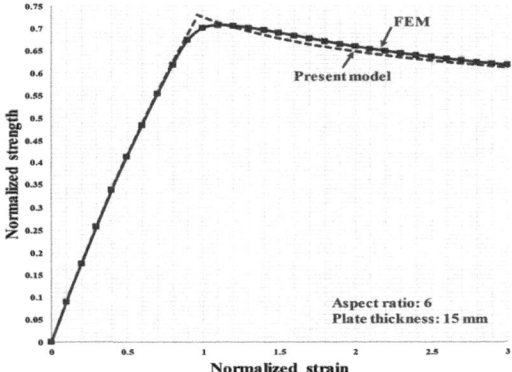

Figure 20. Strength-strain relationship, a/b=6, t_p=15 mm.

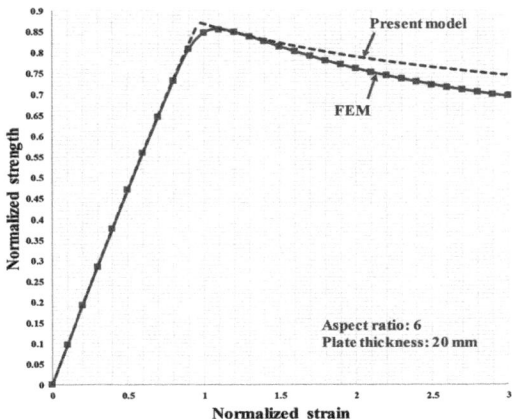

Figure 21. Strength-strain relationship, a/b=6, t_p=20 mm.

As pointed out in Section 4.1 and shown in Figure 15, there is a region between A_1 to A_2 that needs to be better estimated and remains to be a future work to fully match the one predicted by the FEM solution. This is considered to refine the ultimate capacity prediction of the present model and makes it closer to the FEM prediction.

5 CONCLUSIONS

The large deflection behaviour of rectangular plates has been analysed by employing analytical and finite element methods. A new model has been developed in predicting the post-collapse response of the rectangular plate accounting for the roof mode plastic solution. It has been shown that the newly developed model can predict the post-collapse response of the plates showing a reasonable agreement with the finite element solution. However, the presented simple model showed deficiency in defining the region between A_1 to A_2, as shown in Figure 15 that needs further enhancement and remains to be done in future work.

ACKNOWLEDGEMENTS

This work was performed within the Strategic Research Plan of the Centre for Marine Technology and Ocean Engineering (CENTEC), which is financed by Portuguese Foundation for Science and Technology (Fundação para a Ciência e Tecnologia-FCT) under contract UIDB/UIDP/00134/2020.

REFERENCES

ANSYS 2009. Advanced analysis techniques guide, South pointe, 275 technology drive, Canonsburg, PA 15317 Ansys, Inc.

Cui, W., Wang, Y. & Pedersen, P. T., 2002. Strength of ship plates under combined loading. *Marine Structures*, 75–97.

Cui, W. C. & Mansour, A. E. 1999. Generalisation of a simplified method for predicting ultimate compressive strength of ship panels. *Int Shipbuilding Progr*, 46(4479, 291–303.

Fujita, Y., Nomoto, T. & Niho, O. 1977. Ultimate strength of stiffened plates subjected to compression. *J Soc Naval Architects of Japan*.

Okada, H., Oshima, K. & Fukumoto, Y. 1979. Compressive strength of long rectangular plates under hydrostatic pressure. *J. of the Society of Naval Architects of Japan*, 146, 270–280.

Paik, J. K., Kim, J. B. & Seo, J. K., 2008. Methods for ultimate limit state assessment of ships and ship-shaped offshore structures: Part I - Unstiffened plates. *Ocean Engineering*, 35, 261–270.

Paik, J. K. & Pedersen, P. T., 1996. A simplified method for predicting ultimate compressive strength of ship panels. *Int Shipbuilding Progr*, 43(434), 139–157.

Silva, J. E., Garbatov, Y. & Guedes Soares, C. 2013. Ultimate strength assessment of rectangular steel plates subjected to a random localised corrosion degradation *Engineering Structures*, 52, 295–305.

Smith, C. S., Davidson, P. C. & Chapman, J. C. 1988. Strength and stiffness of ship´s plating under in-plane compression and tension. *R.Inst.Nav.Archit.Trans. 130*.

Tekgoz, M., Garbatov, Y. & Guedes Soares, C. 2015. Ultimate strength assessment of welded stiffened plates. *Engineering Structures*, 84, 325–339.

Ueda, Y. & Yao, T. 1985. The influence of complex initial deflection modes on the behaviour and ultimate strength of rectangular plates in compression. *Journal of Construction Steel Research*, 5, 265–302.

Yao, T. & Nikolov, P. I. 1991. Progressive collapse analysis of a ship´s hull under longitudinal bending. *The Society of Naval Architects of Japan*.

Developments in Maritime Technology and Engineering – Guedes Soares & Santos (eds)
© 2021 Copyright the Author(s), ISBN 978-0-367-77376-2

Ultimate strength of stiffened plates subjected to compressive load and spatially distributed mechanical properties

K. Woloszyk
Faculty of Ocean Engineering and Ship Technology, Gdansk Univeristy of Technology, Gdansk, Poland

Y. Garbatov
Centre for Marine Technology and Ocean Engineering (CENTEC), Instituto Superior Técnico, University of Lisbon, Lisbon, Portugal

ABSTRACT: The present study deals with the ultimate strength of stiffened plates subjected to spatially distributed mechanical properties and compressive load. Normally, mean values of mechanical properties based on tensile tests are used to validate the numerical assessment with experimental results. However, mechanical properties may vary within a single specimen. To investigate the impact of that, random fields of yield stress and Young modulus are employed together with the FE method. The variations of mechanical properties are estimated based on tensile tests. Different parameters of the random field are analysed, showing that the structural response will be different when compared with the stiffened plate with constant mechanical properties. Finally, it was revealed that the variation of yield stress has a significant impact, and Young modulus uncertainties play a secondary role. It is concluded, that the spatial variation of mechanical properties need to be considered when validating the numerical assessment with experimental measurements.

1 INTRODUCTION

The ultimate strength of different structural components in intact and degraded conditions has been analysed for a long time employing both experimental and numerical methods. Recently, several experimental studies were carried out on different structural levels, including box girders (Gordo and Guedes Soares, 2008; Saad-Eldeen, *et al.*, 2013), stiffened panels (Estefen, *et al.*, 2016; Saad-Eldeen, *et al.*, 2017), stiffened plates (Garbatov, *et al.*, 2017; Shi, *et al.*, 2017) and plates (Kim, *et al.*, 2009), where the experimental results were also validated using different numerical approaches, including the nonlinear FE method (Paik and Seo, 2009; Shi, *et al.*, 2018; Woloszyk, *et al.*, 2018). Recently ISSC (Czujko, *et al.*, 2018) reported a benchmark study of a FE analysis of the ultimate strength of a box girder compared with experimental data collected from different sources. Apart from the deviations within the numerical results were significant, the deviations with experimental results were also very high. The differences occurred not only in terms of the ultimate strength and also present in the post-collapse regime as well as in the force-displacement relationship. The scatter related to the numerical analyses may originate from different sources, such as the non-ideal modelling of boundary conditions (Woloszyk and Garbatov, 2019a), initial

imperfections existing in real structures (Dow and Smith, 1984) and mechanical properties. In the study presented in (Woloszyk and Garbatov, 2019b), the uncertainty analysis of the ultimate strength of corroded stiffened plates was performed showing the impact of the scatter of the mechanical properties on the resulting ultimate strength. However, in that case, only the mean values of the input material stress-strain relationship parameters deviated.

When one performs a validation of the numerical model within experimental results, the material properties that are coming from tensile tests of standard coupon specimens are typically taken as an input. The hypothesis, which is behind the origin of presented work, states that the deviations of mechanical properties within a single specimen may also impact the collapse behaviour of structural components, such as stiffened plates.

To reflect the non-homogeneity of the mechanical properties within the single structural elements, one needs to identify a proper modelling tool. The random field modelling seems to be a very effective one in this type of problems (Jankowski and Walukiewicz, 1997). The examples of random field applications reflecting different imperfections can be found in (Teixeira and Guedes Soares, 2008; Górski, *et al.*, 2015; Górski and Winkelmann, 2017). In (Woloszyk and Garbatov, 2020), the random fields were used in

DOI: 10.1201/9781003216582-68

order to model the corrosion surfaces of small-scale specimens.

In the presented work, the random field modelling is applied in order to model and analyse the impact of the spatial deviation of the mechanical properties within the stiffened plates made of different thicknesses. Tensile tests of steel coupons are performed in order to quantify the mechanical properties variation within a single plate component (three for each plate). Furthermore, the random fields of both Young modulus and yield stress are generated considering different correlations and used in the FE model. The collapse behaviour of stiffened plates considering constant mechanical properties within the specimen and spatially distributed ones are analysed, and the discrepancy between the two model results are compared and discussed.

2 TENSILE TEST

The mechanical properties can deviate significantly, even for one type of steel grade, when considering different batches of steel plates (Melcher, et al., 2004; Paik, et al., 2017). In that case, the coefficient of variation, CoV of yield stress can reach a level of 10 %. However, it occurs that even for a single plate, the mechanical properties may deviate spatially.

In the present study, the three coupon specimens of a standard size (see Figure 1) were taken from different places of steel sheets (1.25 m x 2.5 m) and analysed. The sheets thicknesses were 5 mm, 6 mm and 8 mm, and there were made from normal strength steel S235. The universal testing machine Zwick-Roell was used to carry out the tensile test and to identify the elasticity modulus, and a mechanical extensometer was adopted. The tests were carried out with accordance to ISO Standard (ISO, 2009).

The results of performed tensile tests are presented in Table 1. As can be noticed from the specimens of a 5 mm and 6 mm thickness, the mechanical properties are subjected to a significant level of COV (the three identical specimens per each thickness were tested). In the case of the yield stress, the highest level of COV is observed for a 6 mm plate, where the COV is about 4 %.

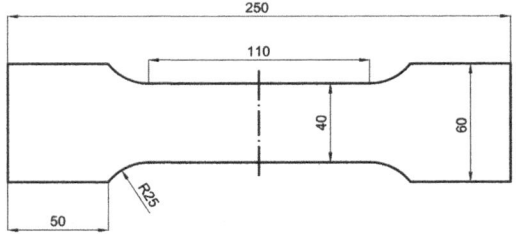

Figure 1. Dimensions of standard coupon specimen.

Table 1. Tensile test descriptors.

Thickness [mm]		Young modulus E [GPa]	Yield stress R_e [MPa]	Ultimate tensile stress R_m [MPa]	Total elongation δ [-]
5	Mean	198.9	263.1	386.5	0.278
	St Dev	9.66	6.42	9.50	0.0079
	COV [-]	0.049	0.024	0.025	0.029
6	Mean	190.5	279.3	404.8	0.267
	St Dev	4.82	10.79	16.84	0.0035
	COV [-]	0.025	0.039	0.042	0.013
8	Mean	196.7	357.5	457.0	0.224
	St Dev	2.42	3.88	0.72	0.0132
	COV [-]	0.012	0.011	0.002	0.059

3 RANDOM FIELD MODELLING

The spatial distribution of mechanical properties may be modelled by employing the random field techniques, which sounds to be most suitable for this purpose, due to the set of an infinite number of spatially correlated random variables. In the case of engineering applications, the random field with a finite number of random variables is needed. The Gaussian random field is defined entirely by its mean $\mu(x)$, variance $\sigma^2(x)$ and autocovariance function $C(x,x')$.

In order to discretise the random field, different methods are available. In the present study, the Karhunen – Loeve expansion (Ghanem and Spanos, 1991) is used, which is employed to generate the random field within a specified mesh density. Other methods can be found in (Ghanem and Spanos, 1991; Li and Der Kiureghian, 1993). However, the Karhunen – Loeve expansion has some useful properties with a comparison to other methods, due to that, it is widely used in engineering applications. According to the Karhunen – Loeve method, the random field can be described as:

$$H(x,\theta) = \mu(x) + \sum_{i=1}^{\infty} \sqrt{\lambda_i} \xi_i(\theta) f_i(x) \qquad (1)$$

where λ_i and $f_i(x)$ are the eigenvalues and eigenvectors of the covariance function $C(x,x')$ and x, x' are two coordinates defined for a specific mesh density. For a random field defined in a 2D surface, each vector consists of two components.

The parameter $\xi_i(\theta)$ is defined as a set of uncorrelated random variables with a mean value and covariance function of:

$$E[\xi_i(\theta)] = 0 \qquad (2)$$

$$E[\xi_i(\theta)\xi_j^{'}(\theta)] = 1 \qquad (3)$$

Assuming that $H(x, \theta)$ is a zero-mean Gaussian process, $\{\xi_1(\theta), \xi_2(\theta), \ldots\}$ is a vector of uncorrelated random variables sampled from a zero-mean normal distribution.

Truncating the series after the nth term from Eq. 1, an approximated solution of $H(x, \theta)$ is obtained:

$$\hat{H}(x, \theta) = \mu(x) + \sum_{i=1}^{N} \sqrt{\lambda_i}\xi_i(\theta)f_i(x) \qquad (4)$$

The corresponding autocovariance function is defined as:

$$\widehat{C_{HH}}(x, x^{'}) = \sum_{i=1}^{N} \sqrt{\lambda_i}f_i(x)f_i(x^{'}) \qquad (5)$$

For constant mean and variance values of the field, the autocovariance function is dependent only from an absolute distance between points x and $x^{'}$, and the field is homogenous. In the presented work, the square exponential autocovariance is used:

$$C(x, x^{'}) = \exp\left(\frac{(x - x^{'})^2}{c_0^2}\right) \qquad (6)$$

where $(x - x^{'})$ is the absolute distance between two points and c_0 is the correlation length (also known as a damping parameter). The spatial variation of the field mostly depends on the correlation length. With the increase of c_0, the correlation is higher, and the field is smoother. When c_0 decreases, the correlation is lower, and the field becomes more irregular. The examples of highly correlated and slightly correlated random fields are presented in Figure 2.

The MatLab software (Mathworks, 2019) is used to generate the random field using one of the available codes as given in (Constantine, 2012) using the random Gaussian fields for a specific mesh density and correlation functions.

4 FE MODELLING

The stiffened plates of a 1.06 m length, a 0.4 m width and with a stiffener of a 0.1 m height are analysed in the present study. The specimen thicknesses are considered the same as analysed sheets of steel.

To generate the FE models, the commercial software ANSYS (ANSYS, 2019) is used with the application of SHELL181 elements. The nonlinear static solver with the use of the Newton – Raphson iterative procedure is applied in order to obtain the structural response of stiffened plates subjected to a uniaxial compressive load. Two types of nonlinearities are considered during the finite element analysis, namely material nonlinearities and geometrical nonlinearities (allowance of large deformations). To find the proper mesh size, mesh convergence studies were performed, and a 0.02 m element size was found to give satisfying results.

With regards to the material model, a bilinear stress-strain material relationship with hardening is employed. One can note that the material properties are subjected to a spatial variation in the present study.

However, based on the sensitivity analysis regarding the ultimate strength of stiffened plates as presented in (Woloszyk, et al., 2018), only the variation of the Young modulus and yield point are considered. The variations of the total elongation and ultimate tensile stress have insignificant influence.

Based on the following assumptions, the statistical characteristics for the Young modulus and yield stress are accounted for as presented in Table 1, as a function of the plate thickness. In the case of the ultimate tensile stress and total elongation, only the mean values as presented in Table 1 are considered, and there are constant across the plate and stiffener.

The considered boundary conditions are presented in Figure 3. The loaded edges of the stiffened plate

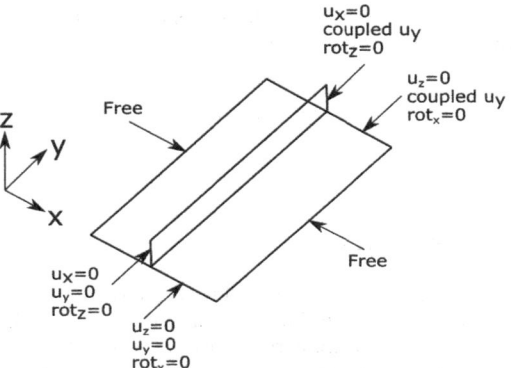

Figure 3. Boundary conditions.

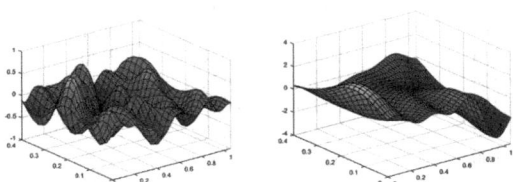

Figure 2. Random field with low (left) and high (right) correlation.

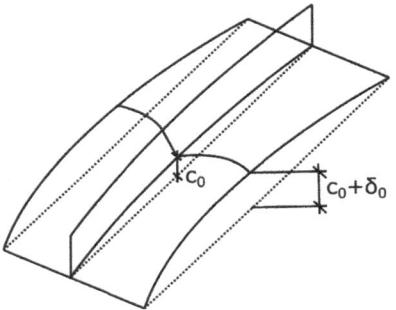

Figure 4. Initial imperfections of stiffened plate.

are fixed, whereas side edges are left free. An incremental displacement is applied to one of the edges in order to obtain the force-displacement response.

In order to evaluate the ultimate strength of the stiffened plates, the initial imperfections need to be adopted in the model. A right way is to follow the Smith approach (Smith, 1977), where the shape, as presented in Figure 4, is considered. The imperfections are coming from the superposition of the global distortions in a longitudinal direction (c_0) and local plate distortions with an imperfection level of (δ_0).

The level of the initial imperfections can be calculated based on the plate slenderness ratio, β:

$$\beta = \frac{b}{t}\left(\frac{Re}{E}\right)^{0,5} \qquad (7)$$

where E is the Young modulus, Re is the yield stress, b is the plate breadth and, t is the plate thickness.

The maximum value of the plate imperfections can be calculated as (Dow and Smith, 1984):

$$\frac{\delta_0}{t} = 0.1\beta^2 \qquad (8)$$

The longitudinal imperfection of the stiffened plate, c_0/l is as well as the stiffener imperfections can be considered equal to:

$$\frac{c_0}{l} = 0.0015 \qquad (9)$$

5 SENSITIVITY ANALYSIS

The generation of random fields accounts for different correlations. This includes a correlation length together for different correlations in the longitudinal and transverse directions of the field. Nevertheless, for one correlation, different realisations of the random field may result in a different structural

response. Due to those reasons, the initial sensitivity analysis is performed in order to analyse the influence of the characteristics of the random field.

A 6 mm plate is chosen for the initial studies, and the considered random field correlation length varies between 0.1 m up to 0.316 m. The minimum correlation length cannot be less than the gauge length of the specimen as presented in Figure 1, because it is the minimum length of the specimen, where the material properties were evaluated. Above the maximum level of the correlation length, it was identified that the random field tends to be more uniform.

For each correlation length, four analyses are performed. For each run, a set of four random fields are generated considering two material properties separately for the plate and stiffener. Based on the analysis of the tensile tests, there was no found any correlation between the Young modulus and yield stress. Due to that, the random fields are generated separately for both of these material properties.

The FE analyses considering different correlation length and realisation are presented in Table 2. The results are compared with the initial reference case, where no deviations of the material properties are considered.

One can notice, that in almost all of the 12 cases, the ultimate strength was reduced with regards to its initial value. For the extreme case, the reduction of 3.22 % was observed (4th run of a correlation length of 0.316 m). There cannot be found an explicit correlation between the reduction of the ultimate strength and the correlation length. However, the extreme reductions of the ultimate strength are observed for somewhat strongly correlated fields. The random fields representing variations of the mechanical properties for the selected cases are presented in Figures 5 to Figure 7. The scales are normalised, i.e. the plotted value is then multiplied by the mean value of the yield stress or Young modulus, respectively.

Table 2. Sensitivity analysis.

Correlation length [m]	Run No	Ultimate force [kN]	Normalised ultimate stress [-]	Difference with initial state [%]
Initial state	-	538.5	0.6427	0.00
0.316	1	532.7	0.6357	-1.08
0.316	2	534.0	0.6373	-0.84
0.316	3	540.4	0.6450	0.35
0.316	4	521.2	0.6220	-3.22
0.224	1	521.3	0.6221	-3.20
0.224	2	531.4	0.6342	-1.32
0.224	3	527.5	0.6296	-2.03
0.224	4	538.9	0.6432	0.07
0.100	1	530.6	0.6332	-1.47
0.100	2	536.1	0.6398	-0.45
0.100	3	531.0	0.6337	-1.39
0.100	4	530.9	0.6336	-1.41

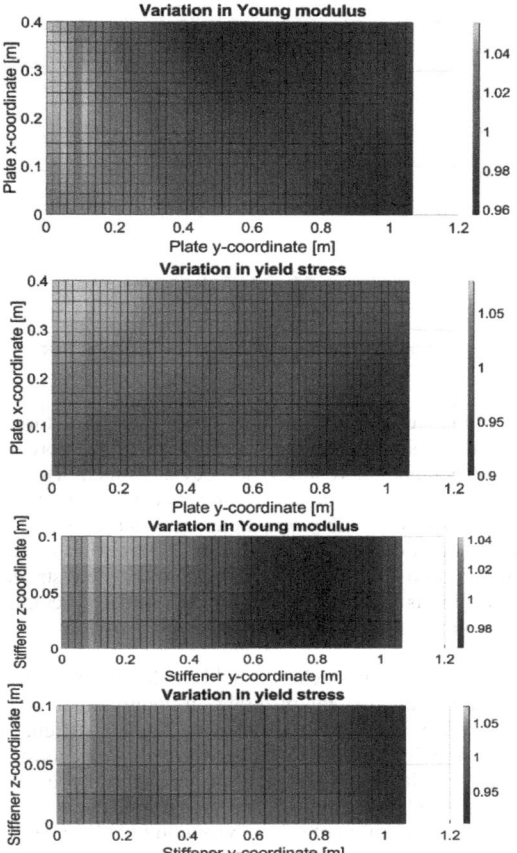

Figure 5. Random fields for second run, correlation length of 0.316 m.

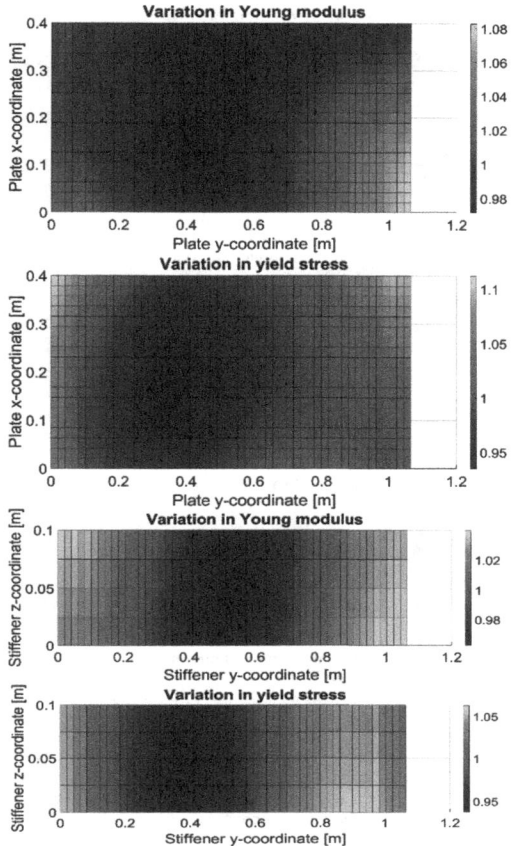

Figure 6. Random fields for fourth run, correlation length of 0.316 m.

Based on the observations from the normal fields with a correlation length of 0.316 m, it could be seen that for the fourth run, the minimum values of material properties are concentrated somewhat near the middle of the plate and stiffener. In the case of the second run, the minimum and maximum values of the material properties are concentrated near the shorter edges. By comparing this with the ultimate strength reductions, it may be concluded that the first case is the weakest one. The middle sections of the stiffened plate are subjected to higher stresses during compressive loading. Thus the reduction of the mechanical properties in that region leads to the highest decrease in the ultimate strength. In the case of a smaller correlation, like presented in Figure 8, there is more than one region of smaller values of the mechanical properties. Due to that, the reduction of the ultimate strength is not very severe, like in the most critical case with a high correlation. However, for a smaller correlation, the ultimate strength reductions are very similar, and there are not so spread. In the case of a higher correlation, some cases show higher ultimate strength in comparison to the initial

case. For these examples, the higher values of the mechanical properties are observed in the mid-sections of the stiffened plate.

For further investigation, the post-collapse shapes for different cases are compared, as shown in Figure 8. It can be noticed, that although the failure mode is similar for each case, which is a local plate buckling mode, followed by stiffener tripping, the position of the collapsed cross-section differs significantly between the cases. In the case of the initial state, the collapsed shape is symmetrical with regards to the middle cross-section. When comparing the shapes with the distributions of the random fields, it can be noted that the positions of the collapsed cross-section occur near the regions of the lower values of the mechanical properties.

For the selected cases, as presented in Figure 9, the force-displacement relationships are plotted. It can be noticed that in the pre-collapse region there are not significant deviation between the curves. In that region, the stresses are below the yield point. This also indicates that the variation of the Young

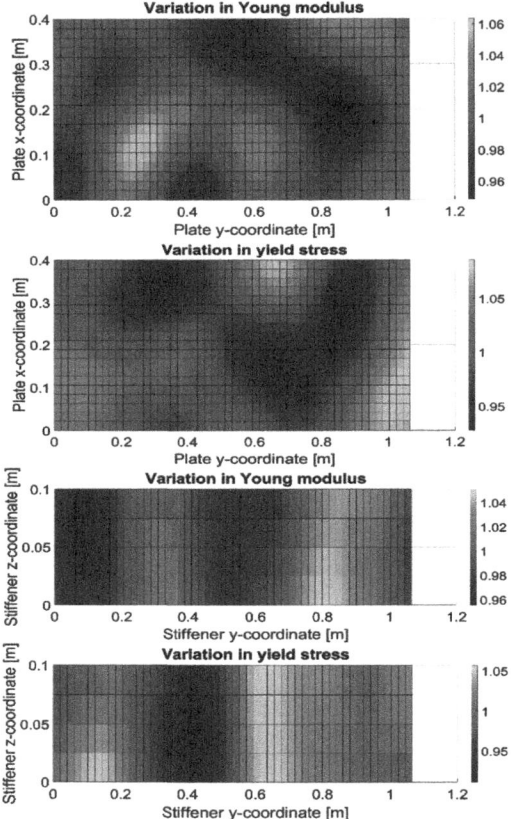

Figure 7. Random fields for first run, correlation length of 0.1 m.

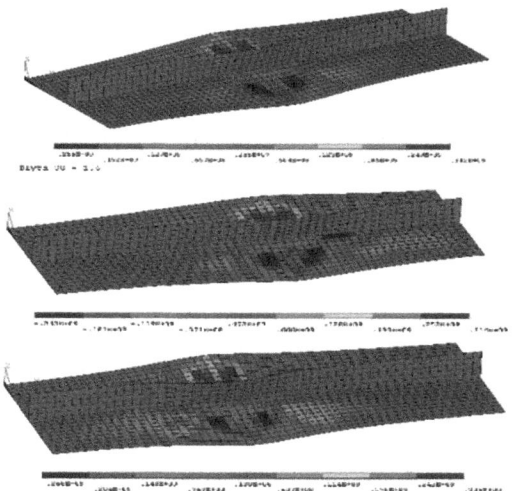

Figure 8. Post collapse shapes for 6 mm specimen – initial case (top), second run of 0.316 m correlation (mid), fourth run of 0.316 m correlation (bottom).

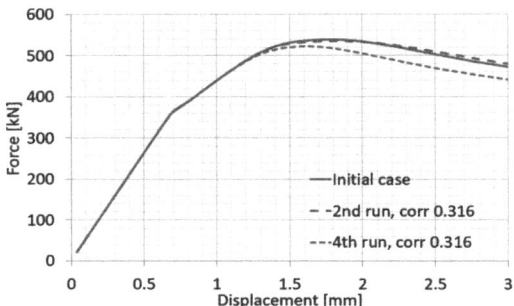

Figure 9. Force – displacement curves for different cases.

modulus does not impact much the structural response, since the initial inclination of the force-displacement curve does not deviate much. However, when the yield point is reached in the middle of the specimen, the deviations are visible, leading to a reduction of the ultimate force. Based on these observations, it can be concluded that the yield stress deviation is the main reason for the ultimate strength reduction of stiffened plates subjected to compressive loadings.

In the case of stochastic field modelling, the correlation can also be different in longitudinal and transverse directions for the plate and or stiffener. The additional analysis is performed, considering a very high correlation between mechanical properties in both longitudinal and transverse directions. The resulting ultimate force was computed for both cases. A higher ultimate strength reduction is obtained for the mechanical properties variations in the longitudinal direction. A reduction of 3.06 % is very close to the maximum one obtained from the previous analysis. In the case of mechanical properties, variating only in the transverse direction, the reduction is almost two times smaller and is about 1.88 %.

6 RESULTS AND DISCUSSION

For most critical random fieldsets, as was found in sensitivity analysis, an analysis is carried out for a 5 mm, 6 mm and 8 mm thickness of the plates, considering different correlation lengths. The mechanical properties and their variations are presented in Table 1, and the results of the analysis are presented in Table 3.

It can be seen, that in the case of a 5 mm and 6 mm thicker plates, the reductions of the ultimate strength with regards to the non-homogenous distribution of mechanical properties is significant, whereas for the 8 mm plate it is rather small. This is caused by different variations of mechanical properties, as presented in Table 1. In the case of an 8 mm plate, the coefficient of variation for both Young modulus and yield stress is around 1 %. In the case of ultimate strength reductions, similar observations

Table 3. Ultimate strength for stiffened plates.

Thickness [mm]	Correlation length [m]	Ultimate force [kN]	Difference with initial state [%]
	Initial	396.5	0.00
5	0.316	387.0	-2.38
	0.224	385.6	-2.75
	0.100	393.7	-0.70
	Initial	538.5	0.00
6	0.316	521.2	-3.22
	0.224	521.3	-3.20
	0.100	530.6	-1.47
	Initial	992.2	0.00
	0.316	987.9	-0.43
8	0.224	989.7	-0.24
	0.100	989.6	-0.26

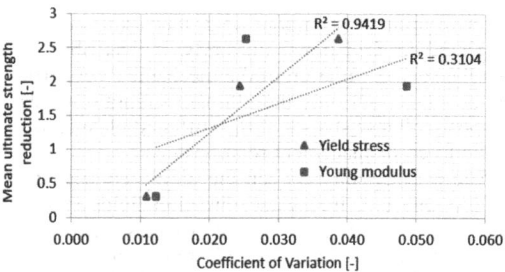

Figure 12. Relation between ultimate strength reduction and mechanical properties uncertainty level.

may be seen, that the higher correlation of the field causes a more significant ultimate strength reduction.

The comparison of the force-displacement curves for the initial and most critical case for 5 mm and 8 mm plates is presented in Figures 10 and 11, respectively. In the case of a 6 mm plate, the curves and discussion were already covered in the previous section. With regards to the 5 mm plate, similar observations can be noted. The initial inclination of the curve is not changed, although the CoV of the Young modulus for that plate reaches the level of 5 %. The differences start to occur near the ultimate strength point region. In

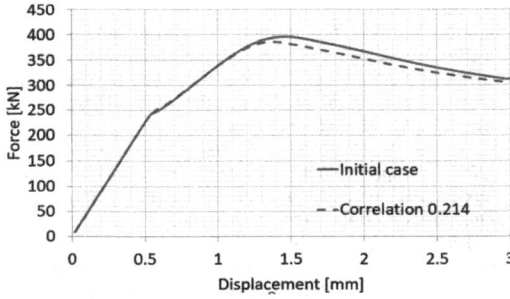

Figure 10. Force – displacement curves for 5 mm plate.

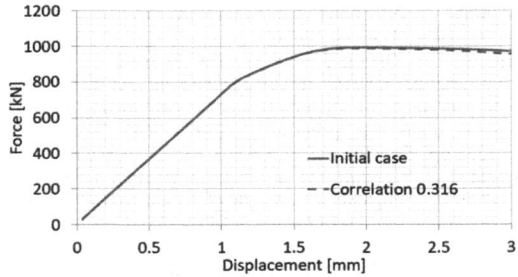

Figure 11. Force – displacement curves for 8 mm plate.

the case of an 8 mm plate, there are very slight differences, due to a very low level of variation of the mechanical properties.

Based on the force-displacement curves, it can be assumed that the deviation in the yield stress is the main reason for the ultimate strength reduction. This hypothesis is furtherly verified. In Figure 12, the mean ultimate strength reductions from Table 3 are compared with both Young modulus, and yield stress coefficients of variation for particular plates and the correlations are plotted.

It can be noted that in the case of the yield stress variation level, the ultimate strength reduction is strongly correlated and the R^2 value is 0.94. However, with the increase of the Young modulus variation level, there is no strict correlation with the ultimate strength reduction. Based on this plots and previous observations regarding the force-displacement curves, it may be concluded that the ultimate strength reduction is mainly caused by the non – homogenous distribution of the yield stress, whereas the Young modulus deviation plays a secondary role.

7 CONCLUSIONS

The work presented here analysed the impact of the non – homogenous distribution of mechanical properties on the ultimate strength of stiffened plates subjected to a compressive load employing the random field methodology coupled with the nonlinear FE analysis. The statistical descriptors of the mechanical properties were obtained based on the tensile test of specimens taken from mild steel sheets of different thicknesses.

In the first place, the present study revealed that non – uniform spatial distribution of the mechanical properties can cause a reduction of the ultimate strength of the stiffened plate up to 3.2 % when compared to the stiffened plate with a mean value uniformly distributed mechanical properties.

The performed sensitivity analysis with regards to the random field correlation length showed that the significant ultimate strength reduction is obtained for a more correlated field. Additionally, depending on the realisation of a random field, the results are also subjected to deviations. Furthermly, it was observed that the spatial variation of the mechanical properties might impact not only the ultimate strength but also the post-collapse shapes of the stiffened plate as well as the force-displacement relationships.

The comparison between the highly correlated in longitudinal and transverse direction fields showed that the second one is the weakest case and a significant ultimate strength reduction is noted when the lower values of the mechanical properties are encountered in the region near the middle of the stiffened plate.

The investigation of the plate with different thicknesses revealed that the ultimate strength reduction is strongly correlated with the COV of the yield stress, whereas the deviations of the Young modulus plays a minor role.

The outcome of the present analysis has shown that the mechanical properties variation within a single stiffened plate causes additional variation with regards to the FE analysis and there is a need to be considered during validation with experimental results.

ACKNOWLEDGEMENTS

This work has been supported by the National Science Centre, Poland (grant No. 2018/31/N/ST8/02380). The ANSYS software used in presented simulations in this paper was available as a part of the partnership cooperation agreement between ANSYS Inc., MESco sp. z o.o., and the Gdansk University of Technology.

REFERENCES

ANSYS (2019) 'Online Manuals, Release 19'.

Constantine, P. (2012) *Random Field Simulation*.

Czujko, J., Bayatfar, A., Smith, M., Xu, M., Wang, D., Lützen, M., Saad-Eldeen, S., Yanagihara, D., Notaro, G., Qian, X., Park, J., Broekhuijsen, J., Benson, S., Pahos, S. and Boulares, J. (2018) 'Committee III.1: Ultimate Strength', in *Proceedings of the 20th International Ship and Offshore Structures Congress (ISSC 2018)*.

Dow, R. S. and Smith, C. S. (1984) 'Effects of localised imperfections on compressive strength of long rectangular plates', *Journal of Constructional Steel Research*. Elsevier, 4(1), pp. 51–76.

Estefen, S. F., Chujutalli, J. H. and Guedes Soares, C. (2016) 'Influence of geometric imperfections on the ultimate strength of the double bottom of a Suezmax tanker', *Engineering Structures*.

Garbatov, Y., Tekgoz, M. and Guedes Soares, C. (2017) 'Experimental and numerical strength assessment of stiffened plates subjected to severe non-uniform corrosion degradation and compressive load', *Ships and Offshore Structures*, 12(4), pp. 461–473.

Ghanem, R. G. and Spanos, P. D. (1991) *Stochastic Finite Elements: A Spectral Approach, Stochastic Finite Elements: A Spectral Approach*. New York, NY: Springer New York.

Gordo, J. and Guedes Soares (2008) 'Experimental evaluation of the behaviour of a mild steel box girder under bending moment', *Ships and Offshore Structures*, 3, pp. 347–358.

Górski, J., Mikulski, T., Oziębło, M. and Winkelmann, K. (2015) 'Effect of geometric imperfections on aluminium silo capacities', *Stahlbau*, 84(1), pp. 52–57.

Górski, J. and Winkelmann, K. (2017) 'Generation of random fields to reflect material and geometric imperfections of plates and shells', in *Shell Structures: Theory and Applications Volume 4*. CRC Press, pp. 537–540.

ISO (2009) 'Metallic materials - Tensile testing - Part 1: Method of test at room temperature.', *International Standard ISO 6892-1*.

Jankowski, R. and Walukiewicz, H. (1997) 'Modelling of two-dimensional random fields', *Probabilistic Engineering Mechanics*, 12(2), pp. 115–121.

Kim, U.-N., Choe, I.-H. and Paik, J. K. (2009) 'Buckling and ultimate strength of perforated plate panels subject to axial compression: experimental and numerical investigations with design formulations', *Ships and Offshore Structures*, 4(4), pp. 337–361.

Li, C. and Der Kiureghian, A. (1993) 'Optimal Discretisation of Random Fields', *Journal of Engineering Mechanics*, 119(6), pp. 1136–1154.

Mathworks (2019) 'MatLab R2019b'.

Melcher, J., Kala, Z., Holický, M., Fajkus, M. and Rozlívka, L. (2004) 'Design characteristics of structural steels based on statistical analysis of metallurgical products', *Journal of Constructional Steel Research*, 60 (3–5), pp. 795–808.

Paik, J. K., Kim, K. J., Lee, J. H., Jung, B. G. and Kim, S. J. (2017) 'Test database of the mechanical properties of mild, high-tensile and stainless steel and aluminium alloy associated with cold temperatures and strain rates', *Ships and Offshore Structures*, 12(sup1), pp. S230–S256.

Paik, J. K. and Seo, J. K. (2009) 'Nonlinear finite element method models for ultimate strength analysis of steel stiffened-plate structures under combined biaxial compression and lateral pressure actions-Part II: Stiffened panels', *Thin-Walled Structures*.

Saad-Eldeen, S., Garbatov, Y. and Guedes Soares, C. (2013) 'Experimental assessment of corroded steel box-girders subjected to uniform bending', *Ships and Offshore Structures*, 8(6), pp. 653–662.

Saad-Eldeen, S., Garbatov, Y. and Guedes Soares, C. (2017) 'Experimental compressive strength analyses of high tensile steel thin-walled stiffened panels with a large lightening opening', *Thin-Walled Structures*.

Shi, X. H., Zhang, J. and Guedes Soares, C. (2017) 'Experimental study on collapse of cracked stiffened plate with initial imperfections under compression', *Thin-Walled Structures*, 114, pp. 39–51.

Shi, X. H., Zhang, J. and Guedes Soares, C. (2018) 'Numerical assessment of experiments on the ultimate strength of stiffened panels with pitting corrosion under compression', *Thin-Walled Structures*, 133, pp. 52–70.

Smith, S. (1977) 'Influence of Local Compressive Failure on Ultimate Longitudinal Strength of a Ship's Hull', *Proc. Int. Sym. on Practical Design in Shipbuilding*, pp. 73–79.

Teixeira, Â. P. and Guedes Soares, C. (2008) 'Ultimate strength of plates with random fields of corrosion', *Structure and Infrastructure Engineering*, 4(5), pp. 363–370.

Woloszyk, K. and Garbatov, Y. (2019a) 'FE analysis of support-specimen interaction of compressive experimental test', in Georgiev, P. and Guedes Soares, C. (eds) *Sustainable Development and Innovations in Marine Technologies*. CRC Press, pp. 423–428.

Woloszyk, K. and Garbatov, Y. (2019b) 'Uncertainty assessment of ultimate strength of corroded stiffened plates subjected to maintenance', in Georgiev, P. and Soares, Guedes (eds) *Sustainable Development and Innovations in Marine Technologies*. CRC Press, pp. 429–436.

Woloszyk, K. and Garbatov, Y. (2020) 'Random field modelling of mechanical behaviour of corroded thin steel plate specimens', *Engineering Structures*, 212, p. 110544.

Woloszyk, K., Kahsin, M. and Garbatov, Y. (2018) 'Numerical assessment of ultimate strength of severe corroded stiffened plates', *Engineering Structures*, 168, pp. 346–354.

Developments in Maritime Technology and Engineering – Guedes Soares & Santos (eds)
© 2021 Copyright the Author(s), ISBN 978-0-367-77376-2

Experimental study on axial compression of misaligned pillars in large passenger ship superstructure

Weiguo Wu & Bin Liu
Green & Smart River-Sea-Going Ship, Cruise and Yacht Research Center, Wuhan University of Technology, Wuhan, China

Pengliang Ren, Jin Gan & Weiguo Tang
School of Transportation, Wuhan University of Technology, Wuhan, China

Yongshui Lin
School of science, Wuhan University of Technology, Wuhan, China

ABSTRACT: This paper takes the typical misaligned pillars in the large passenger ship superstructure as the investigated object. It introduces the misaligned pillars in the large passenger ship and the specific connection between the pillars. The scaled models of the pillars are designed and the ultimate strength experiments are conducted. It analyses the carrying capability characteristics of the pillars in terms of the load-displacement curve, structural failure mode and stress-load curves. An experimental method is provided to study the ultimate strength of misaligned pillars.

1 INTRODUCTION

There are many large-span cabins in the superstructure of large passenger ships, which are mainly used for the auditorium and cinema, etc. These large-span cabins weaken the structural strength of the superstructure greatly. Therefore, it is necessary to arrange pillar structures to improve the structural strength of large-span cabins. In order to facilitate the load transfer between multi-decks, the upper and lower continuous pillars need to be arranged at the same axis (CCS 2018; LR 2018; DNV/GL 2018; BV 2018). However, in the superstructure of large passenger ship, there are many pillars that are not at the same axis because of very strict space requirements at some specific locations. This type of pillars can be called as misaligned pillar. The connection form between the misaligned pillars leads to the bearing capacity loss that is one of the most important difficulties in the structural design of large passenger ships. In this paper, an experimental method is used to study the bearing characteristics of typical misaligned pillars in large passenger ships.

2 EXPERIMENTAL OBJECT

According to the arrangement of pillars in large passenger ships, two typical misaligned pillar structures are selected as the experiment objects. They are divided into two types according to the different pillar combination: (1) circular pillar connected to H-shaped pillar and (2) circular pillar connected to square pillar (see Figure 1). The two types of the pillar connection are different. The connection between circular and H-shaped pillars is more complicated. The connection structure from top to bottom is described as follows: upper pillar, sub-plate, deck, T profile, stiffener, web and lower pillar. The connection structure of circular and square pillars from top to bottom is described as follows: upper pillar, sub-plate, deck, T profile, force transmission plates and lower pillar.

3 DESIGN OF TEST SPECIMENS

The upper and lower pillars are slender. As the processing error of specimens can cause the initial bending deformation of the upper pillar, the structural damage is possible to appear at the foot of the upper pillar in the axial compression. Nevertheless, this type of damage pattern cannot reflect the true picture of the damage of pillar structures. For the misaligned combined pillars, the focus is on the damage of the misaligned connection and the misaligned compression failure of the lower pillar. Thus, in the experiment the upper pillar was shortened and the lower pillar remained to maintain its original length. For the attached deck of pillar structures, it is considered the influence of the range of the attached deck on the

DOI: 10.1201/9781003216582-69

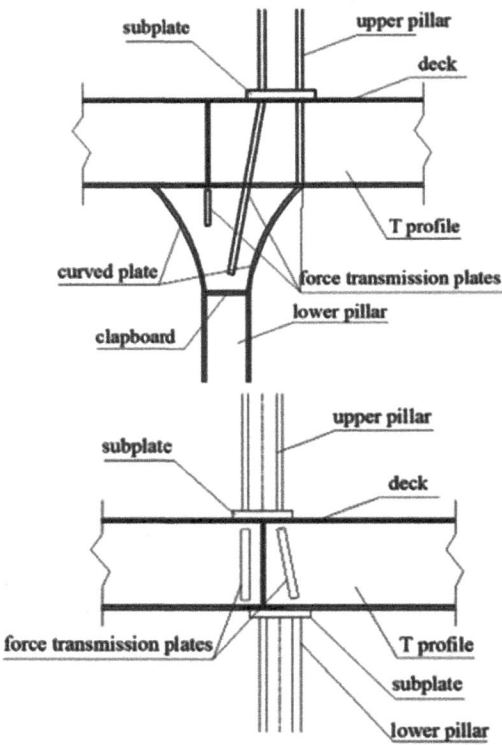

Figure 1. Misaligned pillars and their connection structures.

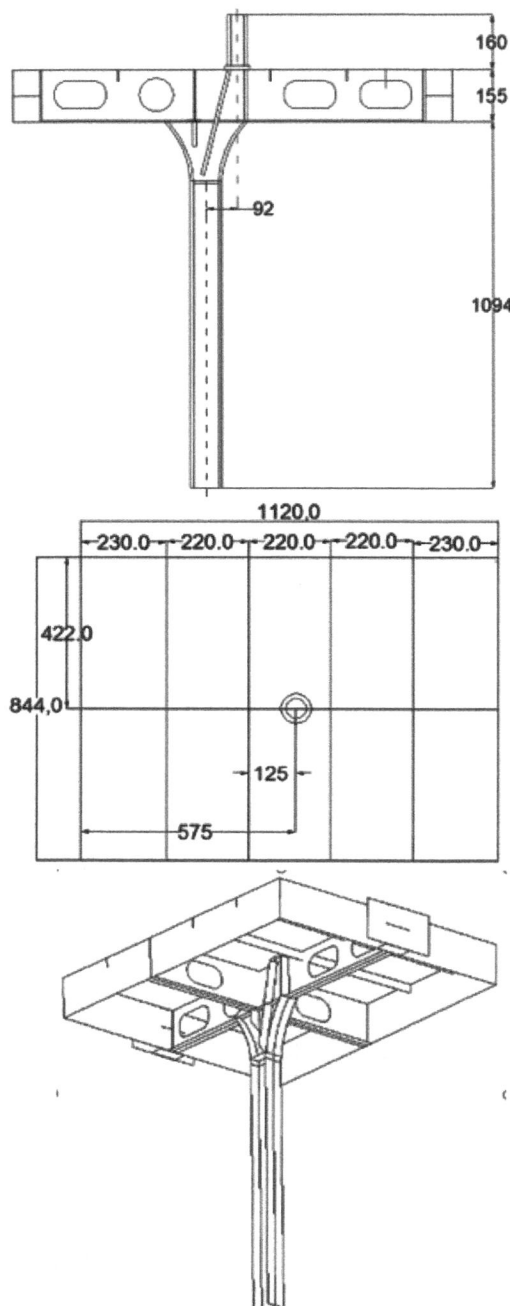

Figure 2. Dimensions of the circular pillar connected to H-shaped pillar.

bearing capacity of the lower pillar. The range of the attached deck was selected to take the distance between two adjacent longitudinal beams of the pillars along the ship breadth and half of the adjacent T-profiles along the ship length. In order to simulate the boundaries of the deck structures, rigid plates are welded around the deck structures.

The specimens are scaled taking the length similarity coefficient $C_L = 1/3$ and the plate thickness similarity coefficient $C_t = 1/2$. According to the scaling law in elastic region, the load relationship between scaled model and prototype structure is 1/6 for achieving the same stress level. However, the ratio of ultimate strength cannot be analysed by the scaling law.

Considering that the actual plate thickness of specimens has to be taken as an integer, the thickness of specimen plate is designed on the premise of ensuring that the structural damage patterns of the specimen and the prototype are consistent. Moreover, as the actual steel plate thickness often has negative tolerance, the actual geometrical size and plate thickness of the specimens need to be measured after manufacturing. The dimensions of the two specimens are shown in Figures 2 and 3, respectively. All components are welded to construct the pillar structures. The misalignment can be defined by the horizontal distance between the axis of upper and lower pillars. Here, the misalignments of the first and second specimens are 92 mm and 53 mm, respectively.

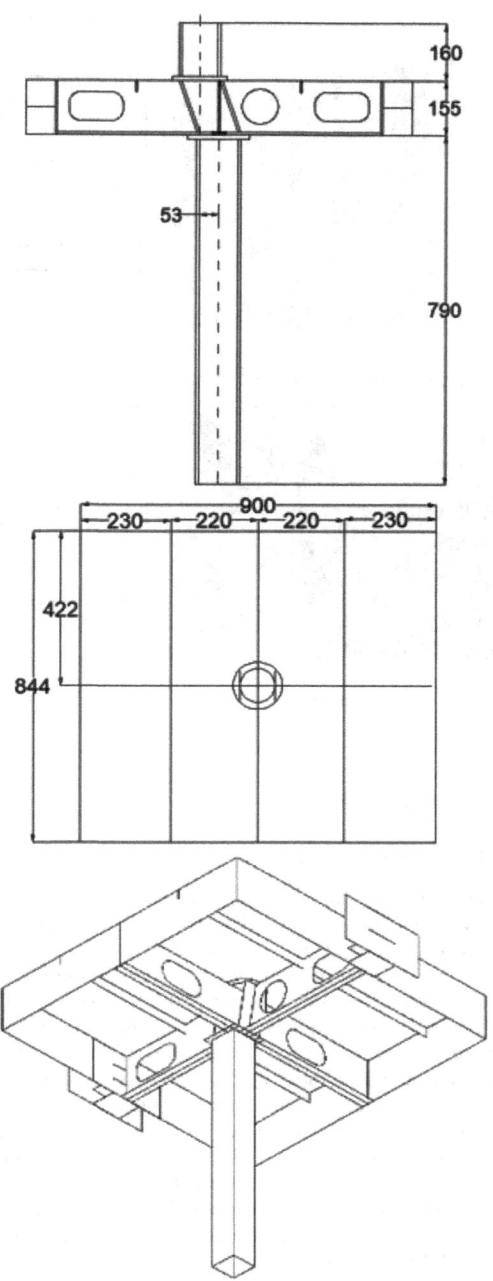

Figure 3. Dimensions of the circular pillar connected to square pillar.

4 EXPERIMENTAL DETAILS

The pillar is the supporting member of deck structures and it mainly suffers the axial load.

Thus, the experiment is designed as the axial compression of the misaligned pillar structures. The two pillar specimens have different bending characteristics. The verticality of lower pillar is adjusted and then the bottom of lower pillar is welded to the structural base. Thus, the initial deflection of the lower pillar is not measured.

Here, the first experiment is performed on the specimen of circular pillar (diameter = 56 mm) connected to H-shaped pillar (H 73 × 66 × 8 × 10 mm), and the second experiment is performed on the specimen of circular pillar (diameter = 91 m) connected to square pillar (80 × 80 × 7 mm). Here, the upper pillars are considered as rigid structures.

4.1 Load condition

In the first experiment, the upper and lower pillars of the specimen have a very large misaligned distance. With the increase of the axial load, the lower pillar (H-shaped) suffers a verge large bending moment. A large hydraulic machine is used to apply the external loading. The hydraulic machine can automatically record the load and displacement during the loading process.

In the second experiment, as the ultimate carrying capacity of the misaligned pillar is large, six hydraulic cylinders are used to apply the loading.

4.2 Arrangement of experimental setup

The test devices include the experimental framework, the loading device, the restraints and the measuring device, as shown in Figures 4 and 5.

The experimental frame is placed vertically and fixed with ground bolts. The loading devices are a large hydraulic machine and hydraulic cylinders, respectively. The restraints are composed of two front and rear chute and fixed seats. The ear plates on the edge of the deck can slide up and down in the chute. The chute adjusts the position of the chute through the screw and the front and back pads.

4.3 Loading scheme

In the first experiment, the loading speed of the large hydraulic machine is set to about 0.5 mm/min to simulate the quasi-static loading. The load is applied three times repeatedly until a displacement of 2 mm to check the linear relationship between loading and stresses in the elastic stage.

In the ultimate experiment, the same loading speed is used and the load-displacement curve is obtained. When the curve shows an unstable increase, the load no longer continues to apply, indicating the collapse of the pillar structures.

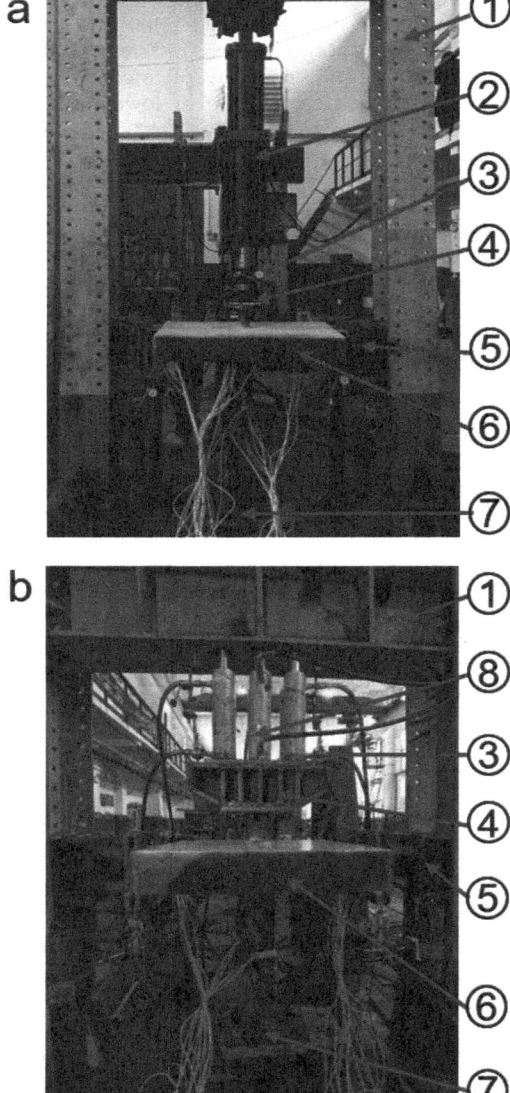

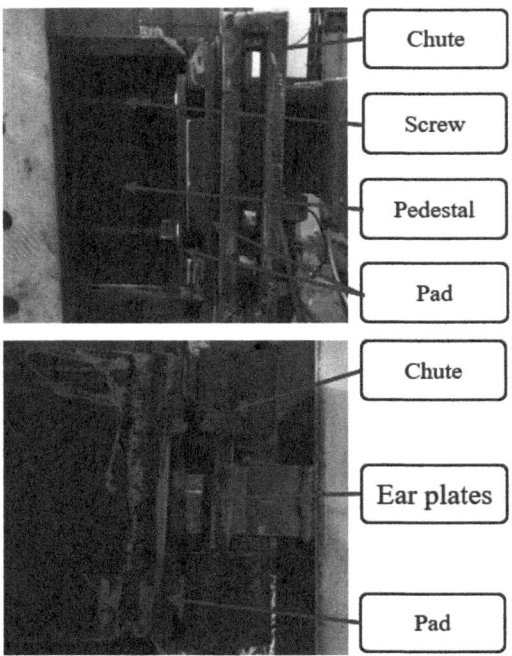

Figure 5. Deck restraints.

Figure 4. Overall schematic diagram of loading system of the (a) first experiment and (b) second experiment. ① Experimental framework, ② Hydraulic machine, ③ Dial indicator, ④ Load connection, ⑤ Restraint, ⑥ Specimen, ⑦ Fixed seats, ⑧ Hydraulic cylinders.

In the second experiment, the loading of a single cylinder is recorded by a load sensor. Three preloads and a ultimate failure load are also performed. In the ultimate experiment, the load of the hydraulic cylinder no longer increases indicating the collapse of the pillar structure.

5 EXPERIMENTAL RESULTS

5.1 *Load-displacement curve*

The experimental load-displacement curves are shown in Figure 6. This displacement indicates the vertical movement of the upper pillar. The ultimate carrying capacity of the pillar structures in the linear elastic phase are 335 kN and 580 kN, respectively. When the load continues to increase, the slope of the curve gradually decreases. It shows that at this moment, the structure begins to fail locally but has not yet reached the ultimate strength. When the slope of the curve decreases to zero, the structure reaches its ultimate limit. The ultimate carrying capacities of the pillar structures are 426.2 kN and 864.6 kN, respectively. The structure collapses completely and the experimental deformations end at 11.5 mm and 22.3 mm, respectively.

5.2 *Failure mode*

The permanent deformation of the misaligned pillar in the first experiment is shown in Figure 7(a). The failure of H-shaped pillar is mainly induced by the bending. The failure is mainly concentrated at the curved plates and fore transmission plates, indicated in Figure 1, above the end of the lower pillar. The deck attached to the misaligned pillar experiences

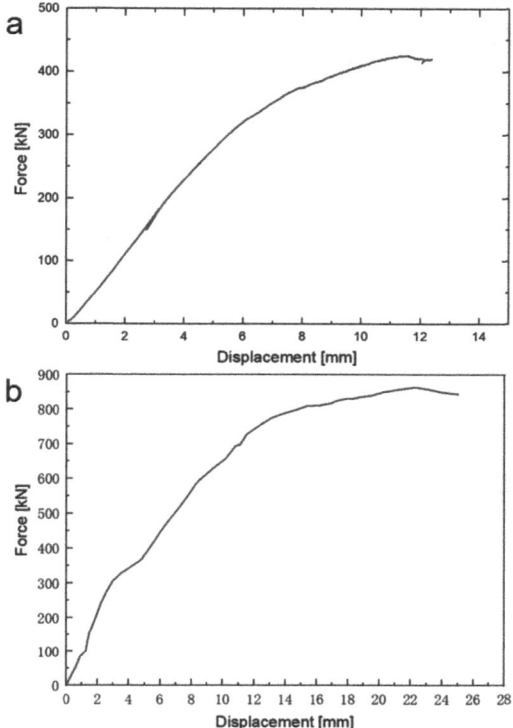

Figure 6. Load-displacement curves for the (a) first and (b) second experiments.

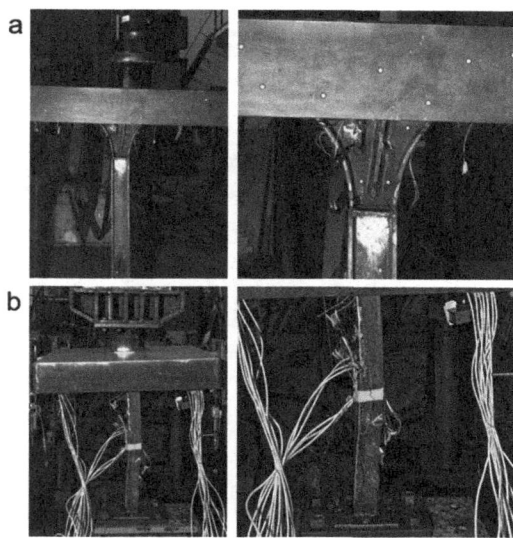

Figure 7. Deformation of pillar (a) round steel-H-shaped steel composite pillar, (b) round steel-H-shaped steel composite pillar.

a significant inclination, which causes structural failure.

The permanent deformation of the misaligned pillar in the second experiment is shown in Figure 7 (b). The failure mode is mainly induced by the bending instability of the square pillar. The main failure is the bending deflection of lower pillar. This mode is similar to the typical instability of the beam under axial compression. At the same time, the deck inclines with the instability of the lower pillar.

5.3 *Stress analysis*

The positions of the strain gauges in the first and second experiments are shown in Figures 8 and 9, respectively. The pillar specimen is composed by three components: deck plate frame, connecting member and lower pillar. Figures 10 and 11 show the von-Mises stresses versus loading curves for the first and second experiments, respectively. The specimen material is AH36 steel and its yield strength is 355 MPa. When the calculated stresses exceed 355 MPa, the stress value will be corrected to 355MPa.

In the first experiment, the stress is mainly concentrated around the connection position between the upper pillar and the deck, for example, large strains are measured by the strain gauges #3, #4, and #10. When reaching the structural ultimate strength, the von-Mises stresses are 177.4 MPa, 138.6 MPa and 150.1MPa at the gauges #3, #4, and #10, respectively. For the connecting members of the upper and lower pillars, the large stresses also occur at the gauges #24, #25, and #31, and their values are 309.6 MPa, 255.5 MPa and 242.6 MPa, respectively. The stresses of #32 and #33 are very small. According to these stress results, it can be obtained that the load transmitted by the upper pillar to the lower pillar is mainly through the #24 and #25 force transmission plates. The structural carrying capacity of #32 and #33 force transmission plates give a relatively small contribution. The stresses on the lower pillar are mainly concentrated at the position of the abrupt change of the pillar structure where the strain gauges #26 to #29 are located. With the increase of the load, the positions of the gauges #27, #29, #28 and #34 reach the material yield limit in sequence.

In the second experiment, the stress of the pillar deck is also mainly concentrated around the connection position of the upper pillar and the deck, such as #3 and #6 strain gauges. When reaching the ultimate carrying capacity of the pillar structure, both the stresses of #3 and #6 reach 355 MPa. The stresses at the T-profile structure position are very large. The measurements of the strain gauges # 9, # 12 and # 15 are 305.1 MPa, 277.3 MPa and 287.2 MPa, respectively when reaching the structural ultimate limit. The lower pillar is the main load-carrying component of the whole structure. As the load continues to increase, the values of strain gauges #21, #22, #23,

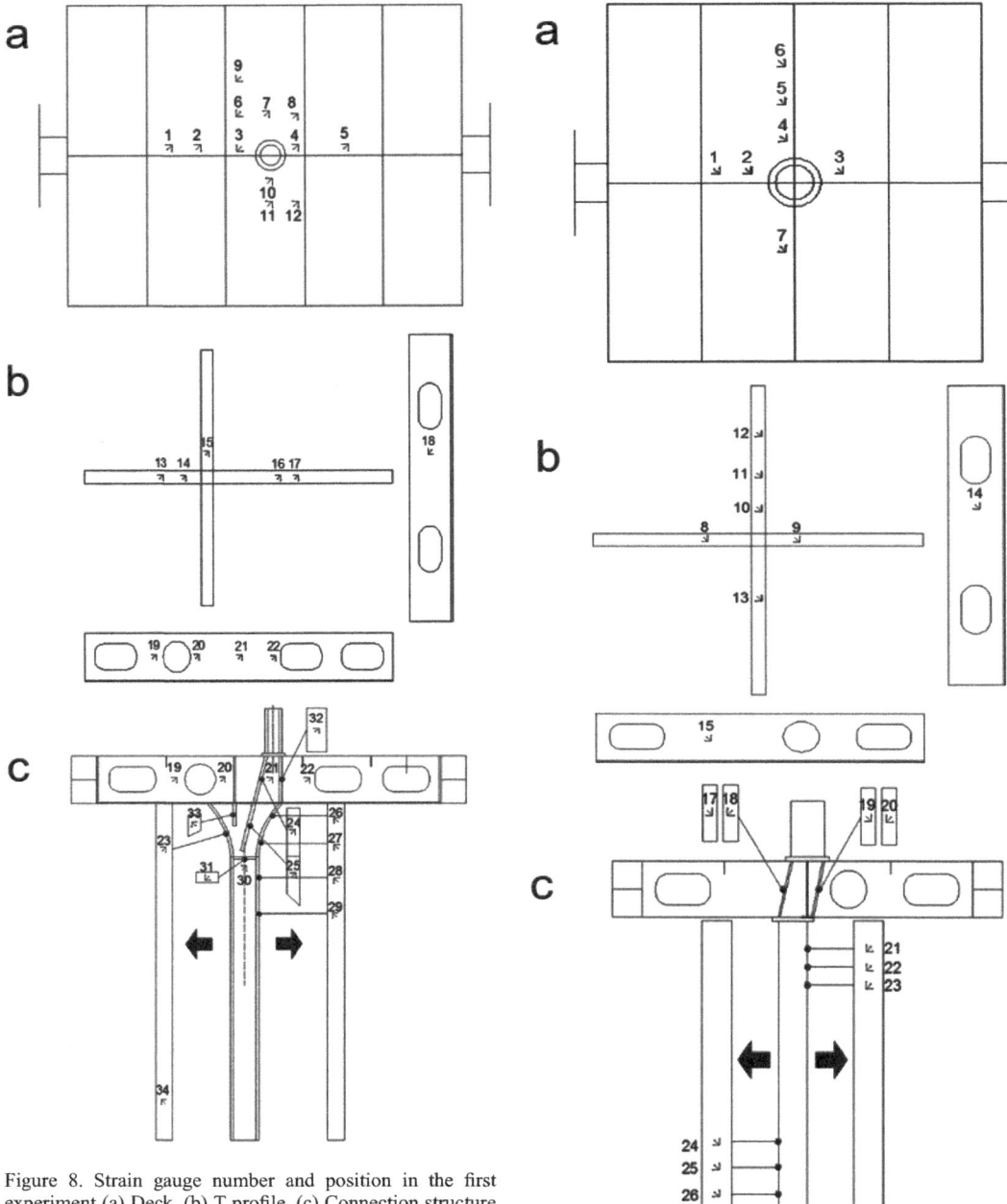

Figure 8. Strain gauge number and position in the first experiment (a) Deck, (b) T profile, (c) Connection structure and pillar.

Figure 9. Strain gauge number and position in the second experiment (a) Deck, (b) T profile, (c) Connection structure and pillar.

#26, #24, and #25 reach 355 MPa (material yield limit) in sequence.

It can be concluded that the stresses of the pillars are mainly concentrated at the connection between the upper pillar and the deck and the upper and lower ends of the lower pillar. In addition, the failure sequence of different components in the misaligned pillars can be obtained. The lower pillars firstly bend at the stress concentration zone and in turn cause the failure of the T-profile web and the partial failure of the deck. Then, the overall structures collapse.

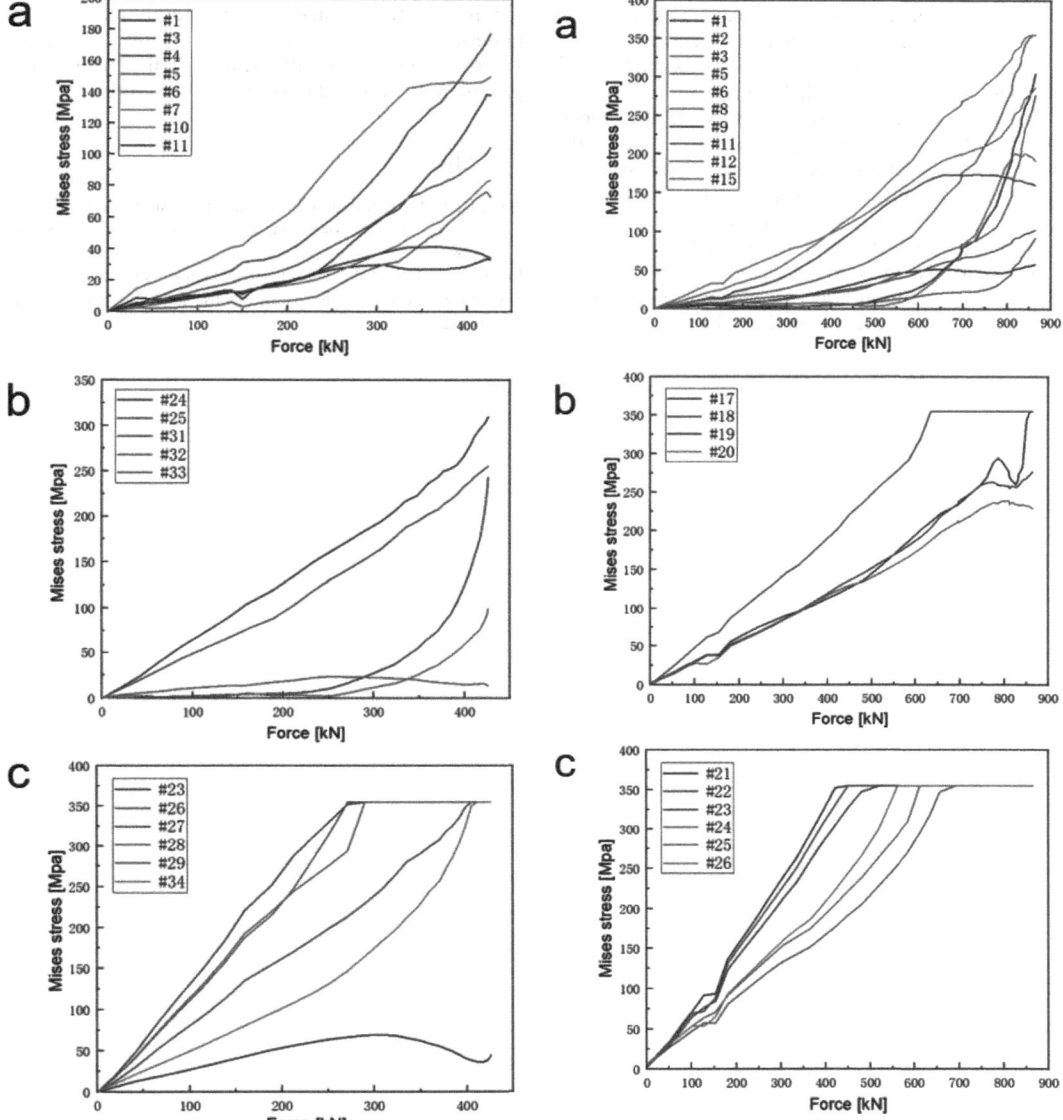

Figure 10. Von-Mises stresses versus external load curves in the first experiment. (a) Deck, (b) T profile, (c) Connection structure and pillar.

Figure 11. Von-Mises stresses versus external load curves in the second experiment. (a) Deck, (b) T profile, (c) Connection structure and pillar.

6 SUMMARY

In this paper, the typical misaligned pillars are selected as the research object. The experimental method is summarised as follows. Firstly, the appropriate scaling is selected to design the pillar specimen. Secondly, the structural details of the misaligned pillars are designed according to the force characteristics of the pillars under axial compression. The experimental load-displacement curve provides the understanding of the load-carrying characteristics of the pillar structures, the maximum capacity of the structure at the linear elastic stage and the maximum capacity of the structure at the ultimate failure.

In view of the structural deformation, the misaligned connection part of the pillar structures under axial compression generates an additional bending moment and this moment causes the incline of the deck attached to the pillar. The overall failure mode of misaligned pillars is mainly caused by the failure of the lower pillar or the connecting member.

The change of the stresses at various components of the pillar specimen shows that the stresses are mainly concentrated at the connection between the upper pillar and the deck and the upper and lower ends of the lower pillar. The failure sequence of the components in the misaligned pillars is obtained. The lower pillar bends firstly at the stress concentration zone, causing the failure of the T-shaped web and localised failure of deck. Eventually, the overall structure collapses.

The axial compression experiments of the misaligned pillars give the bending characteristics of the misaligned pillars, the structural deformation patterns and the key connecting members. It provides a reference for the design of the misaligned pillars in the large passenger ship structures.

REFERENCES

BV. 2018. Rules for the Classification of Steel Ships.
CCS. 2018. Rules for Classification of Sea-going Steel Ships.
DNVGL. 2018. Rules for Classification of Ships.
LR. 2018. Rules and Regulations for the Classification of Ships.

Ship structures - Composites

Developments in Maritime Technology and Engineering – Guedes Soares & Santos (eds)
© 2021 Copyright the Author(s), ISBN 978-0-367-77376-2

Advanced composite materials shafts modelling

E.P. Bilalis & N.G. Tsouvalis

Shipbuilding Technology Laboratory, School of Naval Architecture and Marine Engineering, National Technical University of Athens, Athens, Greece

ABSTRACT: The use of composite materials for the manufacturing of power transmission shafts attracts particular interest. Especially for the marine sector, apart from their high strength and light weight, composite shafts offer the advantages of high fatigue and corrosion resistance. In the context of the present work, advanced finite element models are developed for the simulation of the mechanical behavior of composite shafts. The calibration and validation of the finite element models is pursued by the comparison of their results with experimental data acquired from the industry and literature. These models will consist the base for the future development of the digital twins of composite materials shafts.

1 INTRODUCTION

Composite materials are increasingly becoming the choice for a large number of products and applications, due to their light weight and high strength. The use of composite materials for the manufacturing of power transmission shafts attracts particular interest. Composite materials shafts are hollow and are usually manufactured with the automated filament winding method, which produces high quality end products with very good repeatability. Especially for the marine sector, apart from their high strength and light weight, composite shafts offer the advantages of high fatigue and corrosion resistance. GFRP or CFRP composite shafts have the potential to be 25-80% lighter than the conventional steel shafts of similar specifications. The massive steel shafts on large ships comprise up to 2% (or ~100-200 tons) of the ship's total weight. Ship designers expect a composite shaft to also suppress the transmission of noise from machinery and propellers due to the intrinsic damping properties of composite materials. Hence the acoustic signature of the vessel would be reduced which is important for naval warfare ships. Being non-magnetic, composite shafts will also reduce the magnetic signature of a vessel (Mouritz et al. 2001). Additionally, composite shafts offer the advantages of low bearing loads (or even elimination of bearings) due to their light weight, and of greater flexibility and improved life-cycle cost (Greene 1999).

The international literature includes some works on composite shafts, however in industrial sectors other than marine (Shokrieh et al. 2004, Khoshravan & Paykani 2012, Montagnier & Hochard 2013, Sevkat et al. 2014, James Prasad Rao et al. 2016, Stedile Filho et al. 2018). Studies of marine applications of composite shafts do not exist in the literature and although some commercial applications are available in the market (Figure 1), there are still several open issues, mainly related to the effective and optimum design of these parts, their maintenance and their durability. These issues derive from the general problem of understanding composite material mechanical behavior and their failure modes and mechanisms. Additionally, the definition of the mechanical properties of a composite material which, as the composite shafts, is made with the filament winding method, can be a difficult task due to its special characteristics. These difficulties are multiplied by taking into account the number of parameters influencing their design, such as the torque they carry, their rotational speed, their bearings, their length, etc. In order to overcome these difficulties, it is necessary to combine advanced numerical methods to simulate their behavior together with experimental tests.

Along with the above, a new concept that constantly gains ground in engineering is the *"digital twin"*. It is an evolving digital profile of the past and current mechanical behavior of a structure, which, based on measurements during its operation, helps optimize its performance and predict its future condition. Thus, the digital twin enables specialized maintenance, timely handling of failures, minimization of downtime and extension of the life span of the structure (Malakuti et al. 2018).

In the context of the present work, finite element models are developed for the simulation of the mechanical behavior of composite shafts. The calibration and validation of the finite element models is pursued by the comparison of their results with experimental data acquired from both the industry and literature. These models will consist the base for the future

DOI: 10.1201/9781003216582-70

Figure 1. Carbon fiber reinforced polymer shaft in ship (CENTA, www.centa.info).

optimization of the design of composite materials shafts and the development of their digital twins.

2 EXPERIMENTAL DATA

2.1 Case 1 – GFRP shaft

There is lack of experimental results from torsional tests of composite materials shafts in the literature. Very few papers present experimental results accompanied by material properties and lay up information that are necessary for the modelling of the shaft. Experimental results from the industry are even rarer, due to confidentiality. Despite this, relevant information was obtained by a Greek composites manufacturing company about such a test of a Glass Fiber Reinforced Polymer shaft manufactured by the filament winding method (Bilalis 2016). The main dimensions of the shaft are external diameter = 0.260 m, thickness = 5 mm and length between flanges = 0.692 m. The winding pattern was [±45]₁₂. The layer mechanical properties of the composite material are as follows:

$$E_1 = 37.04 GPa, \ E_2 = 15.04 GPa, \ E_3 = 15.04 \ GPa,$$

$$G_{12} = G_{13} = 5.5 \ GPa, G_{23} = 2.75 \ GPa$$

$$v_{12} = 0.28, v_{23} = 0.3447, v_{13} = 0.28$$

where magnitudes in italics are assumed values, based on classical composites literature (Jones 1975, Christensen 1979).

The shaft had failed at the torque of 61.12 kNm.

2.2 Case 2 – CFRP shafts

Bauchau et al. (1988) presented their study on torsional buckling of graphite/epoxy shafts, which included

Table 1. Lay-up and geometry of specimens.

	Lay-up	L (m)	R (mm)
L#1	15, -15, -45, -15, 15, 45	0.260	41.27
L#2	45, 15, -15, -45, -15, 15	0.260	41.27

analytical and experimental results of a series of cylinders. The torque-twist curves of only two cylinders were presented in this paper (L#1 and L#2), so only these two specimens are analyzed in the context of this work. The lay-up (starting from the innermost ply), length, L, and mean radius, R, of these two cylinders are given in Table 1, whereas the layer mechanical properties considered are as follows:

$$E_1 = 134 GPa, E_2 = 8.5 GPa, E_3 = 8.5 \ GPa,$$

$$G_{12} = G_{13} = 4.6 GPa, \ G_{23} = 4 GPa$$

$$v_{12} = 0.29, v_{23} = 0.40, v_{13} = 0.29$$

where, once more, magnitudes in italics are assumed values, based on classical composites literature.

3 FINITE ELEMENT MODELLING

3.1 Modelling of the GFRP shaft

3.1.1 Layered shell model

The first finite element modelling approach is a layered shell model developed in ANSYS. Layered shell elements are used because the shaft is relatively thin, as its thickness to diameter ratio is 2%. The element type used is SHELL281, the 8-node structural shell element of ANSYS. The developed model implements both eigenvalue buckling analysis and nonlinear buckling analysis. There is a connection between the two, as the modeshapes of the eigenvalue buckling analysis are used as a pattern for the initial imperfections necessary for triggering nonlinear buckling.

For the construction of the element mesh, a convergence analysis was carried out involving four different meshes, from 1080 to 16320 elements and with varying number of integration points through each layer thickness. The difference between the coarser and the finer mesh is 0.09% concerning the critical eigenvalue buckling load, negligible concerning the nonlinear ultimate calculated load and 3.7% concerning the maximum rotation of the free end of the shaft. More integration points through the thickness of the layer offered no benefits in this particular analysis. Considering the results and the duration of the solution, the mesh of 3840 Elements and 3 integration points through the thickness of each layer was chosen.

Multipoint Constraint (MPC184) elements were used for fixing one end of the shaft and for applying torque on the other end. MPCs connected all nodes at the circumference of each end with a master node at the center of the circular end. The torque was applied to the master node of one end, which was allowed only to rotate around the longitudinal axis of symmetry of the shaft. The other end of the shaft was considered fully fixed, by constraining all degrees of freedom of the respective master node.

The expected failure mode of the shaft is rotational buckling. Apart from the eigenvalue buckling analysis which predicts the theoretical buckling load of an ideal linear elastic structure, nonlinear buckling analysis was also carried out because it employs a non-linear, large-deflections, static analysis to predict buckling behaviour. In this analysis, the applied load is gradually increased until a load level is found whereby the structure becomes unstable (i.e. a very small load increase causes very large deflections). The true non-linear nature of this analysis thus permits the modelling of geometric imperfections and material nonlinearities.

Eigenvalue buckling analysis results
In the current analysis the buckling loads of the first 10 modeshapes were calculated and are presented in Table 2. All buckling loads in this table are double, corresponding to identical modeshapes rotated around the longitudinal axis of the cylinder. All modeshapes depict only one crest along the longitudinal direction.

The calculated critical (minimum) eigenvalue buckling load is 62.4 kNm. It is 2% higher than the experimental failure load of the shaft. The results also indicate that, at the critical load, the shaft buckles in

Table 2. Eigenvalue buckling analysis results of the GFRP shaft.

Buckling load (kNm)	Buckling modeshape
	Mode 4
62.4	
-63.2	
	Mode 3
70.1	
-71.1	
	Mode 5
76.1	

a mode 4 modeshape, meaning that 4 crests are formed around the circumference of the shaft.

Nonlinear buckling analysis
Nonlinear buckling analysis consists of two steps. The first is an eigenvalue buckling analysis leading to the generation of the various modeshapes and the second is the nonlinear analysis. The modeshape corresponding to the critical buckling load is scaled by a specific magnitude and is then used in the nonlinear analysis as the initial geometrically imperfect geometry of the shaft, safely assuming that the shaft is going to buckle following this predicted mode shape (Papadakis & Tsouvalis 2016). It is also possible to combine the modeshapes of several eigenvalues to generate the initial imperfections pattern. The size of the geometric imperfections depends on the application and is expressed in this study by the ratio of its maximum value to the shaft's internal diameter, this ratio taken here equal to 0.001. The effect of the size of the initial imperfections has been investigated and is discussed later.

As expected, the solution stopped at some point due to non-convergence. In this case, this happened at a load of 61.06 kNm. Figure 2 depicts the relation between the applied torque and the resulting rotation of the free end (rotation of the corresponding master node). Buckling can be identified in Figure 2 at the point where the curve starts to bend. It can be seen in this figure that the shaft buckles somewhere between 56 kNm and 59 kNm and, therefore, the nonlinear buckling analysis model predicts quite well the experimentally measured buckling load.

Figure 3 shows the maximum stresses at the external surface of the shaft in the direction of the fibers (σ_1) and in the direction normal to the fibers (σ_2). The fiber orientation of the external ply of the shaft is -45°. As expected, the node with the maximum tensile stresses in both directions is situated at mid-length on a crest, and the node with the maximum compressive stresses is situated at mid-length on a trough of the buckled modeshape. Figure 3 shows the typical pattern that stresses follow in buckling. Maximum stresses in the fibers' direction are compressive and when buckling initiates, one of them (on the crest) becomes

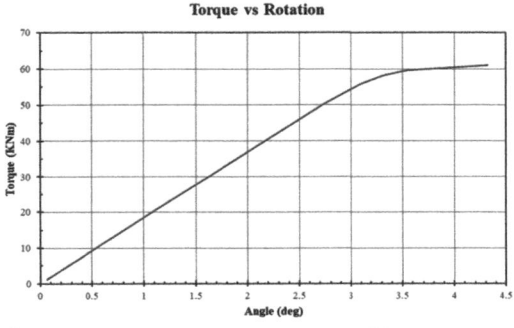

Figure 2. FEM applied torque vs angle of rotation.

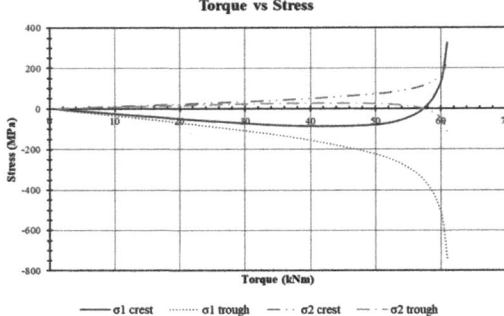

Figure 3. Variation of stresses σ_1 and σ_2 at a crest and a trough on the external surface.

tensile. The stresses in the direction normal to the fibers follow the opposite pattern. It can be also seen in this figure that all stresses present a rapid change after 50 kNm, where buckling initiates. A similar pattern is also presented by the longitudinal and circumferential stresses at exactly the same points.

Figure 4 shows the variation of the same stresses σ_1 and σ_2 at the same nodes as above, but in the internal surface of the shaft. The fiber orientation of the internal ply of the shaft is 45°. Figure 4 shows that, on the internal surface, maximum stresses in the direction of the fibers are tensile, in contrast to the corresponding stresses on the external surface which are compressive. When buckling initiates, the stress developed on a crest reduces its magnitude as it was expected, due to the compression of the internal surface of the shaft at the position of the crest. The stresses in the direction normal to the fibers follow the opposite pattern. Additionally, the magnitudes of these stresses are lower than the ones on the external surface.

3.1.2 *Homogeneous material shell model*

Homogeneous modelling considers the shaft as single-layered, with this single layer having the equivalent mechanical properties of the multilayered composite. This is attempted because filament winding, the manufacturing method of the shaft, does not produce discrete layers, as for example hand layup does. The approach used, transforms a especially orthotropic multilayered material to an equivalent homogeneous orthotropic material. Due to its layup, $[\pm45]_{12}$, the shaft can be considered as especially orthotropic.

The calculation of the equivalent mechanical properties was carried out in accordance with typical procedures of mechanics of composite materials (Jones 1975) and resulted in the following values:

$$E_x = 16.27\text{GPa}, \; E_y = 16.27\text{GPa}, E_z = 16.27\text{GPa},$$

$$G_{xy} = G_{xz} = 11.27\text{GPa}, G_{yz} = 5.64\text{GPa}$$

$$v_{xy} = 0.479, v_{yz} = 0.479, v_{xz} = 0.479$$

where x, y and z are the longitudinal, circumferential and through-thickness directions of the cylinder, respectively.

The calculated material properties were implemented in the corresponding FE shell model. The calculated critical eigenvalue buckling load is 66.3 kNm, 8.5% higher than the experimental failure load and 6.3% higher than the layered shell model result. The critical buckling modeshape is the same to that of the layered shell model.

Figure 5 presents a comparison of the torque vs rotation curves between the two material models, from which it is obvious that the coincidence is very good. Homogeneous shell material modelling did not affect the rotational stiffness of the shaft.

Figure 6 presents the variation of the maximum tensile and compressive stresses in x and y directions, for both the homogeneous and the layered shell GFRP shaft FE model. These stresses were calculated at mid-length, at the position of a crest and a trough, on the external surface of the shaft. A main difference is that all stresses of the layered model start as compressive, whereas this is not true for the homogeneous model. Additionally, there is a symmetry with respect

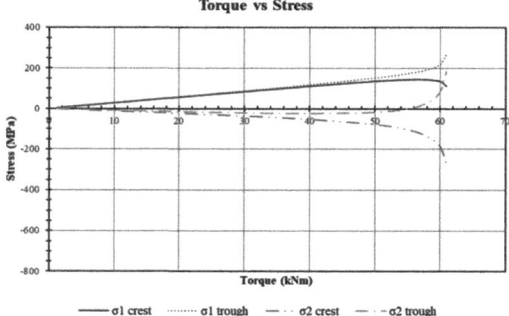

Figure 4. Variation of stresses σ_1 and σ_2 at a crest and a trough on the internal surface.

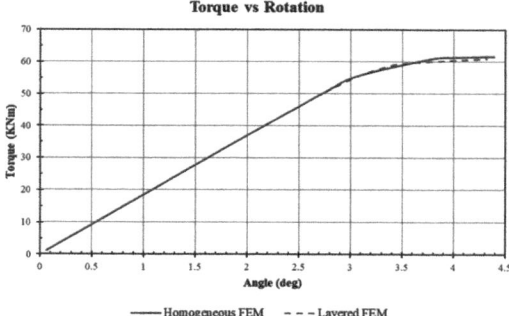

Figure 5. Torque vs rotation - comparison between the homogeneous and layered shell model results.

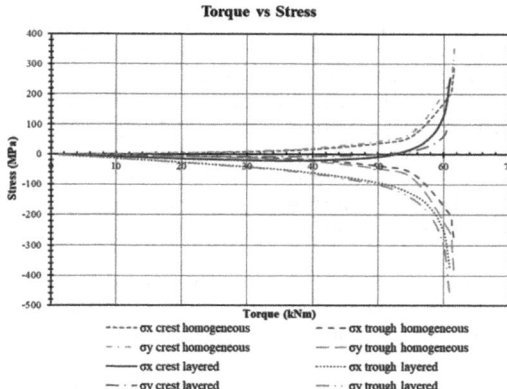

Torque vs Stress

Legend:
- ····· σx crest homogeneous — — σx trough homogeneous
- — · — σy crest homogeneous — — σy trough homogeneous
- ——— σx crest layered ········ σx trough layered
- — · · σy crest layered — · · · σy trough layered

Figure 6. Comparison of the stresses between the layered and the homogeneous shell models.

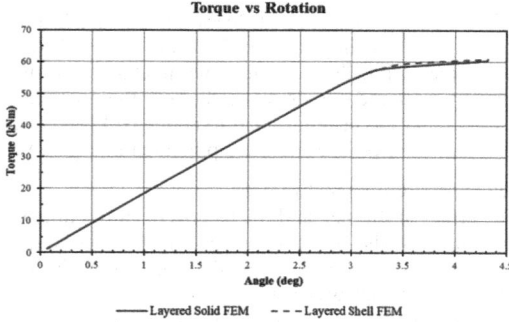

Torque vs Rotation

——— Layered Solid FEM — — — Layered Shell FEM

Figure 7. Torque vs rotation - comparison between the layered solid and layered shell model results.

to the zero value between the magnitudes of the compressive and the tensile stresses of the homogeneous model, as expected. This symmetry does not exist in the layered model due to the orientation of the layers. The existence of this symmetry is an indication that the homogeneous material model represents more accurately the real filament wound shaft.

The comparison of the stresses between the homogeneous and the layered shell model revealed some differences. A main difference is that the stresses of the layered shell model start all as compressive and the ones that remain compressive reach higher magnitudes than the ones that turn tensile. Additionally, the compressive stresses of the layered shell model are about 50 MPa greater than both compressive and tensile stresses of the homogeneous model until 50 kNm torque, and then they become even greater. This occurs due to the orientation of the layers. The small differences in the eigenvalue buckling loads and in the stress range between the two models reveal also that their difference is small, owing to the fact that the large number of layers of the ±45° lay-up considered in the first case is very close to the behavior of a homogeneous material. Finally, a weakness of the homogeneous model is the lack of available strengths for the homogeneous material, so no failure criteria could be introduced.

3.1.3 Layered solid model

The ANSYS element type used in this model is the 20-node structural solid element SOLID186. The layered solid model yielded very similar results to the layered shell model. The critical eigenvalue buckling load is 63.6 kNm, a value 4% greater than the experimental failure load and 1.9% greater than the layered shell model corresponding result. The critical buckling modeshape is also the same as that of the layered shell model. Additionally, the layered solid and layered shell models give almost identical results concerning the rotational stiffness of the shaft (Figure 7) and the

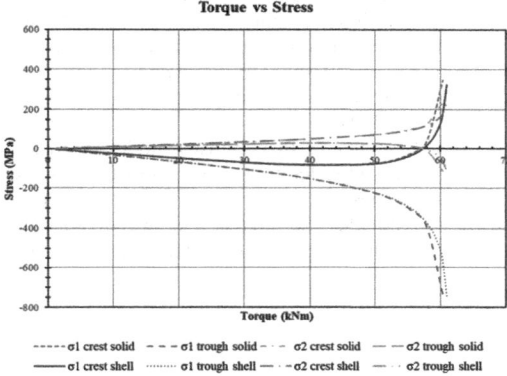

Torque vs Stress

Legend:
- ····· σ1 crest solid — — — σ1 trough solid — — σ2 crest solid — — σ2 trough solid
- ——— σ1 crest shell ········ σ1 trough shell — · — σ2 crest shell — · · · σ2 trough shell

Figure 8. Comparison of the stresses between the layered shell and the layered solid models.

prediction of the stresses (Figure 8). Therefore, for thin-walled composite shaft applications, the extra computing cost of layered solid modelling offers no benefits. For the further investigation of the mechanical behavior of the shaft, the layered shell model is used.

3.1.4 Sensitivity analysis

Effect of material properties and thickness

Various scenarios of modifying specific material properties were investigated, in order to assess their effect on the buckling load and the rotational stiffness of the shaft. In all cases, apart from the modified material properties, all other magnitudes retained their initial values. First, a drastic reduction of modulus E_1 by 50% and of both moduli E_1 and E_2 by 50% is investigated. The first scenario leads to a reduction of the rotational stiffness by 40% and reduced the buckling load to around 45 kNm. The second scenario lead to a reduction of the rotational stiffness by 50% and reduced the buckling load to around 35 kNm. In the sequence, all shear moduli were reduced by 50% and by 90%. These scenarios had no effect on the rotational stiffness of the shaft, but reduced the buckling load to 50 kNm and 32 kNm, respectively. Finally, reducing the thickness of the shaft by 50% reduces dramatically both the rotational stiffness and the buckling load to about 10 kNm. The results of all these scenarios are presented in Figure 9.

Effect of initial imperfections

The effect of the size and the pattern of initial imperfections has also been investigated. As mentioned before, the chosen size of the maximum initial imperfection that was used so far was 0.1% of the internal diameter of the shaft, D_i. The pattern used to generate the initial imperfections was the modeshape of the minimum eigenvalue. The analysis was additionally carried out for 3 more different initial imperfections schemes. The first two used the same modeshape but with maximum sizes of the initial imperfection equal to 0.01% and 1% of the internal diameter. The third scheme used all first 10 modeshapes of the eigenvalue buckling analysis, in order to generate a mixed shape initial imperfection. The size of the maximum initial imperfection in this case was equal to 1% D_i.

Figure 10 indicates that the two models with the smaller initial imperfections have almost the same rotational stiffness. The one with initial imperfection 0.01% D_i buckles somewhere between 60 and 65 kNm, about 10 kNm higher than the one with

0.1% D_i. Increase of the initial imperfection to 1% D_i decreases the rotational stiffness and initiates buckling much earlier, somewhere between 30 and 35 kNm. In the last case that all modeshapes were used for the generation of the initial imperfections, the rotational stiffness showed a slight decrease but was higher than the case with 1% D_i and lower than the other two cases with the critical modeshape imperfection pattern. Buckling started at around 35kNm.

It can be assumed that, within reasonable limits, initial imperfections have a small effect on the rotational stiffness, but have a significant effect on the initiation of buckling, since increasing the initial imperfections decreases notably the buckling load. It must be also emphasized here that, for large initial imperfections, the transition from the linear to the non-linear response of the shaft becomes gradual and, therefore, no clear buckling occurs.

3.1.5 *Progressive damage analysis*

A Progressive damage analysis (PDA) is conducted, based on the layered shell model, in order to predict damage evolution and more accurately assess the near failure behaviour of the shaft. A PDA requires the ultimate strengths of the material of the shaft, which in this case were not known. Consequently, ultimate strengths from the literature were used (Takeyama & Iijima 1988), which, for a unidirectional GFRP laminate, are those shown in Table 3.

ANSYS takes into account four different failure modes, i.e. fiber failure in tension, fiber failure in compression, matrix failure in tension and matrix failure in compression. The Hashin failure criterion was implemented for all failure modes of the composite material and the instant stiffness reduction method was applied. The percentage of degradation of material properties for each failure mode was investigated in order to assess the effect of the various stiffness magnitudes to the final failure load. Five different scenarios were investigated, i.e. reducing by 50, 80 and 90% all stiffnesses, reducing by 90% only the stiffness along the fiber direction and, finally, reducing by 90% only the stiffness normal to the fiber direction (matrix stiffness). The results are presented in Figure 11, from where it can be seen that the strength in the direction normal to the fibers, Y_t, and the corresponding stiffness have

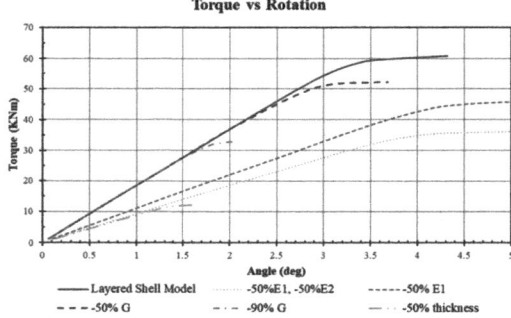

Figure 9. Effect of the material properties and thickness.

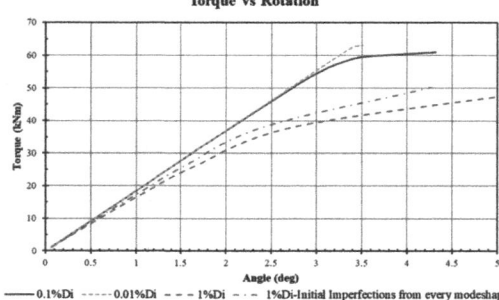

Figure 10. Effect of initial imperfections.

Table 3. Ultimate strengths of UD-GFRP laminate.

Ultimate strengths	MPa
Lo ngitudinal tensile strength, X_t	1200
Longitudinal compressive strength, X_c	800
Transverse tensile strength, Y_t	59
Transverse compressive strength, Y_c	128
In-plane shear strength (0° laminate)	25
In-plane shear strength (90° laminate)	250

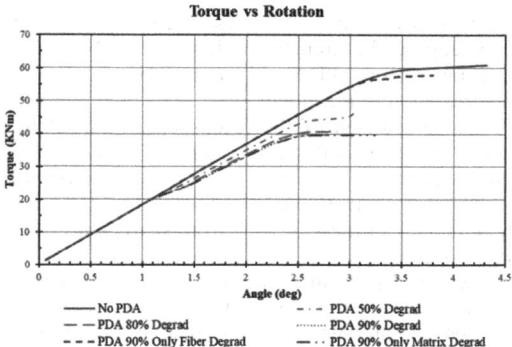

Figure 11. Torque vs rotation - Progressive damage analysis results.

the most significant effect in the reduction of the critical buckling torque. In addition, Figure 11 shows that the PDA analysis closer to the experimental result is that when only the stiffness in the fiber direction is degraded by 90% and the stiffness of the matrix is not degraded. This indicates that the used ultimate transverse tensile strength value and probably the used failure criterion itself do not describe the case of filament wound material accurately, and that matrix failure will occur at a higher load, near the fiber failure load, resulting in the ultimate failure of the shaft. Finally, Figure 11 shows that, when set up correctly, PDA yielded satisfactory results and predicted quite accurately the buckling failure load between 55 kNm and 60 kNm.

3.2 Analysis of the CFRP shafts

For the analysis of the two CFRP shafts described in section 2.2 the developed layered shell FE model is used, taking into consideration all the aforementioned remarks and conclusions. The results of the present eigenvalue and non-linear numerical buckling analyses and their comparison to the experimental values and the analytical predictions of Bauchau et al. (1988) are presented in Table 4 and discussed in the following lines. In this table, CC and SS indicate clamped and simply supported boundary

Table 4. Comparison of experimental and numerical critical buckling torques of CFRP shafts (in Nm).

		Analytical		Numerical	
Shaft	Exp.	CC	SS	Eigenvalue	Nonlinear
L#1	486	541	523	559	500÷520
L#2	655	733	712	767	670÷690

conditions of the shaft, respectively, taken into account by the analytical tools in Bauchau et al. (1988). It is also noted that, as expected, a range of values is given in Table 4 for the non-linear analyses results, since this type of analysis does not end up with a specific critical buckling load. These values were extracted from the corresponding rotational stiffness curves, shown in Figures 12 and 13 for shafts L#1 and L#2, respectively.

Table 4 shows that both eigenvalue and nonlinear buckling analyses are quite close to the analytical predictions of the paper and slightly overestimate the experimental buckling load, the differences being in the order of 16% for the eigenvalue results and 4% for the nonlinear analyses ones. It is noted by Bauchau et al. (1988) that in the experimental set-up the boundary conditions were neither clamped nor simply supported.

Finally, Figures 12 and 13 present the effect of the size of the initial imperfections on the predicted rotational stiffness and the buckling load. The model predicts relatively well the torsional stiffness of shaft L#1 and gives the most accurate prediction of the

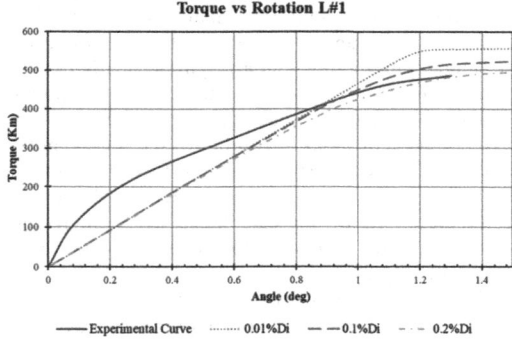

Figure 12. Torque vs Rotation Curves of L#1 for various initial imperfection sizes.

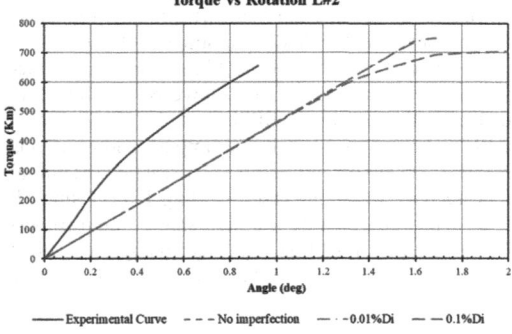

Figure 13. Torque vs Rotation Curves of L#2 for various initial imperfection sizes.

buckling load for an initial imperfection size of 0.1% D_i. However, it seems to underestimate the torsional stiffness of shaft L#2, despite giving a relatively accurate buckling load prediction for an initial imperfection size of 0.1% D_i.

4 CONCLUSIONS

An extensive analysis of the finite element modelling of composite material shafts was presented. The analysis was based on the data of a GFRP shaft test and two CFRP shafts tests acquired from the industry and literature, respectively.

The layered shell model revealed its ability to quite accurately predict the buckling load of the shaft. In the case of the GFRP shaft the eigenvalue buckling analysis and the nonlinear buckling analysis yielded reasonable estimations of the shaft's buckling load, with the latter yielding the more conservative and generally the more accurate of the two. Additionally, the examined stresses showed a logical pattern and reasonable magnitudes.

The homogeneous model of the GFRP shaft yielded almost the same results with the layered shell model concerning rotational stiffness. It also predicted tensile and compressive stresses of almost the same magnitude, which are symmetrically developed about the x-axis. The homogeneous model lacks the ability to calculate stresses in the direction of the fibers, as well as the ability of the direct application of failure criteria due to the lack of material strengths for the homogeneous material. Moreover, the method used for the calculation of the homogeneous properties can only be applied to especially orthotropic lay-ups.

The layered solid model of the GFRP shaft yielded almost identical results with the layered shell model, leading to the conclusion that, for thin-walled layered composite shaft applications, the extra computing cost of layered solid modelling offers no benefits.

The sensitivity analysis showed that the reduction of modulus E_1 has significant effect on the rotational stiffness and the buckling load, whereas the reduction of the shear moduli had insignificant effect on the rotational stiffness and the buckling load. The reduction of the thickness by 50% lowered the buckling load and the rotational stiffness of the shaft significantly. The size of the initial imperfections affects drastically buckling initiation. The smaller the initial imperfection, the higher the buckling load. Using all modeshapes to generate a mixed shape initial imperfection, triggers buckling later than using the critical modeshape initial imperfection with the same magnitude.

Progressive damage analysis requires accurate knowledge of the strengths of the material. However, if set up correctly, PDA will increase the accuracy of the prediction of the model.

The comparison of the results of the model for L#1 and L#2 CFRP shafts evaluated its ability to quite accurately predict the buckling load of a composite shaft. It also tested its ability to predict the torsional stiffness of composite shafts with positive results.

The developed finite element model is a valuable tool able to aid the optimization of the design of composite materials shafts for marine applications and serve as the basis for the development of simulation-based digital twins of composite material shafting systems.

ACKNOWLEDGEMENTS

Special thanks to B&T Composites SA (www.btcomposites.gr) for providing us with the information of the GFRP shaft.

REFERENCES

Bauchau, O.A., Krafchack, T.M., and Hayes, J.F., 1988. Torsional Buckling Analysis and Damage Tolerance of Graphite/Epoxy Shafts. *Journal of Composite Materials*, 22 (3), 258–270.

Bilalis, E., 2016. *Experimental and Numerical Study of Composite Shafts*. Diploma Thesis, National Technical Univeristy of Athens. (http://dx.doi.org/10.26240/heal.ntua.14723)

Christensen, R.M., 1979. *Mechanics of composite materials*. New York: Wiley.

Greene, E., 1999. *Marine Composites*. Annapolis: Eric Greene Associates, Inc.

Rao, B.J.P., Srikanth, D.V., Suresh Kumar, T., and Sreenivasa Rao, L., 2016. Design and analysis of automotive composite propeller shaft using FEA. *Materials Today: Proceedings*, 3 (10), 3673–3679.

Jones, R.M., 1975. *Mechanics of Composite Materials*. Washington, D.C.: Scripta Book Company.

Khoshravan, M.R. and Paykani, A., 2012. Design of a composite drive shaft and its coupling for automotive application. *Journal of Applied Research and Technology*, 10 (6), 826–834.

Malakuti, S., Schlake, J., Grüner, S., Schulz, D., Gitzel, R., Schmitt, J., Platenius-Mohr, M., Vorst, P., and Garrels, K., 2018. Digital twin – a key software component of Industry 4.0. *ABB Review*, 04/2018.

Montagnier, O. and Hochard, C., 2013. Optimisation of hybrid high-modulus/high-strength carbon fibre reinforced plastic composite drive shafts. *Materials and Design*, 46 (October), 88–100.

Mouritz, A.P., Gellert, E., Burchill, P., and Challis, K., 2001. Review of advanced composite structures for naval ships and submarines. *Composite Structures*, 53 (1), 21–24.

Papadakis, A.Z. and Tsouvalis, N.G., 2016. An Experimental and Numerical Study of CFRP Pressure Housings for Deep Sea Environment Research. *In: Proceedings of the Twenty-sixth (2016) International Ocean and Polar Engineering Conference.* 226–233.

Sevkat, E., Tumer, H., Halidun Kelestemur, M., and Dogan, S., 2014. Effect of torsional strain-rate and lay-up sequences on the performance of hybrid composite shafts. *Materials and Design*, 60, 310–319.

Shokrieh, M.M., Hasani, A., and Lessard, L.B., 2004. Shear buckling of a composite drive shaft under torsion. *Composite Structures*, 64 (1), 63–69.

Stedile Filho, P., Almeida, J.H.S., and Amico, S.C., 2018. Carbon/epoxy filament wound composite drive shafts under torsion and compression. *Journal of Composite Materials*, 52 (8), 1103–1111.

Takeyama, H. and Iijima, N., 1988. Machinability of Glass-fiber Reinforced Plastics and Application of Ultrasonic Machining. *CIRP Annals*, 37 (1), 93–96.

Developments in Maritime Technology and Engineering – Guedes Soares (eds)
© 2021 Copyright the Author(s), ISBN 978-0-367-77376-2

Study of a composite pressure hull for point absorber wave energy converter

M. Calvário & C. Guedes Soares
Centre for Marine Technology and Ocean Engineering (CENTEC), Instituto Superior Técnico, Universidade de Lisboa, Lisbon, Portugal

ABSTRACT: This paper presents a buckling and failure analysis of a cylindrical wave energy converter composite buoy. This preliminary study considers a point absorber buoy, made of a glass-fibre reinforced polymer, under hydrostatic lateral pressure. The structure is analysed for the case of simply supported boundary conditions at ends of the structure and in case of cantilever boundary conditions. The buckling analysis is carried out through a finite element analysis and a comparison with an analytical formulation is presented. In addition, the failure of the composite structure is analysed with Tsai-Wu failure criterion. According to the results, the design pressure is determined by the composite failure criterion in both boundary conditions.

1 INTRODUCTION

Wave Energy Converters (WECs) are devices used to extract energy from waves, in which different concepts and prototypes have been studied (Guedes Soares et al 2012), however their development is still in an early stage, as there are no concepts in the market supplying energy on a commercial basis. The high operation and maintenance costs are a barrier to their development although the use of composite materials may play a role to reduce these costs due to their low level of corrosion, low weight and capability to produce complex shapes when compared to steel construction.

These advantages lead to the implementation of composite materials in the structures of a significant number of WECs, e.g. the floats of the Wavestar prototype are made of glass fibre reinforced polymer (GFRP) (Marquis et al. 2010) and Kevlar and rubber composites were used in the Archimedes Wave swing (Marsh 2009). A review of composite materials in wave and tidal energy has been presented in Calvário et al. (2018).

In addition, studies highlighted the role of the slamming phenomenon on axisymmetric buoys (Blommaeart 2009, Blommaeart et al. 2009 and Van Paepegem et al. 2011). Sandwich and monolithic structures were studied and it was concluded that breaking wave slamming is more critical than bottom slamming and composite structures may have large deformation under these phenomena. In spite of not be in accordance with standards, it was claimed that the composite structures can withstand these loads.

The WECs buoys are shell structures which are compressed under the wave loads and a buckling analysis should be performed in order to guarantee

the structural integrity. In Moon et al. (2010) thick-walled filament-wound carbon epoxy cylinders under external hydrostatic pressure were investigated with a finite element analysis (FEA) and experimental results were in agreement with numerical ones. Other studies included optimization procedures to maximize the design pressure of a composite cylindrical shell (Kechum & Guang 2019) and to minimize the buoyancy factor, maximize the deck area and buckling strength factors of a submarine pressure hull (Helal et al. 2019).

Analytical solutions have been proposed for the buckling of thin-walled circular cylinders (NASA 1968), covering isotropic and orthotropic cylinders. Recently, further research have been conducted to assess the behaviour of composite cylindrical shells for different boundary conditions.

Lopatin & Morozov (2012) investigated the buckling of a cantilever circular cylindrical shell under uniform pressure and a comparison was carried out with finite element software, which verified the accuracy of the approach. The same approach was followed in the study of a sandwich cylindrical shell subjected to uniform pressure with both ends fully clamped (Lopatin & Morozov 2015) and in the case of a cylindrical shell subjected to hydrostatic pressure with its ends closed by rigid disks allowing axial displacement (Lopatin & Morozov 2017).

The objective of this paper is to study the buckling behaviour of a point absorber WEC buoy presented in Yang et al. (2018). Due to the aggressive climate where the WEC is deployed it is reasonable to take advantage of a composite construction with inherent corrosion resistance (reducing the maintenance costs)

DOI: 10.1201/9781003216582-71

and its lightness that may reduce the installation costs. The buoy is analysed considering a shell structure, subjected to an external uniform pressure. The results are obtained by an analytical formulation and a comparison with the FEA is presented. Furthermore, the study includes a range of winding angles and is conducted a failure analysis of the composite structure.

The paper is organized in five sections. In Section 2, the WEC buoy is presented, while the formulation for the buckling analysis and composite failure is presented in Section 3. The case study is highlighted in Section 4 and the results are presented and discussed in Section 5. The paper is summarized in the conclusions section.

2 WAVE ENERGY CONVERTER

This study analyses a point absorber WEC (presented in Figure 1), which consists of a buoy made of a long cylindrical tube and an outer rim, anchored to the seabed, in which the production of energy is based on water movement inside the tube (Yang et al. 2018). The buoy (presented in Figure 2) is modelled as closed buoy without water inside.

3 MODELLING

In this section, the methods to determine the design pressure are described. Section 3.1 presents the critical buckling pressure formulation while in section 3.2 discusses the composite failure criterion.

3.1 Buckling pressure

The buckling of a thin walled circular orthotropic composite cylinder is given by (NASA 1968, Vinson & Sierakowski 2008):

$$p_{buck} = \frac{3\left(D_{22} - \frac{B_{22}^2}{A_{22}}\right)}{R^3} \quad (1)$$

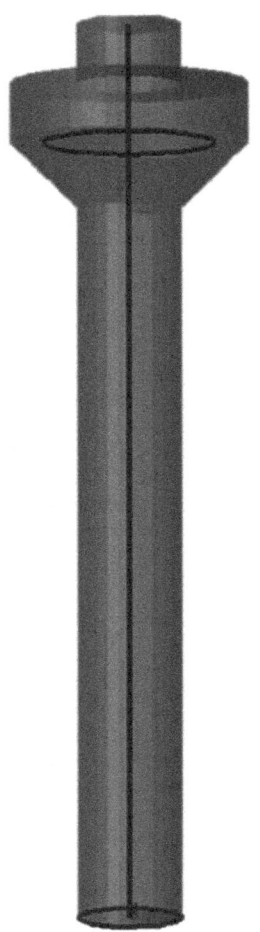

Figure 2. WEC buoy. Adapted from Yang et al. (2018).

where A_{22}, B_{22} and D_{22} = elements of the extensional stiffness matrix $[A]$, bending-extension coupling matrix $[B]$ and the bending stiffness matrix $[D]$; and R = mean radius.

These matrixes are obtained by (Vinson & Sierakowski 2008):

$$[A] = 1\sum_{k=1}^{n}(\bar{Q}_{ij})_k(h_k^1 - h_{k-1}^1)$$

$$[B] = \frac{1}{2}\sum_{k=1}^{n}(\bar{Q}_{ij})_k(h_k^2 - h_{k-1}^2) \quad (2)$$

$$[D] = \frac{1}{3}\sum_{k=1}^{n}(\bar{Q}_{ij})_k(h_k^3 - h_{k-1}^3)$$

where: $\bar{Q}_{ij}$ = Elements of the transformed reduced stiffness matrix; h_i = coordinate location (Figure 3); and N = Number of plies.

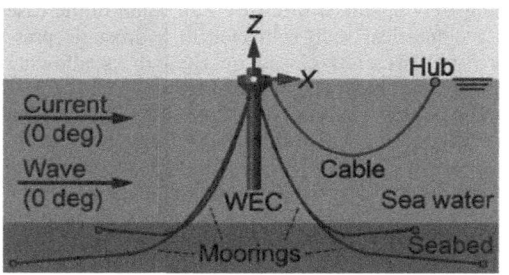

Figure 1. Illustration of the WEC. Retrieved from Yang et al. (2018).

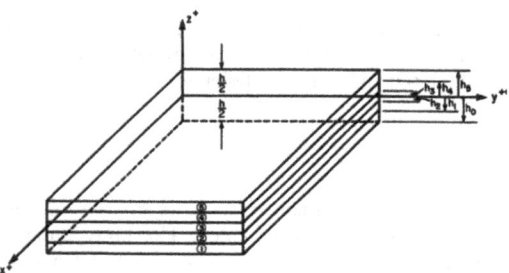

Figure 3. Stacking sequence. Retrieved from Vinson & Sierakowski (2008).

The coefficients $\bar{Q}_{ij}$ are obtained by:

$$\bar{Q}_{11} = Q_{11}m^4 + 2(Q_{12} + 2Q_{16})m^2n^2 + Q_{22}n^4$$

$$\bar{Q}_{12} = (Q_{11} + Q_{22} - 4Q_{66})n^2m^2 + Q_{12}(n^4 + m^4)$$

$$\bar{Q}_{22} = Q_{11}n^4 + 2(Q_{12} + 2Q_{66})n^2m^2 + Q_{22}m^4 \quad (3)$$

$$\bar{Q}_{16} = -mn^3Q_{22} + m^3nQ_{11} - mn(m^2 - n^2)(Q_{12} + 2Q_{66})$$

$$\bar{Q}_{26} = -m^3nQ_{22} + mn^3Q_{11} + mn(m^2 - n^2)(Q_{12} + 2Q_{66})$$

$$\bar{Q}_{66} = (Q_{11} + Q_{22} - 2Q_{12})n^2m^2 + Q_{66}(m^2 - n^4)^2$$

where Q_{ij} = reduced stiffness coefficients; $m = \cos(\theta)$; $n = \sin(\theta)$; and θ = angle lamina.
The coefficients Q_{ij} are obtained by:

$$Q_{11} = \frac{E_1}{1 - \nu_{12}\nu_{21}}$$

$$Q_{12} = \frac{\nu_{12}E_2}{1 - \nu_{12}\nu_{21}} \quad (4)$$

$$Q_{22} = \frac{E_2}{1 - \nu_{12}\nu_{21}}$$

$$Q_{66} = G_{12}$$

where: E_1 = Young's modulus in direction 1; E_2 = Young's modulus in direction 2; ν_{12} = in-plane Poisson's ratio and G_{12} = in-plane shear modulus.
In case of a cantilever circular cylindrical composite shell under uniform lateral pressure, the critical pressure is determined by (Lopatin & Morozov 2012):

$$p_{buck} = \eta_{cr}\frac{D_{22}}{R^3} \quad (5)$$

where η_{cr} = dimensionless buckling coefficient; and R = shell radius.
The coefficient η_n is given by:

$$\eta_{cr} = min(\eta_2, \eta_3, \ldots) \quad (6)$$

$$\eta_n = \frac{a_{11}a_{22}a_{33} - 2a_{12}a_{13}a_{23} - a_{11}a_{23}^2 - a_{22}a_{13}^2 - a_{33}a_{12}^2}{b_{33}(a_{11}a_{22} - a_{12}^2) + b_{22}(a_{11}a_{33} - a_{13}^2) - 2b_{23}(a_{12}a_{13} + a_{11}a_{13})} \quad (7)$$

$$a_{11} = \frac{B_{11}R^2\lambda^4}{D_{22}s} + \frac{B_{33}R^2}{D_{22}}n^2s\sigma\lambda(2 + \sigma\lambda)$$

$$a_{22} = \left(\frac{B_{33}R^2}{D_{22}} + \frac{D_{33}}{D_{22}}\right)\frac{\sigma\lambda}{s}(2 + \sigma\lambda) + \left(\frac{B_{22}R^2}{D_{22}} + 1\right)n^2s$$

$$a_{33} = \left(\frac{B_{33}R^2}{D_{22}} + n^4 + \frac{D_{11}\lambda^4}{D_{22}s^4}\right)s + 2\frac{\sigma\lambda}{s}n^2$$

$$\left[-\frac{D_{12}}{D_{22}}(2 - \sigma\lambda) + 2\frac{D_{33}}{D_{22}}(2 + \sigma\lambda)\right] \quad (8)$$

$$a_{12} = a_{21} = n\sigma\lambda\left[-\frac{B_{12}R^2}{D_{22}}(2 - \sigma\lambda)\right.$$

$$\left. + \frac{B_{33}R^2}{D_{22}}(2 + \sigma\lambda)\right]$$

$$a_{13} = a_{31} = \frac{B_{12}R^2}{D_{22}}\sigma\lambda(2 - \sigma\lambda)$$

$$a_{23} = a_{32} = n\left\{\left(\frac{B_{22}R^2}{D_{22}} + n^2\right)s + \frac{\sigma\lambda}{s}\left[-\frac{D_{12}}{D_{22}}(2 - \sigma\lambda)\right.\right.$$

$$\left.\left. + 2\frac{D_{33}}{D_{22}}(2 - \sigma\lambda)\right]\right\}$$

where $b_{22} = s$; $b_{33} = n^2s$; $b_{23} = b_{32} = ns$; and $s = l/R$.
For a helical filament winding structure, the coefficients B_{ij} and D_{ij} (11, 12, 22, 33) are calculated by (Lopatin & Morozov 2017):

$$B_{ij} = A_{ij}h$$

$$D_{ij} = A_{ij}\frac{h^3}{12}$$

$$A_{11} = \bar{E}_1c^4 + \bar{E}_2s^4 + 2\bar{E}_{12}c^2s^2$$

$$A_{12} = \bar{E}_1\nu_{12} + (\bar{E}_1 + \bar{E}_2 - 2\bar{E}_{12})c^2s^2$$

$$A_{22} = \bar{E}_1 s^4 + \bar{E}_2 c^4 + 2\bar{E}_{12}c^2 s^2$$

$$A_{33} = (\bar{E}_1 + \bar{E}_2 - 2\bar{E}_1 v_{12})c^2 s^2 + G_{12}(c^2 - s^2)$$

$$\bar{E}_1 = \frac{E_1}{1 - v_{12}v_{21}} \qquad (9)$$

$$\bar{E}_2 = \frac{E_2}{1 - v_{12}v_{21}}$$

$$\bar{E}_{12} = \bar{E}_1 v_{12} + 2G_{12}$$

$$E_1 v_{12} = E_2 v_{21}$$

$$c = cos(\theta)$$

$$s = sin(\theta)$$

3.2 Failure criterion

The composite failure analysis, is carried out by the failure index approach, which is expressed by (Barbero 2014):

$$I_f = \frac{stress}{strength} \qquad (10)$$

where: I_F = Index failure. The failure happens when $I_F \geq 1$.

The I_F, defined by the Tsai-Wu failure criterion, is given by

$$I_F = \left[-\frac{B}{2A} + \sqrt{\left(\frac{B}{2A}\right)^2 + \frac{1}{A}} \right]^{-1} \qquad (11)$$

$$A = \frac{\sigma_1^2}{F_{1t}F_{1c}} + \frac{\sigma_2^2}{F_{2t}F_{2c}} + \frac{\sigma_3^2}{F_{3t}F_{3c}} + \frac{\sigma_4^2}{F_4^2} + \frac{\sigma_5^2}{F_5^2} + \frac{\sigma_6^2}{F_6^2}$$

$$+ c_4 \frac{\sigma_2 \sigma_3}{\sqrt{F_{2t}F_{2c}\bar{F}_{3t}\bar{F}_{3c}}} + c_5 \frac{\sigma_1 \sigma_3}{\sqrt{F_{1t}F_{1c}F_{3t}\bar{F}_{3c}}} \qquad (12)$$
$$+ c_6 \frac{\sigma_1 \sigma_2}{\sqrt{F_{1t}F_{1c}F_{2t}\bar{F}_{2c}}}$$

$$B = \left(F_{1t}^{-1} - F_{1c}^{-1}\right)\sigma_1 + \left(F_{2t}^{-1} - F_{2c}^{-1}\right)\sigma_2 \qquad (13)$$
$$+ \left(F_{3t}^{-1} - F_{3c}^{-1}\right)\sigma_3$$

where σ_1, σ_2, σ_3 = Normal stresses along the material directions 1, 2 and 3 respectively; F_{1t} = Longitudinal tensile strength; F_{2t} = Transverse tensile strength; F_{3t} = Transverse- thickness tensile strength; F_{1c} = Longitudinal compressive strength; F_{2c} = Transverse compressive strength; F_{3c} = Transverse-thickness compressive strength; F_4, F_5 =

Interlaminar shear strength in the 2-3 and 1-3 plane respectively; F_6 = In-plane shear strength; and c_1, c_2, c_3 = Tsai-Wu coupling coefficients.

4 CASE STUDY

In this section, the main parameters for the modelling of the system are revealed. The buoy dimensions are presented in Figure 4. It should be noted, that this study just considers the submerged geometry of the buoy.

This study adopts the material properties already studied in the SEEWEC project (Blommaert 2009). The material is glass fibre-polyester resin, in which the elastic material properties are presented in Table 1, while the strength properties are presented in Table 2.

Taking into consideration the lack of data it is assumed that F_{3t}, F_{3c} and F_5 are equal to F_6 (Barbero

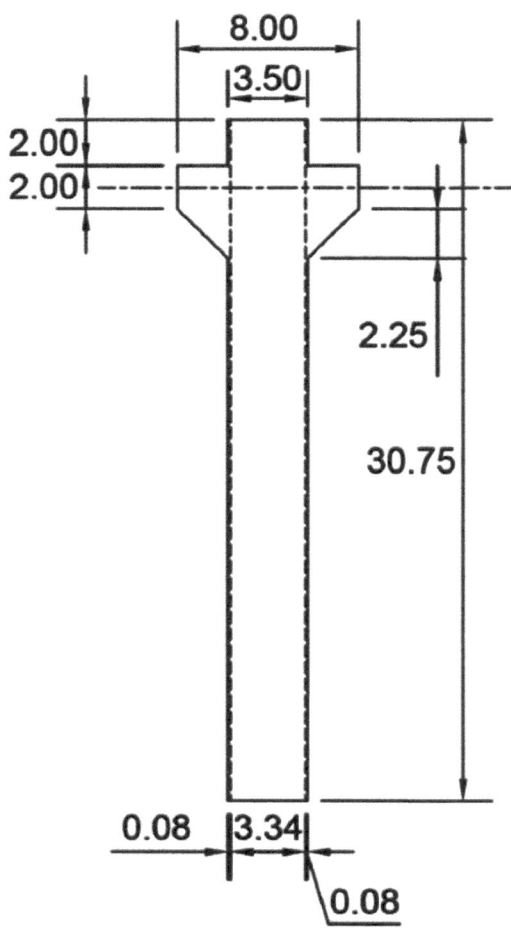

Figure 4. Geometrical parameters of the WEC buoy (units: m). Adapted from Yang et al. (2018).

Table 1. Material elastic properties. Retrieved from Blommaert (2009).

E_1	E_2	G_{12}	G_{13}	G_{23}	v_{12}
[MPa]	[MPa]	[MPa]	[MPa]	[MPa]	
25770	6251.5	4200	4200	2500	0.38

Table 2. Material strength properties. Retrieved from Blommaert (2009).

F_{1t}	F_{1c}	F_{2t}	F_{2c}	F_4	F_6
[MPa]	[MPa]	[MPa]	[MPa]	[MPa]	[MPa]
575.7	419	20.73	72	45.2	71.43

2014). The Tsai-Wu coupling coefficients (c_1,c_2,c_3) are assigned with value of -1.

In addition it is assumed a transversely isotropic material (Barbero, 2014):

$$E_3 = E_2 \qquad (14)$$

$$\nu_{12} = \nu_{13} \qquad (15)$$

$$G_{23} = \frac{E_2}{2(1+\nu_{23})} \qquad (16)$$

where: E_3 = Young's modulus in direction 3; v_{13} = Poisson's ratio direction 1-3; and v_{23} = Interlaminar Poisson's ratio.

The buckling of a shell subjected to uniform lateral pressure is analyzed with simply supported boundary conditions at ends of the structure and with cantilever boundary conditions. The ANSYS software is used to perform the FEA and an illustration of the structure with the two types of boundary conditions and load profile is revealed in Figure 5.

In the first case (Figure 5a) the critical buckling pressure is calculated by Equation (1) and is multiplied by a coefficient of 0.75, in accordance to the recommendations (NASA, 1968). In the problem of a cantilever shell (Figure 5b), the analytical approach of Morozov & Lopatin (2012) is used. In both boundary conditions, the analytical analysis considers a cylinder with a length of 24.5 m and winding angle range between 20 and 90° and each lamina has 1 mm thickness.

5 RESULTS

In this section, a comparative study between the values of the buckling critical obtained by the analytical and

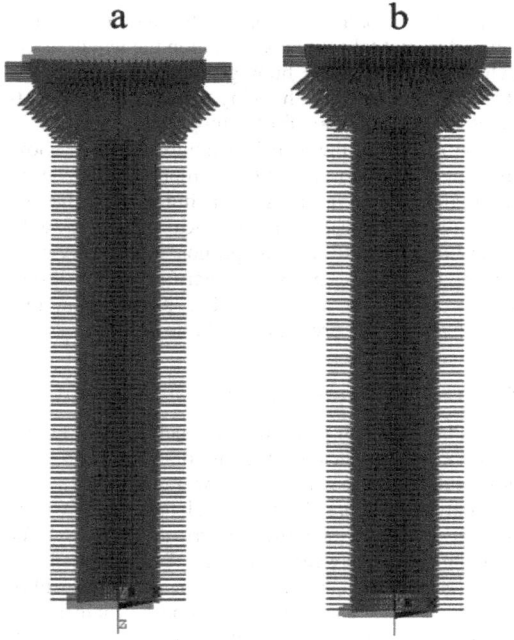

Figure 5. Boundary conditions for the WEC buoy.

numerical approaches is presented, as well as the failure pressure when the index failure is equal to one.

In section 6.1 the case study of simply supported boundary conditions is discussed while in section 6.2 the cantilever condition is highlighted.

5.1 Simply supported boundary conditions

The buckling pressure determined using the analytical formulation (CS$_1$), FEA (CS$_2$) and the failure pressure (CS$_3$) as function of the winding angle θ is presented in Figure 6.

The critical buckling pressure in CS$_1$ is conservative relatively to the results from the FEA, CS$_2$, for all winding angles. The ratio varies from 0.45 (θ=0°) and

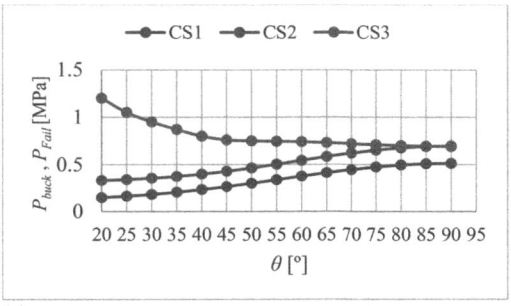

Figure 6. Critical buckling pressure and failure pressure as a function of θ for the simply supported condition.

0.74 (θ=90°). For a long cylinder, the analytical formulation does not take into consideration the length of the cylinder. Moreover, the buoy geometry it is not cylindrical in the entire length, being the analytical procedure an initial estimation for the proposed geometry.

The results of the Tsai-Wu failure criterion show that for the proposed WEC geometry and material properties, the failure pressure is higher than the one from the analytical and numerical approach with exception of θ =90°. For the proposed boundary condition, the failure pressure determines the design pressure (P_D) for a θ =90°. The pressures at this winding angle are presented in Table 3.

5.2 Cantilever beam boundary conditions

A similar approach regarding the buckling and failure pressure was conducted in the case of cantilever boundary conditions. The results are shown in Figure 7.

Analysing Figure 7, the critical buckling pressure in CS_1 is lower than the numerical one (CS_2) for winding angles between 0 and 45°. As mentioned, the buoy geometry it is not cylindrical in the entire length, although the analytical approach obtained reasonable results with a ratio varying from 0.71 to 0.99.

The results of failure pressure (CS_3) are always lower than the results of the numerical analysis and determines the design pressure for this boundary condition. The design pressure is obtained for θ =45° and the pressures for this winding angle are presented in Table 4.

Table 3. Critical buckling pressure, failure pressure and design pressure for the simply supported analysis.

CS	Variable	Unit	Value
1	P_{buck}	[MPa]	0.51
2	P_{buck}	[MPa]	0.70
3	P_{Fail}	[MPa]	0.69
	P_D	[MPa]	0.69

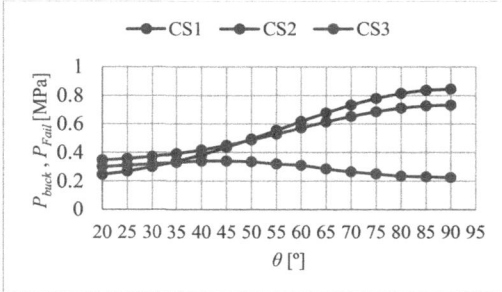

Figure 7. Critical buckling pressure and failure pressure as a function of θ for the cantilever condition.

Table 4. Critical buckling pressure, failure pressure and design pressure for the cantilever supported analysis.

CS	Variable	Unit	Value
1	P_{buck}	[MPa]	0.44
2	P_{buck}	[MPa]	0.45
3	P_{Fail}	[MPa]	0.34
	P_D	[MPa]	0.34

It should be noted that the hydrostatic pressure at the bottom of the structure is approximately 0.27 [MPa] which is relatively close the design pressure in both boundary conditions and this study doesn't take into account, the effect of the dynamic pressure, impacts loads, fatigue sea water aging and manufacturing imperfections. Therefore, for the proposed material (with typical glass fibre-polyester properties) a thick shell (for instance a radius-to-thickness less than 20) should be designed to guarantee the integrity of the structure under the mentioned loads. In addition, reinforce stiffeners should be add to the structure, sandwich construction may improve the structural response in case of impact loads and further research is required in order to determine an optimal layup sequence.

6 CONCLUSIONS

This paper provided a preliminary study regarding the buckling of composite shells for a buoy of point absorber WEC. The structure was analysed with simply supported boundary conditions and with a cantilever condition. In the simply supported case the critical buckling pressure obtained by analytical formulation was more conservative relatively to the results of the FEA while for the cantilever boundary conditions the analytical approach was conservative for winding angles bellow 50°. The results from the analytical approaches are just a first approximation since the geometry of the buoy is not cylindrical in the entire length and these theories don't considerer stress concentrations among other factors.

In both boundary conditions, the design pressure was determined by the failure pressure obtained by the Tsai-Wu criterion. Moreover since the bucking pressure is relatively close to the maximum hydrostatic pressure, is predicted that structure should be thick (for instance a radius-to-thickness less than 20) in order to guarantee the structural integrity under dynamic pressure, impact loads, long time exposition to water and manufacturing imperfections. Complementary studies are required in order to optimize the structure, for instance by means of reinforcement stiffeners, optimal layup sequence and study of sandwich construction.

ACKNOWLEDGEMENTS

The first author has been financed by the Portuguese Foundation for Science and Technology (Fundação para a Ciência e Tecnologia - FCT) and the European Social Fund (Fundo Social Europeu - FCE) under the grant SFRH/BD/137429/2018. This work contributes to the Strategic Research Plan of the Centre for Marine Technology and Ocean Engineering (CENTEC), which is financed by FCT under contract UIDB/UIDP/00134/2020.

REFERENCES

Barbero, E. J. 2014. Finite Element Analysis of Composite Materials using ANSYS. Boca Raton: CRC Press.

Blommaert, C. 2009. Composite Floating 'Point Absorbers' for Wave Energy Converters: Survivability Design, Production Method and Large-Scale Testing (PhD thesis). Gent University.

Blommaert, C., van Paepegem, W., Degrieck, J. 2009. Design of composite material for cost effective large scale production of components for floating offshore structures. *Plastics, Rubber and Composites* 38: 146–152.

Calvário, M., Sutherland, L.S., Guedes Soares, C. A review of the applications composite materials in wave and tidal energy devices. In: C. Guedes Soares & A. P. Teixeira (Eds), *Maritime Transportation and Harvesting Sea Resources*: 695–701. UK: Taylor & Francis Group.

Guedes Soares, C., Bhattacharjee, J., Tello, M., Pietra, L. 2012. Review and classification of wave energy converters. In: C. Guedes Soares, Y. Garbatov, S. Sutulo, T. A. Santos (Eds), *Maritime Engineering and Technology*: 585–594. UK: Taylor & Francis Group.

Helal, M., Huang, H., Wang, D., Fathallah, E. 2019. Numerical Analysis of Sandwich Composite Deep Submarine Pressure Hull Considering Failure Criteria. *Journal of Marine Science and Engineering* 7: 1–22.

Kechun, S. & Guang, P. 2019. Buckling Optimization of Composite Cylinders for Underwater Vehicle Applications Under Tsai-Wu Failure Criterion Constraint. *Journal of Shanghai Jiao Tong University* (*Science*) 24: 534–544.

Lopatin, A.V. & Morozov, E.V. 2012. Buckling of a composite cantilever circular cylindrical shell subjected to uniform external lateral pressure. *Composite Structures* 94: 553–562.

Lopatin, A.V. & Morozov, E.V. 2015. Buckling of the composite sandwich shell with clamped ends under uniform external lateral pressure. *Composite Structures* 122: 209–216.

Lopatin, A.V. & Morozov, E.V. 2017. Buckling of composite cylindrical shells with rigid end disks under hydrostatic pressure. *Composite Structures* 173: 136–143.

Marquis, L., Kramer, M., Frigaard, P. 2010. First Power Production figures from the Wave Star Roshage Wave Energy Converter. In *3rd International Conference on Ocean Energy, Bilbao, 6–8 October 2010.*

Marsh, G., 2009. Wave and tidal power– an emerging new market for composites. *Reinforced Plastics* 53: 20–24.

Moon, C. J., Kim, I. H., Choi, B. H., Kweon, J. H., Choi, J. H. 2010. Buckling of filament-wound composite cylinders subjected to hydrostatic pressure for underwater vehicle applications. *Composite Structures* 92: 2241–2251.

NASA SP-8007. 1968. Buckling of Thin-Walled Circular Cylinders.

Van Paepegem, W., Blommaert, C., De Baere, I., Degrieck, J., De Backer, G., De Rouck, J., Degroote, J., Vierendeels, J., Matthys, S., Taerwe, L. 2011. Slamming Wave Impact of a Composite Buoy for Wave Energy Applications: Design and Large-Scale Testing. *Polymer Composites* 32: 700–713.

Vinson, J. R. & Sierakowski, R. L. 2008. *The Behavior of Structures Composed of Composite Materials.* Dordrecht: Springer.

Yang, S.H., Ringsberg, J.W., Johnson, E. 2018. Parametric study of the dynamic motions and mechanical characteristics of power cables for wave energy converters. *Journal of Marine Science and Technology* 23: 10–29.

Developments in Maritime Technology and Engineering – Guedes Soares & Santos (eds)
© *2021 Copyright the Author(s), ISBN 978-0-367-77376-2*

Fatigue behaviour and damage evolution of glass fiber reinforced composite material

A.K. Haldar, T. Gobikannan, A. Portela & A.J. Comer

Irish Composites Centre (IComp), Bernal Institute (Composites), School of Engineering, University of Limerick, Ireland

ABSTRACT: The proposed extensive use of advanced fiber-reinforced composites in structural applications such as large marine vessels has brought attention to the need to be able to predict the fatigue life of these structures. For design purposes, it is important to be able to predict when fatigue cracks initiate and the duration of the propagation phase before rapid final failure occurs. A fatigue study has been conducted as part of the FIBRESHIP Horizon 2020 funded EU project. This paper addresses one aspect of the fatigue study conducted in FIBRESHIP, focusing on suitable methods to identify crack initiation and monitor the subsequent crack propagation in glass fiber reinforced composite materials (GFRP) under fatigue loading. To correlate the fatigue response with the crack initiation and propagation phases, interrupted fatigue tests were carried out on open-hole GFRP test samples. The reinforcing fibres were predominantly transverse to the loading direction. Fatigue tests were periodically interrupted and stiffness measurements were performed to track the evolution in stiffness as a function of the number of fatigue load cycles. Damage was also monitored using ultrasonic scanning to inspect the progressive failure morphology of the matrix and reinforcing fibres. It was found that the first type of damage that appears in 90° plys under loading is matrix cracking.

1 INTRODUCTION

Composite materials are becoming increasingly popular for use in primary and secondary structural applications within the aerospace, marine, and automotive industries due to their high strength-to-weight ratio, structural efficiency, excellent corrosion and fatigue characteristics in comparison to metal alloys (Mouritz, 2012). Many applications of composites, e.g. aerospace and marine applications, involve repeated cyclic loading, causing fatigue damage within the structure. The fatigue characteristics of composites are not very well understood compared to their quasi-static behaviour.

Notches and holes for bolts and rivets are often required for the assembly of composite structures. Notches such as open-holes act as stress concentration points which experience higher stresses under fatigue loading in composite structures, resulting in significant strength reduction compared to un-notched composite laminates (Haldar & Senthilvelan, 2011). The fatigue damage evolution of open-hole composite specimens has been studied (Nixon & Pearson, 2013). Quasi-isotropic carbon fibre/epoxy open-hole specimens were initially quasi-statically loaded to determine an average static failure load and subsequent samples were cyclically loaded in tension-tension fatigue, at various constant amplitudes. Here, the crack initiation, propagation and failure of the composite was analysed

using a micro CT (computed tomography) scanner. It was found that damage initiates at, and propagates from, the hole edge in the form of matrix cracks and, eventually, delamination causes a significant drop in effective modulus. (Padmaraj, 2019) investigated the damage behavior of glass fibre composites materials under low cyclic fatigue loading. Here, the matrix cracking in the initial stage grew rapidly and, coupled with a corresponding reduction in stiffness, accounted for 60% of the overall damage.

As a part of 'FIBRESHIP' Horizon 2020 funded EU project, significant work has been carried out on the fatigue failure of GFRP. However, limited investigation has been carried out on the crack initiation of composites under FIBERSHIP project. Therefore, this paper seeks to build on the previous data by determining, through the fatigue testing of predominantly transversely reinforced open-hole GFRP composites, the point of matrix crack initiation and propagation on the established S-N curve.

2 MATERIALS AND METHODOLOGY

2.1 *Laminate manufacturing*

Uni-directional (UD) glass laminates were manufactured by the vacuum-assisted liquid resin infusion process. The reinforcement fabric used in this study was Saertex U-E 940 g/m² LEO unidirectional non-crimp

DOI: 10.1201/9781003216582-72

glass fabric. The reinforcement fabric has 90% of the glass fibres aligned with the 0° direction, the remaining glass fibres are oriented in the 90° direction to provide support to the dry reinforcement. The glass fabric layers were arranged in a $[0]_{2S}$ stacking arrangement. A vinyl ester resin (LEO Injection Resin 8500 from BÜFA) was infused into the fabric under ambient temperature and allowed to cure for 24 hours. A post cure of 6 hours at 80° was also performed. Both the fabric and resin are part of the LEO fire retardant composite system. Test coupons were extracted by rotating disc cutter to produce samples with the fiber orientation predominantly transverse to the fatigue load direction (UD 90° GFRP). The nominal sample dimensions of each fatigue test coupon are 300 x 25 x 3 mm and each sample contains an open hole of 6 mm in diameter as per ASTM D 5766.

2.2 Interrupted fatigue loading

Constant load amplitude fatigue tests in accordance with ASTM D3479 were performed using a 300kN Zwick servo-hydraulic fatigue testing machine at the University of Limerick, Ireland. Samples were gripped using hydraulic grips. Two load ranges were chosen to investigate the effect of load range on the number of cycles to crack initiation and propagation. The load ranges were calculated by taking the cyclic maximum load to be either 60% of failure, or 40% of failure. The failure loads were established by performing tensile tests to failure for the composite material system, which is 6.8 kN. Fatigue tests were performed at 4 Hz frequency under tension-tension loading with a 0.5 stress ratio. To correlate the fatigue response with the crack initiation and propagation in the open-hole samples, interrupted fatigue tests were carried out on open-hole GFRP test samples. The tests were interrupted at key stages, where a significant change in the stiffness curve was anticipated. An extensometer mounted on the gauge length of the sample was used to measure the local stiffness across the open hole.

2.3 Non-destructive inspection

The composite laminates were non-destructively inspected before and after the fatigue loading test using through-transmission ultrasonic scanning. The ultrasonic scanning was performed in a water bath using two 2.25 MHz transducers in conjunction with UTwin software (Mistras Group Inc.).

3 RESULTS AND DISCUSSION

Initial fatigue tests were performed on UD 90° GFRP samples to identify the number of cycles until visual whitening (indicative of crack initiation), crack propagation and final failure occurred. Fatigue load data associated with UD 90° open-hole samples is shown in Figure 1. Here, the data within the dotted line circle represents the crack initiation of the samples.

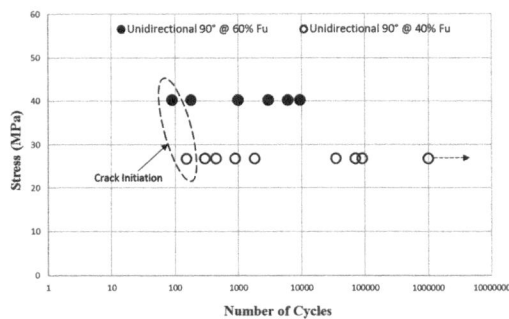

Figure 1. S-N curve to identify the crack initiations and propagation of UD90° samples.

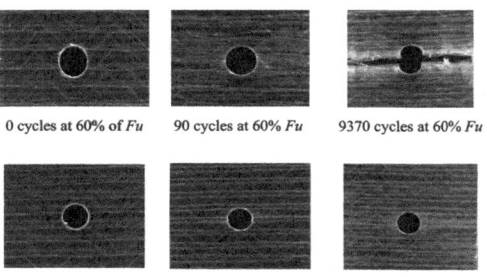

Figure 2. Image of damage propagation of the UD 90° GFRP open-hole samples.

An example of visual inspection of the sample subjected to fatigue loading is also given in Figure 2. Propagation of the crack after the initial whitening was relatively slow and difficult to observe visually. Closer to failure, the whitening grew rapidly, and this ultimately led to sudden failure. The first sign of whitening and crack initiation were visible at 90 and 150 cycles for 60% and 40% of ultimate tensile failure loads, respectively. The test specimens loaded to 60% of the ultimate load experienced sudden failure at 9,370 cycles while samples loaded to 40% of ultimate load did not fail after 1,000,000 cycles.

UD 90° GFRP samples were further investigated by measuring the change in stiffness during fatigue tests conducted at 40% and 60% of the ultimate tensile load. Regarding samples tested at 60%, a significant drop in stiffness was measured in the stiffness curve at an early stage of cyclic load (Figure 3(a)) because the local stress at the hole is greater than the ultimate tensile strength of the material. However, the stiffness only dropped marginally for the 40% sample at crack initiation point, as shown in (Figure 3(b)). Here, micro cracks were identified, and their presence constitutes the initiation of damage. Following this, stiffness dropped continuously until 1,000,000 cycles.

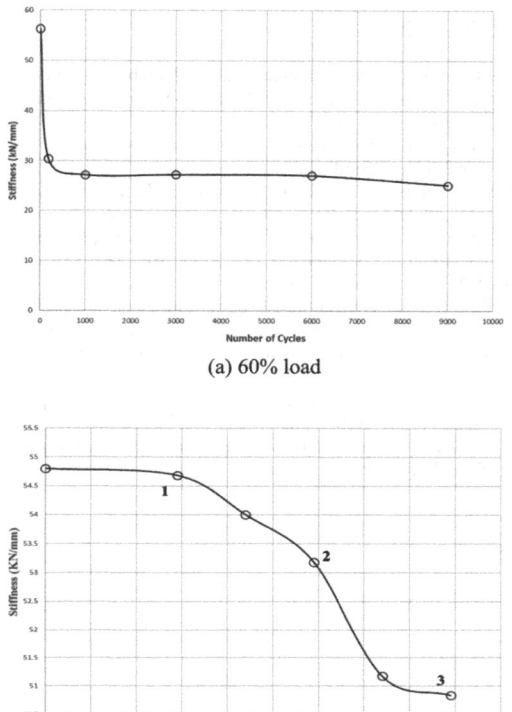

(a) 60% load

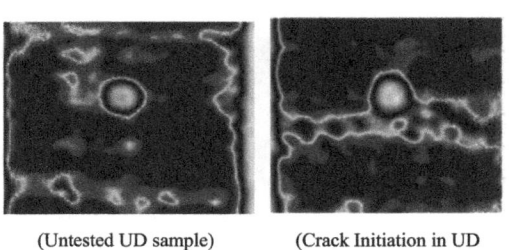

(b) 40% load

Figure 3. Stiffness values of UD 90° GFRP samples subjected to cyclic load.

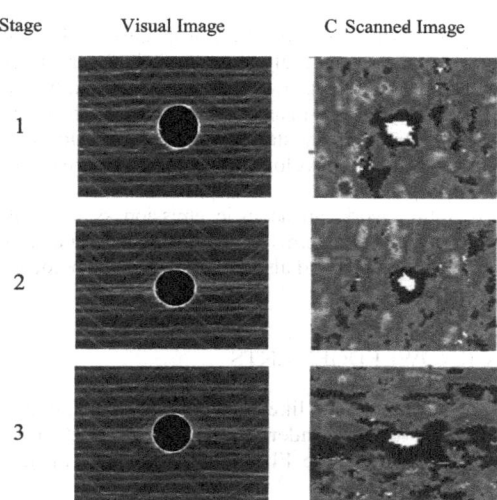

(Untested UD sample)   (Crack Initiation in UD samples subjected to a cyclic load of 60% failure)

Figure 4. C-scan inspection of open-hole composite sample.

More detailed inspections of UD 90° GFRP samples were carried out in the preliminary stages of crack initiation and crack propagation, using a through-transmission ultrasonic scanner to determine the amount of damage. Through-transmission (C-scan) images of the sample revealed the crack initiated as expected at the hole. Examples of C-scan images of

Stage   Visual Image   C Scanned Image

1

2

3

Figure 5. Crack propagation images of UD GFRP sample subjected to cyclic load of 40% tensile failure at key stages at key stages 1, 2 and 3.

the open hole GFRP test samples are given in Figure 4. The images are presented on a scale where red corresponds to a strong receiving ultrasonic signal and blue relates to complete attenuation of the ultrasonic signal. Complete attenuation of the ultrasonic signal is directly correlated to a damaged area due to the formation of cracks.

Figure 5 represents the images corresponding to the key stages 1, 2 and 3, associated with Figure 3(b). Here, the purple colour corresponds to the percentage of damage area of the scanned sample. At the low cyclic fatigue load (40% of tensile failure load), the percentage damage area is ~3%, ~4% and ~14% at the key stages of 150, 300 and 450 cycles, respectively. It is acknowledged that the first type of damage that appears under loading is at ~150 cycles. Micro cracks were visually identified and their presence constitutes the initiation of damage. Subsequently, the greater percentage damage area at 450 cycles is due to the continual matrix cracking.

4 CONCLUSIONS

Fatigue tests were performed on UD 90° GFRP samples to identify the number of fatigue cycles required to initiate cracks. The combination of methods evaluated so far (stiffness measurements, ultrasonic scanning and visual inspection) were successful in helping to establish when crack initiation, in the form of microcracking of the matrix, occurred during the fatigue life of the 90° plies. It was determined that the cracks initiate at a very early stage of the fatigue lifecycle for UD 90°

composite samples. However, propagation of these micro cracks is relatively slow until the later stages where the damage grows exponentially leading to full failure. C-scan inspections also help to determine the preliminary stages of crack initiation and help to show the development of damage area with crack propagation.

In future work, an acoustic emission system will also be evaluated in terms of its effectiveness in detecting crack initiation and also monitoring the subsequent crack propagation.

ACKNOWLEDGEMENTS

The authors would like to acknowledge that this work is funded under the Horizon 2020 programme through the FIBRESHIP project (Project ref: 723360).

REFERENCES

Haldar A.K. and Senthilvelan S, 2011. Notch effect on discontinuous fiber reinforced thermoplastic composites, *Key Eng Mat*, 471, 173–178.

Mouritz, A.P. (2012) *Introduction to aerospace materials*, Cambridge: Woodhead

Nixon-Pearson, O., Hallett, S., Withers, P. and Rouse, J. 2013. 'Damage development in open-hole composite specimens in fatigue. Part 1: Experimental investigation', *Composite Structures*, 106, 882–889.

Nash N.H., Portela A., Bachour C., Manolakis I., Comer A. J., Effect of environmental conditioning on the properties of thermosetting- and thermoplastic-matrix composite materials by resin infusion for marine applications, *Compos Part b*, 177, 2019.

Padmaraj, N., Vijaya, K. and Dayananda, P., 2019. Experimental study on the tension-tension fatigue behaviour of glass/epoxy quasi-isotropic composites, *Journal of King Saud University - Engineering Sciences*.

Developments in Maritime Technology and Engineering – Guedes Soares & Santos (eds)
© 2021 Copyright the Author(s), ISBN 978-0-367-77376-2

Fatigue analysis of a naval composite structure

J. Jurado-Granados, X. Martinez, L. Barbu & D. di Capua
International Center in Numerical Methods in Engineering, CIMNE Universitat Politecnica de Catalunya (UPC),
Barcelona, Spain

ABSTRACT: The use of composite materials in marine industry has many potential benefits, such as lower weight. However, there are still some failure mechanisms, such as fatigue, that are not yet completely understood. This work presents a formulation to simulate the fatigue phenomena, which is based on the Serial/Parallel Rule of Mixtures (S/P RoM) (Rastellini et al. 2008) and a fatigue damage model (Oller et al. 2005) previously developed for metals. The SP-RoM can be understood as a constitutive law manager that provides the elastic and non-linear response of the composite from the constitutive performance of its constituent materials. The fatigue damage model is based on the definition of a reduction function that takes into account the cyclic performance of the material, acting on its strength and stiffness. Both models are used to study a section of a container ship. The results obtained from this analysis validate the capability of the formulation to predict fatigue failure in marine composite structures.

1 INTRODUCTION

The work presented in this manuscript has been developed in the framework of FIBRESHIP. FIBRESHIP is a R&D H2020 project funded by European Commission. Its goal is to enable the construction of the entire hull of large-length ships in FRP materials by overcoming several technical challenges. One of these technical challenges is the prediction of the fatigue phenomena in composite materials. The importance of fatigue in naval structures is crucial, given that such structures are in operation during long time and in hard environments, these structures are subjected under cyclic loads, leading to fatigue phenomena.

Fatigue in composite structures has been studied mainly in the field of wind turbines. In particular, there are many studies about the fatigue life of composite blades (Liu et al. 2019), (Ding et al. 2012). Many of these analysis do not treat fatigue from a constitutive point of view, but using failure criteria based on static criteria, such as Tsai-Hill. The main advantages of these methods are that they are easy to implement in a FE code, they obtain the expected life of the structure and they are cheap in time-computing. On the other hand, such methods could require huge experimental campaigns in order to be reliable, what leads to expensive costs, therefore the reliability of such methods is directly related with the cost of the testing campaign. Although failure criteria can define the failure mechanism of a specific laminate, they do not take into account the interaction between them. Finally, failure criteria methods can not follow the damage evolution in the

material. This last feature is important, given that the initiation of damage does not mean the immediate collapse of the structure. As a consequence, failure criteria are very conservative, what implies higher safety factors and therefore more weight.

Current developments in numerical methods have lead to multi-scale models. These models solve a given problem in a two-scale context: global scale and micro-scale. The global scale, or macro-scale, solves the global response of the composite structure, considering the composite as a homogeneous material. The micro-scale is solved on a Representative Volume Element (RVE) which is defined to characterize the internal micro-structure of the composite. The main advantage of this approach is that it takes into account all micro-structural phenomena and material interactions that occur in the composite. On the other hand, such models are expensive in terms of computation time and they are not suitable for non-advanced users, what makes its use more restricted. Another disadvantage is that they require a deep experimental campaign in order to calibrate the material properties at the micro-level.

In this work is proposed a phenomenological approach in order to capture the fatigue behavior of composite structures. The method obtains the fatigue performance of the composite by means of coupling the fatigue behavior of its constituent materials, fiber and matrix. This procedure is accomplish by means of two formulations, one acts as a constitutive fatigue equation and the other couples the constituent behavior of fiber and matrix. In this work, the fatigue formulation and S/P Mixing theory are introduced. Next, a brief description is made of how

DOI: 10.1201/9781003216582-73

fatigue in composites is characterized. Then, the FE model of the composite structure and material description are introduced. Finally, the results and their conclusions are introduced, explaining the capabilities of the current formulation and methodology.

2 FORMULATION

In this section is described the approach used to obtain the fatigue behavior of the composite by means of the fatigue performance of the constituent materials. This approach is based in the use of S/P Mixing Theory and a constitutive equation for fatigue. The S/P Mixing Theory is basically a constitutive law manager, which obtains the non-linear performance of the composite by solving the non-linear performance of its constituents. The fatigue constitutive model used is a phenomenological approach that follows the strength degradation of each material in function of the number of cycles.

2.1 Serial/Parallel Mixing Theory (S/P RoM)

The classical approaches of Rule of Mixtures (RoM) are based on the definition of a closing equation that relates the stresses and strains of the composite components. Depending on the relations defined, it is possible to define a direct or an inverse rule of mixtures. Both of them were originally developed by Truesdell and Toupin (1960). In order to take into account the non-linear behavior of the composite, Car et al. (2000), (2002) reformulated the classical models formulated by Trusdell using a mechanics of the continuum medium approach. As a consequence, the rule of mixtures became a constitutive law manager, giving that Helmotz free energy of the composite, $^c\Psi$, can be characterized as the sum of the free energies of each constituent material of the composite in proportion of the volumetric participation:

$$^c\Psi(^c\varepsilon, \theta, ^c\beta) = \sum_{i=1}^{n} {}^i k^i \Psi(^i\varepsilon, \theta, ^i\beta) \qquad (1)$$

From the expression of the Helmotz free energy, the expression of the stress of the composite can be obtained as:

$$^c\sigma = \frac{\partial \Psi}{\partial \varepsilon} = \sum_{i=1}^{n} {}^i k \frac{\partial \Psi^i}{\partial \varepsilon^i} = \sum_{i=1}^{n} {}^i k^i \sigma \qquad (2)$$

Later, Rastellini et al. (2008) improved the model with the definition of a serial-parallel rule of mixtures, which can apply the direct and inverse closing equations to different directions of the strain and stress tensors. This approach is the one used on this

work, which is a more realistic model than the classic theory of the rule of mixtures. On one hand, the formulation considers an iso-strain condition for matrix, fiber and composite in the fiber direction (*parallel direction*). On the other hand, for all perpendicular directions to the fiber (*serial directions*), the formulation considers an iso-stress condition. Another advantage of this formulation is its lower computational cost compared with other models which allows take into account the non-linear performance of the composite, such as multi-scale models.

The micro-structure of composite materials gives a different behavior to the composite in function of the loading direction, which implies a highly anisotropic response of the material. Due to the anisotropic performance, there are directions that have a parallel perfromance (iso-strain), while others must be characterized with a serial behaviour (iso-stress).

The equations that define the stress equilibrium and establish the stain compatibility between components are the following:

Parallel behavior:

$$^c\varepsilon_P = {}^f\varepsilon_P = {}^m\varepsilon_P \qquad ^c\sigma_P = k^f \cdot {}^f\sigma_P + k^m \cdot {}^m\sigma_P \qquad (3)$$

Serial behavior:

$$^c\sigma_S = {}^f\sigma_S = {}^m\sigma_S \qquad ^c\varepsilon_S = k^f \cdot {}^f\varepsilon_S + k^m \cdot {}^m\varepsilon_S \qquad (4)$$

where superscripts c, m and f stand for composite, matrix and fiber, respectively, and k^i corresponds to the volume fraction coefficient of each constituent in the composite.

The validation of the current formulation and its capability for predicting different failure mechanisms have been proved in a previous work of the author (Granados et al. 2019). A more deeper understanding of the formulation and its implementation can be obtained in (Rastellini et al. 2008) and (Martínez et al. 2008).

2.2 Fatigue model

In this work, a fatigue characterization of each constituent material is conducted. Consequently, the fatigue behavior of fibers and matrices is monitored and coupled between them in order to obtain the fatigue performance of the composite. The fatigue model uses a phenomenological approach, taking into account the degradation of the mechanical properties of the material due to cyclic loading. In addition, the formulation is able to take into account changes in the stress ratio or stress level applied, allowing the use of variable block loading sequences. The fatigue formulation acts in the current constitutive equation of the material. In the current work, a damage model has been used as a constitutive law, which is defined as:

$$\sigma = (1 - d) \cdot \sigma_0 \qquad (5)$$

Where d is the damage parameter, ranging between 0 and 1. Value 0 means there is no damage in the material, corresponding with no loose of material stiffness. Value 1 means the material is completely damaged, and consequently the has lost its stiffness. This model requires a failure threshold in order to evaluate if damage has started in the material, as well as the evolution of this damage. This can be written as:

$$F^D(\sigma_{i,j}, d) = f^D(\sigma_{i,j}) - K^D(\sigma_{i,j}, d) \leq 0 \qquad (6)$$

where $f^D(\sigma_{i,j})$ is the stress tensor function and $K^D(\sigma_{i,j}, d)$ is the current elastic threshold.

The degradation of the mechanical properties is obtained by modifying the elastic threshold of the constitutive equation of the material. This modification is by means of a reduction function, named $f_{red}(N, \sigma_{max}, R)$, which reduces the failure threshold of the material based on the evolution of the cyclic load applied to it. This strategy allows to use any constitutive equation known and coupling between them, obtaining static non-linear and fatigue behavior. Finally, the constitutive equation (6) can be rewritten of taking into account the fatigue model, such as:

$$f^D(\sigma_{i,j}) - f_{red}(N, \sigma_{max}, R) \cdot K^D(\sigma_{i,j}, d) \leq 0 \qquad (7)$$

The reduction function is defined from material parameters obtained from experimental data of each compounding material. The characterization process to obtain such parameters is based on fitting the experimental curves of each material for a specific stress ratio with the numerical S/N curves given by the formulation. Therefore, the Wholer curves of each constituent material are taken into account in the model to account for the material fatigue behavior.

This work uses the Wohler curves as experimental data to obtain the fatigue models of each component material. S-N curves are defined with an analytic expression that can be found in the owrk by Salomón et al. (2002) and Oller et al. (2005). The material calibration is obtained by fitting the Wohler curves for a basic stress ratio $R = -1$ and extrapolated for the other stress ratio values.

The constitutive equation modified by the fatigue model is an isotropic damage model, as the one shown in equation 5. The stiffness degradation provided by the model can be understood as the growth of small fractures and micro voids on the structure of any material, which describes the typical behavior of brittle materials. For a deeper explanation of this model, the reader will find reference in (Oller 2014).

Finally, the fatigue formulation described is coupled with a step-wise load-advancing strategy, which is described in (Oller et al. 2005) and (Barbu et al. 2015). This consists on two different phases. The first one is defined by load-advance being conducted by small time increments, with the consequent load variation following a cyclic path. The second phase is characterized by load-advance being done with large increments on the number of cycles. The strategy consists of an algorithm that automatically switches from one phase to the other, going repeatedly back and forth between both in accordance with the loading input and the damage increase rate. The main advantage of this strategy is that the model has not to compute the strain-stress state of each GP cycle by cycle, what would lead to a high computing time. For a deeper understanding of the stepwise load advancing strategy algorithm, the author of this paper refers to (Oller et al. 2005, Barbu et al. 2015, Barbu et al. 2019).

3 FATIGUE CHARACTERIZATION OF COMPOSITES

In section 2 has been explained the equations of the SP RoM, which obtains the non-linear performance of the composite by solving the non-linear performance of the constituents. Therefore, a good constitutive calibration of constituents have to be achieved. In the case of the fatigue, the constitutive models used are those that describe the fatigue performance of the fiber and matrix. However, experimental data relates with the mechanical parameters of the laminate, and hence no data of constituents is available. Regarding this topic, in a previous work the author developed a new calibration procedure in order to obtain the material parameters for each constituent from experimental data of UDs loaded at 0° and 90° (Granados et al. 2019). As a result, the numerical failure of the laminate depends only on the geometry, boundary conditions and loads applied in each sample, obtaining the same failure mechanisms found in the testing campaign.

The observations made by different authors (Kawai and Suda 2004) describe a fatigue failure dependence with loading direction in UD laminates. When the laminate is loaded in the fibre direction, fatigue failure is induced by fibre breakage, while if the laminate is loaded in any other direction, failure is produced by matrix failure, either by normal or by shear stresses. Therefore, two conclusions can be suggested: First, the failure mechanism in on-axis laminates is dominated by the strength of fibers. Second, failure mechanism in off-axis laminates is dominated by a combination between transverse and in-plane shear strength of the matrix. As a consequence, the performance of the laminate in each orthogonal direction are driven by one of its compounds, which can be described by their constitutive equation. In this work, the main assumptions of the calibration procedure described in (Granados et al. 2019) and above are applied with the fatigue formulation described.

Consequently, the fatigue model is used as a constitutive equation for each material, and S/P RoM formulation is used as a constitutive equation manager for fiber and matrix, with the aim to couple both performances and take into account the whole fatigue response of the laminate. In particular, the FE model obtains the composite strains. Composite strains are used by the S/P RoM, obtaining the constituent strains and evaluating the constitutive equation for each material. The constitutive equation used is the fatigue model previously explained. Finally, the S/P RoM couples fiber and matrix fatigue performance, obtaining the fatigue behavior of the composite. This idea was previously proved by Barbu et al. (2019) for UD laminates loaded in fiber direction. The fatigue data of each material is obtained by calibration using experimental data from UD0 and UD90. The process to obtain the S/N curves for constituent materials is normalizing the S/N curves of UD0 and UD90, for fiber and matrix respectively. In Figure 1 is shown the normalized S/N curves for fiber and matrix. In this work, the different failure mechanisms arisen in composites are expected to be obtained from the fatigue simulations by the failure of each compounding material. As an example, Figure 2 and 3 show two simulations of the same composite material,

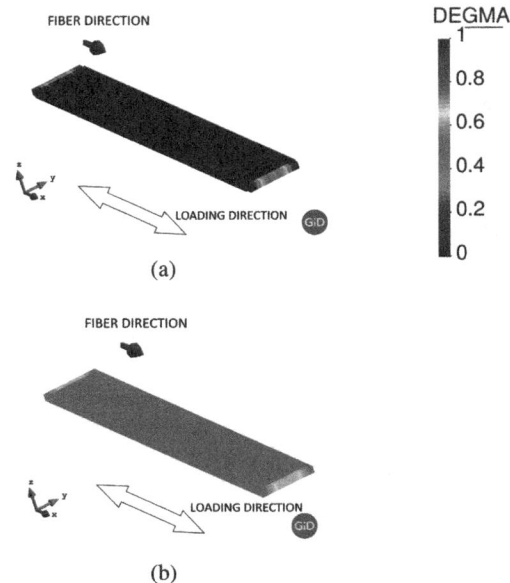

(a)

(b)

Figure 2. Fatigue failure in UD0 laminate. (a) Fiber damage parameter. (b) Matrix damage parameter.

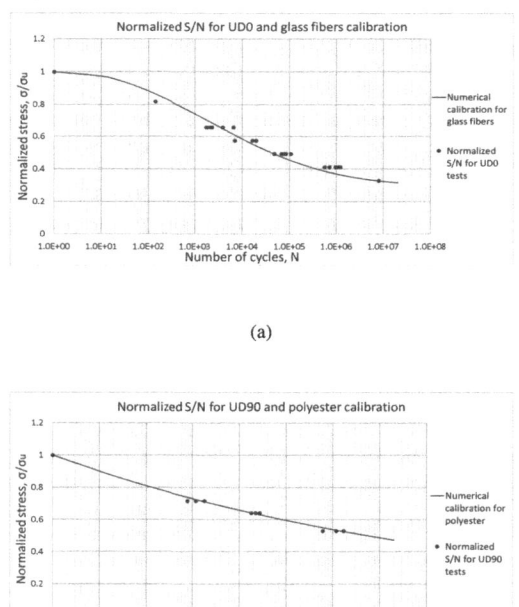

(a)

(b)

Figure 1. Normalized S/N curves. (a) S/N curve for fiber and equivalent to UD0. (b) S/N curve for resin and equivalent to UD90.

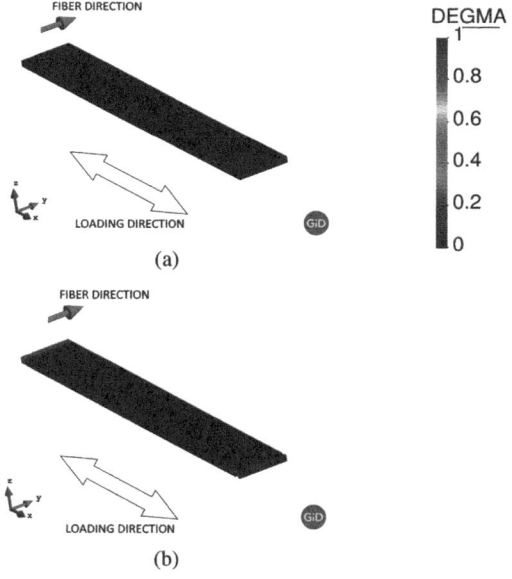

(a)

(b)

Figure 3. Fatigue failure in UD90 laminate. (a) Fiber damage parameter. (b) Matrix damage parameter.

but the only difference is the loading direction. In both figures, material failure is depicted with the damage parameter d, named DEGMA. In Figure 2, a UD laminate loaded in the fiber direction is shown, which failure mechanism is the fiber breakage as it is seen in Figure 2a, although failure in the matrix can be also seen (Figure 2b). On the other hand, in Figure 3, the same material is simulated but the loading direction is perpendicular to the fiber direction. What is observed is that the failure mechanism of the laminate is produced by matrix failure only (Figure 3b), whereas no damage in the fiber was found (Figure 3a). Both behaviors were the ones expected. In conclusion, different fatigue failure mechanisms can be predicted in laminates, independently of their loading direction and without a specific failure predefined.

4 FATIGUE ANALYSIS OF NAVAL STRUCTURES

The fatigue analysis of a composite sub-structure is conducted in order to demonstrate the capabilities of the current formulation. This analysis is able to obtain the failure mechanisms presented in composite structures, such as fiber breakage and matrix failure. In addition, the strength degradation evolution of each constituent material can be monitored, giving useful information to the user of how much strength the materials still have.

The selected composite structure belongs to an early design of a composite ship. This early design is based on an existing steel container ship. The definition of the composite structure has been done to be equivalent in structural stiffness to the steel structure. Given that the material stiffness of steel is different than the stiffness of a composite, a new scantling has been defined using the updated stiffness. Composite stiffness depends on the fiber/matrix system and fiber orientation. A quasi isotropic laminate is used, with a lay-up of $[0°, \pm 45°, 90°]_s$, as it is shown in Figure 4. Each ply corresponds to a UD layer made of glass fiber and polyester resin, which mechanical properties are detailed in Table 1. For each ply, fiber content and ply thickness are equal. The value used for fiber content is $V_f = 54\%$, and the total composite thickness is $35mm$.

Table 1. Mechanical parameters for constituent materials.

Material	E (GPa)	$^t\sigma_u$ (MPa)	$^c\sigma_u$ (MPa)
Glass fiber	68	1900	1900
Polyester	4.7	48.5	125

The cyclic loads applied to the composite substructure are defined as cyclic displacements. A nested process has been followed in order to obtain the value of such cyclic displacements. The nested process connects two models: the composite model of the sub-structure (Figure 5b) and a larger model (Figure 5a). Cyclic pressure loads are applied in the larger model, obtaining the displacements of such structure. These cyclic displacements are used as boundary conditions in the composite model, in order to analyze its fatigue behaviour. The larger model is a previous design of a steel ship, in which a wave load has been applied as cyclic load. The wave load used is a monochromatic wave of 4 meters height and 2 seconds period.

Figure 6 shows the boundary conditions applied in the composite model. The hull and longitudinal girder (blue surfaces in Figure 6a), as well as decks and bulkheads (red surfaces in Figure 6a) have their displacements fixed in all directions. Besides, the displacements applied in the side hull are shown in Figure 6b.

Fatigue simulation has been run with an in-house software: PLCd (CIMNE). In Figure 5b is depicted the FE model used. The model mesh has 25600 linear hexahedral elements.

It must be recalled that currently the design of the ship is not yet completed, therefore the sub-structure analyzed is preliminary and its main purpose is to demonstrate the performance of the fatigue analysis procedure developed.

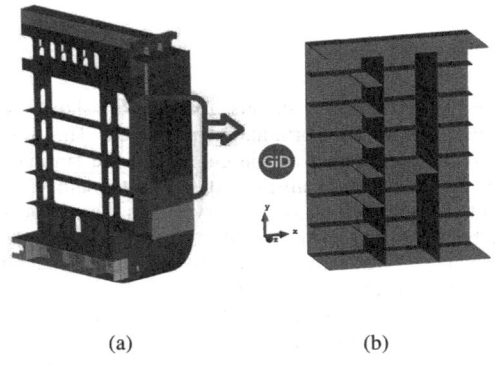

(a)　　　　　　　(b)

Figure 5. Nested process followed to define the FE model for the composite sub-structure. (a) FE steel model. A wave load is applied to obtain the cyclic displacements in the structure. (b) FE composite model. Cyclic displacements are applied to obtain the fatigue performance.

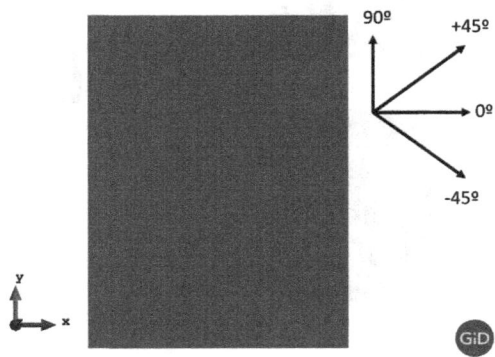

Figure 4. Fiber directions.

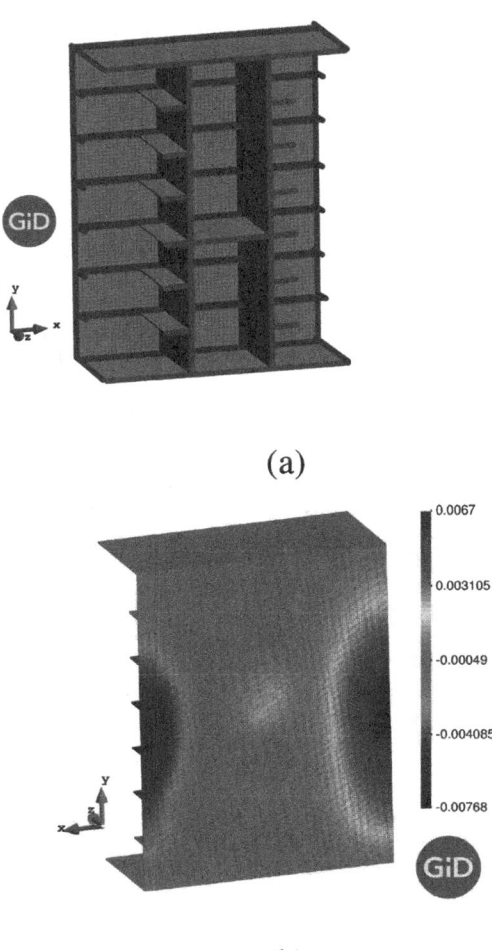

(a)

(b)

Figure 6. Definition of boundary conditions.

5 RESULTS

The results included in this section will show the capabilities of the formulation presented. This formulation can predict the fatigue life of the laminate and its failure mechanisms. As it was explained, in Section 2, the S/P Mixing theory solves the composite behavior by solving the constituent materials performance. Therefore, the results shown are related with the constituent materials.

The results provided by the fatigue analysis are the apparition of damage in the different constituent materials, for a certain number of cycles. The first simulation made, with the loads obtained from the larger model, showed no fatigue failure in the structure, which proves that its design is correct.

A second simulation has been conducted, in which the loads applied to the structure have been

increased by 50%, in order to have a fatigue failure and with the aim of showing the results provided by the model in this case.

Figure 7 shows the region in the panel under study in which fatigue failure takes place at $N = 150.011$ cycles. In Figures 8 and 9, damage parameter in fibers of each ply are depicted (named DEGMA). As it can be seen, only fibers in ply 90° are not suffering damage. On the other hand, the extension of fiber damage in ply −45° is greater than damage in plies 0° and +45°. These damage extension suggests that the main loading direction applied in the structure is close to the fiber −45° direction, followed by 0° and +45° directions. On the contrary, ply 90° is not a critical ply from the point of view of fibers. On the other hand, matrix damage parameter (DEGMA) is shown in Figures 10 and 11. These figures show that the damage area is substantially larger than the area with damaged fibers. In particular, matrix damage in ply 90° has the greatest

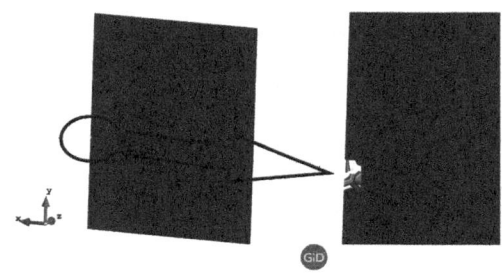

Figure 7. Zone of interest.

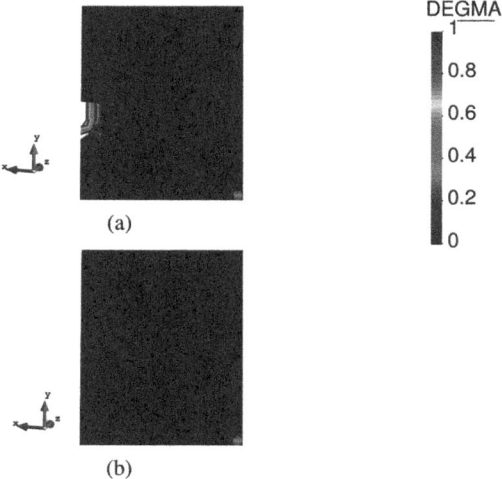

(a)

(b)

Figure 8. Fiber damage. (a) Ply 0°. Medium damage extension in fibers. (b) Ply 90°. No fiber damage appeared.

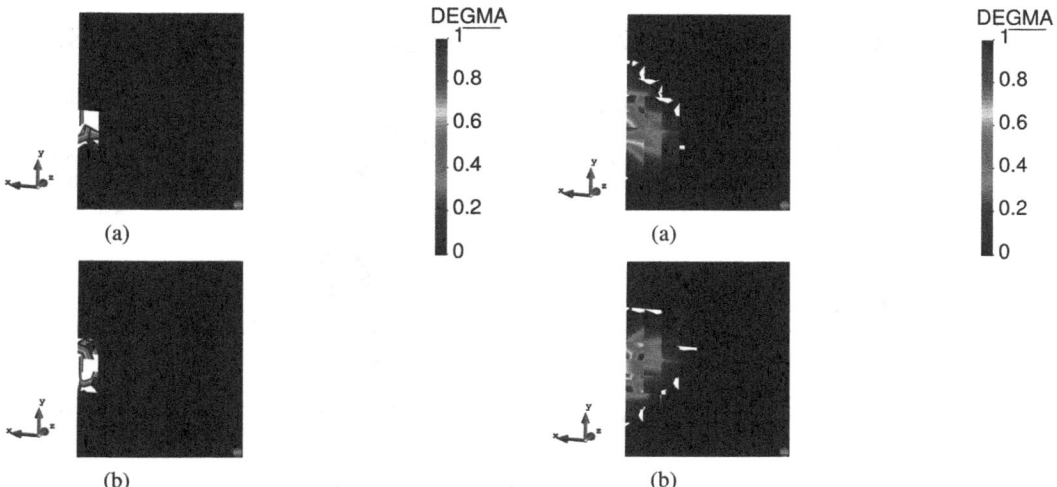

(a)

(b)

Figure 9. Fiber damage. (a) Ply +45°. Medium damage extension in fibers. (b) Ply −45°. Greater fiber damage extension.

Figure 11. (a) Ply +45°. Medium damage extension in matrix. (b) Ply −45°. Medium damage extension in matrix.

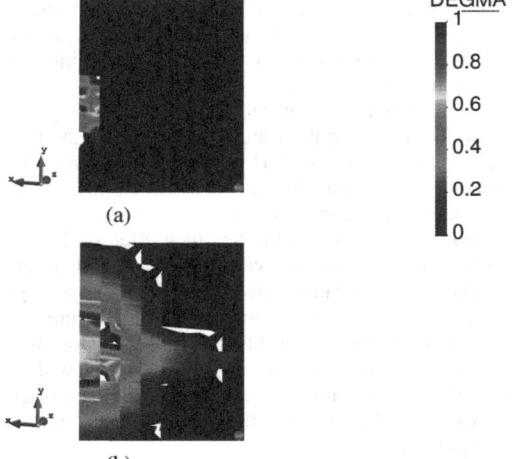

(a)

(b)

Figure 10. Matrix damage for each ply. (a) Ply 0°. Low damage extension in matrix. (b) Ply 90° Huge matrix damage extension.

extension, contrary of what happens in the same ply for fiber damage, as it can be observed in Figures 8b and 10b. The fact that in ply 90° there is no fiber damage but there is a great extension in matrix damage means that fibers are not being loaded, or what is the same, the loading direction is perpendicular to fiber, consequently, in that specific ply the matrix is subjected to higher stresses. Finally, the presence of matrix damage implies the apparition of matrix cracks as well as delamination between plies,

which is one of the main failure mechanisms in composites. As it can be seen, the damage appears in those regions where there are stress concentrations, and therefore, where they are more susceptible to suffer fatigue damage.

Another interesting aspect of the formulation introduced that it is capable of showing the reduction of the material failure stress threhold (RESFA parameter) produced by fatigue loading. Figures 12, 13, 14 and 15 show the value of RESFA for fibers and matrix respectively. The results show that strength reduction in fibers is

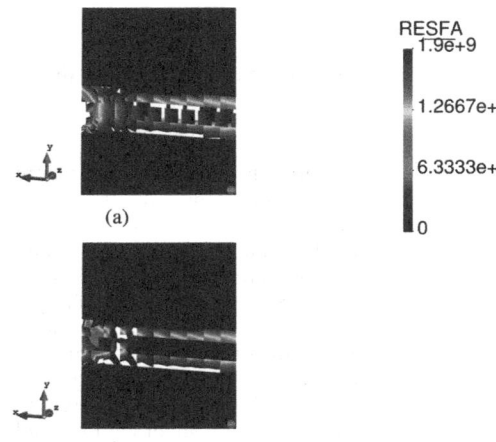

(a)

(b)

Figure 12. Residual strength in fibers. (a) Ply 0°. (b) Ply 90°.

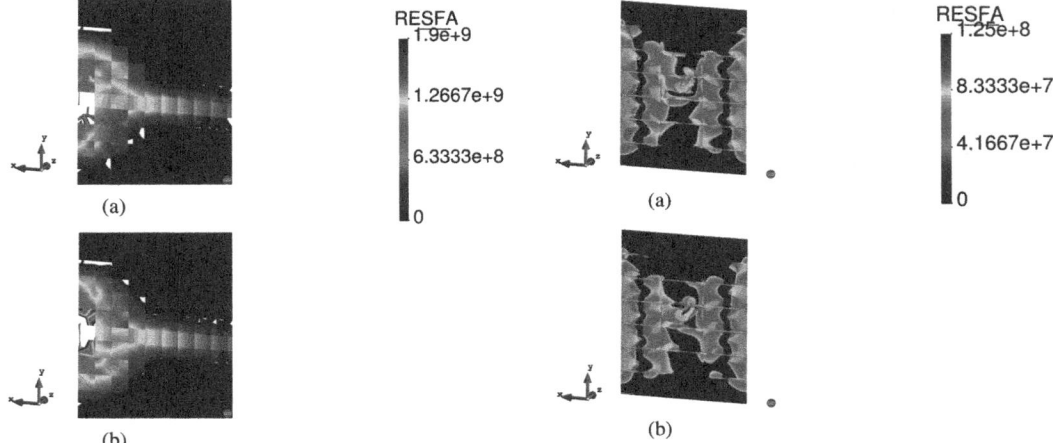

(a)

(b)

Figure 13. Residual strength in fibers. (a) Ply +45°. (b) Ply −45°.

(a)

(b)

Figure 15. Residual strength in matrix. (a) Ply +45°. (b) Ply −45°.

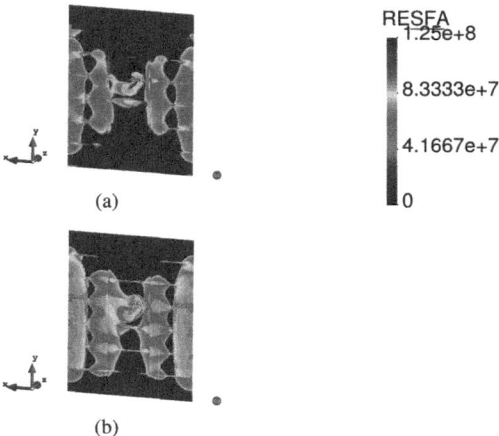

(a)

(b)

Figure 14. Residual strength in matrix. (a) Ply 0°. (b) Ply 90°.

observed locally only in those points where there are stress concentrations. In the other hand, matrix strength degradation is observed more generally in the whole structure. This fact is mainly to the combination of the stress state subjected the structure and the low fatigue strength of polymer resin, in comparison with glass fibers. This feature of the formulation can be used to define safety factors and a design criteria, for instance limiting the fiber strength reduction until a certain level, in order to avoid the collapse of the structure, or limiting the matrix strength reduction to avoid the apparition of delamination.

6 CONCLUSIONS

In this work, a formulation to obtain the fatigue performance of composite structures has been introduced. The formulation is able to follow the fatigue degradation of constituent materials, obtaining the fatigue life of the structure.

The formulations allows to recognize which ply and material is failing in the composite, as well as the failure mechanism triggered. Therefore, a re-design of the structure can be done, in order to avoid fatigue failure. One of the key points for designing structures with composites is the possibility that they offer, of designing the material together with the structure. It means that the engineers can decide prior which are the properties of the material that best fit with the structure that they are designing. With this approach, and formulations such the ones presented in this work, it will be possible to design more reliable composite structures, encouraging the use of this materials in marine applications.

ACKNOWLEDGEMENTS

This work has been supported bu the European Union's Horizon 2020 research and innovation program under grant agreement No. 723360, FibreShip Project.

REFERENCES

Barbu, L., S. Oller, X. Martinez, & A. H. Barbat (2019, jul). High-cycle fatigue constitutive model and a load-advance strategy for the analysis of unidirectional fiber reinforced composites subjected to longitudinal loads. *Composite Structures 220*, 622–641.

Barbu, L. G., S. Oller, X. Martinez, & A. Barbat (2015). High cycle fatigue simulation: A new stepwise load-advancing strategy. *Engineering Structures 97*, 118–129.

Car, E., S. Oller, & E. Oñate (2000, may). An anisotropic elastoplastic constitutive model for large strain analysis of fiber reinforced composite materials. *Computer Methods in Applied Mechanics and Engineering 185*(2–4), 245–277.

Car, E., F. Zalamea, S. Oller, J. Miquel, & E. Oñate (2002, apr). Numerical simulation of fiber reinforced composite materials—two procedures. *International Journal of Solids and Structures 39*(7), 1967–1986.

Ding, C., G. Li, & H. Guo (2012). Fatigue analysis of wind turbine. *Applied Mechanics and Materials 229–231* (August 2018), 621–624.

Granados, J. J., X. Martinez, N. Nash, C. Bachour, I. Manolakis, A. Comer, & D. Di Capua (2019, oct). Numerical and experimental procedure for material calibration using the serial/parallel mixing theory, to analyze different composite failure modes. *Mechanics of Advanced Materials and Structures*, 1–19.

Kawai, M. & H. Suda (2004). Effects of non-negative mean stress on the off-axis fatigue behavior of unidirectional carbon/epoxy composites at room temperature. *Journal of Composite Materials 38*(10), 833–854.

Liu, X., W. Yu, F. Gasco, & J. Goodsell (2019, sep). A unified approach for thermoelastic constitutive modeling of composite structures. *Composites Part B: Engineering 172*, 649–659.

Martínez, X., S. Oller, & E. Barbero (2008). Study of Delamination in Composites by Using the Serial/Parallel Mixing Theory and a Damage Formulation. In *Mechanical Response of Composites. Computational Methods in Applied Sciences*, Chapter Vol. 10, pp. 119–140. Springer, Dordrecht.

Oller, S. (2014). *Nonlinear Dynamics of Structures*. Lecture Notes on Numerical Methods in Engineering and Sciences. Cham: Springer International Publishing.

Oller, S., O. Salomón, & E. Oñate (2005). A continuum mechanics model for mechanical fatigue analysis. *Computational Materials Science 32*(2), 175–195.

Rastellini, F., S. Oller, O. Salomón, & E. Oñate (2008, may). Composite materials non-linear modelling for long fibre-reinforced laminates: Continuum basis, computational aspects and validations. *Computers & Structures 86*(9), 879–896.

Salomón, O., S. Oller, & E. Oñate (2002, dec). Fatigue Analysis of Materials and Structures using a Continuum Damage Model. *International Journal of Forming Processes 5*(2-3-4), 493–503.

Truesdell, C. & R. Toupin (1960). The Classical Field Theories. pp. 226–858. Springer, Berlin, Heidelberg.

Developments in Maritime Technology and Engineering – Guedes Soares & Santos (eds)
© 2021 Copyright the Author(s), ISBN 978-0-367-77376-2

External surface crack growth in offshore steel pipes reinforced with CRS subjected to fatigue bending

Zongchen Li, Xiaoli Jiang & Hans Hopman
Department of Maritime and Transport Technology, Delft University of Technology, Delft, The Netherlands

Ling Zhu
School of Transportation, Wuhan University of Technology, Wuhan, P. R. China

Zhiping Liu
Port Logistic Technology and Equipment Engineering Research of Ministry of Education, Wuhan University of Technology, Wuhan, P. R. China

ABSTRACT: The Composite Repair System (CRS) is an advanced maintenance technique in the pipeline industry. In past a few decades, researchers applied composite materials to repair through-thickness cracks in pipes, which found that the CRS significantly decreased the crack growth rate. However, the investigation on repairing external surface cracked steel pipes is limited in open documents. Understanding the surface crack growth behaviour reinforced with Fibre-Reinforced Polymer (FRP) are of great importance. In this paper, we have experimentally studied the CRS reinforced external surface cracked steel pipes subjected to cyclic bending. In total, nine groups of 27 surface cracked pipes have been tested, containing three different sizes of the initial notch and four different CRS reinforcement schemes. The results show that the CRS has significantly decreased the crack growth rate and prolonged the residual fatigue life. Using more layers of CFRP can evidently facilitate the reinforcement effectiveness, while the surface crack growth is insensitive with the bond length. The inversely diagonal wrapping pattern performs less effective than the other reinforcement schemes. The experimental investigation in this paper is of instructive value to facilitate the further application of CRS in the offshore pipeline industry.

NOMENCLATURE

a	crack depth of surface cracks
a_0	notch depth
a/c	aspect ratio of surface cracks
a/t	normalized crack depth
C	Paris' law constant
D	external diameter of pipes
d	internal diameter of pipes
c	half crack length of surface cracks
c_0	notch length
da/dN	crack growth rate along the depth direction
dc/dN	crack growth rate along the length direction
E_i	elastic modulus
G_{ij}	shear modulus
L	length of the steel pipe
L_c	bond length
L_e	external span of the four-point bending test
L_i	inner span of the four-point bending test

(Continued)

(Cont.)

m	Paris' law constant
N	cyclic index
Nu	Poisson's ratio
R	load ratio
T	tensile strength
t	thickness of the pipe

1 INTRODUCTION

Offshore steel pipes bear dynamic loads long-termly in marine environment. The cyclic bending load, which is generated by wave, current, wind, and 2nd order floater motions, is a dominant load case (Li et al., 2019c). It is commonly applied on critical zones in offshore pipes, such as hang-off zone, sag bend, arch bend and the touch down zone (Li et al.,

DOI: 10.1201/9781003216582-74

2019b). Meanwhile, circumferential external surface cracks, initiating from corrosion pitting or girth weld defects, often appear on the pipe surface in these critical areas (DNV, 2017b, Li et al., 2019c). Under this circumstance, surface cracks might continually propagate to through-thickness cracks, which might eventually result in leakage or collapse (DNV, 2008, DNV, 2017a).

The surface cracked steel pipes with critical size needed to be repaired instantly to avoid further failures, such as oil and gas leakage (API, 2016). The Composite Repair System (CRS) has been recognized as an efficient and advanced repairing technique in the pipeline industry (Li et al., 2019a, Li et al., 2017, Liu et al., 2017). It has been highly valued for the outstanding advantages in terms of effectiveness, time-saving, cost-effective, no secondary damage and ease of installation (Zhao and Zhang, 2007, Li et al., 2018). In recent years, composite materials were applied to repair through-thickness cracked metallic pipes (Woo et al., 2016, Zarrinzadeh et al., 2017b, Achour et al., 2016). The results indicated that composite reinforcement has significantly decreased the crack growth rate. Researchers also indicated that crack-induced debonding would occur in a CRS repaired through-thickness cracked pipes, which affect the CRS reinforcement negatively (Zarrinzadeh et al., 2017a, Li et al., 2019b). Yet, the possible failures of using CRS to repair surface cracked pipes subjected to bending remains unknown. In addition, the surface crack propagates approximately in a semi-elliptical shape, which is more complex than through-thickness crack. Particularly, more concerns are required on the crack growth along the depth direction, since it might trigger the pipe leakage.

At present, the repair of surface cracks in steel pipes using CRS conforms to the guidance of pipe repairing standards based on either the rule of thumb (ASME, 2012) or ultimate strength (BS, 2015). Such repair aims to rehabilitate the load bearing capacity of damaged steel pipes, while the fatigue crack growth rate after reinforcement remains unknown (Li et al., 2019a). To date, there have been not yet well-established methods to predict surface crack growth in pipes reinforced with CRS. This has resulted in a lacking confidence situation which serious restricted the application and development of CRS. In this respect, it is highly demanded to investigate surface crack growth in steel pipes reinforced with CRS comprehensively.

The investigation of using CRS to repair surface cracked metallic pipes has recently been conducted by means of finite element analysis, including circumferential internal surface cracks in pipe subjected to bending (Li et al., 2019a), and longitudinal internal surface cracks in pipe subjected to internal pressure (Chen and Pan, 2013). The results indicated that CRS reinforcement could significantly decrease the crack growth rate. Yet those studied concentrated on the internal surface cracks, thus no contact exists in between the cracked area of the steel substrate and the composite laminates. In addition, the numerical investigations are lacking experimental validations.

To the knowledge of the authors, the investigation of external surface crack growth in steel pipes reinforced with CRS is absent from open documents. As one of the most reliable methods to understand the mechanism of surface crack growth reinforced with CRS, the experimental investigation plays an irreplaceable role.

Given that, in this paper, we conduct an experimental investigation on external surface cracked steel pipes reinforced with the CRS subject to cyclic bending, in order to identify the crack growth behaviour within the CRS reinforcement. In Section 2 the specimen preparation processes are introduced. In Section 3, the test set-up is described. In Section 4, test results are presented and analysed. Finally the conclusions are drawn in Section 5.

2 SPECIMEN PREPARATION

2.1 *Material properties*

The CRS reinforced external surface cracked steel pipe specimen is indicated in Figure 1. Four materials have been used for the specimens: steel substrate, Glass-FRP (GFRP), CFRP, and adhesive. Stainless steel of API 5L X65 for subsea scenarios conforming to API SPEC 5L code (API, 2018) has been used as the steel substrate, which has yield strength of 448

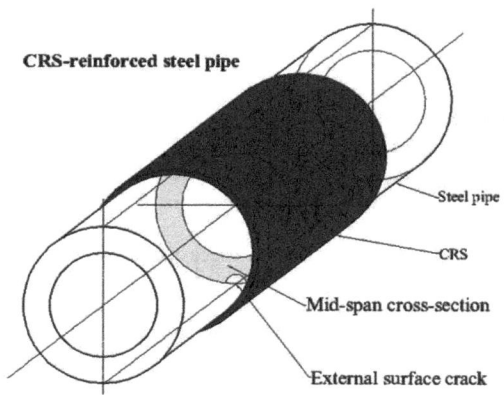

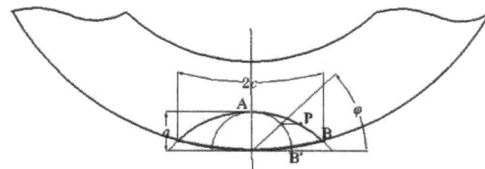

Figure 1. The sketch diagram of the CRS reinforced surface cracked pipe specimens.

MPa, and tensile strength of 530 MPa. In light of the galvanic corrosion between CFRP laminates and steel substrate, one layer of GFRP laminate was adopted as the first layer of the CRS for all reinforced specimens. The GFRP laminate applied the E-glass fibre weave fabric while the CFRP laminate used the Toray™ T700S unidirectional fabric. The adhesive adopted the ZHONGBO™ ER4080 resin epoxy. The detailed material properties are listed in Tables 1 to 3. Note that all material properties are provided by each manufacturer.

Table 1. Material properties of GFRP.

E_1 (Pa)	E_2 (Pa)	T (Pa)	G_{13} (Pa)	G_{23} (Pa)	Nu
72×10^9	72×10^9	1.1×10^9	4.7×10^9	3.5×10^9	0.33

Table 2. Material properties of the CFRP material.

E_1 (Pa)	E_2 (Pa)	T (Pa)	G_{13} (Pa)	G_{23} (Pa)	Nu
230×10^9	25×10^9	4.9×10^9	5.5×10^9	3.9×10^9	0.3

Table 3. Material properties of the resin epoxy.

E (Pa)	T (Pa)	G (Pa)	Nu
2.8×10^9	70×10^6	1.4×10^9	0.35

2.2 Specimen manufacturing

The specimen manufacturing contains three main steps: manufacturing of the notches, the pre-cracking procedure, and the CRS reinforcement, as indicated in Figure 2. Semi-elliptical notches were manufactured in order to guarantee that the surface cracks are semi-elliptical shaped during the fatigue tests, located in the mid-bottom of the steel pipes, orienting circumferentially, as indicated in Figure 3. The notches were made by Micro-Electric Discharging Machining (Micro-EDM) suggested by ASME E2899 in order to achieve

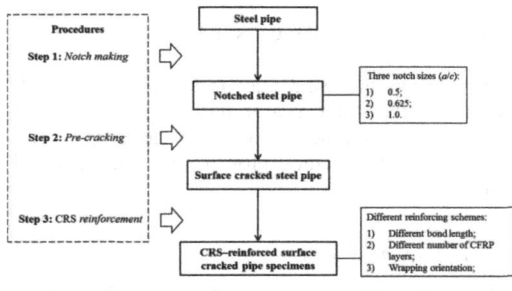

Figure 2. The procedure of specimen preparation.

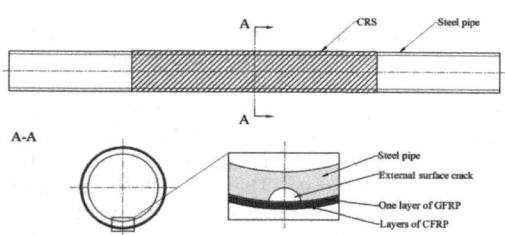

Figure 3. The configuration of CRS reinforced steep pipe specimens.

the user designed notch profile and to avoid the thermal residual stress (ASTM, 2015). The aimed aspect ratio of the notches are 0.5, 0.625, and 1.0, representing the range of common seen shapes of surface cracks in offshore metallic pipes in practice (DNV, 2017a). The width of the notch is controlled as 0.35 mm. The details of the notch size are listed in Table 4.

Before using the FRP laminates to reinforce the steel pipes, a pre-cracking procedure was conducted to generate fatigue surface cracks initiated from the notches (ASTM, 1994). The pre-cracking procedure was conducted on the fatigue machine containing two stages: the first stage adopted 80% yield stress as the load amplitude of the constant amplitude sinusoidal cyclic loading, while the second stage adopted 60% yield stress respectively. Note that both of the two stages were under the load ratio equals to $R = 0.1$. During each stage, bending fatigue load was applied on the specimens until the surface crack initiated from the notch and propagated more than 1.0 mm. Then the size of each surface crack after the pre-cracking procedure was regarded as the initial crack size. The specimens therefore were ready for the CRS reinforcement procedure.

Afterwards, the specimens were reinforced with the FRP laminates on the external surface of the steel pipes by professional workers using hand lay-up technique. The FRP laminates was bonded in the middle of the specimen. The reinforcement procedure contained the surface preparation, cleaning, and composite laminates pasting. The surface preparation

Table 4. Specimen's configuration and details.

Specimen group	L (mm)	D (mm)	L_c (mm)	CFRP wrapping scheme
PE-1	2,000	168.3	\	\
PE-2	2,000	168.3	\	\
PE-3	2,000	168.3	\	\
PE-1-R	2,000	168.3	1,000	L-L-L-H
PE-2-R	2,000	168.3	1,000	L-L-L-H
PE-3-R	2,000	168.3	1,000	L-L-L-H
PE-1-R600	2,000	168.3	600	L-L-L-H
PE-1-R8	2,000	168.3	1,000	L-L-L-H-L-L-L-H
PE-1-R45	2,000	168.3	1,000	Inversely diagonal

663

included rust cleaning and sanding. Note that the surface preparation procedures were selected to maximally meet the practical situation of reinforcing offshore metallic pipes. GFRP was applied for all FRP reinforced specimens as the first layer, in order to prevent the galvanic corrosion between the CFRP and the steel substrate. Then the CFRP laminates were wrapped around the cracked pipes with different reinforcement schemes, i.e., different bond length, orientation, number of layers. Then the CFRP laminates were wrapped and compressed by plastic tapes in order to squeeze redundant resin epoxy and eliminate the bubbles in the interlaminations, as well as to let each laminate bonded tightly. Finally the reinforced specimens were placed at room temperature for solidification of one week, in order to achieve the optimum bond condition.

2.3 Specimens configurations

The length, external diameter and thickness of the steel pipe are 2000 mm, 168.3 mm and 12.7 mm respectively. The thickness of each layer of GFRP and CFRP laminate are 0.35 mm. The parameters of the specimens are shown in Table 4. In total, nine groups of 27 specimens were prepared. Specimens PE-1, PE-2, and PE-are three controlling specimens without FRP reinforcement. The difference between PE-1, PE-2 and PE-3 is the different notch sizes. The specimens PE-1-R to PE-3-R are the corresponding specimens to the specimens of PE-1 to PE-3 respectively, using the default reinforcement schemes: four layers of CFRP using the L-L-L-H wrapping pattern of 1000 mm bond length (L represents longitudinal wrapping, while H represents hoop wrapping). The specimens in specimens PE-1-R600 are reinforced by the same wrapping method with specimens PE-1-R while shorter bond length of 600 mm. Specimens PR-1-R8 applied doubled number of CFRP layers than Specimens R1. The reinforcement scheme of Specimens PR-1-R45 is using the inversely diagonal wrapping method with the same bond length of PE-1-R. Table 4 shows the configuration and details of all specimens. The name of the specimens represents the notch category, CRS reinforcement scheme, CFRP wrapping scheme, and their repetitive number. Take 'PE-1-R(1)' as an example, 'P' means steel pipe, 'E' represents external surface, the '1' stands for the first type of notch, R means reinforcement, and the '(1)' means the No. of the repetitive specimen. Then the specimens were ready for the fatigue test.

3 TEST SET-UP

The fatigue tests were carried out under constant amplitude sinusoidal cyclic loading, generated by MTS Hydraulic Actuator. The schematic of test setup is shown in Figure 4. The load was applied in four-point bending condition to ensure a pure

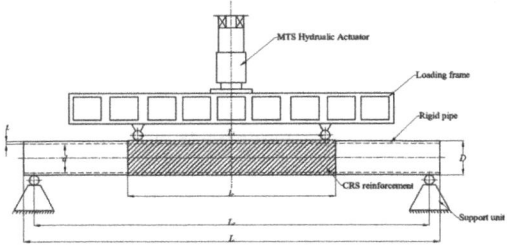

Figure 4. The schematic of the four-point bending test set-up.

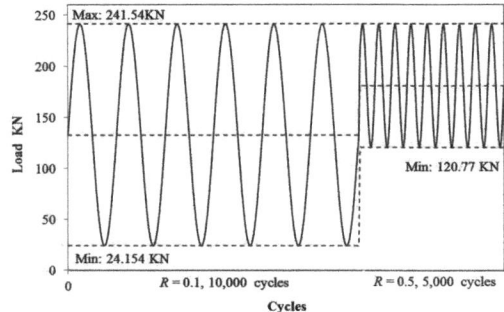

Figure 5. The two-step load spectrum and beach mark generating procedure.

bending statue around the cracked location within the inner span. Note that the inner span L_i is designed more than four times larger than the pipe diameter, in order to eliminate possible negative effects from the loading cells. The fatigue test follows the code of ASTM E647 (ASTM, 1994).

All the fatigue tests were conducted at room temperature and air environment under load control condition. The loading frequency was set as 2.5 Hz. The load ratio R maintained 0.1 for the crack growth process of all tests. The crack growth process was recorded by beach marking technique by means of changing the load ratio R to 0.5 and cycled for 5,000 cycles, as shown in Figure 5. Each test ended automatically once the tensile specimen fractured and trigger the displacement limiter of the fatigue machine.

4 RESULTS AND DISCUSSION

In this section, the test results, including possible failure modes, the strain data around the mid-bottom of the specimens, and the displacement data of the specimens, and surface crack growth, are presented and analysed.

4.1 Failure modes

During the fatigue tests, except the surface crack growth, no failures were observed for all specimens, including interfacial failure, FRP delamination, cohesion failures or FRP rupture. The CRS was perfectly bonded on the steel pipes during the fatigue tests before the crack penetrating the pipe wall. When approaching the end of each test, the crack has already propagated to larger through-thickness cracks, a clearly sound of debonding at the mid-bottom was captured for each CRS reinforced specimen.

4.2 Surface crack growth behaviour and crack growth results

After each test, the cross-section of bending specimens was sampled around the cracked area by the oxy-acetylene cutting. Then the crack growth behaviour was recorded by the beach marking technique using an electronic reading microscope, as shown in Figure 6, which has an accuracy of 0.01 mm. The cycle-index between each two adjacent beach marks in Figure 6 is 10,000, thus the crack growth process versus cyclic relation is obtained. The figures also demonstrate the multiple initiations along the notch fronts, and the surface crack continually propagated as a semi-elliptical shape until the crack penetrated the pipe wall. In addition, the photographs show that by CRS reinforcement, the beach marks is more closeness than the un-reinforced specimens. Especially, the beach marks of specimens PE-1-R8(2) is significantly denser than PE-1(1), which indicates that the crack growth rate was reduced dramatically by using eight layers of CFRP laminates.

4.2.1 CRS reinforcement on the cracked steel pipes with different crack size

It is clearly indicated from Figures 7 to 9 that the experimental results have a good repeatability. The results of crack growth along the depth direction in Figure 7a shows that accounting from N equals to 25,000 cycles, the CRS prolonged the fatigue life from around 65,000 cycles to 110,000 cycles, or

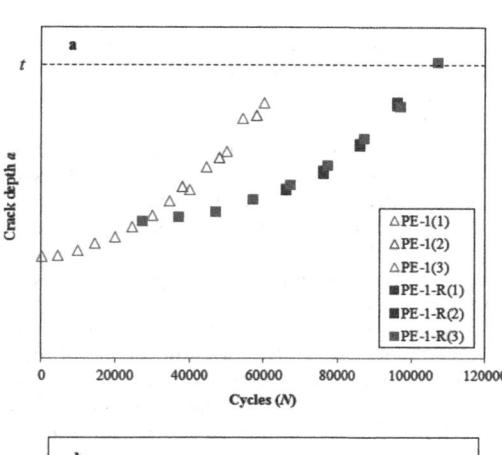

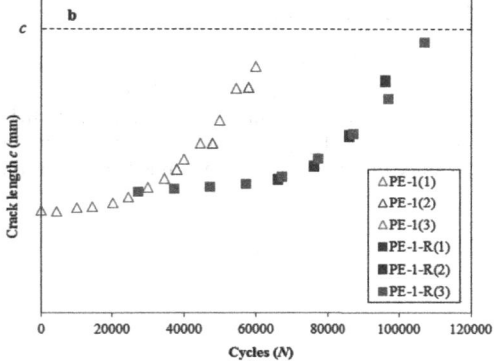

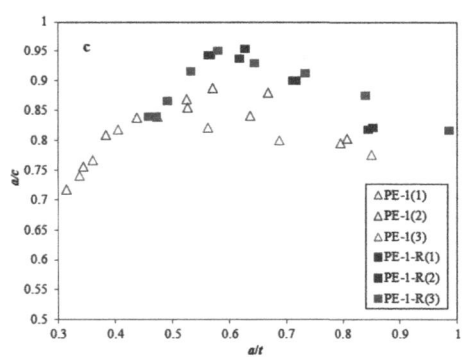

Figure 7. The experimental results of PE-1 and PE-1-R of using the default reinforcement scheme: a) crack growth along depth direction; b) crack growth along length direction; c) aspect ratio variation: a/c versus a/t curves.

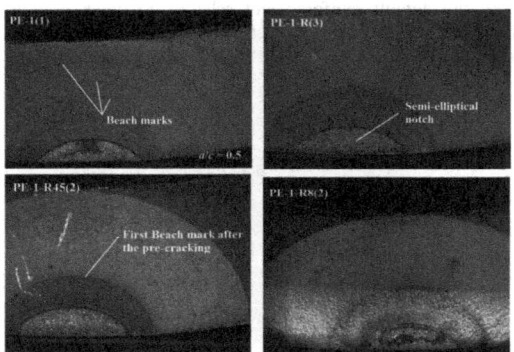

Figure 6. Beach marks on the cross-section of the pipe specimens.

approximately 110%. Figure 7b illustrated that between 30,000 and 60,000 cycles, the cracks barely grew along the length direction. The crack aspect ratio (a/c) variation results in Figure 7c shows that the a/c of the CRS reinforced specimens were always higher than the non-reinforced specimens during the test. Finally, the preferred aspect ratios (a/c when a/t equals to 0.8) of the CRS reinforced specimens are around 0.9, larger than the un-reinforced specimens, which is around 0.8. This indicates that the CRS is more effective on crack growth along the length direction, owing to the crack-bridging effect.

The results in Figure 8 and Figure 9 of using CRS to reinforce the specimens with different crack sizes are similar with the results in Figure 8 in terms of the repeatability, fatigue life prolongation, and aspect ratio variation. Figure 8 indicates that the fatigue life has prolonged from 58,000 cycles to 120,000 cycles, or around 110%, and Figure 9 shows that the fatigue life has prolonged around 120%. Besides, the crack of PE-2-R and PE-3-R barely grew along the length direction before N reached around 60,000 cycles. Thus, the CRS has a similar effect on surface crack with different crack aspect ratios.

4.2.2 *CRS reinforcement on the cracked steel pipes using different reinforcing method*

When using FRP to enhance the strength and stiffness of intact steel pipes, an effective bond length exists: once the bond length has reached a certain value, the effectiveness increases minimal when increasing of the bond length (Kabir et al., 2016). In this section, different reinforcing method including using bond lends, number of bond layers and wrapping pattern are discussed. Besides the default length which is 1000 mm, the less bond length of 600 mm is applied, as shown in Figure 10.

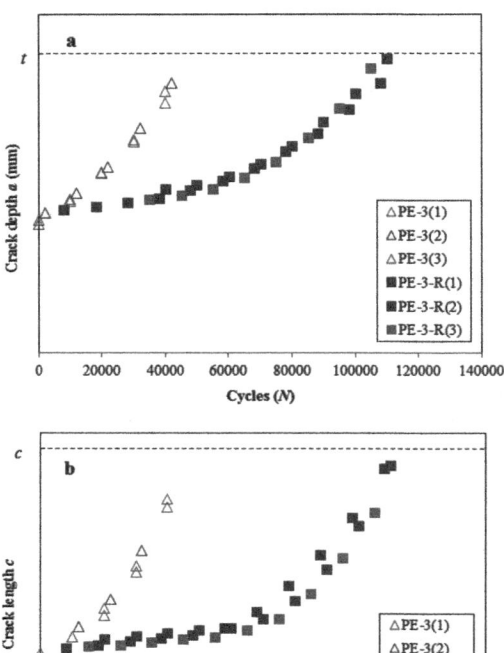

Figure 9. The experimental results of PE-3 and PE-3-R of using the default reinforcement scheme: a) crack growth along depth direction; b) crack growth along length direction.

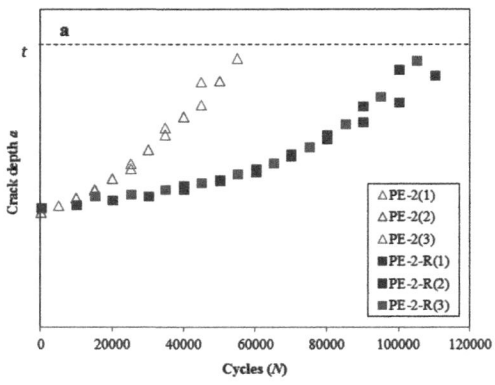

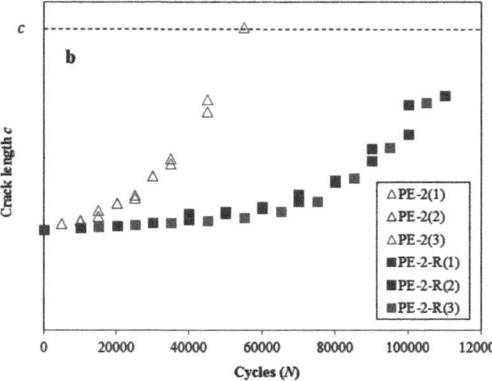

Figure 8. The experimental results of PE-8 and PE-8-R of using the default reinforcement scheme: a) crack growth along depth direction; b) crack growth along length direction.

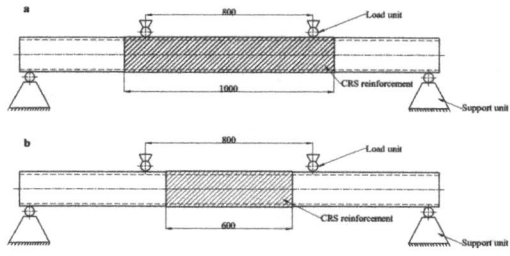

Figure 10. Loading position of specimens using different bond length: a) PE-1-R and b) PE-1-R600.

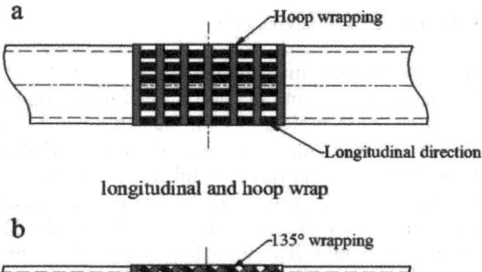

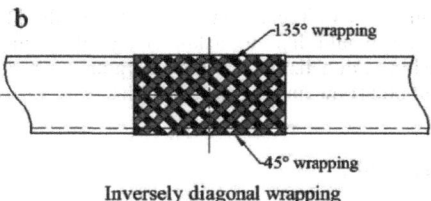

Figure 11. Wrapping scheme: a) the longitudinal and hoop wrapping pattern; b) the inversely diagonal wrapping pattern.

In the previous study, the Stress Intensity Factor (SIF) of internal surface cracked steel pipes reinforced with CRS by using different bond layers has been discussed (Li et al., 2019a). The results indicated that the SIF has staged decreased when applying more layers of CFRP (Li et al., 2019a). Therefore, besides the default number of CFRP laminate which is four layers, eight layers is used by doubling the CFRP layers of PE-1-R (see specimens PE-1-R8 in Table 4).

In light of the practical situation of applying CRS to repair the surface cracked offshore steel pipes, that the wrapping construction will be taking place under water, the longitudinal wrapping in some cases might be difficult to be employed. Thus, an inversely diagonal wrapping pattern, i.e., 45°-135°-45°-135° wrapping, as shown in Figure 11, is applied. The results of surface crack growth of using different reinforcement schemes are shown in Figure 12.

Figures 12 summaries the surface crack growth results of using different reinforcement schemes to repair the specimen PE-1, named as 'PE-1-R', 'PE-1-R600', 'PE-1-R8', and 'PE-1-R45' respectively. It illustrates that there is no evident difference of the crack growth between the specimens using different bond lengths. In this sense, there might be an effective bond length for reinforcing the surface cracked steel pipes as well, which could be less than 600 mm in this case.

It also illustrates that the residual fatigue life prolonged dramatically when applying eight layers of CFRP. Accounting from 25,000 cycles, PE-1-R prolonged the fatigue life from 65,000 to 110,000 cycles of 110%, while the PE-1-R8 prolonged the fatigue life further to 150,000 cycles of around 280%. Figure 12b shows that before N reached around 70,000 cycles, the increments of crack length of specimens PE-1-R8 are minimal,

while the final crack lengths are similar with the specimens of PE-1-R. The aspect ratio variation of PE-1-R8 specimens did not have an evident difference with the PE-1-R specimens, as shown in Figure 12c.

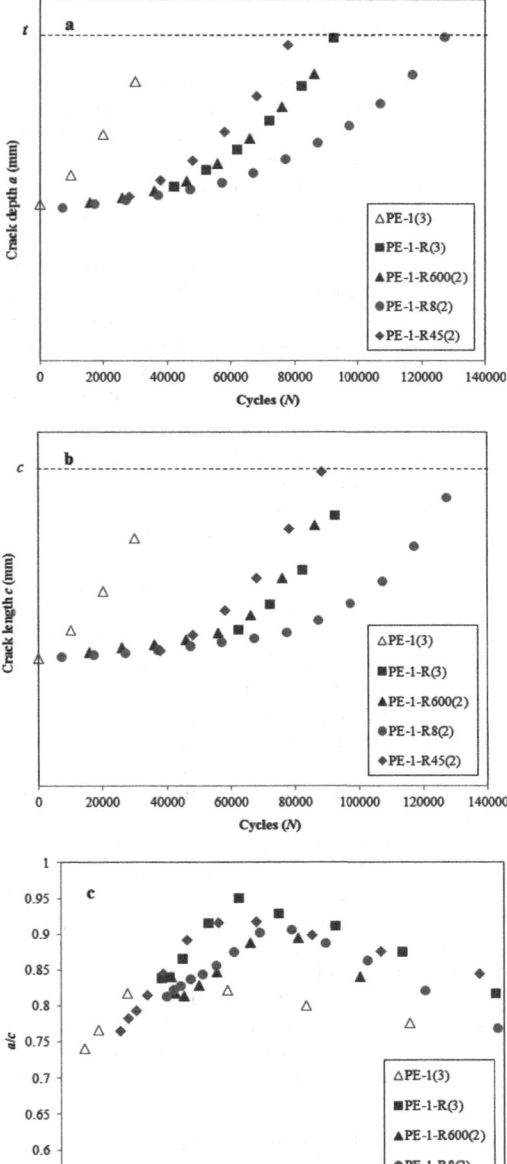

Figure 12. The experimental result comparison of using different reinforcement schemes.

In addition, Figure 12 illustrates that the inversely diagonal wrapping pattern performed less effective than the 'L-L-L-H' wrapping pattern, which prolonged the fatigue life of around 66.7%. Figure 12c shows that there is no evident difference of the aspect ratio variation between the PE-1-R and PE-1-R45 specimens.

5 CONCLUSIONS

Circumferential external surface crack growth is a serious threat to the structural integrity of offshore steel pipes. Critical surface cracks need to be repaired instantly in order to avoid oil and gas leakage. In this paper, CRS reinforcement on external surface cracked steel pipes subjected to bending has been experimental investigated. In total, nine groups of 27 specimens were tested. The effectiveness of CRS reinforcement on surface cracks with different aspect ratio, different reinforcement schemes in terms of bond length, numbers of bond layer, wrapping orientation has been discussed. The crack growth results were presented and analysed. The conclusions are drawn:

- During each fatigue test, before the surface crack propagated to a through-thickness crack, no failures beyond the crack growth in steel pipes were observed: the CRS was perfectly bonded on the steel pipe substrate.
- The CRS has significantly prolonged the residual fatigue life. Applying the default reinforcing scheme, i.e., PE-1-R, prolonged the fatigue life of around 120%, while using eight layers of CFRP laminate have maximally prolonged the fatigue life of approximately 280%. While different from reinforcing internal surface cracks in steel pipes subjected to bending, the CRS performed more efficiently on reducing the crack growth rate along the length direction than along the depth direction, which might owe the crack-bridging effect.
- The surface crack growth is insensitive to the bond lengths which were applied in this paper. It indicates that their might exist an effective bond length which is less than 600 mm. Applying more layers of CFRP laminate can significantly promote the reinforcement effectiveness. The inversely wrapping pattern performed less effective than the other wrapping patterns.

The analysis, design and prediction of external surface crack growth in CRS reinforced steel pipes are being investigated by means of numerical and analytical methods as separate topics. These results will be analysed and presented in future studies.

ACKNOWLEDGEMENTS

The experimental investigation was supported by Overseas Expertise Introduction Project for Discipline Innovation – 111 project of Chinese Ministry of Education and State Administration of Foreign Experts Affair of P. R. China [grant number 444110356] and Department of Maritime and Transport Technology, Delft University of Technology, the Netherlands. The first author would like to acknowledge the China Scholarship Council, P. R. China [grant number 201606950024] for funding his research.

REFERENCES

Achour, A., Albedah, A., Benyahia, F., Bouiadjra, B. A. B. & Ouinas, D. 2016. Analysis of repaired cracks with bonded composite wrap in pipes under bending. *Journal of Pressure Vessel Technology*, 138, 060909.

API 2018. API SPEC 5L:Specification for Line Pipe. WASHINGTON, D.C.: American Petroleum Institute.

API, A. P. I. 2016. 579-1/ASME FFS-1: Fitness-for-service. *The American Society of Mechanical Engineers*.

ASME 2012. Pipeline Transportation Systems for Liquids and Slurries: ASME Code for Pressure Piping, B31.4-2012. American Society of Mechanical Engineers.

ASTM 1994. ASTM E647. Standard Test Method for Measurement of Fatigue Crack Growth Rates.

ASTM, E. 2015. 2899. *Standard Test Method for Measurement of Initiation Toughness in Surface Cracks Under Tension and Bending*, 3.

BS 2015. Petroleum, petrochemical and natural gas industries-Composite repairs for pipework-Qualification and design, installation, testing and inspection. BS EN ISO 24817:2015.: British Standards Institution.

Chen, J. & Pan, H. 2013. Stress intensity factor of semi-elliptical surface crack in a cylinder with hoop wrapped composite layer. *International Journal of Pressure Vessels and Piping*, 110, 77–81.

DNV 2008. Riser Integrity Management. *DNV Recommended Practice DNVGL-RP-F206*.

DNV 2017a. Assessment of flaws in pipeline and riser girth welds. *DNV Recommended Practice DNV-RP-F108*.

DNV 2017b. Assessment of flaws in pipeline and riser girth welds. DNV-RP-F108. Hovik, Norway: DNV GL.

Kabir, M., Fawzia, S., Chan, T., Gamage, J. & Bai, J. 2016. Experimental and numerical investigation of the behaviour of CFRP strengthened CHS beams subjected to bending. *Engineering Structures*, 113, 160–173.

Li, Z., Jiang, X. & Hopman, H. 2019a. Numerical analysis on the SIF of internal surface cracks in steel pipes reinforced with CRS subjected to bending. *Ships and offshore structures*, 1.

Li, Z., Jiang, X. & Hopman, H. Numerical investigation on surface crack growth in steel plates repaired with Carbon Fiber-reinforced Polymer. ASME 2019 38th International Conference on Ocean, Offshore and Arctic Engineering, 2019b. American Society of Mechanical Engineers.

Li, Z., Jiang, X., Hopman, H., Zhu, L. & Liu, Z. 2019c. An investigation on the circumferential surface crack growth in steel pipes subjected to fatigue bending. *Theoretical and Applied Fracture Mechanics*, 105, 102403.

Li, Z., Jiang, X., Liu, Z. & Hopman, H. Internal Surface Crack Growth in Offshore Rigid Pipes Reinforced With CFRP. ASME 2018 37th International Conference on Ocean, Offshore and Arctic Engineering, 2018. American Society of Mechanical Engineers, V004T03A022–V004T03A022.

Li, Z., Jiang, X. & Lodewijks, G. 2017. The latest development of reinforcement techniques on tubular joints. *Progress in the Analysis and Design of Marine Structures*, 783–790.

Liu, Z., Chen, K., Li, Z. & Jiang, X. 2017. Crack monitoring method for an FRP-strengthened steel structure based on an antenna sensor. *Sensors*, 17, 2394.

Woo, K. S., Ahn, J. S. & Yang, S. H. 2016. Cylindrical discrete-layer model for analysis of circumferential cracked pipes with externally bonded composite materials. *Composite Structures*, 143, 317–323.

Zarrinzadeh, H., Kabir, M. & Deylami, A. 2017a. Crack growth and debonding analysis of an aluminum pipe repaired by composite patch under fatigue loading. *Thin-Walled Structures*, 112, 140–148.

Zarrinzadeh, H., Kabir, M. & Deylami, A. 2017b. Experimental and numerical fatigue crack growth of an aluminium pipe repaired by composite patch. *Engineering Structures*, 133, 24–32.

Zhao, X. & Zhang, L. 2007. State-of-the-art review on FRP strengthened steel structures. *Engineering Structures*, 29, 1808–1823.

FIBRESHIP: A great step forward in the design and construction of lightweight large-length vessels

X. Martinez & J. Jurado-Granados
Facultat de Nàutica de Barcelona, UPC, Spain
International Centre for Numerical Methods in Engineering (CIMNE), Barcelona, Spain

A. Jurado
Técnicas y Servicios de Ingeniería S.L. (TSI), Madrid, Spain

J. Garcia
Facultat de Nàutica de Barcelona, UPC, Spain
Compass Ingeniería y Sistemas S.A. (Compass), Barcelona, Spain

ABSTRACT: The application of composite materials in large-length shipbuilding is attracting considerable attention thanks to their significant expected structural weight reduction, resulting in advantages such as bunkering savings, lower greenhouse gas emissions, higher payload cargo capacity, or suppression of corrosion problems, which might produce maintenance costs reduction. The H2020 European research project FIBRESHIP (EU 2020) aims to develop the technologies required to use composites in large length ships. The project covers all aspects required to make reality this new generation of ships, from materials to shipyard requirements; from the numerical tools required for designing fibreships, to the development of guidelines and rules that define how this design should be made. Current work presents the main work conducted in the field of material selection and characterization, numerical tools developments, and shipyard requirements.

1 INTRODUCTION

Seaborne trade is growing and it is expected to continue its growth at a pace of 3% in the upcoming years (UNCTAD, 2018) At the same time, environmental considerations have led the International Maritime Organization (IMO) to establish the target of reducing greenhouse gas emissions by a 50% by 2050, compared with 2008 (UNCTAD, 2018). Under this scenario, and because of the ambitious objective sought, it will be necessary to approach the problem with several strategies, from improvements in the propulsion system, to redefining and/or defining new maritime routes.

One of the approaches that can be followed to reduce the emissions in transportation is by minimizing the weight of the transport vehicle. This approach has been followed in the automobile sector by introducing aluminium and other alloys replacing steel, and in the aeronautical sector by incorporating composite materials in a high percentage of the structure in the latest designed aircrafts. The maritime industry has also embraced the use of composites in order to reduce the vessel weight, but it has only been applied to small crafts. Large length ships, which are responsible of most of the seaborne transport, are not using

composite materials extensively yet, losing an excellent opportunity to reduce greenhouse emissions.

The European research project FIBRESHIP aims to develop the technologies and procedures required to facilitate the construction of large length ships made of composite materials. Eighteen different partners are involved in the project: research institutions (CIMNE, ULIM, VTT and TWI), engineering companies (TSI, COMPASSIS, ATEKNEA and SOERMAR), shipyards (TUCO, iXblue and NAVRON), ship-owners (ANEK LINES, DANAOS, FOINIKAS and IOE) and classification societies (BV, LR and RINA), each one of them bringing their specific expertise to make feasible this new generation of FIBRESHIPS.

1.1 Project objectives and scope of work

The main objective of the FIBRESHIP project is to enable the building of the complete hull and superstructure of large-length seagoing and inland ships in FRP materials. To achieve this goal, it will be necessary to overcome several challenges, some of them listed hereafter:

- Assessment of the adequacy of composites for the construction of fibreships, considering aspects such

DOI: 10.1201/9781003216582-75

as mechanical strength, fatigue, fire-resistance and durability; and development of a selection criteria to find the optimal composite for this purpose.

- Development of specific joint solutions for the connection of large ship sections.
- Development and validation of different computational tools that can assist in the design and functional safety assessment.
- Elaboration of innovative procedures and guidelines for the design of FIBRESHIPS.
- Redesign of the structural configuration of the ship to adapt its strength mechanisms to the new material performance.
- Development and implementation of new production procedures. Shipyards must be adapted to the new fabrication processes.
- Definition of a decision support tool on life-cycle assessment of fibre-based vessels in order to reduce operational costs. This tool is based on inspection methodologies and SHM strategies, among others.
- Development of a cost-benefit calculator of the three targeted vessels in the project as well as a global business model.
- Definition of a FRP adoption roadmap in European shipping market considering end-user's satisfaction.

Furthermore, solving the different technical challenges is not enough, as the market calls for cost efficient solutions, and therefore the results of the project must prove significantly lower life cycle costs in order to ensure market uptake.

Each one of the abovementioned specific objectives is solved in the different tasks in which the project is divided. In particular, it will be conducted specific work related to the material selection and characterization, the development of numerical models to analyse composite materials, the implementation of numerical tools for ship design, the actual design of three different ship vessels (a container ship, a passenger ship and a research vessel) in order to evaluate the structural requirements of fibreships compared to their steel counterparts, the evaluation of shipyard requirements to manufacture this new generation of ships, the definition of new rules and design guidelines specially developed for composite vessels, and the cost-analysis evaluation to assess fibreship feasibility with technologies developed in the project.

Of all these tasks, current paper focus on the main work conducted in the first period of the project, consisting in material selection and characterization, development of numerical tools, and shipyard requirements.

2 MATERIAL SELECTION AND CHARACTERIZATION

2.1 Material selection procedure

The selection of the most promising composite materials to be used in a fibreship has been conducted based on an experimental campaign, using a two-phase approach. In a first phase, seven different matrix systems, reinforced with unidirectional glass fibres, have been tested and analysed. The matrix types considered have been a vinylester, a urethane acrylate, two epoxies, a bio-epoxy, a phenolic resin, and a thermoplastic. The results obtained, together with other relevant information have been weighted in order to classify the different resins and to obtain the two most promising subjects. Over a total punctuation of 110, it has been given a weight of 20 to the mechanical properties of the resin (accounting for flexural strength, stiffness and interlaminar shear strength), a weight of 50 to the manufacturing process (curing infusion temperature, post curing requirements, infusion capability and worldwide knowledge), and a weight of 40 to the material impact (material cost, fire retardancy, health impact for the worker, and possible recyclability). Following this approach, the two most promising resins to be used in a fibreship are vinylesters and epoxies, with a preference for the former one. The second phase consists in the fibre selection. Three different fibres have been tested: carbon, glass and basalt. Despite the better performance is provided by carbon fibres, the cost drives the selection criteria and glass fibres are the ones chosen.

Having decided that the most adequate material to be used in the main structural components of a fibreship is a glass fibre reinforced vinylester, it has to be stated that the selection of this material does not imply that other materials are disregarded. The versatility provided by composites to adapt to specific requirements will result in the use of other composites in different ship sections, if specific properties are required.

2.2 Material characterization

Once selected the main composite system for fibreships, an extensive experimental campaign has been conducted in order to acquire a good knowledge of the material performance, as well as to obtain the material parameters required for the composite simulation.

The experimental campaign has assessed the mechanical behaviour of the glass fibre reinforced vinylester (GFRV) composite selected by conducting the following tests: tensile strength and flexure test, both of them in the fibre direction, $0°$, and perpendicular to fibre direction, $90°$, Inter-Laminar Shear Strength test (ILSS, at $0°$), and fatigue tests at $0°$ and $90°$. All these tests have provided the stiffness and strength of the composite under monotonic and cyclic loads. In the experimental campaign it has also been measured the composite density and the volume fraction of the different composite components.

In parallel, the fire performance of the material has been also characterized conducting the following tests on it: Cone Calorimeter (CC), ThermoGravimetric Analysis (TGA), Micro-scale Combustion Calorimetry (MCC), Dynamic Mechanical Thermal Analysis (DMTA), Differential Scanning Calorimetry (DSC) and Transient Plane Source (TPS).

The material characterization conducted, besides providing the material properties, has defined a systemized procedure to obtain material data, and has also shown that the mechanical properties of GFRV are adequate for its use in a fibreship. On the other hand, the fire performance of the material, with a glass transition temperature around 100°C, forces to use insulation methods in those ship regions that must be fire-proof.

2.3 Connection technology

A fibreship will contain many connection types in terms of the material typologies to be jointed (monolithic to sandwich, monolithic to monolithic, etc.), the connection topology (butt connections, T connections, stiffener reinforcements, etc.), and the connection method (laminated, bonded, etc.). The project has conducted the experimental characterization of some of these connections, in order to assess their mechanical performance and to validate the numerical models developed.

FIBRESHIP project is also testing a new concept of "Dismantling on demand" bonded connection, in order to facilitate the structure dismantling at the end of its span life, or its repair if necessary. This consists in the introduction of a carbon insert inside the bonding material, as shown in Figure 1a. Taking advantage of the conductivity properties of carbon fibres, a current is applied to those in order to heat them. This heat will melt the bonging adhesive breaking apart the connection without the need of any other mechanical action (Figure 1b and 1c).

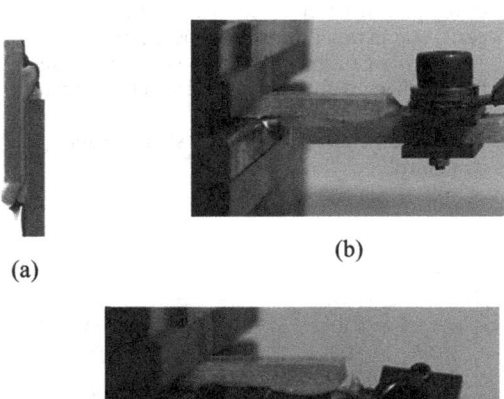

(a)

(b)

(c)

Figure 1. (a) Bonded connection with carbon insert (black line in the middle of the connection). (b) connection at the beginning of the test, when current starts being applied to carbon material. (c) connection failure at the end of the test.

3 NUMERICAL MODELS AND ANALYSIS TOOLS FOR FIBRESHIP DESIGN

3.1 Material models

A homogenization procedure obtains the material properties of the composite from a model that accounts for the internal structure of the material. This model can be either numerical or analytical. The numerical analysis of composites in FIBRESHIP project is conducted using the serial/parallel mixing theory (Jurado et al. 2019, Martinez et al. 2011, Rastellini et al. 2008), an analytical homogenization model specially developed for the analysis of long fibre reinforced composites. This formulation uses the constitutive equations of the composite components, fibre and matrix, to obtain the response of the composite. The model defines an iso-strain relation between the composite constituents in the fibre direction, and an iso-stress relation in the other directions. The final performance of the composite is obtained from the strain-stress state of the components, proportionally to their volumetric participation in the composite.

With this approach, it is possible to obtain the elastic response of the composite for any combination of fibre and matrix, as well as to capture all possible failure modes: fibre failure, transverse matrix cracking, delamination, etc. by just defining the material parameters of the composite constituents. In Figure 2 are compared the experimental and numerical results obtained for the GFRV material tested in the framework of the project. These results prove the capacity of the method developed to predict different material failures.

The serial/parallel mixing theory provides an excellent framework for the fatigue analysis of composites. This is done by modifying the constitutive equation of fibre and matrix introducing a reduction function, f_N, that accounts for the number of cycles, N, the average stress, σ_m, and the reversion factor, R. Equation 1 shows the modified constitutive equation, in which f(σ) is the equivalent stress and K is the stress threshold.

$$f(\sigma) - f_N(N, \sigma_m, R) \cdot K \leq 0 \qquad (1)$$

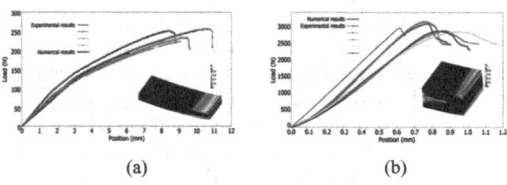

(a)

(b)

Figure 2. Comparison of the numerical and experimental results. (a) Force displacement graph of the Transverse Flexure Test and matrix damage at the final step, and (b) Force-displacement graph for the Interlaminar Shear Strength Test and matrix damage at the final step.

3.2 Analysis tools

In order for fibreships to become a reality, it is necessary to have the technology ready, but also to have the adequate analysis tools for the design of this new generation of ships. With this aim, the project is working in a new version of the software package Tdyn (Compassis, 2020), which will have several new developments to account for the specifies of these vessels. Tdyn is Multiphysics finite element software with seakeeping functionalities, though the SeaFEM package, and a seamless coupling with the structural solver Ramseries. Among the different developments made to the new version of Tdyn for FIBRESHIP project, it is worth to mention:

- Implementation of the serial/parallel mixing theory for a correct mechanical characterization of composite materials, and development of a new Graphical User Interface (GUI) for an easy definition of the composite laminates.
- Development and implementation of a fire and smoke propagation and collapse assessment tool. This couples the computational analysis of fire and smoke propagation, together with a collapse assessment of the FRP structure (based on a thermo-mechanical analysis).
- Development of a full 3D hydro-elasticity solver, consisting in the coupling of a time-domain radiaton/diffraction seakeeping analysis solver (2nd order) with a dynamic FEM structural solver.
- Development of an advanced hull girder model that applies seakeeping analysis tools to a 1D finite element model of the ship. The results obtained in the 1D model can be exported afterwards to a 3D model of the ship for its structural assessment.
- Development of a Structural Health Monitoring system based on the analysis of the natural frequencies of a given structure, and their variation in case of structural damage.

The validity of the numerical tool developed, Tydn, is currently being proved with its use for the design of the different vessel types analysed in the framework of the project.

4 SHIPYARD CONSTRUCTION

The development of fibreships it is not only a technical challenge in terms of materials or design, but also in terms of manufacturing. Most of the current mid- and large shipyards are oriented to construct steel-based vessels while the existing shipyards focused on the building of fibre-based vessels are not prepared for such ships as the ones analysed in the FIBRESHIP project. For all of them, an adaptation to develop this activity of large-length vessel construction should be addressed, including needed new production strategies and building techniques.

In order to assess the shipyard requirements and to evaluate the different problems that can be associated to the construction of fibreships, the project has built a section of one of the vessels designed for the project. The construction of this vessel section has also been used to validate the performance of the materials selected, the manufacturing needs for different connection configurations, and the feasibility to include different non-structural elements in a fibreship such as embedded windows, fire insulation layers, connections, and other elements that will be attached to the structure. Different sections of the demonstrator will be used afterwards to evaluate the fibreship performance in aspects such as noise insulation, fire performance and connection strength. The demonstrator has a dimension of 11m x 11m x 8.6m, and has been constructed at iXblue shipyard. Figure 3 shows a frontal view of it.

The construction of the demonstrator shown in Figure 3, as well as the thorough analysis of the production process required for a large fibreship, has provided the requirements that must be fulfilled by a shipyard prepared to build large length fiberships. Among they, some are specified in the following list:

- Shipyards will have to automatize most processes in order to increase production speed and to reduce the amount of quality controls required.
- Work will have to be divided among shipyards, that will specialize in manufacturing specific components, in order to improve their efficiency
- Fibreships will require a modular construction. This implies logistics challenges in terms of transport and storage.
- New manufacturing procedures will have to be developed in order to construct large composite parts, and in order to connect these parts.
- To build large Fibreships there is a need of qualified workers, with standardized specific formation in composite manufacturing.
- There will be also a need for a standardized quality control.

Figure 3. Fibreship demonstrator.

5 CONCLUSIONS

Fibreships can be a reality. The project has demonstrated that there are materials that can be used for manufacturing fibreships and that these materials can be properly characterized with numerical formulations. The project has also developed numerical tools to conduct a reliable design of fibreships. It has also been shown that shipyards will have to change and, even the ones that already construct with composite materials, will have to adapt some of their manufacturing procedures to overcome such challenge. However, this adaptation is possible and the project has shown the path to do it.

Considering the innovative developments and findings obtained in FIBRESHIP project as well as the evident advantages of composites in shipbuilding, it is possible to continue engaging marine stakeholders and regulatory bodies to enable the design and shipbuilding of large-length vessels by means of new regulations or adaptation of the current ones such as IMO MSC.1/Circ.1574 (IMO, 2017) and SOLAS (IMO, 1974).

Therefore, the project has provided solutions for most of the challenges posed, and has proved that it is possible to find solutions for those that have been detected. Under this scenario, it can be concluded that fibreships can be soon a reality and, with them, it will be possible to reduce the environmental impact of maritime transport, improving its sustainability.

ACKNOWLEDGEMENTS

This work has been supported the European Comission's Horizon 2020 research and innovation program under grant agreement No. 723360 (FIBRESHIP project).

REFERENCES

Compassis. http://www.compassis.com/compass [Accessed: January-8-2020].

European Union's Horizon 2020 research and innovation programme under grant agreement nº 723360 "Engineering, production and life-cycle management for the complete construction of large-length FIBRE-based SHIPs". http://www.fibreship.eu/about/

International Maritime Organization (IMO). 2017. MSC.1/Circ.1574. Interim guidelines for use of fibre reinforced plastic (FRP) elements within ship structures: fire safety issues.

International Maritime Organization (IMO). 1974. International Convention for the Safety of Life at Sea (SOLAS)

Jurado Granados, J. et al. 2019. Numerical and experimental procedure for material calibration using the serial/parallel mixing theory, to analyze different composite failure modes. *Mechanics of Advanced Materials and Structures*. DOI: 10.1080/15376494.2019.1675106

Martinez, X., Rastellini, F., Oller, S, Flores, F., Oñate, E. 2011. Computationally optimized formulation for the simulation of composite materials and delami-nation failures. *Composites Part B: Engineering, vol. 42 (2)*. DOI: 10.1016/j.compositesb.2010.09.013

Rastellini, F., Oller, S., Salomon, O., Oñate, E. 2008. Composite materials non-linear modelling for long fibre-reinforced laminates: Continuum basis, com-putational aspects and validations. *Computers & Structures, vol. 86 (9)*, DOI: 10.1016/j.compstruc.2007.04.009

United Nations Conference of Trade and Development (UNCTAD), 2018. Review of Maritime Transport 2018. New York, USA, ISBN 978-92-1-112928-1.

Developments in Maritime Technology and Engineering – Guedes Soares & Santos (eds)
© 2021 Copyright the Author(s), ISBN 978-0-367-77376-2

Analytical approach for global fatigue of composite-hull vessels

J.P. Tomy, L. Mouton & S. Paboeuf
Composite Materials Section, Department of Expertise, Bureau Veritas Marine & Offshore, Nantes, France

A. Comer, A.K. Haldar & A. Portela
Bernal Institute (Composites), Irish Composites Centre (IComp), School of Engineering, University of Limerick, Ireland

ABSTRACT: Multi-axial strain fatigue is a complex subject that is rather a research topic than applied design subject even in isotropic steel material. In a layered and non-homogeneous structure such as the composite hull of a ship, multi-axial fatigue analysis becomes non-realistic and computationally non-feasible to apply in a preliminary design phase. This study presents a pragmatic approach and the associated results developed with the ambition of having fatigue design tools for design review of hull structural components. Furthermore, the validation of such an analytical characterisation would reduce the number of fatigue tests required to be performed for each new design; thereby reducing the duration and expense of the fatigue testing campaign. The study includes two parts – an analytical part (led by Bureau Veritas) where the fatigue assessment formulations are defined, and an experimental part (led by University of Limerick) used to feed the models, support the hypothesis, and validate the results. The focus is set on defining practical tools which can be applied for the global fatigue analysis of composite-hulled ships.

1 INTRODUCTION

Composite laminates are anisotropic materials, with different structural behaviour along different directions. The commonly used simplification for analytical methods is to consider the material as orthotropic, defining different mechanical properties in three orthogonal axes. While such a simplification is commonly used to characterise the linear, static mechanical behaviour of the material, the application of this analytical methodology to non-linear, multi-axial phenomenon is not prominent in the literature. One such phenomenon is the fatigue behaviour of the composite laminates.

While performing the technical design and analysis for medium-sized composite-hull ships, the structural analysis also includes a global fatigue analysis. Whereas the methodology for global fatigue analysis of metallic ships is well-defined in the classification society rules, the applicability of the same to composite-hull ships requires additional considerations. In addition to the complexity posed by the non-linearity of the fatigue phenomenon, and the complexity in predicting the loads on the ship's hull in a dynamic sea environment, there are additional complexities introduced by the anisotropic behaviour of the composite material, as described in the previous paragraph.

Advances analysis methodologies simulate the progressive evolution of the fatigue damage using solid finite elements (Nishikov & Makeev, 2011). Several scales may be found - representation of the matrix and fibres, or homogeneous properties for each plies. If the former is the most capable to represent the actual physics of fatigue in composite, its cost makes it rather a research or calibration method. Solid homogeneous orthotropic elements are more feasible but still computationally expensive and barely conceivable for global analysis of the ship structure. Furthermore, during design stage, wherein the structural design is still an iterative process, the focus needs to be on developing computationally fast methods, without serious compromise on the accuracy. This is particularly true in an industry where series production is not the norm.

In isotropic steel material, the preliminary fatigue design is performed based on classical S-N curves approach. However, the applicability of the same is limited in the case of multi-axial strain fatigue in a layered and non-homogeneous structure such as a composite laminate. The aim of this paper is to provide a pragmatic approach that can be used in the preliminary design and review of hull structural components. The approach is based on the classical S-N curve approach of metallic hull vessels; but incorporates the anisotropic mechanical behaviour of composite laminates.

DOI: 10.1201/9781003216582-76

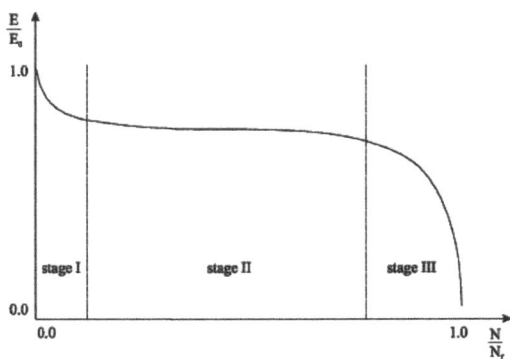

Figure 1. Typical evolution of stiffness over the course of fatigue life of a wide range of fibre-reinforced composite materials.

2 FATIGUE IN COMPOSITES

The evolution of fatigue on composites is not linear in nature. For vast majority of fibre-reinforced composite materials, the fatigue evolution can be divided into three stages, as can be seen in Figure 1 (Paepegem & Degrieck, 2002). An initial phase with a rapid stiffness reduction is followed by an intermediate region with approximately linear reduction in stiffness. The final stage is characterised by stiffness reduction in abrupt steps, ending in specimen fracture. The paper further investigates on how these three stages of stiffness degradation can be modelled.

2.1 Fatigue modelling approaches

Three fatigue modelling approaches for fibre-reinforced polymers can be distinguished (Degrieck & Paepegem, 2001). In an increasing order of complexity, as well as accuracy in representing the fatigue evolution mechanism, these are: (i) fatigue life models, which use S-N curves or Goodman-type diagrams and predict only the final failure of the material; (ii) phenomenological models, which describe the fatigue evolution in terms of an empirical relationship with a phenomena such as stiffness degradation; and (iii) progressive damage models, which use one or more damage variables, such as matrix cracks, to quantify the extent of damage at each iteration.

The classical approach to developing a fatigue damage model is by using the S-N curve approach. This is the most basic fatigue damage model, based on the fatigue life of the material. It does not take into account the actual damage mechanism; the fatigue life of the material is experimentally computed for different stress ranges, and the same is then plotted as an S-N curve which represents the fatigue performance of the material. However, the computations are much simpler and efficient in terms of computational time and effort.

While considering anisotropic materials such as multi-layer composite laminates, the actual mechanism involved is much more complex than that would be for a homogeneous, isotropic material like steel. The interaction between the composite layers can mean that, even if one ply is damaged, the rest of the plies can take extra load. The gradual deterioration in a fatigue damage leads to a continuous redistribution of stresses. Moreover, within a single ply, the fibre and the resin behaviour are entirely different, making it difficult to extend the results from one load case to another.

2.2 Modelling multi-axial fatigue

The fatigue state in composite laminates is a multi-axial phenomenon - even with an orthotropic assumption. Phenomenological models and progressive damage models take into account the fact that the prediction of the evolution of fatigue of the composite structure depends on the 'path' of successive damage states. Using such models allows to identify the state of fatigue during each computational iteration.

However, the procedure requires to simulate each loading cycle numerically. In a layered and non-homogeneous structure such as the composite hull of a ship, multi-axial fatigue analysis using such models become non-realistic and computationally non-feasible to apply. When considering that the minimum fatigue life expected for these application cases fall in the range of 10^9 cycles, such computations become cumbersome, computationally expensive and impractical at the preliminary design stage.

In fatigue design of materials, it is not necessary to understand and follow this 'path' of successive damage states. More importantly, the designer needs to have an idea about the number of cycles that the structure can withstand, when subjected to a sequence of fluctuating stresses. In reality, uncertainty factors which affect the fatigue behaviour are not accounted by any analytical and numerical simulations of the fatigue phenomenon. Some of these uncertainty factors include ageing effects, uncertainty in determining the loads, and uncertainty in manufacturing processes.

Thus, the objective for fatigue design of composite-hull vessels should be to have a methodology that is more analytically developed than the classical S-N curve, but not as computationally intensive as the phenomenological models. The approach proposed within this paper intends to fall in between these two categories.

3 PLY-BY-PLY FATIGUE ANALYSIS METHODOLOGY

Analytical models on composite laminates typically use an orthotropic assumption. Even though the actual state of the material is anisotropic, these models assume that all the mechanical properties can be simplified into three orthogonal planes – parallel to the fibre direction (L-direction), perpendicular to the fibre direction (T-direction) and through the thickness of the laminate (N-direction). While considering unidirectional plies, the properties in the L-direction are dominated by the fibres, and those in the T-direction and N-direction by the matrix.

A ply-by-ply analysis methodology assumes that the behaviour of each ply (or layer) within the laminate can be characterised separately. The individual stresses on each layer depend on the mechanical properties (moduli) of the particular layer, its position in the laminate thickness and the global strain of the laminate. For each ply, the elastic stresses in the L and T directions as well as the shear stress in the L-T plane can be thus obtained. This gives the two-dimensional stress state of the ply, which can subsequently be used to determine its mechanical behaviour. This kind of design philosophy is used for static analysis of composite laminates (for example NR546 BV Rules for Composite Materials (Bureau Veritas, 2018), and the idea is to extend such an approach for cases when the laminate is subjected to fluctuating stresses, i.e. for fatigue loading.

3.1 Still water loads

The loads applied on a vessel does not depend on its material of construction (apart from the lighter displacement of the composite ship). Hence, the loading conditions and the still water loads for fatigue analysis of composite-hull vessels should be the same as that for a steel vessel.

The ship should be considered in the most demanding loading conditions with regard to its cargo loading, ballast and any other variable that has an effect on its displacement. Loading conditions that provide maximum and minimum drafts, still water bending moments or load on the inner structure upon the structural configuration of the ship have to be identified. One such typical example is the alternate cargo loading condition on tankers. Further details can be found in BV Rules for the Classification of Steel Ships – NR467, Part D, Chapter 4, Section 3 (Bureau Veritas, 2020).

The fatigue analysis performed is for the intended service life of the ship. Full loads and ballast conditions are typically selected; specific rules should be used as per the ship type.

3.2 Wave loads

Typically, the hull girder loads are obtained by a summation of the still water loads and wave loads (Bureau Veritas, 2018). Still water loads are induced by the longitudinal distribution of the lightship, which effectively refers to the loading conditions described in Section 3.1. Generally, a fully loaded departure condition corresponds to a sagging condition; and a departure condition with empty tanks corresponds to a hogging condition.

Wave loads are induced by the encountered waves from different heading directions. Classification Society rules enable the designer to assess rapidly the waves encountered by the ship. Usually, this is done for a representative wave with a probability of occurrence of 10^{-8} or 10^{-5}, depending on the rules.

3.3 Laminate loads

From the hull girder loading, the next step is to ascertain the design loads on the structural component under consideration. In steel ships, the preliminary approach is to consider the shear force and bending moment acting on the cross-section under consideration. Thereafter, thin wall hollow sections beam theory is used to calculate the longitudinal stresses and shear stresses in the plating and stiffeners of the considered section.

A similar approach is used for composite-hulls as well. The plating made of laminate is homogenized in order to be able to calculate the inertia of the compo-site hull section. Thus, the strain of each plate and stiffener is calculated. Within this work, this is accomplished using BV MARS software. It should be noted that the approach is only true when the deformation of the section remains linear. In case of an abrupt structural discontinuity (such as the hull-superstructure joint) or in cases where the vertical walls of the ship may deform significantly under shear loads, the hypothesis is not valid anymore. To approach conservatively, two calculations may be done - one with superstructure, and one without.

3.4 Ply-by-ply stresses

From the strain distribution on the laminate, the orthotropic assumption of composite multi-layers can be used to obtain the stresses at each ply. ComposeIT a freeware from Bureau Veritas, enables to rapidly compute the stress in each plies based on its stiffness.

Using the orthotropic assumption, the stresses in the L and T directions (σ_L and σ_T respectively) can be obtained for each ply. The shear stress in the L-T plane (τ_{LT}) can also be obtained using the in-plane shear strain. Such an analysis can be done for various wave load cases, including hogging, sagging, torsion, shear loads and horizontal bending. The difference between the two load cases – wave hogging and wave sagging - would provide the stress range that the ply is subjected to.

It is to be noted that this stress range corresponds only to the characteristic wave case that was considered in Section 3.2. That means, for the particular structural design component under consideration, this value corresponds to the stress range that the component endures with a 10^{-5} (or 10^{-8}) probability of occurrence in its service life. This result then needs to be extended to the entire loading cycle of the ship throughout its service life.

3.5 Loading histogram

To assess the waves encountered by the ship throughout its service life (usually considered as 25 years), rules follow generally a probabilistic approach, with a probability distribution function defining the probability of occurrence of waves with specific wave heights. The total number of encounter cycles is typically in the range of 10^8, corresponding to the service life of 25 years.

The most commonly used probability distribution function to define the encounter waves for a ship in service is the Weibull distribution function (International Association of Classification Societies, 1999). This function is defined by a shape parameter and a characteristic value. The shape parameter is usually defined in the Class rules. The characteristic value is provided by the response of the ship to a characteristic wave corresponding to 10^{-5} probability of occurrence.

The wave loading corresponds to a characteristic wave which corresponds to the wave that is encountered by the ship with a 10^{-5} probability within its service life. The ply-by-ply stresses computed in the previous step, corresponds to the stress range for this characteristic wave. The stress range loading histogram can then be extrapolated for the entire service life using the Weibull law.

For each ply, the probability distribution of stress ranges for the entire service cycle can be thus obtained. For every ply, three different distributions can be obtained – one each for σ_L, σ_T and τ_{LT}.

3.6 S-N curves for unidirectional ply

The stress distribution obtained at the ply level then needs to be compared with the fatigue behaviour of the material, vis-à-vis the S-N curves obtained experimentally for unidirectional (UD) plies or woven roving (RV) plies. A detailed discussion about this campaign is further provided in Section 4.

In order to separate the fatigue behaviour of the anisotropic laminate into orthogonal directions, the fatigue behaviour needs to be separately obtained for the different directions – tensile fatigue in L and T directions and the shear fatigue in L-T plane. The tensile fatigue in L and T directions can be characterised by conducting fatigue experiments on UD 0-degree and UD 90-degree plies respectively. The shear fatigue behaviour is more complicated to ascertain experimentally, and also to combine its effect with the other directions. The fatigue damage calculation proposed within this paper does not consider the effect of shear; but, the framework provided is such that this can be directly included as an additional calculation in a later phase.

3.7 Damage calculation from S-N curves

Once the S-N curves (Section 3.6), and the stress loading histogram (Section 3.5) are established, these can be compared to compute the percentage of

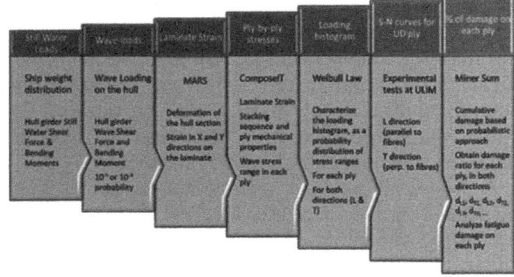

Figure 2. Global fatigue analysis methodology.

fatigue damage in the ply. To achieve this comparison, the Palmgren-Miner rule is used, which assumes accumulation of fatigue damage on the material irrespective of the order in which the loading cycle is applied.

For each histogram of stress range (i), the number of cycles endured within the service life (n_i) is read from the probability distribution function; and the number of endurable stress cycles (N_i) obtained from the S-N curve. The percentage of damage (d) on the ply, according to Palmgren-Miner Rule is then:

$$d = \Sigma \ (n_i/N_i)$$

The percentage of damage is obtained separately for the L-direction (dominated by fibre fatigue) and T-direction (dominated by matrix fatigue). This is done for every ply. Summarising the results would provide an insight about the criticality of fatigue of composite materials for a large composite ship, which is the critical ply for fatigue and also about whether the fatigue accumulation happens more in the stiff fibres (L-direction) or in the matrix (T-direction).

3.8 Global fatigue analysis methodology

A summary of the global fatigue analysis methodology, explained from Sections 3.1 to 3.7, is pictorially represented in Figure 2.

4 EXPERIMENTAL CHARACTERISATION OF S-N CURVES

The experimental campaign was aimed at characterising the fatigue behaviour of each ply in a laminate stacking sequence. It is common that such stacking sequences include the same fibre architecture, but oriented in different directions, in order to have a laminate that can sustain loads in different directions. However, when using a ply-by-ply approach, laminate loads are decomposed to stresses in the directions parallel and perpendicular to the fibre direction in each layer; and thus, can be compared to one fatigue S-N curve of the same fabric, in the relevant orientation. The objective is thus to obtain the S-N curves for the fibre direction (L direction) and perpendicular to the fibre direction (T direction).

To achieve this, glass fibre/vinylester laminate samples consisting of four unidirectional plies ($0°$ and $90°$) were manufactured by vacuum assisted liquid resin infusion. Constant amplitude tension-tension fatigue tests at $R=0.5$, were conducted at the University of Limerick. For each stress range, the number of cycles to failure were observed and then plotted as the S-N curve.

It is important to mention that the fatigue damage recorded experimentally, refers to the final failure of the laminate. Referring back to Figure 1, this represents then end of Stage III of the fatigue evolution. However, the fatigue mechanism in stage III seem to be dominated by non-linear phenomena, such as a transfer to local damage progression wherein the first initial fibre fracture is leading to abrupt collapse of multiple fibres(Paepegem & Degrieck, 2002).

Experimental efforts are in progress to investigate the practicality of restricting the fatigue damage limit to a point within which a linear prediction of the fatigue damage is possible. This is done by attempting to identify the end of stage II of the fatigue life through residual stiffness measurements at different stages of the experiment, and/or detection of damage using, for example, ultrasonic scanning and/or acoustic emission sensors.

5 APPLICATION CASE – FRV DESIGN FOR FIBRESHIP PROJECT

Within the European research project FibreShip (H2020 programme), the technical design and analysis for medium-sized composite-hull ships are carried out. The structural analysis also includes a global fatigue analysis, and the comparison of the behaviour with similar steel-hulled vessels. One such application case is the design of a composite-hull Fishing Research Vessel (FRV).

The proposed global fatigue analysis methodology was applied to the keel plate of the FRV, at the midship section. The demonstrator model of the FRV composite hull is shown in Figure 3, and the location of the laminate is clouded in this figure. The stacking sequence for the keel plate comprises of 45 layers of glass/vinylester rovings and unidirectional plies. It includes fabrics in $0°$, $+45°$, $-45°$ and $90°$ orientation.

The considered ship has a closed section hull (not a U-shape section container carrier) and the hypothesis is to consider head seas only. Not being a cargo ship, the still water loading varies very little compared to the wave loadings. Thus, a single loading condition is considered.

At the time of the study, no finite element results were available, thus, it appeared that the longitudinal bulkheads were quite soft in terms of shearing. Thus, the decks above main deck were not considered in the hull girder calculation, making a conservative approach. The same process could be derived from finite elements calculations including the actual participation of each deck.

5.1 S-N curve results

The glass/vinylester laminates were subjected to constant amplitude fatigue tests, at a stress ratio ($R=\sigma_{min}/\sigma_{max}$) of 0.5, and frequency of 4Hz. The results are plotted in Figure 4 and Figure 5 below.

5.2 Global fatigue analysis results

The global fatigue methodology, proposed in Section 3, was applied to the FRV design. The loading conditions and hull girder loads were computed as per Bureau Veritas (BV) rules (Bureau Veritas, 2018). The laminate loads were obtained using BV MARS software. Ply-by-ply stresses were obtained using the BV tool, ComposeIT. The probability distribution of the stresses and the subsequent fatigue damage calculations were done using the spreadsheet features of Microsoft Excel. It is pertinent to mention that all the tools used in this calculation sequence are available as freeware (Bureau Veritas Marine & Offshore, 2020).

Figure 3. FibreShip demonstrator model of FRV Composite Hull.

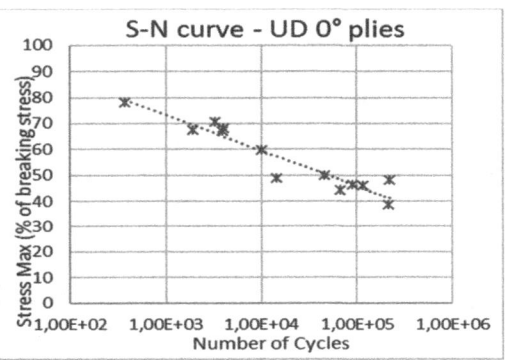

Figure 4. Normalized fatigue S-N curve for UD0° plies.

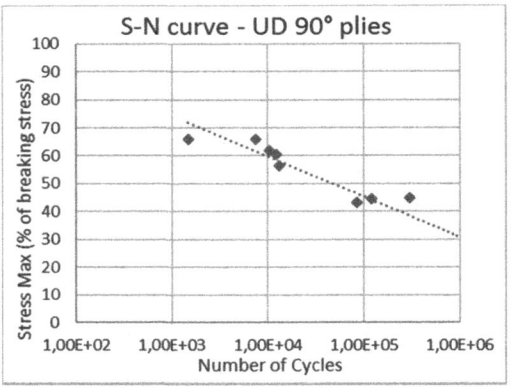

Figure 5. Normalized fatigue S-N curve for UD90° plies.

Fatigue calculations are done for all 45 plies. For the sake of brevity, only representative results are presented here. The loading histograms and comparison to the S-N curve for 0°, 45° and 90° plies are shown in Figure 6 through Figure 11. The cumulative damage for these 3 plies are then computed using the Palmgren-Miner rule and the result is tabulated in Table 1.

The data shown is normalized and plotted with logarithmic scale on both axes; however, the relative magnitudes can still be compared. It can be observed that the most critical damage, in this particular case, occurs for the laminate layers oriented at 90 degrees. The fatigue effect is critical in the T-direction for all the layers, indicating that the multi-axial fatigue is dominated by the fatigue of the matrix.

Another interesting point to note is that, even though the fatigue behaviour (S-N data) is more critical for the matrix (T-direction), the loading histogram is also less critical for these cases. The laminate loads are higher in the 0° plies for the L-direction, and in 90° plies for the T-direction. This is expected, as both these directions correspond to

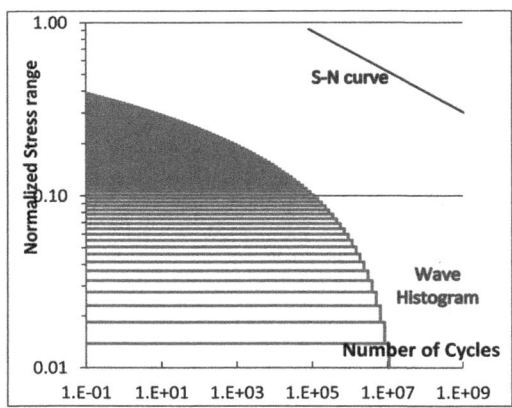

Figure 7. Fatigue damage calculation in L-direction for 45° plies.

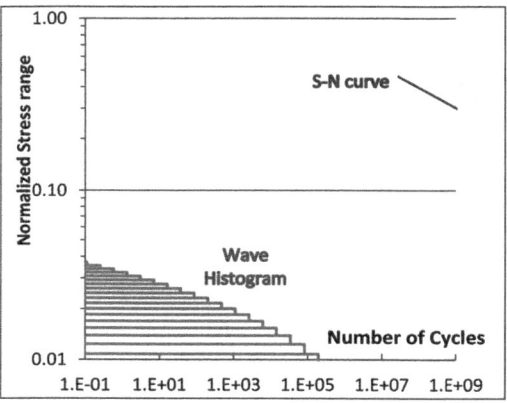

Figure 8. Fatigue damage calculation in L-direction for 90° plies.

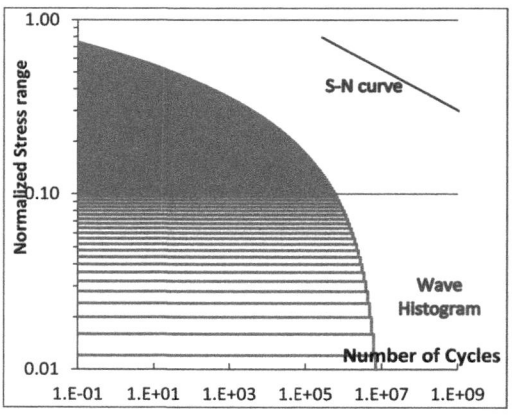

Figure 6. Fatigue damage calculation in L-direction for 0° plies.

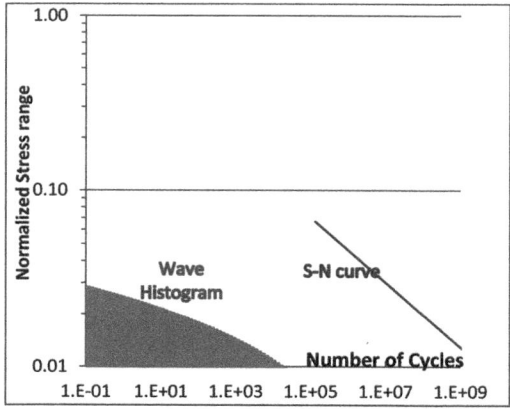

Figure 9. Fatigue damage calculation in T-direction for 0° plies.

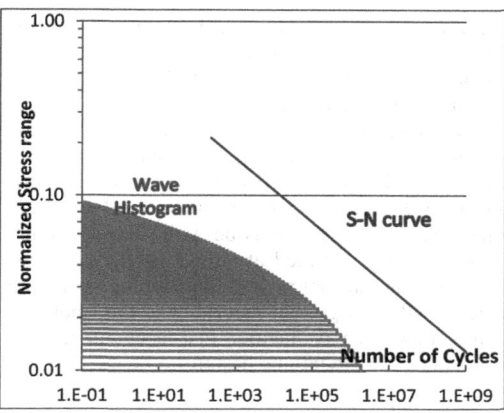

Figure 10. Fatigue damage calculation in T-direction for 45° plies.

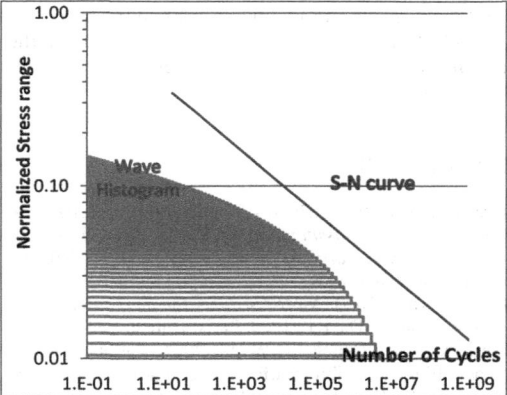

Figure 11. Fatigue damage calculation in T-direction for 90° plies.

Table 1. Cumulative damage for Glass/Vinylester plies.

Material	Orientation	L-Damage	T-Damage
Glass/Vinylester	0°	1.42e-04	1.16e-04
Glass/Vinylester	45°	4.23e-07	6.64e-02
Glass/Vinylester	90°	1.31e-15	8.83e-01

the longitudinal direction on the ship co-ordinate system. Even though the S-N behaviour is critical in the T-direction, the relative magnitude of loads experienced in this direction is also less.

6 LIMITATIONS

(a) *Fatigue Progression Mechanism*: The objective of the global fatigue analysis methodology, as laid out in the end of Section 2, is to provide an analytical framework that accounts for the effects of multi-axial fatigue of composite materials. In order to simplify the computational complexity, the methodology proposed does not consider the entire fatigue progression mechanism, neither through progressive damage models nor through phenomenological models. The framework, however, provides a scientifically based methodology for predicting the fatigue life of the laminates. In practical design, the focus is on identifying weak-points in the design, and not the exact evolution of the fatigue, in order to rectify them accordingly.

(b) *Inter-laminar effects*: Inter-layer fatigue effects are not accounted for. The methodology is aimed at finding the critical plies for fatigue damage, and modify the design accordingly. Design of composite structure is still to be done in order to minimize the inter-laminar stresses. Further work is schedule to produce in the rules the inter-laminar fatigue effect.

(c) *In-plane shear effects*: The methodology currently does not account for the in-plane shear effects. This is because it is difficult to experimentally obtain the fatigue S-N curve for pure shear samples. From the results in Table 1, it is observed that the critical fatigue behaviour is observed in the T-direction of the fibres. The inclusion of in-plane shear calculations would only be an enhancement to the methodology already proposed.

(d) *Effect of stress ratios*: The methodology proposed considers that the S-N behaviour of the material, obtained experimentally for a specific stress ratio, is applicable for all stress ratios. Previous studies have shown that the same is not entirely applicable for non-homogeneous materials such as composite laminates (Kadi & Ellyin, 1994). However, to perform fatigue experiments for multiple stress ratios is time-consuming and impractical. Hence, the methodology should be able to incorporate an equivalence of the S-N behaviour for multiple stress ratios. The most practical idea would be to use the S-N curve for the most conservative stress ratio, separately for tensile and compressive behaviour. The S-N curve data for $R=0.1$ and $R=10$ should provide reasonably conservative estimates of the fatigue behaviour across multiple ranges of stress ratios. Alternatively, techniques that enable to extrapolate the fatigue behaviour from one stress ratio to another can be utilised. One such method, using the Goodman diagram approximation, is under development.

(e) *Compressive fatigue*: In the presented results, the compressive fatigue of the laminate is considered to be the same as the tensile fatigue. This is not entirely correct and compressive fatigue is generally known to be more severe. Techniques to incorporate separate tensile and compressive fatigue behaviour are also in progress.

(f) *Effect of experimental scatter*: It should be noted that the present results do not take into account the experimental scattering in-bound to fatigue testing. This should be taken into consideration by applying methods as defined in standard ISO12107.

7 CONCLUSIONS

The fatigue behaviour of large complex structures is a long-term phenomenon and hence is difficult to predict through simulations. Progressive damage models can indicate how the fatigue damage is evolving over the life of the material, but the designer is more interested in estimating the fatigue life of the structure. It is to this effect that most of the industrial applications for fatigue design utilise fatigue-life models such as an S-N curve approach.

In the classical approach, these methods use the homogeneity of the material to ascertain the fatigue life, i.e. the fatigue behaviour is considered the same in all directions through the material. The same is not applicable for anisotropic materials; and for composite materials, these methods are to be implemented considering the orthotropic assumption.

Within this paper, the classical S-N curve approach has been further developed to make it applicable for orthotropic materials. The fatigue behaviour is separately considered in the orthogonal directions. A ply-by-ply approach for composites is scientifically founded; and the framework proposed, couples this to the classical S-N curve approach. The outcome is to have a global methodology that can be used in the fatigue design of composite-hull ships.

The method, however, needs to be validated further with experimental results. It is expected to be a conservative approach; and hence pragmatic as well. Some effects still need to be developed, such as the impact and combination of in-plane shear fatigue with L and T direction fatigue, consideration of actual compressive fatigue behaviour, extrapolation of results for multiple stress ratios, and the general applicability of linear accumulation of fatigue damage. Such works are in progress and are aimed to enhance the accuracy of this generic methodology.

It shall also be remarked that the failure of a ply in the direction transverse to the fibre does not mean failure of the laminate. However, it may prepare the laminate for further compressive failure through inter-laminar delamination. If this might be acceptable for extreme loading cases, it may be different for fatigue. This is to be considered by engineering judgement. Authors remind that if a progressive model may allow redistribution of stress, the delamination under inter-laminar shear loads, or micro-buckling under compressive loads are still issues to be accounted for.

Future work is aimed at validating the approach to assess the accuracy and developing a design review methodology that incorporates this approach with suitable safety factors that account for the uncertainties in the fatigue design.

ACKNOWLEDGEMENTS

The work was undertaken as part of the FibreShip project. The project has received funding from European Union's Horizon 2020 research and innovation programme under grant agreement N° 723360. The authors would also like to thank the support received from other partners of the Fibre-Ship consortium.

REFERENCES

Bureau Veritas, 2018. *NR600 - Hull Structure and Arrangement for the Classification of Cargo Ships less than 65 m and Non Cargo Ships less than 90 m*. DT R03 ed. Paris: Bureau Veritas.

Bureau Veritas, 2020. *NR467 - Rules for the Classification of Steel Ships*. DT R12 E ed. Paris: Bureau Veritas.

Degrieck, J. & Paepegem, W. V., 2001. Fatigue damage modelling of fibre-reinforced composite materials: review. *Applied Mechanics Review*, 4(54), pp. 279–300.

International Association of Classification Societies, 1999. *IACS Rec. No.56 - Fatigue Assessment of Ship Structures*, s.l.: IACS.

Kadi, H. E. & Ellyin, F., 1994. Effect of stress ratio on the fatigue of unidirectional glass fibre/epoxy composite laminae. *Composites*, 25(10), pp. 917–924.

Nishikov, Y. & Makeev, A., 2011. *Fatigue Life Assessment of Composite Materials*. Jeju Island, Korea, 18th International Conference on Composite Materials.

Paepegem, W. V. & Degrieck, J., 2002. A new coupled approach of residual stiffness and strength for fatigue of fibre-reinforced composites. *International Journal of Fatigue*, 24, pp. 747–762.

Subsea structures - Pipelines

Developments in Maritime Technology and Engineering – Guedes Soares & Santos (eds)
© 2021 Copyright the Author(s), ISBN 978-0-367-77376-2

Reliability assessment of corroded pipelines with different burst strength models

U. Bhardwaj, A.P. Teixeira & C. Guedes Soares
Centre for Marine Technology and Ocean Engineering (CENTEC), Instituto Superior Técnico, Universidade de Lisboa, Lisbon, Portugal

ABSTRACT: This paper assesses the performance of available burst strength prediction models of pipelines. 25 burst strength models for corroded pipelines are analysed and grouped into three main categories based on their formulation. The models' performance is first assessed deterministically by estimating and comparing the burst pressure of corroded pipelines predicted by each model in each category. The model predictions are compared for increasing normalised corrosion defects, taking the predictions of a code strength model as reference. Then, reliability analyses are conducted for limit state functions formulated in terms of the 25 burst strength models by Monte Carlo simulation to check their probabilistic performance and conservatism in each formulation category. Finally, measures of model uncertainty are derived to describe the variability in the structural safety among the models. The objective is to provide information on the available models applicable for integrity management of corroded pipelines.

1 INTRODUCTION

Subsea pipelines being one of the important and the most efficient mode of gas, oil and other hydrocarbon transportation over long distances require attention for its safe operation. The pipelines endure harsh environmental and varying operating conditions that pose several integrity issues. Also, the variabilities in such conditions as well as in geometrical and strength parameters of pipelines lead to various uncertainties in assessing integrity of pipeline.

The random characteristics of several parameters that influence the burst strength of pipelines have motivated many authors to conduct reliability analyses (Caleyo et al. 2002, Hussein et al. 2006, Teixeira et al. 2008, Teixeira et al. 2010). In the reliability studies the failure probabilities are calculated using specific strength models of a failure mode and consider a degradation mechanism like corrosion defects (Barbosa et al. 2017, Teixeira et al. 2019).

Corrosion is a complex phenomenon due to its arbitrary nature, which is difficult to inspect and to repair particularly in subsea pipelines (Teixeira et al. 2010). Out of many failure modes, the failure of corroded pipelines under internal pressure is relatively prominent. Burst strength has been found as the controlling factor for structural design, safe operation and integrity assessment of subsea pipelines (Zhu & Leis 2012). Burst strength of pipelines is a function of material and geometrical properties and defect size. In the last years burst strength models have

been developed through many scientific approaches for accurate prediction of burst strength.

Maintenance of subsea corroded pipelines is arduous and widely depends on the prediction of failure probability or reliability index. This fact makes reliability analysis essential for scheduling inspections and repairs corroded pipelines (Palencia et al. 2019).

Traditional remaining strength assessment models such as ASME B31G (2012) are inherently conservative to consider uncertainties and thus assure safety. Several research studies have compared the accuracy of these models as burst pressure predictors with newer limit state models. The role of strength model thus becomes eminent as a slight variation in the burst pressure prediction may affect significantly the probability of failure.

The differences among these models lie in assumption of flow stress, defect shape, limiting state, failure mechanism, and elastic plastic behaviour. Differences are also due to analytical methods used, empirical fitting of limited data and use of one or few specific materials.

This paper analyses and categorises available burst strength models for corroded pipelines based on their formulations. Their predictions as function of the normalized depth of corrosion with respect to DNV model (DNV-RP-F101 2019) are compared. The limit state equation is defined in terms of model strength prediction and applied operating pressure. Probabilistic properties for basic variable are gathered from literature. Monte Carlo simulation is used

DOI: 10.1201/9781003216582-77

to evaluate the probability of limit state violation and to predict the reliability in each case. Reliability analyses are conducted based on the burst strength models and measures of model uncertainty are derived to describe the variability in the structural safety among the models. The overall objective is of this paper is to assess the influence of the strength model on the safety level and their variability.

2 CODES, METHODS AND APPROACHES OF BURST MODELS OF PIPELINES

2.1 Model based on NG 18 equation (Type I)

In the early 70s, witnessing early failures in pipelines, oil and gas industry formed a working group under the guidance of American Gas Association (AGA) to understand their causes and to develop a practical solution. This led into a well-known project – AGA Natural Gas project (NG-18) that has developed the first and foremost burst strength model (Kiefner et al. 1973; Maxey et al. 1975). The first type of models (Type I) can be defined as the models that have been developed based on NG-18 equation. The Folias, length correction or bulging factor "M" has been derived in this model from linear elastic fracture mechanics of defected area. In addition to this, other empirical models with minor adjustments of the bulging factor or flow stress of the basic NG 18 equation are presented in Appendix A. The models include design code by American Society of Mechanical Engineers (ASME) B31G (ASME B31G 2012), recommended practice by Det Norske Veritas (DNV) (DNV-RP-F101 2019), a modified version of design codes (Mod B31G (Kiefner & Veith 1989)) Chells limit load criteria (Chell, 1990), Sims criteria (Hantz et al. 1993), Canadian Code - CSA Z662-07 (2007) and modifications by Ritchie & Last (1995). It should be noted that the "M" assumes different forms in expressions for different models.

Among all the models the ASME B31G (2012) standard has been the most widely used and accepted model, yet found to be too conservative in prediction. To solve this, a modified version - Mod B31G (Kiefner & Veith 1989) with improved flow stress was conceptualized at the Battelle Memorial Institute (BMI). Later, some other similar approaches resulted in new models such as Sims criteria (SIMS) (Hantz et al. 1993), Canadian Code CSA-Z662 (2007) and SHELL-92 model by Ritchie & Last (1995).

In 1999, DNV and BG technology have developed a very popular recommended practice, DNV-RP-F101, using full-scale experiments and finite element analyses (DNV-RP-F101 2019). Petrobras (Benjamin et al. 2001) modified the DNV model, by minimizing the error between DNV predictions and test data and using especially very long defects. It should be noted that only the DNV part B method is considered here. The distinction among other popular models such as Chell limit load (CLL) (Chell 1990), SIMS (Hantz

et al. 1993), CSA Z662-07 (2007) and SHELL-92 models can be seen in Appendix A.

Some models are not considered in the present study, such as the RSTRENG (Kiefner & Veith 1989) mathematical model, which is congruent with the Modified B31G (Kiefner & Veith 1989) (except the shape of the defect in RSTRENG is river bottom). Similarly, the Linear Pipeline Corrosion Criterion (LPC model) (Fu & Batte 1999) is identical to the DNV-RP-F101 (except the shape of the defect in LPC is only rectangular). Therefore, both RSTRENG & LPC models are not described nor analysed. Also, the Ahammed (1997) version of the modified B31G model is not considered since a little difference exists between the models.

2.2 Model based on PCORRC approach/structure (Type II)

In the year 2000, the American Battelle laboratory under the funding of PCRI (Pipeline Research Council International) has developed the well-known model Pipe CORRosion Criterion (PCORRC) (Stephens & Leis 2000). For the first time, the failure in this model is assumed to be governed by ultimate tensile strength rather than flow stress. The model was developed for moderate to high-toughness pipes (high yield strength to tensile strength ratio). Significantly moderate to high toughness pipes show hardening and geometric instability resulting in the plastic collapse, which does not happen in low toughness pipes. Subsequently, many models are obtained using numerous experimental and numerical simulations, quite in line with the PCORRC approach. The models, either based on the PCORRC (Yeom et al. 2015), (Zhu & Leis 2005) or whose structure is similar to the PCORRC model (Shuai et al. 2017), (Ma et al. 2013), (Zhu & Leis 2005) are categorized hereon as type II.

Generally, the length and depth (axial and radial direction) of the corrosion defect are represented in the models so far. The circumferential extent is far less significant and often ignored in the burst pressure prediction (Terán et al. 2017). Lies & Stephens (1997) have concluded that the width of corrosion has almost no effect on burst strength of pipelines. However, the Shuai et al. (2017) (CUP) model has suggested better accuracy and the importance of the width (circumferential extent) of corrosion. Ma et al. (2013) and Zhu & Leis (2005) (Z&L) have developed their models assuming stress-strain behaviour by a power law. More information about their structure is presented in Appendix A

2.3 Models based on other methods (Type III)

Type III models include entirely new concepts to address burst strength of corroded pipelines or those whose structure does not fit type I and II models, for example, Pipeline requalification criterion (RAM PIPE) (Bea & Xu 1999), Cronin & Pick (2002) (CPS) and Chen et al. (2017) (CHEN) models.

RAMPIPE (Bea & Xu, 1999) and Gajdoš & Šperl 2012) (G&S) models have used the concept of stress concentration in the pipe due to the reduced thickness. Based on fracture mechanics G&S model was basically developed for crack with radial and longitudinal extents. However, they are not explicitly developed for corrosion defect yet, due to analogy in the application; they are incorporated in this study.

In 2002, an iterative assessment method was developed by Cronin & Pick (2002) named as CPS (corroded pipe strength) based on weighted depth difference. This method assumes that the complex shape of the corrosion defect is bounded by two limits. The upper limit implies failure pressure of plain pipe while the lower limit is the burst pressure of infinitely long groove of depth equal to maximum corrosion depth. Moreover, the Ramberg-Osgood strain hardening behaviour was included in this model. Su et al. (2016) (CHLNG) and CUP models (type I) are other models considering Ramberg-Osgood stress-strain relationship, however, the Ramberg-Osgood parameters are not reflected in the model expressions.

Choi et al. (2003) (CHOI) and CHLNG models have utilized small scale experiments and finite element analyses and then dimensionless parameters fitted through regression were included in the formulation of the models. These models express burst pressure as direct function of length of corrosion defect (l) to diameter ratio (D), such relation becomes erroneous when $l > 1.5\ D$.

Netto et al. (2005) have developed a model (NETTO) by fitting a semi-empirical equation for strength reduction prediction based on small scale experiments and non-linear finite element (FEM) analyses. The FEM analyses have been carried out in parallel to the experiments, covering a variety of steel grades (X52, X65, X77) with varying geometrical (t) and defect (d, l, w - width) parameters. Later, Wang & Zarghamee (2014) (W&Z) have modified the NETTO model in two expressions pertaining to pipe diameter using a numerical parametric study of hundred simulations. One of the drawbacks of utilizing these models is that they do not assume flow stress, tensile strength yielding criteria, nor do they consider any material behaviour in their model expressions. However, some utilization schemes for these models are adopted in the present study that will be discussed later.

CLL (Chell, 1990) (type I) and Orynyak (2008) (ORYNYAK) have formulated models by means of analytical and using limiting load instead of maximum allowable stress in the formulation of the model. CLL (Chell 1990) used a global limit load condition while Orynyak (2008) has used local limit load condition as the global limit load sometimes predicts lower values. Orynyak (2008) has developed three models for shapes of corrosion defect. In this study, however, only one model with the most widely used rectangular shape is considered.

There is another way of looking at the models in terms of the number of corrosion defect parameters (d, l and w) utilized in the model. While most of the models use d and l, RAMPIPE and CHEN models have considered only corrosion depth (d), whereas CUP and CHLNG models used all three parameters. More details on the type III models is presented in Appendix A.

3 RELIABILITY ANALYSIS

3.1 Reliability evaluation

A reliability analysis begins by defining the limit state function. A generalized form of limit state function is given by:

$$g(\mathbf{X}) = P_b - P_o \tag{1}$$

where, P_b is burst pressure and P_o is internal operating pressure. The burst pressure P_b may be defined by different mathematical models given as defined in Appendix A.

Based on the limit state function, the probability of failure is the probability of limit state violation given by:

$$P_f = \int_{g(\mathbf{x})<0} f(\mathbf{x})dx \tag{2}$$

where, $\mathbf{X}$ is a vector of random basic variables, including the operating pressure (P_o), tensile strength (σ_t), yield strength (σ_y), thickness (t), Diameter (D), Corrosion depth (d) and length (l). $f_x(\mathbf{X})$ is the joint probability density function of the vector $\mathbf{X}$. Failure is supposed to occur when $g(\mathbf{X})$ is less than or equal to zero. The region may be called failure region, whereas $g(\mathbf{X})$ greater than zero represents safe region. The random vector $\mathbf{X}$ reflects the uncertainty in mathematical modelling, loading, dimensions, and properties of pipelines. To evaluate the failure probability the multidimensional integration of Eq. 2 is required.

The direct calculation of P_f from the integral is rather complicated. There exist many approximation methods to solve such problems, such as FORM/SORM (First Order Reliability Methods/Second Order Reliability Methods) methods and simulation techniques (Melchers & Beck 2018). The present study adopts the well-known Monte Carlo simulation method with 10^6 simulations to calculate the probability of limit state violation. Then, the reliability index (β) can be evaluated from the probability of failure (P_f) as

$$\beta = -\Phi^{-1}(P_f) \tag{3}$$

where, Φ^{-1} is the inverse of normal distribution function with zero mean and unit standard deviation.

3.2 Stochastic models

In pipeline various uncertainties are present such as physical, statistical and human factors that can lead to uncertainties in the values of operating parameters (P_o), dimensional parameters, material strength and defect parameters and model errors. These uncertainties can be described by random variables with specific theoretical probability models.

From literature, the statistical properties of the parameters for modelling an API 5L X 60 pipeline are obtained and represented in Table 1. The table also presents the range of the various parameters considered in analysis. Details on the stochastic models are given in the text below Table 1. Wherever necessary, probability distribution parameters for distributions like Lognormal (LN), Gumbel (G) and Weibull (W) are calculated from their mean values and Std. Dev.

Table 1. Statistical properties of basic random variables.

Parameters	Mean	Std Dev	COV [%]	Distri-bution*
Diameter (mm)	508	0.5	0.1	N
Thickness (mm) [b]	6.35	0.063	1	N
Length of corrosion (mm) [c]	100	50	50	LN
Width of corrosion (mm) [d]	50	25	50	LN
Yeild strength (MPa) [e]	415	33.2	8	LN
Tensile strength (MPa) [f]	520	41.6	8	LN
Operating pressure (MPa) [g]	5.74	0.4	7	G
Depth of corrosion (mm) [h]	0 - 5.2	0 – 0.88	17	W

[a] The COV and distribution for pipe diameter is based on (Caleyo et al. 2002), the outer diameter is suggested to follow the normal distribution with mean equals to nominal values of the actual pipeline diameter and small COV between 2 and 3 % value due to little variability of pipe diameter. This study adopts COV of 1%

[b] [e] [g] [h] The thickness, yield strength, operating pressure and depth of corrosion are assumed to follow distribution and COV in accordance with (Teixeira et al. 2008). Mean value of operating pressure is calculated from characteristic operating pressure (72% of yield pressure). As per EURO code EN 1990 from characteristic value should be 95% upper for load variables (ES 2005).

[c] [d] The distribution for length and width for corrosion are assumed Lognormal as per (Shuai et al. 2017). The COV for length and width used here are comparable to the values suggested in (Shuai et al. 2017).

[f] The tensile strength is given COV of 8% with Lognormal distribution based on the information in (Bhardwaj et al. 2019).

*N-Normal, LN-Lognormal, W-Weibull, G- Gumbel

4 RESULTS AND DISCUSSION

This section presents deterministic and reliability analyses using the 25 models of corroded pipelines. First the models' performance is assessed deterministically by estimating and comparing their burst pressure in each category. Later, reliability analyses are conducted for the 25 models in three categories.

4.1 Deterministic assessment

Figure 1 shows the normalized burst strength predictions for type I models with increasing normalized corrosion depth. At initial level of corrosion all models show lower prediction than DNV, except the Petrobras model. This variation is due to adaptation of yield strength as the main strength criteria.

K shell and ASME B31G strength models exhibit almost linear trends, with K shell having the highest variation in burst pressure followed by Fitnet FFS and NG 18. The minimum variations are observed by CLL and ASME B31G models. All other models show similar trends.

On the other hand type II models (Figure 2) show all similar behaviour with d/t. In this case, differences between models remain almost constant and their predictions are on the average higher than that of the DNV model when varying parametrically the d/t, except for the Mod. PCORRC model. The reason of higher prediction is the use of tensile strength as basic strength criteria in such models and reduced influence of the defect parameters.

As illustrated in Figure 3, Type III models do not follow a common formulation and thus show inconsistent behaviour. Chen and RAMPIPE models show overdependence on d/t while NETTO, CHOI and W & Z models show minimum variation with d/t. The models showing extreme behaviour may not be recommended for strength prediction in general. Further study is required to understand the applicability

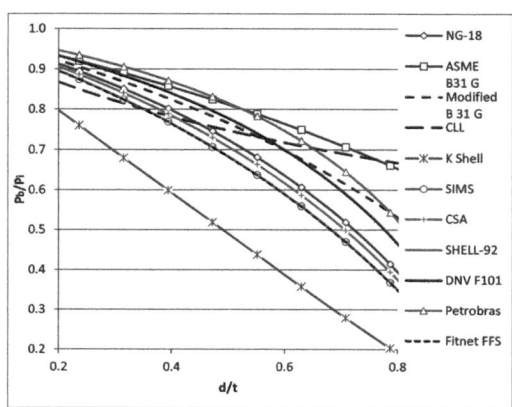

Figure 1. Normalized burst strength of pipes with defects as a function of d/t for type I models.

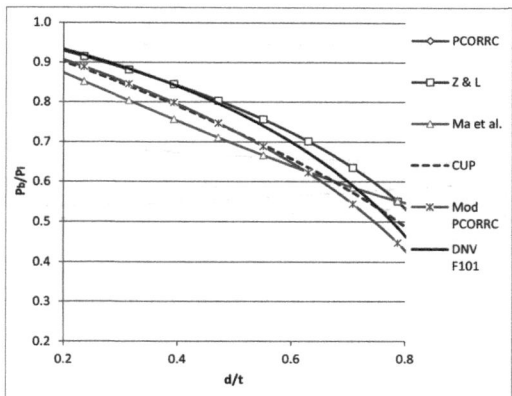

Figure 2. Normalized burst strength of pipes with defects as a function of d/t for type II models.

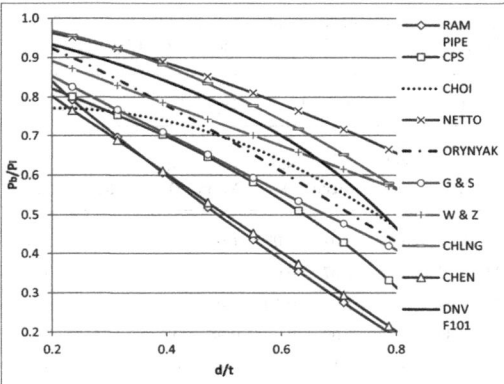

Figure 3. Normalized burst strength of pipes with defects as a function of d/t for type III models.

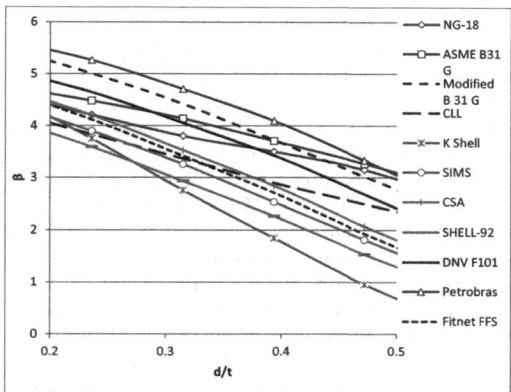

Figure 4. Reliability as a function of d/t for type I models.

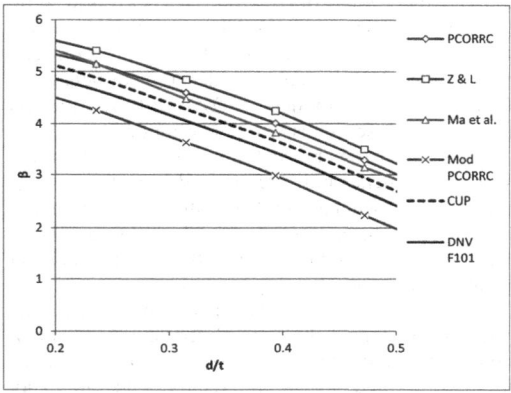

Figure 5. Reliability as a function of d/t for type II models.

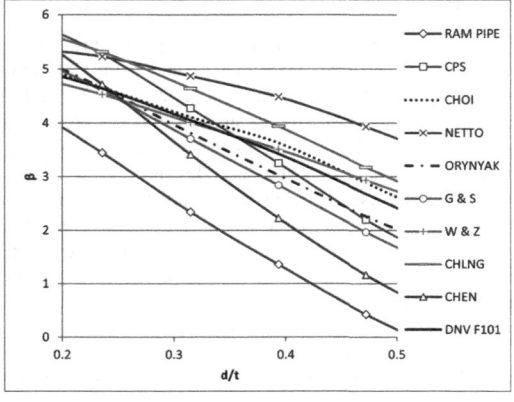

Figure 6. Reliability as a function of d/t for type III models.

of each model. Since, each model is formulated using specific material under special conditions and based on a scientific method, the applicability must be checked within their material grade.

4.2 Reliability assessment

The reliability indices (βs) for increasing normalized depth of corrosion (d/t) are calculated using Monte-Carlo simulation. It is obvious that the reliability index decreases with increase in d/t, however, the relative trends among the 25 models in their category are significantly important and are thus analysed. All models at initial defect level show reliability indices above 3. The reliability index is highly sensitive to the strength model adopted. The results of reliability analysis for different categories of models (Type I, II and III of models) are plotted in Figure 4 to Figure 6.

Normalized parameter of defect depth d/t varies from 0.2 to 0.5.

The reliability indices for type I models with d/t are illustrated in Figure 4. All models tend to show similar decreasing trends, although K shell, Fitnet

FFS, CSA, SIMS and SHELL-92 models show major deviations. K shell with the highest variation in β with d/t while NG-18 shows the lowest reduction in β. DNV model seems to provide the most appropriate trend of reliability estimation with an average behaviour throughout the corrosion range.

Type II models show nearly parallel trends to that DNV model (Figure 5). Z&L and Mod. PCORRC represent upper and lower limits of trends. Almost all models show similar reduction in β with d/t, still Ma et al. model shows the highest reduction.

Figure 6 clearly shows larger variability among the type III model reliability evaluations (with respect to each other model) for different d/t. Some extreme variation is shown by the RAMPIPE model that predicts zero reliability at d/t around 0.5. Similar low reliability at early d/t ratio is shown by CHEN, G & S and CPS models. However, the highest reliability indices are estimated by NETTO followed by CHLNG models. Also, NETTO and W&Z models lead to lowest reduction in β with varying d/t. It can be seen that DNV model provides a mean behaviour with respect to all models. W&Z and CHOI models also follow close trends with DNV model.

The comparisons presented in Figure 4 to Figure 6 reveals the behaviour, trends and sensitivity of β assessed by different models with variation in d/t. The extreme behaviour of models with respect to each other and specifically with DNV model, may not be appropriate for reliability analysis.

The present analysis has also investigated the variation of β with the normalized length of corrosion ($l/\sqrt{Dt}$) and operating pressure (P_o). The results are not presented here but the relative behaviour (not the trends) of models with each other is similar to that presented above with d/t.

4.3 Model uncertainties for reliability prediction

This section assesses the uncertainty on the safety level of the corroded pipes that results from different bursts strength prediction models. The reliability indices predicted by the different strength models are compared to that calculated using DNV model. Table 2 presents the mean and standard deviation (Std. Dev.) of normalized βs estimated from each type of models with respect to β calculated by the DNV model at different d/t. The mean and Std. Dev. of normalized β for all models independently of their type are also evaluated. It is clear that the uncertainties increased appreciably when d/t crosses value 0.4.

Figure 7 shows the trends in the mean of normalized βs presented in Table 2. Almost constant trends are perceived for type I and III models, whereas increasing trends for type II with d/t. Till $d/t = 0.5$, type I and all models together have mean normalized reliability less than 1. Type III models first (till $d/t = 0.35$) shows higher mean normalized reliability then lower normalized reliability.

The standard deviation of normalized reliabilities in each category is shown in Figure 8. Trends of Std.

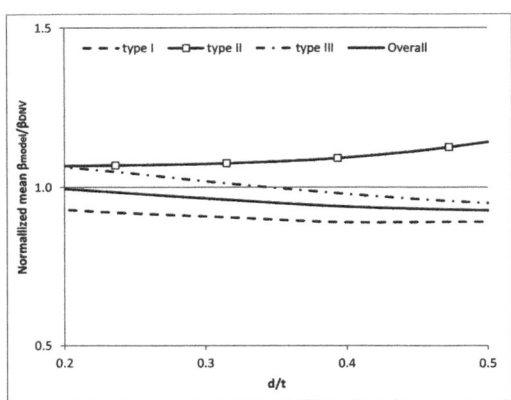

Figure 7. Mean of βs normalised by β calculated by the DNV model at different d/t.

Table 2. Mean and standard deviation of β from models relative to β calculated by DNV model at different d/t.

d/t	Type I Mean	Std Dev	Type II Mean	Std Dev	Type III Mean	Std Dev	All together Mean	Std Dev
0,16	0.94	0.09	1.07	0.07	1.08	0.09	1.01	0.10
0,24	0.92	0.11	1.07	0.08	1.05	0.12	0.98	0.12
0,31	0.91	0.14	1.08	0.10	1.01	0.17	0.96	0.16
0,39	0.89	0.19	1.09	0.12	0.98	0.26	0.94	0.22
0,47	0.89	0.28	1.13	0.16	0.96	0.38	0.93	0.32
0,53	0.89	0.40	1.18	0.21	0.94	0.54	0.92	0.45

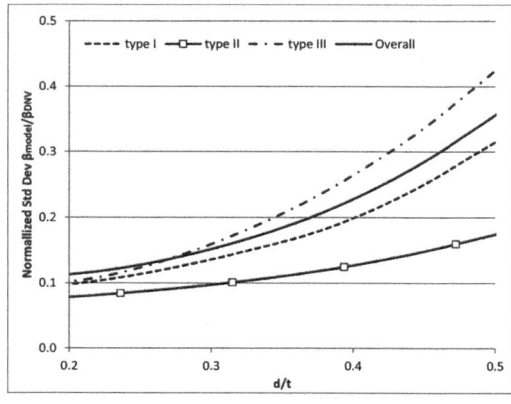

Figure 8. Standard deviation of βs normalised by β calculated by the DNV model at different d/t.

Dev. always increase with d/t. These trends suggest that the uncertainties increase at higher corrosion defects. In particular, after $d/t = 0.5$ the Std. Dev. increases to very high values.

5 CONCLUSIONS

The present study conducts reliability analyses of 25 burst strength models. First, the burst strength models are classified into three main categories as type I, II and III based on their formulation.

The normalized burst strength predictions with varying defect parameter (normalized depth d/t) are compared in each category. It is shown that some models exhibit extreme variations with d/t and therefore it is further required to define applicability limits with respect to material grade, geometrical and defect parameters.

Next, failure limit states are formulated using the 25 models. Stochastic models are adopted for limit state parameters based on literature. The influence of the normalized defect parameter d/t on the reliability in each category of strength models is investigated. Conservativeness in reliability analysis among the models is analysed. DNV model provides almost a mean behaviour of all models. It has been seen that some models exhibit extreme behaviour with respect to each other and especially to the DNV model reliability predictions when varying d/t. It may be deduced that such models with extreme deviations are not appropriate for reliability studies. Finally, measures of model uncertainty are derived to describe the variability in the structural safety among the models. It is seen that uncertainty increases appreciably at higher $(d/t > 0.4)$ in each category.

This paper provides gist of models selection over reliability analysis. Further research will be conducted to analyse models' predictions against burst test data to derive appropriate model uncertainty factors.

ACKNOWLEDGEMENTS

This work contributes to the project "Cementitious cork composites for improved thermal performance of pipelines for ultradeep waters" – SUPBSEAPIPE, with the reference n.° POCI-01-0145-FEDER-031011 funded by European Regional Development Fund (FEDER) through COMPETE2020 - Operational Program Competitiveness and Internationalization (POCI) and with financial support from FCT/MCTES through national funds. This study also contributes to the Strategic Research Plan of the Centre for Marine Technology and Ocean Engineering (CENTEC), which is financed by Portuguese Foundation for Science and Technology (Fundação para a Ciência e Tecnologia - FCT), under contract UIDB/UIDP/00134/2020.

REFERENCES

Ahammed, M. 1997. Prediction of remaining strength of corroded pressurised pipelines. *International Journal of Pressure Vessels and Piping* 71: 213–217.

ASME B31G 2012. Manual for determining the remaining strength of corroded pipelines. A supplement to ANSI/ASME B31 code for pressing piping, American Society of Mechanical Engineers (ASME), An American National Standard, New York.

Barbosa, A.A., Teixeira, A.P. & Guedes Soares, C. 2017. Strength analysis of corroded pipelines subjected to internal pressure and bending moment. In C. Guedes Soares & Y. Garbatov (eds), *Progress in the Analysis and Design of Marine Structures*: Taylor & Francis Group, 803–812.

Bea, R. & Xu, T. 1999. RAM PIPE REQUAL -Risk Assessment and Management (RAM) based guidelines for Requalification of marine pipelines. Berkeley.

Benjamin, A.C., Vieira, R.D., Freire, J.L.F. & De Castro, J.T.P. 2001. Modified equation for the assessment of long corrosion defects. In *Proceedings of the International Conference on Offshore Mechanics and Arctic Engineering*: OMAE2001–4111.

Bhardwaj, U., Teixeira, A.P., Guedes Soares, C., Azad, M.S., Punurai, W. & Asavadorndeja, P. 2019. Reliability assessment of thick high strength pipelines with corrosion defects. *International Journal of Pressure Vessels and Piping* 177: 103982.

Caleyo, F., González, J.L. & Hallen, J.M. 2002. A study on the reliability assessment methodology for pipelines with active corrosion defects. *International Journal of Pressure Vessels and Piping* 79: 77–86.

Chell, G. 1990. Application of the CEGB Failure assessment procedure, R6, to surface flaws, In: *Fracture Mechanics: Twenty First Symposium*: 1074. ASTM STP.

Chen, Z., Yan, S., Ye, H., Shen, X. & Jin, Z. 2017. Effect of the Y/T on the burst pressure for corroded pipelines with high strength. *Journal of Petroleum Science and Engineering* 157: 760–766.

Choi, J.B., Goo, B.K., Kim, J.C., Kim, Y.J. & Kim, W.S. 2003. Development of limit load solutions for corroded gas pipelines. *International Journal of Pressure Vessels and Piping*: 121–128.

Cronin, D.S., & Pick, R.J. 2002. Prediction of the failure pressure for complex corrosion defects. *International Journal of Pressure Vessels and Piping* 79: 279–287.

CSA Z662–07 2007. *Oil and gas pipeline systems, Canadian Standards Association*. Canada.

DNV-RP-F101 2019. *Corroded Pipelines, Recommended Practice*. Det Norske Veritas.

ES 2005. *Eurocode - Basis of structural design*, EN 1990:2002. European Standard, Brussels.

Fu, B. & Batte, A. 1999. *An overview of advanced methods for the assessment of corrosion in linepipe*. Health & Safety Executive, Leicestershire.

Gajdoš, Ľ. & Šperl, M. 2012. Determination of burst pressure of thin-walled pressure vessels. In *18th International Conference Engineering Mechanics*: 323–333. Svratka: Czech Republic.

Hantz, B.F., Sims, J.R., Kenyon, C.T. & Turbak, T.A., 1993. *Fitness for service: Groove like local thin areas on pressure vessels and storage tanks*. American Society of Mechanical Engineers, United States.

Hussein, A., Teixeira, A.P. & Guedes Soares, C. 2006. Reliability assessment of the burst strength of corroded pipelines. In C Guedes Soares & E Zio (eds) *Safety and Reliability for Managing Risk*: Taylor & Francis Group: London. 1467–1474.

Kanninen, M.F., Pagalthivarthi, K. V. & Popelar, C.H. 1992. A theoretical analysis for the residual strength of corroded gas and oil transmission pipelines. In V Chaker (ed.) *Corrosion Forms and Control for Infrastructure*: 183–198. ASTM International: West Conshohocken.

Kiefner, J., Maxey, W., Eiber, R. & Duffy, A. 1973. The failure stress levels of flaws in pressurised cylinders, in: *ASTM STP 536, American Society for Testing and Material. Philadelphia*: 461–481.

Kiefner, J.F. & Veith, P.H. 1989. *A modified criterion for evaluating the remaining strength of corroded pipe.* Pipeline Research Council International, Inc., Battelle Memorial Institute, Columbus.

Koçak, M. 2008. *FITNET European Fitness For Service Network.* Final Technical Report, GTC 2001 43049. GTC 2001 43049, Geesthacht, Germany.

Lies, B.N. & Stephens, D.N.R. 1997. An alternative approach to assess the integrity of corroded line pipe – Part I: current status and Part II: alternative criterion. In *7th International Offshore and Polar Engineering Conference. International Society of Offshore and Polar Engineers* Honolulu: USA.: 624–641.

Ma, B., Shuai, J., Liu, D. & Xu, K. 2013. Assessment on failure pressure of high strength pipeline with corrosion defects. *Engineering Failure Analysis* 32: 209–219.

Maxey, W., Kiefner, J. & Eiber, R. 1975. *Ductile fracture arrest in gas pipelines.* PRCI Catalog No. L32176.

Melchers, R.E. & Beck, A.T. 2018. *Structural Reliability — Analysis and Prediction.* John Wiley & Sons Ltd.: Chichester, UK.

Netto, T.A., Ferraz, U.S. & Estefen, S.F. 2005. The effect of corrosion defects on the burst pressure of pipelines. *Journal of Constructional Steel Research* 61: 1185–1204.

Orynyak, I. V. 2008. Leak and break models of ductile fracture of pressurized pipe with axial defects. In *6th International Pipeline Conference. ASME International, Calgary, Alberta, Canada*, pp. 41–55.

Palencia, O.G., Teixeira, A.P. & Guedes Soares, C. 2019. Safety of Pipelines Subjected to Deterioration Processes Modelled Through Dynamic Bayesian Networks. *Journal of Offshore Mechanics and Arctic Engineering* 141: 011602.

Ritchie, D. & Last, S. 1995. Shell 92 - Burst criteria of corroded pipelines - Defect acceptance criteria. In *EPRG PRC Biennial Joint Technical Meeting on Line Pipe Research*: 32. Cambridge.

Shuai, Y., Shuai, J. & Xu, K. 2017. Probabilistic analysis of corroded pipelines based on a new failure pressure model. *Engineering Failure Analysis* 81: 216–233.

Stephens, D.D.R.D. & Leis, B.N.B. 2000. Development of an alternative criterion for residual strength of corrosion defects in moderate- to high- toughness pipe. In *Third International Pipeline Conference*: 781–792. Calgary, Alta: Canada.

Su, C. liang, Li, X. & Zhou, J. 2016. Failure pressure analysis of corroded moderate-to-high strength pipelines. *China Ocean Engineering* 30: 69–82.

Teixeira, A.P., Guedes Soares, C., Netto, T.A. & Estefen, S. F. 2008. Reliability of pipelines with corrosion defects. *International Journal of Pressure Vessels and Piping* 85: 228–237.

Teixeira, A.P., Palencia, O.G. & Guedes Soares, C. 2019. Reliability analysis of pipelines with local corrosion defects under external pressure. *Journal of Offshore Mechanics and Arctic Engineering* 141(5): 051601.

Teixeira, A.P., Zayed, A. & Guedes Soares, C. 2010. Reliability of pipelines with non-uniform corrosion. *Journal of Ocean Engineering and Technology* 1: 12–30.

Terán, G., Capula-Colindres, S., Velázquez, J.C., Fernández-Cueto, M.J., Angeles-Herrera, D. & Herrera-Hernández, H. 2017. Failure pressure estimations for pipes with combined corrosion defects on the external surface: A comparative study. *International Journal of Electrochemical Science* 12: 10152–10176.

Wang, N. & Zarghamee, M.S. 2014. Evaluating Fitness-for-Service of corroded metal pipelines: Structural reliability bases. *Journal of Pipeline Systems Engineering and Practice* 5: 04013012.

Yeom, K.J., Lee, Y.K., Oh, K.H. & Kim, W.S. 2015. Integrity assessment of a corroded API X70 pipe with a single defect by burst pressure analysis. *Engineering Failure Analysis* 57: 553–561.

Zhu, X.-K. & Leis, B.N. 2005. Influence of Yield-to-Tensile strength ratio on failure assessment of corroded pipelines. *Journal of Pressure Vessel Technology* 127 (4): 436–442.

Zhu, X.K. & Leis, B.N. 2012. Evaluation of burst pressure prediction models for line pipes. *International Journal of Pressure Vessels and Piping* 89: 85–97.

Appendix A.   Burst Strength Models Of Corroded Pipes

	Model name (Ref.)	Type	Model structure
1	NG-18 (Kiefner et al. 1973)	I	$P_b = 2.2(\sigma_y + 68.95)\frac{t}{D}\left[\frac{1-\frac{d}{t}}{1-\frac{d}{tM}}\right]$ $M = \sqrt{1 + 2.51\left(\frac{L}{2\sqrt{Dt}}\right)^2 - 0.54\left(\frac{L}{2\sqrt{Dt}}\right)^4}$
2	ASME B31G (ASME B31G 2012)	I	$P_b = \begin{cases} 2.2\sigma_y\frac{t}{D}\left[\frac{1-\frac{2d}{3t}}{1-\frac{2d}{3tM}}\right] & \frac{L^2}{Dt} < 20 \\ 2.2\sigma_y\frac{t}{D}\left[1-\frac{d}{t}\right] & \frac{L^2}{Dt} \geq 20 \end{cases}$ $M = \sqrt{1 + 0.8\frac{L^2}{Dt}}$
3	Mod. B31G (Kiefner and Veith 1989)	I	$P_b = 2(\sigma_y + 68.95)\frac{t}{D}\left[\frac{1-0.85\frac{d}{t}}{1-0.85\frac{d}{tM}}\right]$ $\begin{cases} M = \sqrt{1 + 0.6275\frac{L^2}{Dt} - 0.003375\left(\frac{L^2}{Dt}\right)^2} & \text{for } \frac{L^2}{Dt} \leq 50 \\ M = 3.3 + 0.032\frac{L^2}{Dt} & \text{for } \frac{L^2}{Dt} \leq 50 \end{cases}$
4	CLL (Chell 1990)	I	$P_b = 2.2\sigma_y\frac{t}{D}\left[1 - \frac{d}{t} + \frac{d}{tM}\right]$ $M = \sqrt{1 + 0.496\frac{L^2}{Dd}}$
5	K Shell (Kanninen et al. 1992)	I	$P_b = 2\sigma_t\frac{t}{D}\left[\frac{1-\frac{d}{t}}{1-\frac{d}{tM}}\right]$ $M = B_1 B_2$ $B_1 = (1 + \eta^4)(\cosh\theta\cdot\sinh\theta + \cos\theta\cdot\sin\theta) + 2\eta^{3/2}(\cosh^2\theta - \cos^2\theta)$ $\quad + 2\eta^2(\cosh\theta\cdot\sinh\theta - \cos\theta\cdot\sin\theta) + 2\eta^{5/2}(\cosh^2\theta - \sin^2\theta)$ $B_2 = \left\{\cosh\theta\cdot\sin\theta + \cos\theta\cdot\sinh\theta + 2\eta^{3/2}\cosh\theta\cdot\cos\theta + \eta^2(\cos\theta\cdot\sinh\theta - \cosh\theta\cdot\sin\theta)\right\}^{-1}$ $\theta = 0.9306\frac{L}{\sqrt{D(t-d)}}$ and $\eta = 1 - \frac{d}{t}$
6	SIMS (Hantz et al. 1993)	I	$P_b = 2.22\sigma_y\frac{t}{D}\left[\frac{1-\frac{d}{t}}{1-\frac{d}{tM}}\right]$ $\begin{cases} \text{for } w > 6d + 0.1D, & M = \sqrt{1 + 2.5\frac{L^2}{Dt}} \\ \text{for } w \leq 6d + 0.1D, & M = \sqrt{1 + 0.8\frac{L^2}{Dt}} \end{cases}$
7	CSA (CSA Z662-07 2007)	I	$P_b = 1.8\sigma_t\frac{t}{D}\left[\frac{1-\frac{d}{t}}{1-\frac{d}{tM}}\right]$ $\begin{cases} M = \sqrt{1 + 0.6275\frac{L^2}{Dt} - 0.003375\left(\frac{L^2}{Dt}\right)^2} & \text{for } \frac{L^2}{Dt} \leq 50 \\ M = 3.3 + 0.032\frac{L^2}{Dt} & \text{for } \frac{L^2}{Dt} \geq 50 \end{cases}$
8	SHELL – 92 (Ritchie & Last 1995)	I	$P_b = 1.8\sigma_t\frac{t}{D}\left[\frac{1-\frac{d}{t}}{1-\frac{d}{tM}}\right]$ $M = \sqrt{1 + 0.805\frac{L^2}{Dt}}$
9	RAM PIPE (Bea & Xu 1999)	III	$P_b = 2.2\sigma_t\frac{(t-d)}{(D-t)}\left[1 + 2\sqrt{\frac{2d}{D}}\right]^{-1}$
10	DNV F101 (DNV-RP-F101 2019)	I	$P_b = 1.8\sigma_t\frac{t}{(D-t)}\left[\frac{1-\frac{d}{t}}{1-\frac{d}{tM}}\right]$ $M = \sqrt{1 + 0.31\frac{L^2}{Dt}}$
11	PCORRC (Stephens & Leis 2000)	II	$P_b = 2\sigma_t\frac{t}{(D-d)}\left[1 - \frac{d}{t}\left\{1 - \exp\left(-0.157\frac{L}{\sqrt{D(t-d)/2}}\right)\right\}\right]$
12	Petrobras (Benjamin et al. 2001)	I	$P_b = 2\sigma_t\frac{t}{(D-t)}\left[\frac{1-\frac{d}{t}}{1-\frac{d}{tM}}\right]$ $M = \sqrt{1 + 0.217\frac{L^2}{Dt} + \frac{1}{1.15\times 10^6}\left(\frac{L^2}{Dt}\right)^2}$
13	CPS (Cronin & Pick 2002)	III	$P_{PP} = 0.9\left(\frac{E\sigma y^{m-1}}{\sqrt{3}am}\right)^{1/m}\frac{4}{\sqrt{3}(D-t)}\frac{t}{\left[exp\left(\frac{t}{2m}\right)\right]^2}$ $P_b = P_{LG} + g(P_{PP} - P_{LG})$ $P_{LG} = \frac{4\sigma_t}{\sqrt{3}(D-t)}(t-d)\exp\left(-\sqrt{\frac{3}{4}}\varepsilon_{crit}\right)$ $g = \frac{4\tan^{-1}\left[exp\left(-\frac{L}{2\sqrt{D(t-d)}}\right)\right]}{\pi}$
14	CHOI (Choi et al. 2003)	III	$P_b = \begin{cases} 1.8\sigma_t\frac{t}{D}\left[C_2\left(\frac{L}{\sqrt{Dt/2}}\right)^2 + C_1\left(\frac{L}{\sqrt{Dt/2}}\right) + C_0\right] & \text{for } \frac{L}{\sqrt{Dt/2}} < 6 \\ 2\sigma_t\frac{t}{D}\left[C_4\left(\frac{L}{\sqrt{Dt/2}}\right) + C_3\right] & \text{for } \frac{L}{\sqrt{Dt/2}} \geq 6 \end{cases}$ $C_0 = 0.06\left(\frac{d}{t}\right)^2 - 0.1035\left(\frac{d}{t}\right) + 1$ $C_1 = -0.6913\left(\frac{d}{t}\right)^2 + 0.4548\left(\frac{d}{t}\right) - 0.1447$ $C_2 = 0.1163\left(\frac{d}{t}\right)^2 - 0.1053\left(\frac{d}{t}\right) + 0.0292$ $C_3 = -0.9847\left(\frac{d}{t}\right) + 1.1101$ $C_4 = 0.0071\left(\frac{d}{t}\right) - 0.0126$
15	Z&L (Zhu &Leis 2005)	II	$P_b = \frac{4\sigma_t}{(\sqrt{3})^{n+1}}\frac{t}{D}\left[1 - \frac{d}{t}\left\{1 - \exp\left(-0.157\frac{L}{\sqrt{D(t-d)/2}}\right)\right\}\right]$

(*Continued*)

695

	Model name (Ref.)	Type	Model structure
16	NETTO (Netto et al. 2005)	III	$\frac{P_b}{P_m} = \left[1 - 0.9435\left(\frac{d}{t}\right)^{1.6}\left(\frac{l}{D}\right)^{0.4}\right]$
17	Fitnet FFS (Koçak 2008)	I	$P_b = 2\sigma_t \frac{1}{D-t}\left(\frac{1}{2}\right)^{\frac{65}{\sigma_y}}\left[\frac{1-\frac{d}{t}}{1-\frac{d}{tM}}\right]$ $M = \sqrt{1 + 0.8\frac{l^2}{Dt}}$
18	ORYNYAK (Orynyak 2008)	III	$P_b = 2\sigma_t \frac{t}{D}\left[\frac{1 + \frac{l^2}{Dt}(1-\frac{d}{t})\frac{d}{t}}{1 + \frac{l^2}{Dt}(\frac{d}{t})}\right]$
19	G&S (Gajdoš & Šperl 2012)	III	$P_b = (\sigma_t + \sigma_y)\frac{t}{D}\left[1 - \frac{\frac{ud}{4t}}{1 + \frac{l}{t}}\right]$
20	Ma et al. (Ma et al. 2013)	II	$P_b = \frac{4\sigma_t}{(\sqrt{3})^{\frac{n+1}{m}}}\frac{t}{D}\left[1 - \frac{d}{t}\left\{1 - 0.7501\exp\left(-0.4174\frac{L}{\sqrt{Dt}}\right)\left(1-\frac{d}{t}\right)^{-0.1151}\right\}\right]$
21	W&Z (Wang & Zarghamee 2014)	III	$\frac{P_b}{P_m} = \begin{cases} 1 - 0.886\left(\frac{d}{t}\right)^1\left(\frac{l}{D}\right)^{0.3} & \text{if } D < 610\ mm \\ 1 - 1.12\left(\frac{d}{t}\right)^{1.15}\left(\frac{l}{D}\right)^{0.3} & \text{if } D \geq 610\ mm \end{cases}$
22	Mod PCORRC (Yeom et al. 2015)	II	$P_b = 1.8\sigma_t\frac{t}{D}\left[1 - \frac{d}{t}\left\{1 - \exp\left(-0.224\frac{L}{\sqrt{D(t-d)/2}}\right)\right\}\right]$
23	CHLNG (Su et al. 2016)	III	$P_b = 2\sigma_t\frac{t}{(D-t)}\left[C_0 + C_1\left(\frac{L}{\sqrt{Dt}}\right) + C_2\left(\frac{L}{\sqrt{Dt}}\right)^2\right]\left[G_0 + G_1\left(\frac{2w}{\pi D}\right) + G_2\left(\frac{2w}{\pi D}\right)^2\right]$ $\begin{cases} for\ L < \sqrt{20Dt} \begin{cases} C_0 = 0.8816 + 0.7942\left(\frac{d}{t}\right) - 0.05329\left(\frac{d}{t}\right)^2 \\ C_1 = 0.03982 - 0.3946\left(\frac{d}{t}\right) - 0.1901\left(\frac{d}{t}\right)^2 \\ C_2 = -0.0044248 + 0.02983\left(\frac{d}{t}\right) + 0.03091\left(\frac{d}{t}\right)^2 \\ G_0 = 1.065 - 0.2992\left(\frac{d}{t}\right) - 0.248\left(\frac{d}{t}\right)^2 \\ G_1 = 0.06604 + 0.7039\left(\frac{d}{t}\right) - 2.027\left(\frac{d}{t}\right)^2 \\ G_2 = -0.000185 - 1.211\left(\frac{d}{t}\right) + 2.356\left(\frac{d}{t}\right)^2 \end{cases} \\ for\ L \geq \sqrt{20Dt} \begin{cases} C_0 = 1.061 - 0.4754\left(\frac{d}{t}\right) - 0.5692\left(\frac{d}{t}\right)^2 \\ C_1 = 0.03102 - 0.1621\left(\frac{d}{t}\right) + 0.1343\left(\frac{d}{t}\right)^2 \\ C_0 = -0.002118 + 0.009434\left(\frac{d}{t}\right) - 0.006719\left(\frac{d}{t}\right)^2 \\ G_0 = G_1 = G_2 = 0 \end{cases} \end{cases}$
24	CUP (Shuai et al. 2017)	II	$P_b = 2\sigma_t\frac{t}{D}\left[1 - \frac{d}{t}\left\{1 - 0.1075\left(1 - \left(\frac{w}{\pi D}\right)^2\right)^6 + 0.8925\exp\left(-0.4103\frac{L}{\sqrt{Dt}}\right)\left(1-\frac{d}{t}\right)^{0.2504}\right\}\right]$
25	CHEN (Chen et al. 2017)	III	$P_b = \frac{2}{\sqrt{3}}\sigma_y\left[1 + \gamma\left(1 - \frac{\sigma_y}{\sigma_t}\right)\right]\frac{f_0 + f_1\left(\frac{d}{t}\right) + f_2\left(\frac{d}{t}\right)^2}{c_0 + c_1\left(\frac{d}{t}\right) + c_2\left(\frac{d}{t}\right)^2}$ $\begin{array}{l} f_0 = 2\lambda - 6\lambda^2 + 8\lambda^3 - 4\lambda^4 \\ f_1 = -2\lambda + 8\lambda^2 - 12\lambda^3 + 8\lambda^4 \\ f_2 = -2\lambda^2 + 4\lambda^3 - 4\lambda^4 \\ c_0 = 1 - 4\lambda + 8\lambda^2 - 8\lambda^3 + 4\lambda^4 \\ c_1 = -4\lambda^3 + 12\lambda^3 - 8\lambda^4 \\ c_2 = -4\lambda^3 + 4\lambda^4 \end{array}$

D-Ma imum diameter of pipe, t ma imum pipe thickness, d ma imum depth of corrosion, L-length of corrosion, w width of corrosion, M-ollias factors, σ_t-Ultimate tensile strength, σ_y-Yield strength, SMYS-Specified minimum Yield strength, SMTS-Specified minimum tensile strength, E-Youngs modulus, ε_{crit}- critical strain, n- strain hardening coefficient (based on stress - starin power law), a&m- Ramberg - Osgood material constant

Developments in Maritime Technology and Engineering – Guedes Soares & Santos (eds)
© 2021 Copyright the Author(s), ISBN 978-0-367-77376-2

Structural integrity of offshore pipelines considering buckling and fracture limit-states

M. Kaveh

Department of Offshore Structures Engineering, Faculty of Marine Science , Petroleum University of Technology, Abadan, Iran

C. Guedes Soares

Centre for Marine Technology and Ocean Engineering (CENTEC), Instituto Superior Técnico, Universidade de Lisboa, Lisbon, Portugal

ABSTRACT: Nonlinear elastic-plastic finite element analyses are performed to investigate the effect of fracture assessment at circumferential girth welds and buckling along the pipes between welded joints at various strains from elastic to large plastic zones. Based on the simulation performed, it was found that while for strains located in the elastic region it was not necessary to consider the buckle mitigating constrains, for large strains of up to 1.5%, there is a need to provide buckle restraints in the middle of the pipe. In addition, it is observed that as the strain increases, the average acceptable defect lengths at girth weld showed decreasing trend. Furthermore, it is concluded that in severe plastic strains larger than 1.5%, structural integrity of pipelines will not be influenced by buckle or fracture limit states and the plastic collapse failure mode will be the main cause of failure.

1 INTRODUCTION

With the ever-rising cost of energy and depletion of most resources, the oil and gas industry is compelled to move into deeper waters and therefore more hostile environments. Reliable pipelines withstanding capable of operating in these environments and the associated extreme loadings are a key factor in such developments.

Offshore pipelines during the installation phase in deep and ultra-deep waters are subjected to different critical loads to the most severe inelastic levels, which induces strains up to 2%, well into the materials plastic range that threaten the integrity of the pipeline structure (Nourpanah & Taheri 2011). It is of very high importance to monitor these relevant load scenarios through a structural integrity assessment procedure.

Local buckling is one of these severe situations that has occurred in offshore pipelines due to excessive bending deformation where the pipeline shaped the S-curve and runs from the installation vessel over stinger structure, down through the water column onto the seabed during the process of deep-water installation. As the pipeline is empty and not subject to any internal pressure, if the external hydrostatic pressure is high enough, the buckle could catastrophically propagate along the pipelines, or even destroy their structural integrity (Gong et al. 2012).

On the other hand, these pipelines are usually composed of a number of short pipes joined by welding and the girth welds may contain weld imperfections of certain size (depth and length) at specific locations in the circumferential direction (Dake at al. 2012). Therefore, in order to form an S-shaped line in over-bend or sag-bend regions, there would be a great potential of pipeline failure due to fracture of weld defects.

There is a number of researchers that have investigated the fracture assessment of offshore pipelines and the effect of different parameters on pipeline response in the process of either installation or operation conditions. For example, (Sharifi et al. 2017, Sharifi et al. 2018) have investigated the effect of crack geometries and mechanical properties mismatching between base and weld metal on fracture failure in their researches.

Extensive studies have been performed in order to introduce a set of closed-form solutions to predict the failure in pipelines. An analytical weld toe magnification factor was proposed by Han et al (2014) for efficient calculation of the stress intensity factors. This factor is applicable to complex geometrical shapes such as cruciform or T-butt joints. Zhao et al (2018) presented a modified reference stress approach under large strain loading exerted to pipeline during reel-lay installation using a derived implicit analytical solution of the nominal stress. In their study, various pipeline geometries and crack dimensions were calculated using the finite element analysis and an empirical formula was determined.

DOI: 10.1201/9781003216582-78

Studies on the effect of different defects on the fatigue life of pipes under cyclic internal pressure were conducted by Pinheiro et al. (2018), leading also to the proposal of an empirical formula. The same set of experimental and numerical results have been used to determine the fatigue reliability of such pipelines (Garbatov & Guedes Soares, 2017).

In addition, literature exist also in the area of buckling behavior of in-service pipelines and proposing simple empirical formulas for predicting buckle behavior of submarine pipelines. From this kind of research, the following can be mentioned. Zhang et al. (2018a, 2018b) have done studies on the relationship between initial imperfection and lateral buckling critical force. Finally, they proposed unified empirical formulas for both common and pipe in pipe systems based on finite element results. Furthermore, a new lateral soil resistance model was proposed by Zhang & Guedes Soares (2019) to describe the lateral resistance force-displacement relation for in-service pipelines. Based on the new lateral soil resistance model, an analytical solution for the lateral buckling of subsea pipelines with initial imperfections is presented.

Pipelines in operation will develop corrosion defects and their effect on collapse and reliability of the pipes has been studied with finite elements by Teixeira et al (2019), while the influence in different design formulations was analyzed in Teixeira et al (2008).

While assessing the integrity of offshore pipelines would not be completed without considering both governing limit-states of fracture and buckling on the tensile and compression sides respectively, there is a lack of research contributing to the structural monitoring of offshore pipelines considering the above-mentioned limit-states simultaneously. Thus, the major novelty of this work is to present a method to control the integrity of pipelines when they undergo high strain loads during installation.

In this paper, limit-state design in accordance with DNV-OS-F-101 (DNV 2013) for submarine pipelines is presented and the governing failure modes are introduced in the second section. In Section 3, geometrical configurations, material properties of pipeline, and loading scenario are illustrated in detail. In section 4, 2D and 3D elastic-plastic finite element modeling and meshing techniques served for analyzing fracture assessments and buckling configurations are presented. Structural integrity of offshore pipelines during installation is investigated in section 5. For doing this purpose, axial and post buckling forces are determined to design buckle mitigating constrains distances along the pipeline. Fracture assessment is conducted afterward through a detailed finite element analysis. Finally, a summary of the results and conclusions are given in the last section.

2 PIPELINE LIMIT STATES

The Limit State Design (LSD) provides cost-effective design strategies, as presented in most major design codes, e.g. DNV-OS-F-101 (DNV2013). Limit state is the condition in which a structure is deemed unsafe beyond it and no longer fulfills the relevant design criteria. According to DNV-OS-F-101, the integrity of offshore pipelines during installation is mostly governed by local buckling and fracture limit states.

DNV-OS-F 101 buckling control criterion is only appropriate for the nominal strains in the elastic region. Therefore, because the pipelines undergo huge plastic strains during the installation phase in deep waters, special precautions should be taken to check the buckling failure.

Therefore, it can be concluded that in order to avoid buckling failure under plastic loads during the installation phase where there is a higher probability of failure corresponding to the serviceability limit state, aids to detect buckle are needed.

On the other hand, pipeline systems shall have adequate resistance against initiation of unstable fracture. Thus, it is necessary to inspect the fracture failure of the welding joint under such extremely large plastic strains. By referring to the DNV-OS-F-101 again, it is indicated that for loads greater than 0.4 % strain, the integrity of the girth welds shall be assessed in accordance with ECA (Engineering Critically Assessment) method that is mentioned in Appendix A. By referring to the above appendix, it can be seen that if the strain exceed 1%, direct ECA method should be used.

In this study, therefore, the finite element method is utilized to ensure pipeline integrity through control fracture limit state in welding joints as well as determining the arrangement of equipment needed to prevent buckling during installation under severe plastic loads.

3 METHODOLOGY

3.1 *Geometry configuration*

2D and 3D nonlinear finite element analyses are performed for buckling and fracture assessments of pipeline respectively. For both analysis, the outer radius of pipeline is 203.2 mm, and the average wall thickness is 20.4 mm.

Two-dimensional analysis has been used first to investigate the pipeline behavior under buckling failure. Because the pipes that are connected to each other at the welding joints are 12 meters in length, buckling analysis for this length has been performed.

No buckling occurs in a direct pipeline, which has no out-of-straightness. Lateral imperfection is introduced to the pipeline to initiate buckle. The imperfection magnitude is taken as 10% of the pipe thickness, as is mentioned in ABAQUS Documents (2018) that in cases of unstable post-buckling response it is usually easiest to approach the analysis by studying the larger imperfection magnitudes, since then the response is smoothest.

After analyzing the pipeline as a whole model, in order to analyze the fracture failure in the tension side, a local analysis is performed using three-dimensional finite element at the girth weld region. The cross-section of the girth weld is shown in Figure 1. As can be seen in Figure 2 a canoe-shape surface crack (with its fillet radius equal to the crack depth) is located at the weldment, which is believed to be a typical weld-defect in offshore pipelines. However, it is worthy to note that the crack shape does not influence the fracture response at the crack center, where the maximum CTOD occurs (Raju & Newman 1982). The crack depth is symbolized as "a", while "2c" represents the crack length. The ratio "c/a" is denoted as the aspect ratio. It is noted that only a segment of the pipe with the length of two times the outer diameter was modeled. It is sufficiently long to capture the strain and stress discontinuity caused by the crack as recommended by Jayadevan et al. (2004).

3.2 Material properties

API 5L Grade X65 is considered as the base metal of the pipe. The mechanical properties of the material used in the pipeline are listed in Table 1.

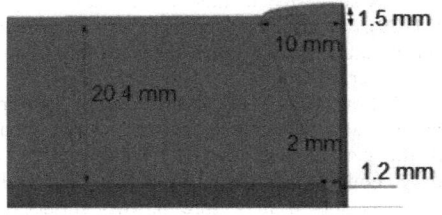

Figure 1. Schematic drawing of the girth welded pipeline cross section.

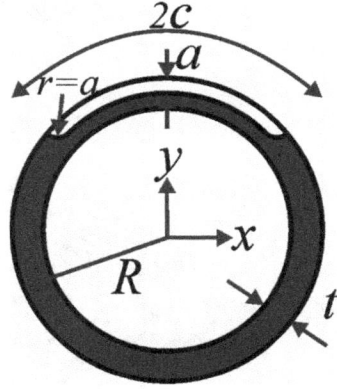

Figure 2. Typical geometry configuration of a canoe shape surface crack.

Table 1. Mechanical properties of materials used in the pipeline.

Material	E (GPa)	υ	YS (MPa)	UTS (MPa)
X65*	207	0.3	545	593

* According to API – 5L

Isotropic strain hardening model is selected to produce uniaxial true stress-strain curves for the carbon steel in all finite element models since it can generate a unique hardening exponent for any given set of Yield Stress (YS), Ultimate Tensile Strength (UTS), and uniform Elongation (uEL). Nonlinear plastic behavior of material is defined as follows:

$$\varepsilon = \frac{\sigma}{E} + \left(0.005 - \frac{YS}{E} \right) \left(\frac{\sigma}{YS} \right)^n \quad (1)$$

where E is Young's modulus, YS is the yield stress at 0.5% strain and n represents the strain hardening defined in equation 2 which is 39.25 in current study.

$$n = \ln \left(\frac{uEL - UTS/YS}{0.005 - YS/E} \right) / \ln \left(\frac{UTS}{YS} \right) \quad (2)$$

The true stress–strain curve for the material is plotted in Figure 3.

In this study, in order to facilitate finite element modeling, a separate material for the welding section is ignored and a conservative assumption of over-match condition for weld metal is assumed in all models as it is proved in the work of Sharifi et al. (2018).

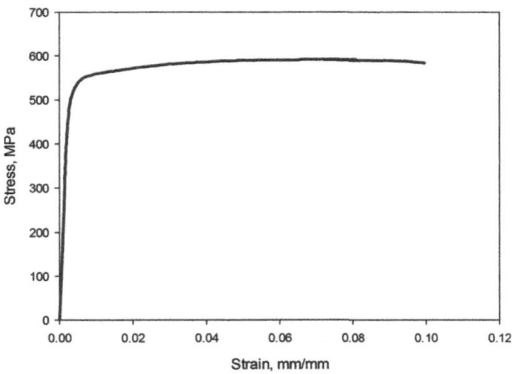

Figure 3. The uniaxial stress - strain curve for API-5L-X65 Carbon steel.

4 FINITE ELEMENT MODELING

The commercial finite element package ABAQUS (2018) was employed in this paper for both 2D and 3D buckling and fracture analysis respectively. The non-linear geometry option in ABAQUS software is activated to include the feature of the nonlinear geometry in both analyses.

Two-dimensional analysis is done using built-in PIPE21H that is a two-node linear pipe element. The letter H at the end of the element name defines hybrid formulation. Hybrid beam elements are used to improve the numerical convergence of the beams where the axial stiffness of the beam is much larger compared to the bending stiffness (ABAQUS 2018). This is useful especially in lateral buckling analysis when axially rigid long beam or pipeline undergoes large rotation when it buckles.

The finite element modeling process of buckling involves following steps:

First, in order to generate imperfection in the pipe, the pipe-buckling modes should be carried out using buckle analysis. The governing mode shape of buckling is illustrated in Figure 4.

In the next step, after the buckling modes are identified, the pipeline installation procedure is simulated. Given the worst-case scenario, the pipeline is assumed empty and therefore, there is no internal pressure in the models. The boundary condition is obtained by fixing one end of the pipeline in all directions that form the condition the pipe is fully restrained and the other end is imposed by displacement in the same way which is indicated the moment effecting the pipe during installation phase.

As in the installation procedure the pipeline in contact with the stinger of installation vessel bent into conformity with a continuous curved, the condition is considered as displacement controlled. Therefore, in order to explore systematically the buckling behavior and produce a practical mitigating lateral buckling solution, a vast range of strains from 0.3% to 2.0% are applied to the pipeline and they are checked for susceptibility to lateral buckling by calculating post-buckling configuration and effective axial forces through the pipeline.

Detailed 3D finite element analysis are done for fracture assessment. For modeling of the crack tip under large plastic strains, Anderson (2005) recommendations of finite radius is utilized. Based on this theory, the initial blunt crack tip should not affect finite element results as long as CTOD value after deformation is at least 5 times the initial value. Therefore, a blunt crack front is modeled with a radius of 0.01 mm (the initial value of CTOD is 0.02 mm). So, finite element analyses with the CTOD larger than 0.1mm are considered to be accurate. Spider web mesh technique is adopted to simulate blunt crack tip region. Figure 5 shows a sample finite element model and a close-up view of the near-tip spider web mesh. The crack tip region was modeled with 10 rows of elements covering in the circumferential and radial directions as suggested in literature (Makmeeking & Parks 1979). In addition, the size of the smallest element around the blunt crack tip is on the order of 0.001 of the crack length.

The J2 flow theory of plasticity with isotropic hardening is adopted to describe the material behavior of the carbon steel and the weld metal. The boundary condition and loading for the model in this study are as shown in Figure 6, two surfaces of

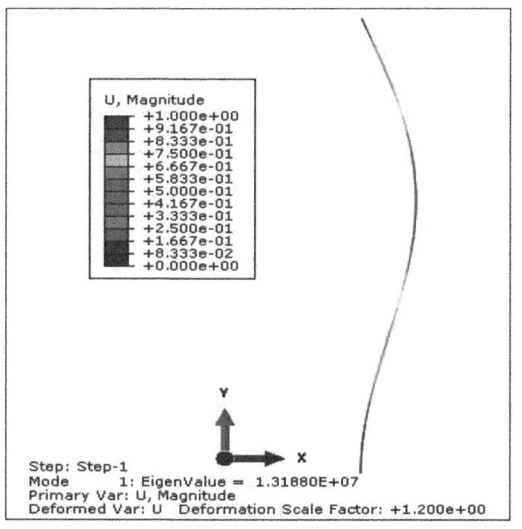

Figure 4. First mode-shape of buckling of an offshore pipeline during installation from a vessel.

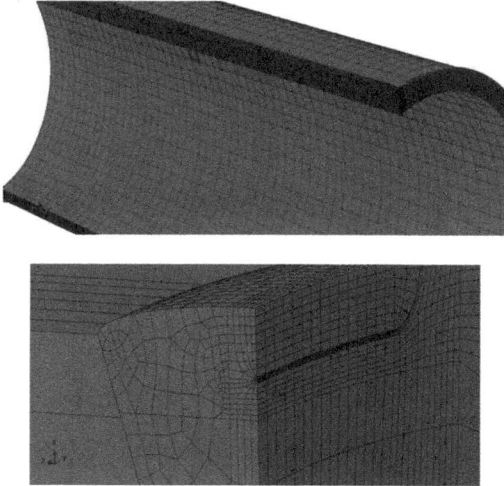

Figure 5. Finite element model and close-up of the near-tip mesh.

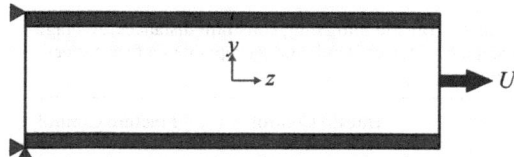

Figure 6. Boundary conditions and loading scenario considered as representative of fracture assessment at girth welds.

which the normal direction is the X axis are not allowed to move along this axis. One end of the pipeline segment is constrained in the Z direction expect the crack surface and the other end is imposed by displacement simulated the force in tension side.

Taking into account that the buckling analysis have to be required to match fracture mechanics assessment, the maximum global strain is assumed to be 2% in both groups of analyses.

5 RESULTS AND DISCUSSION

In this research work, structural integrity of offshore pipelines during installation is investigated through buckling and fracture limit states. For doing this purpose, buckling analysis is performed first to introduce the critical buckle failure load. Axial and post buckling forces are determined afterwards to design buckle mitigating constrains distances along the pipeline. Then, fracture assessment is conducted through a detailed finite element analysis.

According to the research done it is clear that the failure bending strain levels for offshore pipelines were commonly between 1.5 to 2.25%, whereas the local buckling occurred at strain levels in the range of 3.5–4% (Ostby et al. 2005).

Therefore, fracture failure at the welding joints has occurred prior to the collapse of the structure due to buckling. Therefore, in this study, buckling analysis is performed for the distance between the circumferential girth welds. Given the fact that the pipelines are constructed by means of welding 12-meter-long pipes on the installation barge, the length of one pipe is modeled and analysis under various loads ranging from 0.3% to 2.0% strains. Analyzes determine the post buckling configurations and effective axial forces in the pipeline. By comparing these forces with the critical buckling load, design of buckle mitigation constraints for different loading values would be possible. There are some examples of the axial force distribution along the pipe that is shown in Figure 7. It is obvious that when the strains are in the elastic region, the effective axial force raised in the pipe will be less than the critical one and there is no buckling occurring. Whereas, at 0.5% strain where the plastic zone starts the pipe is prone to buckle.

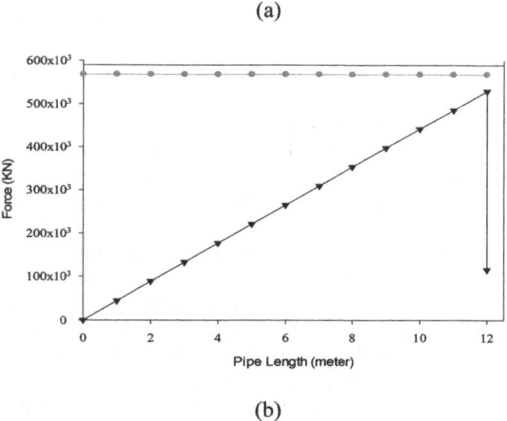

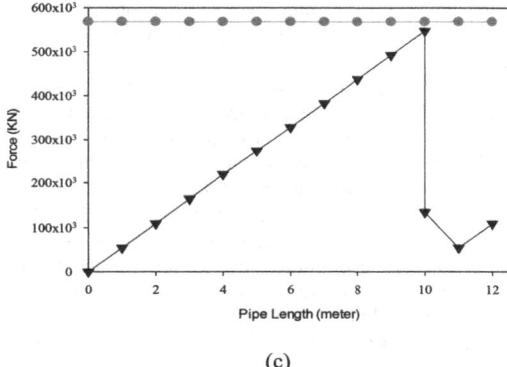

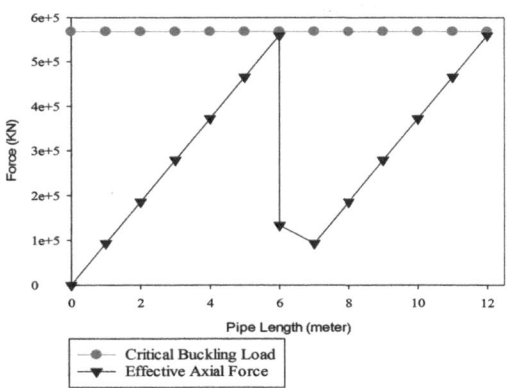

Figure 7. Effective force distribution along the pipeline length in comparison to critical buckling force for: (a) 0.3%, (b) 0.5%, (c) 2.0% strains.

So, buckle mitigating constrains are installed at specified distances from the end of the pipe and cause the axial force to be reduced to post buckling value under the critical buckling load. At heavy plastic strain of 2%, force created in the pipe reached its maximum and constrain should be used at the middle section.

Therefore, it is found that as long as the inserted strain is located in the elastic zone the force generated in the pipe is less than the critical buckling force. Thus, there would probably be no need to consider buckle mitigating constrains. However, by passing the yield point and entering to the plastic area, the cross section buckling capacity is severely reduced but it has reached a constant value for heavy plastic loads, which is indicated that at extreme strains, the failure mode of the structure is shift to plastic collapse and buckling would probably be no longer the dominant failure mode.

For cracks that may exist on girth welds and probably growth on the tension side during installation, the critical sizes include depth and length obtained from direct elastic-plastic finite element. Acceptable defect sizes for 0.5% strain are shown in Figure 8 as an example.

It is observed from Table 2 that with increasing strain values, the acceptable crack depths is decreased which is obviously due to the decrease in crack depth. On the other hand, while the amount of acceptable defect depths remains constant for strains larger than 1.5%, the crack aspect ratio C/a still follows its increasing trend that is a sign of increasing in acceptable crack length in large strains. From this issue, similar to buckling mode, it can be concluded that in severe plastic strains the structure failure will most likely be plastic collapse and the fracture is not critical in this interval. Therefore, evaluation of structural integrity at exerted strains greater than 1.5% should also consider plasticity failure.

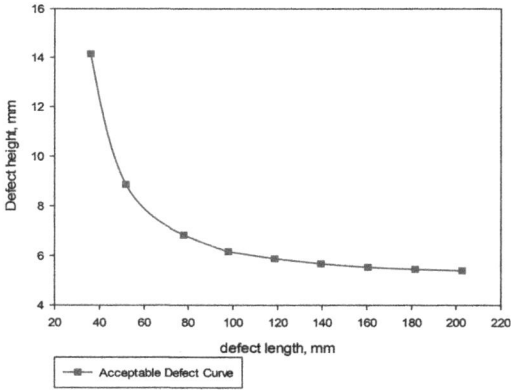

Figure 8. Critical crack sizes curve (crack depth and length) obtained from direct elastic-plastic finite element analysis for 0.5% strain.

Table 2. Buckle mitigating constrain distances, Average Acceptable Defect Depth and Acceptable Defect Aspect Ratios.

Nominal Strain (%)	Buckle Control Constrains Distances (meter)	Fracture Control a* (mm	C/ a**
0.3	12	9	8
0.4	12	8	9
0.5	10	7	10
0.8	6	6	12
1.1	6	5	13
1.5	6	4	14
1.7	6	4	15
2.0	6	4	17

* Average Acceptable Defect Depth
** Average Acceptable Defect Aspect Ratio

6 CONCLUSIONS

In the present study, structural integrity assessment of offshore pipelines subjected to large plastic strains during installation has been performed through elastic–plastic finite element models. The combined effect of buckling and fracture limit states on both tension and compression sides of the pipeline is investigated. It is noteworthy that such a quantitative study with considering combined effect of foresaid parameters has not been performed so far. It is believed that, the models were used would cover many of the practical combinations relevant to offshore pipelines.

Dominant buckle modes and critical buckling force are determined by 2D finite element analysis. Post buckling configuration and effective axial force in pipeline cross section is achieved afterward by installing deliberate horizontal imperfection to trigger buckle at pre-determined location along the pipeline. Therefore, we were able to specify the mitigating constrains distances by comparing the two above-mentioned values for various strains.

The weakest part of the structure is where the two pipes are connected to each other through circumferential girth welds. Because, cracks emergence and growth has its highest probability of occurrence in this section during welding. Thus, Acceptable crack depths and aspect ratios are investigated at girth welds in order to carry out fracture assessment under a spectrum of plastic strains. 3D detailed finite element analysis using solid elements were performed to observe fracture behavior of pipelines in process of deep-water installation. The most important conclusions of the current study are made as follows:

- It was found that for strains located in the plastic region it was not necessary to consider buckle mitigating constrains in order to prevent buckling. This is because the axial force in the pipe is less than the critical buckling load. It is important to note that a 12 m long pipe is considered in the modeling. Because, the fracture failure is more likely to be occurred at the weld joints than buckling phenomena.
- For large strains of up to 1.5%, there is a need to provide buckle restraints in the middle of the pipe. On the other hand, because constrain length will become constant for larger strains, it can be concluded that in severe plastic strains the buckling failure would probably be no longer the dominant mode and the possibility of plastic collapse will be too much.
- By reviewing average acceptable defect lengths at girth welds, it is concluded that as the strain increases, they decrease, which cause in increase of average acceptable defect ratios. However, at strains greater than 1.5%, despite the average acceptable defect lengths remaining constant, the acceptable crack ratio still shows an increasing trend, which means that the crack length should be increased. As it is not a possible true scenario at large plastic strains, it can be deduced that in severe plastic strains more than 1.5%, in both tensile and compressive areas, the predominant mode of structural failure will be plastic collapse.

REFERENCES

Anderson TL. Fracture Mechanics Fundamentals and Applications, 3rd edition, CRC Press, 2005.

Dake, Y., Sridhar, I., Zhongmin, X. & Kumar, S.B., 2012. Fracture capacity of girth welded pipelines with 3D surface cracks subjected to biaxial loading conditions. International Journal of Pressure Vessels and Piping, 92, pp.115–126.

DNV 2013. Offshore Standard DNV-OS-F-101-2013: Submarine Pipeline Systems.

Gong, S., Sun, B., Bao, S. & Bai, Y., 2012. Buckle propagation of offshore pipelines under external pressure. Marine Structures, 29(1),pp.115–130.

Garbatov, Y. & Guedes Soares, C. 2017. Fatigue Reliability of Dented Pipeline based on Limited Experimental Data. International Journal of Pressure Vessels and Piping. 155:15–26.

Han, J.W., Han, D.K. & Han, S.H., 2014. Stress intensity factors for three dimensional weld toe cracks using weld toe magnification factors. Fatigue & Fracture of Engineering Materials & Structures, 37(2),pp.146–156.

Jayadevan, K.R., Østby, E. & Thaulow, C., 2004. Fracture response of pipelines subjected to large plastic deformation under tension. International Journal of Pressure Vessels and Piping, 81(9),pp.771–783.

McMeeking, R. & Parks, D.M., 1979. On criteria for J-dominance of crack-tip fields in large-scale yielding. In Elastic-plastic fracture. ASTM International.

Nourpanah, N. & Taheri, F., 2011. A numerical study on the crack tip constraint of pipelines subject to extreme plastic bending. Engineering Fracture Mechanics, 78(6), pp.1201–1217.

Østby, E., Jayadevan, K.R. & Thaulow, C., 2005. Fracture response of pipelines subject to large plastic deformation under bending. International Journal of Pressure Vessels and Piping, 82(3),pp.201–215.

Pinheiro, B.; Guedes Soares, C., & Pasqualino, I. P. 2019. Generalized expressions for stress concentration factors of pipeline plain dents under cyclic internal pressure. International Journal of Pressure Vessels and Piping. 170: 82–91.

Raju, I.S. & Newman Jr, J.C., 1982. Stress-intensity factors for internal and external surface cracks in cylindrical vessels.

SIMULIA, D.S., 2018. Abaqus 2018. Documentation, Dassault Systèmes, Rhode Island.

Sharifi, S.M.H., Kaveh, M. & Saeidi Googarchin, H., 2017. Investigating The Effect of Crack Geometries and Weld Mismatching In Order to Optimize ECA Analysis of Girth Welded Offshore Pipelines. International Journal of optimization in Civil Engineering, 7(4),pp.645–661.

Sharifi, S.M.H., Kaveh, M. & Saeidi Googarchin, H., 2018. Engineering Critical Assessments of Marine Pipelines with 3D Surface Cracks Considering Weld Mismatch. Journal of Solid Mechanics, 10(2),pp.354–363.

Teixeira, A. P.; Palencia, O. G., & Guedes Soares, C. 2019; Reliability analysis of pipelines with local corrosion defects under external pressure. Journal of Offshore Mechanics and Arctic Engineering. 141: 051601-1-051601-10.

Teixeira, A. P.; Guedes Soares, C.; Netto, T. A., & Estefen, S. F. 2008. Reliability of Pipelines with Corrosion Defects. International Journal of Pressure Vessels and Piping. 85(4):228–237.

Zhang, X., Guedes Soares, C., An, C. & Duan, M., 2018a. An unified formula for the critical force of lateral buckling of imperfect submarine pipelines. Ocean Engineering, 166, pp.324–335.

Zhang, X., Duan, M. & Guedes Soares, C., 2018b. Lateral buckling critical force for submarine pipe-in-pipe pipelines. Applied Ocean Research, 78, pp.99–109.

Zhang, X. & Guedes Soares, C., 2019. Lateral buckling analysis of subsea pipelines on nonlinear foundation. Ocean Engineering, 186, p.106085.

Zhao, X., Xu, L., Jing, H., Zhao, L. & Huang, J., 2018. A modified strain-controlled reference stress approach for submarine pipelines under large-scale plastic strain. Advances in Engineering Software, 119, pp.12–20.

Developments in Maritime Technology and Engineering – Guedes Soares & Santos (eds)
© 2021 Copyright the Author(s), ISBN 978-0-367-77376-2

Numerical study on the effects of initial deflection on ultimate strength of pipeline under external pressure

Ruoxuan Li & C. Guedes Soares
Centre for Marine Technology and Ocean Engineering (CENTEC), Instituto Superior Técnico, Universidade de Lisboa, Lisboa, Portugal

ABSTRACT: The effects of initial deflection are investigated, especially the coupling effect of multiple initial deflections. Numerical simulations with finite elements are conducted. For the pipelines with only one initial deflection, first, the effect axial position of initial imperfection is confirmed to have little influence. Secondly, the effects of the length and amplitude of initial deflection are tested. Then, multiple initial deflections are introduced into the pipelines. First, the effect of distance between two initial deflections with the same size is investigated, resulting in concluding the critical distance, which provides the criteria to judge the coupling effects. Therefore, the equivalent length of the coupled initial deflections is derived. Second, the effect of multiple initial deflections with different properties is investigated. Based on calculation results, an equivalent method is proposed to combined two adjacent initial deflection areas into one.

NOTATION

L	length of pipe
D	diameter of pipe
t	thickness of pipe wall
Ω_0	ovality of pipe cross section, $\Omega_0 = (D_{max} - D_{min})/(D_{max} + D_{min})$
z	the centre position of the initial deflection area in pipe longitudinal direction
w_0	maximum amplitude of initial deflection
λ	length of initial deflection area in pipe longitudinal direction
δ	distance between two adjacent initial deflection area centres
P_{s1}	critical pressure of pipe with only one initial deflection area
P_{s2o}	critical pressure of pipe with two initial deflection areas

1 INTRODUCTION

In the offshore oil and gas industry, pipelines are adopted to transport liquid or gas. In deep sea environment, the pipelines suffer high external pressure. In order to avoid economic loss or environmental pollution, it is important to keep the adequate strength of pipeline after they are installed on the sea bed. The aim of this paper is to investigate the ultimate strength of pipelines under external pressure.

Due to the manufacture and installation procedure, initial deflection would occur onto the pipelines. In most cases, the detail features of initial imperfection of this kind are unknown and random, unless the geometrical property of the pipe is measured individually. Therefore, it is better to start estimating the ultimate strength by a simple way. Fraldi & Guarracino (2013) and Fraldi et al (2011) found that two waves in the circumferential direction usually existed in the pipe initial deflection.

Park & Kyriakides (1996) have investigated the ultimate strength of single pipes, pipe-in-pipe (PIP) and sandwich pipe analytically in 2D cases. A buckling model is proposed, in which four plastic hinges are assumed during the buckling process. For 3D cases, Kyriakides (2002) and Kyriakides & Vogler (2002) introduced artificial dent into the pipes. Then they tested the ultimate strength of the dented pipes experimentally.

Netto et al (2007) have investigated the ultimate strength of pipes under external pressure after corrosion. The method how the corrosion is introduced onto the pipe is by removing material from the pipe wall to reduce the thickness. These type of defects were even used in further reliability studies (Teixeira et al, 2008).

Gong et al (2012) have obtained the buckling propagation pressure of single pipe and PIP experimentally and numerically. An artificial initial deflection is partly introduced into the pipe both in experiment and simulation.

Alrsai et al (2018a, b) have investigated the ultimate strength of pipe-in-pipe under external pressure experimentally and numerically. In their research the

DOI: 10.1201/9781003216582-79

ultimate strength of a single pipe is also obtained. From the results, it could be concluded that the ultimate strength of PIP equals to that of single outer pipe.

So, for the imperfect pipe with single area of initial imperfection the collapse strength is investigated. In this paper, first, a main influential parameter is proposed to affect the ultimate strength of pipe. Then the main task is to obtain the ultimate strength of pipes with multiple areas of initial imperfections. The critical distance between two initial deflection areas is derived to identify whether the two areas are needed to be regarded as one or not.

2 VALIDATION OF THE NUMERICAL SIMULATION

The numerical simulations in this paper are performed in ANSYS. The 2 validation of the FEM simulations is made by comparing the numerical results with the experiments by Alrsai et al (2018a, b). The pipe geometrical parameters and physical properties are shown in Table 1, and the collapse pressure, P_e, obtained in his experiment is 4820kPa.

Only a quarter of the pipe is modelled as shown in Figure 1. Both ends of the pipe are fixed. The nodes in XOZ plane are fixed to avoid deforming in Y direction and the nodes in YOZ plane are fixed to

Table 1. Geometrical parameters and physical properties of pipe in experiment.

Parameter	value
L (mm)	1600
D (mm)	60
t (mm)	2
E (MPa)	66680
σ_0 (MPa)	139
Ω_0 (%)	0.5

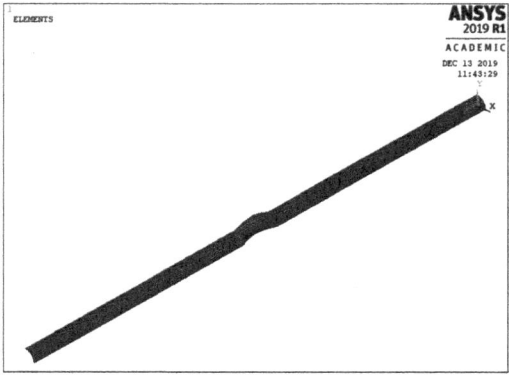

Figure 1. Numerical model of basic case (100 times of actual initial deflection).

avoid deforming in X direction. The mesh divisions are 4, 12 and 320 in radial, circumferential and longitudinal directions, respectively.

The initial deflection introduced into the pipes in this paper is selected to be a cosine mode with half-wave in axial direction and two half-waves in circumferential directions. The maximum amplitude is set to be the same with the value applied in the experiment. The assumption is made according to the potential lead of initial deflection in manufacturing and installation procedures.

The calculation results of deformation and stress distribution of the basic case, which is applied with 20 sub steps are shown in Figure 2. The maximum

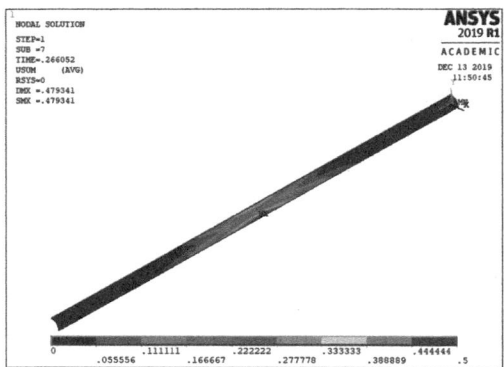

(a) Deformation

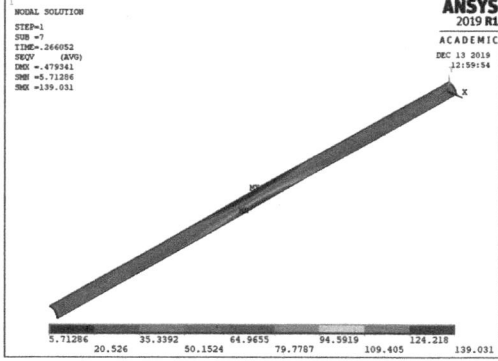

(b) Von Mises stress distribution

Figure 2. Calculation result of basic case at ultimate state: (a) Deformation and (b) Von Mises stress distribution.

Table 2. Comparison results of numerical calculation with different sub steps.

Sub steps	P_{s1}(kPa)	P_e(kPa)	Error*(%)
1	5270.1	4820	9.3
10	5287.4	4820	9.7
20	5321.0	4820	10.4
50	5323.5	4820	10.4

* Error=$(P_{s1}-P_e)/P_e$

deformation is 0.439mm, 0.21*t*. The comparison results of numerical calculations with different options of sub steps are shown in Table 2.

From this comparison, it is confirmed that the numerical calculations are reliable. The scheme with 20 sub steps is selected. Since in this case, the result is acceptable with time efficiency and the same method of parameters setting including time-step and mesh are applied in the following calculations.

3 EFFECTS OF INITIAL DEFLECTION

In this section, the effects of initial deflection are discussed. At first, the pipe scantlings, such as L/D and D/t might have influence. For only one initial deflection, its geometrical parameters are considered including length, λ, and amplitude, w_0. The ultimate strength of pipes under external pressure should be some kind of functional relationship with the above variables. For multiple initial deflection areas of the same size, the distance between two adjacent ones is also a parameter.

However, multiple initial deflection areas with different geometrical properties is another complicated issue. So it is necessary to derive a simplified method based on which random initial deflections could be regarded as one.

The effect of L/D is very similar to the effect of z/L. The reason is that, both ratios reflect the influence of the boundary. The smaller of the value of L/D or z/L means the initial deflection area locates near the boundary. As a result, in this paper the effect of z/L is investigated instead of L/D.

3.1 Effect of λ/D

The length of initial deflection area in pipe longitudinal direction has an influence on the ultimate strength of pipe under external pressure. Larger value of λ/D means more parts of the pipe are influenced by initial deflection and the ultimate strength will decrease, as shown in Table 3 and Figure 3. The calculation results shown in Figure 3 are almost linear. So the ultimate strength of an intact pipe is about 5800kPa, estimated by this result. Additionally, the calculation results of deformation and stress distribution of the case with $\lambda/D=4$ at ultimate state is shown in Figure 4. The maximum deformation is 0.499mm, 0.25*t*.

While the pipe suffers uniform external pressure, the weakest cross section plays the dominant role to

Table 3. Parameter setting and calculation results of relationship between critical pressure and λ/D ratio.

λ(mm)	λ/D	P_{s1} (kPa)
30	0.5	5662.8
60	1	5561.9
120	2	5321.0
150	2.5	5206.6
240	4	4958.2

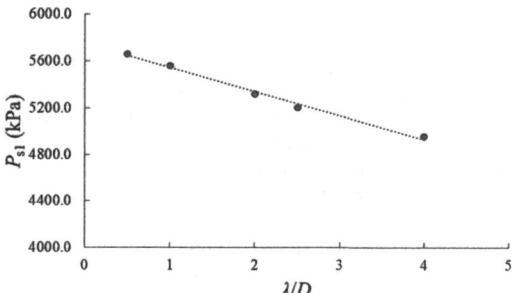

Figure 3. Results comparison of relationship between critical pressure and λ/D ratio.

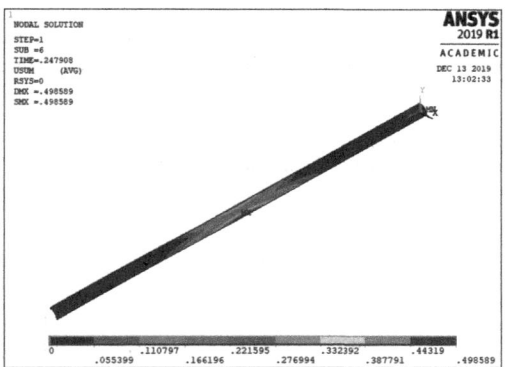

(a) Deformation

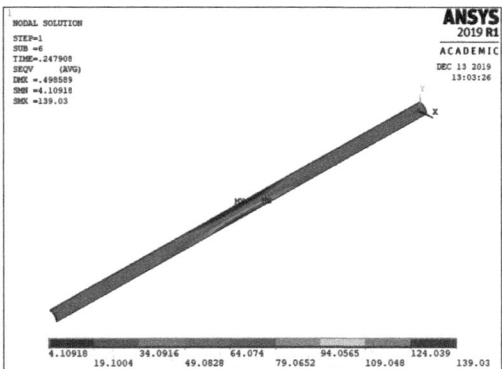

(b) Von Mises stress distribution

Figure 4. Calculation result of case with $\lambda/D=4$ at ultimate state: (a) Deformation and (b) Von Mises stress distribution.

determine the ultimate strength of the structure. In 2D issue, only the weakest cross section is considered, whereas, in 3D issue, the weakest cross section is supported by both sides of it. So large value of λ/D means the weakest cross section has influenced more length of the pipe, and the supports on both sides are weaker. As a result, both sides of the weakest cross section are easier to deform. Therefore, larger value of λ/D leads to lower ultimate strength.

3.2 Effect of w_0/t

As illustrated above, the ultimate strength of pipes under external pressure is sensitive to ovality of cross section. Larger value of Ω_0 will lead to lower value of ultimate strength. So four values of w_0/t are selected here to investigate the effect of w_0/t ratio, as shown in Table 4. The calculation results are shown in Table 4 and Figure 5.

According to the results, the reduction of ultimate strength for small value of w_0/t is too small to make significant difference compared to the intact pipe. So, in order to obtain obvious results, two typical values of w_0/t are selected, 7.5% and 20.0%.

3.3 Effect of z/L

The z/L ratio represents the effect of boundary. So in order to figure out where is the boundary affecting

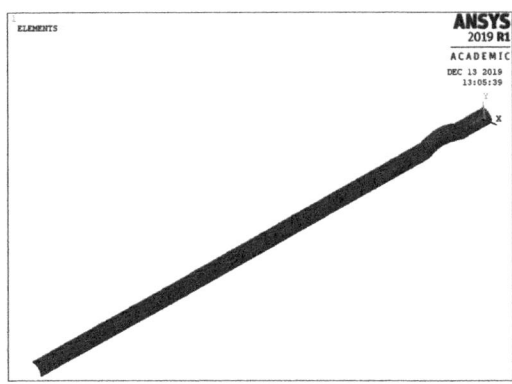

Figure 6. Calculation model of case $z/L=0.1$ (100 times of actual initial deflection).

zone, the initial deflection area is moved along the pipe longitudinal direction and other features of the initial deflection keeps the same. Parameter setting is shown in Table 5. The model of case $z/L =0.1$ is shown in Figure 6.

The comparison results are shown in Table 5, Figure 7 and Figure 8. Figure 7 shows the relationship between critical pressure and z/L ratio, when the z/L ratio exceeds 0.2, the ultimate strength becomes stabile. Figure 8 shows the deformation distribution along the whole pipe at each one's ultimate state. It can be noticed that except the case $z/L =0.1$, the maximum displacement is 0.45mm. In the other cases, the maximum displacement is about 0.48mm.

So it can be concluded from the results that if the center of initial deflection area is in the range of 20%-80% of the total pipe length, the boundaries have very little influence on the ultimate strength of pipe. The weakest situation is that the initial deflection area locates in the mid-span of the pipe, since the middle is the largest distance to the boundary support.

3.4 Effect of D/t

It is common sense that the thinner the pipe is, the lower value of the capacity of the loading the pipe could tolerate. The value of D/t is a typical character

Table 4. Parameter setting the calculation results of relationship between critical pressure and w_0/t ratio.

w_0 (mm)	w_0/t	P_{s1} (kPa)		
		$\lambda/D=1$	$\lambda/D=2$	$\lambda/D=4$
0.02	1.0%	5779.4	5725.6	5674.4
0.05	2.5%	5700.7	5644.5	5468.6
0.15	7.5%	5561.9	5321.0	4958.2
0.4	20.0%	5167.1	4672.6	4121.3

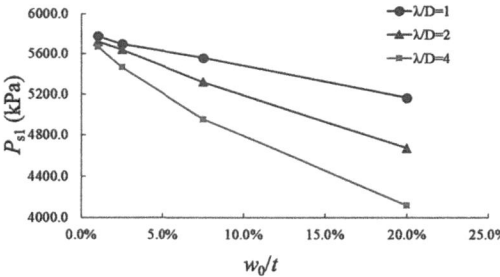

Figure 5. Results comparison of relationship between critical pressure and w_0/t ratio.

Table 5. Parameter setting and calculation results of relationship between critical pressure and z/L ratio.

z/L	P_{s1} (kPa)			
	$w_0/t=7.5\%$ $\lambda/D=2$	$w_0/t=20\%$ $\lambda/D=2$	$w_0/t=7.5\%$ $\lambda/D=4$	$w_0/t=20\%$ $\lambda/D=4$
0.1	5577.0			
0.2	5344.3	4678.4	4975.7	4125.5
0.25	5325.0			
0.3	5322.3			
0.4	5321.3			
0.5	5321.0	4672.6	4958.2	4121.3

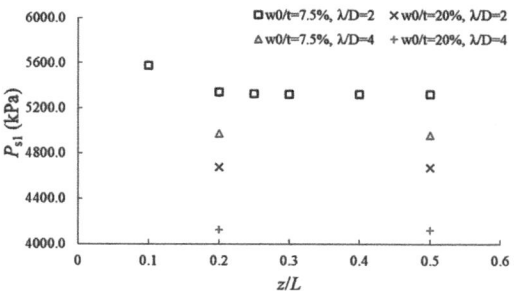

Figure 7. Results comparison of relationship between critical pressure and z/L ratio.

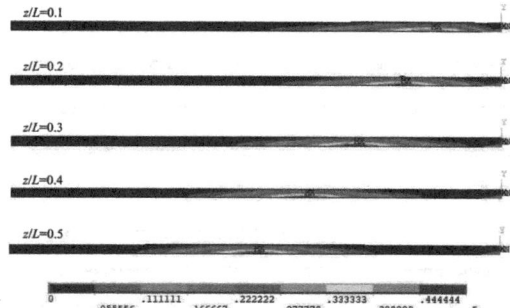

Figure 8. Calculation results of deformation for cases with different ratios of z/L.

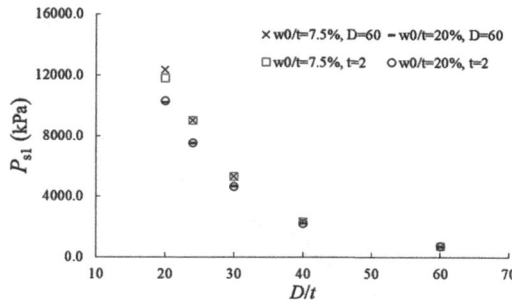

Figure 9. Results comparison of relationship between critical pressure and D/t ratio.

to represent the thickness of the pipe wall. By changing the pipe diameter and thickness separately, the relationship between ultimate strength and D/t is investigated. The detail parameter setting is shown in Table 6. It can be seen that the length of initial deflection area varies with pipe diameter changes, so the value of λ/D keeps to be the same value.

The comparison results are shown in Table 6 and Figure 9. As is illustrated above, the effect of D/t is simple to be understood. Besides, the λ/D ratio instead of λ is one of the key parameters to determine the reduction of pipe ultimate strength.

3.5 Effect of distance between multiple initial deflections, δ/D

In order to determine the equivalence of multiple areas of initial deflections, it is necessary to identify the whether two adjacent areas of initial deflections could be treated as one or not. As has been pointed out above, the length of the initial deflection area has important influence on the ultimate strength of pipelines. So the ultimate strength is investigated by varying the distance between two adjacent initial imperfection areas. At first, two initial imperfection areas with the same geometrical properties are

introduced into the pipe and the two areas are arranged symmetrically about the mid-span of the pipe.

However, in this step, the effect of λ/D should be taken into consideration, so the value of $\lambda/D=1$, 2 and 4 are selected. The parameters setting is shown in Table 7. Figure 10 shows an example of case with $\lambda/D=2$, $\delta/\lambda=3$ and $w_0/t=7.5\%$. The reason why δ is selected to describe the distance between two adjacent initial deflection areas instead of applying δ-λ will be explained later in the result comparison.

The comparison results of ultimate strength of pipes is shown Table 7 and Figure 11. Figure 12, shows the deformation distribution at ultimate state of some cases. In order to explain clearly, only the cases with $w_0/t=20\%$ are given, and the cases with $w_0/t=7.5\%$ almost follow the same tendency.

The critical distance determines whether the two adjacent initial deflection areas could be regarded as one area or two separate areas. Normally, the distance

Table 6. Parameter setting and the calculation results of relationship between critical pressure and D/t ratio.

D (mm)	t (mm)	λ (mm)	D/t	λ/D	P_{s1} (kPa)	
					$w_0/t=7.5$	$w_0/t=20\%$
60	1	120	60	2	701.0	693.9
60	1.5	120	40	2	2334.5	2230.2
60	2	120	30	2	5321.0	4672.6
60	2.5	120	24	2	9045.7	7521.9
60	3	120	20	2	12348.2	10220.7
120	2	240	60	2	738.3	724.0
80	2	160	40	2	2332.1	2223.9
48	2	96	24	2	9041.1	7527.3
40	2	80	20	2	11808.7	10323.4

Table 7. Parameter setting and the calculation results of relationship between critical pressure and δ/D ratio.

λ/D	δ/λ	δ/D	$(\delta-\lambda)/D$	P_{s2o} (kPa)	
				$w_0/t=7.5\%$	$w_0/t=20\%$
1	1	1	0	5329.7	4692.4
1	2	2	1	5368.8	4788.4
1	3	3	2	5387.9	4883.2
1	4	4	3	5449.4	4963.2
1	5	5	4	5476.9	5022.9
1	6	6	5	5514.8	5126.2
1	7	7	6	5492.5	5156.1
1	8	8	7	5542.5	5166.2
1	9	9	8	5536.3	5135.7
2	1	2	0	4984.5	4186.9
2	2	4	2	5158.9	4433.3
2	3	6	4	5239.4	4682.2
2	4	8	6	5300.7	4680.3
4	1	4	0	4736.0	3812.5
4	1.5	6	2	4920.0	4125.4
4	2	8	4	4933.6	4143.9
4	3	12	8	4912.2	4123.4
4	4	16	12	4919.6	4126.2

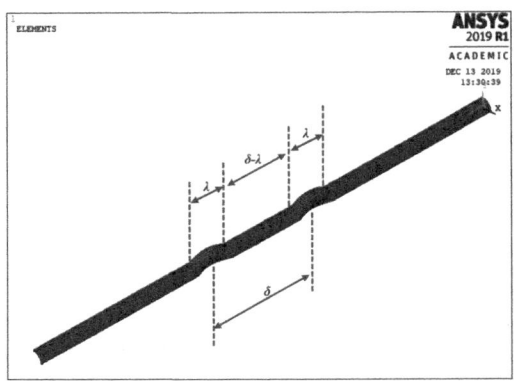

Figure 10. Model detail for the case with $\lambda/D=2$, $\delta/\lambda=3$ and $w_0/t=7.5\%$ (10 time of amplitude).

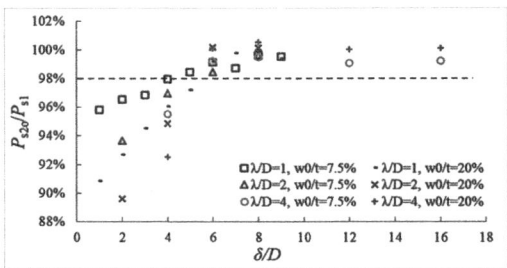

Figure 11. Results comparison of relationship between critical pressure and δ/D ratio.

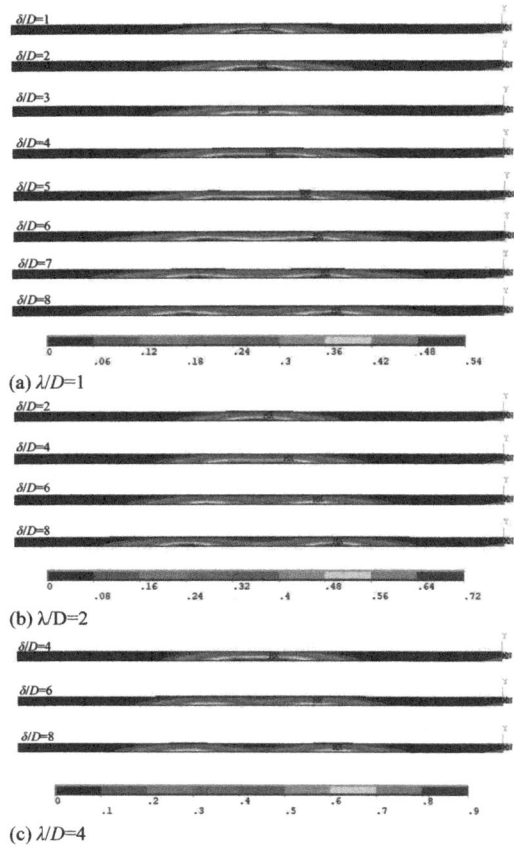

(a) $\lambda/D=1$

(b) $\lambda/D=2$

(c) $\lambda/D=4$

Figure 12. Calculation results of deformation distribution of cases with two initial deflection areas ($w_0/t=20\%$).

between the edges of two initial deflection areas should be the key parameter. So the critical values of $(\delta-\lambda)/.D$ are 5, 4 and 2 for the cases with λ/D ratio of 1, 2 and 4, respectively. So by this method, the critical distance is not unified.

The critical distance could be defined as the distance between two adjacent initial deflection area centers. Whereas, it could be figured out that when the δ/D ratio reaches 6, two peak value points of deformation appear, regardless of the value of λ/D ratio as shown in Figure 12. That means these two initial deflection areas act separately. In other words, when the δ/D ratio is below about 6, the two adjacent initial deflection areas should be treated as one, since coupling effect exists.

The above discussion is based on the principle that the distance between two centers is the key parameter. The reason why the distance between two centers is dominant is also related to the weakest cross section. As is illustrated above, the weakest cross section determines the ultimate strength of the pipe and the effective length is relatively the same value after the weakest cross section is settled. So the distance between the two centres of the initial deflection areas means the distance between two weakest cross sections.

In Figure 11, a dash line is drawn to show the 2% tolerance of error, which means if the difference between two cases is lower than 2%, they could be

regarded to have the same influence on the ultimate strength of pipes under external pressure.

Moreover, it should be noted that, if $\delta/\lambda=1$, it means the two areas of initial deflection are next to each other, and in these cases the ultimate strength of the pipe almost equals to the pipe which has only one initial deflection area with the same area length, for example, P_{s2o} ($\lambda/D=1$, $\delta/\lambda=1$, $w_0/t=7.5$)=5329.7kPa which is very closed to that P_{s1} ($\lambda/D=2$, $w_0/t=7.5$)= 5321.0kPa.

3.6 Equivalence of multiple deflections

As it is confirmed that, if distance between two centres of adjacent initial deflection areas with same size is larger than $6D$, the two areas could be treated separately, this means in terms of ultimate strength that only one of them determines the collapse capacity of the structure. Otherwise, the two initial deflection areas should be regarded as one. If the two initial deflection areas are with different size, it is necessary to provide a method about how to merge the two areas into one.

The parameter setting is shown in Table 8. The arrangement of two initial deflection areas is shown in

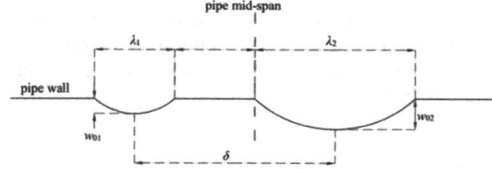

Figure 13. Arrangement of two different initial deflection areas along the pipe longitudinal direction.

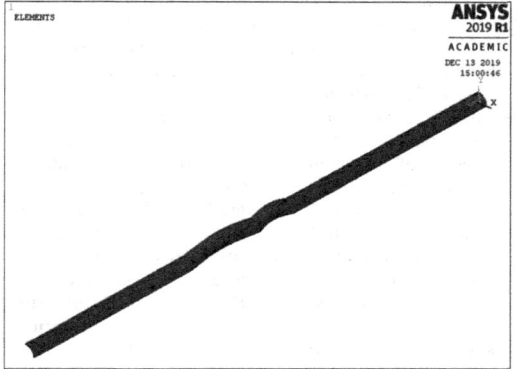

Figure 14. Model of case A-3 (100 times of amplitude).

Table 8. Parameter setting the calculation results of pipes with two different initial deflection areas.

No.	λ_1/D	w_{01}/t	λ_2/D	w_{02}/t	δ/D	P_{s2}(kPa)	P_{s1}(kPa)*	P_{s2}/P_{s1}
A-3	2	7.5%	4	7.5%	3	4803.6	4958.2	96.9%
A-4	2	7.5%	4	7.5%	4	4858.8	4958.2	98.0%
A-5	2	7.5%	4	7.5%	5	4890.0	4958.2	98.6%
A-6	2	7.5%	4	7.5%	6	4916.4	4958.2	99.2%
B-3	2	20%	4	20%	3	3915.0	4121.3	95.0%
B-4	2	20%	4	20%	4	4039.2	4121.3	98.0%
B-5	2	20%	4	20%	5	4104.2	4121.3	99.6%
B-6	2	20%	4	20%	6	4127.4	4121.3	100.1%
C-2	2	7.5%	2	20%	2	4461.0	4672.6	95.5%
C-3	2	7.5%	2	20%	3	4554.2	4672.6	97.5%
C-4	2	7.5%	2	20%	4	4625.0	4672.6	99.0%
C-5	2	7.5%	2	20%	5	4662.4	4672.6	99.8%
C-6	2	7.5%	2	20%	6	4677.0	4672.6	100.1%
D-3	2	20%	4	7.5%	3	4393.8	4672.6	94.0%
D-4	2	20%	4	7.5%	4	4535.6	4672.6	97.1%
D-5	2	20%	4	7.5%	5	4633.0	4672.6	99.2%
D-6	2	20%	4	7.5%	6	4674.0	4672.6	100.0%

* $P_{s1}=MIN[P_{s1}(\lambda_1, w_{01}), P_{s1}(\lambda_2, w_{02})]$

Figure 13, and Figure 14 is an example for case A-3 in Table 8. In the table, the value of P_{s1} is selected to be the smaller value among those of the two initial deflection when they act on the pipe alone separately. For series A, B and C, the weaker initial deflection is that with larger value of λ/D or w_0/t. However, for series D,

the weaker one is the initial deflection with $\lambda/D=2$ and $w_0/t=20\%$, so P_{s1} equals to 4672.6kPa.

An equivalence case is provided for each series. The equivalent case is defined to be a pipe with only one initial deflection area, which length is the sum of the two initial deflection areas, and amplitude is the maximum value among the two initial deflection. Therefore, for series A and C, the initial deflection of equivalent case is $\lambda/D=6$, $w_0/t=7.5\%$ and $\lambda/D=4$, $w_0/t=20\%$, respectively. For series B and D, the initial deflection of equivalent case is the same, to be $\lambda/D=6$, $w_0/t=20\%$. In Figure 15, the comparison results of the relationship between ultimate strength and δ/D ratio of the cases in Table 8 are shown. The spots represent the calculation results of the cases of A, B, C and D series, the solid lines represent the ultimate strength of the pipes only with the weakest initial deflection area and the dotted lines represent the

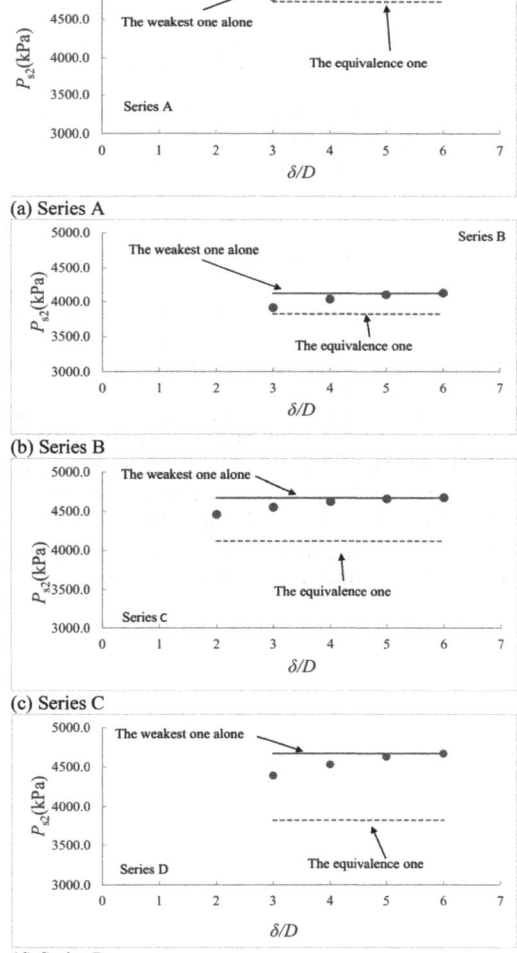

(a) Series A

(b) Series B

(c) Series C

(d) Series D

Figure 15. Comparison results between P_{s2} and δ/D ratio.

ultimate strength of the pipes only with the equivalent initial deflection area. As shown in Figure 15, all the spots are settled between the solid and dotted lines, which means the method to simplify multiple initial deflection areas into one is valid, and this method is conservative in safety side.

4 CONCLUSION

In this paper, the ultimate strength of pipes under external pressure has been investigated. First, the calculation result is compared with that of experiment to confirm the validity of numerical simulation. Secondly, the influence of some effects are investigated, including the pipe D/t ratio, the initial deflection area position z/L ratio, length λ/D ratio and ovality w_0/t ratio. It is concluded that the weakest cross section is the key factor to determine the ultimate strength of pipe under external pressure.

Thirdly, multiple initial deflection areas are introduced into pipe, the coupling effect is investigated. With two areas of same size, the critical distance between two centres of initial deflection areas is derive to be about $6D$. Forth, the above critical distance is verified by two initial deflection areas with different sizes. Moreover, a method is proposed based on which two adjacent initial deflection areas could be regarded as one if they are closed to each other.

ACKNOWLEDGEMENTS

This work was developed in the scope of the project "Cementitious cork composites for improved thermal performance of pipelines for ultradeep waters – SUPBSEAPIPE, with the reference n.° POCI-01-0145-FEDER-031011 funded by European Regional Development Fund (FEDER) through COMPETE2020 - Operational Program Competitiveness and Internationalization (POCI) and with financial support from FCT/MCTES through national funds.

This study contributes to the Strategic Research Plan of the Centre for Marine Technology and Ocean Engineering, which is financed by Portuguese Foundation for Science and Technology (Fundação para a Ciência e Tecnologia - FCT), under contract UIDB/UIDP/00134/2020.

REFERENCES

Alrsai, M., Karampour, H., & Albermani, F. 2018a. On collapse of the inner pipe of a pipe-in-pipe system under external pressure. *Engineering Structures*, *172*, 614–628.

Alrsai, M., Karampour, H., & Albermani, F. 2018b. Numerical study and parametric analysis of the propagation buckling behaviour of subsea pipe-in-pipe systems. *Thin-Walled Structures*, *125*, 119–128.

Fraldi, M., & Guarracino, F. 2013. Towards an accurate assessment of UOE pipes under external pressure: effects of geometric imperfection and material inhomogeneity. *Thin-Walled Structures*, *63*, 147–162.

Fraldi, M., Freeman, R., Slater, S., Walker, A. C., & Guarracino, F. 2011. An improved formulation for the assessment of the capacity load of circular rings and cylindrical shells under external pressure. Part 2. A comparative study with design codes prescriptions, experimental results and numerical simulations. *Thin-Walled Structures*, *49*(9), 1062–1070.

Gong, S., Sun, B., Bao, S., & Bai, Y. 2012. Buckle propagation of offshore pipelines under external pressure. *Marine Structures*, *29*(1), 115–130.

Kyriakides, S. 2002. Buckle propagation in pipe-in-pipe systems.: Part I. Experiments. *International Journal of Solids and Structures*, *39*(2), 351–366.

Kyriakides, S., & Vogler, T. J. 2002. Buckle propagation in pipe-in-pipe systems.: Part II. Analysis. *International journal of solids and structures*, *39*(2), 367–392.

Netto, T. A., Ferraz, U. S., & Botto, A. 2007. On the effect of corrosion defects on the collapse pressure of pipelines. *International Journal of Solids and Structures*, *44* (22-23), 7597–7614.

Park, T. D., & Kyriakides, S. 1996. On the collapse of dented cylinders under external pressure. *International Journal of Mechanical Sciences*, *38*(5), 557–578.

Teixeira, A. P.; Guedes Soares, C.; Netto, T. A., and Estefen, S. F. 2008; Reliability of Pipelines with Corrosion Defects. *International Journal of Pressure Vessels and Piping*. 85(4):228–237.

Developments in Maritime Technology and Engineering – Guedes Soares & Santos (eds)
© 2021 Copyright the Author(s), ISBN 978-0-367-77376-2

Finite element studies of the curvature effect on collapse behaviors of flexible risers subjected to wet collapse

X. Li, X. Jiang & J.J. Hopman
Department of Maritime and Transport Technology, Faculty of Mechanical, Maritime and Material Engineering, Delft University of Technology, Delft, The Netherlands

ABSTRACT: Flexible risers are a multi-layered pipeline device for extracting oil and gas resources from petroleum reservoirs. As offshore hydrocarbon production moves towards ultra-deep water, they have to withstand the huge hydro-static pressure without collapse. The innermost carcass of flexible risers is the main layer for collapse resistance. This layer is designed with sufficient anti-collapse capacity, enabling the risers to operate under the condition where the seawater floods the pipe annuli through the damaged external polymeric layer. In such a flooded annulus condition, however, the anti-collapse capacity of the carcass is susceptible to the pipe curvature, which can lead to the so-called "wet collapse". This paper presented 3D detailed finite element models to investigate the wet collapse behaviors of curved flexible risers. The numerical simulation indicates that the pitch extension of the carcass, and its deformed cross-sectional shape induced by the pipe curvature are the two primary factors for the reduction of wet collapse pressure. This investigation is a part of an ongoing research, aiming to help develop a cost-efficient analytical model for the collapse analyses of flexible risers in design stage. The 3D FE models presented in this work will also be used as tools to verify that developing analytical model.

1 INTRODUCTION

Flexible risers are the device used for transporting hydrocarbon product from sub-sea fields in deep water to floating vessels. A flexible riser is basically a multilayered pipe structure which consists of helical wound metallic strips and tapes, and extruded polymeric layers (Rahmati et al. 2016), as shown in Figure 1. The carcass and pressure armor are two self-interlocking layers wound by metal strips with constant pitches, which take external and internal pressure, separately. The polymeric inner liner in-between is an internal water barrier to contain fluids in the pipe bore. Tensile armors form in pairs to provide tension, bending and torsional resistances. The outer sheath is another water barrier for the seawater outside.

Over the past 40 years, the water depth of the subsea fields for oil and gas production has increased from 125 m (1977, Roncador field) to 2900 m (2016, Stones field) (Vidigal da Silva & Damiens 2016, Moore et al. 2017). Flexible risers are therefore required to withstand huge hydrostatic pressure without collapse. Normally, the external pressure is resisted by all the layers together if the outer sheath is intact. Collapse failure occurs in this situation is called as "dry collapse". However, the outer sheath can be damaged in the ocean environment, leading to a severe flooded annulus situation as illustrated in Figure 2 (Mahé 2015). The anti-collapse capacity of

the flexible riser is designed to resist the hydrostatic pressure in this severe situation, in which the carcass and pressure armor are the main layers for collapse resistance. The collapse failure occurs in such a situation is referred as "wet collapse".

Owing to the structural flexibility, the flexible riser can be bent naturally in the touchdown zone (TDZ) or buoyed regions, as shown in Figure 3 (Anderson & O'Connor 2012). When the flexible riser remains intact, the influence of pipe curvature on the collapse strength is negligible (Gay Neto et al. 2016). Once the outer sheath of the flexible riser is breached, however, a significant reduction of the wet collapse pressure can be caused. The curved collapse tests done by companies such as Technip (Paumier et al. 2009) and Wellstream (Lu et al. 2008, Clevelario et al. 2010) showed that the wet collapse pressure could be reduced up to 20% when the pipe samples were bent to their minimum bending radius (MBR), indicating that wet collapse pressure of flexible risers is susceptible to the pipe curvature.

Although this curvature effect has been evidenced, the estimation of wet collapse pressure for a curved flexible risers remains challenges. Models in numerical analyses are required to be constructed adequately long in order to introduce the pipe curvature and eliminate the end effects (Gay Neto et al. 2012, Axelsson & Skjerve 2014). Moreover, layers such as the carcass, the inner liner and the pressure armor are necessities in

DOI: 10.1201/9781003216582-80

Figure 1. Typical configuration of a flexible riser (NOV 2014).

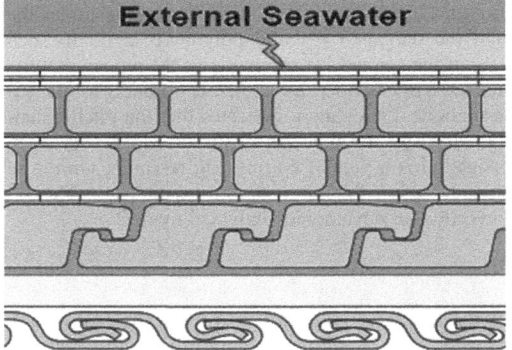

Figure 2. Flooded annulus scenario for the flexible riser with a breached outer sheath (Mahé 2015).

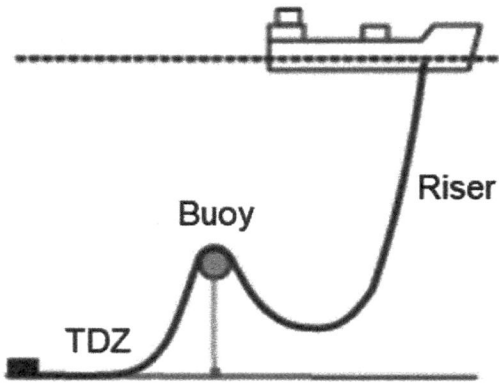

Figure 3. Flexible riser in a curved configuration (Anderson & O'Connor 2012).

the numerical models since they are the main components for wet collapse resistance (Gay Neto et al. 2017). As a result, the numerical models are always cumbersome and time-consuming. Analytical models can be an alternative. Owing to the structural complexity of flexible risers, however, the way in which the pipe curvature affects the wet collapse strength is still unclear (Clevelario et al. 2010, Loureiro Jr & Pasqualino 2012, Edmans 2014). This pose difficulties in the development of analytical approaches.

In view of this, three dimensional finite element models were presented in this paper to gain insights into collapse mechanisms of curved flexible risers. The finite element studies were established to reveal the dominated factors for wet collapse pressure of curved flexible risers, paving the way for the future development of analytical approaches in collapse analyses. As one part of our ongoing research, those FE models established in this paper will also be a verification tool for our developing analytical model. This paper is organized as follows: Section 2 gives a description of the 3D full FE model adopted in this study. In Section 3, case studies were carried out to investigate wet collapse behaviors of flexible risers with increasing pipe curvature. The results of case studies were analyzed in Section 4 to indicate the dominated factors for reducing the wet collapse pressure of flexible risers. The final section concludes the work.

2 MODEL DESCRIPTION

To investigate the wet collapse mechanism of curved flexible risers, a 3D FE model of 4" internal diameter (ID) was constructed using the commercial finite element software Abaqus 6.14. This FE model consists of three layers, the innermost carcass, the inner liner and the pressure armour. The consideration for the layer selection is as follows: the carcass is the major component to provide collapse resistance in a flooded annulus situation; and that wet collapse resistance can be further enhanced by the surround pressure armour since it restrains the radial deformation of the carcass during the collapse process; the inner liner in-between is the barrier for seawater which helps transfer the hydrostatic pressure to the carcass. To exclude the pipe end effects, the length of the model should be at least 5 times its ID by referring to the work of Paumier et al. (Paumier et al. 2009). Therefore, the 3D FE model showed in Figure 4 was built with a length that included a carcass of 32 winding turns, which will be referred as Model-A in the following.

This Model-A was developed from a prototype, as shown in Figure 5, that was presented in (Gay Neto

Figure 4. Model-A for investigating the wet collapse mechanism of curved flexible risers.

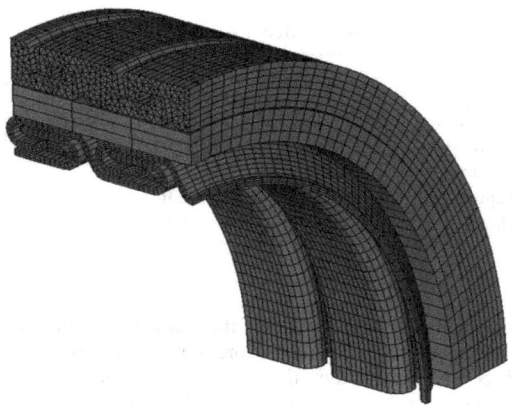

Figure 5. Prototype presented by Gay Neto & Martins (2014).

& Martins 2012, Gay Neto & Martins 2014). The prototype developed in ANSYS 12.0 was a one quarter model of the flexible pipe section, which had a length of two carcass pitches. An initial ovalization of 0.5% was imposed on the carcass in the prototype, which is defined as (API 2014).

$$\Delta = \frac{D_{\max} - D_{\min}}{D_{\max} + D_{\min}} \qquad (1)$$

Where $D_{\max}$ and $D_{\min}$ are maximum and minimum pipe diameter, respectively. Considering the prototype was a symmetric model, the initial ovalization were therefore introduced symmetrically in Model-A, making the entire carcass as well as the liner ovalized in z-axis, as shown in Figure 6. Table 1 lists the basic geometric and material data of the layers within the prototype. Geometric details of interlocking layer profiles and stress-strain curve of the inner liner are given in the source references mentioned above.

Table 1. Geometric and material properties (Gay Neto and Martins 2012, Gay Neto and Martins 2014).

Model	Carcass	Liner	Pressure armor
Internal diameter (mm)	101.6	114.4	124.4
Layer thickness (mm)	6.4	5.0	7.0
Young's Modulus (GPa)	200	-	207
Poisson's Ratio	0.3	0.45	0.3
Tangent Modulus (GPa)	2.02	-	50.00
Yield stress (MPa)	600	-	650

Since the flexible riser is a multi-layered pipe structure, layer contact can occur during the loading process. This was addressed by using a surface-based penalty method. According to the work of Gay Neto & Martins (2014), and Caleyron et al. (2014), a Coulomb friction coefficient of 0.15 was adopted for the tangent sliding between layers while a penalty stiffness factor of 0.1 was used in normal contact. By using the sweep meshing technique, Model-A was meshed as Figure 7. An 8-node linear brick element, C3D8R, was employed as the main element type in the meshing process of all the layers. A small amount of C3D6 elements were adopted to mesh the irregular corners of the pressure armor, which were the 6-node wedge elements.

Loads and boundary conditions were applied to Model-A as follows. The pipe was first bent to a specified curvature. The external pressure was then applied to the outer surface of the liner to compress the riser into collapse. The tips of the model were constrained as two rigid regions to simulate the clamped ends. This was done by creating two reference points (RP) on the model center line to constrain the tip nodes with MPC constraints, as shown in Figure 8. These two RPs were fully fixed in the collapse analysis of straight riser. In curved collapse analysis, the fixed boundary condition on one RP was removed. Instead, a bending moment was applied to this RP to introduce the pipe curvature, making the models bend within YZ plane. Moreover, two lines on the outer surface of the liner were fixed after the model was bent to the specified curvature, as illustrated in Figure 9. This is due to the fact that the external pressure applied onto a curved pipe always restores it to a straight configuration. Therefore, these two lines were fixed after bending to prevent this restoring effect. The nodes on these two

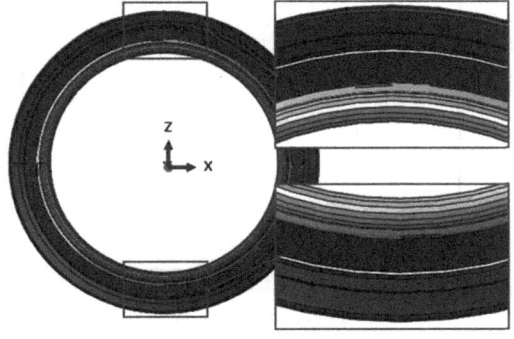

Figure 6. Initial ovalization imposed to the entire carcass and liner.

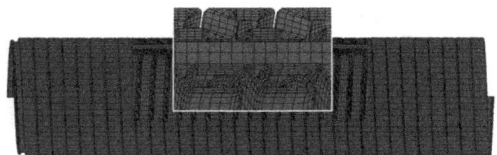

Figure 7. Mesh of Model-A.

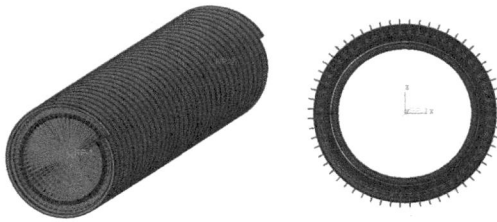

Figure 8. Constraints at the ends (left) and the applied external pressure (right).

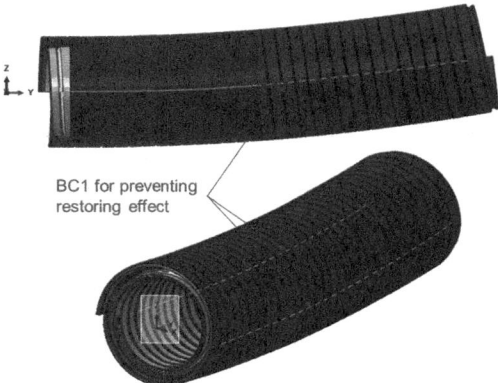

BC1 for preventing restoring effect

Figure 9. BC1 – lines on the outer surface of the inner liner for preventing restoring effect in Model-A.

lines were only allowed to move along x-axis. The line-fixed boundary condition defined here for resisting the restoring effect will be referred as "BC1" in the following sections.

The Riks method was adopted to capture the collapse pressure of the models. To ensure the reliability of Model-A in the wet collapse studies of curved risers, a verification process, as illustrated in Figure 10, was carried out. First, the prototype was reproduced using Abaqus 6.14. After the reproduced model was verified by the prototype, it was then extended to 32 pitches'

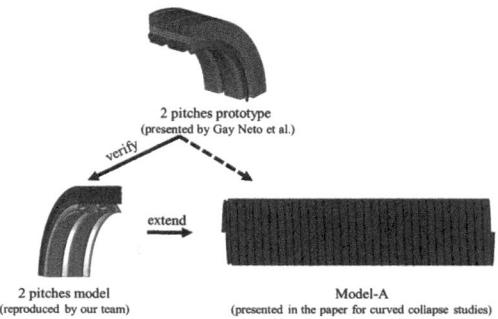

2 pitches prototype
(presented by Gay Neto et al.)

verify

extend

2 pitches model
(reproduced by our team)

Model-A
(presented in the paper for curved collapse studies)

Figure 10. Verification procedure of Model-A.

Table 2. Verification of Model-A.

Model	Pitch number	Collapse pressure (MPa)	Diff (%)
Prototype (Gay Netoand Martins 2014)	2	25.36	–
Reproduced model	2	24.36	3.94
Model-A	32	24.21	4.53

long with full pipe section. If there was not much difference among the collapse pressures given by these three FE models, then the reliability of Model-A was verified. Table 2 lists the wet collapse pressure predicted by the above-mentioned three FE models for a straight riser. The collapse pressure given by reproduced model agrees well with that of the prototype, which performs a difference less than 4%. It indicates that the prototype model was well reproduced in Abaqus. Model-A extended from the reproduced model also shows a good agreement with the prototype and the reproduced model in respect of the collapse pressure prediction. Therefore, this Model-A can be a reliable tool for the following investigation.

3 CASE STUDY

Case studies were carried out to investigate the wet collapse behaviors of curved flexible risers. By applying bending moments to the RP at one end of Model-A, a radius of curvature of 3 meters was imposed. The bending moment for the specified radius of curvature was determined by a bending response study prior to this curved collapse investigation. The curves of external pressure vs. carcass ovalization given by Model-A for curved and straight risers are plotted as Figure 11.

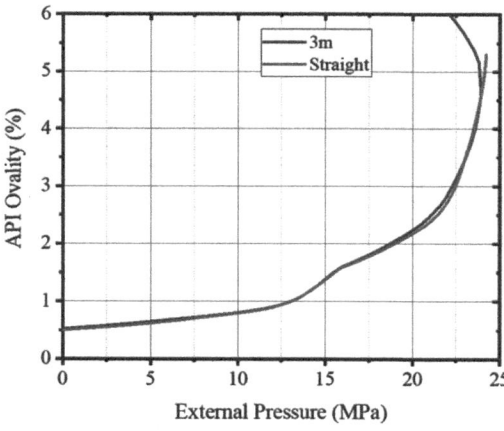

Figure 11. External pressure vs. API ovalization from Model-A for straight and curved flexible risers.

According to the results plotted in Figure 11, it seems that the curvature has little influence on the wet collapse behavior of the curved riser. However, we noted that the cross-sectional shape of the carcass at the collapse moment, as shown in Figure 12, still remained an approximate bi-symmetric shape when it was bent to a radius of curvature of 3 m. On the side of pipe intrados (the section of the pipe which is undergoing compression during the bending process, the opposite section is called extrados), there was an obvious unclosed gap between the layers. This is abnormal since the curvature creates a dis-symmetry between the carcass intrados and extrados, making its extrados less stiff and easier to deformed (Gay Neto et al. 2012, Clevelario et al. 2010). As a result, the external pressure forces the cross section of the carcass to form a symmetric shape rather than a bi-symmetric one. Considering the BC1 for preventing the pipe restoring effect might also restrain the interaction between the carcass and the pressure armor, therefore, an improvement was made to Model-A.

By referring to the work of Gay Neto et al. (2012), a four layers' model called Model-B was built, as shown in Figure 13. This model was improved from Model-A by adding an additional outer sheath. The function of the outer sheath in Model-B was only to maintain the pipe curvature in the collapse process, preventing the pipe model from being restored to straight configuration. BC1 was abandoned in Model-B. Instead, a boundary condition called BC2 was adopted as follows: two middle lines on the outer surface of the outer sheath were only allowed to move along x-axis while the bottom line on the pipe intrados was fully fixed.

Table 3 lists the wet collapse pressure predicted by Model-A and Model-B for risers in straight (∞ represents the straight riser) and curved configurations (a radius of curvature of 3 m). Since the wet collapse is

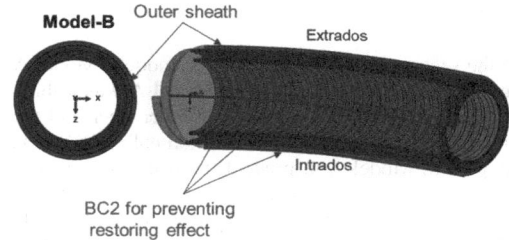

Figure 13. Model-B with an improved boundary condition BC2.

Table 3. Wet collapse pressure predicted by Model-A and Model-B for each radius of curvature.

Radius of curvature (m)		∞	3
Collapse pressure (MPa)	Model-A	24.21	23.92
	Model-B	24.23	22.07

a snap-through buckling issue, the carcass becomes extremely unstable when the external pressure close to its collapse pressure. Self-termination of the calculation were therefore caused in some cases due to the non-convergence issue. For those cases, the critical pressures were read from their last analytical step. The results from Model-A and Model-B are very close for a straight riser, which indicates the newly-added outer sheath has little influence on the wet collapse pressure. Figure 14 plots the curves of external pressure vs. carcass ovalization given by Model-B for straight and curved flexible risers. Discussion of those results are given in the following section.

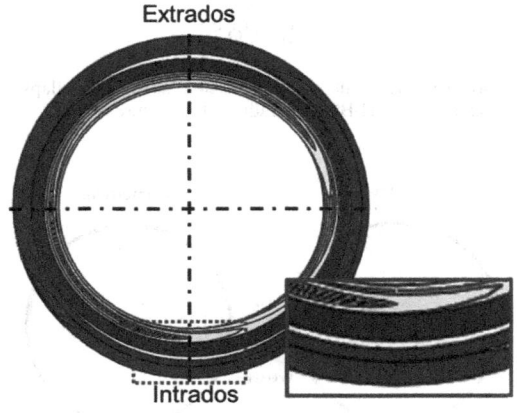

Figure 12. View cut – cross section shape at collapse moment for Model-A with a radius of curvature of 3 m.

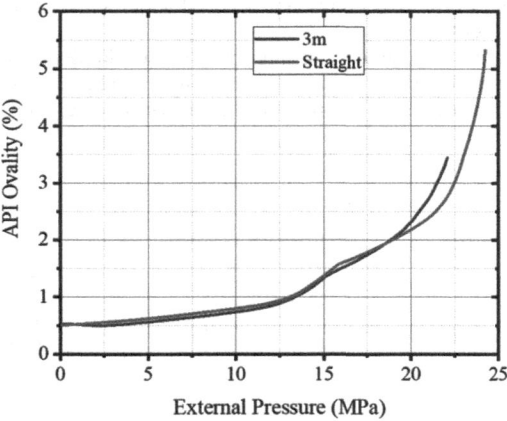

Figure 14. External pressure vs. API ovalization from Model-B for straight and curved flexible risers.

4 DISCUSSION

In the section of case study, two FE models, Model-A and Model-B, were established. Model-A was a three layers model with the carcass, the inner liner and the pressure armor. A boundary condition called BC1 was adopted in Model-A to prevent the restoring effect in curved collapse analyses, which was the nodes on the two lines of inner liner (see Figure 9) that were only allowed to move along x-axis after bending. However, we noted that BC1 had an interference on the collapse behavior of the carcass since the liner it was applied onto was a layer for transferring external pressure to the carcass. Therefore, Model-B was developed with an additional outer sheath, and BC2 replaced BC1 by moving the fixed lines to the outer surface of the outer sheath (see Figure 13).

Both Model-A and Model-B were employed to predict the wet collapse pressure of a 4" ID flexible riser in straight and curved configurations, as listed in Table 3. For the straight riser, the predictions from these two models agree quite well with each other, indicating that the outer sheath in Model-B has little influence on wet collapse pressure. However, they performed different in the curved collapse analyses. In the case studies using Model-A, the curvature did not cause a significant reduction on the wet collapse pressure. For the riser with a radius of curvature of 3 meters, the reduction on its wet collapse pressure was only 1.2% in comparison with the straight one. This reduction was caused by the bending-induced pitch extension of the carcass extrados, as shown in Figure 15. For a straight riser without any external loads, its carcass has a pitch, $P_{neutral}$, in the neutral position. When the riser is bent, the pitch on the carcass extrados is extended. This pitch extension reduces the the superposed area between two carcass profiles within a pitch (Gay Neto & Martins 2012), leading to a decreased radial stiffness on the carcass extrados. As a result, a slight decrease of the wet collapse pressure occurs for the 3m case of Model-A.

As stated above, the application of BC1 in Model-A interfered the layer interaction during the collapse process, leading to incorrect prediction results. Therefore,

Model-B was adopted to study the collapse behavior of curved flexible risers. According to the results listed in Table 3, pipe curvature causes significant reductions on the wet collapse pressure in the case studies using Model-B. Compared to the straight one, there was a 8.9% drop of wet collapse pressure when Model-B was bent to a radius of curvature of 3 meters. By observing the collapse behaviors of the curved Model-B, it was found that the carcass presented a symmetrical cross-sectional shape at its collapse region, as shown in Figure 16. This differs from the one of Model-A, see Figure 12, which performed an approximate bi-symmetrical oval-shaped cross section when it was bent to a radius of curvature of 3 meters.

Figure 17 gives an illustration of these two oval-shaped cross section by treating them as concentric rings. When the external pressure goes up, it forces parts of the carcass to detach from the inner surface of the pressure armor. Those detached portions of

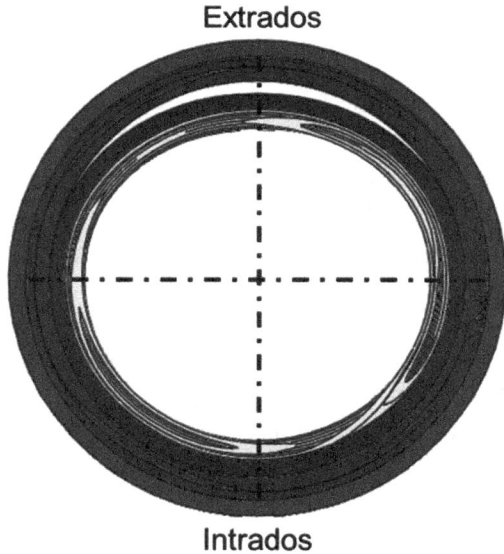

Figure 16. View cut – cross section shape at collapse moment for Model-B with a radius of curvature of 3 m.

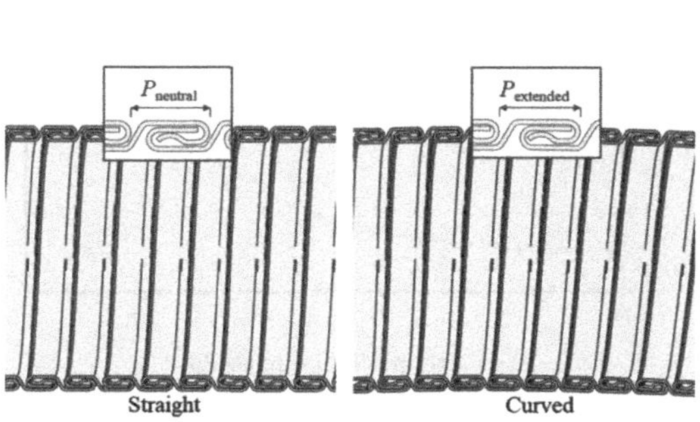

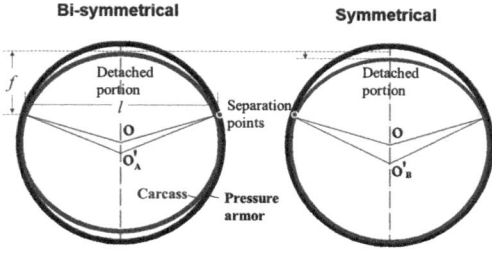

Figure 15. Change of the carcass pitch due to bending.

Figure 17. Bi-symmetrical and symmetrical shapes of the cross section at the collapse moment.

the carcass can be regarded as a circular arch with a new center O' (Li et al. 2019), which has a rise f, the vertical distance between the separation points and crown, and a span l, the horizontal distance between its two separation points (Karnovsky 2011). The bi-symmetrical cross section of the carcass has two identical detached portion at its top and bottom while the symmetrical shape only has one detached portion. If the circumferential length of the carcass is assumed to remain unchanged during the collapse process, its detached arched portion performs a smaller rise-span ratio f/l in the symmetrical oval-shaped cross section. For the arch with a smaller rise-span ratio, a greater circumferential thrust can be triggered by the uniform radial external pressure, making the arch material easier to be plasticized (Timoshenko & Gere 1963, Coccia et al. 2015). As a result, the collapse resistance of the carcass that in a symmetrical cross-sectional shape is reduced.

Since the curvature-induced pitch change causes a dis-symmetry of the radial stiffness on the carcass extrados and intrados, the cross-sectional shape of the carcass is then more close to a symmetrical one at the collapse moment. For the curved flexible riser, its wet collapse pressure is determined by the buckling strength of the detached portion of the carcass extrados. Therefore, the rise-span ratio f/l of this detached portion plays an important role in the collapse resistance. As a result, the wet collapse pressure of the flexible riser is heavily reduced when there is a large pipe curvature.

5 CONCLUSIONS

In this paper, sophisticated 3D FE models were presented to study the wet collapse behaviors of curved flexible risers. A three layers model with a length of 32 pitches was proposed at first, which was made up of the carcass, the inner liner and the pressure armor. To prevent the curved riser from restoring to straight configuration, a boundary condition called BC1 was adopted by fixing two lines on the outer surface of the liner. However, it was then found that BC1 restrained the movement of the carcass while preventing the pipe restoring effect, leading to incorrect prediction of wet collapse pressure for curved risers. In view of this, a model called Model-B was built, which was improved from Model-A by adding an additional outer sheath. The function of this outer sheath was only to maintain the specified curvature in the collapse analysis.

Based on the numerical analyses conducted in this paper, conclusions can be drawn as follows:

(1) Pipe curvature has a significant reduction effect on the wet collapse pressure of flexible risers. When the riser was bent to a radius of curvature of 3 meters, there was a 9% drop of wet collapse pressure in comparison with the straight one according to the results from Model-B.

(2) Curvature-induced pitch extension on the carcass extrados is one factor for reducing the wet collapse pressure. This extension reduces the superposed areas between two carcass profiles within one pitch, making the carcass extrados less stiff and easier to be collapsed.

(3) The deformed cross-sectional shape of the carcass during the collapse process is the dominated factor for the reduction of wet collapse pressure of curved risers. The geometry of the detached portion on carcass extrados is influenced by pipe curvature. Comparing to the straight riser, the curved one has a detached arched portion of a smaller rise-span ratio f/l, which is easier to be collapsed. As a result, the collapse resistance is reduced when the flexible riser is in a curved configuration.

(4) Finite element analyses require very high computational cost for predicting the collapse pressure of flexible risers. Running one job on 2 nodes/32 cores with our HPC took 3~5 days in average. It spent more time if there was a convergence problem. For the collapse analysis in the design stage, this is less efficient to give designers the feedback.

The work presented in this paper is one part of an ongoing research, which aims to develop a cost-efficient analytical model for the collapse analysis of flexible risers in design stage. The factors revealed by the finite element studies are going to be introduced to our developing analytical model. The verification of this analytical model will be conducted by the FE models presented in this work.

ACKNOWLEDGMENTS

This work was supported by the China Scholarship Council [grant number 201606950011].

REFERENCES

Anderson, K. & M. O'Connor (2012). The evolution of Lazy-S flexible riser configuration design for harsh environments. In *Proc. 31st OMAE*, OMAE2012-83404, Rio de Janeiro, Brazil.

API (2014). Recommended practice for flexible pipe - API RP 17B. American Petroleum Institute, 5th ed., Washington.

Axelsson, G. & H. Skjerve (2014). Flexible riser carcass collapse analyses - sensitivity on radial gaps and bending. In *Proc. 33rd OMAE*, OMAE2014-23922, San Francisco, California, USA.

Caleyron, F., M. Guiton, J. Leroy, T. Perdrizet, D. Charliac, P. Estrier, & L. Paumier (2014). A multi-purpose finite element model for flexible risers studies. In *Proc. 33rd OMAE*, OMAE2014-23250, San Francisco, California, USA.

Clevelario, J., F. Pires, G. Falcão, Z. Tan, & T. Sheldrake (2010). Flexible pipe curved collapse behaviour assessment for ultra deepwater developments for the Brazilian pre-salt area. In *Offshore Technology Conference*, OTC-20636, Houston, Texas, USA.

Coccia, S., F. Carlo, & Z. Rinaldi (2015). Collapse displacements for a mechanism of spreading-induced supports in a masonry arch. *Int. J. Adv. Struct. Eng. 7(3)*, 307–320.

Edmans, B. (2014). Finite element studies for assessment of collapse modelling methodologies for unbonded flexible pipes. In *Offshore Technology Conference*, OTC-24815-MS, Kuala Lumpur, Malaysia.

Gay Neto, A. & C. Martins (2012). A comparative wet collapse buckling study for the carcass layer of flexible pipes. *ASME J. Offshore Mech. Arct. Eng. 134*, 031701.

Gay Neto, A. & C. Martins (2014). Flexible pipes: influence of the pressure armor in the wet collapse resistance. *ASME J. Offshore Mech. Arct. Eng. 136 (03)*, 031401.

Gay Neto, A., C. Martins, E. Malta, C. Godinho, T. Neto, & E. Lima (2012). Wet and dry of straight and curved flexible pipes: a 3D FEM modeling. In *Proc. 22nd ISOPE*, Rhodes, Greece, pp. 355–364.

Gay Neto, A., C. Martins, E. Malta, R. Tanaka, & C. Godinho (2016). Simplified finite element models to study the dry collapse of straight and curved flexible pipes. *ASME J. Offshore Mech. Arct. Eng. 138 (2)*, 021701.

Gay Neto, A., C. Martins, E. Malta, R. Tanaka, & C. Godinho (2017). Simplified finite element models to study the wet collapse of straight and curved flexible pipes. *ASME J. Offshore Mech. Arct. Eng. 139 (06)*, 061701.

Karnovsky, I. (2011). *Theory of arched structures: strength, stability, vibration.* xxx: Springer Science and Business Media.

Li, X., X. Jiang, & H. Hopman (2019). An analytical approach for predicting the collapse pressure of the flexible risers with initial ovalization and gap. In *Proc. 38st OMAE*, OMAE2019-95642, Glasgow, United Kingdom.

Loureiro Jr, W. & I. Pasqualino (2012). Numerical-analytical prediction of the collapse of flexible pipes under bending and external pressure. In *Proc. 31st OMAE*, OMAE2012-83476, Rio de Janeiro, Brazil.

Lu, J., F. Ma, Z. Tan, & T. Sheldrake (2008). Bent collapse of an unbonded rough bore flexible pipe. In *Proc. 27th OMAE*, OMAE2008-57063, Estoril, Portugal.

Mahé, A. (2015). Flexible pipe technology for deepwater and gas riser systems. GE Wellstream, AOG Perth Conference.

Moore, B., A. Easton, J. Cabrera, C. Webb, & B. George (2017). Stones development: Turritella FPSO - design and fabrication of the world's deepest producing unit. In *Offshore Technology Conference*, OTC-27663-MS, Houston, Texas, USA.

NOV (2014). Floating production systems: dynamic flexible risers. National Oilwell Varco.

Paumier, L., D. Averbuch, & A. Felix-Henry (2009). Flexible pipe curved collapse resistance calculation. In *Proc. 28th OMAE*, OMAE2009-79117, Honolulu, Hawaii, USA.

Rahmati, M., H. Bahai, & G. Alfano (2016). An accurate and computationally efficient small-scale nonlinear FEA of flexible risers. *Ocean Eng. 121*, 382–391.

Timoshenko, S. & J. Gere (1963). *Theory of elastic stability.* New York, USA: McGraw-Hill.

Vidigal da Silva, J. & A. Damiens (2016). 3000 m water depth flexible pipe configuration portfolio. In *Offshore Technology Conference*, OTC-26933-MS, Houston, Texas, USA.

Developments in Maritime Technology and Engineering – Guedes Soares & Santos (eds)
© *2021 Copyright the Author(s), ISBN 978-0-367-77376-2*

Strength analysis of corroded pipelines in subsea operation condition and heated product transport

M.R. Pacheco & C. Guedes Soares
Centre for Marine Technology and Ocean Engineering (CENTEC), Instituto Superior Técnico, Universidade de Lisboa, Lisbon, Portugal

I.I.T. Riagusoff
Rosenbra Engenharia Brasil Ltda -ROSEN, Brazil

ABSTRACT: Subsea pipelines are subject to internal and external pressures arising from the depth of operation. To ensure adequate flow of transported fluid, many subsea pipelines operate at temperatures above ambient. The coupling of internal pressure and temperature is studied here. The critical load of instability is evaluated according to the increase in internal operating pressure and the external pressure due to the depth of operation. For this operational scenario, integrity and damage levels are allowed, allowing the use or behavior at different integrity levels, assisting in the management of this important asset. The analysis are performed using the finite element method include the geometric and material nonlinearities. Therefore, finite element simulations are developed, allowing to evaluate the influence of temperature on the critical instability loads.

NOMENCLATURE

d	flaw depth
n	hardening exponent
t	pipe wall thickness
w	flaw half-width
D	pipe external diameter
E	elasticity (Young's) modulus
$\grave{E}$	secant stiffness modulus
K	plastic stiffness
L	flaw half-length
M	model length
P	pressure
P_{bs}	critical internal pressure
Sy	tensile test yield strength
Su	tensile test ultimate strength
T	temperature
α	coefficient of thermal dilation
ν	Poison's ratio
ε	strain
ε_θ	circumferential strain
$\bar{\varepsilon}$	equivalent strain
σ	stress
$\bar{\sigma}$	von Mises equivalent stress

1 INTRODUCTION

Analysis of limit force in pipelines is an important step in the pipeline design, investigating various effects such as operating pressure, external pressure (sub-mining pipelines) in addition to the operating temperature. Still, efforts resulting from the installation process or even support on the ground are often evaluated, in cases of submarine pipelines, in these cases additional efforts such as flexion of the pipeline are evaluated, as presented in the study Barbosa et al., (2017). One of the main evaluated effects is lateral buckling, as presented by Zhang et al., (2018) this failure mode as well as others can be intensified by the existence of a flaw that locally compromise the duct, one of the most common flaw being corrosion induced. In the corrosion region, there is intensified tension due to less thickness, which can cause failures or even the local instability of the pipeline. Several studies such as Teixeira et al., (2009 and 2019), evaluate the reliability of operation of corrosion pipelines. This work will evaluate pipelines with and without flaws subjected to mechanical (operating pressure and external pressure - subramine) and thermal conditions.

Some subsea pipelines operate by transporting products with temperatures above 80°C, often allowing depths where the ambient temperature is 4 to 2°C. This condition generates the additional axial use of purely mechanical users caused by external and internal pressure to the pipeline.

Although corrosion is a significant threat for pipelines that operate with heated fluids, the available corrosion assessment methodologies might not be appropriate for this situation. The combination of the

DOI: 10.1201/9781003216582-81

thermal dilation with the axial restraint generates an axial compressive strain. Most assessment methodologies are not prepared to cope with this situation.

Several studies, Roy et al., (1997). Liu et al., (2012), Benjamin (2008), Bjornoy et al., (2000), Smith et al., (2000), Wang et al., (1998), Roberts et al., (1998) and Smith et al, (1996) have been conducted considering a pipeline with a corrosion flaw submitted to axial stress (or load) in addition to pressure. DNV's Corroded Pipeline Assessment Standard, DNV RP-F101 provides a guideline for evaluation of a pipeline with axial load, and recommends the use of the compressive elastic axial stress in the assessment of heated, axially restrained pipelines.

The previous methodologies seem too restrictive to the assessment of a buried and heated pipeline. By heating an axially restrained pipeline, a compressive axial strain is generated. In the elastic regimen this axial strain causes significant axial stress. However, some level of decrease in the axial stress can be expected after yielding, due to the large reduction in the material stiffness and the increase in Poisson's ratio. Since localized yield in the flaws must be allowed in any corrosion assessment methodology, it seems too conservative to use the elastic axial stress in the flaw assessment. Smith & Waldhart (2000) realized that a strain based assessment would be more realistic, but did not develop their proposal into a methodology.

In this article a numerical study of the effects of the temperature in the burst pressure of a pipeline with axial restraint and thermal load is presented. Initially, pipe without defect is studied. Afterwards, simulations for pipes with volumetric flaws were carried out.

It was found that although the thermal effect causes a large compressive axial stress in the elastic regimen, this stress is almost completely relaxed after yielding. No effect of the temperature in the burst pressure was observed in the numerical simulations.

2 NUMERICAL METHODOLOGY

The numerical simulations are carried out by finite element method (FEM) using ANSYS Mechanical 19.0© (2019). The simulations comprise pipes with volumetric flaws and pipes without flaws. The boundary conditions assumed axial restraint with free radial displacement.

These boundary conditions are applied to the cross section of the pipelines, at its two ends, and are the only ones present in the study. It should be noted that in addition to the non-linearity of material in the study, the effect of large deformations was implemented and no stabilization devices (force or energy) are used in the study.

The loading comprises an initial heat of the pipe's material and, afterwards, gradual increase of the pressure until burst. The simulations are performed with solid elements with quadratic interpolation.

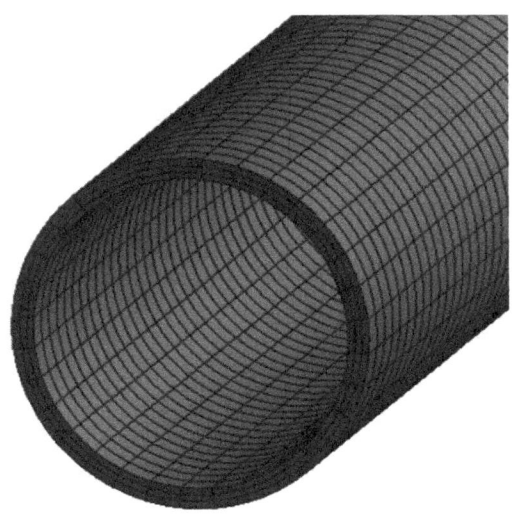

Figure 1. FEM MESH – PIPE without FLAW.

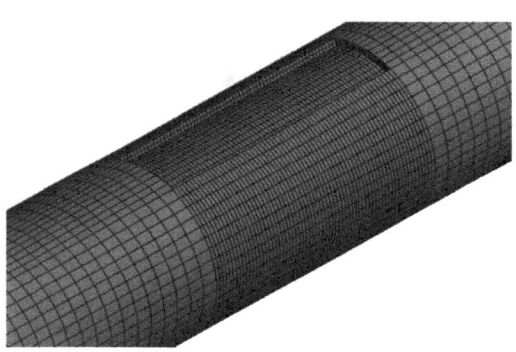

Figure 2. FEM MESH – PIPE with FLAW.

A mesh sensitivity analysis was carried out. Figure 1 illustrates the mesh used for simulating a pipe without flaw and Figure 2 shows a FEM mesh of the pipe with a volumetric flaw.

Four elements were consider along the thickness, an approach that facilitates the maintenance of a structured mesh in the evaluation of the pipelines with and without flaw.

The pipe burst was modelled using Considère's (1885) instability principle. In order to determine the instability pressure, it is necessary to use a numerical methodology capable of evaluating the system behavior close to the instability, and thus safely determine the inflexion point in the pressure-displacement curve.

The most appropriate methodology for instability analysis is the Load Control approach, which incrementally increases the load. However, this approach restricts the numerical method employed, since at the inflexion point of the load-displacement curve, the stiffness matrix has a determinant equal to zero,

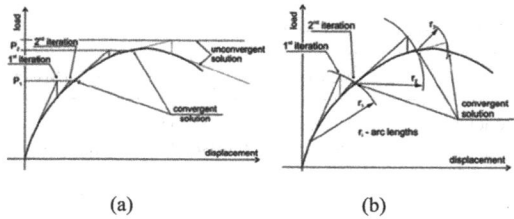

(a)   (b)

Figure 3. Graphic representation of the Newton-Raphson (a) and the Arc-Length (b) load paths.

which causes potential divergence. This characteristic prevents the problem to be solved with the Newton-Raphson method, the most common iterative method for FEM. In fact, it is often assumed that the divergence load in such an analysis is equal to the instability load.

To circumvent this issue, the Arc-Length method, initially developed by Riks (1979), was employed. The Arc-Length method introduces a load factor, λ (-1 < λ < 1), to the Newton-Raphson equations. The load factor enforces a spherical convergence path, enabling a null or negative stiffness matrix, as shown in Figure 3.

The ANSYS software employs a variation of the Riks method, proposed by Chrisfield (1981). The load balance is controlled by Equation (1), where $\left[K_i^t\right]$ is the stiffness matrix for each iteration, $\{F^a\}$ is the total load applied in each sub step, and $\{R_i\}$ is the internal load vector for the nodes.

$$\left[K_i^t\right]\{\Delta u_i\} = \Delta\lambda\{F^a\} - \{R_i\} \qquad (1)$$

The full mathematical details of this algorithm may be found in the ANSYS manual (2019).

Two actual pipeline steel curves X70, are employed in the numerical simulations. The tensile test data was transformed into real stress and logarithmic strain up to Su. The material behavior beyond Su was modeled by Ludwig's power law:

$$\bar{\sigma} = K\bar{\varepsilon}^n \qquad (2)$$

Table 2 presents the parameters utilized for the simulations of pipes with flaw.

Figure 4 presents the X70 tensile curves: the nominal stress vs engineering strain, the real stress (up to Su) vs. logarithmic strain and Ludwig's law vs. logarithmic strain, where the values for K and n were determined using the tensile test data. Table 1 introduces the material parameters found at the tensile test for both steel grades.

Table 1. Material properties.

E[GPa]	ν	S_u [MPa]	S_y [MPa]	K [MPa]	n
215	0,3	713,19	500,1	945,5	0,079

Table 2. Simulation parameters for pipes with and no flaw, API Specification 5L (2018).

Parameter	Symbol	Value	Unit
external diameter	OD	8,625– 219,08	inch -mm
slenderness ratio	D/t	17,25	-
flaw depth	d/t	0.25; 0.75	-
flaw length	L/D	0.1; 1.5	-
flaw half width	w	π/8	rd
model length	M	2286	mm
Temperature Operation	T	80; 120	°C
thermal dilation coeff.	α	1.1 x10^{-6}	°C
material	Steel	X 70	API 5L

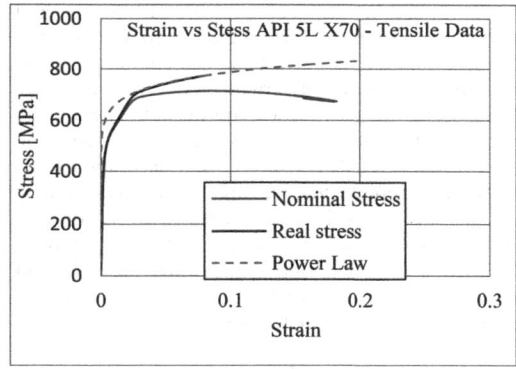

Figure 4. API 5L X70.

3 SIMULATION RESULTS AND DISCUSSION

In this study 6 models were developed and simulated in ANSYS. The simulations were developed with 2 analysis steps, and considering three loads, one thermal and two mechanical (external and internal pressure), show in Table 3.

It is important to note that arc length is applied only from the second step of the simulation, where internal pressure is applied.

Table 4 presents the models developed for numerical simulation. In all cases an external pressure equivalent to 2000 meters depth (22.26 MPa) and an

Table 3. Simulation step loads.

Steps simulations	Termo Load	External pressure	Operating pressure
Step 1	yes	yes	No
Step 2	yes	yes	No

Table 4. Developed models.

	flaw (d/t | L/D)	T [°C]
Model 1	without flaw	120
Model 2	0,25 | 1	120
Model 3	0,75 | 1,5	120
Model 4	without flaw	80
Model 5	0,25 | 1	80
Model 6	0,75 | 1,5	80

external temperature of 4°C is considered. To evaluate the critical pressure, an arbitrary value of internal pressure, much higher than operational, is adopted in this case 160 MPa.

It is important to note that the value of 160MPa was assigned as internal pressure arbitrarily and much higher than the limit pressure. In a simulation that does not use the Arc-Length method, this pressure would cause a convergence error, however, the Arc-Length method allows equilibrium interactions to be maintained and it is essential to observe the representative of the results. This observation must be made by monitoring the force convergence criterion (its decay) and its equivalence in relation to the force by the radial displacement. The target limit pressure is the value immediately after the inflection point of the curve where a decline (or even maintenance) of the force can be observed with increased displacement.

The numerical simulations do not indicate any decrease in the capability of a pipe to withstand pressure to rupture due to the axial compression caused by the thermal dilation of a pipe with complete axial restraint.

The burst pressure obtained by simulation of plastic instability of pipes without defect, and their associated circumferential strain, are presented in Table 5. The simulation results indicate that the temperature does change the strain at instability, but the burst pressure remains virtually unchanged.

The critical instability pressure (P_{bs}), presented in Table 5, corresponds to the maximum internal pressure observed for the instability instant.

Models 3 and 6, which have large defects (75% loss of pipeline wall thickness) still fail in the analysis step where there is only external temperature and pressure.

Table 5. Simulation Results.

	P_{bs} [MPa]	ε_θ	σ_{bs} [MPa]
Model 1	117,36	0,0034783	751,08
Model 2	101,95	0,003632	824,35
Model 3	-	-	-
Model 4	117,33	0,0034749	754,59
Model 5	101,82	0,0036421	827,41
Model 6	-	-	-

Observed as internal pressure increases. In a scenario where the integrity of the duct without external pressure is investigated, an approximate 30% decrease in plastic instability resistance was observed.

The external pressure and temperature are due to the location of the duct installation and are therefore constant. The only variable loading is the internal pressure (instability pressure - P_{bs}), in this work the identification of this limit internal pressure is the objective, being a useful information for the line operators, mainly because it presents internal pressure limit conditions (associated with production) related to different conditions of duct integrity (with and without flaw) and at different temperatures.

Figure 5 shows the von Mises and internal pressure achieved for the 4 simulations performed.

Figure 6 shows the mechanical loading profile (external pressure and internal pressure) through the radial deformation. It is important to note that this pressure does not correspond to the critical pressure of instability, here adopted as internal pressure and previously presented in Figure 5. It is possible to observe that the critical pressure of instability is greater than the maximum pressure shown in Figure 6, this due to the effect of amortization caused by external pressure versus internal pressure.

Figure 7 shows the internal pressure curves for deformation, in which it is possible to observe the inflection point where the pressure starts to fall or

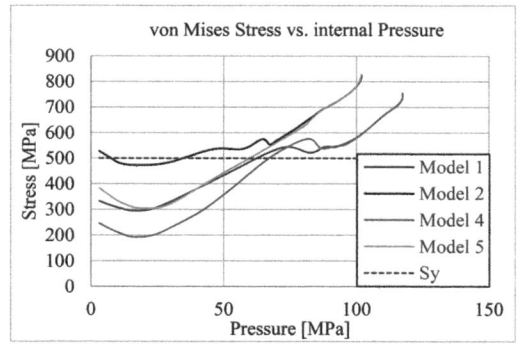

Figure 5. Von Mises Stress vs. Pressure.

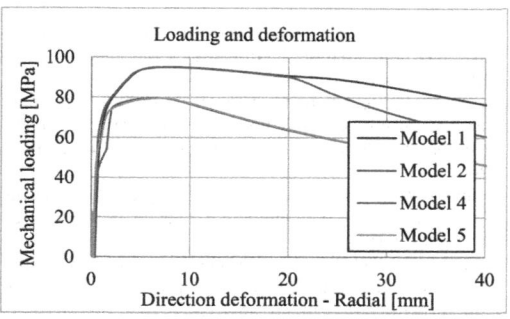

Figure 6. Mechanical loading and deformation -radial [mm].

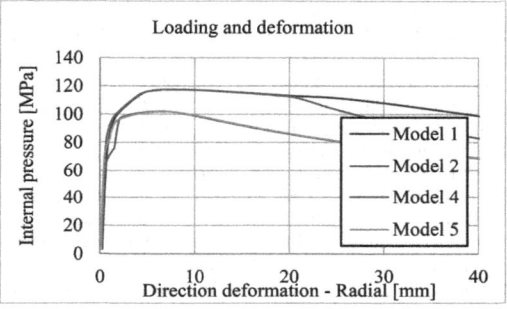

Figure 7. Internal pressure and deformation -radial [mm].

remains constant with the increase in deformation . This point corresponds to the critical internal pressure (P_{bs}) presented in Table 4.

Figure 8 presents the result of von Mises stress for the defective pipeline at 2 evaluated temperatures (80 and 120 ° C), while Figure 9 presents the same results for flaw pipelines (d/t=0,25 and L/D=1). It can be observed that the temperature and external pressure do not impact the critical instability pressure (P_{bs}).

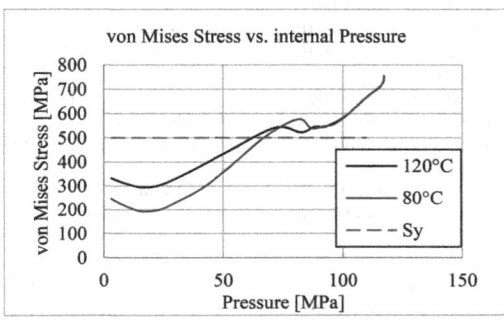

Figure 8. Von Mises Stress vs. Pressure – pipeline without flaw.

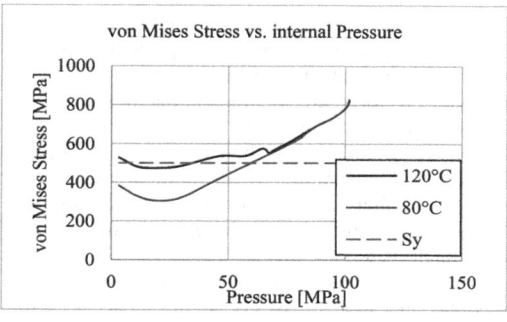

Figure 9. Von Mises Stress vs. Pressure – pipeline flaw. d/t=0,25 and L/D=1.

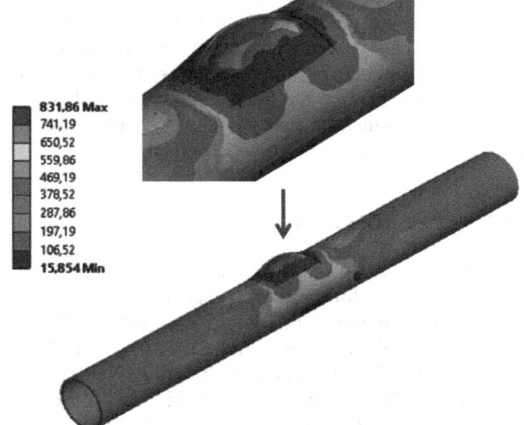

Figure 10. Von Mises Stress [MPa] – pipeline flaw. d/t=0,25 and L/D=1.

Figure 10 shows an image of the stress distribution (von Mises) in the pipeline flaw model. In detail the region of the flaw with critical stress.

When observing the results presented in Figure 10, it is worth mentioning that this is the result of tension found for critical internal pressure (Pbs) for the defective pipeline. like this, pipe burst formulas from API 5C3 can provide nominal burst pressures, the FE models of this paper assess the condition where it is loaded internally and externally, whether it will failure or not will depend on the percentage of metal loss, depending on this percentage it can withstand some metal loss due to inherent project safety factors.

4   CONCLUSIONS

The numerical results used in the test, operating according to external and internal pressure, in conditions with and without defect, are not influenced by the temperature (considering the studied temperature range), which can be observed up to the internal pressure limit, evaluated or that is, the axial thermal expansion efforts are neither impacting or operational limits of the pipeline.

Initially, an increase in the stress was observed as the internal temperature of the fluid increases, this effort due to thermal expansion is axial and compressive as found within the elastic region of the pipeline constituent material.

The increase in internal pressure, up to the limit pressure, raises the stresses beyond the elastic region independently of the stresses of thermal nature. Thus, one can consider that in view of the mechanical stress due to the increasing pressure, the thermal stress is not significant for the pressure that occurs in the plastic region.

In addition, the study can assess the impact of internal pressure on the duct in relation to external pressure, and a significant gain in collapse resistance is observed as internal pressure increases. In a scenario where the integrity of the duct without external pressure is investigated, an approximate 30% decrease in plastic instability resistance was observed.

ACKNOWLEDGMENTS

This work contributes to the project "Cementitious cork composites for improved thermal performance of pipelines for ultradeep waters – SUPBSEAPIPE, with the reference n.° POCI-01-0145-FEDER -031011 funded by European Regional Development Fund (FEDER) through COMPETE2020 - Operational Program Competitiveness and Internationalization (POCI) and with financial support from FCT/ MCTES through national funds.

This study contributes to the Strategic Research Plan of the Centre for Marine Technology and Ocean Engineering, which is financed by Portuguese Foundation for Science and Technology (Fundação para a Ciência e Tecnologia - FCT), under contract UIDB/UIDP/00134/2020.

REFERENCES

ANSYS INC (2019). "ANSYS Mechanical 19.0 - Theory Reference". Manual.

API BULLENTIN 5C3 - Bullentin on Formulas and Calculations for Casing, Tubing, Drill Pipe, and Line Pipe Properties, six edition, October 1, 1994.

API SPECIFICATION 5L, 46th edition, (2018).

Barbosa A.A., Teixeira A.P., C. Guedes Soares, (2017). Strength analysis of corroded pipelines subjected to internal pressure and bending moment. Progress in the Analysis and Design of Marine Structures – Guedes Soares & Garbatov (Eds) Taylor & Francis Group, London, 803–811.

Benjamin A. C., (2008). "Prediction of the failure pressure of corroded pipelines subjected to a longitudinal compressive force superimposed to the pressure loading". International Pipeline Conference 2008, IPC2008-64089.

Bjornoy O.H., Sigurdsson G., Cramer E., (2000). "Residual strength of corroded pipelines, DNV test results", International Offshore and Polar Engineering Conference, 2000.

Considère M, (1885). Mémoires sur l'employ du fer et de l'acier dans les constructions. Annales de Ponts et Chausses; 6 (9):p. 574–775, 1885.

Crisfield, M. A., (1981). "A fast incremental/iterative solution procedure that handles snap-through". Computer and Structures, 1, 1981. p. 55–62.

Cunha S.B., Pacheco M.R., Silva A.B., (2016). Numerical Simulations of Burst of Corroded Pipes with Thermally Induced Compressive Axial. International Pipeline Conference IPC2016, 2016, Calgary, Alberta, Canada.

Cunha, S., (2012). "Comparison and analysis of pipeline failure statistics". International Pipeline Conference 2012, IPC2012-90186.

Davis, P. M., Dubois, J., Gambardella, F., Sanchez-Garcia, E., Uhlig F., (2011). "Performance of European cross-country oil pipelines - Statistical summary of reported spillages in 2010 and since 1971". Report no. 8/11, CONCAWE, http://www.concawe.be.

DNV - Det Norske Veritas. Recommended practice DNV RP-F101 – corroded pipelines, 1999.

Liu J., Chauhan V., Ng P., Wheat S., Hughes C., (2012). "Remaining strength of corroded pipe under secondary (biaxial) loading ". Pipeline Research Council International PRCI Report L 52307, January 2012.

Riks E., (1979). "An Incremental Approach to the Solution of Snapping and Buckling Problems". International Journal of Solids and Structures, 7, p 524–551.

Roberts K. A., Pick R. J., (1998). "Correction for longitudinal stress in assessment of corroded line pipe". International Pipeline Conference 1998, Vol 1 p 553–561.

Roy S., Grigory S., Smith M., Kanninen M. F., Anderson M., (1997) "Numerical simulations of full scale corroded pipe tests with combined loading ", Journal of Pressure Vessel Technology, November 1997, Vol. 119, p 457–466.

Smith M.Q., Grigory S. C., (1996). "New procedures for the residual strength assessment of corroded pipe subjected to combined loads". International Pipeline Conference 1996, Vol 1 p. 387–400.

Smith M.Q., Waldhart C. J., (2000). "Combined loading tests of large diameter corroded pipelines". International Pipeline Conference 2000, Vol 2 p. 769–779.

Teixeira A. P., O. G. Palencia, C. Guedes Soares, (2019). Reliability Analysis of Pipelines with Local Corrosion Defects Under External Pressure. Journal of Offshore Mechanics and Arctic Engineering, ASME Vol. 141/ 051601–1.

Teixeira A.P., C. Guedes Soares, T.A. Netto, S.F. Estefen, (2008). Reliability of pipelines with corrosion defects. International Journal of Pressure Vessels and Piping 85 228–237.

Wang W., Smith M.Q., Popelar C. H., Maple J. A., (1998). "A new rupture model for corroded pipelines under combined loading". International Pipeline Conference 1998, Vol 1 p 563–572.

Zhang X., Duan M., Guedes Soares C., (2018). Lateral buckling critical force for submarine pipe-in-pipe pipelines. Applied Ocean Research 78 99–109.

Developments in Maritime Technology and Engineering – Guedes Soares & Santos (eds)
© 2021 Copyright the Author(s), ISBN 978-0-367-77376-2

Review of the investigation of pipe collapse mechanism based on hyperbaric chambers in Tianjin University

J.X. Yu, M.X. Han & Y. Yu
State Key Laboratory of Hydraulic Engineering Simulation and Safety, Tianjin University, Tianjin, China

J.H. Duan
Offshore Engineering Plan Approval Center of CCS, Tianjin, China

Z.Z. Sun
Powerchina Huadong Engineering Corporation, Zhejiang, China

ABSTRACT: A review of the investigation on pipe buckling, propagation and complicated loading conditions is introduced in the present study. Both numerical simulation and experiments have been used to analyze the pipe buckling mechanism under different loading conditions and corrosion. All of the buckling experiments were conducted by Deep Water Structural Laboratory in Tianjin University. With the upgrading of full-scale and reduced-scale hyperbaric chambers, combined loading conditions including bending-pressure, axial tension-pressure and torsion-pressure can be achieved for testing pipe's load carrying capacity.

1 INTRODUCTION

As an important means for transporting offshore oil and gas, the mechanical behavior of the submarine pipeline during manufacturing, installation, and service had been concerned and studied by lots of scholars. Among them, the problems of pipe buckling, buckling propagation, and buckling crossover were the main concerns of pipelines under the combined action of external pressure, bending moment, and axial force.

Kyriakides and his team at the University of Austin in Texas, USA had focused on the investigation of the buckling mechanism of the offshore pipeline for a long time. Kyriakides (1993) experimentally determined the relationship between pipe collapse pressure and buckling propagation pressure. With the application of ABAQUS software, Xue et al. (2002), and Xue (2006) analyzed the buckling and propagation characteristics of the non-uniform and corroded pipe under external pressure. It was found that the symmetric and anti-symmetric buckling modes were mainly related to the relative thickness and the angle of the non-uniform section of the pipeline. Estefen (1999) evaluated the ultimate strength of the intact pipe under external pressure and bending moment with the consideration of the nonlinear shell theory and the elastoplastic constitutive relationship through reduced scale pipe specimen. Bai & Bai (2005) & Kyriakides & Corona (2007) studied the buckling of pipelines under bending, axial force, and external pressure by tests and numerical simulation. Meanwhile, they

analyzed the effects of diameter-thickness ratio, material properties, initial ovality, and loading path on local buckling of pipelines. Recently, Yu (2019) conducted a sensitivity analysis on the collapse mechanical behavior of thick-walled pipe whose D/t ratio is less than 20.

When local buckling occurs on the pipeline under the deep sea, a large scale buckling propagation will occur along the length of the pipeline under the continuous influence of external pressure. It was first discovered in 1970 by the Buttelle Columbus Laboratory in the United States during model tests. In the following two decades, some experts and scholars carried out a lot of theoretical and experimental research on this phenomenon. Especially in the 1980s, the theoretical model continued to develop with three stages including a discrete ring model, continuous ring model, and the thin-walled cylindrical shell model. In 1975, Palmer (1975) first revealed the pipes buckling propagation phenomenon and established a two-dimensional ring model with four equidistant fixed plastic joints. Based on previous work, J.G. Croll (1985) proposed a more reasonable ring buckling model with the consideration of the strain hardening effect and the axial bending strain energy of the ring. Bhat (1987) further improved the above model and established a moving plastic hinge model, taking into account the axial stresses between the pipe rings. In terms of experiment and numerical simulation, Kyriakides & Netto (2000) simulated quasi-static and dynamic buckling propagation models of the pipeline and founded that dynamic

DOI: 10.1201/9781003216582-82

buckling propagation length was smaller than that under the quasi-static external pressure.

To prevent the phenomenon of large scale buckling propagation, the collapse deformation was usually prevented by buckle arrestors. Park & Kyriakides (1997) carried out quasi-static modeling analysis on the cross-over pressure and cross-over mode of the integral buckling arrestor. Toscano et al. (2008) who comes from the TENARIS Industrial Research Center in Argentina, used ADINA software to simulate the efficiency of integral buckling arrestor on preventing cross-over of pipe propagation compared with other types of buckling arrestors. Lee & Kyriakides (2004) had conducted many quasi-static simulations on the buckling of the integral buckling arrestors to improve its efficiency in preventing pipe buckle.

The research introduced in this paper is mainly based on the full-scale hyperbaric chamber and reduced-scale hyperbaric chamber in the Deep Water Laboratory of Tianjin University. With the expansion of the research content, the functions of the two hyperbaric chambers are also gradually upgraded. The test apparatus will be introduced based on the specific research content below.

2 PIPE BUCKLING INVESTIGATION

As concluded in the standard DNV-OS-F101 (2013), the following formula for calculating the characteristic collapse pressure P_{co} under external pressure is written as follows.

$$
\left.
\begin{aligned}
P_e - P_{\min} &\leq \tfrac{P_{co}(t_1)}{\gamma_m \gamma_{sc}} \\
(P_{co} - P_{el}) \cdot \left(P_{co}^2 - P_p^2 \right) &= P_{co} \cdot P_{el} \cdot P_p \cdot f_0 \cdot \tfrac{D}{t} \\
P_{el} = \tfrac{2E\left(\frac{t}{D}\right)^3}{1-\nu^2}, f_0 &= \tfrac{D_{\max} - D_{\min}}{D} \\
P_p &= 2 \cdot f_y \cdot \alpha_{fab} \cdot \tfrac{t}{D}
\end{aligned}
\right\}
$$
$$(1)$$

Where P_e and P_{min} are the external pressure and minimum internal pressure of the pipe, γ_m and γ_{sc} are material resistance factor and safety class resistance factor respectively; α_{fab} is suggested to be 1 for the seamless pipe; $f_0 \approx 2\Delta_0$ is the initial ovality; D_{max} and D_{min} are the maximum and minimum measured values of the outer diameter, respectively. Equation 1 has been widely applied in the design and evaluation of offshore pipelines at present. It has been found that the ovality in the cross-section and the D/t ratio has a great influence on the collapse pressure, and these two factors are investigated through experiments and numerical simulation in this paper.

2.1 Investigation on the axial length of pipe ovality (Fan 2017)

As Equation 1 neglected the influence of the axial length of ovality, Yu conducted five tests and

Figure 1. Schematic of the scaled hyperbaric chamber.

established a numerical model for investigating this problem. As shown in Figure 1, the experiments were conducted in a reduced-scale hyperbaric chamber at the Institute of Offshore and Naval Engineering of Tianjin University. The test chamber is 1.74m long with a 0.12m inner diameter. For the test pipes, we have artificially imposed a maximum ovality value (Δ_{max}) with a certain straight length (l_s) and have measured its transition length (l_t) in advance, as shown in Figure 2.

$$
\Delta_0 = \frac{D_{\max} - D_{\min}}{D_{\max} + D_{\min}} \tag{2}
$$

The basic geometric parameters of the ovality pipe are shown in Table 1. According to the profile of test specimens, the numerical model is established for comparison.

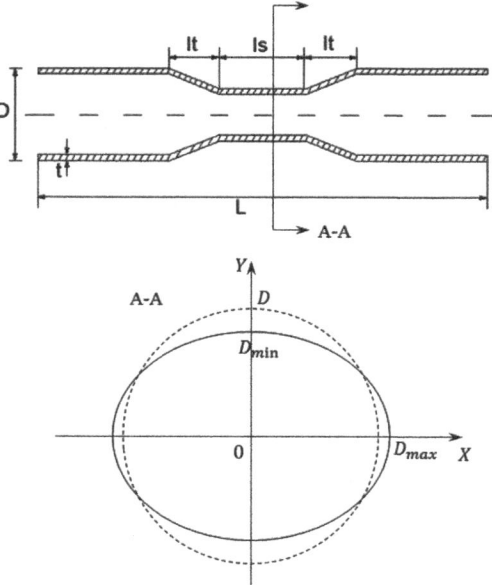

Figure 2. Schematic view of the ovality along the pipe.

Table 1. Geometric characteristics of specimens.

Specimen	D	t	$l_t/2$	l_s	Δ_{max}	
	mm	mm	mm	mm	%	n
SO-1	51.0	3.10	25	30.0	2.03	7.6
SO-2	51.0	3.10	25	17.5	1.76	7.4
SO-3	51.5	3.15	25	47.5	2.51	7.6
SO-4	51.0	3.00	25	65.0	3.76	7.6
SO-5	51.0	3.10	25	80.0	5.62	7.4

As shown in Table 2, the numerical results correspond well with the experimental data and FEM results. Numerical models were developed to conduct a parametric study with the following parameters: a diameter-thickness ratio setting of 17, a total length L fixed at 40D, and a mean value of ovality Δ_0 set to be constant at 0.2%. Figure 3 shows the influence of the straight length (l_s) on the pipe collapse pressure (P_{co}). The collapse pressure gradually decreases as the straight length (l_s) increases, and it becomes constant after l_s exceed 8D.

Table 2. Collapse pressure of ovality pipes.

Specimen	P_{co-EX}	P_{co-TP}	Relative	P_{co-FEM}	Relative error
	MPa	MPa	%	MPa	%
SO-1	29.77	30.27	1.7	29.96	0.63
SO-2	32.36	32.28	-0.2	34.30	5.57
SO-3	28.93	27.62	-0.45	28.35	-2.05
SO-4	23.45	22.95	-2.1	22.73	-3.19
SO-5	24.44	23.36	-4.4	23.52	-3.76

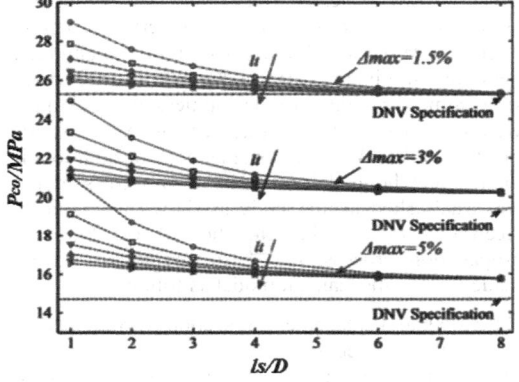

Figure 3. The sensitivity of P_{co} to l_s/D at different Δ_{max}.

In conclusion, the axial length characteristics of maximum ovality had crucial effects on the collapse pressure. A larger error value appeared when axial length parameters of maximum ovality had small values.

2.2 Investigation on the thick-walled pipes buckling (Yu 2019)

Based on the thin-walled theory, the numerical model proposed by Kyriakides & Arikan (1983), Yeh & Kyriakides (1986) & DNV neglected the effect of thickness stretching, which has significant influence in analyzing the collapse behavior of thick-walled tubes as concluded by Yang et al. (1997). Even though, previous studies have obtained beneficial results, until now some of them are still controversial in the applicability, especially for the offshore pipelines with $D/t < 20$ as concluded by Fan et al. (2017), which should be considered as the thick shell. The effects of the disturbing parameters on the collapse pressure are various and complicated. Thus, a 2D model based on the thick shell and nonlinear large deflection theory are established to predict the collapse pressure. It is verified by both the laboratory experiments and the results from other researchers. To conduct sensitivity analysis on the collapse pressure of thick-walled pipes, eight pipe specimens were prepared for verifying the accuracy of the numerical model.

Two full-scale pipes and six reduced-scale pipes were incised from the original pipes and the thickness of them was measured by thickness gauge. A local ovality was pre-inflicted on the specimens and it was measured by the three coordinate joint arms. Meanwhile, the strain gauges were attached to the pipes to record the deformation during collapse experiments. These geometric parameters such as mean diameter, thickness, and ovality are summarized in Table 3.

The full-scale and reduced-scale hyperbaric apparatuses are shown in Figure 4 and Figure 5, respectively. The full-scale one is 11.5 meters in length and 1.25 meters in inner diameter. The reduced-scale one is 3.36 meters in length and 0.24

Table 3. The geometric parameters of the specimen.

Number	D	t	D/t	Δ_0		L
	mm	mm	–	%	mm	
F1	325.0	16.60	19.58	0.34	8000	
F2	273.0	10.20	26.76	0.28	8000	
R1	76.16	6.33	12.04	0.22	2300	
R2	75.70	6.30	12.02	0.26	2300	
R3	75.70	5.97	12.68	0.47	2300	
R4	75.70	6.00	12.62	0.55	2300	
R5	76.00	4.26	17.84	0.43	2300	
R6	76.08	4.21	18.07	0.68	2300	

Figure 4. Schematic of the full-scale hyperbaric chamber (in new campus).

Figure 5. Schematic of the reduced-scale hyperbaric chamber (in new campus).

Table 4. Comparison of the collapse pressure between numerical and experimental results.

No.	D/t	Δ_0 %	E GPa	σ_y MPa	n_h –	P_{co-num} MPa	P_{co-ex} MPa	Error %
F1	19.58	0.34	204	362	13.5	28.73	28.20	-1.88
F2	26.76	0.28	206	271	9.10	14.33	15.24	5.97
R1	12.04	0.22	193	208	6.90	44.45	47.37	6.16
R2	12.02	0.26	193	208	6.9	42.00	43.69	3.87
R3	12.68	0.47	193	208	6.9	36.45	39.31	7.28
R4	12.62	0.55	193	208	6.9	34.85	37.17	6.24
R5	17.84	0.43	193	264	8.3	25.13	26.61	5.56
R6	18.07	0.68	193	264	8.3	23.20	24.15	3.93
R7	19.93	0.62	193	225	7.5	19.11	19.27	0.83
R8	19.91	1.27	193	225	7.5	16.97	17.64	3.80
E1	18.66	0.08	197	241	13	23.67	25.48	7.10
E2	25.66	0.12	191	320	14	16.86	17.31	2.60
E3	28.65	0.15	198	320	14	13.76	13.83	0.51
E4	34.67	0.05	184	272	12	8.66	9.50	8.84
E5	25.45	0.14	199	309	12	17.76	18.26	2.74
E6	19.75	0.02	190	291	10	26.58	25.32	-4.98
E7	13.12	0.04	183	304	9.4	53.13	51.59	-2.99

meters in inner diameter. Both of them consist of a main chamber, a hydraulic loading controlling system, and a data acquisition system.

The comparison of collapse pressure between numerical results (P_{co-num}) and experimental results (P_{co-ex}) is listed in Table 4. Meanwhile, the experimental results obtained by Yeh & Kyriakides (1986) and Madhavan (1988) were also superimposed as E1 ~ E4 and E5 ~ E7. The good agreement of P_{co-num} and P_{co-ex} can be seen in Table 4 and the relative errors of them are no more than 10%.

Verified by the experimental results, sensitivity analysis of the manufacturing imperfections and the material properties on the collapse pressure are also investigated. The main conclusions can be drawn as follows:

(1) Compared with experimental results, the prediction of characteristic collapse pressure given by DNV specification is relatively conservative for thick-walled pipes.

(2) The parametric study shows that thick-walled pipes are more easily affected by manufacturing imperfections and material properties than thin-walled pipes, especially initial ovality (Δ_0), residual stress (σ_R), and hardening factor (n_h). Among the disturbing parameters of interest, initial ovality (Δ_0) is an important influence factor in collapse pressure determination.

Discovery is also found that the collapse pressure of thick-walled pipes is mainly affected by plastic mechanical behavior.

3 PIPE PROPAGATION INVESTIGATION

3.1 Investigation on the propagation theory of pipe (Yu 2014)

To simulate and predict the buckle propagating phenomena of the pipeline, a series of theoretical quasi-static analyses were established in the literature. Relatively speaking, the most successful two-dimensional model is the nonlinear ring model established by Kyriakides & Yeh (1984), and Kyriakides (1993), with the consideration of material hardening effect. Since the membrane effect in the longitudinal direction of pipeline affects the propagating pressure, some three-dimensional numerical simulations were developed with good agreement to the experimental results. With the consideration of both longitudinal and hoop deformations, a ring-truss model was established in this study as shown in Figure 6. According to the characteristics of the buckling propagation process, two basic assumptions were made for this mechanical model as follows.

(1) No axial displacement or section warping occurs for the ring section.

(2) Axial deformation is accounted by longitudinal truss.

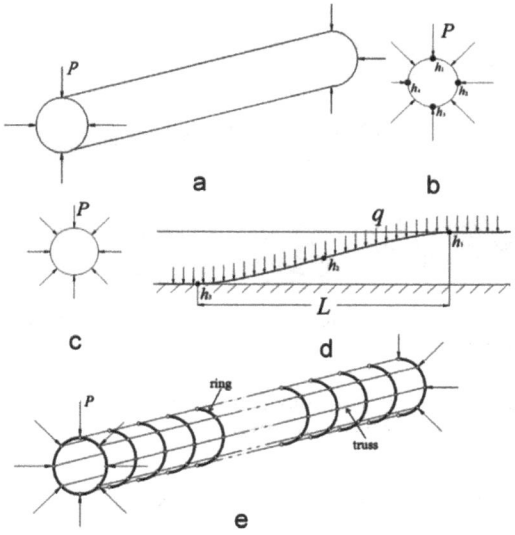

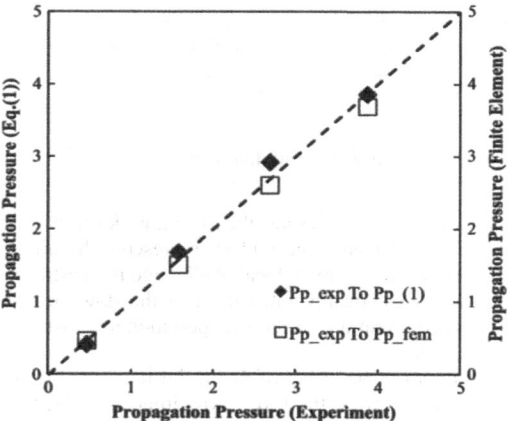

Figure 7. Comparison of propagation pressure determined from experimental and numerical results.

Figure 6. Analysis model: (a) actual situation, (b) four-hinge model, (c) nonlinear ring model, (d) beam-foundation model and (e) ring-truss model.

As the length of the transition zone is essential for the determination of propagation pressure, experiments and numerical simulations are also conducted for correcting the theoretical model. In this study, four types of samples with different diameter to thickness ratio were tested during the experiments. The geometric and material properties of specimens are tabulated in Table 5.

As shown in Figure 7, the results obtained from the full-scale hyperbaric chamber (Figure 8) have good agreement with both the published works and numerical simulation. Based on a lot of numerical calculations, an empirical formula with a modified empirical coefficient is proposed for the determination of the propagation length. The empirical formula is written as follows and the parameter c is modified to 2.15.

$$\left. \begin{array}{l} L/R = c\sqrt{R/t} \\ R = (D - t)/2 \end{array} \right\} \quad (3)$$

Table 5. Properties of the test specimen.

Sample	Material	D mm	t mm	E GPa	σ_0 MPa	σ_y MPa	n mm
1	SS304	51	2	193	313	278	10.1
2	API X65	325	10	206.1	448	400	10.7
3	API X65	406	10	206.1	448	400	10.7
4	Q235B	325	6	206.1	235	203	10.7

Figure 8. Schematic of the full-scaled hyperbaric chamber.

3.2 Investigation on the crossover modes of the pipe integral arrestors (Yu 2017)

Two different cross-over modes of integral buckle arrestors are identified in the published research, i.e., the flattening mode and the flipping mode. However, arrestors are not completely crossed by the flattening mode or the flipping mode according to experimental results (1997). For example, the angle of collapse direction between the upstream and the downstream pipes could be 30° or 60°. Hence this paper explores the reason why different angles of collapse direction appear in the experiments and FE models.

To the best of authors' knowledge, the effect of downstream local ovality is first introduced to analyze the cross-over modes of integral buckle arrestors. Meanwhile, the geometric and material characteristics of pipes and arrestors can also determine the cross-over modes of arrestors. In this study, the test specimens were made with pipe and arrestor which is shown in Figure 9. Meanwhile, the geometric parameters of two full-scale pipes and ten reduced scale pipes are listed in

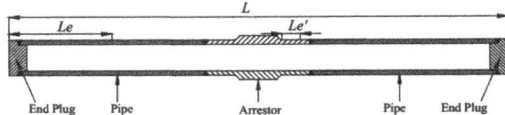

Figure 9. The specimen with four parts.

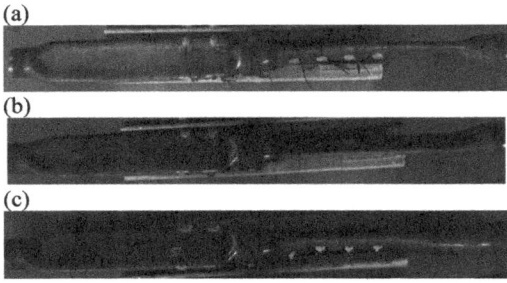

Figure 10. Experimental results after test: (a) S7; (b) S8; (c) S9.

Table 6. Here, Δ_d represents the maximum local ovality of the downstream pipe and θ_d represents the angle between the downstream local ovality and the upstream one. The local ovality introduced to the downstream pipes is common due to the transportation and installation of pipes.

To confirm the effect of the initial local ovality of downstream pipes, the three experiments of S7, S8, and S9 are analyzed as an example whose results are shown in Table 7 and Figure 10.

The arrestors of specimens S7 and S8 were crossed by the flipping mode and the θ'_d of S7 and S8 was equal to their θ_d, respectively. Moreover, the θ'_d of S8 was 60°. Compared with specimen S8, the arrestor of specimen S9, not imposed the local ovality of the downstream pipe, was crossed by the flipping mode. The only difference between specimen S8 and S9 was the local ovality of the downstream pipe. The different results of them showed that the

downstream local ovality had a remarkable effect on the cross-over modes of arrestors.

4 UPGRADE OF THE TEST APPARATUS

Considering the real loading situation of the pipeline in the installation and service period, it usually subjected to the combination of bending, tension, torsion, and external pressure. To analyze the mechanism of the pipe buckling behavior under complicated loading conditions more accurately, our research group has upgraded the existing test apparatus.

First, for the full-scale hyperbaric chamber, we applied a lateral loading vibration device in the side of the chamber in 2019, which can simulate the seismic load and lateral excitation of the current during service time. The schematic diagram of the new vibration equipment is shown in Figure 11.

Secondly, in 2019, our research group upgraded the size and shape of the whole reduced scale chamber which is shown in Figure 12. The combined action of axial force, bending moment, and water pressure can be applied to the test pipe simultaneously. At the same time, the axial force, bending moment, and torque loads can be implemented for cyclic loading to simulate the real loading state when the pipeline is under installation.

Table 6. Geometric characteristics of specimens.

No.	D mm	t mm	Δ_0 %	L_a mm	h mm	Δ_d %	θ_d –	Material
A1	325	6	5	160	14.4	0.2	0°	Q235B
A2	325	6	5	160	15.6	0.0	0°	Q235B
S1	51	2	8	63	4.2	0	0°	SS304
S2	51	2	8	26	4.8	0	0°	SS304
S3a	51	3	8	26	7	0.2	30°	SS304
S3b						0.2	90°	SS304
S4a	51	3	8	26	9	0.2	0°	SS304
S4b						0.3	0°	SS304
S5a	51	3	8	51	9	0.0	0°	SS304
S5b						0.2	0°	SS304
S6a	51	3	8	61	9	0.0	0°	SS304
S6b						0.2	90°	SS304

Table 7. Experimental results of S7, S8, and S9.

No.	D mm	t mm	Δ_0 %	L_a mm	h mm	Δ_d %	θ_d –	Cross-over modes
S7	51	3	8	51	7	0.2	90°	flipping
S8	51	3	8	61	7	0.2	60°	flipping (60°)
S9	51	3	8	61	7	0.0	0°	flipping

Figure 11. Schematic of the upgraded full-scale hyperbaric chamber.

Figure 12. Schematic of the upgraded reduced-scale hyperbaric chamber.

5 CONCLUSIONS

Surrounding the experimental tests, a review of the investigation on pipe buckling, propagation, and the cross-over mechanism of integral arrestors is concluded in this paper. Some main conclusions are summarized according to the previous study as follows.

(1) The present DNV-OS-F101 standard is relatively controversial for calculating the thick-walled pipes' collapse pressure. Even though the length of ovality is neglected by the standard, the experiments and numerical results showed that it can reduce the collapse capacity of the pipe.
(2) A ring-truss model is established to determine the propagation pressure of the pipe with the consideration of both longitudinal and hoop deformation of the pipe. Combing the results of experiments and numerical model, the propagation length is modified with a formulation.
(3) With the verification of full and reduced scale experiments, it is found that the cross-over mechanism of the integral arrestor is affected by the value and angle of the ovality in the downstream of the pipe.
(4) In 2019, our research group upgraded the test apparatus. A lateral loading vibration device was installed beside the full-scale hyperbaric chamber. The cyclic bending, tension, and torque are implemented for the reduced scale chamber.

REFERENCES

Bai, Y. & Bai, Q. 2005. Subsea Pipelines and Risers. Oxford: Elsevier Science Ltd.

Bhat, S.U. & Wierzbicki, T. 1987. On the length of the transition zone in unconfined buckle propagation. Journal of Offshore Mechanics and Arctic Engineering, 109: 155–162.

Croll, J. G. A. 1985. Analysis of buckle propagation in marine pipelines. Construct Steel Research, 5: 103–122.

DNV (Det Norske Veritas). 2013. Offshore standard. DNV-OS-F101: Submarine Pipeline Systems. Høvik: DNV.

Dyau, J. Y. & Kyriakides S. 1993. On the propagation pressure of long cylindrical shells under external pressure. International Journal of Mechanical Sciences, 35 (8):675–713.

Estefen, S. F. 1999. Collapse behavior of intact and damaged deepwater pipelines and the influence of the reeling method of installation. Journal of Constructional Steel Research, 50: 99~114.

Fan, Z. Y. & Yu, J. X. et al. 2017. Effect of axial length parameters of ovality on the collapse pressure of offshore pipelines. Thin-Walled Structures, 116: 19–25.

Kyriakides, S. 1993. Propagating Instabilities in Structures. Advances in Applied Mechanics, 30(114):67–189.

Kyriakides, S. & Arikan, E. 1983. Postbuckling Behavior of Inelastic Inextensional Rings under External Pressure, J. Appl. Mech. 50 (3): 537.

Kyriakides, S. & Corona, E. 2007. Mechanics of Submarine Pipelines. Oxford: Elsevier Science Ltd.

Kyriakides, S. & Netto, T. A. 2000. On the dynamics of propagating buckles in pipelines. International Journal of Solids and Structures. 37: 6843–6867.

Kyriakides, S. & Yeh, M. K. et al. 1984. On the Determination of the Propagation Pressure of Long Circular Tubes. World Journal of Urology, 32(3):691–5.

Lee, L. H. & Kyriakides, S. 2004. On the arresting efficiency of slip-on buckle arrestors for offshore pipelines. International Journal of Mechanical Sciences, 46: 1035–1055.

Madhavan, R.1988. On the Collapse of Long Thick-walled Circular Tubes under Biaxial Loading. PhD dissertation, California Institute of Technology, Pasadena, California.

Palmer, A. C. & Martin, J. H. 1975. Buckle propagation in submarine pipelines. Nature (London), 254: 46~48.

Park, T. D. & Kyriakides. S. 1997. On the performance of integral buckle arrestors for offshore pipelines. International Journal of Mechanical Sciences, 39 (6):643–669.

Smienk, H. & Karjadi, E. et al. 2017. Extending the Limits for Thick Walled Pipe (D/t < 20) for external pressure and combined loading. ASME. International conference on off shore mechanics and arctic engineering, Volume 5B: Pipelines, Risers, and Subsea Systems:V05BT04A002. doi:https://doi.org/10.1115/OMAE2017-61055.

Toscano, R. G. et al. 2008. Collapse arrestors for deepwater pipelines. Computers and structures, 86:728–743.

Xue, J. & Hoo Fatt, M. S. 2002. Buckling of a non-uniform shell subjected to external hydrostatic pressure, Eng. Struct, 24: 1027–1034.

Xue, J. H. 2006. A non-linear finite-element analysis of buckling propagation in subsea Corrode pipelines, Finite Elements in Analysis and Design, 42: 1211–1219.

Yang, C. & Pang, S.S. et al.1997. Buckling analysis of thick-walled composite pipe under external pressure, J. Press. Vessel Technol, 119 (1):111–121.

Yeh, M.K. & Kyriakides, S. 1986. On the collapse of inelastic thick-walled tubes under external pressure, J. Energy Resour. Technol, 108 (1): 35–47.

Yu, J. X. & Duan, J. H. et al. 2017. The cross-over mechanisms of integral buckle arrestors for offshore pipelines. Applied Ocean Research, 76: 236–247.

Yu, J. X. & Han, M. X. et al. 2019. A modified numerical calculation method of collapse pressure for thick-walled offshore pipelines. Applied Ocean Research, 91: 101884.

Yu, J. X. & Sun, Z. Z. et al. 2014. Ring-truss theory on offshore pipelines buckle propagation. Thin-Walled Structures, 85(dec.):313–323.

Shipbuilding and ship repair

Developments in Maritime Technology and Engineering – Guedes Soares & Santos (eds)
© 2021 Copyright the Author(s), ISBN 978-0-367-77376-2

Biofouling control in heat exchangers by statistical techniques

D. Boullosa-Falces & M.A. Gomez-Solaetxe
Department of Nautical Sciences and Marine Systems Engineering, University of the Basque Country UPV/EHU, Portugalete, Spain

S. García, A. Trueba & D. Sanz
Department of Sciences & Techniques of Navigation and Ship Construction, University of Cantabria, Santander, Spain

ABSTRACT: Seawater is an unlimited resource used in the cooling of marine equipment. However, the use of seawater water brings with it an undesirable phenomenon known as biofouling. This adheres to the interior of the heat exchangers, reducing their efficiency and generating production losses due to unplanned shutdowns. This paper proposes a statistical method to predict the growth of biofouling. The inside of the tubes of a heat exchanger cooled by seawater was monitored by statistical techniques, while the growth of the biofouling inside the tubes was allowed. The results showed that the statistical method, prematurely detected the biological adhesion in the heat exchanger, before it reduced its efficiency below the optimal operating conditions. Knowing in advance the evolution of the biofouling growth process will help in the choice and at the moment when a specific antifouling treatment is used and will reduce operation and maintenance costs due to unplanned shutdowns.

1 INTRODUCTION

A major problem in industries that use heat exchanger equipment cooled with seawater in their industrial processes is biological fouling. Biofouling consists of an organic film composed of micro-organisms embedded in a polymer matrix they themselves create (biofilm) that inorganic particles can reach and be retained (salts and/or corrosion products), which are the outcome of other types of fouling undergone in the process. This biofilm composed of micro-organisms (microbial biofouling or microfouling) may give rise to accumulation of macro-organisms (macrobial biofouling or macrofouling) (Eguía et al. 2008).

Biofouling acquires greater relevance when it develops in the marine environment, due to the high biological activity of seawater (García & Trueba 2019). Biofouling on surfaces that are in contact with the seawater implies, acceleration of microbiologically influenced corrosion, reduction in thermal performance, the formation of obstacles to fluid flow, loss of production and increase in the maintenance cost of heat exchanger equipment (Eguía & Trueba 2007; Trueba et al. 2015a).

Fouling on heat transfer surfaces of power plants are a major economic and environmental problem worldwide. Different studies have shown that the total cost of industrialized countries due to fouling in heat exchanger can be more than 0.25% of gross national production. In addition, indirectly fouling also affects the environment due to the increased resistance from fouling will increase power consumption of pumps (Gudmundsson et al. 2008). Furthermore, the costs associated with biofouling in heat exchangers, include operating costs, loss of production due to unscheduled shutdowns and maintenance costs due to the cleaning of heat exchangers. For these reasons it is important to detect and mitigate as soon as possible the biofouling before it reaches levels high enough that the performance of the equipment decreases (Müller-Steinhagen et al. 2011).

In the last decades, various mitigation techniques have been developed to minimize the undesirable phenomenon of biofouling, such us, physical techniques, ultrasound (Bott 2001), ultraviolet light (López-Galindo et al. 2010), reverse flow (Eguía et al. 2008) electromagnetic fields (Trueba et al. 2015b), heat treatment (Rajagopal et al. 2012), chemical techniques include biocide oxidants (Trueba et al. 2013), and non-oxidising (Cloete et al. 1998) and biological techniques (Dobretsov et al. 2013).

In order to implement the appropriate treatment it is necessary to monitor and control the growth of biofouling. Different techniques have been used for this purpose such as acoustic, optical (Withers 1996), x-rays (Ismail et al. 2004), electric resistance (Chen et al. 2004). However, these methods require a intrusive and expensive equipment to control de biofouling process and can even alter it.

DOI: 10.1201/9781003216582-83

The biofouling process generates in a heat exchangers a slow and progressive decrease in the temperature difference between the input and the output. In this case, alternative techniques, such as cumulative sum control charts (CUSUM), can be used. CUSUM charts is a statistical technique wich represents the cumulative sum of the deviations, wich contain information from previous samples (Croarkin et al. 2006). Therefore, they are very efficient in detecting small but progressive changes in the variable being measured (Hawkins & Olwell 2012); In this issue, these charts have been used in the online monitoring of the fuel processing of a marine diesel engine. The effectiveness of the technique was demonstrated by the detection of slow and progressive deviations (Boullosa-Falces et al. 2017). This technique has also been used combined with the slope technique in the detection of biofouling in the district heating industry (Gudmundsson et al. 2008), however a large amount of simulated data was necessary to create the algorithm. Additionally, in the biomedical industry as a surveillance system for continuous monitoring of clinical outcomes, using routinely collected data (Sibanda & Sibanda 2007).

In this study, we proposes used a CUSUM control chart to predict the growth of biofouling in heat exchangers. The growth of the biofouling inside the tubes was allowed until it was maximum, while the inside of the tubes of a heat exchanger cooled by seawater was monitored by CUSUM control chart, which prematurely detected the growth of the biofouling before the heat exchanger worked below its normal operating conditions.

2 MATERIALS AND METHODS

2.1 Heat exchangers

The experimental plant (Figure 1) consisted of eight single-step counter-flow tubular heat exchangers designed and manufactured according to ASME VIII

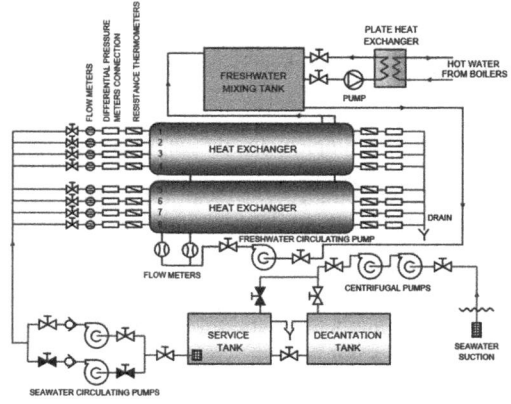

Figure 1. Experimental plant.

(American Society of Mechanical Engineers) and the Tubular Exchanger Manufacturing Association (TEMA) standards. The outer shell was constructed of AISI 304 grade stainless steel with an external diameter of 0.24m (0.02m in thickness). The length of the tubes was 3.1 m, with an inner diameter of 0.010m (0.0015m in thickness); the tubes were linearly arranged and made from AISI 316Ti (N6-class surface roughness).

2.2 Seawater cooling system

The coolant seawater was taken from Santander Bay in norther Spain (43°28′N and 3°48′W) to the laboratory a distance of 250 m by two centrifugal pumps (ITUR AU-M1 1.5/10) in serial operation. There, seawater was subjected to a macro-filtering and decantation process in a 1 m³ tank, after, circulation pumps (Grundfoss CHI 4-50 AWG) pumped the seawater, at 1.9 bar to the tubular bundle in the heat exchanger, through the control instruments, differential pressure meters (PTX 2170-1656) and resistance thermometers (Pt 100). The seawaster flow spped was set at a constant 1 m s^{-1} within the tubes using positive displacement flow meters (Badger Meter M25 PFT-420, Badger Meter Europe, Neuffen, Germany).

2.3 Freshwater heater system

The freshwater circulation rate through the shell-side was established at 17 m³ h^{-1} using a rotameter-type flow metre for eight heat exchangers. The inlet temperatures of the shell-side were kept at a constant temperature of 36 °C using a proportional-integral derivative (PID) automatic control system (OMROM E5CK) that acted on a motorised three-way valve in the primary circuit, where water at 80 °C from the boiler room released heat in a plate heat exchanger (Sedical UFX-26H) to the water from the secondary circuit (Figure 1). The freshwater outlet of the shell-side was pumped by a centrifugal pump (Grundfos MG90SA4-24F 115IP 44CLF) into a mixing tank. The inlet and outlet temperatures of the shell-side were measured using resistance thermometers (Pt 100).

2.4 Data management system

Data management consisted of four multifunctional configurable modules (STEPR model DL01-CPU, STEP Logistica y Control, Barcelona, Spain) that were capable of registering and controlling all of the parameters of the plant. The data were managed with a software program (TCS-01 version 3.0.3) via an RS232 communication port and stored on a personal computer. The monitoring set-up allowed the plant to be controlled in real time and permitted the generation of historic records.

The experiment was completed in 63 days and was conducted during the period of the year when the maximum amount of biological activity occurs (July and August). In each of the eight tubes of the heat exchanger the hydraulic conditions were maintained at a constant flow velocity of 1 m s^{-1} with a constant flow rate of 4.8 1 min^{-1}. During the time that the experiment lasted were acquired 1,486 samples at 1-hour intervals.

The variables selected to develop the study were cooling water inlet temperature (T_{in}) and cooling water outlet temperature (T_{out}). The mean and standard deviation values of each variable are presented in Table 1.

2.5 Biofouling growth monitoring

The growth of biofouling adherence to the internal surface of the heat exchanger of experimental plant was monitored through the heat transfer resistance value (R_f) and CUSUM control graphs. The study focused on tube number one.

2.6 Heat transfer resistance value Rf

The heat transfer resistance variable R_f [m^2 K kW^{-1}], is the sum of the conductive and advective heat transfer resistances (Chen et al. 2016). Conductive heat transfer resistance is the results from the transport of heat from high temperature zone to a lower-temperature zone in a liquid or solid phase by the movement of electrons and molecules, it increases with the addition of insulating layers formed by accumulated biofilm. Advective heat transfer resistance is the result of the transport of heat caused by the movement of the fluid, it decreases with the formation of turbulence in the interface zone due to the existence of biofilm. Heat transfer resistance was calculated using equation (1) (Trueba et al. 2014).

$$R_f = \frac{A_t}{Q\rho c_p \ln\left(\frac{T_{shell} - T_{in}}{T_{shell} - T_{out}}\right)} \quad (1)$$

where A_t is the total surface area in m^2 covered by the biofilm in the tube; Q is the cooling water flow (m^3 s^{-1}); ρ is the seawater density (1,025 kg m^{-3}); c_p is the specific heat at constant pressure (4.18 kJ kg^{-1} K^{-1}); T_{shell} is the freshwater temperature (K); T_{in} is the inlet temperature of the cooling water (K); T_{out} is the outlet temperature of the cooling water (K).

2.7 Cusum statistical technique

CUSUM control chart was used (Montgomery 2009) to monitor ΔT between the cooling water outlet temperature (T_{out}) and the inlet temperature (T_{in}).

The Cusum chart, calculated the deviations of each value with respect to the target value μ_0, distinguishing between deviations positive C$^+$ and negative C$^-$. The statistics C$^+$ and C$^-$ have a form,

$$C_i^+ = max\left[0,\ X_i - (\mu_0 + K) + C_{i-1}^+\right] \quad (2)$$

$$C_i^- = max\left[0, (\mu_0 - K) - X_i + C_{i-1}^-\right] \quad (3)$$

where the starting values are $C_0^+ = C_0^- = 0$

To develop these graphs, the average run length (ARL0) must be defined as the minimum number of samples necessary to minimise the probability of the control chart indicating that the cooler is dirty, when in fact it is not (type I error), or that biofouling has not been detected, when in fact it is present (type II error) (Montgomery et al. 2009). The average value of the ARL0 samples is called μ_0. An ARL0 value of 375 was established for this type of heat exchanger (Boullosa-Falces et al. 2019).

The reference value K, at which the cumulative deviation is considered to be significant. This value determines the greater or lesser sensitivity of the control chart and is selected according to the standard deviation (σ_0), which is determined by means of Equation (4), as follows:

$$K = k \cdot \delta \cdot \sigma_0 \quad (4)$$

where k is a tabulated value and typically accepts a value of 0.5; quantity δ represents the change in the mean of the standard deviations units to be detected which acquired a value of 3 for this type of heat exchanger.

The decision interval H, was determined by the following equation (5):

$$H = h \cdot \delta \cdot \sigma_0 \quad (5)$$

where σ_0 is the standard deviation of the process in control, and h is a tabulated value based on ARL0 and k. The relevant tables can be found in the literature (Hawkins & Olwell 2012).

If C_i^+ or C_i^- exceeds the decision interval H, the process is considered to be out of range.

Table 1. Mean and standard deviation, values (T_{in} and T_{out}).

Measure	Mean(μ)	Standard Deviation (σ)
T_{in} (°C)	21.6	0.6
T_{out} (°C)	26.2	0.5

3 RESULTS

3.1 Monitoring of heat transfer resistance

The evolution of the Rf chart (Figure2) shows three growth phases of biofouling growth clearly differentiated. On the one hand, the induction phase, which appears from the beginning of the experiment to the 700 hours' samples, this is within the first 30 days. Then, the exponential growth phase starts from sample 700 to 1080, which corresponds to days 30 and 45, respectively. Finally, the leveling-off phase started in sample 1080.

3.2 Monitoring of CUSUM statistical technique

The ARLO in control situation of the first 375 samples was established (Boullosa-Falces et al. 2019). Then,1111 samples to be controlled (ARL1) were monitored through the cusum control chart. The mean and standard deviation of the ARL0 in-control situation and the new samples to control ARL1 are presented in Table 2.

The control chart is showed in (Figure 3). In the ordinate axis the values reached by the cusum statistic are shown and in the ordinate axis the hourly samples are presented. The established control limit H, is represented by a horizontal red line. If any value is below the control limit, the chart indicates that the temperature difference between the input and the output is less than that established for optimum operating conditions of the heat exchanger.

The results obtained using the control charts are summarised in Table 3.

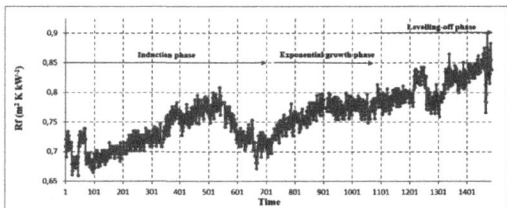

Figure 2. R_f evolution.

Table 2. Mean and standard deviation of ARL0 and ARL1.

		Means (μ)	Standard Deviations (σ)
ΔT (°C)	ARL0	4.62	0.21
	ARL1	4.52	0.24

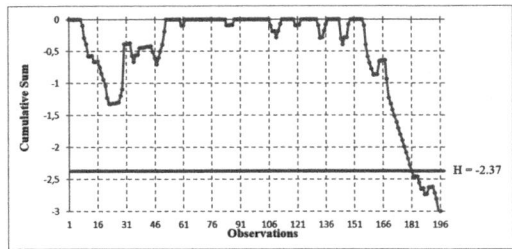

Figure 3. CUSUM chart.

Table 3. CUSUM control graph results.

H	-2.37
K	0.28
ARLO	375
ARL1	181
Number detected sample (ARLO + ARL1)	556
Detection day since last cleaning	23

4 DISCUSSION

The growth of biofouling inside the tubes of the heat exchanger occurs slowly and progressively. As indicated in the state of the art, CUSUM control graphics have been widely used in different processes in the monitoring of variables that have slow and progressive changes. In this work, the Cusum control charts have been used to monitor the growth of biofouling inside on the tubes of heat exchanger and also, to detect an increase greater than that allowed for optimal operation of the heat exchanger with respect to the normal operating conditions.

In an experimental plant formed by two heat exchangers that use seawater as a cooling fluid, ideal conditions for the initiation of biofouling were created. The study focused on tube number 1.

An ARLO of 375 samples was established, which correspond to the optimal operating conditions of the heat exchanger. Subsequently, new samples were monitored, and it was detected where the change occurred with respect to the optimal conditions previously established based on the change in the mean.

The (Figure 3) show the results obtained in the detection of biofouling through the CUSUM control chart. Here, the C⁻ statistic acquires values close to the target value for a long time and tends to take null values up to sample 157, during this time the biofouling has slowly and progressively adhered to the walls of the cooler tubes, remaining heat transfer within the allowed limits. From this moment on, a progressive decrease of the C- statistic begins, which indicates that the amount of biofouling that is adhering to the surface of the cooler tubes is deviating from normal operating conditions, causing a slow decrease in the temperature difference since the last cleaning, to exceed the limit established in sample 181. That is, from day 23, 556 hours' samples, the heat exchanger did not behave according to the normal operating conditions established

The (Figure 2) shows the different phases of biofouling growth as a function of the evolution of the variable Rf. This indicates that until day 30, sample 700, the exponential growth phase does not begin, that is, the cooler is not considered to be dirty enough to have to take preventive measures.

We observe that the CUSUM control chart is efficient in the detection of Biofouling. The algorithm developed so that its implementation is simple and economical through the only analysis of the temperature difference between the input and the output of the cooling water. These variable were acquired with conventional instrumentation available in any installation. However, the monitoring of the Rf variable becomes more complex for any type of user.

The monitoring by the conventional method indicates in a clearer way the evolution of biofouling development in its growth phases (induction, exponential growth and levelling-off). However, the monitoring by means of the CUSUM chart has a more predictive character so it could be used as a complement to the protocols of maintenance and to be able to take preventive actions, before the cooling capacity of the exchanger decreases below a set value for which it was designed, avoiding unplanned stops. Also will help in the choice and at the moment when a specific antifouling treatment is used and will reduce operation and maintenance costs.

5 CONCLUSIONS

This work proposed the CUSUM control charts to predict the growth of biofouling in a heat exchanger. The difference in cooling water temperatures between the inlet and outlet was monitored by means of basic instrumentation, while allowing the growth of biofouling inside the tubes.

The results showed the effectiveness of the CUSUM graphs to detect biological growth prematurely in the heat exchanger with respect to the optimal operating conditions established. This technique of predicting the behavior of biofouling on the walls of the heat exchanger, will help to take preventive actions without having to wait for the cooler to be below normal operating conditions, avoiding unscheduled stops. In addition, the developed algorithm can be used economically and easily, with the basic instrumentation available in any type of heat exchanger.

REFERENCES

Bott T (2001) Potential physical methods for the control of biofouling in water systems. Chem Eng Res Design 79:484–490.

Boullosa-Falces D, Gomez-Solaetxe MA, Sanchez-Varela Z, García S, Trueba A (2019) Validation of CUSUM control chart for biofouling detection in heat exchangers. Appl Therm Eng 152:24–31.

Boullosa-Falces D, Barrena JLL, Lopez-Arraiza A, Menendez J, Solaetxe MAG (2017) Monitoring of fuel oil process of marine diesel engine. Appl Therm Eng 127:517–526.

Chen XD, Li DX, Lin SX, Özkan N (2004) On-line fouling/cleaning detection by measuring electric resistance—equipment development and application to milk fouling detection and chemical cleaning monitoring. J Food Eng 61:181–189.

Chen Y, Sun S, Lai Y, Ma C (2016) Influence of ultrasound to convectional heat transfer with fouling of cooling water. Appl Therm Eng 100:340–347.

Cloete TE, Jacobs L, Brözel VS (1998) The chemical control of biofouling in industrial water systems. Biodegradation 9:23–37.

Croarkin C, Tobias P, Filliben J, Hembree B, Guthrie W (2006) NIST/SEMATECH e-handbook of statistical methods.

Dobretsov S, Abed RM, Teplitski M (2013) Mini-review: Inhibition of biofouling by marine microorganisms. Biofouling 29:423–441.

Eguía E, Trueba A, Río-Calonge B, Girón A, Bielva C (2008) Biofilm control in tubular heat exchangers refrigerated by seawater using flow inversion physical treatment. Int Biodeterior Biodegrad 62:79–87.

Eguía E, Trueba A (2007) Application of marine biotechnology in the production of natural biocides for testing on environmentally innocuous antifouling coatings 4:191–202.

García S, Trueba A (2019) Fouling in Heat Exchangers. In: Anonymous Heat Exchangers, IntechOpen.

Gudmundsson O, Palsson OP, Palsson H, Lalot S (2008) Method to detect fouling in heat exchangers.

Hawkins DM, Olwell DH (2012) Cumulative sum charts and charting for quality improvement. Springer Science & Business Media,.

Ismail B, Ewing D, Chang J, Cotton JS (2004) Development of a non-destructive neutron radiography technique to measure the three-dimensional soot deposition profiles in diesel engine exhaust systems. J Aerosol Sci 35:1275–1288.

López-Galindo C, Casanueva JF, Nebot E (2010) Efficacy of different antifouling treatments for seawater cooling systems. Biofouling 26:923–930.

Montgomery DC (2009) Introduction to statistical quality control. John Wiley & Sons (New York),.

Montgomery DC, Runger GC, Hubele NF (2009) Engineering statistics. John Wiley & Sons,.

Müller-Steinhagen H, Malayeri M, Watkinson A (2011) Heat exchanger fouling: mitigation and cleaning strategies.

Rajagopal S, Jenner HA, Venugopalan VP, Khalanski M (2012) Biofouling control: alternatives to chlorine. In: Anonymous Operational and environmental consequences of large industrial cooling water systems, Springer, pp 227–271.

Sibanda T, Sibanda N (2007) The CUSUM chart method as a tool for continuous monitoring of clinical outcomes using routinely collected data 7:46.

Trueba A, García S, Otero FM, Vega LM, Madariaga E (2015a) Influence of flow velocity on biofilm growth in a tubular heat exchanger-condenser cooled by seawater. Biofouling 31:527–534.

Trueba A, García S, Otero FM, Vega LM, Madariaga E (2015b) The effect of electromagnetic fields on biofouling in a heat exchange system using seawater. Biofouling 31:19–26.

Trueba A, García S, Otero FM (2014) Mitigation of biofouling using electromagnetic fields in tubular heat exchangers–condensers cooled by seawater. Biofouling 30:95–103.

Trueba A, Otero FM, González JA, Vega LM, García S (2013) Study of the activity of quaternary ammonium compounds in the mitigation of biofouling in heat exchangers–condensers cooled by seawater. Biofouling 29:1139–1151.

Withers PM (1996) Ultrasonic, acoustic and optical techniques for the non-invasive detection of fouling in food processing equipment. Trends Food Sci Technol 7:293–298.

Developments in Maritime Technology and Engineering – Guedes Soares & Santos (eds)
© 2021 Copyright the Author(s), ISBN 978-0-367-77376-2

Optimum life-cycle maintenance of fatigue-sensitive structures considering the random effect of ship repair

Jianda Cheng & Yan Liu
Huazhong University of Science and Technology, Wuhan, China

Yiwen Lu
China Ship Development and Design Center, Wuhan, China

ABSTRACT: Ship structures are subjected to fatigue damage during their lifecycle. To maintain the safety level, life-cycle management methods are often adopted to schedule inspection and repair during the lifecycle. Most of the existing life-cycle methods assume that the fatigue crack is restored to the initial as-built state after repair. However, it is often found that cracks propagate again in the repair area. Therefore, it is worthy to examine the random effect of repair for a more reliable maintenance scheme. This paper presents a novel life-cycle framework to address the random effect of repair in life-cycle maintenance framework. The random effect of fatigue damage repair is described by the crack length after the repair. In this paper, we will examine a fatigue detail of a naval vessel with the proposed framework. The inspection schedule is derived by minimizing the expected life-cycle cost and maximizing the expected service life together. Compared with the result of not considering the random effect of repair, our method proves to be more reliable in preparing the structure against lifetime fatigue damage.

1 INTRODUCTION

Structural damage such as fatigue cracking is pervasive among ship structures. Fatigue cracking can lead to serious safety issues and affect the effective service life of ships. Life-cycle management frame-work (Kim et al. 2013, Yang and Frangopol 2018, Zou et al. 2019) provides a rational method to derive maintenance strategies against fatigue damage. Regard-less of the development of life-cycle methods, predicting and modeling fatigue damage are still facing large uncertainties associated with damage occurrence and propagation, operational condition, structural response, maintenance effect, among others (Collette2016). The uncertain effect of crack repair in the life-cycle management of fatigue-sensitive structures is examined in this paper.

Deterioration caused by fatigue is complex process involving a series of uncertain parameters. Therefore, probabilistic methods such as reliability analysis are often adopted to assess fatigue damage degree (Soares & Garbatov 1996). In the lifetime deterioration process, inspection and maintenance are essential to maintain the reliability and safety level of ship structures. In addition to the fatigue damage prediction, decision making of lifetime inspection and maintenance planning has to include considerations of inspection quality, repair criterion, and maintenance effect, and cost (Garbatov and Soares 2001, Kim and

Frangopol 2010, Soliman et al. 2016), among others. Kim et al. (2013) have proposed a framework to optimize inspection planning that can extend the service life and reduce life-cycle cost of deteriorating structures. An optimized inspection time and method can also help to reduce maintenance delay (Kim & Frangopol 2011) and improve fatigue damage predictions (Liu & Frangopol 2018).

The effectiveness of optimized inspection plans relies on appropriate assumptions and approximations of the probabilistic models in the life-cycle maintenance planning frameworks. For example, most of the existing maintenance planning methods adopt the assumption that the crack will be restored to initial as-built state after repair. It is often found that crack propagates again in the repaired area. The effect of crack repair on the fatigue strength has been studied with numerical and experimental methods (Akyel et al. 2017, Abdullah et al. 2012). Ellingwood (2005) indicated that stochastic models of repair area research need. Moreover, the repair effect has not been investigated within life-cycle inspection planning framework.

This paper presents a probabilistic framework for lifetime inspection planning, where the randomness of crack repair is considered with various probability models. The effect of repair on optimum inspection schedule is illustrated with the impact on service life prediction and life-cycle cost of a fatigue detail in ship structure.

DOI: 10.1201/9781003216582-84

2 LIFE-CYCLE INSPECTION AND MAINTENANCE PLANING FRAMEWORK

2.1 Fatigue crack growth model

Prediction of fatigue crack growth often follows a linear elastic fracture mechanics approach, where the ratio of the crack length increment to the increment of the stress cycle is described in the well-known Paris equation (Paris & Erdogan 1963):

$$\frac{da}{dN} = C(\Delta K)^m \qquad (1)$$

Where a is crack length; N is number of cycles; C and m are the material parameters. ΔK is stress intensity factor that is often estimated as (Newman Jr & Raju 1981):

$$\Delta K = S_{re} \cdot G(a)\sqrt{\pi a} \qquad (2)$$

Where S_{re} is stress range and $G(a)$ is geometric correction factor.

Service life t_{life} is usually computed with the cumulative number of cycles required for crack to grow from an initial size a_f:

$$t_{life} = \frac{1}{N_a \cdot C \cdot S_{re}^m} \cdot \int_{a_0}^{a_f} \frac{da}{[G(a)\sqrt{\pi a}]^m} \qquad (3)$$

Where a_0 is initial crack length, and a_f is fatigue crack size associated with fatigue, N_a is the average annual number of stress cycles.

2.2 Probabilistic damage detection

During the service life of the ship, the inspection are scheduled to detect the fatigue damages. Nondestructive inspection methods (NDI) are often used for the fatigue crack detection, such as visual inspection, liquid penetrant inspection, ultrasonic inspection, magnetic particle inspection and acoustic emission inspection. The relation between the fatigue damage degree and the probability of damage detection is described with PoD functions. A widely used lognormal PoD function is described as (Crawshaw & Chambers 2001):

$$PoD = 1 - \Phi\left[\frac{\ln(a) - \lambda}{\beta}\right] \qquad (4)$$

Where Φ is the standard normal cumulative distribution function (CDF); a is cracks size; λ is location parameter; and β is scale parameter of the cumulative

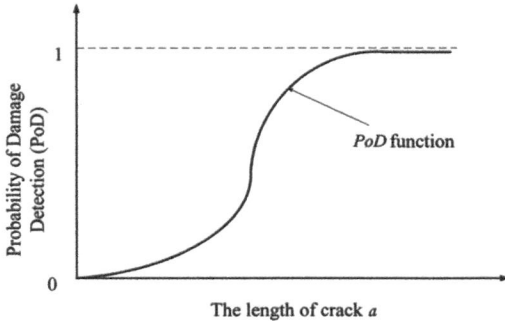

Figure 1. Illustrative PoD function.

lognormal PoD curve. The parameters (λ and β) are determined by the quality of the inspection method. The PoD function is shown in Figure 1.

2.3 Service life and life-cycle maintenance cost

Prediction of service life and life-cycle cost have to include the maintenance activities such as inspection and repair during the lifetime deterioration. Before the first inspection is implemented, the initial service life prediction is t_{life}^0. Repairing a fatigue crack can extend the initial service life as seen in Figure 2. Determine the repair decision has to follow an inspection event subjected to probabilistic detection. If crack is detected, the repair criterion also determine whether a repair is actually implemented.

The possible maintenance events are illustrated in the decision tree model proposed by Kim et al.(2013) shown in Figure 3. The service life and life-cycle cost are associated with respective event in the decision tree. If the inspection is scheduled after the predicted service life from last inspection $t_{insp,i}^{i-1} t_{life}$, no service life extension will be added, and maintenance cost includes the scheduled inspection cost $C_{insp,S}$ only. If the ith inspection fails to detect a crack, the corresponding service life and life-cycle cost is the same.

When the scheduled inspection detect the fatigue damage, a following in-depth inspection (with a cost of $C_{insp,D}$) is implemented to detect the size a growing crack, the decision criterion for crack repair is defined with the threshold crack size C_M. The life-cycle cost of maintenance is the summation of each maintenance cost $C_{life} = \sum_{i=1}^{n} C_{main}^i$, where C_{main}^i is computed based on the scenarios introduced.

The service life is extended only after a actual repair is implemented in the ith inspection as seen in the decision tree model. In this case, the service life after repair $t_{life}^{r,i}$ in i^{th} the inspection is calculated as:

$$t_{life}^r = t_{insp,i} + \frac{1}{C \cdot S_{sr}^m} \cdot \int_{ar}^{a_f} \frac{1}{[G(a)\sqrt{\pi \cdot a}]^m} da \qquad (5)$$

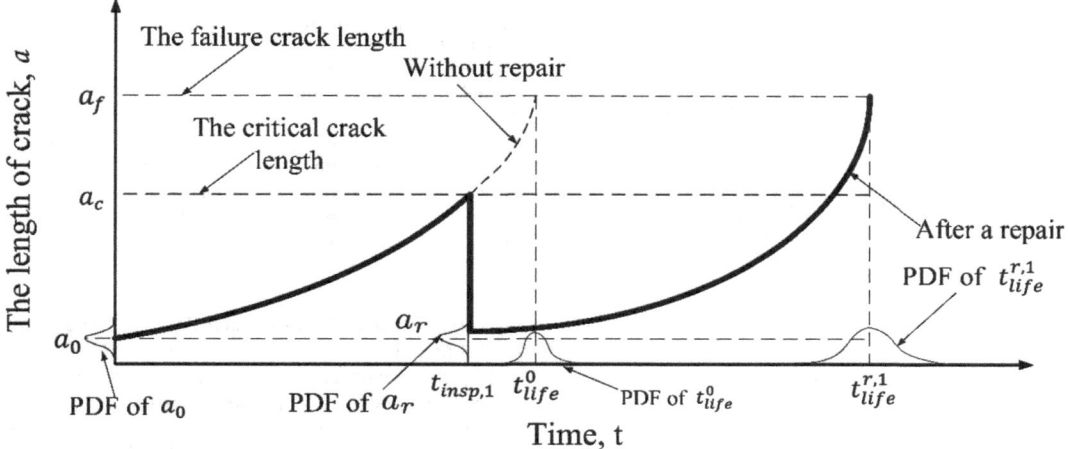

Figure 2. Effect of repair on the fatigue crack growth.

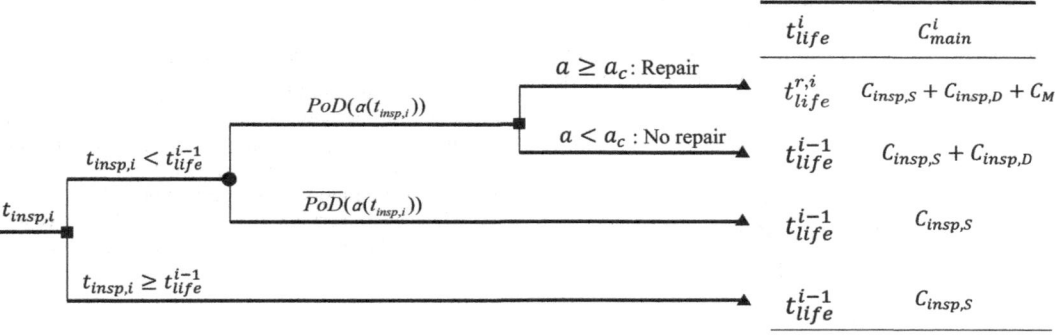

Figure 3. The decision tree to calculate the service life and the life-cycle maintenance cost.

Where t_{insp}^i is the time of ith inspection; a_r is the size of crack after repair. The service life associated with other scenarios are defined with t_{life}^{i-1} as no service life extension is added without repair. For example, before the first inspection time, the service life t0life is calculated with Equation 3.

3 RANDOM EFFECT OF REPAIR

In the previous studies, the commonly used model for maintenance effect is the "as good as new" model. That means the crack is restored to initial as-built condition after repair ($a_r = a_0$). However, the actual condition can be more complicated than this assumption. In practice, it is difficult to ensure that the maintenance model is perfect. The repaired crack size a_r can be deviated from the initial condition.

The potential impact of repair effect is discussed in Zou et al. (2019), where the repaired crack size can return to various states after the intervention. In this paper, we consider three possible cases of repair effect: 1) return to the initial state a_r^0 r; 2) return to

worse than initial state a_r^1; 3) return to random state a_r^2. The three possible effects on crack size a_r after repairing the crack at t_{insp} are illustrated in Figure 4. a_c is the critical crack size and the repair will be applied when the crack size is greater than a_c. a_r^0 is modeled with the same distribution as initial crack size a_0, a_r^1 and a_r^2 are modeled with exponential distribution and uniform distribution, respectively. Considering the repair effect on crack, the new fatigue life after repair t_{rlife}^r is computed as:

$$t_{rlife}^r = \frac{1}{N_a C \cdot S_{re}^m} \int_{a_r}^{a_f} \frac{da}{[G(a)\sqrt{\pi a}]^m} \qquad (6)$$

4 OPTIMUM INSPECTION SCHEDULING

The determination of optimum maintenance strategy is complex process. The maintenance decision

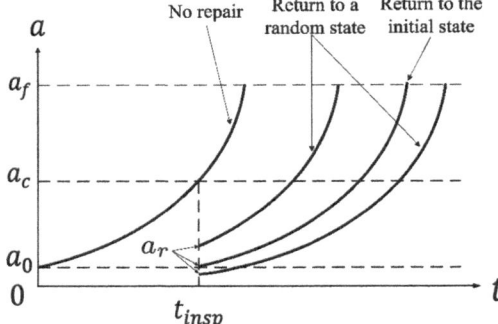

Figure 4. Illustration of repair effect.

affects the effective service life and total maintenance cost, among others. In this paper, the optimum inspection time $t_{insp,i}$ is derived by optimizing two objective functions, the expected service life $E(t_{life})$, and expected life-cycle maintenance cost $E(C_{tota;})$ at the same time. The optimization problem can be expressed as:

$$
\begin{aligned}
&\text{Find}: t_{insp,1}, t_{insp,2,...,}t_{insp,n} \\
&\text{By Max } E(t_{life}) \text{and Min } E\left(C_{tota;}\right) \\
&\text{Given}: \lambda, \beta, a_c, \text{n, PDF of } t_{life}^0
\end{aligned}
\tag{7}
$$

$$\text{S.t. } t_{insp,i} - t_{insp,i-1} 1$$

In the optimization problem, the probabilistic inspection quality is pre-defined with λ and β. The objective functions $E(t_{life})$ and $E(C_{tota;})$ associated with n inspections can be computed according to the decision tree in Section 2.3. Time interval between inspections should be larger than 1 year. PDF of t_{life}^0 is computed with probabilistic fatigue parameters. The multi-objective optimization problem is solved with genetic algorithm. In this work, a python library Geatpyisusedasmulti-objectivegeneticevolutionary algorithm toolbox (Geatpy 2019).

5 ILLUSTRATIVE EXAMPLE

A fatigue detail in a naval ship hull structure is used to illustrate the framework. The location to be analyzed is the joint between longitudinal plate and bottom plate, more information about this fatigue example is provided in Kim et al. (2013). The probabilistic parameters to compute fatigue life are listed in Table 1. The geometry function G(a) in Equation 2 is assumed equal to 1.0. The critical crack length a_c for repair criterion is 2.3mm, the failure crack length a_f is 50mm. The detection method applied in this example is ultrasonic

Table 1. Variables of fatigue example (adopted from Kim et al. (2013)).

Variable	Distribution	Mean	CoV
S_{sr}(ksi)	Weibull	5.8	0.1
N_{cycle}	Lognormal	1.0×10^6	0.2
a_0(in.)	Lognormal	0.0197	0.2
C	Lognormal	1.77×10^{-9}	0.3
M	Deterministic	2.54	-

inspection method, the associated PoD parameters are $\lambda = 0.122$ and $\beta = -0.305$ (Forsyth & Fahr 1998).

In the maintenance cost model used in the example, the cost of scheduled inspection is assumed $C_{insp,S} = \$5000$, the cost of in-depth inspection $C_{insp,D} = \$10,000$, the cost of repair $C_M = \$100,000$. The repair effect on crack is modeled with three possible conditions a_r^0, a_r^1, and a_r^2. The probabilistic distribution and parameters assumed are listed in Table 2.

5.1 Random repair effect on fatigue life t_{life}^r

Considering the possible repair effects on the crack, the impact on the new fatigue life after repair t_{life}^r is investigated first. t_{life}^r can be obtained by the Equation 6 and the results are shown in Figure 5. It can be seen that the random repair effect on trlife is obvious. The mean service life of a_r^2 is 16.51 years greater than the other two, and the service life of a_r^1

Table 2. Investigated cases of repair effect on a_r.

Cases	Mean	CoV	Distribution
a_r^0	0.50mm	0.2	Lognormal
a_r^1	0.75mm	0.67	Exponential
a_r^2	0.50mm	0.58	Uniform

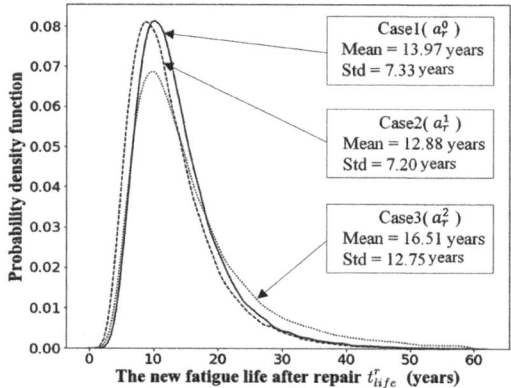

Case1(a_r^0)
Mean = 13.97 years
Std = 7.33 years

Case2(a_r^1)
Mean = 12.88 years
Std = 7.20 years

Case3(a_r^2)
Mean = 16.51 years
Std = 12.75 years

Figure 5. Comparison of repair effect on t_{life}^r.

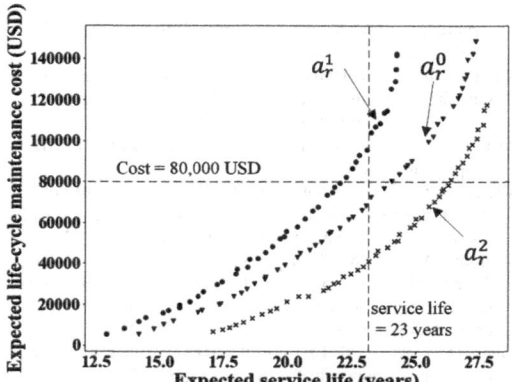

Figure 6. Comparison of repair effect on Pareto solution sets.

is reduced considering the random repair effect. Therefore, the different repair model has definite impact on the fatigue life prediction.

5.2 Comparison of optimum inspection time

The optimization problem is solve with regard to two inspections (n = 2). Generation number of 100 and population number of 50 are used in the multi-objective genetic algorithm to compute the optimum solutions. Three Pareto fronts are plotted in Figure 6 corresponding to respective repair effect considered.

The impact of repair model on the optimization results are substantial by looking at the results. It can be seen that this impact increase as expected service life and life-cycle maintenance cost become larger. A detailed comparison of optimum solutions is shown in Table 3. If the same expected service life (e.g. 23 years) is desired by decision maker, it is advised to schedule the inspection time earlier for the repair models of a_r^0 and a_r^1. In addition, the maintenance planning comes with larger lifetime costs. Similarly, when the same budget is allowed for lifetime maintenance cost (e.g. \$80,000), the repair models of a_r^0 gained less extended service life than a_r^2 model but more than a_r^1.

Table 3. Comparison of optimum inspection schedules.

Cases	$E\left(t_{life}\right)$ (years)	$E\left(C_{life}\right)$ (USD)	$t_{insp,1}$ (years)	$t_{insp,2}$ (years)
a_r^0	23.00	68,342	15.71	31.24
a_r^1	23.00	95,776	12.03	24.18
a_r^2	23.00	35,514	20.91	41.97
a_r^0	24.07	80,000	14.14	29.10
a_r^1	22.65	80,000	13.06	26.76
a_r^2	26.38	80,000	13.95	27.52

6 CONCLUSION

This paper presents life-cycle framework to examine the random effect of repair in optimum life-cycle maintenance scheduling. The framework covers prediction of probabilistic fatigue damage propagation, probabilistic detection which describes the randomness of damage detection in inspection event, evaluation of service life and life-cycle maintenance cost considering maintenance events. The proposed framework was applied to a naval ship subjected to fatigue deterioration. The following conclusions can be drawn:

1. This work focus on examining the effect of crack size (a_r) after repair. The results obtained are limited to the assumption on a_r. More distribution models and parameter choices are needed for comparative study. Meanwhile, it is worthy to look at other effects such as deterioration rate after repair for a more comprehensive study
2. The repair models used in this paper show a significant impact on the optimum results from the life-cycle maintenance scheduling framework. Therefore, it is advised to put careful assumption of repair model in the life-cycle framework for a more reliable prediction of service life and lifecycle cost. The proposed framework can also be applied in other deterioration mechanism such as corrosion

ACKNOWLEDGMENTS

The support from the National Natural Science Foundation of China (Contract Nos. 51839005, 51579109 and51479079) is gratefully acknowledged. The opinions and conclusions presented in this paper are those of the authors and do not necessarily reflect the views of the sponsoring organizations.

REFERENCES

Abdullah, A., M. Mlaki, & A. Eskandari (2012). Strength enhancement of the welded structures by ultrasonic peening. Materials & Design 38, 7–18.

Akyel, A., M. Kolstein, & F. Bijlaard (2017). Fatigue strength of repaired cracks in base material of high strength steels. Journal of Constructional Steel Research 139, 374–384.

Collette, M. (2016). Uncertainty approaches in ship structural performance. Handbook of Uncertainty Quantification, 1–22.

Crawshaw, J. & J. Chambers (2001). A concise course in advanced level statistics: with worked examples. Nelson Thornes.

Ellingwood, B. R. (2005). Risk-informed condition assessment of civil infrastructure: state of practice and research issues. Structure and infrastructure engineering 1(1), 7–18.

Forsyth, D. & A. Fahr (1998). An evaluation of probability of detection statistics. Institute of Aerospace Research, 10–1.

Garbatov, Y. & C. G. Soares (2001). Cost and reliability based strategies for fatigue maintenance planning offloatingstructures.ReliabilityEngineering&System Safety 73 (3), 293–301.

Geatpy (2019). Geatpy 2.2.3 http://geatpy.com/.

Kim, S. & D. M. Frangopol (2010). Cost-based optimum scheduling of inspection and monitoring for fatigue sensitive structures under uncertainty. Journal of Structural Engineering 137(11), 1319–1331.

Kim, S. & D. M. Frangopol (2011). Optimum inspection planning for minimizing fatigue damage detection delay of ship hull structures. International Journal of Fatigue 33(3), 448–459.

Kim, S., D. M. Frangopol, & M. Soliman (2013). Generalized probabilistic framework for optimum inspection and maintenance planning. Journal of Structural Engineering 139(3), 435–447.

Liu, Y. & D. M. Frangopol (2018). Utility and information analysis for optimum inspection of fatigue-sensitive structures. Journal of Structural Engineering 145(2), 04018251.

Newman Jr, J. & I. Raju (1981). An empirical stress intensity factor equation for the surface crack. Engineering fracture mechanics 15(1–2), 185–192.

Paris, P. & F. Erdogan (1963). A critical analysis of crack propagation laws.

Soares, C. G. & Y. Garbatov (1996). Fatigue reliability of the ship hull girder accounting for inspection and repair. Reliability Engineering & System Safety 51(3), 341–351.

Soliman, M., D. M. Frangopol, & A. Mondoro (2016). A probabilistic approach for optimizing inspection, monitoring, and maintenance actions against fatigue of critical ship details. Structural Safety 60, 91–101.

Yang, D. Y. & D. M. Frangopol (2018). Probabilistic optimization framework for inspection/repair planning of fatigue-critical details using dynamic bayesian networks. Computers & Structures 198, 40–50.

Zou, G., A. Gonzalez, K. Banisoleiman, & M. H. Faber (2019). An integrated probabilistic approach for optimum maintenance of fatigue-critical structural components. Marine Structures 68, 102649.

Developments in Maritime Technology and Engineering – Guedes Soares & Santos (eds)
© *2021 Copyright the Author(s), ISBN 978-0-367-77376-2*

Short sea shipping and shipbuilding capacity of the East Mediterranean and Black Sea regions

T. Damyanliev, P. Georgiev, Y. Denev & L. Naydenov
Technical University of Varna, Varna, Bulgaria

Y. Garbatov
Centre for Marine Technology and Ocean Engineering (CENTEC), Instituto Superior Técnico, Universidade de Lisboa, Lisbon, Portugal

I. Atanasova
Varna Maritime Ltd, Varna, Bulgaria

ABSTRACT: The recently released data on the Short Sea Shipping in the EU seas has shown a stable trend in transportation growth. At the same time, the coaster fleet of the East Mediterranean and Black Sea regions (EMBSR) is of considerable age, and the increased freight rates enforce the new orders of multi-purpose ships. The objective of this paper is to analyse the main dimensions of dry cargo ships operating in the EMBSR in the range of 2,000 to 9,500 DWT. The data collected for already built ships is compared to the newly designed ones, accounting for the capital and operational expenditure and required freight rate. Recent investigations demonstrated that the new building of ships in the range of 2,000 to 9,500 DWT is possible to be done in the existing SME ship repair yards in the region. The study investigates the capacity of the shipbuilding industry in the EMBSR. Several conclusions considering the relationships of the deadweight and design governing parameters accounting for the existing production constraints of small and medium enterprises specialised in shipbuilding and repair are also presented.

1 INTRODUCTION

The EU definition on the short sea shipping (SSS) is defined as transportation of cargo and passengers by sea between ports on the enclosed seas bordering, including domestic and international maritime transport, feeder services between the islands, rivers and lakes[1].

Recently published data shows that the yearly amount of SSS goods in the EU is higher than the amount for the year 2009[2]. The increase for the last three years before 2016 is permanent, and the level for 2016 is grater with 2.6 % from the previous year. Despite the relative increase in the amount of transported goods, the current level is still below the ones recorded in the years immediately preceding the economic downturn in 2009 (Figure 1).

The short sea shipping is close to 60 % of the total maritime transport of goods to and from the main EU ports in 2016. The share of short-sea shipping in the entire maritime transport varies considerably between the countries as for example for

Finland is 92.8%, Norway is 87,2%, and Croatia is 81.5%.

In the Black Sea region, the reported short-sea shipping is for Bulgaria, about 80.0%, Romania is 70.0 %, and Turkey is about 68.1%.

Figure 2 presents the distribution of transported goods in the European seas based on the Eurostat.

The total share of the Mediterranean Sea, Black Sea and Atlantic Ocean is approximate as the one of the North and Baltic Sea as can be seen in Figure 2.

The liquid bulk is the dominant type of cargo in the EU short sea shipping. The share of the dry bulk goods in the short-sea transportation of each sea region is more evenly distributed, with a range of 16 % in the Mediterranean Sea to a maximum of 26 % in the Black Sea. After the liquid cargoes, the maximum share in the Mediterranean Sea has the container transportations.

The growth in the SSS is accompanied by the development of the EU's political framework. The "Motorways of the Sea (MoS)" is a concept introducing new intermodal maritime-based logistics chains in Europe.

---

1 http://www.shortsea.info/definition.html
2 ec.europa.eu/eurostat/statistics-explained/index.php/Maritime_transport_statistics_-_short_sea_shipping_of_goods

DOI: 10.1201/9781003216582-85

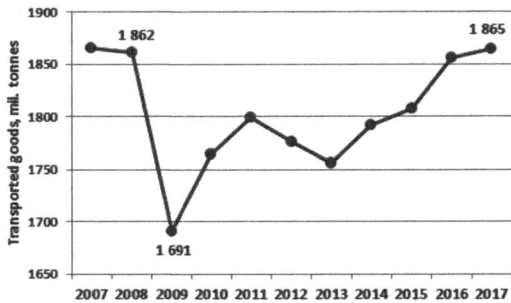

Figure 1. Transported goods by SSS 2007-2017.

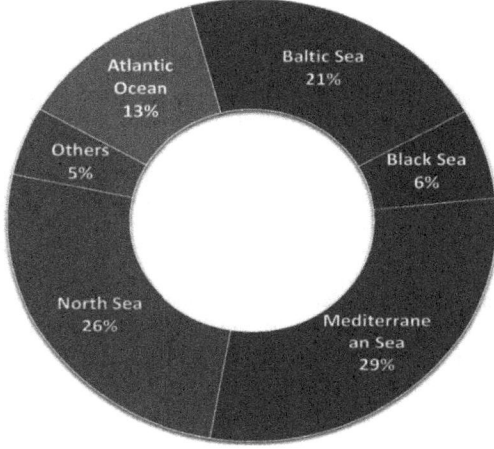

Figure 2. Transported goods in European seas, 2017.

The Detailed Implementation Plan (DIP) for MoS is based on three pillars i.e. environment, integration of maritime transport in the logistics chain and safety, traffic management and human element. The DIP gives development priorities in different areas that belong to the three pillars. The development directions for the ships in Pillar 1 – "Environment" are presented in (Simpson, 2016). According to the DIP the financing of the green shipping priority needs pilot testing of new mechanisms that consider both the capital expenditure (CAPEX) and operational expense (OPEX). They attract the beneficiary stakeholders such as the EU shipyards, engine and equipment manufacturers in the risk-sharing model.

The maritime investments should be considered in the cohesion and other structural funds agenda and to design an overall risk and security management approach in the transport and logistics sector.

Although the problems related to SSS are widely studied, still some long-standing issues have not been resolved (Gustafson et al., 2016). The logistics solutions in the short sea shipping are generally not efficient. One of the main reasons is the high number, more than 15 different organisations,

involved in the transportation chain. As evidence of the low vessel, utilisation is the fact that the bulk and general cargo vessels in the Baltic Sea sail on an average 34% of their sailing time in ballast conditions without a cargo. Furthermore, about half of the ships in dry bulk and general cargo segment operating in the Baltic Sea currently spend at least 40% of their time in ports. Another contributing factor is the slow cargo loading and unloading process during which the vessel remains in the harbour.

A central reason for the lack of technological innovations is the decision-making process in the shipbuilding, only based on the minimisation of the capital expenditure, instead of a more holistic approach. Such a comprehensive approach accounts for all expenses during the life cycle of the ship. Another problem in the design phase, which is the discrepancy between the design speed and the corresponding power and what is the actual need.

In the Baltic Sea, around 70% of vessels sail at a speed that is 15–25% below the design speed (Gustafson et al., 2016). This can be partly explained by the need to sail in icy conditions.

Recent legal developments introduced by the International Maritime Organisation, IMO have an impact on the performance of the SSS (Papadimitriou et al., 2018). The lower sulphur limit in ECAs appears to have a negative effect on the competitiveness of the SSS. Some of the ships are fully exploited in ECAs and are forced to use significantly more expensive fuel.

The Ballast Water Convention (entered in to force on 8 September 2017) also has a negative impact on the SSS. The cost of the new equipment, together with the installation that will keep the ship many days out of the market, required extra cost and loss of profit for the operators from the SSS sector. The requirement to perform ballast water operations at a distance of at least 200 nm from the coast and 200 m depth directly impact of the efficiency of the ships that sail much closer to the coast.

As an alternative mode of the transportation, the SSS offer valuable social benefits including shifting the traffic from the overburdened roads, development of peripheral or isolated regions and islands and a low number (together with aviation) of accidents.

The significant social benefit of the SSS also lies in the contribution to the local shipbuilding industry. Therefore, in Europe, short-sea vessels are primarily ordered by the local shipbuilders who specialise in ship designs of a relatively small ship size. Thus, short sea newbuilding activity offers valuable support to the European shipyards, which face severe financial difficulties. These activities will also reduce unemployment and the same holds true for the ship repair yards.

The next sections of the study will present an overview of the current state of the SSS and the coaster fleet of the East Mediterranean and Black Sea regions. The shipbuilding capacity of the EMBSR is also described, and a comparison of different ship design solutions employing the conceptual design framework "Expert" with already built and operated in the regions ships will be performed. The analysis of a series of newly designed sips, accounting for the CAPEX, OPEX and required freight rate (RFR) will be presented. Finally, several conclusions considering the relationships of the deadweight and design parameters accounting for the existing constraints of the SMEs will also be discussed.

2 SHORT SEA SHIPPING

2.1 European short sea shipping fleet

Twenty years ago, the SSS fleet counted more than 50% from the total fleet with an average age of 20 years as can be seen from Table 1. The dry bulk cargo ships represent the largest European Economic Area (EEA) short sea fleet with almost constant total DWT during the years. From 9,479,320 DWT in 2005 to 11,536,140 DWT at October 2014 (ECSA, 2016). Concerning the age of the ships, in the spring of 2017 more than 35% from the ro-ro ships are over 25-year old (EC, 2018).

Recently, an analysis of the European Coaster Fleet was published in (IDEAS, 2016). The study presented a snapshot of the Black Sea, Mediterranean, Continent and Baltic coaster fleet evaluating the severe problems with the sub-standard shipping, to determine its extent within the coaster fleet.

The data showed that the coaster fleet, defined as 1,000-12,000 DWT, general cargo and bulk carrier vessels, owned by the countries surrounding the Black, Mediterranean, Continent and Baltic seas are over 4,300 active units with a total capacity of 19,788,524 DWT.

Among this fleet, six flag states hold over one million DWT registered ships (see Table 2). The top five classification societies are Germanischer Lloyd (now DNV-GL[3]), Bureau Veritas, Russian Register, Lloyds Register and Registro Italiano Navale (see Table 3).

It should be noted that a significant part of the fleet – 805 vessels with a total of 2,692,502 DWT (over 13,5 per cent), remains unidentified. The possible

Table 1. Data for SSS fleet (OECD, 2001).

		Number (%)	DWT %	Average Age (Years)
EU	Deep Sea	42.7	93.3	14
	Short Sea	57.3	6.7	20
Rest of Europe	Deep Sea	37.3	90.8	13
	Short Sea	62.7	9.2	21
Rest of the World	Deep Sea	31.5	91.1	13
	Short Sea	68.5	8.9	18

Table 2. Flag states of SSS fleet.

Flag state	Total DWT	%
Antigua & Barbuda	2,810,513	14.20%
Russian Federation	2,740,257	13.85%
Netherlands	2,589,966	13.09%
Malta	1,361,415	6.88%
Turkey	1,160,163	5.86%
Panama	1,078,842	5.45%
Others	8,047,368	40.67%
TOTAL	19,788,524	100.0%

Table 3. Classifications societies of SSS fleet.

Classification Society	Total DWT	%
Germanischer Lloyd	4,578,244	23.14%
Bureau Veritas	3,863,651	19.52%
Russian Register of Ships	2,770,517	14.00%
Unidentified	2,692,502	13.61%
Lloyds Register	1,866,422	9.43%
Registro Italiano Navale	701,083	3.54%
Others	3,316,105	16.76%
TOTAL	19,788,524	100%

reasons can be seen as a change of the classification society, and the information is not yet updated in the records, withdrawal of the class of the vessel from the existing classification society. The class renewal survey is overdue, but the records are not adequately updated, or the ship operates or switched to a less recognised class, without access to an international database or does not update its records regularly.

Every year the Paris MoU on the port state control published the "White, Grey and Black (WGB) list" presenting the quality flags and flags with poor performance with high or very high risk. The vessels with undefined class additionally were grouped by the IDEAS according to their flag and age as can be seen in Table 4. One can consider that ships with a total DW about 617,000 t are sub-standard. However, other ships may be added to these as vessels younger than 20 years that are poorly managed, ships registered under White or Grey-listed flag might also belong to

Table 4. Flag and age of vessels with "unidentified" class, (DWT).

Flag	Age > 20 years	Age < 20 years
White of Grey	1,699,771	350,932
Black	617,195	24,604
TOTAL	2,316,966	375,536

---

3 DNV and GL signed a merger contract on 12 September 2013

the sub-standard shipping, and about 1.7 million DWT of the coaster fleet is with unidentified class and are over 20-year old.

Taking into account the WGB list of the Paris MoU, IDEAS, it may be concluded that 485 coasters (about 2.0 million DWT, 10% of the fleet) remain registered under risky flag jurisdictions and low-performing Classification Societies thus bearing a very high risk. A further 866 coasters, (about 2,98 million DWT, 15 % of the fleet) need to be closely monitored. Regardless of their flag state, vessels classified by low-performing or unknown Classification Societies threaten the sustainability of coaster trade is due to fact that they operate under sub-standard or near-sub-standard conditions. They are potentially unsafe and unfit for the international trade and create unfair competition.

It can be argued that the considerable age, high freight rates and the increase in the amount of goods transported by the SSS will result in ordering and construction of new ships.

2.2 Operational cost

Since 2017, the Brokers Market & Trend Information (BMTI) Company launched the first official European Short Sea Index (EUSSIX). The index uses weighted inputs from three sub-regional indexes, including the Northern Europe, Mediterranean Sea and Black Sea-Azov seas. The EUSSIX is published weekly and aggregates freight rates for dry bulk cargoes transported in Europe and adjacent regions with vessels between 1,000-20,000 DWT.

Figure 3 presents the average freight rates for a general short sea cargo shipped from the Azov Sea to Marmara region, June 2017 – June 2018, (BMTI, 2019). The shown cycles in Figure 3 can be explained by the harvesting in the southern Russian and Ukraine.

The Istanbul Freight Index (ISTFIX) is owned and operated by Istanbul Denizcilik Ar Geve Danismanlik[4], which reaches information about the shipping in the EMBSR. The ISTFIX is based on the data for the cargo transportation from the Black Sea to the five other destinations including the Marmara Sea, East, Central and West Mediterranean Sea and Continent Range. The size of ships is ranging from 2,000 to 12,000 DWT, which describes a small coaster ship of dry/general cargo or mini bulkers type.

Figure 4 and Figure 5 show the changes in the IFO-180 and MGO fuel price for the last years. It is clear that during the collapse in the oil price in 2014-2016, a growing supply accumulation, which can be one of the reasons for the increased amount of transported cargo in that period (see Figure 1).

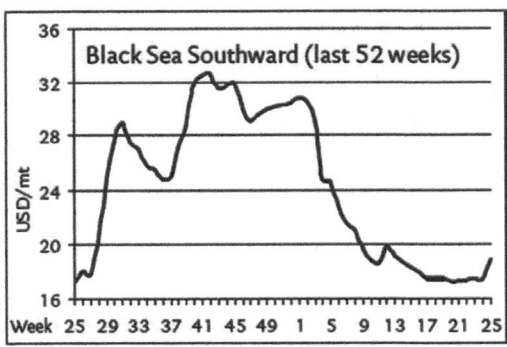

Figure 3. Average freight rates for a general short sea cargo shipped from Azov Sea to Marmara region.

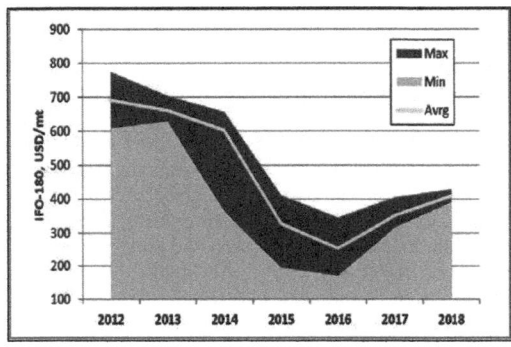

Figure 4. IFO-180 price for the last six years. Own calculations based on ISTFIX.

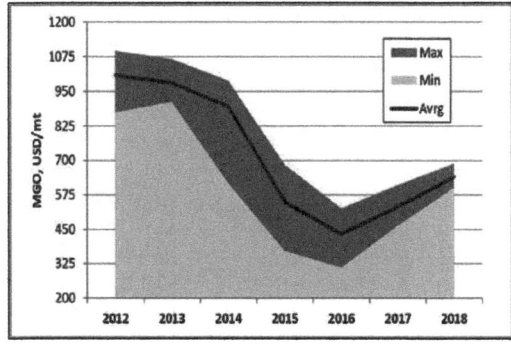

Figure 5. MGO fuel price for last six years. Own calculations based on ISTFIX.

4 http://en.istfix.com/

3 SHIPBUILDING CAPACITY

The shipbuilding potential of the East Mediterranean and the Black Sea regions is evaluated by analysing the shipbuilding facilities of 46 shipyards from 6 countries including Bulgaria represented by 4 shipyards, Greece – 4 shipyards, Romania – 3 shipyards, Russia – 2 shipyards, Turkey - 27 shipyards and Ukraine – 6 shipyards.

The analysed facilities are composed of graving docks, floating docks and slipways. The term slipway comprises a sloped slipway for longitudinal launching, airbags launching, ship lift, rail slipway etc. A total of 117 facilities have been analysed as can be seen from Table 5, and their dimensions are shown in Table 6.

Table 7 to Table 13 present the shipyards with the most significant dimensions of the shipbuilding facilities and lifting capacity of floating docks for each country.

Table 14 shows the primary characteristic of container ships, general cargo carriers, tankers and bulkers typically operating in a short-sea shipping (Tsinker, 2004). The presented dimensions are approximate and serve only to evaluate the capabilities of the shipyard's facilities. The range of different ship types dimensions is shown in Figure 6.

Based on the analysis of the typical dimensions of the ships operating in the SSS can be concluded that existing SMEs have enough capacity in building new ships in EMBSR.

Table 5. Shipyard facilities by country and type.

Country		Graving dock	Floating Dock	Slipway
Bulgaria	BG	3	5	1
Greece	GR	2	7	1
Romania	RO	5	2	1
Russia	RU	-	1	1
Turkey	TR	4	21	48
Ukraine	UA	4	5	6
TOTAL		18	41	58

Table 6. Range of dimensions of the facilities.

Parameter	Graving dock	Floating dock	Slipway
Length, m	370.0	382.0	400.0
	85.0	83.8	92.5
Breadth, m	70.0	58.0	60.0
	12.5	16.0	15.0
Lifting capacity, t	-	50000	-
	-	3000	-

Table 7. Maximum length of graving docks.

Shipyard	Country	Lmax, m
Okean	UA	370
Damen 3	RO	360
Sedef	TR	310
Bulyard 1	BG	240
KRZ- Odesos	BG	240
Perama 1	GR	140

Table 8. Maximum breadth of graving docks.

Shipyard	Country	Bmax, m
Celictrans	TR	70
Okean	UA	60
Zaliv	UA	60
Damen 3	RO	60
Bulyard 1	BG	40
Perama 1	GR	19

Table 9. Maximum length of floating docks.

Shipyard	Country	Lmax, m
Beshiktash 2	TR	382
Elefsis 3	GR	276
Chernomorsk 1	UA	225
KRZ- Odesos 2	BG	219
Azov	RU	200
Constanta 2	RO	180

Table 10. Maximum breadth of floating docks.

Shipyard	Country	Bmax, m
Beshiktash 2	TR	58.0
Nikolayev	UA	40.0
Elefsis 3	GR	40.0
KRZ- Odesos 2	BG	36.0
Constanta 2	RO	32.0
Azov	RU	25.0

Table 11. Maximum lifting capacity of floating docks.

Shipyard	Country	Lifting capacity, t
Tersan 3	TR	50000
Elefsis 3	GR	28000
Chernomorsk 1	UA	20000
KRZ- Odesos 2	BG	20000
Azov	RU	15000

Table 12. Maximum length of slipways.

Shipyard	Country	Lmax, m
Zaliv	UA	400
Tor group 5	TR	360
Vard Tulcea	RO	160
Nevski	RU	150
Terem-Flotski Arsenal	BG	130
Megatechnica	GR	95

Table 13. Maximum breadth of slipways.

Shipyard	Country	Bmax, m
Cemre 3	TR	60
Nikolayev 1	UA	41
Vard Tulcea	RO	27
Megatechnica	GR	25
Nevski	RU	23
Terem- Flotski Arsenal	BG	16

Table 14. Main dimensions of SSS type ship types.

Ship type	DWT, t	LW, t	L, m	B, m
Container ship	6,500	2,500	115	19
	15,000	5,000	180	27
General	1,000	600	65	10
Cargo ship	10,000	3,600	135	20
Tanker	1,000	400	60	9
	10,000	2,000	140	17
Bulker	5,000	2,000	110	16
	10,000	3,000	140	18

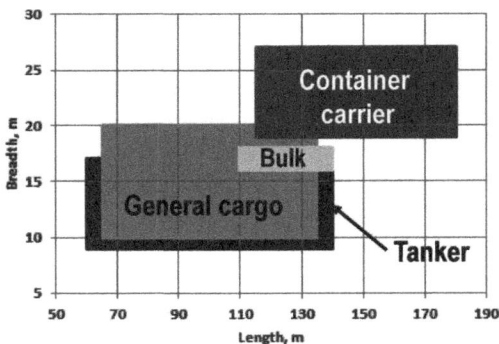

Figure 6. Main dimensions of ships suitable for short sea shipping.

The new design vessels, operating in the SSS conditions, need to satisfy all environmental requirements of the IMO, including the Tier III NOx emission criteria and to achieve significant reductions in CO_2 and particulate matter (PM).

4 DRY CARGO SHIPS

Nowadays, the European maritime industry counts on small and medium-sized enterprises, SME to restore the European shipbuilding industry while ensuring youth employment. In response to this goal, several studies can be highlighted.

One direction is to consider the shipbuilding capacity of the SME shipyards and the demand of the ship-owners for efficient ships operating in the EMBSR. The breadth of the slipway as building limitation was studied in(Atanasova et al., 2018).

The design of series of Multi-Purpose Vessel (MPV) with a DWT range from 5,000 to 10,000 tons subjected to shipbuilding, operational and functional constraints were design by the software tool "Expert" (Damyanliev et al., 2017, Garbatov et al., 2017). The series includes two groups of ships with and without restriction in the breadth due to shipbuilding facility constraints. The analyses of the total ship resistance, intact stability and cargo capacity (Denev et al., 2018) and ship performance taking into account the seakeeping of the ships (Georgiev et al., 2019) showed that some design solutions with a limited breadth leads to better ship performance compared to the vessels of the same deadweight without restrictions. The main conclusion was that ships can be built in small and medium-sized enterprises with constraints of building new ships.

The software tool "Expert" permits ship design optimisation with constraints taking into account CAPEX, OPEX or RFR (Required Freight Rate) as an objective to minimize. The optimal main dimensions of ships with 5,000 - 8,000 DWT at speed = 14,5 kn., stowage factor = 1.45 m^3/t, crew =18 persons and sailing distance of 5,000 nm are presented in Table 15 to Table 18 (i.e Table 15, 16, 17 and 18). These tables include the main dimensions obtained at different objective functions taking into account the specific limitations due to the shipbuilding facilities as analysed in (Denev et al., 2018). In this case, the breadth of the ship is limited to 16 m. Later, the obtained design solutions are compared with the dimensions of already built ships.

Recently, several studies presented the use of regression relationships in defining the main ship dimensions. Ebrahimi et al. (2015), demonstrated the application of the multivariate data analysis in the parametric design of OSVs. The key design characteristics of container ships with a length range of 47.5 to 383 m as a function of DWT and the number of TEU are presented in (Abramowski et al., 2018), where the regression relationships are based on the container ships built from 2005 to 2015 year.

Table 15. Main dimensions of 5,000 DWT MPV.

Objective	Lpp, m	B, m	D, m	d, m	Cb-
No restrictions					
CAPEX	90.09	16.88	8.76	7.04	0.650
OPEX	113.20	15.92	6.93	9.06	0.600
RFR	101.51	16.08	8.88	7.02	0.632
With restrictions					
CAPEX	86.89	16.00	9.03	7.30	0.680
OPEX	121.15	15.90	9.20	7.07	0.574
RFR	88.629	16.00	7.08	8.81	0.690

Table 16. Main dimensions of 6,000 DWT MPV.

Objective	Lpp, m	B, m	D, m	d, m	Cb-
No restrictions					
CAPEX	99.32	16.80	9.34	7.35	0.680
OPEX	131.80	16.50	9.16	6.85	0.627
RFR	116.57	16.43	9.23	7.06	0.652
With restrictions					
CAPEX	104.92	16.00	9.17	7.12	0.700
OPEX	134.90	15.96	9.32	6.88	0.632
RFR	106.60	16.00	8.93	6.88	0.721

Table 17. Main dimensions of 7,000 DWT MPV.

Objective	Lpp, m	B, m	D, m	d, m	Cb-
No restrictions					
CAPEX	108.67	17.06	9.79	7.53	0.680
OPEX	144.22	16.69	9.95	7.44	0.613
RFR	123.24	17.05	9.67	7.29	0.668
With restrictions					
CAPEX	123.37	16.00	9.27	6.87	0.740
OPEX	139.35	16.00	9.40	6.94	0.690
RFR	120.62	16.00	9.03	6.67	0.772

Table 18. Main dimensions of 8,000 DWT MPV.

Objective	Lpp, m	B, m	D, m	d, m	Cb-
No restrictions					
CAPEX	118.58	17.43	10.10	7.59	0.720
OPEX	151.38	17.69	10.46	7.75	0.615
RFR	126.44	17.51	9.98	7.43	0.699
With restrictions					
CAPEX	139.32	16.02	9.42	6.77	0.770
OPEX	141.75	16.00	9.45	6.77	0.766
RFR	135.06	16.00	9.028	6.67	0.772

Different relationships for the preliminary design of Handy size, Medium-Range, Panamax, Post Panamax, Aframax, Suezmax and VLCC tankers built from 2000 to 2018 year may be found in (Cepowski, 2019) and all relationships have been developed concerning deadweight and velocity.

Recently, Santos and Soares (2017) presented a novel methodology for determining the characteristics of a ro-ro ship and the fleet size required for a given short sea shipping route, identifying the vessel characteristics as a function of the capacity.

The present study analyses more than 200 dry cargo ships (MPVs and mini bulk carriers) in the range of 2,100 to 9,300 DWT.

The main dimensions as a function of the DW are presented in Figure 7 to Figure 11, including the optimal design solutions obtained by "Expert" software framework accounting for CAPEX, OPEX and RFR as an objective and for the restricted or non-restricted breadth due to shipbuilding constraints of the SMEs.

Figure 7 and Figure 8 show that the ship length solution obtained by RFR and CAPEX as an objective function is close to the already built ships. The CAPEX leads to a reduced length, while the OPEX increases the length. In the case of a restricted breadth, due to the production constraints of the SMEs, with increasing the DW, the ship length, accounting for CAPEX and RFR, increases too, and the estimated

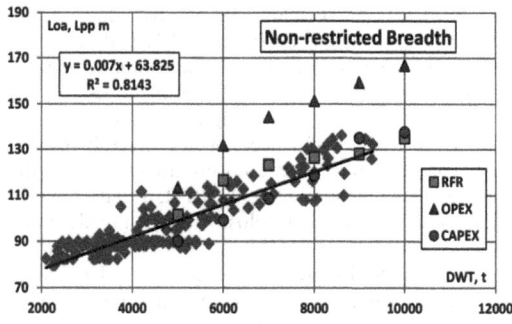

Figure 7. Length as a function of deadweight, coasters.

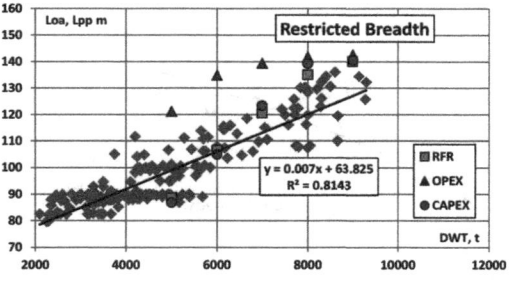

Figure 8. Length as a function of deadweight, coasters.

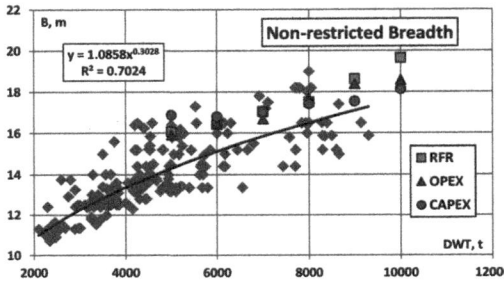

Figure 9. Breadth as a function of deadweight, coasters.

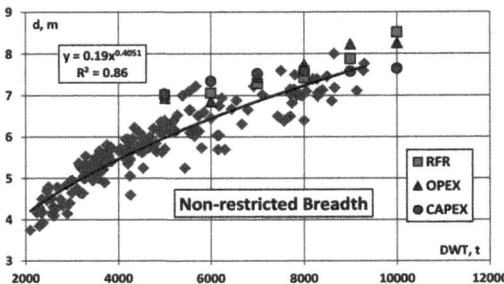

Figure 10. Draught as a function of deadweight, coasters.

values are close. The gradient of increasing the length is more significant than the one observed for already built ships.

Concerning the breadth, the optimal ship design solutions based on the CAPEX, OPEX and RFR are close to the already built ships. The estimated design values are a little bit greater than the ones of the already built ships. The design solutions for the draught are highly influenced by the breadth restrictions. The economic criteria do not affect the draught at the restricted breadth while the RFR, as a design criterion, leads to solutions close to the already built ships

Figure 12 to Figure 14 present the relationship between the gross tonnage (GT), net tonnage (NT) and Cargo capacity versus DWT. A very good match is obtained, and the estimated relations may be used at the preliminary stage of ship design.

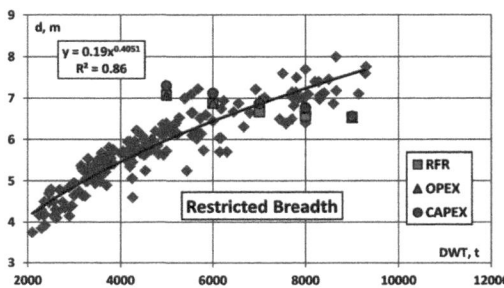

Figure 11. Draught as a function of deadweight, coasters.

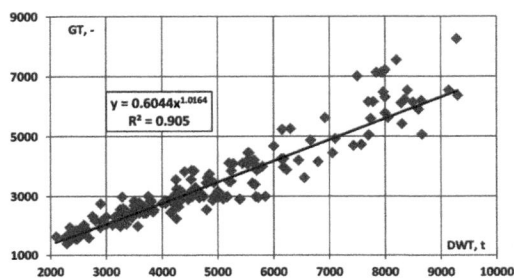

Figure 12. GT as a function of deadweight, coasters.

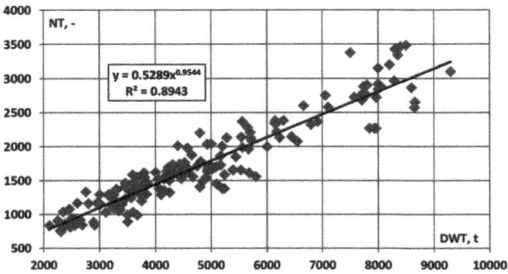

Figure 13. NT as a function of deadweight, coasters.

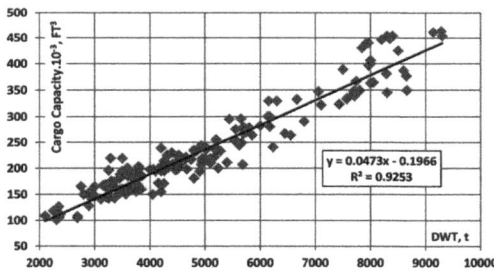

Figure 14. Cargo capacity as a function of deadweight, coasters.

5 CONCLUSIONS

The analysis of the recently published studies showed that the amount of goods transported by SSS in the EU is higher compared to the year 2009 (the worst year) and the increase in the last three years is permanent. The total share of the Mediterranean and the Black Seas is more than one-third of the total amount of transported goods, and transportation in this region has the potential to be increased.

The coastal fleet in the East Mediterranean and Black Sea regions (EMBSR) faces serious challenges related with some long-standing problems of the SSS as for example the legislation and organisation, new environmental requirements of the IMO that will impact the efficiency of the ships and the most essential facts are the considerable age of the

SSS fleet. The IMO includes ballast water treatment, sailing in ECAs and EEDI and EEOI. About 12% of the SSS fleet is over 20-year-old, and 485 from about 4,300 active units (11%) are considered as sub-standard. It can be argued that the considerable age, high freight rates and the increase amount of goods transported by the SSS will lead to new ordering and construction of new-generation coaster ships taking into account all IMO requirements.

The analysis of 117 facilities from six counties of the region has shown that there is enough capacity for the construction in SME shipyards of such new ships.

To respond to the objective of the EC in stimulating the development of the SMEs, the analysis of shipbuilding opportunities, with the consideration of shipbuilding constraints was performed here. To aid the SME in this process, a design tool "Expert" specialised in the conceptual design was employed in defining the optimal design solutions taking in consideration CAPEX, OPEX and RFR. The present study included the design of MPV in the range of 5,000 to 10,000 DWT with and without restriction in the breadth due to the construction limitations of the SMEs. The design solutions were compared with the already built ships, including the ones accounting for restriction in the breadth of the vessel.

REFERENCES

Abramowski, T., Cepowski, T. & Zvolenský, P. 2018. Determination of regression formulas for key design characteristicas of container ships at preliminary design stage. *New Trends in Production Engineering*, Vol.1, 247–257.

Atanasova, I., Damyanliev, T. P., Georgiev, P. & Garbatov, Y. 2018. Analysis of SME ship repair yard capacity in building new ships. *In*: Guedes Soares, C. & Santos, T. A. (eds.) *Progress in Maritime Technology and Engineering*. London: Taylor & Francis Group, 431–438.

BMTI 2019. BMTI Short Sea Report. BMTI Technik & informations Gmbh.

Cepowski, T. 2019. Determination of regression formulas for main tanker dimensions at the preliminary design stage. *Ships and Offshore Structures*, 14, 320–330.

Damyanliev, T. P., Georgiev, P. & Garbatov, Y. 2017. Conceptual ship design framework for designing new commercial ships. *In*: Guedes Soares, C. & Garbatov, Y. (eds.) *Progress in the Analysis and Design of Marine Structures*. London: Taylor & Francis Group, 183–191.

Denev, Y., Georgiev, P. & Garbatov, Y. 2018. Analysis of multipurpose ship performance accounting for SME shipyard building limitations. *In*: Guedes Soares, C. & Santos, T. A. (eds.) *Progress in Maritime Technology and Engineering*. London: Taylor & Francis Group, 165–171.

Ebrahimi, A., Brett, P. O., Gaspar, H. M., Garcia, J. J. & and Kamsvåg, Ø., 2015, Parametric OSV Design Studies – precision and quality assurance via updated statistics, Tokyo, 12th International Marine Design Conference (IMDC).

EC 2018. Detailed analysis of ports and shipping operations. Annex to Motorways of the Sea Detailed Implementation Plan. Brussels: EC – Directorate General for Mobility and Transport.

ECSA 2016. The full potential yet to be unleashed, Brussels: European Community Shipowners' Associations. *ECSA Short Sea Shipping*.

Garbatov, Y., Ventura, M., Georgiev, P., Damyanliev, T. P. & Atanasova, I. 2017. Investment cost estimate accounting for shipbuilding constraints. *In*: Guedes Soares, C. & Teixeira, A. (eds.) *Maritime Transportation and Harvesting of Sea Resources*. London: Taylor&Francis, 913–921.

Georgiev, P., Kirilov, L., Garbatov, Y. & Denev, Y. 2019. Multi-attribute design decision solution of MPV accounting for shipyard building constraints. *In*: Georgiev, P. & Guedes Soares, C. (eds.) *Sustainable development and innovations in marine technologies*. London: Taylor and Frances Group.

Gustafson, M., Nokelainen, T., Tsvetkova, A. & Wikstrom, K. 2016. Revolutionizing short sea shipping. Abo Akademi University.

IDEAS. 2016. *European Coaster Fleet Analysis In View Of Substandard Shipping & Easy Classification*. [Online]. http://denizstrateji.com/:IDEAS. Institute.

OECD 2001. European Conference of Ministers of Transport. Short Sea Shipping in Europe. .

Papadimitriou, S., Koliousis, I. G., Sdoukopoulos, E., Lyridis, D. V., Tsioumas, V. & Stavroulakis, P. J. 2018. *The Dynamics of Short Sea Shipping. New Practices and Trends*, Cham, Switzerland, Palgrave Macmillan.

Santos, T. A. & Soares, C. G. 2017. Methodology for ro-ro ship and fleet sizing with application to short sea shipping. *Maritime Policy & Management*, 44, 859–881.

Simpson, B. 2016. Motorways of the Sea. Detailed Implementation Plan. Brussels: MOVE/B1/2015-201 Project.

Tsinker, G. 2004. *Port Engineering: Planning, Construction, Maintenance, and Security*. John Wiley & Sons, Inc.

Developments in Maritime Technology and Engineering – Guedes Soares & Santos (eds)
© 2021 Copyright the Author(s), ISBN 978-0-367-77376-2

Finite element welding simulation of construction assembly

M. Hashemzadeh, Y. Garbatov & C. Guedes Soares
Centre for Marine Technology and Ocean Engineering (CENTEC), Instituto Superior Técnico, Universidade de Lisboa, Lisbon, Portugal

ABSTRACT: The effect of clamping constraints during the erection process of construction assembly is analysed as concerns the welding residual stresses and distortions. The forced reduction of misalignments and gaps may lead to a considerable amount of pre-stresses in the welded assembly. Once the welding is made, and the assembly is released from the pre-welded fixtures, some stresses will remain due to the initial constraints, additionally to the welding induced residual stresses. The present work describes a numerical investigation of different mounting conditions considering the effect of pre-welding fixture induced forces and time of releasing the constraints. A butt-welded panel is modelled by employing the finite element method, and the transient thermo-mechanical analysis is applied to simulate the welding procedure. The results of simulating the aligning condition, as well as the corresponding releasing of initial constraints, are presented and compared with the butt-welded plate without the pre-stress due to the imposed tolerances.

1 INTRODUCTION

The ship construction is a long process that can differ depending on the size and type of ships. The construction process starts once the design is ended and includes several different steps before the ship is delivered to the shipowner, as can be seen in Figure 1 (Matthews, 1998).

The first step in the manufacturing process is related to the material procurement followed by cutting, shaping and bending of plates and other structural details. Then the details are sent to the fabrication shops where the process of sub assembling is performed. The construction goes through the fabrication of units, surface preparation, painting, outfitting, testing and finally, the delivery of the ready for use ship.

During the ship construction, the welding is the predominant type of joining of structural details, and it is widely used. Independently of the fact that the welding has been proven to be one of best in steel detail joining during the ship construction, the welding induces residual stresses and distortions, which can be additionally magnified during the erection process of construction assemblies (Connor et al., 1987, Masubuchi, 2013). Many studies, based on numerical and experimental approaches, have already been performed in estimating the residual stresses and distortions and the way how to mitigate them.

Thin-walled structures are widely used in the shipbuilding industry. However, this type of structures is subjected to high-level welding-induced distortions compared to the thick-walled structures, as can be seen in Figure 2. Assembling structures with a high level of distortions require the additional capital expenditure in mitigating the misalignment to guarantee good quality of the welding processes (Gumenyuk & Rethmeier, 2013).

In this regard, many studies have been performed on the welding processes of thin-walled structures and the different measures to reduce the resulting residual stresses and distortions during the fabrication process. Gray et al. (2014) showed that the initial imperfections of the welded plate has a significant effect on the post welding distortions.

Conrardy et al. (2006) investigated different fabrication techniques for mitigating the welding-induced distortions in thin-walled welded structures concluding that to reduce the welding-induced distortion, each step of the welding operation need to be precisely controlled keeping the welding governing parameters in the acceptable ranges.

Huang et al. (2004) performed several experimental tests considering different levels of complexity of the welding process to investigate different methods to minimise the distortions and residual stresses of thin-walled welded panels developing a computational tool to control the welding mechanism and its governing parameters.

The Finite Element Method (FEM) employing decoupled thermomechanical analysis is a very sophisticated approach, which has been widely used in simulation and analysing the welding process in many studies (Hashemzadeh et al., 2015, 2016). In this regard, the present work deals with a finite element analysis of different imperfections during the assembling considering the effect of pre-welding fixture induced constraints (Figure 3) and welding-

DOI: 10.1201/9781003216582-86

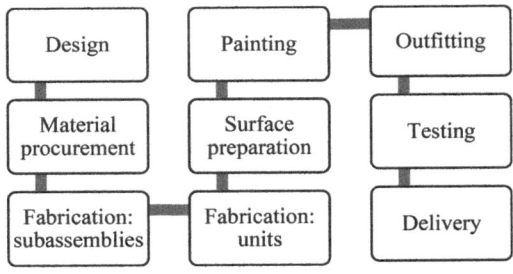

Figure 1. Ship construction process.

Figure 2. Deformed stiffened panel (Chau, 2006).

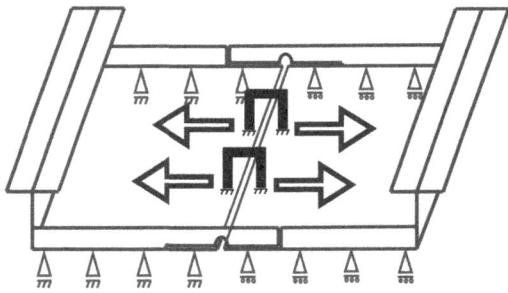

Figure 3. Clearance mitigation using fixtures and clips.

induced residual stresses and distortions. Pre-stress (aligning) conditions, as well as the corresponding releasing of initial constraints, are considered and the impact of them is calculated. Furthermore, a stiffener panel without the pre-stress condition is modelled and those results are used as a reference case study to estimate the effects of pre-stress conditions.

2 WELDING MODELLING

A butt-welded panel is modelled by employing the Finite Element Method, and the transient thermo-mechanical analysis is applied to simulate the welding procedure using commercial software (ANSYS, 2012). The moving heat source model is employed in the welding transient thermal analysis demonstrating to be a very accurate and suitable for welding simulation approach. This approach has been very profoundly studied in many works for example by Hashemzadeh et al. (2014a, 2014b), Hashemzadeh et al. (2018), where the Goldak et al. (1984) double ellipsoidal heat source model is most widely used as a combination of two semi-ellipsoids. Figure 4 shows the double ellipsoid model.

It is assumed that the heat from the welding arc is applied at any given instant of time, t at the point within the first semi-ellipsoid, z = 0, and moving with a constant velocity, v along the x-axis, the rate of the internal welding torch heat is modelled by a double ellipsoidal power density distribution as:

$$q_{sup} = \frac{6\sqrt{3} \cdot Q_{net}}{b \cdot c \cdot \pi \cdot \sqrt{\pi}} \begin{cases} \frac{f_f}{a_f} \exp\left(-\frac{3(x-vt)^2}{a_f^2} - \frac{3y^2}{b^2} - \frac{3z^2}{c^2}\right), x > v \cdot t \\ \frac{f_r}{a_r} \exp\left(-\frac{3(x-vt)^2}{a_r^2} - \frac{3y^2}{b^2} - \frac{3z^2}{c^2}\right), x > v \cdot t \end{cases}$$

(1)

where f_f is the heat input ratio of the front part, equals to 0.6 in the case before the torch passes the welding region, f_r is the heat input proportion in the rear section, 1.4 in the case after the torch passes the welding area, where $f_f + f_r = 2$.

The welding heat input per a unit weld length is $Q_{net} = \eta UI/\vartheta$ in W s/cm, where the current of the direct source is defined as I in A and the voltage between the electrode and the base plane by U. The travelling speed of the electrode is defined as v. The local coordinates of the double ellipsoidal model, aligned with the weld, are defined as x, y, and z. The radius in the y-direction of the weld is defined as b and c is the weld penetration, or it is the radius in the z-direction of the weld. The weld ellipsoid length is determined as a_f and respectively a_r. In the present study, a_f, b and c are equal to 4mm and a_r is equal to 8mm.

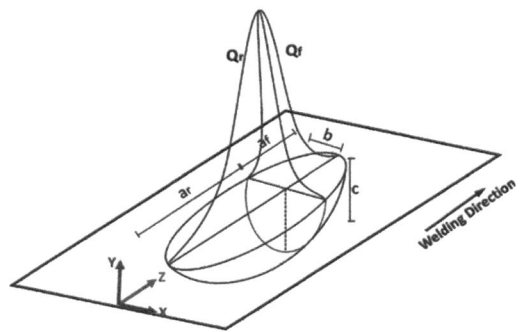

Figure 4. Double-ellipsoidal heat source model.

The steel used in this study is ASTM A36 and the temperature-dependent material properties considering the phase transformation effects (Noda, 1991, Zhu and Chao, 2002) are included in Table 1.

Different mounting conditions considering the effect of pre-welding fixture induced forces, residual stresses and distortions are analysed here, and with this respect, a butt-welded panel is modelled, and transient thermomechanical analysis is performed.

The scantlings of the stiffened panel are presented in Figure 5. The area close to the welding is refined, accounting for the accuracy in the heat source modelling, where 15,606 finite elements are used. The eight-node, 3D brick ANSYS thermal element Solid 70 is used in the thermal analysis. The element Solid 70 can be automatically replaced by the equivalent structural element Solid 185, which is an eight-node, three-dimensional element and has plasticity, hyperelasticity, stress stiffening, creep, massive distortion, and substantial strain capabilities. During the conversion, from thermal to mechanical analysis, the element mesh density remains the same. During the welding simulation and thermal analysis, employing FEM, the convection and radiation of the external areas are considered as the boundary conditions,

defined as h_f [Wm^{-2}] and ε=0.9 (Gery et al., 2005). Other boundary conditions, which are set for the thermal analysis is the environment temperature and in the present study, is assumed to be 25 °C. The total number of load steps in the present work is 77 steps, and the running time for thermo-mechanical analysis is approximately 138 min.

3 RESULTS AND DISCUSSIONS

The results of simulating pre-stress (aligning) condition, as well as the corresponding releasing of initial constraints, are presented and compared with the butt-welded plate without the pre-stress due to the imposed tolerances. The effect of the additional load, originating from the presence of the clips in the processes of aligning of construction assemblies, on the residual welding-induced stresses is investigated.

During the erection process of construction assemblies, angular or linear misalignment may occur, as can be seen in Figure 2 and Figure 3. To mitigate the misalignment between the joining structures, mounting of additional clips is considered here. The objective of the clips is to introduce a load with an opposite sign that allowing the gaps and misalignments between the joining structure to be reduced and good quality of welding to be performed.

It is also assumed that the level of shrinkage and distortions of the assembling parts of the structure are different and need an external load to joint them, which temporarily, during the welding process, can be done by mounting clips as can be seen in Figure 6).

Once the assembled structures are welded, and the clips are removed once the cooling process ended, the welding induced stresses due to the presence of the clips during the welding process are analysed at the time $t > \tau_2$ as can be seen in Figure 7.

The welding-induced residual stresses for both longitudinal and transverse direction are analysed here, considering different levels of clip-induced loads

Table 1. Material properties, ASTM A36 l.

T °C	C J/Kg°C	K W/m°C	α μm/m°C	σy MPa	E GPa
20	450	51	11.2	360	210
100	475	50	11.8	340	195
210	530	49	12.4	320	195
330	560	46	13.1	262	185
420	630	41	13.6	190	168
540	720	38	14.1	145	118
660	830	34	14.6	75	52
780	910	28	14.6	40	12
985	1055	25	14.6	38	11.8
1320	2000	32	14.6	28	10.4
1420	2100	42	14.6	25	10.2
1500	2150	42	14.6	20	10

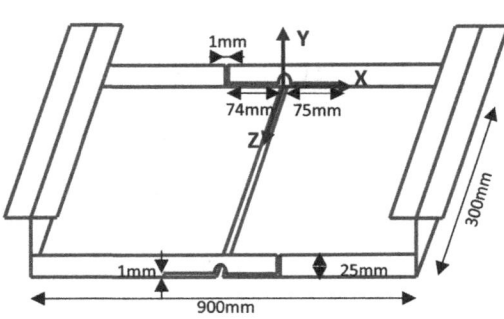

Figure 5. Stiffener panel geometry.

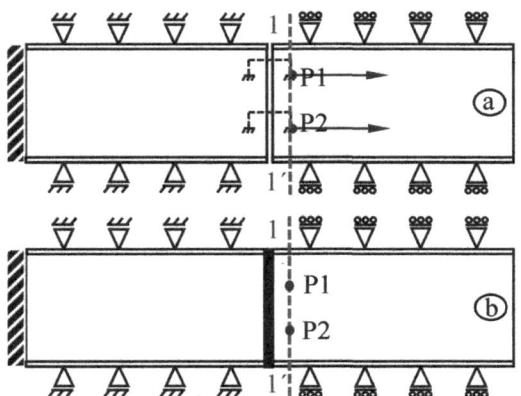

Figure 6. Mounted (upper) and removed clips (lower).

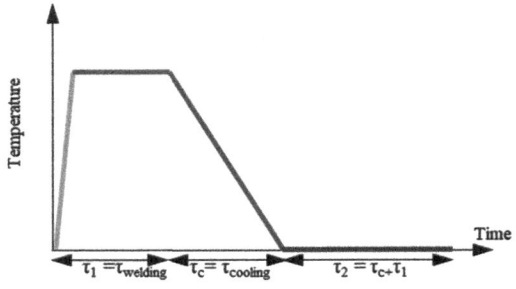

Figure 7. Welding time history.

before (τ_2 in Figure 7) and after the clips are removed from the assembled structure (τ_2 in Figure 7).

Figure 8 and Figure 10 show the longitudinal and transverse residual stress distributions at time τ_c and section 1-1′ as shown in Figure 6a and Figure 7. It can be seen that two peaks are identified in the place where the clips are mounted. Also, away from the location of the clips, it can be observed that by increasing the level of the clips-induced load, there is a reduction in the longitudinal welding-induced residual stresses and an increase in transverse welding-induced residual stress, respectively.

Figure 9 and Figure 11 illustrate the longitudinal and transverse welding-induced residual stresses at the time τ_2 and section 1-1′ (see Figure 6b and Figure 7). It is observed that even after removing the clips, the initial effect of the clips remains at the structure as a stress concentration.

It can be concluded that, applying fixtures (clips) to control and mitigate the initial misalignments cause high stress concentrations in the place where the clips are mounted.

It is also observed that the longitudinal welding-induced residual stresses are more sensitive to the presence of the clips. To analyse more closely, the trend of

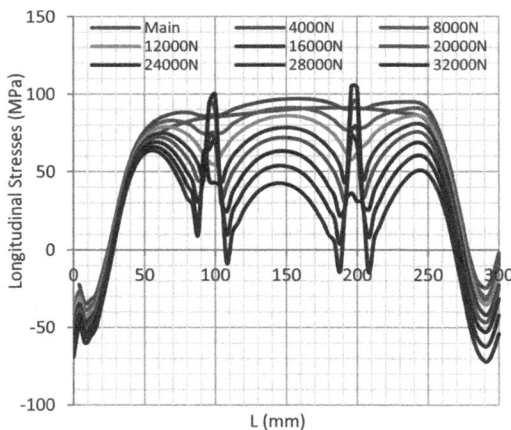

Figure 9. Effects of different levels of clip-induced load, removed clips.

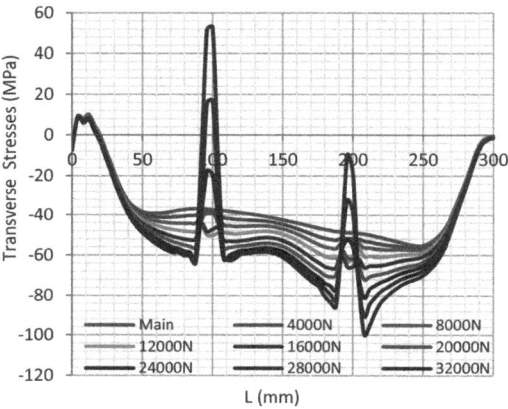

Figure 10. Effects of different levels of clip-induced load.

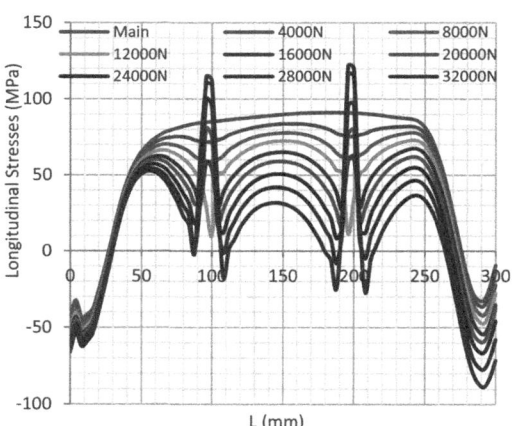

Figure 8. Effects of different levels of clip-induced load.

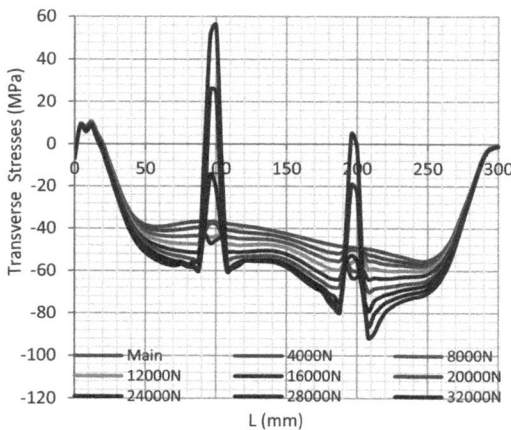

Figure 11. Effects of different levels of clip-induced load, after removing clips.

changing the peak stresses at the location of the clips Figure 8 to Figure 11 are presented and as a function of the level of the clips-induced load in Figure 12.

Figure 12 shows that the peak stress of the longitudinal welding-induced residual stress (S_z) has unstable behaviour, but the peak value of the transverse welding-induced residual stress (S_x) has a more linear trend. Also, it seems that the level of a 24,000 N clips-induced force is the most critical one and above this load, the structure has a different behaviour.

A different number of clips aligning configuration and their effect on the welding-induced residual stresses are studied. Three different cases are analysed, as shown in Figure 13, by considering two, three and four clips. The individual clips-induced load is considered to be 32,000 N. The longitudinal and transverse welding-induced residual stresses are presented for two assembling conditions after welding cooling process ended and after removing the clips.

Figure 13 to Figure 17 show that the number of peak stresses is equal to the number of clips, and they appear at the place, where the clips were mounted. Figure 14 and Figure 15 show the behaviour of the longitudinal welding-induced residual stresses before and after removing the clips, where it can be observed a reduction once the clips are removed.

Figure 16 and Figure 17 show the transverse welding-induced residual stresses, which are increasing in general. The trend of all studied cases looks similar, and it may be concluded that by having the same clips-induced load (associated with the same initial misalignment, distortion and gaps between the assembling parts of the structure), the three clips case study leads to lower the transverse welding-

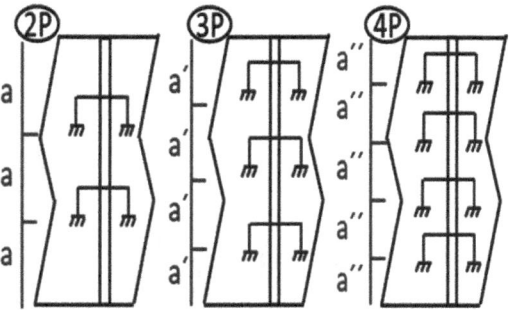

Figure 13. Different configurations of clips.

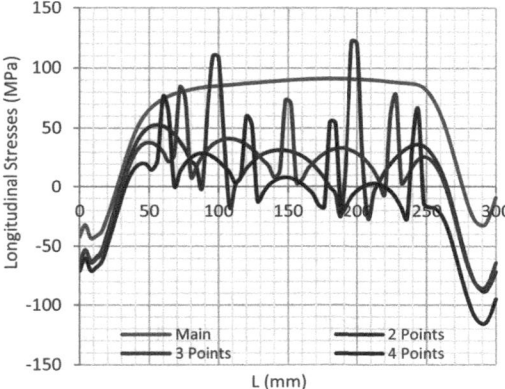

Figure 14. Effects of different number of clips-induced load, longitudinal residual stresses.

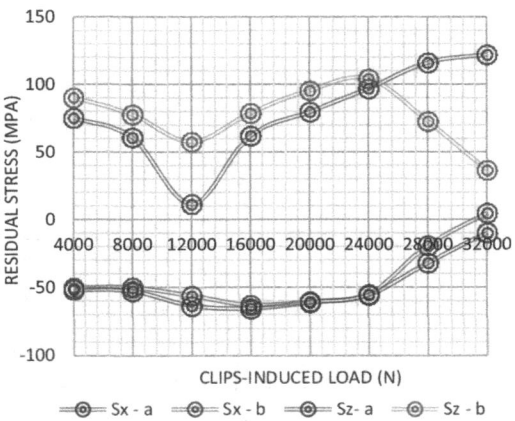

Figure 12. Peak-value welding-induced residual stresses (P2 in Figure 6).

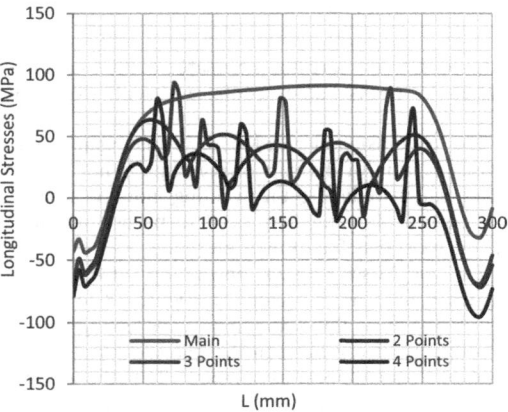

Figure 15. Effects of different number of clips-induced loads, longitudinal residual stresses-removed clips.

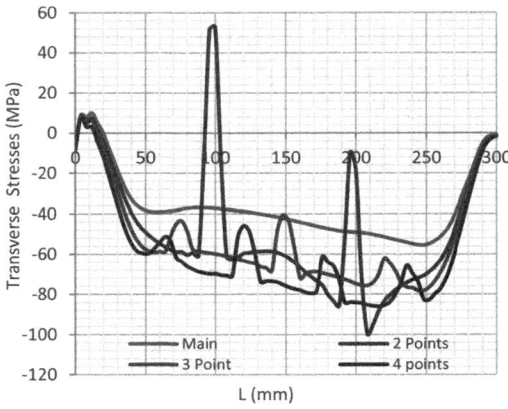

Figure 16. Effects of different number of clips-induced load, transverse residual stresses.

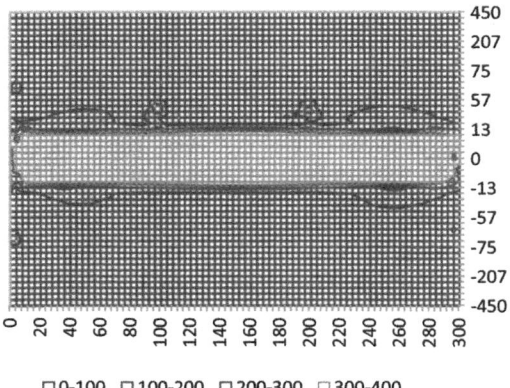

Figure 18. Von Mises stresses, top surface – two clips.

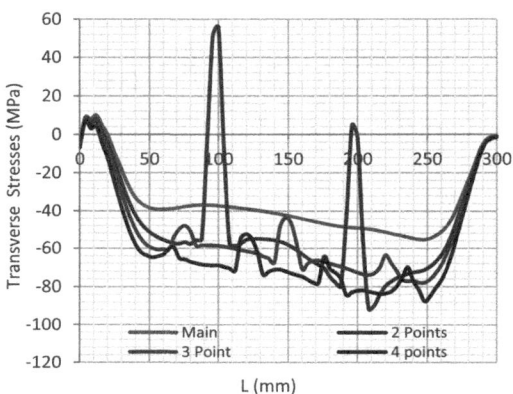

Figure 17. Effects of different number of clips-induced load, transverse residual stresses-removed clips.

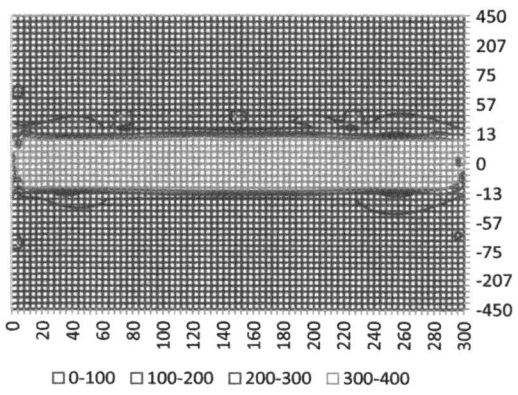

Figure 19. Von Mises stress, top surface – three clips.

induced residual stresses compared with four clips and lower peaks values comparing with two clips case study.

The conclusion derived here is that using more dense location of the supporting clips is not always beneficial in reducing the welding-induced stresses and careful planning may need to be performed to get the most benefits of their presence.

The von Mises stresses, on the top surface of the welded panels, are presented in Figure 18 to Figure 20 for the two, three and four clips configuration, respectively. It can be seen that the welding-induced residual stresses appear asymmetrically to the welding toe line. At the place, where the clips are mounted, a stress concentration is observed, and the other parts of the welded structure are not affected significantly.

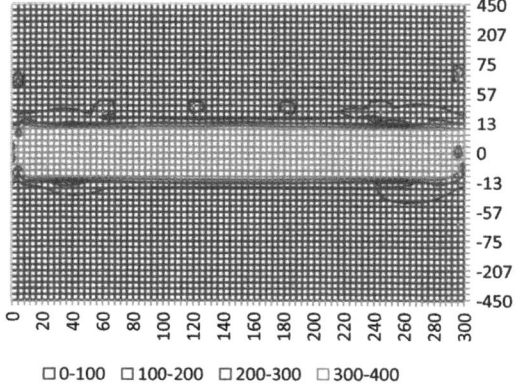

Figure 20. Von Mises stress, top surface – four clips.

4 CONCLUSIONS

The work presented here analysed the effect of clamping constraints as a result of mounting clips, during the erection process of construction assemblies. The principal objective of mounting clips is to reduce misalignments and gaps in performing a good quality of welding, which in the same time may lead to a considerable amount of pre-stresses in the welded assembly. The results of simulating pre-stress (aligning) condition, as well as the corresponding releasing of initial constraints, are presented and compared. It is observed that by increasing the amount of clips-induced load, which represent the scale of the initial imperfection, there is a reduction in the longitudinal welding-induced residual stresses and an increase in the transverse welding-induced residual stress, respectively. After removing the clips, the already induced effect of the clips is not released from the structure. The longitudinal welding-induced residual stress (S_z) has an unstable trend, but the transverse residual (S_x) stress has a more linear one. Having the same clips-induced load, but the different number of clips makes that in the particular case studied here, the three clips configuration leads to a lower transverse welding residual stress compared to other studied cases.

ACKNOWLEDGEMENTS

This work was performed within the Strategic Research Plan of the Centre for Marine Technology and Ocean Engineering (CENTEC), which is financed by Portuguese Foundation for Science and Technology (Fundação para a Ciência e Tecnologia-FCT) under contract UIDB/UIDP/00134/2020.

REFERENCES

Ansys 2012. *Online Manuals. Release*

Chau, T. T. 2006. Welding Effects On Thin Stiffened Panels. *In:* Youtsos, A. G., ed. *Residual Stress and Its Effects on Fatigue and Fracture*, Dordrecht. Springer Netherlands, 117–137.

Connor, L. P., O'Brien, R. L. & Society, A. W., 1987. *Welding Handbook: Welding technology*, American Welding Society.

Conrardy, C., Huang, T. D., Harwig, D., Dong, P., Kvidahl, L., Evans, N. & Treaster, A. 2006. Practical Welding Techniques to Minimize Distortion in Lightweight Ship Structures. *Journal of Ship Production*, 22, 239–247.

Gery, D., Long, H. & Maropoulos, P. 2005. Effects of welding speed, energy input and heat source distribution on temperature variations in butt joint welding. *Journal of Materials Processing Technology*, 167, 393–401.

Goldak, J., Chakravarti, A. & Bibby, M., 1984. A new finite element model for welding heat source. *Metal Transactions B*, 15B, 299–305.

Gray, T., Camilleri, D. & Mcpherson, N. 2014. 2 - Fabrication of stiffened thin-plate structures and the problem of welding distortion. *In:* Gray, T., Camilleri, D. & Mcpherson, N. (eds.) *Control of Welding Distortion in Thin-Plate Fabrication*. Woodhead Publishing, 14–38.

Gumenyuk, A. & Rethmeier, M., 2013. 19 - Developments in hybrid laser-arc welding technology. *In:* Katayama, S. (ed.) *Handbook of Laser Welding Technologies*. Woodhead Publishing, 505–521.

Hashemzadeh, M., Chen, B. Q. & Guedes Soares, C. 2014a. Comparison between Different Heat Sources Types in Thin-plate Welding Simulation. *In:* Guedes Soares, C. & Pena, F. (eds.) *Developments in Maritime Transportation and Exploitation of Sea Resources*. UK: Francis & Taylor Group London 329–335.

Hashemzadeh, M., Chen, B. Q. & Guedes Soares, C. 2014b. Numerical and experimental study on butt weld with dissimilar thickness of thin stainless steel plate. *The International Journal of Advanced Manufacturing Technology*, 78, 319–330.

Hashemzadeh, M., Chen, B. Q. & Guedes Soares, C. 2018. Evaluation of multi-pass welding-induced residual stress using numerical and experimental approaches. *Ships and Offshore Structures*, 13, 847–856.

Hashemzadeh, M., Garbatov, Y. & Guedes Soares, C. 2015. Reduction in Welding Induced Distortions of Butt Welded Plates Subjected to Preventive Measures *In:* Guedes Soares, C. & Shenoi, A. (eds.) *Analysis and Design of Marine Structures*. London, UK: Taylor & Francis Group, 581–588.

Hashemzadeh, M., Garbatov, Y. & Guedes Soares, C. 2016. Reduction in welding induced residual stresses and distortions of butt welded plates subjected to heat treatments. *In:* Guedes Soares, C. & Santos T.A. (eds.) *Maritime Technology and Engineering III*. London, UK: Taylor & Francis Group, 481–488.

Huang, T. D., Dong, P., Decan, L., Harwig, D. & Kumar, R., 2004. Fabrication and Engineering Technology for Lightweight Ship Structures, Part 1: Distortions and Residual Stresses in Panel Fabrication. *Journal of Ship Production*, 20, 43–59.

Masubuchi, K. 2013. *Analysis of welded structures: Residual stresses, distortion, and their consequences*, Elsevier.

Matthews, C., 1998. Ship and boat building and repair: general profile. *Encyclopaedia of occupational health and safety. Fourth edition.* Geneva: International Labour Office, 92, 1–92.

Noda, N. 1991. Thermal stresses in materials with temperature-dependent properties. *Appl. Mech. Rev*, 44, 383–397.

Zhu, X. K. & Chao, Y. J. 2002. Effects of temperature-dependent material properties on welding simulation. *Computers & Structures*, 80, 967–976.

Developments in Maritime Technology and Engineering – Guedes Soares & Santos (eds)
© 2021 Copyright the Author(s), ISBN 978-0-367-77376-2

Outsourcing activities in a shipbuilding cluster: A study via MCDM

P.I.D. Lameira, E.S.P. Loureiro, R.S. Saavedra & T.C.G.M Filgueiras
Federal University of Pará, Belém, Brazil

R.C. Botter
University of São Paulo, São Paulo, Brazil

ABSTRACT: Due to their ability to improve their competitiveness in shipbuilding industry (SI), maritime industrial clusters are in the limelight of logistics research. Companies within a cluster can operate at a higher level of efficiency by using more specialized assets and suppliers, with shorter reaction times than they would be capable of alone. One of important practical advantages of clustering is outsourcing. Through this process, companies and industries carry out nonessential processes through suppliers or outside professionals, so they can focus their occupations on higher value-added professional activities across the value chain. The present study analyzes the activities to be outsourced in the SI, through the selection of criteria and statistical treatment of data, using an AHP approach to the selection of outsourcing activities in the shipbuilding industry. Decision-making elements were used through a comprehensive structure comprising strategic dimensions to evaluate the performance of each sector of the shipbuilding cluster.

1 INTRODUCTION

The ability to improve the competitiveness of industrial clusters is in focus in current scientific research, defined by the interconnection of geographically adjacent related companies operating in a similar commercial follow-up (Porter 1998; Kayvanfar et al. 2018). In the case of the shipbuilding industry (SI), industrial clusters, measured as maritime clusters, in that case, serve as the basis for a reasonable number of business groups. The adoption of shipbuilding clusters is recommended and studied by several authors with the intention of regional development as well.

Clustered companies have the ability to operate at higher levels of efficiency, using more specialized assets and suppliers, with shorter reaction times than they would be able to do in isolation. This capability is performed by the practical advantage of outsourcing, which is natural in a cluster (Sonobe & Otsuka, 2006; Zhang et al., 2017).

This practical advantage translates into the ability to deploy resources external to an internal activity, within an outsourcing regime (Balaji & Brown, 2005). According to Lacity & Willcocks (2013), outsourced activities allow a company to reduce costs, increase its operational efficiency and assist in focusing on the core business, reorganizing critical resources within that company. Ranganathan & Balaji (2007) defined this process as a set of organizational means to effectively exploit internal and external resources (material or labor). Other studies show that logistical performance is also positively affected by outsourcing (Yuan et al., 2020).

Notwithstanding the current trend of using clusters in market relations, the Amazonian shipbuilding industry (SI) has developed based on the use of insourced production where a project is carried out by an individual or department within one. This industry experienced an upward economic curve until mid-2014, with great investment prospects, national government support, industry incentive policies, technology transfer in various regions of the country with the creation of large shipyards. However, the abrupt emergence of very large shipyards, occurred much more as a government plan, than by the real condition and need to absorb all this sudden industrial conglomerate, dependent on the political, economic and social conditions imposed, thus creating situations. extremely unstable business conditions (D'Avila & Bridi, 2017). Given the situations mentioned, it is essential to investigate the implementation of a cluster focused on naval construction and repairs in the northern region of Brazil as a way of evasion and mitigation of financial instability in the regional SI, in order to object the dominant form of insourced production in this area.

It is critical to use a structured approach that considers decision-making elements to select a more effective outsourcing strategy for a given organization, considering short-, medium- and long-term concerns related to outsourcing or not building groups. This research uses a multicriteria decision making (MCDM)

DOI: 10.1201/9781003216582-87

method, through criteria and statistical treatment of data, using an Analytic Hierarchy Process (AHP) approach for the selection of outsourcing activities in the shipbuilding industry taking into consideration variables such as quality of supply, reliability, time-frame for delivery and costs. Decision-making elements were used through a comprehensive structure comprising strategic dimensions to evaluate the performance of each sector of the shipbuilding cluster.

2 BACKGROUND

2.1 Outsourcing in Brazil

According to Dias Ferreira and Barbin Laurindo (2009), the outsourcing process in Brazil has been increasingly incorporated in Brazilian companies. However, from the beginning of the 21st century until mid-2010, the outsourcing process in Brazil did not yet have specific legislation in place regarding the outsourcing process, so as to meet the aspirations of the related society and provide legal certainty, generating possible jobs from that process. However, during this period, a number of judgments - known as summary - were used as a reference for much of the legal action. In this case, these summaries determined that outsourcing in Brazil was only allowed in the middle activities needed to complete the service/product.

In view of this, in 2017 the law 4.302/1998 (Brazil, 1998) was approved, coming with changes from the government of former President of Brazil, Fernando Henrique Cardoso, which took place on March 23, 2017, by the House of Representatives. Deputies, becoming the ordinary law 13,429/2017 (Brazil, 2017), sanctioned by the President Michel Temer and having the endorsement of the Brazilian Supreme Court. This legislation aims to protect the worker of the country, establishing the joint liability, in which the worker can act at the same time contractor and outsourced to pay labor claims. Also, there is the possibility of a service company subcontracting another company to perform the services (the so-called "blocking").

Drawing on the work of Santos (2017), Vasconcellos (2017) and Maeda (2016), current Brazilian outsourcing legislation allows:

- Outsourcing may apply to any business activity, medium or purpose;
- The outsourced company will be responsible for hiring, remunerating and directing workers;
- The contracting company must ensure the safety, hygiene, and health of outsourced workers;
- The duration of temporary work goes from up to three months to up to 180 days, consecutive or not, and there may still be an extension of 90 days, after the termination of the contract, the temporary worker can only provide the same type of service to the company again. after waiting three months.

2.2 Outsourcing and clusters

The idea of division of labor was first expressed by Smith (1776). According to the author, people or companies acting in their own interest, performing tasks that each individual performs better, provide economic growth and more benefits to society. That is, instead of a single individual performing all tasks resulting in a potentially inefficient process, the tasks of a production process should be assigned to specialists. The concept of cluster, in turn, popularized by economist Michel E. Porter in the 1990s, is basically a group of geographically interconnected companies and associated institutions that specialize in a single sphere and united by common interests and complementary (Porter, 2000), being present in several types of production chain in the world and extremely important for the regional development of markets and increase of its international competitiveness.

Industrial clusters have as their main characteristic a high level of competitiveness at the regional or even global level, affecting it on three points: Increasing the productivity of the constituent companies or industries (Girma & Görg, 2004; Görg & Hanley, 2005 Koliousis et al., 2017); increasing the capacity of cluster participants for innovation development and productivity growth (Djoumessi et al., 2019) and; stimulating the formation of new businesses that support innovation and expand the cluster (Stavroulakis & Papadimitriou, 2016).

It is also important to show that the use of clusters does not only generate positive returns. An example of the negative effect of clustering is environmental effects. According to Yoon & Nadvi (2018), clustered companies often generate external waste, increasing pollution costs and the demand for scarce resources. However, this is already a latent concern in studies that investigate the environmental and sustainable aspects of clustering (Daddi et al., 2017; Konstadakopulos, 2008; Varshney et al., 2018a, 2018b; Wang & Wang, 2019; Yoon and Nadvi, 2018). Also, within this area of study, there is some research showing ways to mitigate such effects or even reduce them to insignificant levels (Lund-Thomsen, 2009; Chertow & Park 2019; Baldassarre et al. 2019).

Outsourcing, as far as it is concerned, has been the subject of many studies since its concept was introduced by Smith (1776) to the present day, being debated in various areas of study. Papers in areas of study such as supply management (Brandes et al., 1997; Gilley & Rasheed, 2000), marketing (Wathne & Heide, 2000) and information sciences (Zhang et al., 2019) are examples of research involving this concept.

3 METHODOLOGY

A comprehensive maritime cluster outsourcing group assessment system has been built on the basis of MCDM methods. The general steps of the proposed approach, using data processing and evaluation, consisted of 3 main steps: Data collection and analysis (1), weighting for criteria (2), and net flow and classification of the alternatives (3).

For all cost groups analyzed, responses were obtained from professionals consulted. In the case of alternatives, we also used the answers obtained from the professionals consulted, associated with technical data collected in the research to support the weightings related to the alternatives.

Assuming that outsourcing objectives are based on cost reduction, focus on the core activity and facilitating the management of business processes, the proposed alternatives that seek to meet the objective of the model are: to insource, to outsource with the supply of materials or outsource only the labor. The shipbuilding cost groups for the yard were established as Structure, machinery, electricity and treatment and painting.

According to Macharis et al. (2004), the AHP method is based on 3 principles: Hierarchical construction, priority setting, and logical consistency. First, the hierarchical structure is used to decompose the complex system of the problem into its constituent elements. The structure is built on at least 3 levels: The overall goal or intention at the top; the criteria and subcriteria at the intermediate level and the alternatives considered at the base (Chand et al., 2018; Macharis et al., 2004; Saaty, 2008). Therefore, the equal comparison of the criteria for defining their weights and their reliability is necessary. Then, the relative priorities of each element in the hierarchy are determined by comparing all lower-level elements with the criteria, which need to have a causal correlation.

A 3-tier hierarchical structure was built, where the second tier consists of eleven criteria, namely: Costs, Time, Risk, Flexibility, Focus on Core Activity, Financial Management, Process Management, Quality, Reliability, Productivity and Expansion of the company's core business. The decision criteria were determined through literature review and consultation with experts from the shipbuilding area of Brazil and abroad, who were also responsible for carrying out the selection process of alternatives.

In this paper, the AHP method was used to hierarchically structure the problem and define the criteria weights due to its high level of reliability and easy and intuitive use (Dweiri et al., 2016; Ghimire & Kim, 2018). According to Sennaroglu & Varlik Celebi (2018), AHP fits into the class of subjective weighting methods, due to the judgments of the decision makers, which the pairwise comparison matrices are established.

Criterion weights were calculated according to the procedure presented by Saaty (1987), using paired comparison matrices of the criteria for each group of this research. The consistency ratios of each comparison matrix used was less than 0.1, therefore acceptable, as proposed by Baffoe (2019), Ghimire & Kim (2018) and Saaty (1980).

The criteria "Costs" and "Time" were evaluated through the averages of real values obtained from 3 shipyards in the analyzed region that produce the same type of vessel, where the worst values were established with value "1" and the best with value "9" In the AHP input. The assessment of the remaining criteria is not so simple; therefore, their assessments are based on reviews of academic literature, decision makers' brainstorming, and the responses obtained from the data handled.

The methodology provides a complete ranking of alternatives, from best to worst, where all alternatives within groups are comparable. The results generated by the AHP method are ranked in descending order, that is, the alternative that is closer to 100% is the best option to be used.

4 RESULTS

All groups exposed in this research were considered. The Structures group is fundamental for any shipyard, especially shipyards in the studied region, which specialize in the manufacture of barges. In other words, these shipyards practically only have costs in the Structures group and the Treatment and Painting group. The activities of the group of structures end up being involuntarily treated by everyone in the sector as a core activity. Therefore, the group is regarded as fundamental by shipyards that manufacture any type of vessel.

Thus, the AHP method was applied and through the survey of the criteria, the weighting of the criteria, verification of the consistency of the matrices and definition of the alternatives. In order to elucidate the results obtained, Table 1 shows the final priorities of the alternatives obtained from AHP for the Structures group. This table shows that the most relevant criteria for this group were, costs and time, respectively.

Given the criteria and aspirations proposed for carrying out the activities of this group, according to Table 1, the alternative found was the outsourcing of labor only, considering that the structural part is one of the backbones of the construction process of

Table 1. Pairwise comparison matrix for structure.

Structure	Cost	Time	Flexibility	Core activity focus	Productivity	Priorities
Insourcing	0,35	0,20	0,25	0,67	0,52	35,9%
Outsourcing labor and materials	0,15	0,40	0,43	0,10	0,25	21,1%
Outsourcing labor	0,52	0,41	0,33	0,25	0,25	43,0%

vessels and that the cost for outsourcing including material supply in the region is considerably high.

The matrix assembly of the criteria regarding the group was reliable, precisely because the CI found 0.098 to be less than 0.1.

To facilitate comparisons, Table 2 brings standardized AHP alternative priorities to the machine group, finding "focus on end activity" and "costs" as the most relevant criteria for the group, with 0.34 and 0, 33, respectively. This hierarchy of priorities shows that actually installing and sourcing machines makes it easier to focus on core yard activities, and cost is a deciding factor in choosing to do or outsource.

This alternative "Outsourcing labor and materials" was obtained precisely because of the need for focus on the yard's core activities, with an emphasis on its skills and the ability to reduce costs through outsourcing of labor.

The electricity group has characteristics similar to those mentioned for networks and pipeline. Also depending on the type of vessels, the seasonality applies throughout the working period for construction. In the case of pushers, electricity accounts for approximately 20% of the labor cost, 12% of the total cost, and depending on the type of vessel and propulsion system selected, can reach higher levels, reaching between 20 % and 25% of the cost.

The criteria considered for analysis of the electricity group were: costs; term, focus on end activity; risk; processes management; and reliability.

Table 3 shows the standardized priorities for the alternatives proposed in the Electricity group, as well as the relevance of the criteria. The most relevant criteria were: focus on the end activity; and risk. The focus on end activity can be justified precisely by the phenomenon of seasonal demand for this type of activity throughout the work. The risk has several factors that enable its prioritization, one of them is the labor risk itself due to the need for eventual dismissals, as well as, due to the complexity level of the services demanded in the electrical systems, possible accidents or delays for such reason may compromise the risk. progress of the work.

Outsourcing the installation of electrical systems on vessels is a common practice, albeit partial. The result for the AHP analysis shows that total outsourcing to the electricity group is the best alternative for the case evaluated. It is important to emphasize that the greater the complexity of the deployed systems, the greater the possibility of potential risks and consequently loss of focus on the yard's end activity, so it is critical to verify the supply chain capacity before making the decision to outsource or insource.

The activities related to the treatment and painting of the vessel usually occur at different times of the work, either at the beginning of production through surface treatment and application of the shop primer, or at the end of the work performing finishing painting, and there may also be painting activities. at intermediate times.

The criteria considered most relevant for this group were cost and productivity, respectively with

Table 2. Pairwise comparison matrix for machines.

Machines	Cost	Time	Flexibility	Core activity focus	Reliability	Business Expansion	Priorities
Insourcing	0,32	0,20	0,23	0,19	0,20	0,20	23,6%
Outsourcing labor and materials	0,33	0,66	0,63	0,78	0,71	0,66	59,6%
Outsourcing labor	0,40	0,28	0,28	0,06	0,08	0,28	21,0%

Table 3. Pairwise comparison matrix for electricity.

Electricity	Cost	Time	Core activity focus	Risk	Proc. management	Reliability	Priorities
Insourcing	0,51	0,31	0,10	0,33	0,16	0,11	18,2%
Outsourcing labor and materials	0,21	0,38	0,46	0,41	0,57	0,46	43,8%
Outsourcing labor	0,33	0,38	0,45	0,39	0,32	0,46	37,9%

Table 4. Pairwise comparison matrix for treatment and painting.

Treat. and painting	Cost	Reliability	Core activity focus	Productivity	Proc. management	Quality	Priorities
Insourcing	0,40	0,33	0,43	0,29	0,26	0,30	34,4%
Outsourcing labor and materials	0,27	0,31	0,29	0,40	0,42	0,40	33,1%
Outsourcing labor	0,33	0,35	0,33	0,33	0,30	0,33	32,6%

0.37 and 0.24, according to the standardized priorities presented in Table 4. The result obtained through the proposed analysis, shows the insource as the most viable alternative among those studied.

5 CONCLUSION

After surveying the cost structures of vessels built in the region, in order to increase reliability and validate the proposed analysis, the MCDM method was satisfactory. First, after treatment of the primary data, a uniform sample space was verified and for the first validation using the consistency index, values below 0.1 were reached, within the acceptable limit for criteria matrices, when compared in pairs. It is also noteworthy that the results showed the will of most of the companies consulted, being also in line with global trends.

In addition to the general analysis, a cost group analysis was performed. Vessels with greater complexity, that is, that have more systems, tend to have a higher percentage of outsourcing within the construction process, both labor and materials. Groups that need more assembly technology, such as electricity, in addition to the inherent seasonality of the group, tend to be outsourced at higher rates to reduce costs, meet deadlines, maintain quality and focus on end activity advantages obtained by the coherent selection of activities to be outsourced, based on consistent analysis.

The core activity of businesses that deal with few systems or have disconnected and low complexity processes in their supply chain can be easily defined.

In order to treat the focus on end activity as a criterion for the model, it is essential to define the end activity of a shipbuilding yard, a not so simple task. This criterion is complex to define due to the various systems to be developed for the production of a ship. So, the question about the definition of a yard's end activity is: would it be the structural part, machinery or electrical? After the results obtained in this research, it was concluded that the end activity of a shipyard depends on its competences, the types of vessels it builds and the organization created to meet its demands.

Regarding the outsourcing behavior of a shipyard's production facilities in Brazil, not only of the middle activities but also of the end activities, the change in Brazilian legislation facilitates the shipyards' approach to process and cost control. It is possible today, as long as there is a competent outsourced chain, that shipyards are sites only, as is the case with European yards, which group together service providers. Thus, increased productivity and control are considerations that can considerably increase competitiveness.

From the application of the proposed approach, AHP it was possible to verify that, in the case of shipyards inserted in the Amazon region, most of the studied groups indicated the total outsourcing, labor, and materials, as a more viable alternative. Thus, it can be seen that both subcontracting (labor only) or total outsourcing indicate a gain for production in a general context, whether financial or management. Therefore, the vast majority of cost groups indicated total outsourcing, turning the yard into a contract assembly and concentrator site, catching up with the worldwide trend of manufacturing industries.

In addition, it is complemented that the deep study of the condition of each region is fundamental for the implantation of any organizational structure. It is necessary planning and measures by the private initiative and the public power together to analyze the conditions and propose alternatives that meet the wishes and the regional reality because the regional development is a fundamental element for the evolution of an industrial cluster and feeding of the community.

Thus, this work is relevant, as it proposes ways to structure the SI of the studied region. It also denotes extremely relevant data for the sector and presented an original method, using established techniques, proposing a model for selecting outsourced activities.

REFERENCES

Baffoe, G., 2019. Exploring the utility of Analytic Hierarchy Process (AHP) in ranking livelihood activities for effective and sustainable rural development interventions in developing countries. Evaluation and Program Planning 72, 197–204. 10.1016/j.evalprogplan.2018.10.017.

Balaji, S., Brown, S.A., 2005. Strategic IS Sourcing and Dynamic Capabilities: Bridging the Gap, in: Proceedings of the 38th Annual Hawaii International Conference on System Sciences. Presented at the Proceedings of the 38th Annual Hawaii International Conference on System Sciences, pp. 260b–260b. 10.1109/HICSS.2005.560.

Baldassarre, B., Schepers, M., Bocken, N., Cuppen, E., Korevaar, G., Calabretta, G., 2019. Industrial Symbiosis: towards a design process for eco-industrial clusters by integrating Circular Economy and Industrial Ecology perspectives. Journal of Cleaner Production 216, 446–460. 10.1016/j.jclepro.2019.01.091.

Brandes, H., Lilliecreutz, J., Brege, S., 1997. Outsourcing —success or failure?: Findings from five case studies. European Journal of Purchasing & Supply Management 3, 63–75. 10.1016/S0969-7012(97)00001-4.

Brazil, 2017. Lei no 13.429 de 2017.

Brazil, 1998. Projeto de lei no 4.302 de 1998.

Chand, P., Thakkar, J.J., Ghosh, K.K., 2018. Analysis of supply chain complexity drivers for Indian mining equipment manufacturing companies combining SAP-LAP and AHP. Resources Policy, Sustainable management and exploitation of extractive waste: towards a more efficient resource preservation and waste recycling 59, 389–410. 10.1016/j.resourpol.2018.08.011.

Chertow, M., Park, J., 2019. Reusing Nonhazardous Industrial Waste Across Business Clusters, in: Waste. Elsevier, pp. 353–363. 10.1016/B978-0-12-815060-3.00018-9.

Daddi, T., Nucci, B., Iraldo, F., 2017. Using Life Cycle Assessment (LCA) to measure the environmental benefits of industrial symbiosis in an industrial cluster of SMEs. Journal of Cleaner Production 147, 157–164. 10.1016/j.jclepro.2017.01.090.

D'Avila, A.P.F., Bridi, M.A., 2017. Indústria naval brasileira e a crise recente: o caso do Polo Naval e Offshore de Rio Grande (RS). Cadernos Metrópole 19, 249–268. 10.1590/2236-9996.2017-3810.

Dias Ferreira, A.M., Barbin Laurindo, F.J., 2009. Outsourcing decision-making aspects considered by IT departments in Brazilian companies. International Journal of Production Economics 122, 305–311. 10.1016/j.ijpe.2009.06.024.

Djoumessi, A., Chen, S.-L., Cahoon, S., 2019. Factors influencing innovation in maritime clusters: An empirical study from Australia. Marine Policy 103558. 10.1016/j.marpol.2019.103558.

Dweiri, F., Kumar, S., Khan, S.A., Jain, V., 2016. Designing an integrated AHP based decision support system for supplier selection in automotive industry. Expert Systems with Applications 62, 273–283. 10.1016/j.eswa.2016.06.030.

Ghimire, L.P., Kim, Y., 2018. An analysis on barriers to renewable energy development in the context of Nepal using AHP. Renewable Energy 129, 446–456. 10.1016/j.renene.2018.06.011.

Gilley, K.M., Rasheed, A., 2000. Making More by Doing Less: An Analysis of Outsourcing and its Effects on Firm Performance. Journal of Management 26, 763–790. 10.1177/014920630002600408.

Girma, S., Görg, H., 2004. Outsourcing, foreign ownership, and productivity: evidence from UK establishment-level data. Review of International Economics 12, 817–832.

Görg, H., Hanley, A., 2005. International outsourcing and productivity: evidence from the Irish electronics industry. The North American Journal of Economics and Finance 16, 255–269. 10.1016/j.najef.2004.11.006.

Kayvanfar, V., Moattar Husseini, S.M., Sajadieh, M.S., Karimi, B., 2018. A multi-echelon multi-product stochastic model to supply chain of small-and-medium enterprises in industrial clusters. Computers & Industrial Engineering 115, 69–79. 10.1016/j.cie.2017.11.003.

Koliousis, I.G., Papadimitriou, S., Riza, E., Stavroulakis, P.J., Tsioumas, V., 2017. Strategy, policy, and the formulation of maritime cluster typologies. Marine Policy 86, 31–38. 10.1016/j.marpol.2017.09.010.

Konstadakopulos, D., 2008. Environmental and Resource Degradation Associated with Small-Scale Enterprise Clusters in the Red River Delta of Northern Vietnam. Geographical Research 46, 51–61. 10.1111/j.1745-5871.2007.00491.x.

Lacity, M., Willcocks, L.P., 2013. Outsourcing business processes for innovation. MIT Sloan Management Review 54, 63–69.

Lund-Thomsen, P., 2009. Assessing the Impact of Public–Private Partnerships in the Global South: The Case of the Kasur Tanneries Pollution Control Project. Journal of Business Ethics 90, 57–78. 10.1007/s10551-008-9914-x.

Macharis, C., Springael, J., De Brucker, K., Verbeke, A., 2004. PROMETHEE and AHP: The design of operational synergies in multicriteria analysis. European Journal of Operational Research 153, 307–317. 10.1016/S0377-2217(03)00153-X.

Maeda, P., 2016. Terceirização no Brasil: histórico e perspectivas = Outsourcing in Brazil: history and perspectives. Revista do Tribunal Regional do Trabalho da 15a Região 127–150.

Porter, M.E., 2000. Location, Competition, and Economic Development: Local Clusters in a Global Economy.

Economic Development Quarterly 14, 15–34. 10.1177/089124240001400105.

Porter, M.E., 1998. Clusters and the New Economics of Competition. Harvard Business Review.

Ranganathan, C., Balaji, S., 2007. Critical Capabilities for Offshore Outsourcing of Information Systems. MIS Quarterly Executive 6.

Saaty, R.W., 1987. The analytic hierarchy process—what it is and how it is used. Mathematical Modelling 9, 161–176. 10.1016/0270-0255(87)90473-8.

Saaty, T.L., 2008. Decision making with the analytic hierarchy process. International Journal of Services Sciences 1, 83. 10.1504/IJSSCI.2008.017590.

Saaty, T.L., 1980. The analytic hierarchy process: planning, priority setting, resource allocation. McGraw-Hill International Book Co, New York ; London.

Santos, E.R. dos, 2017. A nova lei da terceirização: Lei n. 13.429/2017: um cheque em branco ao empresariado. Revista eletrônica: acórdãos, sentenças, ementas, artigos e informações 13, 51–59.

Smith, A., 1776. An Inquiry Into the Nature and Causes of the Wealth of Nations. W. Strahan and T. Cadell.

Sonobe, T., .Otsuka, K., 2006. Cluster-Based Industrial Development: An East Asian Model. Springer

Stavroulakis, P.J., Papadimitriou, S., 2016. The strategic factors shaping competitiveness for maritime clusters. Research in Transportation Business & Management 19, 34–41. 10.1016/j.rtbm.2016.03.004.

Varshney, D., Mandade, P., Shastri, Y., 2018a. Optimization of sugarcane bagasse based industrial cluster for economic and environmental benefits, in: Computer Aided Chemical Engineering. Elsevier, pp. 1993–1998. 10.1016/B978-0-444-64241-7.50327-X.

Varshney, D., Mandade, P., Shastri, Y., 2018b. Optimization based design of an industrial cluster for economic and environmental benefits, in: Computer Aided Chemical Engineering. Elsevier, pp. 717–722. 10.1016/B978-0-444-64235-6.50127-3.

Vasconcellos, A.C., 2017. NOVA LEI DA TERCEIRIZAÇÃO: O QUE MUDOU? Revista da Escola Nacional da Inspeção do Trabalho (ENIT) 1, 69–96.

Wang, Y., Wang, J., 2019. Does industrial agglomeration facilitate environmental performance? New evidence from urban China? Journal of Environmental Management 248, 109244. 10.1016/j.jenvman.2019.07.015.

Wathne, K.H., Heide, J.B., 2000. Opportunism in Interfirm Relationships: Forms, Outcomes, and Solutions. Journal of Marketing 64, 36–51. 10.1509/jmkg.64.4.36.18070.

Yoon, S., Nadvi, K., 2018. Industrial clusters and industrial ecology: Building 'eco-collective efficiency' in a South Korean cluster. Geoforum 90, 159–173. 10.1016/j.geoforum.2018.01.013.

Yuan, Y., Chu, Z., Lai, F., Wu, H., 2020. The impact of transaction attributes on logistics outsourcing success: A moderated mediation model. International Journal of Production Economics 219, 54–65. 10.1016/j.ijpe.2019.04.038.

Zhang, X., Yang, J., Thomas, R., 2017. Mechanization outsourcing clusters and division of labor in Chinese agriculture. China Economic Review 43, 184–195. 10.1016/j.chieco.2017.01.012.

Zhang, Y., Xiang, Y., Zhang, L.Y., Yang, L.-X., Zhou, J., 2019. Efficiently and securely outsourcing compressed sensing reconstruction to a cloud. Information Sciences 496, 150–160. 10.1016/j.ins.2019.05.024.

Developments in Maritime Technology and Engineering – Guedes Soares & Santos (eds)
© 2021 Copyright the Author(s), ISBN 978-0-367-77376-2

Economic feasibility analysis for the deployment of a ship repair yard in the Amazon

P.I.D. Lameira, T.C.G.M. Filgueiras, P.P. Souza & H.B. Moraes
Federal University of Pará, Belém, Brazil

R.C. Botter
University of São Paulo, São Paulo, Brazil

ABSTRACT: The Amazonian shipbuilding market, despite having some consolidated shipyards, still does not reach its full potential, having thus, their demands not met due to the limitations of the local shipping industry. Corroborating this problem, the objective of this paper is to analyze economic feasibility for the deployment of a ship repair yard in the northern region of Brazil. First, a SWOT analysis was performed aiming to determine the critical points of the cluster implementation based in the diverse field data collected. Finally, the final analysis was performed using the Net Present Value (NPV) method, the comparison parameter of Internal Return Rate (IRR) and the Minimum Rate of Attractiveness (MRA), and being performed for three possible scenarios. It was concluded that the project to implement a ship repair cluster is completely viable based on all the assumptions considered in the methodology of this research for the northern region of Brazil.

1 INTRODUCTION

The Brazilian naval industry market has experienced ups and downs in recent decades, once again reaching a full rise five years ago, thus contributing to the flow of cargo and passengers around the country, in addition to the significant generation of employment. However, since mid-2014, Brazil has been experiencing a profound political and economic crisis, being the largest economic crisis since 1931, including a contraction of GDP in twelve consecutive quarters and a tripling of the unemployment rate that affects 13 million people. (Dobrovolski et al., 2018)

In view of this, the Brazilian federal government is currently looking for new ways to revive such an important sector, thus creating several tax incentives for the shipbuilding industry such as the REB (Brazilian Special Registry), which provides tax exemption for vessels built in Brazil. Brazil with National Flag and specific financing lines for shipbuilding provided that it has a minimum nationalization index of 60% (Brazil, 1997).

In the northern region of Brazil, due mainly to its strategic position in relation to the main world routes and the region has the greatest potential for inland navigation of the planet, with more than fifty thousand kilometers of navigable stretches (MMA, 2006), it has to be the target of new export routes, thus generating a differentiated demand for inland waterway vessels (barges and river pushers), which

are traditionally built by the Amazonian shipbuilding industry.

In addition, it is of great importance to mention that, with the increase in the number of vessels in the region, the need arises for the installation of ship repair yards in the region following the thought that even the current supply of this area is not sufficient to supply the regional demands causing several vessels to have to perform maintenance services outside the region, generating an economic deficit and reducing the attractiveness of investments to the north of the country.

It is also necessary, of course, to emphasize that there is a need for an economic feasibility analysis for the emergence and implementation of a new business. For Begum et al. (2006), an investment for the company is a disbursement made to generate a future benefit stream, usually over one year, so it is necessary that the business to be implemented be profitable and viable.

Corroborating this problem, the objective of this paper is to carry out an economic feasibility analysis for the installation of a ship repair yard in the Amazon region, having as main parameters the current conditions of the Brazilian shipbuilding industry.

2 FRAMEWORK

2.1 *Amazonian navigation*

According to the Brazilian Ministry of the Environment (MMA, 2006 and MMA, 2006b), the Brazilian

DOI: 10.1201/9781003216582-88

Amazon comprises most of the Amazon River and Tocantins-Araguaia hydrographic regions, with only the former accounting for approximately 44.65% of the total. Brazilian territory, comprising an area of 3.8 million square kilometers with approximately 50 thousand kilometers of rivers, being historically and geographically prone to use the waterway as its main means of transport.

Given the magnitude of the potential of the waterway modal in the Amazon region, navigation is the main mode of transport since the beginning of the occupation process (Gadelha, 2002), making use of a waterway network. This mode continues to be very important today, both in view of the incipience of alternative modes (Dos Santos et al., 2018), due to the lack of great efforts of municipalities in planning and analysis of logistics systems (Marques and Kuwahara, 2009 and Frota, 2008) as well as the fact that rivers are navigable in most of their courses, allowing access to great distances in the Hydrographic Region at a relatively low cost (ANTAQ, 2009).

The study of vessel and shipyard profiles in the Amazon region is extremely scarce when it comes to academic works, we can mention some authors who proposed to work with the subject such as Paula et al. (2019) analyzing from a fuzzy logic the risks to the environment of river vessel sanitation in the Amazon, Soares, and Filho (2015) seeking to increase safety in typical wooden vessels, but not focusing primarily on an efficiency overview of modal. Looking at older works, we can mention Moraes (2001); and Moraes and Vasconcelos (2001) which characterize river passenger boats in a very specific way, but, being already lagged by the time of publication, thus increasing the relevance of this work, which in part aims to characterize the supply and demand of maintenance of vessels in the region.

2.2 *Economic feasibility and SWOT analysis*

To assess competitiveness or even have a mirror of what may happen in a given market, there are several techniques that help in understanding and analyzing the main scenarios and factors that may influence the good or not of a particular business. One is the SWOT analysis (Strengths, Weaknesses, Opportunities, and Threats).

A SWOT analysis (alternatively called a SWOT matrix) and structured planning, the method used to assess the strengths, weaknesses, opportunities, and threats involved in a project or business venture or industry (Hossain et al., 2017). According to Fernandes (2014), the perception that to develop a good strategy requires a lot of knowledge and understanding of the business, the internal and external environments in which the organization is inserted, which in itself justifies the use of SWOT. The intrinsic characteristics of the organization, its strengths, and weaknesses and its extrinsic characteristics, opportunities, and threats of the environment outside the organization form the foundation of the matrix that ultimately represents the result of perceptions of the environments in which the organization is inserted.

This methodology was used in several ways within the academic area, among them we can cite Markovska et al. (2009) who used the method to assess the Macedonian energy system by pointing to the progressive adoption of European Union (EU) standards in energy policy and regulation as the most important achievement in the region's energy sector and Agyekum et al. (2020) turning to the analysis of the use of nuclear energy as a sustainable development option in Ghana.

In addition, the analysis of economic viability is a topic of great importance for the academic environment and essential for any type of enterprise. According to Groppelli and Nikbakht (2010), the relationship between risk and return is the basis on which rational and intelligent investment decisions are made. Moreover, it is essential to use indicators to guide the profitability of the business and its viability. According to Rodrigues et al. (2016), one of the indicators associated with project profitability (gain or wealth creation) and the net present value (NPV), which is defined as the sum of the present values of the estimated flows of an application, calculated from a given and its duration. Other important indices for any economic viability analysis are the Minimum Attractiveness Rate (MAR) and the Internal Rate of Return (IRR). For Kassai (2005), the MAR is the minimum rate to be reached in a given project, otherwise, it must be rejected and is also the rate used to discount cash flows when using the NPV method and the benchmark. for the IRR. Moreover, it is essential to satisfactorily select the indices for the economic viability analysis of a project, so it is necessary to know which index consistently represents the simulations of the scenarios that will be performed.

3 METHODOLOGY

In order to obtain secondary data, a survey was carried out mainly at SINAVAL (National Union of Shipbuilding and Offshore Repair and Construction Industry), at SINCONAPA (Union of Shipbuilding Industry of the State of Pará), at SINDNAVAL (Union of Shipbuilding Industry), Offshore and Amazon Repairs) and in affected organizations from 2006 to 2016; and in updated scientific and regional informative works. Primary data were obtained by applying questionnaires to the actors involved and with expertise in the regional shipbuilding process. The selected questionnaire was of the closed type, which according to Nogueira (2012), despite being more rigid than the open ones, allowed obtaining data for a direct application of statistical data with the aid of computers, which eliminated the need to classify responses that would induce undesirable trends.

A survey was made of the Garra Naval shipyard, located in the district of Icoaraci in the city of Belém - PA, totaling 108 suppliers. For the survey, based on adaptations proposed by the author to the method used by Ruuska et al. (2013), the questionnaires were directly applied in the yard, in two hierarchical levels of the evaluated organization: strategic level and tactical level, represented by directors and technical managers, respectively, where the five attributes were analyzed as follows:

- Product quality;
- Quality in product delivery;
- Price of products;
- Deadline; and
- Meeting deadlines.

Thus, the SWOT analysis was performed, aiming, preliminarily, to determine the strengths, weaknesses, threats, and opportunities of the implementation of a repair yard in the state of Pará, Amazon region. Similarly to that performed by Hossain et al. (2017), this SWOT analysis was performed for the ship repair industry based on field data collected, close observation of the shipbuilding process, supply chain, project management, order book, actions and strategic government plans during the period of 3 years of the elaboration of the research, besides consultations carried out in industries in the studied area.

In the SWOT analysis, the cross-matrix technique was used using the professionals' opinions on the scores, in order to ensure reliability for the matrix. A strength scale from 1 to 5 was used, 1 considered unimportant and 5 extremely important.

In order to elucidate the analyzes and give robustness to the results, we considered the favorability index (FI) to analyze the final results, as proposed by Resende (2015), according to equation 1:

$$FI = \left(\frac{(S + O) - (T + W)}{(S + O) + (T + W)} \right) \qquad (1)$$

where S = Strengh; O = Oportunities; T = Treaths; and W = Weakness.

To reach the factor values, a scoring scale was used according to the response scale, namely:

- 1 represents 0 points;
- 2 represents 2.5 points;
- 3 represents 5.0 points;
- 4 represents 7.5 points;
- 5 represents 10 points.

To elucidate the analyses, factors were classified into appropriate SWOT categories, including strengths, weaknesses, opportunities and threats. The answers are put into categories for the SWOT matrix. The attribution process is based on scoring. Therefore, according to Thamrin and Pamungkas (2017), it was agreed that strengths and opportunities would have positive values and weaknesses and threats of negative values.

Economic viability, in turn, was initially made by NPV, which represents the difference between future cash flows discounted at zero date by the opportunity cost of project capital (discount rate) and initial investment (Adami, 2010), defined by:

$$NPV = -I_0 + \sum_j^n \frac{FC_j}{(1 + i)^j} + RV \qquad (2)$$

where I_0 = Initial investment; FC_j = Cash flow for year j; RV = Residual value of investment; and i = discount rate for investment.

The interpretation of NPV is crucial to verify the prior viability of the business. According to Motta and Calôba (2002) for the economic viability analysis using NPV the following rules must be obeyed:

- NPV > 0 - Then the project must be accepted;
- NPV = 0 - So the project is indifferent;
- NPV < 0 - So the project is not feasible.

Another factor still used as complementary analysis is the Profitability Index (PI). The PI can be defined by:

$$PI = \frac{\frac{FC_1}{(1+i)^1} + \frac{FC_2}{(1+i)^2} + \frac{FC_n}{(1+i)^n}}{FC_0} \qquad (3)$$

where FC_0 = Initial cash flow; FC_n = Cash flows in period n; i = Interest rate/Minimum attractiveness rate.

For Groppelli and Nikbakht (2010) the PI is directly linked with the NPV approach because if the NPV is positive, the PI will be greater than 1 and if the NPV is negative the PI will be less than 1.

Another important factor for economic viability analysis is the IRR. For Adami (2010), the IRR and the discount rate that makes the NPV of an investment project null, that is, the interest rate that makes the investor indifferent as to the application of its resources in design or best available alternative. According to Motta and Calôba (2002) analyzing a scenario as an investment alternative, if the calculated IRR is higher than the market MAR, the alternative deserves to be considered for further analysis, otherwise, it should be rejected.

4 RESULTS

4.1 Supplier characterization

After the application of the questionnaires in the selected yard, we analyzed, from the source of locational supply of each group of systems of the vessel to the assessment of the attributes performed by the actors of different levels (strategic and tactical) of the yard evaluated for each group too. The groups of systems analyzed were:

- Structure - Composed of suppliers of materials focused on the structure of the vessel such as structural plates, structural profiles, castings, and structural tubes, etc.
- Machinery - Composed of suppliers of ship-oriented machinery including propulsion materials, power generation, auxiliary and housing equipment, etc.
- Networks and Pipes - Composed of suppliers of networks and pipes for vessels with pipes for networks, valves, fittings, filters, and flanges, etc.
- Electricity - Composed of suppliers of materials related to the vessel's electricity, such as electrical cables, parts, rails, mounting accessories, electrical panels, electrical panels, demarcators, light fixtures, radar, echo sounder, radios, odometer, autopilot, etc.
- Hull and Deck Accessories - Composed of suppliers of materials related to vessel accessories such as deposits, arches, manholes, fire extinguishers, fixed firefighting system, whistle, smoke detectors, magnetic needle, barometer, anemometer, inclinometer, etc.
- Finishing - Composed of suppliers of materials for finishing vessels such as partitions, linings, floors, wooden doors, fittings, sanitary equipment, furniture, shelves, anchors, moorings, towing and mooring lines, service boats, liferaft. lifebuoys, buoys, life jackets, first aid boxes, pyrotechnic devices, etc.
- Treatment and Painting - Composed of suppliers focused on treatment and painting materials such as shop primer, paints, varnishes, solvents, abrasives, brushes, abrasive blasting, anti-corrosion protection, anodes, etc.

108 suppliers who serve companies in the region were surveyed. In addition, as already mentioned in the methodology, the actors considered in the analysis represent the largest resource-handling yards in the region studied, as well as the largest order book identified, which implies greater reliability of responses.

From the point of view of the origin of the supply, it was possible to verify, for example, that the shipyard studied, being a company focused on the repair of vessels, has a great need for fast service, which implies the use of local suppliers. A particular case is only the machines (50% -50%), exactly because of the lack of spare parts in the local market or even the differentiated price, the latter factor that justifies the high prices applied in the repair industry, which besides being met by local suppliers, often do not use incentives to purchase inputs.

Conducting a general analysis of all groups for the yard evaluated, it can be concluded that:

- In all groups, the attribute "price of products" scored low, between 5 and 8 on average, which shows dissatisfaction with price. The explanation justified by the interviewees was given precisely due to the lack of competitiveness of our industry in relation to the international market and the binding in the nationalization index and incentive policies;
- Regarding the quality of the products supplied to ICN from the Amazon region of Brazil, an average of 7-9 of the evaluated yards were noted. This means that the inputs are of quality in the opinion of respondents, but many emphasize the issue of price to reach this level of quality and that some companies opt for lower quality products to be more competitive in final price;
- In the electricity group, it is possible to verify a slight discrepancy between the delivery time and its fulfillment, exactly because some products are imported and there is difficulty in both ports for receiving, customs and speedy customs activities.

4.2 SWOT analysis

The second stage of this work was the SWOT analysis, initially, a brainstorming was used, a tool commonly used for strategic planning of a business. The technique was used with a total of 22 professionals related to the naval, port and logistics areas, in order to consider the maximum of ideas, solutions and problems relevant to the case. This step aims to develop as many scenarios as possible, with different points of view, so as not to allow relevant factors to be disregarded and the matrix to develop unreliable or only from the author's point of view.

After Brainstorming, Forces, Weaknesses, Opportunities, and Threats were defined. To this end, a list of factors considered relevant to the problem of an ocean and river ship repair yard was developed, based on a future high-capacity offshore port designed for the region and future integration into an industrial cluster.

Below are descriptions of Forces (Table 1), Weaknesses (Table 2), Opportunities (Table 3), and Threats (Table 4):

To obtain the results of internal and external factors, the cross matrix was used. Thus, we arrive at the values of internal factors, strengths, and weaknesses, which are 82.5 and 37.5 points respectively.

For external factors, opportunities and threats, a list of relevant factors and their respective strength scale scores was also developed, reaching 87.5 and 45 points respectively. 7 potential threats and 17 opportunities were raised. It is noteworthy that only one threat was considered with an extremely important force scale, which is the competition with mainly Asian shipyards. As well as, opportunity was also considered only one of extreme importance, the proximity to the main terminals of the world. As very important opportunities also stand out the few competitors in the regional market and the intention of creating a naval pole by various spheres of government.

Table 1. Forces descriptions.

Item	Description	Force Scale
Strategic location	The significant increase in ship size worldwide demands high depths at the world's terminals. Therefore, the location is extremely privileged, with depth for the largest ships in the world, which favors both the port and the repair yard.	5
Possibility of Diversified Service	For the repair yard, there is the provision of dry dikes to service larger vessels, but there is also the provision of transverse careers, which was designed for the repair of small and medium-size vessels,	3
Wide area for possible expansions	Availability of area for expansion, which allows the construction to be modular, in phases to meet projected demands.	3
Modern facilities and equipment	Integration with educational and research institutions is considered as the use of state-of-the-art technology in the yard. As a result, the recent crisis has left industrial parks with state-of-the-art machinery available for purchase at a low cost.	4
Green Port (aligned with the world trend)	The offshore terminal itself has its concept of operating with water-water movements, in addition to planning for the use of renewable energy and sustainable construction. This is a worldwide trend that attracts investors.	4
Less interference in the region's traditional communities	Ship repair organizations will have little impact on traditional communities near the region as their location does not cover any environmental reserves within 20km as well as recognized traditional communities in their vicinity.	2
No dredging or shelter required	The Offshore Terminal is located in a deep and naturally sheltered region, which significantly reduces any intervention required for shelter and dredging, for example.	3
Depth terminal available for the world's largest ships	With the worldwide trend of increasing cargo ship ports, it is sensitive to the need to increase the depths of existing ports, as well as to consider this very important variable in the design of new terminals.	5
Integration with waterways	The very natural vocation of the Amazon region cannot be forgotten, so the proposed shipyards have integration with the waterways, as well as the offshore terminal as	3

(*Continued*)

Table 1. (Cont.)

Item	Description	Force Scale
	well. This point facilitates the planning of operations.	
Presence of a Collaborative Environment	The organization's information needs to be available transparently and to all actors. The very use of technology, integrated with research, facilitates this environment so that all sectors and actors understand the operations and can collaborate with each other.	3
Lower environmental impact on terminal deployment	Because it is an Offshore Terminal, it is also likely to have less environmental impacts on its deployment.	3
Proximity to World's Largest Mineral Deposit	The Carajás mountain range in the state of Pará houses the largest iron ore deposit (approximately 1 million square kilometers) in the world. In addition to iron, it also has abundance in manganese, gold, and nickel.	3
Know-how in the construction of river vessels	The northern region already has extensive experience in the construction of cargo and passenger river vessels. Even with national competitiveness, which facilitates the deployment of the shipbuilding industry.	3
Management model using strategic planning	The business will already be conceived with a rooted management model, that is, strategic planning will be used to service ensuring quality, credibility and adding value to processes. The management model also directly interferes with operational actions, reducing rework and service time.	3

After all completed steps of the cross-SWOT matrix, you can verify the business phase classification. The propitious phase is the development phase, that is, opportunities are favoring the forces, which justifies the need to use strategies for business development. It is noteworthy that this step is fundamental for the selection of the strategy to be applied in the present and future of the business under analysis.

The final step performed in the SWOT analysis was to obtain the favorability index. The favorability index found was 69.31%. In this preliminary analysis, it is possible to verify that the result of the favorability index is the installation of the ship repair yard in the region.

4.3 *Economic feasibility*

To calculate the demand generated by the northern region, the traffic of vessels in the ports of this

Table 2.	Weaknesses descriptions.	

Item	Description	Force Scale
Shortage of skilled labor	The region has an old problem of shortage of skilled labor, especially at the operational level, which makes it difficult to increase productivity and competitiveness.	4
New to the market	Shipyards may suffer directly from being new to the market, as many shipowners may initially lack confidence to invest in new construction or even repair, as possible delays lead to higher expenses and lost profits from vessels.	2
Higher cost than competitors	The so-called "Brazil cost" impacts on raw materials as well as on certain processes. In addition, the tax issue, directly and indirectly, impacts the cost of services provided to organizations located in Brazil.	3
Lack of government support in all spheres	Integration with the Government is essential for the generation of incentives, not only fiscal and financial but also through support for implementation and assistance in the licensing of organizations.	3
No similar modality in the region	There is no similar modality in the region, which hinders the legal and physical implementation of organizations. Therefore, the deployment processes tend to be more time-consuming.	1
Lack of data and specific studies to stimulate business heat	The lack of data creates insecurity on the part of investors and also the public authorities to believe in the implementation of businesses of this magnitude.	2
Lack of land access to Port	Because it is an offshore terminal, there is no land access to the terminal. Therefore, this restricts some operations and, in some ways, makes it difficult to move the terminal.	3
Lack of know-how in the construction and repair of ocean vessels	The region has no know-how in repair or construction of oceanic vessels, which can make it very difficult, especially in the initial phase, to develop the activities of the shipyards. This absence reflects the need for technology transfer and also the medium and long-term training of teams at all hierarchical levels of the organization.	5

Table 3.	Opportunities description.	

Item	Description	Force scale
Availability of Undergraduate, Specialization, and Master of Naval Engineering courses	An excellent opportunity for the cluster is the existence in Pará of an undergraduate degree in naval engineering, specialization courses in the port and shipbuilding areas and even a master's degree in naval engineering. SENAI itself has several technical level courses that also add to the shipbuilding and repair industry.	3
Existence of credit lines of national and international origin destined	There is the merchant navy fund itself (FMM) which has credit available for the construction, expansion, and modernization of industrial units (shipyards), either for construction or repairs, as well as for the vessels themselves.	2
Nearby major terminals in the world (Compared to others in Brazil)	Reinforcing the strategic location regarding the physical and environmental characteristics of the terminal and the shipyard, the terminal location is also a competitive advantage in relation to the proximity of the main terminals in the world, when compared to the other terminals in Brazil.	5
Possibility of offering services to the foreign market	The existing know-how in the manufacture of small and medium cargo vessels, combined with the condition provided by the incentive cluster and aggregate technology, is what can provide competitiveness to the supply of vessels to the international market.	3
In the regional market, there are few competitors	With respect to the repair yard, there are only small businesses to repair small craft, so the medium and large craft market has great potential, especially if it has a competitive price and reliability.	4
Intention of the creation of a naval pole by the class entities	The intention of some associations and federations already exists the intention of creating a so-called "naval pole" that encompasses a Terminal, investment in teaching and research center and shipyards.	4
Benefit of regional teaching and	The studies generated mainly by FINEP and other research incentive agencies have produced technical data, even if outdated today,	2

(Continued)

778

Table 3. (Cont.)

Item	Description	Force scale
research development	but which provides basic information about the conditions of the regional ICRN. Still, the course and undergraduate degree in Naval Engineering from UFPA, created in 2005, which has a maximum grade of recognition by MEC also generated specific manpower and critical mass to develop the naval and port sector.	
Growing industry demand	The construction industry is still riding the wave of commodities for river cargo vessels, and the repair could take advantage of the sheer volume of ships operating in the vicinity, also coming from the recent expansion of the Panama Canal.	4
Huge river fleet mainly due to commodity boom	This opportunity also leverages the possibility of repairs to vessels already in transit in the Amazon region and others to be built.	3
Panama Canal Expansion	Panama's channel expansion allows a volume of vessels traveling through the channel to use the services offered by the proposed cluster.	3
Cluster has a growing trend	The growing amount of exports, mainly of grains, as well as other products, made by Brazil tend to newly developed terminals. This generates a growth trend for the offshore terminal. For the repair market, the very increase in ship traffic and possible competitiveness may bring new demands on the repair industry.	3
High demand from developing tradings in the region (construction and repair)	As already mentioned, the demands of tradings for construction and repair, as well as the possibility of cluster aggregation across the terminal can assist in regional growth and development.	3
Business chain perpetuation through the creation of a management association	The cluster in its conception intends to integrate several organizations to meet the demands and the generated process of horizontalization itself tends to perpetuate the business chain.	3
Possibility of reducing the cost with practice	In the case of the terminal being located in a sheltered area and with high depths in its proposed mooring place, as well as in waterway access, there is a great possibility of reduction in the practical costs, making the terminal even more competitive.	3
Possibility to be a free port	The concept proposed by Moraes (2017) considers the possibility of	3

Table 3. (Cont.)

Item	Description	Force scale
zone (Singapore model)	the Offshore Port becoming a "Free Port", that is, a specific type of special economic zone.	
Possibility of low-cost raw material due to the proximity of the deposits	In the case of an installed processing industry, according to plans already submitted by the local authorities, the raw material will tend to be much cheaper for shipyards that will demand steel and other derived materials to develop their services.	3
Possibility to be the gateway to South America training of teams at all hierarchical levels of the organization.	In the case of the long term development of the cluster, the offshore port by all the characteristics already presented has the capacity to become the gateway of South America	4

Table 4. Threats descriptions.

Item	Description	Force scale
Lack of information from actors and difficulty in collecting information	It is not clear which companies would be involved in the cluster deployment and development process, let alone the information accurately to increase the reliability of the feasibility analysis, which makes estimation even more difficult, either in the planning or implantation.	3
New Competitors Coming	Following the emergence of the cluster, there is a high probability that competitors will emerge from shipyards in the Central and South American region. This fact should be assessed as a threat and preparation for this event is feasible.	2
Resistance from RESEX (Big Mother of Curse)	The onshore environmental reserve closest to the terminal and the cluster itself is the RESEX known as "Big Mother of Curuça" which has 36,678.24 hectares and has already resisted the implementation of projects.	4
Competition with other worldwide yards (China and others)	Other competitors around the world may also face the repair yard, for example. China and some Asian countries generally have the largest market share for international shipowner repair due to the competitive price and know-how acquired.	5
		4

(Continued)

(Continued)

779

Table 4. (Cont.)

Item	Description	Force scale
Slowness in the environmental licensing release process	The very lengthy licensing processes in large projects is already a recurring practice in Brazil and in the region, which implies even reducing the trust and unfeasibility of the project.	
Exchange variation	Foreign exchange variation itself can be an enemy or even benefit the business by bargaining with foreign clients. However, the uncertainty of variation brings risk to the business, especially to construction and repair yards.	3
Need for a specialized supply chain to become the gateway of South America	A specialized supply chain is critical to the maintenance and development of the cluster. This is necessary, as it transcends the shipyard's autonomy only, so it is essential to attract companies from the sector and leading service providers to add logistical support to the cluster	4

region was analyzed. A database was built from the information obtained from ANTAQ data, which quantified the long-haul vessels that docked at seaports during 2018. After the analysis, 792 different vessels were found. The worldwide standard used for forecasting repairs is the age/length parameter for estimating average repair days by vessel class.

Based on the size of the surveyed vessels, it was found that the largest with the largest occurrence is the Cape Size, however, the most frequent are Handymax and Panamax. Thus, to estimate the demand calculation, the parameters of 179 meters in length of the HandyMax class of vessels were used, since the largest class of vessels in the region is Panamax type; and 20.83 years of age, which was based on the UNCTAD study (2019) which indicates the average age of the world fleet of vessels.

Following the above considerations, the number of days of repairs required for this fleet can be estimated. For the purpose of calculating the imminent demand of the repair yard, only 10% of the total demand calculated was used.

For the ship repair yard, the analyzes for a moderate scenario met the following assumptions:

- According to the demand study, only the repairs on Panamax and Handymax class vessels were considered for revenue calculation;
- The only docking mechanisms considered were two dry dikes, sized to suit Cape Size and Handymax vessels, respectively;
- The investment was considered using the maximum participation allowed by the FMM, ie 90%

of the items financed at an interest rate of 3% per year and 10% of the own contribution;
- 330 operating days per year were considered;
- The Minimum Attractiveness Rate - MAR was composed of the opportunity cost represented by SELIC (Special Settlement and Custody System for Federal Securities) of 6.5%, 3% of business risk and 3.87% of liquidity, which is also the estimated inflation for 2019;
- Were sized workshops, equipment and number of employees according to benchmarks used;
- For the residual value, the depreciation of civil facilities and equipment and the appreciation of the land were estimated. The initial depreciation rate was 4% in the first year and 3.5% in the other years. And the appreciation was estimated by the IPCA average of 3.78% per year;
- The period considered for viability was 18 years, given that the total provided by the financial institution and FMM is up to 20 years;
- The social charges were calculated based on the official parameter National System of Costs Survey and Indexes of Construction (SINAPI) for the state of Pará, with reference in October 2018 and totaling approximately 73.1% of the remuneration;
- Demand was well below that projected by studies and class bodies. The Offshore Terminal was considered only as a booster because the demand of the referred terminal has not been projected yet; and
- For the values used in the costs applied in this feasibility study, we used parameters such as SINAPI, Basic Unit Cost (CUB) through the Brazilian Chamber of Construction Industry (CBIC), surveys in regional shipyards carried out during the survey, of the author and budgets made.
- For the dimensioning of the repair yard employees and also the equipment that will compose the workshops, reference indices were used for Cape Size and Panamax class ships, as presented by Butler (2000);

For the sizing of workshops, it is necessary to know the annual productivity of the repair yard. To obtain the number of vessels that can be serviced per year, we considered 330 operating days and an average for the days of repair per ship for each class, assuming the data presented above.

From the data presented above, the number of employees for the Repair Shipyard was dimensioned. Therefore, the workshops were sized, based on the services offered and the number of collaborators calculated. We also estimated the amount of equipment for processes and handling, as well as employees by function.

The CAPEX was developed for the Repair Yard, in line with the minimum values found. It was considered a 30% cash flow over the monthly fixed cost. For the Repair Shipyard OPEX, we used the dimensioning of the workshops, the calculation of labor charges, also considering the health, transportation and meal plan for the employees and the due taxes of the legal entity. Income tax and social contribution were calculated by actual profit, ie, they were computed as a cost only in the months that earned profit. For the maintenance cost, 5% of the equipment acquisition cost was admitted. For insurance, 0.5% of CAPEX was considered, excluding working capital, works, and systems.

To estimate the revenue of the Repair Shipyard, three phases were considered: Implementation, maturation and full phase. The main factors that varied in the 3 phases were: revenues, admitting 25% in the implementation phase, 75% in the maturation phase and 100% in the full phase, as well as the costs were proportionally considered following the same percentages. The implantation was accepted as the first two years, maturation in the following four years and the full phase from the 8th year.

Revenue forecasting was performed. Considering the phases presented, for the value of each service, an average value of 5% of the acquisition value of each vessel was admitted, according to its class, based on the indices proposed by D'almeida (2009).

It was observed that the project has viability, reaching, in the end, an IRR of 47.43% and a Profitability Index of 77.56%. The analysis was also performed for the midpoint of the period, ie, nine years, also showing a positive NPV and an IRR of 42%. The payback of the project was 10 years. The annual result is denoted in Table 5 and the calculated residual value was added to next year's revenue.

Variation in scenarios aims to provide diverse conditions to support the decision maker's gaze. The main purpose of this variation is to guarantee the investor the possibility to consider with different conditions throughout the implementation and development of the business.

For the repair yard evaluated, in the case of the pessimistic scenario, the first stipulated variation was in the scenarios. In the implantation phase, it was considered 15% of the. calculated total capacity, in the 50% maturation phase and in the "full" phase it was considered that the yard could not reach 100% but 80% of the total capacity.

Table 5. Moderate revenue.

9-Years Analysis		
NPV (9 years)	$1.081.534.219,72	Profitability (Above the amount invested)
IRR (9 years)	42,20%	31,06%
18-Years Analysis		
NPV (18 years)	$2.700.484.123,16	Profitability (Above the amount invested)
IRR (18 years)	47,43%	77,56%

Table 6. Pessimistic revenue.

9-Years Analysis		
NPV (9 years)	-$10.201.851,44	Profitability (Above the amount invested)
IRR (9 years)	13,08%	-0,29%
18-Years Analysis		
NPV (18 years)	$1.129.548.052,68	Profitability (Above the amount invested)
IRR (18 years)	26,24%	32,49%

Table 7. Optimistic revenue.

9-Years Analysis		
NPV (9 years)	$1.786.022.244,76	Profitability (Above the amount invested)
IRR (9 years)	80,05%	51,27%
18-Years Analysis		
NPV (18 years)	$3.397.569.301,44	Profitability (Above the amount invested)
IRR (18 years)	80,05%	51,27%

In the case of the pessimistic scenario, the NPV only becomes positive from the 11th year, as shown in Table 6. Despite adverse conditions at the end of the project, an IRR of 26.24% e was obtained. a PI of 32.49%, not so high, but demonstrates the economic viability of the project because even in the worst condition it still has acceptability.

For the optimistic scenario, the same previous variables were modified, being the revenues for the different phases and the mentioned productive capacity, bringing the costs in tow. The optimistic scenario for the repair yard was simulated, considering in the implementation phase the production of 50% of the total, in the 80% maturation phase and in the 100% full phase.

In the optimistic scenario for the repair yard, it could not be different, both the IRR and the PI increased to 81.45% and 98.54%, respectively, at the end of the project. It is possible to analyze that even if it does not reach profitability on the construction site level, the repair site is extremely viable, assuming the adopted boundary conditions. Table 7 presents a summary of the feasibility analysis and the results detailed annually.

5 CONCLUSIONS

All stages of the project were completed and all objectives were achieved, so we concluded first from the SWOT analysis, we realize that the ship repair industry, like any other in the country, suffers from cyclicality, needing to look for mechanisms to increase domestic and even international competitiveness, through solid integration strategies with an incipient supply chain, without mechanisms that guarantee competitiveness in itself, so as to allow the supply to the foreign market to meet the demand disintegrated and without seeking mechanisms to stimulate the development and continuity of the industry.

After characterization of the regional supply chain, preliminary analysis conducted through the SWOT matrix revealed that the prevailing strengths and opportunities suffer from weaknesses and threats as they were driven by the Offshore Terminal considered an integral part of the proposed future project. It was also possible to infer from this initial analysis that the implementation strategy is favorable to development, with a high rate of favorability. Although this analysis is not sufficient to define the viability of the cluster, much less its competitiveness;

Finally, from the economic viability analysis, we tried to get as close as possible to the reality exposed, being considered a repair yard. The scenarios were proposed: moderate, pessimistic and optimistic. After analysis of the scenarios, the businesses were attractive, profitable and, consequently, economically viable.

REFERENCES

Adami, A.C. de O., 2010. Risco e retorno de investimento em citros no Brasil (Doutorado em Economia Aplicada). Universidade de São Paulo, Piracicaba. 10.11606/T.11.2010.tde-24052010-110330.

Agyekum, E.B., Ansah, M.N.S., Afornu, K.B., 2020. Nuclear energy for sustainable development: SWOT analysis on Ghana's nuclear agenda. Energy Reports 6, 107–115. 10.1016/j.egyr.2019.11.163.

ANTAQ, National Waterway Transportation Agency, 2009. Multimodalidade e as Hidrovias no Brasil.

Begum, R.A., Siwar, C., Pereira, J.J., Jaafar, A.H., 2006. A benefit-cost analysis on the economic feasibility of construction waste minimisation: The case of Malaysia. Resources, Conservation and Recycling 48, 86–98. 10.1016/j.resconrec.2006.01.004.

Brazil, 1997. DECRETO no. 2.256, DE 17 DE JUNHO DE 1997.

Butler, D., 2000. Guide to Ship Repair Estimates, 1st ed. Butterworth Heinemann.

D'Almeida, J., 2009. Arquitectura Naval – O Dimensionamento do Navios. Prime Books, Portugal.

Dobrovolski, R., Loyola, R., Rattis, L., Gouveia, S.F., Cardoso, D., Santos-Silva, R., Gonçalves-Souza, D., Bini, L.M., Diniz-Filho, J.A.F., 2018. Science and democracy must orientate Brazil's path to sustainability. Perspectives in Ecology and Conservation 16, 121–124. 10.1016/j.pecon.2018.06.005.

Dos Santos, D.B., Lima, R.D.C., Bassi, R.E., Rodrigues, E. F., Maiellaro, V.R., 2018. A INFRAESTRUTURA NO TRANSPORTE FERROVIÁRIO NO BRASIL. SADSJ 4, 38. 10.24325/issn.2446-5763.v4i10p38-51.

Fernandes, D.R., 2014. Uma Visão Sobre a Análise da Matriz SWOT como Ferramenta para Elaboração da Estratégia. UNOPAR Científica Ciências Jurídicas e Empresariais 13, 42–73. http://dx.doi.org/10.17921/2448-2129.2012v13n2p%25p.

Frota, C.D., 2008. Gestão da Qualidade Aplicada às Empresas Prestadoras do Serviço de Transporte Hidroviário de Passageiros na Amazônia Ocidental: Uma Proposta Prática (Transportation engineering doctorate thesis). Universidade Federal do Rio de Janeiro, Rio de Janeiro, RJ.

Gadelha, R.M.A.F., 2002. Conquista e ocupação da Amazônia: a fronteira Norte do Brasil. Estudos Avançados 16, 63–80. 10.1590/S0103-4014200 2000200 005.

Groppelli, A.A., Nikbakht, E., 2010. Administração financeira. Saraiva, São Paulo (SP).

Hossain, K.A., Zakaria, N.M.G., Sarkar, M.A.R., 2017. SWOT Analysis of China Shipbuilding Industry by Third Eyes. Procedia Engineering 194, 241–246. 10.1016/j.proeng.2017.08.141.

Kassai, J.R., 2005. Retorno de investimento: abordagem matemática e contábil do lucro empresarial. Atlas, São Paulo.

Markovska, N., Taseska, V., Pop-Jordanov, J., 2009. SWOT analyses of the national energy sector for sustainable energy development. Energy 34, 752–756. 10.1016/j.energy.2009.02.006.

Marques, A.O., Kuwahara, N., 2009. Transporte Aquaviário no Amazonas: necessidade de mudanças para a promoção do desenvolvimento econômico e social. Presented at the XXIII Congresso de Ensino e Pesquisa em Transportes – ANPET, Vitória, Espirito Santo.

MMA, Ministry of the Environment of Brazil, 2006a. Caderno da região hidrográfica amazônica. Secretariat of Water Resources of Brazil, Brasília, Brasil.

MMA, Ministry of the Environment of Brazil, 2006b. Caderno da região hidrográfica Tocantins-Araguaia. Secretariat of Water Resources of Brazil, Brasília, Brasil.

Moraes, H.B. de, 2001. Embarcações de Passageiros na Amazônia. Presented at the XVIII Congresso Pan-Americano de Engenharia Naval, Transportes Marítimos, e Engenharia Portuária, IPEN, México.

Moraes, H.B. de, Vasconcelos, J.M., 2001. Análise Qualitativa e Quantitativa dos Tipos e Características das Embarcações Atualmente Empregadas nas Principais Linhas de Transporte de Passageiro e Carg. SUDAN/FADESP, Belém, PA.

Moraes, H.B.D., 2017. Os Portos e as Hidrovias da Amazônia para integração da América do Sul. Presented at the 10o International Seminar On Inland Waterbone Transportation, Belém, PA.

Motta, R.R., Calôba, G.M., 2002. Análise de investimentos tomada de decisão em projetos industriais. Atlas, São Paulo.

Nogueira, R., 2012. Elaboração e análise de questionários: uma revisão da literatura básica e a aplicação dos conceitos a um caso real. UFRJ/COPPEAD, Rio de Janeiro, RJ, Brasil.

Paula, D.L.M. de, Lima, A.C. de M., Vinagre, M.V. de A., Pontes, A.N., Paula, D.L.M. de, Lima, A.C. de M., Vinagre, M.V. de A., Pontes, A.N., 2019. Saneamento nas embarcações fluviais de passageiros na Amazônia: uma análise de risco ao meio ambiente e à saúde por meio da lógica fuzzy. Engenharia Sanitaria e Ambiental 24, 283–294. 10.1590/s1413-41522019150122.

Resende, F.M., 2015. Análise Para Aplicação De Boas Prá-ticas De Gerenciamento De Projetos Em Uma Indústria Metalúrgica Com Produção Customizada (Trabalho de conclusão de curso). Centro Universitário Eurípides de Marília - UNIVEM, Marília SP.

Rodrigues, S., Torabikalaki, R., Faria, F., Cafôfo, N., Chen, X., Ivaki, A.R., Mata-Lima, H., Morgado-Dias, F., 2016. Economic feasibility analysis of small scale PV systems in different countries. Solar Energy 131, 81–95. 10.1016/j.solener.2016.02.019.

Ruuska, I., Ahola, T., Martinsuo, M., Westerholm, T., 2013. Supplier capabilities in large shipbuilding projects. International Journal of Project Management 31, 542–553. 10.1016/j.ijproman.2012.09.017.

Soares, F.J.A., Filho, W. de B.V., 2015. Caracterização Dinâmica de Embarcações Regionais do Amazonas. UNOPAR Científica Ciências Exatas e Tecnológicas 13.

Thamrin, H., Pamungkas, E.W., 2017. A Rule Based SWOT Analysis Application: A Case Study for Indonesian Higher Education Institution. Procedia Computer Science 116, 144–150. 10.1016/j.procs.2017.10.056.

UNCTAD - United Nations Conference on Trade and Development, 2019. Review of Maritime Transport 2018. United Nations.

Developments in Maritime Technology and Engineering – Guedes Soares & Santos (eds)

A simplified method to simulate residual stresses in plates

J.M. Gordo & G. Teixeira
Centre for Marine Technology and Ocean Engineering (CENTEC), Instituto Superior Técnico, Universidade de Lisboa, Lisbon, Portugal

ABSTRACT: Welded structures are subjected to internal residual stress after manufacturing that may affect the structural strength and normally are associated with an increase on initial geometrical imperfections. This study presents a simplified method to generate an adequate representation of residual stresses on Finite Element models for structural analysis of thin-walled structures and other applications. The results obtained shown that the methodology proposed to introduce residual stresses is simple, accurate and efficient on the modulation of post-welding stresses and their pattern, thus it may be used for simulation of the thermal process.

1 INTRODUCTION

Many efforts have been made to simulate the process of welding. Given that the process of welding is a phenomenon using metal in the liquid state, and as such without solid-state stresses, slightly above fusion temperature, the reality is quite difficult to emulate. This phenomenon is usually neglected, and the analysis is concentrated on the heat flux directed to the plate on the welded zone. The heat source, as such, can be of two types: fixed or moving. The first one to be described is the approximation proposed by Rosenthal (1946). The analytical theory of Rosenthal is based on a concentrated heat source. Later, Friedman (1975) presented the Gaussian Distributed heat source, as to express an approximation of the heat flux of the moving heat source. More recently however, a study (Goldak et al. 1985) proposed the usage of a Double-Ellipsoidal moving heat source.

Different heat inputs with different welding speeds have a direct effect on not only the results, but in the simulation and computation as well. Heinze et al. (2012) have studied this phenomenon, which compares different heat inputs with the results obtained.

The *FEA* of the welding process is often defined as a thermo-mechanical coupled analysis. Generally, there are three methods to simulate the welding process. The first method is the most straightforward of the three methods used. It consists in applying directly the temperature field to the plate, with its geometry already defined with structural elements, to calculate directly the distribution of residual stresses and the plate deformation. The temperatures are defined as body loads, and the plate is heated and cooled along time as such.

The two most used methods are known as the direct and indirect methods. The direct method, as the name implies, is characterized by the usage of elements that make possible both a thermal and structural analysis. The indirect method, which is the method more commonly used in simulations of this type, uses a semi-coupled approach. This approach predicts two different steps in the calculations, one in which the heating process is simulated, and elements which only have thermal degrees of freedom are used. This approach was used in studies from Chen (2011), Chen et al. (2011, 2014) to study the deformations welding causes to specimens with 300 mm by 260 mm and 6 mm in thickness.

The geometry used is commonly accepted to be the three-dimensional in general for all simulations. However, it is especially difficult and time-consuming to compute 3D plates with initial imperfections. In these cases, it is common to use a 2D simulation.

Regarding mesh size, when performing an analysis of the deformation due to the heat of the welding process, it is common to use a more refined mesh in the area of the welding chord, and directly near it, as done in the study performed by Chen (2011). For a more refined simulation, the use of dynamic meshes can be applied. In general, dynamic meshing consists of having meshes solidary to the moving heat source. This saves computation time. Runesson & Skyttebol (2007) make a brief case of the methodology used to perform analysis of the welding process in their study.

In modelling the welding process, it is crucial to make the right choice regarding the boundary conditions of the model to be simulated. The boundary conditions affect drastically the outcome of the results. It is of the biggest importance to have the boundary conditions in mind when performing the simulations. With that in mind, when it comes to choosing the boundary conditions, in terms of the structural analysis, it is common for the plates to be simply supported, clamped, or completely free.

DOI: 10.1201/9781003216582-89

A study has been made (Fu et al., 2014) to simulate this effect. The authors compared these boundary conditions to the other cases. It was found that the boundary conditions that restrained the plate have less strain but tend to have more residual stresses. For the clamped case, there are two methods in which the clamping can be simulated. In the study made by Deng et al. (2016), the clamping is simulated by applying a friction coefficient and a force to specific nodes.

The simplest case is to lock the nodes in the areas in which the clamps are attached to simulate clamping, since the movement on the clamping is minimum and can usually be neglected. This approach was used in the study performed by Fu et al. (2014), with the results obtained being very acceptable.

2 RESIDUAL STRESSES APPROACHES

2.1 Stress-strain curve approximation

Gordo & Guedes Soares (1993) proposed a simplified approach to model analytically the distribution and values of the induced residual stresses. In this method, the authors considered a bilinear isotropic elastic-plastic approximation (*BISO*) to represent the material behaviour. This consideration is acceptable in the analysis of structures made of very ductile materials having a yielding plateau. The equation that models this is defined by branches and is the one seen in (1).

$$\phi_e = \begin{cases} -1 \Leftarrow \bar{\varepsilon} < -1 \\ \bar{\varepsilon} \Leftarrow -1 < \bar{\varepsilon} < 1 \\ +1 \Leftarrow \bar{\varepsilon} > 1 \end{cases} \quad (1)$$

where ϕ_e is the stress on the material and $\bar{\varepsilon}$ is the strain normalised by yield strain. This equation considers both compression and tension of the material.

2.2 Residual stress distribution

During welding, the temperature between the sides of the plate, which maintain an even temperature along its width similar to the environment temperature, and the area near the welding, are severely different. According to Masubuchi (1980), the residual stresses that appear on a plate can be of two types:

1. Residual welding stresses that are produced directly by the thermal difference during the welding;
2. Residual stresses caused by external restraint.

Masubuchi & Martin (1965) devised an equation for the distribution of residual stresses in the longitudinal direction along the width of the plate, and is expressed as it can be seen on the equation (2).

$$\sigma_y(x) = \sigma_m \left[1 - \left(\frac{x}{w}\right)^2 \right] e^{-\frac{1}{2}\left(\frac{x}{w}\right)^2} \quad (2)$$

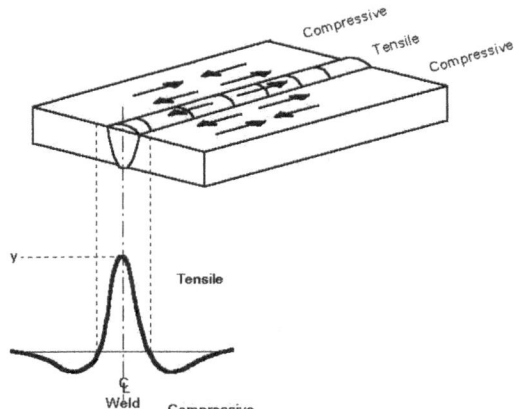

Figure 1. - Residual stresses' distribution parallel to the welding direction.

where σ_y is the residual stress in the longitudinal direction, σ_m is the maximum stress and w is the width of the plate. This equation considers the area of the plate in traction, as well as the area in compression. The value for the σ_m can be as high as the yield stress. The distribution of residual given by equation (2) can be seen in Figure 1.

This distribution has a distinct point in which the stresses change from tensile to compressive. The value of η can be easily obtained by substituting the value for the longitudinal residual stress by zero in the equation and solve for x.

However, it is usual to consider a model, which simplifies the distribution of residual stresses along the width of the plate. With that, it is possible to define the residual stresses in tension as being located in a strip of width ηt, with η being the ratio between the plate's thickness t and the strips' width. The remaining $w - 2\eta t$ width of the plate is in compression.

For the value of η, Guedes Soares & Soreide (1983) state that, although commonly varying from 2 to 8, it is common even to consider values below that. As such, a value of 1.5 was set, and used along the calculations, as well as the geometry for the Finite Element Analysis (*FEA*). Since the model used considers two adjacent plates, the total width of the strip is equal to $2\eta t$. It is in this area that the residual stresses are significantly bigger. These are predominant in the longitudinal direction, and their value can be obtained using equation (3).

$$\sigma_r = \sigma_o \frac{2\eta t}{w - 2\eta t} \quad (3)$$

σ_r and σ_o are the resulting residual and yield stresses, respectively. Timoshenko & Goodier (1982) asserts that a plate experiencing a gradient of temperature

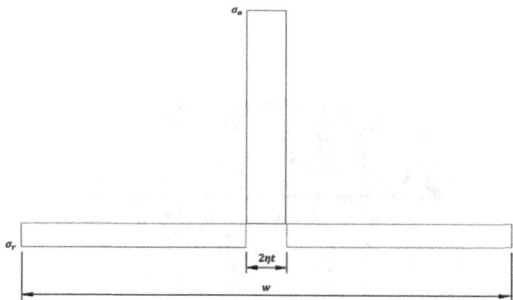

Figure 2. - Theoretical Model for the Distribution of Residual Stresses.

along its width is bound to develop residual stresses. However, for simplicity purposes on performing calculations, a theoretical model has been created. This model features a constant distribution for both the traction residual stresses and compression residual stresses, as shown in Figure 2.

3 MODELLING OF RESIDUAL STRESSES

The FEM model here implemented makes use of a two-dimensional (2D) geometry, where elements of shell type were used. In ANSYS®, elements of the type SHELL281 were used. A quadrangular mesh was used. The mesh is finer in the middle, where the welding toe is located, and coarser in the edges, were low refinement can be considered. The material behaviour is described by the Equation (1), where the Bilinear Isotropic (*BISO*) correction to the elasto-plastic domain was used, with no plastic hardening considered. This behaviour was considered for both traction and compression.

Regarding heat input, a time-dependent heat source was considered, which is uniform along the length of the welding path, to promote speed in the computational process. Various temperatures were considered. Finally, concerning the boundary conditions, two cases were implemented. The first one features a fixed node translation on the transversal direction, applied on the nodes located on the longer edges of the plates. This was considered to promote symmetry, as leaving one edge free would create asymmetry on the resulting geometry. For the first case the shorter edges were left free, whereas in the second case, the shorter edges are allowed only to move as a rigid line, as to simulate a welded plate on these shorter edges.

3.1 Temperature interval

The simulation of the generation of residual stresses in a structure requires to input a temperature variation that creates the tensile residual stresses pattern in the welded region.

At high temperature in this region its material is in compression at yield stress for that temperature; during cooling, the region becomes to behave elastically, reducing the compressive state of stress and, when the reduction of temperature is enough, it becomes to be in tension. Tensile stress increases until the yield stress in tension is achieved. After that reduction in temperature and for elastic-perfectly plastic material the residual stresses pattern remains constant and only the plastic strain increases.

The minimum difference of temperature required to generate the pattern of residual stress at room temperature involves the knowledge of the yield stress and elastic modulus E at the 2 extreme temperature and the coefficient of thermal expansion, α.

A major of this minimum variation of temperature is given by:

$$\Delta T = -\frac{2\sigma_o}{\alpha E} \qquad (4)$$

The steel used in this study was *ASTM* 36 Structural Steel which has yield stress of 370 MPa at 20° C and $\alpha=11.3 \times 10^{-6}$ K^{-1}.

The temperature difference required is ΔT=-312°.

3.2 Implementation of the FEM

In order to validate the feasibility of the two-dimensional (2D) approach for the application of weld-induced residual stresses, a test model was devised.

The dimensions of the plates used were firstly 10 m by 5 m, and later 2 m by 0.8 m. Models with imperfections and with added stiffeners were created by modifying the existing geometry with 2 m by 0.8 m. The thicknesses used were three, of 6, 10 and 20 mm, as these are common thicknesses of plates used in the Naval Industry.

Since this model is to simulate a simplified distribution of residual stresses, the loading is made using only a direct heat input along the full length of the welding path. This heat input is also conducted in the full width of the area of the plate subjected to a traction stress. This width is equal to the tensile strip with $2\eta t$.

Finally, the boundary conditions chosen ensure that the plate has some freedom to move. However, in order to obtain symmetry transversally to the welding path, the longer edges were constrained on the transverse direction, in the x direction. Also, in a second approach, and to simulate the welding on the transversal edges, a constraint of rigid edge in the longitudinal direction was created. Both edges were left free to move.

Regarding the heat source, this was modelled by applying directly the temperature as a body force to the plate. The temperature input was performed using a time-dependent table. In Figure 3 can be seen the temperature distribution, varying as a function of time. As it can be seen in the figure below, the temperature increases to 500° Celsius in 500 seconds, with a decrease to the 20° Celsius.

787

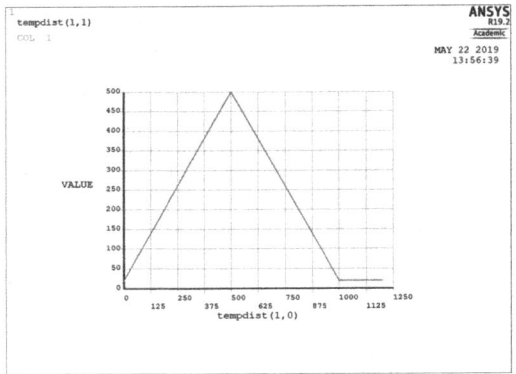

Figure 3. – Time-dependent temperature distribution.

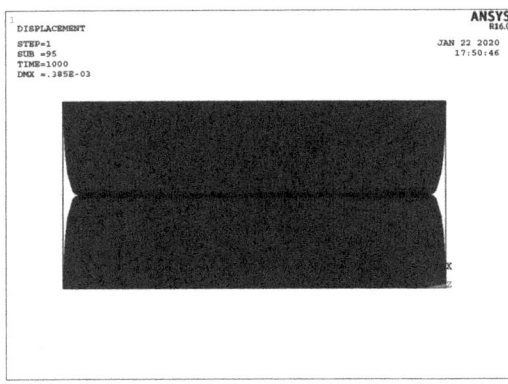

Figure 4. – Mesh and final deformation of a 10mm thick plate with 10x5m after butt welding.

3.2.1 *Butt weld of long plates*

Figure 4 presents the mesh and the deformation of welded plate with 10 m by 5 m with η=3 and t=10mm. It was found during the calibration of the model that the mesh size should be smaller than 1/3 of ηt in order to achieve a good representation of the residual stresses pattern. The resulting distribution of longitudinal residual stresses on the plate is presented in Figure 5.

This simulation of butt weld of 2 long plates shows the shrinkage at the ends of the welding and the residual stress pattern is uniform along the length except in the ends of the welding due to uncompensated thermal loads during the cooling. The level of compressive residual stresses in this butt weld is very low due to the very high ratio between the compressive and tensile width of the plate.

In Figure 6 it is shown the distribution for the longitudinal residual stresses at mid-length of the plate near the welding. The tensile strip is subject to 373 MPa and the compressive region to less than -3 MPa due to the high ratio already mentioned.

The pattern of compressive stresses changes with the longitudinal position as can be seen in Figure 7. In the middle compressive stresses are almost constant due to symmetry and the location been far away from the tops of the part. At half-quarter of the length of the plate the distribution shows an evident shear lag effect in result of unbalanced longitudinal stresses at the tops. The pattern of tensile stresses is very similar in all situations, represented by the almost vertical straight line.

In the middle of the weld the residual stresses are almost constant except near the tops, Figure 8. The tops create unbalanced forces in longitudinal direction that are transform in distortions and shear originating a complex pattern of stress, as shown in Figure 9.

3.2.2 *Butt weld of plates between stiffeners*

Typical plates between stiffeners and frames have different dimensions, been the length of the order of 2.5m and width between 0.5 and 1.0m. In this section it is analysed the case of a plate 2.5 m long and

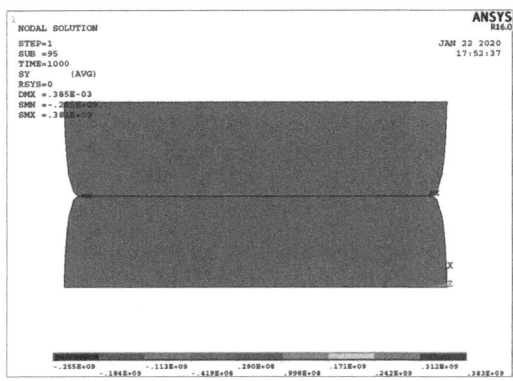

Figure 5. - Distribution of the residual stresses in the longitudinal direction for a long plate with 10 mm thickness.

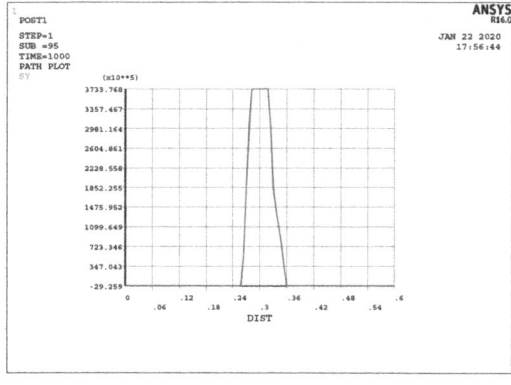

Figure 6. - Distribution of the residual stresses in the middle of a 10x5m plate with 10 mm thickness and η=3. Detail near the welding.

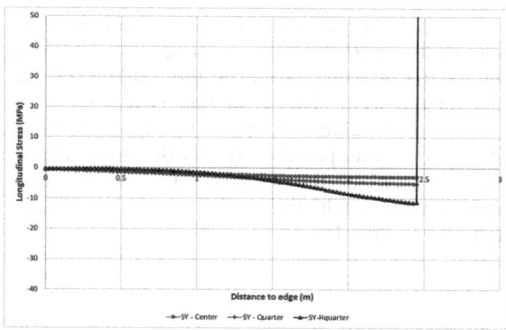

Figure 7. - Distribution of the compressive residual stresses in the middle, quarter and half-quarter of a 10x5m plate with 10 mm thickness and η=3.

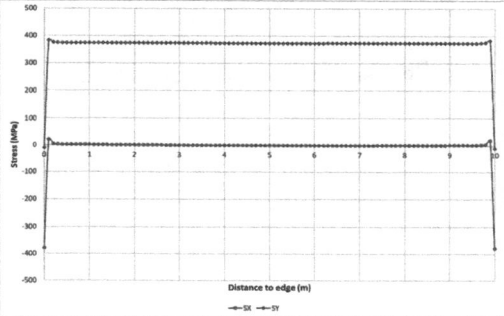

Figure 8. – Longitudinal (Sy) and transversal (Sx) residual stresses along the weld.

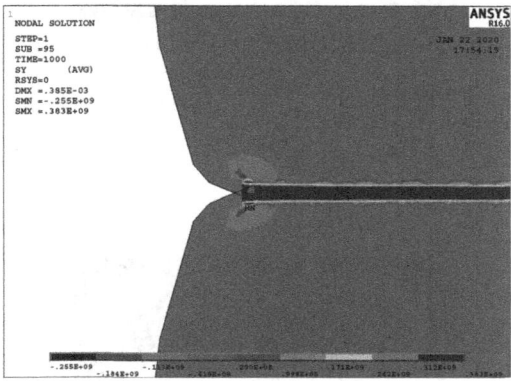

Figure 9. – Detail of longitudinal residual stresses near the top.

1.0 m wide with a welding in the middle of the width.

The distribution of longitudinal residual stresses is in accordance with the theoretical model assumed in codes represented by 3 rectangular regions, 1 in tension in the welding at yield stress and compressive stresses to equilibrate in the rest of the plate, as shown Figure 10.

The same distribution is presented for a path at a quarter and eight-length showing the appearance of some shear lag effect near the top. The distribution of compressive residual stresses in the middle of the plate is almost constant.

In respect to the top effect on the residual stresses it is found the same response as in previous case. A shorter plate was also used to generate the graphic of residual stresses in the middle of the weld along the length of the plate, Figure 11. It can be observed by the overlap near the top that the length do not affect the length of the perturbed region. Apart from that region with typical dimension of 200mm, longitudinal and transversal stresses remain almost constant.

3.2.3 Thickness relevancy

Variation in thickness for 8, 10 and 20mm produced the results seen in Figure 12 that do not affect the residual stress pattern hence the tensile width is kept

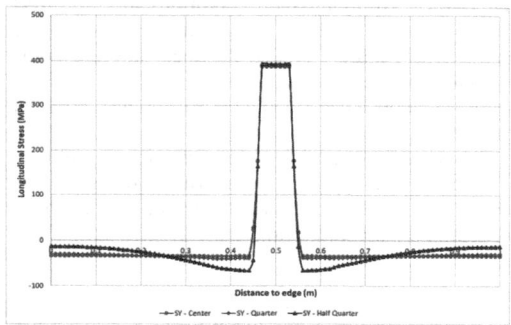

Figure 10. - Distribution of the residual stresses in the middle, quarter and half-quarter of a 10x5m plate with 10 mm thickness and η=1.5.

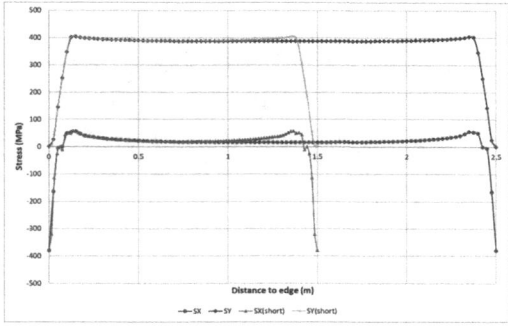

Figure 11. – Longitudinal (Sy) and transversal (Sx) residual stresses along the weld of a 2.5 and 1.5 m long plate.

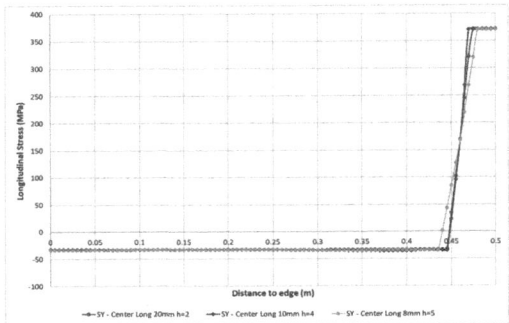

Figure 12. Comparison between the distributions of longitudinal residual stresses for different thicknesses.

constant. In order to guaranty the same width of the tensile zone η parameter changes to 5, 4 and 2, respectively for plate 8, 10 and 20mm thick.

3.2.4 Boundary conditions and imperfections

In continuous reinforced panels the boundary conditions (BC) tend to be better represent by constraint condition of the edges due to the rigidity of stiffeners and frames. Applying such conditions the top effect mentioned previously disappears, as may be observed in Figure 13. Also the shear lag due to the proximity of the tops disappears.

The compressive residual stresses are 10% of yield stress.

This model has an initial imperfections amplitude of 4mm before applying the thermal process, introducing displacements on the nodes of the flat plate according to sinusoidal function:

$$\omega = 0.004 \, sin\frac{\pi x}{1.0} \, sin\frac{\pi y}{2.5} \qquad (5)$$

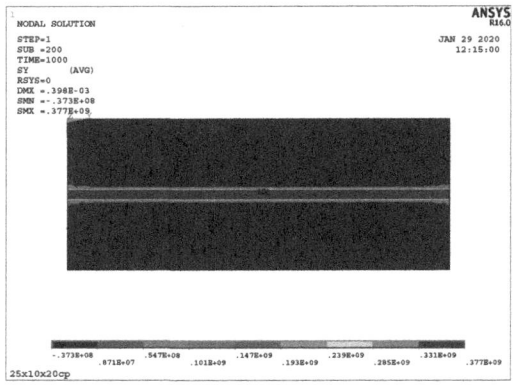

Figure 13. Distribution of the residual stresses in the longitudinal direction for a plate with 2.5x1.0x0.020 m with η=2.

After the thermal process to generate residual stresses the amplitude was reduced by 0.155mm due to the tensile forces in the middle of the plate on the weld. This reduction is less than 4% of the initial value, but at 500°, before cooling, the variation was positive by 5% due to compressive stresses in the middle. The shape also changes very much during the whole process, as shown in Figure 14.

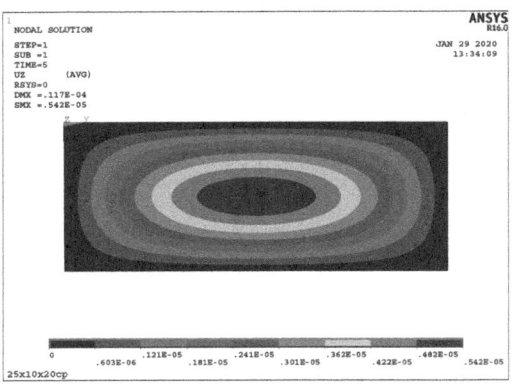

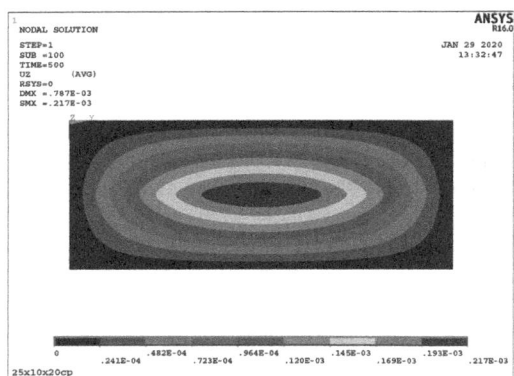

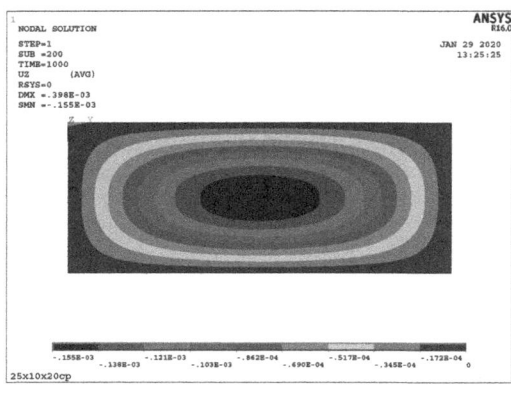

Figure 14. – Variation of out-of-plane deformations for the plate with 4mm amplitude of imperfections; Temperature 25° (top), 500° (middle) and 20° (down).

3.3 Stiffened plates

Stiffened models were also created, with the addition of formerly one and later two stiffeners. The model with two stiffeners was created by mirroring the model with one stiffener along the longer edge, creating a model constituted by a plate with 2 m by 1.6 m and two stiffeners separated by a distance of 800 mm.

It can be noticed that the distribution resembles the theoretical approach. Symmetry is also present, as it would be expected.

Regarding the results for a stiffened model, the resulting distribution of longitudinal residual stresses in the plating is very similar to the ones presented for unstiffened plates with one weld. Figure 15 presents the internal state of stresses before cooling and after it. As can be noticed the induced deformations reverse direction from hot weld to cold weld due to reverse of internal stresses in the toe.

The pattern of residual stresses in the web can be seen in Figure 16. The linear reduction on the compressive stress from the toe is according to theory and is a result of the bending of the stiffened plate.

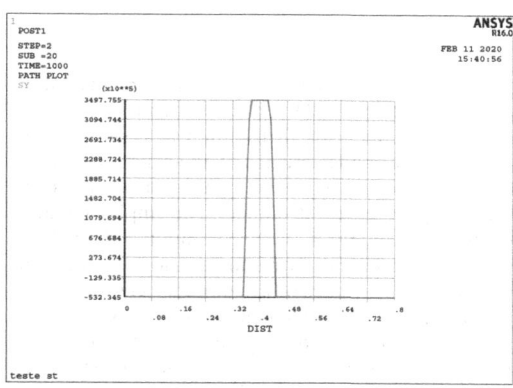

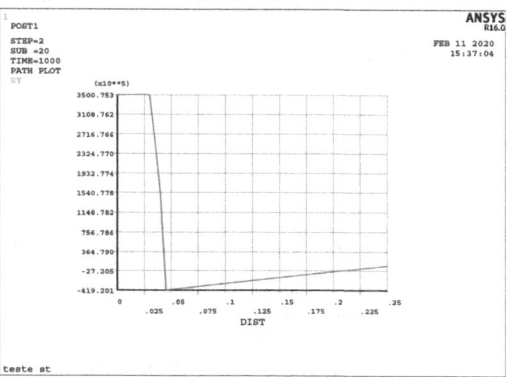

Figure 16. - Distribution of residual stresses across the width of the plate at mid-length (up) and along the web of the stiffener (down).

In Figure 17 can also be seen the distribution across the width of the panel for the model with two stiffeners.

These results are important to acknowledge that the addition of a stiffener to the geometry of the model does not influence the existing stiffener.

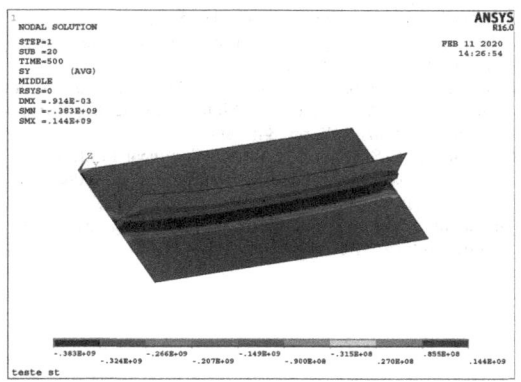

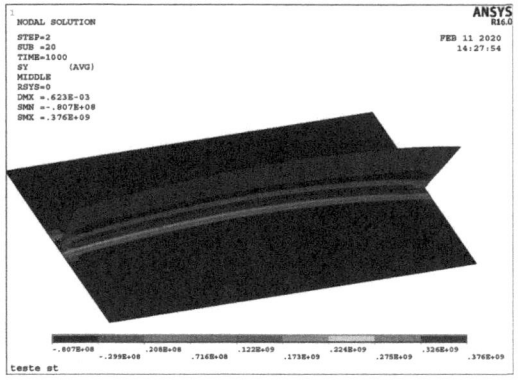

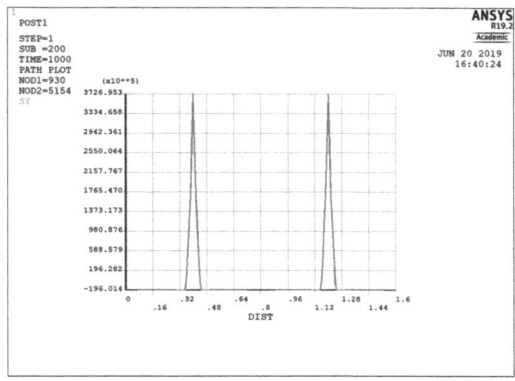

Figure 15. – Internal stresses in longitudinal direction and associated deformations for a perfect flat panel of 2x0.8m with welding strip at 500°C (up) and after cooling (down).

Figure 17. - Distribution of the longitudinal residual stresses on the plate with two stiffeners.

4 CONCLUSIONS

Firstly, it can be concluded that the implementation of a FEM model in 2D is capable of producing reasonable results for the distribution of residual stresses. With the model created it is easier to perform simulations faster and have significant results using different geometries at choice. The implementation of simple geometries and heat inputs allowed for a significant amount of runs to be made, for different thicknesses and modes of initial imperfection.

The successful implementation of a simple yet effective model of inducing residual stresses along several geometries is definitely a major achievement because it allow performed structural analysis in a simplest way.

Regarding the uniform heat input, it was proven to achieve adequate results for the validation in question to be performed.

REFERENCES

Chen, B. 2011. *Prediction of Heating Induced Temperature Fields and Distortions in Steel Plates*. MSc thesis, IST, Portugal.

Chen, B., Adak, M. & Guedes Soares, C. 2011. 'Thermo-Mechanical Analysis of the Effects of Weld Parameters in Ship Plates During Welding Process', in ICSOT: *Technological Innovations in Shipbuilding*, RINA, UK, pp. 23–30. doi: 10.13140/2.1.1759.1680.

Chen BQ, Hashemzadeh M, Guedes Soares C. 2014. Numerical and experimental studies on temperature and distortion patterns in butt-welded plates. *International Journal of Advanced Manufacturing Technology*, 72, pp. 1121–1131.

Deng, D., Liu, X., He, J. et al. 2016. Investigating the influence of external restraint on welding distortion in thin-plate bead-on joint by means of numerical simulation and experiment. *Int J Adv Manuf Technol.* 82, 1049–1062 https://doi.org/10.1007/s00170-015-7413-7.

Friedman, E. 1975. 'Thermo-Mechanical Analysis of the Welding Process Using the Finite Element Method', *J. Pressure Vessel Tech*, 973(3), pp. 206–213.

Fu, G. et al. 2014. 'Effect of Boundary Conditions on Residual Stress and Distortion in T-joint Welds', *Journal of Constructional Steel Research*. 102: 121–135. doi: 10.1016/j.jcsr.2014.07.008.

Goldak, J. A., Chakravarti, A. & Bibby, M. J. 1985. *A Double Ellipsoid Finite Element Model for Welding and Heat Sources*, IIW Document.

Gordo, J. M. & Guedes Soares, C. 1993. Approximate load shortening curves for stiffened plates under uniaxial compression". Edited by D. Faulkner, M. J. Cowling A. Incecik & P. K. Das. *Integrity of Offshore Structures - 5*; Glasgow. Warley, U.K.: EMAS; pp. 189–211.

Guedes Soares, C. & Soreide, T. H. 1983. 'Behaviour and Design of Stiffened Plates Under Predominantly Compressive Loads', *International Shipbuilding Progress*, 30(341): 13–27.

Heinze, C., Schwenk, C. & Rethmeier, M. 2012. 'Effect of heat source configuration on the result quality of numerical calculation of welding-induced distortion', *Simulation Modelling Practice and Theory*, 20(1), pp. 112–123. doi: 10.1016/j.simpat.2011.09.004.

Masubuchi, K. 1980. *Analysis of Welded Structures*. D. W. Hopk. Pergamon Press.

Masubuchi, K. & Martin, D.C. 1965. 'Investigation of Residual Stresses by Use of Hydrogen Cracking', *Welding Journal*, 196, 40, 553–563.

Rosenthal, D. 1946. 'The Theory of Moving Sources of Heat and Its Applications to Metal Treatments', *Transactions of ASME*, 68, pp. 849–866.

Runesson, K. & Skyttebol, A. 2007. 'Nonlinear Finite Element Analysis and Applications to Welded Structures', in *Comprehensive Structural Integrity*. pp. 255–320.

Timoshenko, S. P. and Goodier, J. N. 1982. *Theory of Elasticity*. 20th Ed. McGraw-Hill.

Developments in Maritime Technology and Engineering – Guedes Soares & Santos (eds)
© 2021 Copyright the Author(s), ISBN 978-0-367-77376-2

Research on modular building system for accommodation cabins of large luxury cruise

J.H. Xu, C.X. Pan, J. Jiang & L. Huang
Wuhan University of Technology, Wuhan, Hubei Province, P. R. China

ABSTRACT: This paper is primarily supposed to establish the integration mode of "design - construction" technology for accommodation cabins of large luxury cruise based on the modular system, which is discussed from the following two aspects: modular decomposition and reorganization; modular design and construction. According to the standards for the design of accommodation cabins of large luxury cruise, the overall system and operation process, combining the design and the construction process of cabin units, are constructed through modular, typed and assembled analysis of the unit components. Based on boards, profiles, auxiliary materials, equipment and toilet unit modules, this paper carried out a classified study on the material components, module integration and connection modes of the cruise cabin by adopting technical methods such as three-dimensional graphic, collaborative design and simulation construction. The paper focuses on the module integration mode and construction logic of cabin unit system, trying to provide a direction of technical research for further improving the building system for accommodation cabins of China's luxury cruises.

1 INTRODUCTION

With the development of China's economy in recent years, there has been a rapid growth in the market of large luxury cruise involving areas of design, R&D, construction, etc. China is actively seeking cooperation with international markets. As an important interior part of large luxury cruise, the accommodation cabins account for about 50% of the total superstructure area, with their total number reaching more than 2,000. At present, the modular building process of "overall design - factory assembly - on-site installation" is internationally adopted for the construction of accommodation cabins. Although a small number of Chinese companies handling interior finishing materials for ships have mastered this method, their comfort standards, construction technology and material performance are far from those in foreign countries due to the variation of service object standards, and the biggest gap is intensively seen in the overall modular design and construction of cabins as well as their working efficiency.

Therefore, based on the communication, study and investigation with well-known foreign companies, this paper attempts to discuss the standard requirements, module composition, process flow and other aspects of the modular building system for cabins of large luxury cruise, while constructing the overall system and operation process combining the design and the construction of cabin units, as a basic research for independent research and development of cabins in China.

2 DEVELOPMENT OF THE ACCOMMODATION CABINS OF LARGE LUXURY CRUISE

Since the industrial society, the prefabricated modular building of accommodation cabins has become an important embodiment of the modular design and construction of luxury cruise. The United States took the lead in proposing the idea of modular shipbuilding in the 1930s. Since then, Baltic countries represented by Germany have made outstanding achievements in the field of modular shipbuilding. At present, International renowned shipbuilding company have monopolized the global field integrating the design and construction of luxury cabins all by virtue of advanced manufacturing technology. On the basis of self-organized production lines, these companies each adopt the process technology to accomplish the assembly of 2,500 standard cabins of the overall ship within two months with an average efficiency of 40-50 standard cabins per day. As a highly reflection of modular construction of luxury cruise cabins, such high efficiency of construction represents the integration of high standard, high quality and high operation in terms of material supply, performance technology, design capability, aesthetic perception and construction organization, which contributes to veritable high-tech ship products.

In China, the modular design and construction of accommodation cabins for luxury cruise are still in the exploratory stage, there still have a distance with the

DOI: 10.1201/9781003216582-90

old Nordic shipping agencies in terms of efficiency and standards, with a gap mainly reflected in aspects of work coordination, construction period and efficiency, full industrialization and standardization, and accuracy of materials and technical equipment. Therefore, China basically does not have a share in the global mainstream market of luxury cruise construction. Nevertheless, China's huge market potential and its rapidly rising design and construction capabilities are pushing for the rapid localization and systematization of its modular building of accommodation cabins of luxury cruises.

3 DESIGN STANDARD FOR ACCOMMODATION CABINS OF LARGE LUXURY CRUISE

3.1 *Lightweight requirement*

The lightweight of luxury cruise cabins is mainly realized by adopting novel boards featured by light weight and high strength (Li, 2014). Under the condition of satisfying the structural security check, the lighter the cabin is, the better. And it is already a great progress when the weight is lessened by 5%. In the market, rock wool boards with various joints are used as the main boards with a thickness of 25-50mm to meet the different requirements for wall boards in the cabin. By comparing various indexes (Table 1) of traditional rock wool board and aluminum honeycomb board, it can be found that, the aluminum honeycomb board can not only ensure the overall stiffness, but also effectively reduce the weight, although with a higher cost.

3.2 *Comfort requirement*

The comfort of a luxury cruise cabin is mainly reflected in the physical environment: humidity and temperature; vibration and sound; and color and lighting.

First of all, in terms of humidity and temperature environment, the standards are set according to the optimal temperature and humidity in physiological sense, rather than the somatosensory changes of tourists' subjective feelings. Physiologists believe that when the indoor temperature is too high, it will affect the function of temperature regulation in the human body, and would cause effervescence, vasodilation, pulse acceleration and acceleration of heart rate due to poor heat dissipation. If the indoor temperature is kept above 25°C in winter, people will feel tired, dizzy, mentally retarded and suffer from poor memory. In case that the indoor temperature is too low, the metabolic function of the human body will decline; the pulse and respiration will slow down; the subcutaneous blood vessels will contract; the skin will be excessively tense; the resistance of respiratory mucosa will weaken; and respiratory diseases will be easily induced. According to this, scientists set the lower limit temperature of "cold tolerance" at 11°C and the upper limit temperature of "heat tolerance" at 32°C. Based on the humidity research, the upper limit value of relative humidity should not exceed 80%, with lower limit value not be less than 30%, for achieving optimal indoor dry and wet temperature (Table 2). In addition, with the impact of the global epidemic, attention has been increasingly paid to safe and healthy gaseous environment, ventilation and the regulation of humidity in accommodation cabins (Liu, 2011).

Secondly, in terms of vibration and sound environment, the allowable noise values for the first-class and second-class cabin are 45dB and 50dB respectively according to the reference value of the noise standard for ocean-going ships. Necessary sound insulation and noise reduction measures should be taken for accommodation cabins, and special sound-absorbing materials and structures are supposed to be used at noise sources such as doors and windows (Liu et al, 2018) In addition, the limitation of vibration value is also an important standard for the design of luxury cruise cabins. Under normal gravity, the natural resonant frequency of a person as an elastomer is maintained in a balanced state. When the vibration acceleration of the hull changes, it would greatly disturb the

Table 1. Comparison of Rock Wool Board and Aluminum Honeycomb Board.

Type of Boards	Thickness mm	Weight (kg/m²)	Density (kg/m)	Bending stiffness (N·m²)
Rock wool board	25	13.0	120	1.15×10^6
Aluminum honey-comb Board	20	6.5	70	1.15×10^6

* Source: A Dissertation for the Degree of M. Eng *Research on the Design of Shock-Resistance Cabin Module*

Table 2. The most comfortable Indoor Temperature and Humidity.

	Optimal temperature	Optimal humidity
In winter	18 - 25°C	30% - 80%
In summer	23 - 28°C	30% - 60%
Equipped with air conditioner	19 - 24°C	40% - 50%
With influence on people's thinking activities	18°C	40% - 60%

* Source: Summarized by the Research Group through research

tourists' attention and rest, thus affecting their living comfort.

Finally, in terms of color and lighting environment, the objects that make up indoor colors can be classified into the following four categories from the perspective of color design: indoor building components, indoor equipment, indoor exhibits, and indoor textiles. With different shapes, scales, decorations, materials and textures, these four kinds of color objects can provide a combination of colors by interacting with furniture with different dimensions, states and textures, e.g., furniture walls combined by furniture and walls, soft furniture integrated by textiles and furniture, as well as wallpaper and wall cloth involving walls and dyed fabrics (Jiang, 2003). Regarding lighting, the international standards for the lighting design of luxury cruise cabins are also clearly stipulated (Table 3). In addition to meeting the illuminance standard, accommodation cabins also require proper and non-monotonous illuminance, as well as directional and diffusive light to enhance visual effect. There could also be a combination of centralized and diffusive lighting, which provides harmonious colors and aesthetic feeling.

3.3 Artistic requirement

In the artistic aspect, it requires to follow certain aesthetic rules and apply various modeling elements, such as form, color and texture, for proper combination and creation. The high quality of luxury cruise cabins is not only reflected in the design of decoration that beautifies the accommodation cabins, but also in the unity of art, technology and craft, which is also the essence of modular design and construction.

The artistic elements of accommodation cabins of luxury cruise are mainly reflected in three aspects: the craft aesthetics of exquisite and concise interface construction; the material aesthetics of comfortable coating furniture with texture; and the art aesthetics of the soft decorative surface with distinctive theme (Figure 1).

Table 3. Suggestions on Lighting Labeling of US and JPN Ships.

Places	US (lx)	JPN (1) (lx)	JPN (2) (lx)	Illumination rate	
				Direct lighting	Semi-indirect lighting
Living room	40 - 50	60 - 100	70 - 100	0.5	0.4
Toilet	45	40	30- 70	0.5	

* Source: Summarized by the Research Group through research

Figure 1. Artistic Experience of Living in Luxury Cruise Cabin.

* Source: Summarized by the Research Group through research

4 MODULE SYSTEM FOR ACCOMMODATION CABINS OF LARGE LUXURY CRUISE

As for prefabricated modular accommodation cabin, it is subject to a process that adopts standardized functional module units to construct a system with independent living function through a combination based on functional analysis and system decomposition. It mainly includes basic modules such as enclosure structure (bulkhead ceiling), connecting profiles, doors and windows, equipment, furniture, toilet units, etc. (Figure 2). For the modular system, importance should be attached to the modular coordination of all functional module units and the mode of connection between the functional modules.

For the modulus of the cabin's functional module, it can be designed based on a set of applicable values obtained through statistical analysis and investigation data with a consideration of the behavior scale of the Chinese. On the premise that the structural requirements for the ship are satisfied, the dimension value of the accommodation cabin should be the coordinated dimension of the overall modular shape of the ship's superstructure, and the rib spacing (generally 700-800mm) shall be considered as much as possible, with the coordinated dimension suggested by the Standard being adopted.

According to China's Standard for "Coordination of Dimensions of Ship's Accommodation Cabins", the preferred order of standard modulus is 300, 100 and 50, and the coordination dimensions are multiples

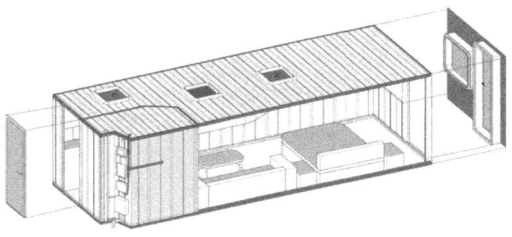

Figure 2. Large and Luxury Accommodation Cabin (Balcony Room).

* Source: Provided by the Research Group

thereof. The per capita area is usually over 6m²according to the specification. With regard to the overall size of the accommodation cabin, the statistical results show that the optimum size of the clear height suitable for the Chinese is 2100mm and the minimum endurance height is 1950 mm. The width of the cabin is taken as an integral multiple of the rib spacing. If coordinated in line with the dimension of the ship's ribs, the appropriate clear width of the accommodation cabin will be 2400mm and 2800 mm. The length of the cabin is generally 6-8m, and the dimensions shall be coordinated according to the basic requirements of toilet units, furniture modules, etc.

Secondly, the form of connection between the functional modules can greatly affect the construction difficulty, the stability and impact resistance of the cabin's structure, so it is hard to handle in modular design and construction.

It should be noted that, it is not easy to standardize and serialize the accommodation cabins of luxury cruise as a whole. Apart from considering the basic physiological characteristics, social features and national customs of human activities, it is also necessary to coordinate with the overall hull structure. Therefore, the overall layout of the cabin space is also the premise and key of modularization. If the accommodation cabin is arranged by the remaining space after completing the physical design of superstructure and the layout of structural enclosure wall, it is undoubtedly hard to achieve satisfactory results of modular standardization.

4.1 Module of wallboard and profile structure for accommodation cabin

As an important part of the accommodation cabin, the wallboard constitutes the module of the cabin's enclosure structure. It is made by composite boards and matching connecting profiles, and has various modes of connection with profiles (Figure 3).

The wallboard is mainly composed of rock wool board (which can be replaced by aluminum honeycomb sandwich panel) and roof board. The modulus of the overall material of the wallboard is in accordance with the specification of 550mm width,, and a ceiling panel of 300mm width is formed on this basis. There are also square wallboards of 600mm×600mm, with board thicknesses of 25mm, 30mm or 50 mm. The profiles for connecting boards include angle steel, channel steel, I-shaped, Z-shaped and Y-shaped materials, etc. The complexity of the ceiling wallboard module lies in its combination with electrical, ventilating or sound equipment and in the arrangement of lamp sockets, pipelines, cable ports and ventilation openings while achieving seamless connection of wallboards as well as aligned interfaces pleasing to the eye. As mentioned earlier, the surface coating of cabin wallboard is the key point affecting comfort level, and it determines the fire resistance, durability and artistry in combination, so this technology will also be the key China's future research and development.

The profile, usually galvanized steel plate, is a material that increases stiffness and structural support through malformation. It is not only an important mechanical component, but also a structural aesthetic component, so its modular design is the focus of process design. Particularly the ceiling profile, which is applied to the connection between the bulkhead and the ceiling, is the focus of vision in the cabin since it is the intersection line between two perpendicular interfaces.

4.2 Furniture module for accommodation cabin

The standardized and intensive design of furniture modules, mainly including doors, beds, sanitary units, sofas, desks, wardrobes, etc., is a major embodiment of cabin comfort. And the key points of furniture module design and the research on product design also lie in suitable ergonomic dimensions (Table 4), reasonable and efficient storage capacity, as well as the elements of safety, beauty and stability.

In addition, the study found that the mode of connection between furniture and bulkhead is also an essential aspect that affects the stability and impact resistance of the cabin's overall structure. Different connection modes will lead to distinct structural performance. In order to reduce the impact response of the cabin more efficiently, the cabin and the deck are generally separated by vibration isolation devices. The door opening of the cabin is designed according to the

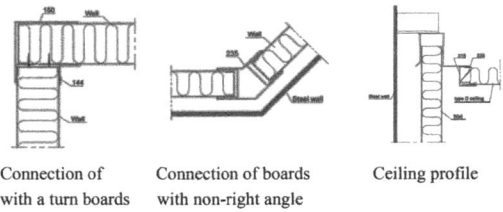

| Connection of with a turn boards | Connection of boards with non-right angle | Ceiling profile |

Figure 3. Modes of Connection between Boards and Profiles.

* Source: Provided by the Research Group

Table 4. Static Space of Furniture Modules in Cruise Accommodation Cabin.

Type of furniture modules	Size of static space (mm)
Double bed module	2000×1700
Sanitary unit module	1900×1400
Sofa module	1900×800
Desk module	1300×550
Wardrobe module	1300×550

* Source: Summarized by the Research Group through research

minimum standard size of the door in the construction. Lamp sockets, pipelines, cable ports and ventilation openings are arranged at the ceiling and bulkhead of the cabin. To facilitate the arrangement of cables, the lamp sockets are generally set at the center of the ceiling. Since the deck stiffness is large relative to the stiffness of cabin top, the opening position at the bulkhead should be properly close to the cabin bottom.

The furniture module that affects the cabin stiffness consists of two parts: bed and wardrobe. Structural damage can easily occur at the joint between the bed and the side wall, and there is a higher probability for the outer plate unit of the cabin module to deform. For prevention, it is necessary to enhance the deformation resistance of the outer plate structure of the cabin by changing the mode of connection between the bed and the side wall and reducing the span at the fixed end of the plate unit. Angle steel and bed side walls already exist at the connection between the wardrobe and bulkhead and ceiling, so such connection has little influence on the stiffness of the model (Guan, 2011).

4.3 Toilet unit module

The basic structure of the sanitary unit consists of steel frames or other frames with equivalent strength, steel plate chassis, wallboards, ceilings and internal equipment (Figure 4). The toilet unit is an important area for equipment integration, and the interface between water supply and drainage equipment and electrical equipment is the focus of construction inspection. The crucial part of the design lies in overall design, i.e., the sanitary ware, accessories, hardware, doors, etc. inside

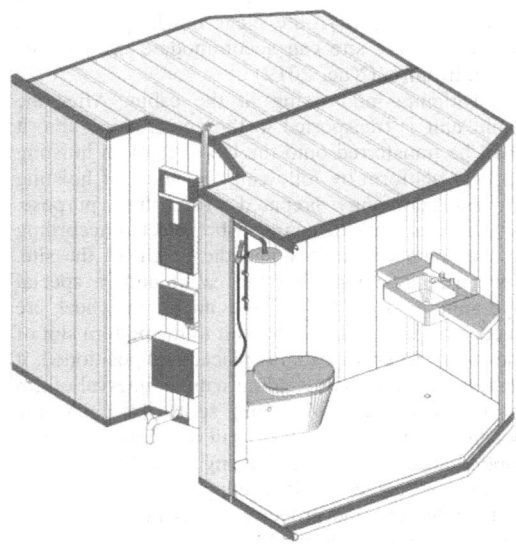

Figure 4. Exploded View of the Basic Structure of Toilet Unit.

* Source: Provided by the Research Group

the integral sanitary unit are completely installed; the pipe system butt joint and electrical junction box are externally reserved; lifting earrings or brackets are installed at the top; and adjusting screws are installed at the bottom. The whole sanitary unit is prefabricated and processed in the factory, and it only needs to be transported to the site for integral hoisting, and can be used when connected to the water source and power supply.

The toilet unit can be fast installed with simple operation process, and it is not easy to leak with unified standards for its eternal settings, but there are requirements of precise design and accurate installation. During installation, the position and size of a toilet module shall be accurately found to ensure its consistency with the position and size of others including the rib, and the error of upper and lower piping interfaces shall be avoided (Fan, 2019). The pipeline part of the sanitary unit is divided into water inlet and drainage systems. During design and positioning, the distance between the unit and the cabin wall shall be calculated with the interface shall be connected properly. The door frame of the sanitary unit shall be connected well with the cabin's internal enclosure wall. At the same time, a maintenance space is supposed to be reserved behind the toilet bowl, and it should avoid the strong cross beam on the reverse side of the deck to prevent failed connection of water pipes. In addition, the maintenance space is required to avoid the dead angle of the steel enclosure wall as far as possible to facilitate the installation and later maintenance of the working staff (Zhou, 2012).

5 INSTALLATION PROCESS FOR ACCOMMODATION CABIN OF LARGE LUXURY CRUISE

The modular construction of accommodation cabin of large luxury cruise typically shows the integration of assembly design and construction. Such integration involves not only the design of modular scheme, but also the overall planning of material decomposition and procurement as well as later assembly. The process approach, construction organization and technical specifications included in the preliminary design are the key links of the scheme design, while the detailed design and production design in the later stage are closely related to the material suppliers. Internationally wellknown companies engaged in cabin design and construction all follow their technical guidance for standardized operation (including technical operation manual and declaration and inspection process), and they all have material supply database as the support for design work. Each of the above links will be the research focus of modular building of luxury cruise cabins in China.

The modular process for the installation of accommodation cabins of large luxury cruise can be generally divided into three stages: design and procurement; platform assembly; and on-board installation. The entire process can be mastered through investigation and research, but there are still many detailed measures to be further explored in module design, precision control and process assurance. Meanwhile, the whole process shall be controlled by qualified workers, whose skills play a key role in the installation process.

5.1 Design and procurement

The difficulty in the phase of design and procurement lies not in the preliminary design of cabin scheme itself, but in the control of material supply and the overall plan of the platform assembly project. Regarding the expression of design drawings, the investigation found that the drawing standards are different among many design companies. It is mainly reflected in the adoption of module number and index system, which are also the embodiment of their respective technical styles.

The connecting work designed for later phase is mainly to prepare the cabin construction process and provide the material purchase order, which corresponds to the material supply database and needs a scientific and detailed work plan. The design department should start drawing up the purchase plan, providing purchase instructions and purchase lists two months in advance of the construction. The purchase process shall be subject to quality control and factory acceptance test (FAT) or site acceptance test (SAT). As the key to the procurement process, FAT or SAT should be implemented according to the acceptance outline, which must include the specification requirements of this functional module.

5.2 Platform assembly

During construction, most companies engaged in cabin design and construction will build cabin assembly workshops and production lines in the vicinity of the ship construction area to minimize transportation costs and losses. The fundamental of the whole assembly line is to control the levelness and precision of the assembly platform (the error is within 2mm), as well as the assembly efficiency and process quality of each functional module.

The precision control of assembly platform is the core of the installation. In fact, the platform itself is an essential component that requires multi-disciplinary design. An ideal assembly platform should be equipped with precise positioning lines, tracks that can move and rotate, and industrial integrated mechanical devices that are convenient to operate. However, at present, most shipbuilding plants in China still adopt the production process streamline and the belt-shaped assembly platform with divided sections, which not only reduces the

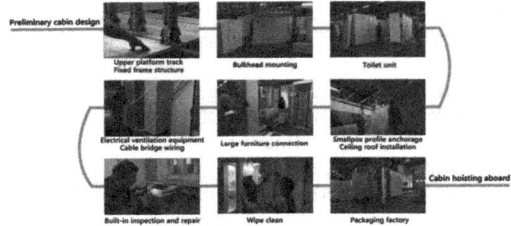

Figure 5. Process of Platform Assembly of Modular Cabins.

* Source: Provided by the Research Group

work efficiency but also easily leads to large errors. Therefore, the assembly process of functional modules is a decisive guarantee for process quality (Figure 5).

5.3 On-board installation

In the process of on-board installation, important nodes include the work of positioning and fixing in the cabin, equipment connection and debugging, and bulk furniture layout. Besides, preparations must be done before hoisting onto the board. For example, the number of cabins produced should match the number of cabins hoisted with a plan made for linkage. Tools on the construction site should be prepared, such as the lengthened fork teeth of the forklift, the hoisting cage for prefabricated modular cabin unit, as well as corresponding transfer trolley and pry bar The on-board site should be cleaned and tidied up, and obstacles on the route shall be removed to ensure that the on-route clear height pushed and pulled into the site can accommodate the passage of cabin units (Xue, 2019).

Positioning and fixing in the cabin. After the cabin unit is transported to the construction site, it shall be transferred onto the board through hoisting cage or platform by following the guide of hoisting process drawing prepared for this purpose. A telescopic forklift can also be selected according to the crane's capability and the height of the site. For pushing and pulling the cabin unit, a special hydraulic booster trolley and an angle wheel are needed to push and pull the unit to the bottom slot of the cabin. After the cabin is placed and positioned, it shall be fixed by spot welding with an interval of 2m for the distribution of welding spots. After that, the position of the window box shall be marked and perforated according to the drawing. The floor is often treated by cement self-leveling, and the Carpet, being ruled out and properly cut, is pulled and leveled by way of brushing glue on the floor.

Equipment linking and debugging. The equipment mainly includes air conditioning, water intake and drainage system, electrical and ventilating parts, as well as lighting and sound equipment. The general control interface of the equipment is basically concentrated in the maintenance triangle area between the two cabins. And it is even feasible that the junction boxes are shared by the two cabins to reduce the number of junction boxes for lightweight purpose. The module of the air conditioning unit is installed on the back of the ceiling panel of the cabin and fixed with a suitable supporting structure, with air duct and electric wires connected according to relevant drawings. The cold and hot water pipes are respectively connected with corresponding valve joints. The drainage system is separated into grey water (reclaimed water) and black water (sewage) drainage systems. The main power supply circuit of the power receiving box is distributed into light circuit, signal cable, emergency power supply, etc. In the phase of final debugging, it requires procedures such as a pressure test for the pipeline, an inspection on ventilation interface and electrical system, and a check on the status of insulation and fire work.

After the furniture in bulk is arranged, there should be a check on the paint state, iron outfitting state, door and window glass, etc., and it will be submitted to the ship owner for inspection after a thorough cleaning with soft cloth. And in the process of on-site construction, it is especially necessary to avoid damage and adopt necessary protective measures for the affected areas of construction.

6 CONCLUSION

To sum up, there are three key points in the modular building system for accommodation cabins of luxury cruise: (1) The application and control of materials. Most of China's high-end materials have not been domestically produced, and the artistic design of cabins also needs a support of customized coating materials and facing materials from abroad. Only when the material system is fully controlled, can the modular design of cabins and factory assembly be carried out smoothly. Therefore, it is the premise to construct a set of applicable material supply database as the basis; (2) Precision and process control of the platform assembly. It involves not only the control of materials and operating methods, but also the proficiency of skilled workers. Precision and process control is the guarantee of luxury cabin's quality. Real-time monitoring should be carried out on the data of a number of key nodes, with continuous improvement and correction of process operation manual as

essential means to handle it; and (3) Research and development of technical tools in the cabin building process. Such tools include cabin assembly platform, push-pull forklift for getting into the cabin, equipment debugging and testing tools, levelness and precision measuring tools, etc. As a guarantee for the quality of luxury accommodation cabins, these technical tools can be developed in cooperation with institutions specialized in mechanical automation.

This paper has constructed a modular building system for accommodation cabins of large luxury cruise, with a summary of the above three points based on the composition of each functional module unit and the process of design, assembly and on-site installation. However, it cannot cover the modular cabin building process on the whole, nor can it go deep into the operation details. This conclusion of the main points in this paper aims to provide a direction of technical research for further improving the building system for accommodation cabins of China's luxury cruises.

REFERENCES

Byun, L.-S. (2006). Peculiarity of interior design materials for accommodation areas of cruise ships: A state-of-the-art review. J. Taylor &; Francis Group. 1(3),5–8.

Fan, Z. (2019). Built-in modular technology and its application research of the ship. J. Science and Technology & Innovation. (13): 123-124, 2–3.

Guan, Y. (2011). Research on the Design of Shock-Resistance Cabin Module. D. Harbin Engineering University, 46–50.

Jiang, Z., (2003). Modelling and cabin design of the ship. M. Harbin Engineering University Press, 112–135.

Li, P. (2014). A Study on Simulation Analysis Method of Lightweight Bulkhead. D. Ocean University of China, 16–18.

Liu, H. (2011). An Investigation and Analysis of Indoor Environment in Air-Conditioned Chinese Ship Vessel Cabins. J.SAGE Publications. 20(3),5–6.

Liu, H., Lian, Z., Gong, Z., Wang, Y. and Yu, G. (2018). Thermal comfort, vibration, and noise in Chinese ship cabin environment in winter time. J. Elsevier Ltd. 135,4–5.

Luo, X., Yang, Y., Ge, Z., Wen, X. and Guan, F. (2015). Maintainability-based facility layout optimum design of ship cabin. J. Taylor & Francis. 53(3),4–6.

Xue, W. (2019). Application analysis of prefabricated modular cabin unit technology in passenger ship construction. J. Mechanical and electrical technology. (04): 74-76,2–3.

Zhou, X. (2012). Design and location for wet units. J. SHIP & BOAT.23 (05):72-75, 2–4.

Developments in Maritime Technology and Engineering – Guedes Soares & Santos (eds)
© 2021 Copyright the Author(s), ISBN 978-0-367-77376-2

A measurement of shipbuilding productivity

P. Zuniga Roque & J.M. Gordo
Centre for Marine Technology and Ocean Engineering (CENTEC), Instituto Superior Técnico, Universidade de Lisboa, Lisbon, Portugal

ABSTRACT: The present work studies the concept of productivity in shipbuilding and how it should be measured. The existing metrics, shipbuilding process and shipyard organization were studies in order to choose the most adequate metrics which would allow the measuring of a shipyard productivity in a systematic and holistic way. This is achieved by gathering the man-hours spent in each ship organized by cost centre and using Compensated Gross Tonnages as the measure of output from the shipyard. Data was gathered for thirty ships built in the same European yard organized by cost centre. From the data collected it was found that the ratio of hours spent in outfitting to the hours spent in structures is proportional to the complexity of the ship. There was also opportunity to study the work reduction resulting from building ships in series and the shares of labour for ships series and across ship types.

1 INTRODUCTION

In the current times shipbuilding faces several challenges, on one hand shipyards must compete in a fierce international market while on the other hand shipyards, which are usually traditional and conservative, must adapt to a quickly changing market and technologies. From the International Maritime Organization (IMO) 2020 sulphur cap and the target to reduce the Green House Gas Emissions by at least 50% by 2050 (compared to 2008), to the Internet of Things, Building Information Modelling (BIM) which allow the yards to optimize and monitor its shipbuilding process in a systematic and automatic way.

This has driven and shaped the shipbuilding market, on one hand we are now seeing shipyards adapting to newer requirements by owners to provide environmentally friendly ships, be it either by using scrubbers, alternate fuels such as LNG or, in some cases by opting for an emission free electric vessel. While on the other hand, yards have to keep up with technological advances and improve their efficiency in order to remain competitive. Some yards have also adapted to newer profitable markets, namely some yards will also produce steel structures for the offshore market, such as the offshore wind. Andritsos et al. (2000), Sharma et al. (2010) and Yoshihiko et al. (2011) are a few examples of the efforts being made to improve and modernize the shipbuilding industry.

For a yard seeking to improve their productivity and efficiency the first challenge is the definition of these two terms in the shipbuilding industry and finding different methods to quantify them.

Independently of the various forms of quantifying a shipyard's efficiency, it is necessary to have a holistic understanding of not only the assets but of all the steps involved in the process of building a ship, the technologies involved and general shipyard organization. When choosing the metrics to use, the facility to obtain the necessary data should also be considered. While thorough data can be used, such as done by Guofu et al. (2017) and Sulaiman et al. (2017), which provide insightful and important information on the shipbuilding process, the metrics chosen should be readily available from the data which is often accounted for by the yard cost centres.

Only then, and after choosing econometric indicators related to production, one can identify possible bottlenecks and indicate a course of action to improve the efficiency.

Inevitably, most indicators end up analysing the selling price and the cost of its produced vessels. The major costs involved in building a ship can be more easily understood once we decompose them in two main partitions which are labour and materials (where materials and intermediate products can represent can to up to 70% of the total ship cost, as seen in Lamb, 2003). While material costs should be similar in every country (not always the case), labour is not, and is where the yard is presented with greater change of improvement.

Lastly the suggested metrics where gathered for a case study consisting of 30 ships built in the same yard. Which allowed the study of the weight of four main cost centres (Hull, OTF, Support and Project) on different ship types (Container ships and Chemical Tankers) and showed the limitations of the

DOI: 10.1201/9781003216582-91

Compensated Gross Tonnage (CGT) system as a measure of output.

2 SHIPYARD ORGANIZATION AND SHIPBUILDING PROCESS

In order to identify the factors that influence the productivity of the yard it is first needed to understand the process of shipbuilding.

2.1 Shipbuilding process

The shipbuilding processes a shipyard adopts will be dependent on their production strategy. The shipbuilding industry is very characteristic and most of the shipyards will build several ships at a time with significant variation among them but will try to make use of standardization to implement the gains of mass production. This production strategy is known as Group Technology (Lamb, 1986), as presented in Figure 1.

In this strategy the yard will establish a work breakdown structure for the ships it builds and will group identical products (products which suffer the same processes) into intermediate products families, as in Figure 2. By grouping those intermediate products and fabricating them for several ships at the same time the yard manages to take some of the gains from mass production while allowing for variety among ships.

To allow the grouping of similar products into families a good coding and classification system in essential. Classification separated products through similarities (properties, shape, processes among others) by use of a code system. Lamb (1986) proposed a seventeen-digit shipbuilding classification and coding system. Pal (2015) analysed different work breakdown systems as well as coding systems identifying three as the most relevant: the SWBS (Ship Work Breakdown Structure); PWBS (Product Work Breakdown Structure) and the SFI (Senter for Forskningsdrevet Innovasjon) system. Of those SFI is widely used in both project and shipyards, namely, to assign costs to cost centres, as shown in Table 1.

There are also concurrent activities which, while not contributing directly to production, are essential for a proper organization and operation of a yard.

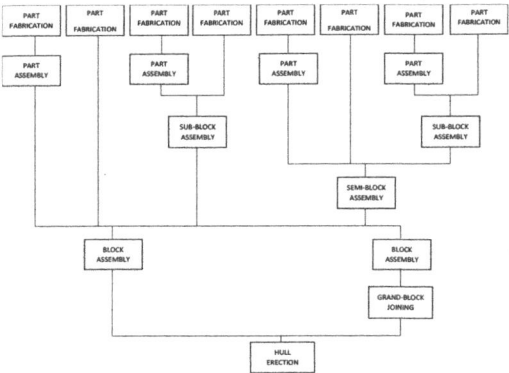

Figure 2. Manufacturing levels for the hull construction method.

Table 1. SFI system main groups (Pal, 2015).

SFI Group	Description
000	(reserved)
100	Ship General
200	Hull
300	Equipment for Cargo
400	Ship Equipment
500	Equipment for Crew and Passengers
600	Machinery Main Components
700	Systems for Machinery Main Components
800	Ship Common Systems
900	(reserved)

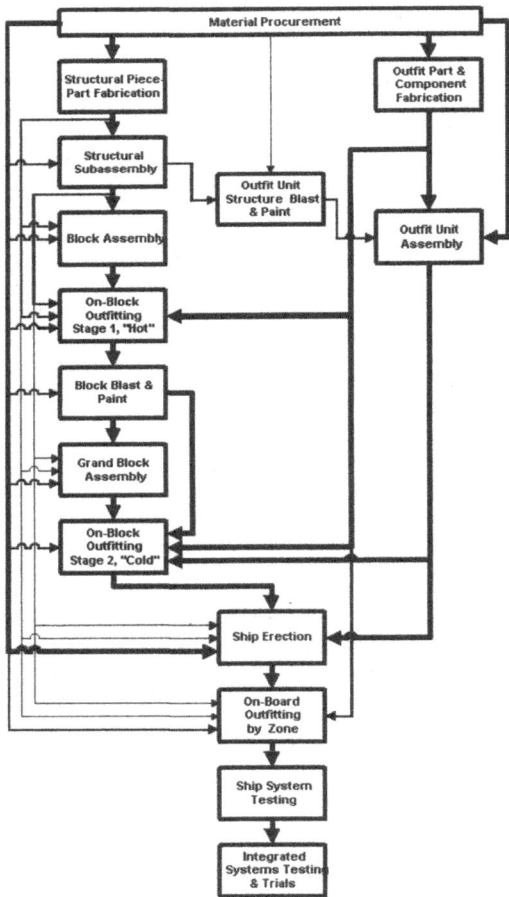

Figure 1. Shipyard Material and Workflow for a shipyard employing Group Technology (Lamb, 2004).

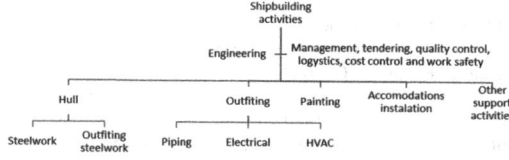

Figure 3. Shipbuilding activities.

The activities involved in the process of building a ship can be divided in production activities, support activities and engineering, production activities includes hull work (steelwork), outfitting (which includes piping, electrical and HVAC) and painting, while support activities are not directly involved in production but are still essential to support and provide the information needed to produce the ship, as systematised in Figure 3. Associated with each of those activities there will also be a cost centre, where the hours and resources spent for that activity are registered.

2.2 Measuring shipbuilding productivity

Productivity is defined as a measure of the efficiency of converting inputs into outputs. In shipbuilding there are several metrics used to measure a shipyard productivity, the choice of the metric to use will depend on its purpose.

Krishnan (2012) analyses the productivity measurement system for shipbuilding, mentioning the difficulty of calculating total productivity and defines some of the usages of productivity as being benchmark performance, value of comparison, measurement of production capacity, resource utilization and measure profitability. The intention of this work is to study a shipyard productivity in order to provide a benchmark performance value, hence, the choice of the metrics should reflect this.

2.2.1 Inputs for productivity measurement

Krishnan (2012) focuses on inputs such as labour, ship launching, shop floor area and total shipyard area. Pires et al. (2009) also present production cost, building time and quality as basic criteria to evaluate the performance of a shipyard from the competitiveness point of view, while capacity (total area, erection area, capacity for moving blocks), industrial environment and technology as indicators and influencing factors.

Inputs are often divided into five main types (Coelli et al. 2005): energy, Material, Purchased Services, Capital and Labour. The first three types are often aggregated into one single input.

2.2.1.1 Labour inputs

Labour inputs are one of the major input categories and measures the human work employed to produce the output.

Some of the most common ways to measure labour are the number of employed persons, the number of hours of labour (MH – man hours), and number of full-time equivalent employees.

Worked hours is the preferred metric (OECD, 2001), since it does account for hours paid by not worked, due to illness, leave among others. A well-organized yard will keep a registry of MH per cost centre for each ship, this is the ideal source of labour as it is the most complete and detailed, it allows not only to calculate the yard productivity but also allows to study the results from each cost centre.

2.2.1.2 Capital inputs

The capital of a shipyard is comprised of all the assets it owns. On a shipyard the most relevant capital would be those which contribute for production, productive assets. In these categories we will find the heavy and machinery of each workshop as the principal productive assets, as well as the area of the yard (Pires et al. 2009).

Ideally capital would be measured using the PIM (Perpetual Inventory Method). However, for this method requires the time series of investment expenditures on the yard assets, which might not always be available. Coelli et al. (2005) presents the following alternative measures of capital:

– Replacement value
– Sale price
– Physical measure
– Depreciated capital stock

From the alternative measures of capital two option stand out as the ones for which information is more readily available. Those are; physical measures and the depreciated capital stock.

For the physical measures it would be required to make an inventory of the main machinery used in the yard (heavy machinery) and the area of the yard. The differences between machinery quality and category should be accounted; the main equipment's could be categorized depending on their capabilities, however this would lead either to only a few categories being used, to maintain a simple approach, which would lead to a significant decrease in differentiation, or too many categories being considered which would lead to an exhaustive list of equipment's being created which, due to the variability among yards, would lead to results difficult to compare.

For these reasons the depreciated capital stock of the yard is a preferable method, since the majority of yards will either publish annual financial reports or keep track of their depreciated capital stock for finances purposes.

2.2.1.3 Energy, materials and purchased services

Materials and equipment's can account for most of the cost of a ship (up to 70% of the total ship cost, Jiang et al. 2011).

However, in this study there was no opportunity to develop the study of the materials cost which per se would be worthy of an individual study. The price of steel depends on the location of the yard, transport costs and, when applicable, import taxes. Yards in China and Europe will purchase steel at different prices, which can make the yard which buys steel cheaper appear more efficient, while it might only be more competitive, but not necessarily more efficient.

The services of painting, interiors, insulation, cleaning, HVAC, Scaffolding and others which include both labour and materials should also be considered in this category (energy, materials and purchased services inputs). The remaining subcontracted labour, which does not include materials, should be included as labour. In the cases where no man hours are known for that service, then the price must be converted to man hours worked by using the maritime industry worker average hour price.

2.2.2 *Outputs*

The output of a shipyard are the ships it produces, however the number of ships produced, by itself, is not an adequate metric as it does not account neither for the complexity nor size of each ship. Compensated Gross Tonnage (CGT) is the recommended measure of output of a shipyard as it accounts for the complexity and size of the ship. This metric has been used in OECD studies as well as by Lamb *et al.* (2001), Pires *et al.* (2009) and Krishnan (2012), among others.

Initially super yachts and naval vessels were not included in the original CGT coefficients; however, recent works have been done to include them. Hopman *et al.* (2010) proposed a factor A = 278 and B = 0.58 for super yachts and Craggs *et al.* (2004) presents a formula to calculate the CGT of naval vessels by calculating

Table 2. CGT coefficients (OECD, 2007; Hopman *et al.*, 2010).

Ship Type	A	B
Oil tankers (double hull)	48	0.57
Chemical tankers	84	0.55
Bulk carriers	29	0.61
Combined Carriers	33	0.62
General cargo ships	27	0.64
Reefers	27	0.68
Full container	19	0.68
Ro ro vessels	32	0.63
Car carriers	15	0.7
LPG carriers	62	0.57
LNG carriers	32	0.68
Ferries	20	0.71
Passenger ships	49	0.67
Fishing vessels	24	0.71
NCCV	46	0.62
Mega Yacht	278	0.58

Table 3. Customer factor.

Customer Factor	Characteristic
1.00	Normal commercial contract
1.06	Naval auxiliaries for Ministry of Defence and typical export combatants
1.12	Combatants built for Ministry of Defence and demanding export customer

a base CGT coefficient, dependent on the outfit weight to lightship ratio and a customer factor which represent the extra work required for naval vessels.

The formula for calculating the CGT can be seen in Eq. 1, while the corresponding coefficients are shown on Table 2.

$$cgt = A * gt^B \qquad (1)$$

To calculate the CGT of a naval vessel Eq. 2 should be used instead, where the base CGT coefficient is calculated using Eq. 3 and the client factor is obtained from Table 3.

$$cgt = gt * BC * CF \qquad (2)$$

where gt is the gross tonnage; BC is the base CGT coefficient for naval vessels; CF is the customer factor, as presented in Table 3 according to Craggs *et al.* (2004). The base CGT coefficient is given by:

$$BC = 44.65 \text{x} (\text{Outfit weight/Lightship})^{3.19} \qquad (3)$$

3 CASE STUDY

For this case study data was gathered for thirty ships, thirteen chemical tankers and seventeen containerships, built in the same European shipyard. The ships where built in five distinct series. Series A, B and C comprised of eight, three and two chemical tankers, respectfully, while series D and E comprised of thirteen and four containerships respectfully. The series A ships had stainless steel cargo hold, series B had painted holds and series C had icebreaking capacity.

3.2 *Data collected*

The data gathered consists on the man hours registered in the yards' custom cost centres for each ship, as well as each ship GT and type.

The cost centres are organized into four main groups: Structures, Outfitting, Support and Project. Structures included the hours spent on fabricating, (cutting, welding) and assembly of the ship structures plus steel outfitting. Outfitting included piping, mechanical works, electricity, on board outfitting and insulation. Support included painting, cleaning, interiors, scaffolding, transportation, quality and others support activities. Project included the hours spent in engineering for each ship.

Table 4 compares each ship to the most expensive one in terms of man-hours for every main cost centre. Table 5 presents the composition of every main cost centre in percentage of the total cost in MH for each ship.

3.3 Subcontracted services

On the data gathered from the yard there where man-hours from subcontracted services included. Some of those services included both the labour and materials necessary for the service, those services where Interiors, insulation, painting, cleaning, scaffolding and HVAC, for those the shipyard estimated the man-hours spent, however, due to the uncertainties regarding those estimations those hours where excluded in

Table 4. Man-hours in % to maximum MH by cost centre.

ID [Series /Nº]	Man hours / maximum MH in the series for the group (%)					
	Hull	OTF	Support	Project	T. Prod.	Total
SERIES A - CHEMICAL TANKERS (SS TANKS)						
A1	100.00%	100.00%	100.00%	100.00%	100.00%	100.00%
A2	95.64%	83.78%	79.86%	7.73%	89.75%	75.48%
A3	87.64%	75.71%	75.67%	10.67%	81.94%	69.54%
A4	83.98%	71.48%	70.11%	3.03%	77.93%	64.90%
A5	81.73%	67.40%	70.65%	4.41%	75.10%	62.80%
A6	77.64%	66.79%	73.94%	1.72%	72.88%	60.50%
A7	72.36%	63.67%	71.00%	1.62%	68.64%	56.98%
A8	71.38%	62.14%	72.79%	1.24%	67.58%	56.04%
SERIES B - CHEMICAL TANKERS (PAINTED TANKS)						
B1	97.56%	96.64%	100.00%	100.00%	97.61%	100.00%
B2	93.82%	90.02%	83.07%	14.18%	92.07%	78.32%
B3	100.00%	100.00%	97.74%	22.64%	100.00%	86.51%
SERIES C - CHEMICAL TANKERS (ICEBREAKERS)						
C1	100.00%	100.00%	100.00%	100.00%	100.00%	100.00%
C2	98.10%	88.28%	85.71%	7.57%	94.08%	81.72%
SERIES D - CONTAINERSHIP (HEAVY LIFT)						
D1	100.00%	100.00%	72.01%	100.00%	100.00%	100.00%
D2	90.69%	80.71%	67.54%	18.33%	88.01%	77.29%
D3	86.80%	78.61%	60.53%	8.22%	84.31%	72.60%
D4	86.87%	77.89%	66.17%	6.68%	84.59%	72.59%
D5	85.00%	76.11%	66.36%	10.81%	82.86%	71.77%
D6	82.52%	71.87%	57.24%	5.95%	79.31%	68.02%
D7	85.92%	73.20%	59.07%	7.06%	82.07%	70.53%
D8	84.67%	88.27%	100.00%	55.89%	88.71%	83.68%
D9	77.13%	83.09%	91.43%	7.93%	81.60%	70.26%
D10	74.92%	76.47%	81.83%	6.62%	77.51%	66.60%
D11	75.35%	70.03%	83.00%	3.41%	76.05%	64.87%
D12	78.62%	71.30%	73.51%	13.62%	77.84%	67.95%
D13	78.49%	68.89%	73.96%	2.30%	77.10%	65.58%
SERIES E - CONTAINERSHIP						
E1	94.77%	94.92%	100.00%	100.00%	95.19%	100.00%
E2	98.05%	89.73%	93.21%	16.65%	95.73%	89.30%
E3	96.21%	88.92%	94.77%	5.77%	94.38%	86.61%
E4	100.00%	100.00%	99.69%	6.29%	100.00%	91.79%

Table 5. Man-hours in % to total production MH in series by cost centre.

ID [Series/ Nº]	Man hours / total production MH for the ship (%)					
	Hull	OTF	Support	Project	T. Prod.	Total
SERIES A - CHEMICAL TANKERS (SS TANKS)						
A1	52.27%	41.95%	5.78%	21.06%	100.00%	121.06%
A2	55.70%	39.16%	5.14%	1.81%	100.00%	101.81%
A3	55.90%	38.76%	5.34%	2.74%	100.00%	102.74%
A4	56.32%	38.48%	5.20%	0.82%	100.00%	100.82%
A5	56.91%	37.65%	5.44%	1.24%	100.00%	101.24%
A6	55.69%	38.45%	5.86%	0.50%	100.00%	100.50%
A7	55.10%	38.92%	5.98%	0.50%	100.00%	100.50%
A8	55.20%	38.57%	6.22%	0.39%	100.00%	100.39%
SERIES B - CHEMICAL TANKERS (PAINTED TANKS)						
B1	63.05%	29.74%	7.21%	24.95%	100.00%	124.95%
B2	64.28%	29.37%	6.35%	3.75%	100.00%	103.75%
B3	63.08%	30.04%	6.88%	5.51%	100.00%	105.51%
SERIES C - CHEMICAL TANKERS (ICEBREAKERS)						
C1	60.88%	32.32%	6.80%	16.67%	100.00%	116.67%
C2	63.48%	30.32%	6.19%	1.34%	100.00%	101.34%
SERIES D - CONTAINERSHIP (HEAVY LIFT)						
D1	65.91%	28.53%	5.56%	18.19%	100.00%	118.19%
D2	67.91%	26.16%	5.92%	3.79%	100.00%	103.79%
D3	67.86%	26.60%	5.54%	1.77%	100.00%	101.77%
D4	67.69%	26.27%	6.04%	1.44%	100.00%	101.44%
D5	67.61%	26.21%	6.18%	2.37%	100.00%	102.37%
D6	68.57%	25.86%	5.57%	1.36%	100.00%	101.36%
D7	69.00%	25.45%	5.56%	1.57%	100.00%	101.57%
D8	62.91%	28.39%	8.70%	11.46%	100.00%	111.46%
D9	62.30%	29.05%	8.65%	1.77%	100.00%	101.77%
D10	63.70%	28.15%	8.15%	1.55%	100.00%	101.55%
D11	65.30%	26.27%	8.43%	0.82%	100.00%	100.82%
D12	66.57%	26.14%	7.29%	3.18%	100.00%	103.18%
D13	67.10%	25.49%	7.41%	0.54%	100.00%	100.54%
SERIES E - CONTAINERSHIP						
E1	68.63%	23.95%	7.42%	15.50%	100.00%	115.50%
E2	70.61%	22.51%	6.88%	2.57%	100.00%	102.57%
E3	70.28%	22.63%	7.09%	0.90%	100.00%	100.90%
E4	68.94%	24.02%	7.04%	0.93%	100.00%	100.93%

Table 6. Man-hours subcontracted in % to total production MH by ship.

Series	Type	AVG [%]	σ	RSD [%]
A	CT	16	0.01	7
B	CT	17	0.01	9
C	CT	20	0.01	5
E	C	2	0.00	5

this study, except for the final shipyard productivity calculation, where their exclusion would show an erroneous higher productivity.

The share of subcontracted labour was, nevertheless, studied for all series, except series D, whose data was not trustworthy. In Table 6 it is shown the ratio average value to maximum value in series and the coefficient of variation (COV) for each series.

It was found that the values for subcontracted serviced remained fairly constant along the series, with an average relative standard deviation of 10%.

3.4 Total production man-hours

By plotting the total man-hours per ship along the series, it becomes evident that series A and D have a steady increase in efficiency until a limit efficiency is reached, Figure 4. Series A peak efficiency is reached on the 8th ship while series D construction in series was interrupted on the 8th ship due to changes in that ship project. For series B, C and E, in Figure 5, the number of ships built were not enough to notice improvements in building ships in series.

The improvement seen follows a logarithmic decrease in total production man-hours. The regression obtained for series A and D are shown in Table 7 and compared with the regression shown on OECD (2007).

Series A gives a regression very close to the one suggested on OECD (2007), with an R^2 of 0.996. The division of series D into two subseries, D1-6

Table 7. Efficiency in a shipbuilding series.

Series	f(x)	R^2
A	$f(x) = -0.1571 \cdot \ln x + 1.000$	0.996
OECD (2007)	$f(x) = -0.1483 \cdot \ln x + 0.972$	0.972
D	$f(x) = -0.073 \cdot \ln x + 0.958$	0.748
D1-D6	$f(x) = -0.104 \cdot \ln x + 0.921$	0.921
D8-D13	$f(x) = -0.066 \cdot \ln x + 0.873$	0.844

and D8-13, yields better results, obtaining a regression with R^2 0.921 and 0.844 respectfully.

From the decomposition of hours spent in each ship by cost centre the following conclusion were reached: the share of each cost centre is maintained virtually the same along the series, showing that the decrease in man-hours has affected all cost centres proportionately, as presented from Figure 7 to Figure 11.

It was also found that the ratio of OTF to Hull was identical between ship types, 0.55 for chemical tankers and 0.37 for containerships, those two also represented the majority of the hours spent per ship, composing 85-90% of the total production man-hours. An higher ratio of OTF to Hull appear to represent a more complex ship, as chemical tankers are more complex ships than containerships they have an higher ratio (this is also reflected on the CGT coefficients, that are higher for chemical tankers), this also goes in accordance to Craggs et al. (2004) which uses this ratio to obtain the base CGT coefficient for naval vessels, and as seen in equation 3 the higher this ratio the higher the base CGT coefficient and thus the higher is the ship complexity.

Support activities share is constant and independent of ship type, ranging from 5% to 7%. Project man-hours ranged from 3% to 10% of the total hours spent. This share remained fairly constant at 3-4% with the exception of series B and C, as in Figure 6.

Contrary to the other cost centres, project man-hours decrease along the series is not constant, a steep decrease is noted from the first to the second ship, showing an average decrease of 86%, after which will remain relatively constant and at a residual value of 2.1%, on average, of the total project man-hours spent on the first ship, as may be estimated from Table 8 and Table 9.

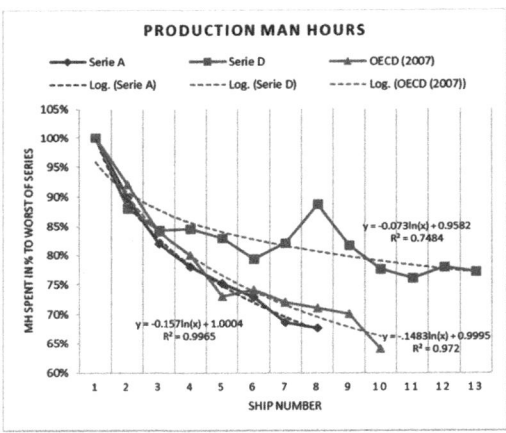

Figure 4. Evolution of man-hours required per ship in series A series D and OECD (2007).

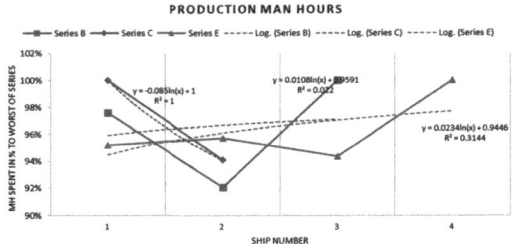

Figure 5. Evolution of man-hours required per ship in series B, C and E.

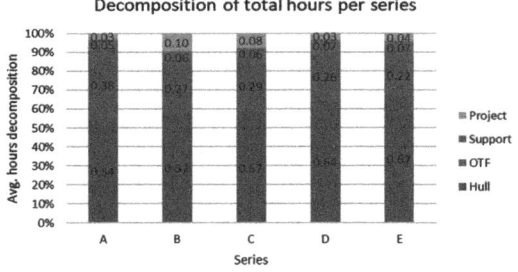

Figure 6. Average share of the 4 main areas in the total hours spent, by ship series.

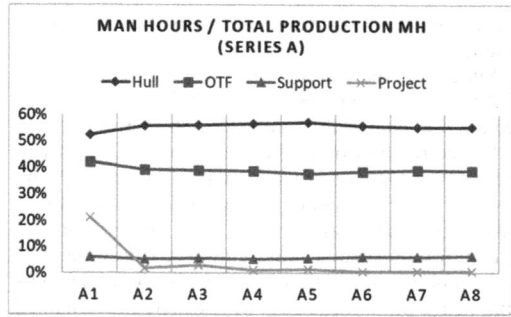

Figure 7. Evolution of MH/TP(MH) in series A.

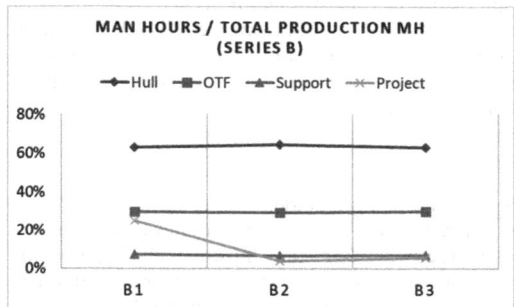

Figure 8. Evolution of MH/TP(MH) in series B.

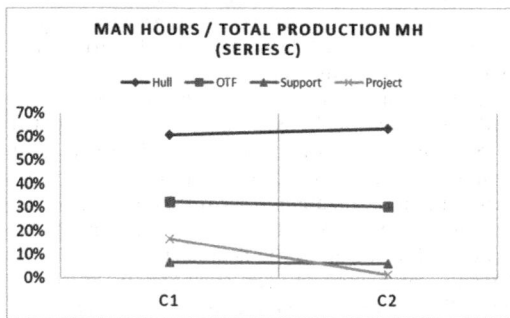

Figure 9. Evolution of MH/TP(MH) in series C.

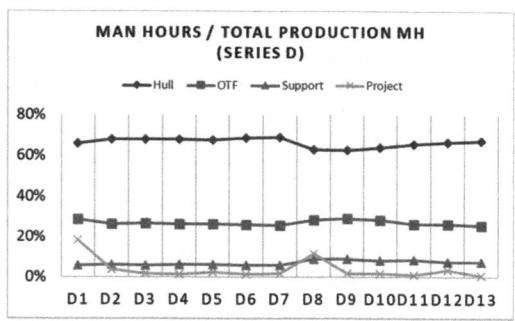

Figure 10. Evolution of MH/TP(MH) in series D.

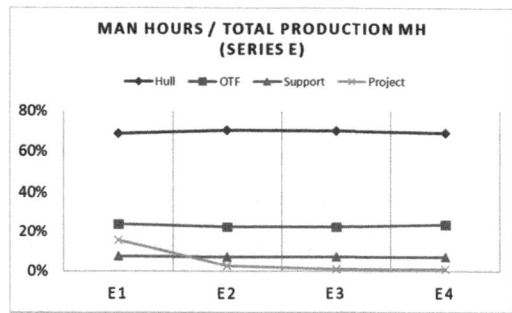

Figure 11. Evolution of MH/TP(MH) in series E.

Table 8. Project hours in % to total production hours.

Project MH/Production MH (%)			
Series	1st project	2nd project	Avg. rest
Comparison by ship series			
A (CT)	21.1	1.8	1.1
B (CT)	25.0	3.8	5.5
C (CT)	16.7	1.3	-
D (C)	18.2	3.8	2.6
E (C)	15.5	2.6	0.9
Avg.	19.3	2.7	2.5
Comparsion by ship type			
Chemical tanker	20.89	2.30	3.30
Containership	16.85	3.18	1.77
Dif. CT to C [%]	+19%	-38%	+46%

3.5 Shipyard's productivity

Using Eq. 1 and the factors in Table 2 the CGT of each ship was found. The inverse of the productivity was calculated by the ratio MH/CGT for each ship. This provides the average amount of MH spent to produce one unit of CGT, which can be later be used for productivity control and for tendering. To avoid getting a mistakenly low value the estimated MH spent in subcontracted services were included. If the average price of MH (in €/MH) was known, as well as the depreciated capital stock it would also be possible to calculate the average cost to produce one unit of CGT (excluding materials costs).

In Figure 12 the inverse of the productivity was plotted for each ship. The three lowest points shown correspond to the chemical tankers with painted tanks, which presented, in average, a 46% lower MH/CGT ratio when compared with the tankers with stainless

Table 9. Drop in project hours, from 1st to 2nd ship.

Series	Ship Type	Drop 1st to 2nd (%)	Avg. after second ship (%)
A	CT	91.4	1.1
B	CT	85	4.6
C	CT	92	1.3
D	CT	79.2	1.8
E	CT	83.4	1.5
	Avg.	86.2	2.1

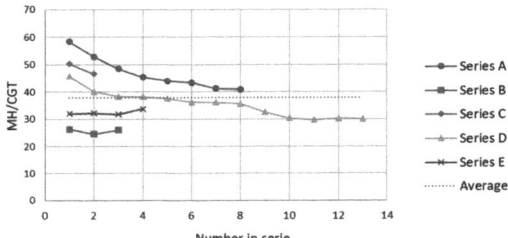

Figure 12. Case study of shipyard productivity in MH/CGT.

steel tanks. This shows a fragility of the CGT system, that it does not account for difference in complexity inside the same category. Chemical tankers achieved an average productivity of 40.3 MH/CGT, while containerships required a lower 33.9 MH/CGT, thus giving the shipyard an average productivity of 37.7 MH/CGT. In theory, by using the appropriate CGT coefficients the productivity should be identical across ship types, however as shown this was not the case.

4 CONCLUSIONS

By realizing this work, it was possible to conclude that the CGT coefficients still do not account for the differences in complexity for ships of the same type. In this study it was observed that chemical tankers with painted holds required 46% less MH/CGT than tankers with stainless steel holds, but it still would have the same CGT factors for being the same ship type.

The average productivity obtained for the chemical tankers and for the containerships was also different, of 40.3 and 33.9 MH/CGT respectfully, with containerships requiring approximately 15% less man-hours to produce one CGT. This shows that the CGT system still does not provide ideal coefficients for use in micro-economic analysis, for macro-economic analysis (which were the intended analysis when the CGT was created) this effect should be diluted due to the vast amount of ships being built,

and it would be expectable to obtain a more similar result.

Subcontracted labour accounted for 7.6% to 10.1% of the total MH spent and were found to remain moderately the same along the series.

Structures and outfitting were found to represent the greatest amount of work, with structures ranging from 54% to 67% and outfitting from 22% to 38%, it was also found that a higher ratio of outfitting to structure represents a more complex ship (chemical tankers obtained a ratio of 0.55 while containerships obtained a ratio of 0.37). Support hours remained virtually constant among the series ranging from 5% to 7%.

Project man-hours represented 3% to 10% of the total man-hours, in average, however, were found to be reduced 82% to 92% from the 1st to the 2nd ship, after which would remain a relatively small portion of the total man-hours showing only slight gains along the series.

4.1 Further work

Despite not being included in this study it is becoming increasingly frequent for yards to produce other steel structures other than ships, namely it is becoming more frequent for yards to also build offshore wind contractions, such as transition pillars. The offshore wind market is highly competitive and provides great opportunities to yards. In the offshore wind market, it is also essential for a yard to be highly productive, so that it can complete a project as fast as possible, while maintaining the strict quality required in this market.

As a continuation of this work data from more yards should be gathered, with man-hours divided by Structures, Outfitting, Support and Project cost centres. To account for different production departments and rationalization in the different yards, CGT (output) of the yard would be a function of the labour (MH) as well as the depreciated capital stock (€) and purchased services (€). The inclusion of those three inputs should address the mentioned issues, as a yard with higher automation would require less MH/CGT but would have a significantly bigger depreciated capital stock. With enough data points it would also become possible to establish production frontier using either data envelope analysis or stochastic analysis.

With data for more ships gathered from more yards it would also be interesting to further study the CGT system, the variation found among ships of the same type and study a solution to this issue.

In further works it would also be interesting to study yards involved in the offshore wind and compare its profitability of this market against the shipbuilding one as well as the compatibility and challenges to a yard organization engaged both in offshore wind and shipbuilding.

REFERENCES

Andritsos, F., & Prat, J. (2000). The Automation and Integration of Production Processes in Shipbuilding, State-of-the-A. European Commission Joint Research Centre.

Coelli, T., Rao, D., O'Donnell, C. & Battese, G. (2005). *An Introduction to Efficiency and Productivity Analysis*. 2nd Edition. Springer.

Craggs, J., Damien, B., Brian, T. & Hamish, B. (2004). Naval Compensated Gross Coefficients and Shipyard Learning. *Journal of Ship Production*, Vol.20, No.2, pp. 107–113.

Guofu, S., Xiaobing, L., Yizhuang, X., & Yao Nailong (2017). Measurement and Evaluation Model of Shipbuilding Production Efficiency. *International Journal of Economic Behavior and Organization*, Vol. 5, Issue 6, pp. 149–161.

Hopman, J., Pruyn, J. & Hekkenberg, R. (2010). *Determination of the Compensated Gross Tonnage Factors for Super Yatchs*. TU Delft.

Jiang, L., Strandenes, S. (2011). Assessing the Cost Competitiveness of China's Shipbuilding Industry. *Maritime Economics & Logistics*. Palgrave Macmillan

Krishnan, S. (2012). A Scientific Approach to Measure Shipbuilding Productivity. Maritime Affairs: *Journal of the National Maritime Foundation of India*, Vol. 8, Issue 1, pp. 136–149.

Lamb, T. (1986). Engineering for Ship Production. *Ship Production Committee, Education and Training Panel (SP-9)*, SNAME.

Lamb, T. & Hellesoy, A. (2001). A Shipbuilding Productivity Predictor. *Ship Production Symposium*, SNAME, 13-15, Ypsilanti, Michigan.

Lamb, T. (2003). Discussion of "Methodology Used to Calculate Naval Compensated Gross Tonnage Factors" by John Craggs, Damien Bloor, Brian Tanner, and Hamish Bullen. *Journal of Ship Production*, Vol.19, No.1, pp.29–30.

Lamb, T. (2004). *Ship Design and Construction*. The Society of Naval Architects and Marine Engineers.

OECD (2001). Measuring Productivity, Measurement of Aggregate and Industry-Level Productivity Growth, *OECD Manual*. OECD.

OECD (2007). *Compensated Gross TON (CGT) System*.

Pal, M. (2015). Ship Work Breakdown Structures Through Different Ship Lifecycle Stages. *International Conference on Computer Applications in Shipbuilding*, Bremen, Germany.

Pires, F., Lamb, T. & Souza, C. (2009). Shipbuilding Performance Benchmarking. *International Journal of Business Performance Management*, Vol. 11, Issue 3, pp. 216–235.

Sharma, R. & Kim, T. (2010). Development of a Logic-Based Product Life-Cycle Management (LBPLM) System for Shipbuilding Industry-Conceptual Development. *Journal of Ship Production and Design*, Vol.26, No.4, November 2010, pp. 1–21.

Sulaiman, Sasono, E., Susilo, S., & Suharto (2017). Factors Affecting Shipbuilding Productivity. *International Journal of Civil Engineering and Technology (IJCIET), Vol. 8, Issue 7, pp. 961-975.*

Yoshihiko, T., Morinobu, I. & Hiroyuki, S. (2011). "IHIMU- a" A Fully Automated Steel Plate Bending System for Shipbuilding. *IHI Engineering Review*, Vol.44, No.1, 2011.

Author Index

Proceedings in marine technology and ocean engineering

1. Advances in Renewable Energies Offshore
   Edited by C. Guedes Soares
   ISBN: 978-1-138-58535-5 (Hbk + Multimedia)
   ISBN: 978-0-429-50532-4 (eBook)

2. Trends in the Analysis and Design of Marine Structures
   Edited by Josko Parunov & C. Guedes Soares
   ISBN: 978-0-367-27809-0 (Hbk + Multimedia)
   ISBN: 978-0-429-29887-5 (eBook)

3. Sustainable Development and Innovations in Marine Technologies, 2019
   Edited by P. Georgiev and C. Guedes Soares
   ISBN: 978-0-367-40951-7 (Hbk + Multimedia)
   ISBN: 978-0-367-81008-5 (eBook)

4. Developments in the Collision and Grounding of Ships and Offshore Structures, 2019
   Edited by C. Guedes Soares
   ISBN: 978-0-367-43313-0 (Hbk)
   ISBN: 978-1-003-00242-0 (eBook)

5. Developments in Renewable Energies Offshore, 2021
   Edited by C. Guedes Soares
   ISBN: 9780367681319 (Hbk)
   ISBN: 9781003134572 (eBook)

6. Developments in Maritime Technology and Engineering, 2021*
   Edited by C. Guedes Soares and T.A. Santos
   ISBN: 9780367773748 (Hbk)
   ISBN: 9781003171072 (eBook)

7. Developments in the Analysis and Design of Marine Structures, 2021
   Edited by J. Amdahl and C. Guedes Soares (Ed.).
   ISBN: 9781032136653 (Hbk)
   ISBN: 9781003230373 (eBook)